AUTOMOBILE ENGINEERING

Vol. II

Automobile Engines
including
Electrical/Electronic Equipment

Dr. KIRPAL SINGH
B.Sc. Engg. (Mech) Hons., M. Sc. (Engg.), Ph. D.
Adv. Dip. (German), Gert. E.T.T. (Russian)
M.I.E. (India), M.I.S.T.E., M.S.A.E. (India)
Professor and Head (Retd.)
Mechanical Engineering Department
Punjab Engineering College, Chandigarh
(now Deemed University)

STANDARD PUBLISHERS DISTRIBUTORS
1705-B, NAI SARAK, POST BOX No.: 1066, DELHI-110006
#: 4581/15 G.F., ANSARI ROAD, DARYAGANJ, NEW DELHI-110002
Phones: 91-011-23262700, 91-011-23285798, Fax: 91-011-23243180
www.standardpublishers.com email: stpub@vsnl.com
www.engineeringbooks.co.in www.standardpublications.in standard1705@gmail.com

Published by:
A.K. Jain
For Standard Publishers Distributors,
1705-B, Nai Sarak, Delhi-110006.

1st Edition	:	1971
2nd Edition	:	1975
3rd Edition	:	1979
4th Edition	:	1983
5th Edition	:	1986
6th Edition	:	1989
7th Edition	:	1994
8th Edition	:	2000
9th Edition	:	2004
10th Edition	:	2007
11th Edition	:	2009
12th Edition	:	2011
13th Edition	:	2014
14th Edition	:	2018
Reprint	:	2019
Reprint	:	2021

Cover—Engine for 2018 Chevrolet Tahoe car.

Price : Rs. 450-00

ISBN : 978-81-8014-251-2

Laser Typesetting by:
Ajesh Bhargave, Delhi.

Printed by: Giriraj Offset Press.

SHINING STARS ON THE AUTOMOTIVE SKYLINE

Free

DR. KIRPAL SINGH

B.Sc. Engg. (Mech.) Hons., M.Sc. (Engg.), Ph.D.,
Ad. Dip. (German), Cert. E.T.T. (Russian),
M.I.E. (India), M.I.S.T.E., M.S.A.E.
Professor and Head (Retd.)
Mechanical Engineering Department
Punjab Engineering College, Chandigarh

STANDARD PUBLISHERS DISTRIBUTORS

1705-B, NAI SARAK, POST BOX NO.: 1066, DELHI-110006
Phones: 011-23262700, 011-23285798, Fax: 011-23243180
www.standardpublishers.com email: stpub@vsnl.com
www.standardpublishers.in www.standardpublications.in

Published by:
A.K. Jain
For Standard Publishers Distributors
1705-B, Nai Sarak, Delhi-110006.

1st Edition, 2018

ISBN: 978-81-8014-252-9

Laser Typesetting by:
Ajesh Bhargave Laser Printers, Delhi.

Printed by:
Giriraj Offset Press, Delhi.

SHINING STARS
ON
THE AUTOMOTIVE SKYLINE

Automobiles are influencing our life styles more and more as the importance of transporting men and material from one place to another, both nearby as well as distant, is gradually increasing in the highly competitive world. Universities, engineering colleges and technical institutes, worldwide, are busy producing researchers, engineers and technicians for the automotive industry, which is working under neck-to-neck competition, to improve the existing techniques and devise the new ones to make the automobiles more safe, comfortable, affordable and beautiful.

It is in this context that here a sight is focused on the horizon of Automotive education and industry to see and show to the readers, the beautiful shining starts thereon. The main objective is to induce in the budding engineers, a curiosity and thereby to motivate them to devote them fully to the automotive fields so as to make them competent and perfectionists to make further improvements therein. To help in this further, the 'Shining Stars' here include also information about the facilities for higher education worldwide, in the automotive field.

The skyline stars where the sight has been focused briefly are:

1. Main automotive P.G. education facilities worldwide.
2. Some futuristic developments (Technical advances presented at the SAE Automotive World Congress, 2017).
3. Vehicle technologies one may see in the near future.
4. Main auto engine plants in European Union.

MAIN AUTOMOTIVE P.G. EDUCATION FACILITIES WORLDWIDE

S.No.	*Country/Location*	*University/College/School*	*Degrees Available*
1.	Australia/Victoria	Deakin University, Faculty of Science and Technology	ME. Mechanical Engineering (Automotive Engineering)
2.	Australia/Melbourne	RMIT University	ME International Automotive Engineering
3.	Canada/Oshawa, Ontario	University of Ontario, Institute of Technology, Faculty of Engineering	M.S. Automotive Engineering
4.	China/Beijing	Tsinghua University, School of Mechanical Engineering	MS, Ph.D., Vehicle Engineering
5.	China/Changchun	Jilian University, College of Automotive Engineering	ME, MS Vehicle Engineering MS, Ph.D, Vehicle Body Engineering
6.	China Changsha City	Changsha University of Technology and Science, School of Automobile and Mechanical Engineering	MS Vehicle Engineering

S.No.	*Country/Location*	*University/College/School*	*Degrees Available*
7.	China/Chongqing	Chongqing University, College of Mechanical Engineering	ME, MS, Ph.D. Automotive Engineering
8.	China/Dalian City	Dalian University of Technology, Faculty of Vehicle Engineering and Mechanics	MS. Ph.D. Vehicle Engineering
9.	China/Guangzhou	South China University, School of Mechanical and Automotive Engineering	MS, Ph.D. Vehicle Engineering
10.	China/Shanghai	Shanghai Jiao Tong University; School of Mechanical and Power Energy Engineering	ME Automobile Engineering Ph.D., Vehicle Engineering
11.	China/Shanghai	Tongji Universtiy, Collage of Automotive Engineering	MS Vehicle Engineering, Ph.D., Mechanical Engineering with sub-field of Vehicle Engineering
12.	China/Xi'an	Chang'an University, School of Automobile	ME Vehicle Engineering
13.	France/Paris	IFP School, Engineering Department	MS Powertrain Engineering
14.	France/Paris	IFP School, Engineering Department	MS Vehicles and Sustainable Transport, Specialisation: Electrification of Automotive Propulsion
15.	France/Paris	Paris Tech, Engineering Department	MS Mobility and Electric Vehicles
16.	Germany/Braunschweig	Braunschweig University of Technology, Faculty of Mechanical Engineering	MS Automotive Technology

S.No.	*Country/Location*	*University/College/School*	*Degrees Available*
17.	Germany/Berlin	Berlin Institute of Technology. School V-Mechanical Engineering and Transport Systems	MS Vehicle Engineering
18.	Germany/Cologne	Cologne University of Applied Sciences, Faculty of Automotive Systems and Production	MS Automotive Mechatronics
19.	Germany/Ingolstadt	Ingolstadt University of Applied Sciences, Faculty of Mechanical Engineering	MS International Automotive Engineering
20.	Gerrmany/Kaiserslautern	Universitaet Karlruhe (TH), Graduate School of Commercial Vehicle Technology	MS, Ph.D, Commercial Vehicle Technology
21.	Germany/Munich	Technical University of Munich, Faculty of Mechanical Engineering	MS Automotive and Combustion Engine Technology
22.	India/Banashankari. Bangalore	PES University	M.Tech. Automotive Engineering
23.	India/Chennai	Anna University Chennai, Madras Institute of Technology	ME Automobile Engineering
24.	India/Chennai	VELS University	ME Automobile Engineering

S.No.	Country/Location	University/College/School	Degrees Available
25.	India/Coimbatore, Tamil Nadu	PSG College of Technology	ME Automobile Engineering
26.	India/Chandigar	Chandigarh University	M.E. Automobile Engineering
27.	India/Gujarat	Gujarat Technological University	ME. Automobile Engineering
28.	India/Karnataka	Visvesvaraya Technological University	ME Automotive Engineering
29.	India/Kelambakkam, Channai	Hindustan University	M.Tech. Automobile Engineering
30.	India/Dehradun, Uttrakhand	University of Petroleum and Energy Studies	M.Tech Automotive Electronics
31.	India/Kolkata	Jadavpur University, Faculty of Engineering and Technology	ME Automobile Engineering
32.	India/Lohegaon, Pune	Jeenkya DY Patil University, School of Engineering	M.Tech Automotive Product Engineering
33.	India/Mayunga West, Mumbai	VJ Technolgical Institute	M.Tech. Automobile Engineering
34.	India/Noida	Amity University, Amity School of Engineering and Technologies	M.Tech. Automobile Engineering
35.	India/Tamil Nadu	Bharath University, School of Automotive Technology	MTech Automobile Engineering

S.No.	*Country/Location*	*University/College/School*	*Degrees Available*
36.	India/Vollore, (TN) Pune	VIT University and ARAI (jointly)	M Tech Automotive Engineering
37.	Italy/Modena and Raggio Emilia	University of Modena and Reggio Emilia	MS Automotive Engineering
38.	Japan/Takyo	Tokyo City University, Faulty of Engineering	MS, PhD. Internal Combustion Engines
39.	Malaysia	University Putra Malaysia (UPM), Vehicle Engineering	D Eng. Automotive Engineering MSc. Automotive Engineering (By Research) Ph.D. Automotive Engineering (By Research)
40.	Malaysia	University of Malaya, Vehicle Engineering	Ph.D Battery Charge (By Research) Ph.D. Electric Vehicle System (By Research) Ph.D. Electric Motor Drives and Machine Design (By Research) M.Phil Electric Vehicle System (By Research)

S.No.	*Country/Location*	*University/College/School*	*Degrees Available*
41.	Malaysia	Universiti Teknikal Malaysia Melaka, (UTEM). Vehicle Engineering	Ph.D, M.Sc. Mechanical Engineering-Automotive Engineering (By research) M.Mech Engg. (Automotive) (By Mixed Mode), M.Mech. Engg. (Automotive) (By Taught Course)
42.	Malaysia	Universiti Teknologi Malaysia, (UTM), Vehicle Engineering	ME, Ph.D. Automotive Engineering (By Research)
43.	Malaysia	University Teknologi Mara (UiTM). Vehicle Engineering	MSc. Ph.D, Mechanical Engineering (Automotive Engines) (By Research)
44.	Mexico/Monterrey	Monterrey Institute of Technology, Group de Ingenieria y Ciencias	MS Automotive Engineering
45.	South Koree/Seoul	Kookmin University, School of Mechanical Engineering	MS Automotive Engineering
46.	South Korea/Ulsan	University of Ulsan. College of Engineering	MS, Ph.D Mechanical and Automotive Engineering

S.No.	*Country/Location*	*University/College/School*	*Degrees Available*
47.	Sweden/Gothenburg	Chalmers University of Technology, School of Machine and Vehicle Systems	MS Automotive Engineering
48.	Sweden/LinKoping	LinKoping University, Faculty of Engineering and Computer Science	MS Intelligent Transport Systems
49.	UK/Bath	University of Bath, Faculty of Engineering and Design	MSc Automotive Engineering
50.	UK/Birmingham	Birmingham City University, Faculty of Technology, Engineering and the Environment	MSc Automotive Engineering MSc Automotive Calibration and Control
51.	UK/Bradford	University of Bradford, School of Engineering. Design and Technology	M. Eng. Mechanical and Automotive Engineering
52.	UK/Coventry	Coventry University, Faculty of Engineering and Computing	M. Eng. Automotive Engineering MA Automotive Design MSc. Automotive Engineering MSc. Engine Design and Development MSc. Vehicle Dynamics MSc. Vehicle Interiors MSc. Vehicle Structures and Safety

S.No.	*Country/Location*	*University/College/School*	*Degrees Available*
53.	UK/Cranfield	Cranfield University, School of Engineering	MSc Automotive Product Engineering MSc Automotive Technology Management MSc. Driver Behaviour and Education MSc. Integrated Vehicle Health Management MSc. Military Vehicle Technology MSc/M Tech. Motorsport Engineering and Management
54.	UK/Hertfordshire	University of Hertfordshire, School of Engineering and Technology	M. Eng. Automotive Engineering with Motorsport M. Eng. Automotive Engineering MSc. Automotive Engineering
55.	UK/Huddersfield	University of Huddersfield, Computer and Engineering	M Eng. Automotive Engineering MSc. Automotive System Design and Analysis

S.No.	*Country/Location*	*University/College/School*	*Degrees Available*
56.	UK/Leeds	University of Leeds, School of Mechanical Engineering	M Eng. Automotive Engineering MSc Automotive Engineering
57.	UK/Leicestershire	Loughborough University, Faculty of Engineering	MSc Automotive Systems Engineering
58.	UK/London	Kingston University, Faculty of Engineering	MSc Automotive Engineering MSc Automotive Engineering (Commercial Vehicle Technology) MSc Automotive Engineering (Motorcycle Technology)
59.	UK/Nottingham	University of Nottingham, Faculty of Engineering	ME Mechanical Engineering Automotive Engineering Specialization
60.	UK/Oxford	Oxford Brookes University, School of Technology	MSc Automotive Engineering MSc Motorsport Engineering
61.	UK/Sussex	University of Sussex, School of Engineering and Design	MSc Automotive Engineering
62.	UK/West London	Brunel University, School of Engineering and Design	M Eng.Motorsport Engineering MSc Automotive and Motorsport Engineering

S.No.	*Country/Location*	*University/College/School*	*Degrees Available*
63.	USA/Chicago, IL	Illinois Institute of Technology, Dept. of Electrical and Computer Engineering	MS Electrical Engineering Hybrid and Plug-in hybrid Electric Vehicle focus
64.	USA/Clemson, SC	Clemson University, Department of Automotive Engineering	MS, Ph.D Automotive Engineering
65.	USA/Daytona Beach, FL	Embry-Riddle Aeronautical Society, Dept. of Mech. Engineering	MS Mechanical Engineering, Hybrid Vehicle focus
66.	USA/Dearborn, MI	University of Michigan-Dearborn, College of Engineering and Computer Science	MS, Ph.D. Automotive Systems
67.	USA/Detroit, MI	Wayne State University, College of Engineeering	MS Electric-Drive Vehicle Engineering
68.	USA/East Lansing, MI	Michigan State University, College of Engineering	MS, Ph.D Mechanical Engineering MS, Ph.D Engineering Mechanics
69.	USA/Flint, MI	Kettering University, Mechanical Engineering Department	MS Engineering-Automotive Systems
70.	USA/Houghton, MI	Michigan Tech, College of Engineering	MS. Advanced Graduate Certificate Hybrid Electric Vehicle Engineering
71.	USA/Houghton, MI	Michigan Technological University, College of Engineering	M Eng. Hybrid Electric Drive Vehicle Engineering

S.No.	*Country/Locction*	*University/College/School*	*Degrees Available*
72.	USA/Michigan	Michigan Tech, College of Engineering	ME With emphasis on Hybrid Electric Vehicle Engineering
73.	USA/San Lu s Obispo, CA	California Polytechnic State University, College of Engineering	MS, Mechanical Engineering
74.	USA/Southfield, MI	Lawrence Technological Instutute, College of Engineering	MS Automotive Engineering

SOME FUTURISTIC DEVELOPMENTS

Some of the prominent automotive technical advances concepts presented at the SAE Automotive World Congress 2017 are:

1. One-cycle engine

Wisconsin US based 'Monolith Engines' presented an engine block for such on engine which they claim to have developed. The block was cast using 3D-printed sand-casting moulds made by Tooling and Equipment International. Inside this casting is a long cylinder in which run two opposed pistons, each of which is double sided, so that a combustion chamber each is formed between the pistons and at each outer end of the long cylinder. A pair of crankshafts connect to the centre of each double piston, thus providing a symmetrical power take-off, thereby making the pistons remain centered in the cylinder which minimizes friction. There are no valves and the fuel-injected engine does breathing through ports. The engine is named 'one-cycle' due to the fact that the pistons are always driven in each direction, never being interfered with another piston's work cycle. The engine is supposed to be mainly used as range- extender for hybrid electric vehicles.

2. Variable displacement supercharger

The concept was introduced by Minnesota, USA based Hansen Engine Technoliges. Herein a Lysholm type positive displacement twin-screw supercharger is employed, which has also a sliding window in one side of the housing. When that window is open, no pressure is generated. However, intake air is compressed when the window is closed.

The system uses a typical throttle to control air flow into the engine. When the accelerator pedal is about one-third of the way down, the throttle is wide open. The next one-third of the pedal travel causes the housing window to gradually close, thereby compressing the intake air and hence increasing the intake air flow. Pressing the accelerator pedal further would trigger higher engine rpm which would result in a typical turbo or blower opening its waste gate. However, here the window also starts opening back up, which results in higher efficiency and superior full-range engine response of a supercharger. This is being further tested on converted turbo engines.

3. Active shock absorber

Developed by German supplier Rausch & Pausch GmbH (RaPa), this active shocker is expected to reduce or eliminate the need for anti-roll bars. Instead of working on centrally pumped and distributed hydraulic oil or air, here each corner has its own fast-acting, 48 volt electrically operated hydraulic pump, which can switch directions at 5Hz, which is about double the typical frequency of car body motions (usually about 2-3 Hz). The shock absorber inside is just like a typical one, but the quicker flow of oil, a wheel can be placed down into a pot hole sensed by forward-looking cameras and then also pull it back out of that hole. Moreover, when driven passively, the gear oil pump can actually recover electrical energy from the suspension.

4. CT scanner for vehicles

Caresoft Global Inc. of Illinois, USA has developed a new machine for reverse-engineering competitor vehicles. So far, this is being done by buying the competing vehicle, tearing it down and analyzing the same for its engineering design, processes, materials and cost to build, etc. After scanning a Tesla model XP90D for three weeks, the company obtained sufficient data for making geometric models of most of its constituent parts. However, for part which have close tolerances, the same must be disassembled and rescanned for greater accuracy; although after doing all that a complete CAD model of the vehicle is obtained.

The scanner can also sense density of parts and hence can distinguish between ferrous metals from aluminium, plastic, etc. However, it cannot determine the constituents of alloys, due to which it is still required to do destructive analysis of many parts to produce a computer model for crash-test analysis. However, as claimed by the company, there is no other technology as yet, which can produce a complete model of the complete wiring harness without taking anything apart.

There is a big limitatum, however. As the machine requires a 9000 kW X-ray machine (compared to 100 kW for ordinary medical CT scanners), it has to be housed in a big bunker lined with lead. Further, the cost of scanning is quite high as yet, which the company hopes to decrease in the very near future.

5. Electric axle

The 'e-AAM' axle produced by American Axle manufacturing (AAM) Co. of USA is connected to an electric motor with a single-speed gear reduction of between 9:1 and 11.1, alongwith a epicyclic (planetary) gear set which enables three modes of torque-vectoring to the outside road-wheel:

1. Neutral, which disconnects the electric motor at high speeds or when it is not required.
2. Open-differential mode.
3. Torque-vectoring mode, in which rotating the motor in one direction or the other biases the torque to the outside wheel on left or right turns.

However, during torque-vectoring, the e-AAM axle does not, itself, provide any propulsion. The system is not in production as yet.

VEHICLE TECHNOLOGIES WE MAY SEE IN THE NEAR FUTURE

1. V2V AND V2I Communication technologies

A Vehicle to Vehicle (V2V) communication technology is being tested by automotive manufacturers like Ford, which would help reduce the road-accidents. The vehicles would be able to 'speak' to one another (without any action or even knowledge of the occupants) providing important information to stay safe and steer clear of the danger. For instance, imagine you approach an intersection where another car in a perpendicular road is coming with an intention to jump red light. Obviously you would not see it earlier. But if that car and your car are both equipped with V2V technology, your car would automatically get a signal from the other car that it is crossing your path and warning you of the potential collision or even hits the brakes of either or both the cars automatically to avoid the accident. According to a study by the NHTSA (US), this technology may be able to reduce the vehicle crashes on the road by 79 percent.

This technology works by using wireless signals for sending information to and fro between vehicles, regarding their location, speed and direction. This information is then communicated to the vehicles around it so that the vehicles are able to keep safe distances from each other. Algorithms are being developed to process the information from the vehicles, to determine the best measure to evade the onconing collision between the vehicles, if any.

Vehicle to Infrastructure (V2I), which is also being perfected, would allow the vehicles to communicate with road infrastructure such as road signs or traffic signals and would provide information about safety matters, to vehicles. The technology would also enable the vehicle to interact with traffic management system and access the best available routes.

2. Air bags as active safety measure

Mercedes is developing a new method to change the air bags from a *passive* safety measure to a part of an *active* safety system.

Mercedes is experimenting with air bags which deploy underneath the vehicle that helps it to stop before the crash. The deployment is initiated when sensors determine that a crash is imminent. The bags themselves have a friction coating on them which helps the vehicle to slow down and can double the stopping power of the vehicle. The active bags also lift the vehicle upto eight centimeters which counters the dipping motion of the vehicle during hard braking, improves bumper-to-bumper contact and helps prevent passengers from sliding under seat belts during a collision.

This type of active air bag technology uses only existing safety systems in the vehicle. However, it is not available so far, on any production models.

3. Biometric vehicle access

With this technology in the very near future, motorists would be able to unlock their vehicles without any key or key-fab, simply using their finger prints or eyeball. All the main automotive manufacturers are already working on finger print readers and retina scanners to identify the vehicle owners so that the vehicle door can be opened simply by placing the designated finger on the door handle or by looking through a window.

4. Reconfigurable vehicle

Automotive engineers are now working on the concept to recon figurate small sized SUVs and Trucks. The idea is to take an SUV with light-weight body panels and advanced motors which retract the roof and side glass into the lower body panels, and add Chrysler minivan stow-and-go seats– and you get a truck. Similarly you can change the truck back into an SUV.

The concept is in the design stage as yet.

5. Active health monitoring of the driver

Ford Motor Co. has previewed the idea of seat belt or steering wheel sensors which track the vital statistics of a vehicle driver. The technology, not only can alert the driver and passengers in the event of the health of the driver being seriously affected due to e.g., a heart attack or some other such medical emergency, but the vehicle would also be able to pull over onto the side of the road, shut down and call paramedics. This technology is capable of saving many lives.

6. Comprehensive vehicle tracking

This technology would allow insurance companies to track all the movements of the car. Some U.S. state governments are even considering the legislation which would authorize the vehicle insurers to charge fees depending upon the distance over which a person drives in a year. Although the move is being opposed by motorists, but insurance companies in U.S. are pushing hard for comprehensive vehicle tracking.

7. Energy-storing body panels

Although hybrid vehicles save a lot of energy, yet the weight of the batteries even in case of lithium-ion type, remains a drag. This has led a group of nine European Auto manufacturers to start researching and testing body panels which can store energy through regenerative braking, etc. and charge faster than the conventional batteries currently available. The body panels being tested are made of polymer fibre and carbon resin. Such panels are strong enough for automotive use and suitable also for being moulded to form panels. Such panels are expected to reduce the vehicle weight by upto 15 percent, by decreasing the size of the hybrid batteries.

Toyota is also working on light-weight energy-storing panels that would actually capture solar energy and store the same in the light-weight panels.

8. Driver-override technology

Available in the market today are vehicles that would apply brakes, if required to prevent a crash, if the driver fails to do so. However, it is expected that by about 2022, the vehicles would apply brakes if the sensors so determine, *even if* the driver ıloors the accelerator pedal, which means complete disregard of the driver. Not only that, the vehicles would also be able to 'talk' to other vehicles in the area, without even the knowledge of the drivers involved, enabling two or more vehicles to make decisions independent of their drivers and to do as they consider right, on the road.

9. Transparent car pillars

Strength considerations make it necessary to have large car pillars. However, they make it difficult sometimes to check the sonoundings.

Toyota's latest U.S. patent is about a device which makes the A-pillars in a car completely transparent. This has been done using many mirrors, which makes the light bend around a vehicle pillar and see its other side.

The technology to achieve the same already exists, making use of cameras and other costly components. However, since Toyota's method does not use any such devices, it is seen as more practical and less costly solution.

It is not known so far when Toyota would implement this technology in their vehicles.

MAIN AUTO ENGINE PLANTS IN EUROPEAN UNION

S.No.	*Manufacturer/Group*	*Country/Location*
1.	BMW	1. Hams Hall, UK
		2. Munich. Germany
		3. Steyr. Austria
2.	Daimler	1. Affalterbach, Germany
		2. Berlin-Marienfelde, Germany
		3. Gaggenau, Germany
		4. Koelleda, Germany
		5. Sabes, Romania
		6. Untertuerkheim, Germany
3.	Fiat Group Automobiles	1. Bielsko-Biala, Poland
		2. Cento, Italy
		3. Foggia, Italy
		4. Maranello, Italy
		5. Mirafiori, Italy
		6. Pratola Serra, Italy
		7. Termoli, Italy
		8. Verrone, Italy
4.	Ford of Europe	1. Bordeause, France
		2. Bridgend, UK
		3. Cologne, Germany
		4. Dagenham, UK
		5. Valencia, Spain
5.	General Motors	1. Aspern, Austria
		2. Kaiserslautern, Germany
		3. Ruesselsheim, Germany
		4. Tychy, Poland

6. Honda	1. Swindon, U.K.
7. Magna Powertrain	1. Ilz, Austria
	2. Lannach, Austria
	3. Roitzsch, Germany
8. PSA/Peugeot-Citroen	1. Douvrin, France
	2. Herimoncourt, France
	3. Metz, France
	4. Tremery, France
	5. Valenciennes, France
9. Renault-Nissan	1. Barcelona, Spain
	2. Cleon, France
	3. Ruitz, France
	4. Seville, Spain
	5. Sunderland, UK
	6. Togliatti, Russia
	7. Valladolid, Spain
10. Toyota	1. Deeside, UK
	2. Jelez-Laskowice, Poland
	3. Walbrzych, Poland
11. Volkswagen	1. Chemnitz, Germany
	2. Crewe, UK
	3. Gyor, Hungary
	4. Kaluga, Russia
	5. Kassel, Germany
	6. Prat, Spain
	7. Salzgitter, Germany
	8. Zuffenhausen, Germany
12. ZF Friedrichshafen	1. Brandenburg, Germany
	2. Eger, Hungary
	3. Saarbruecken, Germany

SELECTED
GLOBAL AUTOMOTIVE UPDATES
(Vol. 2)

Free

DR. KIRPAL SINGH

B.Sc. Engg. (Mech.) Hons., M.Sc. (Engg.), Ph.D.,
Ad. Dip. (German), Cert. E.T.T. (Russian),
M.I.E. (India), M.I.S.T.E., M.S.A.E.
Professor and Head (Retd.)
Mechanical Engineering Department
Punjab Engineering College, Chandigarh

STANDARD PUBLISHERS DISTRIBUTORS

1705-B, NAI SARAK, POST BOX NO.: 1066, DELHI-110006
Phones: 011-23262700, 011-23285798, Fax: 011-23243180
www.standardpublishers.com email: stpub@vsnl.com
www.standardpublishers.in www.standardpublications.in

Published by:
A.K. Jain
For Standard Publishers Distributors
1705-B, Nai Sarak, Delhi-110006.

1st Edition, 2014
2nd Edition, 2018

ISBN: 978-81-8014-253-6

Laser Typesetting by:
Ajesh Bhargave Laser Printers, Delhi.
Printed by:
Prabhat Offset Press, Delhi.

ASIA

INDIA

October 2017

1. Tata Motors' shelved project, two-seater sports coupe 'Racemo' bags a German Design Award. 'Racemo' appeared first time in public at the 87th International Motor Show. The award won was in the Gold category for 'International Excellence' in 'Exceptional Product Design'.
2. A subsidiary of SAIC Motor Corporation of China, 'MG Motor India' announced to start production at its new plant in Halol by 2019. In the first phase, they have scheduled to produce about 75,000 cars for Indian and South Asian markets.
3. Delhi government's transport department started sending registration certificates of vehicles through Speed Post, from October 1, 2017.
4. Mahindra and Mahindra announced its plan to roll out electric UVs and Crossovers. Besides it would also start producing petrol cars when BS VI emission norms are enforced in the country.
5. British premium motor cycle manufacturer 'Triumph Motorcycles' would start assembling 90 percent of its products in India at its Manesar plant by July 2018.
6. American motorcycle manufacturer 'Harley Davidson' announced its plans to start 'Harley Davidson University' at Delhi in India. It would be mainly a skill centre to train its employees and dealerships staff, apart from courses on management, finance, branding, etc.

Sept. 2017

1. Tata Motors wins the tender to supply 10,000 electric vehicles to government of India in two phases.
2. Suzuki to set up lithium-ion batteries plant in Gujarat, in

partnership with Denso and Toshiba. The plant is scheduled to start production in 2020.

August 2017

1. Government notifies the amendments to rules governing the use of cylinders carrying gas under pressure, which would pave the way for LNG refueling stations and help start commercial service by LNG vehicles.

March 2017

1. Supreme Court of India banned the sale of Bharat Stage III vehicles after March 31, 2017.

October 2016

1. In a recent notification, the road transport ministry has amended Form 22 under the Central Motor Vehicles Act, which is used to provide the initial certification of compliance with pollution standards, safety standards of components quality and road-worthiness certificate for all vehicles.
2. Mahindra-Ssangyong announced the development of new range of petrol engines.

CHINA

October 2017

1. 'Volvo Cars' and its owner 'Geely Holding' announced in Shanghai, China, to pledge 750 million US dollars to support Polestar's initial product and brand development, as a standalone electrified performance car manufacturer. For this a new manufacturing facility would be established in Chengdu, China.
2. American electric car maker Tesla has struck a deal with the Chinese government to establish a manufacturing facility at Shanghai in China, with a choice to accept any of the two options- Either to set up a joint venture with a Chinese partner, sharing their technology and profits, or to build a wholly owned facility in Shanghai's free-trade zone, but subject to 25 percent import tariff.

September 2017

1. On 9th September 2017, Xin Guobin, vice minister of industry and information technology announced in an

automotive conference in Tianjin, China, that the Government is structuring a long term plan to phase out vehicles powered by fossil fuels.

August 2017

1. 'SAIC-Volkswagen Automotive Co.', the joint venture of Volkswagen Group and SAIC Motor Co., announce fresh investment to renovate its powertrain plant in Shanghai, China, to manufacture engines for plug-in hybrid crossovers, which would hit the market in 2018.
2. Volvo Car and Geely Holding (Chinese car group) announced the formation of two companies to develop more rapidly, the next generation electrified vehicle technology:
   - (*i*) GV Automobile Technology (Ningbo) Co. Ltd. would be one company with 50/50 share holding and headquartered in China with a subsidiary in Gothenberg, Sweden.
   - (*ii*) A separate 'LYNK & Co.' responsible for 'LYNK & Co.' car line would be formed, jointly owned by Volvo Cars, Geely Holding and Geely Auto.

April 2017

1. ZF announced the expansion of its Engineering Centre in Songjiang district in Shanghai. The expansion is expected to be completed in 2022.
2. Volvo announced to build its first fully electric car in China, which would be based on Volvo's Compact Modular Architecture for smaller cars and is expected to be available in 2019.
3. A new development centre for engine components was opened by Thyssenkrupp, in Dalian in China, on the grounds of its existing production site in Dalian, which is in operation since 2005.

JAPAN

September 2017

1. Mazda, Toyota and Denso announced the establishing of a joint venture for the development of basic structural technologies for electric vehicles, under the name 'EV CA Spirit Co.'

August 2017

1. Mazda announced its new long-term plan for new technologies, named 'Sustainable Zoom-Zoom 2030'. In the first phase of this vision, would be the next-generation engine called SkyActiv-X, which would be the world's first commercial petrol engine, using compression ignition. This is scheduled to debut in 2019.

   Compared to current 'SkyActiv-G' engines, the 'SkyActive-X' engines are expected to use between 20 and 30 percent less fuel and coupled with supercharger, the technology would produce 10-30 percent more torque than the current naturally aspirated SkyActive range.

July 2017

1. Hitachi Automotive Systems Ltd. And Honda Motor Co. Ltd. announced the establishment of a joint venture company on the premises of Hitachi Automotive Systems in Hitachinaka-shi, Ibaraki Prefecture. The joint company would develop, manufacture and market motors for electric vehicles.

February 2017

1. Toyota Motor Corporation announced the commercial availability of new smaller catalytic converter which uses 20 percent less precious metal in approximately 20 percent less volume, still maintaining the same exhaust gas purification performance.

November 2016

1. AVL announced the opening of its new technical centre in Kawasaki, which is going to serve as a technology hub for the Japanese automotive industry.

October 2016

1. Toyota announced to start sale of Fuel cell buses in Japan under the Toyota brand from early 2017. Toyota plans to introduce over 100 FC buses mainly in the Tokyo area, ahead of the Tokyo 2020 Olympic and Paralympic Games.

SOUTH KOREA

August 2017

1. Hyundai exhibits the near-production-ready version of its

new hydrogen-powered SUV, at a special previous event held in Seoul. The official launch for the same would be in early 2018.

The new SUV previews Hyundai Motor's second commercially produced hydrogen model, using its fourth generation fuel cell.

MALAYSIA

August 2017

1. Mazda Motor Corporation announced that Mazda Malaysia Sdn. Bhd. Has unveiled a locally made all-new CX-5, whose production started in this month only.

EUROPE

October 2017

1. Renowned auto-component Tier 1 supplier, GKN opened a new innovation centre in Abingdon, Oxfordshire, UK, to develop state-of-art vehicle technologies and systems for the electric racing car.
2. Dutch chipmaker NXP launches a new generation of automotive processors with better performance, lower power consumption and reduced software development effort.
3. Oxford, the University city, announced plans to ban petrol and diesel cars from its centre, thereby becoming first UK city for such a ban to curb pollution.
4. Paris authorities announced plans to ban all combustion-engine cars by 2030.

September 2017

1. Horiba MIRA announce the setting up of a global skills centre at MIRA Technology Park, in partnership with North Warwickshire and South Leicestershire College, Coventry University, Loughborough University and University of Leicester. The site is scheduled to open in September 2018.
2. DMS alongwith Mahle, has developed a high-heat-resistant plastic for use in charge air cooler applications, which would enable the industry to replace metal end caps, resulting in better performance at lower system cost.

July 2017

1. Continental has developed on all-new electronically switchable engine mount, which uses an electromechanical shift actuator which is suitable for the latest generation vehicles. The actuator adjusts the hydraulic properties of the engine so that the dynamic rigidity of the mount during idling is less than the static rigidity, which helps to decrease significantly the transmission of structure-born noise Besides, the new engine mount is also about 10% lighter than standard variants.
2. BorgWarner delivered its already proven waste gate turbocharger for Hond'a new three-cylinder 1.0 litre direct injection petrol engine, initially available for 'Civic' in Asia and Europe. The compact turbocharger improves engine efficiency and boosts performance, helping the engine to output 95 kW (127 hp) with quick engine response. The unit is also claimed to improve fuel economy while providing 'excellent' torque characteristics over the entire engine speed range.
3. Volvo Cars announced that all the car models to be launched by it from 2019 would have an electric motor. Volvo would launch five fully electric cars between 2019 and 2021, three out of which will be Volvo models and two will be Volvo Cars' performance car arm, Polestar models of high-performance electrified cars. All these five models would also have optionally available range of petrol and diesel plug-in hybrid and mild-hybrid 48-volt options.

June 2017

1. Aston Martin has confirmed that its all-electric 'Rapid E' will enter production in 2019. Rapid E is the first all-electric model from Aston Martin.
2. A cylinder liner foundry was opened by Federal-Mogul Powertrain, at its Naberezhnye Chelny plant in Russia, which is its joint venture with Kazaz already operating since 2009.

May 2017

1. Volkswagen unveiled an apprentice-led project at the worthersee GTI fans meeting in Austria. The project adds a 48-V mild-hybrid drive system to the traditional GTI hatch.

April 2017

1. Continental unveiled clean diesel technologies at the Vienna Motor Symposium 2017, which reduce NOx emissions by 60%.

March 2017

1. Ford unveiled at 2017 Geneva Motor Show, its all-new 3-cylinder, 1.5 litre EcoBoost engine for its next-generation Fiesta ST. The new engine uses technology including variable cam timing for optimized performance and fuel efficiency.
2. Toyota unveiled its new 1.5 litre VVT-i engine at 2017 Geneva Motor Show. The 4-cylnder petrol engine powers the latest Yaris model.

December 2016

1. UPM's tests on Finnish wood-based diesel fuel both in laboratory conditions as well as in traffic, clearly demonstrated that this renewable diesel, named 'Bio Verno' works similarly as the best diesel fuels. Laboratory tests were conducted at VTT Technical Research Centre of Finland (VTT) and field tests in Helsinki region bus traffic.
2. BMW Group, Volkswagen Group, Daimler AG and Ford Motor Co. signed a memorandum of understanding to create the highest-powered charging network in Europe. The network will be based on Combined Charging System standard technology.
3. Tremec opened a new development and manufacturing facility at its site in Zedelgem, Belgium.

AMERICA

USA.

October 2017

1. Cummins announces to acquire battery pack developer 'Brammo', for getting an entry point into the growing electric vehicle market.
2. US Postal Service announced their working with University of Michigan to develop and build a self-driving mail truck called 'Autonomous Rural Delivery Vehicle'. It proposes to pilot the vehicle on 10 routes through rural America in 2019

and if it succeeds, it would roll out such vehicles on 28,000 rural routes by 2025.

3. Nissan announced that its 'ProPILOT Assist' technology would be available on its compact SUV, the 2018 'Rogue' SL grade. 'ProPILOT Assist' is stated to be the foundation for the future autonomous vehicle, helping drivers to maintain lane control, navigate stop-and-go traffic, maintain a set vehicle speed and maintain a set distance to the vehicle ahead– all with a simple two-button operation.

May 2017

1. Ford announced to add a new transmission to expand its lineup of fuel-efficient power trains at its Livonia Transmission Plant. It would be completed in 2019.

April 2017

1. Eaton and Cummins announced an agreement to form a 50-50 joint venture for automated transmissions for heavy duty and medium duty commercial vehicles. The joint venture would be called 'Eaton Cummins Automated Transmission Technologies'.

January 2017

1. Infinity 3.0 liter twin-turbo V6 engine named to Wards 10 Best Engines list for 2017.

November 2016

1. Ford announced that its award-winning 1.0 litre EcoBoost petrol engine would be offered with cylinder deactivation technology from 2018, to further reduce CO_2 emissions and improved fuel efficiency.

October 2016

1. Audi of America announced that new 2.0 liter TFSI 4-cylnder engine would now be offered on 2017 Audi Q7 models.

PREFACE TO THE FOURTEENTH EDITION

The book has been thoroughly revised for the fourteenth edition. Care has been taken of even the minor details and new topics have been added in a number of chapters to update the same.

Chapters on 'Introduction' and 'Petrol engine fuel supply systems' have been reorganised thoroughly due to huge expansion of technology and have been divided into two chapters each.

As the market is being flooded with numerous models from various manufacturers, data has been further included/revised about the engines of many latest vehicles, such as Audi A3; BMW 1-Series, X5; Ford Ecosport, Figo, Endeavour; GM Chevrolet Enjoy; Honda CR-V 2.0, Jazz, BR-V; Hyundai Tucson, Elantra, Grand i10, Creta, i20 Elite, Xcent; Mahindra KUV 100 K8, Nuvosport N8, XUV500, TUV 300; Maruti Ertiga, Celerio, S-Cross, Baleno, Ciaz; Mercedes A180, B180, S 500, GL 350 CDI; Nissan Datsun Go, Redigo S, Terrano; Renault Duster AWD, Lodgy DCI 110, Kwid; Skoda Superb (3rd Gen), Octavia; Tata Nano Twist, Bolt XT, Tiago; Toyota Innova MT/AT; Volkswagen Jetta Highline, Vento; Volvo S60; Citroen C1, C3, C4, C5; Peugeot 508, 3008, 5008 (1.2 & 1.6 litres); Rolls Royce Phantom, Ghost Series II; Vauxhall New Viva 1.0i, 1.4i and Porsche 718 Boxter S.

To keep the size of the book manageable, technical specifications of many old vehicles have been phased out. The book was updated last in May 2017.

Many Appendices at the end of the book have been updated/revised. Further all types of questions at the end of each chapter have been supplemented/updated. Exclusive use of S.I. system has been made for all the numerical problems in the book. The above improvements have resulted into the addition of sizeable matter.

Since 1971, when the first edition of this volume was brought out, automotive technology has undergone tremendous change. Since then many techniques have become obsolete and were replaced by the newer innovations. However, for the sake of continuity, wider perspective and better understanding of the subject, many of such old techniques have been retained in the book, alongwith the newer ones.

It is hoped that the revised book will be more useful to not only the students of the engineering institutes, but also to the serving engineers in automotive field. The book is also expected to serve the students aspiring for the membership of professional bodies through examinations.

However, as the man is never perfect, many errors and omissions may have been left still. The author will be highly indebted for any suggestions to improve the book further.

In the end, the author thanks various manufacturers who have provided illustrations of their products for the book. The individual acknowledgments are made alongwith such illustrations in the book.

Chandigarh, June 2017

Dr. KIRPAL SINGH

PREFACE TO THE FIRST EDITION

Automobile Industry is one of the most important industries. In India, though presently in the infancy stage, it is bound to become one of the major industries, in the immediate future. The book "Automobile Engineering" is an endeavour to help the engineering students in acquiring a thorough knowledge of the automotive details.

"Automobile Engineering" has been divided into two volumes. Complete chassis is dealt with in volume I and this volume is concerned with the automotive engines. Various engine components alongwith the main circuits have been explained thoroughly. Though the last chapter on "lighting System and Accessories" does not form a part of the engine, yet it has been included in this volume to keep the entire electrical systems together.

Each chapter contains selected questions at the end, both for the written as well as viva voce examinations. Special care has been taken that the diagrams are easy to reproduce, without losing, however, any of the important details. Data regarding premier Indian automobiles has been given in each chapter to enable the reader to understand the applicability of the things he has studies in theory.

Any suggestions for the improvement of the book will be welcome.

Chandigarh, March 1971

Dr. KIRPAL SINGH

CONTENTS

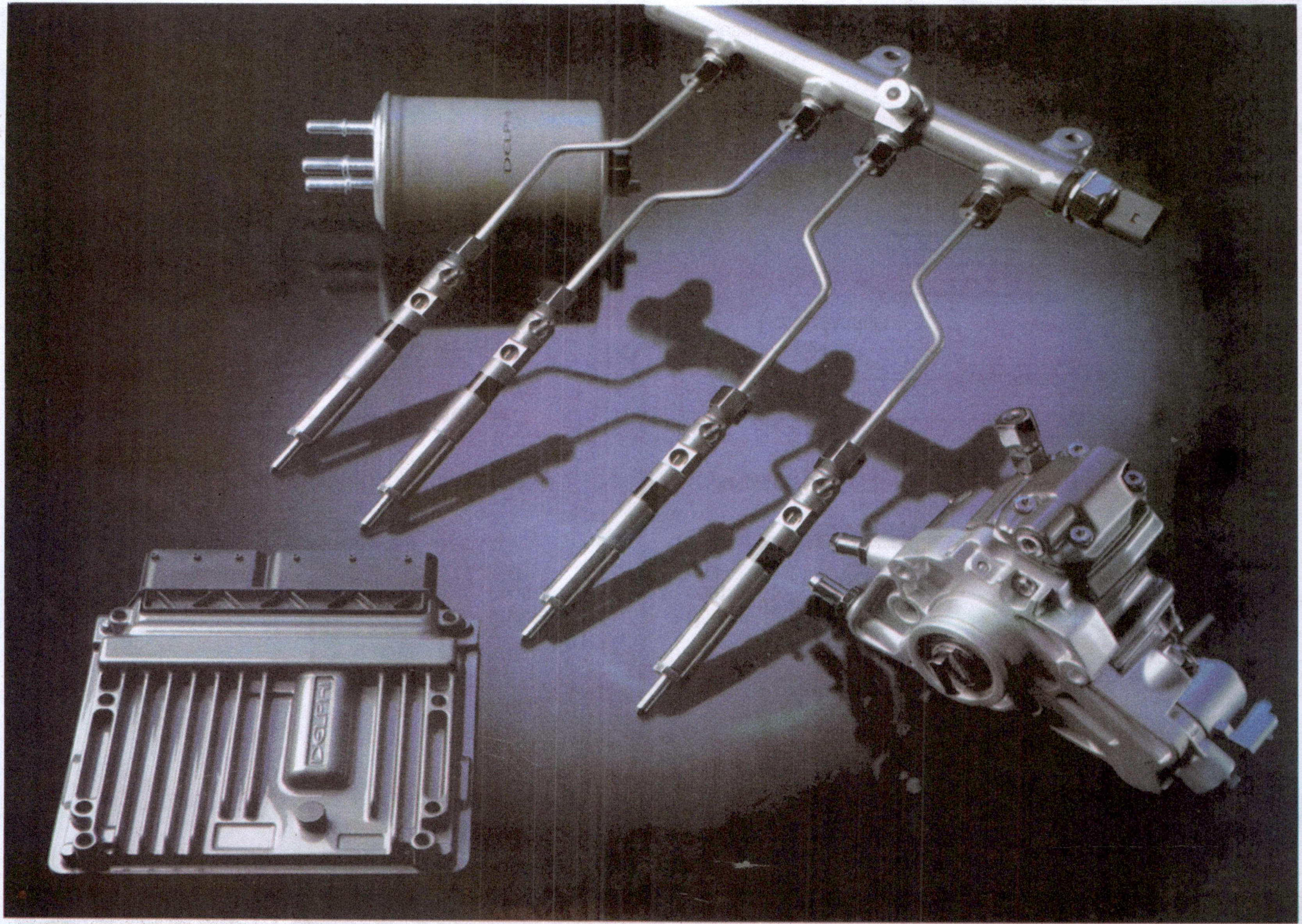

Diesal Common Rail Injection System (Courtesy - Delphi TVS Diesel System Ltd., India)

Design of the SCR catalytic converter with Denoxtronic PC/LD

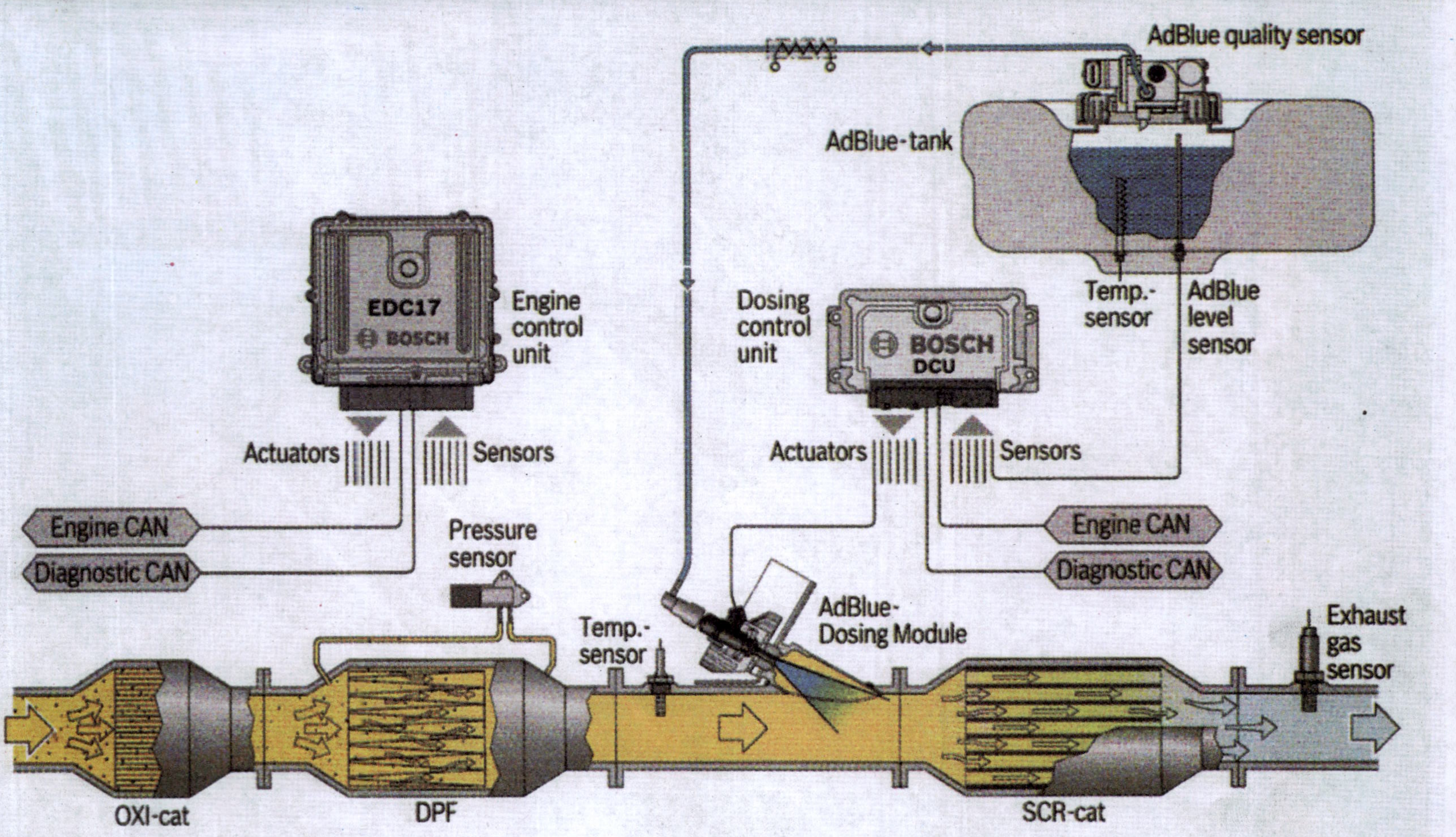

(Courtesy - Robert Bosch Gmph Diesel Systems, Germany)

Introduction–I

1

1. COMPONENTS OF AN AUTOMOBILE

The main components of an automobile are:

1. *The basic structure.* It consists of the frame, the suspension system, axles, wheels and tyres.
2. *The engine.* It provides the motive power for all the various functions which the vehicle or any part of it, may be required to perform. The engine for automotive use is of internal combustion type.
3. *The transmission system.* It consists of a clutch, a gear box giving four, five or even more different ratios of torque output to torque input, a propeller shaft to transmit the torque output from the gear box to the rear axle and a differential gear to distribute the final torque equally between the driving wheels.
4. *The auxiliaries.* The principal one out of these is the electrical system.
5. *The controls.* They consist of steering system and the brakes.
6. *The superstructure.* In those cases where frameless construction is not adopted there must be superstructure, *i.e.*, the body.

Out of these the basic structure, the transmission system and the controls have been dealt in volume 1. The present volume would cover the engine along with the auxiliaries (excluding lighting system and accessories which have been covered in volume 1).

2. ENGINE SYSTEMS

Various systems of an automobile engine are:

1. *Cooling system:* This prevents the engine from overheating. It may be of air cooling or water cooling types, the former being more commonly used in smaller engines of two-wheelers, while the later types of cooling systems are predominantly used in engines of the rest of the automobiles.
2. *Fuel system:* This includes storage tank for the fuel, piping work for supply to the engine and arrangement for mixing with air and spraying into the engine cylinder.
3. *Exhaust system:* Its function is to vent exhaust gases with least back pressure and also to reduce engine noise with the help of muffler. These days converters are also fitted in the exhaust pipe to reduce the harmful constituents in the exhaust gases.
4. *Lubrication system:* This reduces friction to decrease wear of moving parts. Relatively less viscous lubricating oils are used in engine, whereas heavier oils and greases are used in the transmission and wheel bearings.
5. *Ignition system:* Its function is to supply high voltage surge at the desired instant and of adequate strength to produce a spark in the engine. Earlier electro-mechanical systems were used, which have given way to electronic systems in modern engines.

6. *Electrical system:* It consists of a storage battery, charging system and starting system. The battery supplies electricity for starting the engine, providing energy for spark and for all the electrical devices in the vehicle, *e.g.*, lighting, heating, visiting, music system, etc. The charging system consists mainly of an alternator and is used to continually charge the battery to keep it charged at all times. The starting system consists of a starting motor and a device to disconnect the same from the engine as soon as the engine is started.

3. BASIC ENGINE TERMINOLOGY

A number of basic terms are used to describe and compare engines. A few commonly used terms are explained here.

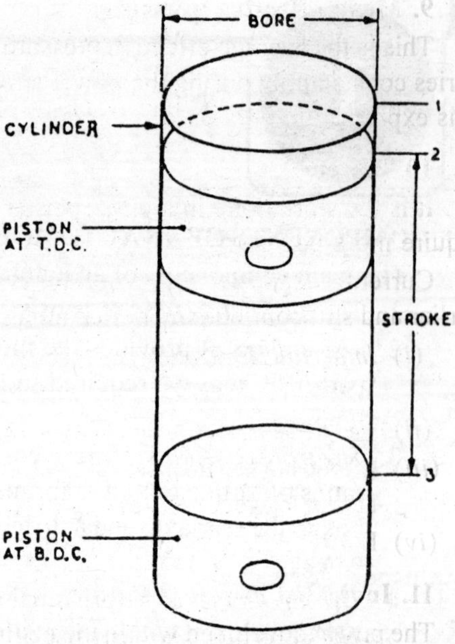

Fig. 1.1. Engine Terminology

1. Top dead centre (T.D.C.)

The refers to the position of the crankshaft when the piston is in its topmost position *i.e.*, the position closest to the cylinder head (position 2 in Fig. 1.1).

2. Bottom dead centre (B.D.C.)

This refers to the position of the crankshaft when the Piston is in its lowest position, *i.e.*, the position farthest from the cylinder head (position 3 in Fig. 1.1).

3. Bore

Diameter of the engine cylinder is referred to as the bore.

4. Stroke

Distance travelled by the piston in moving from T.D.C. to the B.D.C. (Fig. 1.1), is called 'stroke'.

5. Clearance volume

The volume of cylinder (including the combustion chamber) above the piston when it is in the T.D.C. position (volume between '1' and '2' in Fig. 1.1) is referred to as 'clearance volume' (Vc).

6. Piston displacement

This is the volume swept by the piston in moving from T.D.C. to B.D.C. This is also called 'swept volume'. If '*d*' is the cylinder bore and 'S' the stroke, the piston displacement, Vs is given by

$$V_s = \frac{\pi}{4} d^2 \cdot S \quad ...(1.1)$$

7. Engine Capacity

This is a total piston displacement or the swept volume of all the cylinder. If '*n*' is the number of cylin-ders and V_S is the piston displacement, then 'engine displacement' or the engine capacity' V_d, is given by

$$V_d = V_S \cdot n \quad ...(1.2)$$

8. Compression ratio

This indicates the extent to which the charge in the engine is compressed. This is calculated as the ratio of the volume above the piston at B.D.C. to the volume above the piston at T.D.C. If *r* is the compression ratio, then

$$r = \frac{V_S + V_C}{V_C} \quad \text{...(1.3)}$$

For petrol engines, compression ratios are about 8 to 10, whereas for diesel engines, there vary from 15 to 24.

9. Mean effective pressure

This is the average effective pressure throughout the whole power stroke. In fact the cylinder pressure varies considerably during the power stroke. Thus it is more helpful to refer to the mean pressure instead. It is expressed in bars or kilo Pascals (1 bar = 100 kPa).

10. Power

It is the work done in a given period of time. Doing the same amount of work in a lesser time would require more power.

Currently, standard unit for power is kilowatt (kW), although the following units have also been used earlier and still continue to be used alongwith kW:

(*i*) *Imperial Horse Power,* exactly 550 foot-pounds per second, approximately equivalent to 745.7 Watts.

(*ii*) *Metric Horse Power,* exactly 75 kgf-m per second, approximately equivalent to 735.5 Watts.

(*iii*) Pferdestarke (PS), which is German translation for Horse Power and is equivalent to one Metric Horse Power.

(*iv*) For rating electric motors, one Horse Power is equal to 746 Watts.

11. Indicated Power (I.P.)

The power developed within the engine cylinders is called indicated power. This is calculated from the area of the engine indicator diagram. It is usually expressed in kilowatts (kW).

12. Brake Power (B.P.)

This is the actual power delivered at the crankshaft. It is obtained by deducting various power losses in the engine from the indicated power. It is measured with a dynamometer and is expressed in kilowatts (kW). It is always less than the Indicated Power, due to frictional and pumping losses in the cylinders and the reciprocating mechanism.

Brake Power is what would keep the vehicle running at any speed once you have accelerated.

13. Engine torque

It is the force of rotation acting about the crankshaft axis at any given instant of time.

It is given by,

$$T = Fr \quad \text{...(1.4)}$$

where 'T' is the engine torque (Nm), 'F' is the force applied to the crank (N) and 'r' is the effective crank-radius (m). As the value of 'r' varies during the power stroke, the torque on the power-stroke is continually varying. Moreover, there is no torque delivered during the three idle strokes. Therefore, the engine manufacturers always quote the average value of torque throughout the engine cycle.

Engine torque goes through the vehicle transmission system, to the road wheels and is responsible for rotation of the latter and hence for pulling of the vehicle.

Torque provides the vehicle acceleration and is directly proportional to the engine power.

Thus a high torque, low brake power engine would accelerate well, but would not be able to maintain a high speed. On the other hand, a high brake power but low torque engine would accelerate slowly, but will be able to better hold the speed once it is accelerated.

4. ENGINE CYCLES

The two commonly used cycles on which the I.C. engines operate are:

1. Otto cycle
2. Diesel cycle

1. Otto Cycle. Theoretical Otto-cycle is shown in Fig. 1.2. The cycle was originally introduced by Bean de Rochas, but Otto was the first to succeed in working an engine on this cycle. The cycle consists

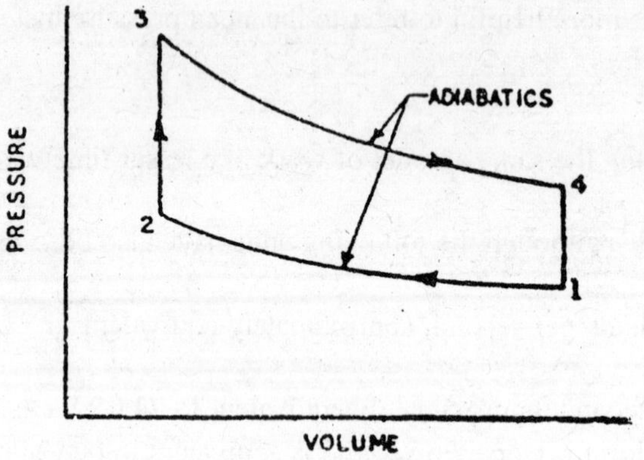

Fig. 1.2. Theoretical Otto Cycle-*p-v* diagram

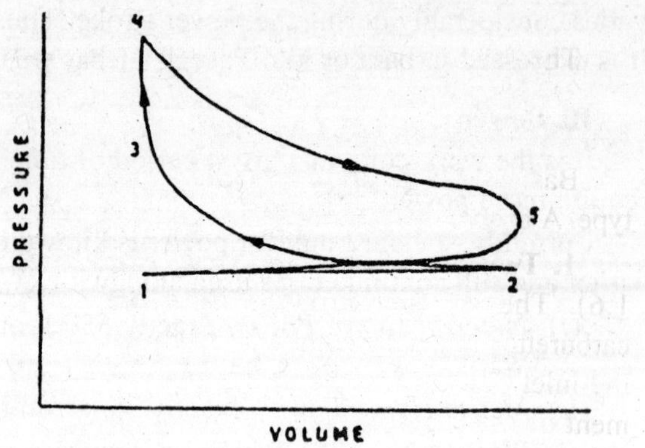

Fig. 1.3. Actual Otto Cycle *p-v* diagram

of two adiabatics and two constant volume lines. At (1) cylinder is full of air. Process (1)—(2) represents adiabatic compression of the air. The combustion takes place at constant volume, from (2) to (3) after which the air expands adiabatically from (3) to (4).

The air standard effective, η, of this cycle is given by

where r = compression ratio

γ = ratio of specific heats.

The actual *p-v* diagram of the cycle differs slightly from the theoretical one and is shown in Fig. 1.3. (1)—(2) is the intake process, (2)—(3) compression and (3)—(4) the combustion process. (4)—(5) represents expansion and (5)—(1) is the exhaust process.

2. Diesel Cycle. Theoretical diesel cycle *p-v* diagram is shown in Fig. 1.4. The popular diesel engine works on this cycle. This is similar to Otto cycle, except that the combustion here takes place at constant

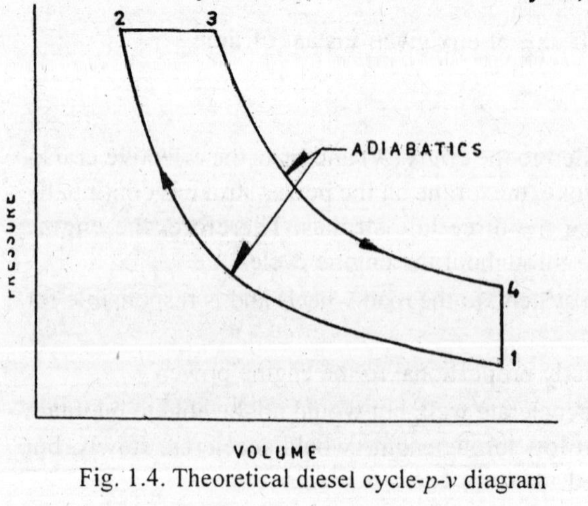

Fig. 1.4. Theoretical diesel cycle-*p-v* diagram

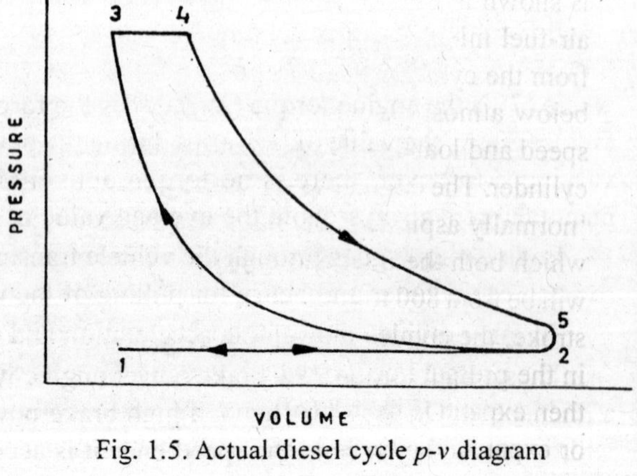

Fig. 1.5. Actual diesel cycle *p-v* diagram

pressure. The air standard efficiency, η, of this cycle is given as:

$$\eta = 1 - \frac{r^{1-\gamma}}{r}\left(\frac{\rho^{\gamma-1}}{\rho - 1}\right)$$

where ρ = volume ratio at cut off

$$= \frac{\text{Volume at (3)}}{\text{Volume at (1)}}$$

The actual *p-v* diagram of a diesel engine is slightly different and is shown in Fig. 1.5.

5. WORKING OF I.C. ENGINE

Basically there are 2 types of internal combustion engines, *viz.*, the two-stroke type and the 4-stroke type. A brief description of these is given below:

1. Two stroke petrol engine (Fig. 1.6). The air fuel mixture from the carburettor enters the cranks case through the inlet port during the upward movement of the piston [Fig. 1.6 (*a*)]. At the same time the mixture in the cylinder is compressed, which is ignited when the piston is just at T.D.C. The combustion takes place and the piston moves imparting motion to the crankshaft. During the down- ward movement of the piston the mixture in the crankshaft is compressed and pushed into the cylinder through the transfer port, which pushes out the exhaust gases through the exhaust port, at the same time filling the cylinder with a new charge [Fig. 1.6 (*b*)]. This process is called cross-flow scavenging. Thus the whole cycle is completed in two strokes, *i.e.* one revolution of the crankshaft.

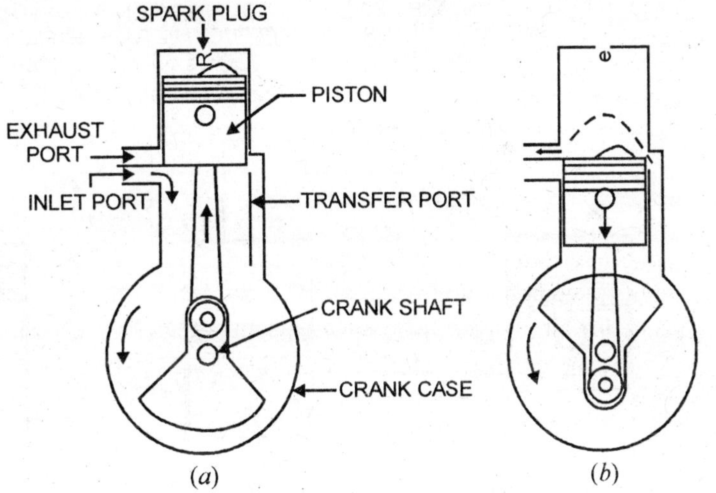

Fig. 1.6. Two-stroke engine

2. Four stroke petrol engine (Fig. 1.7). The cycle of events that take place in a 4-stroke petrol engine is shown in Fig. 1.7, a simplified version of which is given in Fig. 1.8. Fig. 1.8 (*a*) shows the suction of air-fuel mixture in the cylinder during the downward movement of the piston. The piston moving away from the cylinder head creates a pressure reduction or depression reaching maximum of 30 kPa (0.3 bar) below atmospheric pressure at one third of the piston stroke. The actual depression depends upon the speed and load on the engine. This depression is responsible for sucking the air-fuel mixture in the engine cylinder. The engines in which air-fuel mixture is inducted only by the depression in the cylinder are called 'normally aspirated' or 'naturally aspirated' engines. In Fig. 1.8 (*b*) is shown the compression stroke in which both the inlet and exhaust valves are closed and at the end of which the typical cylinder pressures will be from 800 to 1300 kPa (8 to 13 bars) with the engine running under load. Towards the end of combustion stroke, the combustion of the charge ignited by a spark plug, occurs. This generates heat and a rapid rise in the pressure occurs which may reach as high as 6000 kPa (60 bars) under full-load. The burning gases then expand [Fig. 1.8 (*c*)] pushing the piston away from the cylinder head. This is called the power-stroke or expansion-stroke at the end of which the pressure may drop to as low as 400 kPa (4 bars). At the end

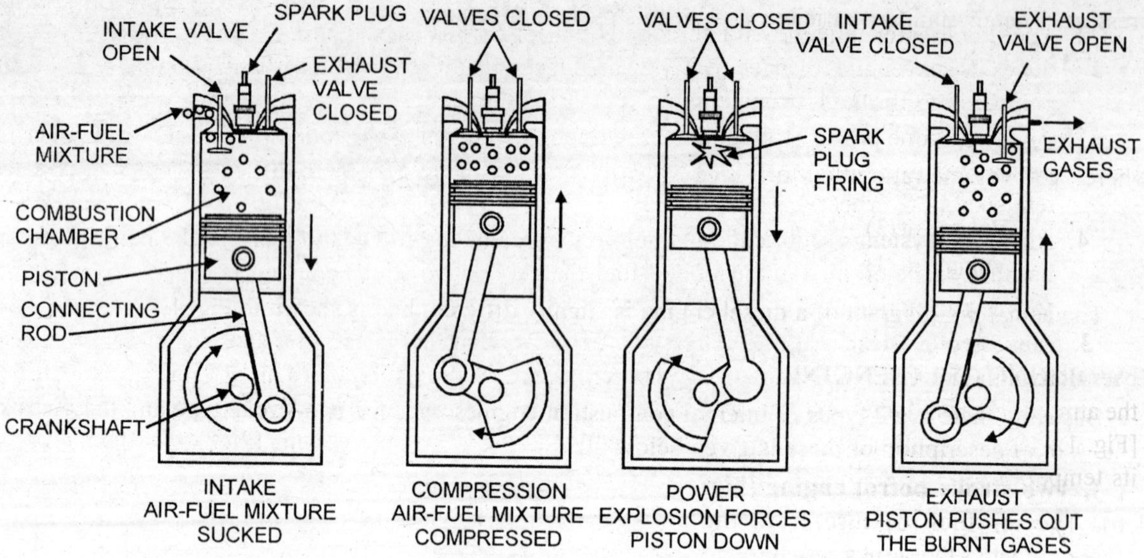

Fig. 1.7. Cycle of events in a 4-stroke petrol engine

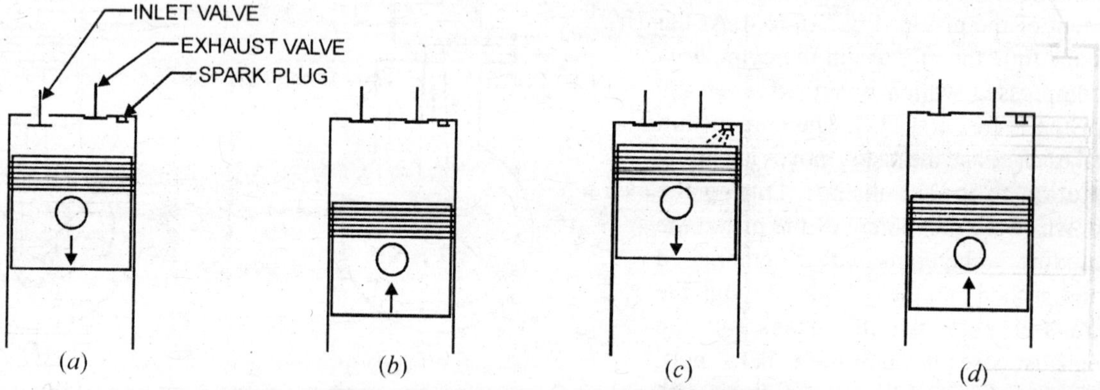

Fig. 1.8. Cycle of events in a four-stroke petrol engine

of the power stroke the inlet valve remains closed but the exhaust valve opens, the piston moves towards the cylinder head expelling most of the burnt gases to the atmosphere [Fig. 1.8 (*d*)]. At the end of this stroke, called the exhaust stroke, the cylinder pressure may drop down to atmospheric or even less. In this way, the entire cycle is completed in four strokes, *i.e.*, two crankshaft revolutions.

Comparison of 2-stroke and 4-stroke engines. Two-stroke engine possesses the following advantages over the four-stroke type:

1. Because of its valveless construction, the two-stroke engine is simpler, more compact, lighter and consequently cheaper to manufacture.
2. Due to twice as many power strokes per revolution, compared to the four stroke type, it delivers smoother torque. However, at lower speeds, this may not happen due to poor scavenging which may give rise to 'four-stroking' or irregular firing.
3. The petroil system of lubrication, which has to be used in case of 2-stroke engines, is simpler.

However, the two-stroke engine has got certain disadvantages also.

1. The 'scavenging efficiency' of the two-stroke engine is poor, *i.e.*, it is not possible to completely purge the exhaust gases from the cylinder and fill it completely with the fresh intake mixture. This results in decreased volumetric efficiency.

2. There is greater tendency for the deposition of carbon on the exhaust ports, which further reduces its scavenging efficiency. This would mean that it requires decarbonization more frequently compared to the four-stroke engine.
3. As there is one power stroke in one revolution compared to the four stroke engine which has one power stroke in two revolutions, the cooling is comparatively more difficult. Thus, it has greater tendency to overheat.
4. As petroil system of lubrication is not very efficient and it tends to increase the harmful exhaust because of the oil mist in the air and fuel charge, the two-stroke engine has become obsolete now for the car engines.

3. Four Stroke Diesel Engine. The cycle of events that take place in a 4-stroke diesel engine has been shown in Fig. 1.9. Unlike the petrol engine which draws in a mixture of air and petrol, only air from the atmosphere is inducted into the diesel engine cylinder, through the inlet valve, during the suction stroke [Fig. 1.9 (*a*)]. This air is then compressed to about 1/22 of its original volume [Fig. 1.9 (*b*)] which raises its temperature.

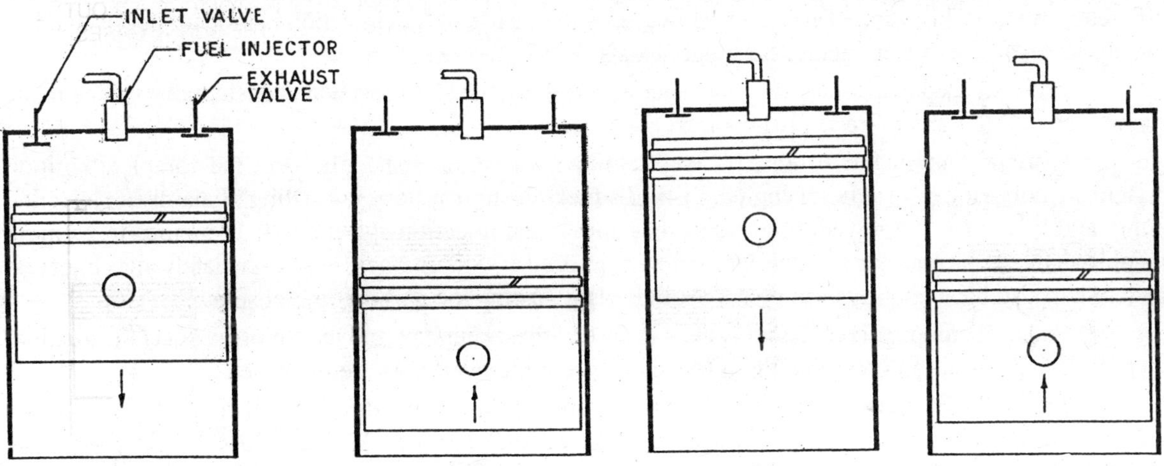

Fig. 1.9. Cycle of events in a 4-stroke diesel engine

The average automotive diesel engine has a compression ratio of 22:1 and a cranking compression pressure of 2.8 MPa. For every 7 kPa of compression pressure, the increase of air temperature in the cylinder would be approximately 1.1°C. Therefore, the compression pressure of 2.8 MPa would raise the cylinder temperature to about 440°C. The ignition temperature of diesel fuel is about 400°C.

When the piston is at the top dead centre, the fuel is injected into the cylinder. The raised temperature of the compressed air is sufficient to ignite the fuel. There is an increase in pressure which drives the piston down giving working stroke [Fig. 1.9 (*c*)]. The exhaust valve opens at the end of the working stroke and the exhaust gases are driven out as the piston ascends [Fig. 1.9 (*d*)].

The combustion in a diesel engine occurs in four stages. These stages are going on simultaneously in different parts of the combustion chamber during injection of fuel.

(*i*) Delay period, in which the fuel injected into the cylinder changes from liquid to vapour phase.
(*ii*) Uncontrolled burning of the fuel already injected into the combustion chamber.
(*iii*) Controlled burning as the injector continues to feed fuel into the combustion chamber.
(*iv*) Afterburning, which is the period for which the fuel remaining in the combustion chamber after the injection stops, continues to burn.

The amount of fuel injected into the combustion chamber determines the power developed in a diesel engine. Increasing the amount of fuel injected increases the engine speed or torque. The maximum amount of fuel depends upon the amount of air inducted in the intake stroke. However, the fuel injected is limited by a governor, which is usually set to provide fuel so as to result in maximum enrichment of about 10:1 by weight.

Comparison of Petrol and Diesel Engines

1. In petrol engine, a mixture of air and petrol is drawn into the cylinder and compressed, whereas in case of diesel engine, air alone is compressed.

2. The compression ratios used in petrol engines are limited to the range 9:1 to 11:1, whereas automotive diesel engines usually employ compression ratios from 15:1 to 24:1. This results in higher thermal efficiency of the diesel engine. Thus a saving of fuel occurs; in diesel engines fuel consumption per horse power output on this account is less by 30 to 40 per cent as compared to petrol engines. In latest designs of diesel engines, the fuel consumption is claimed to be only half as in case of petrol engines of comparable size.

3. There being no throttle valve in the air supply passage of diesel engine, it operates at a higher volumetric efficiency, which results in fuel economy. However, with the use of fuel injection in modern petrol engines, this disadvantage of petrol engines no longer exists.

4. Diesel fuel used in C.I. engines is cheaper and less volatile. This results in considerable saving in the running cost. The risk of fire is also comparatively less.

5. Important components of a petrol engine, like carburettor, spark plug and the complete ignition system are not required in a diesel engine. Instead a fuel injection system, consisting of mainly an injection pump and injector is employed. Experience has shown the injection system of C.I. engines to be more reliable than the carburettor and the ignition system used in S.I. engines. However, lately fuel injection system has also been employed in petrol engines also, making the carburettors obsolete.

6. Because of comparatively lesser speeds in case of diesel engines and their more robust construction, there is reduced wear and tear and the intervals between the overhauls are increased.

7. Compared with petrol engines which often run with stoichiometric air/fuel ratio (14.7:1), diesel engines use very lean air/fuel ratios, which for full load vary between 17:1 and 29:1; while at idling or under no load, this ratio can exceed 145:1.

8. Higher combustion efficiency of diesel engines means less carbon monoxide and hydrocarbons; so ordinarily there is no need of special emission control devices for these, but till recently NO_x emission was a problem is diesel engines as in petrol engines. However, with rapid technical advances, NO_x and particulate emissions in diesel engines have been reduced by 85% since nineteen nineties. Green house gas (CO_2) emissions have also been reduced significantly. Particulate filters have also been developed to limit the particulate matter in the engine exhaust.

Moreover, due to better fuel efficiency, light-duty diesel engines now use 30-40% less fuel than petrol engines of similar power under similar circumstances.

9. With its smooth combustion, a diesel engine offers a high and more stable torque over a wide range for easier operation. In modern common rail direct injection diesel engines, more torque is available even at low engine speeds than the petrol engines of similar size and similar circumstances.

10. The initial cost of a diesel engine is high on account of mainly two factors; firstly, because the engine has to be made more robust owing to the higher pressure involved compared to a petrol engine, and secondly, due to high precision involved in the manufacture of fuel injection pump. However, this higher initial cost is fairly compensated by reduced running costs, the break even point is reached after a running of about one lakh kilometers or even less.

11. As in case of diesel engines only fresh air is drawn and very small time is available for the fuel

injected to form a uniform mixture, about 20% excess air at full load than required for chemically correct mixture has to be inducted so that less CO_2 gas is expelled with the exhaust. Besides CO and HC emission are also less. However, they do tend to generate higher NO_x and PM (Particulate Matter) than petrol engines. Higher NO_x emissions are caused by higher temperature and excess oxygen of the lean combustion process in diesel engines. Higher PM levels result due to incomplete combustion caused by low oxygen levels around fuel droplets. This occurs inspite of the excess air since diesel fuel vapourizes less readily than petrol resulting in the incomplete air/fuel mixing in the injection system. For the same reason, for the same power output under similar conditions, a diesel engine will have to have a cylinder capacity at least 20% greater than the equivalent petrol engine.

12. Another disadvantage of diesel engine was that it ran less smoothly and was more noisy than the petrol engine. However now with the use of high pressure common rail direct injection, the noise level in diesel engines has vastly deceased, even acceptable for luxury class automobiles.

13. Earlier, a governor in a diesel engine was a must to control critical speeds like idling and maximum B.P. However, electronic control has now made this redundant.

14. In diesel engines sometimes, especially during a cold start in winter, the cranking speed may not be sufficient to raise the temperature of compressed air to the self-ignition value. For this reason heater plugs or other starting aids have to be employed.

Example 1.1. *An engine has a swept volume of 300 cm³. If the compression ratio is 8.5 : 1, determine the clearance volume.*

Solution. Swept volume, V_s = 300 cm³

Compression ratio, r = 8.5 : 1

From Eq. (1.3), $8.5 = \dfrac{300 + V_c}{V_c}$ where V_c is the clearance volume.

$$V_c = \mathbf{40\,cm^3}$$

Example 1.2. *An auto engine has a bore of 90 m, stroke of 85 mm and compression ratio of 8.5. The engine is rebored to 2 mm oversize. Calculate (i) original capacity of the engine (ii) percentage increase in capacity due to reboring (iii) compression ratio after reboring.*

Solution.

(*i*) Original capacity, $V_s = \dfrac{\pi}{4}(90)^2 \times 85 = 540747\,\text{mm}^3 = \mathbf{540.747\,cm^2}$

(*ii*) Capacity after reboring, $V_s' = \dfrac{\pi}{4}(92)^2 \times 85 = 565047\,\text{mm}^3 = \mathbf{565.047\,cm^2}$

∴ Percentage increase in capacity $= \dfrac{V'_s - V_s}{V_s} = \dfrac{565.047 - 540.747}{540.747} \times 100 = \mathbf{4.49\%}$

(*iii*) Let V_c = clearance volume of engine in cm³. Thus from Eq. (1.3),

$$8.5 = \frac{540.747 + V_c}{V_c}$$

∴ $$Vc = 72.1\,\text{cm}^3$$

∴ After reboring $$r = \frac{565.047 + 72.1}{72.1} = \mathbf{8.84}$$

6. FIRING ORDER

If in multi-cylinder engines, various cylinders are made to fire one after the other, interference between adjacent cylinders would occur, since these may have overlapping periods which prevents proper distribution of the fresh charge in the inlet manifold. Likewise, overlapping of the exhaust periods in case of adjacent or nearby cylinders would increase exhaust back-pressure which may prevent the combustion products escaping from the cylinders. Therefore, to obtain best engine performance in multi cylinder engines, the occurrence of power strokes in various cylinders is not kept directly one after the other. For example, in 4-cylinder engines it is not kept as 1-2-3-4, but is provided as 1-3-4-2 or 1-4-3-2. Similarly firing order of a 6-cylinder in-line engines is 1-5-3-6-2-4 and in a V-8 engine it is 1-5-4-2-6-3-7-8 (wherein, cylinder nos. 1 and 5 are the front ones). Proper firing order also helps in suppressing certain critical torsional vibrations.

The occurrence of various events in the 5-cylinder engine having firing order as 1-3-4-2 is shown below:

Crank shaft degrees	*Cylinder*			
0	*1*	*2*	*3*	*4*
180	Power stroke	Exhaust	Compression	Suction
360	Exhaust	Suction	Power stroke	Compression
540	Suction	Compression	Exhaust	Power stroke
720	Compression	Power stroke	Suction	Exhaust

The driver must know the correct firing order of the engine so that he is able to connect the H.T. wires correctly to various spark plugs.

7. BASIC ENGINE DESIGN CONSIDERATIONS

7. 1. Effect of Engine Size

The two parameters involved are the diameter, *d* and stroke, *L* of the engine. The effect of doubling each of this is discussed below:

1. The piston head area $A = \frac{\pi}{4}(d)^2$. Therefore, doubling the cylinder diameter would increase the piston head areas (or the cylinder area) four times which means that piston thrust will become four times for the same mean effective gas pressure. This would further mean that for a given piston speed and mean effective gas pressure, doubling the cylinder diameter would increase the power four times.
2. Doubling the stroke length for the same cylinder diameter would double the power.
3. If, for a given crankshaft speed, the stroke is doubled, the piston speed would also be doubled. Or in other words if the piston speed is to remain constant even after doubling the stroke, the crankshaft speed need to be reduced to half.
4. Doubling the cylinder diameter increases the piston thrust four times, while doubling the stroke doubles the crank-throw. Since torque is proportional both to the piston thrust and the crank-throw, doubling both the cylinder diameter and the piston stroke would increase the torque 4 × 2 *i.e.* 8 times.

7.2. Effect of Flywheel Size

In a four-stroke engine, energy is supplied to crankshaft only during the power stroke. During the remaining three strokes, some of this energy is absorbed to overcome the frictional and pumping losses, besides inertia loads due to reversing of the direction of motion of the piston assembly. Due to this there would be a substantial fluctuation in the crankshaft speed in one cycle.

The flywheel on the crankshaft absorbs the excess energy during the power stroke and transfers back some of this energy to the crankshaft during the remaining strokes. Thus flywheel reduces the fluctuation of crankshaft speed during one cycle. However, it never completely eliminates this fluctuation. A cyclic torque curve for a single cylinder engine is shown in Fig. 1.10. The mean torque may be defined as the steady torque which would provide torque load equivalent to the actual fluctuating torque load during one cycle. Thus the area of torque diagram above the mean torque line represents the excess energy stored in the flywheel resulting in increase of its speed, while the area of torque diagram below the mean torque curve represents the energy drawn from the flywheel, causing reduction in the crankshaft speed.

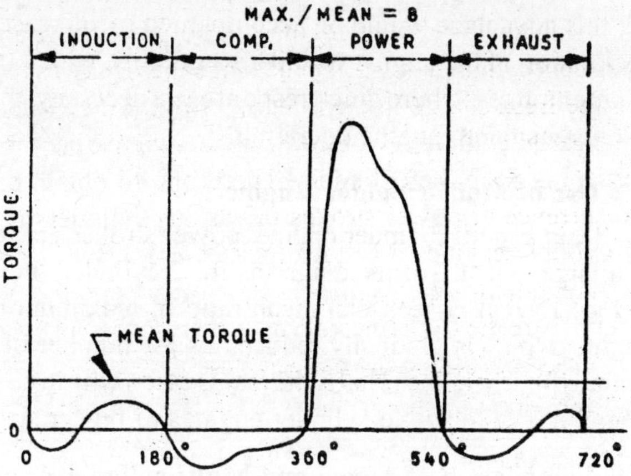

Fig. 1.10. Single cylinder torque diagram

As the size of the flywheel determines the energy absorbed or released by it, it would determine the

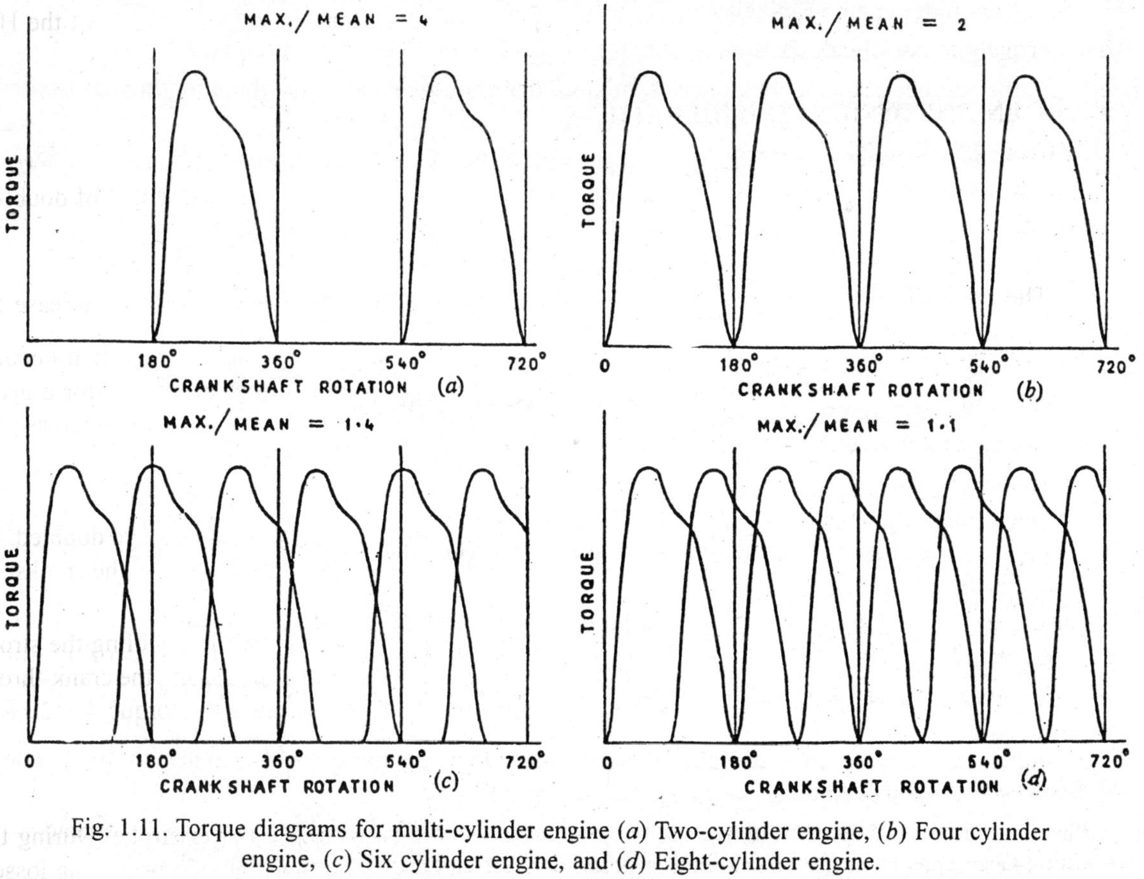

Fig. 1.11. Torque diagrams for multi-cylinder engine (*a*) Two-cylinder engine, (*b*) Four cylinder engine, (*c*) Six cylinder engine, and (*d*) Eight-cylinder engine.

amount of speed variation during a cycle. A large flywheel would reduce the speed fluctuation to minimum, but this advantage would be accompanied by increase of flywheel inertia due to which the acceleration or declaration of the engine would also be slow, which in other words means sluggish response. Thus in case of automobiles, where quick response is a necessity, there is a limit to which flywheel size may be increased to obtain smooth engine operation.

7.3. Use of Multi-cylinder Engine

If, in a multi-cylinder engine, power strokes are staggered, the limitation of flywheel can be overcome to a large extent. Torque diagrams for 2 cylinder, 4 cylinder, 6 cylinder and 8 cylinder engines are shown in Fig. 1.11. It is seen that mean ratio of maximum to mean torque is about 8 in case of single cylinder engine, which is gradually reduced as the number of cylinder is increased and is about 1.1 in case of an eight-cylinder engine. This is the main reason why in case of automobiles, multi-cylinder engines are employed in preference to single cylinder engines of bigger size.

7.4. Comparison of Single and Multi-cylinder Engines

1. Power output is more uniform in case of multi-cylinder engines due to increase in frequency of power impulses with the increase in the number of cylinders.
2. A multi-cylinder engine of comparable size would have smaller cylinders, which means the greater surface to volume ratio leading to more cooling. Therefore higher compression ratios can be used without the occurrence of detonation.
3. Due to lighter reciprocating components and smaller flywheel, acceleration response in case of a multi-cylinder engine is quicker.
4. As the number of cylinders increases, torsional vibrations become a problem.
5. As the number of cylinders increases, mixture distribution in case of carburettor engines becomes difficult.
6. With more number of cylinders, the cost of providing exactly duplicate components in various cylinders becomes excessive.

8. TYPES OF ENGINES

The internal combustion engines for the automotive use may be classified according to the following considerations:

1. Engine cycle
2. Number of strokes
3. Fuel used
4. Type of ignition
5. Number and arrangement of cylinders
6. Valve arrangement
7. Type of cooling.

An automotive I.C. engine may run on Otto cycle or diesel cycle. It may also be classified as a 2-stroke or a 4-stroke engine. The fuel on which the engine runs may also be a basis of its class, *e.g.*, a petrol engine, diesel engine, gas engine etc.

The charge in the engine cylinder after compression requires to be ignited. Two types of ignition systems are in use which classify the engines. These are, the spark ignition and the compression ignition. Correspondingly, the engines are named as spark ignition (S.I.) engines or the compression ignition (C.I.) engines.

The automotive engines may also be classified, based upon the number and arrangement of cylinders, which has been discussed in detail in the subsequent article.

Based upon the arrangement of valves, the engines are classified as side valve type, overhead valve type etc. These types have been explained in detail in the chapter on engine construction.

The engines are broadly divided into two categories depending upon the method of cooling, *viz.*, air cooled engines and the liquid cooled engines. These have been discussed in detail later in Chapter 5.

9. NUMBER AND ARRANGEMENT OF ENGINE CYLINDERS

Based upon the number and arrangement of cylinder, the automobile engines may be classified as:

1. Single cylinder engines
2. Two cylinder engines
   (*a*) In-line vertical type
   (*b*) V-type
   (*c*) Opposed type
3. Three cylinder engines
4. Four cylinder engines
   (*a*) In-line vertical type
   (*b*) V-type
   (*c*) Opposed type
5. Six and eight cylinder engines
6. Radial engines

These will now be discussed in detail.

1. Single cylinder engines. A single cylinder engine has a power stroke every 720° of crankshaft rotation for a four-stroke engine. These are used generally for scooters and motor cycles. It is seen that the maximum size of the single cylinder engine is restricted to about 500-600 c.c., because of the higher unbalance forces, which become difficult to be balanced. Further the weight of the fly wheel required becomes excessive for higher engine sizes.

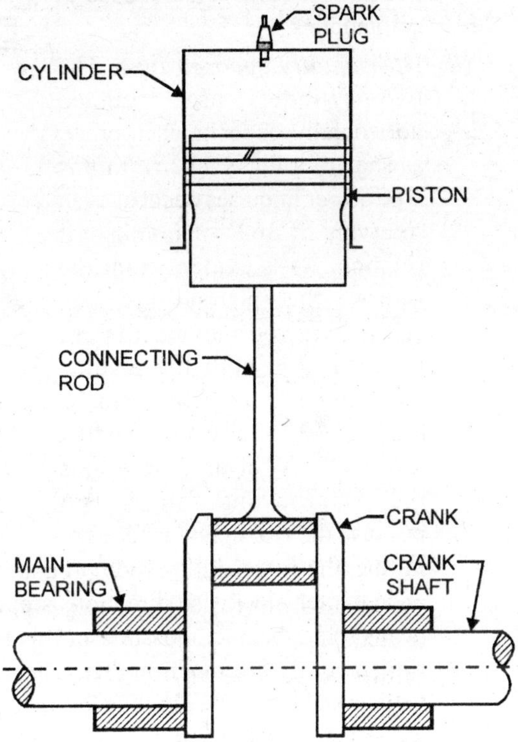

Fig. 1.12. Single cylinder engine

2. Two cylinder engines. This type of engine, apart from providing more power, gives more uniform torque and balancing possible is also better as compared to single cylinder engines. However, in practice two cylinder engines are rarely employed for automotive use. Fig. 1.13 shows three different types of 2-cylinder engines.

(*a*) *In-line type cylinders placed side by side.* This type has two single cylinders placed side by side vertically so that their pistons are in phase. Such an engine will have a power impulse every 360 degrees of crankshaft rotation.

(*b*) *In-line, cylinders 180° out of phase.* In this the two single cylinders are in effect, placed side by side vertically, so that their cranks are 180° out of phase. This type provides good balancing, but the disadvantage is in unequal firing intervals: the spark takes place at 0°, 180°, 720°, 900° and so on.

(*c*) *V-type.* In this two cylinders are placed with their centre lines at some angle to each other. Generally this angle is kept 60°, but in some instances angles from 40° to 90° have also been used. Its main advantage is that it can be made more compact, *e.g.* same crankcase as used in a single cylinder engine may be used here also. Further short crankshaft also means more rigidity, due to which the engine runs more smoothly at higher speeds. Besides, overall length and height of the engine are decreased. The weight/volume ratio is lower compared to an in-line engine. However, there are uneven intervals between power impulses (which take place every 270° and 450° of crankshaft rotation), cost of manufacture is high and balancing more difficult. For these reasons it is not used in automobiles.

(*d*) *Horizontally opposed type.* The two cylinders are arranged horizontally opposite each other, as shown in Fig. 1.13 (*d*). In this type power impulses occur at even intervals of 360° of crankshaft rotation. As is evident from the figure, the piston and the connecting rod movements are identical, *i.e.* both the pistons approach TDC and BDC positions together. This causes the engine to be perfectly balanced, as regards the reciprocating forces. But because the two cylinders are not in line, the forces in the rods produce a rocking couple, which acts alternately in the clockwise and counter-clockwise directions, depending upon the movement of the pistons. Because of the near-perfect balance possible in this type, the highest possible speeds at which the engine can run smoothly are much higher. The engine centre of gravity is also low which contributes to vehicle stability. The obvious disadvan- tage of this type is that the length of the engine increases too much, which has to be placed in the transverse direction in the vehicle. This hinders proper streamlining of the car.

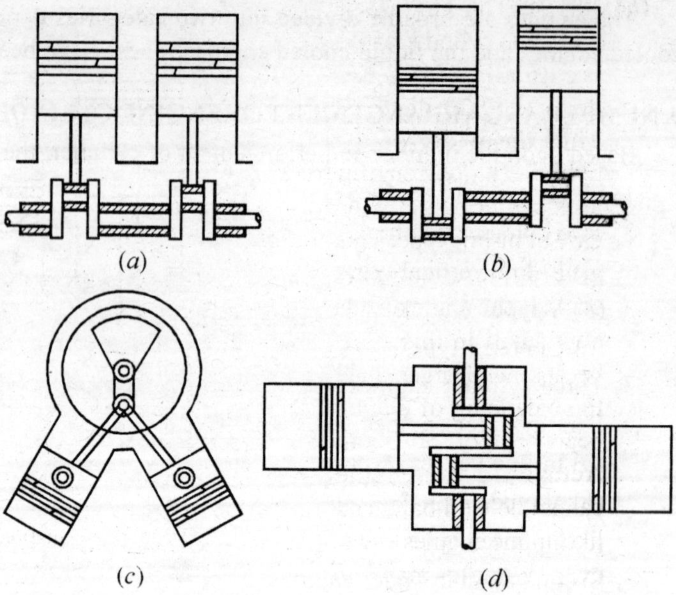

Fig. 1.13. Types of 2-cylinder engines.
(*a*) In line type, cylinders, side by side; (*b*) In line type, cylinders 180° out of phase; (*c*) V-type; (*d*) Opposed type

3. Three-cylinder engines. In three-cylinder in-line engine the power impulse occurs every 240° of crankshaft rotation. These are dynamically balanced, but there remain some unbalanced rocking forces. However, the three cylinders smooth out the cyclic torque adequately so that this has become a good competitor of the more popular four-cylinder in-line engine for small cars. Its other advantages are reduced weight, length and drag besides improved fuel consumption. Maruti 800 vehicles in India employ this engine.

4. Four cylinder engines. In this type, the torque obtained, as compared to a single cylinder engine, is much more uniform because two working strokes per revolution are obtained. Further, the balancing is also better. Apart from this, the maintenance is also easier as compared to the engines with larger number number of cylinders.

(*a*) *In-line vertical type.* This is perhaps the most popular engine for ordinary cars of medium size (from 0.75 to 2.0 litres). Fig. 1.14 shows this type of engine. It has a power impulse every 180° of crankshaft rotation.

The four cylinders are all in a line. The second and third cylinders are in phase, while cylinders 1 and 4 are also in phase, but in the direction opposite to 2 and 3. Thus the reciprocating forces are also nearly balanced. Firing order is 1-3-4-2 or 1-2-4-3. Regard- ing crankshaft torsional wind-up and uneven breathing spacing between adjacent cylinders, both these firing orders have equal merits and demerits.

(*b*) *V-type*. The angle between two cylinder banks is usually from 15 degrees to 20 degrees. Thus the width of the cylinder block remains almost the same as in case of in-line type. This give a more compact engine. The angle of V is kept small in this case, because with larger angle the balancing of engine becomes more difficult. Actually a V-4 engine cannot be well balanced like in-line engines whatever V-angle may be adopted. Therefore, to keep the vibration level within limits in case of V-4 engines, an extra counter-weighted crankshaft is used.

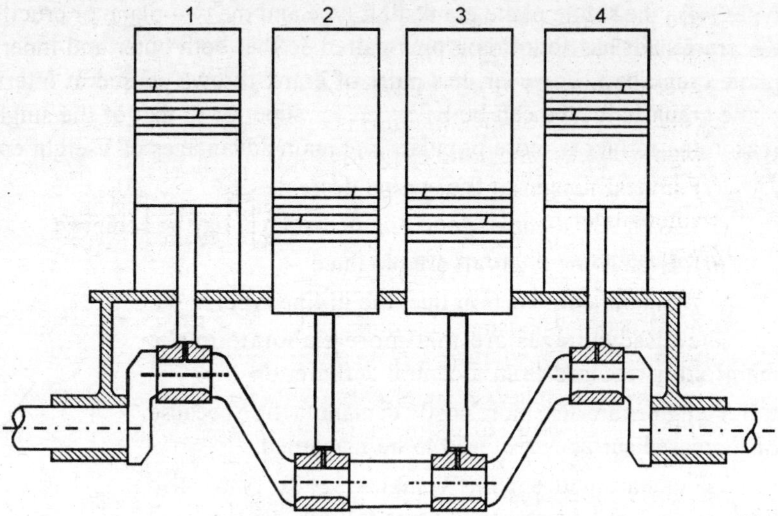

Fig. 1.14. In-line vertical type 4-cylinder engine

(*c*) *Horizontally Opposed flat type.* This type is similar in construction to the 2-cylinder opposed type. Rather it has advantage over the 2-cylinder type that in this perfect balance is possible. Even the rocking couple is also balanced. Its dynamic balance is even somewhat better than the in-line four cylinder engine, although smoothness of torque is the same in the two cases. Thus it may be operated at much higher speeds. Due to this reason, this type provides very compact engine. Its low flat shape makes it particularly suitable for rear-mounted engines. The only disadvantage, as already mentioned, is its excessive length which leaves little space for servicing of cylinder head. The famous Volkswagen car has used this type of engine, which is air cooled. To provide the necessary draft of cooling air, a special centrifugal fan is provided.

The manufacturer of Subaru automobiles, Fuji Heavy Industries Ltd. celebrated the golden anniversary of its horizontally-opposed engine called 'Boxer' in 2016. The Boxer engine is so called because the movement of the engine's pistons resembles the movement of a boxer's fists. Now (2016), every vehicle sold by Subaru (excluding OEM-supplied mini vehicles in Japan) is fitted with a Boxer engine.

5. Six and eight cylinder engines. For higher h.p. and smoother torque, the six and eight cylinder engines have been used. Like the four-cylinder engines, they have also been designed in the form of in-line, V-type and opposed type configurations. Whereas out of the six-cylinder engine types, the in-line type is most widely used, in case of eight-cylinder engines, it is the V-type which is almost universally in use.

In six-cylinder in-line engine, there is a power impulse every 120° of crankshaft rotation. Thus, the crankshaft consists of six cranks arranged in pairs in three planes. For heavy diesel engines, there are seven main journal bearings to support the crankshaft, whereas petrol engines usually have five or even four main bearings. Commonly employed firing order is 1-5-3-6-2-4, whereas equally suitable firing order also used in 1- 4-2-6-3-5.This engine has smoothness of torque and excellent dynamic balance. for this reason its use is quite common for engine sizes larger than 2.5 litres, provided length of engine is not the main consideration.

The angle between cylinder banks in V-engines is generally kept 90°. V-eight type engine is of two types, *viz.*, the single plane crankshaft type and the two-plane or cruciform crankshaft type. In the former, the crankshaft has four crank pins paired so that both outer and inner crank pins are in phase. The two-plane crankshaft, however, has pairs of crank throws phased at intervals of 90°. Out of these, the two-plane crankshaft type can be balanced far superior to that of the single plane crankshaft type, for which reason the former is more popular. The main advantages of V-eight engines are:

(*i*) Overall length of the engine decreases.

(*ii*) With 90° V-angle, good engine balance may be achieved.

(*iii*) Even firing intervals are obtained.

(*iv*) Simple construction than the in-line type.

The disadvantages are that more elaborate intake manifold is required from a central carburettor.

V-engines are also more costly to manufacture because of increased surfaces required to be machined.

V-8 engines with smaller V-angles, say 20°, have also been made. The advantages of narrow V are less torsional vibrations and increased stiffness. However, the valve-operating mechanism is complicated.

6. **Radial Engines.** In this case the various cylinder are arranged in radial directions. about the crankshaft which is placed at the centre (Fig. 1.15). The number of cylinders generally used is odd, say 5, 7, 9 etc., to obtain uniform firing intervals.

This type of engine is compact and gives higher B.H.P./ weight ratio. The construction is also simpler. However, this engine has a very large frontal area, which creates difficulty in streamlining the vehicle shape.

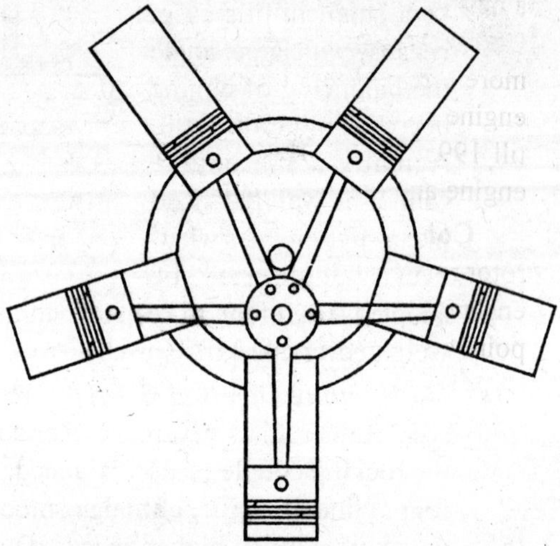

Fig. 1.15. Radial engine

10. NUMBERING OF ENGINE CYLINDERS

In case of in-line engines, cylinders are numbered from the front to the rear, number one cylinder being at the front [Fig. 1.16 (*a*)]. For numbering of V engines, two different approaches have been adopted and both are in vogue. In one system (adopted by Ford Motor Co.) one bank cylinders are numbered one to four for V-8 engines, with number one at the front, whereas the other bank cylinders are numbered five to eight with number five at the front [Fig. 1.16 (*b*)]. In the second system (used by General Motors), however, the cylinders are numbered in the same order in which the connecting rods are attached to the crankshaft [Fig. 1.16 (*c*)] starting with number one at the front of the crankshaft. As with all parts of the automobile, the left and the right banks are as viewed from the driver's position.

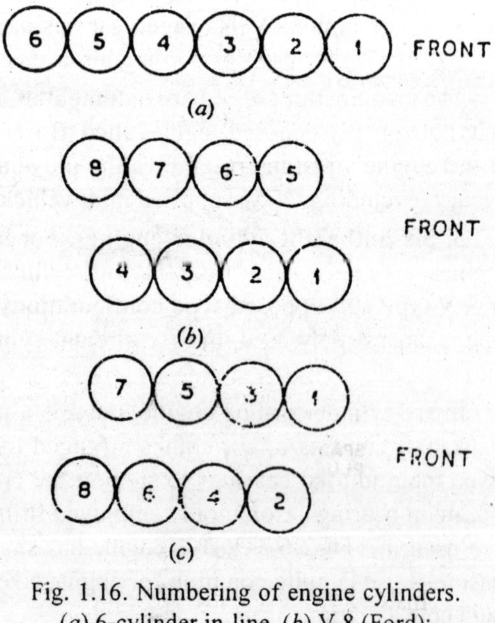

Fig. 1.16. Numbering of engine cylinders. (*a*) 6-cylinder in-line, (*b*) V-8 (Ford); (*c*) V-8 (General Motors)

11. SQUARE ENGINE

This is an engine having stroke equal to its bore. In the ordinary engines the stroke/bore ratio is more than 1.

For the same crankshaft speeds, the square engines have lower piston speeds than the corresponding engines of larger stroke. They develop more power and consume less fuel as compared to the longer stroke engines. The examples are the Vauxhall and the Velox square engines.

Engines with stroke/bore ratio less than 1 are called *oversquare,* while with stroke/bore ratio more than 1 are called *undersquare.*

12. WANKEL ENGINE

This engine works on ordinary Otto cycle, but differs from the reciprocating I.C. engine in that the piston in this case undergoes rotary instead of reciprocating motion. This design of Felix Wankel has provided a new challenge to the existing reciprocating engine. This has been developed by N.S.U. of Germany and Curtiss Wright Co. of U.S.A. for commercial purposes. However. research is still going on to make it more efficient and to solve the sealing problem more effectively. Mazda of Japan introduced its Wankel engine sports car, RX-7 in 1979, which was the most successful rotary engine-powered car and was sold till 1995. Mazda R-8, a new car from Mazda has rotary engine called the Renesis, which is a two-rotor engine and has been named the international engine of the year 2003.

Construction. Fig. 1.17 shows in simplified from the construction of the Wankel engine Three-lobed rotor rotates in the housing, which is epitrochoidal in shape. This combination of housing and rotor shapes ensures that they remain in contact with each other throughout the rotation. Epitrocnoid is the locus of a point on the radius of a circle, which rolls along the circumference of another circle.

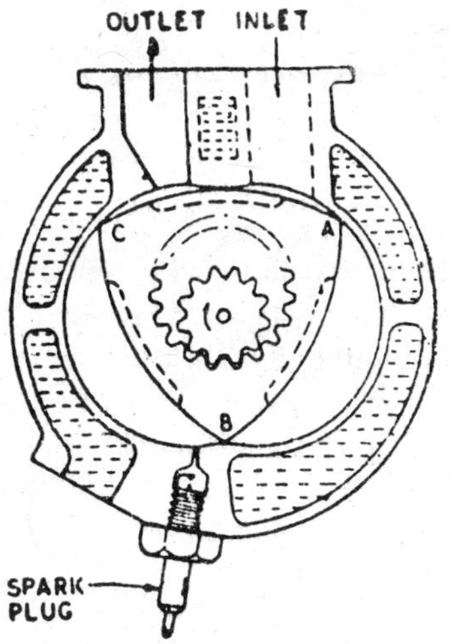

Fig. 1.17. Wankel Engine

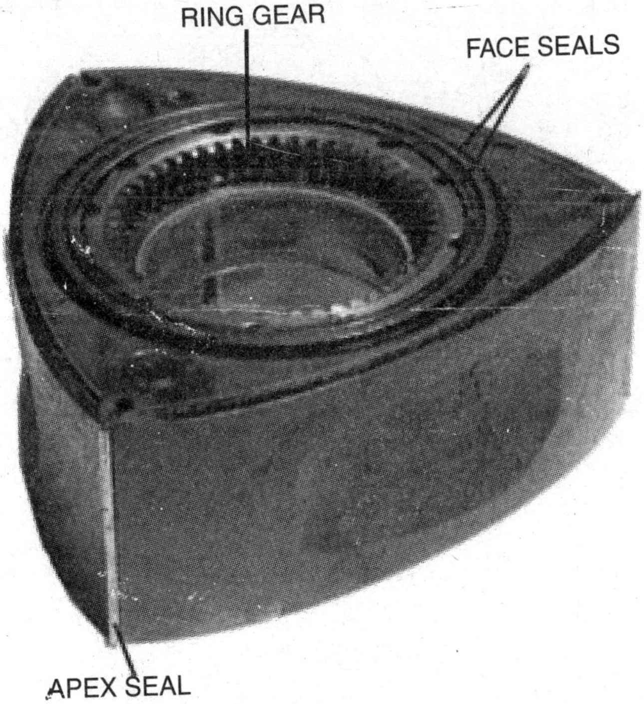

Fig. 1.18. Rotor of Wankel Engine

Instead of valves, there are the inlet and the exhaust ports. The housing is surrounded by water jackets for cooling. The rotor is having internal teeth cut into the centre of one side (Fig. 1.18) and rotates about the pinion in an epicyclic manner. The pinion is fixed to the housing. However, the rotor is mounted on an

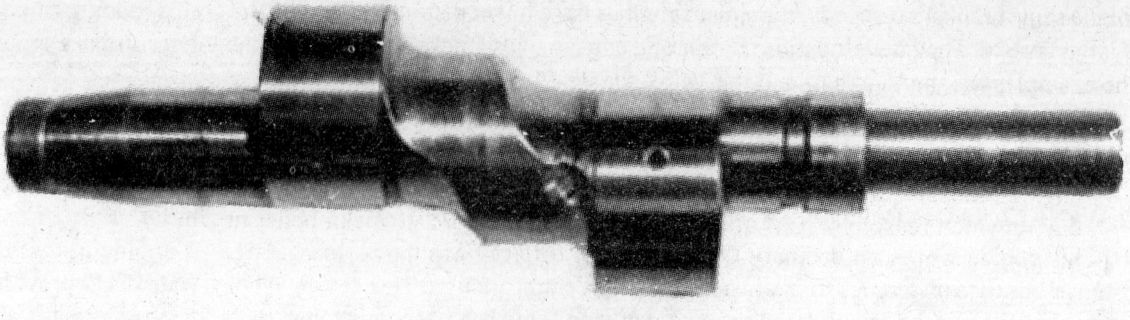

Fig. 1.19. Output shaft of a two-rotor engine

output shaft having eccentric lobes, one for each rotor (Fig. 1.19). Each rotor is mounted over one of these lobes. The lobes provide the eccentricity and act like crankshafts in a reciprocating engine, causing the shaft to spin, thus providing torque.

Operation. To understand the operation of this engine, consider Fig. 1.20 (*a*) to (*d*) showing various positions of the rotor in sequence. In Fig. 1.20 (*a*), it is seen that referring to side CA, the induction process is going to start (stage 1), which gets completed in stage (4) in Fig. 1.20 (*d*). Next consider side AB of the rotor in Fig. 1.20 (*a*), where the compression process just starts and it gets completed through stages (5),

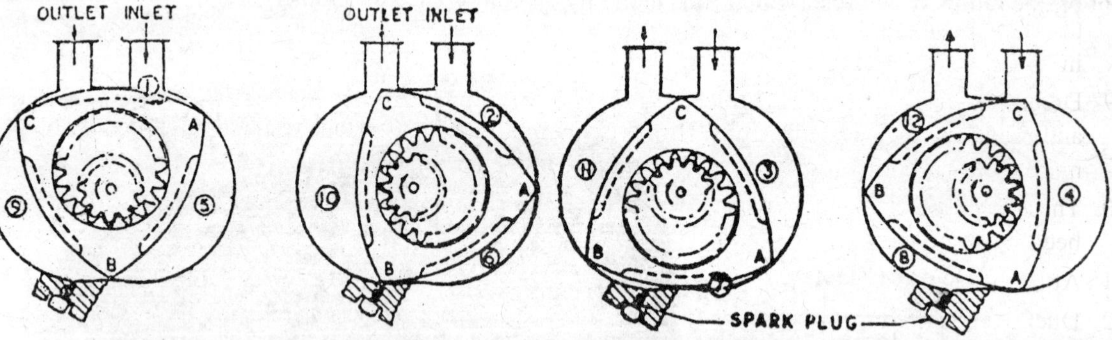

(*a*) (*b*) (*c*) (*d*)

Fig. 1.20. Different position showing the operation of a Wankel Engine

(6) and (7). At stage (8), the spark plug fires and the expansion process starts, which proceeds through stages (9) and (10) (side BC). This is followed by the exhaust process in stages (11) and (12). It is thus observed that one complete cycle of operations is completed in one revolution of the rotor. It may, however, be appreciated that similar processes are also taking place at the same time on the other rotor sides and hence it may be easily visualized that one rotor revolution competes three Otto cycles, which is thrice as much compared to a two-stroke engine and six times compared to a 4-stroke engine. Herein lies the secret of the compact size of the Wankel Engine.

Wankel engine can have more than one rotor also. Units with a maximum of four rotors have been produced. However, the most common has been the twin rotor Mazda engine. The principle of working remains the same through, irrespective of the number of rotors used.

Comparison with Conventional Engines

1. For the same power output Wankel engine is smaller in size and weight. Weights as low as 400 gm per kW brake power have been attained.
2. The balancing is easier, because Wankel engine does not contain any reciprocating parts, due to

which reasons it runs very smoothly. The delivery of power is also smoother. It can be seen that in a Wankel engine, the working stroke lasts for 270 degrees of the output shaft rotation, whereas in the reciprocating engine, the working stroke lasts only 180 degrees in two-revolutions, *i.e.* only 90 degrees of the each revolution of crankshaft.

3. The construction is simpler in that number of working parts is far less as compared to the conventional engines. For example, instead of the valves and the valve-operating mechanism, there are only ports, due to which reasons it runs quietly. Fewer moving parts also mean better reliability.
4. Volumetric efficiency of the Wankel engine is higher and the pumping loss lower, due to the absence of choking valves.
5. On account of the quasi-overlap of the power strokes, the Wankel engine is very quick to react to throttle changes and is able to deliver quickly, a surge of power whenever required, especially at higher speeds.
6. Wankel engine housing is made of aluminium having higher coefficient of thermal expansion than cast iron or steel, which ensures that even a severely overheated Wankel engine cannot seize, contrary to in a reciprocating piston engine.
7. In Wankel engine, there is more time to complete combustion due to 50% longer stroke duration compared to a reciprocating engine. This makes it more suitable for direct injection.
8. In the Wankel engine, the rotor spins at one-third the speed of the output shaft, which means that the main moving part of the Wankel engine moves much slower than in a piston engine. This results in lesser wear of the rotor and better reliability.
9. Due to large cooling surface around the combustion chamber, it can operate on low-octane petrol and produces low *NOx* emissions. However, the same cooling surface results in the production of more hydrocarbon emissions and also in more fuel consumption.
10. The main obstacle in the development of the Wankel engine was the sealing problem, which has been largely overcome by now.
11. At lower speeds the torque produced is less as compared to the conventional engine.
12. Due to the low compression ratio and long shape of the combustion chamber, the thermodynamic efficiency is reduced which results in more fuel consumption compared to a piston engine.
13. The trailing side of the Wankel engine's combustion chamber develops a squeeze stream which pushes back the flame front. Due to this, in case of conventional one or two spark plug systems, the flame front is prevented from propagating to the trailing side, especially at mid and high engine speeds. This results in more carbon divide and unburnt hydrocarbons in the Wankel's exhaust. Mazda 26 B engine avoided this by using 3-plug system. 'Renesis' avoided this with a side exhaust, due to which the unburnt mixture cannot escape.
14. Frequent spark plug changes are required because of ignition troubles experienced with ordinary coil ignition system. However, this has been remedied by using transistorised ignition.
15. As the number of Wankel engines produced in much less compared to the piston engines, production costs are higher.

Mazda's Renesis hydrogen rotary engine (2006) which is the world's first rotary hydrogen engine, maintains a natural driving feeling similar to that of internal combustion engines. This engine employs direct injection with electronically-controlled hydrogen gas injectors. In this, air is drawn from a side port and hydrogen is injected directly into the intake chamber with an electrically-controlled injector installed on the top of the rotor housing. Moreover, a gas injector premixing is installed on the intake pipe which enables it to use direct injection and premixing, combined, depending on driving conditions, which results in optimal hydrogen combustion.

The engine has a dual-fuel system which allows it to run on both hydrogen as well as petrol. When the system runs out of hydrogen fuel, it automatically switches to petrol. Alternatively, the driver can also shift from hydrogen to petrol manually. When it runs in the petrol mode, fuel is supplied from the same petrol injector as in the standard petrol engine.

For emission control, a three-way catalytic converter along with lean burn and Exhaust Gas Recirculation (EGR) are employed. The catalytic converter is the same as used in a standard petrol engine. NOx is mainly reduced by lean burn at low engine speeds and by EGR and catalytic converter at high engine speeds.

The advantages of this engine are:

1. It gives out zero CO_2 emissions.
2. Performance and reliability same as that of the petrol rotary engines.
3. Having a dual-fuel capability, it is highly convenient for long distance journeys and in areas with no hydrogen fuel supply.

Mazda Premacy Hydrogen RE Hybrid uses the RX-8 Hydrogen RE's duel fuel system plus a synchronous motor of 110 kW and Lithium ion battery, to make it a hybrid. It gives 40% more power and better acceleration than the plain RX-8 Hydrogen RE. The 200 km range is also double.

13. ATKINSON CYCLE AND MILLER CYCLE ENGINES

Atkinson cycle engine was developed by a British engineer James Atkinson at around the same time as German engineer Nikolaus A. Otto's famous Otto cycle engine. Atkinson cycle engine was quite similar to Otto cycle engine. The main difference was that in this the compression stroke was shorter than the expansion stroke; as a result compression ratio was decreased and expansion ratio was increased. This was possible by means of a special complex linkage between the piston and the flywheel, quite different from simple connecting rod and crankshaft of conventional engines. The advantages of such an engine were:

1. Decreased compression ratio made the engine less prone to detonation.
2. Increased expansion ratio meant longer power stroke and reduced heat wasted in exhaust, leading to greater engine efficiency and hence lesser fuel consumption.
3. Due to special linkage, all the four strokes in the Atkinson cycle engine were completed in one rotation of the flywheel, which meant that valves could be operated directly from the flywheel shaft, obviating the necessity to have a separate camshaft for this purpose.

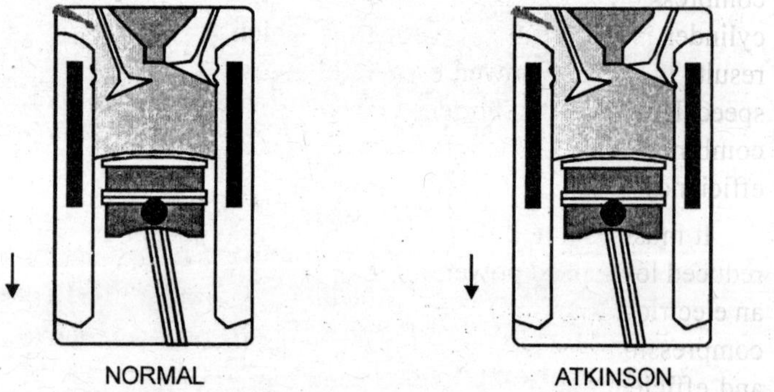

(*a*) Intake- No difference

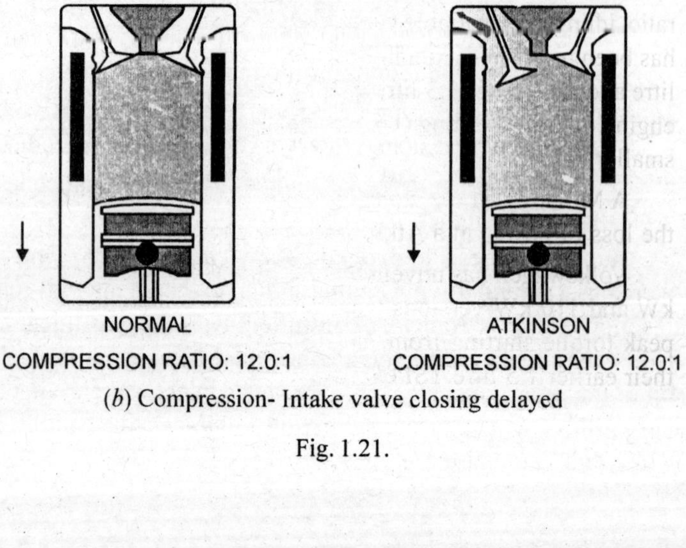

(*b*) Compression- Intake valve closing delayed

Fig. 1.21.

However, the above advantages were also accompanied by following disadvantages:

1. The complexity of the piston-flywheel linkage.
2. Increased pumping and frictional losses resulting in loss of power.

Due to above reasons, the original Atkinson cycle engine could not compete with the Otto cycle engine end was soon obsolete.

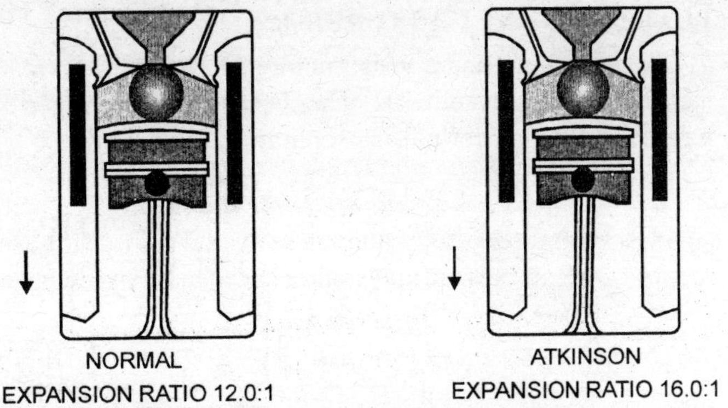

(*c*) Power- Expansion ratio increased

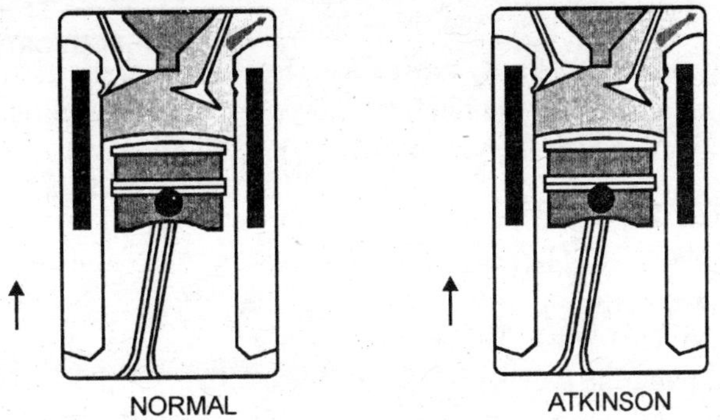

(*d*) Exhaust- No difference

Fig. 1.21. Comparison of Atkinson and Otto cycles

Although original Atkinson cycle engine failed, yet its air cycle is being widely used presently, due to the fact that now instead of the complex linkage, it uses the variable valve timing, which is far cheaper and easier. In a modern Atkinson engine, intake valve is held open a little too long so that when the piston moves up for compression, some of the air-fuel mixture can escape back into the intake manifold, thus lowering the compression ratio. This also means that the cylinder is never filled completely, which results in reduced power, especially at low speed. However, the shortened intake stroke combined with full-length expansion stroke squeezes more work out of even reduced fuel, increasing efficiency. Fig. 1.21 (*a*) to (*d*) illustrated the differences between a modern Atkinson cycle and Otto cycle.

It must be noted that even modern Atkinson cycle engine is not suitable for common cars due to its reduced low-speed power output. However, since in a hybrid car, the low-speed operation is carried on by an electric motor, ralter than the I.C. engine, Atkinson cycle engine is ideally suited. In most such engines, compression ratio is kept at highest possible value from detonation considerations, so as to optimize power and efficiency.

Atkinson cycle, with many further improvements such as reshaped intake port, higher compression ratio, idling stop, variable value Timing intelligent Electric (VVT-iE) and a cooled Exhaust Gas Recirculation, has been used by Toyota in their two modern extra fuel-efficient small petrol engines, *i.e.*, a 3-cylinder 1.0 litre and 4-cylinder 1.3 litre (2015), which is being deployed in their new hybrid cars. The smaller 1.0 litre engine delivers 78 mpg (US) in the Toyota Aygo, giving 30% improvement, compression ratio used in the smaller engine is 11.5, while in 1.3 litre engine, it is kept higher, *i.e.*, 13.5.

A Millar cycle engine is basically an Atkinson cycle engine with a super charger, which makes up for the loss of power in a Atkinson cycle engine, at low speeds.

Volkswagen has unveiled it 1.5 litre Miller cycle engine at the 2016 Vienna Motor Symposium, the 96 kW and 110 kW versions were released in the market in late 2016. VW claims that its new engine gives peak torque starting from just 1300 rpm and also improves fuel economy by litre/100 km compared to their earlier 1.5 litre TSI engine.

14. AUTOMOTIVE GAS TURBINE

A gas turbine unit consist mainly of 3 members, viz., the compressor, the combustion chamber and the turbine. For automotive use it has been found advantageous to have instead two turbines *viz.*, the compressor turbine and the power turbine separately.

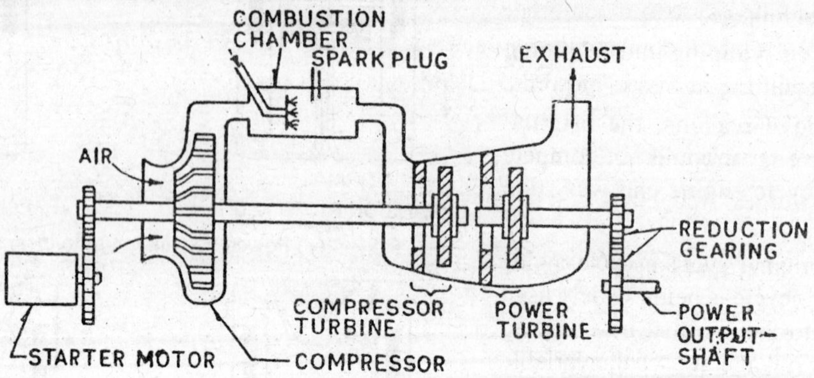

Fig. 1.22. Principle of operation of a gas turbine for automotive use

Ref. Fig. 1.22, the air from the atmosphere is drawn into the compressor which may be centrifugal or axial in design. The air gets compressed in the compressor, whereafter it enters the combustion chamber. Fuel is sprayed into the combustion chamber. The chamber is also fitted with a spark plug, which ignites the air-fuel mixture. The hot combustion gases under pressure then pass through blades of the compressor turbine and the exhaust from the compressor turbine enters the power turbine where it expands to give power. The compressor turbine is so designed that it produces power which is just sufficient to run the compressor. The motive power is generated by the power turbine. As the turbine speeds for efficient operation are very high, a reduction gear has to be used. For starting, an electric motor run by batteries is used.

To increase the thermodynamic efficiency of the turbine unit a regenerative heat exchanger is used (Fig. 1.23). The heat exchanger serves to utilize the heat in the exhaust gases by transferring the same to the compressed air entering the combustion chamber.

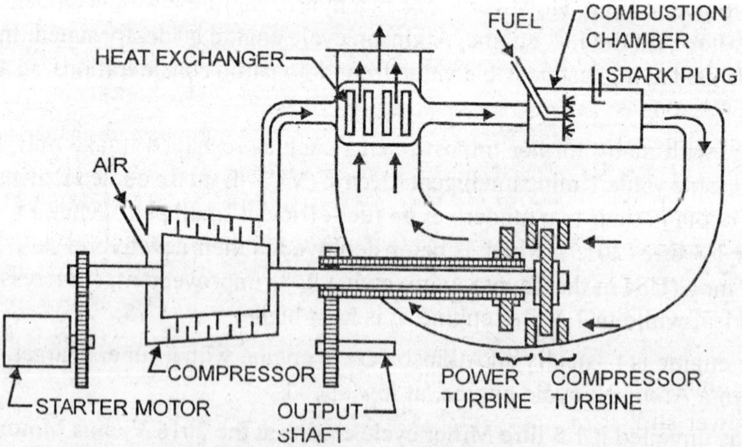

Fig. 1.23. Use of heat exchanger with automotive gas turbine

Advantages of gas turbine. The gas turbine possesses many advantages over the reciprocating engines in current automotive use, because of which it is finding increasing use as a power plant for commercial vehicles. These advantages are:

(*i*) Less wear because of absence of rubbing parts like pistons in I.C. engines.

(*ii*) Better control of vibrations due to absence of reciprocating parts.

(*iii*) Any distillate fuel can be used.

(*iv*) Starting is very easy.

(*v*) No cooling system is required.

(*vi*) Lubricating oil consumption is far less, about 2% only of the corresponding I.C. engine.

(*vii*) Simpler transmission required. The clutch is not required and the gear box is needed only the reversing.

(*viii*) More compact size.

Disadvantages. Accompanying the above advantages of the gas turbine, there are many disadvantages also:

(*i*) Very high fuel consumption (about double that of corresponding I.C. engine).

(*ii*) The very high speeds at which the turbines have to run introduce complications regarding reduction gears and controls.

(*iii*) Absence of engine braking and acceleration.

(*iv*) Very large quantities of exhaust gases may be a nuisance in the cities.

(*v*) For higher thermal efficiency, very high operating temperatures are required, which necessitates the use of non-metallic components, say, of ceramics, which are otherwise brittle.

15. HOMOGENEOUS CHARGE COMPRESSION IGNITION (HCCI) ENGINE

15.1. Introduction

This is an internal combustion engine which combines some of the combustion strategies of both S.I. and C.I. engines to get high fuel efficiencies of C.I. engines (diesels), but without the NO_x and soot emissions they produce. In such an engine, well-mixed fuel and oxidizer (air) are compressed to the point of auto-ignition. Thus it has the characteristics of combustion of both the spark ignition (petrol) as well as the compression ignition (diesel) engines.

An HCCI engine is based on the four-stroke Otto cycle. On the intake stroke, fuel is injected into each cylinder's combustion chamber by means of fuel injectors mounted directly in the cylinder head. The air induction is independent, from the air intake plenum. Towards the end of the intake stroke, air and fuel have been fully introduced and mixed in the combustion chamber. Now, as the piston starts moving up during the compression stroke, heat begins to build in the combustion chamber and towards the end of this stroke sufficient heat accumulates to cause spontaneous combustion of the fuel/air mixture, *without any spark* (as in a C.I. engine), forcing the piston down for power stroke.

HCCI engine employs a homogenous, *i.e.* well-mixed and a very lean (high proportion of air to fuel) mixture which is compressed until it ignites automatically, *i.e.*, without any spark. Unlike (SI or CI) engines, here the ignition occurs at several places at a time [Fig. 1.24 (*c*)], which makes the air/fuel mixture burn almost spontaneously, without any initiator of combustion and any flame propagation, which eliminates heterogeneous air/fuel mixture regions. The entire fuel/air mixture is burnt simultaneously producing more power but using much less fuel and releasing far less emissions.

After the power stroke, piston reverses direction to start the exhaust stroke. But the exhaust valves are closed before the piston reaches T.D.C., *i.e.*, before all the exhaust gases are evacuated. This heat of the trapped exhaust gases is preserved and a small quantity of fuel is injected into the combustion chamber for a pre-charge before the next intake stroke begins, which helps to control combustion temperatures and emissions.

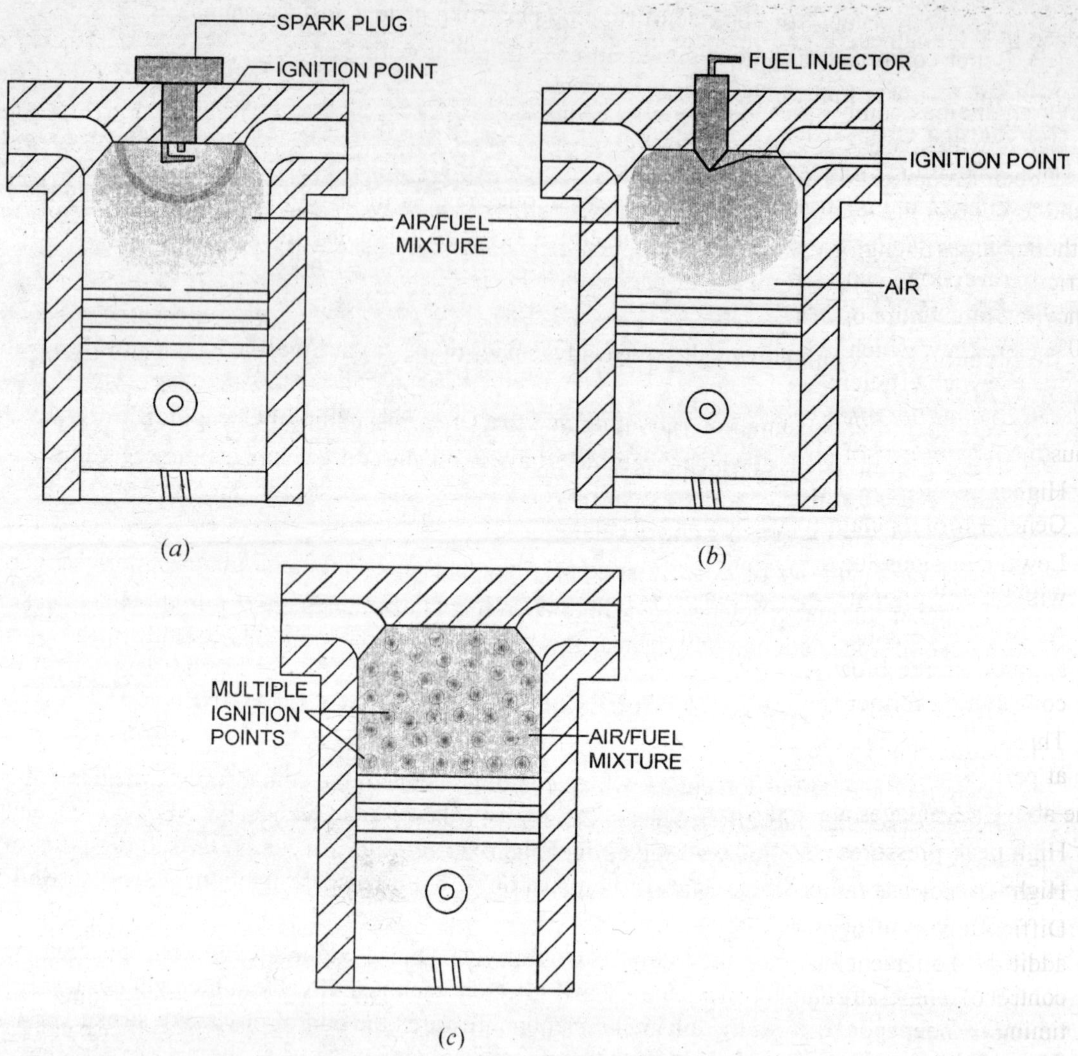

Fig. 1.24. Comparison of combustion in various types of I.C. engines, (*a*) S.I. engine, (*b*) C.I. engine, (*c*) HCCI engine

15.2. Comparison with SI & CI engines

The S.I. engine has relatively clear exhaust, but is very inefficient at part-loads since the only way to vary the engine torque is to open or close the butterfly valves which regulate the amount of air sucked in before it is mixed with fuel and fed to cylinders. Thus at part loads, it is this 'throttling' effect which wastes so much energy, leading to poor engine efficiency. Of course, the positive point is that air/fuel mixture is well-mixed before combustion starts. However, there is no 'throttling' in case of C.I. (diesel) engine; the torque- variation is achieved simply by varying the amount of fuel injected into the engine. Thus at part-load, it does not have the S.I. engine's throttling losses. Further higher compression ratios can be used in a C.I. engine, resulting in greater efficiency, while the same cannot be done in case of S.I. engines because of detonation considerations.

On the other hand in case of C.I. engine, higher compression also results in higher temperatures giving rise to increased *NOx* emissions. Moreover, since the air and fuel are not completely mixed before ignition takes place in C.I. engines, there is a lot of unburnt fuel going into the exhaust in the form of particulate matter.

HCCI engine has combustion temperatures lower than either an S.I. (petrol) engine or C.I. (diesel) engine, due to which very small amounts of NO_x are produced. Moreover, due to thorough mixing of the fuel and air before combustion, there is almost no particulate matter going in the exhaust. However, emissions are found to contain higher amounts of HC and CO, which have to be treated with a catalytic converter.

Although a petrol engine can achieve upto 30% efficiency when the throttle is wide open and there is no restriction to air flow, the efficiency at part load is far less due to throttling of air flow, the average efficiency over the entire operating range being about 17%. An HCCI engine on the other hand can have upto 40% efficiency, which is a big improvement. In diesel engines, however, there is no throttling of the charge due to which efficiency is already high, which is also increased further by high compression ratio. *As such HCCI does not offer much improvement in efficiency over a diesel engine.*

Thus the *advantages* of an HCCI engine can be summarised as:

1. Higher efficiency compared to S.I. engines, due to lean combustion & no throttling of the charge. General Motors claims upto 15 percent fuel savings, while meeting current emission standards.
2. Lower emissions due to homogeneous mixing of charge. In particular, NO_x levels are almost negligible without any after-treatment (due to lower combustion temperatures).
3. As in HCCI engine, diesel efficiency is approached without high compression ratios of a diesel engine, engine blocks, heads and other components can be much lighter and less expensive, contributing further to fuel savings of the vehicle due to lesser weight.
4. Throttle-less induction eliminates frictional pumping losses occurring in conventional S.I. engines at part loads and lower speeds.

The above advantages are also accompanied by following disadvantages:

1. High peak pressures, requiring stronger engine construction.
2. High heat release rates.
3. Difficulty of control. Control software should be able to handle weather and altitude worldwide, in addition to different road and driver conditions. Flameless combustion requires much more complex control of temperature and mixture composition, compared to traditional methods of adjusting spark timing and intensity. Besides, fuel injection, valve timing and lift need also highly precise control. For all this, reliable and accurate cylinder pressure sensors are required.
4. Limited power range.
5. Higher pre-catalyst emissions of HC and CO.

15.3. Current scenario

The main problems due to which *no production car has been produced to date with an HCCI engine,* are:

1. It is very difficult to control HCCI over a vide range of speeds and loads, since unlike petrol or diesel engines it has no external trigger like a spark plug or fuel injector to initiate the combustion and to control its duration. However, supercomputers and simulation software developed or being further developed are helping the engineers a lot in developing a practical solution to this. Some of the approaches being researched upon are:

(*i*) Variable compression ratio.
(*ii*) Variable induction temperature.
(*iii*) Variable exhaust gas percentage in EGR system.
(*iv*) Variable valve actuation.

2. Still the manufactures have not been able to find a catalyst which works as good as in case of a petrol engine, but also at much lower temperatures and also at a reasonable cost.

The manufacturers are, however, already coming up with *concept hybrid cars* with HCCI, wherein the engine runs in HCCI mode only at part-load/speed conditions and as an S.I. engine at full or near-full load/speed conditions. Two prominent engines launched in 2007 are:

1. General motors have show-cased 'Saturn Aura' and 'Opel Vectra', each equipped with a 2.2 litre HCCI engine. The engine operated in HCCI mode upto 55 mph (88 kph) and partial loads. GM uses spark ignition when starting a cold HCCI engine to produce the heat required for warming the exhaust catalyst and to enable HCCI operation. Also, at higher speeds or heavier loads, the engine runs with normal spark ignition, since HCCI engine runs best at relatively constant, partial-load conditions.
2. Mercedes-Benz presented at Frankfurt Motor Show 2007, their 'F 700' concept car featuring their 'Dies Otto' engine which is also a combination HCCI/SI design, but also has a turbo-charger. This is a 1.8 litre, 4 cylinder engine which is claimed to give a performance level of a similar current S-class car with a 3.5 litre naturally aspirated V6 petrol engine or a 3.0 litre V6 turbo diesel. Its fuel consumption is only 5.3 litres of petrol per 100 km and CO_2 emissions of only 127 g/km, very low for a vehicle of this class.

16. GASOLINE DIRECT INJECTION COMPRESSION IGNITION (GDCI) ENGINE

This is a US Department of Energy funded project undertaken by Delphi and Hyundai with an aim to develop a practical four-cylinder IC unit using RON 87 regular gasoline (petrol) which may be installed in a vehicle, rather than a single-cylinder lab prototype. Keeping in mind the work done by General Motors, Volkswagen and Mercedes Benz on HCCI engine, the GDCI team adopted a different approach as follows:

1. Instead of homogeneous discharge compression ignition process, a new partially premixed combustion process without switching of combustion mode (as in HCCI) was adopted.
2. 1.8 litre I.C. unit was designed with both a supercharger and a turbocharger. The clutched Eaton supercharger was placed on the intake side for low speed boost, which cuts off at about 2500 rpm. The Borg Warner turbocharger was on the exhaust side for high-speed boost. Viariable geometry design of the turbocharger enables creation of back pressure and internal EGR when required. Each of the supercharger or the turbocharger can produce independently, maximum boost of 2.5 bar.
3. No spark plugs were used in clear contrast to HCCI engine, since here no big-bang combustion is used.
4. As here the fuel is injected quite late and under high compression ratio (14.85:1), there are high temperature in the cylinder and a lot of EGR, partially premixed compression ignition delays the combustion and controls the burning very effectively compared to the HCCI.
5. GDCI engine also used a Delphi electric cam timing system with brushless DC motors which is much faster and accurate than the hydraulic camshaft phasers used in HCCI engines.
6. GDCI engine also employs a cam-in-cam design which makes the exhaust valve open a second time during intake stroke due to which any left-over exhaust in the exhaust port is sucked into the combustion chamber, adding rich, unburned hydrocarbon. This helps to get, on the next compression stroke, very high temperatures to enable autoignition to occur.

Compared with modern diesel engines, a GDCI engine uses much lower injection pressures (About 500 bar compared to 2800 bar or higher). However, in comparison to modern direct injection petrol engines, this engine (due to high-compression super charged design) has to use very strong steel pistons with a peak in crown and made out of special alloy for this application.

It is reported that installed on a 2009 Hyundai Sonata and employing an automatic transmission with fewer forward speeds than the current transmissions (since the torque curve cannot stand more number of speeds), GDCI engine gave 25% better fuel economy compared to Hyundai's mass-produced MPI petrol engine.

Typical diesel exhaust problems *viz.*, particulates and NOx emissions are not an issue here due to low combustion temperatures. However the lean mixtures reduce the effectiveness of the standard three way catalytic converter to address which a second oxidation catalyst is fitted to decrease carbon monoxide and hydrocarbon emission.

17. SPECIAL CHARACTERISTICS OF SPORTS CAR ENGINES

In the case of the sports car engines, the principal aim is not to design the same from the economy point of view, but the efficient high-power characteristics and the reliability have to be attained.

Such engines have to be designed specifically and often contain many refined systems which are normally not provided in ordinary cars. Their important characteristics are:

1. Generally these engines are water-cooled, 4-strokes engines with often 4 but also in many cases 6 to 8 cylinders.
2. The double overhead camshaft is used for valve actuation to get reliable action at high speeds during continuous operation.
3. To provide adequate fuel supply both in quality and in quantity, multiple carburettors were used. In many cases, to improve volumetric efficient of the engine, the supercharges have been employed. However, now electrically controlled fuel injection system has been resorted to, which has been found to be highly adequate are reliable.
4. To ensure ignition-reliability of high order, two spark plugs are used, which also improve the knocking characteristics of the engine. Further, far better characteristics of an electronic ignition system, preferably the capacitor discharge type, makes it almost universal in use in sports cars.
5. To reduce weight and increase capability, the metals with low specific weight and high thermal conductivity, *e.g.* aluminium and copper have been used in place of steel and cast iron.

18. R.A.C. RATING

The formula adopted by the Royal Automobile Club of England has been used for taxation purposes. It gives the horse power of an engine as :

$$h.p. = \frac{(\text{Diameter of cylinder in inches})^2 \times (\text{Number of cylinders})}{2.5}$$

In deriving this formula, the conditions prevailing at that time were considered. These were :

Piston speed = 100 ft/min.

Mean effective pressure = 90 *p.s.i.*

Mechanical Efficiency = 75 per cent.

Due to improved design and better materials, all these factors have changed by now and it is seen that the actual horse power of an engine today is about five to six times its R.A.C. rating.

19. S.A.E. RATING

The Society of Automotive Engineers, U.S.A. have specified the method of measuring the power output of an engine, for standardization purposes. The engine is run without the generator, air cleaner, cooling fan, etc. However, the standard water circulating pump and fuel pump are fitted to the engine. The horse power of the engine is measured with a dynamometer and the measured value is then corrected for the standard temperature, pressure and humidity conditions prescribed.

20. DIN RATING

In this DIN (Deutsch Industries Normale of Germany) method, the horse power of the engine is measured with all the accessories, *e.g.* generator, air cleaner, cooling fan, etc. fitted on the engine. The measured value is corrected to the standard conditions of temperature, pressure and humidity.

21. SUPPORTING OF ENGINE

The engine, together with the clutch and gear box, is mounted on the chassis frame at three points. This arrangement has greater mounting flexibility than the 'four point' mounting system (now obsolete) due to which engine torque fluctuations induced by the power impulses from each cylinder, are absorbed more effectively. The engine support mountings are made of rubber, which is highly resilient and dampens noise. The rubber mountings are placed in the inclined position so as to load them partly in shear and partly in compression.

REFERENCES

1. Sellnau, M., Moore, W., Sinnamon, J., Hoyer, K, et al., "GDCI Multi-Cylinder Engine for High Fuel Efficiency and Low Emissions", SAE Int. J. Engines 8 (2): 2015, doi: 10.4271/2015-01/0834.
2. McGraw, Jim; 'One step beyond', Engine Technology International.com, March 2014.

THEORY QUESTIONS

1. Outline major components of an automobile and explain the functions of each.
2. Enlist various systems of an automotive engine and explain the function of each.
3. Describe with p-v diagrams the two commonly used cycles for automotive internal combustion engines.
4. Describe clearly the working of a four-stroke internal combustion (*i*) petrol engine, (*ii*) diesel engine. Discuss in detail the comparison between these two types.
5. What is a two-stroke engine? Describe its working and compare it with the four-stroke type.
6. What do you understand from the team 'firing order'? Discuss its significance with an example of a four-cylinder engine.
7. Explain clearly the effect of varying the engine dimensions on the engine output.
8. Explain the use of flywheel in an engine with a neat diagram
9. Discuss in detail why higher output automotive engines are made multicylinder and not single cylinder type. Explain also the limitation of increasing the number of cylinders in an engine.
10. What different factors are considered in classifying the engines? Discuss the same in detail giving appropriate examples.
11. What is an opposed type of engine? How is it compared with the in-line type of engine?
12. Write notes on:

    (*a*) Square Engine (*b*) R.A.C. Rating

    (*c*) S.A.E. Rating (*d*) DIN rating.
13. Explain with the help of suitable sketches, the construction of Wankel engine and discuss its working. How do you compare it with conventional engines?
14. Describe in detail the following engines stating clearly their advantages, disadvantages and typical examples where used:

(*a*) Atkinson cycle engine (*b*) Miller cycle engine

15. Explain briefly the construction of an automotive gas turbine. What are its advantages and disadvantages?
16. What is an HCCI engine? How does it compare with S.I. and C.I. engines? Discuss in detail its merits and demerits, describing also the current scenario regarding its application in automobile engines.
17. What is a GDCI engine. Discuss in detail how it differs from a HCCI engine.
18. Discuss briefly special characteristics required in case of sports car engines.

QUESTIONS FOR VIVA

1. What do the terms T.D.C. and B.D.C. stand for?
2. Define 'piston stroke'.
3. What is clearance volume?
4. What is 'piston displacement' in an engine?
5. How do you define compression ratio? What values of compression ratio are in general use for automobile engine?
6. What is mean effective pressure?
7. Define I.P. and B.P. State their units.
8. What do you understand from the term 'engine torque'? How is it measured?
9. What are the units of engine torque?
10. Differentiate between 'power' and 'torque'.
11. What are the factors which form the basis for engine classification?
12. What is an Otto cycle?
13. How does Diesel cycle differ from Otto cycle?
14. What is the air standard efficiency of (*i*) an Otto cycle (*ii*) a diesel cycle?
15. How do you compare a two-stroke engine with a four-stroke one?
16. On a two-stroke engine, is there a power stroke in each revolution of crankshaft?
17. What do the abbreviations, S.I. and C.I. (with reference to I.C. engines) stand for?
18. What is the difference between internal and external combustion engines?
19. What compression ratios are used for automotive diesel engines?
20. How is the fuel ignited in a diesel engine?
21. What is approximate cranking compression pressure in an automotive diesel engine?
22. What is approximate cylinder temperature after compression in a diesel engine?
23. What is ignition temperature of diesel fuel?
24. Name the four stages of combustion in a diesel engine.
25. What determines the power developed in a given diesel engine?
26. What limits the amount of fuel injected in a diesel engine?
27. State the approximate maximum air-fuel ratio employed in diesel engines.
28. State the reason for higher volumetric efficiency of diesel engines compared to petrol engines.
29. Which is more efficient out of the petrol and the diesel engines? Why?
30. Out of petrol and diesel engines, which has usually more reliable fuel supply system?
31. Why is a diesel engine also called C.I. engine?
32. What is approximately the consumption of fuel in a diesel engine as compared to that in an equivalent petrol engine?
33. Out of the diesel and the petrol engines, which involves less risk of catching fire? Why?
34. Which is more costly: a petrol or a diesel engine? Why?
35. If the diameter of an I.C. engine is doubled, how will it affect the power output?

36. How will any change in stroke length affect the power output of an engine?
37. If both the diameter as well as the stroke length of an I.C. engine are doubled, how will it affect (*i*) power output, (*ii*) torque produced?
38. What is the function of an engine flywheel?
39. Why does the increase in number of cylinders makes the engine power output more uniform?
40. Why is the acceleration response in a multicylinder engine quicker compared to a single cylinder engine of the comparable size?
41. State any three problems that arise if number of cylinders in an engine is more.
42. Ordinarily what is the maximum size of a single-cylinder engine?
43. What is the main disadvantage of a 2-cylinder in-line type engine as compared to the opposed type?
44. What is an opposed type engine?
45. Which can be balanced better: a single cylinder engine or a multicylinder engine?
46. What is the main advantage of V-type engine?
47. What is the interval between power impulses in a V-8 engine?
48. What is a radial engine?
49. How are the cylinders numbered in case of (*i*) in-line engines, (*ii*) V-engines?
50. What is 'firing order'? What are the correct firing orders for 4- cylinder and 6-cylinder in-line engines?
51. How are the left and the right banks of a V-engine identified?
52. What is a square engine?
53. Theoretically, how many times a Wankel engine produces power as compared to a conventional 4-stroke engine of comparable size?
54. Name any car with a rotary engine.
55. What are the main advantages of a rotary engine?
56. What is an Atkinson cycle?
57. State advantages of an Atkinson cycle engine. What are its disadvantages?
58. What is a Miller cycle engine? Give typical example of such an engine.
59. How does the automotive gas turbine compare in contract to conventional engine.
60. What is an HCCI engine?
61. State the main advantages of HCCI engine.
62. Why both a supercharger as well as a turbocharger are used in the GDCI project engine?
63. Why no spark plug was used in GDCI engine?
64. How much petrol injection pressure is used in a GDCI engine?
65. Why do you generally use double overhead camshaft in case of sports car engines?
66. What is the advantage of using two spark plugs in an engine?
67. What do the terms 'R.A.C.' and 'S.A.E.' signify?
68. Is R.A.C. rating applicable to the modern automobile engines also?
69. Which out of the '3-point' or '4-point' mounting system is used for automobile engines? Why?

MULTIPLE CHOICE QUESTIONS

Mark the correct answers:

1. The cross-sectional area of one cylinder of an engine multiplied by its stroke is called:
   (*a*) clearance volume (*b*) combustion volume
   (*c*) swept volume (*d*) engine capacity
2. The unit of brake power is:
   (*a*) kilo Newton (*b*) kilo Watt
   (*c*) kilo Pascal (*d*) kilo Newton-meter

3. One bar is equal to
   (*a*) one kPa (*b*) 10 kPa
   (*c*) 100 kPa (*d*) 1000 kpa
4. Maximum combustion pressure in a car petrol engine is about
   (*a*) 6 kpa (*b*) 60 kPa
   (*c*) 6 MPa (*d*) 60 MPa
5. Doubling the cylinder diameter of an engine could give power output, which is '*x*' times the original value, where '*x*' is
   (*a*) 2 (*b*) 3
   (*c*) 4 (*d*) 8
6. The fluctuation of engine speed during a cycle depends upon
   (*a*) mass of flywheel (*b*) mass of crankshaft
   (*c*) speed of flywheel (*d*) power output
7. Can engineers are usually
   (*a*) Single cylinder type (*b*) Twin-cylinder type
   (*c*) four-cylinder in-line type (*d*) four-cylinder V-type
8. In a six-cylinder engine, the power impulse occurs after '*x*' degrees of crankshaft rotation, where '*x*' is
   (*a*) 90 (*b*) 120
   (*c*) 240 (*d*) 360
9. Firing order of a 6-cylinder in-line engine is usually
   (*a*) 1-5-3-6-2-4 (*b*) 1-6-3-5-2-4
   (*c*) 1-4-3-2-6-5 (*d*) 1-3-6-2-4-5
10. A 'square engine' is an engine having
    (*a*) cylinders of square cross-section (*b*) combustion chambers of square section
    (*c*) bore and stroke equal (*d*) none of these.
11. Compared to a HCCI engine, the fuel injection pressure in case of GDCI engine is about
    (*a*) five times (*b*) ten times
    (*c*) one fifth (*d*) one tenth

[**Key:** 1. (*c*), 2. (*b*), 3. (*c*), 4. (*c*), 5. (*b*), 6. (*a*), 7. (*c*), 8. (*b*), 9. (*a*), 10. (*c*), 11. (*c*)]

2 Introduction–II

1. ELECTRIC VEHICLE

It is a vehicle which is powered totally or partially, by electric power from batteries charged through the electrical network. These are also called sometimes, 'Electric drive vehicles'.

Types: These may be categorised as:

1. Battery Electric Vehicles, also called 'All-Electric vehicles' or 'Pure electric vehicles'.
2. Hybrid electric vehicles.
3. Plug-in hybrid electric vehicles.

These would now be described in detail.

2. BATTERY ELECTRIC VEHICLES (BEVs)

In the battery- or pure electric car, the engine is replaced by an electric motor which gets electrical energy from a set of batteries. The main *advantages* of an electric vehicle are:

1. **Energy-efficiency.** Electric vehicles have about thrice the 'grid (tank) to wheels' efficiency compared to the I.C. engine vehicles, converting about 59% to 62% of the electrical energy from the grid in comparison to just 17% to 21% of the energy in the petrol tank to power at the wheels.
2. **Environment friendly.** No exhaust fumes, no fuel evaporation, no fuel refining and no fuel transport to petrol pumps (fuel stations). Due to this reason, battery-operated electric vehicles are sometimes referred to as "zero emission vehicles". Even considering the fact that the electricity used for recharging batteries comes from power stations which themselves are a source of pollution, the net pollution caused by an electric vehicle is far less. According to Electric Vehicle Association of Canada (EVAC), even electric vehicles recharged from coal-powered electric plants reduce the carbon emission by 50 percent, whereas electric vehicles recharged from cleaner electric generation plants, *e.g.*, hydro-or nuclear plants, reduce the carbon emissions to less than one percent of those produced by vehicles powered by I.C. engines.
3. **Rapid acceleration.**
4. **Noise-free operation.**
5. **No loss of power in idling.** Electric motor does not need to idle at stops, thus avoiding the power loss during idling in case of conventional vehicles.
6. **Easy to drive.** No clutch operation or gear shifts are required. Full power is available as soon as you step on the pedal.
7. **High reliability.** Due to absence of numerous engine parts, such as pistons, connecting rods, cams, valves, clutch, flywheel, starter motor, distributor, spark plugs, head gaskets, oil pump, valve tappets, etc., electric car is highly reliable.

8. **Easy maintenance.** Due to very small number of moving parts in the electric motor, the maintenance is much easier.

However, there are many *disadvantages* also:

1. The range is limited because of the need to recharge the batteries. Electric vehicles using lead-acid batteries have a range of about 125 km, whereas Nickel-Metal-Hydride (NiMH) batteries perform better with a range of about 190 km. The maximum range is however, obtained with lithium-ion batteries, which is about 350 km.
2. The top speed is quite low (about 60 km per hour).
3. Fully recharging the battery pack can take 4 to 8 hours. Even fast charging to 80% of full value needs 30 minutes.
4. Battery packs take up a considerable part of the vehicle space and also its weight-carrying capacity.
5. Life of batteries is quite short, which means very high replacement costs. Thus it is found that it is much more expensive per kilometer run compared to the conventional petrol engine vehicle. However, with the rapid development of batteries, their cost is coming down and the life is increasing rapidly.

A general layout of an electric vehicle is given in Fig. 2.1.

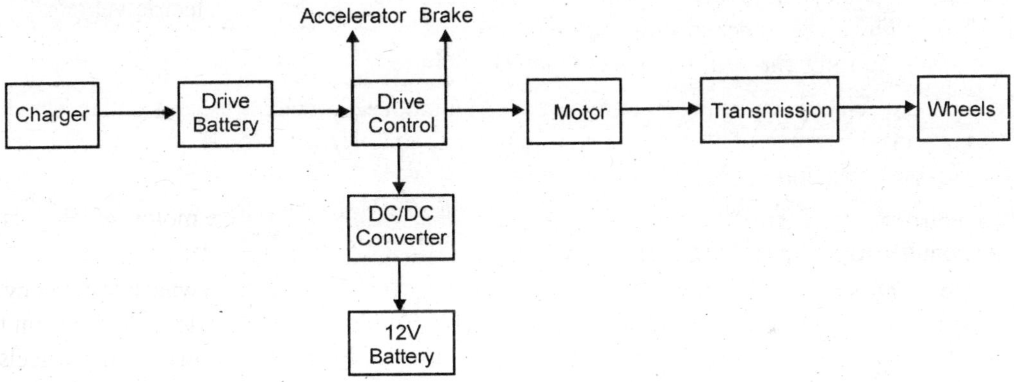

Fig. 2.1. Layout of an electric car

The battery used is the lead-acid type, whose only disadvantage is relatively low specific energy (27-33) Wh/kg). Sodium-sulphur type with specific energy 79-81 Wh/kg is a good alternative. However, specific cost being very high, its use is as yet limited. With further research, lithium-based batteries may be a still better alternative, with the specific energy about 90 Wh/kg.

The drive controller takes power from the batteries and delivers to the motor. The drive controller consists of two potentiometers (variable resistors) operated by accelerator pedal or brake pedal. Depending upon the position of the accelerator pedal, signal from potentiometers tells the drive controller how much power is to be delivered to the motor. Two potentiometers are used for safety. The drive controller reads signals from both these potentiometers, but shall operate only if the two signals are equal. This takes care of an eventuality when the potentiometer fails when in full-on position. The drive controller takes in about 300 volts DC from the battery pack and converts it into a maximum of 240 volts three-phase AC to send the same to the motor.

There are many types of drive motors to choose from. A DC shunt wound motor of about 50 kW rating is a common choice for smaller vehicles. It is also suitable for regenerative braking by increasing the field strength at appropriate time. However, AC motors, with their merits, are likely to become more popular in future.

Apart from drive batteries, 12 V/24 V battery is also required for the rest of the vehicle electrical system which may be charged when required from the drive batteries from a DC/DC converter.

An example of electric car is EV-1 introduced by General Motor in 1996. It has an aluminium spaceframe chassis (40% lighter than steel) with composite body panels. It has an electronically regulated top speed of 120 km/h. It can reach 96 km/h from a standing start in less than 9 seconds. Its electrical power train consists of a 103 kW three-phase AC induction motor and an integral, single speed, dual-reduction gear set driving the front wheels. The battery pack uses 26, 12 V lead-acid batteries giving a range of 112 km per charge in urban conditions and 144 km on the open road. However, later using the nickel-metal-hydride (NiMH) batteries, the range increased to 224 km in the city and 252 km on highways.

'Leaf' 2013 from Nissan uses a 80 kW synchronous electric motor and 24 kWh Lithium-ion battery, giving it a driving range of 120 kms on full charge. EPA (US) fuel economy is equivalent of 55.5/43.9 km/h in city/highway respectively. 2013 model Honda 'Fit' (U.S. model) employs a 20 kWh Li-ion battery which can be charged in 3 hours at 240 V or in 15 hours at 120 V. It employs an AC synchronous permanent magnet motor giving 92 kW (Sports mode), 75 kW (Normal mode) and 47 kW (Economy mode) all at 3695 to 10320 rpm. The torque available is 260 Nm at 0 to 3056 rpm. Its EPA mileage rating is equivalent of 56.7/45.1/50.7 (City/Highway/Combined) km/h. It gives a range of 131 km on full charge. Toyota Scion iQ electric vehicle, 2013 (US) model gives highest EPA fuel economy figures, which are 59.3 km/h in city and 52 km/h on highway. Its range on full charge is however lower at 80 km. It uses 12 kWh Lithium ion battery which can be fully charged in 3 hours from 240V outlet.

In India the only electric car being manufactured is Mahindra e20, which was launched in March 2013. The car has city driving range of 100 km, after which the Li-ion batteries would need to be recharged. It employs a 3-phase induction motor giving an output of 19 kW at 3750 rpm and torque of 53 Nm at 0 to 3400 rpm. The batteries take 5 hours to be charged from a 220 V, 15 Amp socket.

Hyundai launched its new 'Kia Ray' EV in Korea, based on Kia's City car in 2012. It employs a 50 kW, 167 Nm electric motor and 16.4 kWh Li-ion polymer, pouch type battery pack from 'LG Chem' designed for 10 years' life. It has a range of 138 km and top speed of 130 km/h. it can be fully charged in 6 hours from a Korean 220 V socket and in only 25 minutes by a quick charger.

As in 2016, fully charge ranges of various typical pure electric vehicles are:

BMW i3/Chevrolet spark/Fiat 500e	– 130 km
Kia Soul	– 150 km
Mercedes B-class	– 136 km
Nissan Leaf	– 170 km
Tesla Model S	– 425 km
Tesla Model X	– 400 km
Tesla Model 3	– 320 km
Volkswagen E-Golf	– 133 km

Recently (July 2016) Mercedes-Benz unveiled a 26 tonne electric-powered urban eTruck, which is based on a heavy-duty, three-axle short-radius Mercedes-Benz distribution truck, with the conventional drive train replaced by electrically driven rear axle with electric motors adjacent to the wheel hubs. Each motor is of 125 kW and torque 11 kNm. The power is supplied by three lithium-ion battery models of total capacity 212 kWh, which give a range of about 200 km. Charging is done with a 100 kW Combined Charging System Type 2 connector, standardized throughout Europe.

3. HYBRID ELECTRIC VEHICLES

3.1. Introduction

These are powered by an I.C. engine (which runs on conventional or alternative fuel) and an electric motor which uses energy stored in a battery pack. The batteries in hybrid vehicles are not plugged in for charging, but are charged through regenerative braking and by the I.C. engine. Most hybrids employ advanced technologies.

Internal combustion engines produce appreciable emissions and are also less efficient at part loads. On the other hand electric drives have zero emissions, but also very limited range. It is thus logical to combine the best aspects of both and the result is a hybrid vehicle. Optimum strategy would then be to use electric drive during slow moving city traffic, for acceleration and for hill climbing and I.C. engine at cruising speeds on highways. This would also result in reduced pollution in cities, alongwith improved mileage.

The engine on a conventional car is sized for the peak power requirement, which is seldom required in actual practice. The hybrid car uses a much smaller engine, whose size is kept closer to the *average* power requirement rather than the *peak* power. A smaller engine is always more efficient due to the reason that it would run at its optimum capacity most of the time as compared to a bigger engine running at part load most of the time.

Main advantages of a hybrid vehicles are:

1. *Electric motor Drive or Assist.* Since electric motor provides power to help the engine in accelerating or moving up the gradient, a smaller, more efficient I.C. engine may be used. In addition, at low speeds, I.C. engine is completely shut-off, the complete power required being provided by the electric motor, which is more efficient than I.C. engine at low speeds.
2. *Automatic start/stop.* This is provided by the integrated starter-generator so that the engine stops automatically when the vehicle comes to a stop, *e.g.*, at traffic lights; the engine restarts on turning the lights green, the driver simply presses the accelerator pedal. In this way loss of energy during idling is reduced.
3. *Provides regenerative braking to capture energy during braking.* Regenerative braking recaptures the energy ordinarily lost during braking or coasting. It utilizes the forward motion of the road wheels to turn the rotor of the motor in reverse direction, which results in the motor acting as generator, thus generating the electricity while slowing down the vehicle.

The switching on or off of power sources is computer-controlled with the help of an ECU and various sensors. Recharging of the battery is also automatic and is normally not required to be done externally. This is mainly achieved with regenerative braking.

3.2. Types of hybrid systems

A. The most common types of hybrid systems are the series and the parallel. Most hybrids currently are of the parallel type, which can be further subdivided into the 'mild' and the 'full'. The different types are differentiated by:

(*i*) The connection between the electric motor and the I.C. engine.

(*ii*) The situations when each drive system is in operation, and

(*iii*) The proportion of power provided by each propulsion system.

Series system (Fig. 2.2) in which engine driven generator is employed to recharge the batteries, while

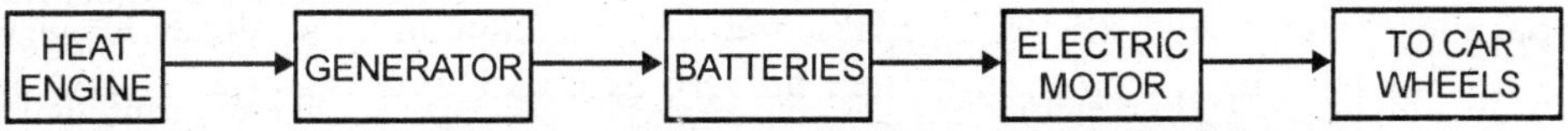

Fig. 2.2. Series system of hybrid car

the electric motor drives the car. In this way the range of the car is extended. However, this system involves significant mechanical-electrical-mechanical conversion losses and is thus very inefficient. The advantages of this type are:

(*i*) The engine is never required to idle, thus reducing exhaust emissions.

(*ii*) The engine drives a generator to run at optimum performance.

(*iii*) Gear box is unusually not required.

These can operate without recharging as long as there is liquid fuel in the tank.

The disadvantages are:

(*i*) The system being inherently less efficient as already explained, it requires larger and thus heavier battery pack than the parallel type.

(*ii*) The engine works hard to maintain battery charge as the system is not running in parallel.

Examples of series hybrids are the Renault Kangoo Elect, Toyota's Coaster light-duty passenger bus, Daimler Chrysler's Orion hybrid bus and Fisker Karma.

Parallel system (Fig. 2.3) in which heat engine and the electric motor are both directly connected to the car wheels which makes it possible for both of them to supply power to the road wheels at the same time. Two possible parallel configurations are shown in Fig. 2.3. It may be noted that in either of these it is necessary to use a gear box to synchronise the engine and the motor. Thus not only does it increase the car range but also supplements the electric motor to produce increased power when required. The recharging of batteries, however, has to be done only at home when the car is not in use.

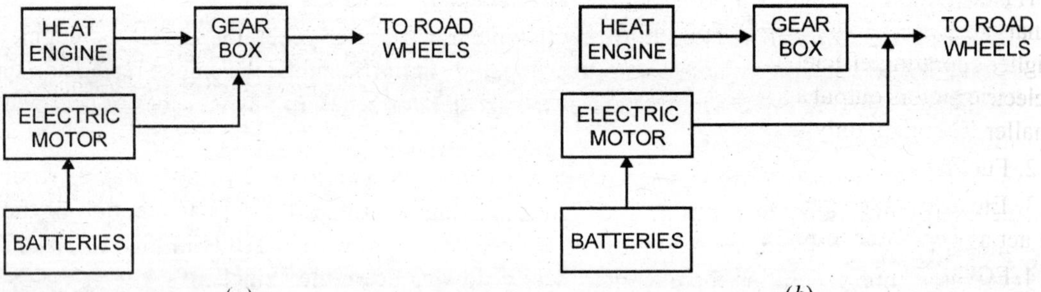

Fig. 2.3. Parallel systems of hybrid car

Parallel hybrids can be programmed to use the electric motor to substitute for the I.C. engine at lower power demands as well as to substantially increase the power available to a smaller I.C. engine, both of which increase the fuel economy substantially compared to an ordinary vehicle run only by an IC engine. The advantages of the parallel type are:

(*i*) Both the engine and the motor supply power simultaneously, due to which the vehicle has more power.

(*ii*) Most parallel hybrid vehicles do not require separate generator for recharging since the motor regenerates the batteries.

(*iii*) Because the power is directly carried to the road wheels, it is more efficient.

Examples of parallel hybrids are Honda's Insight, Civic & Accord Chevrolet Malibu and Toyota Prius.

Series-parallel system which allows the hybrid vehicle to operate in either the series or the parallel mode. Such systems have been employed in their modern hybrids by Ford, Lexus, Nissan and Toyota, which are sometimes referred as series-parallel system with power split.

B. Based on the degree of hybridization, hybrid systems can also be divided into full hybrid, mild hybrid and micro hybrid.

Full hybrid. In this the engine and the motor can operate the vehicle separately on their own or together. Toyota Prius and Ford Escape are examples of this type. A prominent feature of this type is the large battery and motor which enable the vehicle to be driven in pure electric mode for sometime. Such hybrids employ split drive system which allows flexibility and balances the drive from each propulsion source. An ECU controls the hybrid operation depending upon the circumstance. For example, the engine is shut down when the vehicle is stationary, the electric drive launches the vehicle into motion (charge depleting zone) and the engine is used only when extra power is required or for recharging the propulsion battery (charge sustaining zone).

Other features available with the I.C. engine are both stop-start and regenerative brake functions. Examples are Toyota Prius, Ford Escape and Volvo 7900.

Mild hybrid. In this the engine provides the main source of power and the small electric motor only provides additional power whenever required, for example, during acceleration (assist mode) and during regenerative braking. Obviously, such a vehicle cannot be driven in pure electric mode and is operated only in charge-sustaining zone. Honda 'Civic' and 'Insight' are examples of this type.

Micro hybrid. Such a vehicle uses a start-stop system where regenerative braking technology helps to stop an i.c. engine when the vehicle comes to a stop (*e.g.* at red lights) and to restart it when the driver starts to accelerate (on turning lights green). This type has the smallest battery size out of these three types of hybrid systems. This type of system is being used on Mahindra Scorpio M2DI and Bolero SLX.

3.3. Main components

1. Heat engine (Petrol/diesel). IC engines installed in hybrid vehicles are usually smaller than those in ordinary cars, so that they are able to run at higher throttle even at part loads and/or lower speeds, resulting in higher operating efficiency and lower emissions. Moreover, since more often their purpose is to augment the electric motors output and regenerative braking also serves to recapture braking energy to charge batteries, a smaller IC engine only is required.

2. Fuel tank.

3. Electric motor-generator, which acts as a motor when electrical energy is supplied to it and can also act as generator to return energy of the moving vehicle to the batteries by regenerative braking.

4. ECU.

5. Batteries to store energy for electric motor, which can input energy into batteries as well as draw energy from them.

6. Transmission, which may be manual or continuously variable type.

3.4. Advantages of hybrid vehicles

1. Combining the best points of the electrical energy and IC engine for vehicle propulsion.

2. Lower emissions and better mileage. For example, average petrol consumption for 2012 Honda Civic hybrid is estimated at about 13 kmpl whereas a conventional 2012 Civic consumes about 18 kmpl. Regarding emission, these are zero at the tailpipe. However, life cycle emissions for an electric vehicle depend on the sources of electricity used to charge it.

3. Smaller engine means lesser weight.

4. Reduction of dependency on fossil fuels.

5. Lesser routine maintenance due to the following, for the electrical system:

(*i*) Battery, motor and the related electronics do not need regular maintenance.

(*ii*) Brake wear is less due to regenerative braking.

(*iii*) Far lesser number of moving parts compared to a conventional engine.

Disadvantages

1. Higher initial cost.
2. Significant cost of the battery pack which has to be replaced every few years.
3. More weight due to battery pack.
4. Lower acceleration than the cars running with IC engines only.

3.5. Examples

An example of a real hybrid electric car is 'Durango' launched recently by the Daimler Chrysler. The Dodge Powerbox version of the Durangr shall use CNG for the 186 kW, V6 internal combustion engine with Nickel/Metal-hydride (NiMH) batteries. The hybrid is expected to achieve a 9.0 litre/100 km fuel consumption.

'Tino' from Japan is another example of a hybrid electric car. This is parallel type, employing either the electric motor or the IC engine or the two together depending upon operating conditions. A second motor/generator starts the IC engine and when driven by the engine, recharges batteries. The engine is cut off when the vehicle comes to a stop, except when the air conditioner is in use. The vehicle always starts off on electric power drawn by a Hitachi three phase AC synchronous motor producing 17kW between 1390 and 5600 rpm and 155 Nm between 0 and 700 rpm, operating on a 345 V system. The petrol IC engine takes over as the primary propulsion source at about 25 km/h and above. The engine is DOHC, 16-valve inline 4-cylinder type with 1769 cm^3 capacity and 12.0:1 compression ratio. Its output is 74 kW at 5200 rpm and 141 Nm at 4000 rpm. The urban fuel consumption is 4.3 litres/100 km. It uses lithium-ion battery pack.

Another hybrid example is Honda Insight using 1 litre 3-cylinder engine which produces 50 kW at 5700 rpm. With the electric motor running alongwith engine, 54.5 kW BP is produced at 5700 rpm, which means the contribution of motor at 5700 rpm is only 4.5 kW. However, the real effect of motor is seen at lower engine speeds; it produces 10 kW at 3000 rpm.

Further, engine alone produces a maximum torque of 90 Nm at 4800 rpm, whereas with electric motor also running, the torque produced at 2000 rpm is 124 Nm, which is about 50% more than the engine peak torque.

Various functions which the electric motor on the Honda Insight performs are:

(*i*) To provide extra power during acceleration or hill climbing.
(*ii*) Regenerative braking to recharge batteries during braking.
(*iii*) Starting the engine, thus not requiring a starter.

However, by itself the motor cannot move the vehicle from start for which the engine must be running. Like Insight, the Toyota Prius is also a parallel hybrid car which employs specially designed power-train, called the Toyota Hybrid system, which imparts also some advantages of a series hybrid.

Prius uses a 1.5 litre petrol engine producing 57 kW at 5000 rpm whereas the electric motor is rated at 50 kW from 1040 to 5600 rpm, a real flat power curve. The electric motor also gives torque of 400 Nm from 0 to 400 rpm, which is quite sufficient to move the vehicle from start without the help of the petrol engine.

Some Top rated 2017 model US Hybrid cars are:

1. Chevrolet Bolt
2. Toyota Avalon Hybrid
3. Chevrolet Malibu Hybrid
4. Honda Accord Hybrid
5. Toyota Camry Hybrid
6. Toyota Prius
7. Chevrolet Volt
8. Ford Fusion Hybrid
9. Hyundai Ioniq
10. Kia Optima Hyrbid

3.6. Diesel hybrid

Diesel engine being move efficient then petrol engine, it would seem to be more logical to make a hybrid with diesel engine, but till recently, all the hybrids have been petrol-electric. Till 2010, there were no production diesel hybrids in existence. The reason for this is that diesel engines are basically slow speed engines. Their higher and constant torque at lower rpm makes them suitable for trucks and even for ordinary non-hybrid cars, but not so good for a hybrid, which basically is meant to avail the relative strengths of the electric motor (at low-speed and for the urban stop-and-go type driving) and the petrol engine (when overtaking and cruising at high speed as on an express way). In addition, tail pipe emissions in diesel engines, being different from those in petrol engines, require in addition to a catalytic converter, also a particulate trap, urea injection, etc., all of which would make a diesel engine hybrid more expensive.

The situation, however, changed in 2010, when Peugeot unveiled the world's first production diesel hybrid 'Peugeot 3008 Hybrid 4' which employs a 163 bhp 2.0 litre turbo diesel engine driving front wheels and an electric motor powered by NiMH batteries which drive the rear axle with a total system output of 200 hp. The car can run at a maximum speed of 130 mph (210 kmph). The car can be driven either in front drive or rear drive or all wheel drive modes. It also has four operating modes, *viz.*, *Auto Mode* (where only electronics handles change between engine and motor), *ZEV mode* (car uses only electric motor for a maximum of 3.7 km), *4WD mode* (both power trains operating together) and Sports mode (More power, at the cost of efficiency).

4. PLUG-IN HYBRID

A plug-in hybrid electric vehicle (PHEV) is a hybrid vehicle with batteries which can be recharged from ordinary household electric power plug.

U.S. 'Energy Independence and Security Act of 2007' defines a plug-in electric vehicle as a vehicle which

(*i*) draws motive power from a battery with a capacity of at least 4 kWh.

(*ii*) can be recharged from an external source of electricity for motive power, and

(*iii*) is a light, medium or heavy-duty motor vehicle or non-road vehicle.

Thus it is different from regular production hybrids, which cannot be charged by electricity direct from the grid. Plug-in hybrids have also been defined similarly by the Institute of Electrical and Electronics Engineers (IEEE), with an additional requirement that PHEV can drive at least 16 km in all-electric mode, consuming no engine fuel.

All-electric range of a PHEV is designated as PHEV [*miles*] or PHEV [*kilometers*] km. The number in brackets are the miles or kilometers which the vehicle can travel only in all-electric mode. For example, PHEV 30 km hybrid can travel 30 km without using its engine.

Operating modes

There are a number of *modes* in which the PHEVs can operate. These modes manage the vehicle battery discharge strategy, due to which they directly affect the type and the size of the battery required. Various operating modes are:

1. *Charge-depleting mode.* In this the vehicle operates only on electricity until the battery charge level is depleted to a predetermined level at which the IC engine would be engaged. A pure electric vehicle can operate only in this mode.
2. *Charge-sustaining mode.* In this mode, the vehicle operates by combining optimally the power from both the engine as well as battery, such that the battery charge always remains above a pre-determined level. Ordinary production hybrids operate in this mode. In case of a plug-in hybrid, it can switch auto-matically into this mode as soon as it has exhausted its all-electric charge-depleting mode.

3. *Blended mode.* This mode is applied in case of vehicles not having enough electric power to sustain high speeds, without the help of I.C. engine of the vehicle. Thus, this is a type of charge-depleting mode. This mode is used in vehicles such as Renault Kangoo and some conversions of Toyota Prius.
4. *Mixed mode.* The term is used for a trip in which the combinations of the above modes are applied. For example, a PHEV 32 km may start a trip with 8 km of low-speed charge-depleting mode, then enter an expressway and operate in blended mode for 32 km, which consumes 16 km of all-electric range at twice the fuel economy. After covering this 32 km the driver exits expressway and drives further 8 km without running the engine, thus utilizing 8 + 16 + 8 = 32 km, *i.e.*, its full all-electric range. After this he can switch on to the charge-sustaining mode for another 20 km to reach his destination. Such a trip is termed a mixed mode.

Advantages

1. PHEVs are more efficient than conventional hybrids, due to their IC engine operating more often closer to its maximum efficiency.
2. Significant reduction of green house gas (CO_2) emissions due to higher efficiency of the drive train. 15% reduction in CO_2 emissions is expected, compared to ordinary production hybrid.

Disadvantages

1. Additional cost, weight and size due to larger batteries compared to a conventional hybrid.
2. Electric outlets have to be provided at public places for charging since all people do not keep their vehicles in garages.

Model availability

As in 2016, prominent models of PHEVs available, with their all-electric ranges are, Audi A3 E-Tron (16 miles), BMW i8 (25 miles), BMW X5 xdrive 40e, SUV (13 miles), Chevy Volt sedan (53 miles), Ford C-Max Energi, Wagon/Van (20 miles), Ford Fusion Energi, sedan (20 miles), Porsche Cayenne SE, SUV (14 miles), Toyota Prius, sedan (11 miles), Mercedes C 350, sedan (20 miles), Volvo XC90 T8, SUV (17 miles), Porsche 918 Spyder, Sports Coupe (12 miles) and McLaren P1, sports coupe (19 miles). Besides, as example, the I.C. engine only mileage for MBW X5 xdrive 40e is 24 mpg. Likewise examples of combined range of some PHEVs are, Cadillac ELR, coupe-340 miles and Ford C-Max Energi, Wagon/Van-620 miles.

5. BATTERY REQUIREMENTS FOR ELECTRIC VEHICLES

Ordinary hybrids, *i.e.*, Hybrid Electric Vehicles (HEVs) require high power in short pulses, from 1.0 to 1.5 kWh, which means that the batteries must be able to provide many shallow charging cycles.

On the other hand, PHEVs need a larger battery which can provide energy in all-electric (charge-depleting) mode for a defined distance. For instance, a PHEV would need 10 kWh energy for a 40 miles (64 km) range and 5 k Wh for a 10 miles (16 km) range. Besides this energy, the battery also must provide higher continuous power discharge-deep charging cycles similar to that of an electric vehicle. In addition, the batteries should also be able to provide many shallow cycles for the blended/mixed modes.

Batteries for EV (Electric Vehicle) should be able to produce about 40 kWh of continuous-discharge energy and be capable of one charge per week.

Most manufacturers of plug-in vehicles are offering 8 year/1 lakh mile battery warranties. Tests and Simulation results from USA's National Renewable Energy Laboratory indicate a battery life of 12 to 15 years in moderate climates and 8 to 12 years in extreme climates. However, battery prices are expected to decrease with improvement in battery technology and increase in production volume.

6. HYDRAULIC HYBRID VEHICLES (HHV)

These are the vehicles which use a regular I.C. engine and a hydraulic drive to rotate the wheels. The hydraulic system consists of:

1. Working fluid.
2. Reservoir.
3. High pressure hydraulic fluid vessels called 'accumulators'.
4. Hydraulic drive pump-motors.

The accumulators are used to store pressurized fluid. Like the regenerative braking in the electric hybrid vehicles during which the kinetic energy of the vehicle is used to charge the battery, here also, the pump extracts the kinetic energy during braking to pump the working fluid from the reservoir to the accumulator, thus pressurizing the working fluid, leading to energy storage. However, acting as a motor, the hydraulic drive uses the pressurized fluid to rotate the wheels.

Types

These are of two types:

1. *Parallel hybrids.* In these the engine and the hydraulic system both interact mechanically with the wheels. The hydraulic pump-motor is a part of the drive shaft or the differential. The pump-motor is typically installed between the engine and the gear box or between the gear box and the differential. The pump-motor provides assistance to the engine during acceleration and recaptures energy under braking which would otherwise be lost as heat in the conventional brakes. However, in such a system, the pump-motor may or may not be able to drive the vehicle along when the engine is off.
2. *Series hybrids.* In these, engine is not directly connected to the wheels. Instead, a pump-motor, acting as a motor, uses pressurized fluid from an accumulators to drive the wheels. The I.C. engine is only connected to a pump and is so set that it runs only in its most efficient power range, to maintain the optimum hydraulic pressure in the accumulator. The traction motor must be able to supply all the torque required for the maximum acceleration, with the engine running or even when stopped.

Advantages

1. Simple construction.
2. Highly efficient, recovering upto 75% of the vehicle's kinetic energy (in series type) as compared to 25% for electric hybrids. The main factors contributing to increased efficiency are:
   (*a*) Regenerative braking.
   (*b*) Engine shutting off when not required (in series type). When needed, the engine is activated by a hybrid controller. As a result, engine use is nearly reduced to half in stop-and-go city driving.
   (*c*) Optimum engine control. Since the engine is not mechanically connected to the drive wheels, a hybrid controller can easily command the engine to operate only at its optimum efficiency.
3. Cost-effective due to lower manufacturing cost, reduced brake maintenance and highly increased fuel efficiency.

Limitations

1. Hydraulic hybrid vehicles are useful only in stop and go driving urban roads, since it is here that they run more efficiently. On highways there is very less regenerative braking (which recovers energy) and the I.C. engine also runs at its optimum efficiency, due to which there is not much utility to have hydraulic hybrid.
2. Noise and size are other main challenges *due to which no passenger cars have so far been designed with such a drive.*

Applications

Due to the reasons listed above, the use of HHVs has been quite limited to cities, for delivery vans and trucks, where they have been quite successful. American EPA is already running such vehicles since 2005 in partnership with UPS and Eaton Corporation.

7. WHEEL MOTORS

A wheel motor is an electric motor which is installed into a wheel and drives it directly.

The outside of a wheel with in-wheel motor looks like an ordinary vehicle wheel. However, on taking off the wheel from the vehicle, the main parts of the in-wheel electric motor system become visible. The small area there contains the braking system, an active suspension system and the electric motor which actually drives the wheel. The active suspension system as used in the Mickelin's 'Active wheel System' is electrically operated and can correct pitching and rolling motions in 3/1000 ths of a second automatically. Some in-wheel motor designs also have regenerative braking.

Automobile Engineers are expecting these to replace the I.C. engine for motive power shortly because of certain advantages, although a small engine may still be required to run the generator for charging batteries. Wheel motor is a critical element for making fuel efficient all-wheel drive and advanced technology vehicles at a reasonable cost.

Various advantages of wheel motors are:

1. Since the power in case of in-wheel electric motor goes directly to the wheel, the efficiency of transmissions is 90 to 95 percent.
2. Much higher starting torque. These can easily generate 700 Nm.
3. Power shafts and axles become redundant, which means simpler construction, lesser weight and saving of space. A hybrid may retain the conventional IC power train, with All-electric and Hybrid-electric vehicles needing space only for batteries.
4. Because of independent motors at all wheels, it is simpler and natural to have all-wheel drive at no extra cost.
5. No mechanical brakes are required as the braking is done by the wheel motors acting as very efficient electric generators which return almost all the vehicle kinetic energy back to the battery system.
6. Each wheel motor has to be precisely controlled electronically. Thus modern control systems, *e.g.*, ABS, traction control, stability control, etc., would be included as standard out fits, with these motors, ABS can also be applied to acceleration since the motor can smoothly control torque delivery to or from the road. Thus pressing the brake or the accelerator pedal hard would only result in controlled maximum torque giving the shortest possible braking or acceleration time. Traction control dynamically distributes torque when any wheel is in skid management mode. The wheels are in constant communication with each other deciding almost 1000 times in a second, how much torque share is optimum for the current driving conditions. Therefore, the need for crude differential gears is completely eliminated, with the advantage of independent automatic control of the drive torque to two or even all four wheels, without the cost, complexity and packaging problems of controlled differentials and drive shafts.
7. The only parts in wheel motors subject to wear are the wheel bearings, which means longer life and no deterioration in the overall vehicle efficiency with time.

The above advantages are also accompanied by the following disadvantages:

1. Since the wheels are not sprung, additional mass of wheel motors would increase the unsprung weight due to which ride would become bumpier. To remedy this, suspension system would have to be redesigned which may increase its complexity and/or cost.

2. To avoid under increase in the unsprung weight, a single bigger wheel motor cannot be used as in an ordinary electric car and one smaller motoı at each wheel has to be used, which makes the system more complex.

Vehicles with wheel motors have three modes, selectable by the driver:

1. Eco mode which is all-electric and is for frequent start-stop driving encountered in town traffic.
2. Normal mode in which engine starts when the battery drops to 50% state of charge.
3. Sport mode for enhanced performance.

As in 2008, there have been only demonstration vehicle and concept cars. One example is, PML Flightlink's hybrid 'MINI', which is a conversion of BMW Mini, wherein they discarded the engine, the disc brakes, the wheels and the gear box which were replaced by four electric wheels, a lithium polymer battery, a large ultra capacitor, a very small I.C. engine with a very small generator, an energy management system and a display module. Siemens VDO are also working to make a concept car based on wheel motors and named 'eCorner' which would use 'drive-by-wire' technology. The Company expects that during the next decade, 'eCorner' would integrate all systems into the wheel with the exception of electronic shock absorbers and electronic steering. Other examples of in-wheel motor vehicles are, 'Mitsubishi CT' concept and 'Venturi Volage' concept, employing Michelin's 'Active wheel'.

Fig. 2.4. Protean Electric in-wheel Motor Details

Michigan-Based 'Protean Electric' is building a manufacturing center in 'Liyang' in China, which would have an initial capacity to produce 50,000 motors annually and is scheduled to start pre-production in early 2013. Protean design has stationary permanent magnet at the center of the motor while the rotor is on the outside, making it easier to attach the wheel directly. Inverters and power electronics are housed between the two, simplifying and reducing the weight of the system. Each motor is made up of 4 to 8 parallel sub motors, so that in case of failure of one, the unit can operate till the same is serviced. As per Protean Electric, their in-wheel motor with micro inverters has a mass of 31 kg and generates 81 kW power and 800 Nm torque, which are sufficient to be used in hybrid, PHEV or as the only source of traction drive for EV applications.

Other manufacturers of wheel motors are:

1. E-Traction (Netherlands)
2. Tepse (Taiwan)
3. Evens Electric (Australia)
4. Sim Drive (South Korea)
5. Aisin Aw Co. Ltd. (Japan)
6. Mitsubishi Motors (Japan)
7. Siemens (Germany)
8. Michelin (USA)

9. PML Flightlink (USA)
10. Volvo cars (Sweden)

8. FUEL CELL VEHICLE

The working of fuel cell is essentially similar to that of a battery. It relies on chemical reactions at electrodes to convert chemical energy into electrical energy. It combines electrochemically, the oxygen from the air with hydrogen from a hydrocarbon fuel to produce electricity. However, unlike battery, a fuel cell stores the reactants outside the electrodes. They are fed to the cell as 'fuel' for the reaction.

Fig. 2.5 shows the schematic diagram of a fuel cell. The hydrogen is the input at the anode, while oxygen enters the cathode of the cell. The Hydrogen atom is separated into a proton and an electron due to the presence of a catalyst. These take different paths to the cathode. The proton passes through the electrolyte, while the electron stream is utilised to generate current which can be used before it joins the cathode and forms a molecule of water with the protons and the oxygen present there. A fuel cell system can use hydrogen from any hydrocarbon fuel, although a fuel reformer may be required to some cases. Following types of fuel cells have been developed:

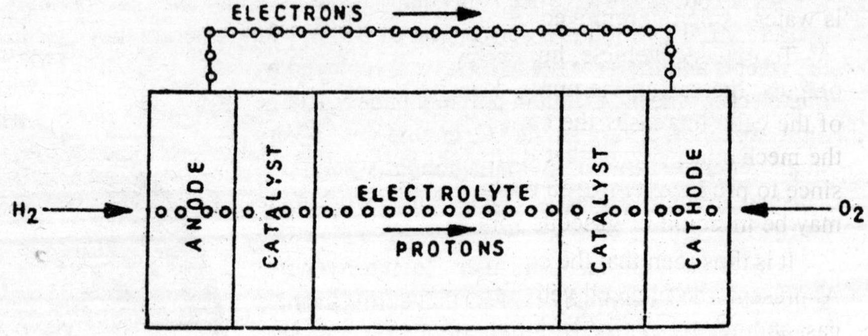

Fig. 2.5. Schematic diagram of a fuel cell

1. Alkaline
2. Molten carbonates
3. Phosphoric acid based
4. Proton Exchange Membrane (PEM)
5. Solid Oxide

Out of the above only PEM types have been found suitable for automotive use. Another name for these cells is the Polymer Electrolyte Membrane, abbreviated also as PEM. The working of this cell is made more clear by Fig. 2.6, wherein hydrogen fuel is passed through field flow plates to the anode on one side of the fuel cell, while the air is channeled to the cathode on the other side. Hydrogen can be supplied to the fuel cell directly or may be obtained from natural gas, methanol or petroleum using a fuel-processor, which converts the hydrocarbons

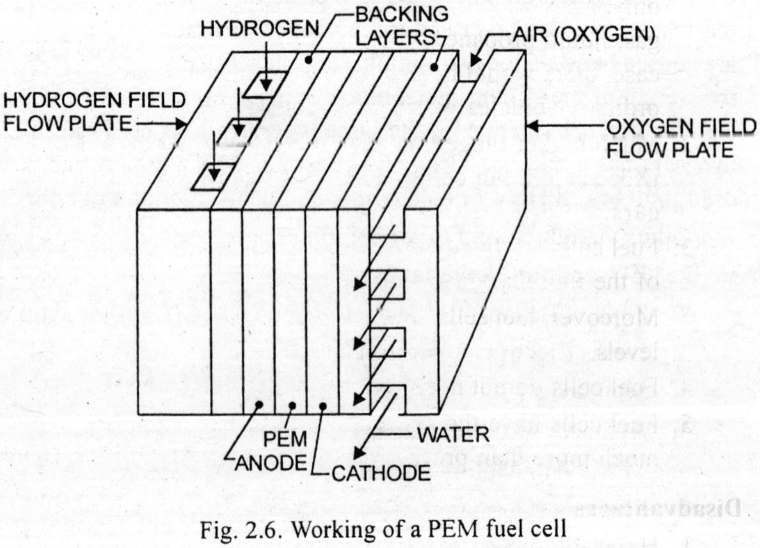

Fig. 2.6. Working of a PEM fuel cell

into hydrogen and carbon dioxide by means of a chemical reaction aided by a catalyst. There is a catalyst at the anode which causes the hydrogen to split into protons (positive hydrogen ions) and electrons. The catalyst is usually made of platinum powder very thinly coated onto carbon paper or cloth. It is rough and porous which helps the maximum surface area of the platinum to be exposed to the hydrogen or oxygen. The platinum-coated side of the catalyst faces the PEM.

Polymer Electrolyte Membrane permits only the protons to pass through it to travel further to cathode, whereas electrons have to travel through the external circuit, creating an electrical current. At the cathode, the protons and the electrons combine with oxygen to form water and heat. Thus the only exhaust emission is water, which is removed by the air stream.

To power a vehicle, hundreds of fuel cells are stacked together in series, combining their individual outputs. Increasing the number of cells in a stack increases the voltage while increasing the surface area of the cells increases the current. The stack sends electricity to the electric motor. The motor produces the mechanical power that turns the wheels to propell the vehicle. A fuel tank stores hydrogen as a fuel since to produce hydrogen on board safely and economically is not possible as yet. The hydrogen stored may be in liquid or gaseous form.

It is thus seen that the only emission in case of a fuel cell would be water, which means zero pollution. At present, the fuel cell vehicles being developed are using methanol which can be produced from natural gas, abundantly available. In the long term, it may be obtained from renewable sources like organic waste or timber residues. Moreover, if we are able to get hydrogen from water by electrolysis economically and safely on board, it would be possible to run vehicles on water only. Extensive research is being done in this direction. Hy Power Fuel Inc., U.S.A. has recently (2006) announced that they have equipped a Volkswagen GTi with its H_2 Reactor (H_2R) hydrogen system that can produce sufficient hydrogen on board and on demand, to power the vehicle using only water. This H_2R employs electrolysis to convert water into hydrogen and oxygen gases, which are then used to power the engine. This reactor is claimed to be the most efficient design requiring 1 Wh to produce 1 liter of hydrogen, which is almost twice efficient compared to other competing technologies. However, these are only preliminary laboratory results which will require much improvement to make it a reality for a production vehicle.

Advantages

1. Fuel cell cars have zero or low emissions. A fuel cell vehicle running on pure hydrogen produces only water vapour. Fuel cell vehicles can also be run on a variety of fossil fuels— petrol, natural gas, liquid propane, etc., as well as on renewable fuels— methanol, ethanol. However, in such a case, there would be CO_2 and other trace emissions, which would also be much less compared to ordinary vehicles with I.C. engine.
2. Much higher range compared to pure battery vehicles. For example, the range of Hyundai FCV, iX35, is 590 km compared to 210 km of the Nissan Leaf, which is one of the best pure electric cars.
3. Fuel cells are highly efficient. Depending upon the fuel used they can capture 40%-60% or more of the fuel-energy to run the car.
   Moreover, fuel cells, quite contrary to the I.C. engines, deliver maximum efficiency at lower power levels.
4. Fuel cells do not need recharging, unlike batteries.
5. Fuel cells have the potential to have life expectancy equal to that of the vehicle itself, which is much more than propulsion batteries used in electric (pure/hybrid) vehicles.

Disadvantages

1. Heat build-up in the fuel cell due to internal chemical reaction as well as the fuel and internal water

flows need to be controlled and managed carefully.

2. High cost of the membrane and the catalysts.
3. Problematic ability of the fuel cell to supply power at low ambient temperatures. Most fuel cell designs work at higher temperatures to achieve reasonable operating efficiencies. To obtain similar efficiencies at lower temperatures would require large quantities of expensive catalysts like platinum.
4. Low cell voltage 0.6-0.7 volt means that the stack needs a large number of cells to obtain a normal operating voltage to power the drive train motor.
5. Unlike a primary cell, a fuel cell cannot accept regenerative braking loads since it can generate electrical energy but cannot store such energy.
6. Fuel cells have a low dynamic range and slow transient response, which would result in unacceptable lag in responding to calls for power by the driver, to remedy which power boost from either a battery or super capacitors is required.
7. Electrodes of fuel cell are prone to contamination.
8. Fuel cell vehicles use heavy tanks to store hydrogen at very high pressure, say 35 MPa, which increases appreciably the vehicle weight.
9. As yet, there are no common fuel cell stack and fuel cell system which can be employed by all companies in their fuel cell electric vehicles, which further increases the cost to the customer.
10. As yet there are no networks of hydrogen filling stations.
11. Current manufacturing cost for fuel cell electric cars are about twice than that for lithium-ion battery models. The cost may, however, come down with the increase in production of fuel cell cars.

Examples

A few examples of the fuel cell based demonstration vehicles produced by various manufactures are given here.

1. GM introduced the 'Hy-wire', the world's first *drivable* fuel cell and drive-by wire concept vehicle, at the Paris Motor show in september 2002. All of Hy-wire's propulsion and control systems are contained within an 11 inch thick skateboard-like chassis. There is no engine to see, no dashboard, no pedals—merely a single driver-control unit featuring hand controls for driving and stopping.

'Sequal' is the recent fuel cell vehicle from GM, unveiled in 2005 at the North American International Auto Show in Detroit. It has 300-mile (480 km) range and can achieve 0 to 100 kmph in 10 seconds.

'HundroGen 4' is the successor of the fuel cell vehicle `HydroGen 3' developed by GM/Opel and unveiled in 2007 at the International Auto Show in Frankfurt, Germany.

In January 2017, G.M. and Honda have announced a joint venture to build hydrogen fuel cell systems to power electric vehicles. They have already been working since 2013 together to develop a next-generation fuel system.

2. Ford has been working on fuel cell vehicle development and technology demonstration since 2005. 30 Ford Focus FCVs were fabricated and operated as demonstration and testing vehicles for about a decade. In January 2013, Ford formed a partnership with Daimler AG and Nissan Motor Co. for accelerating the commercialization of fuel cell vehicle technology. This collaboration is expected to lead to the launch of affordable, mass-market FCVs soon, hopefully in 4-5 years.

3. Honda's 2017 'Clarity' fuel cell car is a production vehicle, having range of 366 miles (586 km). Hydrogen is stored in two aluminium-lined, composite-reinforced cylindrical tanks (31 gallons and 6 gallons). To aid fuel cell deliver fuel power on critical times, such as at start (fuel cell takes a few seconds to start peak generation), during acceleration or going up a gradient, an air-cooled 1.7 kWh lithium-ion battery pack is also installed under the front seats.

4. Toyota 'Mirai' was unveiled at the November 2014 Los Angeles Auto Show. Cumulative Mirai sales by mid-February 2017 totaled 2840 in Japan, USA, Europe and UAE. According to U.S. E.P.A., the 2016 model Mirai had a total range of 502 km (312 miles).

5. Hyundai Motors of S. Korea also unveiled a fuel cell car model based on Santa FE SUV in March 2001. It had a range of 160 km and maximum speed of 124 km/h. Their third generation FCV (Fuel Cell Vehicle) is based on Tucson, wherein Hyundai's own fuel-cell stack is used which supplies electrical energy to the front drive traction motor of 100 kW output giving 160 km/h maximum vehicle speed. However, this is only a demonstration vehicle.

From there Hyundai moved on to make 1000 compact SUVs made in 2015, based on their iX35 SUV, on 'pre-commercial production' basis. In iX35, the fuel, compressed hydrogen, is stored in two gas cylinders at a pressure of 700 bar— a 104 litre tank behind the rear axle and a smaller 40 litre unit in front of the rear axle. Fuel cell is located under the bonnet and hybrid battery packs of 24 kW are placed under the vehicle, which are used mainly to assist the fuel cell stack when the power demand is highest, e.g., while accelerating or when moving up a gradient. The driving range claimed is 590 km. Next, Hyundai plans to start commercial production of 10,000 vehicles annually on global basis.

The *Freedom* CAR (Cooperative Automotive Research) *and Fuel Partnership* is a research collabora tion among the U.S. Department of Energy, three Detroit auto-manufacturers and five energy companies. Its aim is to develop technology that would allow U.S. auto-manufacturers to decide by 2015 whether hydrogen-powered vehicles could be manufactured on a large scale. To achieve this programme partners are seeking safe, cost-effective methods to produce, deliver, dispense and store hydrogen for vehicles. Research is also sponsored to reduce the size, weight and cost of vehicle components to increase fuel efficiency.

During the first review of the programme, it was found that the most difficult long-term challenge was solving the problem of hydrogen storage in vehicles, which (whether gas or liquid) takes up more space than petrol thus needing large, heavy fuel tanks and more frequent refueling. The review report has suggested extending these studies till 2030-2035 to account for the probable emergence of better hydrogen fuel systems and to make sure that most critical factors in their production and delivery are understood.

Recently (April 2016), Nissan Motor Co. of Japan unveiled in Brazil, World's first Solid Oxide Fuel Cell (SOFT)-powered prototype fuel-cell vehicle which runs on bio-ethanol electric power The 'e-Bio Fuel-cell' prototype vehicle runs on 100 percent ethanol (or ethanol-blended water) to charge a 24kWh battery, enabling a cruising range of more than 600 km. Its carbon-neutral emissions are completely clean. Besides quick acceleration and silent driving, running costs are also low, alongwith a long driving range. Bio-ethanol fuels are mainly sourced from sugarcane and corn, and are widely available in North and South America. Due to the easy availability of ethanol and low combustibility of ethanol-blended water, their distribution is very easy, even off-the-shelf from retail stores.

9. SOLAR CARS

Solar cars run only on solar power from the sun. These cars use photovoltaic cells to convert sun light into energy. Photovoltaic cells are the components in solar panels which convert the energy from the sun into electricity. These cells are made up of light-absorbing semiconductors, usually silicon. The electricity thus produced from sun rays is used to power the electric motor which turns the car wheels. Solar vehicles tend to be flat and wide so as to maximize the surface areas, still retaining the aerodynamic shape. Generally, the maximum speed of solar cars is from 65 to 95 km/hr.

Toyota Prius (2010) was the first solar model to offer a solar-powered cooling system which keeps the interior of the car cooled, even when parked. This smart feature automatically activates as soon as the car interior crosses 68°F.

Ford's 'C-Max Solar Energi' concept debuted in January 2014 at the Consumer Electronics Show in Las Vegas, USA. It was a collaborative project of Ford, Sun Power Corp based in San Jose, California and Atlanta-based Georgia Institute of Technology. Solar Energy is basically a plug-in electric-solar hybrid. Like other plug-in hybrids, it has both petrol (gasoline) engine as well as electric motors. Total range is 994 km (621 miles) which includes 34 km (21 miles) of all-electric range. In case sun is not there, 'C-Max Solar Energi' can be charged from the electric grid like other plug-in electric vehicles.

The roof of the Solar Energi is covered with 1.5 sq. m of photovoltaic cells designed by Sun Power. These cells have metal backing for greater support. The silica cells are much thinner than conventional solar cells making then stronger and more flexible, which helps the car-top in maintaining a streamlined curve. Although these solar cells harvest 50 percent more energy than conventional cells, yet it is still about 320 watts in one day, which is just not sufficient to charge Solar Energi. To help in this, Georgia institute designed a solar concentrator on the car roof, made from acrylic Fresnel lenses to direct sun light to the solar cells while boosting the impact of the sunlight by a factor of eight. Fresnel is a compact lens originally developed for use in light houses. The system tracks the sun as it moves from east to west, drawing sufficient power from the sun through the concentrator equivalent to a 4-hour battery charge (8 kWh) in one day.

According to the estimates of Ford, Solar Energi would reduce car owner's CO_2 emissions by 4.4 tons. Further, the car would be useful where electricity supplies are unreliable or expensive. Of course, the solar car would not be any good for areas with rainy climates where sun is not seen for a major part of the year.

Solar Energi is being tested in the real world to determine its feasibility. *As on now (2017), there is no company in the world which is making a solar production car.*

10. eGALLON

'eGallon' is the term coined by U.S. Department of energy for the cost to drive an electric vehicle the same distance one could go on one gallon of gasoline (petrol). It is calculated as:

$$\text{eGallon (\$/gal)} = FE \times EC \times EP$$

where,

FE = Average comparable passenger car adjusted combined fuel economy in miles/gallon.

EC = Average electricity consumption of the top 5 selling PEVs (Plug Electric Vehicles) in the U.S. (kWh/mile)

and EP = Average U.S. electricity price ($/kWh).

11. MILES PER GALLON EQUIVALENT

'MPGe' is a term used in U.S.A. to compare efficiencies of alternative types of vehicle motive power plants, e.g., I.C. engine, electric motor, etc. This is done using a conversion determined by the Argoma National Laboratory of the U.S. Department of Energy, which determined that one gallon of regular gasoline (petrol) contains the equivalent of 33.7 kWh of electricity.

Thus, to determine the fuel consumption of a vehicle with only I.C. engine, fuel tank is filled completely and the vehicle is driven up a specific distance after which the fuel tank is filled to full capacity again, which gives the fuel consumed to cover the specific distance. Dividing this distance by the amount of fuel gives MPG Miles Per Gallon (or likewise, kilometers per litre).

In the same way, a pure electric vehicle is also driven a specific distance after the battery is fully charged after which it is again charged fully. The electricity consumed in kWh to recharge the battery to full charge, divided by 33.7 converts it into '*equivalent gallons*'. Dividing the distance travelled by the equivalent gallons would give 'Miles Per Gallon Equivalent' (MPGe).

Hybrid vehicles would then have two ratings, *viz.*, MPGe rating which gives the efficiency when running with a charged battery, and the traditional MPG rating when driving only with IC engine after the battery is depleted.

REFERENCES

1. Keith Read, "Fuel for thought", Electric & Hybrid Vehicle Technology International, July 2013, pp 32-39.
2. Steve Williams, "Hub motor dynamics", Vehicle Dynamics International, Annual Showcase 2011.

THEORY QUESTIONS

1. Describe briefly the layout and operation of an electric vehicle. Explain also its advantages and disadvantages compared to a conventional vehicle powered by a petrol engine. State salient features of an electric vehicle.
2. What is a hybrid vehicle? Discuss clearly various hybrid systems used for automotive purposes alongwith the merits and demerits of each. Describe also the main components of a hybrid vehicle.
3. What is a plug-in hybrid? Describe various modes in which it can operate, stating also their relative advantages and disadvantages.
4. Describe in detail the battery requirements for various electrical/ hybrid vehicles.
5. Discuss in detail various advantages, disadvantages and modes of operation of automotive wheel motors.
6. What is a fuel cell? Describe its working and explain advantages. Give typical examples of vehicles based on fuel cells.
7. Describe briefly, the main features of any concept solar car.

QUESTIONS FOR VIVA

1. What is an electric vehicle?
2. Name various types of electric vehicles.
3. State advantages of an electric vehicle.
4. How long does it take to charge an electric car?
5. What is approximate driving range of an electric car?
6. What is approximate driving speed of an electric car?
7. How long does an electric vehicle battery last?
8. What is regenerative braking?
9. Which type of batteries are used in an electric vehicle?
10. Which type of electric motor is usually employed in an electric vehicle?
11. What does the drive controller of an electric car consist of?
12. What is a hybrid vehicle?
13. What are 'series' and 'parallel' hybrid systems?
14. Which type of hybrid system is more popular?
15. Enlist main components of a hybrid electric car.
16. Give two typical examples of a modern hybrid electric car.
17. What is a 'series-parallel hybrid'? Give two practical examples.
18. What is a (*i*) Mild hybrid, (*ii*) Full hybrid?
19. What is a plug-in hybrid?
20. Name operating modes of a plug-in hybrid.
21. Give two typical examples of a plug-in hybrid.
22. Compare plug-in hybrids with conventional hybrids.
23. What is a micro-hybrid vehicle?

24. What is a wheel motor?
25. Give two examples of typical vehicles with wheel motors.
26. What is a fuel cell?
27. Name the fuel in a fuel cell?
28. What are the reactants in a fuel cell?
29. What is the cell voltage in case of a fuel cell stack?
30. How is a fuel cell different from an ordinary battery cell?
31. Name various type of fuel cells.
32. Which type of fuel cells is suitable for automotive application?
33. Name the emission products in a fuel cell.
34. Name a few vehicles based on fuel cell.
34. Name a few vehicles based on fuel cell.
35. What is a hydraulic hybrid vehicle?
36. Name the two type of hydraulic hybrid vehicles.
37. State the advantages of a hydraulic hybrid vehicle.
38. What are the limitations of a hydraulic hybrid vehicle which restrict their practical utility?
39. What is 'Solar Energi'?

MULTIPLE CHOICE QUESTIONS

Mark the correct answers:

1. Out of the full-mild- , and micro-hybrids, the largest battery is used in case of:
   (*a*) Full hybrid (*b*) Mild hybrid
   (*c*) Micro hybrid (*d*) Both mild- and micro-hybrids
2. Out of the full- , mild- and micro-hybrids, the smallest battery is used in case of:
   (*a*) Full hybrid (*b*) Mild hybrid
   (*c*) Micro hybrid (*d*) Both full-and micro-hybrids
3. Which of the following statements in case of fuel cells is correct?
   (*a*) Cell voltage is 1.6 volts
   (*b*) Fuel cell cannot accept regenerative braking.
   (*c*) Fuel cell cars have high emissions.
   (*d*) Fuel cell life is very low.
4. 'Solar Energi' is a
   (*a*) Pure solar car (*b*) Solar electric hybrid car
   (*c*) Solar-electric plug-in hybrid car (*d*) Micro-hybrid car
5. Out of trial range of 994 km of 'Solar Energi' concept, pure electric range is about
   (*a*) 35 km (*b*) 65 km
   (*c*) 95 km (*d*) 125 km

[**Key:** 1. (*a*), 2. (*c*), 3. (*b*), 4. (*c*), 5. (*a*)]

Constructional Details–I

3

1. INTRODUCTION

The main parts of an automotive engine are:

1. Cylinder block and crankcase	2. Cylinder head
3. Sump or oil pan	4. Manifolds
5. Gaskets	6. Cylinders
7. Pistons	8. Piston rings
9. Connecting rods	10. Piston pins
11. Crankshaft	12. Main bearings
13. Valves and valve actuating mechanisms	14. Mufflers.

Out of these, components from S. No. 1 to 10 will be discussed in detail in this chapter, whereas remaining components shall be taken up for discussion in the next chapter.

2. CYLINDER BLOCK AND CRANKCASE

The basic framework of the engine is formed by the cylinder block. It houses the engine cylinders, which serve as bearings and guides for the pistons reciprocating in them. Around the cylinders, there are passages for the circulation of coolant. At the top of the cylinders, these passages end either just below the block's machined joint face, which is then called 'closed deck' or at the level of the joint face, which is then known as 'open deck'. In case of a closed deck, circulation of the coolant is provided by vertical drilled holes mating with the corresponding holes in the cylinder head. This is preferred over the open deck on account of its greater reliability to provide leak proof joint, although the later is relatively easier to produce.

Cylinder block also carries lubrication oil to various components through drilled passages called oil galleries. During manufacture, the oil holes are drilled and rarely cast. In case of a curved passage, intersecting drilled holes are used. The ends of the drilled holes are sealed by plugs. At the lower end, crankcase is cast integral with the block. The function of the crankcase is to provide support for the main journals and bearings of the crankshaft, rigidly maintaining the alignment of their axes of rotation under various engine loads. Previously, crankcase was cast separately and attached to the block, but now the integral or the monoblock construction is preferred. Crankcase is shaped simply like a box having no bottom, its roof being formed by the lower deck of the cylinder block. The bottom of the crankcase walls is flanged to strengthen the casing and to provide a machined joint face for the sump to be attached. Crankshaft is supported in the crankcase through a number of bearings called main bearings. The construction of the crankcase has to be such as to provide very high rigidity, because it must provide reactions for the heavy forces set up due to gas pressures in the cylinders. Fig. 3.1 depicts the bottom view of cylinder block and crankcase showing main bearing housings to support the crankshaft.

At the top of the cylinder block is attached the cylinder head. Besides, other parts like timing gear, water pump, ignition distributor, flywheel, fuel pump, etc. are also attached to it. Mating surfaces of the block are carefully machined to provide perfect sealing surface for attaching these parts. The function of the camshaft is to phase the opening and closing of the inlet and the exhaust valves for each cylinder. Camshaft may be mounted in the cylinder block or head or even above the cylinder head. It is always placed parallel to the crankshaft. When in the cylinder block, it is mounted to one side of the cylinders either low down, just above the crankshaft or high, slightly below the cylinder head (Fig. 3.1). The camshaft is supported in usually three bearings force fit in the cylinder block bore. Main bearing bores and camshaft bearing bores are align-bored to allow these shafts to rotate freely. The centre lines of all cylinders must be at exact right angle to the crankshaft centre line to avoid side stresses on pistons, connecting rods and cylinders, thus ensuring free movement of pistons. Likewise, valve lifter bores are also machined at right angle to the camshaft bearing bores.

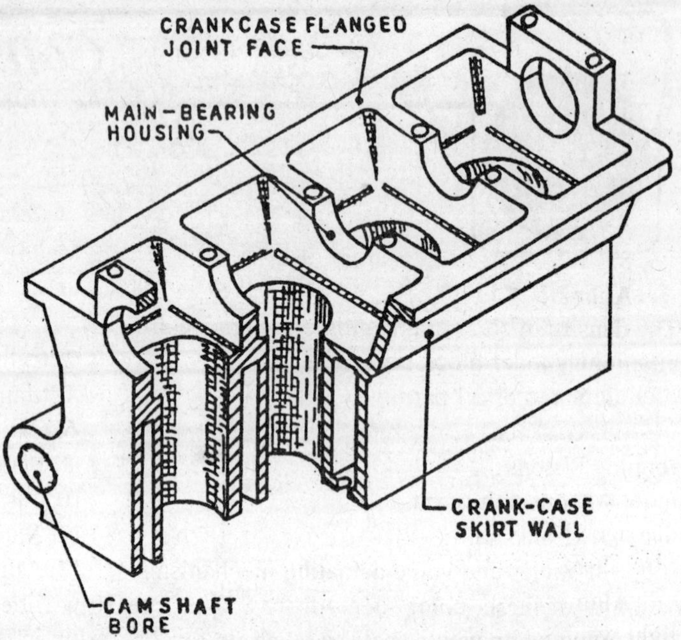

Fig. 3.1. Crankshaft support in the crankcase

Cylinders in the block are machined and honed to a very accurate finish for good lubrication and proper ring seating. Cylinder walls are finished to 1 micron finish with a fine rotating hone. An intersecting visible crosshatch scratch pattern is produced by the hone form.

The cylinder block is basically a casting product. Both monoblock and individual cylinder casting techniques have been tried but the former has proved more useful due to the following reasons and is, therefore, being employed universally.

1. The problem of having good water sealed joints is reduced, because of reduction in their quantity.
2. More positive water circulation is achieved.
3. The manufacturing operations, like machining the joint faces, cylinder boring etc. are simplified.
4. Closer cylinder spacing is achieved, thus reducing engine size.
5. An assembly far rigid than in the individual cylinder construction is achieved, thus reducing tendency to vibrate.

The individual cylinder construction, on the order hand, has its own advantages:

1. Replacement of a single cylinder casting is less costly than replacing a multi-cylinder casting.
2. Because of lesser weight of individual cylinders, handling during repairs is easy.
3. In case of air cooled cylinders having fins around them to increase the heat transfer area, it is easier to adopt individual cylinder construction than to cast en-block.

The materials used for cylinder block are grey cast iron and aluminium alloys. Grey cast iron is so called because of its grey appearance when fractured (compared to white appearance of white cast iron fracture). In this the carbon is present as flakes of graphite, which makes it more wear and corrosion resistant, apart from its better mach inability. The aluminium alloys used for cylinder block are the aluminium-silicon types which have good casting properties, are corrosion-resistant and retain their strength at moderate temperatures encountered in the engine block. Out of these two materials, cast iron is the one which is mainly used because of the following advantages:

1. It is a good foundry material.
2. It has high mach inability.
3. It does not wrap under the high temperatures and pressures developed in the cylinders.
4. Due to its slightly porous nature, it retains better the lubricating oil film.
5. It does not wear too much.
6. It has sound-damping properties.
7. It has a low value of coefficient of thermal expansion.
8. It is relatively cheap.

Aluminium alloys have been used as alternative cylinder block material chiefly owing to their lightness. The density of aluminium is about one third that of cast iron. However, considering the lesser strength of aluminium which necessitates the use of thicker sections to carry same load, the saving in weight in case of aluminium alloys is only about 50% compared to cast iron.

Besides this, aluminium has a higher thermal conductivity than cast iron. This results in cooler engine running, thereby making it possible to use higher compression ratios. However, it has the disadvantages of more wear, to obviate which cast iron liners have to be used. Besides this, the threads in aluminium are damaged easily in case of any mishandling. Further, since it cannot stand high temperatures unlike cast iron, there are more chances of damage occurring in case of any loss of coolant.

Aluminium alloy cylinder blocks with chromium plated cylinder bores have also been used to obtain light weight and also less wearing surface.

The grey cast iron for the cylinder block usually has the composition : 3.5% carbon, 2.5% silicon and

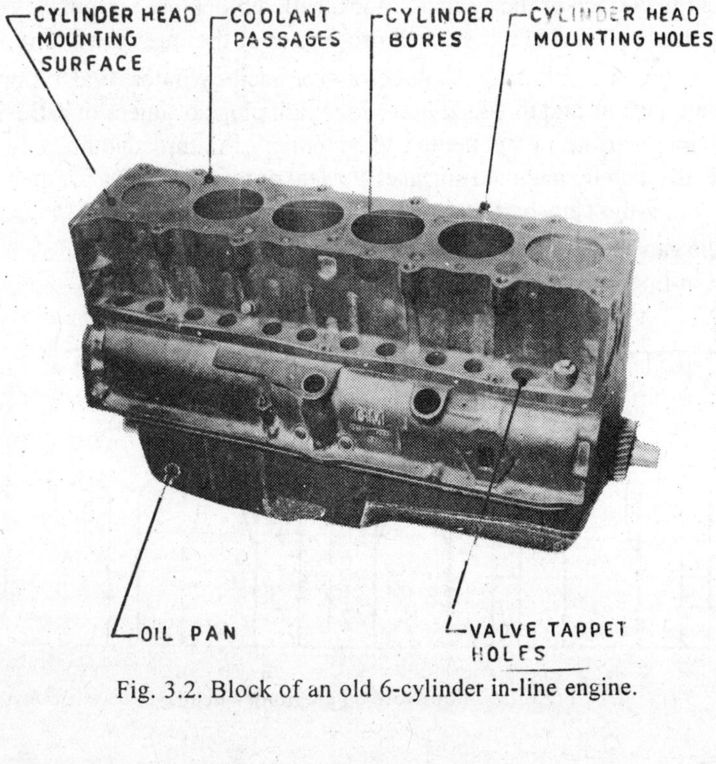

Fig. 3.2. Block of an old 6-cylinder in-line engine.

0.65% manganese. Carbon serves to provide graphite which improves lubrication, silicon provides wear resistance by forming pearlitic structure, while manganese increases the strength and toughness. A typical aluminium alloy for cylinder block contains 11% silicon, 0.5% manganese and 0.4% magnesium. Silicon reduces expansion and increases strength and wear resistance, while manganese and magnesium improve strength of aluminium structure.

Cylinder blocks for diesel engines are similar, in general, to petrol engines, except that the former are heavier in construction. The increased strength and rigidity are required on account of much higher combustion pressures produced in diesel engines. This causes greater loads on cylinders, crankshafts, main bearing supports and caps resulting in their heavier construction.

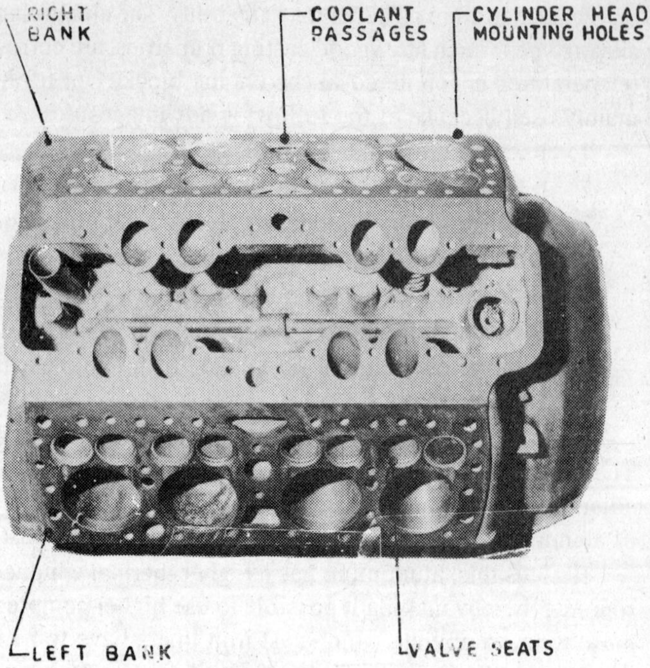

Fig. 3.3. Block of a V-8 engine.

Fig. 3.2 shows cylinder block for a 6-cylinder in-line engine. The engine block for V-8 engine is given in Fig. 3.3.

3. CYLINDER HEAD

It is attached to the top surface of the cylinder block, called block *deck*, by means of studs fixed to the block. The gaskets are used to provide a tight leak proof joint at the interface of the head and the block.

The cylinder head forms a combustion chamber above each cylinder. It also contains valve guides, valve seats, ports, coolant jackets and threaded holes for spark plugs or injections. Besides, there are other threaded holes for attaching various components and a number of unthreaded holes for attaching the head to the block. There are also some machined surfaces for leakproof joints. On overhead camshaft engines, the provisions for mounting the camshaft and related parts also exists.

Depending upon the valve and port layout, the cylinder head may be classified as loop-flow type, offset cross-flow type or the in-line cross-flow type. In the loop-flow type (Fig. 3.4), the inlet and the exhaust

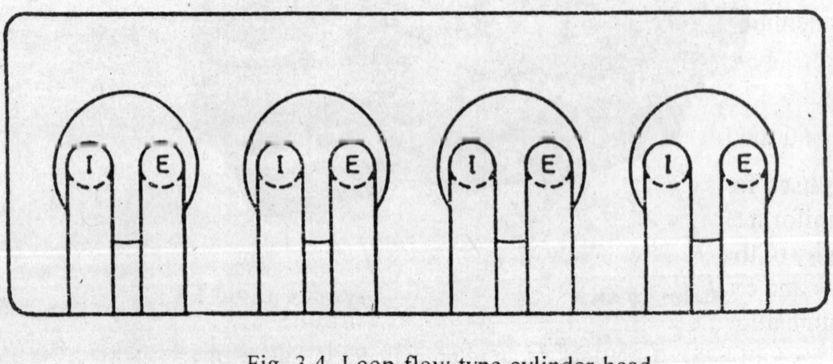

Fig. 3.4. Loop flow type cylinder head.

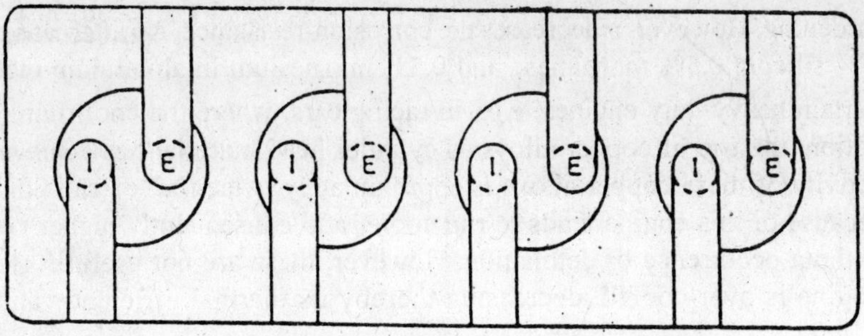

Fig. 3.5. Offset cross-flow type cylinder head.

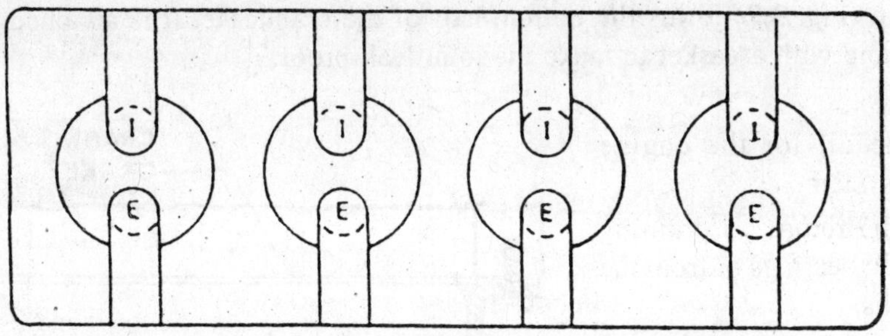

Fig. 3.6. In-line cross-flow type cylinder head.

manifolds are on the same side, which facilitates preheating of the intake air. When, however, with the same valve arrangement, the inlet and the exhaust manifolds are placed on different sides of the cylinder head, it is called offset cross-flow type (Fig. 3.5). This type gives lower exhaust valve temperatures. In the in-line cross-flow type (Fig. 3.6), the valves are positioned transversely and usually inclined to each other, while the inlet and the exhaust manifolds are on different sides of the cylinder head. This arrangement gives better performance, but is costlier.

The head is almost universally cast en-block even for individual cylinder constructions for the cylinder block.

The cylinder heads cast integral with cylinder blocks have also been produced in very few cases and that too in racing engines, which obviates the necessity of a gas tight joint. But this advantages of the integral construction is not very important, because use of gaskets gives a reasonably good gas tight joint. Further the detachable head types have many other advantages over the integral construction:

1. From production point of view, the cylinder block casting with open bore for the detachable type head, is much more simplified.
2. Operations like decarbonizing and valve grinding are simplified.
3. The compression ratio can be changed slightly by changing the thickness of the gasket used between the block and the head.

The materials used for cylinder heads are generally cast iron and aluminium alloy. Apart from weight reduction, more uniform temperature is maintained in case of aluminium cylinder heads because of greater thermal conductivity of the aluminium alloy. Quite often, aluminium alloy head is also used along with cast iron cylinder block and crankcase. Grey cast iron for cylinder head is the same as that used for the cylinder block, but aluminium alloys are usually different, a typical one containing 3% copper, 5% silicon and 0.5% manganese in a matrix of aluminium. Copper increases the hardness and strength of aluminium with time on account of age-hardening. However, it decreases the corrosion-resistance. An alternative alloy commonly used consists of 4.5% silicon, 0.5% manganese and 0.5% magnesium in aluminium matrix.

However, for certain heavy duty engines, *e.g.*, in racing cars, where the cool running of the engines is a major consideration, the use of copper alloys as cylinder head material has achieved much success. The thermal conductivity of these copper alloys is approximately twice that of aluminium and six times that of cast iron. Because of this engine tends to run cooler and consequently higher compression ratios can be employed without occurrence of detonation. However, these are not useful for ordinary engines, because then the engine is over-cooled, decreasing thereby its thermal efficiency and increasing fuel consumption.

4. OIL PAN

Oil pan or sump (Fig. 3.7) forms the bottom half of the crankcase. It is attached to the crankcase through set screws and with a gasket to make the joint leak-proof.

Its functions are:

1. To store the oil for the engine lubricating system.
2. To collect the return oil draining from the main bearings or from the cylinder walls.
3. To serve as a container in which any impurities or foreign matter, *e.g.*, liquid fuel, condensed water, blow-by gases, sludge, metal particles etc., can settle down.
4. To enable the hot churned up lubricating oil to settle for a while before being circulated.

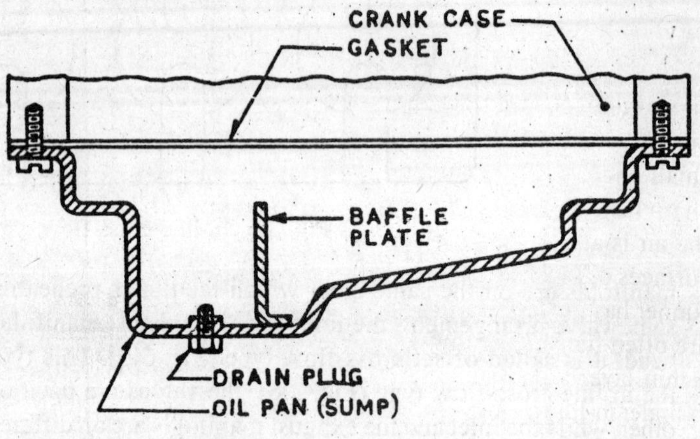

Fig. 3.7. Oil pan or sump.

5. To provide for cooling of the hot oil in the sump by transfer of heat to the outsider air stream.

The sump has a shallow downward slope at one end which merges with deep, narrow reservoir at the other end. This is done to ensue that under all conditions of vehicle running, there will be oil in the reservoir where oil pump is mounted. Inside the sump, baffles are usually provided to reduce oil surging during running of the automobile. At the bottom of the oil sump a drain plug is provided to drain out the dirty oil at the time of oil replacement.

Generally, the sump is made of pressed steel sheet since it is not expected to have much rigidity. However, in some cases, sump of aluminium alloy casting is used, which has adequate stiffness and rigidity. This also provides better oil cooling on account of its higher thermal conductivity. Moreover, with such a sump, particularly when provided with outside ribs, resonant vibration noise is avoided. However, cast sump cannot withstand shocks, which may cause cracks. On the other hand pressed steel sheet sump in such a situation may be dented, but will not crack.

5. MANIFOLDS

There are separate sets of pipes attached to the cylinder head which carry the air-fuel mixture and the exhaust gases. These are called manifolds.

5.1. Inlet manifold

The *inlet manifold* carries the air-fuel mixture from the carburettor to the cylinders. The shape and

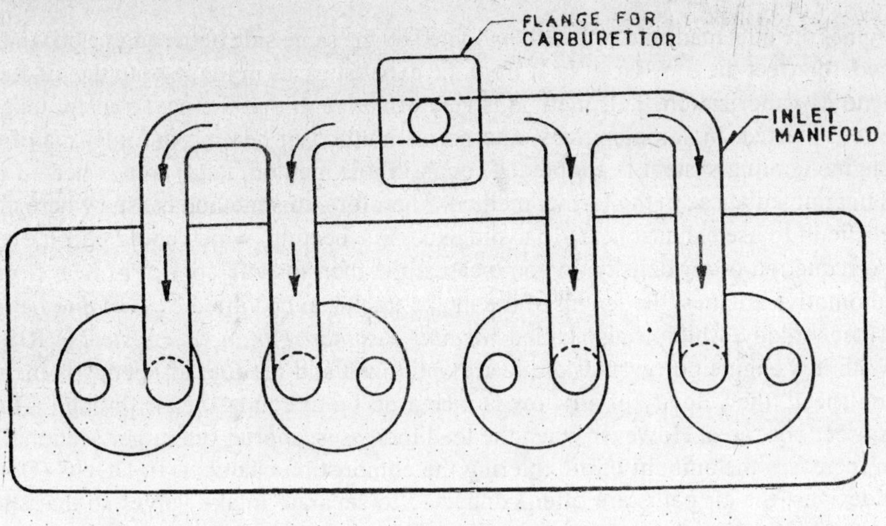

Fig. 3.8. Inlet manifold for 4-cylinder in-line engine with single carburettor.

size of the inlet manifold must be such as to prohibit the formation of fuel droplets without restricting the air flow. The manifold must be large enough to allow sufficient flow for maximum power. Yet it has to be small enough to maintain adequate velocities for keeping the fuel droplets in suspension in air. Further, sharp bends in the inlet manifold tend to increase fuel separation. On account of its relatively lesser inertia, the air is able to travel through turns more quickly compared to the heavier fuel droplets. Rough interior surfaces of the manifold runners will increase the resistance to the mixture flow. Although a circular section runner has the largest cross-sectional area for its wall surface area, yet passenger car inlet manifold runners are often flat so that any liquid fuel dropping out of the charge may spread evenly on the manifold floor in a thin layer and evaporate quickly. Ribs and guide vanes are usually employed in the manifold for a 4-cylinder in-line engine fed by a single carburettor, while the inlet manifold for the engine fed by two carburettors is shown in Fig. 3.9. Various manifold passages are also called *runners*. Intake manifolds on V-8 engines are usually designed with runners on two levels and are fitted between the two heads. Cylinders, in the sequence of their firing order, are fed successively, alternatively from the upper and the lower runners.

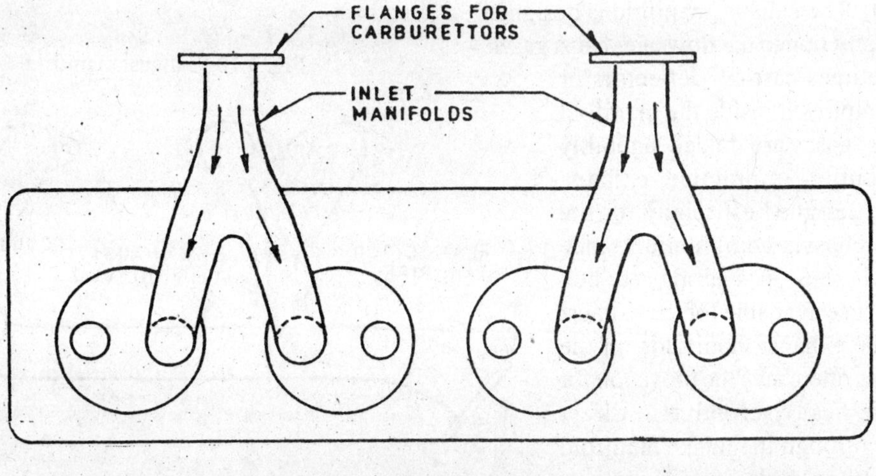

Fig. 3.9. Inlet manifold for 4-cylinder in-line engine with two carburettors.

On in-line engines the inlet manifold is usually mounted on the same side of the engine on which exhaust manifold is located, to effect the transfer of heat from the exhaust gases in the later to the air-fuel mixture in the former, to aid its vaporization. This method is very effective in that the heat transfer takes place as soon as the engine is started. In some engines, alternative method of heating the inlet manifold by hot water from the engine cooling system is adopted. Though in this method, large area is heated uniformly, yet the heating is not immediate as in the former method. Therefore, this method is used where mechanical design makes it difficult to use exhaust heat. To avoid excessive heating, which could become a source of power loss due to reduction of air density, in some cases, the thermostatic control of heat flow is used.

In modern automotive engines, the length of the intake track can be varied by *Variable Length Intake Manifold* (VLIM) technology, which is also called *Variable Resonance Induction System* (VRIS) by some manufacturers. With this engine power and torque are optimized and the fuel efficiency is increased.

At low engine speed, the velocity of air flow is increased by directing the air through a longer path with limited cross-sectional area. However when the load increases, shorter but larger sectional area path is opened, which increases the amount of air entering the combustion chamber. In DOHC (Double Over Head Camshaft) designs, the air paths are often connected to separate intake valves so that shorter path can be excluded by de-activating the concerned intake valve itself.

Some examples of automotive engines where this technology is employed are:

1. Ford DSI (Dual-Stage Intake) on their Duratec 2.5 and 3.0 litre V65 and 2.0 litre Zetec four.
2. Mazda VICS (Variable Intake Control System) in Mazda K V6 engines.
3. Mitsubishi Cyclone is employed in their 2 litre I44G63 engines.
4. Porsche Vario Ram is used on 964, 993, 996 and Boxter engines.
5. Toyota T-VIS (Toyota-Variable Induction System) is employed in their 3S-GE engines.
6. Volvo VVIS (Volvo Variable Induction System) is used on their B52 engines.

5.2. Exhaust manifold

The *exhaust manifold* is the set of pipes carrying exhaust gases from the cylinder head to the exhaust system (Figs. 3.10 and 3.11). It is generally made of cast iron so that it is able to withstand the heat of the exhaust gases. A smoother flow of exhaust gases is provided by the two-in-one manifold (Fig. 3.11) as compared to the simple one (Fig. 3.10). The exhaust manifold is designed to minimise restriction to the flow of exhaust gases. For this sometimes cast-rib deflectors or dividers are employed inside the manifold so that exhaust gases are guided smoothly towards the outlet. In practice, exhaust manifolds are designed especially for the given engine-chassis combination since chassis front suspension, steering gear box etc. limit the space available for the exhaust manifolds. The exhaust manifolds on the modern engines often have the provision for a thermostatic heat riser valve to direct exhaust gases through the intake manifold cross-over passage during warming up.

Fig. 3.10. Exhaust manifold.

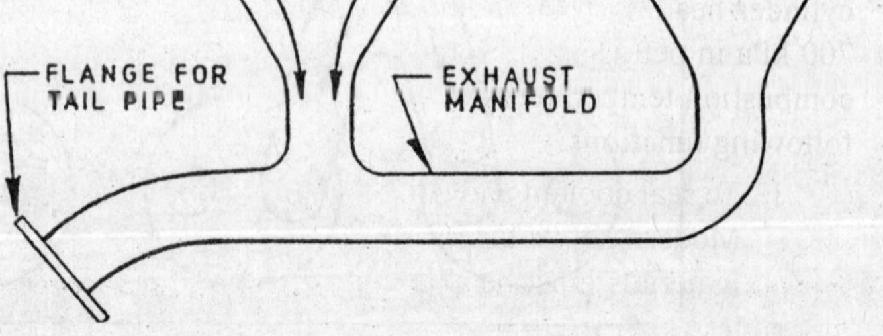

Fig. 3.11. Two-in-one exhaust manifold.

6. GASKETS

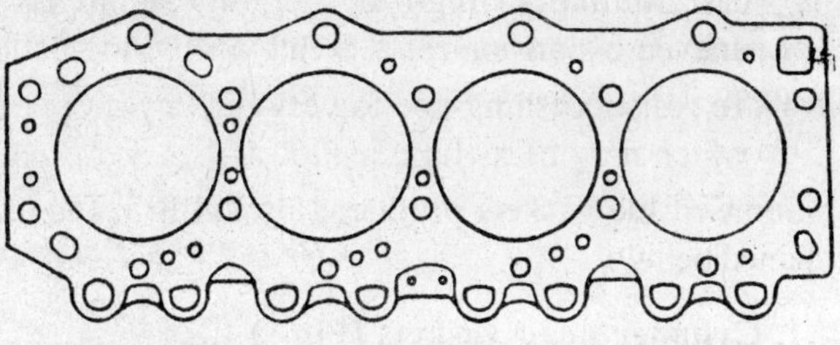

Fig. 3.12. A typical cylinder head gasket.

Gaskets are used to provide a tight fitting joint between two surfaces: for example, the joint between cylinder head and block, between crank-case and oil pan, between the cylinder block and manifolds or coolant pump.

Main requirements of gaskets are:

1. **Conformity.** The gaskets must be able to conform to the mating surfaces which may have roughness or warpage.
2. **Impermeability.** The gaskets must be impermeable to the fluid it is expected to seal.
3. **Resistance to chemical attack.** The gaskets should be resistant to the chemicals with which it may come into contact, *e.g.* a cylinder head gasket should be resistant to the fuel, products of combustion, coolant and engine oil.
4. **Resistance to operating conditions.** As an example, the cylinder head gasket should be resistant to widely fluctuating and pulsating internal pressures and temperatures and exposure to flame.
5. **Provision of apertures.** The gaskets must have apertures for any studs, bolts, opening etc. For instance, the cylinder head gasket should have apertures for cylinders, coolant and lubrication passages, studs and bolts and valve push rods.

Various materials used for different engine gaskets are:

1. **Cork.** It is one of the oldest gasket materials obtained from the bark of oak and cork pine trees. Use of cork gaskets is limited to lightly loaded joints with uneven surfaces, *e.g.* rocker covers, oil pans etc. However, a cork gasket is highly impermeable and conforms easily.
2. **Asbestos.** Fibres of asbestos alone or alongwith fibres of cellulose are bonded together; the binder used determines the gasket properties. Some such gaskets swell to seal the joint. However, such swelling gaskets cannot be used where high pressures are present. Generally, asbestos gaskets, instead of cork gaskets, are used for joints under high pressures. However, these are not as conformable as cork; which necessitates better parting surface smoothness in case of asbestos gaskets. Such gaskets are no longer in use due to asbestos being injurious to human health.
3. **Rubber.** Moulded oil-resistant synthetic rubber gaskets are frequently used in oil pan corner joints and on the end joints of intake manifold.

It is important to remember that each time a joint is assembled, a new gasket has to be used, since after one time use, a gasket loses most of its sealing properties. Quite commonly, the labour cost of installing a gasket would be rather high, compared to the cost of the gasket itself.

The most critical gasket on an engine is the cylinder head gasket, which seals the joint between the cylinder head and the engine block deck. Such a gasket has to seal combustion pressures of upto about 700 kPa in petrol engines and 2000 kPa in turbocharged diesel engines. Besides, it also has to withstand combustion temperatures of more than 1100°C. Besides this, a cylinder head gasket also has to perform following functions:

1. To seal coolant and hot, thin lubricating oil flowing under pressure between the block and the head. Modern coolants and lubricating oils tend to cling to surfaces and soak into gaskets. Hence gasket materials must be selected carefully.
2. To resist the forces tending to scuff gasket surfaces and preventing proper sealing.
3. To accommodate the uneven expansion/contraction rates in case of bi-metal (aluminum head and

cast iron block) engines, since the aluminum expands almost double than cast iron due to which uneven expansion rates create a shearing action in the gaskets.

4. To resist crushing due to very high cylinder damping forces unevenly distributed across the head, which may be as high as 900 kN.

Some of the gaskets produced by Fel-Pro. Inc. of U.S.A. are explained below:

1. Cylinder head gaskets (Fig. 3.13)

Embossed steel or shim type is generally used on I-head engines.

Metal sandwich type is made of either copper and asbestos or steel and asbestos; mainly used on passenger cars, trucks, etc.

Shimbestos is a combination of thin steel heat shield on one side and a highly resilient, specially treated metal reinforced asbestos on the top of side.

Felbestos is made of perforated steel sand- wiched between two treated asbestos sheets.

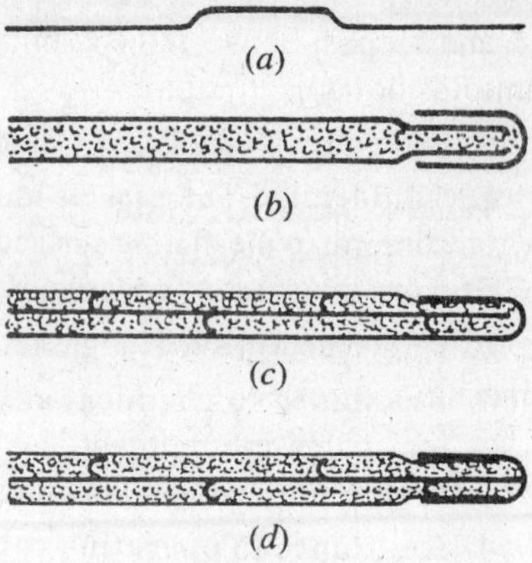

Fig. 3.13. Cylinder head gaskets: (*a*) Embossed steel type, (*b*) Metal Sandwich type, (*c*) Shimbestos, (*d*) Felbestos.

2. Oil Pan Gaskets

These are available in three different materials:

(*a*) **Cord** has the advantage of high compressibility but is subject to shrinkage and expansion.

(*b*) **Felcoid** is an improvement on cork. It is highly compressible and able to withstand considerable bending and twisting without the danger of breakage. It is less subject to shrinkage and expansion.

(*c*) **Felcoprene** is a further improvement. It is a synthetic rubber compound that is highly resilient and compressible and is not affected by oils and greases. It is not subject to expansion or shrinkage and can withstand rough handling.

3. Manifold Gaskets

Three types are available—the *metal encased* asbestos; *felbestos* which is perforated steel base with asbestos mechanically bonded to one or both sides, and *metal embossed* shim gaskets. Metal encased types are more resistant to burn out but are more expensive.

4. Pump Gaskets

A number of materials are available for pump gaskets such as asbestos, karropak, felcoid, etc. They are treated to withstand oil, water, petrol and anti- freeze liquids. This treatment also provides softness, flexibility and tensile strength.

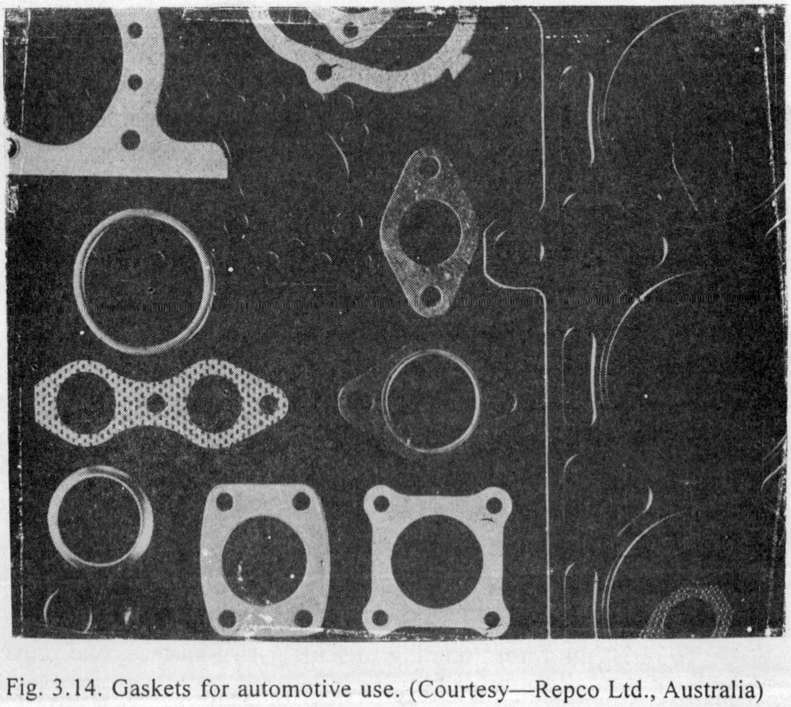

Fig. 3.14. Gaskets for automotive use. (Courtesy—Repco Ltd., Australia)

Karropak is a high quality vegetable fibre, while *Felcoid* is a combination of fibre and cork granules that is more compressible and resilent.

Fig. 3.14 shows some gaskets for cylinder heads, mani- folds, oil pans, tuning covers, carburettors, etc.

Dana Corporation of U.S.A. also makes the automotive gaskets and markets them under the trade name 'Victocor'. Their important products are:

1. Cylinder head gaskets

Victocor (Fig. 3.15) has a steel core which gives it excellent strength. The steel core is mechanically clinched to a facing material which is resistant to temperature, oil and coolant and prevents separation and shape changes under extreme engine operating conditions. There is a further coating of silicon-Teflon on the facing material which provides the combined sealing and non-stick feature required in particular in bimetallic engines.

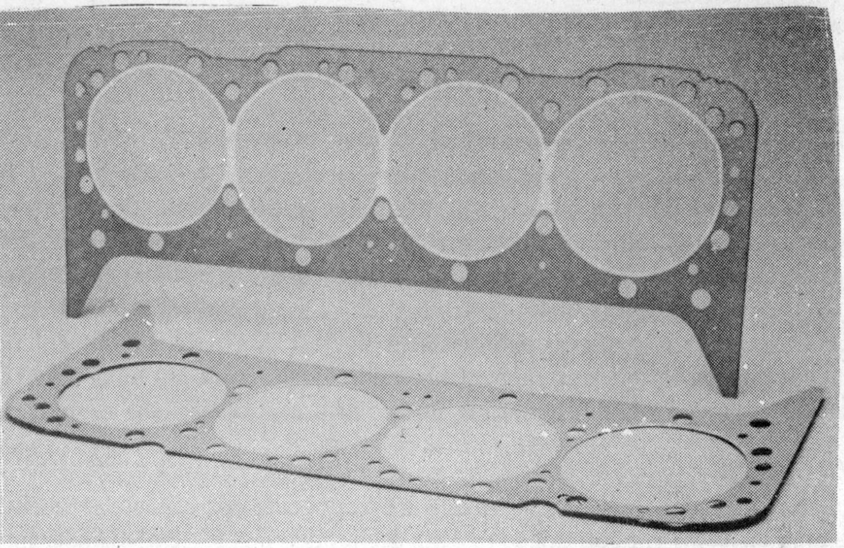

(*a*)

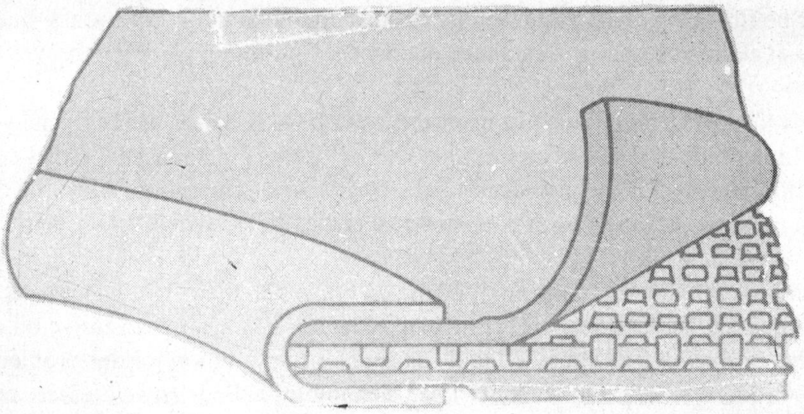

(*b*)

Fig. 3.15. Victocor cylinder head gaskets (*a*) gaskets, (*b*) composition. (Courtesy—Dana Corp., U.S.A.)

Silicor (Fig. 3.16) is the gasket for engines where radial strength is important. In this the core is solid steel for improved strength. A composition facing materials is chemically bonded to each side of the core to provide outstanding conformability. A red Teflon micro-seal coating on the facing material acts as both a releasing agent and a lubricant.

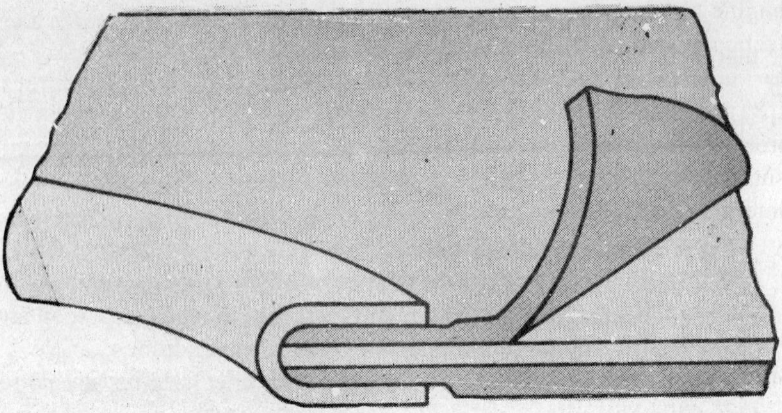

Fig. 3.16. Silicor cylinder head gasket.
(Courtesy--Dana Corp., U.S.A.)

Ultraseal 2000 series gaskets retain the pierced core design of Victocor but adds a thicker core material and a higher density, black, asbestos-free facing material. It combines the merits of Victocor and Silicor.

2. Other gaskets

Tuff cork gaskets are laminated gaskets with two outer layers of cork composition bonded to a core of steel. These provide better torque retention than standard cork gaskets because of steel core. These are resistant to hot oil and coolant. These are used for oil pan, valve cover, timing cover, water pump, water outlet, differential cover, fuel pump, etc.

Cork-rubber gaskets consist of cork particles held together by a synthetic rubber binder and therefore provide all the advantages of cork and rubber. It compresses like cork but holds its shape like rubber. It does not shrink, crack or became brittle during storage or use. It is resistant to oils, fuels and solvents.

7. CYLINDER LINERS

The problem of cylinder wear in the engines is a very acute one especially when cylinder block is made from aluminium alloy. The solution to this has been found in the use of cylinder liners, which can be replaced when these are worn out. Liners are also used to restore to its original bore size a cylinder block which has been rebored beyond allowable limits. They are made in the form of barrels from special alloy iron containing silicon, manganese, nickel and chromium. These are cast centrifugally.

The liners may be further hardened by nitriding or chromium plating. For nitriding, liners are exposed to ammonia vapour at about 500 degree C and then quenched. Chromium plating improves their resistance to wear and corrosion.

Aluminium alloy liners with chromium plating on the inside have also been used especially in combination with aluminium cylinder blocks. The use of aluminium alloy results in increased thermal efficiency due to better heat conduction. Further, as the pistons used are invariably of aluminium alloy, the relative thermal expansion between the liner and the piston does not take place, due to which larger cold clearance is not needed.

In modern higher output engines with small interbore bridges in their light weight aluminium blocks, cylinder bore distortion and potential for affecting performance and causing excessive oil consumption is of major concern. Recently (2012), Federal-Mogul has developed a solution for this problem in downsized petrol engines, which is a dual-material cylinder liner technology, called *'Hybrid Liner'* by the company.

The 'Hybrid Liner', produced at Federal-Mogul's Friedberg (Germany) factory, consists of a conventional cast-iron sleeve coated on the outside with on AlSi 12 alloy with a melting point below that of the aluminum engine block, using a proprietary Federal-Mogul process. This process employs wire-arc thermal spray technology which ensures that the coverage, thickness and bonding strength are uniform over the complete height and circumference of each liner. The outer layer surface topography mechanically locks the liner into the aluminium block casting, structurally and thermally both, unlike alternative technologies. The cast iron sleeve provides good tribological properties and is compatible with piston-ring packs. The coating surface roughness is also controlled to provide undercuts and micro-porosities required for forming a strong intermetallic transition zone between the coating and the cast aluminium block material.

The advantages claimed for the Hybrid Liner are:

1. Compared to alternative technologies, it reduces bore distortion in a running engine by two-thirds. Maximum second-order bore distortion in conventional cast-in liners is seen as about 11%. Cylindrical distortion under operating loads is claimed to be upto three times better, which results in upto 40% reduction in oil consumption.
2. Use of this liner also results in upto 30% higher heat transfer rates, which can reduce cylinder wall temperature by upto 40°C.
3. Increased dynamic strength claimed for this liner. Federal-Mogul's cyclic pulsing pressure test showed the survival of this liner at pressures up to 200 bar, whereas a standard liner block design cracked at 100 bar.
4. With this liner, it is claimed that a 3 mm wall thickness between cylinder bores is sufficient on pressure-die-cast blocks, which means reduced space between adjacent cylinders and hence improvement in overall engine packaging by OEMs.
5. With a water channel drilled between two cylinders, the Hybrid Liner technology is also claimed to limit the crack formation, allowing coolant to penetrate only upto the liner coating. In convention liners, coolant may seep down the joint between the liner and the casting, reaching the crankcase and contaminating the lubricating oil.

The cylinder liners are of two types, the dry and the wet type. These are discussed below in detail:

7.1. Dry Liners

As shown in Fig. 3.17, this type of liner is made in the shape of a barrel flange at the top which keeps it into position. The entire outer surface bears against the cylinder block casting and hence has to be machined very accurately both from the inside and the outside. It is put in position by shrinking the liner. This introduces some stresses due to shrinkage and hence the linear bore has to be machined accurately again after the liner has been put into the cylinder casting.

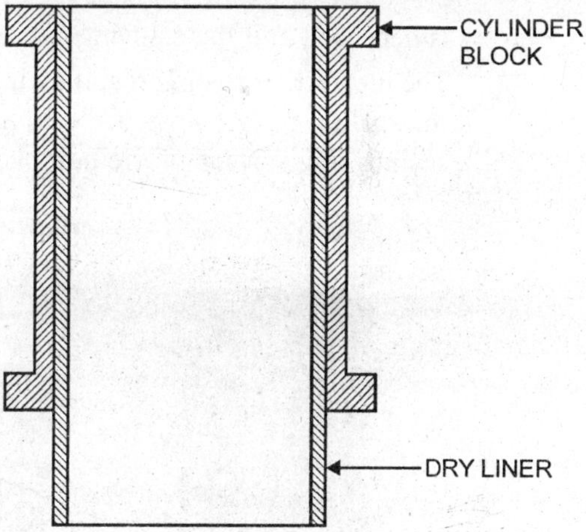

Fig. 3.17. Dry liner

Too loose a liner will result in poor heat dissipation because of absence of a good contact with the cylinder block. This will result in higher operating temperatures. If the lubrication is also deficient, it may cause scuffing. Too tight a liner is even worse than the too loose case. It produces distortion of cylinder block, liner cracking, hot spots and scuffing. Even if a correct liner is fitted in a cylinder block which itself is badly distorted, it will result in poor sealing action of rings if the liner is thin because then it will also tend to adopt the shape of the distorted block in which it is fitted. Even if the liner is thick enough to resist change of shape, there will be some hot spots which will lead to scuffing on the inner surface of the liner.

7.2. Wet Liners

This type of liner is shown in Fig. 3.18. It is in direct contact with cooling water on the outside and hence the entire outer surface does not require very accurate machining. But the water-tight joints have to be provided. This is done as shown in Fig. 3.18. At the top, the liner is provided with a flange which fits into the groove in the cylinder block. At the bottom either the block or the liner is provided with grooves, generally three in number. The middle groove is left empty and in the top and bottom ones are inserted packing rings, made of synthetic rubber. Arrangements for drainage are provided from the middle groove, for any water that may leak through the upper ring. However, the number of grooves will vary depending upon the design.

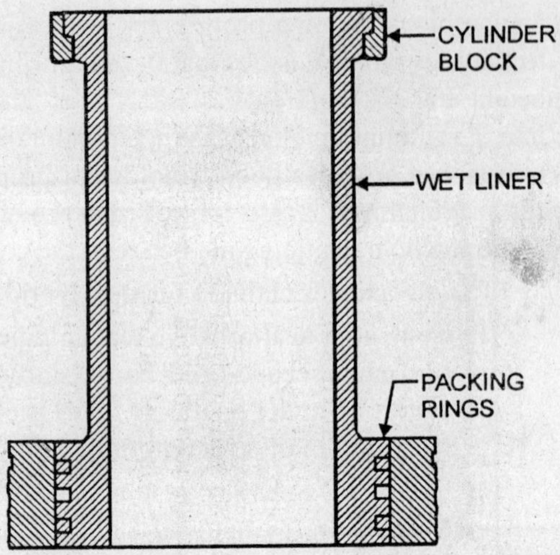

Fig. 3.18. Wet liner

Wet liners are sometimes coated with aluminium on the outside, which makes them corrosion resistant.

The main troubles experienced with a wet type liner are:

(*a*) **Breaking of flange.** This may be caused by wrong tightening sequence of cylinder head bolts or their excessive tightening, uneven counterbore in the block to receive the liner, or the worn out counterbore seat at the inner edge causing the seat to tilt downward. All the factors listed here are such as to result in excessive stressing of the liner flange.

(*b*) **Scuffing near sealing ring area.** While installing, the rubber sealing rings at the lower deck may get locally twisted or rolled. On twisting, the rubber becomes hard and this is accelerated under high temperature of the cylinder. Due to this localized hardening of the rings, very high pressures are exerted against the cylinder liner resulting in its distortion so that running clearance is deceased and the scuffing occurs near the affected area.

(*c*) The ineffective sealing resulting into leakage of cooling water into the crankcase is caused when the sealing rubber rings are damaged or lower deck is either eroded or pitted or unclean at the sealing ring surface of the bore. Excessive wear of the block at the lower deck area, apart from

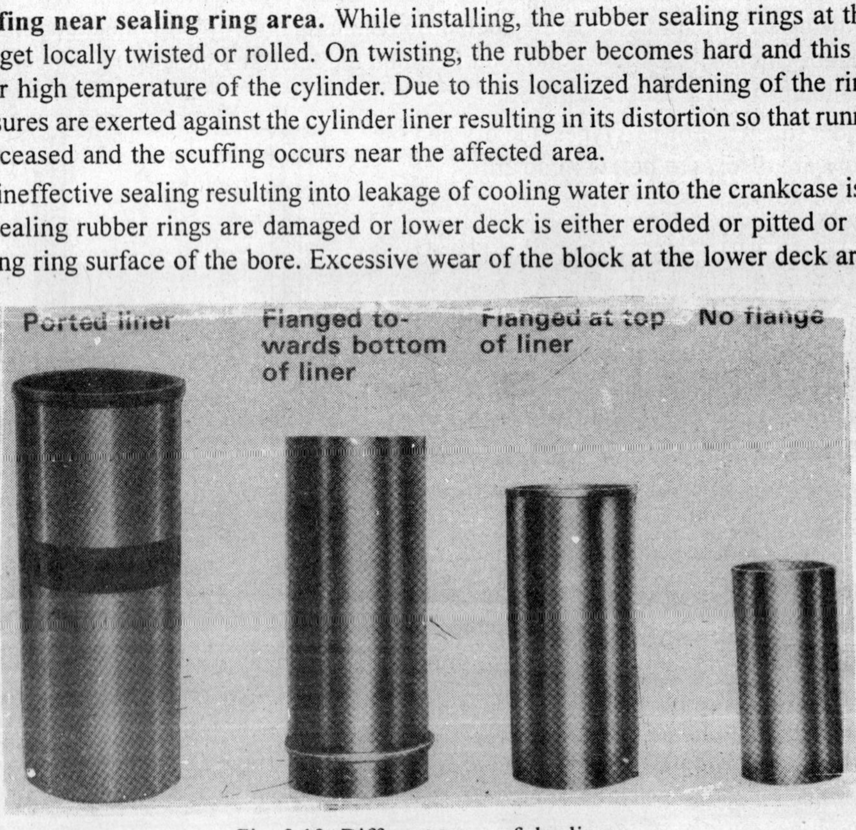

Fig. 3.19. Different types of dry liners.

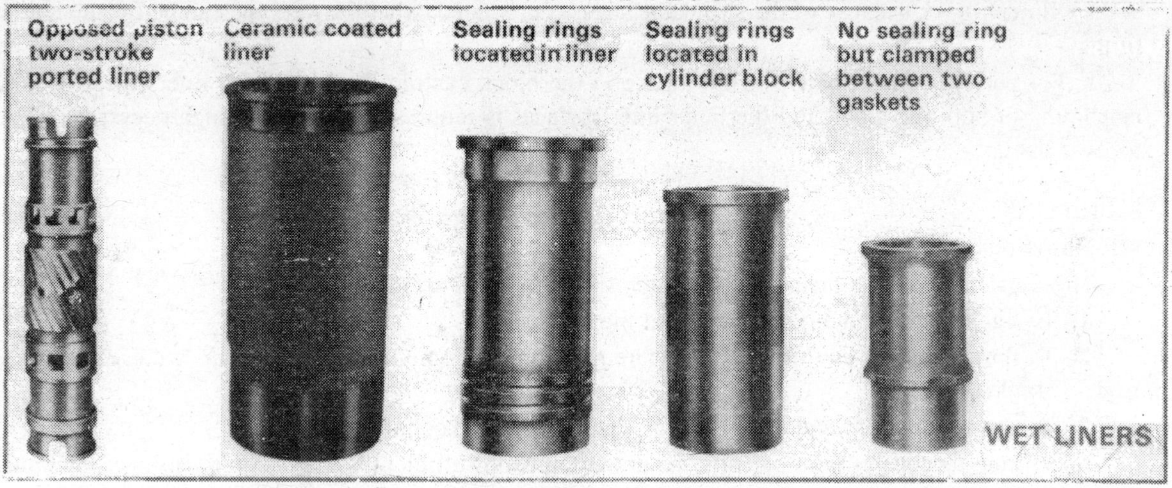

Fig. 3.20. Various types of wet liners.

causing directly, an ineffective seal, also allows the liner to swing which allows foreign particles to become embedded in the sealing rings and decrease their effectiveness.

Dry and wet liners manufactured by Wellworthy Limited, England are shown in Figs. 3.19 and 3.20 and the complete assembly (Sealed Power Corporation, USA) is shown in Fig. 3.21.

Fig. 3.21. Piston-liner assembly.

7.3. Comparison of Dry and Wet Liners

1. Dry type liners may be provided either in the original design or even afterwards, whereas the wet type have to be included in the original cylinder design.

2. A leak proof joint between the cylinder casting and the liner has to be provided in case of wet pipe, whereas there is no such requirement in the case of dry liners.

3. A cylinder block with dry liners is generally more robust than a block with wet liners.

4. In case of a cylinder block with siamesed adjacent cylinder bores (*i.e.* without any coolant passage between the adjacent bores for reducing the overall cylinder length), only dry liners can be used.

5. In case of wet liners, the casing of cylinder block is very much simplified.

6. Better cylinder cooling is ensured in the case of wet liners because the coolant is in direct contact with the liner in this case. Besides better outside surface finish and more uniform wall thickness of the wet liner improves both the heat conduction and the uniformity of cylinder cooling.

7. Dry type cannot be finished finally before they are fitted

to the cylinder block because of the shrinkage stresses produced, whereas wet type can be finished before fitting.

8. For perfect contact between the liner and the block casting in case of dry liners, very accurate machining of both the block and the outer liner surfaces is required, whereas no such necessity is there for wet liners.

8. PISTONS

8.1. Functions

The functions which a piston is called upon to perform in an I.C. engine are:

1. To transmit the force of explosion to the crankshaft.
2. To form a seal so that the high pressure gases in the combustion chamber do not escape into the crankcase.
3. To serve as a guide and a bearing for small end of the connecting rod.

Apart from its capability to perform the above functions efficiently, the piston must have some other desirable characteristics:

1. It should be silent in operation both during warming up and the normal running.
2. The design should be such that the seizure does not occur.
3. It should offer sufficient resistance to corrosion due to some products of combustion, *e.g.* sulphur dioxide.
4. It should have the shortest possible length so as to decrease overall engine size.
5. It should be lighter in weight so that inertia forces created by its reciprocating motion are minimum.
6. Its material should have a high thermal conductivity for efficient heat transfer so that higher compression ratios may be used without the occurrence of detonation.
7. It must have a long life.

8.2. Constructional Features and Materials

A typical piston is shown in Fig. 3.22. The top of the piston is called head or crown. Generally, low cost, low-performance engines have flat head as shown. In some such pistons which come quite close to the valves, the head is provided valve relief (Fig. 3.23). Pistons used in some high powered engines may have a raised dome, which is used to increase the compression ratio as well as to control combustion (Fig. 3.24). In some other engines, the pistons may be specially dished to from a desired shape of the combustion chamber, jointly with cylinder head (Fig. 3.25 and 3.26). In case of a piston containing part of the combustion chamber in its crown, compression ratio can be controlled very accurately, but the disadvantage is that in this case much larger amount of heat has to be dissipated through the piston and the rings.

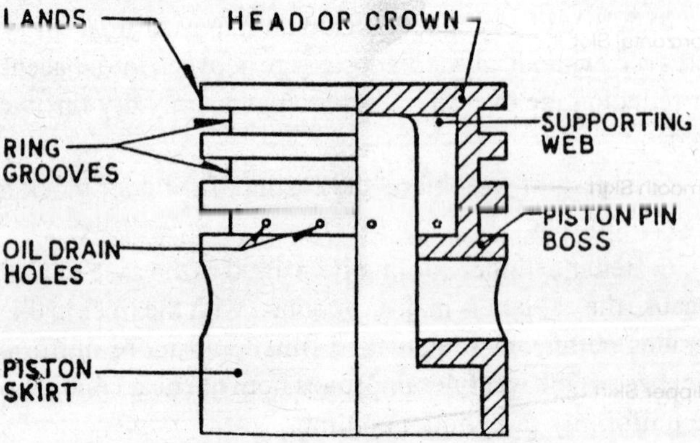

Fig. 3.22. I.C. Engine Piston.

Towards the top of the piston a few grooves are cut to house the piston rings. Properly designed ring grooves have a small degree of vertical uplift, which compensates for uneven temperature rise as the piston reaches operating temperature. Ring groove should must also be smooth since any roughness or waviness would cause poor ring seal and would lead to micro-welding wherein rings attach themselves

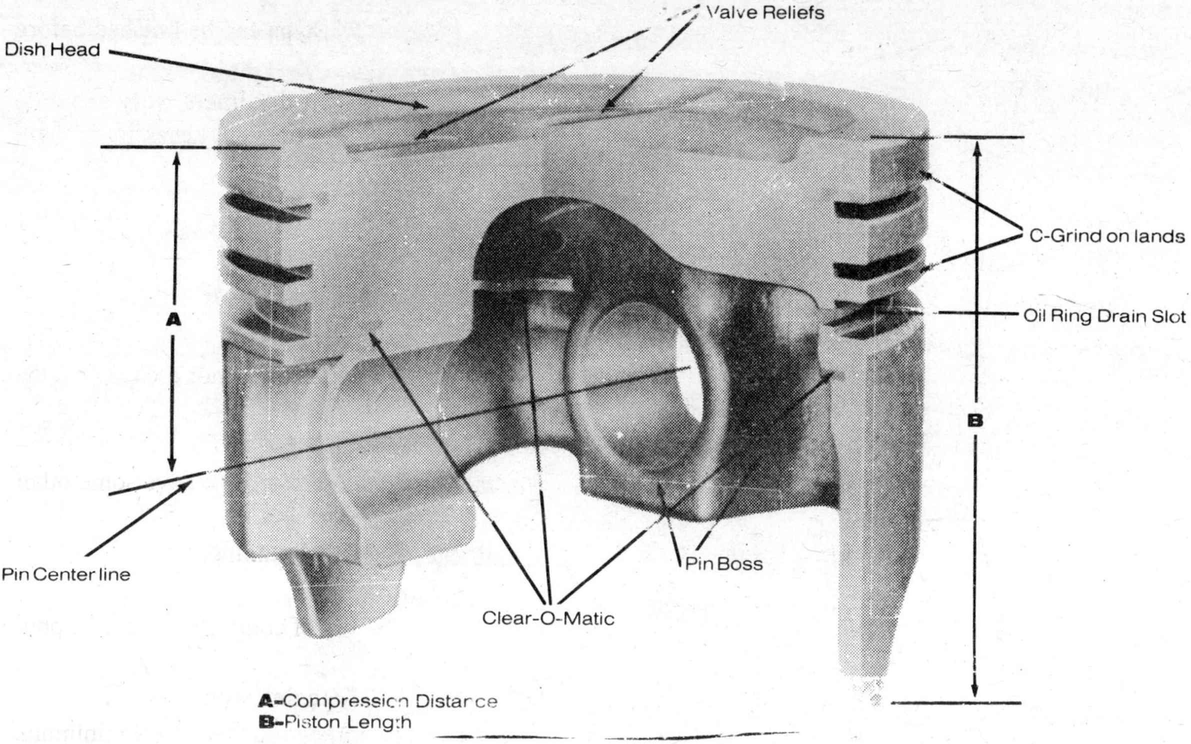

Fig. 3.23. Typical petrol engine piston (Courtesy—Zollner Corporation, U.S.A.)

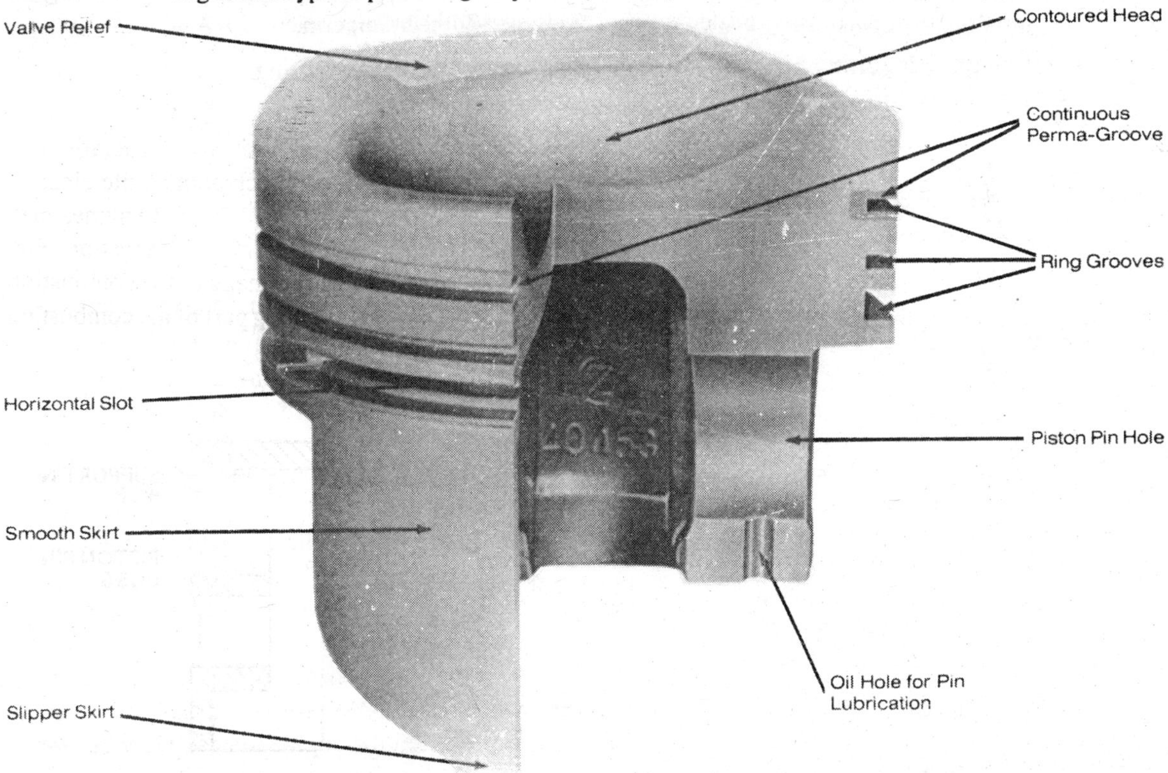

Fig. 3.24. Typical petrol engine piston. (Courtesy—Zollner Corporation, U.S.A.)

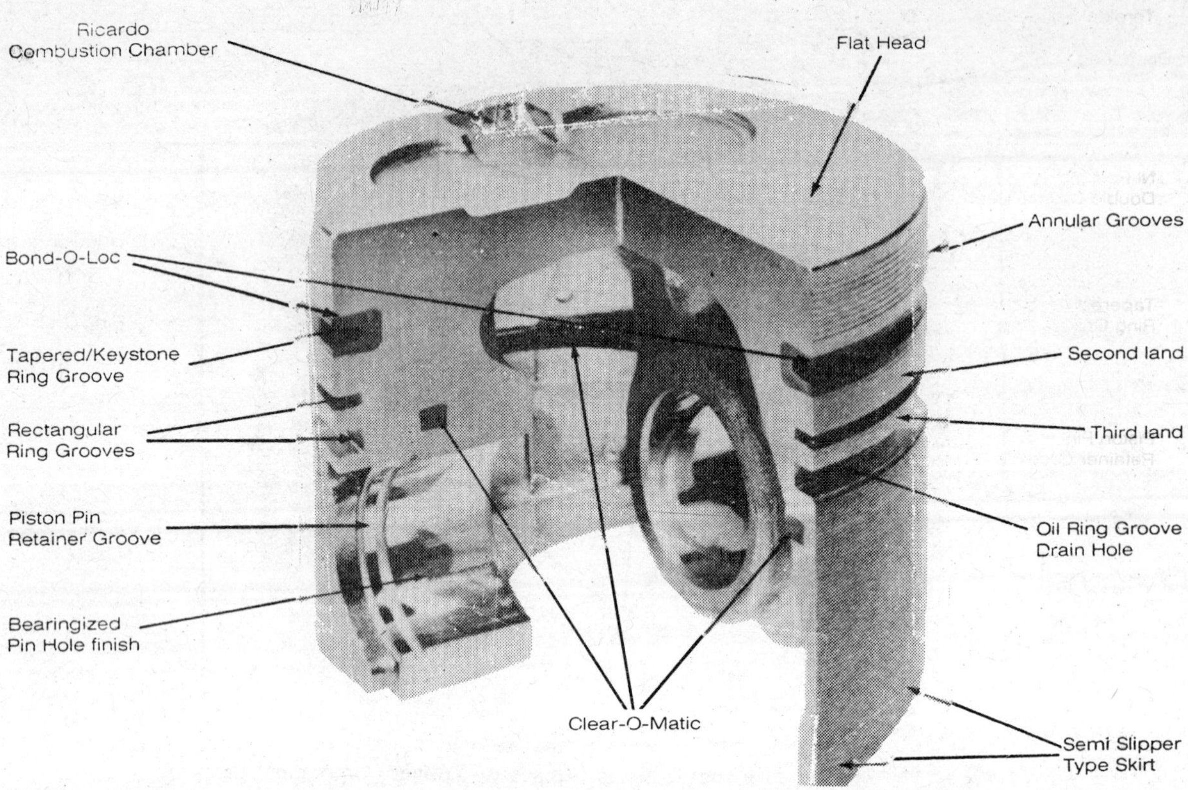

Fig. 3.25. Typical diesel engine piston. (Courtesy—Zollner Corporation, U.S.A.)

momentarily to high spots on the ring groove. To avoid groove distortion and ring land breakage due to fatigue, the vertical and horizontal surfaces should meet each other at a small radius, thus avoiding sharp corners for the fatigue crack to initiate.

The bands left between the grooves are known as 'lands'. These lands support the rings against the gas pressure and guide them so that they may flex freely in the radial direction. The supporting webs transmit the force of explosion directly from the crown to the piston pin bosses, thereby relieving the ring groove portion of large loads and thus preventing deformation of the ring grooves. The part of the piston below the rings is called 'skirt'. Its function is to form a guide suitable for absorbing side-thrust due to gas pressure. The side thrust is produced on account of the inclination of the connecting rod with the cylinder axis. The skirt is provided with bosses on the inside to support the piston pin. It must be of sufficient length to resist tilting of the piston under load. It is kept quite close fitting in the cylinder, but even then it is separated from the cylinder walls by means of lubricating oil film, for smooth running. The combustion pressure from the piston crown is transmitted to the connecting rod through webs inside the piston. The bosses form a bearing surface for the rocking motion of the connecting rod. The thick-sectioned webs also form heat paths from the piston crown to the gudgeon-pin bosses and the skirt and thus have to be designed so as to avoid expansion problems.

The distance between the axis of the piston pin and the top of the piston crown is called 'compression height' and determines the compression ratio for a given engine. Thus for the same engine, a piston with lesser compression height would give lesser compression ratio and *vice versa.*

The material used for piston at one time was cast iron which has good wearing qualities. However, it had a big disadvantage due to its high specific weight, which subjected the engine to heavy inertia forces while the piston changes direction at the dead-centres.

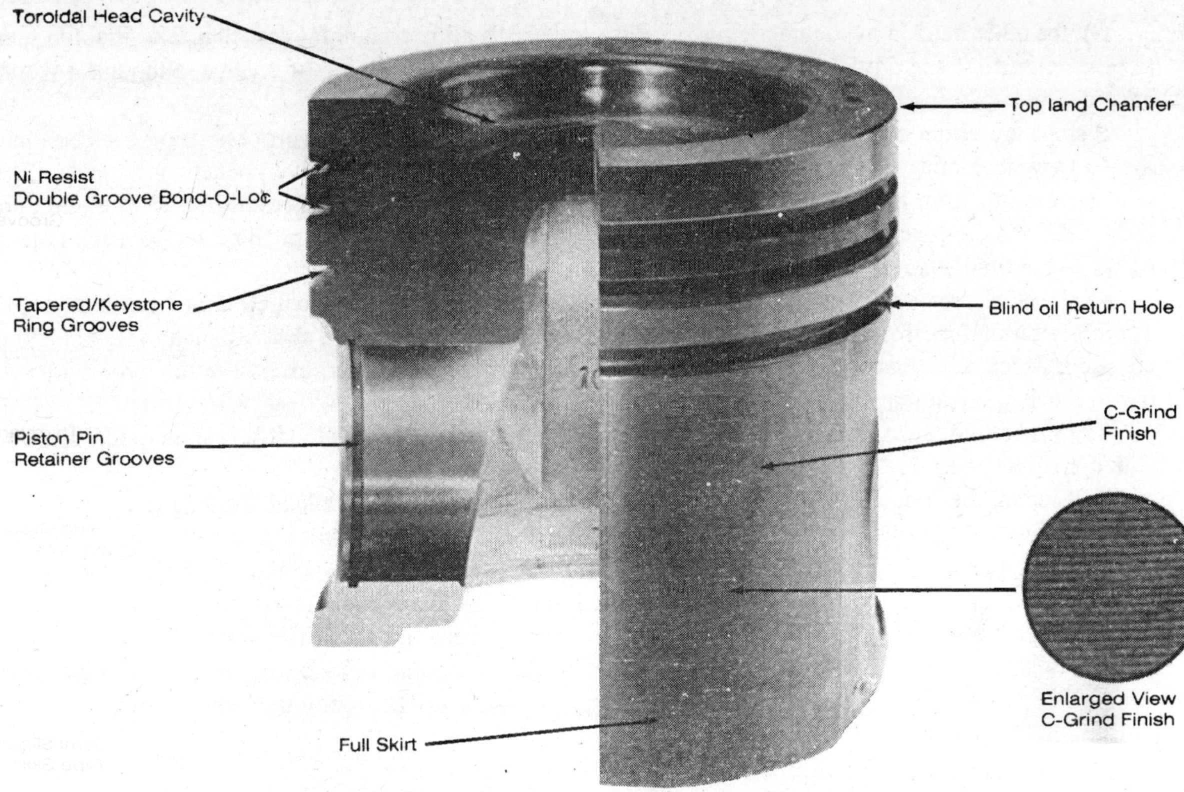

Fig. 3.26. Typical diesel engine piston. (Courtesy—Zollner Corporation, U.S.A.)

Aluminium has two distinct advantages. Firstly, aluminium is as much as three times lighter than cast iron, which makes it desirable from inertia point of view. Secondly, it possesses a higher thermal conductivity, which cause it to run cool. But the aluminium alloy has its own disadvantages, too. It is not as strong as cast iron and hence thicker sections have to be used, as a result of which the weight of piston is increased. Further, aluminium alloy is relatively soft, as a result of which fine particles in the lubricating oil become embedded in it. Aluminium piston with fine particles embedded in it causes a sort of grinding or abrasion of the cylinder walls, thus shortening cylinder life. Another important drawback of using aluminium pistons for cast iron cylinders is their unequal coefficients of expansion which causes engine slap; because if the "cold clearance" is kept just sufficient, there is danger of seizure at operating temperatures and if the cold clearance is kept large, the engine knocks or slaps when cold.

As the technology developed, aluminium alloy containing silicon replaced cast iron as piston material. *Y*-alloy was the best known early aluminium alloy. It is seen that as the percentage of silicon in the alloy increases, its coefficient of expansion decreases. However, silicon is bonded to aluminium at the molecular level, *i.e.*, maximum amount of silicon remains in solution with aluminium only upto about 12%. If silicon content increases beyond this level (called eutectic), the excess silicon remains free, *i.e.*, it retains its distinct granular form, which reduces its strength and toughness. Presently two alloys with silicon content 12% (eutectic) and 22% (hyper-eutectic) are being used, which have coefficient of expansion of 0.000021 per °C and 0.0000175 per °C respectively compared to 0.0000245 per °C for aluminium.

In fact, the main reason of adding silicon to the aluminum alloy for making piston was only to reduce the piston slap. Another advantage of adding silicon is that the piston becomes harder and thus less liable to scuffing and wear. However, higher silicon content also makes the alloy brittle which may result in the piston developing cracks if the detonation or pre-ignition occurs. That is why the silicon has to be restricted.

On the other hand, a hypoeutectic piston alloy, *e.g.*, 2618 alloy containing less than 2% silicon is able to bear detonation and any damage due to impact, such as due to dropping of a valve, although it would have decreased strength and hardness.

Pistons are either cast or forged. Castings are made by gravity die cast process using 5-piece inner dies, a two-piece outer and usually a head chill which can form any cast valve pockets and bowls. Core pins are used to form pin holes. On the other hand, forgings are made from a solid billet of aluminium alloy, which is pre-heated and then pressed into shape in a sequence of two or three stages, each getting closer to the final shape.

Cast pistons are quite strong and light, but cannot take much impact due to their lesser tensile strength. Therefore for high performance engines, say, producing 75 kW per litre and above, usually forged pistons are used, which are about three times stronger than the cast ones. However, excessive cost prohibits their use in stock applications where cast pistons can be used.

The problem of cold slapping, as described above, has been overcome by different methods, some of which are given below:

1. Keeping the heat away from lower part of the piston as far as possible. This is done by:

(*a*) cutting horizontal slot in the piston on the thrust and non-thrust sides, just below the oil control ring (Fig. 3.27). Thus the skirt does not become very hot and consequently it does not expand quite so much. In some designs the circumferential slots are made in the oil control ring groove and these slots end in inclined slots extending downwards (Fig. 3.28). These elongated slots provide additional heat barriers and so reduce even more the amount of heat reaching the working faces of the skirt. Moreover, the drooping ends makes the skirt flexible in the upper region.

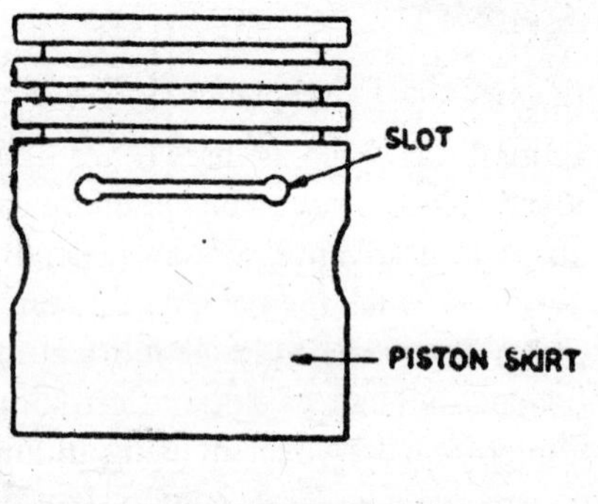

Fig. 3.27. Piston with horizontal slot.

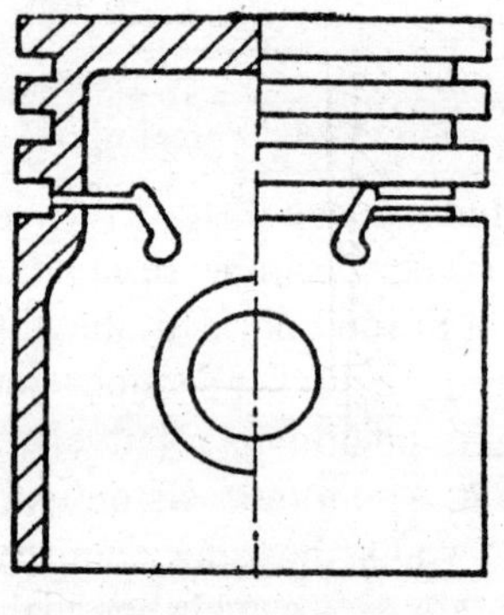

Fig. 3.28. Inclined slot in oil ring groove.

(*b*) making a *heat dam* (Fig. 3.29). It consists of a groove cut near the top of the piston. This reduces the path of heat travel from the piston crown to the skirt. The skirt, therefore, runs cooler and does not expand much.

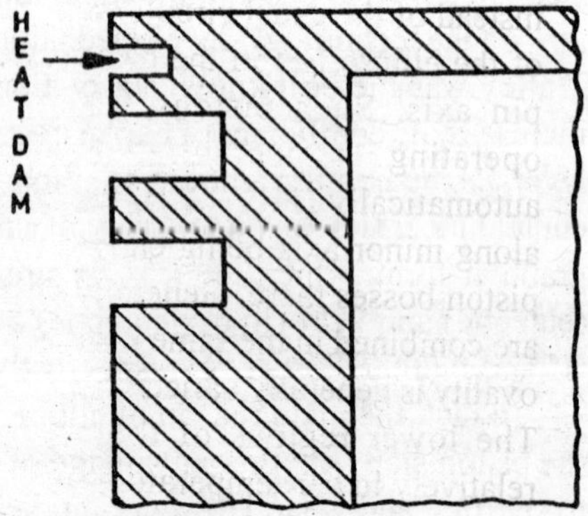

Fig. 3.29. Heat Dam.

2. Use of vertical or T-slots. (Fig. 3.30 and 2.31). Vertical or T-slots on the non-thrust side of the piston were earlier used quite commonly. These slots allow the piston skirt to expand without increase of diameter. However, the mechanical strength is decreased on account of slot. Moreover, with the slot the skirt tends to collapse inwards without elastic recovery. As a result the diameter is reduced permanently which increases the piston slap, instead of decreasing it. For these reasons, fully split skirts are no longer used. However, as a compromise, semi-split skirts in which the slot goes only about

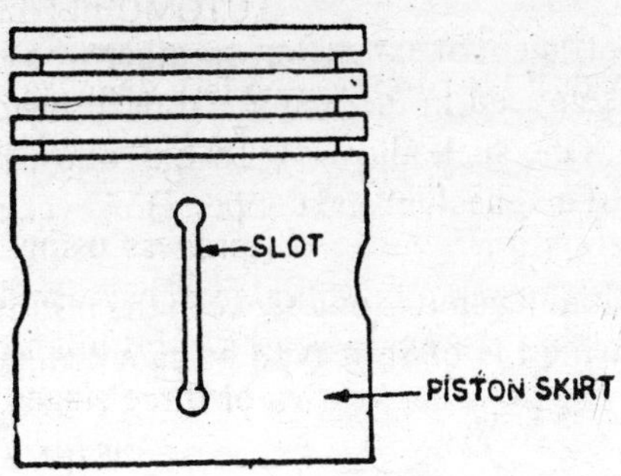

Fig. 3.30. Piston with vertical slot.

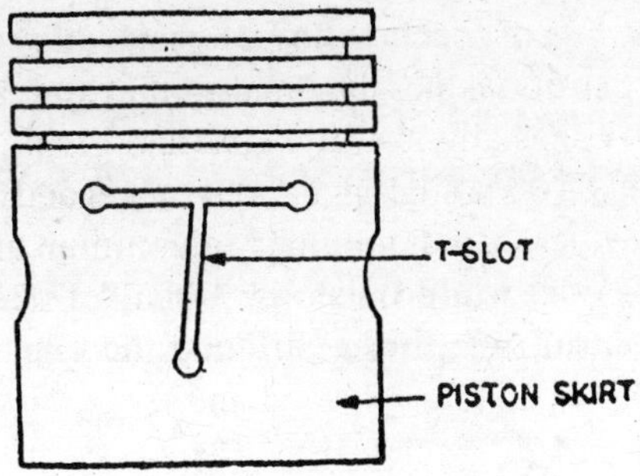

Fig. 3.31. Piston with T-slot.

halfway up with blunting holes at the ends to avoid stress concentration, have been used in pistons of some light duty engines. Heavy duty pistons never use any such slots.

3. Taper Pistons. The pistons are sometimes turned taper, the crown side being smaller in diameter than the skirt end (Fig. 3.32). As higher temperatures occur towards the crown, that side expands more than the skirt, due to which the piston diameter becomes uniform under running conditions.

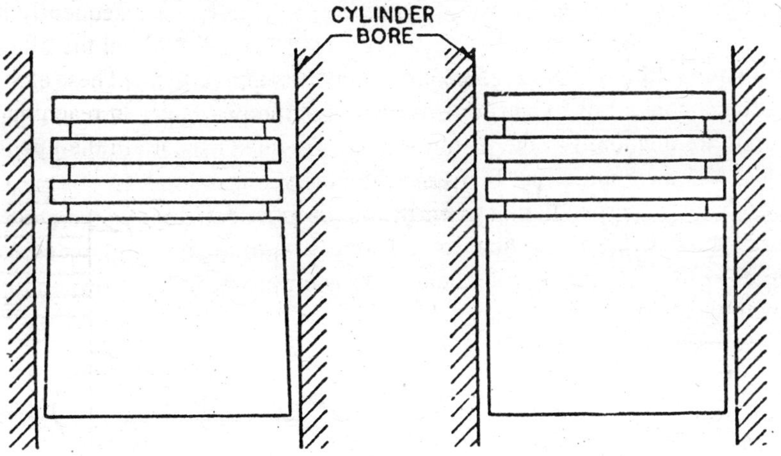

(*a*) (*b*)

Fig. 3.32. Taper piston. (*a*) cold, (*b*) hot.

4. Cam ground pistons. The pistons are cam ground such that they have elliptical section instead of the usual circular one. The minor dia of the ellipse lies in the direction of the piston pin axis. Such pistons, after expanding at operating temperature become circular automatically (Fig. 3.33), the more expansion along minor axis being caused by the metal of piston bosses there. Generally taper and ovality are combined in the same piston. The amount of ovality is generally varied along the skirt height. The lower regions of the skirt operating at relatively lower temperatures, the amount of ovality is kept maximum at the piston-pin-boss level and is reduced gradually towards the bottom of the skirt.

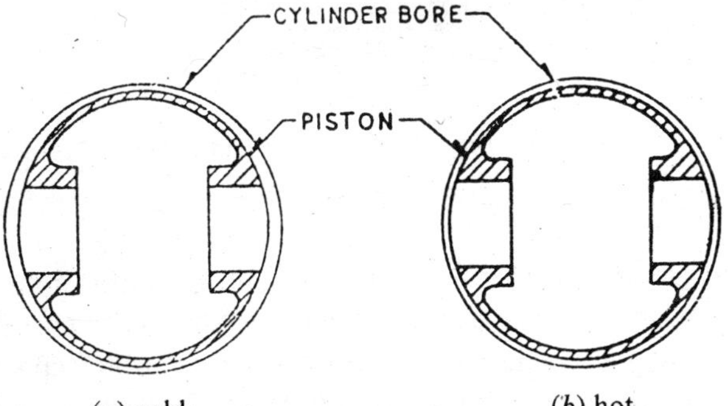

(*a*) cold (*b*) hot

Fig. 2.33. Effect of piston ovality.

5. Use of special alloys. Special alloys having low coefficient of expansion or rather whose coefficient of expansion is nearly equal to that for cast iron have been used in the manufacture of pistons, without split or specially shaped skirts and giving no piston slap. One such alloy is "Lo-Ex" alloy. It is an alloy having 12-15% silicon, 1.5-3% nickel and about 1% each of magnesium and copper. However, such pistons are costlier than the ordinary aluminium alloy piston.

6. Wire wound pistons. A band of steel wire under initial tension is put between the piston pin and the oil control ring, thus restricting the expansion of skirt which is of split type or cam ground type (Fig. 3.34).

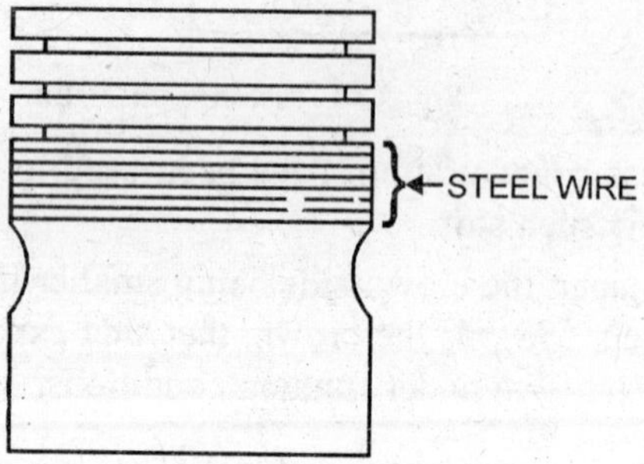

Fig. 3.34. Wire wound piston.

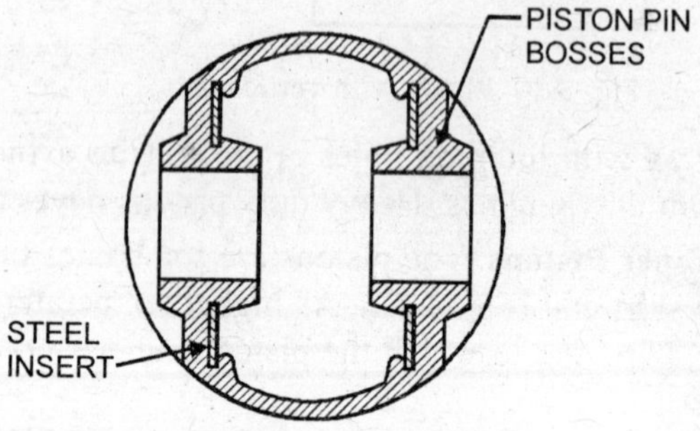

Fig. 3.35. Autothermic piston.

7. Autothermic Pistons. Referring to Fig. 3.35 in which the section of an autothermic piston is shown, it is seen that the piston contains low expansion steel insert at the piston pin bosses. These inserts are so moulded that their ends are anchored in the piston skirt as shown. There is no chemical or metallurgical bond between the steel and the aluminium, there is only a mechanical bond. In this case pistons are cam ground with the major diameter perpendicular to the piston pin axis. At higher temperatures during running the bimetallic action of the inserts causes them to bend outward, thus causing the piston to expand along the piston pin, whereas in the direction perpendicular to piston pin there is corresponding contraction of the piston due to metallic action.

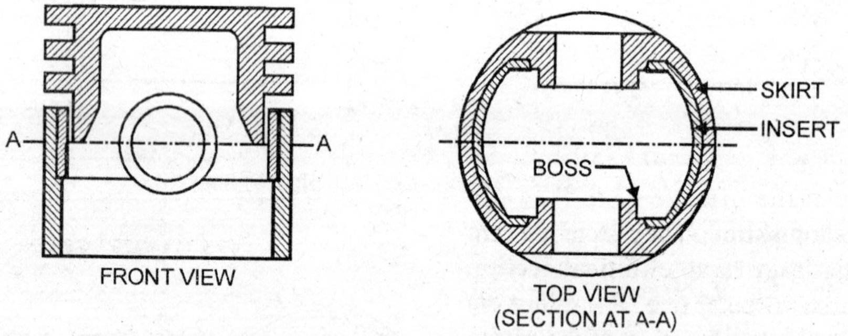

Fig. 3.36. Alternative method of skirt expansion control with steel inserts.

Another such arrangement which provides even better expansion control enabling smaller skirt to cylinder wall clearances to be employed is shown in Fig. 3.36. In this steel inserts are employed on the thrust and non-thrust sides of the piston skirt. There is clearance between the inserts on the inside and the upper piston region, whereas the outer faces of the steel inserts are cast with the aluminium alloy. The gap between the steel inserts and the upper piston region interrupts the heat flow to the skirt and keeps its temperature and hence expansion lower. Moreover, these steel inserts having lower coefficient of expansion than the aluminium alloy, prevent the skirt from expanding as much as it would if made of only aluminium alloy. Besides these inserts strengthen the skirt, still maintaining the flexibility between the upper part of the piston and the piston pin bosses. A Wellworthy piston with steel insert is shown in Fig. 3.37.

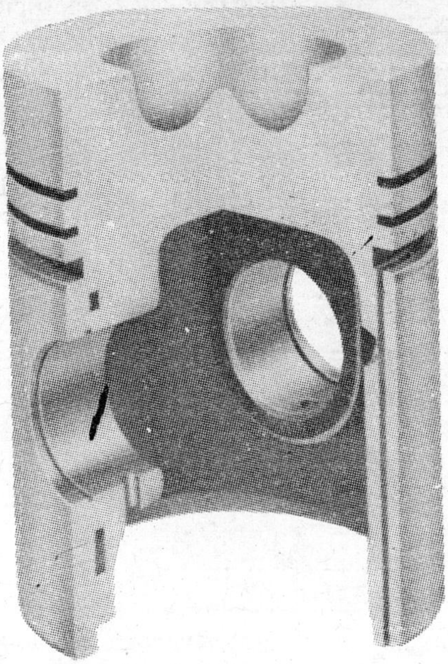
Fig. 3.37. Wellworthy steel insert piston.

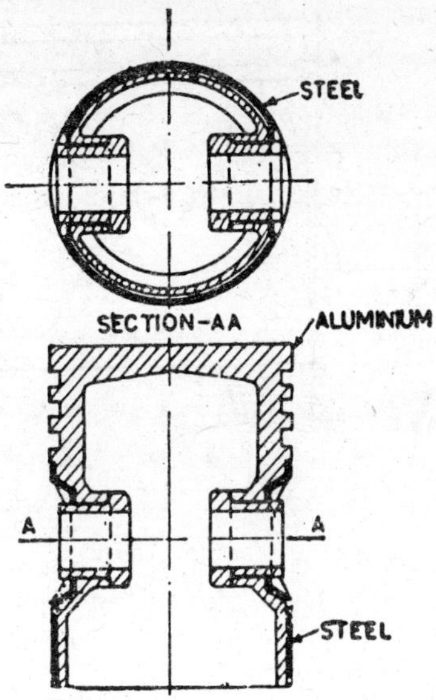

Fig. 3.38. Bi-metal piston.

8. Bi-metal Piston. These pistons are made from both steel and aluminium (Fig. 3.38). The skirt is formed by steei and the aluminium alloy cast inside it forms piston head and piston pin bosses. As the coefficient of thermal expansion for steel is quite small, the piston will not expand much and hence smaller cold clearances can be maintained.

9. Offset Piston. Another method which is sometimes used to eliminate slapping tendency of pistons is to offset them from the cylinder centre line (Fig. 3.39) towards the major thrust side. The piston pin hole is not symmetrically placed as usual, but is slightly offset as shown in figure.

As the piston approaches top dead centre during compression stroke the offset causes it to tilt slightiy so that its top skirt surface on the minor thrust side is placed against the cylinder well. [Fig. 3.40 (*a*)]. Simultaneously the bottom surface of the piston skirt on the major thrust side is also placed against the cylinder wall. When the top dead centre is passed, the inclination of the connecting rod forces the piston against the major thrust surface [Fig. 3.40 (*b*)]. However, since the lower surface of the skirt is already in contact with the cylinder wall because of the offset, as explained above, the piston skirt very smoothly wipes into full wall contact, thus avoiding piston slap. One such piston as used in Daimler-Benz engine 123.920 is shown in Fig. 3.41.

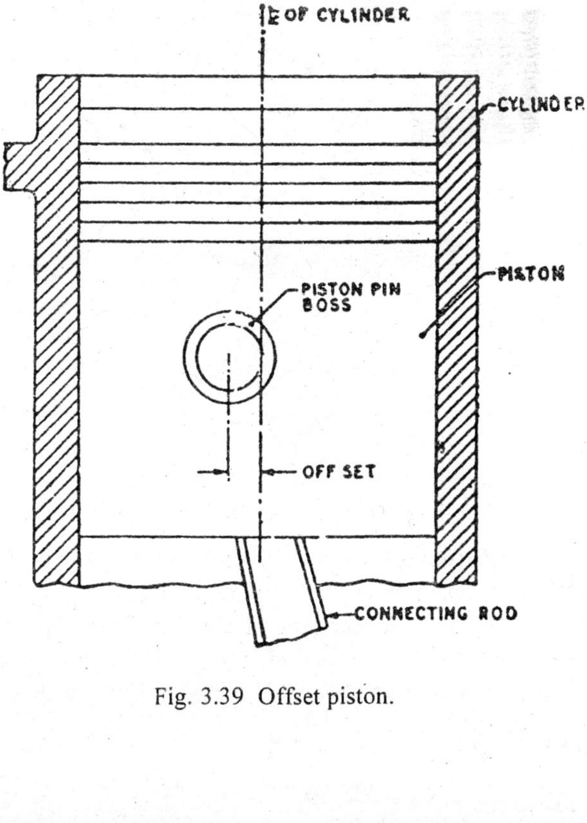

Fig. 3.39 Offset piston.

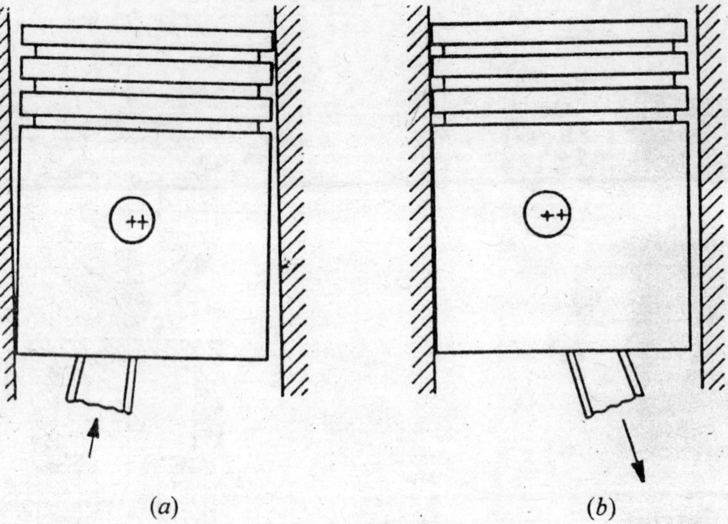

Fig. 3.40. Effect of piston pin offset. (*a*) Piston moving up (just before T.D.C.), (*b*) Piston moving down in power stroke (after T.D.C.)

If the piston offset is located towards the minor thrust side, a higher mechanical advantage is obtained. Moreover, less piston to cylinder friction is caused. However, it will not decrease the piston slap. Therefore, in racing car engines, the pistons are offset towards minor thrust side since noise and life are not as important as performance in such engines.

Apart from the special constructions of pistons mentioned above, solely for the purpose of avoiding piston slap, there exist other piston types, too, having their own advantages. Some of these are given below:

Fig. 3.41. Piston for Daimler-Benz engine 123.920.
(Courtesy—Daimler-Benz Aktiengesselschaft, Germany)

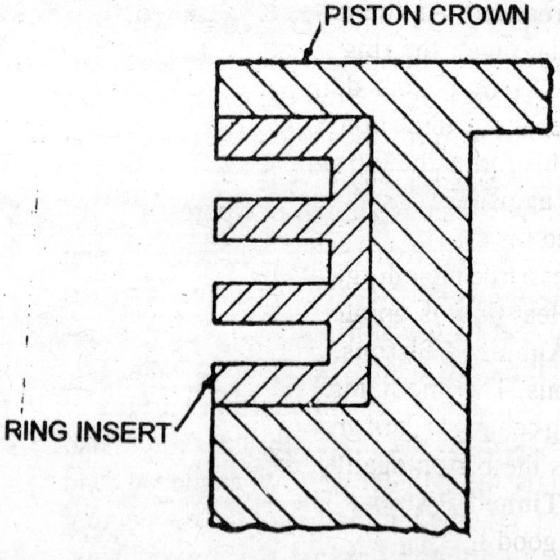

Fig. 3.42. Inserted ring carrier.

1. Pistons with inserted ring carrier (Fig. 3.42). The fuels having high sulphur content from sulphur dioxide or sulphur trioxide on combustion, which with water also formed alongside form sulphurous and sulphuric acids respectively, these having corrosive action on the piston rings particularly, because these acids are formed on cylinder walls with which the rings come into contact. To avoid this special ring inserts made from austinitic form of cast iron have been used, which resist the corrosive action of the

Fig. 3.43. A V-8 engine 240 B.H.P. Truck Piston (Wellworthy).

3.44. A1-fin heavy duty piston with twin inserts (Wellworthy)

acids produced. As the co-efficient of thermal expansion of the austinitic iron and aluminium alloy are nearly the same, the ring insert does not work loose.

The A1-fin pistons developed by Wellworthy Limited, England have been very popular. Such a piston embodies an austinitic iron member molecularly bonded to the aluminium body. This forms the top ring carrier, the ring groove being machined into its as usual (Fig. 3.43). For more heavy ratings, the carrier is provided for the second groove also (Fig. 3.44). The bonding process produces a true metallurgical union of the two metals ensuring a perfect bond. Heat flow is uninterrupted.

2. Anodized Pistons. Anodizing improves the bearing properties of pistons. The anodizing is done by the sulphuric acid process and resulting coating is dark grey in colour. The very hard aluminium coating protects the piston against scoring.

3. Tinned Pistons. Tin deposits of about 0.007 mm thickness ensure good lubrication during starting operations of the piston and this avoids wearing out of the piston. The tin layer gradually wears off during subsequent running of the piston.

4. Oil cooled pistons. The pistons are sometimes oil cooled when they are meant for high and medium speed diesel engines. Oil cooling reduces the piston temperatures and is introduced to satisfy strength requirements in highly loaded regions, and to reduce the temperature at the top ring to prevent carbon formation in the groove and to achieve sufficient oil film thickness on the ring periphery.

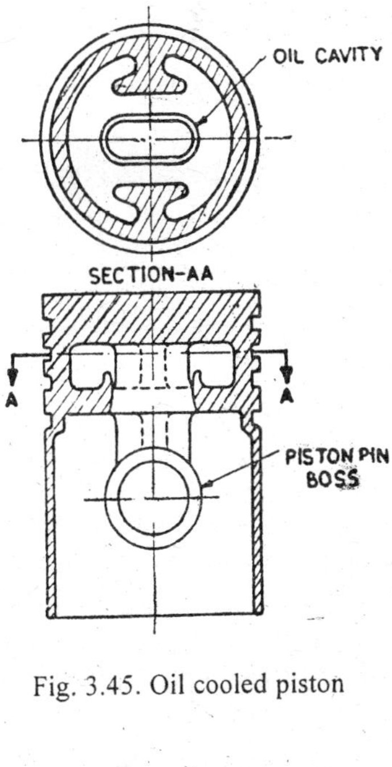

Fig. 3.45. Oil cooled piston

Fig. 3.45 shows such a piston, wherein a cavity is provided underneath the piston crown. The oil from a connecting rod nozzle is directed against the inside of piston crown and then it falls in the cavity. The reciprocating motion of the piston then throws the oil constantly onto the underneath of the crown, keeping it cool. Alternatively, the oil may be supplied from a standing jet in the crankcase. The effectiveness of oil cooling is largely determined by the under crown surface area and the distance to the bowl edge, and is generally suitable for moderate levels of turbocharging.

The most effective form of cooling is obtained with pistons containing enclosed galleries. The gallery surface area is large and is suitably positioned in the crown. The cooling effect is still increased by running the gallery partially filled with oil. The gallery is formed by the soluble core technique.

A few open and closed gallery oil cooled pistons developed by Wellworthy Limited, England are shown in Figs. 3.46 and 2.47.

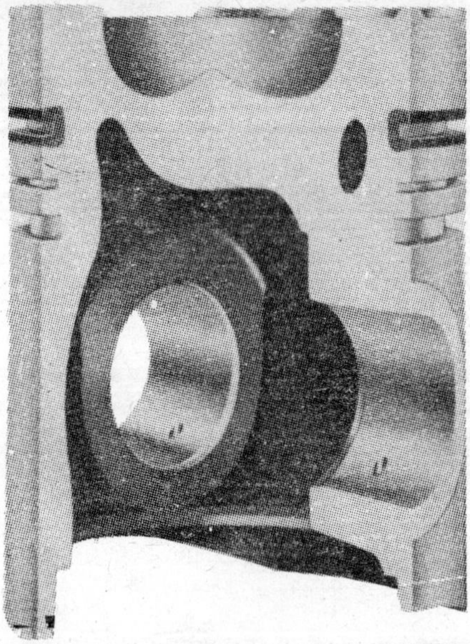

Fig. 3.46. Oil cooled piston with open gallery

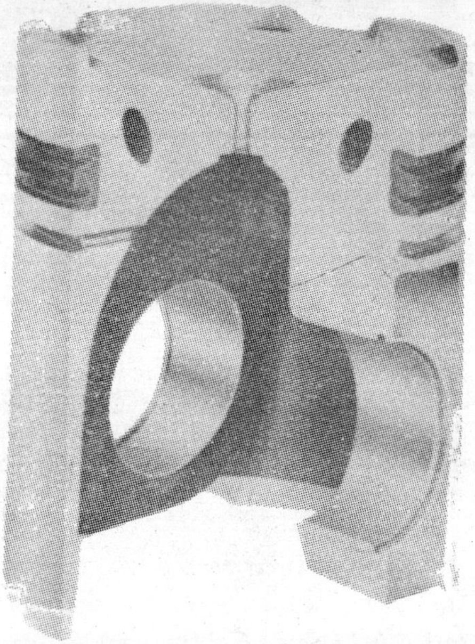

Fig. 3.47. Oil cooled piston with closed gallery

A Mahle aluminium piston with cooled ring carrier and suitable for automotive diesel engines is shown in Fig. 3.48. The cooled ring carrier consists of a Niresist ring carrier to which a thin austenitic steel sheet with intake and outlet openings is welded. This combined insert is cast into the piston and brings the heat transfer to the cooling oil much nearer the combustion chamber and the first ring groove, which makes the cooling much more effective than the usual salt core cooling channels.

The piston shown also has pin bore bushes and aluminium oxide fibres infiltrated to reinforce the thermally-stressed combustion chamber bowl. Mahle's ADC (Advanced Diesel Coating) process produces a very fine-grained structure resulting in high fatigue-resistance in the bowl rim.

5. Two-piece pistons. For applications where engine temperatures exceed the working range for oil cooled aluminium alloy pistons, two-piece pistons are used. In these, the steel crown which resists high temperatures is bolted to aluminium or S.G. iron skirt (Fig. 3.49 and 3.50). The main advantage of this type of piston is that different materials can be used for the crown and the body which better suit their functional requirements. Further a large oil gallery for oil cooling is formed between the two parts. Two-piece pistons are most commonly used in engines with bore size exceeding 180 mm and operating at b.m.e.p. ratings of about 20 bar.

Fig. 3.48. Mahle Aluminium piston with cooled ring carrier.
(*Courtesy*—Mahle GmbH, Germany)

6. Composite insulated (heat shielded) piston. Dana Engine products division (Perfect Circle) U.S.A. have developed a cool running piston by using a stainless steel cap bonded to an aluminium piston (Fig. 3.51). The piston consists of a stainless steel cap on the crown of otherwise cast aluminium piston body and skirt. The insulating barrier is the air between the cap and the piston body.

The stamped steel cap has a wire mesh bonded to its underside by a sintering process. This cap is placed into a mould. As aluminium is poured into the mould, it is entrapped in the wire mesh and this provides a solid mechanical lock. There is also some diffusion between the aluminium and the stainless steel cap with wire mesh providing a metallurgical bond as well. But the main bond is mechanical. The geometry of the cap can be formed to most of the crown shapes commonly found in aluminium pistons for current generation diesels.

A comparative idea of the temperature reductions possible can be had from Fig. 3.52. wherein temperatures above each line are for straight aluminium pistons while the numbers below each line are for a heat shielded piston.

7. Squeeze cast pistons. The process 'squeeze casting' is so called because high pressure is applied

Fig. 3.49. A 185 mm two-piece piston with forged steel crown and forged aluminium body. (Courtesy—Wellworthy Limited, England)

Fig. 3.50. A two-piece piston having a steel head secured to the aluminium body with inclined bolts (Courtesy—Wellworthy Limited, England)

to the casting during solidifications, which eliminates inherent micro-porosity in aluminium castings giving increased fatigue strength. It also permits the introduction of ceramic or metal fibres to provide local strengthening or increased wear resistance. Al-fin ring carriers and controlled expansion skirt inserts can also be provided additionally.

Fig. 3.51. Composite insulated piston. (Courtesy--Dana Engine Products Division, U.S.A.)

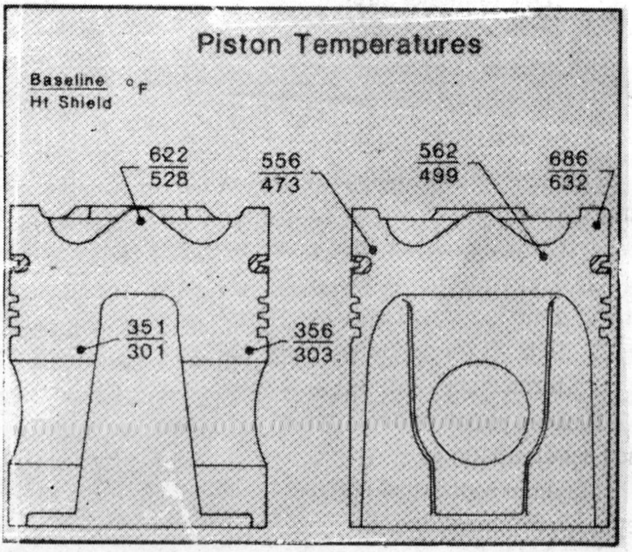

Fig. 3.52. Comparison of piston temperatures in case of the composite insulated and the straight aluminium pistons. (Courtesy—Dana Engine Products Division, U.S.A.)

8. A Econoguide piston. This is a noval method developed by Wellworthy Limited to reduce skirt friction (Fig. 3.53). The skirt contact area is reduced by about 75% compared to conventional pistons and consists of raised pads which are specially shaped to assist the creation of hydrodynamic lubrication. The pad-profile may be machined by a numerically controlled machine.

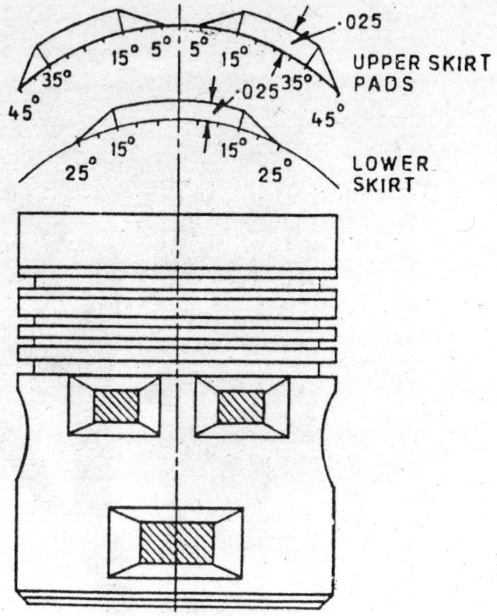

Fig. 3.53. AEconoguide piston.
(Courtesy—Wellworthy Limited, England)

9. Steel pistons. These pistons have definite advantages over the conventional aluminium alloy pistons, particularly with reference to the modern engines having more severe operating conditions. These are:

(*i*) More durable, *i.e.*, longer life.

(*ii*) Having lower coefficient of thermal expansion than aluminium, being approximately same as that of the gray cast iron block, minimum clearance between the piston and the cylinder walls would be required since the same would not be affected by the change of temperature.

(*iii*) Due to higher strength, the height of the steel piston is reduced for the same piston size. A reduction of about 30 percent in the piston height has been achieved in practice, which results in:

- Reduction of friction between the piston and the cylinder walls, leading to saving of fuel ultimately.
- Lower piston height means lower cylinder height, which fistly, beings down the vehicle's center of gravity, thus providing stability and secondly, increases the distance between the engine and the hood, helping to protect pedestrians.

Due to higher strength, the skirt area can also be reduced, which would again result in reduction of friction resulting in reduction of fuel consumption.

(*iv*) The higher strength of steel also makes it possible to reduce the thickness of the piston walls, leading to improvement in the rate of heat transfer from the piston crown to the cylinder walls. Consequently the temperatures in the bowl rim are remain lower.

Fig. 3.54 shows Mahle's 'MonoXcomp' steel piston manufactured by Mahle, Germany. It consists of three main parts:

1. Crown–A screw thread and an anti-fatigue shaft are moulded directly to it.
2. Thrust sleeve–It contains mainly the nut thread and an elastic shapeable part. Appropriate contact zone is formed on it, to transmit power to the piston skirt.
3. Skirt–It forms the main contact zone between the piston and the cylinder walls. It has an integrally formed anti-fatigue coller and the interior is deformable like a spring washer.

Besides above, a cooling gallery circulating the cooling oil under the piston crown further improves the heat dissipation.

Magnum Monosteel piston, developed by Federal-Mogul for heavy duty truck market, consists of two separate forgings, *viz.*, the top or crown forging and the lower or skirt forging. The two forgings are friction welded.

Magnum has an aperture in the skirt which reduces skirt drag and reciprocating mass. Although, magnum skirt design reduces the skirt area by 40 percent, yet the smaller skirt does not decrease the piston stability or guidance. Magnum skirt contacts the cylinder bore on both the upper and the lower bands simultaneously,

Fig. 3.54. Mahle's "MonoXcomp" piston for Commercial Vehicle Diesel Engines
(*Courtesy*—Mahie GmbH, Germany)

achieving the same guidance capability as a full length skirt and also an improved protective oil film on the upper skirt band.

Federal-Mogul is also developing a smaller-bore Light Duty Vehicle (LDV) Monosteel, which would have similar architecture, but may use thinner wall castings which are more efficient than forgings. Moreover it may also make use of a patented induction welded process wherein induction coils are used to provide the welding heat. Induction coils permit thinner design sections, closer to aluminium or articulated piston weights. Such pistons are expected to reach market by 2014/15.

Steel piston patented by KS Kolbenschmidt Pierburg consists of a one-piece forging having enclosed cooling gallery. The design is claimed to have clear advantages in fuel consumption and emission levels without compromising functional performance.

8.3. Dura Bowl piston technology

This technology of Federal Mogul enables the production of aluminum pistons with better geometry, higher performance and greater durability in heavily boosted diesel engines. In a modern diesel engine, temperatures can exceed 400°C and pressure over 200 bar. Under these conditions, aluminium expands eight times as much as free primary silicon particles which decreases the piston strength and repeated mechanical loads on the piston may cause failure from the corners of the silicon particles. The Dura Bowl technology strengthens the piston crown by locally remelting the alloy around the bowl, which improves the fatigue strength over that region. During the process, the remelted area cools down much faster than the original cast condition, producing silicon particles one-tenth of their previous size, which increases the alloy's strength and durability. Consequently, this technology helps the manufacturers in achieving piston combustion bowl geometries which help the air-fuel mixture to burn more efficiently, resulting in improved fuel economy and emissions. Piston mass can also be reduced, achieving still improved engine performance and durability.

It is claimed that this technology improves the piston's fatigue resistance four to eight times under thermomechanical cyclic loading and also a four-fold increase in the life of a cast piston with improved bowl rim geometries.

Daimler has recently (2012) started using Dura Bowl pistons in the new Mercedes-Benz OM642 V6 3-litre and the current OM 651 2.2 litre diesel engines.

8.4. Piston coating

Federal-Mogul has developed a piston-coating, called 'EcoTough', which provides lower friction losses and lower wear under challenging conditions of modern engines. EcoTough-coated pistons reduce fuel consumption by 0.8 percent and also decrease CO_2 Emissions.

EcoTough is a polymer coating containing finely and uniformly dispersed solid lubricants including graphite and molybdenum disulfide (MoS_2). The coating is also reinforced with carbon fibres. It is applied to the piston skirt by a screen-printing process, just as in case of standard coatings. Ordinary piston coatings work mostly based on graphite. However, EcoTough improves the tribological properties of the piston during operation. The mix of solid lubricants in this coating's polymer matrix limits friction under severe boundary lubrication conditions, often found in modern highly efficient engines. The graphite fibres make the coating more wear resistant without increasing friction. Other advantages of this coating are:

1. Simple pretreatment of the piston surface.
2. Coating process is very reliable and stable.
3. For more tough engine operating conditions, thicker layer of coating way be applied, which option is not available in case of other coatings.

8.5. Piston Selection

Pistons are selected by comparing material, manufacturing process used for making the piston, design, compression ratio, the cost and most important, on the intended use of the engine. It is relatively easier in case of a racing car engine since the conditions under which the piston has to operate, and the fuel to be used are known. In case of the road vehicle, the piston selection has to consider the quality of the fuel available, level of performance expected from the vehicle and the possibility of later additions, e.g., turbo/ super chargers, nitrous oxide systems. Any condition which would increase cylinder pressure and the probability of detonation occurring has also to be considered. The detailed discussion on each of these criteria is as follows:

1. Manufacturing process. Piston may be cast or forged. Modern hypereutectic alloys, with improved thermal properties, greater hardness and increased resistance to scuffing allow tight bore clearances which minimizes noise on a cold engine start, which, alongwith lower cost, are main advantages of cast pistons compared to the forged ones. Thus these are suitable for road vehicles and may also work well in moderate supercharged engines. On the other hand forged pistons are particularly suitable for performance engines used in racing cars. Forging removes porosity and increases ductility in the metal and allows the piston to run cooler than a comparable cast one. Main advantages of the forged pistons are minimum piston weight without compromising strength, better resistance to high cylinder pressures and skirt loads present in racing car engines and also more likely to withstand limited detonation and valve-piston contact which may occur during a race. Besides they are also suitable for engines with high compression ratio (11:1 and more), high boost superchargers, nitrous oxide or under conditions just approaching detonation.

2. Piston design. Most important areas for piston design are the head design, piston to valve and cylinder head clearances, piston skirt design and bore clearances. All these are discussed further in detail.

(*i*) *Piston head design* depends on the desired compression ratio, the shape and volume of the cylinder head and the desired number, size and location of valve reliefs. Dome on a piston interferes with flame travel and airflow within the cylinder, but may have to be used in case of large volume heads, it being the only way to achieve therein, a desired compression ratio. Generally a flat head piston with a smaller chamber head is more desirable.

(*ii*) *Piston to valves and cylinder head clearances.* Minimum clearance between piston and valves should be 0.100" (2.5 mm). The minimum clearance between piston and cylinder head should be 0.040" (1.0 mm) in case of steel connecting rods while additional 0.010"-0.020" (0.25-0.50 mm) clearance is required in case of aluminium rods due to their higher coefficients of thermal expansion.

(*iii*) *Piston skirt design and bore clearances.* Piston skirt strength and bore clearances depend on material, manufacturing process, skirt cross-section shape, oil ring groove drain back design. Ordinary road vehicle engine pistons, whether cast or forged, use slots to return oil scrapped from the cylinder walls; this makes the skirt more flexible and allows tighter cold bore clearances. On the other hand a high performance engine piston, *e.g.*, the racing car engine piston, would use drilled holes for the return of oil. Such a design results in higher structural strength for the skirt, but needs greater bore clearances due to decreased flexibility of the skirt and increased amount of heat transfer from the cylinder head.

3. Compression ratio. Excessively high compression ratio would cause detonation which must be avoided. A golden rule is to use slightly lower ratio rather than the one which is too high. Generally road vehicles should have engines (petrol) with compression ratio between 9 and 10 while racing car petrol engines can be built with a maximum ratio of 12.5 to 13.25 depending on the type of car, track size and the quality of fuel used.

4. Piston weight. To keep the inertia forces minimum, piston with minimum possible weight should be selected.

5. Cost. Cast pistons are cheaper than the forged ones due to which the same should be used for engines of ordinary road vehicles.

8.6. Piston Failures

1. Piston Scuffing. This occurs when due to excessive heat the piston expands and becomes tight in the cylinder. As a result the lubricant is squeezed out from the cylinder walls causing metal to metal contact. The main reasons for the piston scuffing or scoring are:

(*i*) Insufficient lubrication of cylinder walls. The lubrication system may be inspected and suitably serviced.

(*ii*) Overloading the engine. This depends on driving habits and the conditions.

(*iii*) Detonation resulting in high engine

Fig. 3.55. Scored Piston.
(Courtesy—Sealed Power Corporation, U.S.A.)

temperatures. Suitable measures may be taken to prevent detonation (refer Art. 6, Chapter 9).

(*iv*) Inefficient cooling system, which may be inspected and rectified.

(*v*) Leakage of cooling water in the cylinders causing lubricant film breakdown.

(*vi*) Piston pin may be too tight in either the piston bosses or the connecting rod bush. This will put a restraint on free expansion or contraction of the piston due to increase or decrease in is temperature.

Fig. 3.56. Piston scuffing at 45° of bosses. (Courtesy—Sealed Power Corporation, U.S.A.)

Fig. 3.57. Piston with broken ring land. (Courtesy—Sealed Power Corporation, U.S.A.)

Thus it will not come back to its natural shape when the engine is stopped. On restarting, therefore, with the increase of temperature the piston will not return to circular shape, but to some other distorted shape, which will result in scuffing and scoring in the areas of the piston skirt located about 45° from the bosses (Fig. 3.56).

2. Burnt Piston. This may be chiefly on account of detonation or preignition. The burning due to detonation is generally at a point farthest from the spark plug where the hot end gases rapidly release their energy. In case of preignition, however, the burning is usually near the centre of the piston head. Remedial measures should be taken to avoid the recurrence of these.

3. Damage to ring land. This may be on account of:

(*i*) Excessive ring groove side clearance.

(*ii*) Detonation or preignition.

(*iii*) Ring not compressed properly while installing.

(*iv*) Attempt to take out piston-rod assembly without first removing cylinder ridge.

Fig. 3.58. Piston with damaged boss. (Courtesy—Sealed Power Corporation, U.S.A.)

(*v*) Leakage of water into the cylinder. The source of leakage may be identified and properly rectified.

4. Damaged piston boss and circlip groove. This may be due to any or more of the following:

(*i*) Bent connecting rod, which produces a lateral rocking movement on the piston as well as the pin causing wear of circlip grooves. The sharp end thrust of the pin against the circlips, then removes them completely, damaging the bosses also.

(*ii*) Crankpins that are tapered or out of parallel with the crankshaft journals will produce the similar effect as described above in (*i*).

(*iii*) Too much end play in the crankshaft will also produce lateral rocking motion of the piston pin.

(*iv*) The circlips may be installed loose. This usually happens because of overcompressing the same while installing, so as to cross yield point causing a loose ring.

Fig. 3.59. Worn out circlip groove.
(Courtesy—Auto Piston Mfg. Co. Ltd. India)

9. PISTONS RINGS

9.1. Definition

These are the engine components which form the dynamic seal between the piston and the cylinder wall.

9.2. Functions

The piston rings in I.C. engine have to perform the following functions:

1. To form a seal for the high pressure gases from the combustion chamber against leak into the crank case.
2. To provide easy passage for heat flow from the piston crown to the cylinder walls.
3. To maintain sufficient lubricating oil on cylinder walls throughout the entire length of the piston travel, minimizing the ring and cylinder wear, and at the same time, control the thickness of the oil film so that satisfactory oil control is maintained. The oil is not to be allowed to go up into the combustion chamber where eventually it would burn to leave carbon deposits.

9.3. Construction

The construction of a piston ring and the nomenclature of its various parts is shown in Fig. 3.60. The ring is generally cast individually and machined carefully so

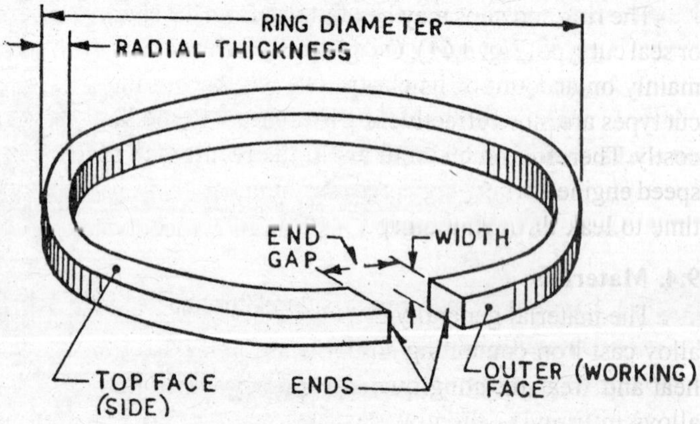

Fig. 3.60. Ring construction.

that when in position, it is able to exert uniform pressure against the cylinder walls. A gap has to be cut at the ends so that while inserting the ring onto the piston, it can be expanded, slipped over the piston head and released into the ring groove. Further, the gap is almost closed when the piston is inside the cylinder, due to which the ring is able to exert pressure on cylinder walls, which is a must for sealing purpose. In the installed condition, ring must be circular so as to match the cylinder. Rings can be designed to have a uniform contact pressure against the cylinder around the entire circumferences of the ring or to have higher or lower pressure at the gaps. Complex mathematical calculations are required to determine the special non-circular cam shapes which a ring requires in the free state for obtaining the ideal contact pressure patterns. Further, any circumferential expansion of the ring at higher operating temperatures may also be accommodated by the end gap. Some differential expansion of the ring with respect to the cylinder is always likely to occur inspite of the equal coefficients of cylinder and ring materials due to the fact that the ring is always operating at higher temperatures than the cylinder walls; that is why direction of heat flow is from the rings to the walls.

The sealing action of the top ring is due to the fact that the high pressure in the combustion chamber presses the top ring tightly on the base of the piston ring groove, thus sealing the ring. However some leakage does take place through the end gap of the top compression ring. This leakage is useful in that it provides the pressure for sealing action of the second piston ring, where sealing action takes place in the same way as in case of the top compression ring described above.

The amount of end gap should, however, be determined cautiously. Excessive end gap would result in blow-by and scuffing of the rings. On the other hand, lesser clearance would cause the ring ends to butt at higher temperatures, resulting in excessive and non-uniform pressure on the cylinder walls causing excessive wear.

In practice, the end gap for the top compression ring is controlled from 0.07 to 0.12 mm per 25 mm of bore size. The second ring end gap should be 0.12 to 0.25 mm *wider* than the top ring gap to prevent gas build-up between the top and the second rings, which could have caused the top ring to lift off the bottom of the piston ring groove and lose contact with the sealing surfaces and also inhibit the ring's ability to transfer heat from the piston. Moreover, with the end gap of the second ring being more than that of the top ring, the gas pressure would now be directed downwards towards the sump, due to which any lubricating oil that may have collected in the ring pack areas would go down with it.

Since the steel has lower coefficient of thermal expansion than cast iron, somewhat smaller end gaps are used for steel rings compared to the rings made from cast iron.

'Total Seal' of USA also manufactures their patented *gapless rings*, which though costlier, are claimed to reduce gas leakage by 10 to 20 percent compared to a conventional ring.

The ring end gaps may be either straight butt type or tapered or seal cut type (Fig. 3.61). Out of these butt type is most common mainly on account of its cheapness. The tapered and the seal-cut types are more effective in preventing leakage, but are more costly. Therefore, such joints are used only in case of some low-speed engines, where high pressure combustion gases have more time to leak through the gap.

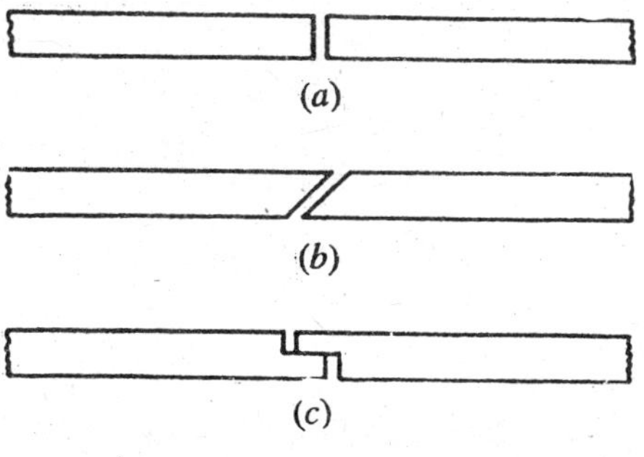

Fig. 3.61. Types of piston ring end gaps. (*a*) Butt type, (*b*) Tapered type, (*c*) Seal cut type.

9.4. Materials

The material generally used for piston rings is fine-grained alloy cast iron containing silicon and manganese. In has good heat and wear resisting qualities. There are two types of iron alloys in use viz., the gray cast iron and the ductile iron. Till recently, gray cast iron has been used extensively for automotive piston rings. Uncoated gray iron is compatible with cast iron cylinder walls and does not scuff. But it is brittle, due to which special ring expanders have to be used for installing the rings on the piston.

With the reduced ring thickness in modern engines, ductile iron has replaced the gray iron. Ductile iron is also called 'nodular' iron because of the nodular or rounded shape of its grains. It has got higher tensile strength, which is about 4 times that in case of gray cast iron (Table 3.1). Due to this, it is now possible to use rings of very small thicknesses, say about 1 mm. However, it is more expensive and also not as compatible with cast iron cylinder walls as the gray cast iron. Ductile iron ring also tends to scuff unless faced with chrome or molybdenum. Although, ductile iron is harder, yet not necessarily more were-resistant, for which chromium or molybdenum coating has to be used.

TABLE 3.1. GENERAL PISTON RING MATERIAL PROPERTIES

Material	*Hardness (HRC)*	*Tensile Strength (MPa)*	*Fatigue Strength (MPa)*
Gray cast iron	22-23	310	210
Ductile iron	38-40	1240	600
Steel	44-53	1650	960

Chromium-plated rings are usually used as the top ring, which is subjected to the highest working temperatures and the corrosive action of the combustion products. However, for heavy duty pistons, other compression rings may also be chromium-plated. Chromium plated rings have resulted in considerable saving in the cylinder bore life. Although chromium itself is very hard due to which it is normally expected to wear the cylinder walls rapidly, yet it does the trick because of very fine finish of its coating on the ring. Chromium plating further helps the rings to resist scuffing because it is difficult to be welded to cast iron cylinder. However, these rings should not be used when the cylinder bore itself is lined with chromium or any such hard material.

The rings are provided generally, a porous phosphate coating to reduce the scoring of the surfaces during running in. These coatings are formed by immersing the ring in a bath of phosphoric acid and manganese, which deposits a layer of iron manganese phosphate on the ring surface. The porous surface has cavities for the worn particles and also acts as oil reservoir, which remains even after the coating has worn away. Thus the rubbing friction between the rings and cylinder walls is reduced.

Rings with molybdenum-filled face have also been introduced recently. The molybdenum surface has larger oil carrying capacity. It, therefore, provides better cylinder wall lubrication with resultant longer engine life. The higher melting point of molybdenum (2620°C) enables the ring to stand higher temperatures than other ring metals and thus resist scuffing. A more recent development is thermo chemically treated chromium (TC chrome) especially suited for top ring application where lubrication is marginal.

Steel rings are stronger than ductile iron, due to which they can withstand high speed pounding and high loads experienced in modern downsized petrol engines and in high compression turbocharged and supercharged diesel engines Generally, these have been used as top compression rings. In fact these are ideal for applications involving higher combustion temperatures, higher compression loads and tougher emission standards. Narrow widths of rings are possible due to higher strength and this also means lighter ring which provides a more effective seal against the bottom of the ring groove. Generally speaking, their advantages are:

1. Higher breaking resistance.
2. Better heat resistance.
3. Higher resistance to mechanical stress.
4. Less ring side wear.

5. Less groove side wear.
6. Longer service life.

Like ductile iron, steel rings are also not compatible with cast iron cylinder walls, due to which reason they must be applied some suitable coating like chrome or molybdenum or gas nitriding. Steel rings are made from pre-formed steel wire.

Machining required to finish a steel ring is much less compared to that needed for finishing a gray iron or ductile iron ring which makes them actually less expensive, especially when made in large quantity.

Alloy steels have also been used as ring material. Stainless steel oil rings resist pitting and corrosion to remain clean and do not clog with carbon as quickly as other rings. Further, these resist excessive tension loss at engine operating temperatures.

In general, practically the steel and ductile rings are considered interchangeable for replacement purposes. However, one should never substitute a gray cast iron ring for a ductile iron or steel top compression rings.

Typical piston ring materials alongwith their compositions and properties are gives in Table 3.2.

TABLE 3.2. TYPICAL WELLWORTHY PISTON RING MATERIALS

S. No.	*B.S. Symbol*	*Material Type and Structural Characteristics*	*Composition*	*Min. Properties (See note below)*					
				RTS	*B/R*	*M/S*	E_n	*HB*	*Thermal set*
1.	U 1	Centrifugally cast, Medium phosphorus, pearlitic grey iron. Free ferrite 10% max.	C 3.5 max., Si 1.8/2.5, P 0.4/0.65 Cr 0.3/0.5	247	338	396	107	230/295	12.5
2.	U 3	Centrifugally cast. Semi malleablised carbidic alloy iron. Pearlitic with free carbides and temper carbon.	C 2.8/3.1, Si 1.9/2.5, P 0.3/0.45, Cr 0.75/1.15, Mo 0.8/1.0	402	550	643	138	269/305	6.5
3.	U 0	Statically sand cast. Medium phosphorus pearlitic grey iron. Type A Graphite. Free ferrite 5% max.	C 3.0/3.3, Si 1.4/2.2, P 0.35/0.55, Cr 0.4 max	247	338	396	96.5	200/245	10.5
4.	U 0	Individually cast. Medium phosphorus pearlitic grey iron. Type A/B Graphite. Free ferrite 10% max.	C 3.4/3.9, Si 2.1/2.9, P 0.4/0.8, Cr 0.4 max	275	375	443	82.9	230/295	20.0
5.	U 5	Centrifugally cast. Malleablised fine pearlitic or martensitic low phosphorus iron with temper carbon. High strength and ductility.	C 3.0/3.3, Si 1.0/1.4, P 0.1 max, Cr 0.07/0.3	586	803	975	145	230/300	12.0
6.	U 6	Centrifugally cast. Martensitic spherical graphite iron. No free ferrite.	C 3.0/3.65, Si 1.7/2.7 P 0.1 max, S 0.03 max, Mo 0.35/0.65, Cu 0.35/0.65, Mg. 0.03/0.07, Ni 0.8/1.7	850	1160	1354	159	272/437	18.0

Note: RTS = British Standard Ring Tensile Strength (MN/m^2)
BS = Breaking Strength (MN/m^2)
M/R = Modulus of Rupture (MN/m^2)
E_n = Nominal Modulus of Elasticity (GN/m^2)
HB = Hardness Brinell (P/D^2 = 30)
Thermal set = % gap collpase when ring closed to nominal cylinder dia for 6 hrs at 350°C.

9.5. Number of Rings

The number of rings to be used on a piston varies depending upon the requirements. Earlier two to four compression rings and one to oil control rings were used, but with modern design trends of decreased car and engine heights, the number of rings is restricted to usually three, out of which one is the oil control ring.

A minimum of two compression rings are required because of the high pressure difference between the combustion chamber and the crankcase at the beginning of the power stroke. This difference may be as high as 70 atmospheres. A single piston ring cannot take such high pressure, which necessitates the use of at least two compression rings, which divide the pressure between themselves. Increasing the number of rings (which is restricted by the maximum piston height) also reduces the design pressure between the rings and the cylinder walls which results in decreased wear and consequently increased life.

9.6. Types of Rings

The piston rings are, generally speaking, of two types:

1. Compression rings
2. Oil control rings

The top compression ring (*i.e.* the ring nearest to the combustion chamber) has to do the hard work of gas sealing and transfer of heat from the piston crown to the cylinder walls. The second compression rings perform a double role. They seal and transfer heat, and also assist the oil rings in controlling oil.

How the compression rings perform their primary function of sealing the compression charge and combustion pressure is made clear by means of Fig. 3.62. When the piston is moving up during compression stroke, the pressure in the combustion chamber acts on top of the compression ring which forces the ring against the cylinder wall [Fig. 3.62 (*a*)]. This provides a very effective seal without leakage. However, for this the surface finish of the ring, groove and cylinder wall must be of adequate quality. A similar situation exists during the power stroke in which the piston actuates downward [Fig. 3.62 (*b*)]. In this case, the upper ring face is flattened against the top of the ring groove.

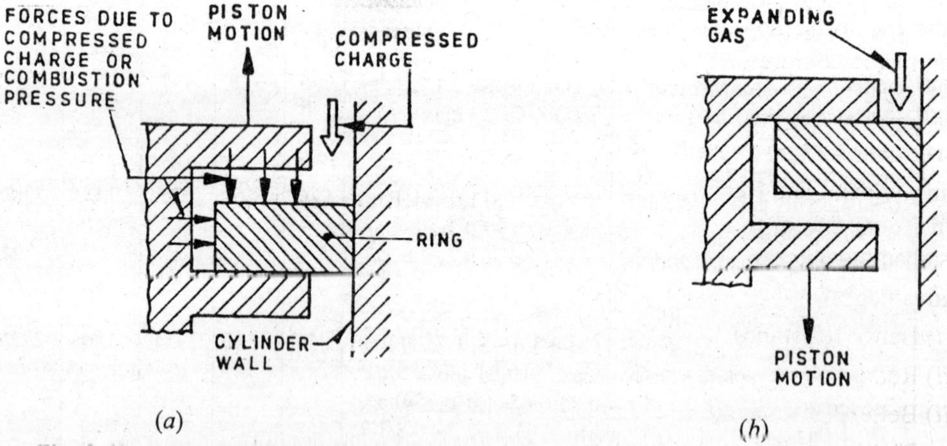

Fig. 3.62. Action of compression ring when the piston is moving (*a*) up, (*b*) down.

The second compression ring may be dispensed with in performance engines. Doing so would allow the piston height to be shorter, resulting in weight savings and consequently reduction in inertia forces. *Formula one engines have only one compression ring.*

The function of the oil control rings is evident from their name. To perform this function effectively, they must prevent excessive amounts of oil from passing:

(*i*) between the ring face and the cylinder wall.

(*ii*) through the ring end gap, and

(*iii*) around behind the ring.

To understand the action of the oil control ring, consider Fig. 3.63. When the piston is moving up, the lower face of the piston ring groove is pressed against the lower ring face which makes the outer ring face to slide over the oil on the cylinder wall. While doing so it scrapes some oil in front of it, which goes to the

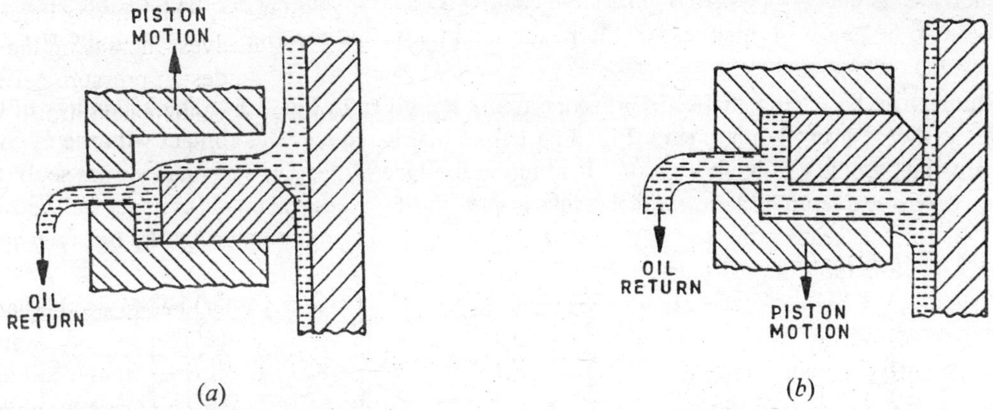

Fig. 3.63. Action of the oil control ring when the piston is moving (*a*) up, (*b*) down.

clearance between the ring and the groove face and thence through a number of holes in the groove, to the sump. Similarly when the piston moves down, the sharp edge of the oil control ring scrapes the oil on the cylinder wall, the excess oil returning to the sump through the holes in the groove. The oil holes in the oil ring groove must be adequate to ensure free flow of all the scraped oil, otherwise some pressure may build-up in the groove which would cause the oil to move up to the combustion chamber side and ultimately get burnt there.

Whether the piston moves up or down, in either case sufficient oil is allowed to be metered for satisfactory lubrication of the compression rings.

9.7. Design considerations and trends

I. Compression Rings

1. Ring width. The trend is toward reducing ring width. During the past 50 years, the ring widths have been reduced to even less than about 1.0 mm. Coupled with advantages of reduced ring width are also it disadvantages:

(*a*) Advantages:

(*i*) Better resistance to ring scuffing.

(*ii*) Reduced ring width means lower piston height and consequently lower engine height.

(*iii*) Better resistance to ring flutter.

(*iv*) Problems of ring inertia are reduced. Ring inertia is the tendency of the ring to continue its motion as the piston changes direction. The problem is particularly severe at high speeds.

(*b*) Disadvantages:

(*i*) Machining very narrow grooves in the piston accurately is quite difficult.

(*ii*) Rings with too much reduced width and without a satisfactory thickness/width ratio become unstable in the ring grooves.

2. Position. The latest trend is to position the rings closest to the top of the piston. Today, the land width between the top ring groove and the piston crown is about 3 mm in some engines compared to more than double a decade ago. Moving the ring up reduces the depth of the crevice above the top compression ring, which reduces the amount of air-fuel mixture entering this crevice during compression stroke. Since such air-fuel mixture can not burn completely on ignition, positioning the top ring closer to the top saves fuel and improves emission.

However, moving the ring closer to the top has also a disadvantage. The top ring is subjected to greater heat, which means necessity to provide suitable coating on the top ring. More over, piston strength is also reduced due to decrease in thickness of the piston land above the top ring, increasing the risk of piston failure.

3. Shape. The basic ring shapes in present day use are given in Fig. 3.64. The advantage of the taper face [Fig. 3.64 (*b*)] over the plain ring [Fig. 3.64 (*a*)] is that is reduces the contact with the cylinder wall to a narrow line, which results in high unit loading on the face of the ring to enhance ring sealing. It also provides a downward scraping action which results in carrying the excess oil on the cylinder walls downwards, ahead of it. For this reason, the top edge of the compression ring must never be the *only portion* of the ring face to be touching the cylinder walls. In case that happens, the oil would be scraped upwards, resulting in very high rate of oil consumption.

Machining of the inside upper corner makes it a torsional twist ring [Fig. 3.64 (*c*)]. The internal forces are changed due to machining, so that when it is installed in the cylinder, it turns about its axis so as to provide a line contract between the ring and its groove and also contact with the cylinder wall at its lower edge, which contributes towards a better control of passage of both the combustion gases and the engine oil. The scraper type torsional twist ring [Fig. 3.64 (*d*)] functions like the bevelled torsional-twist ring discussed above. However, due to its narrow face, it has the advantage of the higher unit face loading in the untwisted position. The taper-face, torsional twist ring [Fig. 3.64 (*e*)] is a combination of the taper face and torsional-twist designs. Keystone rings [Fig. 3.64 (*f*)] have inclined side faces and operate in grooves of similar geometry. Relative movement between the ring and the groove in the transverse direction discourages the build up of carbon deposits and therefore, prevents ring stick. Rings of this form are most commonly used for the top ring in turbocharged engines.

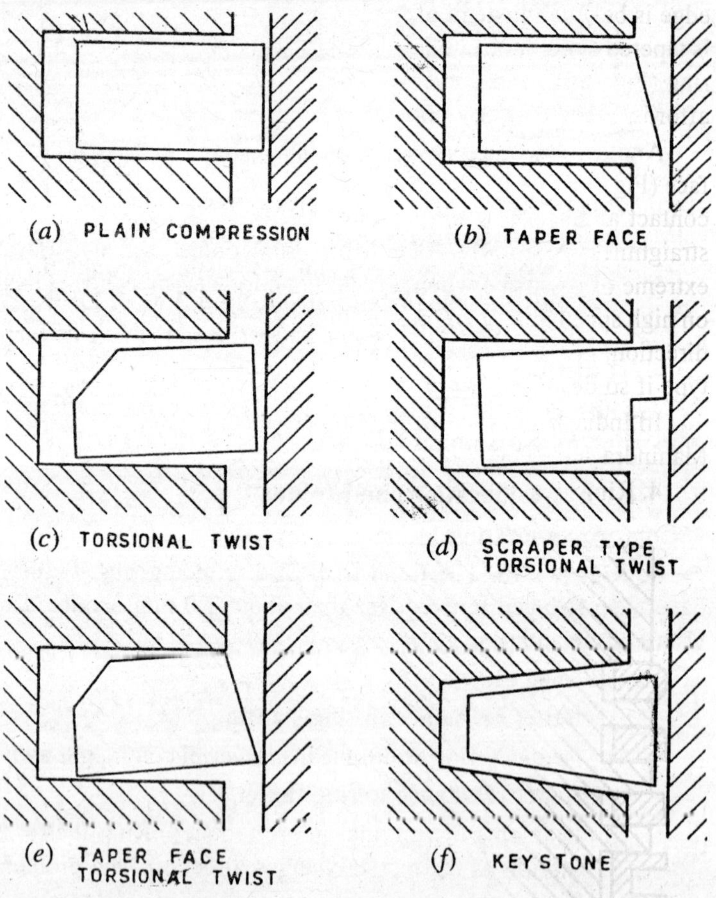

Fig. 3.64. Compression rings.

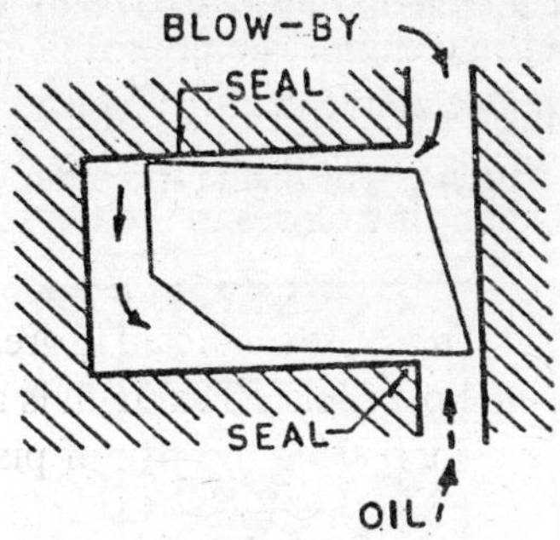

Fig. 3.65. Action of reverse twist ring. (Courtesy—Sealed Power Corporation, U.S.A.)

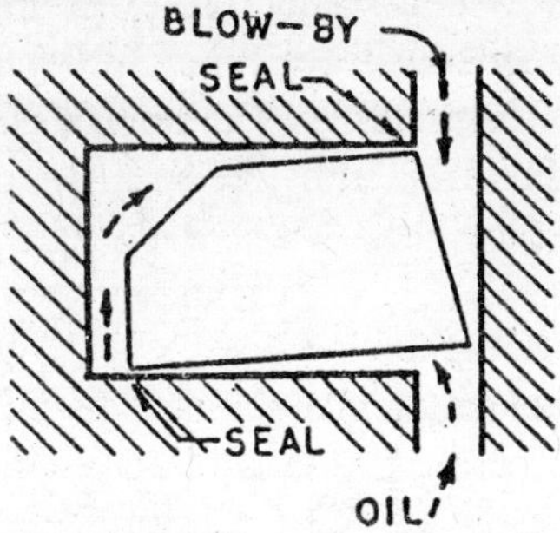

Fig. 3.66. Action of positive twist ring. (Courtesy—Sealed Power Corporation, U.S.A.)

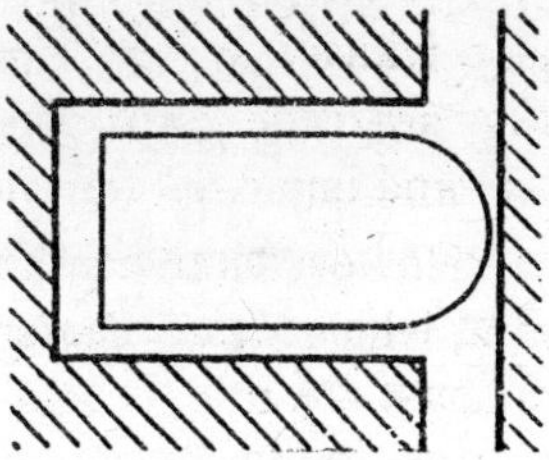

Fig. 3.67. Radius face ring. (Courtesy—Sealed Power Corporation, U.S.A.)

A recent modification of the torsional-twist ring is the reverse-twist ring (Fig. 3.65) whose lower inner edge is bevelled instead of the upper inner edge in case of torsional twist ring. Also the outer face of ring is tapered to compensate for the twist which is in opposite direction to that in case of normal positive twist ring (Fig. 3.66). The reversed twist ring offers better oil control than a normal twist ring, but the latter affords better blow-by control.

Another recent improvement in compression ring design, which has become quite popular is the radius-face (Fig. 3.67). The advantage of the radius shape is that there is high unit loading due to the narrow-line contact and secondly the contact with the cylinder wall remains intact even if the piston ring groove straightness is slightly off. Further, this contact is not lost even when the ring changes direction at the extreme of its stroke when a rocking action can take lace. Thus, it results n good-oil and blow by control on high speed output engines. Moreover, the wear at the top of the cylinder wall where the top ring change direction, is also reduced. The radial face ring may be bevelled to make it positive twist or reverse twist type if so desired.

In India radial face ring has been used for the first time as top ringing the diesel engine DP-4.90 of Mahindra Jeep MM-540 DP (Fig. 3.68).

4. Rings for worn cylinders. An ordinary piston ring meant for correctly bored cylinder will not work

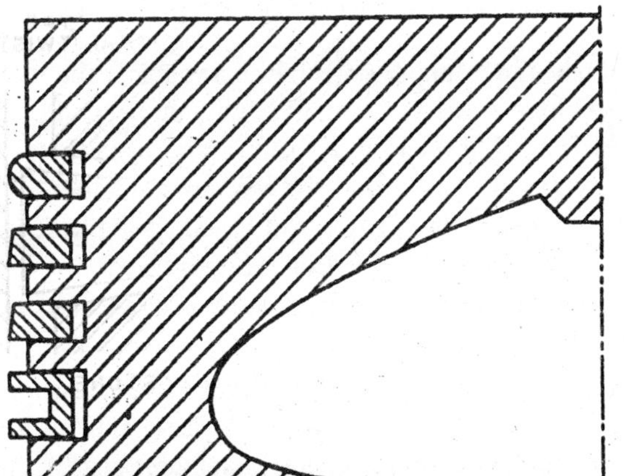

Fig. 3.68. Rings used in diesel jeep MM-540 DP. (Courtesy—Mahindra and Mahindra Ltd., India)

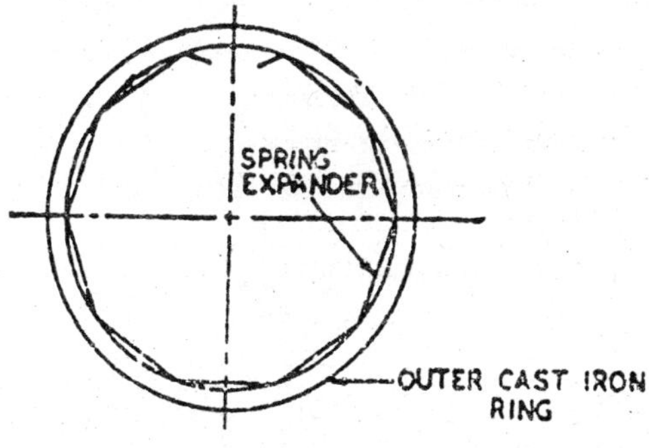

Fig. 3.69. Spring expander piston ring.

efficiently if fitted in a bore which is worn oval. In such cases, spring expander piston ring may be employed. They may be two-piece, three- piece, or four-piece type. Spring expanders are made of spring steel with crimps spaced uniformly along the circumference. The outer cast iron ring exerts only a part of the total pressure on the walls, the rest being contributed by the spring expander which is put inside the outer cast iron one (Fig. 3.69). In three-piece type, apart from the cast iron ring and the spring expander, a spiral steel side rail consisting of two turns of thin flat steel is located below the cast iron ring. In four-piece type, there are two such side rails placed one below and the other above the cast iron ring.

II. Oil control rings

Commonly used oil control rings are shown in Fig. 3.70. The bevelled ring [Fig. 3.70 (*a*)] is installed with the bevel side towards the top. This is the most simple of oil control rings. In the stepped scraper ring [Fig. 3.70 (*b*)], due to the decreased width of the outer face the pressure between the ring and the cylinder wall is increased, resulting in better scrapping. The slotted scraper ring, [Fig. 3.70 (*c*)] has a recess on the outer face, which results in the formation of narrow ring lands providing a higher radial pressure against the cylinder walls and also two-stage scraping action. The oil holes or slots are also provided between this recess and the inner ring face, which facilitates adequate flow of excess oil to the inside of the piston and back to the sump. In the 'delayed action scraper ring' [Fig. 3.70 (*d*)] there is a central raised land, which wears out during the run-in period until both the outer lands control the scraping. In the 'double action scraper ring' [Fig. 3.70 (*e*)] there are two scraper rings in the same groove, with an expander at the back to increase the radial pressure against the cylinder wall. The 'composite rail scraper ring' [Fig. 3.70 (*f*)] is highly useful for worn out cylinders having ovality or taper in the bores. There are a number of steel rails which are chrome-plated and have rounded ends, a spacer in the form of a crimped spring to keep the rails apart and an expander. The rounded ends provide a smooth scraping action reducing wear of cylinder wall, the spacer pushes the rails against the sides of the ring groove and the expander provides additional radial force between the ring and the cylinder wall.

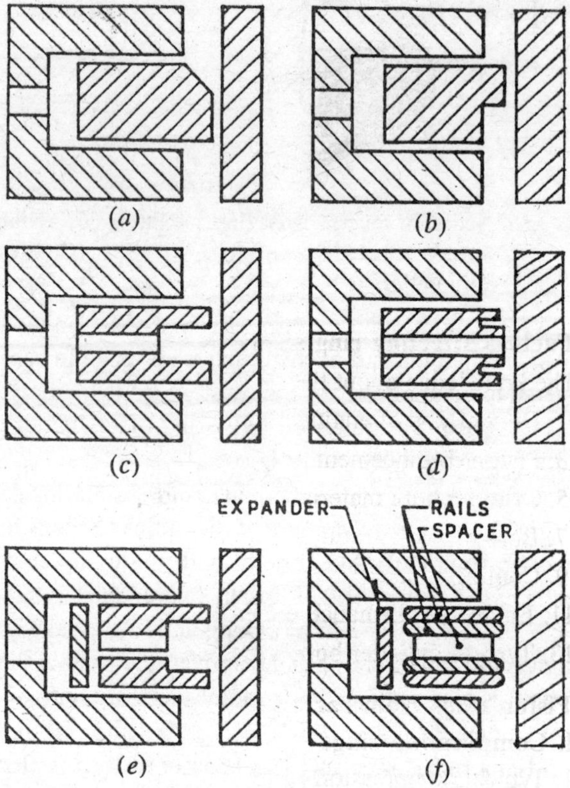

Fig. 3.70. Oil control rings. (*a*) bevelled ring, (*b*) stepped scraper, (*c*) slotted scraper, (*d*) delayed action scraper, (*e*) double action scraper, (*f*) composite rail scraper.

In later designs, expanders were combined with spacers. Such a construction is light having less inertia, is open so that oil can easily flow to return to the crankcase, provides very good oil control and has a long service life. One such ring is shown in Fig. 3.71. It consists of a stainless steel expander spacer that is used to exert uniform outward pressure on two steel rails. The narrow width of steel rails and the tension of the stainless steel expander results in high unit pressure loading. The steel rails provide very good side sealing because the expander exerts side pressure to the rails as well as outward pressure to the rails.

Fig. 3.71. Multi-piece oil control ring.
(Courtesy—Sealed Power Corporation, U.S.A.)

9.8. Factors affecting ring selection

The knowledge about the following factors is necessary to select rings for a particular engine:

1. Dimensions of engine block
2. Piston design
3. Piston displacement
4. Piston speed
5. Cylinder bore material
6. Carburation
7. Bore to stroke ratio
8. Compression ratio
9. Cooling capacity
10. Crankcase pressures
11. Engine performance expected
12. Horsepower requirement
13. Type of cylinder bore lubrication
14. Peak manifold vacuum.

9.9. Piston rings from various manufactures

I. Compression rings

1. Typical compression rings manufactured by *Sealed Power Corporation, U.S.A.*

(*i*) TF-10U. This is a chrome-plated and tapered-face compression ring. The tapered face provides a narrow contact with the cylinder wall to give high initial unit pressure and instant seating.

(*ii*) BT-10U. This is a chrome-plated, inside-beveled and tapered face compression ring. The inside bevel gives a torsional twist to the ring when compressed and aids in preventing 'blow-by'.

(*iii*) TFK-10UA. This is a chrome-plated tapered face ductile iron compression ring with face grooves, to be used for heavy duty applications.

(*iv*) MIB-10Y. This is an inside-bevel type ring with molybdenum filling in the face. Molybdenum provides long services life with reduced wear.

(*v*) TF-30. Combined with the expander spring, this spring can readily

Fig. 3.72. Sealed Power compression ring TF-10U.

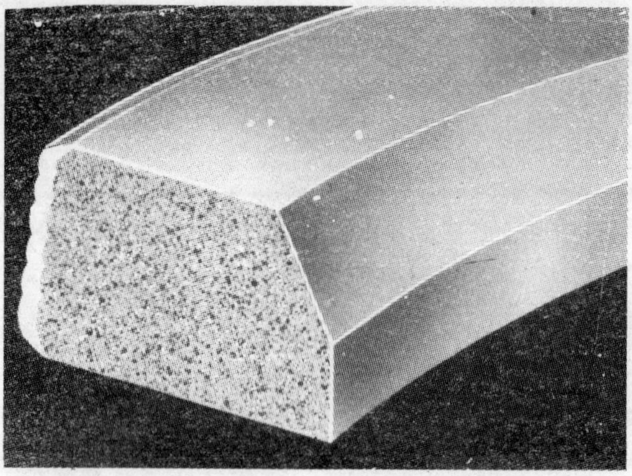

Fig. 3.73. Sealed Power compression ring BT-10U.

Fig. 3.74. Sealed Power compression ring TFK-10UA.

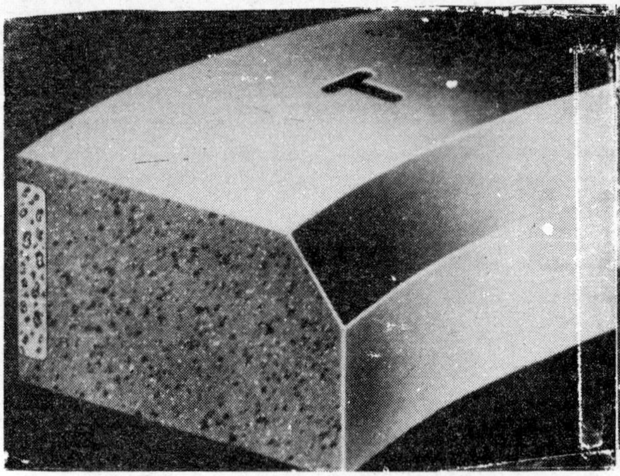

Fig. 3.75. Sealed Power compression ring MIB-10Y.

Fig. 3.76. Sealed Power compression ring TF-30.

conform to cylinder bore irregularities. It is used mainly as compression ring in the second groove.

2. Compression rings manufactured by *Perfect Circle Division of Dana Corporation, U.S.A.*

(*i*) *Chrome 200 cm* (Fig. 3.77). The upper inner corner of this compression ring has a counter bore to produce torsional twist resulting in a positive lower outer edge face contact with the cylinder. May be plated or surface treated as specified.

Fig. 3.77. Perfect circle chrome 200 cm ring.

Fig. 3.78. Perfect circle molly filled 75 cm ring.

(*ii*) *Moly filled 75 cm* (Fig. 3.78). This ring is designed as a top compression ring for a piston groove located very close to the combustion chamber. Only a narrow ring band may separate the ring face from the combustion chamber. Applications include high r.p.m. two-stroke engines.

Fig. 3.79. Perfect circle chrome grooved cm ring.

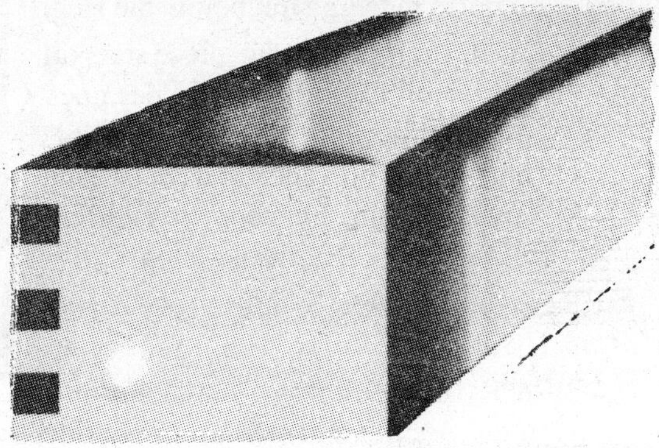

Fig. 3.80. Perfect circle B cm ring.

(*iii*) *Chrome grooved cm* (Fig. 3.79). This ring is especially designed to resist scuffing in high output engines. It may have one or more face grooves.

(*iv*) *B cm* (Fig. 3.80). Especially suitable for use in heavy duty engines, the B cm ring is identified by its narrow Ferrox-filled grooves that minimize scuffing.

II. Oil control rings

Typical oil control rings manufactured by reputed American and British concerns have been discussed below:

1. Rings manufactured by *Sealed Power Corporation, U.S.A.*

(*i*) HD-20. This is a ventilated oil ring with narrow tapered lands. The taper exerts a scraping action on the downward stroke making the ring efficient in controlling oil.

(*ii*) MCL-40 UW. This is a cast iron, chrome-plated ring with stainless steel coiled wire expander, to provide tension. This type of expander provides high unit pressure and better conformability to cylinder deformation, thus allowing the ring to function independently of groove depth, thereby improving the oil control as compared to non-spring-backed rings. Such rings are, therefore preferred on heavy duty engines.

Fig. 3.81. Sealed Power oil ring HD-20.

(*iii*) DLS-40 UW. In this type, a stainless steel wire expander, located above the slots, applies equal pressure to both lands, retains tension and assures comformability without any regard to groove depth. It is used for heavy duty operation.

(*iv*) GB-50. This three-piece oil ring, which is specially designed for some diesel engines, has two scraper lands to provide double cylinder wall contact for positive oil control. The spring is not slotted in this and therefore, drainage is provided by oil holes in the piston below the ring.

(*v*) CS-50 U. This is a four-piece steel oil ring which uses a specially designed spring to achieve side-sealing. This construction prevents oil from passing around the ring, but does not interfere with normal oil control action.

Fig. 3.82. Sealed Power oil ring MCL-40 UW.

2. Rings manufactured by *Muskagon Piston Ring Co., U.S.A.*

(*i*) MPR-160. The chrome plated side rails in this are flexible enough to provide maximum comformability. The spacer is circumferential and the tension comes from the abutment of its ends.

(*ii*) MPR-260. This is similar to MPR-160 described above, except the addition of another member called 'Gaplatch' which prevents expander overlap.

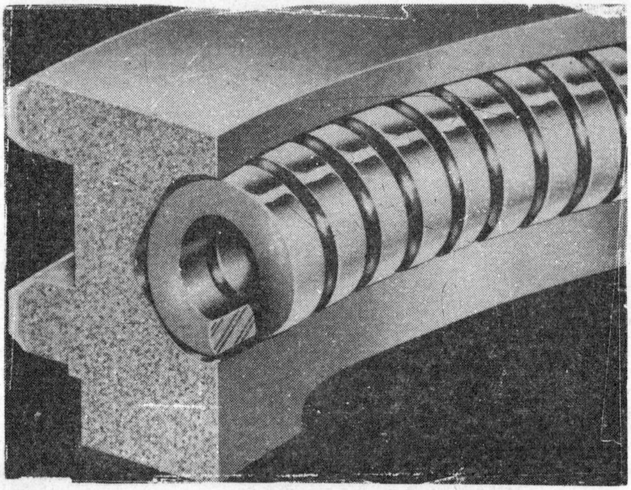

Fig. 3.83. Sealed Power oil ring DLS-40 UW.

3. Rings manufactured by *Wellworthy Ltd. England.*

(*i*) *Flexoil.* This type of ring is equipped with polygonal flat strip steel expanders bearing on the bottom of the piston groove. These rings are ideal for use in chrome-plated bores.

(*ii*) *Conformable oil ring.* Having independence of groove depth, these rings have high confor-mability but required sufficient back clearance for ventilation. Loading is achieved by circum-ferential abutment of the expander which may be either axially crimped or a coil spring.

Fig. 3.84. Sealed Power oil ring GB-50.

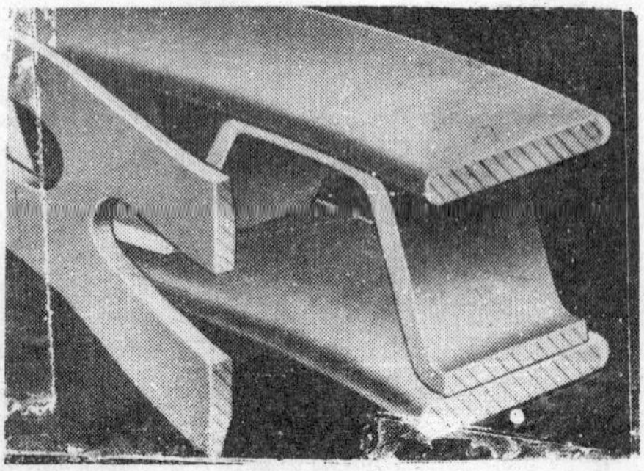

Fig. 3.85. Sealed Power oil ring GS-50 U.

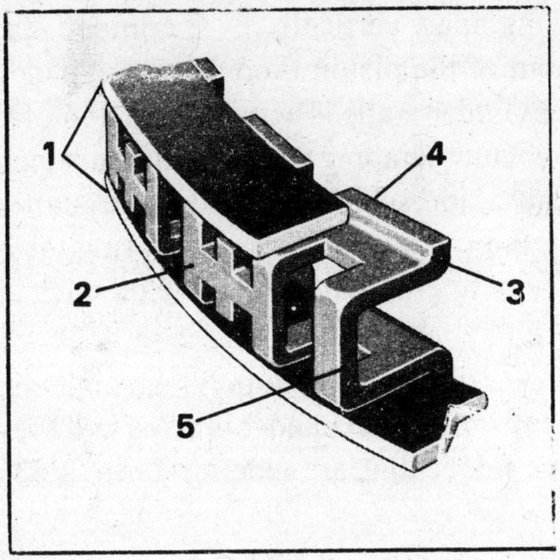

Fig. 3.86. Muskagon oil control ring MPR-160.

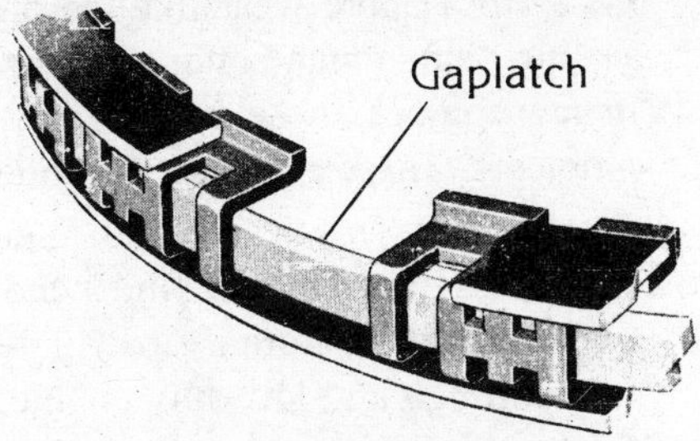

Fig. 3.87. Muskagon oil control ring MPR-260.

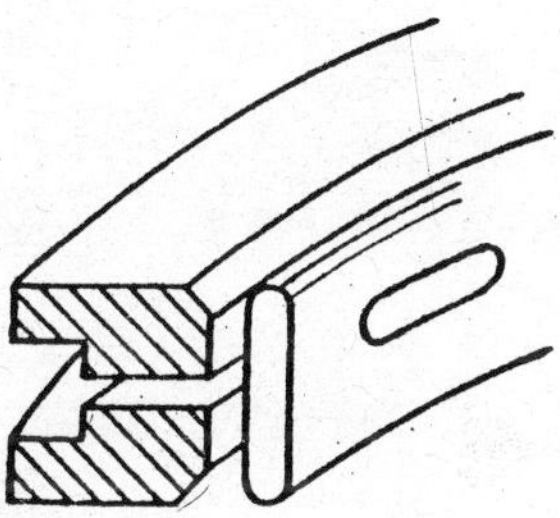
Fig. 3.88. Wellworthy 'Flexoil' oil control ring.

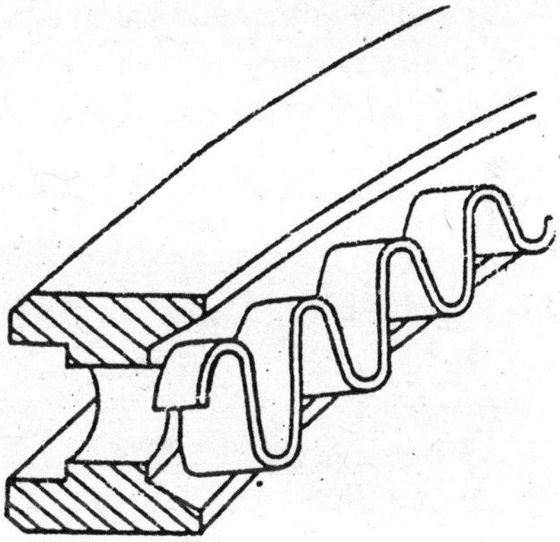
Fig. 3.89. Wellworthy conformable oil control ring.

(*iii*) *Dualcrom.* This is a two-piece cast iron ring, in which peripheries of both parts are hard chromed, thus preventing the breakdown of the control edge. By having two parts the upper one of which is ventilated by cutways, two lines of contact are established which are maintained in spite of groove wear. A single piece standard ring, upon rocking in a worn groove presents only one edge to the cylinder wall.

(*iv*) *Standard Duaflex.* This ring is all made of steel and consists of a number of rails having reduced

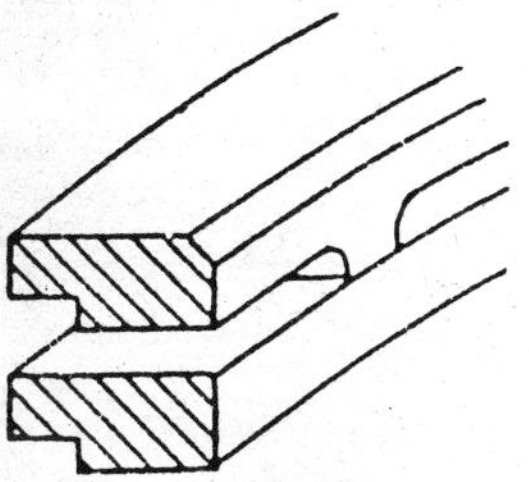
Fig. 3.90. Wellworthy 'Dualcrom' oil control ring.

Fig. 3.91. Wellworthy 'Standard Duaflex' oil control ring.
1. Crimped spring, 2. Expander, 3. Rails.

edges which are hard chromed. The rails are sprung apart vertically by a crimped spring so that they make a seal against the top and the bottom of the piston grooves. the expander ring springs them outwardly to make a seal against the cylinder walls. The wiping action of the rail peripheries squeezes the oil down the cylinder walls without scraping the bores, thus not increasing the cylinder bore wear like the ordinary rings. Thus, side sealing as well as circumferential sealing being much better than conventional rings, these are best suited for re-ring jobs. Even in new engines, these are accepted as standard fitment to eliminate excessive oil consumption problems. These rings are dependent on groove depth.

(*v*) *Duaflex 100.* This is similar to 'Standard Duaflex' described above, but with an abutment type expander. These rings are independent of groove depth and the expander is fixed and prevented from rotation. Conformity to cylinder bore and groove side sealing are superior. These are suitable for both high and low duty engines.

4. Rings manufactured by *Perfect Circle Division of Dana Corporation, U.S.A.*

(*i*) *Chrome 98 oil* (Fig. 3.93). This three piece 85 oil ring has an improved type circumferentially expansible spacer and chrome-plated steel rails. Performance is not affected by groove depth. Provides side seal as well as face seal.

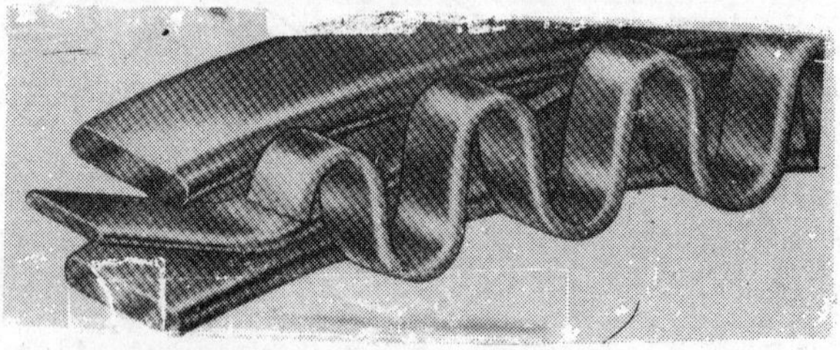

Fig. 3.92. Wellworthy "Duaflex-100" oil control ring.

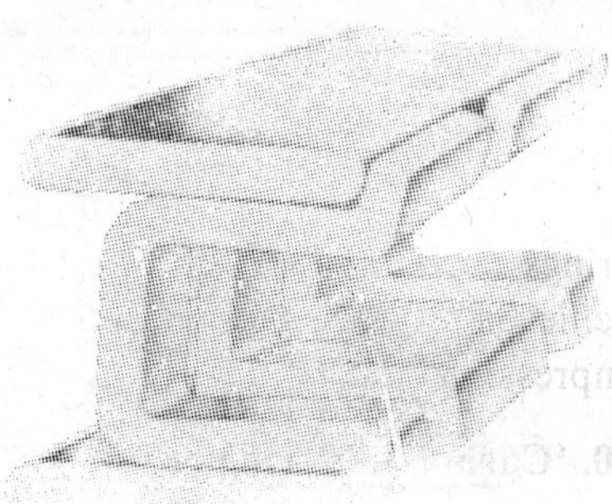

Fig. 3.93. Perfect Circle chrome 98 oil ring.

(*ii*) *Key 85 oil* (Fig. 3.94). This basic type 85 oil ring has been given a keystone design to provide maximum freedom from ring sticking May be chrome plated.

Fig. 3.94. Perfect Circle key 85 oil ring.

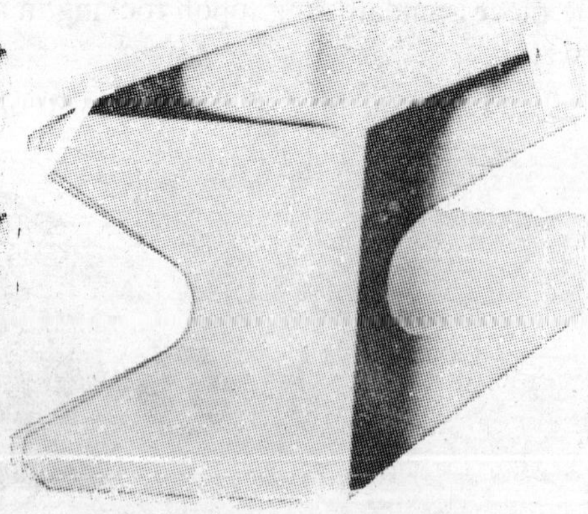

Fig. 3.95. Perfect Circle chrome 85 oil ring.

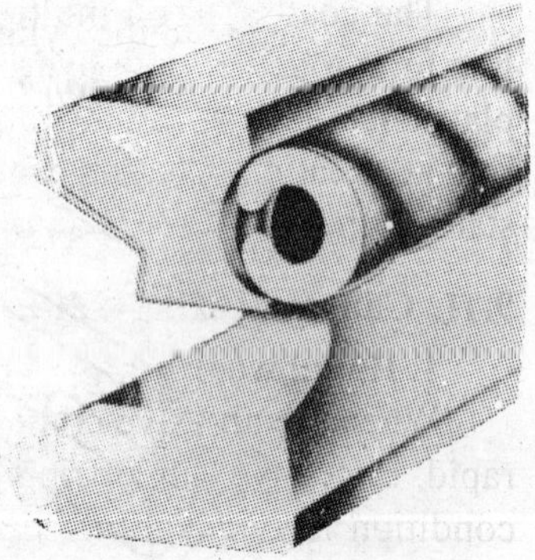

Fig. 3.96. Perfect Circle chrome OS-89 oil ring.

(*iii*) *Chrome 85 oil* Fig. 3.95). Thick hard chrome plate provides long wear and the wide V slot minimizes clogging.

(*iv*) *Chrome OS-89 oil* (Fig. 3.96). This ring has the advantage of maximum conformability and unit pressure independent of groove depth. In addition it has maximum ventilation to prevent clogging since the coil expander is at the side away from the ventilation slots.

5. Federal Mogul's LKZ oil control ring.

The company claims that this ring (Fig. 3.97) is equally suitable for diesel and petrol engines, which reduces oil consumption by up to 50% and friction by up to 15%. It is also claimed that unlike conventional

Fig. 3.97. Federal Mogul's LKZ oil control ring

oil rings, such rings also provide consistently low oil consumption throughout the whole engine life, which it achieves by reducing build-up of carbon on critical piston parts such as the area immediately above the compression ring.

9.10. 'CarboGlide' ring coating

Third generation of coating for petrol engines developed by Federal-Mogul is claimed to reduce friction by upto 20% as compared to nitride or other commonly used coatings, thereby resulting in fuel economy and CO_2 emission improvement. Other advantages are:

1. Due to high wear resistance of the coating it is expected to last for full engine life and can be used for modern high output petrol engines with turbo-charging or direct injection.
2. It also protects the cylinder surface from scuffing and scoring, particularly under most critical lubrication conditions, due to its high physical and chemical stability.

The coating has a multiplayer amorphous microstructure containing carbon, deposited in diamond-like form together with hydrogen and tungsten. A special process based on the combined physical and the plasma-assisted chemical vapour depositions has been developed by the company specifically for piston-ring application. Coating of thickness upto 10 microns which is more than 3 times that of industry's latest state-of-the art DLC (Diamond Like Coating) can be applied using this process.

9.11. Causes of Ring Failures

1. Rapid wear

When the abrasive content in the intake air is very high and particles are also very hard, wear is very rapid. Excessive ring or ring groove-wear is caused by the scraping action of abrasive in the engine. This condition is evident from the excessive oil consumption and poor engine performance. It is found that rapid abrasive wear is the type of ring failure most commonly experienced.

2. Scuffing

Scuffing or scoring is caused by break- down of protective lubricating oil film on the cylinder wall.

This allows metal to metal contact of the ring faces on the cylinder wall producing heat due to friction causing surface welding. The material from the ring face gets transferred onto the cylinder wall causing ring scuffing (Fig. 3.98), which gets worse with further additional running. The factors pro- moting scuffing may be listed as:

(*a*) Overheated engine—This may be due to:
- (*i*) Cooling system, which may be restricted, clogged, or otherwise inefficient.
- (*ii*) Leakage of coolant or defective thermostat valve.
- (*iii*) Incorrect ignition timing, weak air-fuel mixture causing detonation.

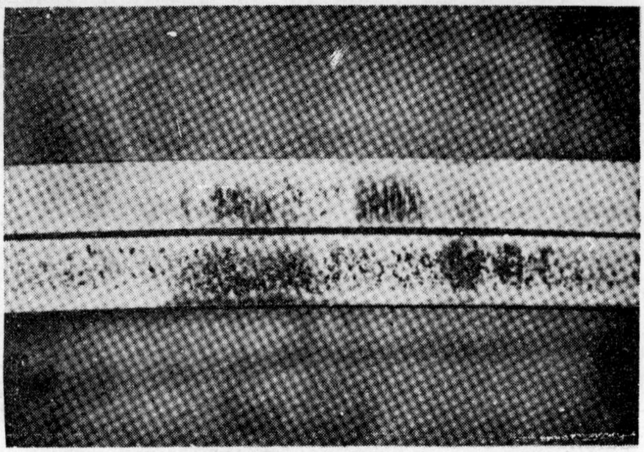

Fig. 3.98. Ring scuffing.
(Courtesy—Sealed Power Corporation, U.S.A.)

(*b*) Lack of or deficient lubrication—This may be caused by:
- (*i*) Less bearing clearances.
- (*ii*) Low oil pressure.
- (*iii*) Water leaking in cylinders causing breakdown of lubrication film.
- (*iv*) Defective oil pump.
- (*v*) Dilution of oil in crankcase.
- (*vi*) Low oil level.

(*c*) Insufficient clearance in cylinder bore. Any or more of the following may be the sources of the trouble:
- (*i*) Wrong size piston or piston rings.
- (*ii*) Distorted cylinder bores.
- (*iii*) Wrong size cylinder bore.
- (*iv*) Cylinder liner distorted or loose in the cylinder block.

(*d*) Improper cylinder finish which may be due to honing not having been done after boring.

3. Ring breakage

Sometimes the ring breaks during service. This may be on account of overstressing due to shock loading, fatigue, etc. The following are the main causes of such overstressing:
- (*i*) Use of ring of incorrect size.
- (*ii*) Ring sticking due to carbon deposits.
- (*iii*) Detonation or preignition.
- (*iv*) Excessive ring side clearance.
- (*v*) Uneven wear of ring grooves.
- (*vi*) Insufficient end gap of rings.
- (*vii*) Excessive ring flutter cased by overspeeding.

10. CONNECTING ROD

The function of the connecting rod is to convert the reciprocating motion of the piston into the rotary motion of the crankshaft.

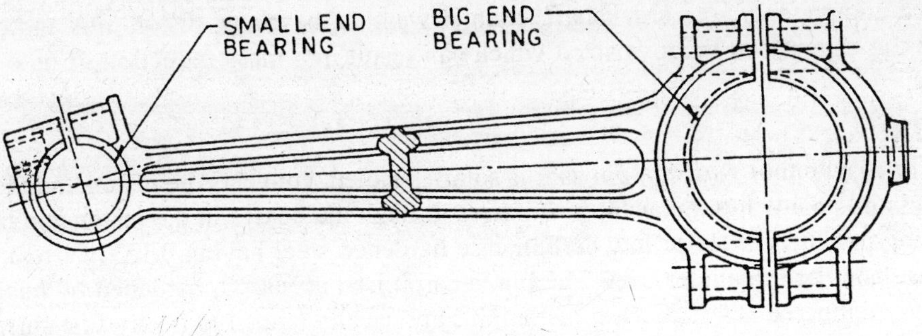

Fig. 3.99. Connecting rod.

A typical connecting rod is shown in Fig. 3.99. To reduce the obliquity of the rod with the cylinder axis, its length should be kept as large as possible. Reduced obliquity decreases the oscillating angular motion of the rod about its small end, thereby decreasing the piston side-thrust and improving the reciprocating balance of the engine. However, on account of the current practice to reduce the overall height of the engine, the length of the connecting rod has been decreased from 4 to 4.5 times the crank throw to about 3.5 times. This has been made possible on account of the improved design. A combination of axial and bending stresses act on the rod in operation. The axial stresses are produced due to cylinder gas pressure (compressive only) and the inertia force arising on account of reciprocating motion (both tensile as well as compressive), whereas bending stresses are caused due to the centrifugal effects. To provide the maximum rigidity with minimum weight, the main cross-section of the connecting rod is made an I-section. This I-section is made to blend smoothly into the two rod-ends, called the small end and the big end. The small end of the rod has either a 'solid' eye or a 'split' eye; this end holding the piston pin. The big end works on the cranks pin and is always split. In some connecting rods, a hole is drilled between two ends for carrying lubricating oil from the big end to the small end for lubrication of piston pin and the piston.

Fig. 3.100. Connecting rod.

The connecting rods are generally made by drop forging of steel or duralumin. However, with the progress of technology, the connecting rods these days are also cast from malleable or spheroidal graphite cast iron. In the later method, a closer weight tolerance can be maintained compared to the forging method. In general, forged connecting rods are compact and lighter which is an advantage from inertia view point, whereas cast connecting rods are comparatively cheaper, but on account of lesser strength their use is limited to small and medium size petrol engines.

Connecting rod is a main component which affects the mass of the rest of the cranktrain. A lighter connecting rod would permit reduced ancillary component mass. Whereas the basic design has remained the same, new materials and more accurate calculation techniques are now being used to significantly reduce the mass without compromising the durability. One latest example is Mahle's ULWC (Ultra Light Weight Connecting rod), developed recently (2011). The material used is 46MnVS6, which is a fine-grained ferritic-pearlitic micro alloy forged steel having mean fatigue strength of 496 MPa at R = – 2.5. This has about 20% higher strength than that of premium 3% copper powdered forged alloys.

The ULWC design for Chrysler 3.5 L V6 engine resulted in 27% mass reduction, down from 548 g to 400 g. This mass reduction has significant effects on reciprocating mass and reliability. Due to this reduction, other associated components viz., crankshaft, counter weights, bearings, piston pins and other critical components in the engine can be downsized which can result in a mass reduction of upto 2 kg in a 6-cylinder engine.

11. PISTON PIN

Piston pin or *wrist pin* or *gudgeon pin* as it is so often called, connects the piston and the connecting rod. For lightness it is made in tubular form. It passes through the bosses in the piston and the small end of the connecting rod. It is made of low carbon case hardened steel having 0.15% carbon, 0.3 silicon, 0.5% manganese and the remainder iron. The pin is carburised at 900°C, hardened by quenching from 780°C and finally tempered at 150°C. The piston pins are usually lapped to a very fine surface finish of about 0.1 ìm, without which it is likely to fail very early due to fatigue caused by surface irregularities. Piston pin operating clearances are generally kept about 7.5 ìm, the larger clearance would result in more noise and decreased life.

Piston pin–connecting rod connections are of three types:

1. Piston pin is fastened to the piston by set screws through the piston bosses. It has a bearing on the connecting rod small end (Fig. 3.101).

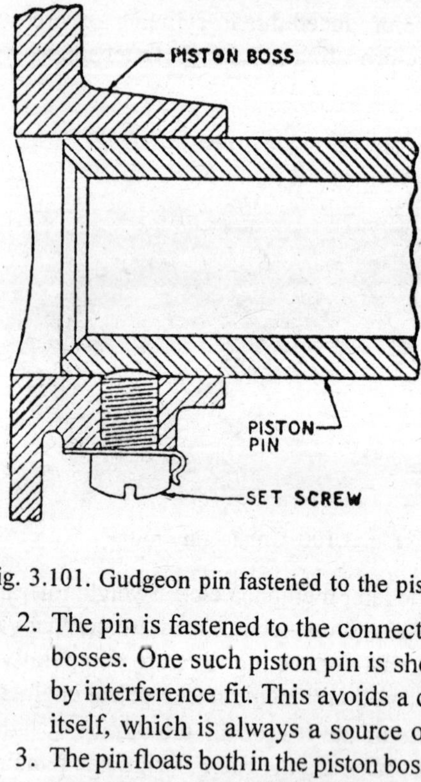

Fig. 3.101. Gudgeon pin fastened to the piston.

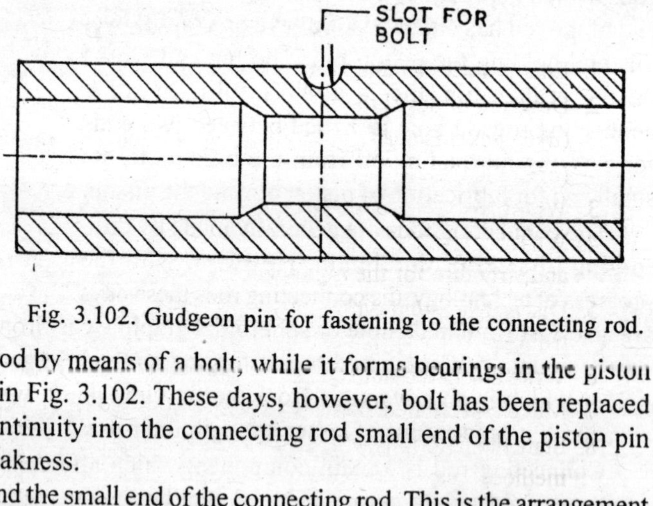

Fig. 3.102. Gudgeon pin for fastening to the connecting rod.

2. The pin is fastened to the connecting rod by means of a bolt, while it forms bearings in the piston bosses. One such piston pin is shown in Fig. 3.102. These days, however, bolt has been replaced by interference fit. This avoids a discontinuity into the connecting rod small end of the piston pin itself, which is always a source of weakness.
3. The pin floats both in the piston bosses and the small end of the connecting rod. This is the arrangement most commonly used. To prevent end movement, circlips are used as shown in Fig. 3.103. Alternatively, piston pin may be prevented from sliding by using spherical end-pads made of aluminium, brass or bronze.

In heavy duty engines, semi-floating pins as described at serial nos. 1 and 2 above, may freeze in their bearings. This does not happen in case of pins floating both in the piston bosses and the connecting rod small end, because of the double swivel action of such pins.

A piston-connecting rod assembly is shown in Fig. 3.104.

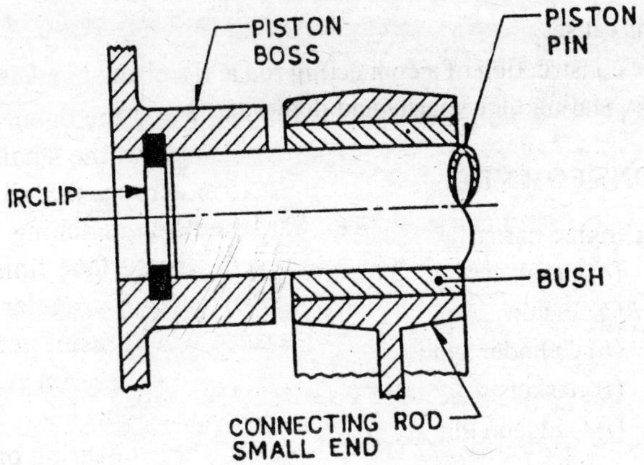

Fig. 3.103. Floating gudgeon pin.

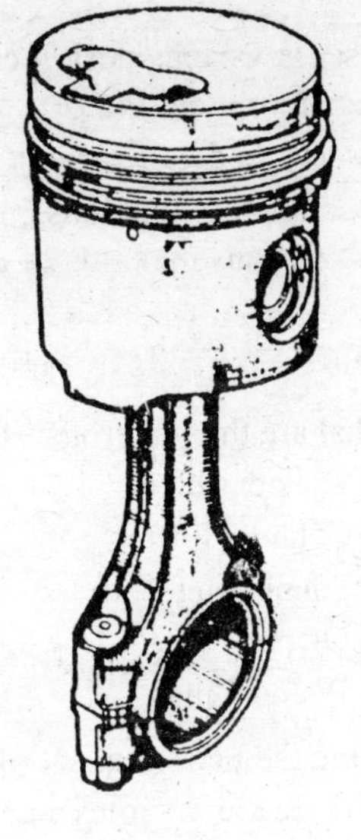

Fig. 3.104. A piston-connecting rod assembly.

REFERENCES

1. Next-generation piston technologies, Engine Technology International.com, January 2012, pp 40-41.
2. Emmerich Ottliczky, Marco Voigt, Hans-Joachim Weimar, Eberhard Weiss; Steel pistons for passenger car diesel engines, ATZ autotechnology, May 2011, Vol. 11, pp 38-42.
3. Piston coating optimization, Engine Technology International.com, April 2011, pp 70-71.

THEORY QUESTIONS

1. Enumerate the main parts of an automobile engine.
2. Describe in detail the function, construction and materials used for the
   (*a*) Cylinder Block (*b*) Cylinder Head
   (*c*) Crank Case (*d*) Oil Pan
3. What are the functions of engine inlet and exhaust manifolds? Describe their materials and construction.
4. What is a gasket? State its function. Discuss various requirements of the gaskets. Describe the materials and structure for the main gaskets used in automotive practice.
5. Discuss different types of cylinder liners. What are their comparative advantages? Discuss various liner troubles.
6. Write a detailed note on 'Hybrid liner'.
7. Make a sectioned sketch of a petrol engine piston and name its various parts.
8. State the functions which a piston in a automobile engine cylinder is required to perform. Discuss various methods used to avoid piston-slap.
9. Describe in detail various types of automotive pistons.
10. Write brief reports on 'Magnum Monosteel Piston' and 'DuraBowl' piston technology.
11. Describe salient features of any modern piston coating.
12. Discuss in detail various factors considered for selecting piston for an automotive engine.
13. Discuss in detail various types of piston failures. Give suggestions to avoid the same.
14. What are the functions which the piston rings are required to perform? With the help of suitable sketches explain the constructional features of various types of piston rings.
15. Explain with the help of simple sketches the working of the compression and the oil control rings.

16. Discuss various design considerations for piston rings. What factors would you consider for selecting rings for an engine?
17. Discuss in detail various ring failure and their causes.
18. Describe the functions, materials used and the construction of a connecting rod and a piston pin. Discuss also various methods of connecting these two, stating their merits and demerits.

QUESTIONS FOR VIVA

1. What are the materials used for the following engine parts:
   (*a*) Cylinder block
   (*c*) Crank case
   (*e*) Cylinder liner
   (*g*) Piston
   (*i*) Connecting Rod
   (*b*) Cylinder head
   (*d*) Oil pan
   (*f*) Gaskets
   (*h*) Piston Rings
   (*j*) Gudgeon Pin
2. Name the parts that are attached to the engine block.
3. How are the oil holes made in a cylinder block?
4. Why is monoblock construction for cylinder blocks preferred over the individual cylinder design?
5. Out of the open and the closed decks which is preferred in case of cylinder block?
6. Why should a material for cylinder block or head have higher thermal conductivity? What happens if this factor is overdone?
7. Why is grey cast iron used for making cylinder blocks?
8. Name any other material apart from cast iron that is used for cylinder blocks. State its advantage and disadvantage compared to cast iron.
9. What should be the order of the cylinder bore finish and why?
10. Why are the bores of cylinder blocks made of aluminium alloy sometimes plated with chromium?
11. Where is the crankshaft mounted in the cylinder block?
12. What is the function of camshaft?
13. Why should the centre lines of all the cylinders be at right angles to the crankshaft axis?
14. Where is the camshaft mounted on the cylinder head?
15. Why should valve lifter bores in the cylinder block be at right angles to the camshaft bore?
16. What is the function of a cylinder head?
17. Which material is usually used for cylinder head?
18. State the relative merits and demerits of the detachable and the integral types of cylinder heads.
19. Name the three classifications of cylinder head based on the valve and port layout.
20. What is the function of crankcase?
21. State the functions of an oil sump.
22. Why is the sump formed into a deep, shallow reservoir?
23. Why are the baffles provided in an oil sump?
24. State the relative merits and demerits of using steel sheet and aluminium alloy casting for the oil sump.
25. What causes fuel separation in the inlet manifold?
26. Where is the inlet manifold mounted on the engine?
27. How is the inlet manifold arranged to ensure pre-heating of the air-fuel mixture?
28. What for the abbreviations (*i*) VLIM, (*ii*) VLIS stand?
29. Give three examples of automotive engines where VLIM technology is used.
30. State main design consideration for an exhaust manifold.
31. Which material is used for exhaust manifold?

32. What is the function of a gasket?
33. State various requirements of engine gaskets?
34. Which is the most critical gasket on an engine?
35. Name different materials for engine gaskets.
36. Name two companies making engine gaskets.
37. Why should a gasket be used only once?
38. What is the function of a cylinder liner?
39. State any advantages of using aluminium alloy as liner material.
40. Which type of cylinder liner ensures more efficient cooling?
41. Why the dry liner must be finished finally after it is fitted to the cylinder block, whereas there is no such compulsion in the case of wet liner?
42. What happens if a cylinder liner is (*i*) too loose, (*ii*) too tight in the cylinder block?
43. What may be the cause of defective sealing in case of wet liners?
44. what will be the result if the rubber in wet liner gets twisted while installing?
45. What is a hybrid liner?
46. What are the main functions of a piston?
47. How does the piston head shape affect engine performance?
48. What are (*i*) skirt, (*ii*) land, (*iii*) crown, in a piston?
49. Define the term 'compression height' in a piston and state its function.
50. What is the function of piston skirt?
51. State the advantages and disadvantages of using aluminium alloy as the piston material.
52. State the two advantages of adding silicon to the aluminium alloy for making pistons.
53. Name the best known aluminium alloys for automative pistons.
54. What are 'hypoeutectic' and 'hypereutectic' piston alloys?
55. What is 'piston slap' and how it is avoided?
56. What is a 'heat dam' in a piston?
57. Why are the following used in a piston:
    (*a*) T-Slot
    (*c*) Oval Skirt
    (*e*) Inserted Ring Carriers.
    (*b*) Horizontal Slot
    (*d*) Wire winding
58. What is the composition of 'Lo-Ex' alloy? Where is it used?
59. What is an 'Al-fin' piston?
60. What is the advantage of a cast steel piston?
61. What advantages are obtained by anodising or tinning the pistons?
62. Why are the pistons cooled in some engines?
63. State the advantage of a two-piece piston?
64. What is the function of oil gallery in some pistons?
65. What is a composite heat-shielded piston? What is its advantage?
66. What is a 'squeeze cast piston'? What advantage is claimed for the same?
67. What is 'Aeconoguide' piston?
68. Enlist various advantages of the 'Eco Tough' piston coating.
69. Name various factors for piston selection.
70. Name various important areas for piston design.
71. What is 'piston scuffing'? State any three causes for the same.
72. What is the main cause of a burnt piston?
73. State any two reasons for a damaged ring land.

74. What are the functions of piston rings?
75. Which type of ring end gap is most commonly used?
76. Which type of ring end gap should be used in slow speed engines?
77. What is a gapless ring?
78. State the advantages and disadvantages of a steel ring.
79. Why is there a trend of placing the top ring closest to the piston top in modern engines?
80. State any two materials used for piston rings.
81. What advantage is obtained by having phosphate coating over the piston rings?
82. How is phosphate coating of piston rings done?
83. State the advantage of chrome-plating the piston rings.
84. What is the advantage of a molybdenum-filled face piston rings?
85. What is a 'TC Chrome' piston ring?
86. What advantage is obtained by using stainless steel for piston rings?
87. How many rings are usually there on a piston of a automotive engine?
88. Why are a minimum of two compression rings required on a piston?
89. What functions are performed by a compression ring?
90. What advantage is obtained by having a tapered external face for a piston ring?
91. What do you understand from the term 'blow-by' in case of engine cylinders?
92. Why should the ring end gaps in piston be staggered?
93. Which ring wears most, the top or the bottom one, and why?
94. What is a 'positive twist ring'? What is its advantage?
95. What is the advantage of reverse twist ring over positive twist ring?
96. What is a 'keystone ring'? What advantage is obtained by its use?
97. In which type of engines is the keystone ring commonly used?
98. Name any four types of a oil control rings. State the advantage of each.
99. What is the function of a spring expander?
100. What are the 'spacer' and 'expander' in an oil control ring?
101. What advantages are claimed for a multi-piece (or a composite rail) scraper ring?
102. Name any two Indian and two foreign manufacturers of automobile pistons and rings.
103. State the advantages claimed for 'CarboGlide' ring coating.
104. What is the function of a connecting rod?
105. Why should the connecting rod be lighter yet strong?
106. Name any two materials used for connecting rods.
107. Compare the forged and the cast connecting rods.
108. Why do some connecting rods have hole drilled from the small end to the big end bearings?
109. Which type of arrangement is mostly used for piston-connecting rod connection?
110. Why is a 'full-floating' piston pin so-called? What is its advantages?
111. What is a 'circlip'? Where is it used?

MULTIPLE CHOICE QUESTIONS

Mark the correct answers:

1. The materials used for cylinder block are
   (*a*) cast iron and steel (*b*) cast iron and aluminium alloy
   (*c*) steel and aluminium alloy (*d*) brass and steel
2. The cylinder bores of aluminium alloy cylinder blocks are usually plated with

(*a*) steel (*b*) brass
(*c*) chromium (*d*) molybdenum

3. The angle between the cylinder axes and the crankshaft centre line in an engine is
(*a*) 60° (*b*) 90°
(*c*) 120° (*d*) 180°
4. Most difficult gasket sealing problem occurs at the
(*a*) head (*b*) oil pan
(*c*) timing cover (*d*) intake manifold
5. Each time a joint is assembled, it is recommended to replace
(*a*) cork gaskets (*b*) asbestos gaskets
(*c*) rubber gaskets (*d*) all gaskets
6. Flat bottomed inlet manifolds are commonly used in car engine to
(*a*) allow space for two runner levels
(*b*) increase fuel evaporation in the manifold
(*c*) provide uniform air-fuel mixture to various cylinders
(*d*) maximise the cross-sectional area for the given wall surface area of the manifold
7. Camshaft in an engine is always mounted
(*a*) parallel to the crankshaft (*b*) perpendicular to the crankshaft
(*c*) inclined to the crankshaft (*d*) none of these
8. The oil pan in an engine may be made of
(*a*) steel or aluminium (*b*) steel or cast iron
(*c*) cast iron or brass (*d*) cast iron or zinc
9. A dish in the piston head is employed to
(*a*) decrease piston weight (*b*) provide valve clearance
(*c*) control compression ratio (*d*) adjust piston displacement
10. The largest diameter of a camground piston is
(*a*) along piston pin axis (*b*) at the piston land
(*c*) at 45° to the piston pin axis (*d*) at 90° to the piston pin axis
11. The piston-skirt clearance at room temperature is about
(*a*) 0.004 mm (*b*) 0.04 mm
(*c*) 0.4 mm (*d*) 1 mm
12. Piston pins on some engines are offset to the
(*a*) right side (*b*) left side
(*c*) major thrust side (*d*) minor thrust side
13. The purpose of piston rings is to control
(*a*) combustion pressures (*b*) cylinder wall lubrication
(*c*) oil consumption (*d*) all of the above
14. In autothermic pistons steel inserts are used to
(*a*) reduce piston expansion (*b*) control the direction of expansion
(*c*) reduce piston weight (*d*) make the piston stronger
15. The minimum number of compression rings in an automotive engine is
(*a*) one (*b*) two
(*c*) three (*d*) four
16. compression rings are generally made of
(*a*) low carbon steel (*b*) cast iron
(*c*) aluminium (*d*) chromium

17. The uppermost ring on a piston is usually plated with
    (*a*) steel (*b*) cast iron
    (*c*) aluminium (*d*) chromium
18. On the compression stroke the rings are pressed against
    (*a*) top of groove (*b*) bottom of groove
    (*c*) inner side of groove (*d*) all of these
19. The primary purpose of a ring expander is to
    (*a*) make up for cylinder wear (*b*) increase static ring tension
    (*c*) decrease dynamic ring tension (*d*) reduce ring vibrations
20. Connecting rod connects the crankshaft and the
    (*a*) cylinder head (*b*) cylinder block
    (*c*) piston (*d*) camshaft
21. The piston pins in the modern automobile engines are usually
    (*a*) semi-floating
    (*b*) fully-floating
    (*c*) three-quarter floating
    (*d*) fixed to both piston and the connecting rod end

[**Key:** 1. (*b*), 2. (*c*), 3. (*b*), 4. (*a*), 5. (*d*), 6. (*b*), 7. (*a*), 8. (*a*), 9. (*c*), 10. (*a*), 11. (*b*), 12. (*b*), 13. (*d*), 14. (*b*), 15. (*b*), 16. (*b*), 17. (*d*), 18. (*b*), 19. (*a*), 20. (*c*), 21. (*b*)]

Constructional Details-II

4

1. CRANKSHAFTS

Crankshaft is the engine component from which the power is taken. It receives the power from the connecting rods in the designated sequence for onward transmission to the clutch and subsequently to the wheels.

The crankshaft assembly includes the crankshaft and bearings, the flywheel, vibration damper, sprocket or gear to drive camshaft and oil seals at the front and rear.

1.1. Construction

Crankshaft is a shaft consisting of the following major parts:

1. Main journals
2. Crank pins
3. Crank webs
4. Counter weights
5. Oil holes

A simplified sketch of the crankshaft for a 4 cylinder in-line engine is shown in Fig. 4.1.

Main journals are supported in main bearings in the crankcase. These form the axis for the rotation of the crankshaft. Their number is always one more or one less than the number of cylinders. The crankpins are the journals for the connecting rod big end bearings and are supported by the crankwebs. The crankwebs

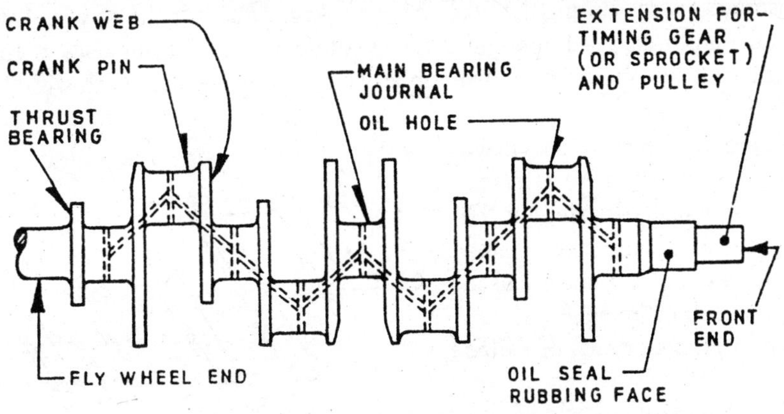

Fig. 4.1. Crankshaft.

should be adequately strong to withstand the twisting and the bending loads. However, increasing their size excessively would increase the inertia which would tend to wind and unwind the crankshaft when the engine is running.

The distance between the axis of the main journals and the crankpin centre lines is exactly one half of the engine stroke and is called the 'crank-throw', which determines the crankshaft turning effort. Oil holes are drilled from main journals to the crankpins through crankwebs to provide lubrication of big end bearings. Main bearings are lubricated from oil galleries in the cylinder block.

When the engine is running, the centrifugal forces acting at each crankpin due to rotation of both the crankshaft as well as the big end of the connecting rod tend to bend and distort the crankshaft. To counter this tendency, counterweights are either formed as integral part of the crank web or attached separately, on the side opposite to the crankpin (Fig. 4.2). As the separate weights can be made to overlap the crankwebs, more mass can be concentrated at a smaller radius, thus preventing the inertia from being excessive. However, in that case very high accuracy is also needed while attaching them, otherwise it would result in unbalanced crankshaft inspite of the counterweights.

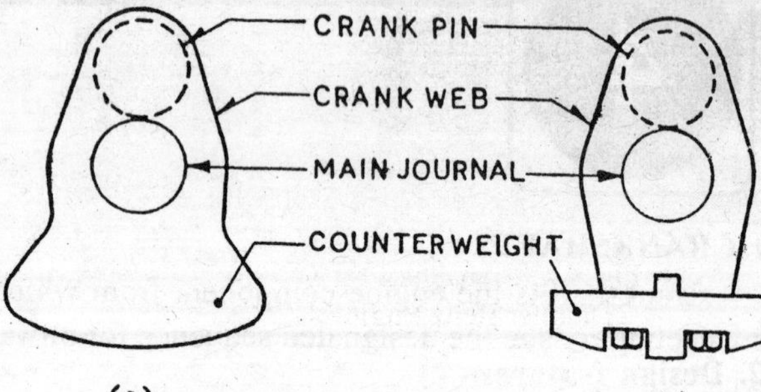

Fig. 4.2. Counter weights.
(*a*) integral, (*b*) attached separately.

On one of the main bearing journals usually near the flywheel end, thrust bearing is located so as to support the loads in the direction of the shaft axis. Such loads may arise due to clutch release forces, forces in the helical gear valve- timing gear train when accelerating or deceleration, forces in the gear train when driving various auxiliary components, *e.g.*, the oil pump, water pump, supercharger, etc. or forces in the helical gear drive on front wheel driven cars with integral engine and gear box.

On the front of the crankshaft are mounted:

1. Timing gear or sprocket which drives the camshaft.
2. Vibration damper.
3. Pulley for driving the water pump, fan and the generator.

On the rear end is mounted the flywheel, which serves as energy reservoir. Though there is one working stroke in two revolutions of the crankshaft, the flywheel absorbs excess energy during the power stroke and gives out the absorbed energy during the other three strokes. Thus it is the flywheel which keeps the crankshaft rotating at the uniform speed throughout inspite of uneven power impulses of engine cylinders. The flywheel also has a ring gear mounted on its periphery having teeth which mesh with the pinion of the starting drive to start the engine.

The flywheel also provides the frictional drive surface for a clutch-type mechanical drive. Besides, the timing marks are also usually located on the flywheel. These marks indicate the TDC piston position in degrees for cylinder no. 1 and are used to determine ignition or injection timing. Flywheels are made from cast iron alloys or cast or rolled steel.

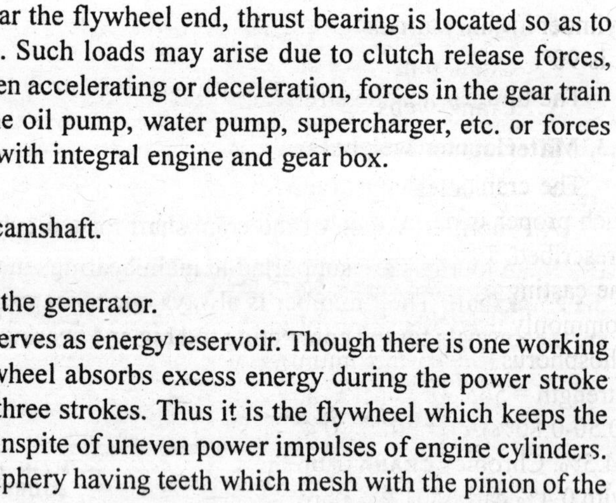

Fig. 4.3. Crankshaft for a 4-cylinder in-line engine

The crankshafts are generally of two types, *viz.*, one piece and built up. In the built up construction the crank pins and journals are bolted to crank arms, which also serve as flywheels. One-piece construction is almost universally used for automotive crankshafts.

The crankshafts for the 4-cylinder and 6-cylinder in-line engines are shown in Figs. 4.3 and 4.4. Fig. 4.5 depicts a crankshaft for V-8 engine.

Fig. 4.4. Crankshaft for 6-cylinder Daimler Benz Engine 123.920.

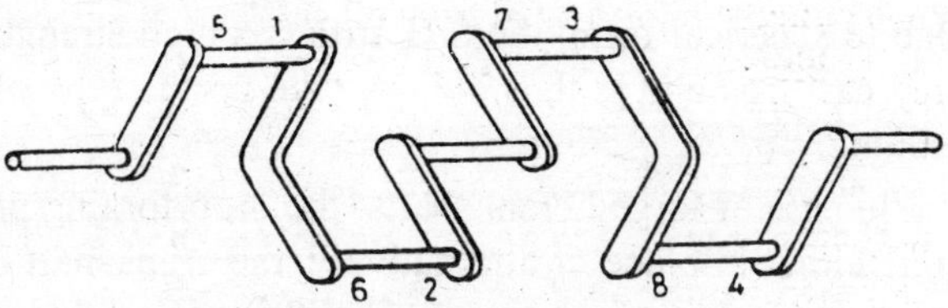

Fig. 4.5. Crankshaft for a V-8 engine.

1.2. Design features

For car engines, which are petrol engines in majority of the cases, the crank pin length is at least 30% of its diameter, which itself is usually not less than 60% of the cylinder core. The thickness of the crankweb is usually about 20% of the cylinder bore size, the main journal diameter is bigger than the crank pin diameter and is usually 75% of the cylinder bore whereas its length is about half its diameter.

The angles between crankthrows are selected from the consideration of smooth power output. In 4-cylinder in-line engines the angle is 90°, whereas for 6-cylinder in-line engine, it is 120° and for V-8 engines it is 90°.

The ultimate tensile strength of the crankshaft material is about 600 MPa.

1.3. Materials and manufacture

The crankshaft must be adequately strong, tough, hard and should possess high fatigue strength. As such proper material has to be selected and suitable fabrication processes and heat treatment have to be prescribed. Earlier the crankshafts were only forged, but with the improvements in foundry techniques, the casting of crankshafts has also become quite common. S.A.E. carbon steels 1045 and 3140 are the commonly used steels for forged crankshafts. S.A.E. 1045 contains carbon (0.45%), manganese (0.75%), phosphorus (0.04% maximum), and sulphur (0.05% maximum) and has Mechanical properties, *viz.*, Tensile Strength = 565 – 625 MPa, and elongation = 12%, S.A.E. 3140 contains carbon (0.40 -0.50%), manganese (0.50-0.80%), chromium (0.45-0.75%) & nickel (1.00-1.50%), with tensile strength 690 MPa and elongation 24.5%. Chrome-vanadium and chrome molybdenum steels have also been used. A typical example is steel with 0.4% carbon, 1.2% chromium and 0.3% molybdenum. Chromium forms hard carbides which harden the steel. A small amount of molybdenum improves the impact strength and machinability. Another typical steel used for heavy duty diesel engine forged crankshaft has 0.3% carbon, 2.5% nickel, 0.65% chromium and 0.55% molybdenum. Here nickel refines the grain and prevents grain growth during heat treatment due to which strength and toughness are improved. However, its tendency to convert the carbon into its soft graphite form is compensated by chromium which forms hard wear-resisting carbides.

In case of cast crankshafts, both cast steel as well as S.G. iron have been used. A typical cast steel for crankshafts has the following chemical composition:

Carbon	1.35—1.6	percent
Chromium	0.05—0.5	percent
Silicon	0.85—11	percent

Manganese	0.6—0.8	percent
Copper	1.5—2.0	percent
Phosphorus	0.10	percent (max.)
Sulphur	0.9	percent (max.)

The Spheroidal Graphite type or commonly called S.G. iron is a high strength cast iron in which the carbon present is in the form of spherical modules of graphite (compared to flakes in the grey cast iron). Due to spherical form, the S.G. iron has high strength, improved ductility and larger toughness than the grey cast iron and is thus able to take up the type of stresses that are imposed on the crankshaft while running.

In case of forged crankshafts, the drop forging process with closed dies is used. The forging may be done either in-place or alternatively, the crankshaft is forged in one plane and then wound to place the crank throws at desired angles. The blank so produced is then heat treated to remove residual stresses and then machined to final dimensions. The machining includes mainly the rough turning, final grinding and final lapping of main journals and crankpins.

The cast crankshaft is manufactured by the shell- moulding process, in which a thin shell-like mould is made from the sand and synthetic resin. This mould is then used for further casting. This method has the advantage of high accuracy in terms of closer tolerances which reduces the machining cost. Of course, the cast crankshaft has also to be heat treated and machined as in case of forged ones.

The crankshafts produced by the forging process are very dense and tough with grains parallel to the principal directions. The cast crankshafts have a uniform and random structure throughout. They are slightly less dense compared to forged crankshafts, but are also cheaper due to reduced machining. Due to these reasons, both the types are in use. However due to slightly lower toughness and fatigue strength, larger sections and more number of main journals are used in case of the cast type.

2. VIBRATION DAMPERS

Vibration damper is also called a *harmonic balancer.* It is mounted on the front end of the crankshaft. In consists of a damper flywheel-cum-crankshaft pulley and a driving flange connected together with a rubber ring in between (Fig. 4.6).

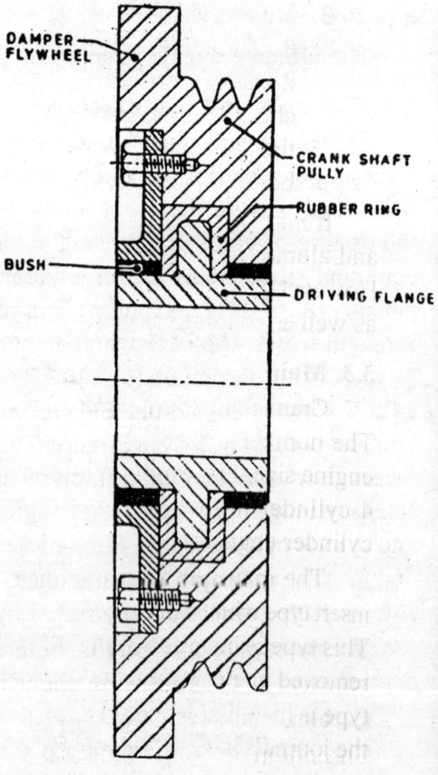

Fig. 4.6. Vibration damper.

During the power stroke in an engine cylinder, a force is applied to its crankpin, which tends to twist the crankshaft. This force disappears during the other strokes, while during the next power stroke it again reappears due to which the crankshaft is untwisted and again twisted. The same cycle is repeated at every crankpin on the engine crankshaft. This twisting and untwisting of crankshaft sets up torsional vibrations, which may become amplified if their frequency becomes equal to the natural frequency of the vibrating parts due to resonance.

The vibration damper reduces the torsional vibrations by means of the dragging effect produced by the inertia of damper flywheel. On account of this dragging effect the rubber ring is also flexed which tends to avoid the vibrations and keep the crankshaft speed uniform. Engines with belt-driven crankshafts do not need vibration damper. The belt acts as a damper.

3. ENGINE BEARINGS

3.1. Requirements

1. Engine bearings should support the operating loads with minimum friction using lubricant.

2. They should have a long life. Bearing metals tend to fatigue with use, which shows as fine cranks in the bearing surface. Harder bearing materials have longer fatigue life. However, softer bearing materials are less costly and therefore are used in light duty bearings.

3. The bearings should have adequate *conformability,* which may be defined as the ability of a bearing to flow slightly so as to match the small variations in the shaft. Such a bearing would conform to the shaft during running-in period.

4. The bearings should have *score resistance* that prevents the bearing materials from seizing to the shaft during boundary lubrication.

5. Corrosion resistance is another property which is required in engine bearings which are subject to acids formed due to the oil additives breaking down and combining with moisture.

3.2. Bearing materials

A modern engine bearing consists of bearing proper and a backing. The backing is usually of steel which provides support against the shaft load. The three bearing materials in use are:

1. *Babbit.* It may be lead or tin based and contains small quantities of copper and antimoney. It is the oldest bearing material for automobile engines and is still used in applications where soft material shaft under moderate loads and speeds is employed.
2. *Copper-lead.* It is stronger and costlier than babbit. However, this material is easily corroded by acid accumulated in the engine oil.
3. *Aluminium.* Alloyed with small quantities of tin and copper, aluminium is the latest bearing material for automobile engines. Aluminium bearings are used for high load and speed conditions. However, it is more expensive due to which such bearings are sometimes used alongwith bearings of some cheaper material. For example, the lower shell of main bearing is more heavily loaded compared to the upper shell. As such aluminium bearing may be used for the lower shell while babbit metal may be used for the upper shell on the same bearing journal.

It may be noted that quite often an *overlay* or third layer of babbit is used on top of the copper-lead and aluminium bearings. Due to overlay, which is 0.01 to 0.02 mm thick, such bearings are expensive. However, the overlay cushions the journal during the running-in period and thus increases the life of bearing as well as journal.

3.3. Main Bearings

Crankshaft is supported in main bearings which are located in the lower portion of the engine block. The number of bearings depends upon the number of cylinders in the engine, their arrangement and on engine size. For example in the Hindustan Ambassador car engine which is 4-cylinder in line type, there are three main bearings, whereas V-type 8-cylinder engines generally have five main bearings.

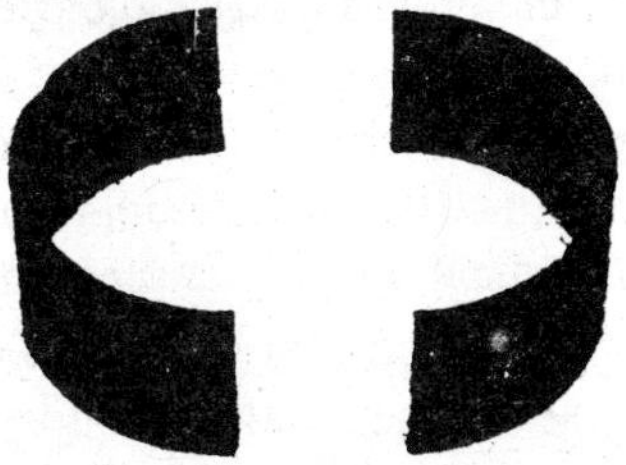
Fig. 4.7. Main bearing.

The main bearings used are mainly of two types. One is the precision insert type which is available both in finished as well as semi-finished condition. This type is installed directly in the bearing saddle cores with the crankshaft removed or even when the crankshaft is in position in the engine. The finished type is installed when the crankshaft journals are of the standard size. When the journals have worn out and are ground to some undersize so that available standard size bearings do not give the bearings clearance within the permissible limits, the semi-finished bearings are used which are rebored to the desired size. The second type of main bearings available are the cast type in which the bearing metal is sprayed directly onto the saddle bores and the inside of the bearing caps after which they are line-bored to the desired dimensions. A number of shims on each side of the bearing, within it and the cap are also provided in the cast type for adjustment of clearances. Taking out a number of shims reduces the bearings clearance.

To prevent the crankshaft from moving to and fro in the endwise direction, one of the bearings is made a thrust bearing *i.e.*, it is provided with flanges on both sides, which fit accurately with the sides of the crankweb on which it is installed.

Fig. 4.8. Flange type main bearing.
(Courtesy—Vandervell Products Ltd., U.S.A.)

Modern bearing of the insert type usually has a back of steel on which there is lining of suitable bearing material. For example, Sealed Power bearings are of the following types:

1. **Standard duty.** Babbit lining with steel or bronze back. Babbit metals or white metals are basically either antimony-tin or antimony-lead alloys. They have very good conformability, embeddability and corrosion resistance, but a low fatigue strength. A typical composition for a tin-based babbit metal is antimoney 7.5%, lead 0.2%, arsenic 0.1%, copper 3.0% and tin remainder.
2. **Medium duty.** Copper alloy lining with steel back. Copper-based alloys are harder and have a higher fatigue strength than babbit metals, but less conformability, embedability and corrosion resistance.
3. **Heavy duty.** Tin flash coating, babbit overplate, brass barrier plate, copper alloy lining and steel back.
4. **Extra heavy duty.** Tin flash plate, babbit overplate, aluminium alloy lining and steel back. A typical aluminium alloy used for turbo-charged heavy duty diesel engines contains 11% silicon, 1% copper and the remainder aluminium. These alloys do not suffer from corrosion as is the case with copper-based alloys.

4. VALVE TIMING

Theoretically the inlet valve should open at top dead centre and close at bottom dead centre. Similarly the exhaust valve should open instantaneously at bottom dead centre and should remain open till the piston reaches top dead centre, where it must close. But in practice, due to inertia the valves cannot open instantaneously and also the air-fuel mixture has its own inertia; consequently more time has to be allowed for the intake and exhaust processes to complete. The inlet valve is generally opened about 0–15° before T.D.C. (Called *valve lead*) and closed after about 60° of B.D.C. (called *valve lag*). Similarly the exhaust valve is opened before about 60° of B.D.C. and closed after 20° of T.D.C. The condition when both the inlet and the exhaust valves are open simultaneously during so many degrees of crankshaft rotation, is called *valve overlap*. The diagram representing this opening and closing of the valves is known as *valve timing diagram.* A typical diagram is shown in Fig. 4.9.

The inlet valve is made to open early to make use of the depression formed by the outgoing exhaust gases in their wake, due to which the fresh charge in the induction port is drawn in the direction of the escaping exhaust gases. In this way the fresh charge sweeps out the remaining combustion products and occupies the combustion chamber space. In fact, if the inlet valve opening lead is excessive, some of the

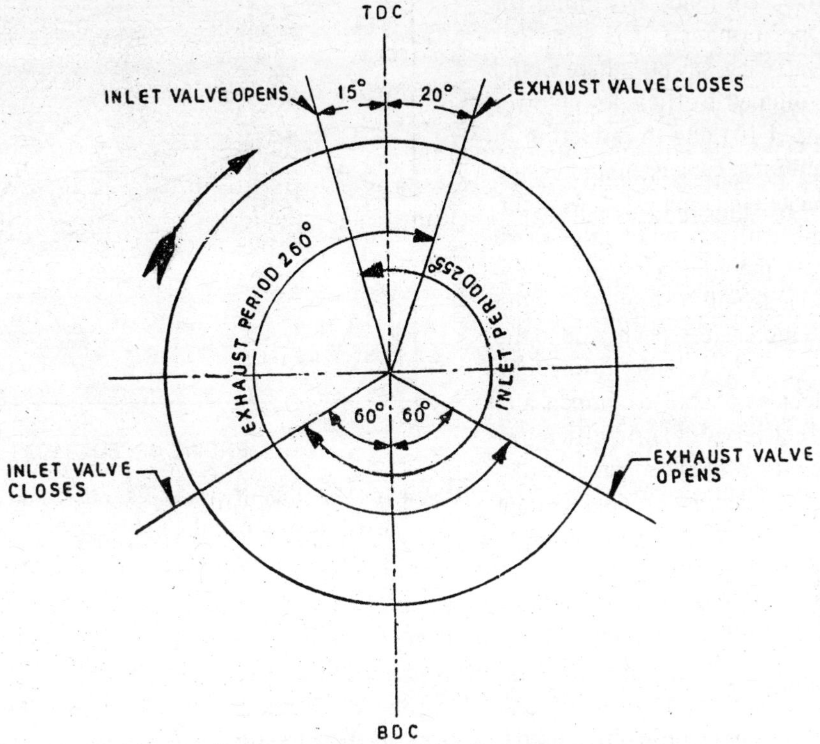

Fig. 4.9. Valve timing diagram.

exhaust gases will be pushed back through the inlet valve into the induction manifold. This may occur particularly under part-throttle conditions, when the average depression in the induction manifold is far higher than in the cylinder, as the piston is approaching TDC towards the end of the exhaust stroke.

The internal EGR (Exhaust Gas Recirculation) as above will be detrimental to full load engine performance as it takes up space which could otherwise be taken by fresh charge. However, this EGR may be beneficial at part load conditions in terms of efficiency and emissions.

The later opening of the intake valve, on the other hand, would restrict the entry of air/fuel mixture from the manifold which can cause lowering of the cylinder pressure as the piston starts to descend after TDC, consequently resulting in EGR, if the exhaust valve is still open as the exhaust gases would be drawn back into the cylinder. The effect of this EGR would be same as discussed above in case of early opening of the intake valve. However, in case the exhaust valve is closed, late opening of intake valve would not directly influence the amount of fresh charge in the cylinder, due to which it would not be particularly significant.

In fact, the opening of the intake valve is timed generally by full-load optimisation, to avoid internal EGR. Typical timing is about 0 to 10° before TDC, which results in fairly symmetrical valve overlap around TDC.

The effect of delaying the closure of the inlet valve until after BDC, when the piston has already started its compression stroke, is to improve the cylinder volumetric efficiency in the medium to high engine speed range. However, at lower engine speeds there is insufficient density and momentum of air (in C.I. engine) or air and petrol mixture (in S.I. engine), to pile into the cylinder to boost the mass of the charge.

One measure of the torque an engine can produce is the volumetric efficiency variation curve over the engine's speed range, due to which inlet and exhaust tuning of the valve timing for the highest possible overall volumetric efficiency is a must for optimum engine performance. It is seen that with a large inlet valve lag, say 80°, there would be a lowering of volumetric efficiency at lower engine speed (Fig. 4.10) due to reduction in charge velocity ramming, causing an increasing amount of flow back to the inlet valve, to occur. On the other hand, with a smaller inlet valve closing lag, say 40°, the volumetric efficiency at lower engine speeds will be highest, falling gradually with the increasing speed, due to the progressive decrease in the cylinder filling time which the ram effect is not able to compensate. Thus the most effective volumetric efficiency improvement over the mid-speed range of an engine is achieved by selecting a moderate inlet valve closing lag.

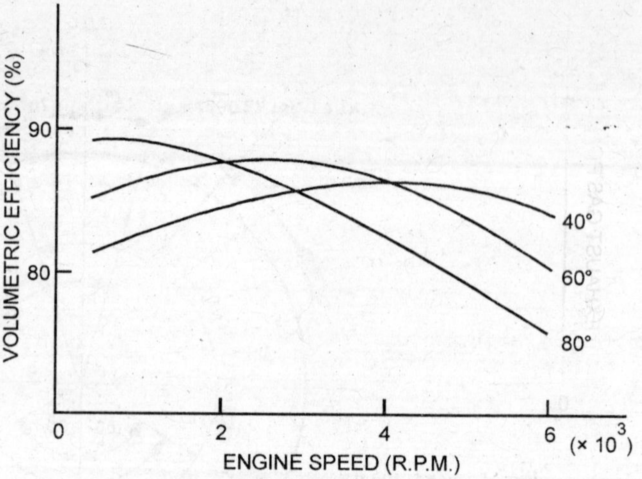

Fig. 4.10. Variation of volumetric efficiency with the engine speed at various inlet valve closing lag valves.

A typical timing for the closing of intake valve is in the range 50-60° after BDC which results from a compromise between high and low speed requirements. At lower engine speeds, a tendency to flow back the mixture into the intake manifold just before closing would be present, whereas at higher speeds, there would still be inflow of oil/fuel mixture when the intake valve closes.

The timing of exhaust valve opening depends upon two conflicting considerations. On the one hand, it is desirable for the pressure in the cylinder to drop to its lowest possible value, *i.e.* the exhaust back pressure, before the piston starts rising, *i.e.*, before BDC so as to minimise the work done by the piston in expelling the exhaust before the intake of a fresh charge. On the other hand it would be desirable not to open the exhaust valve before the piston reaches BDC so as to extract maximum work from the expansion of gas in the cylinder. Moreover, in actual valve train, the inertia of its components causes the valve to lift relatively slowly, so that it becomes necessary to open the exhaust valve some time before BDC. A typical exhaust valve opening is about 50-60° before BDC. The optimum timing for exhaust valve opening depends upon the engine speed and load and the pressure of gases in the cylinder. Under part load conditions, lower cylinder pressure takes less time to escape through the valve, contrary to the full load conditions, where engine cylinder pressure is higher. Thus, it is better to delay the exhaust valve opening under part load and advance it under full load conditions. However, in a conventional valve train, this cannot be changed due to which reason ideal timing is seldom achieved.

The delay in closing the exhaust valve until the piston has moved past TDC encourages the fresh charge to enter the combustion chamber even when the inlet valve has partially opened and piston speed is also very low, due to the partial vacuum created in the exhaust port by the out-rushing exhaust gases. However, excessive closing lag of the exhaust valve will cause the fresh charge to chase the last of the exhaust gases across the combustion chamber and even out of the exhaust port. It is seen that for full load operation, it is desirable to retain minimum possible quantity of exhaust gas in the cylinder so that maximum volume of fresh air/fuel mixture can enter during the intake stroke, which means exhaust valve should close at or shortly after TDC. On the other hand, for part-load operation it is desirable to retain some of the exhaust gases so that lesser fresh air/fuel mixture is drawn. This is done by delaying further the exhaust valve closing after TDC, which increases the level of internal EGR with a corresponding reduction in exhaust emissions. However, there is a limit to maximum permissible EGR, beyond which combustion becomes unstable. Typical exhaust valve closing timings are in the range 5-15° after TDC, wherein internal EGR is almost eliminated so as not to affect adversely the full-load performance.

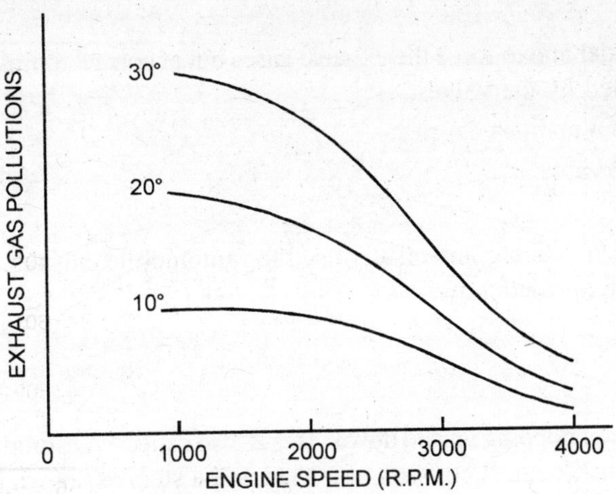

Fig. 4.11. Effect of excessive lag in the exhaust valve closing.

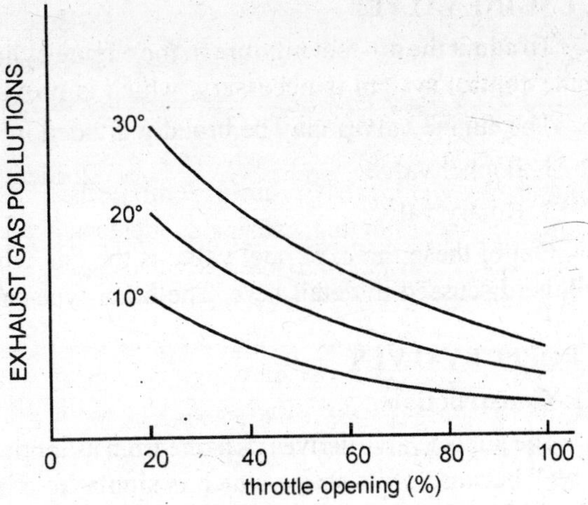

Fig. 4.12. Effect of excessive lead in the inlet valve opening.

In an ideal situation, the valve overlap should allow the leaving exhaust gases to draw the fresh charge into the cylinder without any of the charge passing straight into the exhaust system. In such a situation, the exhaust gases in the combustion chamber at TDC are replaced by the intake charge, which now exceeds the quantity of charge drawn into the cylinder only by the swept volume of the piston. However, it must be noted that the above mentioned advantages of valve overlap, *i.e.*, delaying the exhaust closing and opening the inlet valve earlier are pronounced in the mid to upper speed range of the engine. Also, these are accompanied by increase in the pollutants (hydrocarbons and carbon monoxide) in the exhaust, which is more marked at lower engine speeds (Figs. 4.11 and 4.12). Accordingly, representative values of valve lead and lag ranges for various types of auto engines are as follows:

Inlet valve opening before TDC	10° - 30°
Inlet valve closing after BDC	40° - 75°
Exhaust valve opening before BDC	40° - 75°
Exhaust valve closing after TDC	10° - 30°
Valve overlap at TDC	20° - 60°
Valve overlap at BDC	80° - 150°

5. MAXIMUM VALVE OPENING

Maximum valve opening depends directly on 'Valve Peak Lift', on which directly depend the ability of the air/fuel mixture to flow into the cylinder and the exhaust to leave the cylinder. Due to this, valve peak lift directly affects the engine performance. Thus, maximum power output of the engine would generally be increased by increasing the peak valve lift. However, excessive value of valve peak lift would cause the low-speed performance to deteriorate. Moreover, for maximum torque at a given speed, lift should be kept as low as possible, but again limited by the point where the intake of fresh charge becomes restricted. However, part-load fuel economy will improve by the intake charge restriction by valve lift, since it decreases the necessity of throttling by the engine throttle, thereby reducing pumping losses. Inspite of this advantage, lift amount is not decided by part-load considerations since this would severely limit the maximum engine power output. Thus, the design value of valve lift is usually selected by striking a compromise between low-speed and high-speed full-load considerations and is kept usually in the range 8-10 mm.

6. ENGINE VALVES

To admit the air-fuel mixture in the engine cylinder and to force the exhaust gases out at correct timings, some control system is necessary, which is provided by the valves.

The engine valves may be broadly divided into 3 main categories:

1. Poppet valve
2. Sleeve valve
3. Rotary valve

Out of these three, poppet valve is the one which is being universally used for automobile engines. It will be discussed in detail here. The other types are obsolete now.

7. POPPET VALVES

7.1. Construction

The poppet valve derives its name from its motion of popping up and down. This is also called "mushroom valve" because of its shape which is similar to a mushroom. It consists of a head and a stem as shown in Fig. 4.13. It possesses certain advantages over the other valve types because of which it is extensively used in the automotive engines:

1. Simplicity of construction.
2. These are self-centering.
3. These are free to rotate about the stem-axis to new position.
4. Maintenance of sealing efficiency is relatively easier in their case.

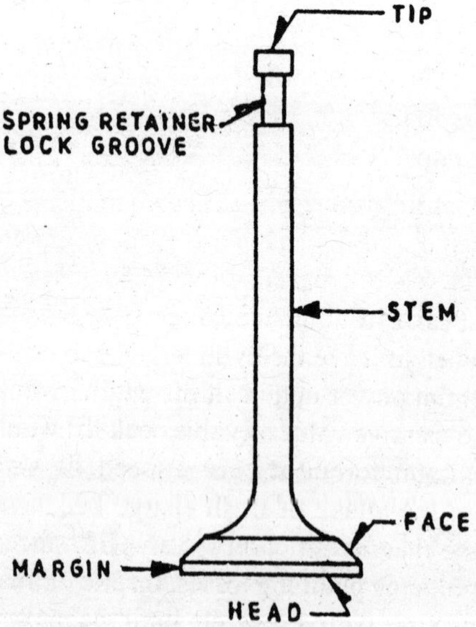

Fig. 4.13. Poppet valve

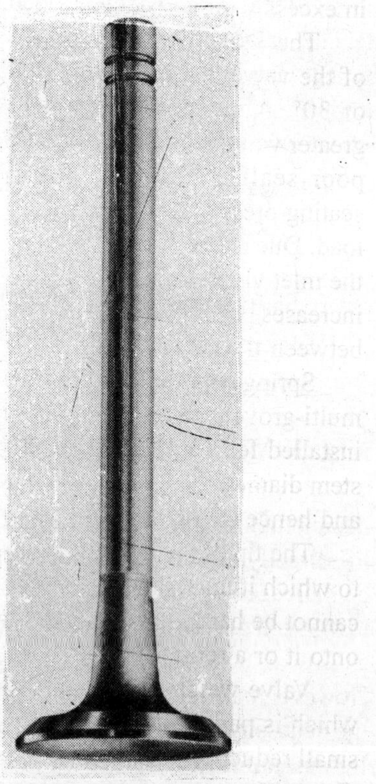

Fig. 4.14. Conventional poppet valve.
(Courtesy—Sealed Power Corporation, U.S.A.)

Valve head shape and size are critical to engine performance; key parameters are head diameter and seat angle. Valves with a dished area on the combustion chamber side of the head are somewhat lighter, but this lightness results in some decrease of compression ratio due to the increased chamber volume. Valve head diameter has a direct effect on airflow and consequently, engine power. Ideally the valve should be large enough so as to provide for a smooth air flow without any restriction. However, excessively large valve would lessen low-end throttle response and torque. Power size selection is a compromise between low speed performance and top end power and ultimately depends on the intended use of the engine.

Generally inlet valves are larger than the exhaust valves, because speed of incoming air-fuel mixture is less than the velocity of exhaust gases which leave under pressure. Further because of pressure, the density of exhaust gases is also comparatively high. Moreover, smaller exhaust valve is also preferred because of shorter path of heat flow in this case and consequent reduced thermal loading. Generally inlet valves and exhaust valves are 45% and 38% of the cylinder bore respectively. Further, to improve heat transfer to the cylinder head, the stem diameter of the exhaust valve is generally 10 to 15% greater than that o the inlet valve. Moreover, the valve lift in both inlet and exhaust valves should be at least equal to 25% of the valve head diameter which would provide the annular valve-opening area to the port throat area. If the valve lift is less, the volumetric efficiency of the engine will be decreased. On the other hand if it is excessive, the inertia of the valve actuating mechanism would be unduly large resulting in excessive noise and wear.

The valve face angle (with the plane of the valve head) is generally kept 45° or 30°. A smaller face angle provides greater valve opening for a given lift, but poor sealing because of the reduced seating pressure for a given valve spring load. Due to this reason in some engines, the inlet valve face angle may be kept 30° or 45° whereas the exhaust valve face angle is only 45°, as this increases its heat dissipation. In some cases, a further differential angle of about 1/2 deg to 1 deg is provided between the valve and its seating (Fig. 4.15), which results in better sealing conditions.

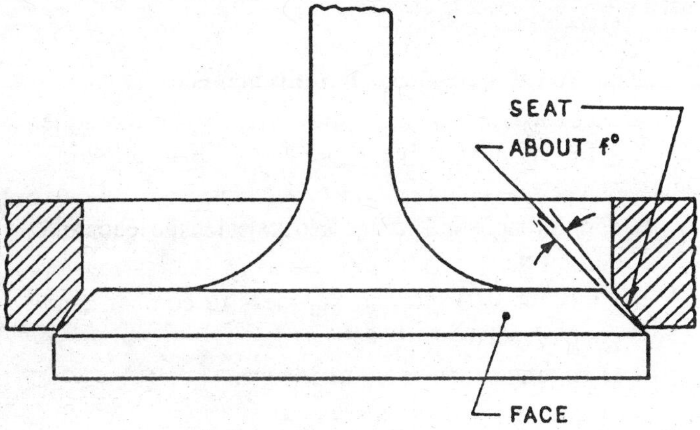

Fig. 4.15. Valve face-seating angle

Spring retainer lock groove is usually the single rectangular groove type as shown in Fig. 4.13. However, multi-grove type locks have also been used in some engines, particularly where valve rotators have been installed for keeping the seating surfaces clean of debris, thus increasing valve longevity. For very small stem diameter valves, sometimes single radius type grooves have also been used which avoids the corners and hence stress concentration in case of rectangular grooves.

The tip of the valve stem has to withstand continuously the dynamic contact with the rocker arm, due to which it must be quite hard. Non-stainless steel alloys are usually hardenable, but stainless steel valves cannot be hardened enough to meet this requirement due to which these must use either a hard tip welded onto it or a removable lash cap.

Valve weight is a crucial factor in valves of high speed engines, due to the inertia forces involved, which is particularly important in case of intake valves due to their large size and hence bigger mass. A small reduction in weight may be achieved by undercutting the area below the valve-guide or by reducing the entire stem diameter. However, titanium alloys, though costly, result in appreciable weight reduction. Titanium is 40% lighter than steel and maintains its strength at high temperatures. However, it is quite costly, thus limiting its use to high performance high speed engines and mostly the inlet valves.

7.2. Operating Conditions

Exhaust valves operate under relatively more severe conditions on account of higher temperatures involved. An exhaust valve is subjected to:

1. Longitudinal cyclic stresses due to the return spring load and the inertia response of the valve assembly.
2. Thermal stresses in the circumferential and longitudinal directions due to the large temperature gradient from the centre of the head to its periphery and from the crown to the stem. A typical variation of temperature in an exhaust valve is given in Fig. 4.16.
3. Creep conditions due to operation at very high temperatures, particularly in case of valve head.
4. Corrosion conditions.

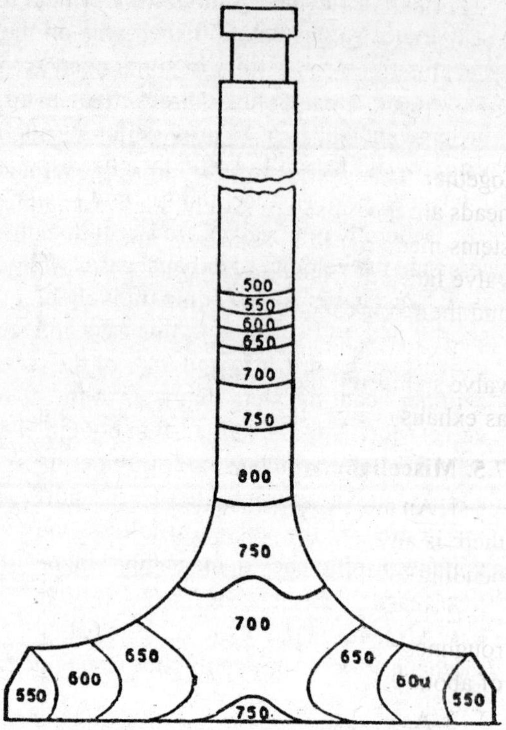

Fig. 4.16. Temperature distribution in degree celsius in an exhaust valve.

7.3. Exhaust Valve Material Requirements

On account of operating conditions described above the material for exhaust valve should have the following requirements:

1. High strength and hardness to resist tensile loads and stem wear.
2. High hot strength and hardness to combat head cupping ad wear of seats.
3. High fatigue and creep resistance.
4. Adequate corrosion resistance.
5. Least coefficient of thermal expansion to avoid excessive thermal stresses in the head.
6. High thermal conductivity for better heat dissipation.

7.4. Materials

The materials used for inlet and exhaust valves are generally different because of the different operating conditions to which these are subjected.

Silico-chrome steel containing about 0.4% carbon, 0.5% nickel, 0.5% manganese, 3.5% silicon and 8% chromium is the material generally used for inlet valves. For early exhaust valves, molybdenum was added to it. The more recent materials for exhaust valves are the austinitic steels and precipitation hardening steel. A typical austinitic steel is the '21-12', which contains 0.25% carbon, 1.5% manganese, 1% silicon, 12% nikel and 21% chromium. Another improved austinitic steel '21-4N' contains 0.5% carbon, 9% manganese, 0.25% silicon, 4% nickel, 21% chromium and 0.4% nitrogen. A typical precipitation hardening steel has the following composition:

Carbon	0.4 — 0.5	percent
Chromium	23.0 — 24.0	percent
Nickel	4.5 — 5.0	percent
Molybdenum	2.5 — 3.0	percent
Phosphorus	0.035	percent
Sulphur	0.035	percent

For exhaust valves of modern heavy duty engines, 'nimonic alloys' are used. A typical such alloy '80 A' consists of 0.05% carbon, 1% manganese, 0.6% silicon, 20% chromium, 2% cobalt, 2.5% titanium, 1.2% aluminium, 5% iron and the remainder nickel. These alloys have higher hardness, fatigue and creep strengths and corrosion resistance. However, these are very costly.

For heavy duty applications, the exhaust valves are sometimes made in two parts and then welded together. This is done only when a single material cannot satisfy all the stringent requirements. The valve heads are then made of special high strength corrosion resistant and creep resistant alloy and welded on to stems made out of a wear-resistant alloy. Alternatively, hard alloy such as 'stellite' may be welded to the valve face and valve tip. Stellite is a cobalt-base alloy having 1.8% carbon, 9% tungsten, 29% chromium and the remainder cobalt.

To avoid corrosion, the valve is given an aluminium coating. Aluminium oxide formed separates the valve stem from the cast iron seat to keep the face metal from sticking. The stems of both the inlet as well as exhaust valves may be chromium plated to make them wear resistant.

7.5. Miscellaneous Considerations

1. An adequately designed valve with proper material can also fail due to local stress concentrations if there is any unevenness around the valve-seat interface on account of distortion of valve heads or seats, bending of valve stem or trapping of carbon particles between the valve and the seat.

2. Excessive surface finish of the valve stem will result in loss of lubricating oil film, while excessive roughness of the stem would increase the guide wear. A thin layer of chromium giving the surface finish of about 0.5 ìm would provide the optimum condition.

3. As engine thermal efficiency is increased with increase of compression ratio, low valve temperatures would result in case of higher compression ratio.

4. Arranging the inlet and the exhaust ports in the cylinder head alternately would increase the transfer of heat from the exhaust to the inlet valves, compared to the case when the like valves are placed together. This would result in decreased exhaust valve temperatures. However, this would also complicate the design of the inlet and the exhaust manifolds in case both are to be on the same side of the engine.

7.6. Valve cooling

As already described, the exhaust valve temperatures in modern engines reach very high values of the order of 750°C. In heavy duty engines, it may be still higher. Therefore, cooling of exhaust valves becomes very important. To do this cooling water jackets are arranged as near the valves as possible. In many cases, nozzles are directed towards the hot spot caused by the exhaust valve.

In heavy-duty engines, sodium cooled valves are used. Sodium is a high conductivity metal which melts at 105°C. Thus sodium is in liquid state at the operating temperatures. Sodium is filled up to about 40 percent volume of the hollow stem of the valve (Fig. 4.17). When the valve is operating, the liquid sodium moves up and down in the hollow stem. When it goes up it absorbs heat from the hot valve head and on coming back down it gives the same heat to the stem, from where it goes via the valve guide, to the cylinder block and from there to the cooling water circulating in water jackets. This arrangement cools the valve by about 100°C.

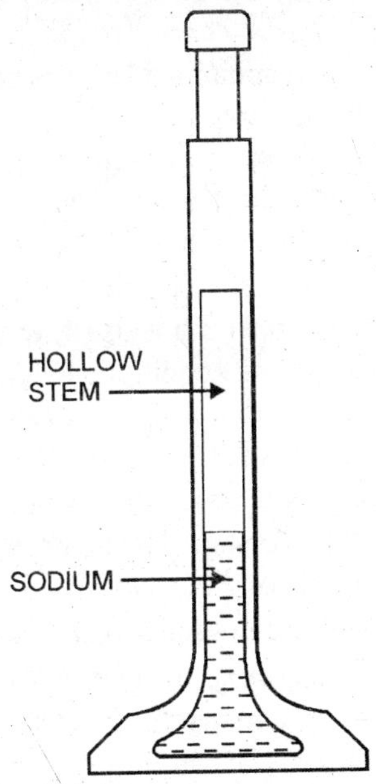

Fig. 4.17. Sodium cooled valve.

7.7. Valve Seats

The valve seats must be faced very accurately, so that there is complete contact between the valve and the valve seat when the former closes. Valve seat face is thus ground to the same angle to which the valve face is ground. This may have any value from 30° to 45°.

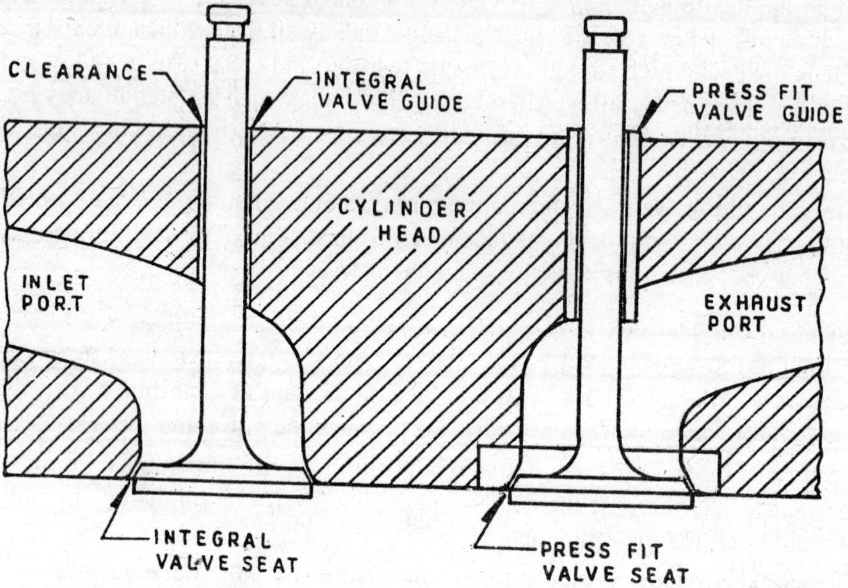

Fig. 4.18. Types of valve seats and guides

For cylinder blocks or heads made of grey iron, the inlet valve seats are directly machined on the cylinder blocks or heads as the case may be because working conditions are not severe. These are called *integral seats*. However, where aluminium blocks or heads are used, separate valve seat inserts are employed even for inlet valves. For the exhaust valves, always the separate valves seat inserts are used, the operating conditions being very severe (Fig. 4.18). Insert seats are also used as salvage procedure when badly damaged integral seats are reconditioned.

Fig. 4.19. Valve seats inserts
(Courtesy—Dana Corporation, U.S.A.)

Valve seat inserts are simply rings made of alloy steel consisting of chromium, silicon, tungsten or cobalt with a conical seat on one of the inside edges (Fig. 4.19). These are force-fitted in the recesses machined in the cylinder head. When worn, these inserts can be easily replaced.

7.8. Valve guides

The stem of the poppet valve needs a guide for the alignment of its up and down motion so that the face of the valve is maintained in a central position with respect to the valve seat while opening and closing. The simplest valve guide is a reamed hole in the cylinder block or head in which the valve stem moves. This type of integral valve guide is cheaper to provide, reduces weight and the heat transfer is also better.

However, the modern tendency is to provide separate valve guides, which are inserted into the holes in the cylinder block or head as the case may be. These are particularly required where the valve stem and cylinder head materials are not compatible. These inserts are of cylindrical shape and they are made from pearlitic cast iron having minimum hardness of the order of 220 Brinell Hardness Number. Sometimes alloy irons containing nickel and chromium are also used. Silicon, aluminium or bronze guide offer maximum resistance to fatigue, corrosion and heat. Both the types of valve guides are shown in Fig. 4.18.

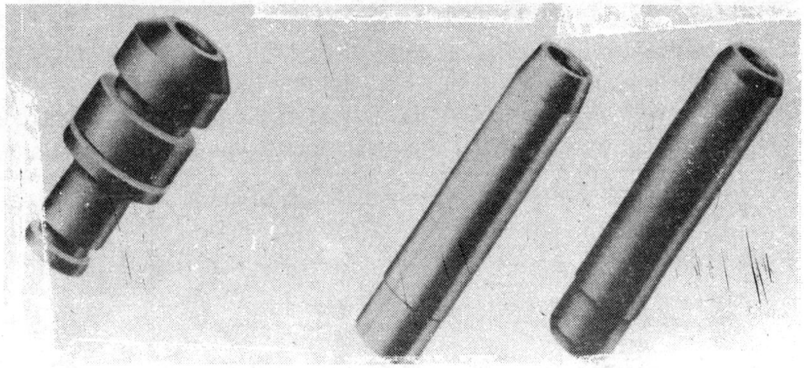

Fig. 4.20. Various types of valve guides.
(Courtesy—Sealed Power Corporation, U.S.A.)

Pictorial views of various types of valve guides are shown in Fig. 4.20.

For optimum sliding between the valve stem and the guide, the operating clearance should be just sufficient to maintain a lubricant film around the stem. Excessive clearance would hinder the conduction of heat transfer from the stem to the guide, Moreover, it would also produce a side rocking motion whenever the rocker arm contacts the valve tip, ultimately causing the bell-mouth wear of the guide at its ends and localized wear of the valve stem where the tilting stem contacts the guide bore.

7.9. Valve Springs

Helical springs are used to keep the valve in constant contact with the tappet and the tappet with the cam. Since the spring is subject to compressive loads, it is ground flat at each end to ensure even distribution of pressure. The coil ends are also placed diametrically opposed to avoid the bending tendency of the spring under compression. The arrangement for the retention of the springs is simple. A ring split into two halves with internal projection to fit into the valve spring retaining groove and the outer surface tapered is employed (Fig. 4.21). Over the split ring, another ring is inverted which supports the spring.

The valve springs are subject to heavy service. These are, therefore, made from high grade spring steel wire, the materials being generally hard-drawn carbon steel or chrome-vanadium steel. Valve springs are often shot peened to make them fatigue resistant.

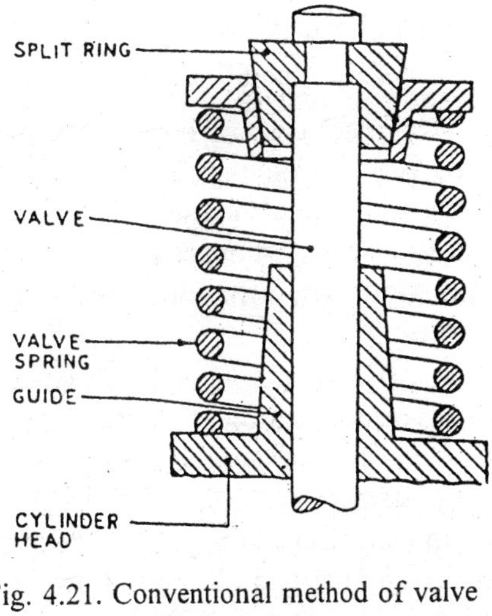

Fig. 4.21. Conventional method of valve spring retaining

8. VALVE ACTUATING MECHANISMS

In all the valve actuating mechanisms *i.e.* valve trains, a cam driven at half the crankshaft speed is used to operate each valve, inlet or exhaust. However, there are different methods of operating the valves from the cam. These may be broadly divided into two types viz., mechanisms with side camshaft and the mechanisms with overhead camshaft. These will now be explained further.

1. Mechanisms with side camshaft

In these, the camshaft is on the side of the engine and the valves are operated either directly by the cams or through the push rods and rocker arms. These may be further classified as:

(*a*) *Double row side valve mechanism (T-head)*. This is the oldest type of valve actuating mechanism and is shown in Fig. 4.22. In this the inlet and the exhaust valves are operated by separate camshafts, which makes the mechanism complicated. Moreover, the shape of the combustion chamber provides poor combustion and low engine performance, due to which this type of mechanism is obsolete.

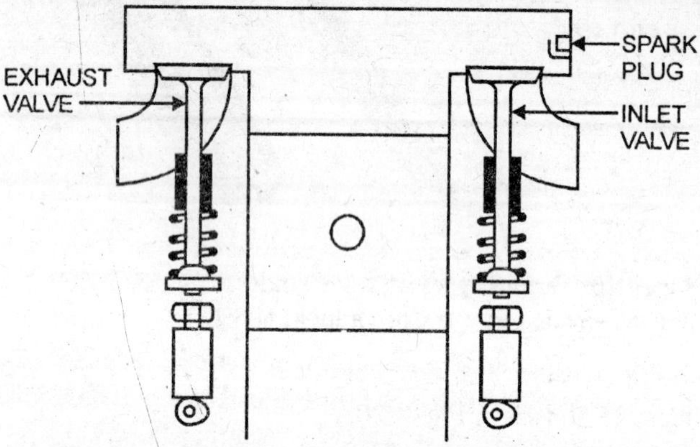

Fig. 4.22. Double row side valve-mechanism

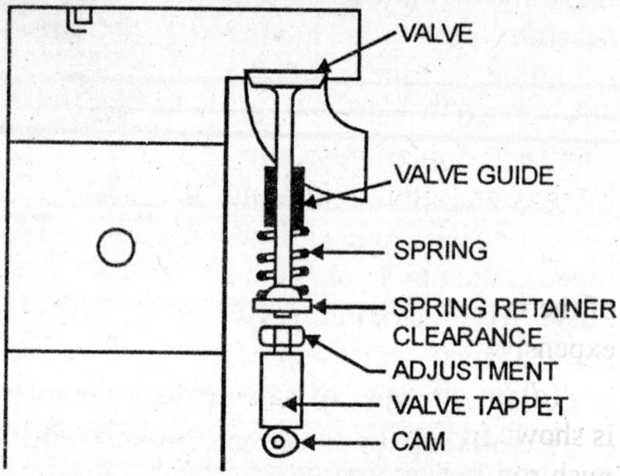

Fig. 4.23. Single row side valve mechanism

(*b*) *Single row side valve mechanism (L-head)*. In this, the inlet and the exhaust valves are all arranged in a single row and operated from the same camshaft (Fig. 4.23). This method was once quite popular on account of the following advantages:

(*i*) Low engine height (*ii*) Low production cost

(*iii*) Quiet operation (*iv*) Ease of lubrication

This mechanism is, however, no more in use, because it was found to be very inefficient on account of the complicated shape of the combustion chamber which is more prone to detonation. There were also restrictions of space on the size of inlet valves that would be used. Moreover, difficulties were experienced in cooling the exhaust valves.

(*c*) *Overhead inlet and side exhaust valve mechanism (F-head)*. This is a combination of the two systems described above. Overhead valve mechanism is used for the inlet valve operation and the side valve mechanism for the exhaust valve. It is used in F-head engines. This mechanism is simpler than the overhead camshaft operated types described later and allows the use of larger inlet valves, but larger valves being heavier, there is also a limitation on the maximum speed

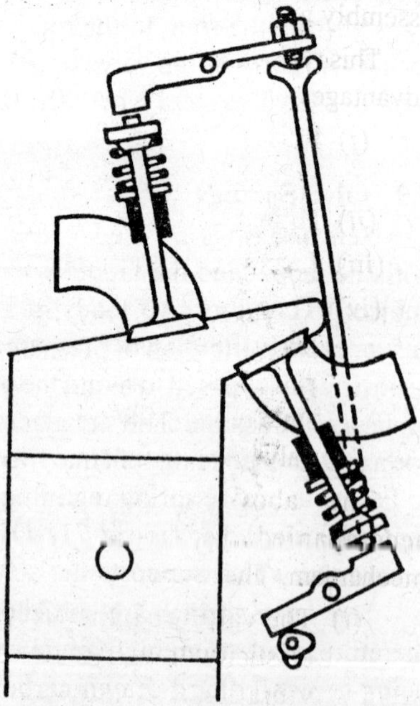

Fig. 4.24. Overhead inlet and side exhaust valve mechanism.

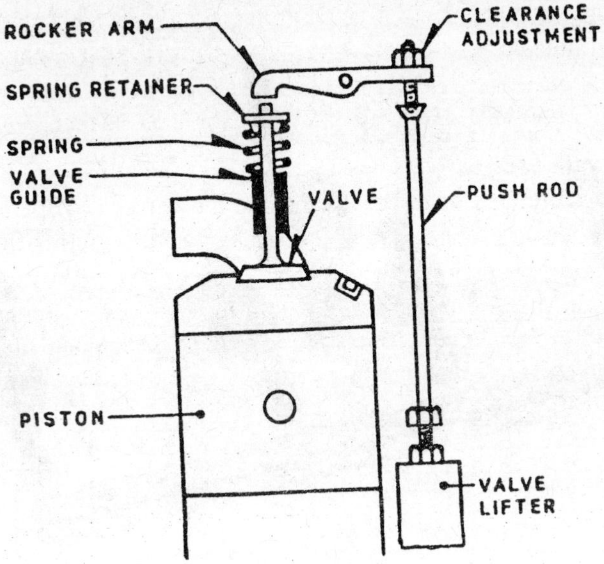

Fig. 4.25. Single row overhead valve mechanism.

of the engine that could be allowed. F-head engines were found to be less efficient and were also more expensive due to which these have also become obsolete.

(d) Single row overhead valve mechanism (I-head). This type is used quite extensively these days and is shown in Fig. 4.25 and 3.26. The cam operates the valve lifter which in turn actuates the push rod. The push rod further operates the rocker arm, which actuates the valve. The details of the rocker assembly are made clear in Fig. 4.27.

This type of mechanism is having the following advantages:

(*i*) Higher volumetric efficiency than the side valve design.

(*ii*) Higher compression ratios can be used.

(*iii*) Leaner air-fuel mixtures can be burnt.

(*iv*) The rocker-arm leverage makes it possible to impact desired cam profile lift multiplication to the system and hence use smaller cam lobes compared to the side valve mechanism.

The above advantages are, however, accompanied by some drawbacks of the mechanism. These are:

(*i*) The valve operation, on account of the elasticity of the system and the resulting vibrations, is not very precise while accelerating at high engine speeds.

(*ii*) Larger valve lifter clearance are required.

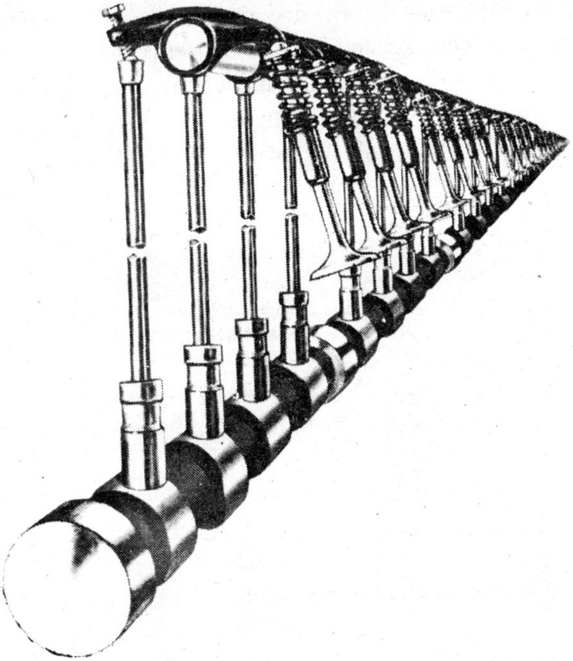

Fig. 4.26. Single row overhead valve mechanism. (Courtesy—Sealed Power Corporation, U.S.A.)

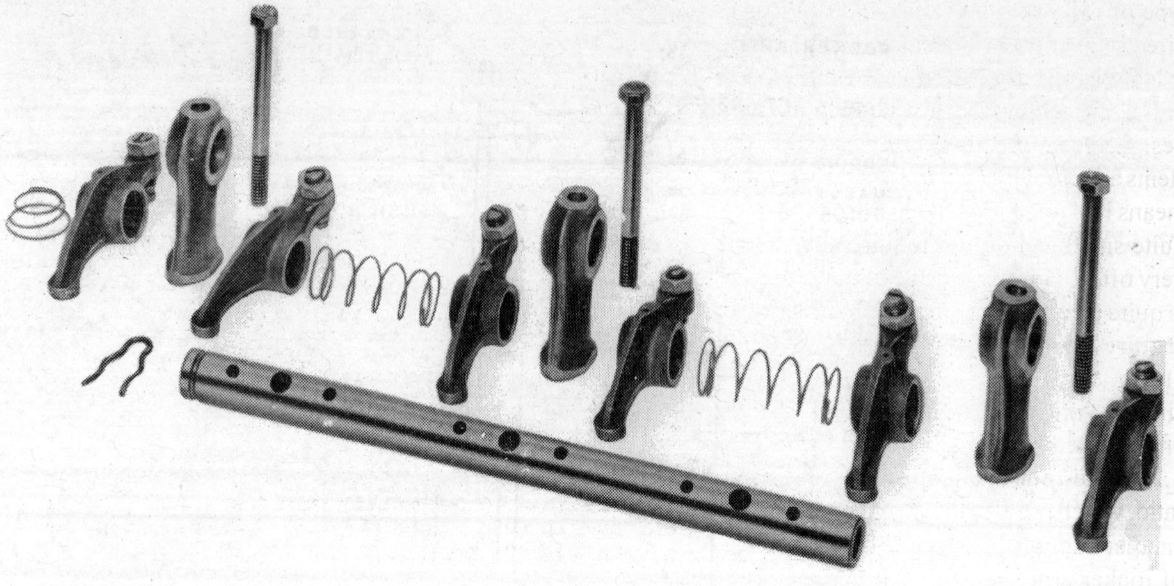

Fig. 4.27. Details of rocker assembly.
(Courtesy—Sealed Power Corporation, U.S.A.)

(*iii*) Noisy operation.

(*iv*) Greater maintenance required due to more wear at more joints.

2. Mechanisms with overhead camshaft

The valve operating mechanisms with overhead single or double camshafts are highly efficient. Overhead camshaft means that the distance between the cam and the valves is much shorter due to which the valve response is quicker and valve adjustment can be more accurate. Moreover, due to fewer valve train components compared to the side camshaft engine, this system permits higher engine speeds. However, with these considerably more lubricating oil is needed to flood the cam profiles as compared to the overhead valves operated by side camshafts. Moreover, they have the disadvantage of higher initial costs. Fig. 4.28 shows single row valves operated by a single overhead camshaft and an inverted bucket type follower. With this

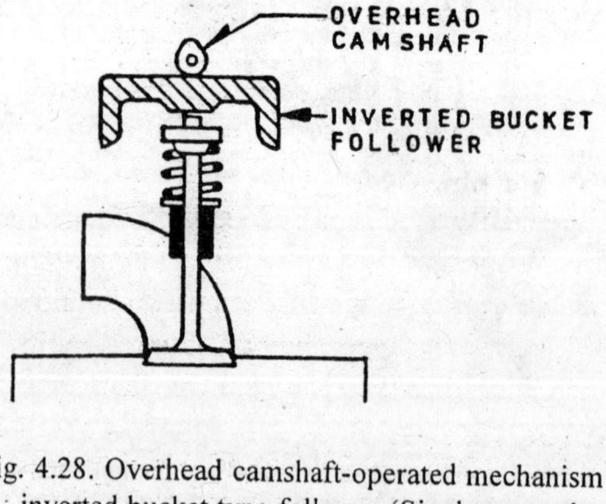

Fig. 4.28. Overhead camshaft-operated mechanism with inverted bucket type follower (Single row valves).

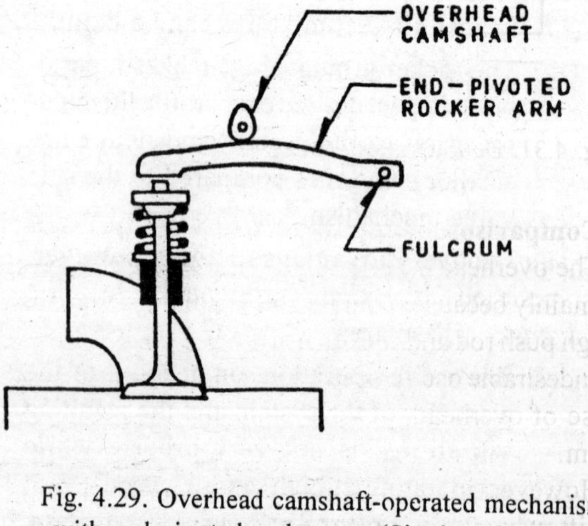

Fig. 4.29. Overhead camshaft-operated mechanism with end-pivoted rocker arm (Single row valves).

type of follower, the camshaft is arranged directly over the valve stems. This type of mechanism is direct and very rigid so that valve movement follows precisely the designed cam-profile lift. Moreover, valve stems are not subjected to side-thrust which means less wear. Tappet clearances are also quite small and do not require adjustment very often. However, drive to the camshaft is quite complicated, positive lubrication is required and adjustment of valve lifter clearance is relatively more difficult. A similar valve-operating mechanism with end-pivoted rocker arm is shown in Fig. 4.29. The rocker arm provides leverage ratio, which enables the designer to provide smaller cam profile. Moreover, the inertia of rocker arm follower is less compared to the sliding bucket type described earlier and adjustment of tappets is easy. However, due to the elastic bending of the rocker arm, the stiffness of the system and hence precision of valve operation is decreased, a side-thrust is produced to the valve stem and guide and more wear and noise occur. In Fig. 4.30 is depicted a mechanism for inlet and exhaust valves in separate rows, but operated by a single overhead camshaft with inverted bucket type follower and the pivoted rocker arm. However, quite often the double-row valves are operated by two separate overhead camshafts as shown in Fig. 4.31 and 3.32.

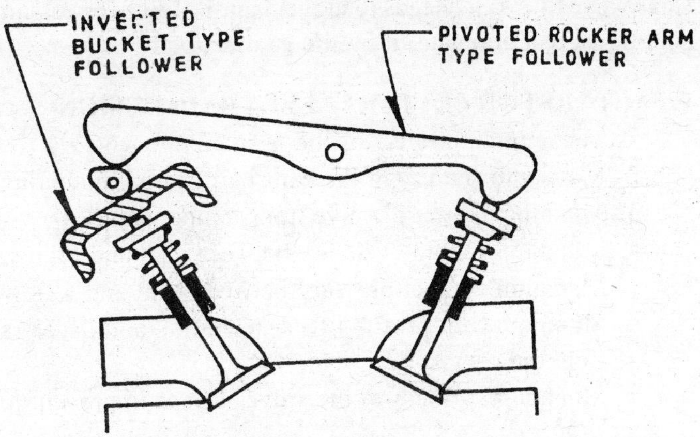

Fig. 4.30. Overhead camshaft-operated mechanism (double row valves), inlet valve operated by inverted bucket type follower and exhaust valve by pivoted rocker arm.

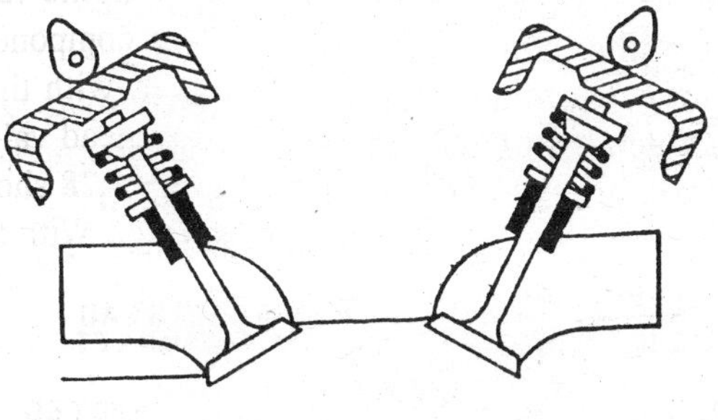

Fig. 4.31. Double overhead camshaft-operated mechanism with inverted bucket type followers

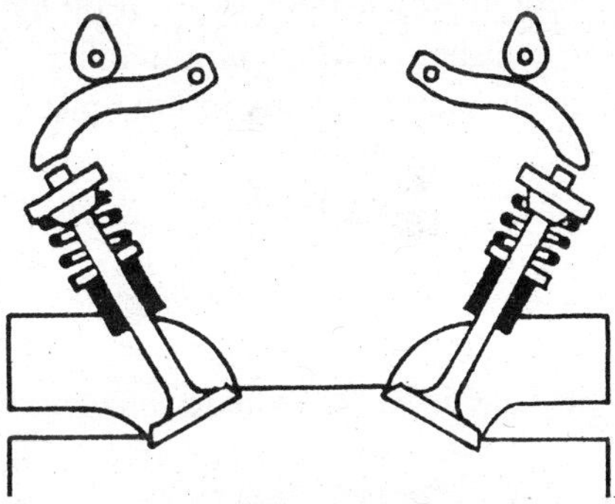

Fig. 4.32. Double overhead camshaft-operated mechanism with separate rocker arms.

Comparison of the Side Camshaft and the Overhead Camshaft Mechanisms

The overhead camshaft type valve actuating mechanisms is generally preferred over the side camshaft type mainly because of its greater rigidity since the camshaft directly operates the valve instead of operating through push rod and rocker arm. Due to this the valve is opened and closed quickly with decreased vibrations and undesirable oscillation. This means in case of high-lift, high-acceleration cam profile, the valve operation in case of overhead camshaft is much more precise and smooth than in case of the side camshaft valve system.

However, in case of overhead camshaft with inverted bucket follower, the valve lift is equal to the cam lift, whereas in case of the side camshaft, the valve lift can be adjusted by suitable design, for a given cam lift. This means that in case of overhead camshaft, the cam size has to be relatively larger for the same valve lift, which leads to higher cam-to-follower velocities and relative rubbing velocities resulting in side-thrust reaction caused by the cam action.

9. FACTORS FOR BETTER VALVE TRAIN DESIGN

1. Maximum stiffeners of the train components *e.g.*, cam, follower, push rod, etc.
2. Maximum rigidity of the camshaft bearing mounting.
3. Minimum inertia of valve train components.
4. Minimum rubbing velocity between cam and follower.
5. Minimum contact pressure between cam and follower.
6. Minimum cam profile lift, acceleration and deceleration so that similar movement is transferred from the cam to the valve.
7. Minimum stiffness of the valve springs to prevent jumping of valve at high speeds.

10. VALVE TRAIN COMPONENT DETAILS

The main components of the valve actuating mechanisms are now discussed in detail. The discussion applies to all types of mechanisms.

10.1. Camshaft

A camshaft is a shaft with a cam for each intake and exhaust valve. Each cam has a high spot called cam lobe which controls the valve opening. The camshaft provides a means of actuating the opening and controlling the opening and closing intervals of the inlet and the exhaust valves relative to each piston position along its respective stroke. It also provides a drive for the ignition distributor and the mechanical fuel pump.

A cam is a rotating member which imparts reciprocating motion to another member, called follower, in a plane at right angles to the cam-axis. The cam profile is appropriately shaped to provide a smooth rise and fall for the predicted motion of the follower, which is then transmitted to the individual valves.

The camshaft consists of a number of cams at suitable angular positions for operating the valves at approximate timings relative to the piston movement and in a sequence according to the selected firing order (Fig. 4.33). There are two lobes on the camshaft for each cylinder of the engine; one to operate the intake valve and the other to operate the exhaust valve. A number of integral bearing journals support the shaft in bearings. Camshaft bearing journals are always larger than the cam lobes so that the camshaft may be installed in the engine through the cam bearings. To provide room for lubrication and metal expansion, a clearance of 0.05 to 0.125 mm is usually provided between the bearing journals and the bores. The camshaft can be located in either cylinder block or the cylinder head.

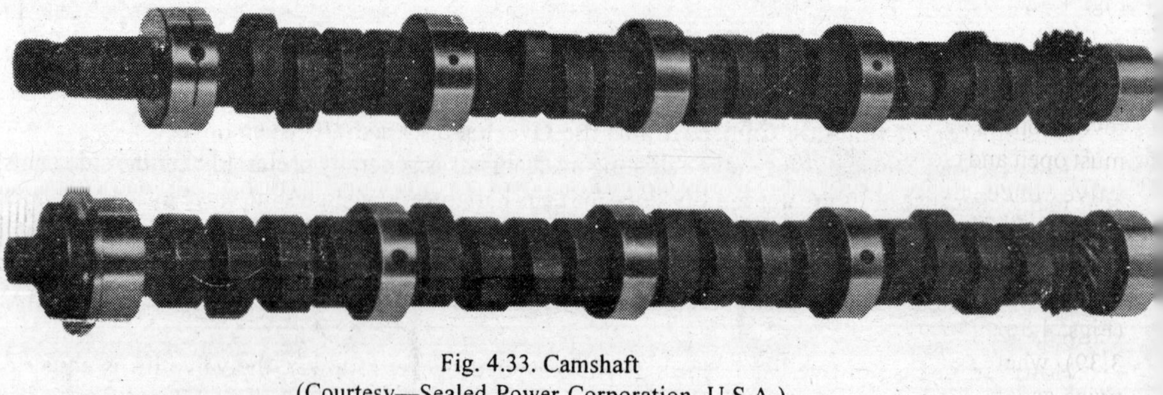

Fig. 4.33. Camshaft
(Courtesy—Sealed Power Corporation, U.S.A.)

Usually there is an integral spiral toothed gear on the camshaft to drive the distributor and the oil pump. The fuel pump is operated from an integral eccentric or a bolt-on eccentric. In some cases, however, distributors are driven directly from the camshaft end. Endwise movement of the camshaft is limited by a thrust plate between the front bearing journal and the drive gear or the sprocket. This thrust plate is bolted to the engine. However, in some engines, instead of the thrust plate, the same is achieved by the tapered spiral teeth of the distributor oil pump drive.

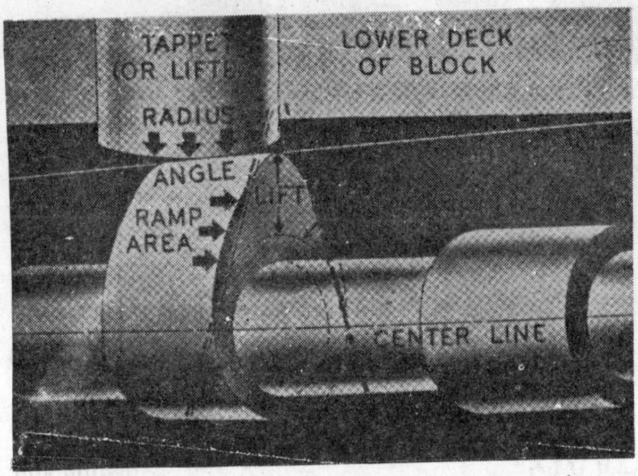

Fig. 4.34. Taper cam operating spherical based tappet. (Courtesy—Sealed Power Corporation, U.S.A.)

Fig. 4.35. Camshaft supported in bearings. (Courtesy—Daimler-Benz Aktiengesselschaft, Germany)

The camshaft is forged from alloy steel or cast from hardenable cast iron and is case hardened. A typical cast iron alloy for a camshaft would consist of 3.3% carbon, 2% silicon, 0.65% manganese, 0.5% chromium, 0.25% molybdenum and the remainder iron. After chilling this, the surface hardness attained is about 500 to 600 DPN. In modern engines, cam lobes are ground with a slight taper across the face (Fig. 4.34). The tappets used with such camshafts have spherical base and are slightly offset from the cam face. This provides tappet rotation and a wear pattern preventing edge loading which is a major cause of failure. Camshaft is supported in a number of bearings, one of which is shown in Fig. 4.35.

10.2. Camshaft drive

One four-stroke cycle is completed in two revolutions of the crankshaft. Therefore, each cam must open and close its respective inlet or exhaust valve once every two revolutions of the crankshaft. That is why, the camshaft is made to rotate at half the crankshaft speed. The drive from crankshaft to the camshaft may be either chain drive (Figs. 4.36 and 3.37) or gear drive (Fig. 4.38 and 3.39), where camshaft gear or sprocket wheel is twice as large as the crankshaft gear or sprocket

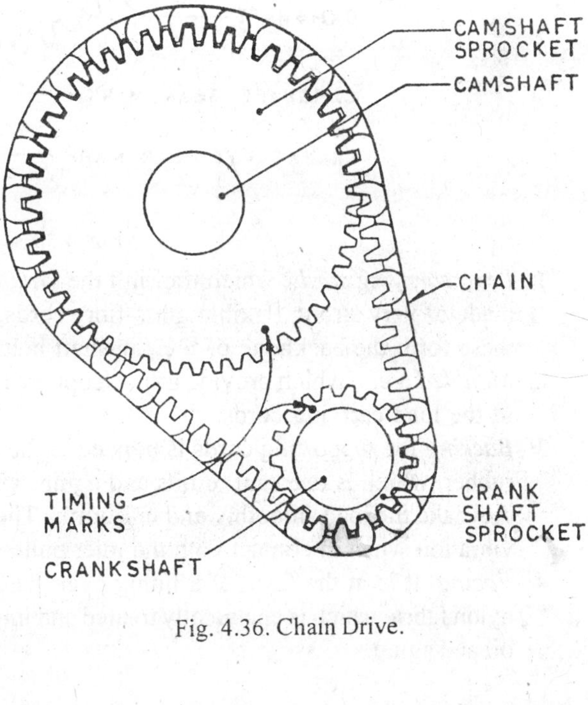

Fig. 4.36. Chain Drive.

wheel. There are *timing marks* on the sprockets or gears of the camshaft and the crankshaft to ensure correct valve timing. In a chain drive a separate idler gear has to be provided. Moreover, a long chain drive tends to whip to avoid which a tensioning device (Fig. 4.40) is used, which may be a roller or spring-loaded steel strip or a rubber pad attached to a spring-loaded piston. Gear drives, on the other hand, need no tensioner, but are noisy and are suitable only for camshafts mounted close to the crankshafts. Obviously gear is not suitable for overhead camshafts. The latest type of drive is however, by means of a toothed rubber belt, which is made of rubber mouled onto a non-stretching cord (Fig. 4.41). Toothed belts consist of four basic components:

Fig. 4.37. Perfect circle chain drive assembly. (Courtesy—Dana Corporation, U.S.A.)

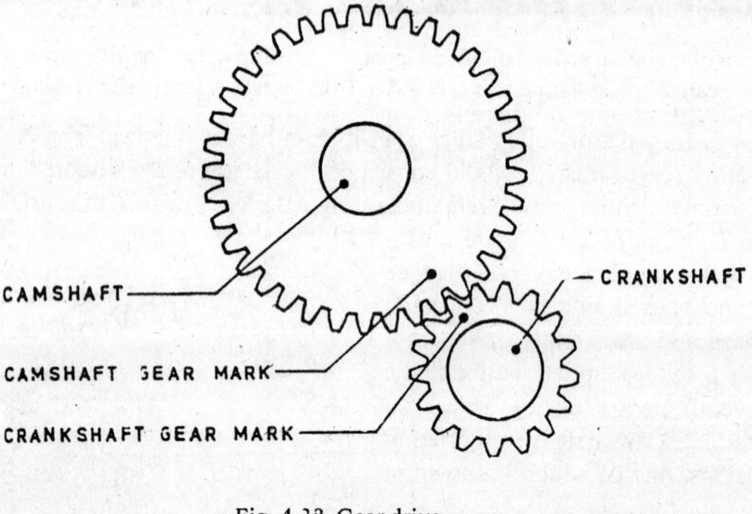

Fig. 4.38. Gear drive.

1. *Load carrying cords,* which transmit the torque load from the drive to the driven pulley. These are made of very strong flexible glass-fibre cords. Due to their flexibility, stability and compactness, these form the backbone of the camshaft belt drive.
2. *Moulded teeth,* which provide grip. Neoprene rubber teeth mouldings are bonded to the inside layer of the load carrying cords.
3. *Backing* for supporting cords is bonded to the outside layer of the cords. It is made from synthetic rubber, which is resistant to oils and aging. This backing positions and supports the cords so that they take the load smoothly and uniformly. The outer surface of the backing is smooth and reduces vibration when in contact with the idler pulley.
4. *Facing.* It is in the form of a lining over the toothed side of the belt. It is made of tough woven nylon fabric which is chemically treated and impregnated with synthetic rubber resistant to abrasion, oil and aging.

Fig. 4.39. Perfect circle gear drive assembly.
(Courtesy—Dana Corporation, U.S.A.)

Fig. 4.40. Perfect circle tensioners.
(Courtesy—Dana Corporation, U.S.A.)

The toothed belt drives have the following advantages due to which these are becoming increasingly popular for both petrol as well as diesel engines with overhead camshaft:

1. Silent drive due to absence of metal-to-metal contact as in a chain drive.
2. No belt chatter or vibration due to absence of chordal rise and fall of the pitch line, creep or slippage.
3. Low heat built-up on account of very low friction between the belt and the pulley and due to very low tension.
4. Negligible hysterisis loss because the belt is quite thin and flexible.

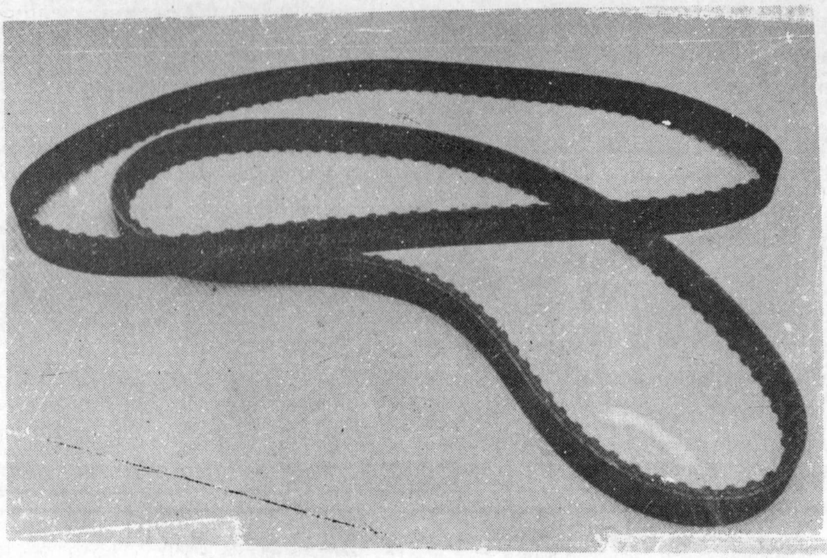

Fig. 4.41. Perfect circle timing belt.
(Courtesy—Dana Corporation, U.S.A.)

5. No lubrication is required due to absence of metal-to-metal contact and low belt tension. Also no periodic adjustments are required on account of low radial pulley bearing load. Only maintenance required is the periodic tensioning checks.
6. Cost of the drive is relatively low.

However, the life expectancy of a toothed belt is about one-third of the chain drive for conventional belts. But, with newer materials and processes, this gap in the life span has been bridged. However, there is no warning indication when a belt is about to break. It must be remembered that in case of sudden failure of the belt, valves in the cylinder head may hit the piston crowns, necessitating the replacement of both. There is also some possibility of the belt teeth jumping over the engaging teeth of the pulley due to instantaneous belt loads caused by a cold start. In such a case, the valve, ignition or injection timing would change. Fig. 4.42 shows such a drive for double overhead camshafts.

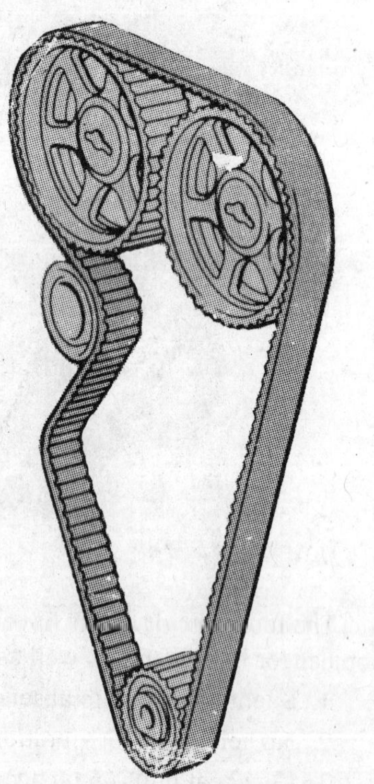

Fig. 4.42. Toothed drive belt for double overhead camshafts.
(Courtesy—Rover Powertrain, U.K.)

Timing gears are made of cast iron, steel, aluminium or laminated fibre. Chains are made with links of alloy steel with ground and case hardened pins. Sprockets are made of nylon, aluminium, steel or iron.

It is, however, seen that if the valve timings are adjusted automatically depending upon the engine working conditions, it results in increased volumetic efficiency and reduced pumping losses. This can be done by varying cam timing by means of a *variable cam phaser* (VCP). It is used to shift the intake cam, the exhaust cam or both on dual overhead camshaft engines. This helps in increasing engine efficiency, improving idle stability and producing more power and torque. Fuel consumption and hydrocarbon emissions are also reduced. A modern variable cam phaser replaces the standard pulley, sprocket or gear. It enables the angular relationship of the pulley, sprocket or gear to the camshaft to be changed when the engine is running, depending on the operating engine parameters. The angular position, *i.e.*, phase relationship is controlled by the internal rotor/stater pack of the VCP as per commands from the engine powertrain control module.

10.3. Valve tappet (valve lifter or cam follower)

The valve tappet *i.e.* cam follower follows the shape of the cam lobe on the camshaft. Thus it provides a method of converting the angular movement of the cam into a reciprocating motion in the valve train which is directly proportional to the amount the cam lobe deviates from the base circle.

This is placed in between the push rod and the cam. so that sideways thrust of the cam is not exerted on the push rod. Generally tappet is placed slightly eccentric with respect to the cam so that it (tappet) keeps on rotating slightly in every stroke which makes it wear uniformly.

It is very necessary that the valves close and open properly and maintain gas tightness. For this purpose some clearance must be allowed between the cam and the tappet to allow for expansion due to heat. This valve train clearance is called *valve lash.* This clearance should not be excessive, otherwise it will cause noise or premature failure. The most common method of doing this in case of a solid or hollow valve tappet is by means of an adjusting screw and lock-nut. Such tappet is shown in Fig. 4.44. Similar to the camshaft, such a tappet is also made from a chilled low-alloy cast iron consisting of iron, carbon, silicon manganese and chromium.

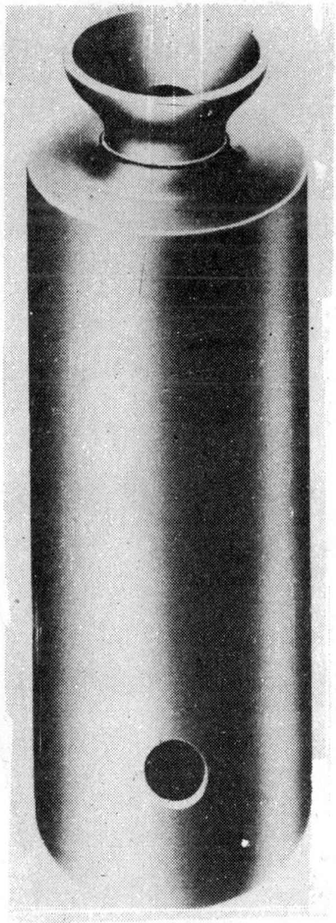

Fig. 4.43. Fixed type valve tappet.
(Courtesy—Sealed Power Corporation, U.S.A.)

Fig. 4.44. Adjustable type valve tappet.
(Courtesy—Sealed Power Corporation, U.S.A.)

Another type of tappet which does not need any adjustment to be provided is the hydraulic tappet shown in Figs. 4.45 to 3.48. To understand working of this type, consider Fig. 4.47. It consists of a body containing groove and holes which is loose fitted in a hole in the cylinder block. A lubricant oil passage in the cylinder block coincides with the outer groove on the body of the tappet through which the oil can enter inside it. Inside the body, is a plunger with a groove and hole on its outside. At the top of the plunger is tightly fitted a cap containing a seat for the lower end of the push rod. A circlip is fitted in a groove at the top of the body to prevent the various components in the body from coming out when the push rod is not seated on its seat. In the bottom of the plunger, there is an oil feed hole which is controlled with a ball valve, which is spring-loaded and retained in position by means of a cup shaped retainer having slots on its sides.

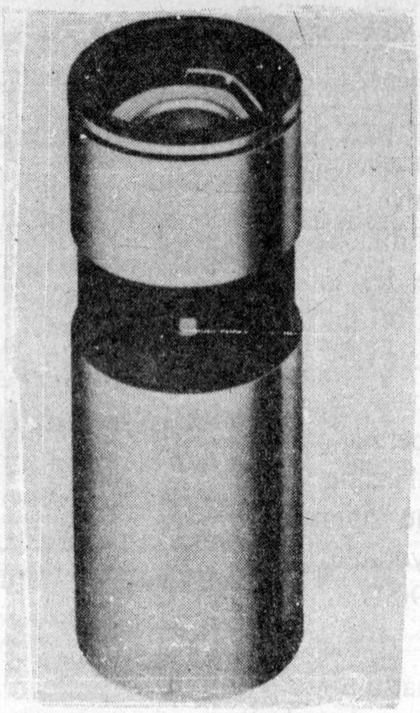

Fig. 4.45. Hydraulic tappet. (Courtesy—Sealed Power Corporation, U.S.A.)

To understand the operation of this tappet, imagine the cam operating it to be in such a position that the tappet body is in its lowermost position (valve closed position). In this situation, the plunger spring pushes the plunger from where the spring force is transmitted via the push rod to the valve, thus taking up any clearance in the valve gear. The spring force is so designed that it is just sufficient to take up any clearance and retain the valve on its seat. Simultaneously, as the spring opens up in this position, the volume of the chamber below the plunger increases, the oil pressure thus drops, the ball valve comes off its seat and more oil enters the chamber through the feed hole in the plunger.

As the cam rotates to lift the body up, the body moves up relative to the plunger, thereby increasing pressure in the chamber below, which causes the ball valve to close the feed hole. The force due to cam

Fig. 4.46. Exploded view of a hydraulic tappet. (Courtesy—Sealed Power Corporation, U.S.A.)

lifting is then transmitted through the column of oil under pressure in the chamber, to the plunger and thence to the push rod and valve stem, opening the valve. Any shocks are thus absorbed as the valve now is supported by a column of oil. The body and plunger fit is so designed that a small quantity of oil leaks through the space in between when the oil is under pressure at the time of valve opening. This helps in changing the oil inside continuously. Fig. 4.48 is the operational diagram of a hydraulic tappet in Mercedes Benz 190 D 2.5 engine.

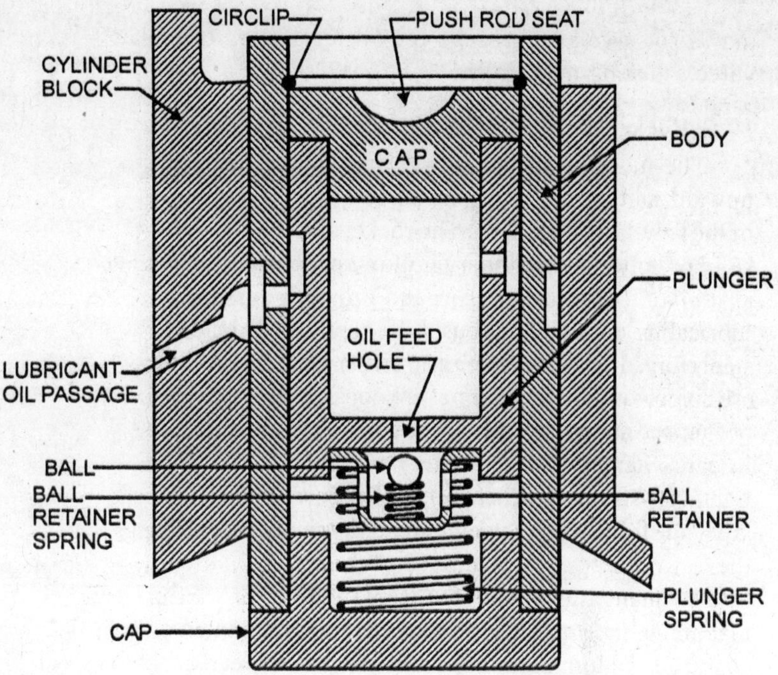

Fig. 4.47. Hydraulic tappet construction (in the 'valve closed' position)

It may be seen from the above that the use of a hydraulic tappet gives following advantages:

1. Extra valve train lash is taken up when the engine is cold and during the warm up period or even otherwise due to wear or incorrect adjustment. Thus it eliminates tappet noise in the valve train.
2. As the initial tappet clearance is not critical with a hydraulic tappet, it need never be adjusted. It automatically maintains the camshaft's designed valve-timing periods during normal engine operation.

10.4. Push Rod

It serves to transmit the reciprocating motion of the valve lifter to the rocker arm. Since the valve lifter moves in a straight line whereas the rocker arm end moves in an arc about its pivot, to provide compatibility of the two, the push rod forms part of ball and socket joints on both ends. Push rods are made of carbon-manganese steel. A typical push rod steel contains 0.35% carbon, 0.2% silicon and 1.5% manganese. After hardening and tempering, a hardness of about 250 B.H.N. is obtained. The push rod may be either solid or hollow. A hollow push rod is lighter, resulting in decrease of inertia forces. Further, it

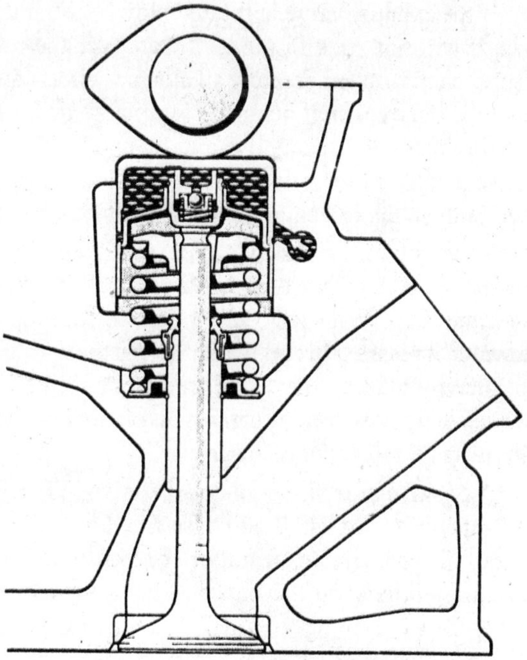

Fig. 4.48. Operational diagram of hydraulic tappet on Mercedes- Benz 190 D 2.5 engine. (Courtesy—Mercedes Benz Aktiengesselschaft, Germany)

may also serve as a passage for oil lubrication of the valve actuating mechanism.

10.5. Rocker Arm and Rocker Shaft

The function of the rocker arm is to reverse the upward motion of the push rod to downward motion of the valve and vice versa. The rocker arm may be either solid or hollow. A stationary hollow rocker shaft serves as a pivot to the rocker arms and provides passage for lubricating oil simultaneously. Rocker arm is made of steel (forged or stamped) or iron (cast). Cast rocker arms are comparatively cheaper but are not as strong as forged or stamped ones. However, these give satisfactory service in cars. Stamped rocker arms have been found to be light, very strong, yet cheapest of all types. Rocker shafts are made from hollow steel tubing. A typical material for these would consist of 0.55% carbon, 0.2% silicon, 0.65% manganese and the remainder iron. After machining the shaft is case-hardened. It is mounted on cast iron or aluminium pedestals placed between each pair of rocker arms.

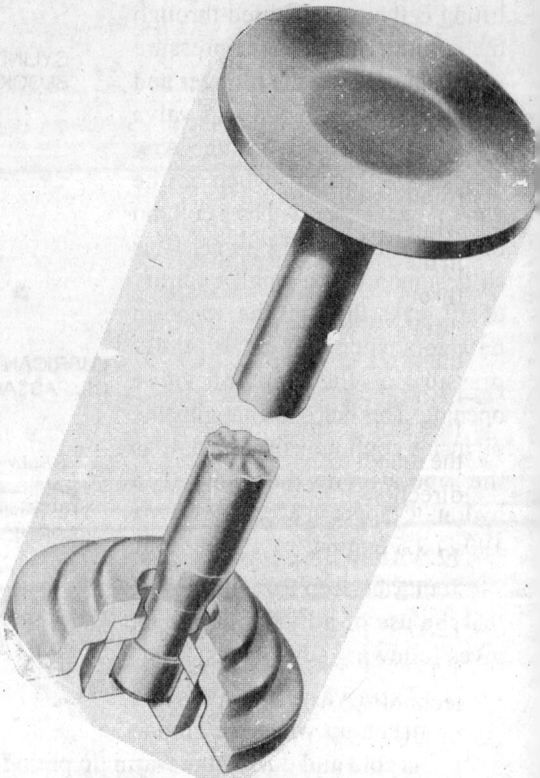

Fig. 4.49. Free type valve rotator.
(Courtesy—Sealed Power Corporation, U.S.A.)

11. VALVE ROTATORS

If the exhaust valve is rotated slowly about its own axis, it will not stick in guides. Moreover the rotation results in uniform temperature all along its circumference, due to which reason it becomes less liable to warping. Thus slight rotation of the valve ensures even wear distribution on the seat, valve and tip and helps in removing small particles between the valve and the seat. Consequently better valve seating and longer valve life are obtained. On some engines, valve rotators are used both on the inlet and the exhaust valves, while in others these are installed only on the exhaust valves.

Valve rotators are of two types, *i.e.* free type and the positive type. The free type valve rotator is shown in Fig. 4.49 and 3.50. When the valve lifter moves up to open the valve it presses tip cup which moves the spring retainer up, thereby making the valve free from spring pressure. Because it is now free, it can rotate on account of engine vibration and turbulence of gases.

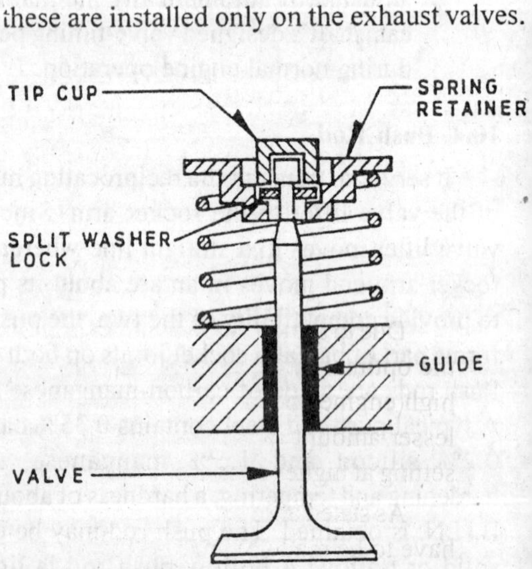

Fig. 4.50. Free type valve rotator.

A positive type valve rotator is shown in Fig. 4.51 and 4.52. It consists of a ball cup base fixed outside the valve guide. There are a number of grooves arranged circumferentially on the ball cup base. In each of these groves there is a spring-loaded ball on top of which is placed a spring washer. The top ends of these washers touch the rotator cover which supports the valve retaining spring. When the valve is closed as is shown in Fig. 4.51 and 3.52 (*a*), the ends of spring washers are just touching the ball cup base and the rotator cover so that it is just pressing the balls, with a very small force due to which reason, the balls are pushed by the balls at the ends of their races which have got higher base level towards the ball ends. On the other hand, when the valve tip is pressed to open up, the rotator cover is also pressed through the valve spring. The valve rotator in turn presses the upper end of the spring washer which presses the balls against their springs to move down in their races. Since the rotator cover, connected to the valve through the valve retainer, is also connected to the balls through the spring washer, this movement of the balls causes the valve to rotate in the counterclockwise direction as shown in Fig. 4.52 (*b*). On closing the valve, the balls become almost free so that their springs move them up the incline causing the rotator cover and hence the valve to rotate in the clockwise direction.

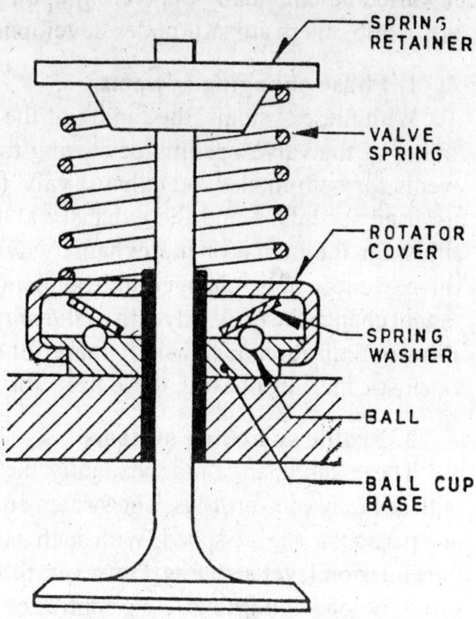

Fig. 4.51. Positive type valve rotator.

12. VARIABLE VALVE TIMING (VVT) TECHNOLOGIES

12.1. Types

VVT is a generic term for an automobile engine technology which allows the *lift* or *duration* or *timing* (some or all) of the intake or/ and exhaust valves to be changed when the engine is working.

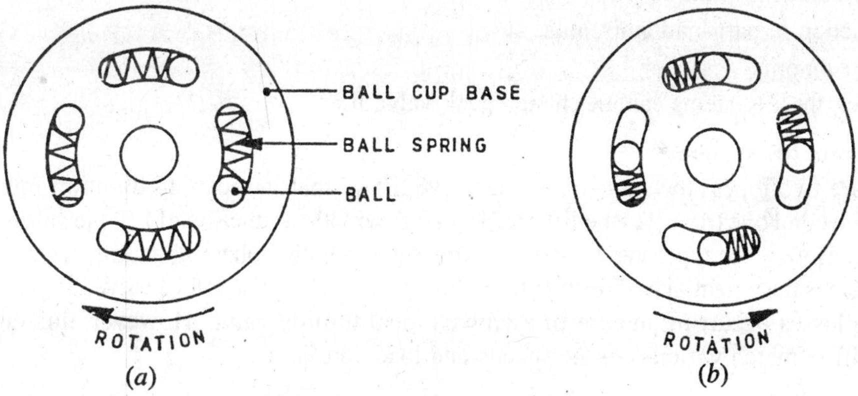

Fig. 4.52. Details of positive type valve rotator.
(*a*) Valve closed, (*b*) Valve open.

Due to the variation in behaviour of the air-fuel mixture before combustion and the gases after combustion, the optimal valve timing, lift and duration settings at low engine speeds are quite different from those at high engine speeds. Any optimal setting of the valve lift, duration and timing at low speeds would cause lesser amount of air and fuel at higher speeds, resulting in loss of engine power output. Similarly, valve setting at higher engine speeds would result in difficult idling and very rough engine operation at low speeds.

As stated above and already discussed in detail in Art 4, for an engine with fixed valve timing, compromises have to be made between emissions, high/low speed torque and full/part-load efficiency. To avoid making such compromises, the following technologies are gradually becoming common, although still limited to high-end cars due to cost considerations. These technologies basically enable the valve timing and lift to be varied automatically depending upon the engine operating conditions. Some of these are currently in use, while others are still under development.

1. Phase changing systems

With these systems, the timing of the camshaft is changed relative to the crankshaft which causes the timing of the valve opening or closing to be advanced or retarded. With a single camshaft, all the valve events for both intake and exhaust valves are shifted by the same amount, whereas if separate camshafts are used for intake and the exhaust valves, such a phase changing system can change the valve events either for the intake or the exhaust valves. But, with two camshafts and two phase- changing systems valve events can be changed independently for the intake and the exhaust valves. However, such systems cannot change the peak valve lift or the *duration* of the valve events. More advanced phase-changing systems, allow optimisation of camshaft phase for different engine speeds and loads, and also permit the phasing of exhaust camshaft phasing to be used for internal EGR control.

2. Profile switching systems

These can change independently the valve-event timing and the valve lift. There are two camshafts with separate cam-profiles. The system enables the valve-operation to be switched between these camshafts at a particular engine speed. With such a capability very high power output can be obtained, while keeping the emission level also low. However, these are not suitable for optimising valve-timing parameters under different load conditions, *e.g.*, control of internal EGR.

3. Variable event-timing systems

Such systems can change both the phase and duration of the valve events and can be optimised for different engine speed and load conditions due to which it is possible to obtain:

1. Increase in full-load torque.
2. Reduction in part-load emissions.
3. Fuel economy.

However, these systems cannot change peak valve lift.

4. Variable lift systems

These are two types: the first scales the valve lift so that its opening duration remains unchanged, whereas the second one truncates the lift profile so that with the reduction of lift, the valve opening duration also reduces. Both these types can be used in combination with a phase changing system. Main advantage of such system is the potential of throttling the cylinder by reducing the intake valve lift, resulting in reduction of pumping losses occurring in case of a conventional throttle valve. However, this needs very precise control of lift to match various engine speeds and load conditions.

5. Electronic valve-actuation systems

Due to their precise control, these have the maximum potential for optimising the valve events. Besides, all the valve events including the lift can be directly controlled.

12.2. Applications

The first VVT system in production cars was used by Alfa Romeo in 1980. This system was mechanical. In 1986 Nissan developed NVCS (Nissan Valve-timing Control System), which focused mainly at low and medium engine speeds since engine would be rarely running at the extremely high speeds. This system worked electronically and was used in their VG 30 DE engine which was employed first in their car 300 ZX/300 ZR model in 1987. Next, Honda brought in their VTEC system in 1989, which gave the engine the ability to operate on two completely different cam profiles. One profile was meant for operating the valves at low engine speeds, while the second was for operation at high engine speeds. The first VTEC engine

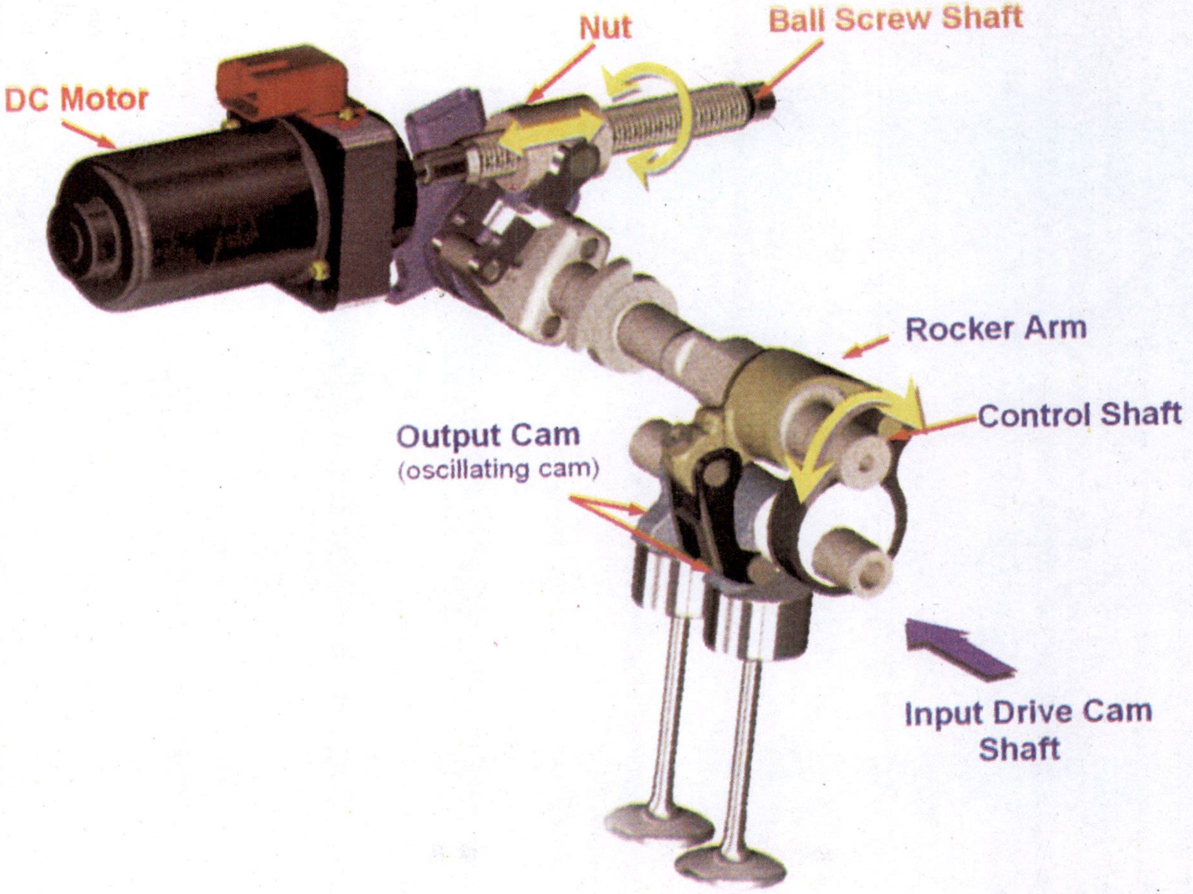

Fig. 4.53: VVEL System for 2008 model Infiniti G37

was Honda's B16A, which was installed in their Civic hatchback. VTEC technology has been used in Honda city car in India and is discussed separately in detail in the next Article.

Another VVT system (VANOS) similar to Nissan's NVCS, was introduced by BMW in 1992. Later in 1998, Double VANOS system was introduced, which is the first VVT system which could provide electronically controlled, continuous timing variation for both the intake and exhaust valves. The latest is the *Valvetronic* system introduced by BMW in 2001. In addition to continuous variation of timing for both the intake and the exhaust valves, this system can also continuously vary intake valve lift, from almost closed to fully open positions, and everywhere in between, the primary means of controlling engine output is transferred from throttle plate to the intake valve train. Reducing the intake duration instead of throttling, pumping losses are reduced, improving fuel economy; while reducing the valve lift asymmetrically on 4-valve engines, swirl is generated in the cylinder, leading to better mixing of air/fuel mixture leading again to fuel economy. However, the throttle plate is not removed, but defaults to fully open position once the engine is running. When the engine is first started, the throttle closes partially to create initial vacuum needed for some engine functions such as emission control. When the engine reads the operating speed, a vacuum pump run by the side exhaust camshaft provides a vacuum source and the throttle plate goes to the fully open position. Besides, the throttle plate also serves as an emergency backup in case of valvetronic system failure.

Valvetronic was not earlier used in direct-injection engines. New generation compact Valvetronic was, however, used in 2009 in BMW's direct-injection and turbo-charged engines such as N55 B30 in MBW5 Series Gran Turismo car. The 2012 BMW M5 vehicle with updated S63 engine has been installed Valvetronic and the 2013 BMW 650 I Gran coupe is also to have Valvetronic-added N63 engine. Thus, such a precise control of intake valves allows the intake charge to be controlled entirely by the intake valves, eliminating the necessity for throttle valve and reducing significantly the pumping loss, which, it is claimed, results in 10% increase in power and fuel economy.

Nissan also introduced its new generation VVEL (Variable Valve Event and Lift) first in US market in late 2007 and later in 2008 on Infiniti G 37 coupe having Nissan VQ 37VHR engine. 2008 model *Infiniti G coupe* employs a variable intake valve lift system (VVEL) with a range of 0.9 to 12 mm, alongwith a wide-band (65°) variable intake timing. The VVEL system is shown in Fig. 4.53. It uses a computer-controlled dc electric motor with ball screw shaft to turn a control shaft in either clockwise or anti-clockwise direction as required by the running conditions. This turning of the control shaft changes the operating relationship between the rocker arm and oscillating output cam, thereby varying the lift. Drive is from an input camshaft through a rocker arm. Intake valve timing is hydraulically controlled by a computer-controlled cam phaser. Exhaust valve timing is however, fixed.

The VVEL improves fuel efficiency most effectively in the low-to-medium operating speed range. Thus it is best suited for the multicylinder and higher displacement engines which typically operate within that range. For such engines, Nissan is now combining VVEL technology with the direct-injection system.

Toyota's *'Valvematic'* system also provides continuous adjustment of valve timing and lift and improves fuel efficiency by using the valve control ratter than the throttle plate control for regulating air/fuel mixture. This system was introduced in 2007 in the Noah and later in 2009 in the ZR engine family used on 'Avensis'. It is simpler in design compared to valvetronic and VVEL.

Borg Warner's CTA (Cam Torque Actuated) variable cam timing technology for upgraded Ford Duratec 3.0 litre V-6 engine, debuted in the 2009 Ford Escape. Unlike traditional cam phasing methods which usually use engine oil pressure to rotate the camshaft, this technology, claimed to be industry-first, utilized the existing torsional energy in the valve trains to achieve this, similar to a hydraulic ratchet. Such cam-phasers operate more quickly at a wider range of engine speeds and temperatures than the traditional ones actuated by oil pressure. This results in improved fuel economy, increased power output and reduced emissions.

In 2013, Borg Warner Morse TEC have introduced CTA phasers with MPL (Mid Position Lock) technology, which employs a hydraulic circuit to passively move the system, allowing increased range of camshaft positioning and having a default stop at an intermediate position within expanded range of travel. With a mid-range start position, advanced combustion strategies, such as LIVC (Live Intake Valve Closing) during part load can be employed without any risk of combustion instability during restart, allowing for additional fuel economy and emission improvement. CTA phasers with MPL lock are claimed to result in fuel economy up to 4% compared with conventional VCT systems.

In 2012, GM also announced their IVLC (Inlet Valve Lift Control) to be used in their new 2.5 litre engine (which is the Gen-III of their Ecotec engine) for the first time, in their 2014 Chevrolet Impala. IVLC provides variable valve lift in addition to continuous variable timing.

The PVLC system has two roller followers built into the rocker arm and a two-section cam lobe—, one profile and follower for low lift and the other profile and follower for the high lift A spring-loaded sliding piston in an oil cylinder in the rocker arm, is actuated by an oil pressure feed from an electronically regulated control valve through a dual-path lash adjuster. When oil pressure is applied, the piston moves in to latch only the low-lift follower to the rocker arm. On the other hand when oil pressure is relieved, the piston moves outward to lock the high-lift follower to the rocker arm. Low lift is applied for most of the operating conditions while high lift is provided for high speed operation and high acceleration. 4.0 mm valve lift is provided during low-lift operation and 10.5 mm lift is there during high lift operation.

13. VTEC TECHNOLOGY

VTEC stands for Variable valve Timing and lift Electronic Control system.

In racing engines, inlet valve is opened earlier and exhaust valve closed later. Large amount of overlap between the inlet and the exhaust valves is especially beneficial at high speeds as this increases the total time available to clear out the exhaust. Earlier opening the intake valve also increases the time for sucking the fuel-air mixture. All this results in maximum power at high speeds [Fig. 4.54 (*a*)].

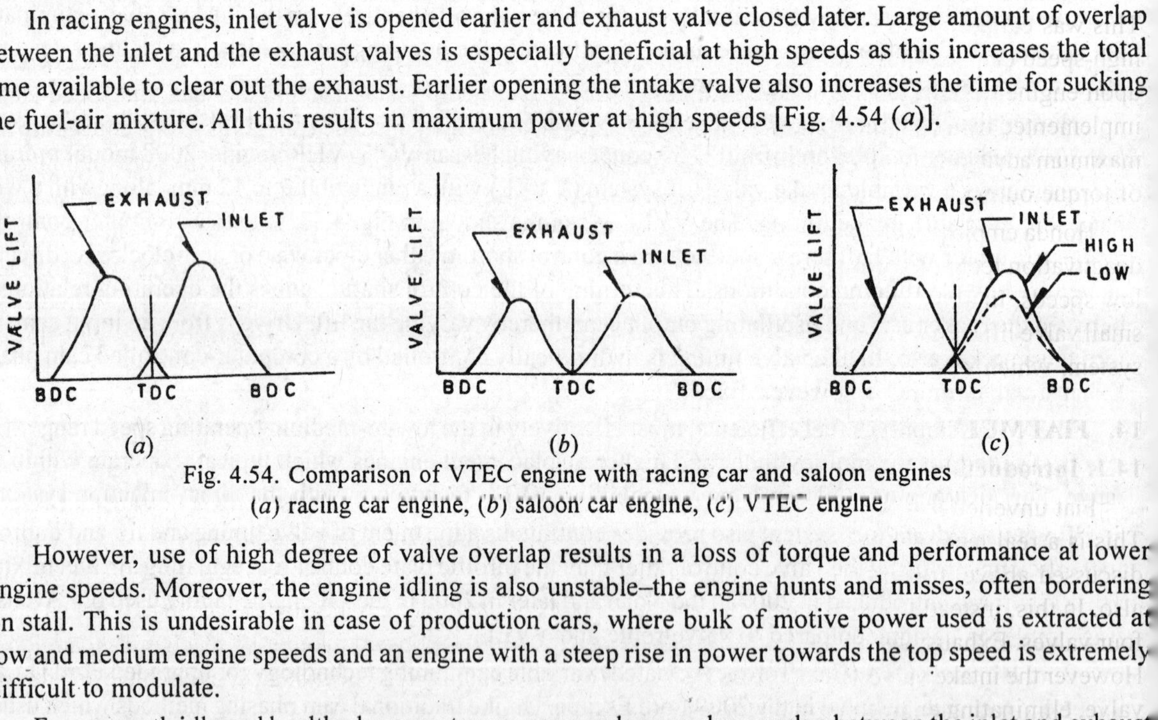

Fig. 4.54. Comparison of VTEC engine with racing car and saloon engines
(*a*) racing car engine, (*b*) saloon car engine, (*c*) VTEC engine

However, use of high degree of valve overlap results in a loss of torque and performance at lower engine speeds. Moreover, the engine idling is also unstable–the engine hunts and misses, often bordering on stall. This is undesirable in case of production cars, where bulk of motive power used is extracted at low and medium engine speeds and an engine with a steep rise in power towards the top speed is extremely difficult to modulate.

For a smooth idle and healthy low rpm torque you need zero valve overlap between the inlet and exhaust valves. With zero overlap, compression is maintained much longer as there is no valve open for the compression to leak out [Fig. 4.54 (*b*)]. Thus a compromise has to be maintained between low rpm torque, smooth engine idle on one side and maximum top-end power on the other.

The Honda VTEC system achieves this compromise by using variable valve timing and lift that is actuated by a separate set of cam lobes. As a result, the vehicle in question has zero valve overlap between the two inlet and exhaust valves at low rpm and high degree of overlap at high rpm [Fig. 4.54 (*c*)].

The VTEC used in Indian 'Honda City' is **SOHC type** that varies timing and lift of only the intake valves. The reason is that in the SOHC engine, the spark plugs have to be placed at an angle to clear the camshaft and also the spark plug tubes are situated between the two exhaust valves, making VTEC on the exhaust impossible. The Electronic Control Unit of the engine does this on the basis of engine speed, load, temperature and road speed of the vehicle. This type of system is economical. The **DOHC type** of VTEC varies the valve timing and lift of both the intake and the exhaust valves for maximum power and effect. This is used in Honda's sports cars, like S2000.

Next version of VTEC was **VTEC-E**, in which at low engine speeds, only one of the two intake valves is allowed to open, increasing the swirl of air/fuel mixture in the cylinder, allowing a very lean mixture to be used. As the engine speed increases, both vales are required to supply sufficient mixture, which is done by a sliding pin as in the regular SOHC/DOHC types, which connects the two valves together at the desired speed. This type has been used in Honda Civic HX in 1996-2000 USA models. Honda also combined the features of SOHC, VTEC and VTEC-E types, called **3-stage VTEC**. In this at low speeds, only one intake valve is used; at medium speeds, both are used while at high speeds, the engine switches to the high-speed cam profile as in regular VTEC. This results in improved both low-speed economy and high-speed efficiency and power.

A further improvement of VTEC was employed by Honda in their 2002 K-series 4-cylinder engines. This was called **i-VTEC**, in which the valve lift and duration are still limited to distinct low-speed and high-speed cam profiles, but now the intake camshaft can advance between 25 and 50 degrees (depending upon engine design) during operation. Phasing depends on combination of engine load and speed and is implemented by a computer-controlled oil driven adjustable cam gear. The cam is fully retarded at idle and maximum advanced at full throttle and low speeds. Obvious result of this improvement is still better optimization of torque output, particularly at low and medium engine speeds.

Honda employed **i-VTEC V6** on their 'Odyssey' in 2004 in which, instead of cam-phasing, their cylinder deactivation technology was used which closes the valves on one bank of three cylinders at light load and low speeds. In a recent version of i-VTEC used in Honda Civic's R-series 4-cylinder engine, there is very small valve lift at low speeds and light loads, alongwith large throttle openings, using a drive-by-wire throttle system, which reduces pumping losses, thereby improving fuel economy.

14. FIAT MULTIAIR TECHNOLOGY

14.1. Introduction

Fiat unveiled its new air management technology named 'MultiAir' at the 2009 Geneva Motor Show. This is a real technological breakthorough and claimed to be much better than any of the VVT systems discussed above. It can be incorporated not only in new engines, it may be fitted in any existing engine also. In this, instead of twin camshafts of a four valves per cylinder engine, a single camshaft opens up all four valves. Exhaust valves are not variable and are opened in the usual manner by mechanical cam lobes. However the intake valves are operated by electro-hydraulic actuation. The system does not have any throttle valve, eliminating engine pumping losses.

14.2. Working

The MultiAir system consists of:

1. A hydraulic-mechanical actuator fitted inside the engine.
2. Hardware and software electronic components located into PCM (Powertrain Control Module), which manage the movement of the intake valve.

The main component for variable valve actuation is the variable valve actuation module which is bolted to the top of the camshaft bearing housing above the intake valves. The intake lobes on the camshaft operate hydraulic pumping elements which provide high pressure oil in the hydraulic chamber for opening the intake valves. The relationship between the camshaft lobe and the intake valves is controlled by a solenoid-valve which is normally open. When this solenoid valve is closed, the oil in the hydraulic chamber, being incompressible, behaves like a solid body due to which the intake valves are lifted as per the shape of the mechanical intake cam. However, when the solenoid valve is open, the hydraulic chamber and intake valves are decoupled, the intake valves do not follow the intake cam and close under the valve spring action. Thus, by controlling solenoid valve opening and closing times, optimum valve-opening schedules can be obtained for different engine operating conditions.

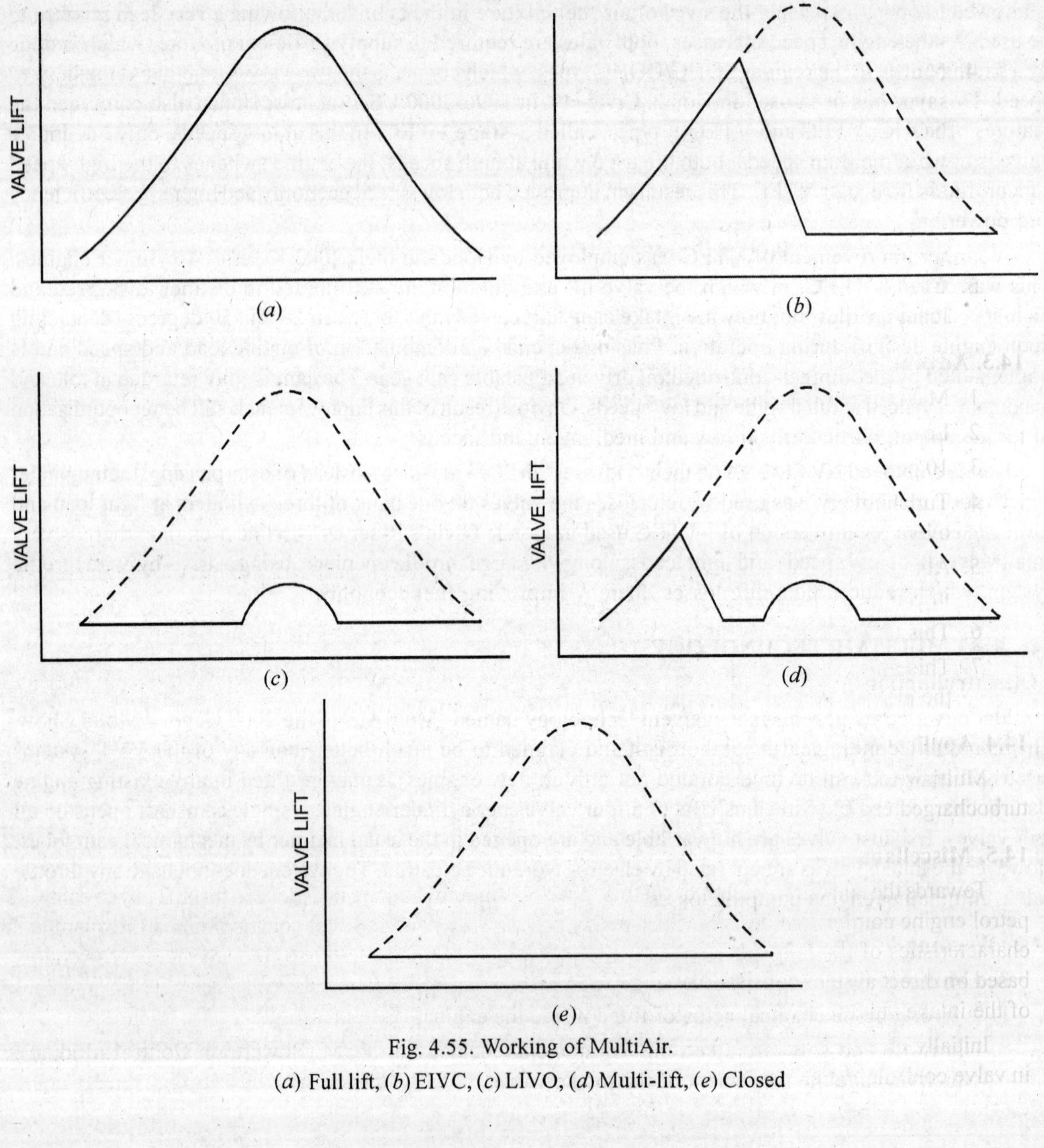

Fig. 4.55. Working of MultiAir.

(*a*) Full lift, (*b*) EIVC, (*c*) LIVO, (*d*) Multi-lift, (*e*) Closed

1. **Full lift:** For maximum power at higher speeds and loads, solenoid valve is always closed which gives full valve opening directly by the mechanical cam which is designed specifically to maximize power at high engine speeds. This mode would be rarely used in everyday driving.
2. **Early Intake Valve Closing (EIVC):** For low speed torque, the solenoid valve is opened near the end of the cam profile, resulting in early closing of inlet valves. This eliminates unwanted backflow into the manifold and maximizes the air mass trapped in the cylinders. The exact timing and lift can be infinitely varied to meet driver requirements.
3. **Late Intake Valve Opening (LIVO):** In this mode, the camshaft lobe lift is *not* transferred to intake valves in the beginning of the lift duration cycle causing the intake valves to open late. The exact valve lift timing can be varied infinitely within the full profile of the camshaft lobe. The lift profile would follow the camshaft lobe profile for the time interval during which hydraulic link is complete. LIVO provides lower emissions and higher efficiency at *lower loads or idle* conditions.
4. **Multi-lift:** The second and the last actuation modes as described above can be combined in the same intake stroke, thus making it a so called *multi-lift mode*. In this the hydraulic connection between the camshaft lobe and intake valves is closed early and then re-opened later in the cycle, which results in *longer duration, but smaller amount* of valve lift. This causes higher velocity of air flow into the cylinder over a longer period of time which produces increased turbulence and thus, higher rate of combustion. This mode may be used in mixed driving of acceleration and deceleration at moderate engine speeds.
5. **Closed:** In this mode the intake valves are kept fully closed by means of springs and the camshaft lobes are not operative at all.

14.3. Advantages

1. Maximum power increased by upto 10 percent.
2. Low speed torque improved by upto 15 percent.
3. 10 percent reduction in fuel consumption and CO_2 emissions due to elimination of pumping losses.
4. Turbocharged downsized engines using Multiair technology can achieve upto 25 percent fuel economy over conventional naturally aspirated engines with similar performance.
5. About 30 percent reduction in unburnt hydrocarbons and CO_2 emissions and upto 60 percent reduction in NOx emissions.
6. This technology may be applied to all I.C. engines, irrespective of the fuel used.
7. This technology can be adapted even for diesel engines to reduce NOx emissions and to increase the effectiveness of the particulate filters.

14.4. Application

Multiair technology is being used in the Fiat's 16-valve 1.4 litre family of naturally aspirated and turbocharged engines and the first car with such an engine is 'Alfa NiTo.

14.5. Miscellaneous

Towards the end of the twentieth century, the engine manufacturers had realized that the key to controlling petrol engine combustion and therefore performance, emissions and fuel consumption is the quantity and characteristics of the fresh air charge in the cylinders. A fundamental breakthrough in air mass control is based on direct air charge metering at the cylinder inlet ports by advanced electronic actuation and control of the intake valves, while maintaining a constant natural upstream pressure.

Initially, electromagnetic actuation concept was favoured due to very high flexibility and dynamic response in valve control, but it was soon realized that this concept is not fail-safe and absorbs high energy and that these drawbacks could not be overcome completely. This led to shifting of emphasis on electromechanical concepts based on valve lift variation and camshaft phasers, some of which have bean discussed as VVT in previous articles. The main limitation of these systems is low flexibility in valve-opening schedules and not so good a dynamic response, besides a lot of power consumption.

Fiat Group selected the electro-hydraulic actuation on account of its relative simplicity, low power requirements, intrinsic fail-safe nature and low cost.

15. BORGWARNER'S VEMB

Valve Event Modulated Boost (VEMB), divides the engine exhaust into two phases with the early stage (called blow-down) going directly to the turbocharger while the later stage (called scavenge) being configured to bypass the turbine and go straight to the exhaust catalyst. A phaser-controlled concentric type camshaft is used, the alternate lobes on the same achieving the variable phase change between the blow down and the scavenge stages and opens the selected exhaust valves in such a way that various stages pass through the bifurcated exhaust port with the desired phase difference. This phasing isolates the backpressure required to extract high turbine work from the engine from the displacement phase of the exhaust stroke. This results in significant reduction in exhaust pumping work and the residual-gas fraction.

The boost obtained can be controlled by appropriately phasing the two exhaust valves relative to each other. Cooled EGR is used at the scavenge port to burn the excess hydrocarbons, which are concentrated in that location. Various advantages claimed for VEMB are:

1. 2 percent gain in fuel economy at 2000 bar,
2. Due to elimination of typical high back pressures, compression ratios higher by at least one can be used with consequential 2% fuel economy occurring additionally.

16. CAMLESS ENGINES

In a camless engine, each engine valve is operated individually by an actuator. There is no camshaft. The advantages claimed for such an engine are:

1. Infinitely variable valve timing which results in improvements in engine performance, fuel efficiency, exhaust emissions.
2. Cylinder deactivation is possible.
3. Less oil required for lubrication due to absence of traditional camshaft valve system.
4. Cold start wear of the valve train hardware is also minimal.
5. Higher overall valve train efficiency due to elimination of frictional losses of camshaft mechanism and the weight of the mechanism.

Chinese automaker, Qoros debuted a camless 'FreeValve' engine in a ''Qoros 3' concept car at the Beijing Auto show in April 2016. Lightweight, low-cost, low-friction Free Valve engine concept functions without camshafts and allows valves to open and close almost instantaneously.

In the Free Valve engine, the valves are opened and closed by a system consisting of electronic actuators with pneumatic and hydraulic controls. An electrical pulse is sent to the valve solenoid, which causes the valve to open. Air and hydraulic pressure jointly stabilize the valve-opening operation, damping any oscillations caused by the solenoid. In addition, valve lift is also determined by the air and the hydraulic pressures. On removing the electrical signal from the solenoid, the valve is closed by the hydraulic pressure. Various advantages claimed for this engine are:

1. A much more efficient fuel-air mixture provided on account of almost instantaneous complete opening and closing of valves.
2. Free Valve's 2 intake and 2 exhaust valves can be controlled individually and slightly differing designs

of exhaust ports of each cylinder, permits 100 percent of the combustion residuals to be scavenged from the cylinders.

3. Simpler construction. Complete camshaft assembly including the timing assembly, gears, chain/belt and cover are eliminated. Besides only one valve per cylinder feeds the turbocharger which means exhaust outlet pressures can be controlled more precisely than with a Watergate, which is eliminated alongwith its associated controls and plumbing. Further, the pre-catalytic converter can also be done away with. On cold start, this engine's exhaust valves can be opened so as to by-pass the turbocharger, resulting in heating-up the catalytic converter quickly.
4. Smaller and lighter engine. A reduction of 20 kg in the Free Valve Qoros engine is claimed.
5. 30% increase in volumetric efficiency over single-and dual over head cam engines of the same displacement.
6. Compared to the 'Qoros 3' hatchback with camshaft engine, 47% increase in maximum power output, 45% increase in maximum torque and 15% reduction in fuel consumption was achieved with the 1.6 litre turbocharged camless engine, as claimed by the company.
7. Problem of slipping/breaking timing belts is also eliminated since there is no such belt in the camless engine. In case of the failure of the valve-control system, the engine simply stops running and the valves remain in the fully closed position, pistons.
8. Less costly than comparable camshaft engines due to:
   (*i*) elimination of many components.
   (*ii*) saving in installation of the eliminated parts.
   (*iii*) use of port-injection by Free Valve engine, which is itself cheaper than direct injection. Moreover, less expensive fuel pumps and regulators can be used in this due to lower injection pressures required in case of port injection.

17. VALVE TROUBLES

The main troubles occurring in the engine valve (poppet type) are discussed below, along with their possible causes and steps to avoid their recurrence.

1. Burning of valve face. This may be caused due to:

(*i*) Poor valve seating that allows the high temperature and high pressure gases to leak between the valve and the seat. The causes of poor seating may be:

(*a*) Insufficient valve lash, which may be reset.

(*b*) Distortion of the valve seat, which may be checked and refaced if required.

(*c*) Hard carbon deposits caught between the valve and the valve seat; these deposits being caused by loosening from the combustion chamber. Positive valve rotators help to avoid this defect.

(*d*) Sticking valve stem due to hard carbon and varnish deposits arising on account of breakdown of the fuel

Fig. 4.56. Burnt valve face.
(Courtesy—Sealed Power Corporation, U.S.A.)

and oil on the hot valve. Because of this valve is not able to seat property. The remedy is to clean up the valve stem and the guide.

(*e*) Excessive run-out of the valve stem, which may be checked; if necessary, the valve guides and seats have to be reconditioned alongwith the straightening of the valve stem or replacing the same if required.

(*ii*) Lean mixture; the carburettor should be correctly tuned.

(*iii*) Detonation and preignition; the carburettor should be reset properly.

(*iv*) Incorrect valve timing, which may be checked and reset.

2. Valve head deformation. This may happen due to mechanical overload caused due to any contamination wedged in between the valve head and the valve seat or on account of thermal overload caused due to:

1. Abnormal combustion.
2. Lesser valve clearance.
3. Attempt to get more power by tuning.

1. Check and remove any contamination from the valve seat/face.
2. Check and rectify the fuel injection system.
3. Reset the valve clearance correctly.

3. Necking of valve stem. This defect may be caused due to leakage of gases at the valve seat. These high speed, high temperature gases are also corrosive. So they tend to erode and corrode the valve stem. The rate of corrosion almost doubles for every 14°C rise of temperature. These erosion and corrosion effects combine to cause necking of the valve stem. The defective valve should be replaced and the seat should be properly examined and reground as to make a proper contact with the new valve.

Fig. 4.57. Valve stem necking.
(Courtesy—Sealed Power Corporation, U.S.A.)

4. Valve stem scuffing. This may be due to geometrical irregularities or over-speeding. Various specific causes and their remedies are:

1. Misalignment of the valve guide and the seat, correct realignment should be done for rectification.
2. Bent or slanted valve, which may need replacement.
3. A loose valve seat insert also causes misalignment of the valve guide. Install the seat insert as given in the Manual.
4. Overspeeding can cause breakdown of the tribological system, *i.e.*, breakdown of the lubricating oil film between the valve stem and guide, resulting in overheating and thereby, the scuffing.

5. Distortion of the valve stem. Probable causes and their remedies are:

1. Incorrect valve clearance leading to valve impacts. The clearance may be reset correctly.
2. Incorrect valve timing, which may be reset as given in the Manual.
3. Jumping of the timing belt or chain due to faulty clamping device, which may be replaced.

Besides above, overspeeding of the engine may also result in insufficient return speed of the valve springs due to which the piston and the valves may collide. Thus always avoid over-speeding to prevent valve stem distortion.

6. Valve face wear. The probable causes for this are :

(*i*) Weak springs, which may be checked and replaced if found defective.

(*ii*) Excessive guide wear causing too much clearance between the valve stem and the guide. The guide will have to be replaced in this case.

(*iii*) Valve seat may be too narrow, causing high stresses. The narrow seat may be reground to make it wider.

(*iv*) Leakage at seat, which may necessitate regrinding and matching it properly with the valve face to provide a good contact.

7. Valve seat wear. This may occur due to following reasons:

1. Misalignment of the valve guide and the seat, which may be realigned correctly.
2. Lesser valve clearance, which may be reset as per specification.
3. Knocking, the cause(s) for which may be thoroughly investigated and remedied.
4. Excessive mechanical loading of the valve seats due to spurious valve springs or camshafts. If found so, the same have to be replaced.

8. Valve stem and guide wear. When the valve stem and the guide wear giving excessive clearance and the bell-mouth shape at the guide outlet (Fig. 4.58), the more probable cause is the lack or deficiency of lubrication of the valve guide, which may be checked and properly set.

Fig. 4.58. Valve stem and Guide wear.
(Courtesy—Sealed Power Corporation, U.S.A.)

9. Valve cracking or breakage. This may be due to any or more of the following causes:

(*i*) Excessive lash in the valve mechanism, which should be checked and reset.

(*ii*) Excessive stem to guide clearance. Replace the guide.

(*iii*) Loose seat insert, which may be properly checked and replaced if found defective.

Fig. 4.59. Valve face cracking
(Courtesy—Sealed Power Corporation, U.S.A.)

Fig. 4.60. Broken valve
(Courtesy—Sealed Power Corporation, U.S.A.)

(*iv*) Defective valve springs. Both weak or broken and unduly stiff springs will cause this defect. Check them and replace if necessary.

(*v*) Excessive vibrations and loading due to overspeeding. This requires only change in driving habits.

10. Noisy valve operation. This may be due to:

(*i*) wrongly adjusted tappet, the noise may be adjusted properly.

(*ii*) (in case of hydraulic tappet) the noise may be due to dirt having accumulated inside the tappet; it may be dissembled, cleaned and reassembled.

(*iii*) sticking valves due to gumming deposits. The valve train has to be dismantled cleaned and properly reassembled.

(*iv*) improper rocker arm adjustment, which may be readjusted according to specifications.

(*v*) worn out components of valve train. The engine train may be inspected and defective parts suitable reconditioned or replaced.

18. ROTATING CYLINDER VALVE ENGINE

U.K. based RCV Engines has unveiled in February 2006, a prototype 125 cc scooter engine Rotating Cylinder Valve (RCV), which is claimed to be ideally suited for motor cycle/scooter engines of upto 250cc size. It gives emission levels and fuel consumption of a four-stroke engine, with the performance and manufacturing cost of the two-stroke type. It can run on alternative fuels, *e.g.*, kerosene, bio-fuels, etc. Besides, it is suitable for advanced technologies like variable valve timing, variable compression ratio and super-charging, which can further improve performance and emission levels.

While most components of this engine are similar to those of a conventional four-stroke engine, the main difference is in the cylinder which is mounted on bearings and rotates at exactly half the crankshaft speed and is driven from the crankshaft either through a gear train or toothed belt drive system. A single large cross-sectional area port and combustion chamber is combined within the cylinder, the entrance to which forms the rotating cylindrical valve together with the openings of the external intake and exhaust ports.

Inspite of many advantages claimed, this system has not found any application in any production model so far.

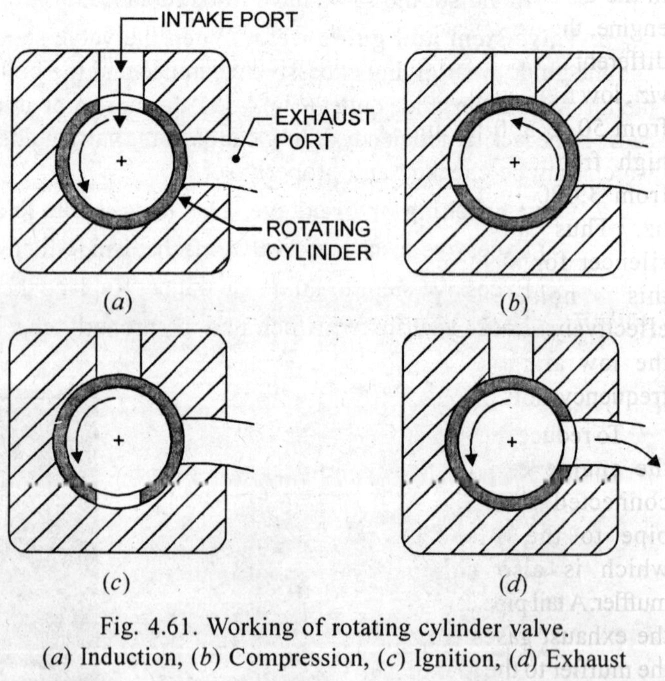

Fig. 4.61. Working of rotating cylinder valve.
(*a*) Induction, (*b*) Compression, (*c*) Ignition, (*d*) Exhaust

19. EXHAUST SYSTEM

Fig. 4.62 shows exhaust system for a modern car engine. The exhaust pipe from the engine leads to a catalytic converter. The treated gas from the converter goes to muffler, to reduce noise, from which the exhaust goes to the atmosphere through the tail pipe. The whole unit is connected to the car by means of a number of clamps and hangers.

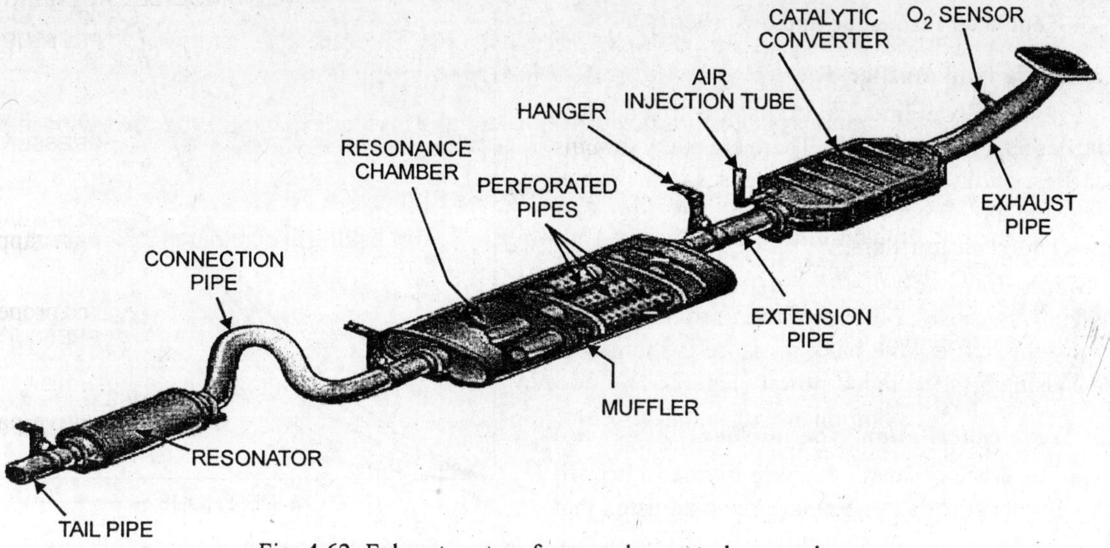

Fig. 4.62. Exhaust system for a modern petrol car engine

20. MUFFLERS

When the exhaust valve opens, high-pressure exhaust gas is released, which causes a pressure wave in the air causing an explosion. Since high pressure gases are released rapidly one after the other in an engine, the explosions occurring very fast combine together to form a steady noise. This noise consists of different notes of various frequencies. The predominant notes out of these have been found in two groups, *viz.*, low frequency notes from 50 to 500 hz and high frequency notes from 3,000 to 10,000 hz. Thus any ideal silencer for absorbing this noise must effectively reduce both the low and the high frequency notes.

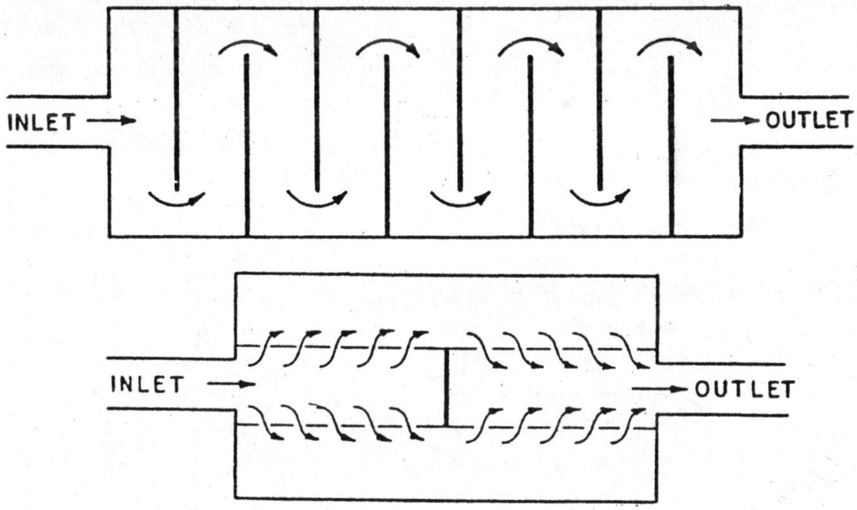

Fig. 4.63. Baffle type mufflers.

To reduce the noise, the engine exhaust is connected via exhaust pipe to the silencer, which is also called muffler. A tail pipe carries the exhaust gases from the muffler to the rear or side of the vehicle near the near wheel. The mufflers used are of the various types.

20.1. Types of Mufflers

1. Baffle type
2. Wave cancellation type
3. Resonance type
4. Absorber type

5. Combined resonance and absorber type

20.2. Baffle type muffler

It is generally cylindrical in shape with a number of baffles spot welded inside. There are many designs of baffles, but the principle in all cases is the same, *i.e.* closing any direct passage for the gas. Fig. 4.63 shows two such mufflers.

Major drawback of the mufflers is their low efficiency. Because of the restriction provided to the flow by the baffles, the back pressure is increased, thus causing loss in engine power (Fig. 4.64).

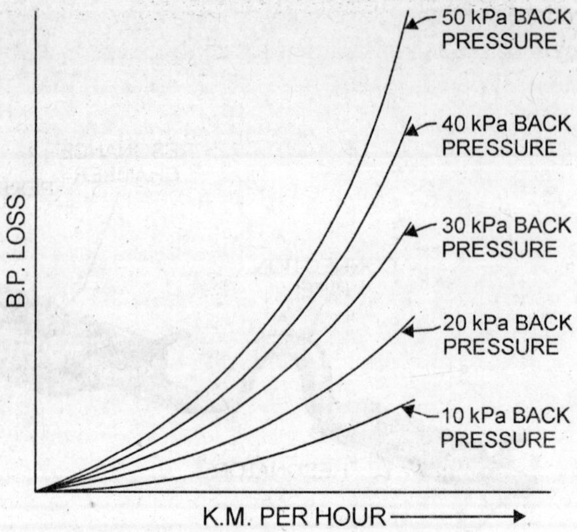

Fig. 4.64. Variation of engine B.P. loss with back pressure.

20.3. Wave cancellation type muffler

In this type, exhaust gases are divided into two parts. The lengths of these paths are so adjusted that after they come out of the muffler, the crests of one wave coincide with the troughs of the second wave, thus cancelling each other and reducing the noise to zero theoretically. This is achieved if the lengths of the two passages differ by half the wavelength. In practice this type of muffler does not eliminate noise completely, because this is possible only at one frequency for which muffler is designed, whereas the noise is a combination of different frequencies. However, appreciable noise reduction is achieved.

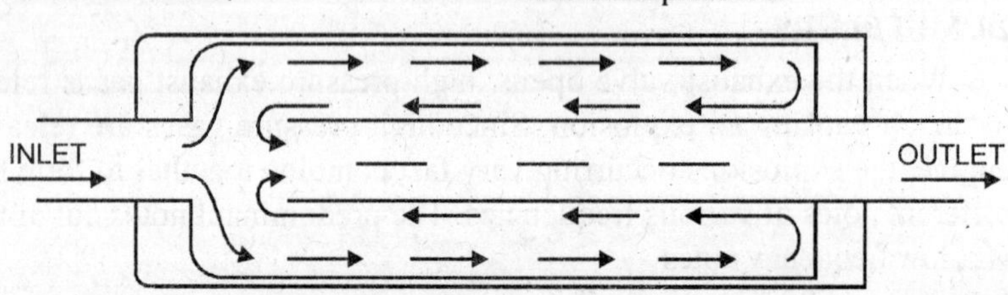

Fig. 4.65. Wave cancellation type muffler.

One such type of muffler is shown in Fig. 4.65. In this the resistance to the main gas flow is very small as compared to the baffle type.

20.4. Resonance type muffler

These are also called Helmhotlz type, after the person who originated the idea. It consists of a number of Helmhotlz resonators in series, through which a pipe containing access ports passes (Fig. 4.66). The exhaust gases flow through this type and thus experience no resistance. Series of resonators eliminate the fundamental and higher harmonics of the engine noise.

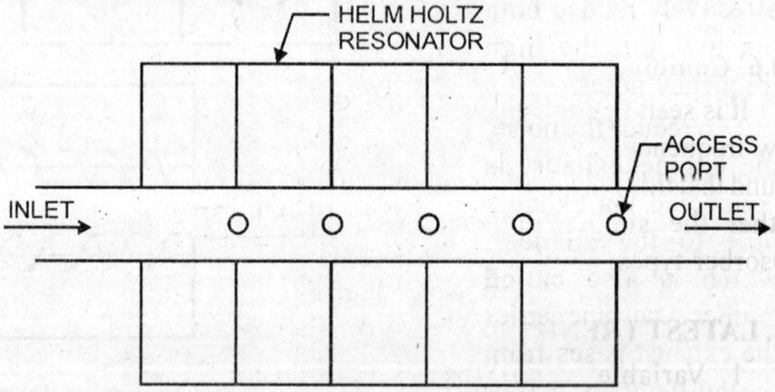

Fig. 4.66. Resonance type muffler.

20.5. Absorber type muffler

The sound absorbing material, usually fibre glass, is placed in this case around the perforated tube through which the exhaust gases pass. The sound is reduced by conversion into heat by the sound absorbing material. Such mufflers are commonly known as glass-packs and are quite common in performance and racing cars due to their little resistance to flow. These essentially attenuate higher frequency sounds and produce the deep, throaty sound usually associated with a high performance engine. A common design is a cylindrical can with a perforated tube which is surrounded by packing material. These mufflers may be straight-through type (Fig. 4.67) or reverse flow type (Fig. 4.68).

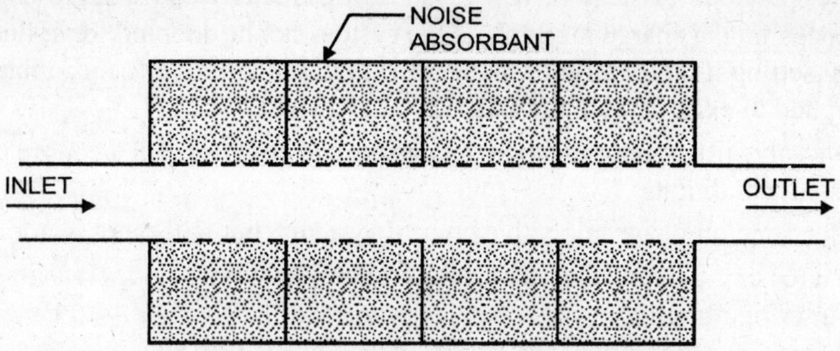

Fig. 4.67. Straight through absorber muffler.

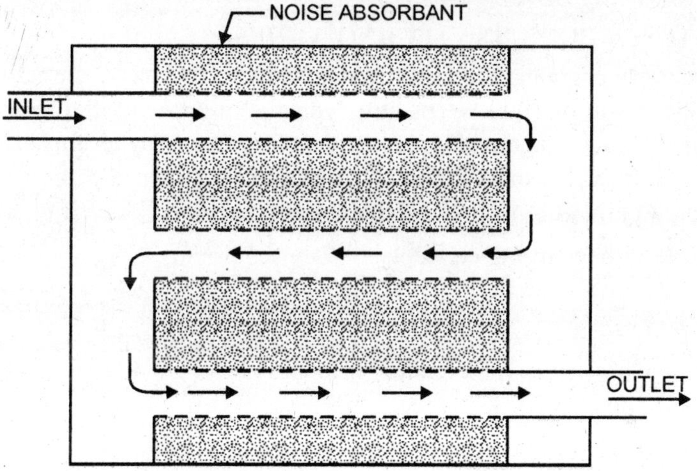

Fig. 4.68. Reserve flow absorber type muffler.

20.6. Combined resonance and absorber type muffler

It is seen that the absorber type muffler, has a drawback in that it is not efficient in reducing noise of low frequency. To obviate this defect, this is combined with a resonant chamber (Fig. 4.69). It has been found that this type is more efficient than either the simple resonance or the absorber types.

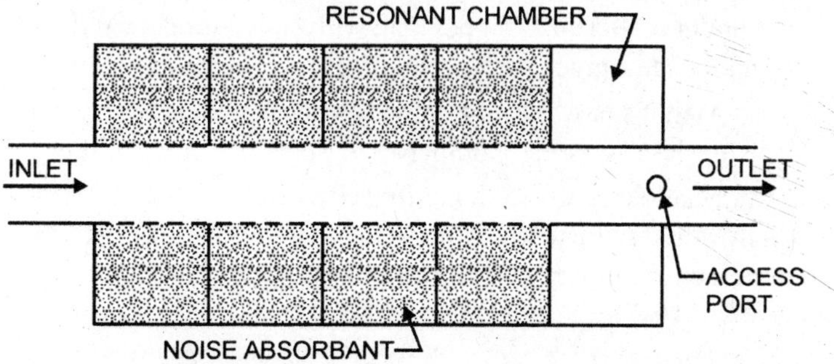

Fig. 4.69. Combined resonance and absorber type muffler.

21. LATEST TRENDS

1. Variable valve lift is being increasingly used. There are also less expensive alternatives, such as GM's Twin Port system which disables one of the two intake ports on its Opel 14 simulating the effects of a two-stage variable lift system for less cost.

2. Variable cam timing is also the favourite of the present engine designers. Delphi Corp. is moving from its original spline-type devices to vane-type cam phasers. The best situation is to combine the variable valve lift and cam phasing.
3. Mechanically operated two-stage valve lift systems are much cheaper and provide significant benefits in the low-speed intake charge motion and the system can be adopted for cylinder deactivation. At the lower lift setting it offers air motion benefits that will improve fuel economy. Moreover, it can be easily applied to existing cylinder heads.
4. Fully variable valve lift systems are more expensive and complex, but are more precise and effective under almost all conditions.
5. Work is being done on electronic valve control systems, but still there is a long way to go.
6. Gradually increasing power density of auto engines is putting an increasing strain on bearings. Although puttering aluminium onto bearings increase their capacity to carry loads, yet is quite expensive. Dana Corp. has developed a coating, called RP 100, which imparts similar benefits, yet is cheaper.

22. MAHLE'S EFFICIENCY-BOOSTING TECHNOLOGIES

1. Mahle's demonstrator engine at Frankfurt Motor Show, 2007 is a 1.2 litre 3-cylinder petrol engine which produces specific power of 120 kW per litre, typical of a current 2.5 litre engine. This 50 percent reduction in the engine capacity would result in more than 20 percent higher fuel efficiency. This has been achieved by a number of different technologies consisting of:

(*i*) Reduced mass and internal friction (pistons, bearings, etc.).
(*ii*) Two alternative turbocharging concepts.
(*iii*) Forged pistons.
(*iv*) DOHC.
(*v*) Direct injection with injection-tip cooling.
(*vi*) Air-water inter-cooler.
(*vii*) Cooled high-load exhaust gas recirculation.

2. The piston-ring friction has been reduced by coating the piston pins with a diamond-like carbon coating, cylinder inside surfaces with Nikasil and the piston rings with PVD (Physical Vapour Deposition).

3. Mahle's light-weight valves have a hollow body. The Valve cone and the disc are produced using a multistage metal-forming process and stem is made of steel pipe. Individual components are joined using high-precision laser welding, which reduces the mechanical finishing by 25 percent compared to conventional forged-steel valves. Hollow design allows valve cooling by filling sodium. Such valves can stand temperatures 50-100°C higher than with conventional valves, apart from reduced mass and increased strength in high-speed ranges while reducing wear and noise.

4. Mahle's new petrol-engine 'Evotec' pistons are lighter due to use of narrow walls on the side where there are less stresses. Compared to standard pistons, these asymmetric ones are 20 percent lighter.

5. Mahle expects to make first diesel-passenger-car forged steel piston applications by 2012 to 2015, which would withstand higher loads and temperatures compared to aluminium pistons for which load and temperature limits are 20 MPa (200 bar) and 400°C.

TABLE 4.1. COMPARATIVE ENGINE DATA OF SOME OLD INDIAN AUTOMOBILES

S. No.	*Item*	*Hindustan Ambassador Mark II*	*Maruti 800*	*Fiat 1100*	*Jeep Universal*	*Dodge/ Fargo Model 89M4 (Petrol Type)*	*Ashok Leyland Comet Passenger*	*Tata Diesel Vehicles 1210*	*Swaraj Mazda (T-3500 series SL engine)*
1.	Type	I-Head	I-Head S.O.H.C.	I-Head	F-Head	L-Head	I-Head	I-Head	I-Head
2.	No. of cylinders	4	3	4	4	6	6	6	4
3.	Firing order	1, 3, 4, 2	1, 3, 2	1, 3, 4, 2	1, 3, 4, 2	1, 5, 3, 6, 2, 4	1, 5, 3, 6, 2, 4	1, 5, 3, 6, 2, 4	1, 3, 4, 2
4.	Bore, mm	73.02	68.5	68	79.3	87.2	103.88	92	100
5.	Stroke, mm	88.9	72.0	75	111.12	114.3	120.56	120	110
6.	Cubic capacity, c.c.	1489	796	1089	2199	4106.56	6075	4788	3455
7.	Compres sion ratio	7.2 : 1	8.7 : 1	7 : 1	7.4 : 1	6.6 : 1	16 : 1	17 : 1	18 : 1
8.	B.H.P.	50 at 2000 R.P.M.		43 at 4800 R.P.M.	75 at 4300 R.P.M.	110 at 3600 R.P.M.	110 at 2400 R.P.M.	112 at 2800 R.P.M.	86.5 at 3000 R.P.M.
9.	Gross torque	11.04 kgm at 3000 R.P.M.		7.25 kgm. at 3200 R.P.M.	15.7 kgm. at 2000 R.P.M.	26.95 kgm at 1400 R.P.M.	37.6 kgm at 1600 R.P.M.	30 kgm at 1800 to 2000 R.P.M	N.A.
10.	Cylinder Liners	Originally no separate liner fitted	Originally no separate liner fitted	Originally no separate liner fitted	Originally no separate liner fitted	Originally no separate liner fitted	Prefinished dry, press fit, flange located type	Dry type, press fit	Dry & loose
11.	Piston	Aluminium alloy concave crown, split skirt	Aluminium alloy with cavity	Aluminium alloy oval tapered, split skirt	Aluminium alloy cam ground, T-slot	Aluminium alloy split skirt	Aluminium alloy spheroidal cavity	Aluminium alloy, with cavity; cam and taper ground	Aluminium alloy with cavity
12.	Compres sion Rings	3	2	1	2	2	3	3	2
13.	Oil Control Rings	1	1	2	1	2	2	1	1
14.	Gudgeon pin connection	Clamped in connecting rod	Floating	Floating	Clamped in connecting rod	Fully floating	Fully floating	Fully floating	Floating
15.	No. of main bearings	3	4	3	3	5	7	7	5
16.	No. of camshaft bearings	3	4	3	3	3	3	4	4
17.	Camshaft drive	Chain	Belt	Chain	Single helical gear	Chain	Single helical gear	Single helical gear	Gear

S. No.	*Item*	*Hindustan Ambassador Mark II*	*Maruti 800*	*Fiat 1100*	*Jeep Universal*	*Dodge/ Fargo Model 89M4 (Petrol Type)*	*Ashok Leyland Comet Passenger*	*Tata Diesel Vehicles 1210*	*Swaraj Mazda (T-3500 series SL engine)*
18.	Valve type	Poppet	Poppet	Poppet	Poppet	Poppet	Poppet	Proppet	Poppet
19.	Valve mechanism	Overhead	Single overhead camshaft	Overhead	Overhead inlet and side exhaust	Side	Overhead	Overhead	Overhead
20.	Valve seat	Originally no separate valve seat insert is fitted	Originally no separate valve seat insert is fitted	Cadmium steel insert	Eatonite E.M.S. 58 insert for exhaust	Cadimium steel insert	Replaceable on both inlet and exhaust	Replaceable on both inlet and exhaust	Separate seat insert
21.	Valve seat angle	45°	45°	45°	45°	45°	30°	45°	Inlet 45°, Exhaust 30°
22.	Valve timing	Inlet open 5° B.T.D.C., closes 5° A.B.D.C. Exhaust opens 40° B.B.D.C., closes 10° A.T.D.C.	N.A.	Inlet open 16° B.T.D.C., closes 56° A.B.D.C. Exhaust opens 56° B.B.D.C., closes 16° A.T.D.C.	Inlet open 13° B.T.D.C., closes 45° A.B.D.C. Exhaust opens 47° B.B.D.C., closes 20° A.T.D.C.	Inlet open 12° B.T.D.C., closes 50° A.B.D.C. Exhaust opens 51° B.B.D.C., closes 60° A.T.D.C.	Inlet open 10° B.T.D.C., closes 50° A.B.D.C. Exhaust opens 46° B.B.D.C., closes 14° A.T.D.C. closes 14° 402 A.T.D.C.	Inlet open 34° 562 B.T.D.C., closes 49°442 A.B.D.C. Exhaust opens 59°322 B.B.D.C., closes 14° A.T.D.C.	Inlet open 19° B.T.D.C., closes 45° A.B.D.C. Exhaust opens 52° B.B.D.C.,

THEORY QUESTIONS

1. Describe clearly the function and construction of a crankshaft. Discuss also the materials and methods of manufacture for these.
2. Explain the function of a variation damper. Discuss its constructional details with the help of a neat sketch.
3. Explain different requirements of engine bearings.
4. What do you understand from the term valve timing? Describe the valve timing diagram in detail.
5. Explain fully the construction of a poppet valve with the help of a sketch.
6. Discuss thoroughly the operating conditions to which the exhaust valve is subjected and consequently the requirements of materials for the same.
7. Describe suitable materials used for the inlet and the exhaust valves of automobile engines.
8. Discuss how sodium metal is used to cool an exhaust valve. Illustrate your answer with a neat sketch.
9. Discuss in detail the construction and materials for valve seats and guides.
10. Describe with a neat sketch how a valve spring is retained by a split ring. Explain also the function of this spring.
11. Describe with the help of neat sketches, various types of valve actuating mechanisms for automotive engines.
12. make a detailed comparison of valve operating mechanisms using rocker arm with the one using overhead camshaft.
13. With the help of a neat sketch discuss in detail the constructional features and working of a valve-actuating

mechanism widely used in modern car engines.

14. Describe various factors for better design of valve train.
15. Discuss various types of drives for operating engine camshafts.
16. Discuss the construction and working of a hydraulic valve lifter. What advantage does it possess over the simple conventional type?
17. Describe in detail various types of variable valve timing technologies. Discuss also their practical applications.
18. What is Fiat's 'Multiair technology'? Discusses its working, advantages and application.
19. Write a brief note on camless engines.
20. Discuss various valve troubles, their causes and the steps to avoid their recurrence.
21. Write a detailed note on rotating cylinder valve engine.
22. With the help of suitable diagrams, explain fully various types of mufflers.
23. Discuss Mahle's various technologies being developed to increase engine efficiency.

QUESTIONS FOR VIVA

1. What is the function of crankshaft?
2. Name the main parts of a crankshaft.
3. What is the function of main journals?
4. What is 'crank-throw'?
5. What determines the turning effort of a crankshaft?
6. What is a crank web?
7. State the type of load on the crankshaft web.
8. What results if the crankwebs are made unduly large?
9. State the necessity of counterweights in a crankshaft.
10. What are the advantage and the disadvantage of attaching separate counterweight compared to integral type?
11. Name the components that are mounted at the front end of a crankshaft.
12. Which component is mounted on the rear end of a crankshaft?
13. What is the function of a thrust bearing on a crankshaft?
14. What are the approximate relations of the crankpin and the main journal diameters with the cylinder bore?
15. Apart from energy balancing, what other purposes are served by the engine flywheel?
16. State the materials used for engine flywheel.
17. How are the crankshafts manufactured?
18. State the properties required for the crankshaft material.
19. State any one material for each of the forged and the cast crankshafts. Give chemical composition of such materials.
20. Compare briefly the forged and the cast crankshafts.
21. What is the function of a vibration damper?
22. State various requirements of engine bearings.
23. Name the three materials used for engine bearings.
24. What is the composition of babbit metal?
25. State the merits and demerits of aluminium bearings.
26. Why sometimes bearing shells of different materials are used on the same bearing journal?
27. What corrodes an engine bearing running in engine oil?
28. What is a main bearing?
29. How many main bearings are there in a 4-cylinder automobile engine?
30. Name the two types of main bearings.

31. What is a white metal? Give typical composition of such a metal.
32. How do you compare copper-based alloys with babbit metals for use as main bearing?
33. What is the main disadvantage of a copper-based alloy to be used for bearings?
34. What is a valve-timing diagram?
35. Theoretically an inlet valve should open exactly at T.D.C. and close at B.D.C. Similarly an exhaust valve should open at B.D.C, and close at T.D.C. But in practice, this is not done. Why?
36. What is valve overlap? State its advantages.
37. What are the disadvantages of excessive valve overlap?
38. State the two considerations on which the timing the exhaust valve opening depends.
39 Why is a poppet valve so-called?
40. A poppet valve is sometimes called a mushroom valve. What is the reason behind this?
41. Name different types of cylinder valves.
42. State the advantage of a poppet-valve.
43. Why are inlet valves generally kept larger than the exhaust valves?
44. What happens if the valve lift is less than its optimum value?
45. What is the likely result if the valve lift is excessive?
46. What are the commonly used valve face angles for the inlet valves?
47. State the value of the commonly used face angle for exhaust valves.
48. What purpose is served by slightly varying the face angles of a valve and its seat?
49. What is the effect of making the combustion side of the valve like a dish, on the compression ratio?
50. How does valve head diameter affect the engine performance?
51. Why is the valve weight an important factor for designing valves of high-speed engines?
52. State various loads and conditions to which exhaust valve is subjected.
53. State the requirements of exhaust valve material.
54. Which material is used for inlet valves? State its chemical composition.
55. What are 'austinitic' and 'precipitation hardening' steels? Where are these used?
56. Why are exhaust valves sometimes made of two different materials welded together?
57. Give the composition of a 'nimonic alloy'. What are the advantages and disadvantage of such an alloy for use as exhaust valve?
58. What is a 'stellite' alloy? Where is it used in automotive engines?
59. Why are the engine valves sometimes coated with aluminium?
60. What is the use of chrome-plating the stems of engine valves?
61. What happens to the valve if carbon particles are trapped between the valve and its seat?
62. What is the effect of excessive surface finish of the valve stem?
63. What happens if the surface of the valve stem is rough?
64. Why sodium cooling is used only for exhaust valves?
65. What are the highest exhaust valve temperatures reached in modern automobile engines?
66. At what temperature does the sodium metal melt?
67. What are the common valve seat angles?
68. Why separate valve seats are being used in cylinders?
69. What material is used for separate valve guides?
70. What is the function of a valve guide?
71. Why are separate valve guides provided sometimes?
72. Why are the valve springs sometimes shot-peened?
73. Name various valve-actuating mechanisms.
74. Which type of valve-actuating mechanism is used most commonly in modern automobiles? What are its advantages?

75. Which type of valve operating mechanism is the cheapest and also gives the lowest engine height?
76. Which valve-actuating mechanisms have become obsolete and why?
77. What are the relative advantages and disadvantages of a side camshaft and an overhead camshaft for valve actuation?
78. What is a cam?
79. What is a follower?
80. What is the function of a camshaft? Where is it located on the engine?
81. Ordinarily how many cams would be there on a camshaft for a 4- cylinder engine?
82. Why must the bearing journals on the camshaft be always larger than the cam lobes?
83. How does a fuel pump get drive from the camshaft?
84. How is the endwise movement of the camshaft limited?
85. Give the chemical composition of a typical material for camshafts.
86. Why are faces of cam lobes in modern engines ground taper?
87. Why are the tappets used in modern engines slightly offset from the cam face?
88. The valves are opened by the camshaft. What is the agency for closing them?
89. Name the three types of camshaft drives.
90. What is the function of a tensioning device in a camshaft drive?
91. Which type of camshaft drive does not need lubrication?
92. What is a timing gear? Of which material is it made?
93. Which materials are used for chain links and sprockets of camshaft drives?
94. State the components of a toothed belt.
95. What are the advantages of toothed belt drive? State also its disadvantages.
96. What is the function of a valve tappet? Where is it placed in the valve train?
97. What is the 'valve lash?' How is it adjusted? What happens if it becomes excessive?
98. Which material is used for valve tappets?
99. What is the advantage of a hydraulic tappet?
100. Is the small oil leakage between the plunger and the body in the hydraulic valve lifter intentional? If so, why is it provided?
101. What additional purpose is served by making push rods hollow?
102. How do the stamped rocker arms compare with cast ones?
103. Give the typical composition of a material for rocker shafts.
104. What is the necessity of using valve rotators?
105. What is a 'free type' valve rotator?
106. What is a 'positive type' valve rotator?
107. What do you understand from terms (*i*) VVT, (*ii*) VVEL, (*iii*) VTEC, (*iv*) IVLC, (*vi*) VEMB?
108. Name different variable valve timing technologies.
109. Name various variable-lift systems of valves.
110. What is the function of 'MultiAir' system?
111. Name different modes of 'MultiAir' system.
112. What is the effect of early closing of inlet valve?
113. What is the effect of late opening of inlet valve?
114. State the advantages claimed for a camless engine.
115. What two types of camless valve systems are being developed currently?
116. Name common valve troubles.
117. State any two reasons for each of the following:
    (*a*) Burnt valve face   (*b*) Necking of valve stem

(*c*) Valve face wear (*d*) Valve cracking
(*e*) Noisy valve operation. (*f*) Valve head deformation.
(*g*) Valve stem scuffing. (*h*) Distortion of the valve stem.

118. What would be the effect on engine performance if the valve springs are too weak?
119. Why do we use a muffler with an automotive engine, when it decreases the engine output?
120. Name different types of silencers.
121. What is the principle of wave cancellation type of silencer?
122. Name any suitable material for filling in an absorber type muffler.
123. Name various technologies used for Mahle's downsized demonstrator engine.

MULTIPLE CHOICE QUESTIONS

Mark the correct answers:

1. The counterweights on a crankshaft are located opposite the
 (*a*) main bearings (*b*) big-end bearings
 (*c*) small-end bearings (*d*) vibration damper
2. On the front end of a crankshaft is mounted
 (*a*) timing gear (*b*) vibration damper
 (*c*) fan pulley (*d*) all of these
3. On the rear end of a crankshaft mounted
 (*a*) flywheel (*b*) counterweight
 (*c*) vibration damper (*d*) timing sprocket
4. The ratio of main journal diameter and the cylinder bore in a car engine is approximately
 (*a*) 0.1 (*b*) 0.25
 (*c*) 0.5 (*d*) 0.75
5. In a six-cylinder car engine the angle between the successive crankthrows is
 (*a*) 60° (*b*) 90°
 (*c*) 120° (*d*) 180°
6. The vibration damper on a crankshaft reduces the
 (*a*) longitudinal vibrations (*b*) transverse vibrations
 (*c*) torsional vibrations (*d*) all of these
7. The number of main bearings in a 4-cylinder car engine is usually
 (*a*) 2 (*c*) 3
 (*d*) 4 (*d*) 6
8. The most commonly used valve in an automobile engine is
 (*a*) poppet valve (*b*) sleeve valve
 (*c*) rotary valve (*d*) none of these
9. Exhaust valve face angle is generally
 (*a*) 30° (*b*) 45°
 (*c*) 60° (*d*) 75°
10. The material used for inlet valve is usually
 (*a*) silico-chrome steel (*b*) austinitic steel
 (*c*) precipitation-hardening steel (*d*) nimonic alloy
11. Most commonly used valve-actuating mechanism in modern cars is
 (*a*) side valve mechanism
 (*b*) overhead valve mechanism

(*c*) overhead inlet and side exhaust valve mechanism
(*d*) none of these

12. The engine valves are closed by
(*a*) crankshaft (*b*) camshaft
(*c*) timing device (*d*) valve springs
13. The camshaft controls
(*a*) valve opening (*b*) valve closing
(*c*) valve timing (*d*) all of these
14. How many times in a minute does each valve on a four stroke engine running at 2000 RPM open and close?
(*a*) 1000 (*b*) 2000
(*c*) 4000 (*d*) 6000
15. The camshaft drive which does not require lubrication is
(*a*) chain drive (*b*) gear drive
(*c*) toothed rubber belt (*d*) none of these
16. The push rod is located between the
(*a*) valve and valve spring (*b*) valve and rocker arm
(*c*) tappet and rocker arm (*d*) none of these
17. A says that rocker arms reverse reciprocating motion while *B* says that rocker arms convert rotating motion to reciprocating motion. Who is right?
(*a*) A (*b*) B
(*c*) Both are right (*d*) Both are wrong
18. A asserts that intake valves are cooled while they are open. However, *B* states that exhaust valves are heated while they are open. Who is right?
(*a*) A (*b*) B
(*c*) Both are right (*d*) Both are wrong
19. The exhaust valve usually starts opening
(*a*) at BDC (*b*) at TDC
(*c*) before BDC (*d*) before TDC
20. Valve overlap occurs between
(*a*) intake and compression strokes (*b*) compression and power strokes
(*c*) power and exhaust strokes (*d*) exhaust and intake strokes.

[**Key:** 1. (*b*), 2. (*d*), 3. (*a*), 4. (*d*), 5. (*c*), 6. (*c*), 7. (*b*), 8. (*a*), 9. (*b*), 10. (*a*), 11. (*b*), 12. (*d*), 13. (*d*), 14. (*a*), 15. (*c*), 16. (*c*), 17. (*a*), 18. (*c*), 19. (*c*), 20. (*d*)]

5 Engine Service

1. INTRODUCTION

Before the engine is disassembled for service, it is desirable to ascertain whether the overhauling is really necessary or not. Generally the overhauling is required in the following events:

1. *Power loss due to poor engine compression.* Loss of power also results due to other factors such as slipping clutch, dragging brakes or defective fuel system. In such cases, the power loss can be remedied simply by rectification of the concerned system. In case power loss is not due to such other factors, the compression pressure may be measured and comparing the value with those given in the shop manual, it may be decided whether the engine requires overhauling or not.
2. *Excessive consumption of lubricating oil,* which should be determined accurately.
3. *Lubricating oil pressure too low,* provided the same is not due to external leakage, less oil in the sump, defective oil pump or faulty oil pressure gauge.
4. *Mechanical failures,* such as excessive clearance in main or big end bearings or in the valve tappets, piston slap or some broken component indicate that the overhauling is required. However, if excessive noise is due to defective ignition or injection or faults in generator, water pump, etc., the engine overhauling is not necessary.

The servicing procedure for the following engine parts has been given in this chapter.

1. Cylinder head
2. Valves and valve mechanism
3. Piston-connecting rod assembly
4. Cylinder
5. Crankshaft and main bearings.

The methods of servicing for various engine systems, *viz*, cooling, lubrication, fuel, electrical, have been described in their respective chapters.

2. ENGINE REMOVAL

Before deciding finally whether it is necessary to remove the engine from the chassis, it is a good practice to clean the engine externally thoroughly so as to remove all surface grime. This makes it easier to visible inspect the engine and to work on it and also helps to keep the dirt out of it while disassembling. While cleaning, take special care not to damage electronic and fuel injection components. Cover any part that can be damaged with a plastic bag. Special attention should be paid to any physical damage, leaks at gaskets, external contamination etc. If it is found necessary to remove the engine the same will have to be lifted after removing components that come in the way of its removal and also those that are likely to be damaged while removing. First of all the battery is removed and the oil and coolant are drained off. The hood of the engine compartment is then removed. All coolant hoses are disconnected, the radiator mounting bolts are removed and the radiator is lifted from the engine compartment. If the vehicle is air conditioned, the compressor is then disconnected from the engine and removed. All air-conditioning lines should be plugged to prevent the dirt and moisture going inside. All wires, tubing and controls connecting the engine to the automobile are removed and tagged. It is better that spark plugs, distributor, carburettor, alternator and fan are then removed to avoid damage during removal of engine. After that the drive shaft and the exhaust pipes are disconnected. On some vehicles it may be necessary to remove the steering linkage also to make adequate space for engine removal. A hoist is then moved near the vehicle engine compartment and its sling connected to the manifold or a lifting eye bolt on the engine top. The mounting bolts are then removed and the hoist operated to lift the engine. The engine is then swung clear and lowered onto the engine stand on the blocks placed on the floor and the hoist sling is disconnected.

3. CYLINDER HEAD

3.1. Removing cylinder head

If various components, *e.g.*, the air cleaner, carburettor, head cover, spark plugs, ignition coil, distributor, etc., were not removed earlier, remove them from the cylinder now in proper sequence. The exact procedure for removing should be as given in the shop manual. It is important to remember, however, that an aluminium cylinder head must be allowed to cool completely before removing the same, failing which the same may get distorted.

On L-head engines, the manifolds need not be removed, but in I-head engines, these have to be removed in some cases while in others there is no such need. In case the gasket is sticking, the same may be loosened with the help of a flat blade, taking care that parting surfaces are not damaged. The parting surfaces may then be inspected for leakage. In case the manifold is not removed and the carburettor is also left mounted on it, the same must be covered with a clean cloth before proceeding further. After removal of manifolds, remove the rocker arm assemblies and keep them in proper sequence so that abnormalities can be pinpointed and also these can be replaced afterwards at the same places from where removed. After the rocker arm assemblies have been removed, check the rocker area for any sludge, excessive build-up of which may indicate a poor oil change schedule which may, later, be corroborated from similar wear patterns on other components.

After disconnecting the required components, loosen the head bolts or nuts. Starting from the centre, gradually outwards loosen slightly, all the bolts and nuts. Repeat the loosening operation all over again after which loosen them completely one by one and take them off. The purpose of all this is to avoid uneven stresses on the head. After removing all the head nuts (or bolts) the head can be lifted up. It happens sometimes that at this stage, the head does not separate from the block easily. In such a case do not try to separate the head by inserting screw driver or any other sharp

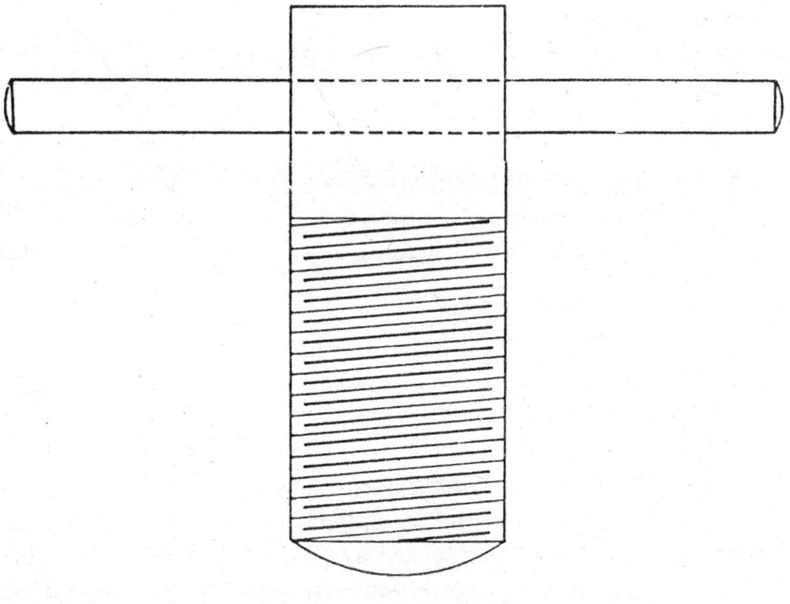

Fig. 5.1. Lifting screw.

object in the joint, because this is likely to damage the machined surface of the head or the bloc. Instead, tap each side of the head with a wooden hammer or with an ordinary hammer interposing a block of wood in-between hammer and the head. This will facilitate breaking of joint. Use may also be made of the lifting screws (Fig. 5.1). Two of these may be inserted in the extreme spark plug holes and the head lifted vertically.

In case of overhead valve engines, now remove the timing cover, after which camshaft should be removed. While doing so take care to avoid dragging lobes over the bearing surfaces, which may damage both the lobes as well as bearings. After removing the camshaft, examine the same visually for any signs of wear. In case either the camshafts or the lifters are worn, both will have to be replaced. Never install old valve lifters with a new camshaft or a used camshaft with new valve lifters.

3.2. Cleaning and Inspection

The first thing to do after the head is lifted off is to observe the nature of carbon deposits which would indicate the cause of excessive oil consumption or overheating if present. The combustion chamber of the engine which has been running under normal conditions would be coated with a layer of light coloured but hard carbon. In case of combustion chamber which has been running too hot, the carbon layer would be thin and of lighter colour. The carbon deposits in the combustion chamber which has been running cold, would be black, thick and dry. The excessive oil entering the combustion chamber causes heavy wet carbon deposits. A careful examination would reveal the exact entry point for the oil.

Next step is to clean the head. Before proceeding to do this cover all the water ports, etc. with a clean cloth. Then remove the carbon from the combustion chamber, valve seats, etc. with the help of the carbon scraper. It must be ensured that no metal is removed during scraping. Stud holes and valve guides may be cleaned with a wire brush fitted in an electric drill. This arrangement, however, should not be used for cleaning pistons. Remove all traces of carbon dust by blowing out with air under pressure. In case compressor is not at hand, use may be made of a tyre pump. Subsequently clean the complete head with kerosene and let it dry.

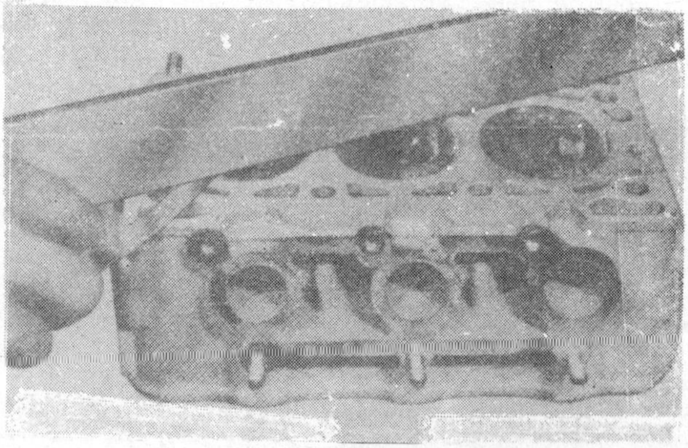

(*a*)

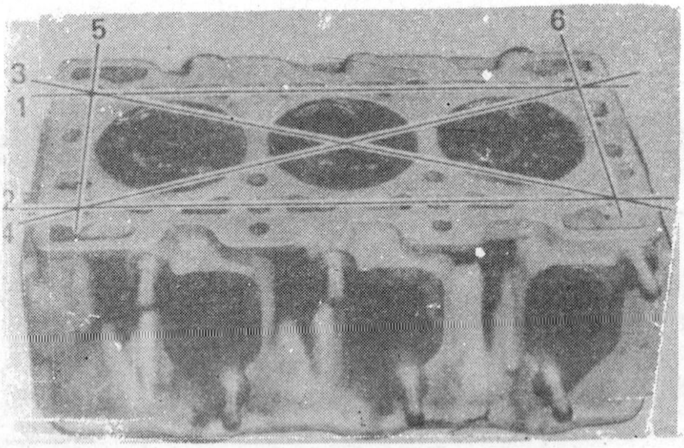

(*b*)

Fig. 5.2. Checking flatness of gasketed surface of cylinder head.
(Courtesy—Maruti Udyog Ltd.)

After cleaning, inspect the cylinder head visually. Look out for warpage, damaged machined surfaces and cracks. Warpage will misalign the seats and guides when the head is installed and tightened on the block. The warpage may be checked by means of a straight metal rule placed against machined surfaces [Fig. 5.2 (*a*)] in a number of directions [Fig. 5.2 (*b*)]. The head surface has to be ground or milled flat if a thickness gauge of more than 0.025 mm per cylinder fits between the straight metal rule and head. To detect small cracks, wet the head with kerosene with a clean cloth and strike again lightly. The cracks will show up clearly. Badly burnt or poor seating valves are clearly visible in the head.

3.3. Refitting the cylinder head

Replacing the cylinder head after servicing the engine is a reverse process of removal. The main precaution to be observed is that the nuts should be tightened in proper sequence and in steps to avoid warpage of the head. The nuts must also be tightened to the specified torque value with a torque spanner (Fig. 5.3). For instance the sequence specified in an engine of Ashok Leyland vehicles is given in Fig. 5.4. The nuts in this case should be tightened to a torque of about 150 Nm.

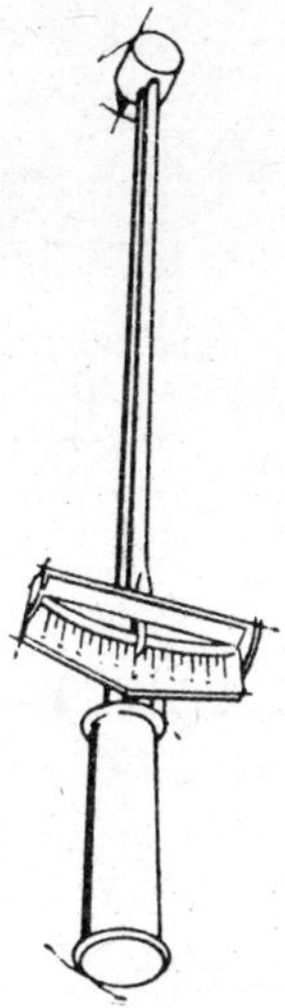

Fig. 5.3. Torque spanner.

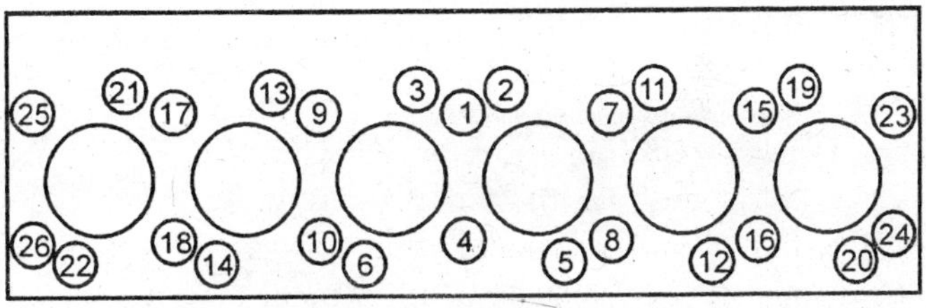

Fig. 5.4. Sequence of tightening cylinder head nuts in an engine of Ashok Leyland vehicles.

4. INSTALLING CYLINDER HEAD GASKETS

Various steps involved are as follows:

1. Allow the engine to cool to room temperature before disassembling.
2. Loosen the bolts and remove them in reverse torque sequence.
3. Clean the mating surfaces thoroughly. Scraper and wire brush can be used on cast iron surfaces, while on aluminium use a non-metallic scraper.
4. Inspect the old gasket, which would provide clue to any engine problems, which must be rectified before replacing the gasket.
5. Check for warpage and distortion as explained in the Art. 3.2. Resurface the head or block if the surface flatness exceeds the specifications of the manufacturer. However, excessive removal of surface metal should be avoided to prevent any possible valve/piston interference.
6. Check the new gasket for proper fit on the head and the block. The gasket should meet the most recent OEM recommendations for the replacement use.
7. Check the finish of the mating surfaces. Too smooth surfaces would not allow the gasket to grip properly to provide durable combustion seal. On the other hand if the mating surfaces are too rough, the gasket would not conform to the surface. A micro inch finish of 60 to 100 RA is preferred for cast iron and 50 to 60 RA for aluminium.
8. Clean and prepare bolt holes and check for correct bolt lengths. If the bolts enter the coolant passages,

coat their threads with a non-hardening sealer. This would prevent coolant seepage around the bolt threads.

9. Tighten the head bolts in the proper sequence with the specified torque.
10. Run the engine after complete reassembly and check for any fluid and combustion gas leaks.
11. In case the cylinder head gasket used needs retorquing, run the engine to normal operating temperature for 10 to 15 minutes and tighten the cylinder head bolts again with the specified torque. In case of a cast iron engine, retorquing can be done while the engine is still warm. However if the head or block is of aluminium, allow the engine to cool to room temperature before retorquing.

5. VALVES AND VALVE MECHANISM

5.1. Valve Removal

Before proceeding to remove valves, certain precautions have to be taken. Firstly mark the sequence of the valves with a dot punch or number punch, if not already marked. This is very important because the valves must be fitted back to their original positions after servicing. Note the difference in size of the inlet and exhaust valves, if any. It is better to place the valves along with other components of the valve assembly in proper sequence so that, it is easier to locate and assemble while fitting. Secondly, it is advisable to plug any holes near the valve chests with pieces of clean cloth so that any component does not fall down through the hole into the sump while the valves are being removed.

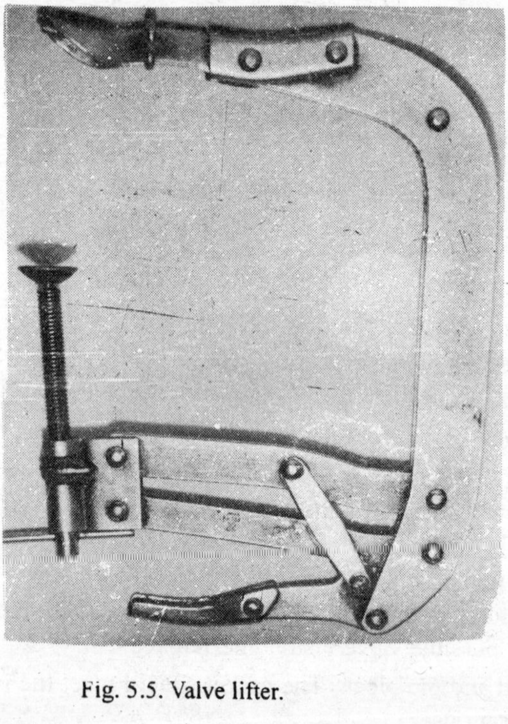

Fig. 5.5. Valve lifter.

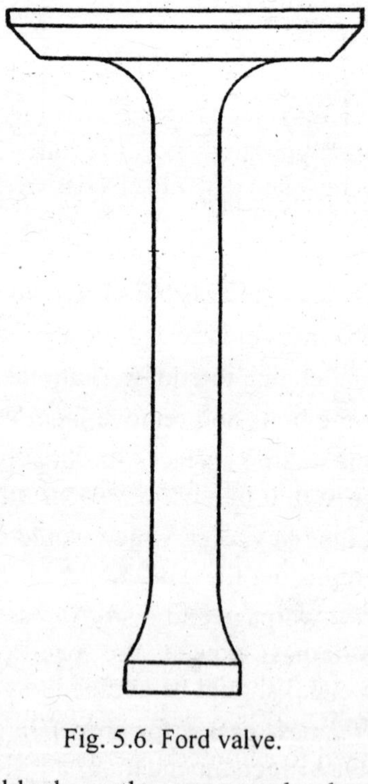

Fig. 5.6. Ford valve.

The valves may be removed from the cylinder head or block, as the case may be, by means of valve lifters. One such valve lifter is shown in Fig. 5.5. The function of the valve lifter is to compress the spring and remove the spring retainer lock after which the valve may be taken out. In the earlier type Ford engines, the valves were not provided with any spring retainer groove (Fig. 5.6). Such valves are fitted in the split guides (Fig. 5.7).

A valve of this type is removed by means of a bar lifter, which is inserted through the spring and above the flange of the valve guide and is pushed down sufficiently so as to release the valve guide retainer and to take it out. The valve assembly, along with the guide, is then taken out from above as a whole and various components dissembled later.

If during valve removal process it is found that a valve is sticking in its guide, it may be due to its tip having a ridge around it. In such a case do not drive the valve through the guide; doing so may score or crack the valve guide or head. Simply raise the valve stem and file off the excess metal till the stem slides easily through the guide.

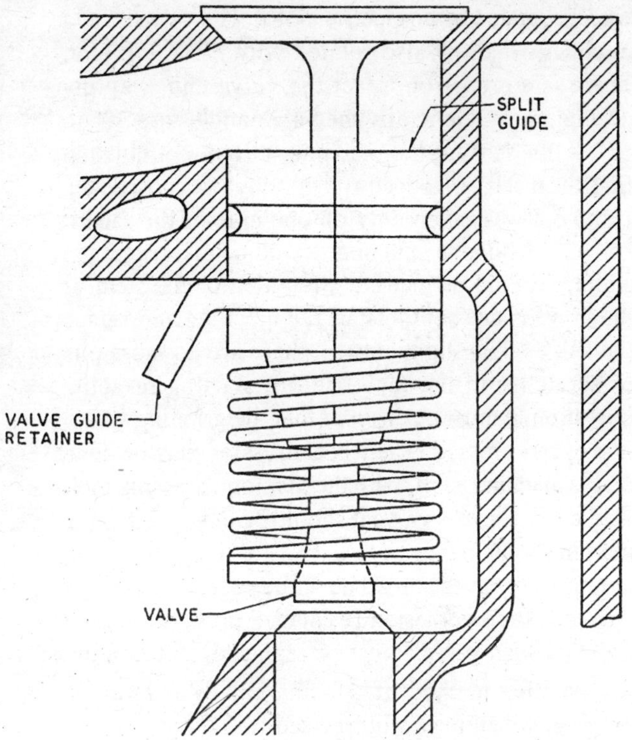

Fig. 5.7. Ford valve fitted in split guide.

After disassembly, clean the valves, guides, springs and other parts of the valve actuating mechanism. Then inspect various parts to determine if they require any service or replacement after which the desired action may be taken.

5.2. Valves

Carbon can be removed from the valves by scraping or even chipping whenever required, though care must be taken to avoid any damage to the valve while chipping. A wire brush mounted in an electric drill may be used to remove carbon from the valves and the valve guides. Valve stems may also be cleaned safely with a fine abrasive cloth. The valve is rotated in a lathe chuck and the abrasive cloth held around the stem. After scraping, chipping or abrasion, wash the valves in kerosene and dry by wiping with a clean cloth.

Subsequent to cleaning, inspect the valves carefully and replace them if—

1. they are burnt, cracked, distorted or badly pitted.
2. they are worn to the extent that the margin has been reduced to a thin edge.
3. they are excessively bent so that the valve stem is not true and square relative to the head. The radial runout may be checked with a dial gauge and V-block (Fig. 5.8).
4. the valve stems are worn out excessively at the ends due to striking of the rocker arm end intermittenly in operation. If the grinding of the metal to smoothen the end surface is excessive (*e.g.*, more than 0.5 mm for Maruti cars), the valve has to be replaced.
5. the stem clearances is 0.03 mm or more.

Fig. 5.8. Checking radial runout of valve.
(Courtesy—Maruti Udyog Ltd.)

If the valves do not need replacement, the same should be refaced before reinstalling them in the engine. This may be done on a special valve refacing machine (Fig. 5.9). The machine consists of a grinding wheel operated by an electric motor. The valve is held in chuck which is also rotated with electric motor. There is a provision to set the valve chuck at any desired angle. Generally the valve angles are 30° to 45°. After starting the machine, a light cut is taken first, the machine stopped and the valve face inspected. If the cut was taken only on one side of the face it means the valve is bent and should be discarded. No attempt should be made at straightening the stem. If the valve is not bent, restart the machine and reface the valve taking light cuts till there are no more pits or scratches in the face. Before starting refacing operation, it must be ensured that the grinding wheel is properly dressed. The wheel dressing may be done on the machine itself, with the diamond dressing tool. While refacing the valve it should be ensured that the margin should never be less than 1.5 mm. Moreover, aluminized valves should never be ground since they will lose their corrosion resistance properties when ground. Such valves must be replaced, if it is found that they require refacing.

Fig. 5.9. Valve refacing machine.

Another method of refacing valves and valves seats, which has become obsolete now, is the valve lapping, which may still be used if the special refacing or grinding machines are not available. After thoroughly cleaning the valves, valve seats and guides, a small quantity of the special lapping compound which is in the form of a paste, is applied on the valve face and valve is slipped into the guide hole. The lapping stick which has a hollow rubber casing at its end, is pressed lightly onto the valve so that the hollow rubber casing gets attached to the valve top because of the vacuum thus introduced inside it (casing) (Fig. 5.10). The valve is now given reciprocating rotary motion with the palms of the hands on the stick. While doing so, the valve is lifted off its seat every now and then so as to spread the grinding paste properly evenly on the valve face and seat. Sometimes, to facilitate this lifting up, a small spring is kept in between the valve seat and valve. Then for

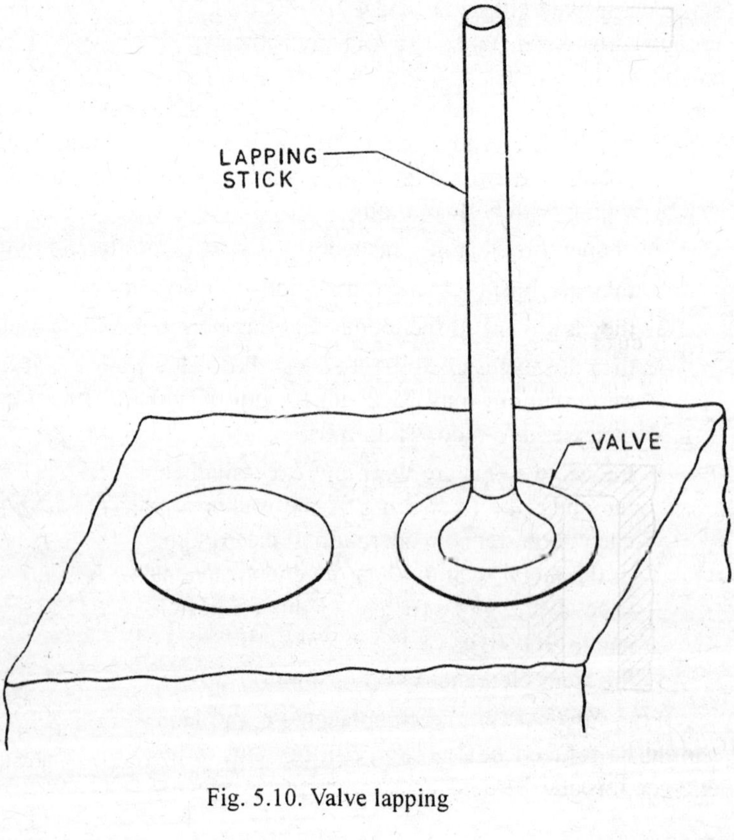

Fig. 5.10. Valve lapping

lifting up, you simply have to relieve the stick, of the slight vertical pressure which you put all the time during lapping operation. The process is continued till pit free smooth surface is obtained.

5.3. Valve seats

The valve seats must provide a tight seal joint with the valve closed on it. For this it is required that the valve must be concentric with the seat and both must have the same angle. Apart from tight seal, the better contact of the valve and the seat is also conducive to better cooling of the valve, which is quite important in case of exhaust valves in particular. From this viewpoint the wider the contact, better it is. But in practice it is very difficult to maintain a very wide tight seal joint. As a compromise the width of the valve seat is generally kept about 2 mm.

Valve seats which have become pitted or worn out due to constant use must be ground. In case of the integral seat which is generally cast iron, this may be done with a hand seat cutter, whereas in case of insert type which is made of harder metal, valve seat grinding is done with special grinders using abrasive stones. Fig. 5.11 shows a hand seat cutter fitted onto the valve seat to be refaced. The cutter of proper angle must be selected, which is fitted to the valve guide through a centering pilot. The tool is rotated by hand by means of the sliding handle provided at the top. An electric type grinder is shown in Fig. 5.12. It consists of a driver (electric motor type), grinding wheel holder, grinding wheels, dressing fixture and a number of pilots to suit various sizes of valve guides. Three types of grinding wheels are generally provided, *viz.*, for the integral type cast iron seats, for the harder special alloy steel inserts and for finishing the seat. The suitable grinding wheel is selected for initial grinding and refaced on the dressing fixture to the desired

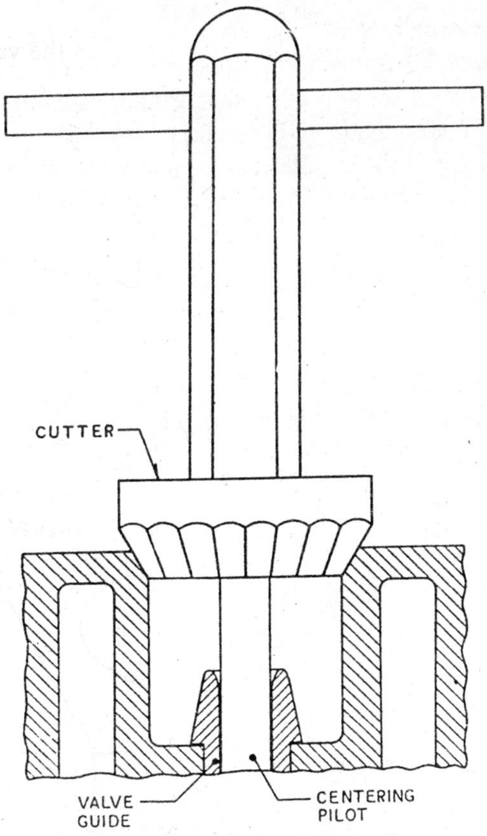

Fig. 5.11. Hand seat cutter (installed on the seat).

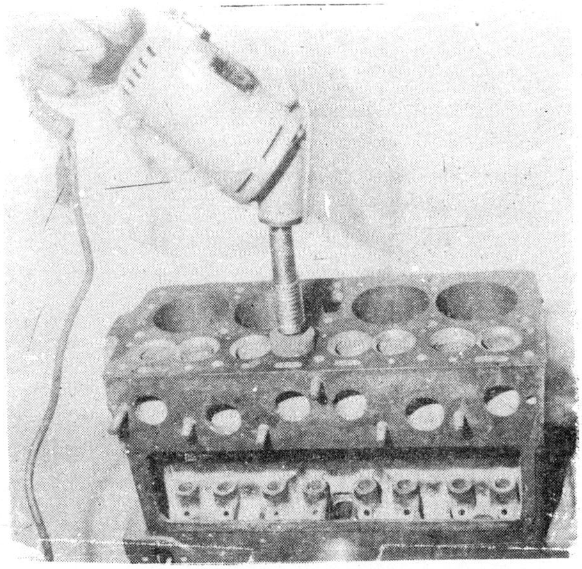

Fig. 5.12. Valve seat grinder.

angle after which it is mounted on the wheel holder. Pilot of suitable size is then selected and inserted in the guide hole which has already been thoroughly cleaned and serviced. The wheel holder with wheel is then supported on the pilot and the driver placed on the holder. Care should be taken that the weight of the driver does not act on the grinding wheel. The driver switch is then turned on, which rotates the wheel to grind the seat. It takes only a few seconds to grind the integral type seat, whereas for the special alloy inset types, it takes slightly longer. The grinding wheel is then taken off the holder and instead a finishing wheel suitably dressed, is fixed on the holder which is placed on the pilot. The seat is then finally finished with the driver.

If the valve seats become wider due to a number of grindings, the same may be narrowed down to the desired width by grinding with wheel of lower angle (30° for 45° seat) at the top and with wheel of higher angle (60° for 45° seat) at the bottom as shown in Fig. 5.13.

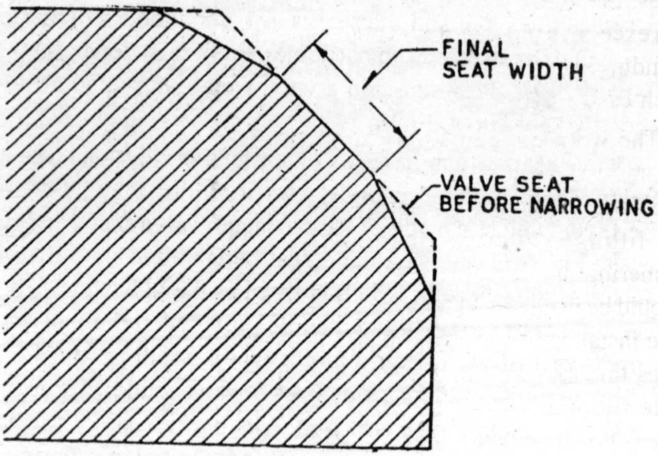

Fig. 5.13. Narrowing down of valve seat.

In case of special alloy steel inserts, if the seat wear is too much and further narrowing down as described above is not possible or the seat is otherwise badly damaged, the insert may be replaced. In very bad cases of integral seats also, suitable insert may be fitted. To remove the insert, holes are drilled on the opposite side throughout the thickness of the insert. Then the insert is removed with a chisel and a hammer, taking care not to damage the counterbore. The counterbore is now checked. If it is found in good condition new insert may be fitted by chilling it in ice for 15 minutes and then pressing it in the counterbore so that it touches the bottom. If however, the counterbore is found damaged, the same may be rebored oversize and an oversize insert may be fitted. After fitting the insert, it should be finish ground to proper angle.

Subsequent to valve refacing, seat grinding and guide servicing the valve seats may be tested for proper contact with the valve. Coat the valve face lightly with prussian blue and press it on valve seat. Give half a turn to the valve and then lift it to examine the seat. If the valve seat has become evenly coated with prussian blue the contact is alright, otherwise not.

5.4. Valve guides

Valve guides must be inspected and suitably serviced if required before proceeding with any work on valves and valve seats, because any attempt to true the valve face or the seat when the guide is worn out, will be in vain.

The guides may be cleaned by a wire brush fitted in an electric drill or a special cleaner whose blades are adjustable. After cleaning the valve and the guide both, the stem clearance in the guide is checked. This can be done with a dial gauge which is installed at any suitable place on the engine head near the valve (Fig. 5.14). The dial gauge is so installed that its stem touches the stem of the valve which has been raised a little by turning the camshaft. The

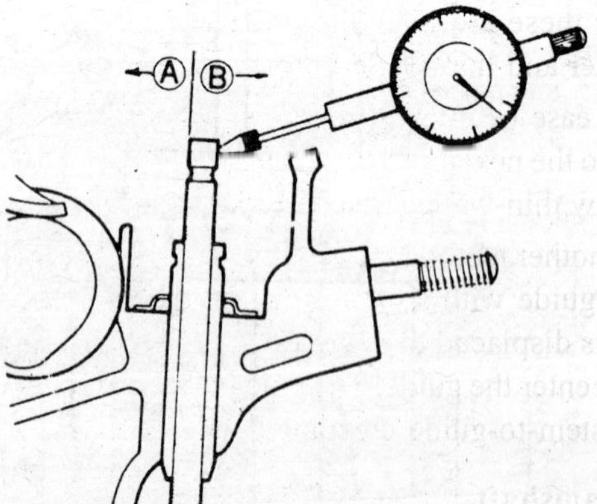

Fig. 5.14. Determination of valve stem clearance.
(Courtesy—Maruti Udyog Ltd.)

valve is then moved to-and-fro in the guide and dial gauge reading noted to register the maximum difference (A—B), which is the stem clearance. If this is more than 0.125 mm for intake valves or 0.150 mm for exhaust valves, the guide should be replaced. Apart from exce sive clearance, the guide may also be unduly worn bell-mouthed at the top, in which case again it has to be replaced.

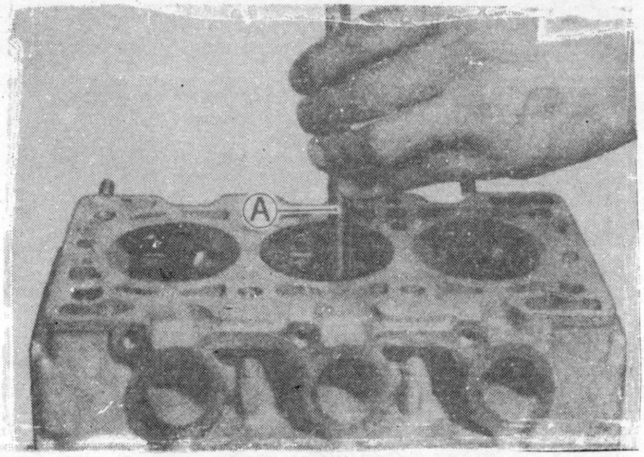

Fig. 5.15. Taking out valve guide with the guide remover (A). (Courtesy—Maruti Udyog Ltd.)

The worn out guide may be driven out by hammering with a proper drift or with a special tool called guide remover (Fig. 5.15). For fitting the new guide, however, hammering should not be resorted to; instead it should be installed with the help of a special guide installer (Fig. 5.16). In case of I-head valves this may be done in an arbor press. While fitting it should be ensured that the guide is properly positioned, *i.e.*, its top end is at a fixed distance from the machined surface of the cylinder block or head, which is specified by the manufacturer. After installation, the guide is reamed to proper size. Valves with standard size stems can then be used.

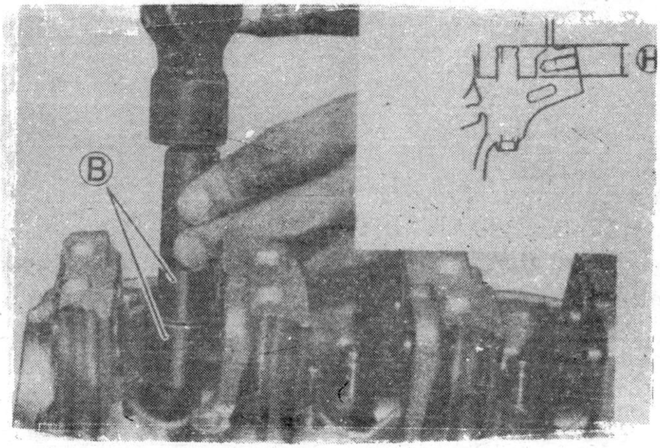

Fig. 5.16. Installing out valve guide with the guide installer (B). (Courtesy—Maruti Udyog Ltd.)

Special care has to be taken while replacing values guides on cylinder heads made of aluminium. Aluminium being soft, can be easily torn out while removing or installing the valve guide. To avoid this, sometimes the worn guides are not removed. Rather these are machined to some larger diameter and thin-walled bronze inserts are installed, with which standard valves may be used.

In case of integral guides (*i.e.* only bore in the cylinder block or head), one simply has to ream the guide to the next oversize an the valves with oversize stem fitted. Alternatively, guide holes may be machined and new thin-walled inserts of standard diameter installed.

Another method to recondition valve guides is *knurling*. Knurling produces a spiral groove on the inside of the guide with an appropriate knurling tool. The knurling tool does not remove any metal; rather the metal is displaced to decrease the diameter of the guide. Successively larger tools are used until the valve cannot enter the guide. The guide is then reamed to restore the guide to original diameter and provide correct valve stem-to-guide clearance.

5.5. Camshaft

Failure of camshaft is indicated by excessive wear of cam lobes, although camshaft in modern engines should normally run for the engine life of the engine and need not be replaced. The same may, however, be checked for straightness when the engine is dismantled, by placing it on V blocks or holding it between

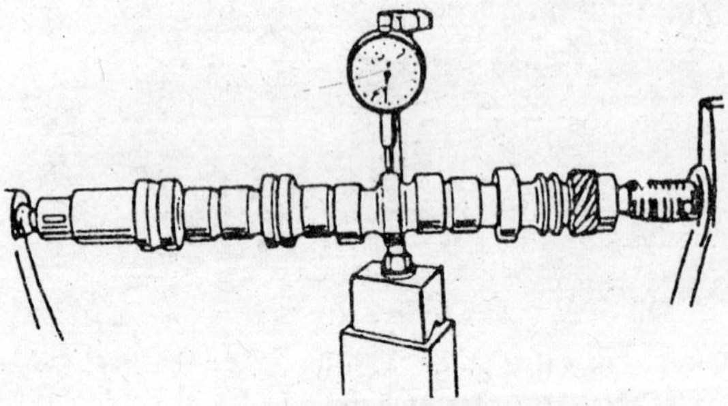

Fig. 5.17. Checking camshaft for straightness.
(Courtesy—Maruti Udyog Ltd.)

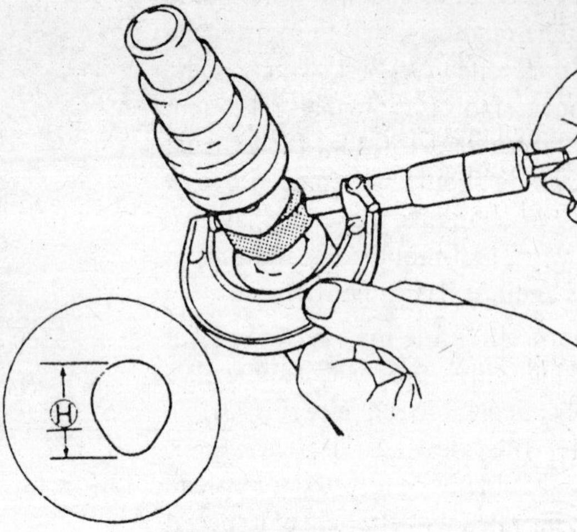

Fig. 5.18. Measuring cam height.
(Courtesy—Maruti Udyog Ltd.)

centres (Fig. 5.17) and using a dial indicator. In many cases a bent camshaft may be straightened in a hydraulic press.

The wear of the cams should also be measured. It the height H of each cam (Fig. 5.18) is less than the prescribed limit, replace the camshaft. The thrust clearance may also be measured and if found excessive, the thrust plate or the camshaft has to be replaced.

Camshaft runs at half the crankshaft speed, due to which reason its bearings have a longer life. The bearing clearance may be found out from the difference in the internal diameter of the bearing and the diameter of the corresponding camshaft journal which can be measured with a telescopic cylinder gauge (Fig. 5.19) and micrometer calliper (Fig. 5.20) respectively. The camshaft bearing clearance generally allowed is up to 0.15 mm. If the measured clearance exceeds this, the bearing will need replacement. The worn out bearings can be removed from the cylinder block by means of a special puller and the new installed using a replacing tool. While replacing bearings, care should be taken that no foreign particles are left in the oil passage connected to the bearings and oil holes in the bearings are in line with the corresponding oil holes in the cylinder block.

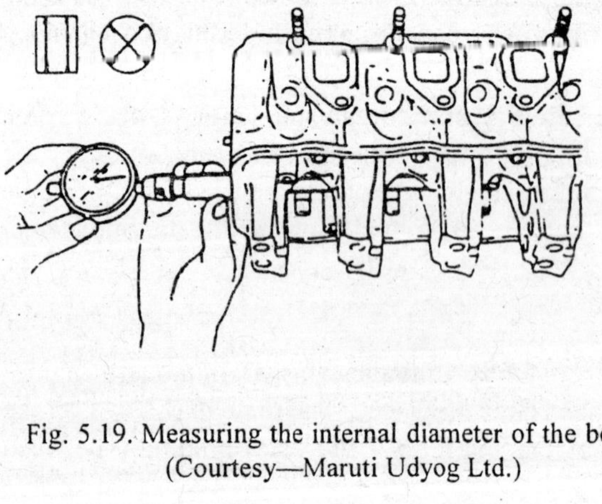

Fig. 5.19. Measuring the internal diameter of the bearing.
(Courtesy—Maruti Udyog Ltd.)

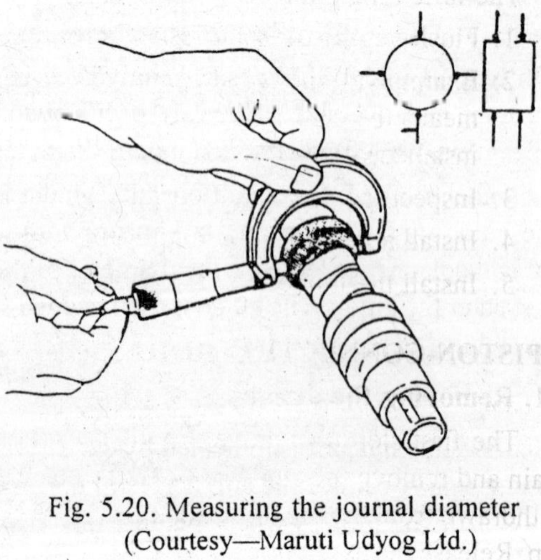

Fig. 5.20. Measuring the journal diameter
(Courtesy—Maruti Udyog Ltd.)

5.6. Tappets

In case the valve operation becomes noisy, the most probable cause may be faulty tappets causing excessive lash in the valve train. Inspect the tappet for any cracks and replace the same if necessary. To check the oil clearance between the tappets and their guides, measure outer diameter of each tappet (Fig. 5.21) and inner diameter of the corresponding tappet guide (Fig. 5.22). The difference of the two gives the amount of oil clearance, which should not exceed about 0.2 mm.

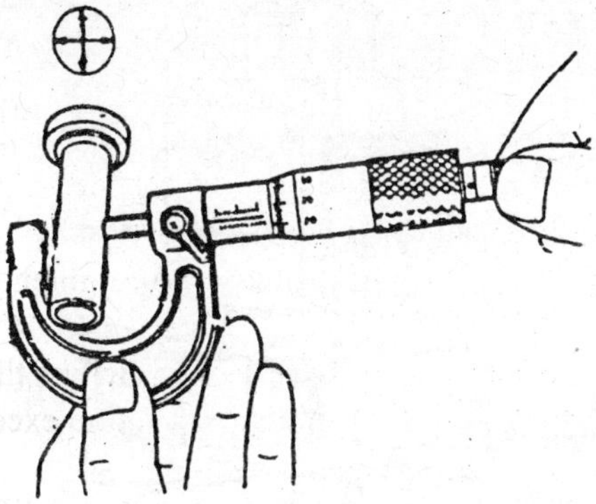

Fig. 5.21. Measuring outer diameter of the tappet (Courtesy—Swaraj Mazda Ltd.)

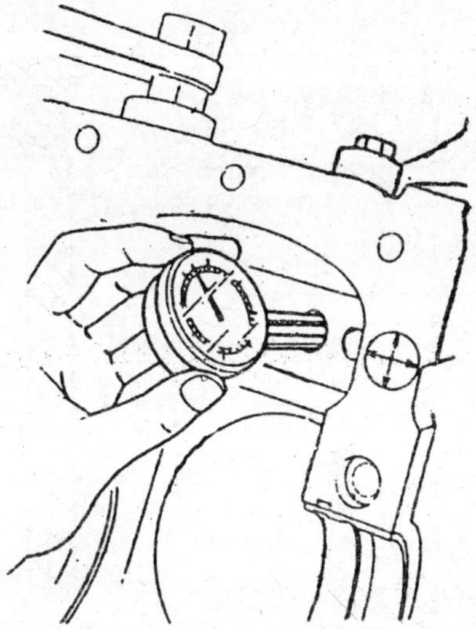

Fig. 5.22. Measuring inner diameter of the tappet guide

The hydraulic tappets are more liable to dirt which enters the tappet through the oil gallery in the form of carbon, sludge, etc. These particles of dirt get stuck between the ball check valve and its seat or between plunger and body and cause noise. The other reasons of tappet malfunction are excessive wear of plunger and body, gummy deposits restricting free plunger movement, air in the oil or faulty adjustment. The other fault may be the excessive wear on the tappet-foot.

The following procedure should be followed for changing hydraulic tappets:

1. Flush out the oil galleries and clean the oil pan.
2. Examine all the tappets removed. In case they are found to have worn concave at the lifter face, it means the camshaft is also worn and the same must be replaced first because the new tappets are installed.
3. Inspect the entire valve train. Repair or replace any defective component, as required.
4. Install a new oil filter and fill the oil pan with lubricating oil of correct specifications.
5. Install the tappets.

6. PISTON-CONNECTION ROD ASSEMBLY

6.1. Removing the assembly

The first step to remove the piston-connecting rod assembly is to remove the cylinder head. Then drain and remove the sump and oil strainer. The piston-connecting rod assembly is usually designed to be withdrawn from the top of the cylinder block. Unlock and remove the big-end bolts and remove the bearing cap. Release the assemblies from other cylinders. It is very important that the same assembly as withdrawn from a particular cylinder must be refitted back to it. To facilitate this piston and connecting rods and their big end caps should be marked with the number of the cylinder from which they are removed. Also they should be placed together side by side in proper sequence, so that it is easier to locate them when required.

There is an important precaution which must be observed. After removing the cylinder head, various cylinders should be inspected for the ridge at the top which shows the upper limit of piston travel in the cylinder (Fig. 5.23). This ridge if found must be removed *before* proceeding further, because otherwise the rings are liable to break once the attempt is made to take out the piston-rod assembly from the top. The

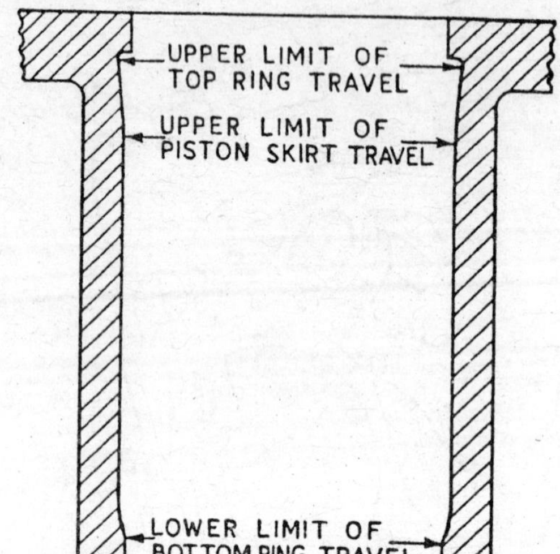

Fig. 5.23. Ridge formation.

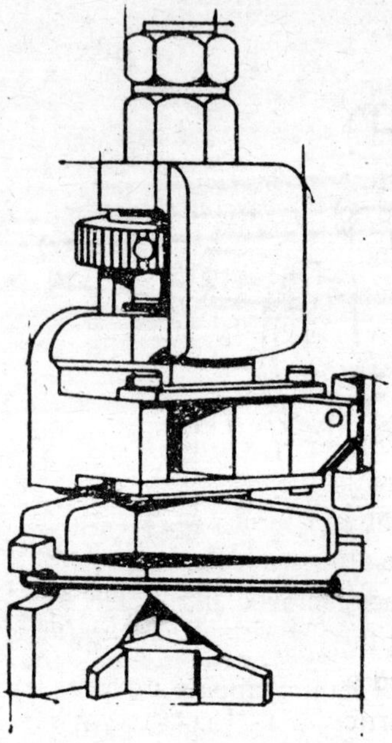

Fig. 5.24. Ridge Reamer.
(Courtesy—Muskegon Piston Ring Co., U.S.A.)

ridge is removed by means of a special ridge reamer (Fig. 5.24). However, care should be taken not to remove any excess metal from the cylinder. Further before starting ridge removal, rotate the crankshaft so that piston of the cylinder being worked upon is at the bottom of the stroke. In this position stuff clean cloth in the cylinder and install ridge reamer. Cover other cylinders and valve ports etc., so that the metal cuttings do not fall in them. Then start the reamer, which rotates to cut away the ridge. The procedure is then repeated for other cylinders.

6.2. Dismantling the assembly

This is done in the following steps:

1. Removing rings from the piston
2. Removing connecting rod from the piston

The rings are removed from the piston by means of the ring expander (Fig. 5.25). This facilitates uniform expansion of the ring avoiding stress concentration at any point so that the cast iron brittle ring is not broken on this account. If, however, the ring remover is not available, a thin strip *e.g.*, a worn out feeler gauge about 0.50 mm size or a hacksaw blade may be used to remove rings. The strip is inserted below the ring at its end and placed in between the ring and the piston. The strip is rotated around the piston with small pressure applied at the ring end so that it rests on the land above the ring groove. From there the ring is removed from the top. The topmost ring is removed first and other rings later in sequence. No attempt should be made ever to take out the ring from the skirt side.

The method of disconnecting the connecting rod from the piston depends upon the manner in which the rod is assembled with the piston. The circlips are removed first from the piston bosses. If it is free on the piston pin which itself is shrink-fit in the piston bosses the piston is heated in boiling water till it expands when the pin is pushed out of its bosses, while the piston is hot. This method is adopted for the piston in Ashok Leyland Engines. In case, however, the connecting rod is attached to the piston pin by means of a clamp bolt, as is the case in Hindustan Ambassador car engine, special holding plugs are inserted in the ends of the piston pin. The pin is then held in a vice and the clamp bolt is opened and removed completely. The assembly is then removed from the vice and the piston pin is pushed out.

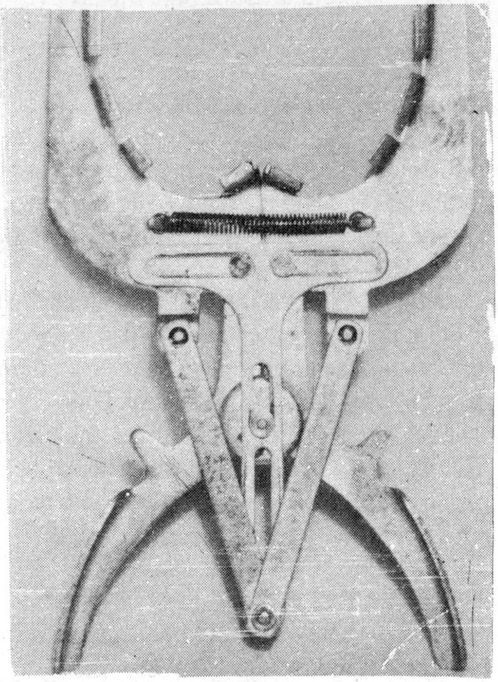

Fig. 5.25. Ring expander.

6.3. Pistons

Before the inspection of the pistons can be carried out, it is very important that they are cleaned thoroughly, so as to remove all traces of carbon or other deposits. With the rings already removed, the pistons are cleaned on both sides, inner and the outer. Piston crown and the inner side can be cleaned with a carbon scraper, whereas ring grooves are cleaned with a special groove cleaner (Fig. 5.26). Particular care is needed, however, in cleaning the piston skirt as the carbon scrapper is not recommended for this. The pistons may, therefore, be soaked in some suitable cleaning solution (which does not adversely affect the piston material) so as to soften the carbon deposits after which a stiff wire brush is used to remove the deposits. Any blocked oil holes in oil ring grooves may be cleaned with a drill. However, the size of the drill bit used must be so selected as not to enlarge the holes.

Fig. 5.26. Piston groove cleaner.

After cleaning thoroughly, the piston is inspected for pitting, scratches, cracks, scuffing, collapsed skirt, fit in the cylinder bore etc. A piston found badly pitted, cracked or otherwise damaged should be replaced. Scuffing of piston occurs when a newly installed piston is loaded heavily without sufficient lubrication due to which it expands and takes up the small clearance between itself and the cylinder bore. In case of heavily scuffed piston, replacement is the only alternative available.

For checking whether the piston has collapsed skirt or not, measure the diameter of the skirt at its top and bottom ends. If the bottom dimension is less than the top, it means the skirt is collapsed. This however, can be remedied by means of skirt expanders, shot blasting against the inner surface of the thrust sides of the piston, peening the skirt by a mechanical hammer at high speeds or knurlizing.

The wear of the ring grooves is checked with a new piston ring and feeler gauge. The new piston ring is installed or placed with its outer side anywhere in the piston ring groove and clearance between the new ring and the side of the ring groove is measured with feeler gauge (Fig. 5.27). If the clearance is more than 0.15 mm, the piston would need replacement. In some cases, it is also possible to remachine the grooves to next oversize and to install oversize rings.

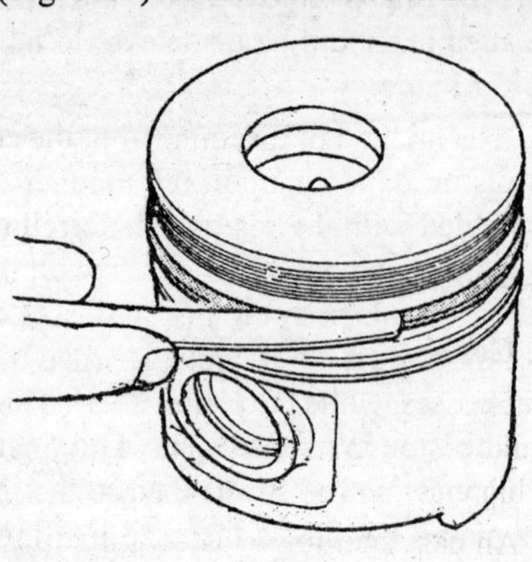
Fig. 5.27. Measuring piston ring groove clearance (Courtesy—Swaraj, Mazda Ltd.)

The fit of the piston in the cylinder bore can be tested by placing a feeler gauge on the thrust face of the cylinder and inserting the piston upside down. A spring balance is attached to the outer end of the feeler gauge and is pulled. The force on the spring scale is noted while the gauge is being pulled up. For satisfactory clearance this pull should be about 30 to 50 N. (The exact amount is specified by the manufacturer for which the shop manual of the vehicle may be consulted). If the pull is more, repeat with a thinner gauge and if it is less, try with a thicker gauge. After a number of trials, the gauge will be found to give the pull within the specified range. The thickness of this gauge gives the clearance. The shop manual of the vehicle should be then consulted to find whether with the clearance thus determined, the piston needs replacement or not. The oversize pistons generally available are in steps of 0.010 inch (0.25 mm) above the standard size.

6.4. Piston Rings

The piston rings should be examined for uneven wear, loss of tension, scratches, etc. However, as a rule, new rings are always fitted once rings are removed from the piston. In case the engine has to be dismantled at low mileage, due to some other defect, the rings should not be removed from the piston; then they can be retained. While fitting new rings, select the proper ring for the piston to be used, *i.e.*, standard size rings should be used for a standard size piston and oversize rings for the oversize piston. However, before fitting new rings, the following checks may be made:

1. *Side clearance.* Place the new ring with its outer side in the piston ring groove and roll it throughout the circumference. The ring should not stick, neither should it be excessively free. The side clearance may be checked using a feeler gauge placed between the new ring and the ring groove side and it should be within limits specified by the manufacturer, which are generally from 0.025 to 0.100 mm for all grooves except the top one for which both the limits are higher, *i.e.*, from 0.040 to 0.150 mm.
2. *End clearance.* The clearance or the gap between the ring ends when fitted in the cylinder should be more than the minimum specified, which is generally taken as 0.001 mm per mm of circumference. If the gap is less than this amount, the ring ends will come together, butt against each other and will ultimately break at higher temperatures.

To check the end clearance, the ring is placed in the cylinder bore slightly above the lowest point of ring travel. Then the piston in the inverted position is placed on the ring and the ring moved down till it comes to the position of lowest ring travel. By so doing, the ring becomes square with the cylinder sides. The point of lowest ring travel is selected because it is here that the wear and hence the cylinder diameter is minimum in the region of ring travel. The end clearance is then measured with a thickness gauge. If the clearance is found to be less than the specified one, the same may be rectified by filing in a special ring filing tool (Fig. 5.28). In case of newly rebored cylinders, however, the cylinder diameter will be uniform and, therefore, the end clearance may be checked at any convenient point.

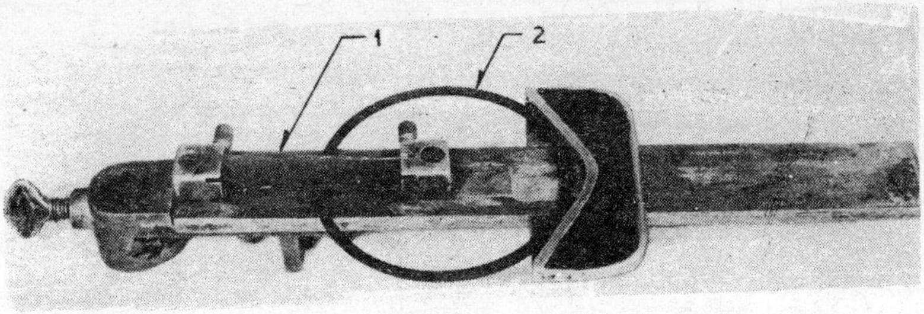

Fig. 5.28. Ring filing tool. 1. File, 2. Ring

6.5. Connecting Rod

The connecting rods are mainly inspected for alignment and wear bearings. Apart from this oil holes should be checked and ensured that they are not blocked. The studs and nuts for connecting the caps to the big ends should also be examined for fitness.

An easy method of checking alignment of the connecting rod is to remove oil pan, crank the engine and observe the big end. If alignment is correct, it should remain centered on the crankpin. If on rotating the crankshaft, the big end moves to and fro on the pin, it must be due to misalignment. Special fixtures for checking the alignment of connecting rods are also available. A connecting rod which is not correctly aligned should be removed and either straightened or replaced. If allowed to run in the misaligned condition, it would cause a rapid damage to the piston and the cylinder.

The clearance between the connecting rod big end bearings and crank pins must be within specified limits, otherwise oil will be thrown on the cylinder walls in much larger quantities than it could be controlled by the piston oil-scraper rings. The same would therefore, go to the combustion chamber and burn there. The result would be thus excessive consumption of lubricating oil and significant reduction of period between engine overhauls.

The bearing clearances can be measured by any of the following methods:

1. With internal and external micrometers.
2. With feeler gauge.
3. With plastigage.

1. With internal and external micrometers. The internal diameters of the connecting rod bearing is measured with an internal micrometer. Before measurement, however, the cap should be tightened on the big end with the torque specified for tightening ordinarily. The diameter should be measured at a number of points to determine the maximum value. The diameter of the crank pin is found out at a number of points by means of the external micrometer. Half the difference in the internal diameter of the big end bearing and the external diameter of the crank pin gives the clearance.

2. With feeler gauge. Dip a feeler gauge about 6 mm wide and of any small thickness, say 0.05 mm thick in lubricating oil, place it between the bearing and the crankpin and tighten the nuts on the bearing cap with the specified torque. Rotate the big end about the pin. If any additional drag is felt than before, it means the clearance is less than the thickness of the feeler gauge, another gauge of smaller thickness may then be tried. If however, no additional drag is felt, the clearance is certainly more than the gauge thickness and this necessitates another trial with a thicker gauge. This is continued till the consecutive two sizes of feeler gauge indicate no drag and slight additional drag respectively. The mean of the two thicknesses can then be taken as the bearing clearance.

3. With plastigage. 'Plastigage' is a special gauge manufactured by Perfect Circle Corporation, U.S.A. to measure bearing clearance. It consists of a fine thread of plastic material which changes shape under pressure, and a gauge graduated to indicate the bearing clearance. The gauges are available in different ranges.

The gauge is placed in the bearing cap in the cross direction which is then fitted into the big end at the specified torque. The nuts are then loosened and the cap removed. The plastic thread has by now changed its size. The gauge scale is then placed along it to read out the bearing clearance (Fig. 5.29). Two precautions must be observed while using plastigage. Firstly, the gauge plastic thread is soluble in lubricating oil. The bearing, therefore, must be thoroughly cleaned before using plastigage. Secondly the pin and the bearing should not be given relative rotary motion which will distort the gauge thread and will give wrong reading.

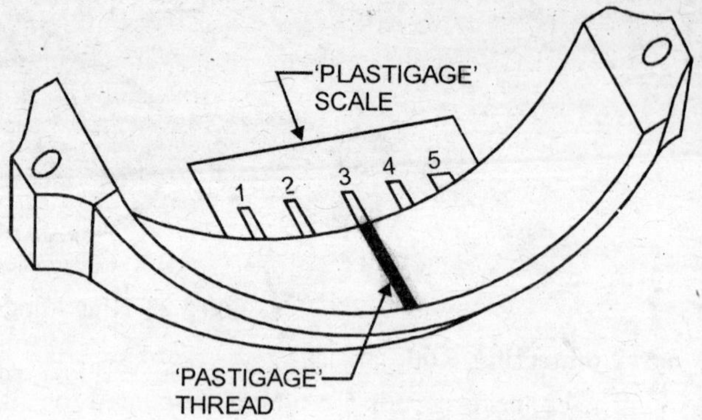

Fig. 5.29. Measuring clearance with a plastigage.

If the bearing clearance measured with any of the methods described above, exceeds the specified value which is generally 0.04 mm, the bearing has either to be replaced if it is of insert type, or adjusted/replaced if it is of direct bonded type. The adjustment in the later type is provided for by means of shims which are inserted around the studs between the cap and the big end. The clearance can be reduced suitably by removing one or more shims on both sides.

The wear (ovality and taper) of the crank pin determines the size of the replacement bearing. If they are within limits, usually 0.04 mm for ovality and 0.02 mm for taper, standard size bearing is fitted. However, if the wear exceeds specified limits, the crank pin has to be ground undersize to make it perfectly round and uniform. Accordingly, suitable undersize bearing will have to be fitted in this case.

In case a rapid wear is suspected, it is always advisable to look out for the source of trouble and rectify the same before undertaking the replacement. The main sources of bearing failure may be lack of lubrication, presence of dirt in the bearing, fatigue failure, misalignment of the connecting rod, tapered crankpin, etc.

6.6. Piston pin

Piston pins should be inspected for wear, scratching etc. The clearance between the piston pin and the connecting rod small end bush or the piston bosses should not exceed the limits prescribed by the manufacturer, which is generally 0.025 mm. The clearance may be measured as in case of connecting rod big end bearing clearance already discussed. If the clearance is found to be excessive, the piston bosses and small end bush have to be reamed and honed oversize and a new pin installed. For reaming, the reamer is inserted in the hole, and clamped in vice while the piston is rotated by hand. For reaming piston bosses some guide should be used so that the two holes are properly aligned.

6.7. Reassembly

Reassembly is generally the reverse process of dismantling. However, the engine shop manual should be consulted for exact procedure. In case of Ashok Leyland engine, *e.g.*, the piston has to be heated in hot water to about 100°C when the bosses expand sufficiently to accommodate the piston pin.

For cars generally the piston pins are thumb pressed into the piston bosses and the small end bush and the circlips installed on both sides. The following precautions should be observed when assembling pins:

1. Pins supplied in sets with the pistons should not be interchanged.
2. Circlips should always be inserted and removed with the help of a special round nose plier. Care should be taken not to damage the right angle corner of both the circlips and the grooves.
3. Circlips should neither be expanded, nor contracted or twisted with violent force, otherwise yield

point of the metal may be crossed and the circlips will reduce their tension. This may lead to damage of circlip grooves and the piston bosses.

4. After assembling the piston and the connecting rod, the rings are fitted. This is done with the help of a ring expander as discussed already while describing dismantling. While fitting, expand the ring expander just enough to slip the ring onto the piston. Over-expanding may result in the ring deformation. The lowermost ring is fitted first of all, after which the other rings are fitted in sequence, while the topmost ring is replaced last of all. The marked side of each ring should always be kept at the top failing which, the oil consumption would be increased tremendously. After installing the rings, distribute their end gaps as shown in Fig. 5.30.

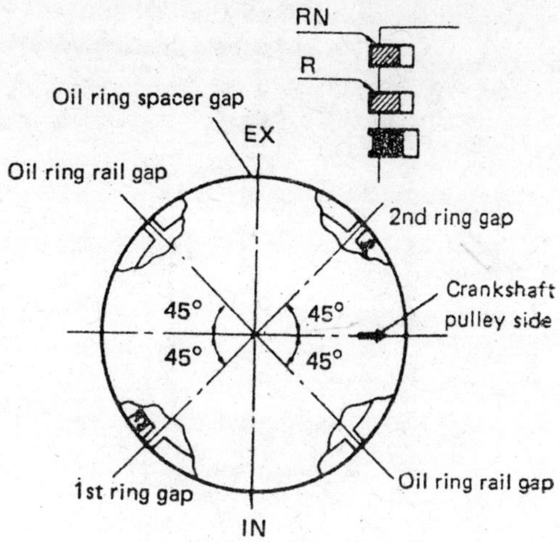

Fig. 5.30. Distribution of ring end gaps. (Courtesy—Maruti Udyog Ltd.)

7. CYLINDER BLOCK

The cylinder bores get worn with time and tend to become oval shaped. Further the wear is not uniform even on the same side from top to bottom. The maximum wear occurs at the upper limit of ring travel.

The cylinder bore should be measured to determine ovality and taper. This can be done using either an inside micrometer or a dial gauge. The diameter is measured both at the top and the bottom of cylinder in two directions in each case *viz.*, in the longitudinal direction of the cylinder block and in the direction perpendicular to it. In each case the measurement is made by slightly probing the position so as to give maximum reading. The difference in readings at top and at bottom gives the ovality there and the difference in the top and bottom readings in the same direction gives the taper in that direction.

After determining the ovality and the taper, shop manual is consulted to find out whether reboring the cylinder is necessary or honing only will be sufficient. Generally the maximum permissible ovality is specified as 0.01 mm and the taper as 0.25 mm. Depending upon the wear of the cylinder bore next oversize is selected so as to remove all ovality and taper. The oversizes generally specified are 0.010, 0.020, 0.030, 0.040, 0.050, 0.060 inch (0.25 mm, 0.50 mm, in case of Maruti 800 car). Corresponding oversize piston has also to be fitted in such a case.

Cylinder bores with less than 0.01 mm ovality need not be rebored. They can be rectified by honing only. However, if the ovality exceeds 0.01 mm, reboring has to be done first after which it is preferred if the bore is finished by honing. It must be noted that all the cylinders have to be rebored or honed to the same oversize, irrespective of the wear of any individual cylinder. This means the oversize is calculated on the base of the *maximum* reading in any cylinder.

Reboring is done on special reboring machines, like the one shown in Fig. 5.31. The top surface of the cylinder block is thoroughly cleaned and the boring bar set on one of the cylinder bores. The bar is centered there and clamped properly. The cutter is then set to the desired depth of cut and the machine started. The cutter will bore progressively to the other end, and the machine stops when the boring is complete. Cylinder is bored 0.0125 mm to .05 mm less than the required size. After reboring, the cylinder size and wall finish are completed by means of honing. A cylinder hone is shown in Fig. 5.32.

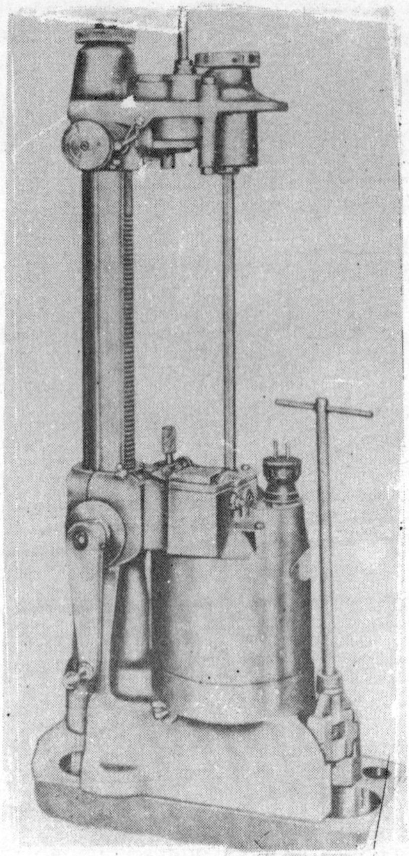

Fig. 5.31. Van Norman 944S Perfecto Boring Bar.

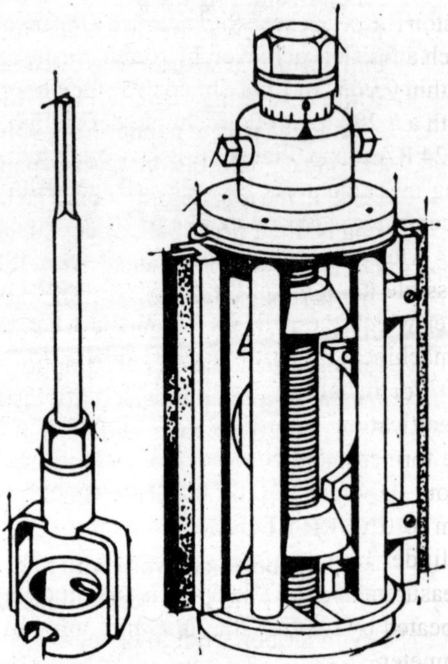

Fig. 5.32. Cylinder hone.
(Courtesy—Muskegon Piston Ring Co., U.S.A.)

Before starting reboring operation it is very important that suitable steps are taken so as to prevent the entry of metal of abrasive particles in the oil galleries. It is a good practice to wrap insulation tapes on crankpins and to cover other parts as is possible. After the job is complete, again the cylinder must be washed thoroughly to remove all abrasive particles. In case the crankshaft is in the block, walls are oiled thoroughly and wiped with clean cloth. This is repeated a number of times till a clean cloth is not stained by abrasives. If, however, the cylinder block is bare, cleaning may be done by thoroughly scrubbing the cylinder walls with a brush using suitable detergent (soap) and water.

A boring bar leaves a pattern on the cylinder wall. Therefore, a cylinder hone has to be used after reboring, so as to provide correct finish to the bore. A proper finish would provide flat areas or plateaus on the surface, which act as a bearing surface on which an oil film can form to separate the rings from the wall. A cylinder hone generally consists of two or three stones. The hone rotates at a selected speed and is moved up and down the bore, removing some metal from the bore while rotating with outward pressure. To control the temperature and to flush out the honed metallic and abrasive residue, honing oil flows over the stones. Honing stones are classified by their grit size. Honing also leaves many small crisscross grooves in the cylinder walls that serve as lubricant reservoirs. While honing with conventional abrasives, allowance must be kept for wear for the stone, otherwise a tapered bore would result. Diamond honing stones wear minimal, due to which they generally produce straighter bores. Likewise, a honing machine with variable speed stroking and which can dwell in the bore while maintaining the same loading would give better results.

Similarly, using a coolant which is compatible with the honing stones used would also result in improved bore geometry.

Surface finish to be achieved for road engines is usually 15 to 20 microinch RA (Roughness Average), while it is smoother, say 10 to 15 microinch RA in case of performance engines. A 'profilometer' is used to measure the surface finish in a freshly honed cylinder. However, regardless of the RA number, most piston rings seat best with a *plateau finished* cylinders, which is equivalent to a broken-in cylinder. Generally such a finish is achieved by rough boring a cylinder to within 0.005" of final dimension or rough honing to within 0.003" of final dimension, then honing with # 220 grit stones to the last 0.001" and finally, finishing with a # 280 to #400 grit stone, depending on the application. In such a case, the finish would be in the 22 to 24 RA range. Plateau finishing with # 600 grit stones would improve the RA number to below 20. Many ring manufacturers specify a surface finish of 15 to 20 RA for molybdenum-faced rings while chrome-faced rings can tolerate above 20 RA finish.

After reboring has been done on a cylinder block a number of times, it is found that it is no longer possible to rebore any further. In such cases dry or wet liners may be fitted (depending upon the design), after which the cylinder dimensions are reduced to the original standard values. Some liners have to be remachined after fitting while others do not need such a process.

For installing dry liners, the required interference fit has to be noted first, as per manufacturer's specifications. In general, aluminium blocks have more expansion than the cast iron blocks, due to which the former need more interference fit than the later. For water cooled cast iron blocks, interference fit of about 0.04 mm to 0.05 mm is adequate, whereas for aluminium blocks, the same should be about 0.05 mm to 0.10 mm. To determine the fit, average outside diameter of the liner and inside diameter of the cylinder bore are measured with the help of accurate gauges. To determine the average diameters, measurements are taken at the top, middle and bottom of the liner/bore and then these measurements are repeated by rotating the liner/bore through 90°. The mean value of all these measurements is the average diameter.

Earlier liners were fitted by pressing in an arbor press, so that a gradual uniform force is applied. However, these days it is preferred to install the liners either by chilling the same or by preheating the block. Placing the liner in a household deep freezer would shrink the outer diameter typically by about 0.05 mm. Alternatively packing them with dry ice or spraying them with some liquid nitrogen would shrink them more, say, about 0.07 mm. This procedure would permit the liner to simply drop in the cylinder bore without the necessity to apply any force, thereby risking the distortion of the sleeve. Similarly, preheating the block would open up the bore making installation easier. In case a liner fitted earlier has to be removed first, the same may be done by using a special puller for this purpose.

After installation of the cylinder liners, the cylinder bores are finished as described above.

During reconditioning of the engine block, old core and oil gallery plugs are usually replaced with new plugs or correct size and type. While doing so, plug bores should also be inspected for any damage which would hinder proper sealing. If found so, the same should be rebored and oversize plug may be fitted. It is always better to coat the plug or bore with a nonhardening oil resistant sealing compound in case of oil galleries or a water-resistant sealing compound in case of cooling jackets.

8. REINSTALLING THE ASSEMBLY IN THE CYLINDER

After reboring of the cylinder block and servicing the piston connecting rod assemblies, the same have to be reinstalled in the block. Care must be taken to ensure that the same assembly as removed from a particular cylinder for servicing is replaced in the same cylinder. Further the pistons are generally marked at the top to indicate which side should be laced in front. In case they are not marked, the split side of the pistons should be so placed that it is not on the thrust side, which is the right side of the engine having clockwise rotation viewed from front (Fig. 5.33).

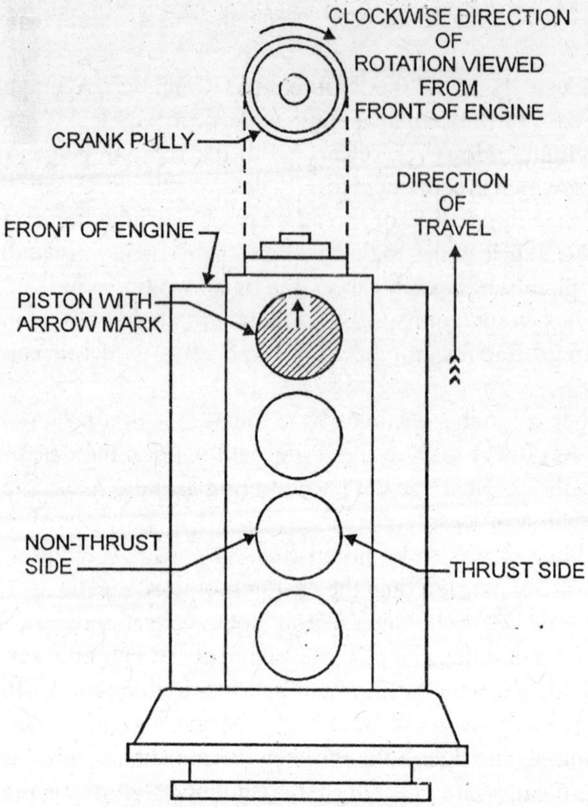

Fig. 5.33. Location of thrust side.

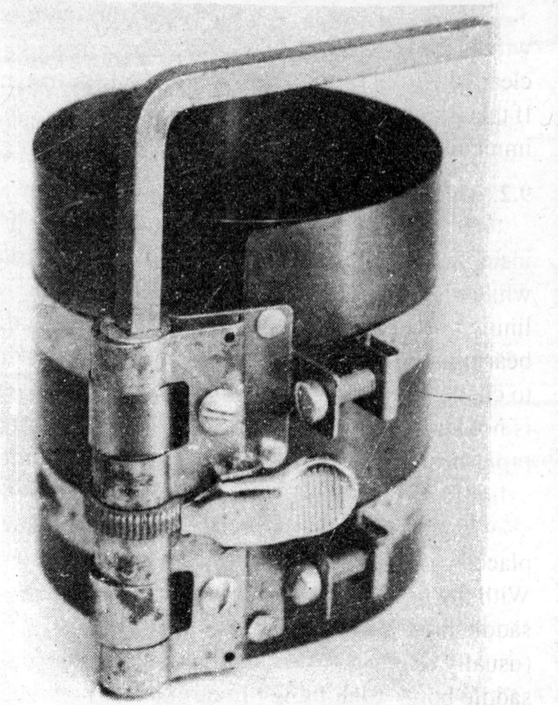

Fig. 5.34. Ring compressor.

The assembly is inserted from the top with connecting rod bearing cap removed and is installed with the help of a ring compressor (Fig. 5.34). The piston rings and groves are lubricated and the ring compressor slid over the piston. The compressor is then tightened so as to compress the rings completely. The whole assembly is then placed on the cylinder bore and light stroke are given on the piston crown with a wooden hammer, which drives the piston in. The big end bearing cap is then installed and tightened properly.

9. CRANKSHAFT AND MAIN BEARINGS

9.1. Crankshaft

The crankshaft should be inspected for alignment and for wear of journal and the crankpin. For checking the alignment, the crankshaft is supported at its end journals on V-blocks and the alignment is checked by means of a dial gauge whose base is fixed at some suitable place. Instead of V-blocks, the crankshaft may also be supported on lathe centres. But in this case care must be taken to ensure that the centres and the centre holes should not be defective, otherwise the readings obtained will be wrong. While placing the crankshaft on V-blocks or lathe centres as explained above, there is likelihood of the self weight of the shaft causing bending. To avoid this, the crankshaft may be set on end. The permissible limit of bending generally allowed is 0.08 mm. In case excessive bend is found, the crankshaft may be straightened in a press.

The wear of the journal and the crank pin may be determined by measuring diameter at a number of places, with a micrometer. The ovality and the taper is then determined. If either of these is found to be more than 0.04 mm, the journals or pins, as the case may be, have to be ground or machined to next undersize. If ovality and taper are within permissible limits, then light scratches, etc. can be honed and then polished with a fine polishing paper or cloth. Polishing should be done in the direction opposite to that of crankshaft rotation.

The oil passages in the crankshaft should also be inspected. If blocked, they may be cleaned with a wire brush. If the crankshaft has been reground or polished it is very important that it should be thoroughly cleaned before reinstallation. A suitable solvent and wire brush should be used to clear out all oil passages. If this is not done, any abrasive left in oil line will cause a very rapid wear of bearings. Lubricate the passage immediately after cleaning.

9.2. Main bearings

The main bearings should be inspected for wear. The bearing clearances may be determined by using inside and outside micrometer or feeler gauge or plastigage the details of which have already been given while explaining servicing of connecting rod big end bearings (Art. 6.5). If the clearances exceed the specified limits, the bearings have to be changed (in case of insert type) or remetalled (in case of cast type). The bearing may be changed either with the shaft removed or in position. It is not a good practice, however, to change the bearings without removing the crankshaft, because this way the condition of the saddle bore is not known. In case a new bearing is installed in the saddle bore which is too much oval and tapered, a rapid bearing failure would result. For remetalling, however, the crankshaft has to be taken out of the cylinder block.

To check the saddle bore, valve gear and other necessary components are removed and the engine placed upside down. The oil pan is removed and the bearing cap nuts loosened to remove the crankshaft. With the bearings removed and the caps tightened again to the specified torque value, the diameter of the saddle bore is measured at different places. In case the ovality of the bore is less than the specified value (usually 0.05 mm), the standard bearings can be installed. However, if the ovality exceeds this value, the saddle bores have to be line-bored to the original diameter. This is made possible by accurately filing the cap machined surfaces by an amount which exceeds the ovality and then reboring. The original size bearing may then be installed.

Sometimes when the journals have been ground undersize, fitting new bearings as above may not reduce the bearing clearance to within permissible limits. In that case the semifinished bearings are installed which are rebored after installation, to the desired diameter.

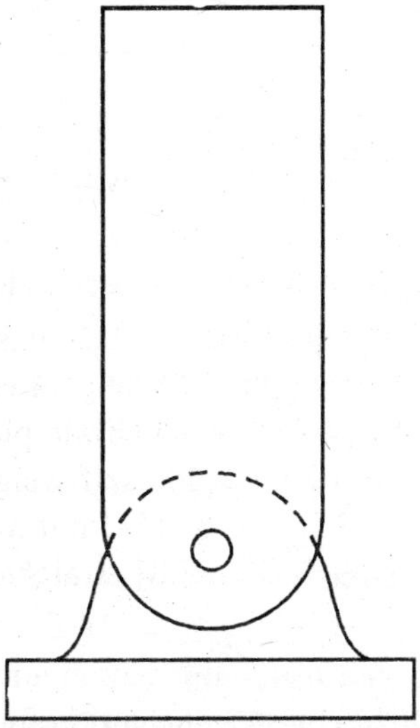

Fig. 5.35. Special tool for bearing installation.

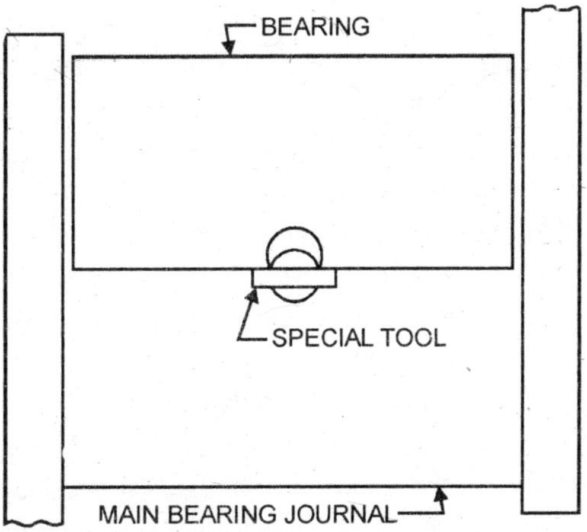

Fig. 5.36. Use of special tool for bearing installation.

For installation of bearings without removing the crankshaft, a special tool (Fig. 5.35) may be used. It is inserted in the oil hole in the journal (Fig. 5.36) and the bearing shell is placed with its hole matching the tool stem. The crankshaft is then rotated which causes the tool to carry away the bearing shell underneath the shaft. The second shell-half is placed over the shaft, the cap installed and the nuts tightened to specified torque value.

10. ENGINE REASSEMBLY

The major steps involved in the engine reassembly after the overhauling are:

1. Final and thorough cleaning of all the engine parts to remove any dust or dirt.
2. The crankshaft is thoroughly coated with the recommended lubricant and carefully placed.
3. The rear main oil seal is installed. The crankshaft and the new main bearings are lubricated. The crankshaft is installed and bearing caps fixed to the standard torque values.
4. Timing gears, sprockets, chains, etc. are aligned and installed. The backlash in the timing gear or chain is checked and ensured so that the same is within limits. New oil seal is then installed in the timing cover. The timing cover is then installed using a new gasket. Before fixing the timing cover, the crankshaft damper is also installed.
5. Piston-connecting rod assemblies are installed.
6. Oil pump, cleaned and lubricated, is installed with new gasket and seals.
7. The oil pan is installed with new gasket.
8. Cylinder head is assembled and installed. The valve lifters and push rods are then placed, followed by rocker arms, in correct locations.
9. Intake manifolds are fixed into the cylinder block, using new gaskets. The water pump and outlet neck are then installed.
10. In case the engine is to be fitted to automobile chassis directly, the other parts like flywheel, clutch, gear box (transmission) housing, fuel pump, carburettor, distributor, alternator, power accessories, fan belt etc. are installed. Before starting the engine, it is pre-lubricated by cranking without spark plugs until oil pressure builds up. However, if the engine is to be stored for future use, all opening should be sealed and the entire engine packed properly in a plastic cover.

11. PRECAUTIONS

During engine service following general precautions must be followed for safety.

1. Hot engine parts should be handled with care.
2. Taking out lubricating oil from the hot engine must be done with care; the same may be hot enough to cause burns.
3. Be careful while removing radiator cap of a hot engine. Cooling system is pressurised and sudden removal of cap may cause the coolant to boil and overflow causing burn injuries on hand and face.
4. There are many heavy engine parts. Use proper equipment and methods to handle such parts. Negligence on this account may result in personal injury or in the part being damaged.
5. Engine mating surfaces are precision machined. Such surfaces must be handled with utmost care to avoid damage.
6. Wrong connections may cause damage to electrical system. Required precautions should be taken to avoid this.
7. Utmost care should be exercised to maintain cleanliness of not only the engine parts, but also the entire service area because a very small amount of dirt may damage the friction surfaces severely or may even cause complete engine failure.

THEORY QUESTIONS

1. Describe the circumstances which necessitate the servicing of an automotive engine.
2. Discuss in detail various steps for installing cylinder head gasket.
3. Give in detail the procedure for servicing the valve system in an engine.
4. Discuss how you will remove, service and reinstall the piston-connecting rod assembly. Mention important precautions to be observed.
5. How do you measure the amount of wear in case of connecting rod big end bearings? Discuss various methods.
6. How are the ovality and the taper of a cylinder bore checked? Discuss the procedure for repair in case the cylinder wear is found to be more than the permissible limits.
7. Discuss in detail the inspection and repair procedures for the engine crankshafts.
8. How do you check the wear of main bearings? Explain in detail the procedure of replacing the main bearings. Mention also the important precautions.
9. Discuss the general precautions to be observed during servicing of an engine.

QUESTIONS FOR VIVA

1. Why should you cover the carburettor before decarburising?
2. Why it it necessary that cylinder head nuts or bolts should be tightened and loosened in proper sequence?
3. Why should you use a torque spanner specifically for tightening cylinder head nuts?
4. What do you do in case you find the cylinder head jammed over the block after you have removed all nuts for removing the head?
5. Why is it a must to allow the aluminium cylinder head cool completely before its removal?
6. What ia s head lifter screw? How is it used?
7. How do you remove carbon from inside the valve guide?
8. How do you detect small cracks in the cylinder head or block casting?
9. Why should each valve be marked properly before being removed for service?
10. In what make of valves a split guide is provided?
11. What should be done in case a valve is found to be cracked, badly pitted or with a bent stem?
12. How will you proceed when you find no more margin is left on the valves you have removed from the engine for servicing?
13. What is the approximate permissible valve stem clearance in the guide?
14. Why is it necessary that the valve and the valve seat are concentric and have the same angle?
15. What is the disadvantage of a valve seat which is too narrow?
16. Why don't we keep a very wide valve seat?
17. How do you narrow down a valve seat which has become too wide due to a number of previous grindings?
18. How do you check whether there is proper contact between the valve and the valve seat?
19. Should the valve or the guide be serviced first? Why?
20. Out of the valve seat and the guide which should be trued up first and why?
21. How do you check the alignment of a camshaft?
22. How do you straighten out a bent camshaft?
23. Why do the camshaft bearings have a longer life than the main bearings?
24. What is the approximate permissible camshaft bearing clearance?
25. What will happen if the abrasive particles during servicing operation enter the oil holes and the engine is assembled without removing them?
26. From which side of the cylinder block is the piston-connecting rod assembly taken out for servicing?

27. Why should the piston-connecting rod assemblies be marked with the number of the respective cylinders from where they are removed?
28. How do you remove the ridge at the cylinder top?
29. Why should the cylinder ridge be removed before removing pistons?
30. Why should the rings be removed preferably with a special tool called ring remover?
31. What is the cause of a scuffed piston?
32. How should hard carbon be removed from pistons?
33. How do you check the wear of the ring grooves?
34. What will happen if the end clearance of rings is less than the specified value?
35. How will you check the end clearance of a piston ring?
36. At what point of the cylinder bore should the end clearance of a ring be checked?
37. Without using any special fixture, how will you determine whether connecting rod is properly aligned or not?
38. What should be done with a connecting rod which is found to be misaligned?
39. What happens if a misaligned connecting rod is allowed to run in the engine?
40. What happens if the big end bearing clearances exceed the permissible limits?
41. Name various methods of measuring bearing clearances.
42. Why is it necessary that bearing caps be tightened to the specified torque value before measuring the bearing clearances?
43. What is a 'Plastigage'?
44. What are the precautions to be observed while using a 'Plastigage?'
45. What is the approximate maximum permissible value for the bearing clearance?
46. What is the function of shims in bearings?
47. Name the main sources of bearing failure.
48. Where in the cylinder bore does the maximum wear occur?
49. What over sizes are generally available for pistons?
50. Due to some accident, only one cylinder bore has been damaged and it is found that it needs reboring to next oversize, whereas other cylinder bores are perfectly alright. Should only the effected cylinder be rebored and fitted with oversize piston or should all the cylinders be rebored oversize?
51. What is honing?
52. Why is honing necessary after reboring?
53. Why is thorough cleaning of a cylinder block a must after reboring and honing?
54. What does a cylinder hone consist of?
55. When should a cylinder be honed?
56. How are the honing stones classified?
57. How should a honed cylinder be cleaned?
58. In case of a cylinder block brought for servicing it is found that the same had been rebored already a number of times and so it was no longer possible to rebore it further. What is the remedy?
59. How do you install a new cylinder liner?
60. What is interference fit? Why is it necessary while installing a new dry liner?
61. What is the approximate range of interference fit for installing dry liners?
62. How do you instal a dry liner with the required interference fit?
63. While reinstalling the piston-connecting rod assembly in the cylinder, on which side should you keep the split side of the piston skirt?
64. What is the function of a ring compressor?
65. How do you check the alignment of a crankshaft?

66. What is the disadvantage of changing the main bearings without removing the crankshaft from engine?
67. Why should you check the saddle bore before installing a new bearing?
68. How are the worn out bearings again bored to the original standard size?

MULTIPLE CHOICE QUESTIONS

Mark the correct answers:

1. The engine requires overhauling in case of
   (*a*) poor compression (*b*) excessive consumption of lubricating oil
   (*c*) mechanical failure (*d*) all of the above
2. The carbon from the cylinder head is removed with
   (*a*) water (*b*) caustic soda
   (*c*) scraper (*d*) soap
3. The warpage of cylinder head is checked with a
   (*a*) metal rule (*b*) vernier calliper
   (*c*) micrometer (*d*) dial gauge
4. The cylinder head nuts should be tightened to a torque of about
   (*a*) 15 Nm (*b*) 50 Nm
   (*c*) 150 Nm (*d*) 1500 Nm
5. Aluminized valves are never ground; they have to be replaced because
   (*a*) aluminium is a costly material
   (*b*) aluminium surface is difficult to be ground
   (*c*) grinding such valves will make them corrosion-prone
   (*d*) none of these.
6. The valves have to be replaced if the metal to be removed for smoothening their end surfaces is more than
   (*a*) 0.05 mm (*b*) 0.5 mm
   (*c*) 1.5 mm (*d*) 5 mm
7. The width of the valve seat should not be kept more than
   (*a*) 0.2 mm (*b*) 0.5 mm
   (*c*) 1.2 mm (*d*) 5 mm
8. While lapping a valve, the lapping compound is applied to its
   (*a*) face (*b*) stem
   (*c*) guide (*d*) tip
9. The stem clearance in the valve guide is checked by means of a
   (*a*) metal rule (*b*) vernier calliper
   (*c*) micrometer (*d*) dial gauge
10. Maximum permissible bearing clearance for the camshaft is usually
    (*a*) 0.05 mm (*b*) 0.15 mm
    (*c*) 0.5 mm (*d*) 1.5 mm
11. The bearing clearance is never measured with a
    (*a*) vernier calliper (*b*) micrometer
    (*c*) fealer gauge (*d*) plastigage
12. Maximum permissible ovality in a cylinder is usually
    (*a*) 0.01 mm (*b*) 0.1 mm
    (*c*) 0.5 mm (*d*) 1.0 mm
13. The cylinder bore oversize for reboring is calculated on the basis of

(*a*) minimum ovality in any cylinder
(*b*) maximum ovality in any cylinder
(*c*) average value of ovality in the cylinder
(*d*) size of the piston available

14. For checking the alignment of the crankshaft, the best method is to place it
(*a*) on V-blocks (*b*) on its ends
(*c*) on lathe centres (*d*) on a plane smooth table

15. The crankshaft bending should not generally exceed
(*a*) 0.008 mm (*b*) 0.08 mm
(*c*) 0.8 mm (*d*) 8 mm.

[**Key:** 1. (*d*), 2. (*c*), 3. (*a*), 4. (*c*), 5. (*c*), 6. (*b*), 7. (*c*), 8. (*a*), 9. (*d*), 10. (*a*), 11. (*a*), 12. (*a*), 13. (*b*), 14. (*b*), 15. (*b*)]

Cooling System

6

1. NECESSITY

All the heat produced by the combustion of fuel in the engine cylinders is not converted into useful power at the crankshaft. A typical distribution for the fuel energy at full engine load is given below:

Useful work at the crank shaft	= 35 percent
Loss to the cylinders walls	= 25 percent
Loss in exhaust gases	= 30 percent
Loss in friction, radiation, misfiring, etc.	= 10 percent

Besides, there is appreciable pumping loss of about 15% of the total input at idling, which vanishes at full load.

It is seen that the quantity of heat given to the cylinder walls is considerable and if this heat is not removed from the cylinders it would result in the preignition of the charge. In addition, the lubricant would also burn away, thereby causing the seizing of the piston. Excess heating will also damage the cylinder material.

Keeping the above factors in view, it is observed that suitable means must be provided to dissipate the excess heat from the cylinder walls, so as to maintain the temperature below certain limits.

However, cooling beyond optimum limits is not desirable, because it decreases the overall efficiency due to the following reasons:

1. Thermal efficiency is decreased due to more loss of heat to the cylinder walls.
2. The vaporization of the fuel is less; this results in fall of combustion efficiency.
3. Low temperatures increase the viscosity of lubricant and hence more piston friction is encountered, thus decreasing the mechanical efficiency.

Though more cooling improves the volumetric efficiency, yet the factors mentioned above result in the decrease of overall efficiency.

Thus it may be observed that only sufficient cooling is desirable and any deviation from the optimum limits will result in the deterioration of the engine performance.

2. METHODS OF COOLING

Various methods used for cooling of automobile engines are:

1. Air cooling 2. Water cooling

3. AIR COOLING

The basic principle involved in this method is to have current of air flowing continuously over the heated metal surface from where the heat is to be removed. The heat dissipated depends upon following factors:

(*a*) Surface area of metal into contact with air.

(*b*) Mass flow rate of air.

(*c*) Temperature difference between the heated surface and air.

(*d*) Conductivity of metal.

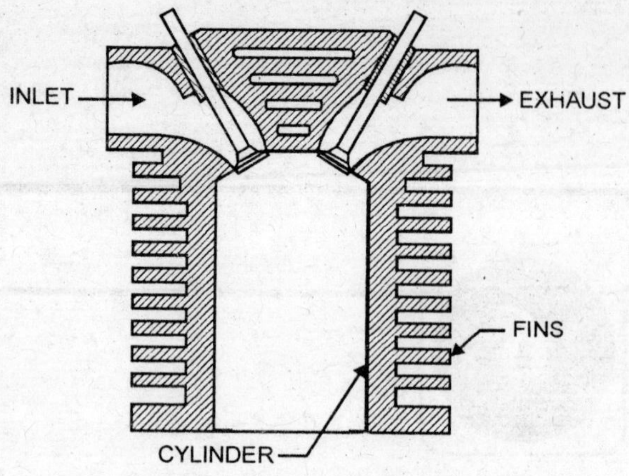

Fig. 6.1. Cylinder with cast fins.

Thus for an effective cooling the surface area of the metal which is in contact with the air should be increased. This is done by using fins over the cylinder barrels. These fins are either cast as an integral part of the cylinder or separate finned barrels are inserted over the cylinder barrel. Sometimes, particularly in the case of aero engines, the fins are machined from the forged cylinder blanks. Fig. 6.1 shows a cylinder with fins over the cylinder barrel.

To increase the contact area still further, baffles are used sometimes. Fig. 6.2 shows a few types.

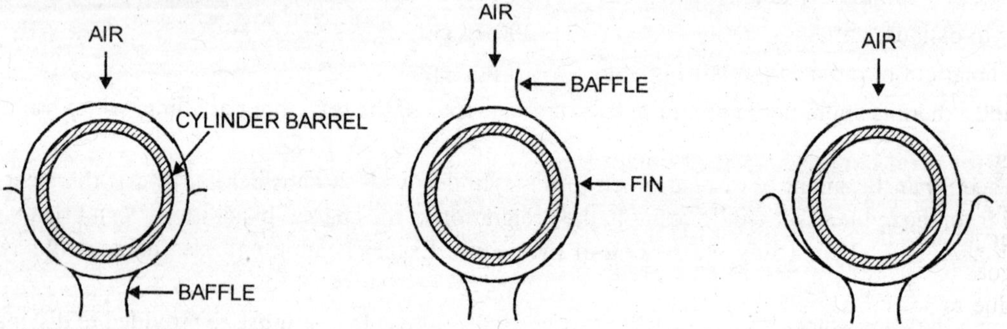

(*a*) Normal baffle (*b*) Complete baffle (*c*) Short baffle

Fig. 6.2. Different tapes of baffles.

Use of copper and steel alloys has also been made to improve heat transfer because of their better thermal conductivity.

Advantages:

1. Air cooled engines are lighter because of the absence of the radiator, the cooling jackets and the coolant.
2. They can be operated in extreme climates, where the water may freeze.
3. In certain areas where there is scarcity of cooling water, the air cooled engine is an advantage.
4. Maintenance is easier because the problem of leakage is not there.
5. Air cooled engines get warmed up earlier than the water cooled engines.

Disadvantages:

1. It is not easy to maintain even cooling all around the cylinder, so that the distortion of the cylinders takes place. This defect has been remedied sometimes by using fins parallel to the cylinder axis (Fig. 6.3). This is also helpful where a number of cylinders in a row are to the cooled. However, this increases the overall engine length.
2. As the coefficient of heat transfer for air is less than that for water, there is less efficient cooling

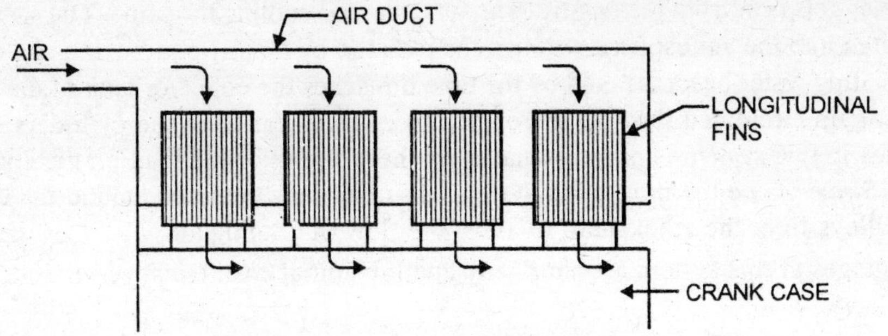

Fig. 6.3. Use of longitudinal fins.

in this case and as a result the highest useful compression ratio is lesser in the case of cooled engines than in the water cooled ones.

3. The fan used is very bulky and absorbs a considerable portion of the engine power (about 5%) to drive it.
4. Air cooled engines are more noisy, because of the absence of cooling water which acts as sound insulator.
5. Some engine components may become inaccessible easily due to the guiding baffles and cooling, which makes the maintenance difficult.
6. The cooling fins around the cylinders may vibrate under certain conditions due to which noise level would be considerably enhanced.

4. WATER COOLING (LIQUID COOLING)

In water cooling system, the cooling medium used is water. In this, the engine cylinders are surrounded by water jackets through which the cooling water flows. Heat flows from the cylinder walls into water which goes to the radiator where it loses its heat to the air. Usually some antifreeze is added to the cooling water, due to which it is often referred to as *coolant* and, the cooling system then also called the 'liquid cooling system'. Both these terms have been used in this chapter, often meaning same unless the context requires otherwise.

Water cooling systems are of two types:

1. Thermosyphon system
2. Pump circulation system

1. Thermosyphon system. A very simple system, which was used in many early automobiles upto about nineteen forty, is obsolete now. It consists of a radiator connected to the engine through flexible hoses. In this system, circulation of water is obtained from the difference in densities of the hot and the cold regions of cooling water. The

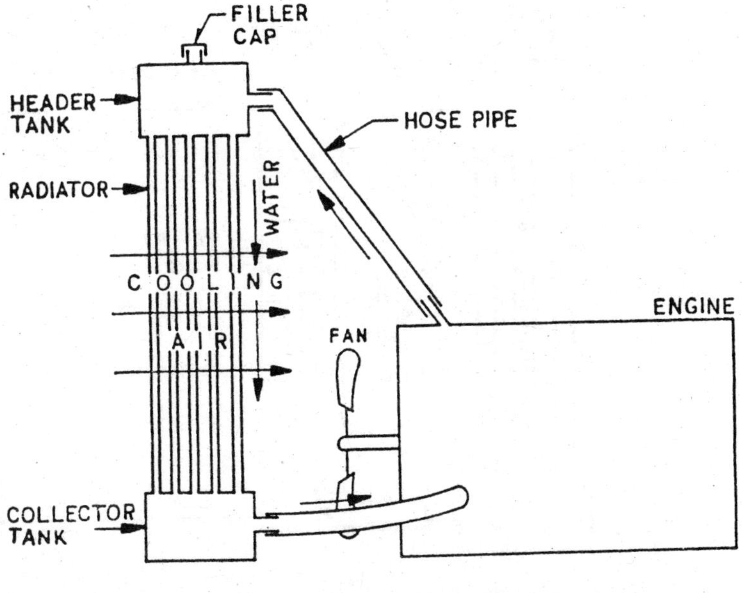

Fig. 6.4. Thermosyphon system of cooling

circulating water gets heat from the engine cylinders, thereby cooling the same. The same heat in the water is then dissipated into the atmosphere, through the radiator, by mainly conduction and convention. Therefore, the circulating water becomes cold by the time it reaches the collector tank of the radiator. The same water is then again circulated through the engine to collect heat from the cylinders. The rate at which water circulates in this system is proportional to the heat output or the load on the engine and not to the engine speed. Some of the thermosyphon systems also had fans mounted behind the radiator and driven by belt and pulleys from the crankshaft, to assist the flow of cooling air.

The advantages of this system are simplicity and low initial cost. However these are accompanied by many disadvantages also:

1. As the circulation of coolant is maintained by natural conventional only, the cooling is rather slow. Therefore, to have adequate cooling, the capacity of the system has to be large.
2. Due to the quantity of coolant being large, it takes, more time for the engine to reach the operating temperature.
3. Radiator header tank must be located higher than the top of the cylinder coolant jackets, which is no more possible with the modern body styles.
4. Certain minimum level of coolant must be maintained in the system. If the coolant falls below that level, continuity of flow would break and the system would consequently fail.

2. Pump circulation system. This system is similar to thermosyphon system described above with the only differences that a pump is used for the circulation of coolant and a thermostat is employed to control the flow of coolant (Fig. 6.5). The pump is driven by means of a belt from the engine crankshaft. The drive for the fan is also obtained from the same belt that drives the pump and the generator.

This system has the following advantages over the thermosyphon system:

1. Circulation of coolant is proportional to both load and speed.
2. Circulation of coolant is positive and hence more efficient due to which the smaller water jackets can be used resulting in overall decrease of engine size.

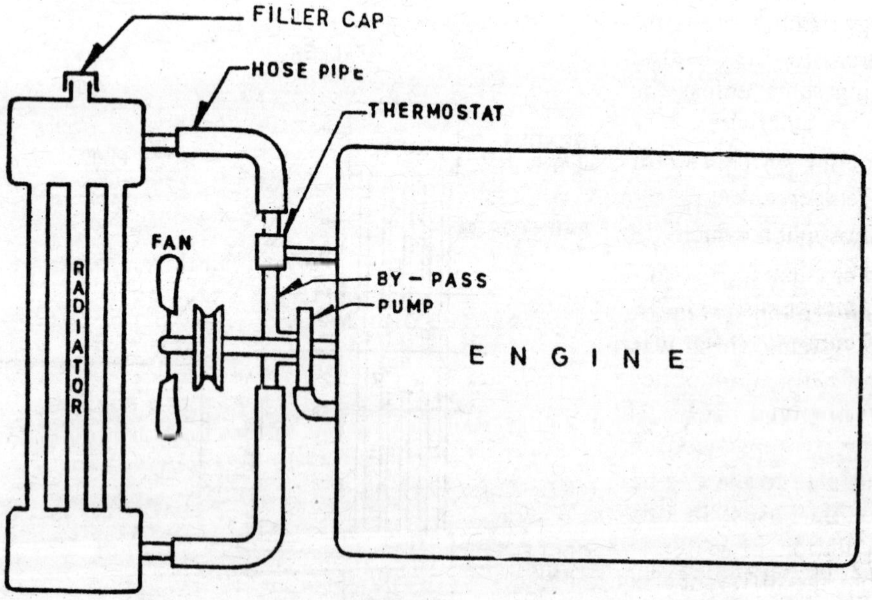

Fig. 6.5. Pump circulation system.

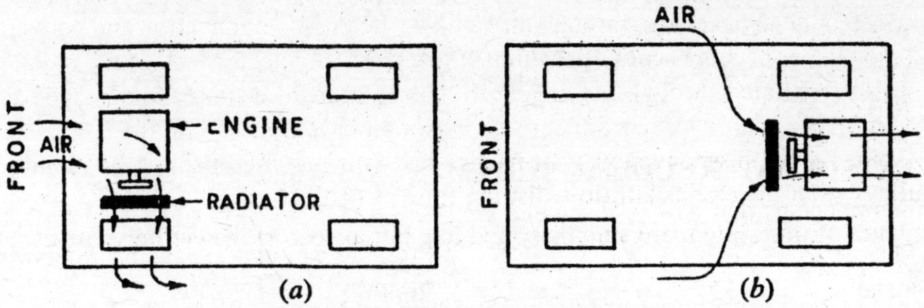

Fig. 6.6. Unconventional positioning of radiator.
(*a*) Side (with transverse engine), (*b*) Rear (with rear engine).

3. Unlike in thermosyphon system, it is not necessary to place the radiator header (upper) tank at above the engine level.
4. Even the radiator need not be placed in the front. It can be placed on the side [Fig. 6.6 (*a*)] or in the rear [Fig. 6.6 (*b*)], if the design conditions so require.

However, this is more complicated and costlier than the thermosyphon system.

Example 6.1. *A water cooled, four cylinder engine with cast iron cylinder walls, bore 80 mm, stroke 110 mm and wall thickness 7 mm is running at 3500 r.p.m. At this speed, 10 percent of the energy input is being transferred to the cylinder walls. The consumption of fuel, whose calorific valve is 42,000 kJ/kg is 300 gm/minute. Assume thermal conductivity of cast iron as 168 kJ/m hr.°C.*

Calculate:

(*i*) *Temperature drop through cylinder walls.*

(*ii*) *Temperature drop from the cylinder walls to cooling water which is being circulated at a speed of 5 cm/sec. Assume film coefficient of heat transfer at this speed as 37800 kJ/m² hr°C.*

Solution.

(*i*) Energy input $= .3 \times 42{,}000 = 12600$ kJ D minute

∴ Hat transfer, $Q = 0.1 \times 12600 = 1260$ kJ D minute

$$= \frac{12060 \times 60}{4 \times \pi (80 \times 10^{-3}) \times (110 \times 10^{-3})} = 683 \times 10^3 \text{ kJ/m}^2\text{hr}$$

Let T = Temperature drop through cylinder walls.

k = Thermal conductivity of wall material.

= 168 kJ/m hr.°C (given)

x = Thickness of cylinder wall

$= 7 \text{ mm} = 7 \times 10^{-3}$m

Then, $$Q = \frac{k}{x} \Delta T$$

∴ $$\Delta T = Q \cdot \frac{x}{k} = \frac{(683 \times 10^3) \times (7 \times 10^{-3})}{168} = 28.46°\text{C}$$

(*ii*) Let $T2$ be the temperature drop from cylinder walls to cooling water

h = film coefficient of heat transfer

= 37800 kJ/m³.hr.°C (given)

Then $Q = h \times \Delta T'$

$$\Delta T' = \frac{Q}{h} = \frac{683 \times 10^3}{37800} = 18.07°C$$

5. COMPONENTS OF WATER COOLING SYSTEM

The main components that will be discussed are:

1. Radiator
2. Pressure cap and expansion reservoir
3. Thermostat
4. Pump
5. Fan

6. RADIATOR

The function of the radiator is to ensure close contact of the hot coolant coming out of the engine with outside air, so as to ensure high rates of heat transfer from the coolant to air. A radiator consists of an upper (or header) tank, core and the lower (or collector) tank (Fig. 6.7). Besides, an overflow pipe in

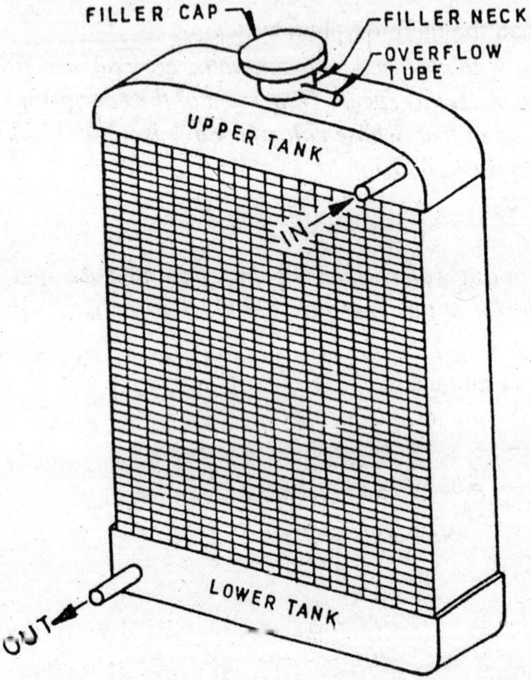

Fig. 6.7. Radiator.

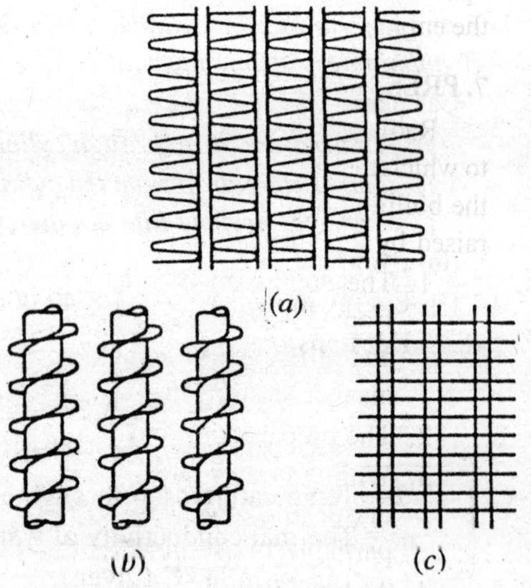

Fig. 6.8. Tubular type radiator cores. (*a*) with surpentine fins, (*b*) with spiral fins, (*c*) with plate fins.

the header tank and drain pipe in the lower tank are provided. Hot coolant from the engine enters the radiator at the top and is cooled by the cross-flow of air, while flowing down the radiator. The coolant collects in the collector tank from where it is pumped to the engine for cooling.

There are two basic types of radiator cores, *viz.*, tubular type and cellular type. In the former, it is the coolant that flows through tubes and air passes around them, while in the cellular type the air passes through the tubes and coolant flows in the spaces in between them. Out of these, tubular type cores are the most commonly used which are further classified depending upon the shape of the fins around the tubes (Fig. 6.8), which are meant to increase the area for heat transfer from coolant to the cooling air.

Both the core tubes as well as the fins are made from thinnest possible material. Tubes are made from 0.1 mm to 0.3 mm sheet, whereas fins are made from about 0.1 mm thick material.

The materials used for radiators should be resistant to corrosion, possess higher thermal conductivity and form easily, apart from having adequate strength. Copper and yellow brass are the widely used materials for radiators. Aluminium is also used from weight and cost considerations. Some late model radiators have plastic tanks with aluminium cores.

The size of the radiator must be adequate to remove the heat which is approximately equal to the heat energy utilized for producing power in the engine. Alternatively, the radiator size is matched to the displacement volume of the engine. The air conditioned vehicle would require a larger radiator due to extra heat load on account of the compressor. Besides, it is also ensured that maximum cooling is attained with minimum air resistance. Thus frontal area of the radiator is kept minimum, which may be achieved by making the core thicker and accommodating more core material into the same volume without increasing the air resistance.

For heavy duty applications, radiators shutters are also sometimes used. These are automatically controlled by means of the compressed air taken from the brake system. Shutter control mechanism is installed with the upper hose of the cooling system. Depending upon the coolant temperature coming from the engine, which itself depends upon the engine temperature, the shutter control mechanism causes the shutter to open or close. Thus, for example, the shutter remains closed during starting and it gradually opens up as the engine attains its working temperature, thus allowing the fresh air to come into contact with the radiator.

7. PRESSURE CAP AND EXPANSION RESERVOIR

Radiator filler neck in modern use is covered with a pressure cap, which forms an air tight joint due to which the coolant is maintained at some pressure higher than the atmosphere. Due to this higher pressure the boiling point of coolant is raised. For approximately, 10 kPa increase in pressure, the boiling point is raised by 2.5°C. The following advantages thus result from the use of pressure cap:

1. The engine can operate at higher temperature without boiling the coolant. As the rate of heat transfer from the cooling system to the atmosphere depends on the difference between coolant temperature and atmospheric temperature, this will result in additional heat transfer to the atmosphere. This means that for the same engine, a smaller radiator can be used.
2. The preparation of air-fuel mixture is improved at the higher operating temperatures.
3. With sealed cap, loss of coolant due to evaporation or surging is prevented, which makes the system particularly useful in deserts.
4. At high altitudes, the atmospheric pressure is low which causes the coolant to boil at a lower temperature. Thus under severe working conditions at high altitudes, *e.g.*, during steep climb, the coolant in the radiator may boil if the system is open to atmosphere. With pressure cap, a higher pressure is maintained inside, irrespective of any change in the atmospheric pressure, thus avoiding overheating while driving at higher altitudes and thus maintaining the cooling efficiency of the system.

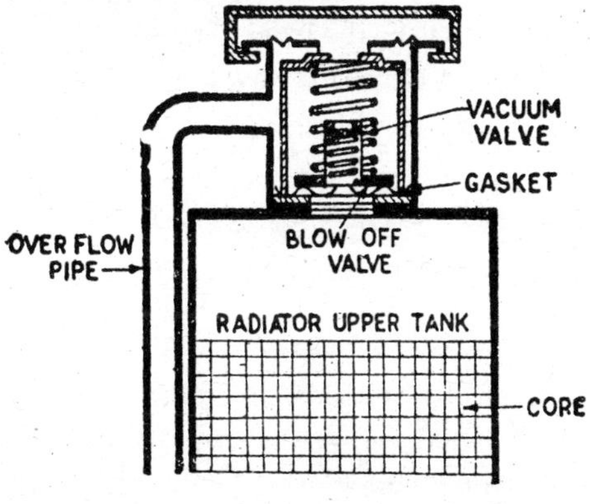

Fig. 6.9. Pressure cap.

A pressure cap is shown in Fig. 6.9. It contains a

pressure valve and a vacuum valve. When due to severe working conditions, the coolant starts boiling and vaporises, the pressure in the system exceeds a certain predetermined value (50-100 kPa), the pressure blow-off valve opens releasing the excess pressure to the atmosphere through the overflow pipe. On the other hand if due to any reason (*e.g.* sudden cooling of the hot radiator), a vacuum is created inside, the vacuum valve operates to avoid collapse of the radiator. This valve is usually set to operate when the vacuum exceeds about 5 kPa. An actual pressure cap is shown in Fig. 6.10.

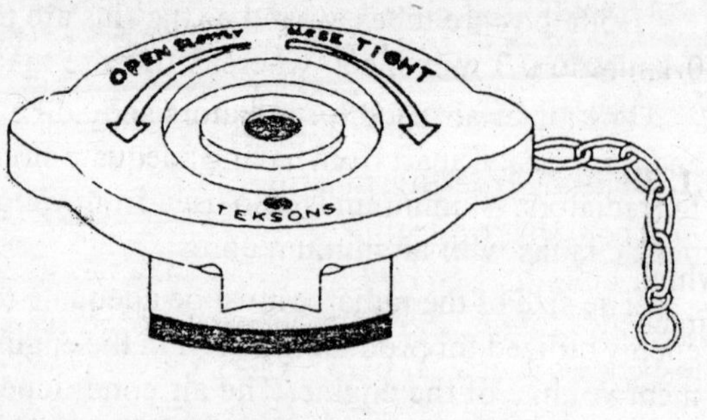

Fig. 6.10. Pressure cap.

It is important to remember that a pressure cap should never be opened when the radiator is still hot, because on removing the cap, the pressure in the cooling system will suddenly drop, causing the boiling point of the coolant to be decreased all of a sudden. This causes the coolant to start boiling immediately, which may spill over and burn anyone standing nearby severely. Further, in case of loss of cap, the replacement cap must be of the same pressure rating as the original one.

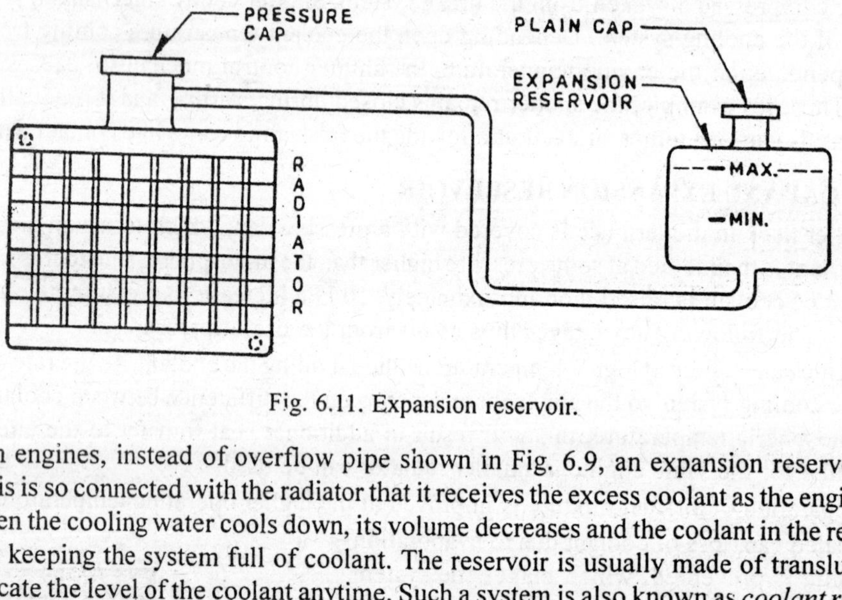

Fig. 6.11. Expansion reservoir.

In modern engines, instead of overflow pipe shown in Fig. 6.9, an expansion reservoir is provided (Fig. 6.11). This is so connected with the radiator that it receives the excess coolant as the engine temperature increases. When the cooling water cools down, its volume decreases and the coolant in the reservoir returns to the radiator keeping the system full of coolant. The reservoir is usually made of translucent plastic so that it can indicate the level of the coolant anytime. Such a system is also known as *coolant recovery system* and it has the following advantages:

1. There is no loss of coolant due to overflow on account of expansion.
2. As air does not enter the cooling system with this arrangement, corrosion of the cooling jackets and passages and deterioration of antifreeze is reduced appreciably.
3. Relatively smaller upper tank may be used with the radiator.

8. THERMOSTAT

It is already discussed in Article 1 that only optimum cooling of the engine is desirable and overcooling results in deterioration of engine efficiency. To keep a rigid control over the cooling, therefore, a thermostat is used which automatically keeps the cooling water temperature at a predetermined value. Moreover,it also helps the engine to reach the operating temperature as soon as possible after starting as the engines are designed to operate most efficiently over a small temperature range of 80° to 100°C.

Two types of thermostats are used in automobiles:

1. Bellows or aneroid type
2. Wax or hydrostatic type

These will now be described in detail.

8.1. Bellows type thermostat

It consists of metallic bellows particularly filled with some volatile liquid like acetone, alcohol or ether which boils between 70-85°C. A valve is attached to one end of the bellows, while to the other end is attached a frame which fits into the cooling passage. The thermostat is fitted in the coolant hose pipe at the engine outlet (Fig. 6.13). When the engine after start is warming up, it is desired that the cooling system

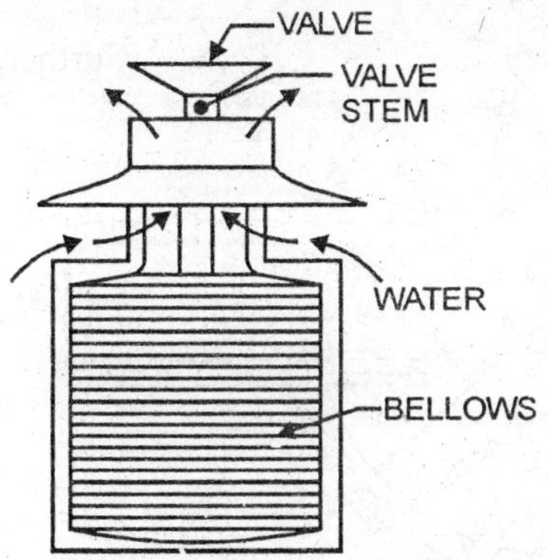

Fig. 6.12. Bellows type Thermostat

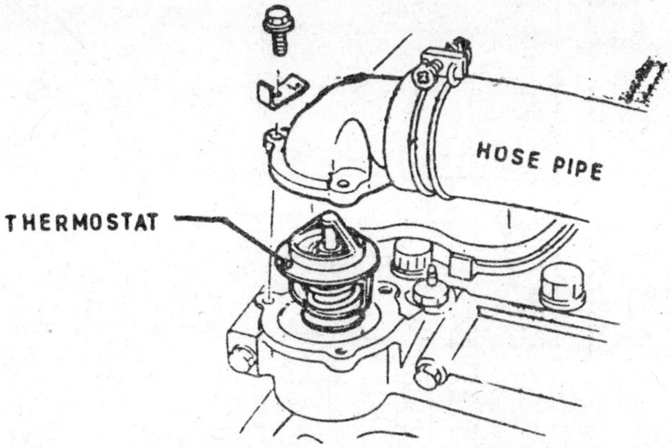

Fig. 6.13. Thermostat valve positioned on the engine.

should not operate so that the engine warms up early. During this period, the thermostat valve remains closed, because the liquid inside as yet has not changed its state and, therefore, does not exert any pressure on the valve. As the thermostat valve is closed with the coolant pump running, to avoid excessive pressure build-up, a part of the held-up coolant is made to circulate back through a by-pass to the pump inlet.

But as the coolant temperature reaches a predetermined value, (about 80°C) the liquid inside the thermostat is converted into vapour which exerts a pressure on the valve, which begins to open, so that the water circulation through the radiator starts. The valve then opens gradually further as the water temperature rises, until it is fully open at about 90-95°C. Thus the thermostat controls the flow of water through the radiator according to the engine cooling requirements.

In this type of the thermostat, the valve movement depends upon the difference between the vapour pressure in the bellows at any given temperature and the cooling system pressure.

8.2. Wax Thermostat

A wax-element thermostat manufactured by the Dole valve Co., Illinois, U.S.A. is shown in Fig. 6.14. As the coolant is heated, it transmits its heat to the copper-loaded wax having high coefficient of volumetric thermal expansion (0.28% per °C) which expands so that the rubber plug contracts against the plunger and exerts a force on it upwards so that it moves vertically. This movement of the plunger opens a valve in the thermostat (not shown) to allow coolant to flow through the radiator.

This type of thermostat, in contrast to the bellows type, is not sensitive to the pressure variations, as discussed above. Thus it is more reliable to operate within the specified temperature range, due to which reason, it is being increasingly used.

8.3. Two-stage thermostat

The construction of a two stage thermostat is similar to ordinary single stage thermostats described above, except that in this there is an additional subvalve, which starts opening at lower temperature (say 79°C), thereby allowing limited circulation of the coolant during warm-up period, thus preventing peak temperature hot spots. The main valve starts opening at some higher temperature, say 88°C and both valves are fully open at 100°C. Both the valves are fully closed below 79°C, providing quicker warm-up of the engine. Thus a two-stage thermostat provides more precise control of engine temperature by reducing peak temperature in critical areas.

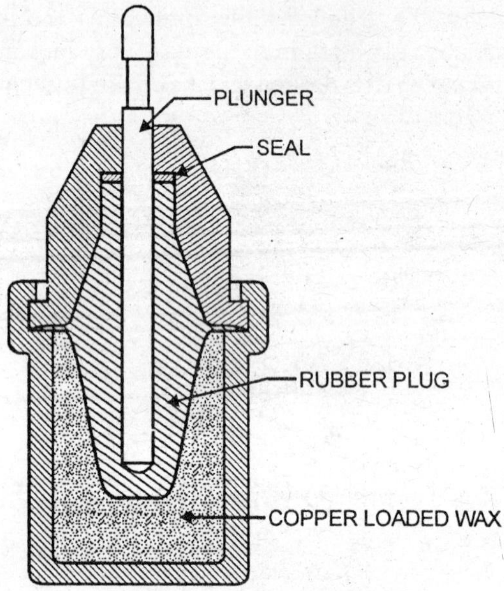

Fig. 6.14. Dole Thermostat.

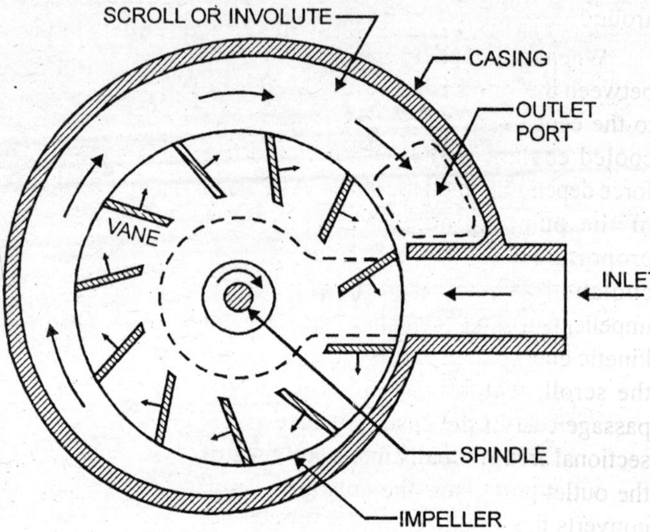

Fig. 6.15. Coolant pump

9. COOLANT PUMP

A coolant pump is a necessity for the forced circulation type of engine cooling system. The pump is mounted at the front end of the engine and is driven from the crankshaft by means of a V- belt. Centrifugal

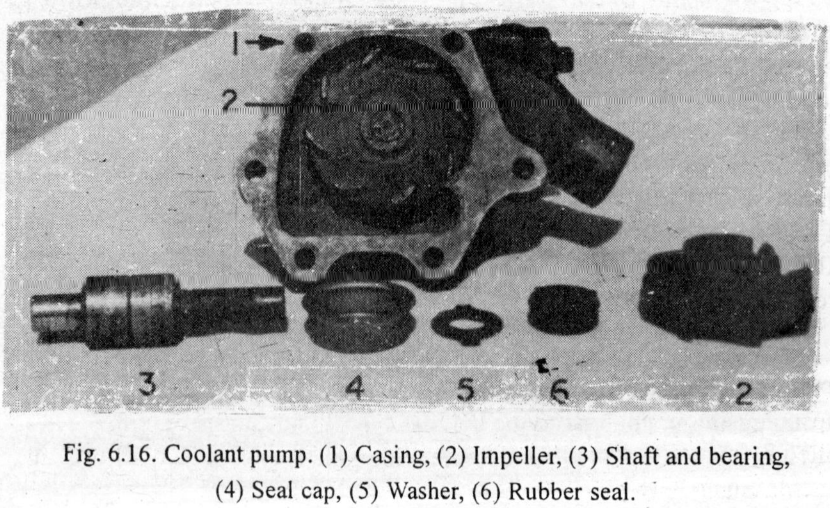

Fig. 6.16. Coolant pump. (1) Casing, (2) Impeller, (3) Shaft and bearing, (4) Seal cap, (5) Washer, (6) Rubber seal.

type pump is the one which is used for this purpose (Fig. 6.15). The coolant from the radiator enters the pump at the centre where inlet is located. The flow of the coolant depends upon the pump speed which is proportional to engine speed. This is desirable since at higher engine speeds more heat is developed which requires more cooling. The scroll is connected to the front of the engine to direct the coolant into the block. For V-engine the coolant pump has two outlets, one for each bank of cylinders. Fig. 6.16 shows a photograph depicting various components of a coolant pump. The main parts of the pump are a casing (1) and a shaft-mounted impeller (2) having a number of vanes. The impeller shaft is mounted on bearing (3) while the seal (6) serves to prevent the leakage of coolant around the shaft.

When the impeller rotates, the coolant between the vanes is thrown outward due to the centrifugal force, thus forcing the cooled coolant at the periphery, with a force depending upon the speed of rotation of the pump spindle, which itself is proportional to the engine speed. This coolant leaving the periphery of the impeller tangentially and having maximum kinetic energy then enters the involute or the scroll, which is a smoothly curved passage cast in the casing, whose cross-sectional area gradually increases towards the outlet port. Thus the enlarging scroll converts the kinetic energy of coolant to pressure energy. In this way a coolant pressure is created at the pump outlet that forces the coolant through the cooling system.

Fig. 6.17. Coolant pump (cast iron)

Different types of coolant pumps manufactured by Sealed Power Corporation. U.S.A. are shown in Figs. 6.17 to 6.19.

The present trend is to make coolant pumps smaller and lighter. Light alloys are increasingly being used for pump bodies while use of synthetic resins is on increase for impellers. The shaft-bearing design in which shaft and bearings are combined in one unit has resulted in more compact design.

Fig. 6.18. Coolant pump (aluminium)

9.1. Thermo-responsive Coolant Pump

M/s Thompson Products Inc. (U.S.A.) have devised a pump with which the circulation of coolant is directly controlled by means of metallic vanes on the impeller, instead of by the use of thermostat valve. Bimetallic elements are provided on the periphery of the impeller. These elements control the apertures on

Fig. 6.19. Coolant pump (heavy duty) for trucks, buses etc.

the periphery of the impeller. Under low temperature conditions the elements close the apertures and there is no pumping action. However, as the coolant temperature increases the elements deflect to open the apertures and the pump discharges coolant into circulation.

9.2. Smart Coolant pump

In conventional cooling pumps, as discussed above, the rate of coolant flow varies in proportion to the engine speed, because of the pump being driven directly by means of a belt. Due to this, the cooling system is always over-designed at normal operation because it has to meet full-load requirements at low engine speeds. Thus it is not flexible to actual load and driving conditions. This results in the engine heat-up times being longer than actually required, thus resul- ting in lower engine temperatures below optimum values, which causes high friction losses and low exhaust temperature.

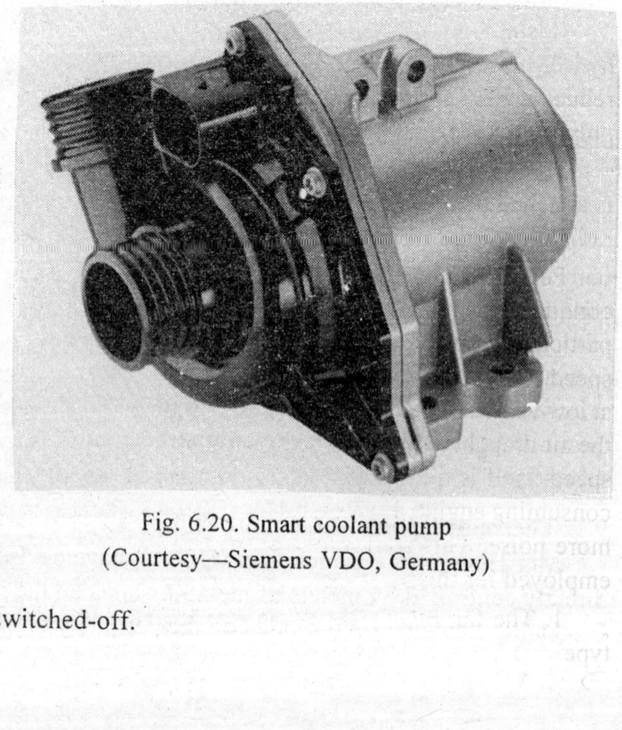

Fig. 6.20. Smart coolant pump
(Courtesy—Siemens VDO, Germany)

Siemens VDO's Smart coolant pump uses an electronically commutated 12V D.C. motor. This pump can be easily controlled by the engine management system, so that precise engine temperatures can be maintained by varying the rate of coolant flow. This is done exactly according to actual physical requirement, independent of the engine speed. This results in higher fuel combustion efficiency and improved exhaust emissions. Moreover, the pump can be installed almost anywhere, even detached from the engine block. Besides, it can be operated even with the engine switched-off.

10. FAN

When the vehicle is going at high speed with light load, the natural draft of air passing through the radiator may be sufficient for cooling of the engine, but when the vehicle is moving under heavy load and at a slow speed *e.g.* when driving uphill, the natural draft is certainly insufficient to produce the desired cooling. This explains why fan is a necessary part of the engine cooling system.

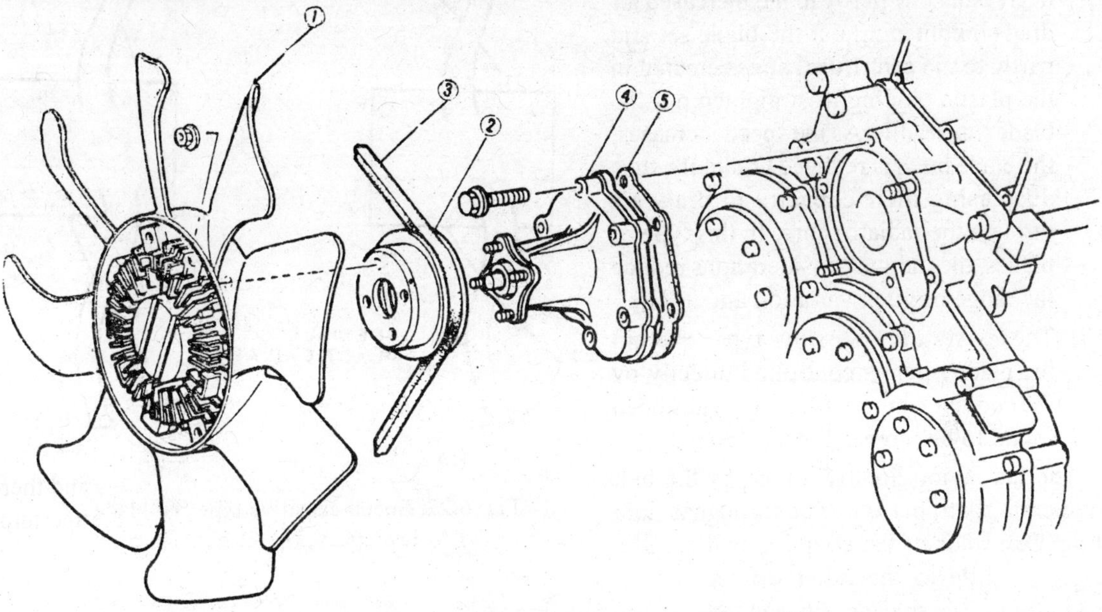

Fig. 6.21. Water pump and cooling fan
1. Cooling fan, 2. Water pump pulley, 3. Alternator drive belt, 4. Water pump, 5. Gasket
(Courtesy—Swaraj Mazda Ltd.)

It is mounted behind the radiator on the same shaft on which the water pump is mounted. It is driven by a V-belt from the crankshaft pulley. It may have four to seven blades, sometimes spaced unevenly to reduce noise. It is generally made of sheet metal, but these days moulded plastic materials *e.g.*, nylon or polypropylene are also being used for making fans.

Sometimes the fan is mounted inside a thin plastic or metal housing around its periphery. This housing is attached behind and against the radiator and is called *fan shroud.* It allows fan to pull more air past the radiator.

For efficient and economical running, it is required that the fan must give adequate air flow at all the conditions of vehicle load and speed. More flow than the minimum necessary for effecting cooling at any particular time is simply uneconomical. Thus the commonly used method of running the fan at one constant speed ratio with the engine is not desirable. If, for example, the fan is designed to give adequate air flow at low vehicle speeds, say, when going uphill when the air flow due to vehicle speed is very small, obviously the air draught at high vehicle speeds will be much more than the desired, when the air flow due to vehicle speed itself is quite high. Thus a fan that is always running when the engine runs, will be unnecessarily consuming engine power which has been estimated as much as 5 percent of the engine B.P. and producing more noise. This is clearly a wastage and must be avoided. The following methods are currently being employed for this:

1. The fan blades are of variable pitch type, which may be speed-sensitive or temperature-sensitive type.

(*i*) *The speed-sensitive* type consist of flexible one-piece plastic-moulded fan-blade assemblies (Fig. 6.22). The blades are set to have maximum pitch when stationary. As these are rotated, their pitch starts decreasing due partly to the increased air drag straightening out the blade set and partly to the centrifugal stress created in the plastic tending to straighten out the blade tips readily. As the speed increases, the blade pitch is reduced gradually, thus decreasing their capacity to draw air through the radiator core. In this type of blades, the fatigue considerations reduce the range of pitch variation and its life.

(*ii*) The *temperature-sensitive* type, wherein the blade pitch is controlled directly by the coolant temperature, which has been found to be more efficient design.

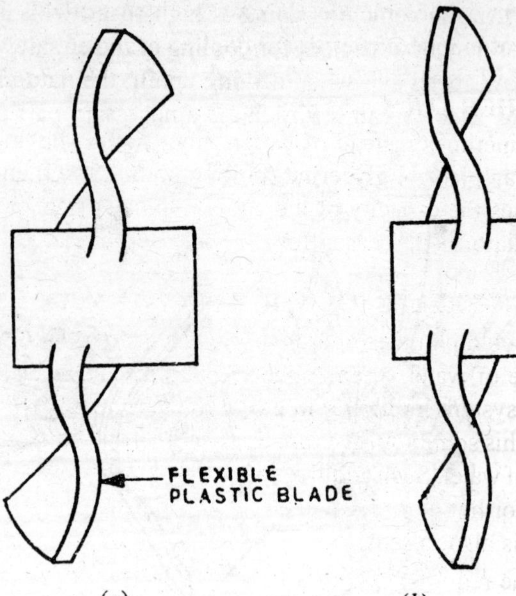

Fig. 6.22. Speed-sensitive type fan blades (*a*) At low speed, (*b*) At high speed.

2. The fan is not directly driven by the belt from the crankshaft, but is driven through a fluid coupling. One rotor of the coupling is driven by the engine, while to the other one is attached the fan. The spacing between the rotors is controlled by the temperature of the cooling system, which changes the slip of the coupling thus changing the fan speed according to the requirements.

Alternatively the fluid coupling is filled with a various fluid, *viz.*, the silicon fluid. The shearing action caused by the difference in the speeds of the two rotors transmits the torque from the drive pulley to the fan. As the engine speed increases, the torque required to transmit the drive to the fan blades also increases. Actually, the fan speed does increase, at first directly with the engine speed and then at a lower rate due to the increase in viscous drag till the limit of the torque capacity for the drive is reached. At this point, fan speed is maximum which is not exceeded even with the further increase in engine speed.

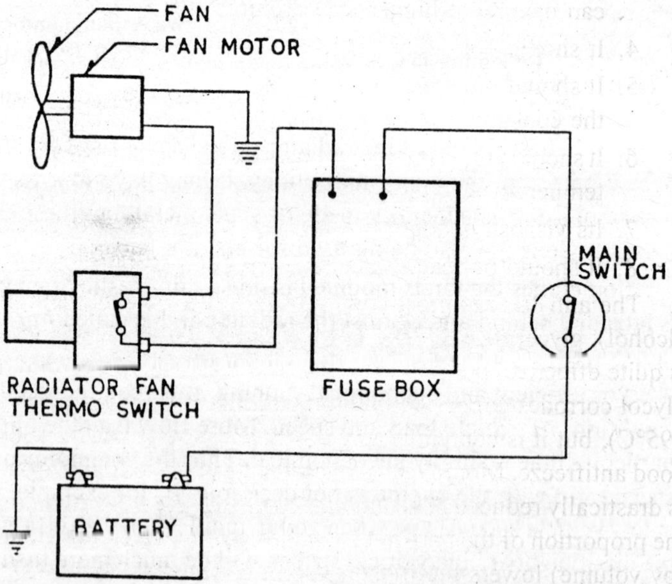

Fig. 6.23. Electric circuit for the fan motor.

3. The fan is driven by means of a separate electric motor which is supplied with power directly from the electric circuit of the engine (Fig. 6.23). This system has been used on Maruti cars. A thermostat switch is placed at an appropriate place in the cooling system and depending upon the cooling system temperature

it operates to switch on or off the fan motor. it has been found that under ordinary conditions, only about 5 percent of the time the fan motor remains in 'on' position, while 95 percent of the time it is 'off'. The saving of engine power thus achieved may, therefore,be well imagined.

11. OTHER COOLANTS

Sometimes instead of water, other liquids having higher boiling points are used for engine cooling. The examples are glycerine (boiling point 290°C) and ethylene glycol (b.p. 195°C). Higher boiling point increases the capacity of the coolant to carry heat. Consequently weight of the coolant and hence that of the radiator is decreased for the same engine output.

12. ANTI-FREEZE SOLUTIONS

In cold climates, there is always a danger that the water in the cooling system may get frozen. As the volume of water when converted into ice increases by about 10%, this may result in the damage of the entire system including in extreme cases the bursting of the radiator core and the cylinder jackets. To avoid this some additives are used, which when mixed with water in suitable proportions, lower the freezing point of water. Such additives are called antifreezes and the solution thus formed becomes antifreeze solution. Even for hot climates the use of antifreeze has the incidental advantage that boiling point of the cooling water is also raised.

The requirements of an anti-freeze may be enumerated as:

1. It should be thoroughly miscible with water and should prevent the freezing of the coolant down to the lowest ambient temperatures.
2. It should not have any corrosive action on system components, especially the radiator hose pipes.
3. Its boiling point should be high so that there is minimum loss due to evaporation and the coolant can operate at higher temperatures.
4. It should not deposit any foreign matter in the jackets, hose pipes or radiator core.
5. It should have high specific heat capacity so as to be comparable to the specific heat capacity of the coolant.
6. It should not be inflammable and its flash point should be higher than the maximum possible operating temperature.
7. Its viscosity should not be excessive so that the circulation is not affected.
8. It should be readily miscible with corrosion inhibitor and anti- scaling compounds.

The anti-freezes most commonly used are wood alcohol (methyl alcohol), denatured alcohol (ethyl alcohol), glycerine, ethylene glycol etc. Each of these has its own advantages and disadvantages. Alcohol is quite effective, but it is very much volatile and due to this reason evaporation losses are high. Ethylene glycol corrodes copper, aluminium and tin-lead solder alloys. The glycerine is less volatile (boiling point 195°C), but it is comparatively costly and also it attacks rubber hose pipes. Calcium chloride is another good antifreeze. Moreover, with chromates like sodium chromate added to it, the corrosion of most metals is drastically reduced. The amount by which the freezing point of cooling water is lowered, depends upon the proportion of the anti-freeze in the cooling water. For example, a 50% concentration of ethylene glycol (by volume) lowers the freezing point of water to about –37°C.

Antifreeze solutions usually spoil the finish of the paint work. Therefore, adequate precaution should be taken while topping up the cooling system with antifreeze solution.

12.1. Corrosion inhibitors

These are the chemicals which prevent the corrosion of the metallic parts of the cooling system and constitute about 10% of the antifreeze. In addition these also provide water-softening and thereby prevent the scale formation caused by mineral deposits. Commonly used corrosion-inhibitors are silicates and chromates.

In practice, antifreezes and corrosion inhibitors start losing their effectiveness with time. That is why, the manufacturers recommended change of coolant periodically, say, every one or two years.

13. TEMPERATURE GAUGES

These are generally provided on the engine as optional accessory. This seems to be only due to economy reasons, because the installation of the temperature gauge on the dashboard helps the driver to avert serious consequences, as he has before him the temperature indication all the time. Supposing the thermostat gets stuck up and does not open, the water circulation will stop and the engine temperatures will go very high. However, if the temperature gauge is provided, the driver can stop the engine atonce and remedy the situation, before the engine is overheated and some damage has already occurred.

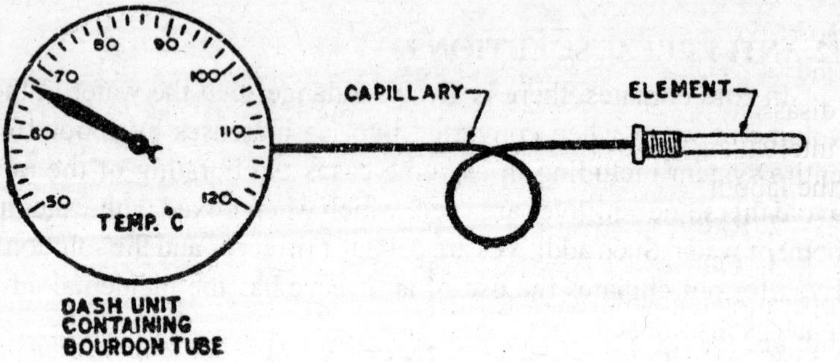

Fig. 6.24. Bourdon tube type temperature gauge.

The temperature gauges are of two types:

1. Bourdon tube type.
2. Electrically operated type.

1. Bourdon tube type. This has a Bourdon tube inside, which is connected by a capillary tube to the element, containing some volatile liquid at suitable temperature and which is inserted in the cooling water circuit at an appropriate point, generally on the engine side of the thermo- stat. As the temperature of cooling water increases, the liquid in the element evaporates and exerts its pressure in the capillary, which is further transmitted to the Bourdon tube. Due to this pressure, the Bourdon tube tries to straighten out and thus moves a pointer attached to it, to show higher temperature on the scale.

2. Electrically operated type. This contains an element made of such a material that its electrical resistance decreases with increase of temperature. The element is connected to the coils inside the dash unit as shown in Fig. 6.25 (*b*) while Fig. 6.25 (*a*) depicts entire electrical circuit. The gauge element is inserted into the coolant at some appropriate place. As the cooling water temperature in- creases, the resistance of element decreases, which causes more current to flow in the coil (2), thus

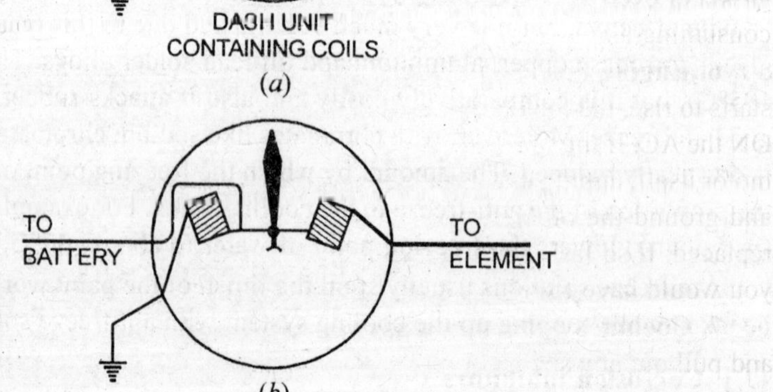

Fig. 6.25. Electrically operated type temperature gauge.

increasing the e.m.f. built up there. The pull of the coil (2) on the armature carrying indicator pointer, therefore, increases and the pointer moves to show the higher temperature.

14. COOLING SYSTEM PREVENTIVE MAINTENANCE PROGRAMME

NARSA (National Automotive Radiators Service Association) of USA recommends a seven-point programme for this purpose, to be undertaken at least once in two years. It is briefly described as follows:

1. Visual inspection of all system components, including the belts and hoses. There should not be any cracked or split hoses or any bulging or swelling at their ends. In case of any sign of such a problem, replace the hose with the correct part number for the year, make and model of the vehicle.

On most older vehicles, the water pump is driven by a V-belt on the front of the engine which is easier to inspect and replace if found cracked or otherwise damaged. However, since on some late model vehicles, the water pump may be driven by the timing belt, which is inside the engine and would require partial disassembly of the engine to inspect, it is always better and ultimately economical to replace it at the specified interval and also to replace the water pump at the same time the belt in replaced since almost 90 percent of the labour to replace water pump is already done when the timing belt is replaced.

2. Radiator pressure cap test. The cap should be able to hold the rated pressure for at least 2 minutes.

3. Thermostat check for proper opening and closing. Start the engine and bring it to the normal operating temperature. If it takes unduly long time for the engine to warm up or the heater to start delivering hot air, the thermostat may be stuck in open position. However, if the engine does heat u quickly, shut it off and touch the inlet and the outlet hoses of the radiator. If one hole is hot and the other is cold, the thermostat may be stuck closed. To confirm the suspicion, remove the thermostat and test by placing in a pot of water being heated. The opening temperature would be indicated by a thermometer, which should match the specified value.

4. Pressure test to identify external leaks. This test is performed only when the cooling system is cool enough to allow you to remove the pressure cap safely. The cooling system should be full of coolant. For the test, remove the pressure cap and attach in its place the cooling system pressure tester. Pump the tester to the specified pressure (usually given on the cap) and begin to look for leaks. Also watch the gauge on the pressure tester to see if it loses pressure. If the pressure drops more than a few pounds in two minutes, there is likely a leak somewhere which may be hidden. For easier determination of the leak source it is better to place the vehicle on a lift and look over with a shop light.

5. Internal leak test. If you are losing coolant, without any external leaks, the head gasket might have been damaged. This can be seen with the help of a combustion leak test on the radiator, using a block tester, which performs a chemical test on vapours in the radiator. If the blue tester fluid in the tester container turns yellow during the test, the exhaust gases are present in the radiator, which points to a blown head gasket or even a cracked cylinder block or head. Replacing the gasket or cylinder head/block is quite time-consuming and costly.

6. Engine fan test. If you find that soon after the vehicle comes to a stop, the engine temperature starts to rise, the fan operation is the first suspect. If the fan is not turning when the engine is hot, just turn ON the AC. If the fan starts turning, the temperature sensor in the fan circuit may be faulty. To test the fan motor itself, unplug the wire connector from the fan and instead, connect 12-volt battery to one fan terminal and ground the other one. If the fan motor starts to turn, the motor is OK. It not, the same has to be replaced. If so far, everything is OK, the trouble may be in the fan relay or fan control module for which you would have to refer the repair manual for the particular vehicle, make, model and year.

7. Cooling system power flush and refill. The power flushing is done to remove all the old coolant and pull out any sediment and scale along with it. Power flushing is done with a special machine for this purpose. The thermostat is removed. And the machine is connected with hoses, between the radiator outlet and the engine coolant inlet. Water (sometimes, a cleaning agent) is pumped through the cooling circuit in the direction reverse to the normal flow, which causes any scale to loosen and flow out. When clear water starts coming out, the hose is reconnected, a new thermostat is installed and the appropriate amount of water and antifreeze is filled in and brought up to the specified level.

15. INTELLIGENT COOLING SYSTEM

Dana's Intelligent Cooling technology enables the engine cooling system to operate at more accurate temperature with increased uniformity and greater control. Such an intelligent cooling system consists of:

1. High-performance electric water pump instead of an engine-speed-related water pump in a conventional system.
2. Multi-port, proportional flow control valve instead of thermostat.
3. Speed-controllable cooling fan.
4. Cylinder head gasket with embedded sensors.

When the engine is warming up after a cold-start, the pump and valve, being independent of the engine, zero to very low coolant flow can be achieved so as to maximise the rate at which the engine warms up. The advantages of this include improvement in thermal efficiency, improvement in fuel economy and reduced warm-up time.

Besides, the pump, being electric can be mounted anywhere at a suitable location, providing greater flexibility in engine compartment configuration designing. Moreover, even after the engine is shut off, the cooling system can monitor the engine temperature and keep the system below set limits. Even in case of extreme conditions, the system can maintain the set temperature limits.

16. COOLING SYSTEM TROUBLE SHOOTING

The common faults occurring in the cooling system are described below alongwith their possible causes and suitable remedies.

1. Overheating. This may be caused due to any one or more of the following reasons. However, it must be borne in mind that the overheating is caused not only due to defects in the cooling system. Rather the same may be caused by many factors completely different, *i.e.*, pertaining to other systems also, *e.g.*, too much retarded ignition timing, weak air-fuel mixture etc.

(*i*) Accumulation of rust or scale in the coolant jackets and the radiator. Rust and scale are both bad thermal conductors and thus the rate of heat transfer is decreased. This may be remedied by use of suitable chemicals (*e.g.* ordinary washing soda)and reverse flushing.

Reverse flushing is done by circulating water under pressure in the direction opposite to the normal. This is done with the help of special air pressure gun which forces the water through. Flushing of the engine and the radiator is done separately. The effect of the chemical is to loosen the scale and the reverse flushing then washes it away. An important precaution is to use only warm water for flushing a warm engine. In some cases the reverse flushing may cause damage to the engine. So before resorting to this, the manufacturer's manual must be studied.

(*ii*) Defective hose or faulty connections may be another source of overheating. Defective hose may become a source of the hose material accumulating in the radiator passages or engine water jackets while the faulty connections will cause leaking of coolant thus reducing the capacity of the cooling system to extract heat. The remedy is to replace the defective hose and to tighten the various connections, so that there is no more any leakage.

(*iii*) Sometimes the engine block may also be cracked, causing loss of coolant. This can be tested by plugging coolant inlet and outlet and pumping in the water under pressure. If the block is found cracked, the same may be suitably repaired.

(*iv*) Defective coolant pump reduces the volume of the coolant to be circulated and thus may be the cause of overheating. There is no definite method to test the coolant pump. However, if the upper hose is opened out and hand is placed over the water jacket outlet while the engine is running, then the feeling of some pressure (which one knows by experience) by the hand indicates that coolant pump is functioning normally. If it is not functioning properly, open it out and look for the defects, which may be corrected.

(*v*) A defective thermostat that may be sticking in the closed position will not allow the coolant to go to the radiator and be cooled there. As a result the engine will be overheated. The thermostat valve may be tested by taking it out and placing it in a hot water bath in which the thermometer is also placed (Fig. 6.26). If the valve opens and closes at the same temperatures as specified in the manufacturer's manual, it is alright; otherwise it is defective. The remedy in this case is to replace it.

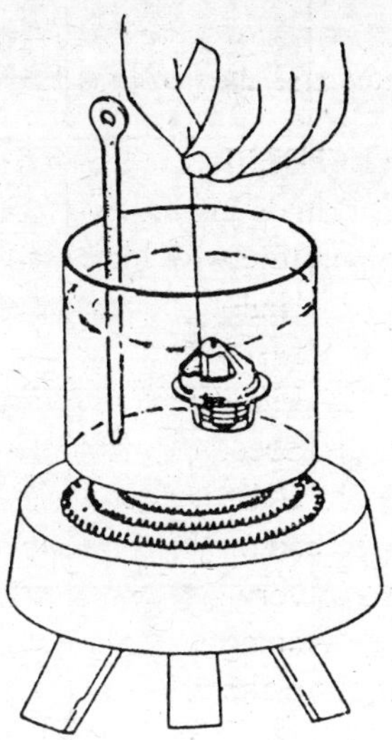

Fig. 6.26. Testing the thermostat.
(Courtesy—Maruti Udyog Ltd.)

(*vi*) Fan belt which is slipping may be another source of engine overheating. The slipping may be either due to the belt being lose or by the oiling of the belt or any of the pulleys. A loose belt may be tested by pressing it at the centre with hand. If the belt sags by more than 10-15 mm with a force of 30-40 N, it is definitely loose and needs tightening, which is done by shifting the generator and retightening its belts. In MM-540 DP jeep the belt is tightened by screwing to bring index (3) in front of face (*a*) of flange marking (Fig. 6.27). The oiled belt or pulleys should be cleaned with a piece of cloth. If the belt is otherwise defective, it should be replaced atone.

(*vii*) The air passages in the radiator may be blocked. These may be opened up by blowing air under pressure from the fan side of the radiator. The fins of the radiator may also be damaged so as to

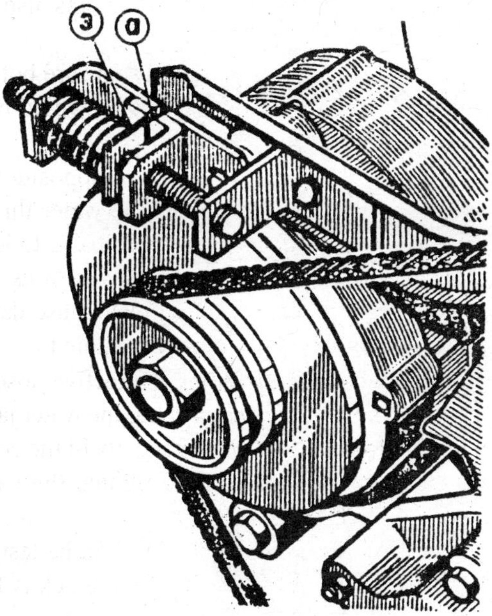

Fig. 6.27. Tightening fan belt.
(Courtesy—Mahindra and Mahindra Ltd.)

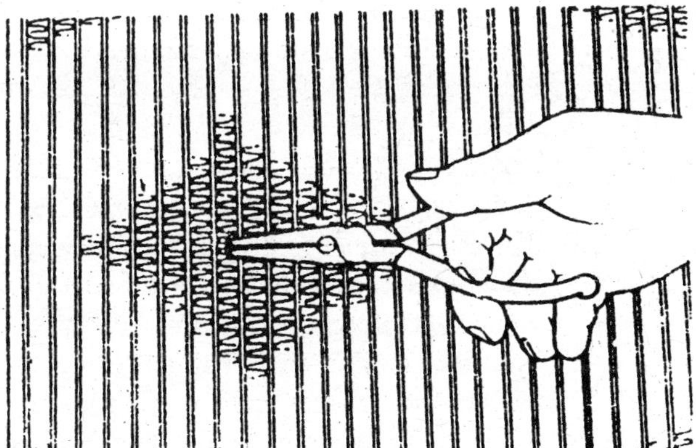

Fig. 6.28. Correcting the radiator fins.
(Courtesy—Swaraj Mazda Ltd.)

TABLE 6.1. COMPARATIVE COOLING SYSTEM DATA OF SOME INDIAN AUTOMOBILES

S. No.	Item	*Maruti 800 dor Mark II*	*Hindustan Ambassa-*	*Fiat 1100*	*Jeep Universal*	*Dodge/ Fargo Model 89M (Petrol type) 4*	*Ashok Leyland Comet Passenger*	*Standard 20*	*Tata Diesel Vehicles 1210*	*Swaraj Mazda*
1.	Type	Water cooling sealed type	Water cooling	Water cooling sealed type	Water cooling	Water cooling	Water cooling	Water cooling	Water cooling	Water cooling
2.	Capacity (litres)	3.6	7.96	4.5	10	9	25	7.41	18	13.5
3.	Thermostat	Wax pallet type, starts to open at 87°C, fully open at 95°C	Bellow type	Bellow type	Starts to open at 74°C Fully open at 86°C	Bellows type	Starts to open at 74°C Fully open at 85°C	Bellows type	Bellows/ wax type: starts to open at 82°C (80°C for wax type) fully open at 92°C (87°C for wax type)	Wax type starts to open at 82°C fully open at 95°C
4.	Fan blades	4	4	4	4	4	4	6	4	8
5.	Recom-mended anti-freeze	Special anti-freeze 'Maruti	Ethylene glycol Colden'	Special anti freeze FIAT mixture, or natural glycerine or denatured alcohol	Methyl alcohol and ethylene glycol	Special anti freeze FIAT mixture, or natural glycerine or denatured alcohol	Ethylene glycol	No particular recommen-dation	Castrol anti-freeze recommended	Any ethylene glycol based

obstruct the flow of air. The same may be corrected with pliers or screw driver (Fig. 6.28).

(*viii*) Loss of coolant by evaporation may also be a cause of overheating, which may be corrected by filling the radiator to the correct level.

(*ix*) Radiator fan thermostat switch, if provided, may be defective. This can be tested by immersing the switch in hot water and testing its continuity (Fig. 6.29). If the switch conducts when the hot water temperature rises to 86-90°C and it does not when the hot water temperature falls down to 81-85°C, it is in good condition. If found defective, the switch should be replaced.

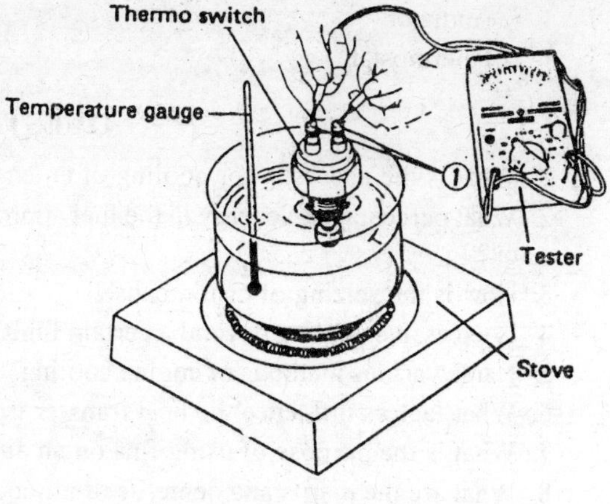

Fig. 6.29. Testing the radiator thermostat switch. (Courtesy—Maruti Udyog Ltd.)

2. Slow warm up (or overcooling). When the engine takes unduly more time to warm, defect may be also due to the thermostat valve being stuck open. It may be removed and tested. If found defective, the same has to be replaced.

3. Noisy coolant pump. If the coolant pump is making too much noise, the reason may be either defective seal or defective bearing. The only remedy is the replacement of the faulty components.

THEORY QUESTIONS

1. Name different methods of engine cooling. Explain in detail the air cooling method.
2. Discuss in detail the water cooling system for automotive engines.
3. What are the main components of water cooling system for an automotive engine? Describe these in detail.
4. Discuss the pump circulation cooling system thoroughly and compare the same with the thermosyphon system.
5. With the help of a neat sketch show the construction of a radiator. Discuss different types of radiator cores used in modern car engines?
6. What is pressure cooling system? Illustrate your answer with a clear sketch of the pressure sealed cap. Explain the advantages of such a system.
7. Describe in detail the coolant recovery system. Illustrate your answer with a neat sketch.
8. What is the necessity of using thermostat in the engine cooling system? Discuss the construction and working of any thermostat.
9. Which type of coolant circulation pump is used in an engine cooling system? Describe its construction and explain the working.
10. Why is a fan required in the engine cooling system? Discuss various methods of controlling the speed of fan.
11. Write a comprehensive note on liquid cooling and explain its advantage over the water cooling system.
12. Explain why it becomes essential to use anti-freezes when the atmospheric temperature falls below zero. Discuss the requirements which are taken into account while selecting a suitable antifreeze and describe various commonly used antifreeze compounds.
13. With the help of suitable diagrams, explain various types of temperature gauges.
14. Describe in detail any preventive maintenance programme for the engine cooling system.
15. Discuss the common faults occurring in the engine cooling system and also suggest suitable remedies.

16. Describe in detail how you would test:
    (*a*) radiator
    (*b*) thermostat

QUESTIONS FOR VIVA

1. What is the necessity for cooling of an engine?
2. What percentage of energy in the fuel approximately is absorbed by the cylinder walls and the lubricating oil?
3. How is the seizing of piston caused?
4. Is excessive cooling beyond a certain limit desirable? If not, why?
5. Name various methods of engine cooling.
6. What factors influence the heat transfer from engine to the surrounding air?
7. What is the purpose of using fins on an air cooled engine cylinder?
8. What are the merits and demerits of air cooled engines?
9. It is said that ordinary air cooled cylinders are subjected to distortion. It is true?
10. What is the magnitude of power absorbed by the fan in air cooled engines?
11. Which types of engines generally employ air cooling system?
12. Out of thermosyphon and pump circulation systems of water cooling which is better and why?
13. In the thermosyphon system, why should the header tank of radiator be placed at a level higher than the engine?
14. In case of thermosyphon system can be radiator be placed at the rear? Why?
15. Which out of the thermosyphon and the pump circulation systems result in lower engine weight?
16. Name the main components of water cooling system.
17. Where is the inlet for cooling water in the engine?
18. Where is the coolant outlet in the engine located?
19. What is the function of a radiator in cooling system?
20. What is tubular type of radiator core?
21. Name any three tubular type radiator cores.
22. What properties should the radiator material have?
23. Name the commonly used materials for radiator cores.
24. What materials are used for the radiator tanks?
25. Why should the radiator have minimum frontal area?
26. What is a pressure cap?
27. Which valves does a pressure cap have?
28. What is the advantage of using a pressure cap on the radiator filler neck?
29. What is the approximate relationship between the increase in pressure and raise of boiling point of water?
30. What advantages are obtained when engine operates at higher temperatures?
31. How is pressure-sealed cooling useful at higher altitudes?
32. What is the advantage of blow-off valve in the pressure cap?
33. When does the vacuum valve in the pressure cap operate?
34. Why should a pressure cap be not opened with the radiator still hot?
35. What is an expansion reservoir?
36. What is the purpose of using translucent material for the expansion reservoir?
37. What is a coolant recovery system?
38. State the advantages of the coolant recovery system?
39. What is the necessity of using a thermostat in the cooling system of an engine?

40. Name the two types of commonly used thermostats.
41. Which liquids are usually used in the bellows types thermostat?
42. What is the purpose of a thermostat bypass?
43. At what temperature does the thermostat start opening?
44. At what temperature approximately is the thermostat fully open?
45. What is a two-stage thermostat? What is its function?
46. What is the function of coolant pump?
47. Where is coolant pump mounted on the engine?
48. How is the drive for coolant pump obtained?
49. Which type of pump is used for circulating coolant?
50. Name the main parts of a coolant pump.
51. How is the water pressure created at the outlet of the coolant pump?
52. What materials are used for coolant pump casing and impeller?
53. What is a thermo-responsive coolant pump?
54. What is the necessity of fan in the water cooling system?
55. Where is the fan mounted?
56. How is the drive for fan obtained?
57. How many blades does the fan usually have?
58. Why are the fan blades sometimes spaced unevenly?
59. What materials are commonly used for fan blades?
60. What is fan-shroud?
61. Why it is not economical to run the fan continuously when the engine is running?
62. Which type of fan control has been used in Maruti car?
63. Is there any advantage in using glycerine over water as coolant?
64. Approximately how much is the volume change when water is cooled to ice?
65. What is an anti-freeze?
66. Name a few commonly used anti-freeze compounds.
67. What is a corrosion inhibitor?
68. Name any two corrosion inhibitors.
69. Why should the body paint work be protected from a spill of anti- freeze solution?
70. Enumerate different type of temperature gauges.
71. What is the advantage of using a temperature gauge in the cooling system of an engine?
72. What may be the possible causes of engine overheating?
73. What is reverse flushing of radiator? How is it done?
74. How does a defective coolant pump affect the performance of cooling system?
75. What happens if the thermostat in the cooling system gets stuck?
76. What happens if the fan belt gets loose?
77. What is the effect on the cooling system if the belt or any of the pulleys gets oiled?
78. How do you clean up the blocked radiator passages?

MULTIPLE CHOICE QUESTIONS

Mark the correct answers:

1. Percentage of fuel energy lost to the cylinder walls in an automobile engine is approximately.
   (*a*) 1 per cent (*b*) 10 per cent
   (*c*) 30 per cent (*d*) 80 per cent

2. Sequence of coolant circulation is:
   (*a*) pump-radiator-block-head (*b*) pump-block-head-radiator
   (*c*) pump-block-radiator-head (*d*) pump-radiator-head-block
3. On leaving the engine the coolant goes to
   (*a*) pump inlet (*b*) header tank
   (*c*) collector tank (*d*) none of the above
4. The radiator core is made of
   (*a*) brass (*b*) steel
   (*c*) cast iron (*d*) plastic
5. A pressure cap contains a
   (*a*) pressure valve (*b*) thermostat valve
   (*c*) blow-off valve (*d*) pressure and vacuum valve
6. The purpose of the thermostat is to keep the engine
   (*a*) hot (*b*) cool
   (*c*) at desired temperature (*d*) at desired pressure
7. The thermostat valve starts to open at about
   (*a*) 90°C (*b*) 80°C
   (*c*) 50°C (*d*) 20°C
8. Coolant pumps are of
   (*a*) vane type (*b*) reciprocating type
   (*c*) centrifugal type (*d*) all the above types
9. The purpose of the fan is to
   (*a*) increase flow of coolant (*b*) draw air through the radiator
   (*c*) cool the engine by blowing air over it (*d*) provide drive to the coolant pump
10. Cooling fans are driven by
    (*a*) electricity and belts (*b*) gears and belts
    (*c*) chains and gears (*d*) by all of the above
11. The fan in the Maruti car is controlled
    (*a*) electrically (*b*) mechanically
    (*c*) hydraulically (*d*) magnetically
12. Engine overheating may result due to
    (*a*) radiator pressure cap stuck closed (*b*) thermostat stuck open
    (*c*) broken fan belt (*d*) excess coolant in the system

[**Key:** 1. (*c*), 2. (*b*), 3. (*b*), 4. (*a*), 5. (*d*), 6. (*c*), 7. (*b*), 8. (*c*), 9. (*b*), 10. (*a*), 11. (*a*), 12. (*c*)]

7 Lubrication and Lubricants

1. INTRODUCTION

Lubrication circuit is one of the most important ones in the engine. The engine cannot run smoothly for more than a few minutes without the lubricating oil.

Whenever two metallic surfaces move over each other under direct contact, *dry or solid* friction is produced. This is due to the irregularities on the two surfaces interlocking each other. The dry friction thus created produces a lot of heat and results in wear of the metal surface.

However, when a film of some lubricating oil is interposed between two surfaces so that the two are not in actual physical contact with each other, the only resistance to motion remains the resistance of the oil itself. This type of friction is termed as *viscous or fluid* friction. This state, from lubrication point of view, is called *hydrodynamic lubrication*. It is observed in practice that the coefficients of friction obtained in viscous friction *i.e.,* hydrodynamic lubrication are far less as compared to those prevailing in dry friction. Thus it is the viscous friction which is most desirable and every effort is made to keep the viscous frictions conditions in the running engine. Engine bearings and pistons mostly operate in the hydrodynamic lubrication region.

A third type of friction, however, also exists, *i.e., greasy or boundary* friction which state is also termed *boundary lubrication* from lubrication view point. In this some lubricant is there but that is not sufficient to cause the complete separation of the two surfaces *i.e.,* only their high spots touch. In this the friction coefficients are undoubtedly higher than in case of viscous friction, but nevertheless they are far less than the coefficients in dry friction. Boundary lubrication may exist in piston rings and valve train.

Another type of lubrication, called *elasto-hydrodynamic lubrication* occurs under high loads, most commonly with hydrocarbon-based oils. In this case, the pressure developed in the lubricant is sufficiently high to deform elastically the metal surfaces on either side of the oil film. This occurs on account of increase in oil viscosity significantly with the increase in pressure. Occasionally, the engine valves and the piston rings operate in this region.

2. OBJECTS OF LUBRICATION

The *primary* objects of lubrication are:

1. To reduce friction between moving parts to its minimum value so that power loss is minimised.
2. To reduce wear of the moving parts as far as possible.

Apart from these primary objects the lubrication also serves other important purposes, which may be called *secondary*. These are:

1. *To provide cooling effect.* The lubricating oil takes heat from the hot moving parts during its circulation and delivers it to the surrounding air through the crank case.

2. *To provide cushioning effect.* The lubricating oil serves also as a good cushion against the shocks present in the engine. For example, instant combustion of the fuel in the combustion chamber produces a sudden pressure rise in the cylinder and the shock goes to the bearings through the piston, gudgeon pin and the connecting rod. This shock is then absorbed by the layer of oil present in the main bearings.
3. *To provide cleaning action.* Apart from the objects to be achieved by lubrication as described above, the lubricating oil serves another useful purpose in providing a cleaning action. During its circulation it dissolves many impurities, *e.g.*, carbon particles. This oil may be purified by filtration.
4. *To provide a sealing action.* The lubricating oil also helps the piston rings to maintain an effective seal against the high pressure gases in the cylinder from leaking out toward the crank case side.

3. REQUIREMENTS OF LUBRICANTS

1. Viscosity

In simply language, the viscosity may be considered as the resistance of the lubricating oil to flow. It is this property alone due to which the two bearing surfaces are kept apart, *i.e.*, hydrodynamic lubrication is maintained.

The viscosity of the lubricating oil should be just sufficient to ensure hydrodynamic lubrication. A higher value than this would be of no use since it will involve higher power losses due to the increased oil resistance.

The viscosity of the lubricating oil, at the time of starting the engine, should be low, otherwise if the viscosity is very high, the engine may not start. On the other hand with the engine running, the oil viscosity decreases due to increase of temperature, which is contrary to what is desirable, since at all operating temperatures, the viscosity should not fall below the minimum value required to maintain hydrodynamic lubrication. Therefore, the viscosity should ideally remain the same at all temperatures. However, the viscosity of all the oils, does vary with temperature and, therefore, the oil with minimum variation is preferred. *Viscosity Index* (VI) is an indication of the change of viscosity of oil with temperature. A high viscosity index means less change of oil viscosity with temperature rise. Petroleum lubricating oils general have viscosity index from 100 to 110, which may be increased to 120 to 130 by means of additives.

Viscosity is the most important property of lubricants and they are mostly selected on the basis of their viscosity and viscosity index.

2. Physical Stability

The lubricating oil must be stable physically at the lowest and the highest temperatures encountered in practice. There should not be any separation of solids at the lower temperatures and at higher temperatures it should not vapourise beyond a certain limit.

3. Chemical Stability

At higher temperatures the oil should remain chemically stable. There should not be any tendency for oxide formation; many of the oxidation products being sticky substances clog the lines and cause faulty piston rings and valve action.

The oil should also not decompose at high temperatures to form carbon. Carbon thus formed lowers rings efficiency, thereby reducing engine compression. The spark plugs and the valves also do not function efficiency due to sticking of carbon particles.

4. Resistance Against Corrosion

The oil should not have any tendency to corrode the pipe lines, crank case and other engine parts with which it comes into contact.

5. Pour Point

The minimum temperature at which the oil will pour is called its pour point. Obviously, since the oil will not be able to flow below the pour point, it cannot be used below this temperature for lubrication. Thus, the pour point of the oil should be less than the lowest temperature encountered in the engine.

6. Flash Point

The flash point of the oil should be sufficiently high so as to avoid flashing of oil vapours at the temperatures occurring in common use. A flash point higher than the minimum desired value will not serve any useful purpose.

7. Cleanliness

The oil should be sufficiently clean and stable itself so that the crank case and oil lines are kept clean. Further, it must contain agents, called detergents, which remove the impurities from the engine parts during oil circulation. These impurities may either be filtered out or removed with the change of oil at periodic intervals.

8. Resistance Against Extreme Pressure

In modern automobile engines, lubricating oil is subjected to very high pressures, particularly in bearings and the valve actuating mechanisms. This would cause the oil to be driven out. The lubricating oil should have sufficient resistance against this tendency. Some oils can cling to the metal surfaces under high pressures better than the other oils and thus may reduce the friction in greater magnitude. Such oils are said to be more 'oily' than the others and this property of soils to adhere to a metal surface under high pressures is called 'oiliness'.

4. TYPES OF LUBRICANTS

According to their origin, the lubricating oils may be classified as:

1. Animal Oils

These are obtained from the animal fat. These are not at all suitable for automotive engine lubrication, because they are exidized easily and become gummy after scme use.

2. Vegetable Oils

These are produced from seeds, fruit of plants, trees etc. Like animal oils, they are also easily exidized and become gummy. However, castor oil, which is a vegetable oil once found extensive use in automobile industry because of its high viscosity and high film strength.

3. Mineral Oils

Mineral oils derived from petroleum are perhaps the most widely used in automobiles. The advantages they possess over the animal or the vegetable oils are:

(*i*) Greater chemical stability at higher temperatures.

(*ii*) Less tendency to form emulsions with water.

(*iii*) More plentiful and cheaper.

Chiefly the mineral oils contains hydrocarbons. According to their molecular structures they may be further sub-classified into paraffins, naphthenes, aromatics and olefins. Paraffins are saturated hydro-carbons with straight or branched chains. Naphthenes are saturated hydrocarbons with ring structure, aromatics are hydrocarbons with benzene nucleus and olefins are unsaturated hydrocarbons with double bonds. In practice the mineral lubricating oil is a mixture of paraffins, naphthenes and aromatics.

4. Synthetic Lubricants

The examples of synthetic lubricants are polyorganosiloxanes or silicon fluids, polyglycol ethers and aliphatic diester oils.

Following advantages are claimed for these over the minerals oils:

(*i*) Higher viscosity index.

(*ii*) Reduced lacquer formation.

(*iii*)Reduced loss due to evaporation.

(*iv*) Considerably less oil consumption.

(*v*) Less engine deposits.

(*vi*) Less frequent changes of lubricating oil.

(*vii*) Increased fuel economy.

However, their high cost limits their use.

5. Greases

Grease is a lubricant which is formed when the liquid lubricating oil is made solid or semi-solid with the help of special thickening agents such as lithium, calcium, aluminium, sodium, et6c. Chemically, they are suspensions of metallic soaps (calcium or sodium soaps of fatty acids) dispersed in a lubricating oil. Greases find use in automobiles at places where retention of liquid lubricants is difficult and where high temperatures are encountered, *e.g.* in axles.

Greases are classified according to the purpose for which they are used and the kind of soap base. The examples of the former type are, chassis grease, multi-purpose grease etc., while for the later type, the examples are sodium base greases, calcium base greases, lithium base greases etc. Sodium base greases can withstand sufficiently high temperatures. Calcium base greases are used for wheel bearings and chassis lubrication. Aluminium base greases are not suitable for high temperatures and are useful on chassis, transmissions and chains. Lithium base greases are suitable for use on wheel bearings, universal joints, brake cables etc. and they can withstand also high temperatures. These are commonly referred to as multipurpose greases.

Like lubricating oils, in greases also, additives are used to improve their properties.

The hardness or penetration of greases is determined as per NLGI (National Lubricating Grease Institute) ratings, which in increasing order are 000,00,0,1,2,3,4,5 and 6.

6. Solid Lubricants

Graphite has been used as a cylinder lubricant, both in powder and colloidal form. It is stable at high temperatures and has low coefficient of friction. The special feature of graphite is that it fill the pores of the metal and makes it friction-free.

5. ENGINE OIL RATINGS

5.1. Viscosity rating

The Society of Automotive Engineers (SAE) has classified the lubricating oils according to their viscosity and service ratings. Since Sept., 1980, the viscosity ratings for engine oil were OW, 5W, 10W, 20W, 25W, 10, 15, 20, 30, 40, 50, 60. However, currently, the high temperature S.A.E. viscosity ratings are 8, 12, 16, 20, 30, 40, 50 and 60. Earlier high temperature ratings were in steps of 10 and low temperature ratings in steps of 5. Due to The introduction of new oil for high temperature for new specially designed engines with lower hydrodynamic friction, resulting in lower fuel consumption, new viscosity standards had to be formulated. On April 2, 2013, SAE introduced a new such viscosity grade SAE 16. The earlier practice of steps of 5 or 10 was discontinued to avoid confusion with SAE 15W, if they called it as SAE 15. Thus for clarity then and also in future, they introduced steps of 4 for the lower range of high temperature ratings (upto SAE 20). This also increased the precision since now there are 2 additional steps of 12 and 16 in between 10 and 20, whereas earlier there was none. 'SAE J 300' is the SAE standard for viscosity classification of engine oils.

The oils with *W* are for winter use and their viscosity tests are made at 0°F (–18 °C), while those without *W* are for use in summer months and the viscosity tests in their cases are done at 210°F (99°C).

These days *multigrade oils* are also available. In their case change of viscosity with temperature is minimum. For example, SAE 10W/30 oil has the same viscosity as SAE 10W oil at 0°F (–18°C) and as SAE 30 oil at 210°F (99°C) (Fig. 7.1). The advantage of such oils is that during cold starting there is reduced power loss on account of less friction than would be there if ordinary single grade oil suitable for engine operating temperatures is used. Further this also avoids the necessity of using different oils for winter and summer months.

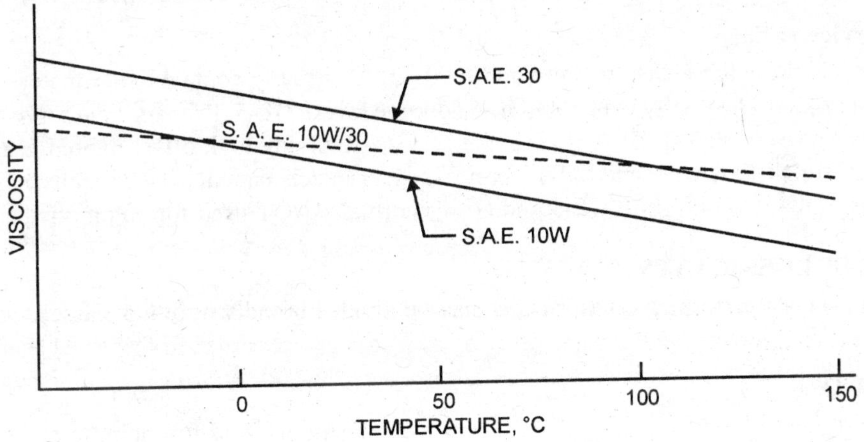

Fig. 7.1. Viscosity behaviour of multigrade oils.

5.2. Service Rating

American Petroleum Institute (A.P.I.) has provided additional service classifications for lubricating oils. The S.A.E. and the A.P.I. classifications are used simultaneously.

The classification of lubricating oils according to service conditions, however, is different for petrol and diesel engines. First letter S stands for service ratings for petrol engine oils, *viz.,* SA, SB, SC, SD, SE, SF, SG, SH, SJ, SL, SM and SN. API intentionally omitted OI and OK from the sequence of ratings. Oils with SA designations are for mildest conditions. The other designations SB, SC, etc. are in order of increasing severity of service conditions. Currently engine oils upto SH ratings are hardly used for automobile engines on account of the gradually increasing severity of service conditions with improvements in engine design. Grade 'SN' was introduced in October 2010 designed to provide high temperature deposit protection for pistons, more stringent sludge control and seal compatibility. API SN with Resource Conserving matches ILSAC GF-5 by combing API SN performance with improved fuel economy, turbocharger protection, emission control system compatibility and protection of engine operating on ethane-containing fuels up to E85.

In the same way, there are separate service ratings for diesel-engine oils. In increasing order of severity of service conditions, these are CA, CB, CC, CD, CE, CF, CG, CI, CJ and CK. Ratings upto CG are obsolete and only CH, CI and CJ are current. Rating CJ-4 was introduced in 2006 for high speed, four stroke engines designed to meet 2007 model year on-highway exhaust emission standards. CJ-4 oils are effective at sustaining emission control system durability where particulate filters and other advanced after-treatment systems are used. Optimum protection is provided for control of catalyst poisoning, particulate filter blocking, engine wear, piston deposits, low-and high-temperature stability, soot handling properties, oxidative thickening, foaming and viscosity loss due to shear.

API Service Rating 'CK-4' describes oils for use in high-speed four-stroke diesel engines designed to meet 2017 model year on-highway and Tier 4, *non-road* exhaust emission standard as well as for previous model year diesel engines. These oils are especially effective at sustaining emission control system durability where particulate filters and other advanced after-treatment systems are used.

6. TRANSMISSION (GEAR) OIL RATINGS

6.1. SAE Viscosity ratings

These are 70W, 75W, 80W, 85W, 80, 85, 90, 110, 140, 190 and 250. Letter 'W' indicates low temperature oil for which the viscosity is determined at maximum temperature of –55, –40, – 26 and –12°C for 70W, 75W, 80W and 85W respectively. For higher temperatures, viscosity determined is at 100°C.

6.2. API Service ratings

For gear oils these are GL-1 (without additive), GL-2 (usually contains fatty materials), GL-3 (with mild EP additive), GL-4 (usually with 50% GL-5 additive level), GL-5 (Primary field service recommended for passenger cars and trucks), GL-6 (for severe service involving high offset hypoid gears, often for oil used in limited slip differentials) and MT-1 (for non-synchronized manual transmissions in buses and trucks at a higher level than GL-4, GL-2, GL-3 and GL-6; normally NOT used for automotive applications).

7. TESTING OF LUBRICANTS

The various tests performed on lubricants may be divided broadly into two classes, *viz.*, physical and chemical tests. These are:

Physical Tests

1. Viscosity test
2. Flash point and fire point test
3. Loss due to evaporation
4. Cold test or pour point test
5. Specific gravity test

Chemical Tests

1. Acid value test
2. Sponification value test
3. Insoluble reside test
4. Moisture and emulsification test

7.1. Viscosity Tests

The apparatus used for determination of viscosity is called the viscometers. Some important viscometers are:

1. Redwood viscometer
2. Saybolt viscometer
3. Engler viscometer

Apart from these, less time-consuming tests have been evolved for viscosity measurements recently, *e.g.*, the Cold Cranking Simulator test. The various units used for viscosity measurements are:

(*i*) Saybolt universal seconds (SUS)

(*ii*) Centistokes (for kinematic viscosity)

(*iii*) Centipoises (for absolute viscosity)

7.2. Flash Point and Fire Point Test

Flash point of a lubricating oil is the minimum temperature at which it gives off sufficient vapour so as to form an explosive mixture with air. If the oil is heated further a stage will reach when it will began to burn continuously on applying a flame to it. This temperature is called the fire point. Flash point and fire point are determined with Abel's apparatus.

7.3. Loss Due to Evaporation

When the lubricants are subjected to high temperature, they are liable to vaporize to a certain extent. Loss due to evaporation is determined by Arch Butt apparatus. Air is passed at a rate of 1/2 to 1 litre per minute for half an hour over the lubricant which is kept in a definite quantity in a weighted platinum dish and is heated with burners. The hot air heats the dish and the lubricant volatilizes. From this the loss of weight is calculated.

7.4. Cold Test or Pour Point Test

The pour point is the temperature at which at which a lubricant just ceases to flow when cooled under standard conditions. The lower the pour point, the better is the lubricant. It is determined with the help of pour point apparatus.

7.5. Specific Gravity Test

The specific gravity of a lubricating oil is generally calculated as:

$$\text{Specific gravity } 60°/60°\text{F} = \frac{\text{Density of oil at } 60°\text{F}}{\text{Density of water at } 60°\text{F}}$$

The American Petroleum Institute (A.P.I.) has its own system of calculating specific gravity.

$$\text{Degree A.P.I.} = \frac{141.5}{\text{Specific gravity } 60°/60°\text{F}}$$

Specific gravity can be determined by any of the following methods;

(*i*) Archimedes' Principle

(*ii*) Specific gravity bottle

(*iii*) Hydrometers

Out of these, hydrometers, though not very accurate, are used extensively for commercial determination of specific gravity because of the quickness and ease with which they can be used.

7.6. Acid Value Test

Acid value of an oil is the number of milligrams of potassium hydroxide (KOH) required to neutralize one gram of the oil.

To determine the acid value, about 5-10 gm of the oil is taken in a conical flask. About 40-50 c.c. of neutral alcohol is added to it and the flask is then given a rotary motion so as to break the emulsion.It is then allowed to stand when after some time two distinct layers separate out. The upper layer is of alcohol and the lower one of the oil, but the acid content of the oil has now gone over to the upper alcohol layer, which is then removed and titrated against N/10 KOH, using phenolphthalein as indicator.

One litre of normal KOH, contains 56 gm of KOH. From this and the results of titration, number of milligrams of KOH required to neutralize the free acid in one gm of oil may be calculated.

7.7. Sponification Value Test

Sponification value of an oil is the number of milligrams of KOH required to sponify completely one gm. of the oil.

Determination of sponification value helps in the detection of adulteration. Sponification value actually refers to the small amounts of compounds containing oxygen, which are not removed during refining process. The procedure for the sponification value test is given below:

Weigh out accurately about 5 gm. of the oil specimen into a conic flask. Pipette out into it, 50 c.c. of approximately N/2 alcoholic KOH. Into an exactly similar flask, which is empty, pipette out another 50 c.c. of the same alcoholic KOH. Connect both the flasks to water cooled condensers and reflux briskly for about 1/2 to 1 hour. Then wash the mouths of condensers and the flasks and titrate the contents of each flask against standard N/2 hydrochloric acid using phenolpathalein as indicator. Note the volume of acid used (titre) in each case.

The oils contain a small amount of free fatty acids and a large amount of glyceryled esters of fatty acids besides a small amount of unsponifiable matter. When alcoholic KOH is added to the oil, the free acid directly reacts with part of alkali and converts it into soap. When the contents are refluxed the ester decomposes forming soap.

The amount of KOH consumed is equal to difference between experimental and blank titres. One titre of *N* hydrochloric acid corresponds to 56 gm. of KOH. From this the number of milligrams of KOH required for the sponification of one gm. of oil may be calculated.

7.8. Insoluble Residue Test

Mineral oils contain some suspended carbon particles, which cause much harm to the machinery by way of wear of parts. To test the presence of insoluble particles in the oil, a known weight of it is dissolved in some suitable solvent, *e.g.* spirit, carbon tetrachloride etc. The solution is filtered and the residue washed with pure solvent, dried and weighed. From this percentage of insoluble residue in the oil is calculated.

7.9. Moisture and Emulsification Test

When a lubricating oil mixes thoroughly with water, an intimate mixture of the two known as emulsion is formed. The lesser the emulsification tendency, the better the oil is. A good lubricant should be such that the emulsion if formed, should break off quickly.

The tendency to de-emulsify is quantitatively determined by A.S.T.M. emulsion test. An emulsion of oil and water is made under standard conditions. The time in seconds, required for the constituents to separate from this emulsion is known as the emulsion number.

8. OIL ADDITIVES

These are the chemical substances which are added to the lubricating oil either to reinforce some of its natural properties or to provide it with certain new properties which it does not possess originally. Oil additives are classified according to the property of the oil which they reinforce or add. The important additives in current use are:

1. *Oxidation inhibitors.* These inhibit the formation of gums and varnishes by preventing the oxidation of the oil at the engine operating conditions. These are usually organic compounds such as amines, sulphides or phenols with metals like tin, zinc or barium.

2. *Corrosion inhibitors.* Combustion in the engine cylinder produces a number of by-products including acids and water which would corrode the engine parts such as bearings. Cold engine, rich fuel mixture, poorly tuned engine, cold-weather driving etc. are the causes which increase the amount of acids produced. Corrosion inhibitors prevent or at least reduce the formation of acids which would cause bearing corrosion. These consists of oxidation inhibi- tors with the addition of metal salts of thiophosphoric acid and sulphurized waxes.

3. *Detergents and Dispersants.* Additives which handle high-temperature deposits such as gums and varnishes are called detergents. Those which deal with low temperature deposits such as cold sludge, are called dispersants.

Detergents and dispersants act in a similar manner and may be conceived as tadpole-like with a head and a tail. The heads attach themselves to particles while the tails keep the particles separated (Fig. 7.2). The result is that any dirt (paint, varnish or sludge) is kept finely dispersed and stays in the oil as fine suspension and is removed when the engine oil is changed. Thus various engine parts, *e.g.*, piston rings, main bearings, oil galleries, connecting rod etc. remain clean. The detergent additives are the polymers and polyalyenyl succinimides.

However, the additives in a two-stroke engine oil act in a different manner. These convert the particles of carbon into a crumbled powdered form which is easily scavenged through the exhaust port.

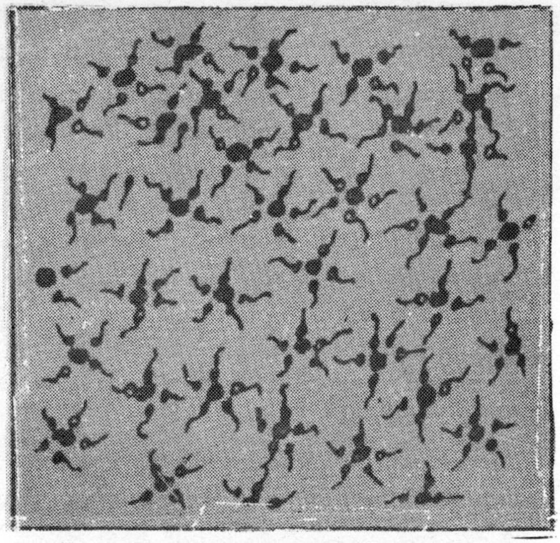

Detergent-dispersant molecules attach themselves to particles, keep them apart and hold them in suspension

Fig. 7.2. Action of detergents and dispersants. (Courtesy—Castrol India Limited)

4. *Anti-scuff additives.* There is more likelihood of scuffing during engine running-in period and after engine overhaul. Antiscuff additives help to polish moving parts such as pistons, cylinder walls, cams etc. and thereby prevent their tendency of scuffing.

5. *Viscosity Index Improvers*. These are the additives which do not allow or at least minimise the decrease of oil viscosity with the increase in temperature.

In case of ordinary lubricating oil, these are mainly hydrocarbon molecules whose internal energy increases with increase in temperature so that they move faster and further apart and thus become thinner. The additives are usually polymers such as from acriloid plastics. These have a long chain molecules which are coiled up when cold so that they can then move as freely as the hydrocarbon molecules. However, on heating, the polymer molecules unwind and hinder the movement of the hydrocarbon molecules, thus compensating for the decrease of oil viscosity with temperature. (Fig. 7.3).

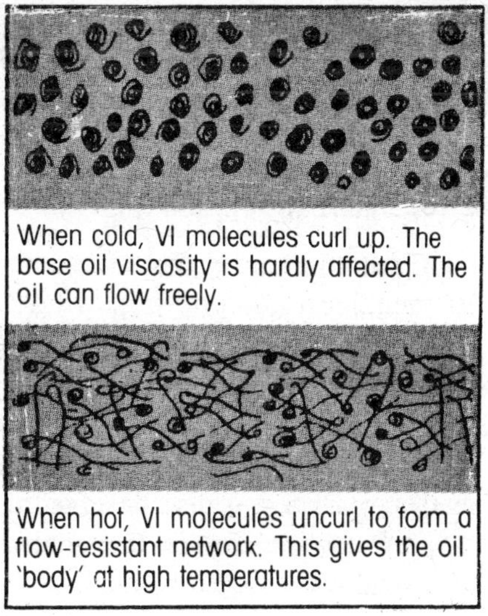

Fig. 7.3. Action of viscosity index improvers. (Courtesy—Castrol India Ltd.)

6. *Anti foaming additives.* The lubricating oils, in general, have a tendency to foam due to engine vibrations which give rise to churning of oil in the sump. Heat and churning mix the oil with air to produce form. If allowed to foam, the oil cannot properly lubricate the engine bearings, apart from loss by overflowing. Anti-foaming additives are available which suppress this foaming tendency of oils. Polyorganosiloxanes are the most common anti-foaming additives.

7. *Extreme Pressure (E.P.) additives.* These cater for more difficult conditions of lubrication, *e.g.*, the one arising between the highly stressed cams and valve tappets. Under severe operating conditions, local temperatures can go upto 160°C and pressures can develop upto about 1400 MPa. At such extreme temperatures and pressures the oil film between moving parts is likely to break down and allow metal-to-metal contact with disastrous results. These additives provide the extra film strength by forming a chemical film. Polymeric materials such as polyisobutene form such additives.

8. *Pour point depressants.* These serve to lower the pour point of the oil by coating the wax crystals in the oil so that they would not stick together and thus facilitate oil flow.

9. *Others.* Apart from above a larger variety of additives are available. Some of these are, rust inhibitors, water repellents, emulsifiers, dyes, odour controllers, etc.

Note. 1. In general, diesel engines tend to produce more acid than petrol engines. Therefore, they require special oil containing additives like a strong alkalizing agent to neutralize acidity and a high performance detergent that prevents carbon from sticking to high-temperature engine parts.

2. Transmission gears are not subjected to very high forces, but since they are located near the engine and the oil is subjected to churning due to gear motion, the same should contain viscosity index improvers and anti-foaming additives.

3. Differential gears are subjected to high forces and side- slipping. As such the lubricating oil for them should have extreme pressure additives.

9. EFFECT OF ENGINE CONDITIONS ON LUBRICATING OIL

The severe engine conditions of temperature and pressure to which the lubricating oil is subjected, cause its deterioration in many ways, which are discussed here.

1. Sludge formation

Sludge is mushy material composed of oil, water and other combustion products. It readily clogs the oil lines and galleries. It is caused by the condensation of water in the engine crankcase, which forms an emulsion with the oil, dirt and the other matter. Obviously this occurs in the case of engine which seldom runs hot to drive away the water vapour out of the crankcase. A vehicle, which runs mostly in traffic, running slow and stopping intermittently is thus more prone to sludge formation. This can be remedied at least partially in case of such a vehicle by adjusting its cooling system so that engine remains at a reasonably high temperature.

2. Lacquer formation

Lacquer or varnish is formed when the oil gets oxidised due to high temperatures. The high speed, high load, high temperature driving increases varnish formation. Lacquer is often responsible for sticking valves and clogged piston rings. To avoid this use suitable additives and change the oil regularly.

3. Oil dilution

Oil dilution is caused by the leakage of gasoline past the piston into the crankcase, or by the condensation of water vapour in the crankcase. This happens often when the engine is cold *i.e.* at the time of starting and warming up. In winter, choke is used frequently to start the engine. The effect of using choke is to introduce a small additional quantity of gasoline in the cylinder, which contributes towards oil dilution. This it is seen that choke should be applied only when it becomes a must, as for example, in very cold weather. Further it is also observed that the dilution occurs mostly under starting and idling conditions only. It follows, therefore, that engine should not be raced heavily just after starting because there is no lubrication at the time.

Oil dilution may give sometimes a faulty dipstick reading. For example, if the dipstick shows satisfactory oil level when put into the sump when just the engine has started, the same will be misleading because the condensed water vapour is also present there in the sump. As the engine warms up and comes to normal conditions, the water will vaporize and the indicated oil level will therefore fall. And this will be actual oil level.

Oil dilution reduces the lubricating property of the oil. Thus, if exceeded beyond limits, it will cause metal to metal contact and hence very rapid wear of engine parts.

4. Carbon deposition

Carbon formation occurs normally on the combustion chamber walls, pistons, valve stems and piston rings. It results due to incomplete combustion of fuel, particularly so at the time of starting and idling when richer fuel is supplied.

Carbon is a bad conductor of heat and therefore dissipation of heat from the engine will be less. The result is higher combustion chamber temperatures, which encourage oxidation of the oil and thicken out the deposits further. Moreover, the tendency to detonate is increased.

The sticking of rings in the grooves causes blow-by of hot gases and the consumption of oil is increased.

10. CRITICAL LUBRICATION CONDITIONS FOR AUTOMOBILE ENGINES

1. Stationary vehicle

During periods of inactivity, such as parking or garaging poor quality oil may drain away from metal surfaces, leaving the engine starved of lubrication. Moreover, under these conditions, internal rusting is bound to occur. Thus starting up an engine after a long period of inactivity is bound to be disastrous due to direct metal to metal contact and layers of abrasive rust being scraped into the oil. Thus the lubricating oil must contain suitable additives to counter such a condition.

2. Short distance journeys

To circulate oil completely in the engine may take upto five minutes. Moreover, some cars may have to travel about 15 km before the engine attains its normal working temperature. Thus in case of short journeys the engine temperatures would be lower due to which any water formed due to combustion of fuel will remain in the engine to combine with acids and corrode the cylinder walls. Thus the lubricating oil must contain corrosion inhibitors and dispersants to take care of lower temperatures during short journeys.

3. Traffic jams

During traffic jams the automobile will have to move very slowly, down to say 10 km/h. This will result in extremely high local temperatures in the engine critical parts and the oil will tend to thin out. Due to high temperatures, varnish and carbon will accumulate around the piston rings and combustion ash will coat the engine valves and spark plugs. Thus oxidation inhibitors and detergents are required to be present in the engine oil to counter such a condition.

4. Highway travel

At very high speeds (beyond 100 km/h), local temperatures and pressures between the piston and the cylinder walls may reach extremely high values. Thus to provide effective lubrication under these conditions, oxidation inhibitors and extreme pressure additives must be present in the lubricating oil.

11. CONSUMPTION OF LUBRICATING OIL

The various sources of oil losses in the engine are:

1. *Combustion.* The oil works up past the piston rings into the combustion chamber, where it is burnt up. This cannot be avoided altogether since some of the oil has to go past the piston rings to lubricate the upper portion of the piston and the cylinder walls.

2. *Loss through leakage.* There may be some loss of oil through leakage at the faulty crankshaft bearing seals, at the defective joint between the cylinder block (or crankcase) and the oil pan, at the loose or faulty drain plug, or due to some crack in the crankcase. However, with proper care such losses can be reduced to negligible.

3. *Loss through crankcase ventilation.* Some of the oil is carried away from the crank case as vapour or mist or alongwith the exhaust gases.

4. *Loss on account of wear of engine parts:*

(*i*) When the cylinder bore, the piston skirt or the rings wear out, more oil is pumped past the piston rings which causes increased oil consumption. Actually, higher the pressure intensity between the piston rings and the cylinder walls, the lower the oil consumption. This is achieved by putting steel expander inside the ring so that even when the ring is worn out, though the pressure due to ring itself decreases, yet the pressure due to steel expander is not decreased, thus maintaining a higher pressure intensity between the rings and cylinder walls.

(*ii*) Worn out big end bearings mean excessive clearances causing too much oil to be thrown on cylinder walls.

(*iii*) Worn out rocker shaft or rocker bushes may also be a source of excessive oil consumption.

5. *Loss due to excessive vehicle speed.* The oil consumption increases with increase in speed of the engine (Fig. 7.4). There are several reasons why the oil consumption should increase at high engine speeds:

(*i*) High speed means high engine temperature and consequently reduced oil viscosity which means the oil would be thinner and, therefore, will flow with relative ease. This means more of oil will pass up the piston rings and burn into the combustion chamber. Further thinner oil will be susceptible to easy leakage.

(*ii*) At high speeds there is more centrifugal effect, as a result of which more oil is fed to the oil lines and more is sprayed on the cylinder walls. This causes more oil consumption.

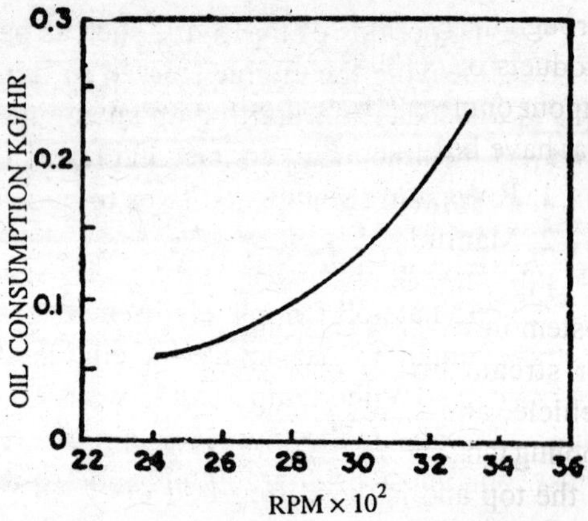

Fig. 7.4. Effect of speed on oil consumption.

(*iii*) 'Ring shimmy' is caused at high engine speed which pumps large amounts of lubricating oil past the rings into the combustion chamber.

(*iv*) At higher speeds more oil is lost in the form of mist through crankcase ventilation.

In an automobile engine working under normal conditions, average consumption of lubricating oil is estimated at about 3000-4000 km. per litre. The consumption increases as the pistons and the cylinders wear out.

12. CRANKCASE VENTILATION

Although, the piston and the rings are designed to form a gas-tight seal between the piston and the cylinder walls, yet the products of combustion do leak from the combustion chamber into the crankcase. These contain mainly unburnt air-fuel mixture, partially burnt hydrocarbons, nitrogen oxides, water vapour, carbon monoxide and carbon dioxide, but may also consist of traces of acid due to sulphur present in the fuel.

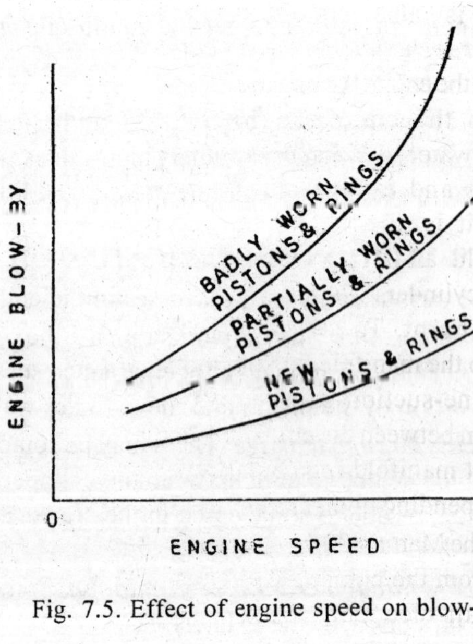

Fig. 7.5. Effect of engine speed on blow-by

Piston blow-by increases with engine speed and with

wear of cylinder bore, pistons and rings (Fig. 7.5). The effects of piston blow-by are:

1. The air-fuel mixture, partially burnt and fully burnt vapour fumes will condense and contaminate the lubricating oil in the crankcase, which may result in the deterioration of performance of the engine lubricating oil and also in the production of sludge by the oxidation of the oil.

2. High concentration of combustible air-fuel mixture in the crankcase, which may cause an explosion there.

3. Corrosion of engine parts in case of fuels containing sulphur.

Crankcase ventilation was developed as a remedial measure against this. A steady flow of air is carried through the engine crankcase which carries away the products of combustion alongwith it in the form of vapour or mist. The methods of crankcase ventilation that have been developed so far are:

1. Road draft system.
2. Manifold suction system.

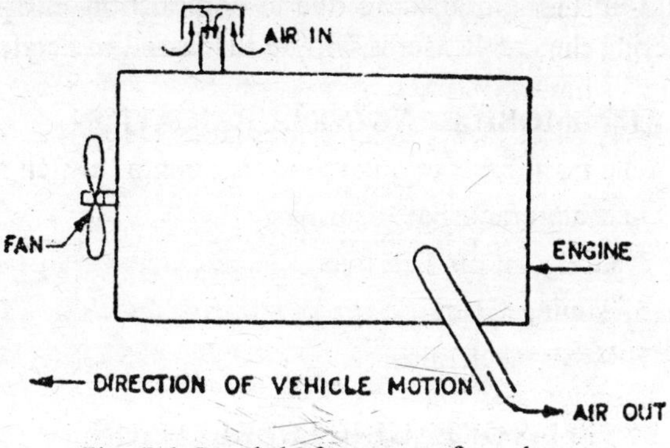

Fig. 7.6. Road draft system of crankcase

Fig. 7.6 shows a road draft system in which use is made of the air stream due to motion of the vehicle, and supplemented by the cooling fan. The air enters the engine at the top and is then circulated in the crankcase where it carries the mist, vapours, etc. after which it comes out through the outlet and goes into the atmosphere.

In the manifold suction system, which is also called *positive crankcase ventilation system,* filtered air from the engine air intake is drawn through the crankcase where it gathers water vapours, unburnt fuel vapours and exhaust gases after which it is directed to the inlet manifold and from there to the engine cylinders where any unburnt fuel is burnt. To avoid large air flows to the manifold during idling, an engine-suction-operated valve placed in between the crankcase and the inlet manifold regulates the air flow depending upon engine speed.

In the Maruti 800 engine blowby gases from the crankcase are taken to the air cleaner through an oil

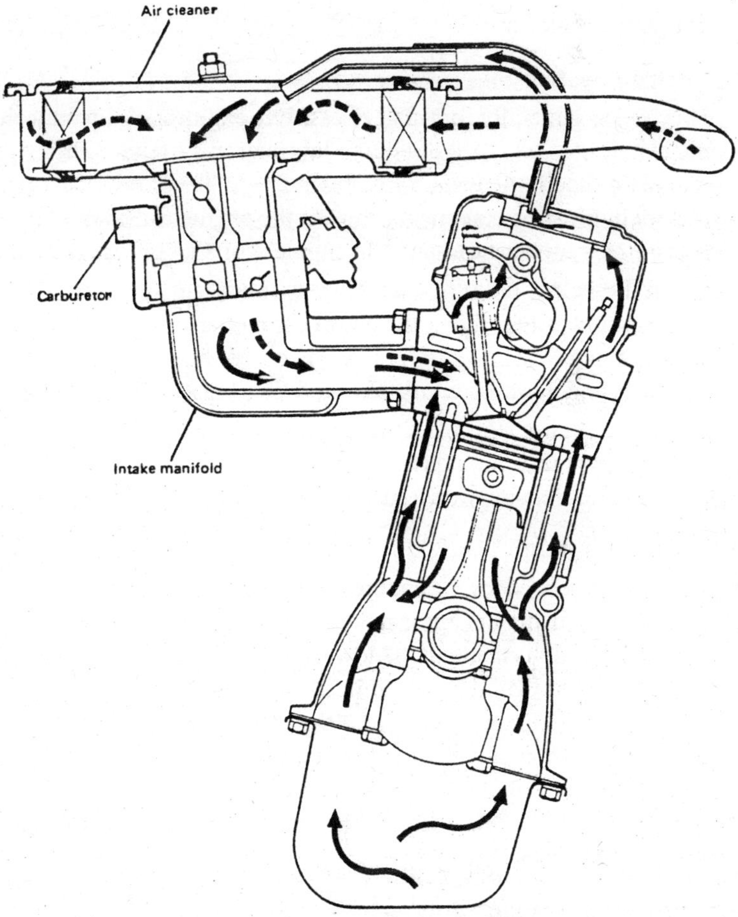

Fig. 7.7. Blowby gas recycling system.
(Courtesy—Maruti, Udyog Ltd.)

separator which removes the oil particles accompanying the gases. In the air cleaner the blow-by gases mix with the fresh air intake and the combined air is passed on to the carburettor (Fig. 7.7).

The main advantage of the manifold suction system is that being a closed type of system this does not cause pollution of atmosphere, which occurs in the road draft system. Moreover, it prevents pressure build-up in the crankcase due to combustion- pressure blow-by and also prevents backfire gases from entering the crankcase. It will be discussed in detail in Chapter 20.

13. AUTOMOBILE ENGINE LUBRICATION

The main parts of an automotive engine which require lubrication are:

1. Main crankshaft bearings
2. Big end bearings
3. Gudgeon pin bearings
4. Piston rings and cylinder walls
5. Timing Gears
6. Camshaft and camshaft bearings
7. Valve mechanism
8. Electrical equipment

14. SYSTEMS OF ENGINE LUBRICATION

The various systems adopted for the lubrication of automobile engine are:

1. Petroil System
2. Splash System
3. Pressure System
4. Dry-Sump System

14.1. Petroil System

This is used generally for small two-stroke engines, *e.g.* in majority of scooter and motor cycle engines. It is the simplest of all types of engine lubrication systems. Certain amount of the lubricating oil is mixed with the petrol itself, the usual ratio being 2% to 3% of oil. If it is less, there is danger of oil starvation or insufficient lubrication causing damage to the engine; if however it is more, there will be excessive carbon deposits in the cylinder head and the engine will also give dark smoke.

When the petrol mixture enters the crankcase, due to high temperatures there, the petrol component vaporizes leaving a thin film of lubricating oil on the crankcase, cylinder walls, crankshaft and bearings.

The main requirements of lubricating oil for two-stroke engine are that it should readily mix with petrol and burn without leaving much residue.

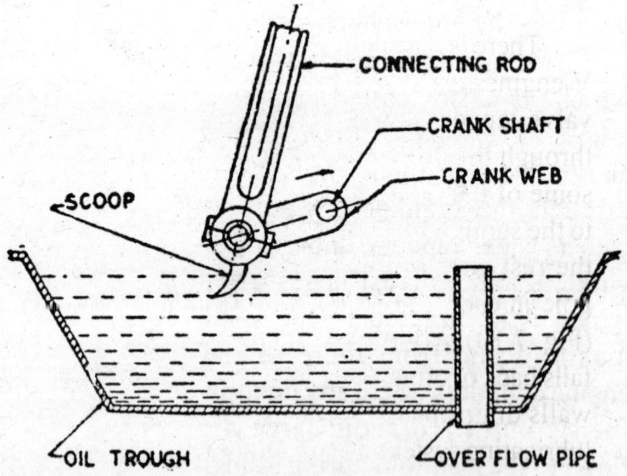

Fig. 7.8. Splash system lubrication.

14.2. Splash System

This was employed for the engines of early motor cycles. It is one of the cheapest methods of engine lubrication. A scoop is made in the lowest part of the connecting rod and the oil is stored in the oil trough (Fig. 7.8), it being pumped there from the crankcase oil sump. When the engine runs, the scoop causes the oil to splash on the cylinder walls each time it passes through its B.D.C. position. This affects the lubrication of engine walls, gudgeon pin, main crankshaft bearings, big end bearings etc.

14.3. Pressure System

This system is used almost universally in modern car engines. The splash system though cheaper, is not suitable for automobile engines because of the absence of positive lubrication. In the pressure system (Fig. 7.9), an oil pump takes the oil from the wet sump through a strainer and delivers it through a filter to the main oil gallery at a pressure of 200 to 400 kPa. The oil pressure is controlled by means of a pressure-relief valve, situated in the filter unit or the pump housing.

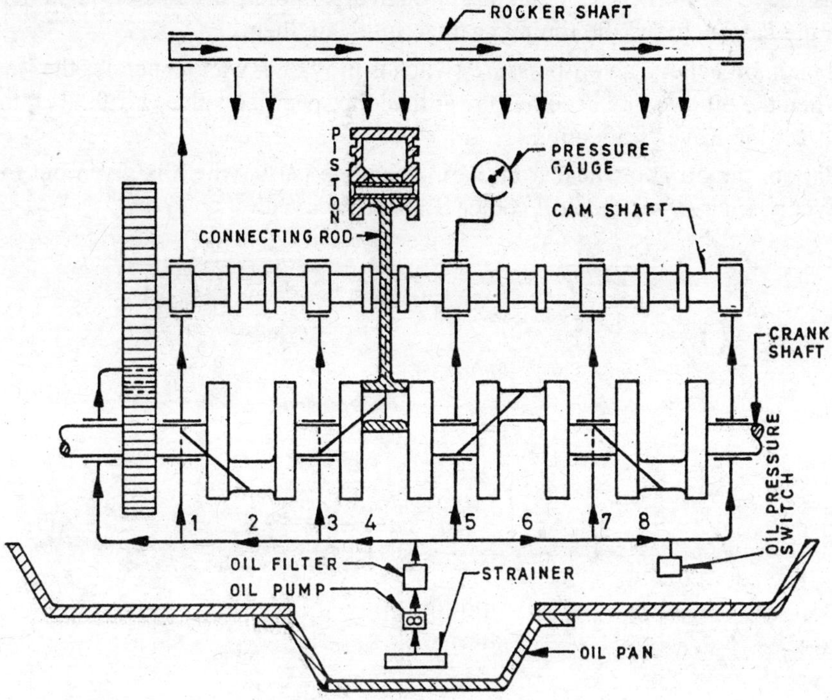

Fig. 7.9. Pressure system of lubrication.

There is one main gallery in case of inline engines, whereas V-engines use two main galleries or one main gallery and two valve tappet galle- ries. From the main gallery, the oil goes through the drilled passages to the main bearings from where some of the oil after lubricating the main bearings falls back to the sump, some is splashed to lubricate cylinder walls, while the rest goes through a hole to the crankpins from where a hole in each connecting rod web, leads it to the gudgeon pin (Fig. 7.10). After lubricating the gudgeon pin bearings the oil falls back or effects ring lubrication. The oil that falls on cylinder walls drains back into the oil pan and is recirculated through lubricating system. However, in light duty engines, the simplest and most commonly used method of cylinder and piston lubrication is by the big end splash. For the camshaft and timing gears, lubricating oil is led through separate oil lines from the oil gallery to each camshaft bearing through a pressure-reducing valve. To lubricate the timing gears and the sprocket chain, sometimes a directed oil jet is employed. The valve tappets are lubricated by connecting the main oil gallery to the tappet

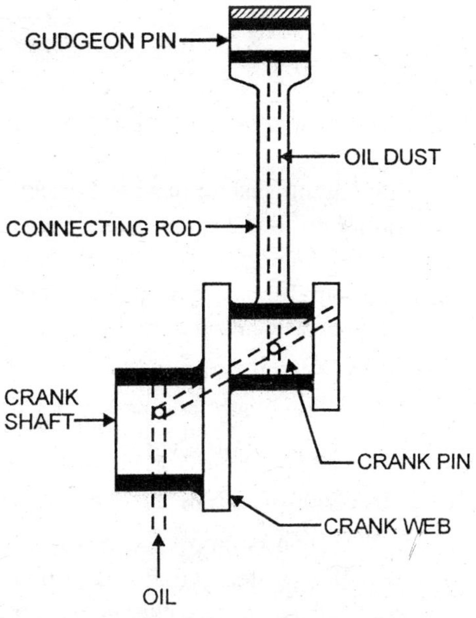

Fig. 7.10. Small end lubrication

guide surfaces through drilled holes. In case of overhead valve mechanism, the rocker arms are sometimes mounted on hollow shafts which carry oil under pressure from the oil gallery. These hollow shafts feed oil for the lubrication of rocker arms. In some other designs, the push rods are made hollow and the oil under pressure is made to flow up to pro- vide lubrication of various parts of the valve train. The oil that seeps from the valve train as above is returned to the oil pan through drain holes, which are sometimes so placed that the oil drains on the camshaft and the timing gears to lubricate them.

At the end of the main oil gallery, an oil-pressure switch is provided, which operates the warning light on the dash board. When the oil pressure becomes lower than the specified value, say 0.3 bar, the warning light comes on, to attract the driver's attention.

During its circulation, the oil gains heat from various engine parts, which is given out to the sump walls. In some heavy duty engines separate oil cooler is also employed.

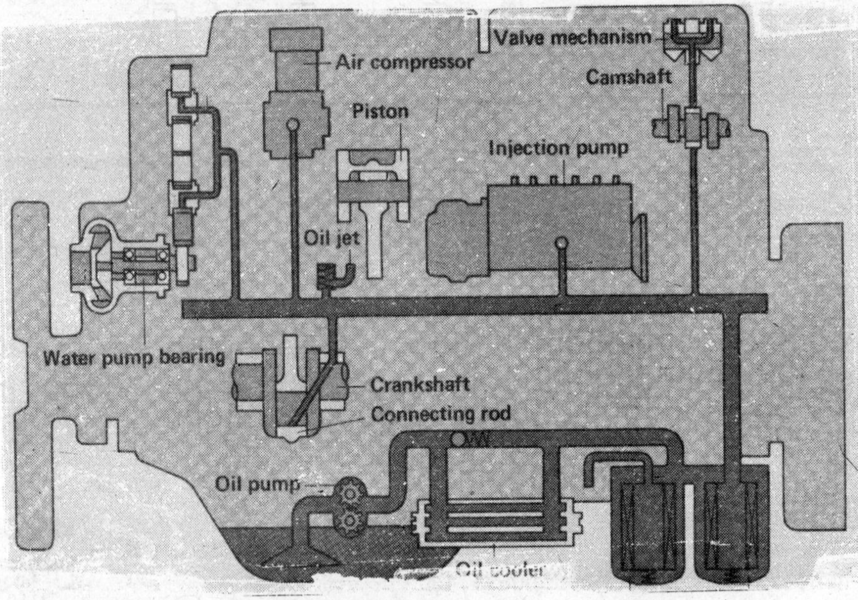

Fig. 7.11. Nissan Diesel pressure lubricating system

Lubricating system used in Nissan Diesel trucks is shown in Fig. 7.11. An oil pump force-feeds the lubricating oil to the crank shaft, valve mechanisms, timing gear chain, injection pump, air compressor and water pump bearings. The pistons are lubricated and cooled by an old jet system. Two oil filters, a full-flow and a by-pass type, provide double protection to extend engine life. Heat absorbed by the circulating oil is removed by an oil cooler.

The advantages of a wet sump system are:

1. It is less costly compared to a dry sump system.

2. Checking of oil level is easy.

14.4. Dry-sump system

This system is employed in some racing car engines and for situations where the vehicle has to be operated at very steep angles. If ordinary wet sump pressure system of lubrication is used in such cases, the situations may arise when there is no oil at the place where oil pump is installed. To avoid such instances, dry sump system is used (Fig. 7.12), wherein two pumps, instead of one, are used. The scavenge pump

A is installed in the crankcase portion which is the lowest. It pumps oil to a separate reservoir *B*, from where the pressure pump *C* pumps the oil through filter *D*, to the cylinder bearings; a full pressure system of lubrication is employed. The oil pressure is maintained at 400-500 kPa for the main and big end bearings while about 50-100 kPa pressure is used for timing gears and cam-shaft bearings etc.

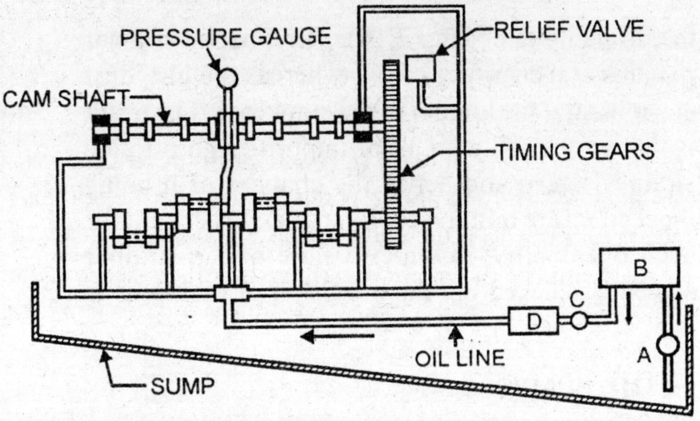

Fig. 7.12. Dry sump system

Obviously, in this system, the sump can be smaller in size and can be placed anywhere in the vehicle.

Usually, it is placed low down so that center of gravity is lower. This system is thus more reliable, but is more complex and costlier due to additional pump and high pressure oil lines, due to which it is not used in common vehicles.

14.5. Pre-Lubrication System

Daimler-Benz of Germany have devised a system by means of which lubrication of the engine is ensured before it actually starts. The arrangement is shown in Fig. 7.13. *A* is the main oil pump which is operated by the engine and it delivers oil to the oil gallery through filter *B*. A secondary lubricating pump *H* is provided which is operated by an electric motor. The oil from *H* is divided into two branches. one branch goes through relief valve and filter *B* to the oil gallery and the other leads the oil through filter *C* and non return valve *D* to the fuel control device *G*. The main oil gallery is further connected through pressure switch *F* and valve *E* to the fuel control device *G*.

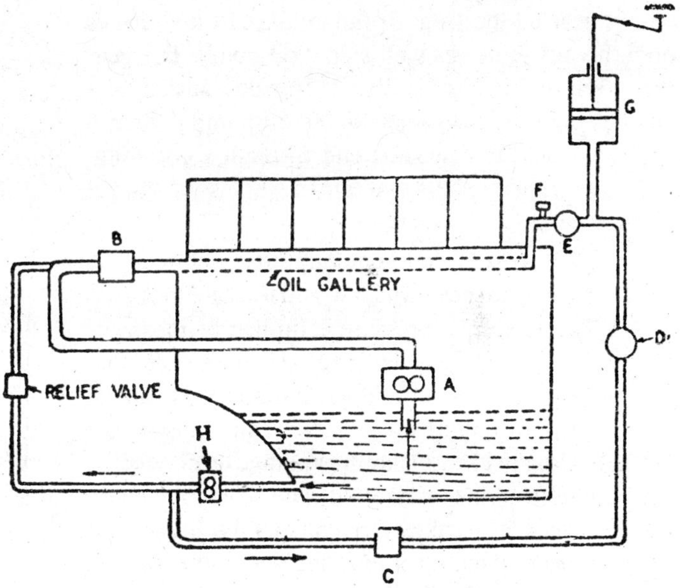

Fig. 7.13. Pre-lubrication system

When the pump *H* is started, in the pre-lubrication stage the oil flows through filter *B* to oil gallery. The switch *F* is opened at 0.5 atmosphere and the oil goes through *E* to device *G*. Simultaneously oil from pump H is also flowing through filter *C* and valve *D* to device *G*. When the pressure in the circuit reaches 1.5 atmosphere, the fuel control device is operated and the fuel is delivered to the engine causing it so start.

In this system double safety is provided. If filter *C* is blocked the device *G* can still be operated by direct circuit through pressure switch *F* and valve *E*.

15. OIL STRAINERS

Oil strainer is attached at the inlet of the oil pump to guard it against the entry of grit etc. The strainer is made of ordinary wire mesh screen. A good practice is to install a floating strainer, which is hinged to the oil pump inlet (Fig. 7.14). The floating strainer remains at the surface of oil, whereas the girt, dust etc. remain at the bottom of the crankcase. The result is that very small amount of impurities goes to the strainer screen and hence the chances of it being blocked up are minimised. Even then to account for such eventuality, in which whole of the strainer screen is blocked up, the provision of a by-pass is kept as shown in Fig. 7.14.

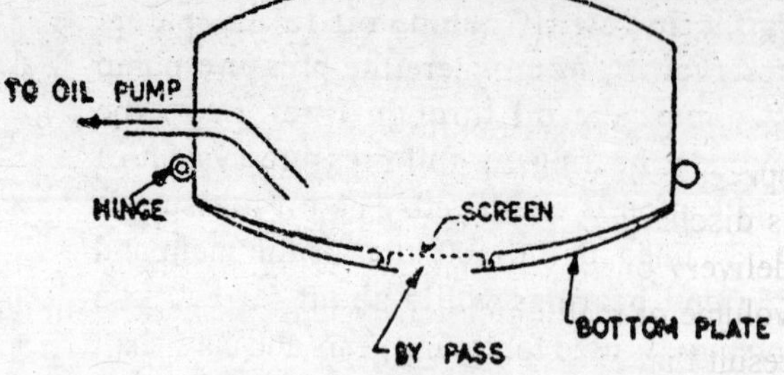

Fig. 7.14. Oil strainer.

16. OIL PUMPS

Next to oil strainer in the lubrication system sequence, comes the pump. Its function is to supply oil under pressure to the various engine parts. The oil pump is generally located inside the crank case below the oil level. However, instances exist where it has been mounted outside the crank case above the oil level. The pump is usually driven from the end of the distributor shaft, which gets its drive from the camshaft through a skew gear (Fig. 7.15) with 1:1 gear ratio, if a low- mounted camshaft is employed. However in case of engines with high-mounted camshaft, the skew gear on a relatively shorter distributor extension shaft meshes with a gear on the front of the crankshaft and drives both the distributor as well as the oil pump. The gear reduction in this case is 2 to 1. In some automobile engines mounted transversely, the oil pump is driven directly from the camshaft end through a coupling. Since no separate shaft is required, this is a compact arrangement.

The oil pressure in the engine increases with the increase in engine speed which would increase the pump speed. The maximum pressure is limited by means of a pressure relief valve and this valve is so selected that the pump will deliver sufficient lubricating oil to all the engine parts. Minimum oil pressure required is almost 100 kPa. Usually 15 to 30 litres per minute oil circulation is enough for engine lubrication. The size of the pump is so selected that necessary volume of oil at the desired pressure is supplied by it at moderate engine speeds without the occurrence of cavitation.

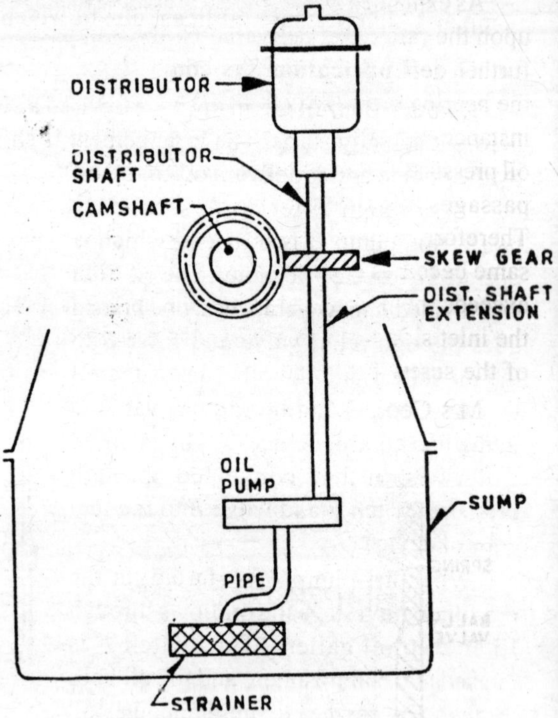

Fig. 7.15. Drive for oil pump.

Further, with wear of bearings and other engine parts, more oil will start leaking which may result in loss of engine pressure. The size of the pump should be sufficient to maintain the desired pressure with reasonable amount of wear. The different types of pumps used for engine lubrication are:

1. Gear pump
2. Crescent type gear pump
3. Rotor pump
4. Plunger pump
5. Vane pump.

16.1. Gear Pump (Fig. 7.16)

This is the type once almost universally used in the automotive engines. Its construction is very simple in that it consists of two spur, or for quieter running, helical gears only which are in mesh with each other. One gear is mounted on a stub shaft and is driven only whereas the other gear is the driving gear, itself being driven directly by the cam shaft through the same gear which drives the distributor shaft. The oil is transported from the inlet to the outlet side in gear spaces between the gear teeth as shown in figure and is discharged through the outlet port. Continuous delivery of the oil from the outlet port exceeds the volume of the lubrication system passages, with the result that excess discharged oil then causes the oil pressure in the system and rate of circulation of oil to be increased.

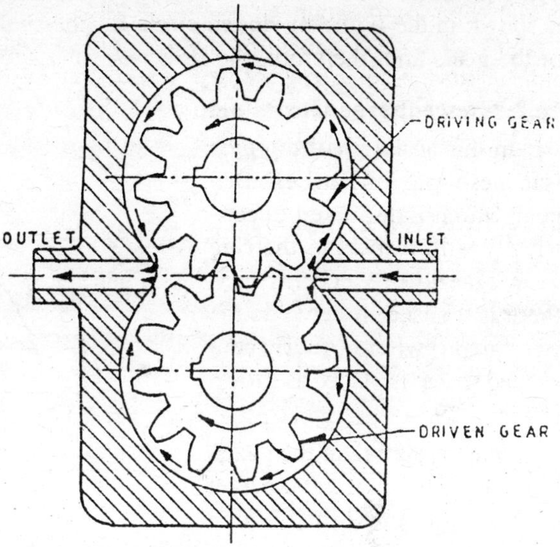

Fig. 7.16. Gear pump.

The pump is always submerged in oil in the crankcase and, therefore no priming is necessary. The pump delivers oil at about 300-400 kPa.

As explained above, the oil pressure would depend upon the rate of oil delivered by the pump, which is further dependent upon the pump speed as well as the bearing clearances in the lubrication system. For instance, lesser bearing clearance would mean higher oil pressure. Likewise any restriction in the lubrication passages would also increase the oil pressure. Therefore, a pressure relief valve is often provided in many oil pumps to relieve the pressure when the same becomes excessive due to high engine speeds or the clogged oil lines (Fig. 7.17). When the pressure exceeds the prescribed value, the spring is compressed, ball is lifted off its seat and the oil passes back to the inlet side through the bypass provided. The pressure at which valve opens may be adjusted by means of the screw provided.

M/s Georg Wiggermann and Water Reiners of Germany have patented a gear pump which provides

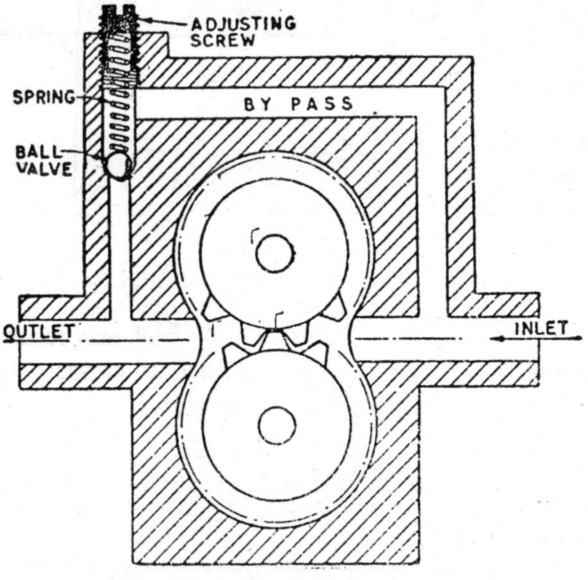

Fig. 7.17. Gear pump with pressure relief valve.

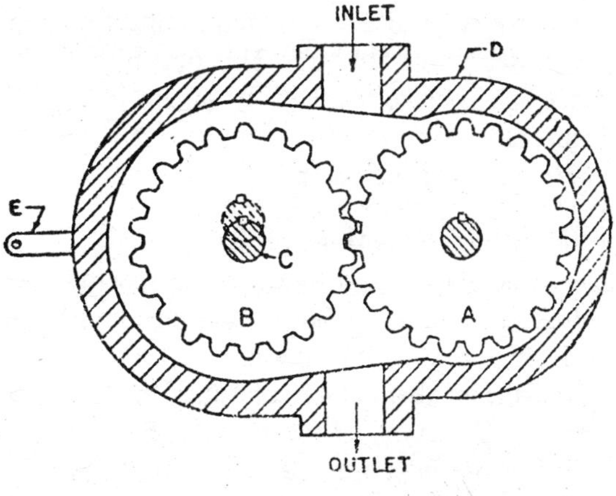

Fig. 7.18. Variable flow gear pump.

infinitely variable flow between full and half capacities. Fig. 7.18 explains the construction of such a pump. *A* and *B* are the meshing gears placed in the body *D*. Gear *A* is mounted on the driving shaft, while gear *B* is mounted on dead axle *C* whose position can be changed by means of lever *E*. One such changed position is shown in the figure by dotted circle for the shaft. This adjustment alters distances separating the axes of the gears and, therefore, the flow.

16.2. Crescent type Gear Pump

In this an internal ring gear is in mesh with a driving external gear which is mounted eccentrically with respect to the ring gear. Due to this eccentricity, there is a space in between the two gears where a crescent shaped spacer is placed as shown in Fig. 7.19.

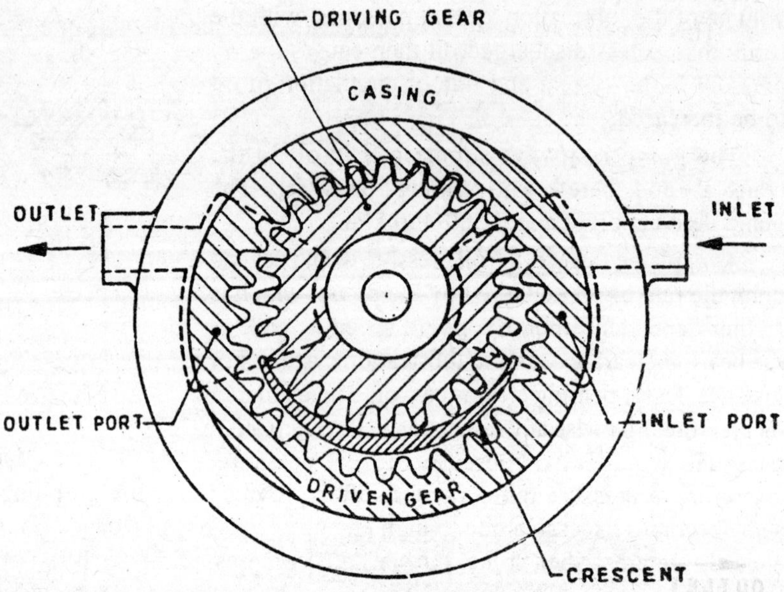

Fig. 7.19. Crescent type gear pump.

Due to rotation of the gears, a suction is produced at the inlet end of the crescent, which draws in oil. This oil is then trapped in tooth spaces between the crescent and the driving gear and the crescent and the tooth spaces of the internal driven gear and is thus carried towards to outlet side of the crescent where it is discharged into the outlet port. With continuous running the spaces between the gear teeth will be filled with oil after which the extra oil will cause the oil pressure to increase. The oil pressure will subsequently depend upon the oil supplied by the pump and the oil escaping from the engine bearings. This type of pump is very suitable for small compact engines where camshaft is mounted high. This occupies a very small space and is driven directly by means of a gear on the crankshaft, which besides driving, also serves as the internal gear of the pump.

16.3. Rotor Pump (Fig. 7.20)

This type of oil pump is now widely used in automobiles. It is similar to the gear pump except that in this two gears mesh internally. A is the external rotor having the number of lobes one more than on the internal rotor B. The axes of rotation of the two rotors are different which causes the size of the spaces between them to vary. Rotor B gets its drive from the engine and causes rotor *A* to rotate along with it. The oil enters the

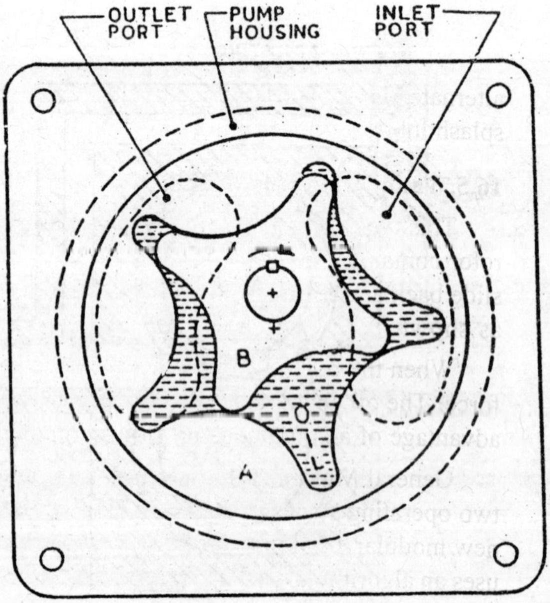

Fig. 7.20. Rotor pump.

pump through inlet port, as the rotor lobes are moving out of mesh. The oil is then transported from inlet to the outlet of the pump in the spaces between rotor lobes. The rotation reduces the clearance between the lobes and the oil is discharged under pressure as the lobes of the rotors move into mesh at the outlet port. This type of pump is about 25% more efficient and compact than the gear type pump. This is also quieter running since since there are comparatively lesser teeth in mesh for each revolution. Because of these advantages, its use is on increase in the automobile engines.

16.4. Plunger Pump (Fig. 7.21)

This is reciprocating type of pump in which the oil gets communicated from the inlet to the outlet by

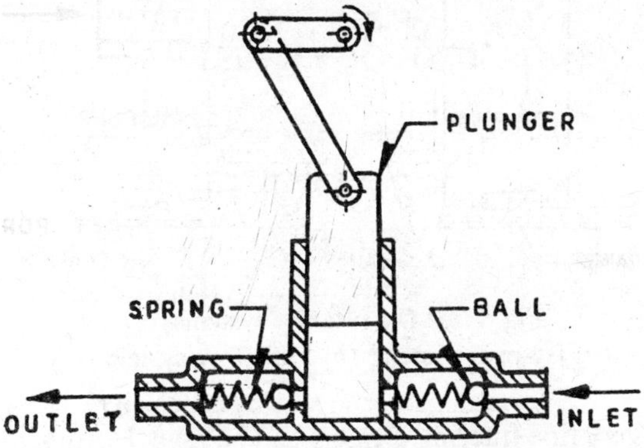

Fig. 7.21. Plunger pump

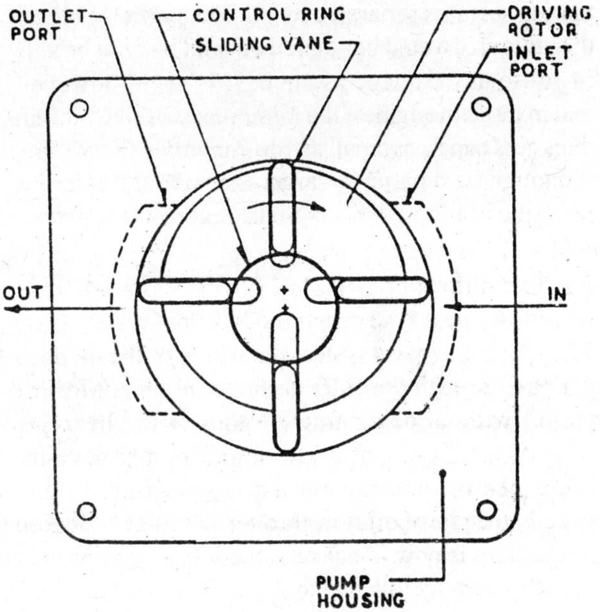

Fig. 7.22. Vane pump

alternate suction and pressure created by a reciprocating plunger. This was once used in engines using splash lubrication for filling the oil troughs, but is obsolete now.

16.5. Vane Pump (Fig. 7.22)

This type consists of a cylindrical housing, within which is eccentrically mounted a driven rotor. The rotor contains a number of vanes in the rotor slots equally spaced around its periphery. These vanes can slide back and forth into these slots. The movement of sliding vanes is guided by means of a control ring as shown.

When the pump is operating, the vanes are pressed outwards against the housing by their centrifugal force. The oil enters the inlet port and is swept by the vanes to the outlet port. This type of pump has an advantage of a continuous oil flow compared to the pulsating oil flow in case of gear pump.

General Motors is to use a new computer-controlled variable displacement vane type oil pump with two operating stages in their Gen-III Ecotec 2.5 litre naturally aspirated in-line four cylinder engine with new modular 14 architecture to be employed in their 2014 Chevrolet Impala. In this pump, the computer uses an algorithm based on engine load and coolant temperature for calculating oil temperature and adjusting the oil pressure accordingly. It is to replace the greater type oil pump in the older Gen-II 2.4 L engine.

17. OIL FILTERS

As discussed earlier, the lubricating oil with use, is deteriorated resulting in the formation of sludge, lacquer and carbon. Further, it is contaminated by various by-products of combustion of fuel, *viz.*, water, acids and unburnt fuel. In addition to these the fine particles of metal due to wear, especially during the running-in period and particles of rust formed in the engine are the other impurities presents in the oil. It is therefore, absolutely necessary to remove these impurities to avoid permanent damage to any or more runnings parts of the engine.

It has been found from experience that impurity particles of 5 to 15 microns diameter are most damaging, because the oil film size is bigger than this due to which smaller particles never come into contact with the mating surfaces separated by oil. The particles bigger than this size would be filtered out and will not be able to cause any damage as such. It is seen, however, that in modern engines, the minimum working oil film thickness can be as small as 1 to 2 microns. Therefore, if enough hard particles larger than about 3 microns reach the bearings, they may cause excessive and rapid wear.

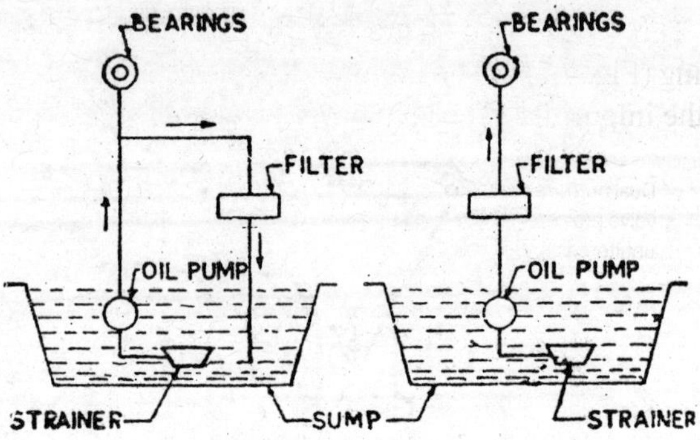

Fig. 7.23. Oil filtering systems
(*a*). By-pass system, (*b*) Full flow system

Two different types of oil filter systems are there *viz.*, the by-pass system and the full flow system (Fig. 7.23). In the by-pass system the whole of the oil does not pass through the filter at the same time. Most of the oil without being filtered goes to the bearings whereas the rest (about 10%) passing through the filter is cleaned out and is returned to the sump. In this system, the rate of oil flow through the filter is slow so that very fine filtering element can be used. However, this system is now obsolete since it has been found statistically impossible to clean the whole oil content of the system by this method. On the other hand full-flow system provides a more certain filteration process.

In the full-flow system, however, all the oil which goes to the bearings must pass through the filter first. Thus, if any time the filter is blocked in the full flow system the oil flow would be stopped and the bearings would be starved. To avoid this to happen, a spring-loaded relief valve is incorporated in the filter, which bypasses an emergency supply of unfiltered oil to the bearings, thus saving them from oil starvation and consequent damage. It is designed to open at 40 to 100 kPa, the actual value depending on the engine model and the normal pressure drop across the filter. This relief valve is in addition to the main relief valve (Fig. 7.24). However, since the entire oil has to be filtered, the rate of flow in this system has to be quite high due to which, the filtering elements used are comparatively coarser.

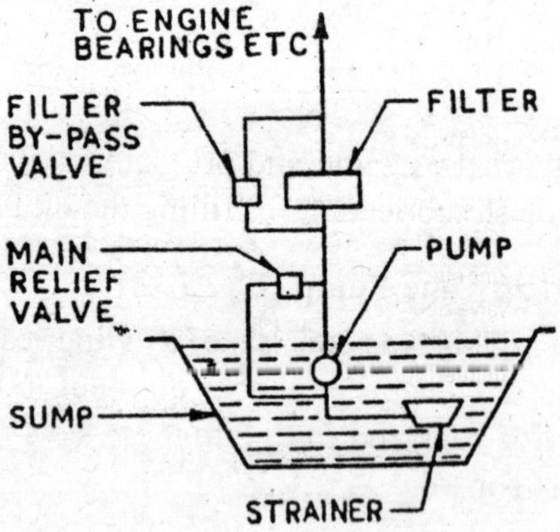

Fig. 7.24. Full-flow filteration system.

An oil filter must have the following characteristics:

1. High filteration efficiency.

2. Less pressure loss. Excessive pressure loss occurs due to high resistance in the filter element, preventing oil from reaching all the engine parts requiring lubrication.

3. Long life.

Oil filters are basically of two types, *viz.*, the primary and the secondary. A primary filter is also known as *surface filter* since the impurities are retained on the outer surface of the filter. It uses coarse filter material and in installed on the inlet side of the oil pump. The secondary filter is installed on the outlet side of the oil pump and since the impurities in this are retained progressively along the depth of the filter material, it is also called *depth filter.*

Primary filter is more commonly known as strainer and is formed from wire gauze of 40-60 ìm. The secondary filters used in the automobile engines are of various kinds, the important ones out of them being:

1. Cartridge type.
2. Edge type.
3. Centrifugal type.

17.1. Cartridge type oil filter

This type is used on most automobile engines and it consists of a filtering element placed in the metallic casing (Fig. 7.25). The impure lubricating oil is made to pass through the filtering element, which takes up all the impurities. The element is given a pleated form to maximize the surface area of the filter for a given

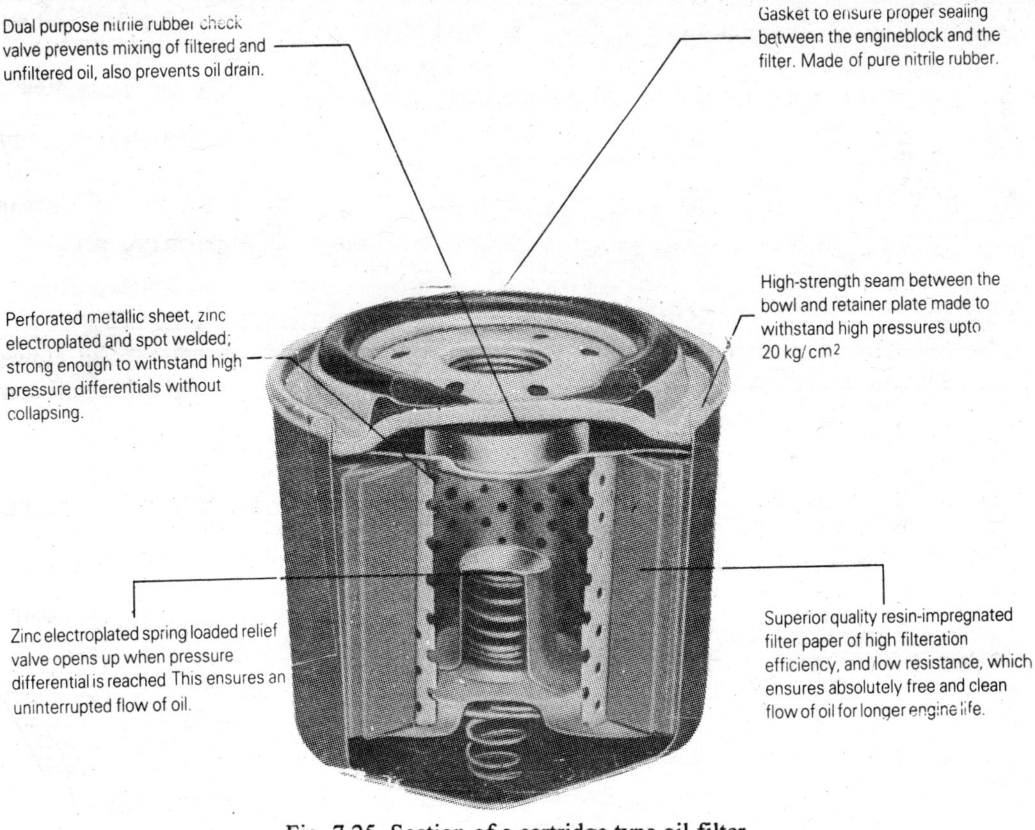

Fig. 7.25. Section of a cartridge type oil filter.
(Courtesy—Elofic Industries Pvt. Ltd., India)

size of element. Currently filter elements with fine pores have been employed which has made it practicable to arrest particles of size down to within the region of 5 microns. In the filter shown in Fig. 7.26 the oil enters the filter at the top and passes through the filter elements as shown by arrows. The pure oil then goes to the porous metallic tube from where it goes to the outlet for circulation. A drain plug is also provided as shown.

The filtering elements of two types are available, *i.e.*, the ones which can be cleaned and those which have to be replaced after certain intervals, say, 10000 km.

Commonly used materials for filtering are wire gauze, cotton, felt, paper, plastic -impregnated paper etc. The filtering element must let the oil pass through without must resistance, but should prevent the undesirable particles from entering the oil galleries. The resistance to flow offered by a new filter element is about 0.3 to 0.5 bar. However, with use the resistance increases due to clogging of the filter element. The maximum permissible resistance is about 2 bars.

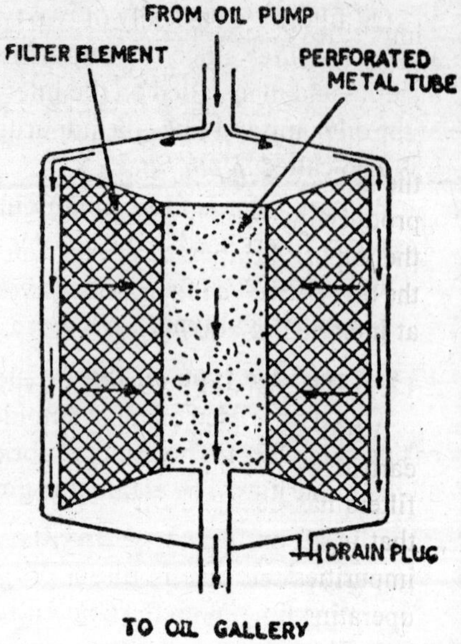

Fig. 7.26. Cartridge type oil filter

17.2. Edge type oil filter

It is also called the *stack type.* In this the oil is made to pass through a number of closely spaced discs (Fig. 7.27). The alternate discs are mounted over a central spindle, while the discs in between these are attached to a separate fixed spindle as shown. The clearance between two successive discs is a few microns. The oil is made to flow through the spaces between these discs and because of the very small spaces involved the impurities are left on the disc peripheries from where these are periodically removed by simply operating the central knob. This may be done ether manually or the knob may be connected to the clutch system and operated periodically by means of clutch action.

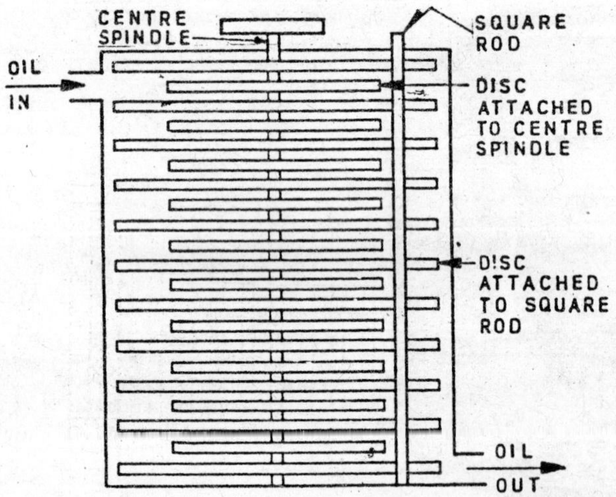

Fig. 7.27. Edge type oil filter

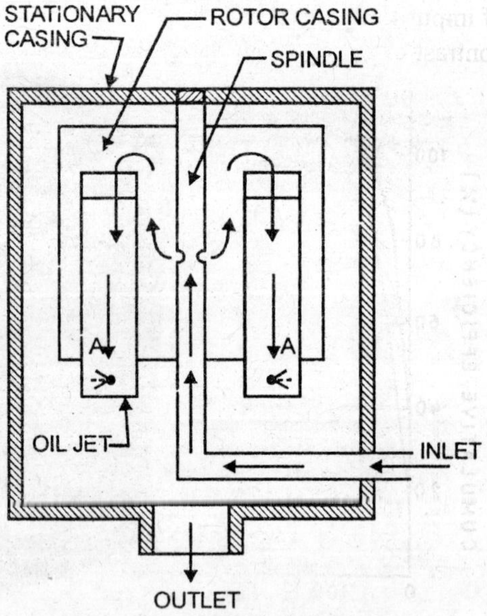

Fig. 7.28. Centrifugal type oil filter

17.3. Centrifugal type oil filter

In this the impure or dirty oil from the engine enters the hollow central spindle having holes around its periphery as shown in Fig. 7.28. The dirty oil comes out of these holes and fills the rotor casing after which it passes down the tubes A at the ends of which jets are attached. The oil under pressure passes through these jets, the reaction of which gives the motion to the rotor casing in the opposite direction so that it starts rotating. The oil impinges on the outer stationary casing under heavy pressures, where the impurities are retained and clean oil falls below from where it is taken out. The working speed of the rotor is usually between 2000 and 7000 rpm, depending upon the oil pressure circulating the oil. Force of upto 2000 times the gravity are produced which are quite effective in throwing the impurities, including very fine ones, out of the oil system. The filter walls have to be cleaned at intervals of about 70,000 km.

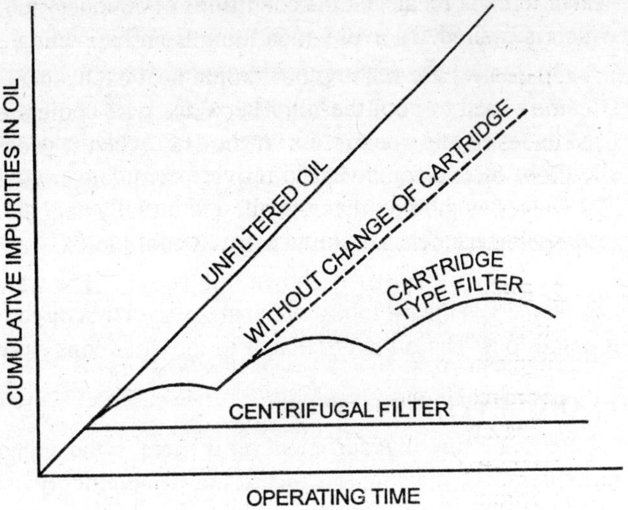

Fig. 7.29. Variation of cumulative dirt in oil in different filters.

17.4. Performance curves of filters

A comparison of the performance of the cartridge type and the centrifugal type of oil filters has been shown in Fig. 7.29. It is seen that in an unfiltered system, the accumulated impurities increase in direct proportion to the operating time. With the conventional cartridge type filter, the accumulation rises first with the unfiltered curve and then decreases to the first filter cartridge change. After changing the cartridge, rate of accumulation again rises and then falls upto the second cartridge change point and this continues further. If the cartridge is not changed, the accumulation of impurities continues to increase at about the same rate as unfiltered system. However, it is seen that in contrast to the cartridge type, the rate of accumulation in the centrifugal filter is similar to the unfiltered

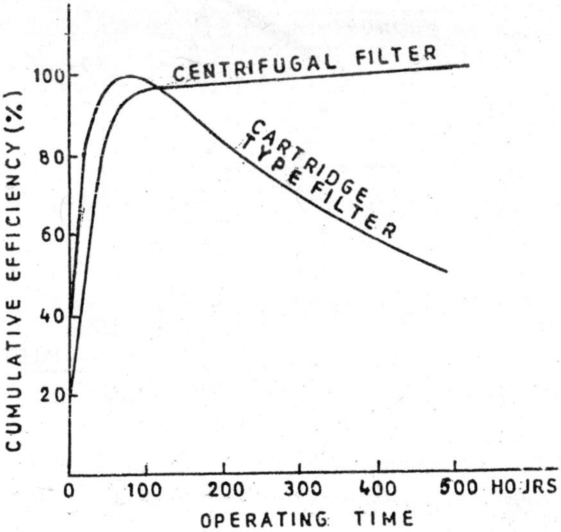

Fig. 7.30. Comparison of performance of the cartridge type and the centrifugal type filters.

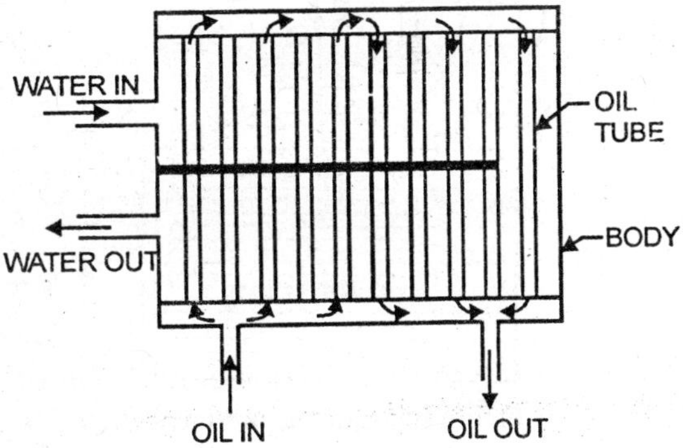

Fig. 7.31. Oil cooler.

system initially, but then later it levels off once the oil circulation through it reaches the minimum design level. Fig. 7.30 shows the variation of cumulative efficiency curves of the two types of filters.

18. OIL COOLERS

In all heavy duty engines the engine temperature and hence the temperature of oil becomes quite high. As the viscosity of lubricating oil decreases with temperature rise, at higher temperatures the oil film in the bearings might break and the conditions of boundary lubrication may be created instead of fluid lubrication which is desired. To avoid such thing from happening, the heavy engines are provided with oil coolers.

Oil coolers are nothing but simple heat exchangers. Either the cold water from the radiator or the air stream is used to cool the oil. The water type coolers are, however, more common because they can be used as reversible coolers, *i.e.* at the start when it is desirable that the oil should not be cooled, rather it should be heated somewhat to provide complete circulation in the lubrication system as fast as possible. The water being hotter than the oil, will initially heat the oil and when higher temperatures are reached, the reverse will happen, *i.e.* the water will cool the oil.

One such oil cooler is shown in Fig. 7.31. The water enters the cooler as shown by means of arrows and then cools the oil tubes which are placed in a direction perpendicular to the flow of water, after which it comes out. The oil is circulated in the directions shown in the figure.

19. OIL LEVEL INDICATORS

To make sure that sufficient oil is there in the engine crankcase some means must be provided for its measurement. The simplest method most commonly used is the 'dip stick' (Fig. 7.32). It is a long stick with a handle at one end for holding. The marks are provided on the dip stick indicating full, half full or quarter full crankcase. The critical mark is also provided below which the level should not fall. It is the duty of the driver to check the oil level at regular intervals, particularly before undertaking any long journey.

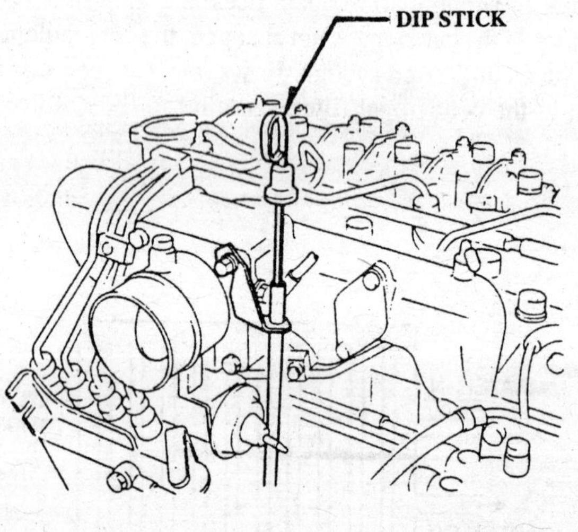

Fig. 7.32. Dipstick.

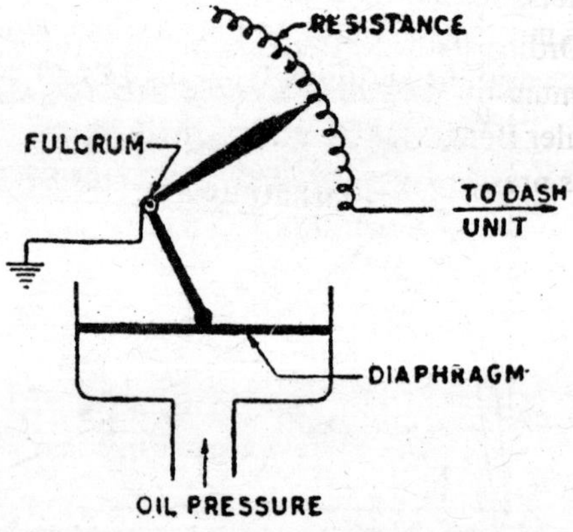

Fig. 7.33. Principle of electrically operated oil pressure gauge.

Another type of oil level indicator is based upon the float system. To the float inside the crankcase, is attached suitable indicator to indicate the oil level outside the crankcase.

20. OIL PRESSURE GAUGES

If any leakage occurs at any part of the lubrication system, the pressure in the entire system would fall, reducing consequently the oil supply to various bearings as a result of which they are bound to starve and be damaged. Thus it is very important for the driver to keep a watch on the oil pressure and any reduction in the system pressure must come into his notice.

The gauges in use are of two types, the Bourdon type and the electrically operated type. The working principle of these gauges is the same virtually as has already been explained for the respective temperature gauges in the cooling system (Art. 13, Chapter 6). Only instead of thermal element in case of electrically operated type, here we have a linear resistance, which is controlled by the sliding contact attached to a diaphragm, whose deflection is proportional to the oil pressure (Fig. 7.33).

21. OIL PRESSURE WARNING LIGHT

A warning light is provided on the dashboard which lights up when the oil pressure is lower than the specified value. This light is controlled by an oil pressure switch mounted in the main oil gallery (Fig. 7.34). It consists of a diaphragm with a projection at the centre. When the oil pressure is above the prescribed

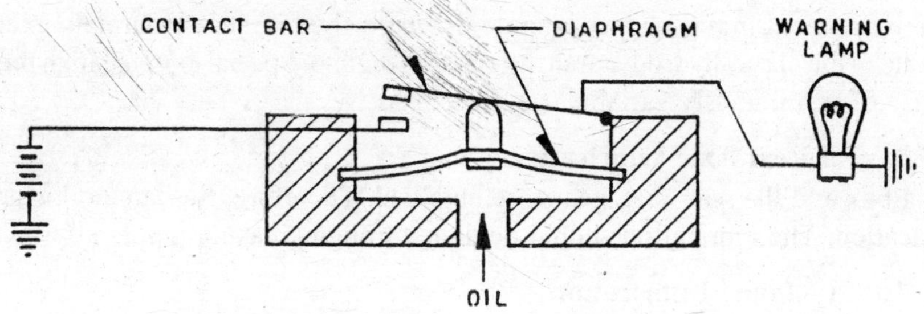

Fig. 7.34. Oil pressure warning light and switch.

value, the projection keeps the contact bar pushed up, thereby separating the contacts so that the warning light is off. However, when the oil pressure falls, the diaphragm also moves down thereby making the contacts, so that the warning light comes on.

Ordinary warning lamps are sensitive to only one oil pressure irrespective of engine speed, though the minimum oil pressure required depends upon the engine speed. The remedy has been provided by M/s Daimler Benz AG (Germany) in the form of a 4 stage diaphragm-operated switch (Fig. 7.35). When the oil exerts pressure on the diaphragm, the projection attached to it presses the uppermost contact bar, breaking

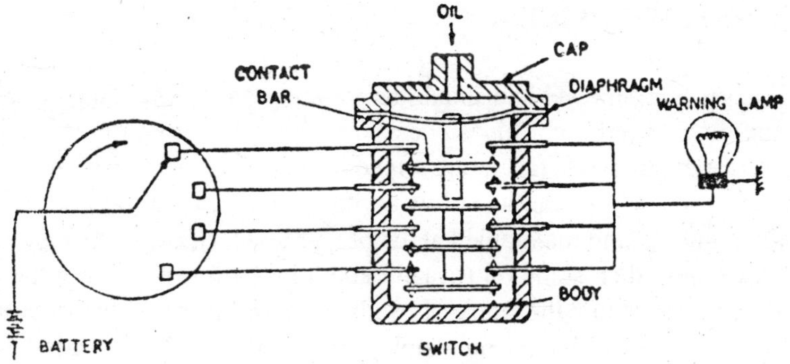

Fig. 7.35. Oil pressure warning system.

the contacts between the two projections. Further movement causes a projection from the center of the first contact bar to deflect a second bar and so on, upto the fourth stage. Each of these four stags is brought singly into circuit with the warning lamp by a selector switch operated in conjunction with the speedometer. The warning light in this way lights up only when the oil pressure falls below the value corresponding to the minimum engine speed.

22. CHASSIS LUBRICATION

Different lubricants and methods of lubrication are employed for chassis lubrication. The manufacturer supplies complete information about the location of lubrication points, manner of lubrication, type of lubricant and the period after which the lubrication must be done, etc. in the form of lubrication charts. A brief general guideline, however, about the lubrication of only a few main chassis components is given here.

1. Gear Box Lubrication

Gear box is filled with lubricating oil upto the specified level, which may be checked with a dipstick. The oil level should be checked after every 1500 km. and replenished, if necessary. After first 800 km and subsequently every 10,000 km, the gear box should be drained, flushed and refilled with the fresh lubricant. For moderate temperatures (above 10°C), generally S.A.E. 90 oil or hypoid 90 oil is used. For heavy duty applications, S.A.E. 140 oil is used. The exact specification for a particular vehicle is given by the manufacturer.

Lubrication environment in a gear box is quite difficult; the pressures routinely exceed 10^9 Pascals and the gears operate in the elasto-hydrodynamic lubrication regime. Special extreme high pressure lubricants are required in case of continuously variable transmissions.

2. Steering Box and Rear Axle Lubrication

In the steering box and the rear axle, gears are involved. Therefore, the same oil is generally used as for gear box lubrication. The lubrication should be done through greasing nipple after every 20,000 km.

3. Constant Velocity Joints Lubrication

Greases are commonly used to lubricate C.V. joints which connect the driving wheels of the vehicle with the live axles, while allowing the suspension to move up and down.

4. Lubrication of Wheel Hubs

Special grease is used for wheel hubs. For example, Mobihub grease is specified in case of Hindustan Ambassador car and Leyland Specification G grease for Ashok Leyland vehicles.

5. General Chassis Lubrication

S.A.E. 140 oil or L.M. Specification G grease is specified for Ashok Leyland vehicles and Mobilgrease No. 4 is prescribed for Ambassador cars. For frequency of lubrication, the manual should be consulted.

23. LUBRICATION SYSTEM SERVICE

1. Oil level

For checking engine oil level, the vehicle should be standing on a level plane. Moreover, it is ideal to check it as the first thing in the morning. This is checked with the help of a dip stick. There is a mark on the dip stick to indicate proper level. It is taken out, wiped clean, inserted in the pump and finally taken out to see the wet portion on it. If the dip stick is not wet upto the mark, more oil has to be added. Oil level should be checked daily regularly and maintained at the proper level. However, it should never be checked when the engine is hot, because then some oil (usually about half a litre) is still in the oil galleries and it would take some time to drain back into the crankcase sump. If oil is measured in this condition, the reading could be erroneous, *i.e.*, the oil level would be visibly low and if more oil is added to being it upto the correct minimum level, the same would become excessive once the engine is cooled and the oil from the galleries has returned to the crankcase. In case oil level is too high, the following consequences may result:

1. The crankcase, being too close to the oil surface, will cause foaming of oil so that air enters the lubricating system which would cause inconsistent oil pressure.

2. The big ends of the connecting rods will cause excessive oil splashing than which the piston rings can control, due to which some of the oil would enter the combustion chamber and get burnt. This would result in blue exhaust smoke and excessive oil consumption. Moreover, excessive oil reaching the piston ring grooves would cause the sticking of rings.

3. Excessive built-up of pressure in the crankcase may damage the crankshaft seals. Even the cork gaskets may be damaged so that the leakage will still remain even if the oil level is corrected.

On the other hand, decrease of oil level below the specified level may expose the oil strainer, due to which the air will be drawn into the lubricating system alongwith the oil and may result in insufficient or interrupted lubrication of the engine parts. This effect of exposing the strainer will be more prominent on slopes and turns. Moreover, lower oil level means lesser quantity of oil which may become overheated in the engine due to which its lubrication properties will get deteriorated earlier. Thus it is seen that fall of oil level below the correct level is more dangerous as it may lead to oil starvation of the engine parts.

2. Oil change

While checking oil level with the dip stick, put a few drops of oil on your first finger and then rub the same with thumb. With practice, this procedure tells you about the dirt etc. present in the oil, and also whether the oil is sufficiently thick or not. Apart from this the colour of the oil will also indicate its quality. If the oil is too dark and thin, the same has to be changed. Usually oil is changed after 10000 km. intervals or earlier depending upon the conditions of operation. However, manufacturer's instructions should be followed for this.

For changing the oil, warm up the engine and drain while it is still warm. Now change 'the oil filter'. It is a good practice to soak the new filter with the new oil before installing so that it does not absorb the oil later and thus result in lowering of the oil level. Finally ensure that the drain plug is tight and refill the new lubricating oil of suitable specifications as recommended by the manufacturer. Check the oil level with dip stick as described above.

3. Checking the oil pump

The points to be tested in a gear pump are the clearance between the gear teeth, stub shaft wear, bush and oil relief valve. The clearances are measured with a feeler gauge. If the clearance between the gear teeth is more than 0.5 mm, the gears have to be replaced. Similarly stub shaft is replaced if the wear thereon is more than 0.5 mm. The bush in the drive gear or the drive shaft has to be discarded if the clearance there has exceeded 0.1 mm. In the relief valve, the spring is to be tested for stiffness and if found not according to design specifications, is to be replaced.

4. Checking of the oil filter

Open the oil filter and inspect the element. If the same is found clogged, the same may be cleaned and reused, but the better practice is to discard clogged element and replace it with a new one. To check whether the filter is really clogged, touch the same after running the engine for some time. If filter is hot to touch, it means it is clean.

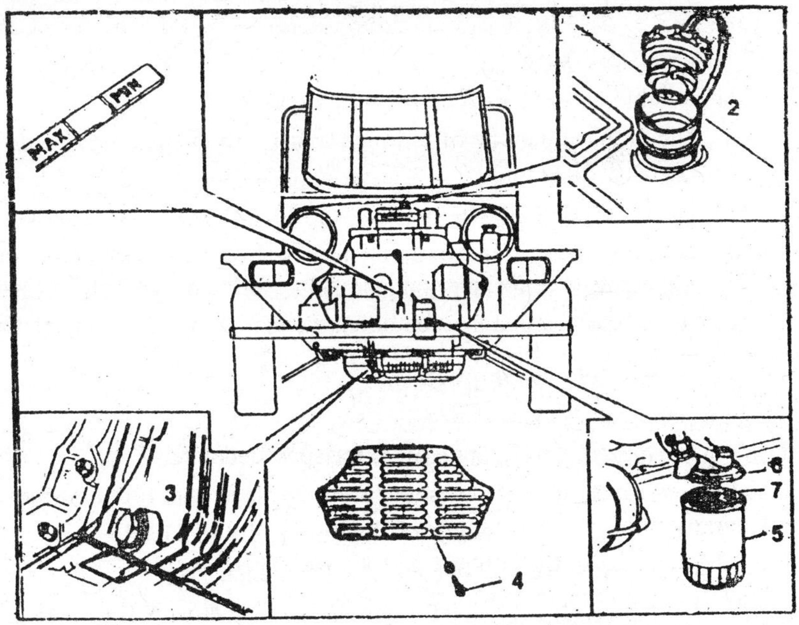

Fig. 7.36. Service points in 'Moke' engine.
1. Dipstick, 2. Oil filter plug, 3. Oil drain plug, 4. Grill screws, 5. Filter canister, 6. Filter head, 7. Seal.
(Courtesy—Leyland Motor Corporation of Australia Ltd.)

However, if it is cold, this indicates that no oil passed through it, because of clogging. Generally this is changed after 10,000 km. However, again it depends upon the conditions under which the vehicle has been running.

How all the above operations are done in Australia's 'Moke' vehicle is made clear by means of Fig. 7.36, which is self explanatory.

24. LUBRICATION SYSTEM TROUBLE SHOOTING

The commonly occurring troubles in the lubricating system are unduly low or excessively high oil pressures. Their possible causes are discussed here and the remedies suggested.

1. Low oil pressure

Low oil pressure may be caused due to any or more of the following:

(*a*) Low oil level, which may be checked by the dipstick. Fill the crankcase to the correct level.

(*b*) Weak relief valve spring may also cause the ball to lift at lower pressure than prescribed, thus lowering the system pressure. The remedy is to replace the defective spring.

(*c*) The oil pump may have worn out. The same may be checked and the defective part replaced.

(*d*) Defective oil line or loose joirts in the oil line may cause leakage and thus reduce the pressure. Loose joints should be tightened and defective lines replaced.

(*e*) The oil may have been diluted with constant use, which has to be changed.

(*f*) Oil filter may be clogged, which should be replaced.

(*g*) Worn out crankshaft or camshaft bearings will pass more oil than the oil pump is capable of delivering. The only remedy is their replacement.

(*h*) The pressure gauge indicating low oil pressure may be defective, which may be checked and the defective parts repaired or replaced.

2. High oil pressure

The possible causes of excessive high pressure are:

(*a*) The oil used may be of too high a viscosity, which may be drained out and the oil with correct viscosity filled in.

(*b*) The oil lines, galleries may be clogged, which may be cleaned.

(*c*) The relief valve in the oil pump may be stuck in the closed position. The valve should be inspected and repaired accordingly. If the valve spring is too stiff, the same may be changed.

(*d*) Oil pressure gauge may be defective which may be inspected and repaired.

3. Excessive oil consumption

This may be possibly due to:

(*a*) Too high oil level. Drain oil to the specified level.

(*b*) Low viscosity oil level being used. Replace with oil of correct specification.

(*c*) P.C.V. valve stuck closed, which has to be replaced.

(*d*) Defective valve seal, which should be changed.

(*e*) Worn out valve stem and/or guide. The defective part in such a situation will have to be changed.

(*f*) Worn piston, cylinder liner or piston rings. The only remedy in such a case is the replacement.

(*g*) Piston rings may be broken, which should be replaced.

(*h*) Incorrect piston ring gap, which may be measured and ring replaced if necessary.

(*i*) Piston rings may be sticking in the grooves. If so, measures ring side clearance and repair as necessary.

(*j*) Compression rings may be installed upside down which may be rectified.

(*k*) Piston ring gaps may not be properly staggered, which may be corrected.

(*l*) Excessive clearance in main or camshaft or connecting rod bearings. The clearances in such case may be measured and repaired as required.

(*m*) External leakage of oil which may be caused due to:

(*i*) Defective gasket between the crankcase and oil pan.

(*ii*) Defective oil pump gasket.

(*iii*) Oil filter cap missing or leaking.

(*iv*) Defective oil filter gasket.

(*v*) Oil pan rear oil seal broken or otherwise defective.

(*vi*) Defective timing case cover oil seal.

(*vii*) Oil pan drain plug loose or having stripped threads.

(*viii*) Rear oil gallery plug loose.

(*ix*) Rear camshaft plug loose or otherwise defective.

(*x*) Cracked oil pan.

All the above situations have to be examined and defective parts repaired or replaced.

TABLE 7.1. COMPARATIVE LUBRICATION SYSTEM DATA OF SOME INDIAN AUTOMOBILES

S. No.	*Item*	*Hindustan Ambassador mark II*	*Maruti 800*	*Fiat 1100*	*Jeep Universal*	*Dodge/Fargo Model 89 M4 (Petrol type)*	*Ashok Leyland Comet passenger*	*Tata Diesel Vehicles 1210*	*Swaraj Mazda*
1.	Type	Wet sump, pressure type	Wet sump, pressure type	Wet sump, pressure type	Wet sump, pressure type	Wet sump, pressure type	Wet sump, pressure type	Wet sump, pressure type	Wet sump, pressure type
2.	Oil pump	Eccentric type	Internal gear type	Grear type	Internal rotor type	Rotor type	Gear type	Gear type	Rotor (trochoid type)
3.	Oil pressure (normal running) kPa	350	340-400 at 3000 rpm	250	250	316-387	420	At max. speed, new engine = 800 old engine = 250	650
4.	Oil filter	Full flow with felt element or paper element	Full flow, cartridge type	(*i*) By pass cartridge type (*ii*) Full-flow screen type	By-pass cartridge type	Sealed type	Full flow, cloth element	Full flow, paper cartridge	Full flow, paper filter, cartridge type

THEORY QUESTIONS

1. Explain the terms (*a*) dry friction (*b*) viscous friction, (*c*) boundary friction. Which out of these is most desirable and why?
2. What are the objects of lubricating an engine? Explain different methods of engine lubrication.

3. What are the characteristics which are considered while selecting a particular lubricant for an automobile engine?
4. Discuss different types of lubricants. Explain the terms "blending" and "compounding".
5. Explain in detail different tests to which lubricants are subjected. How do you determine the viscosity of a lubricating oil?
6. What do you know about oil additives? Explain briefly the commonly used oil additives.
7. Discuss in detail the effect of different engine conditions on the lubricating oil.
8. Discuss in detail the factors on which the consumption of lubricating oil in an engine depends.
9. Explain clearly pressure system of lubrication for an automotive engine.
10. Make a neat sketch showing the components of a dry sump method of engine lubrication. Explain its working.
11. What is crankcase ventilation? How is it achieved in an automobile engine?
12. With the help of a neat sketch, explain the construction of an oil strainer.
13. Explain the construction and working of different types of oil pumps used in the automotive engines.
14. Explain why it is necessary to have filter in the engine lubrication system. Describe the bye-pass and the full flow oil filtration systems.
15. Discuss the different types of oil filters in automotive use. Illustrate your answer with the help of neat sketches. Describe also their performance curves.
16. What is the object of using oil cooler? Explain briefly the construction and working of a water type of oil cooler.
17. What type of oil pressure gauges are used in the lubricating system of engine? Explain each of these gauges with the help of suitable diagrams.
18. Discuss the construction and working of an oil pressure warning lamp switch.
19. Explain how would you service the lubricating system of an automotive engine.
20. Discuss the common troubles occurring in the lubricating system of an automotive engine. Enlist the possible causes and suggest suitable remedies.

QUESTIONS FOR VIVA

1. What is the necessity of engine lubrication?
2. What is hydrodynamic lubrication?
3. What is boundary lubrication?
4. What is viscous lubrication?
5. What is elasto-hydrodynamic lubrication?
6. What are the primary objects of lubrication?
7. What are the secondary objects of lubrication?
8. How does lubricating oil provide cushioning effect in the engine?
9. What is viscosity?
10. What is viscosity index? What is its importance?
11. What are the disadvantages of a very high and very low viscosity lubricating oil?
12. Which oil property is taken as the main criterion for the its selection as engine lubricant?
13. Define 'oiliness' of a lubricating oil.
14. From which source are the commonly used lubricating oils in automobiles derived?
15. State the advantages of a mineral lubricant?
16. Why are the animal and vegetable oils not suitable for automotive use?
17. Give any example of a synthetic lubricant.
18. State the advantages of synthetic lubricating oils.

19. What is a grease?
20. Where do we have to necessarily use a grease rather than oil for lubrication?
21. Give examples of greases for automotive use.
22. Which solid lubricant is most commonly used in automobiles?
23. State the advantages of graphite as a lubricant.
24. What is the significance of S.A.E. viscosity ratings?
25. What do the following stand for:
    (*a*) A.P.I.,
    (*b*) S.A.E.?
26. What is a multigrade oil?
27. State different S.A.E. viscosity ratings for oils.
28. Classify the lubricating oils according to service conditions.
29. Enumerate different tests for lubricants.
30. What is flash point? Why is it important?
31. What is 'pour point' of oil? What is its significance?
32. What is 'acid value' of an oil?
33. What is the significance of acid value test for lubricating oils?
34. Define 'sponification value' of an oil.
35. What use is made of the results of sponification value test for lubricating oils?
36. What is an oil additive?
37. State the nature of oxidation inhibitors.
38. State the factors which promote the formation of acids in the engine cylinders.
39. What is function of a detergent? How does it achieve this?
40. What is an anti-scuff additive? How does it perform its function?
41. How do the viscosity index improvers achieve their object?
42. Why do we require an 'anti-foaming' additive?
43. Give some example of anti-foaming additives.
44. Why are the extreme pressure additives required?
45. How is gum formed by a lubricating oil?
46. How is the sludge formed in the oil lines?
47. State the engine conditions that promote the formation of varnish.
48. What are the possible causes of oil dilution?
49. When should the dipstick reading be taken: when the engine has just started, or when it has warmed up? why?
50. What is the effect of oil dilution on engine performance?
51. Which engine surfaces are most prone to carbon deposit formation?
52. What is the cause of carbon deposition?
53. It is said that the engine, immediately after starting, should not be raced. Why?
54. State various sources of oil loss in the engine.
55. "The oil consumption decreases with increase in engine speed." True or false?
56. What is the average consumption of lubrication oil in an automobile engine?
57. State the effects of piston blow-by.
58. What is crankcase ventilation?
59. What is positive crankcase ventilation?
60. Give example of any Indian car employing positive crankcase ventilation.
61. Name the various parts of an automobile engine which require lubrication.

62. Name various methods of engine lubrication.
63. In what types of vehicles, the use is made of the petroil system of lubrication?
64. Which type of lubricating system is commonly used in modern automobiles?
65. What are the approximate pressures encountered in a pressure system of lubrication?
66. How is the piston pin lubricated?
67. How are the timing gears lubricated?
68. How are the valve trains lubricated?
69. What are the 'wet sump' and the 'dry sump' lubrication systems?
70. In which type of vehicles is the dry sump system used?
71. What are the merits and demerits of dry sump method of lubrication over the wet sump system?
72. What is pre-lubrication?
73. Where is the oil pump usually located?
74. From where is the drive for oil pump obtained?
75. How is the oil pressure at delivery in a gear pump increased?
76. Why do we provide relief valve in the lubrication pump?
77. How does an oil pump differ in construction from a coolant pump?
78. Name different types of oil pumps.
79. Which type of oil pump is being increasingly used in automobiles?
80. State the factors on which the rate of oil delivered by a pump depends.
81. State the advantages of a crescent type pump. For which type of engines is this pump particularly suitable?
82. State the advantages of a vane pump.
83. Of which size impurity particles are most damaging? Why?
84. Name a few materials used for oil filter elements.
85. Approximately how much resistance to oil flow is offered by a new filter?
86. What is the maximum resistance to oil flow offered by the oil filter that can be tolerated?
87. How do you compare the by-pass and the full-flow systems of oil filteration?
88. Why do we provide a relief valve in the full-flow oil filteration system?
89. Compared to the by-pass system, do we require a finer or a coarser filter element in case of full flow system?
90. What are the primary and the secondary filters?
91. What is a surface filter? Where is it installed?
92. What is a depth filter? Where is it installed?
93. What is the life of the filter element in a cartridge type oil filter?
94. After how much interval do the walls of a centrifugal type oil filter normally require cleaning?
95. Which type of oil cooler is more efficient and why?
96. What is a 'dip stick'?
97. How do you determine the level of oil present in the sump of an engine?
98. Name different types of oil pressure gauges.
99. Where in the engine is the oil pressure switch mounted?
100. How is the oil pressure warning light operated?
101. State the specification of a gear box oil.
102. Give the specification of oil for steering box and for rear axle lubrication.
103. Which lubricant is used for wheel hubs?
104. What is the approximate period after which oil is changed in the modern cars?
105. While checking a gear pump, what are the points to be noted?

106. When will you order a gear change in the gear pump?
107. What happens if the relief valve spring of a gear pump, while servicing, gets changed with some other spring of the same size, but lower stiffness?
108. While inspecting the lubricating system of an engine after the same has been run for some time, you find that filter is not at all hot to touch. What do you infer from this?
109. Generally, what is the suggested period for filter change?
110. What will be the effect of worn out crankshaft or camshaft bearings on engine lubrication?
111. What happens if the oil level in the sump is:
   (*i*) too low,
   (*ii*) too high?

MULTIPLE CHOICE QUESTIONS

Mark the correct answers:

1. The friction that occurs between the layers of oil in an oil film is called
   (*a*) viscous friction (*b*) solid friction
   (*c*) boundary friction (*d*) greasy friction
2. The primary function of lubrication is to
   (*a*) provide cooling effect (*b*) provide sealing action
   (*c*) provide cleaning action (*d*) reduce wear
3. The most important characteristic of a lubricating oil is its
   (*a*) viscosity (*b*) physical stability
   (*c*) chemical stability (*d*) resistance against corrosion
4. Most commonly used lubricants in automobiles are the
   (*a*) animals oils (*b*) mineral oils
   (*c*) vegetable oils (*d*) synthetic oils
5. The most commonly used lubrication system in automobiles is the
   (*a*) petroil system (*b*) splash system
   (*c*) pressure system (*d*) none of the above
6. The approximate oil pressure in the lubrication system in modern cars is
   (*a*) 40-50 kPa (*b*) 200-400 kPa
   (*c*) 1-5 MPa (*d*) more than 10 MPa
7. Maximum oil pressure in the lubrication system is controlled by
   (*a*) oil filter (*b*) pump rotor
   (*c*) pressure relief valve (*d*) pressure switch
8. The lubrication system in all modern cars has
   (*a*) gear type pump (*b*) rotor type pump
   (*c*) oil filter (*d*) oil cooler
9. From the oil pump the oil directly to
   (*a*) oil gallery (*b*) oil strainer
   (*c*) oil filter (*d*) main bearings
10. End play of gears of an oil pump may be checked by using
   (*a*) inside caliper (*b*) outside caliper
   (*c*) vernier caliper and feeler gauge (*d*) straight edge and feeler gauge

[**Key:** 1. (*a*), 2. (*d*), 3. (*a*), 4. (*b*), 5. (*c*), 6. (*b*), 7. (*c*), 8. (*c*), 9. (*c*), 10. (*d*)]

8 Fuels

1. PETROLEUM

The main source of engine fuels is crude petroleum, which is a dark viscous liquid. The crude oil occurs in underground deposits in typical geological formations as shown in Fig. 8.1.

The oil taken out from these deposits is a mixture of many hydrocarbons and is further processed to yield gasoline which is also called petrol and is the most widely used fuel for automobiles throughout the world.

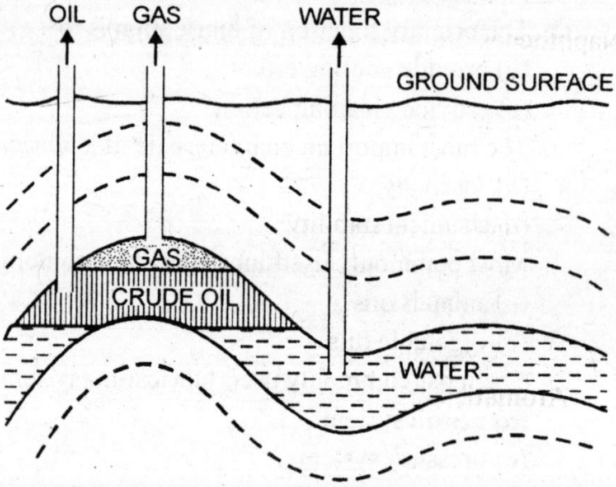

Fig. 8.1. Typical oil field geological structure.

2. HYDROCARBONS

The hydrocarbons are broadly classified into four categories:

1. Paraffins
2. Olefins
3. Naphthenes
4. Aromatics

Paraffins are saturated hydrocarbons having the formula C_nH_{2n+2}. The lower paraffins are gases, the higher being liquids and still higher, solids. The examples of paraffins are:

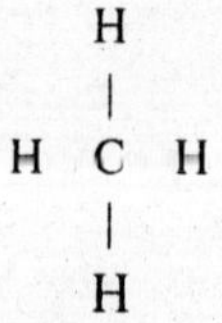
Methane

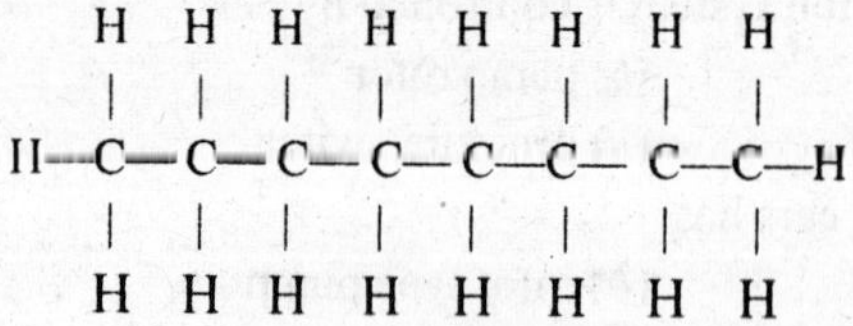
n-octane

Paraffins with open chain structures are called *normal paraffins* while those with branched chain structure are *iso-paraffins*.

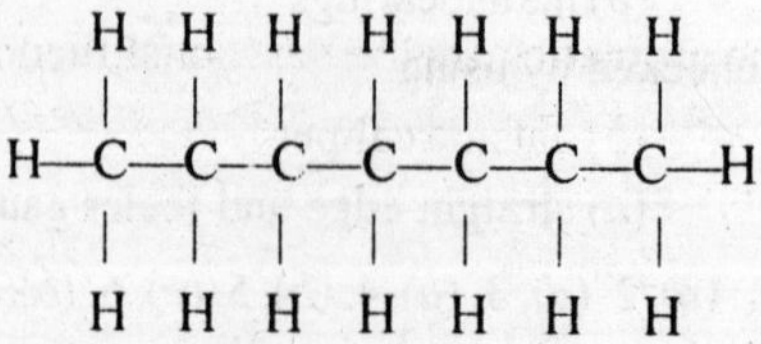
Normal Heptane

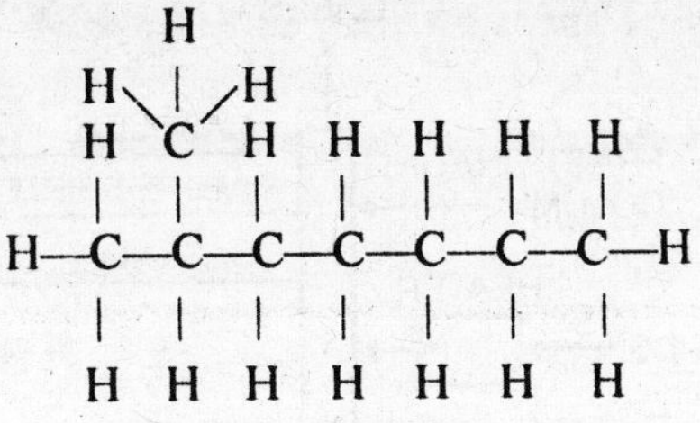

Isooctane (2-methyl heptane)

Olefins are the unsaturated hydrocarbons with the formula C_nH_{2n}. They have double bond between some of the carbon atoms. The examples are:

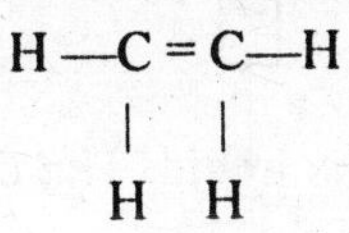

Ethene

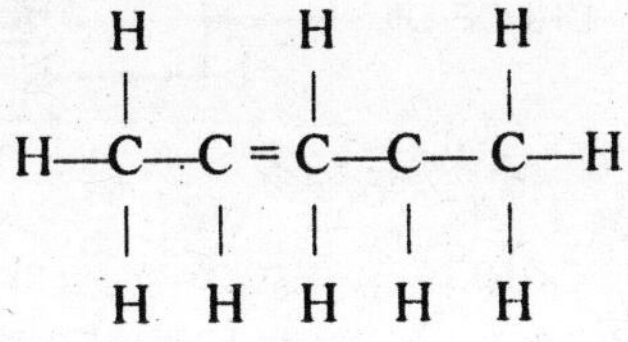

3-Pentane

Naphthenes are also called *Cycloparaffins*. They have a ring type structure.

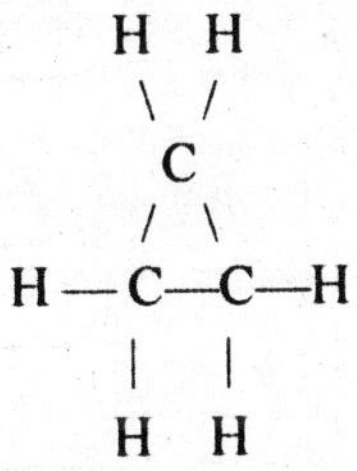

Cyclopropane

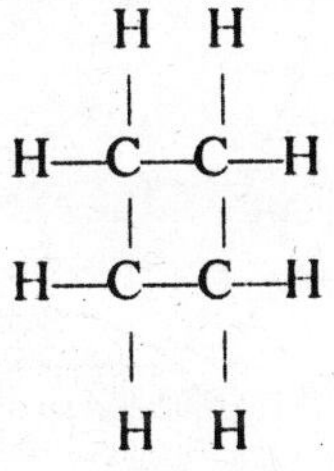

Cyclobutane

Aromatics are the hydrocarbons based upon benzene structure.

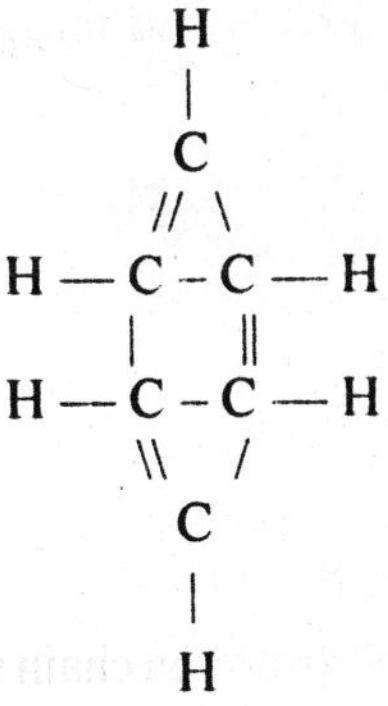

Benzene

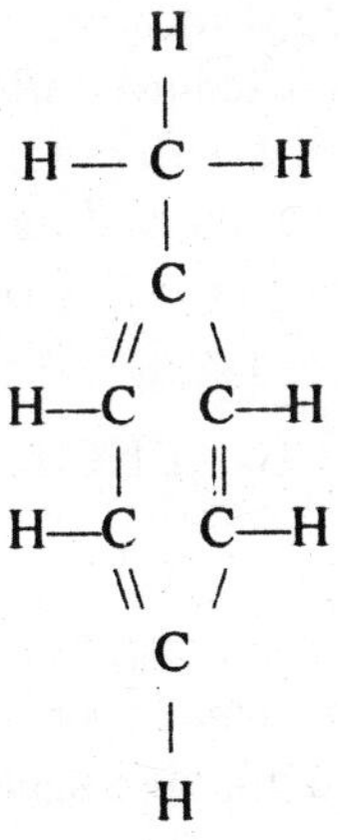

Toluene

As regards the anti-knock characteristics, the aromatics are the best constituents of a crude oil. However, other disadvantages limit their use in large proportions. These have, comparatively high freezing and boiling points. Further they have a corrosive action on rubber of the fuel system.

3. REFINING OF CRUDE OIL

Crude oil distillation is done in fractionating column (Fig. 8.2). Crude oil is heated to about 400°C and is passed through fractionating column. Different fractions depending upon their boiling points are taken out at different places in the column as shown. The unvaporized crude oil is again passed through a heater and vacuum to obtain heavy gas oil and lubricating oils.

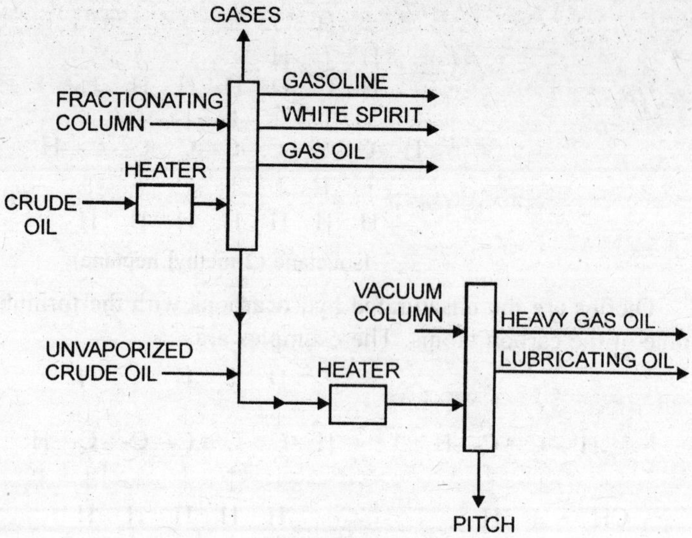

Fig. 8.2. Distillation of crude oil.

The fraction obtained in the fractionating column in the range of about 70°C to 120°C is called the *straight run gasoline,* which may be used as fuel for automobiles, but it has got poor antiknock rating (octane no. = 50). Further the straight run gasoline obtained in this way is very small in quantity to meet the large demands and is thus hardly economical to produce gasolines by this method. The other processes which obviate these disadvantages and have been used extensively are:

1. Cracking
   (*a*) Thermal Cracking
   (*b*) Catalytic Cracking
2. Reforming
3. Polymerization
4. Alkylation

3.1. Thermal Cracking

This is a process by means of which larger, heavier hydrocarbon molecules are broken into smaller lighter fractions suitable for use as gasoline (petrol). This is achieved with application of heat and pressure. Gasolines of about 75 Octane number have been obtained with pressure of 1-3 MPa and temperatures around 500°C. For example, *n*-decane breaks up into *n*-hexane and butene.

$$CH_3—CH_2—CH_2—CH_2—CH_2—CH_2—CH_2—CH_2—CH_2—CH_3$$

$$\xrightarrow{\text{Cracking}} CH_3—CH_2—CH_2—CH_2—CH_2—CH_3 + CH_2 = CHCH_2—CH_3$$

3.2. Catalytic Cracking

This is an improved form of thermal cracking, wherein a catalyst is used to promote cracking. The catalytic cracking thus involves less severe conditions of temperature and pressure. Also comparatively better grades of gasoline are obtained. The catalysts used are manufactured aluminium silicates or some naturally occurring clays.

3.3. Reforming

This process is used for improving the antiknock qualities of fuel by changing the molecular structure of the straight run gasolines. Like cracking, reforming process may be either thermal or catalytic. It is carried out in the presence of hydrogen. The straight chain paraffins and naphthenes are converted into aromatics which have much better anti-knock values. Typical examples of reforming process are:

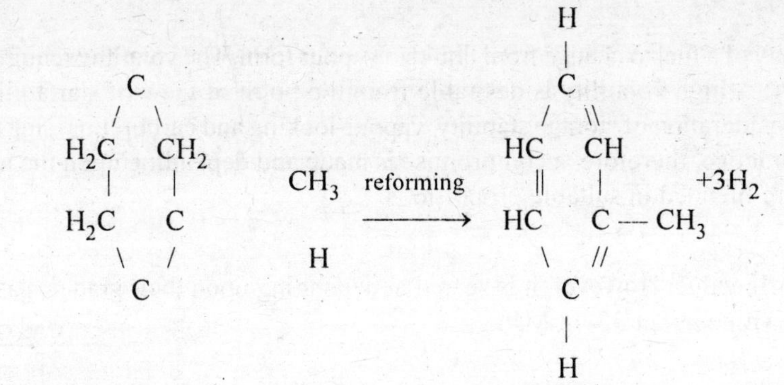

Methyl cyclohexane Toluene

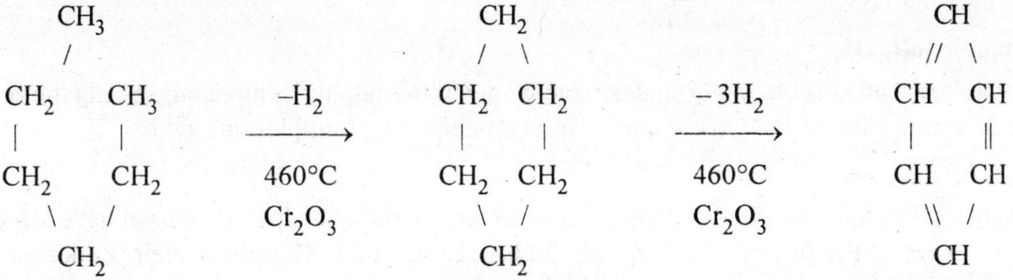

n-hexane Cyclohexane Benzene

3.4. Polymerization

This is the process which is used for combining two or more lighter hydrocarbons into higher hydrocarbons of suitable octane rating. Production of iso-octane from isobutylene is a typical example of polymerization. It is carried out in the presence of hydrogen.

3.5. Alkylation

Alkylation means substitution of alkyl group for a hydrogen atom in a hydrocarbon. With this method, fuels of octane ratings upto 100 have been produced. The products formed are called alkylates. The typical example of alkylation process is:

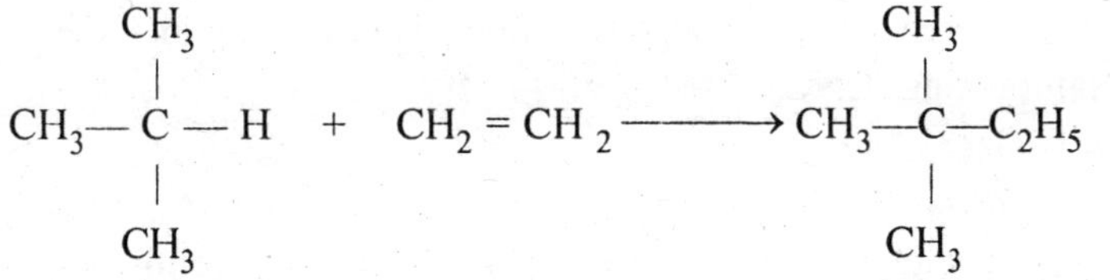

isobutane ethylene 2, 2, Dimethyl butane

Alkylation may be either thermal wherein high temperatures and pressures are used, or catalytic which may be performed at lower temperatures, but in the presence of a catalyst. The catalysts most frequently used are concentrated sulphuric acid or hydrochloric acid.

4. DESIRABLE PROPERTIES OF ENGINE FUELS

1. Knock Rating

The fuel should have high anti-knock value to make use of the higher compression ratios. However, in case of an existing engine nothing is grained by using a fuel of higher knock rating than the minimum required by the engine to avoid knocking during the whole range of its operating conditions.

2. Volatility

Volatility indicates the tendency of a fuel to change from liquid to vapour form. The volatility requirements of engine fuels are quite contrary. A high volatility is desirable from the point of view of startability and good distribution, whereas the consideration of storage stability, vapour-locking and carburettor icing dictate the use of low volatile fuels. In practice, therefore, a compromise is made and depending upon the use, the high and low volatile fractions are blended in suitable proportions.

3. Calorific Value

The fuel should a high calorific value. However, it is seen that depending upon their grades, gasolines have approximately the calorific value about 38-48 MJ/kg.

4. Gum Content

A good fuel should have the minimum gum content. Gum is formed by the polymerization of some unsaturated hydrocarbons.

5. Sulphur Content

A high sulphur content is obviously undesirable. Some of the sulphur compounds restrict the action of tetraethyl lead while some of them form corrosive compounds, *e.g.*, sulphurous acid.

6. Aromatic Content

As already explained the aromatics, though, are high knock-rating hydrocarbons, yet have a corrosive action on the rubber of the fuel system and have high freezing point. Therefore, their proportion in the engine fuel should not exceed beyond a certain limit.

5. TYPES OF AUTO ENGINE FUELS

These can be classified as:

1. Conventional fuels

These are obtained from petroleum after processing. Main such fuels are;

(*i*) Petrol (*ii*) Diesel

2. Alternative fuels

These are fuels other than petrol or diesel and are being increasingly used since these are cleaner than conventional fuels. Moreover, many of these are available within the country and as such their use saves foreign exchange as well as increased security of availability. These fuels includes:

(*i*) Compressed Natural Gas (CNG)
(*ii*) Liquefied Natural Gas (LNG)
(*iii*) Liquefied Petroleum Gas (LPG)
(*iv*) Alcohol fuels such as methyl alcohol (methanol), ethyl alcohol (ehthanol) either in pure form or as mixture with petrol. For example M85/E85 are mixtures of 85% methanol/ethanol with 15% petrol by volume.
(*v*) Electricity (including solar energy)
(*vi*) Hydrogen
(*vii*) Bio-fuels, which are derived from biological materials like soya bean, rapeseed, jatropha, etc.

Alternative fuels came into prominence in nineteen seventees due to the falling reserves of petroleum in the world and consequential fears of same running short of requirements. But, now with many new reserves having been found out and the companies opening vast new reserves using 'hydraulic fracking' technology, the prices of conventional fuels have considerably fallen. However, the importance of alternative fuels has not decreased, albeit due to different reasons, *i.e.*, due to increasingly restrictive emissions (including

CO_2) regulations globally. Moreover, natural gas is also available abundantly- to below petrol prices on an energy-equivalent basis.

Research and Advanced Engineering group at Ford has observed that if a low-octane fuel is used in an engine, modern sensors would retard spark to prevent knock, a condition most common when the engine is running at high load and high torque both. This seems to be quite convenient, but retarding spark results in decreased efficiency and torque. However, if higher-octane fuel is used, spark would be automatically advanced, resulting in higher efficiency and torque. The Ford group found out that all the techniques which engine makers are currently using to increase fuel efficiency [Gasoline Turbocharged Direct Injection (GTDI), downsizing, cylinder deactivation and hybridization] would gain from the higher-octane fuels combined with higher compression ratio.

Now, higher octane rating can be obtained either through reforming low-octane components such as naphtha into petrol (which increase cost due to decrease of petrol yield) or alternatively blending the refinery-petrol with ethanol (an alternative fuel), which is being preferred now due to easy availability of ethanol from crude oil or bio-sources or even conversion from natural gas. Technology has also been developed now to convert natural gas (another alternative fuel) into diesel using gas-to-liquid (GTL) technologies. However, it is found that converting natural gas to ethanol for octane-enhancing requires less energy than GTL. Besides, relatively smaller projects are viable for the former alternative.

Main properties of important conventional as well as alternative fuels are given in the succeeding articles.

6. PROPERTIES OF CONVENTIONAL FUELS

6.1. Petrol (Gasoline)

Petrol is basically a mixture of hydrocarbons. As described already, it is manufactured by crude distillation of petroleum and refining it further by other processes. It does not have a fixed composition. The constituents very depending upon the origin of petroleum. The important properties of gasoline are given below:

1. Volatility

Volatility of petrol is indicated by the distillation test (ASTM D 86). Another method by which the volatility is determined is the Reid Vapour Pressure (RVP), which is a measure of petrol vapour pressure at 38°C under specified conditions (ASTM D 323). Mostly, it is an indication of the butane content of petrol and can vary from about 75 to 70 kPa at 38°C.

Volatility affects the following engine/vehicle conditions:

1. Less volatility means difficult starting and stumbling during warm-up. Excessive volatility would result in vapour-lock.
2. Less volatility would cause the droplets of liquid fuel to break down the oil film on the cylinder wall, causing scuffing or scoring. Eventually these droplets enter the crankcase oil and cause its dilution, besides forming sludge, gum, varnish.
3. Less volatility would also result in uneven distribution of fuel to various cylinders.

2. Specific Gravity

It lies in the range 0.7 to 0.78. Sometimes instead of specific gravity, A.P.I. index of gravity is used.

$$\text{Degrees A.P.I.} = \frac{143.5}{\text{S.P.Gravity at } 15.5°\text{F}} - 131.5$$

3. Calorific Value

It is abut 45 MJ/kg.

4. Specific Heat

It is nearly 2.1 kJ/kg°C.

5. Octane number

In India the minimum research octane number (RON) is 87.

The detailed specifications of Indian motor gasoline are given in Appendix–A.

Various additives which are mixed with petrol to achieve desirable properties are:

1. *Oxidation inhibitors.* These are meant to avoid reaction of some components of petrol with each other and with oxygen and thus help in controlling gum and deposit formation during storage.
2. *Rust inhibitors.* These are added to protect components of the fuel supply system from rusting.
3. *Tetraethyl lead.* This was added to increase the octance rating and is kept at 0.56 gm/litre of load in leaded petrol and 0.013 gm/litre of lead in unleaded petrol in India.
4. *Metal deactivators.* These are meant to inhibit reactions between the petrol and metals in the fuel supply system.
5. *Detergents.* As the name indicates, these keep the injector nozzles and carburettor jets clean and thus prevent their clogging.

Petrol has low overall NO_x emissions, while CO emissions are highest among all fuels. PM emissions are negligible.

Use of catalytic converters now compulsory in India has necessitated the use of unleaded petrol to avoid damage to the catalyst used in the converter. As stated above, the maximum lead content in such a petrol is limited to 0.013 gm/litre for Bharat Stage II (Euro II) and to 0.005 gm/litre for Bharat Stage III (Euro III) fuel specifications. However, to compensate for decrease in the octane rating due to decrease in lead content, other chemicals of the required octane quality are blended with the unleaded petrol which do not damage the catalyst in the converter and at the same time keep the octane number at 87 or higher.

6.2. Diesel Oil

Diesel oil is also obtained from petroleum and is of a higher boiling range, *i.e.* about 180°C to 300°C. It is very widely used because of its cheapness and better thermal efficiency. Some of the important properties of high speed diesel oil used for automobiles are:

1. Volatility

It is very less volatile as compared to gasoline. Excessive or too low volatility promotes smoking, carbon deposits and dilution of oil.

2. Specific Gravity

It has higher specific gravity than gasoline *i.e.* from 0.82 to 0.92.

3. Calorific Value

It is about 50 MJ/kg.

4. Specific Heat

It is about 1.9 kJ/kg°C.

5. Cetane number

For Indian HSD, minimum cetane number specified is 48.

6. Sulphur content

Sulphur in diesel fuel causes deposits in combustion chamber, corrodes the exhaust system and wears the pistons, rings and cylinders. Indian high speed diesel oil contains a maximum of 1% sulphur, which is considered high as per international standards, where winter grade diesels often have less than 0.2% sulphur.

7. Pour point

For winter operation, pour point becomes a critical parameter. Wax crystals start forming at a temperature even 5 to 8°C above the pour point. These wax crystals plug the fuel filters. To improve the fluid flow at low temperatures, additives called flow improvers are often added to the diesel fuel. For Indian HSD, pour point is 6°C for winter and 18°C for summer.

Detailed specifications of Indian high speed diesel fuel are given in Appendix - B.

Engines running on diesel emit relatively less CO_2 and negligible HC. However, overall NO_x and PM emitted are more than in case of petrol engines.

7. ALTERNATIVE FUELS

7.1. Alcohols

Alcohols have been used as engine fuels for long. Both ethanol (C_2H_5OH) as well as methanol (CH_3OH) are now used as automobile fuels.

Alcohols have high anti-knock rating, due to which they, especially methanol, have been used in racing car engines. Besides, methanol is also safer than petrol in the event of accidental fire, since it burns cooler. However, its disadvantages are:

1. Flame of methanol fire is difficult to be seen in daylight, due to which it may not be detected immediately.
2. Methanol contains about half the energy than petrol per litre, which means lesser kilometers per litre, due to which the fuel tank of larger size has to be used to keep the same range as that with petrol.

Ethanol or ethyl alcohol can be made chemically from petroleum or by fermenting sugar derived from vegetable matter, the main sources being sugar cane, sugar beet, cereals and maize (which conflicts with their use as food). The next-generation processes to get sugar by hydrolysis from waste straw, sawdust, grass, some cellulose sources are now under development. Ethanol also has its disadvantages:

1. Corrosion of rubber, magnesium and aluminium parts occurs at higher ethanol percentages.
2. Ethanol is electrically conducting, whereas petrol is insulator effectively. This can cause problems with some electric fuel pumps and fuel tank sensors.

A cost-benefit analysis by the RAND Corporation has found that both the diesel engines and petrol-electric hybrids have significant benefit over vehicles running on E 85 (85% ethanol with 15% diesel/ petrol). The study found that the high cost and resource-intensive means of producing and transporting E 85 makes it less beneficial to society than using conventional petrol. The operating cost of an E 85 vehicle is also higher than a similar petrol vehicle due to less energy per unit volume in case of E 85.

However, the alcohol blends with 10-15% ethanol and 85-90% petrol (E 10/15) are being increasingly used due to following advantages;

1. Ethanol keeps the engine clean since it burns completely and at a slightly lower temperature than petrol, which means longer spark plug life and lesser combustion deposits. Being oxygenate (its molecules contain oxygen), the oxygen atoms inside ethanol aid the oxygen molecules in the air to help ethanol burn more completely. This extra oxygen in the ethanol also help the petrol to burn better when mixed with ethanol.
2. Better combustion also results in fewer harmful emissions. As per U.S. Department of Energy, E 10 can reduce carbon monoxide emissions from older vehicles by upto 25 percent. Emissions of smog-producing hydrocarbons, soot particles and toxic chemicals are also decreased.
3. It improves the engine starting in cold weather.
4. Improves flame visibility in daylight.

However, pure ethanol is less volatile than petrol due to which it becomes difficult to ignite at lower temperatures, which means, starting a cold engine during winter becomes difficult. Petrol blending with ethanol solves this problem. Moreover, ethanol is expensive to produce and with the sugar cane production varying from year to year, its availability and price are always uncertain.

Many petrol engines can use petrol with upto 10% ethanol. For higher concentrations, specially designed engines are necessary, which are usually 'flexible fuel engines' which can detect the level of ethanol concentration and then adjust their tuning accordingly. Besides E10 and E15, E85 is also used for light duty vehicles and E95 for heavy duty vehicles, *e.g.*, trucks and buses. However in such cases,

1. Alcohol being more corrosive than petrol, vehicles for using E85 have to use special lubricants and selected components made from advance anti-corrosive materials.
2. Although E95 is good fuel for buses and trucks, yet it cannot be used in standard, diesel engines. It is expected that a newer fuel 'E-diesel', which contains 15% ethanol, 80% diesel and 5% additive to keep the components remain mixed together, would be able to reduce the exhaust emissions significantly in case of heavy vehicles.

7.2. Compressed Natural Gas (CNG)

Natural gas, like petroleum comes from underground reserves. It consists, mainly (about 95 percent) of methane (CH_4). The remaining 5 percent comprises butane, propane, ethane with small amount of water vapour. Simple chemical structure of methane makes possible its nearly complete combustion, releasing lesser emissions.

For ease in transportation, storage and automotive use, the natural gas is compressed to pressures upto 250 bar (3600 psi) and stored in high-pressure cylinders. That is why it is called Compressed Natural Gas.

CNG posesses many advantages when used as automotive fuel. These are:

1. Vary low emission. Due to its combustion at lower temperatures, less NOx are generated. Further, due to its relatively higher hydrogen content, it produces less CO_2 and more H_2O than petrol or diesel. The exhaust gas is odourless and contains no particles. Compared to petrol, CNG combustion produces about 25% less CO_2. However due to high levels of methane, it gives highest HC emissions.
2. Saving a gram of CO_2 with a CNG power train is many times less expensive than with the plug-in hybrids.
3. It requires no additives and its production does not need complicated refining processes.
4. CNG vehicles place no load on world's scanty petroleum resources. Its main component, methane, can also be prepared from organic substances.
5. CNG has very high knock resistance, its octane number being 130 ROZ compared to maximum 100 RON for best petrol. This makes it highly suitable for supercharging and higher compression.
6. CNG offers lowest running cost.
7. Large reserves of material gas means steady and low prices. While there are forecasts about the shortage of crude oil in future, there are no such forecasts about the natural gas resources. The currently known sources of natural gas are said to be enough for the next 250 years.

The above advantages are accompanied by the following disadvantages:

1. There being no lubrication, engine wear and tear is more.
2. Volumetric efficiency is less than comparable engines using liquid fuels.
3. If methane in the natural gas is allowed to escape into the atmosphere, it may produce 25 times the green house effect compared to that in case of CO_2.

There are 3 types of CNG vehicles:

1. Dedicated vehicles, which run only on CNG.
2. Bi-fuel vehicles operate on CNG, but can also use petrol as a reserve fuel. The engine can run on either fuel but not on both simultaneously.
3. Dual fuel engines use a small amount of diesel as a pilot source of ignition. CNG is the primary fuel which is mixed with the intake air as the bulk fuel.

The examples of vehicles running on CNG are 'Honda Civic CGX' and 'Ford Crown Victoria' (USA) and Fiat Panda, Fiat Punto, Lancia Ypsilon, Fiat Fiorino/Qubo (Europe).

7.3. Liquefied Natural Gas (LNG)

LNG means natural gas in liquid form. It is made by refrigerating natural gas to minus 161°C. It is a clear, colourless, odourless liquid. During the liquefaction process, natural gas is cooled below its boiling point removing most of the water vapour, butane, propane etc., so that LNG is more than 98 percent methane (CH_4).

LNG is much more dense than CNG, which means more energy per unit volume. The implication is the higher vehicle range for the same size of fuel cylinder.

As discussed earlier, simple molecular structure of methane is responsible for lesser emissions in a LNG engine.

7.4. Liquefied Petroleum Gas (LPG)

The liquefied petroleum gas is also an automotive fuel, consisting mainly of butane and propane. However, LPG is formed naturally, interspersed with deposits of petroleum and natural gas. Natural gas contains LPG, water vapour and other impurities which have to be removed before transpoıting in pipelines. LPG is also produced by refining of heavier fractions of crude oil or in the fuel synthesis process. LPG is also produced in fuel-synthesis processes using biomass or electricity and CO_2 as feedstock.

LPG is also called 'Auto gas' when used in cars. Its wordwide use as automotive alternative fuel comes only ext to ethanol. It is mainly used in spark ignition engines, either as only LPG engines or as bi-fuel engines capable of running on either fuel at a time. It is also being used as a secondary fuel, being injected into compression ignition (diesel) engines for heavy vehicles.

It is stored in liquid form in special cylinders at a pressure of about 100 psi (700 kPa) and the engine is provided with a special fuel system. The advantages of LPG are:

1. Better mixing with air and improved distribution, which means lesser tail pipe emissions. Due to its relatively higher hydrogen content it produces less CO_2 and more water than petrol or diesel per unit of energy produced. Compared to equivalent petrol cars, there are 15% less tail pipe emissions. Nitrogen oxide and carbon monoxide emissions are 20% and 60% less respectively. LPG engine supply system being tightly sealed, there are no evaporative emissions, due to which there is lesser tendency to form smog.
2. It is cheaper than petrol.
3. No need of a fuel pump.
4. No carbon deposits.
5. No crank case dilution because of vapour form. This means lesser oil consumption.
6. High octane rating (about 105).
7. Less engine wear.
8. Due to extremely low sulphur content, LPG does not contribute much to acid rain.
9. Reasonably easy to modify a petrol car to run on LPG.

The disadvantages are:

1. Special fuel system has to be provided.
2. Heavy pressure cylinders increase the vehicle weight unnecessarily.
3. Harder to start in winter.
4. Like petrol it can form explosive mixtures with air. LPG being slightly heavier than air, it may form a continuous stream stretching a considerable distance from a leak, which is likely to result in a flashback explosion on contacting any source of ignition.
5. Distribution network not as widespread as that for petrol and diesel.
6. High octane rating of LPG goes waste while using in petrol engines which have low compression rations.

7.5. Bio-diesel

This is an alternative (or additive) to standard diesel fuel, and is made from biological sources instead of petroleum. It is made mainly from organic vegetable oils by a production process, called trans-esterification. It is usually mixed in practice with standard diesel; the most common blend being B 20 which is 20% bio-diesel mixed with 80% standard diesel. Pure bio-diesel is referred to as B 100. According to U.S. National Bio-diesel Board (NSB), the technical definition of bio-diesel is "a fuel comprised of mono-alkyl esters of long chain fatty acids derived from vegetable oils or animal fats, designated B 100, and meeting the requirements of ASTM D 6751." Specifications of Indian Bio-diesel are given in Appendix C.

Advantages

1. It has lesser emissions. Table 8.1 gives average bio-diesel emissions compared to standard diesel fuel.

Further, B 100 reduces carbon dioxide emissions also by 78%. Besides, biodiesel can also reduce the carcinogenic properties of diesel fuel by 94 percent.

2. It is biodegradable. It has been found that its degradation rate is four times that of conventional diesel fuel. Practically it means that is case of a spill-over, the cleanup would be easier.

3. Biodiesel is made from plant materials, which are available in supplies which can never be exhausted. The sun supplies the energy contained in plants through photosynthesis, which is a natural process going on automatically. Moreover, this energy content still remains even after the processing of plants into other materials. Bio-diesel is made in U.S.A. mainly from soyabean oil. In India it can be made from jatropha plant which is available in abundance. Thus, it can save our precious foreign exchange in importing costly crude petroleum.

TABLE 8.1

Constituent	*B 100*	*B 20*
Total unburnt hydrocarbons	– 67%	– 20%
Carbon monoxide	– 48%	– 12%
Particulate matter	– 47%	– 12%
NO_x	+ 10%	+ 2%
Sulphates	– 100%	– 20%

4. Bio-diesel also assists in the process of engine lubrication. Acting as a solvent, it helps to loosen deposits from the inside of the engine that could have clogged the lubrication oil passages. On its own, pure bio-diesel does not leave any deposit. Thus it results in longer engine life.

5. It also safer and non-toxic. Having higher flash point than conventional diesel oil, accidental fires are less likely. This makes it easier to store and transport.

Disadvantages

1. It increases NO_x emissions which contribute to formation of smog.

2. Although bio-diesel being solvent helps in better lubrication, yet the loosening deposits from the conventional diesel engine component may clog the fuel filter. This factor is more obvious when bio-diesel replaces the conventional diesel in the same old engine when earlier deposits may loosen up all of a sudden. It is therefore, advised to change fuel filter shortly after change over to bio-diesel.

Besides, bio-diesel also breaks down rubber components.

3. In some engines, slight decrease in power and increase in fuel consumption has also been noticed. On an average, about 10% reduction in power occurs, which means about 1.1 litres of bio-diesel would be equivalent to 1 litre of conventional diesel oil.

4. Higher cost than conventional diesel oil.

5. Low yield to land ratio affects food production security.

6. Far less Availability. In India, only recently a company has been formed to manufacture bio-diesel. Even in U.S.A., it is not produced in all the states and is available only at the designated outlets. According to U.S.A.'s N.B.B., there are about 100 fleets using bio-diesel in that country.

7.6. Hydrogen

It is the simplest element and occurs in abundance in the universe. It is considered ideal fuel because of its advantages:

1. It is the cleanest fuel which burns to give energy and leaves only water as a product of combustion.
2. It does not produce any greenhouse gases, *e.g.*, CO_2.
3. It is safe to manufacture.

However, its use is still restricted to space vehicles and concept cars due to the following dis-advantages:

1. Very high cost of manufacture due to the large amount of energy required to make it by separating hydrogen and oxygen in water.
2. It is highly flammable and can ignite easily in low concentrations, which means a small leak in transport on storage can be a big public safety hazard.
3. Very low energy density.

However new technologies such as hydrogen I.C. engines and onboard reforming for fuel cells are being developed to decrease the energy required for separating it from water and/or carbon fuels and also to enhance safety in use.

7.7. AUDI E-FUELS

Since 2013, Audi has been doing research with selected partners to develop these third generation renewable fuels. The work so far indicates that these fuels have the following advantages:

1. Unlike fossil fuels, the composition of which varies depending on their place of origin, e-fuels are absolutely pure fuels, which do not contain any olefins or aromatic hydrocarbons. As a result, their combustion is cleaner and there are lesser emissions.
2. No biomass is required to produce e-fuels and they can be produced in regions which are not suitable for agriculture. Thus they do not compete with the production of food or animal feed, unlike biodiesel, for example.
3. Comprehensive life cycle analysis shows tremendous potential of these fuels for reducing CO_2 emissions. In fact these fuels are produced using CO_2 as a new material.

Only 4 elements are needed for the production of e-fuels— water, CO_2, sun light and some specific microorganisms. Instead of using photosynthesis to form new cells, these microorganisms produce fuel continuously, requiring sun light, CO_2, salt water/waste water. They do not need clean drinking water or any farm land.

Audi is specifically researching e-gas, e-gasoline (petrol), e-diesel and e-ethanol. A brief description of each is given below:

1. **Audi e-gas.** It is synthetic methane gas produced in Germany at two facilities, in Werite and the new Viessmann facility. The gas produced is used to run Audi A3 Sportsback g-tron cars, which can be considered the most eco-friendly form of long-distance mobility, since the production of e-gas requires as much CO_2 as would be later released by Audi A3 g-tron. Range of such a car running on pure Audi e-gas is about 400 km.
2. **Audi e-gasoline (petrol).** Audi has started its research to produce non-fossil 'e-benzin' in early 2014 in partnership with French biotech company Global Bio-energies.
3. **Audi-diesel.** Audi is involved in this since 2011, in partnership with US biotech company Joule. Audi e-diesel has been found to have high cetane number and excellent ignition performance. Its chemical composition is such that it can be blended unlimited with fossil diesel. The latest facility to produce diesel fuel from water, CO_2 and green electricity, is a pilot plant in Dresden, Germany set up in November 2014 in partnership with Clime works and Sun fire. This facility produced the first batch of high-quality diesel in April 2015. Daily output of 'blue crude' from this plant in the first phase is around 160 litres 80% of which can be converted into synthetic diesel.
4. **Audi e-ethanol.** In partnership with Joule, Audi optimized in 2007, a process to produce either ethanol or alkanes (main component of diesel fuel) from CO_2, sun light and certain micro organisms, where CO_2 required can be easily got from the exhaust flow of a factory, biogas plant or even from the air. E-ethanol produced for the first time in this manner in Hobbs, New Mexico, was found to be similar to the classic bio-ethanol. This e-ethanol can also be mixed with fossil gasoline to form, *e.g.*, either E10 gasoline or E85 ethanol fuel (85% ethanol, 15% gasoline).

   Testing of e-ethanol in the pressure chamber and in the glass engine in January 2014 confirmed that this fuel has similar or even better properties than the bio-ethanol.

8. COMPARISON OF CONVENTIOALAND ALTERNATIVE FUELS

In favour of conventional fuels

1. Higher energy-density compared to alternative fuels.
2. Well established production infrastructure. There are 7000 oil rigs and more than 700 refineries world wide producing 80-85 million bbl of refined products, mostly petrol and diesel per day.
3. Distribution network well established.
4. With petroleum industry experts continuously innovating, more and more oil reserves are being found out and exploited profitably, *e.g.*, extracting the oil profitably from miles under the sea or separating it from once stubbornly tight and heavy oil formations with improved oil-recovery techniques, such as hydraulic fracturing. Thus there is no risk of running short of petroleum reserves in the near future.
5. Reasonable price level.

In favour of alternative fuels

1. In case of spurt in demand, alternative fuels may fill the gap, may be for a short time till the petroleum industry sets up the increased capacity. Already, International Energy Administration in 2012 projected

worldwide demand of petroleum rising by about 15% by 2035 with the passenger car production doubting globally by then.

2. Alternative fuels would be mainly useful in controlling pollution locally within congested cities worldwide.
3. With the oil production and refining happening at large distances across continents, there are apprehensions of serious disruption in the supply chains of crude oil and the refined products, due to earthquakes, hurricanes, wars or civil disturbances in the critical areas. In such a scenario, the price of petrol and diesel may rise to a level so that alternative fuels become economical.
4. Considering only the energy cost, electric vehicles are cheaper than their petrol or diesel-based vehicles. However, the cost of batteries including replacement plays spoilsport. If and when there is a technological breakthrough so that the batteries last longer, charge faster and cost less, it is sure to see universal shift from petroleum fuels to electricity as an alternative fuel, which may come from batteries or fuel cells.

9. AUTO FUEL SAFETY CONSIDERATIONS

1. Petrol and diesel are poisonous and inflammable. As such, take care never to swallow and to avoid accidental fire.
2. Ethanol is denatured alcohol, which means poison has been added to stop people from drinking the same. It is also inflammable.
3. Methanol, in addition to being inflammable, is much more toxic. Even a small amount can make one very sick or even kill. Direct contact with skin can cause injury. It is also difficult to detect flame in case of accidental fire.
4. CNG is also inflammable. As it is stored under high pressure, accidental release can cause extensive damage. It is odourless, due to which odorants must be added to detect leaks and spills.
5. As LNG is cooled cryogenically, body contact with it can cause frostbite. Further, odorants cannot be added to LNG. So methane gas detectors have to be installed to detect leak for avoiding fire.
6. Though pressure is less than with CNG, yet strong fuel tank is required. Moreover, odorants have to be added to detect leak.
7. Bio-diesel, though coming from vegetable sources, is also not safe to drink. It is also inflammable, requiring care to avoid accidental fire.

REFERENCES

1. Bruce Morey, "Alternative fuels face challenges", Automotive Engineering; March 18, 2014; pp 18-27.
2. Winterhagen Johannes, "VW bets big on CNG", autonewseurope.com, August 2013, p. 16.
3. Antony Ingram, "Audi tests cleaner e-fuels with glass engine", www.motorauthority.com/news/1090087-audi-tests-cleaner-e-fuels-with-glass-engine, Jan 2014.
4. Bruce Morey, "The case for alternative fuels", articles. Sae.org/12358/, August 2013.
5. Keefe, Griffin & Graham, 'The Benefits and Costs of New Fuels and Engines for Cars and Light Trucks, Rand Corporation, 2007.

THEORY QUESTIONS

1. What is petroleum? Write a comprehensive note on hydrocarbons.
2. Discuss how the crude petroleum oil is refined.

3. Write notes on:
4. What are the properties which are sought for in a good engine fuel? Explain briefly.
5. What is petrol? Discuss its properties.
6. Apart from petrol, there are other fuels also which are fit for automotive use. Describe them briefly.
7. Describe on detail various types of fuels for automotive engines. Give specific examples in each case.
8. What is an alternative fuel? Describe various types of alternative fuels for auto engines. Discuss also the merits and demerits of each such fuel described.
9. How is bio-diesel obtained? Discuss its advantages and disadvantages compared to ordinary diesel oil.
10. Write a detailed note on Audi e-fuels, stating also their advantages.
11. Compare in detail the merits and demerits of the conventional and alternative fuels.
12. Describe various precautions necessary while storing and using auto fuels.

QUESTIONS FOR VIVA

1. What is the main source of engine fuels?
2. What are hydrocarbons? Enumerate different hydrocarbon types.
3. What is a branched chain structure?
4. How do you distinguish between normal and iso-paraffins?
5. Which type of hydrocarbons show best anti-knock characteristics?
6. What are the factors prohibiting the extensive use of aromatics as engine fuels?
7. What is fractional distillation?
8. Name different processes for refining crude petroleum oil.
9. What is straight run gasoline?
10. What are the catalysts used for catalytic cracking?
11. What is reforming?
12. How do you define alkylation?
13. What is polymerization?
14. Name the various desirable properties of fuels.
15. What is the effect of fuel volatility on auto engine working?
16. Classify fuels for auto engines.
17. What is an alternative fuel?
18. Name any six alternative fuels for auto engines.
19. What is the disadvantage of retarding engine spark?
20. Name the main techniques currently being used for increasing fuel efficiency in an engine?
21. Enlist various types of CNG vehicles.
22. What is the maximum percentage of ethanol which can be blended with petrol, to be used in an engine without making any changes in it?
23. Sate the advantages of blending alcohol with petrol.
24. What is bio-diesel? State its advantages and limitations.
25. What is gasoline?
26. What is Reid vapour pressure? What does it indicate?
27. State the octane rating of petrol used in Indian automobiles.
28. What is the purpose of a fuel-additive?
29. What has necessitated the use of lead-free petrol in Indian?
30. Is the octane rating of lead-free petrol in India less than that of the leaded petrol? If not, how is the decrease of TEL compensated? If yes, how can the same be used in old car engines designed considering the leaded petrol rating only?

31. Name various engine fuels, apart from gasoline.
32. State the cetane number of Indian high speed diesel.
33. What is the effect of excessive sulphur in diesel oil?
34. What is pour point of a fuel? State its significance.
35. What are the merits and demerits of liquefied petroleum gas as engine fuel?
36. State the advantages and limitations of hydrogen as automotive fuel.

MULTIPLE CHOICE QUESTIONS

Mark the correct answers:

1. The best hydrocarbons from detonation viewpoint are:
   (*a*) paraffins (*b*) olefins
   (*c*) naphthenes (*d*) aromatics
2. The process of changing the molecular structure of the straight- run gasolines is called
   (*a*) cracking (*b*) reforming
   (*c*) refining (*d*) none of these
3. The process in which hydrocarbons are decomposed into smaller hydrocarbons is called
   (*a*) cracking (*b*) reforming
   (*c*) polymerization (*d*) alkylation
5. Blending refinery petrol with ethanol results in
   (*a*) decrease in octane rating
   (*b*) increase in octane rating
   (*c*) no effect on octane rating
   (*d*) decrease/increase in octane rating depending of blending quantity.
5. The calorific value of gasoline is about
   (*a*) 45 J/kg (*b*) 45 kJ/kg
   (*c*) 45 MJ/kg (*d*) 45 GJ/kg
6. The calorific valve of alcohol is
   (*a*) less than that of gasoline (*b*) equals to that of gasoline
   (*c*) more than that of gasoline (*d*) depends upon type of the engine where used
7. The octone number of iso-octane is
   (*a*) 0 (*b*) 10
   (*c*) 80 (*d*) 100

[Key: 1. (*d*), 2. (*b*), 3. (*a*), 4. (*b*), 5. (*c*), 6. (*a*), 7. (*d*)]

9 Combustion and Combustion Chambers

1. COMBUSTION IN SPARK IGNITION ENGINES

In S.I. engines, a flame starts from the spark plug points and spreads continuously outward from the ignition point. The outward boundary of this flame is called the *flame front.* The combustion is called normal when this flame front travels to the end of the combustion chamber without any abrupt change. However, this flame front com-presses the air-fuel mixture ahead of it, raising both its pressure as well as temperature, due to which the mixture ahead ignites even before the flame front reaches there. This phenomenon is called *auto-ignition*, which may give rise to a phenomenon called detonation, discussed later in detail.

Combustion process in S.I. engines may be divided into three phases (Fig. 9.1). However, no clear distinction may be made between these phases as no abrupt change is noticed.

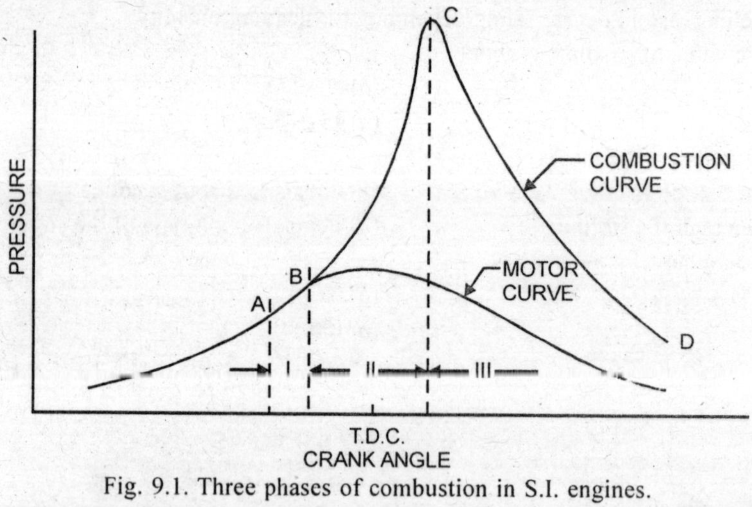

Fig. 9.1. Three phases of combustion in S.I. engines.

Phase I, also called *ignition lag* or the *delay period* is the duration between the occurrence of the spark at the spark plug and the deviation of the combustion curve from the motor curve. This is shown as AB in the Fig. 9.1. This period tends to be very nearly constant *in time.* The factors influencing this phase are:

1. *Fuel*—Higher the self-ignition temperature of the fuel, longer is the ignition lag.
2. *Mixture ratio*—Minimum ignition time occurs for a mixture of relative fuel-air ratio, F_R, about 1.2 (Fig. 9.2). Relative fuel-air ratio is the fuel-air ratio of the actual mixture divided by the fuel-air ratio of the chemically correct mixture. Ignition lag increases as F_R increases or decreases from 1.2. The rate of increase of ignition lag, however, is relatively more in case of weaker mixtures.

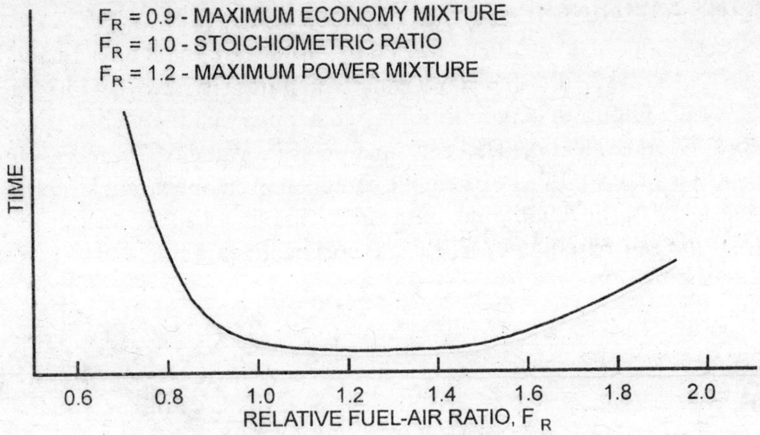

Fig. 9.2. Effect of mixture strength on ignition lag.

3. *Initial temperature*—Ignition lag increases rapidly with increase in temperature of the air-fuel mixture.
4. Pressure does affect the ignition lag, but only marginally.
5. The temperature of the flame between the spark plug electrodes (Fig. 9.3).

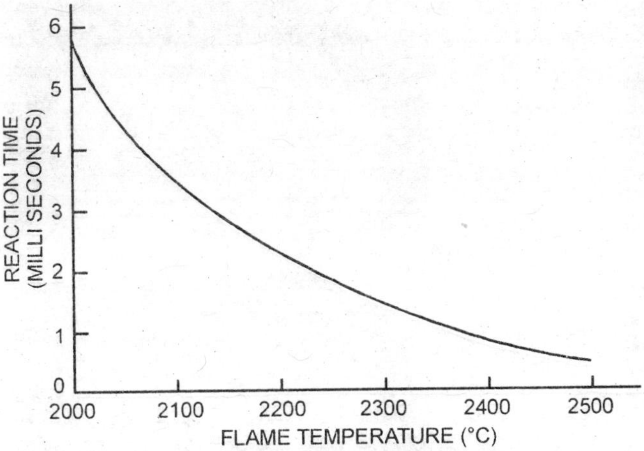

Fig. 9.3. Effect of flame temperature on rate of burning.

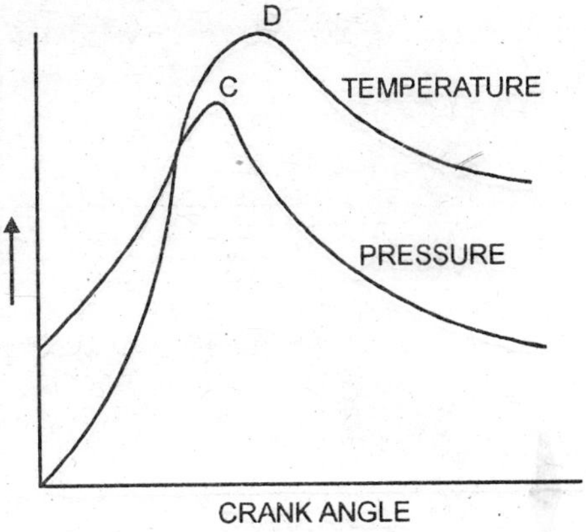

Fig. 9.4. Change of pressure and temperature during second and third phases of combustion.

Phase II, shows as BC in Fig. 9.1 starts when the combustion curve deviates from the motor curve and lasts till the maximum pressure is reached. This phase is also called *rapid combustion phase.* It may be noted that mean temperature of gases in the engine cylinder continues to increase beyond the point of maximum pressure (Fig. 9.4). The time required for this phase depends mainly on the intensity of the turbulence or state of agitation of the air-fuel mixture. The duration of this period is approximately constant *in terms of the crank-angle movement.*

Phase III, shown as CD in Fig. 9.1 occurs between points of maximum pressure and maximum temperature. After the flame front has reached the cylinder walls, there will be about 25% of the charge still not completely burnt. At this stage it becomes difficult for the remaining oxygen in the charge to react with petrol vapour so that the rate of combustion slows down; this condition is known as *after burning.* At the sametime, there will be heat liberated due to chemical interaction caused by reassociation of the combustion products throughout the expansion stroke.

2. DETONATION IN S.I. ENGINES

When the air fuel mixture in an engine is ignited by the spark plug, it does not burn all at once. On the other hand, a wave front is formed at the spark plug which travels further with a certain flame velocity, gradually compressing the unburnt charge, till it may so happen that the whole of unburnt charge is burnt all at once in the form of an explosion. This sets up a pressure wave which travels inside the combustion chamber to and fro, striking the walls. The gas in the combustion chamber will be compressed and expanded alternately by the pressure wave till equilibrium is restored. This disturbance forces the walls of the combustion chamber to vibrate at the same frequency as the gas. A peculiar ringing sound is thus produced and the

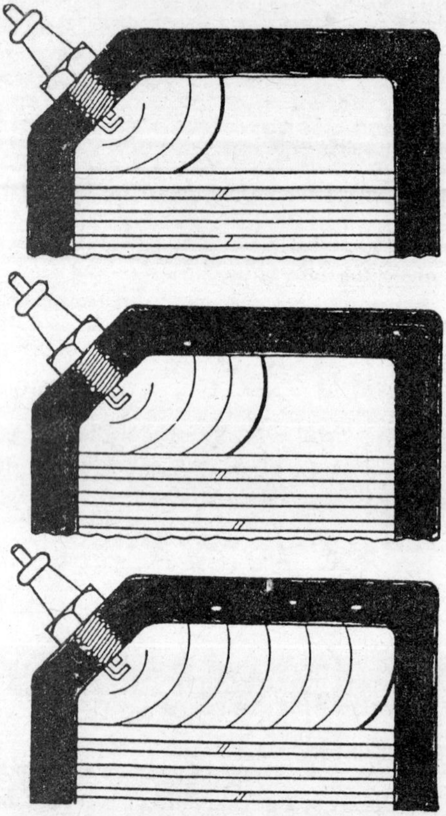

Fig. 9.5. Progressive stages of normal combustion.

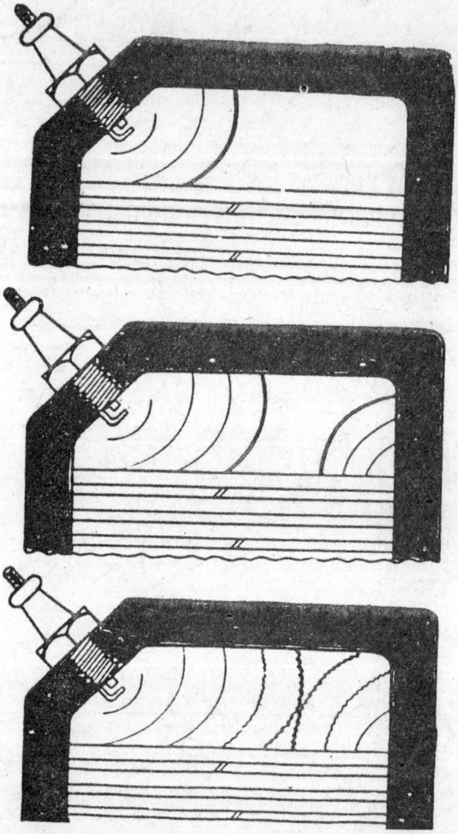

Fig. 9.6. Progressive stages of detonation

phenomenon is known as detonation or knocking Fig. 9.5 and 8.6 describe the progressive stages of normal combustion and detonation respectively. The pressure wave is shown in Fig. 9.7.

The average vibration frequency of the detonation is about 500 cycles/second, whereas the velocities of the shockwaves propagating through the burnt and unburnt charge reach the range 1000 to 1200 metres/ second.

3. THEORIES OF DETONATION

1. Ricardo's Theory

According to this theory detonation is caused by the sudden instantaneous ignition of the unburnt charge, when the rise in temperature exceeds the rate of heat dissipation, thus subjecting the unburnt charge to the temperature and pressure sufficient to cause its ignition.

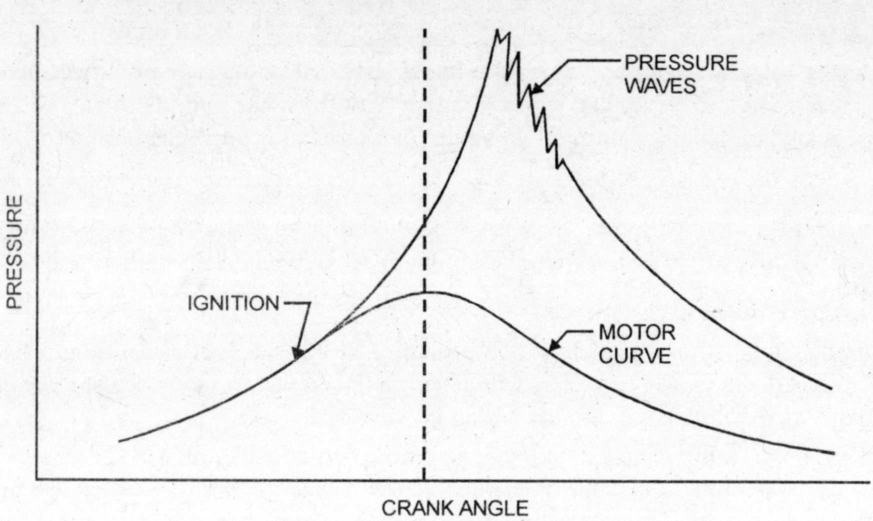

Fig. 9.7. Detonation pressure waves.

2. Callender's Theory

This theory states that the detonation is caused due to the presence of peroxide substance formed in nuclear drops of unvaporised fuel, which acts as foci for the sudden ignition of the unburnt charge.

3. Lewis' Theory

The "chain reaction" theory states that the advancing wave-front projects into the unburnt gas, hot reaction products rich in energy. Each of these reaction products starts a reaction chain which initiates combustion ahead of flame front. As the wave front can project reactant molecules at velocities considerably higher than its own, preflame-front combustion can proceed at very high velocities, forming a detonation wave.

4. FACTORS INFLUENCING DETONATION

The factors affecting detonation may be broadly divided into four categories:

1. Factors concerning engine.
2. Factors concerning fuel.
3. Factors concerning air.
4. Mixture strength.

4.1. Factors Concerning Engine

1. Compression Ratio

Increase in compression ratio results in rise of end gas temperature and pressure, thus reducing reaction time for a particular fuel-air ratio and consequently increasing the tendency to detonate. H.U.C.R. *i.e.* the highest useful compression ratio is the maximum engine compression ratio which can be used without occurrence of detonation. Any compression ratio higher than this would result in detonation being caused.

2. Engine Dimensions

For a fixed compression ratio and engine speed, increase in cylinder size increases the tendency to knock because the flame travel is large in them. That is why S.I. engines must have relatively smaller cylinder diameters; the bore size being limited to about 150 mm in practice.

3. Throttle Opening

Part throttle openings serve to prevent knocking, because due to part throttle the engine pressure at the start of compression stroke, and consequently the maximum pressure during combustion is reduced.

4. Spark Advance

Retarted spark helps to decrease the engine knock, because comparatively lower maximum engine pressures are produced. Over-advanced ignition timing may cause the cylinder peak pressure to occur at TDC, thereby causing higher than normal maximum pressures to occur, thus promoting detonation.

5. Spark Plug Location and Heat range

For reduced detonation location of the spark plug should be such that the length of the flame travel is minimum. Use of spark plug of hotter than the optimum range would also cause the detonation.

6. Combustion Chamber Design

The design of combustion chamber which promotes turbulence can cause more rapid combustion and hence decreases the tendency to knock. Moreover, in the end gas regions, the surface to volume ratio should be large so as to provide maximum cooling in the critical zone.

The optimum combustion chamber shape from detonation consideration is the hemispherical with a spark plug located at the centre. This type of chamber allows faster combustion, so that less time is available for the wave front to ignite the mixture ahead, thus avoiding detonation.

7. Cooling system

Defective cooling system design or operation due to slipping water pump, stuck thermostat etc. tends to increase the knocking tendency. Overheating may be caused by a low coolant level, a slipping fan clutch, a stuck closed thermostat, a slipping water pump, too small a fan or even a missing fan-shroud. Build-up of lime-deposits or steam pockets (arising from trapped air pockets) in the water jackets may also cause the engine to overheat.

8. Carburization

Excess carburization of combustion chamber, valves and piston crown promotes detonation due to two reasons. Firstly, these carbon deposits can increase compression ratio to the point where detonation starts. Such deposits can be especially thick if the engine consumes more oil due to worn out valve guides and seals, worn or broken piston rings and/or cylinder wear. Secondly, these deposits also have an insulating effect, decreasing the rate of heat transfer from the combustion chamber, which results in increase of combustion temperatures, thus contributing to the tendency to detonation.

9. Engine speed

At higher engine speed, there is less time for the flame front to overheat the end-charge so that end mixture may be relatively smaller or even non-existent. Further, the volumetric efficiency also decreases at higher speeds so that the conditions to cause detonation are relatively less severe. On the other hand, the shorter time available at higher speeds, for the heat of combustion to dissipate, increases the cylinder temperature and hence tendency to detonate.

However, the combined effect of all the above considerations favours reducing the detonating tendency at higher engine speeds.

10. Valve timing

The valve timing which increases the volumetric efficiency, *i.e.*, fills the cylinder with more air/fuel mixture, causes higher cylinder pressure and hence increased tendency to detonate.

11. Turbocharging

Turbocharging also forces additional air/fuel mixture in the engine cylinder leading to higher combustion pressures, consequently increasing the knock-tendency. That is why turbo wastegate is provided in turbocharged engines which bleeds off boost-pressure in response of rising intake manifold pressure. In most of the modern engines, the operation of the wastegate is regulated by a computer-controlled solenoid. However, a malfunction in the manifold pressure sensor, the control-solenoid, the wastegate itself or a leak in the vacuum connections between these components can result in excessive turbo-boost, thereby causing detonation.

12. EGR (Exhaust Gas Recirculation)

The purpose of EGR is to reduce the NOx pollutants by circulating small amounts of exhaust gas into the intake manifold through the EGR valve which results in diluting the air/fuel mixture slightly, resulting in a cooling effect on combustion temperatures and hence in decreasing the formation of NOx as well as in the tendency to detonate.

In case the EGR valve is defective or is otherwise not opening due to loose or wrong vacuum-hose connections, a defective vacuum control valve or solenoid, the cooling effect would not be produced, resulting in higher combustion temperatures under load, thus increasing chances of detonation to occur.

13. Knock Sensor

Many modern engines are provided with a knock sensor which produces a voltage signal corresponding to the critical frequency vibrations characteristically produced by detonation (typically 6-8 kHz), signaling the computer to momentarily retard the ignition timing until the detonation stops.

In case of 'check-engine' light coming ON, vehicle's on-board computer system may be checked for the trouble code corresponding to a bad knock sensor, e.g. 42 or 43 for GM, 25for Ford. A knock sensor can also be checked by rapping a wrench on the manifold close to the sensor (but not the sensor itself). If the timing does not retard, the sensor may be defective or the problem may lie with the electronic circuit to control the spark timing. For further diagnosis, the appropriate service manual for the vehicle would have to be consulted and followed.

4.2. Factors concerning fuel

The experiments have established certain relations between the chemical structure of molecule of a fuel and its knock rating.

1. Paraffins

(*a*) As the length of straight chain increases, the octane number decreases. For example, octane number of pentane is 60 whereas for normal heptane it is zero.

(*b*) Branched chain paraffins have higher octane numbers than their straight chain counter parts. For example, 2-methyl butane which has the same molecular formula as the normal pentane (octane number 60) has much higher octane value (90).

2. Olefins

The olefins have higher anti-knock values than corresponding paraffins. For instance, normal pentene has octane number 80 as against 60 for normal pentane.

3. Naphthenes

These also have better anti-knock characteristics than the corresponding paraffins. Cyclohexane, for instance, has octane number 85, while for normal hexane it is 25 only.

4. Aromatics

Of all the hydrocarbons, aromatics have the best anti-knock qualities. Benzene, for example, has octane number 95-108.

4.3. Factors Concerning Air

1. Temperature

Higher air temperature increases the tendency to knock.

2. Density

The density of air decreases at high altitudes, which reduces the tendency to detonate.

3. Humidity

The tendency to knock is decreased with the increase in humidity of the air entering the engine.

4.4. Mixture Strength

The chemically correct mixture is most liable to detonate. Tendency to knock decreases if the mixture becomes either weaker or richer than the chemically correct one.

5. EFFECTS OF DETONATION

The main effects of detonation are:

1. Inefficient combustion leading to higher output of undesirable emissions. *Carbon monoxide* levels increase significantly with detonation. This happens due to decrease in volumetric efficiency which occurs on account of higher combustion chamber wall temperatures. *Hydrocarbons* emissions are not much affected. Although due to detonation there is incomplete combustion in general, yet, heavy detonation also causes significant boundary layer to be removed from the cylinder walls. Since this layer contains higher concentrations of hydrocarbons which, during detonation, are mixed with the rest of cylinder gases and burnet, this compensates for incomplete combustion of the bulk air/fuel mixture and this may even lower the hydrocarbon emissions sometimes. *Nitrogen oxides* emissions, however, increase significantly due to high combustion temperatures during detonation.

2. Loss of power. Medium to heavy detonation levels reduce the engine power output, due to higher proportion of the chemical energy contained in the fuel being lost on account of inefficient combustion, leaving less energy to be converted into useful work.

3. Mechanical engine damage/ failure. On first inspection, the surface appears as having been sandblasted Further analysis reveals the type of damage being similar to that due to hydraulic cavitation. This damage occurs due to excessively high pressures waves and increased heat transfer rates during detonation. Local pressures of over 180 bar (18 MPa) have been recorded in test engines, at frequencies of upto 250 kHz.

Common failure occurs due to erosion of the piston head at the position where last part of the end gases burn. In extreme cases, the aluminium alloy piston may even melt due to excessively high cylinder temperature. Moreover, the aluminium alloy continues to age when subjected to high temperatures and high pressure detonation waves, which makes the alloy brittle and the piston may, then, break catastrophically.

4. Noise. Very high noise levels are a natural result of high pressures detonation waves. This noise is much louder than other vehicle noises and is a definite nuisance.

6. PREVENTION OF DETONATION

The different methods used for the prevention of detonation are:

1. Anti-knock agents.
2. Cooling of the charge.
3. Reducing the time factor.

These will now the discussed in detail.

6.1. Anti-knock Agents

These are the compounds which are soluble in the fuels and decrease their tendency to detonate. The exact mechanism by which this is done is not known so far. The anti-knock agent that has been most commonly used is tetraethyl lead (TEL). Very small quantities, of the order of 5 parts of TEL in 10,000 parts by volume of fuel are needed. It is seen that beyond a certain maximum, further addition of TEL has no effect on the knock rating of fuel (Fig. 9.8). TEL is poisonous and, therefore, sufficient care must be taken while using leaded fuels.

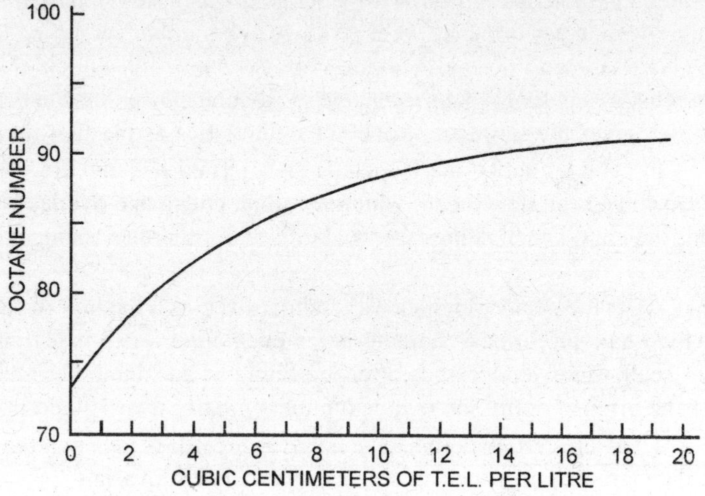

Fig. 9.8. Effect of TEL on Octane Number.

In practice, however, use of TEL causes lead deposition and for this purpose it is never used alone. It is used in the form of "ethyl fluid" which consists of 2/3 by volume of TEL and 1/3 of ethylene dibromide, which converts the lead formed during combustion into lead bromide, which is a gas at the cylinder temperatures and hence goes out alongwith exhaust gases, thus avoiding deposition.

The other anti-knock agents are tetramethyl lead and methylcyclo-pentadienyl manganese tricarbonyl.

Leaded gasoline, however, usually forms deposits that plug the catalytic converter for the treatment of exhaust gases to reduce the pollutants therein. Therefore, if unleaded gasoline is specified, leaded gasoline should not be used. Likewise, for engines with fuel-injection, gasoline with detergent additives should be used.

6.2. Cooling of the charge

The cooling of the charge before induction into engine cylinder results in decrease of knocking tendency. The simplest method is to inject a volatile fluid, *i.e.*, methyl alcohol into the charge. Internal coolants have also been used for this purpose.

6.3. Reducing the time factor

The time factor may be reduced by injecting the fuel in different stages into the air. The method has been successfully used with fuels upto octane number 100 in engines with compression ratio 10.

7. FUEL RATING

1. Highest Useful Compression Ratio (H.U.C.R.)

This system of fuel rating was introduced by Ricardo. He found that if the compression ratio is increased for a particular fuel, a stage comes when the detonation starts. This compression ratio at which the detonation starts is known as highest useful compression ratio. As it was found to be different for different engines, the tests were made on a special engine devised by Ricardo known as Ricardo Engine, where compression ratio could be varied continuously.

2. Octane Number

This method of fuel rating was introduced by Co-operative Fuel Research (C.F.R.) Committee. In this two hydrocarbons, one with worst anti-knock characteristics (normal heptane) and the other with best anti-knock characteristics (isooctane) were selected. These were given arbitrary values, zero for normal heptane and 100 for isoctane. Any fuel which gave the anti-knock performance similar to normal heptane in the specially designed C.F.R. engine was given octane number zero, while the fuel which showed anti-knock characteristics similar to isooctane was given octane number as 100. The other fuels which had intermediate performance were given octane number according to the percentage of isooctane in a mixture of isooctane and normal heptane with the same anti-knock characteristics as the fuel. Thus the octane number of any fuel is the percentage of isooctane *by volume* in the mixture of isooctane and normal heptane which gives the same anti-knock characteristics as the fuel under standard test conditions.

In actual practice the isooctane and normal heptane are very costly so that the *secondary reference fuels* have been developed, which are quite cheap and available in bulk. These secondary fuels are tested and their anti-knock ratings are standardized. Further fuel testing is carried out with the help of these secondary fuels.

After the octane number was standardized as a method of fuel rating, many hydrocarbons were found which had anti-knock characteristics even superior to isooctane. Thus to compare their performances, the scale was extended and correspondingly octane numbers greater than 100 were given, though according to the original definition octane numbers greater than 100 do not mean anything.

There are certain points to be noted as regards octane number. As the knock rating is the most important characteristics of a fuel, this property has been invariably connected with best performing fuels. But it must be remembered that octane numbers are simply arbitrary numbers having absolutely no direct relation with the engine performance. Further a high octane fuel, as is generally supposed, does not mean necessarily more power for a given engine, so long as the fuel does not detonate in the engine. However, a new engine design can be made for the higher octane number fuel, so that it can withstand the loading incident with higher compression ratios which can be used with this fuel and this engine will definitely give more power. But for a given engine, increasing the fuel octane number beyond the value which is just necessary to prevent detonation, does not entail any use.

Further it may be noted that octane number has absolutely no relation with the actual percentage of isooctane in the given fuel. For example, we talk of octane number of benzene, toluene, hexane, etc. which are independent compounds having no isooctane.

Two methods of finding octane ratings have been standardized by C.F.R. Committee. These are:

1. C.F.R. Research Method
2. C.F.R. Motor Method

Both these methods use similar type of laboratory single cylinder engine, which is provided with a variable head to change compression ratio and a knock meter to indicate knock intensity.

The Research method measures the octane rating under relatively mild operating conditions and is considered similar to the the detonation tendency of a fuel when the engine is accelerating from low speed in top gear with the throttle wide open, under medium load. In this the conditions maintained are: inlet air temperature = 52°C, coolant temperature = 100°C, engine speed 600 R.P.M. and ignition advance setting = 13° BTDC. The Motor method, on the other hand involves relatively severe operating conditions. It is considered similar to the detonation tendency of a fuel when the engine is driven at medium speed in top gear with throttle wide-open under heavy load. The conditions maintained in this are: inlet air temperature = 50°C, coolant temperature = 100°C, engine speed = 900 R.P.M., ignition advance setting = 19 to 26° BTDC. As these methods involve different test conditions, there is always a difference between the octane numbers obtained with these methods. This difference is known as *fuel sensitivety*. The octane number found for the same fuel with Motor Method (MON) is lower than the one found with Research Method (RON) which is less severe. Both these ratings are needed to predict the performance of petrol in actual road use. However, their relative importance depends on the type of engine and the operating conditions.

Usually, for trucks and buses which operate at high temperatures and maximum throttle and for cars with automatic transmissions, more stringent motor method is considered more significant. Likewise in cars with manual transmission, the research method is ralied upon more at full throttle.

Sometimes octance number is determined by a test car under full throttle acceleration from 16 to 80 kmph. However, since it is relatively very inconvenient and too costly to perform such a test to determine what is termed as 'road octane number', usually these are avoided and instead road octane number is approximated as average of the research and the motor octane numbers.

$$\text{Road Octane Number} = \frac{\text{RON} + \text{MON}}{2}$$

8. PREIGNITION

This is a phenomenon very distinct from the usual combustion knock. It is caused on account of the early ignition of the cylinder charge due to some hot spot present in the cylinder. This hot spot may be the heated spark plug itself or the exhaust valve or any other highly carbonized part of the combustion chamber. This phenomenon, however, is not very common in automobile engines.

The early ignition caused by preignition extends the total time the burnt gases remain in the cylinder, thereby increasing the heat transfer to the chamber walls, due to which the self- ignition temperature will occur earlier and earlier on each successive compression stroke. As a result the occurrence of peak cylinder pressure will gradually advance to T.D.C. and will coincide with the cylinder temperature peak. The combined effect of the increased combustion time and rising peak cylinder pressure and temperature will cause the self-ignition point to creep further ahead of T.D.C. so that now negative work will be done in compressing the combustion products. Thus pre-ignition in a single cylinder engine will result in continuous reduction in speed and power output. Moreover, if allowed to persist, the self-ignition temperature will be reached so early that incoming air-fuel charge will ignite at the inlet itself, causing the flame to propagate back through the intake port and manifold.

In a multi-cylinder engine, there is another undesirable effect of preignition, when it occurs only in one or a few cylindes only. In this condition, the unaffected cylinders will continue to develop full power and speed, thus dragging the piston in affected cylinders to and fro until the increased heat generated causes the seizure of pistons and rings of the preigniting cylinders.

Preignition can be readily distinguished from combustion knock by a simple test. If the ignition is closed and the engine continues to run for some time, it is preignition; if the engine stops it is detonation. Apart from this, the damage to the piston by preignition occurs in the form of molten piston crown, whereas the pistons damaged by detonation have the appearance of sand blasted and not melted. Fig. 9.9 gives progressive stages

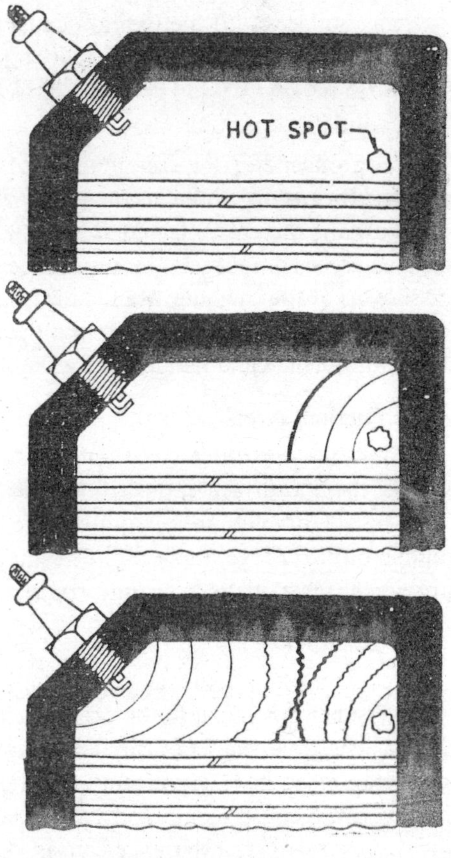

Fig. 9.9. Progressive stages of knocking due to preignition.

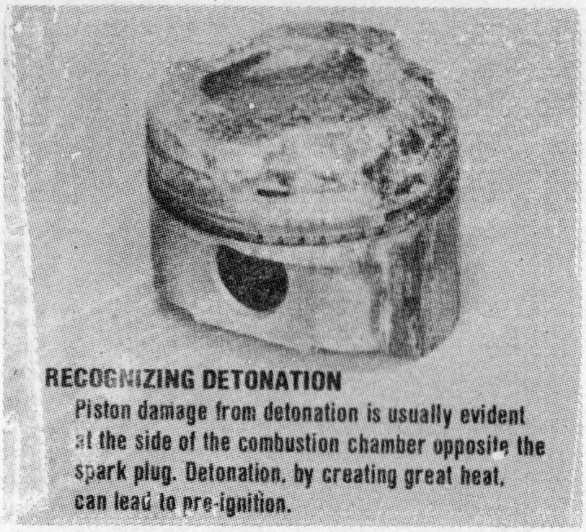

Fig. 9.10. Piston damaged due to detonation.
(Courtesy—Zollner Corporation, U.S.A.)

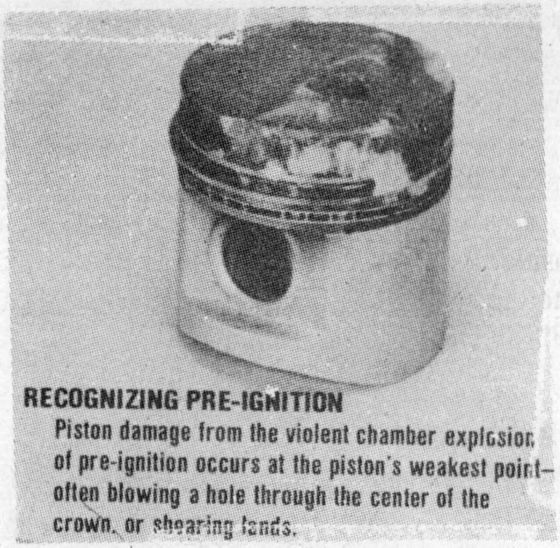

Fig. 9.11. Piston damaged due to preignition.
(Courtesy—Zollner Corporation, U.S.A.)

of preignition, which can be compared with Figs. 9.5 and 9.6. Figs. 9.10 and 9.11 show the pistons damaged due to detonation and preignition respectively.

In a similar way if the ignition caused is late, the phenomenon is termed as *post ignition.*

9. TERMS RELATING TO RATE OF COMBUSTION

1. Squish

It is produced by the compression of the charge trapped between the parallel portions of the cylinder head and the piston crown, as the piston approaches TDC. Due to this compression, the air (or the air-fuel mixture, as is the case), is squeezed at a high velocity towards the much enlarged combustion chamber space situated either in the piston (as shown) or the cylinder head. This highly compressed air or air-fuel mixture suddenly expanding into the larger space, increases the charge turbulence and hence the fuel mixing.

Fig. 9.12. Squish

2. Quench Area

It is defined as the area of flat surface of the piston crown that opposes the flat surface of the cylinder head as the piston approaches TDC. As the quench area is quite large as compared to the small volume of the charge, there will be a large amount of heat transfer through the metal walls, *causing a rapid cooling or quenching effect.*

3. Turbulence

It consists of randomly dispersed vortices of different sizes which become superimposed into the charge (air or air-fuel mixture) flow stream. These vortices are carried alongwith the flow stream and form breakways having concentric spiral motion. The whirling of these vortices cause interaction with adjacent vortices which increases the fuel mixing and heat transfer rate. In other words, turbulence causes very rapid movement of the charge, which results in better mixing of the air-fuel mixture and consequently better combustion.

10. CONSIDERATIONS AFFECTING COMBUSTION CHAMBER DESIGN

Important considerations for better design of combustion chamber and different methods to achieve the same are discussed briefly here.

1. Swirl

It is the rotational flow of charge within the cylinder, about its axis. Induction swirl is created by so positioning the induction (*i.e.* inlet) port passage to one side of the cylinder axis so that the flow discharges into the cylinder tangentially (Fig. 9.13). This may be done in a number of ways as follows:

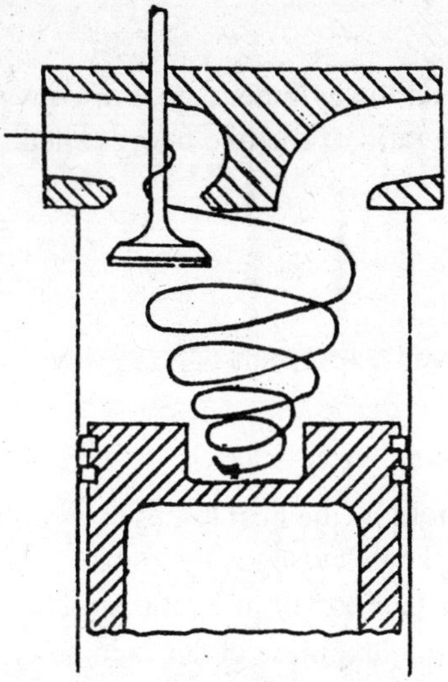

Fig. 9.13. Induction swirl

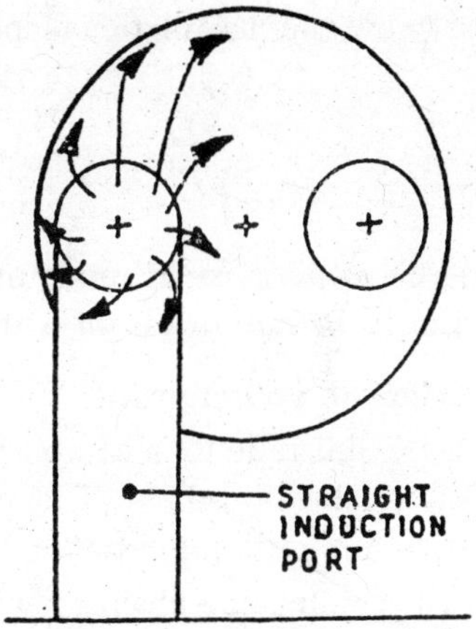

Fig. 9.14. Directed straight port

(*i*) Straight induction port (Fig. 9.14). A straight and parallel port passage takes the charge to the valve opening in the desired tangential direction towards the cylinder wall where it is deflected. The downwards and the circumferential motions combine to give a spiral motion about the cylinder axis.

(*ii*) Deflector wall port (Fig. 9.15). In this the port passage is slightly curved and tapered due to which

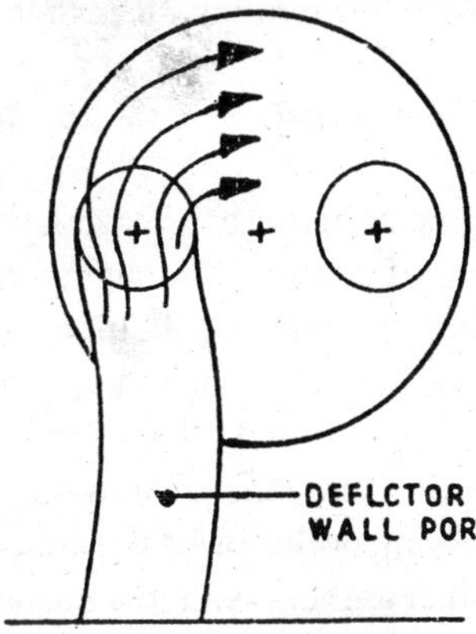

Fig. 9.15. Deflector wall port

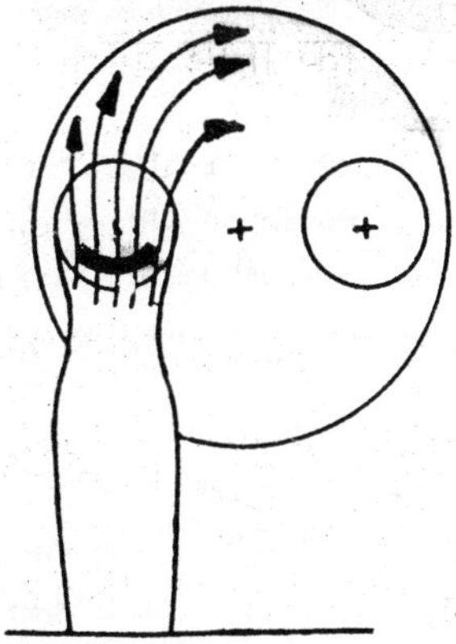

Fig. 9.16. Masked valve port

the charge follows a circular flow path towards the wall opening and then discharges into the cylinder in the downward spiraling motion about the axis of the cylinder.

(*iii*) Masked valve port (Fig. 9.16). In this the valve seat is partially closed in such a way that the charge is made to discharge into the cylinder with a downward spiral motion about the cylinder axis.

(*iv*) Helical port. In this the swirl is produced above the valve seats, within the port. This swirl flow is then guided by the walls of the port passage which make it spiral in the helical shaped passage above the valve seat. Thereafter, when it is ejected into the cylinder, it continues to have its downward rotational motion.

2. Swirl ratio

Cylinder air swirl is the angular rotational speed of air about the cylinder axis. For convenience it can be related to the angular rotational speed of the crankshaft, the ratio of the two being called as swirl ratio, *i.e.*,

$$\text{Swirl ratio} = \frac{\text{air rotational speed}}{\text{crankshaft rotational speed}}$$

With helical ports, swirl ratios of 5 at TDC can be achieved with a flat piston crown, which can be increased to 15 in case of piston with bowl.

3. Surface to volume ratio

The larger this ratio for a combustion chamber, the higher will be the heat losses and cooler will be the walls of the combustion chamber, due to which higher will be the exhaust gas hydrocarbon concentration. Therefore, to minimise the formation of hydrocarbons, the surface to volume ratio should be minimum.

(*i*) The combustion chamber with the least surface area containing a given volume is the spherical type. Thus a double hemispherical chamber has the least surface to volume ratio.

(*ii*) Chambers with small quench area have small surface to volume ratios.

(*iii*) Increasing compression ratio increases the surface to volume ratio.

(*iv*) For a given cylinder capacity, lesser the number of cylinders and larger their size, the lower will be the overall surface to volume ratio for the engine.

(*v*) A long stroke-to-bore ratio results in a smaller surface to volume ratio than a short stroke-to-bore ratio.

When the stoke length and the bore diameter are equal, the engine's stroke to bore ratio is said to be *square* whereas it will be *oversquare* or *undersquare* if the stroke to bore ratio is less than or greater than 1 respectively.

11. COMBUSTION CHAMBERS FOR S.I. ENGINES

Two most important considerations in the design of combustion chambers are the production of turbulence and the location of spark plug. High turbulence improves the rate of combustion chamber, while suitable location of the spark plug can shorten the path of flame travel. Both these factors help to avoid detonation. Another consideration that has become quite important these days is the surface area of the combustion chamber. The charge near the chamber walls, from 0.15 mm to 0.5 mm thick, does not burn because the cold chamber walls cool the same to a temperature below its ignition temperature and thus the fuel in this charge goes in the exhaust as unburned hydrocarbon emissions causing pollution of the atmosphere, besides wastage of fuel. Thus effort is always made to design combustion chambers with low surface area/volume ratio, for less emissions of unburned hydrocarbons. Most prominent types of the combustion chamber for gasoline engines discussed here are:

1. Side-valve type
2. Wedge type
3. Inverted bath tub type
4. Flat-head type
5. Hemispherical type
6. Stratified charge type
7. Multi-valve type
8. Split level type
9. Twin spark plug type

1. Side valve type

The side valve gear used with this type of combustion chamber is simpler than the overhead valve gear used alongwith the other types of combustion chambers described later, due to which reason this is also the cheapest type. However, this type of combustion chamber is less efficient from combustion point of view and was used for lower compression ratios only, say upto about 7:1. On account of this reason this type is no longer in use.

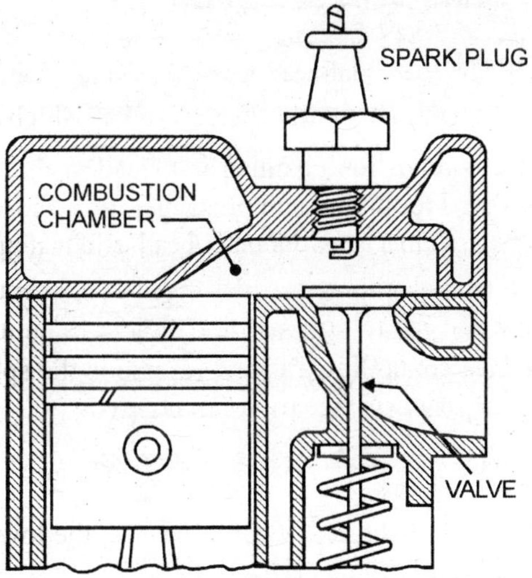

Fig. 9.17. Side valve type combustion chamber

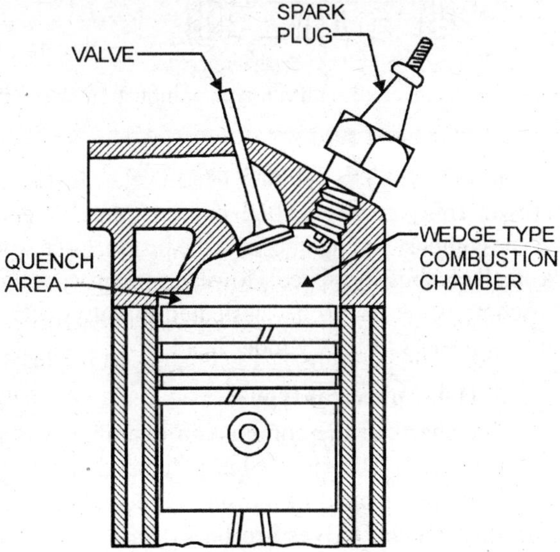

Fig. 9.18. Wedge type combustion chamber

2. Wedge type

In this type, the single row valves are tilted to accommodate the sloping roof of the chamber. The spark plug is located on the thick side of the wedge., midway between the valves. In this, the rate of burning of fuel is controlled which produces smooth and uniform combustion. As the piston reaches the end of the compression stroke, the piston comes near the low flat surface area of the combustion chamber, called the quench area. The gases squeezed from the quench area cause turbulence, which completely mixes the air and fuel in the charge. This results in fast and smooth burning. The spark plug is placed in the end away from the quench area where turbulence is very high. As the gases in the end area *viz.,* the quench area are squeezed very thin (less than 2.5 mm) when the piston is at the top dead centre, they are cooled, which prevents their burning abnormally and thus detonation is avoided. Such turbulent combustion chambers as the wedge type are left without machining, which reduces the cost of production.

3. Inverted bath tub type

This is another form of wedge type chamber and is shown in Fig. 9.19. In this turbulence is promoted and the flame spreads rapidly due to which reason tendency to detonate is decreased.

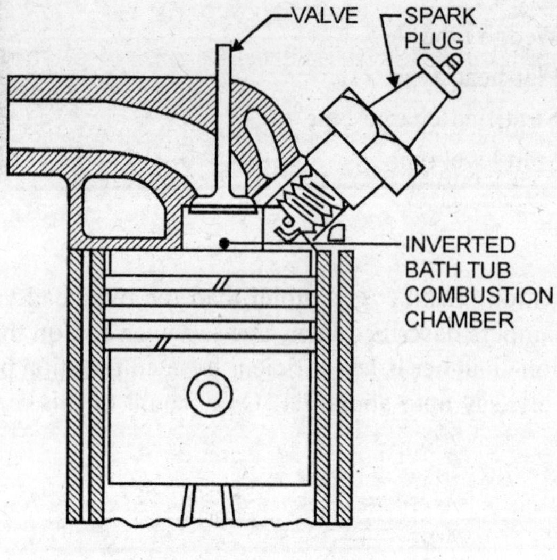

Fig. 9.19. Inverted bath tub type combustion chamber

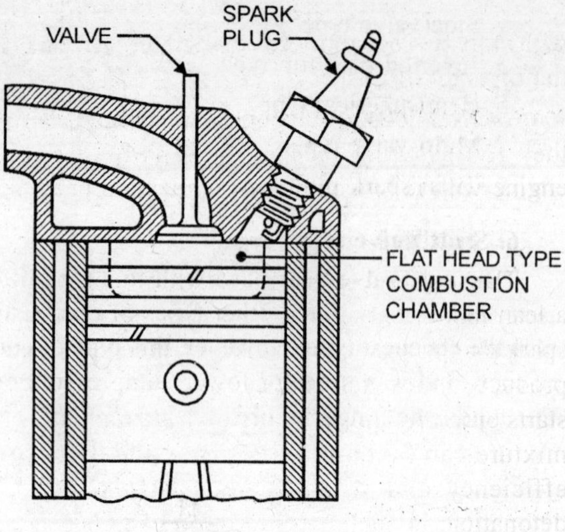

Fig. 9.20. Flat-head type combustion chamber

4. Flat-head type

In this type the cylinder head is flat. Instead, the entire combustion chamber is placed in the piston crown. This type affords high combustion efficiency and is suited for higher compression ratios. Moreover, as the combustion chamber is in the piston, large valves may be used in a large diameter head, still maintaining a small combustion space. However, in this case the heat of combustion has to be dissipated mainly through the piston instead of the cylinder head.

5. Hemispherical type

The shape of the combustion chamber is close to hemispherical. The spark plug is located centrally and the valves are placed in the slanted position. The inlet and the exhaust valves are on different sides of the chamber, requiring two camshafts to be used.

This type of combustion chamber is efficient and compact. Though there is no turbulence in this design, due to which the combustion rate is slower than in the wedge type turbulent combustion chamber, yet the flame travel is short in this, which means better performance against detonation. Moreover, the flow of incoming and outgoing gases is without obstruction which imparts to it a high volumetric efficiency. However, its main disadvantage is its high production cost on account of the fact the separate valve operating mechanisms have to be provided for the inlet and the exhaust valves. Another disadvantage is that in this rapidly burning charge produces a high rate of pressure rise which does produce some engine roughness and noise when the engine is run under medium and heavy loads at low speeds. Thus this type of combustion chamber is best suited for racing cars which are seldom run at low speeds.

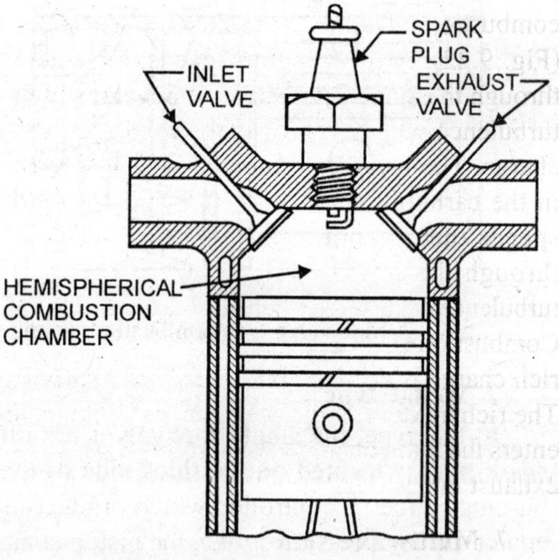

Fig. 9.21. Hemispherical type combustion chamber

This type of combustion chamber is particularly suitable for small, high speed automotive engines. Compared to the wedge type combustion chamber of the same size, it produces more power and torque with more fuel economy. In such small engines, quite often aluminium heads are used which further improves the thermal efficiency.

Such combustion chambers are completely machined, which increases their production cost. Moreover, the valve actuating mechanism in their case is more complex and hence costlier than that in case of an engine with the wedge type combustion chamber.

6. Stratified charge type

The stratified charge means that some of the layers have a rich air-fuel mixture, while the others have a lean mixture. The overall charge has a very lean air-fuel mixture compared to ordinary engines. The spark plug in such combustion chambers is located in the region of rich strata. After ignition, the hot flame produced in the rich strata subjects the remaining lean mixture to heat and pressure, due to which it also starts burning. Thus a relatively leaner air-fuel mixture can be burnt with high combustion efficiency and without the occurrence of detonation, in such a combustion chamber. It is claimed that with such a combustion chamber using direct injection can allow a petrol engine to run with air-fuel ratios as lean as 150:1.

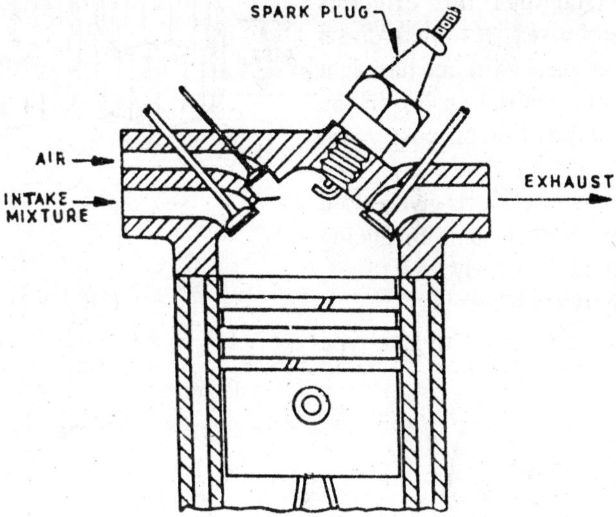

Fig. 9.22. M.C.A. combustion chamber

An example of an engine with such a combustion chamber is the Mitsubishi Clean Air (MCA) system in which air entres the combustion chamber through two inlet valves (Fig. 9.22). At low engine speeds, air enters through the small inlet valve. It develops high turbulence which improves the mixing of the charge. The location of the air intake passage in the carburettor is so designed that at high speeds there is not any significant air flow through the smaller valve and only normal turbulence occurs. Another example of the stratified charge engine is the Honda Compound Vertex Controlled Combustion (CVCC) engine, in which the second intake valve is located in a prechamber, in which the rich charge is drawn, while the lean mixture is drawn in the main chamber through the main inlet valve. The rich mixture in the prechamber is ignited by the spark plug. The burning charge from the prechamber enters the main chamber and ignites the lean charge there. This results in efficient combustion and reduced exhaust emissions.

7. Multi-valve type

Comparison of valve opening and hence the amount of intake air-fuel mixture in case of 2-valve and 3-valve combustion chambers is shown in Fig. 9.23. In both cases, the size of the chamber is same. In case of 2-valve chamber the total opening of inlet valve is $\pi \times 35 \times L$ (where L is the valve lift), whereas in case of 3-valve chamber, the total inlet valve opening is $2\pi \times 28 \times L$ *i.e.* $56\ \pi L$, which is obviously greater than the former. The total inlet valve opening can be increased still further by having two inlet and two outlet valves (Fig. 9.24). A combination chamber with four valves has a pentroof design, wherein each pair of valves is in line (Fig. 9.25) or hemispherical design, wherein each valve is inclined to the others. The spark plug in such cases is placed at the centre of the chamber to have the shortest path of flame travel.

The four-valve design optimises the volume of the mixture that can be taken in, ignited and exhausted.

The behaviour of the mixture before and during the split second of combustion is critical to achieve fast and active burning. The four-valve design creates a barell shaped swirl [Fig. 9.25 (*a*)] which produces a turbulence highly favourable to good combustion efficiency. By keeping the angle of pent roof combustion chamber at a low 45 degrees, this effect is optimised. It breaks down the barrel swirl into turbulent 'mini swirls' earlier in the combustion process [Fig. 9.25 (*b*)]. As the piston rises, these mini swirls give a more thorough and uniform mixing of fuel and air and faster, more even burning.

Thus a multivalve type

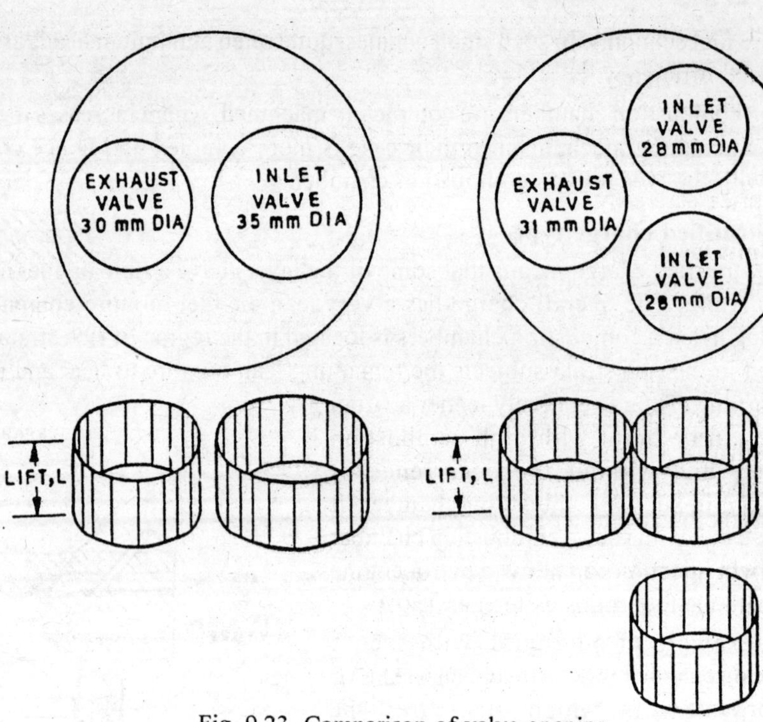

Fig. 9.23. Comparison of valve opening.
(*a*) 2-valve combustion chamber, (*b*) 3-valve combustion chamber.

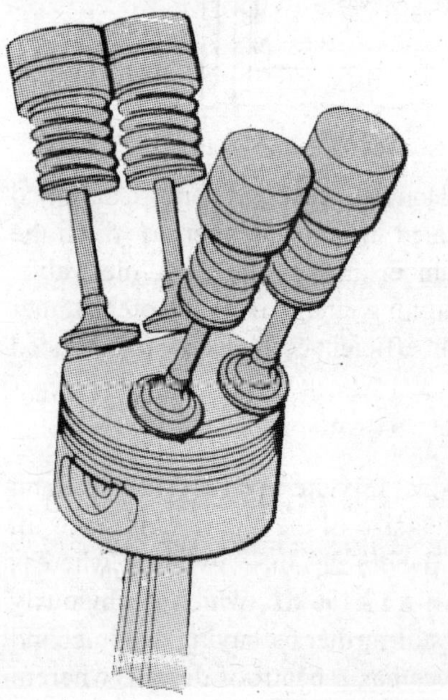

Fig. 9.24. Four-valve design.

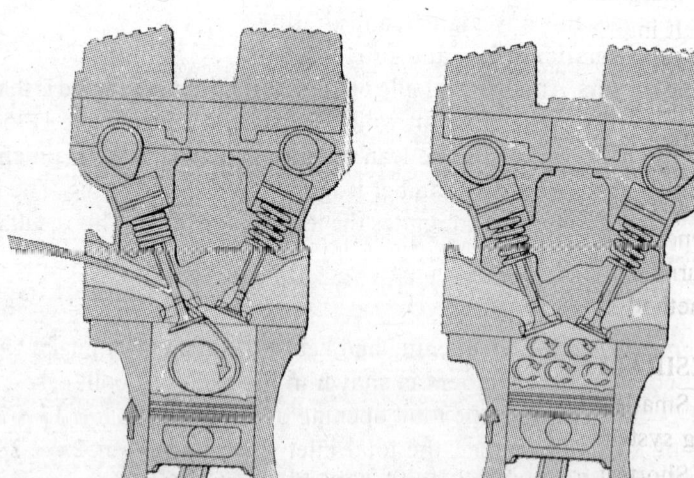

(*a*) (*b*)

Fig. 9.25. Pent roof combustion chamber
(Courtesy—Rover Powertrain, U.K.)

of combustion chamber provides larger total valve and port capacity for any given cylinder bore size. The resulting advantages include higher volumetric efficiency as a result of increased flow rate, better combustion, reduced emissions, better performance and more fuel economy. The disadvantage is the high cost and more complex construction on account of separate camshafts for intake and exhaust valves, which limits its use to engines of high price cars only.

8. Split level type

In this type of combustion chamber [Fig. 9.26), the exhaust valve head is deeply recessed in a circular chamber, whereas the inlet valve head is sunk just below the flat face of the cylinder head. The walls of the deep exhaust valve are steep. This type of chamber is claimed to operate satisfactorily with a 97 RON petrol with a 16:1 compression ratio and 92 RON petrol with a 12:1 compression ratio, without detonation, on account of the elimination of most end-gases before they become overheated.

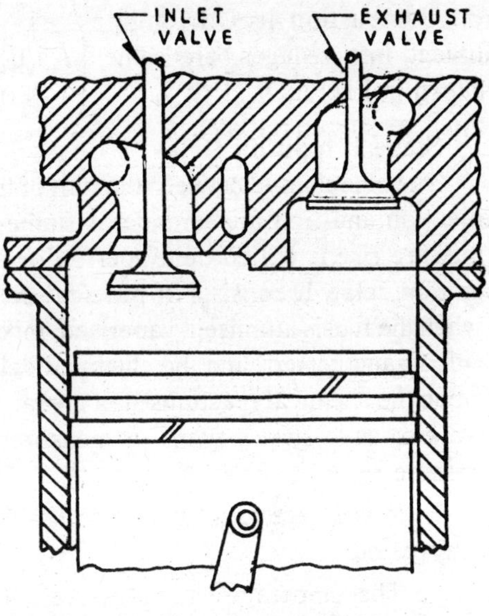

Fig. 9.26. Split level combustion chamber.

9. Twin spark plug type

In a conventional two-valve cylinder head, the single spark plug is usually located between, but to one side of, the inlet and the exhaust valves. The flame path in this case is quite long. By incorporating two spark plugs in the cylinder head and locating them on either side of the valves, the distance the flame travels now before reaching the cylinder walls is reduced. The twin-ignition combustion chambers have the following advantages as compared to single-ignition types:

1. It helps to reduce rough and harsh running, particularly under light loads.
2. It helps to develop more torque at lower speeds. However, at higher speeds this benefit becomes marginal.
3. It improves the fuel consumption under part loads.
4. It enables more exhaust gas recirculation to be utilized to reduce exhaust emission.

A further improvement in the twin spark plug type combustion chamber is to fire two sparking plugs at different times which ensures that the mixture burns thoroughly whatever the prevailing conditions. At lower engine speeds, the spark plug nearest the inlet valve is fired much earlier than usual while the plug nearest the exhaust valve is fired later than usual. At higher speeds, the spark plugs are fired simultaneously. The benefit of controlling the timing of each plug individually is that it increases cylinder pressure as the fuel burns and also allows a higher compression ratio to be used without the occurrence of detonation. This method has been used in Honda's Jazz engine introduced in 2001.

12. DESIRABLE FACTORS FOR COMBUSTION CHAMBERS IN S.I. ENGINES

1. Smallest possible ratio of surface area of chamber to its volume. This minimizes heat loss to the cooling system.

2. Shortest travel distance for the flame front.

3. Sufficient swirl of the incoming mixture to get a uniform mixture rapidly. However, excessive swirl may cause excessive heat loss.

4. Adequate cooling of the exhaust valve.

5. Provision for the cooling of spark plug by the incoming fresh charge to avoid preignition under wide throttle conditions.

6. Adequate sizes and numbers of inlet and exhaust valves for higher volumetric efficiency at higher speeds and to expel the exhaust gases completely, respectively.

13. COMBUSTION IN COMPRESSION IGNITION ENGINES

Combustion in C.I. engines may be divided into 4 stages. The same have been shown in Fig. 9.27.

Stage I. Ignition Delay

The time elapsed between the start of fuel injection and first appearance of flame or pressure rise is called delay period or the ignition delay. It consists of physical delay (when the fuel is atomised, vaporised, mixed with air and heated) and the chemical delay (when the chemical reactions start and accelerate so as to start a flame or measurable pressure rise).

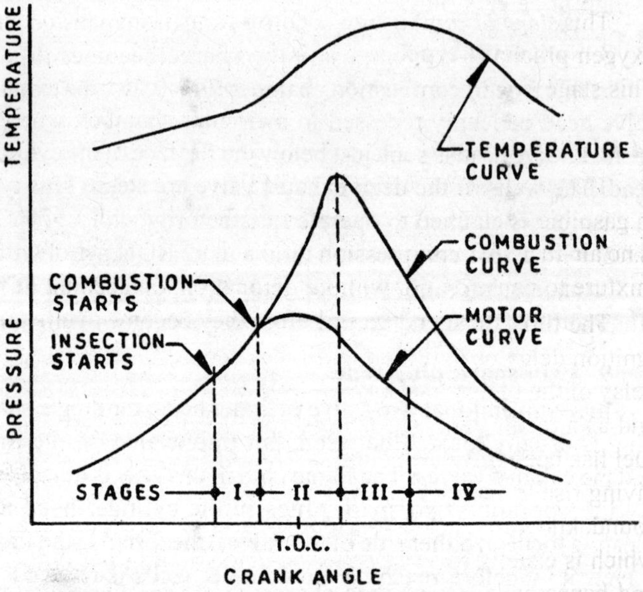

Fig. 9.27. Combustion in C.I. engines

Following factors influence the ignition delay period:

1. The temperature and pressure of the air in the cylinder at the moment fuel is injected.
2. The amount of turbulence present at the time of injection.
3. The amount of penetration and atomization of the fuel spray.
4. The nature of the fuel.

Stage II. Rapid or uncontrolled combustion

By the time the delay period is over, high temperatures of the order of about 450°C to 550°C already exist in the combustion chamber. Therefore once the combustion starts, it proceeds very fast due also to the existence of multiple ignition points. This stage starts at the end of the ignition delay and continues till the maximum pressure is reached.

The rapid pressure rise and peak pressure during this period is influenced by the duration of the delay period. A large ignition delay causes a very high rate of pressure rise, whereas a small delay results in a more gradual rate of pressure increase. Thus for example, if the ignition delay period is small, only small amount of fuel will have entered the cylinder before the combustion starts. Accordingly, the energy released by this small amount of fuel will only provide a smaller rate of pressure rise during the following uncontrolled combustion period, which would mean smooth combustion. On the other hand, a longer delay period would result in rough running.

Stage III. Controlled combustion

This is a stage of slow combustion which continues till the maximum temperature in the combustion chamber in reached. At the end of the rapid combustion, *i.e.*, rapid pressure rise phase, the late fuel droplets injected enter a charge which is far higher heated than the initial injection droplets. Therefore, these droplets at the end of the second phase ignite almost as soon as they come out of the nozzle, due to which the available oxygen in that region is reduced, consequently slowing down the speed of burning and thus raising the cylinder pressure at a much reduced rate. Therefore, this stage is mechanically controlled by the output characteristics of the fuel injection pump, which can extend or shorten the duration the fuel delivery. As such the duration of this stage directly depends upon the amount of additional fuel supplied in excess of that required for no-load running.

Stage IV. Afterburning

This stage of combustion occurs during the expansion stroke. During this state the availability of unused oxygen gradually decreases and the charge becomes diluted by the formation of combustion products. This stage continues from the point of maximum cycle temperature to about 70-80 degrees after T.D.C.

14. DIESEL KNOCK

This phenomenon which occurs in the case of diesel engines is similar to the combustion knock experienced in gasoline engines. However, the cause of occurrence is distinctly different in the two cases. Since there is no air-fuel mixture in a diesel engine; the air alone cannot be compressed by the flame front of the unburnt mixture to cause detonation. Likewise air alone cannot be preignited even if any hot spot is present.

The time between the fuel injection in the combustion chamber and its actual burning is called the ignition delay or ignition lag. In practice, actual injection of the fuel also takes some time. If the ignition delay of the fuel is less, a small amount of fuel only has been injected before the first droplets start to burn and as a result the combustion is smooth. If however, ignition delay is long, much larger quantity of the fuel has been injected before the first droplets burn and this may result in too high a rate of combustion, giving rise to very high rates of pressure rise and also higher peak pressures, causing a peculiar thudding sound, known as diesel knock. It is thus seen that the main cause of diesel knock is prolonged delay period, which is exactly in contrast to the detonation in petrol engines. The factors prolonging the ignition delay and hence encouraging the occurrence of diesel knock are:

1. Low compression ratio.
2. Low combustion pressure due to worn out pistons, rings and/or valves.
3. Inadequate atomisation of fuel spray, due to worn out fuel injector, blocked nozzle holes or insufficient spring load on the injector needle.
4. Over-advanced injection.

15. CETANE NUMBER

Cetane number in the case of diesel fuels is similar to the octane number for gasolines. Earlier it was defined as the percentage of cetane in a mixture of cetane and α-methylnaphthalene, which has the same diesel knock characteristics as the fuel under test. Cetane ($C_{16}H_{34}$) (*n*-hexadecane) is a straight chain paraffin and assigned to the top of the scale, *i.e.*, cetane number of 100, whereas alpha methylnaphthalene ($C_{11}H_{10}$) was given cetane number zero. However, recently, alpha methylnaphalene has been replaced by a more stable compound, *heptamethylnonane*, which having a slightly better knockrating, is given cetane number of 15.

Cetane number, like the octane number, is determined in a standard single cylinder pre-chamber type variable compression ratio engine under the following standard conditions:

Intake temperature	65°C
Jacket temperature	100°C
Speed, rpm	900
Injection timing	13° BTDC
Injection pressure	103.5 bar

Cetane number expresses the auto-ignition quality of the diesel fuel. A fuel with shorter ignition delay period will have higher cetane number and vice versa. The cetane number affects cold staring, warm up, roughness, acceleration, deposits and exhaust smoke density in a diesel engine. Most diesel fuels for high speed engines have cetane numbers from 45 to 55.

16. DIESEL INDEX

Cetane number tests are very expensive to conduct. Therefore, recourse is often taken to adopt simpler and cheaper methods. One such method is the 'diesel index method' which is not very exact, but nonetheless, a useful indicator for estimating ignition quality.

Diesel index uses certain physical characteristics of the fuel, *viz.*, the aniline point and the A.P.I. gravity.

$$\text{Diesel index (DI)} = \frac{\text{Aniline point (°F)} \times \text{API gravity (deg)}}{100}$$

Aniline point of a fuel is the temperature at which equal parts of the fuel and pure aniline dissolve in each other. Likewise, a high API gravity means a low specific gravity, indicating a high paraffinic content, corresponding to gocd ignition quality.

17. COMPARISON BETWEEN DETONATION AND DIESEL KNOCK

1. Detonation in S.I. engine is caused by the auto-ignition of the *end gas towards the end* of combustion. whereas knocking in C.I. engines is due to the auto-ignition of the *first charge at the start* of the combustion (Fig. 9.28). Thus, to avoid detonation, auto ignition of the end-gas has to be prevented, whereas to avoid diesel knock earlier possible auto-ignition should occur.

2. In S.I. engines, the air-fuel mixture that self-ignites is homogeneous due to which rate of pressure rise and hence intensity of detonation is substantial compared to C.I. engines, where there is no homogeneous air-fuel mixture present at the time of self-ignition.

3. Fuels found good to prevent deto- nation have longer ignition delay time, which prove to be poor preventing diesel knock, which means the fuels having higher octane rating have poor cetane rating and *vice-versa.* Actually, an approximate inverse relationship exists between the octane number (ON) and the cetane number (CN), which has been found to be accurate within ± 5 percent. The relationship is:

$$CN = 60 - ON/2$$

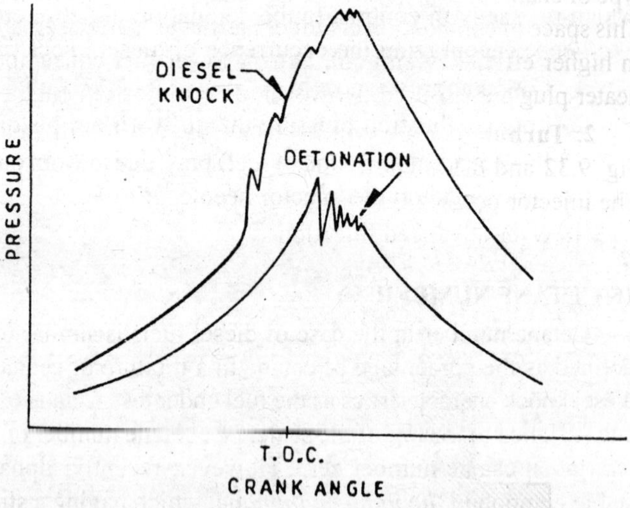

Fig. 9.28. Comparison of detonation and diesel knock.

4. Compression ratio has to be limited in case of S.I. engines, beyond which detonation would occur. However, in case of C.I. engines, higher the compression ratio, lesser the ignition delay and hence lesser possibility of diesel knock occurring.

5. Larger cylinder size promotes detonation, whereas diesel knock is reduced with the same. That is the reason why engines with cylinder diameters larger than 250 mm are almost always of C.I. type.

18. COMBUSTION CHAMBERS FOR C.I. ENGINES

The combustion chambers used for automative diesel engines are mainly of the following three categories:

1. Direct injection type or open type.
2. Turbulent or swirl type.
3. Pre-chamber type.

1. Direct injection type combustion chamber. In engines with this type of combustion chamber, there in no separate combustion space in the head, the fuel being directly sprayed into the cylinder. The

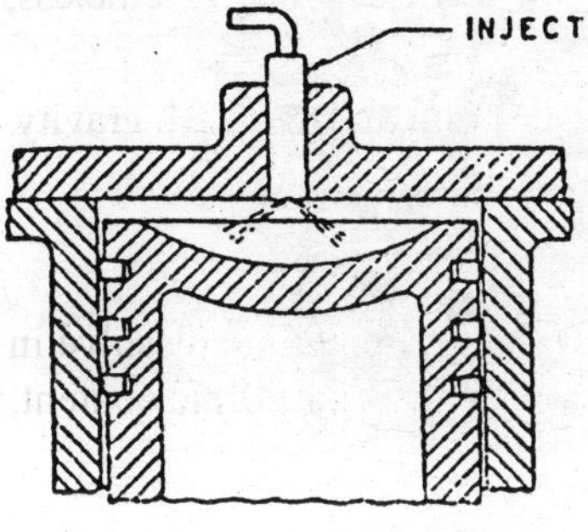

Fig. 9.29. Direct injection type combustion chamber.

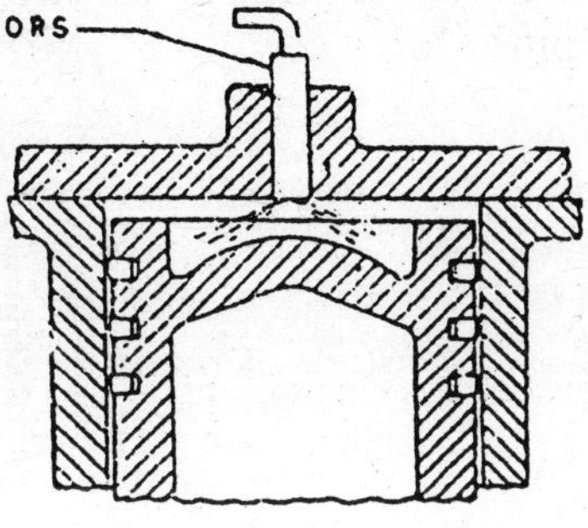

Fig. 9.30. Direct injection type combustion chamber.

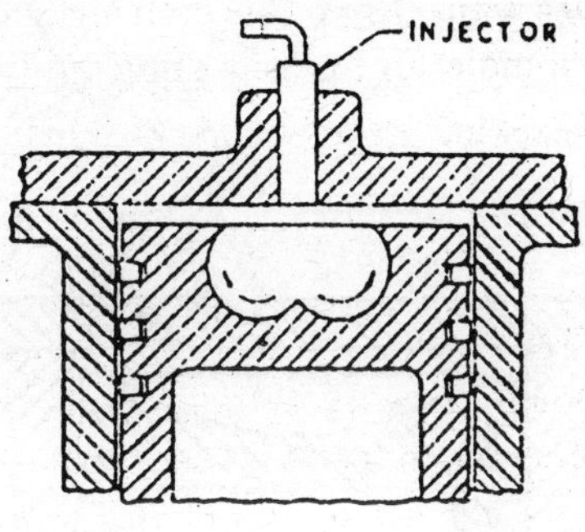

Fig. 9.31. Direct injection type combustion chamber (torroidal)

piston crown contains a depression, which may be bowl shaped or torroidal (Fig. 9.29 to 8.31). The space between the head and the piston crown forms the combustion chamber. The injector nozzles used for this type of chamber are generally multi-hole type, injecting the fuel at relatively high pressure (about 30 MPa). This space of combustion chamber has lower 'surface area/volume' ratio, thus reducing heat losses resulting in higher efficiency and consequent lower fuel consumption. Further, no auxiliary starting devices, like heater plug are required. However, direct injection engines are generally more rough in running.

2. Turbulent type combustion chamber. Two such types of combustion chambers are shown in Fig. 9.32 and 8.33. These are designed in such a way as to introduce turbulence of the compressed air. The injector nozzles generally are of single-hole type. In the vortex type (Fig. 9.32), the air at the end of

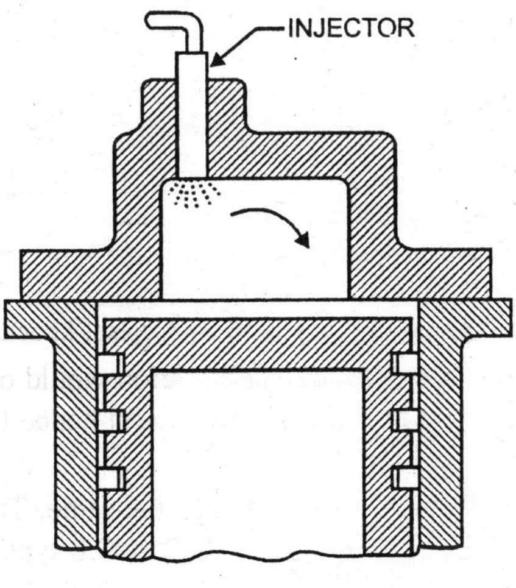

Fig. 9.32. Turbulent type combustion chamber (Vortex).

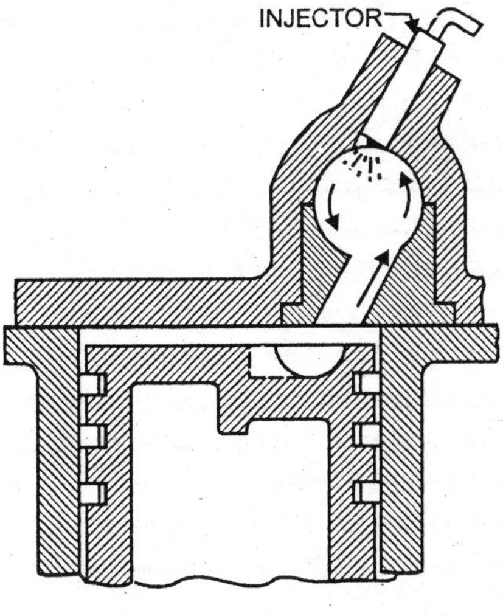

Fig. 9.33. Turbulent type combustion chamber (Comet).

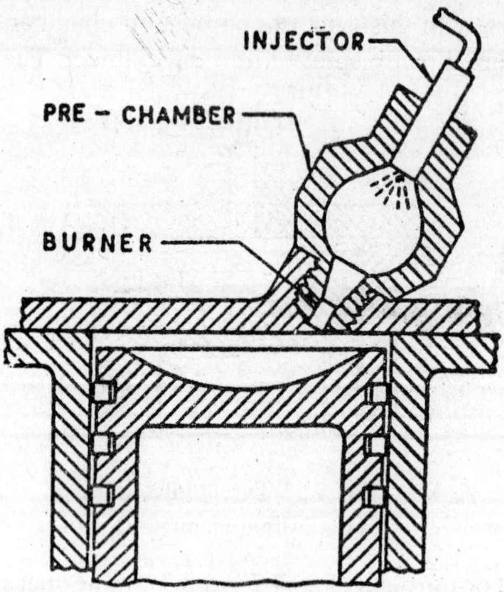

Fig. 9.34. Pre-chamber.

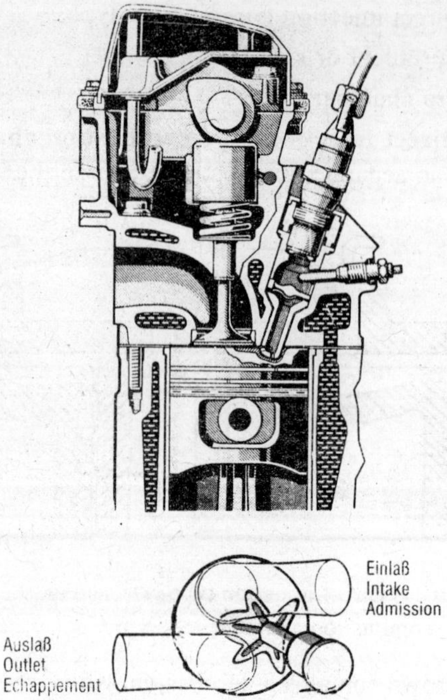

Fig. 9.35. Pre-chamber for Mercedes Benz OM602 Engine. (Courtesy—Mercedes Benz Aktiengesselschaft, Germany)

the compression stroke is pushed in the cylindrical chamber in the head, when the fuel is injected. The comet type (Fig. 9.33) employs a spherical combustion chamber which is, connected tangentially through a narrow passage called throat, to a cavity in the piston crown. The compressed air enters the spherical chamber tangentially through the throat creating turbulence when the fuel is injected by the nozzle.

3. Pre-chamber or ante-chamber. In this type, the combustion space is located both in the cylinder as well as in the head. About 40 per cent of the combustion space is in the shape of pre-chamber whereas the main chamber is formed between the cylinder head and the cavity in the piston crown. The pre-chamber is connected with the main space through a restricted passage called burner.

The compressed air is forced into the pre-chamber through the burner when at the end of the compression stroke the fuel is injected. This causes the charge in the pre-chamber to ignite. A gush of gas escapes the pre-chamber, into the main chamber and completes its combustion along with the air there. The advantages of this type are the high turbulence, freedom from high maximum pressure and smooth running. There are many disadvantages, however. The engines using this types of combustion chamber are difficult to start without heater plugs which are a necessity, require high compression ratios and involve heat loss to the burner. Figs. 9.35 and 9.36 show the pre-chamber design used on OM 602 engine for Mercedes Benz 190 D2.5 car. This new Mercedes diesel prechamber 1989 design reduces particulate emissions by about 40 per cent on average. In this the injection nozzle is off-centre and 7^0 more inclined than the previous model. Also with appropriate shape and position of the underside of the ball pin, is created a moderate air swirl into which the fuel is injected in parallel flow. The glow plug is thus positioned in the downward current. This improves the combustion sequence considerably.

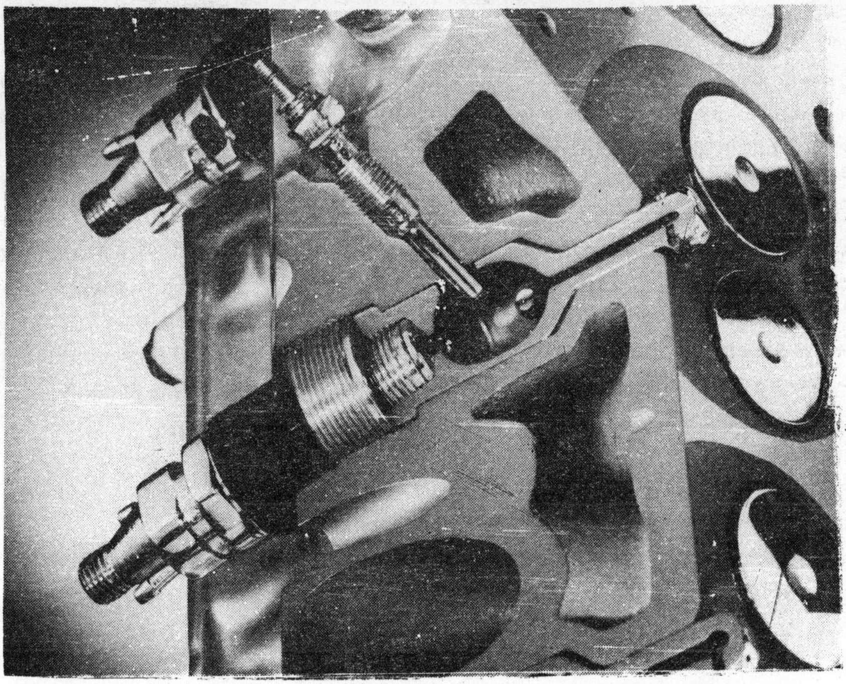

Fig. 9.36. Details of Pre-chamber design of Mercedes Benz OM 602 Engine.
(Courtesy—Mercedes Benz Aktiengesselschaft, Germany)

THEORY QUESTIONS

1. State various conditions affecting combustion chamber design and discuss the same in detail.
2. Explain clearly various technical terms concerning rate of combustion in internal combustion engines.
3. Discuss in detail the phenomenon of combustion in spark ignition engines.
4. How do you define detonation? Describe briefly the various theories of detonation.
5. Give a detailed account of the factors that influence detonation in an engine.
6. Discuss the various steps which should be taken to prevent detonation in I.C. engines.
7. What are the different methods of fuel rating? Explain briefly.
8. What is octane rating? Discuss its significance for a particular fuel to be used in a given engine.
9. Describe phenomenon of preignition in spark ignition engines and compare the same with detonation.
10. Describe clearly how preignition differs from detonation. Use simple sketches to explain this.
11. Describe, with the help of sketches, various shapes of combustion chambers for petrol engines. Explain the particular features of each type.
12. Explain clearly various desirable factors for combustion in S.I. engines.
13. Discuss the phenomenon of combustion in compression ignition engines.
14. Explain the term diesel knock. Describe how it is produced and compare the same with detonation in petrol engines.
15. Write brief note on cetane number of fuels.
16. Explain the term 'diesel index' and discuss its utility.
17. Discuss in detail various types of combustion chambers for automotive diesel engines.

QUESTIONS FOR VIVA

1. What is a flame front?
2. What is auto-ignition?
3. In how many stages does the combustion occurs in S.I. engines?
4. What is ignition lag? State the factors influencing it.
5. What is detonation?
6. State various factors that influence detonation.
7. "As the length of straight chain increases in the case of paraffins the octane number decreases." Is the statement true?
8. Which type of hydrocarbons have the poorest anti-knock characteristics?
9. Which is the optimum combustion chamber shape from detonation consideration?
10. What are the effects of detonation in an I.C. engine?
11. What are the methods used for preventing detonation?
12. What does T.E.L. stand for?
13. What is ethyl fluid?
14. What is the proportion of T.E.L. in fuel that is used in practice?
15. Why has the use of T.E.L. been discontinued in modern gasolines?
16. What does H.U.C.R. stand for?
17. What is Ricardo Engine?
18. What does the abbreviation C.F.R. stand for?
19. What is octane number?
20. How do you explain octane numbers greater than 100?
21. What are secondary fuels?
22. Is the octane number of a fuel indicative of its quality?
23. "A high octane fuel does not necessarily mean more power." True or false?
24. Benzene is pure hydrocarbon. How is it then that you talk of octane number of benzene?
25. What are the standard C.F.R. methods of octane rating of fuels?
26. Out of Research and Motor methods of octane rating, which is more severe?
27. Define the 'sensitivity' of a fuel.
28. What is pre-ignition?
29. How do you distinguish pre-ignition from detonation?
30. What is post-ignition?
31. What is 'squish'? State its function.
32. What is 'quench area'? What does it do?
33. Define 'turbulence'. Why is desirable to introduce turbulence in a combustion chamber?
34. What is swirl? What is its function?
35. Define the term 'swirl ratio'.
36. Why should the surface to volume ratio of a combustion chamber be minimum?
37. Which type of combustion chamber has the least surface to volume ratio?
38. What is a square engine?
39. What is a combustion chamber?
40. State two main considerations in the design of combustion chamber?
41. Name various types of combustion chambers in use.
42. In which type of combustion chamber there is no turbulence?
43. Why is the side valve type of combustion chamber obsolete?

44. What are the advantages and disadvantages of a hemispherical combustion chamber?
45. Why in a hemispherical combustion chamber is the spark plug located at the centre?
46. What is the function of quench area in a wedge type combustion chamber?
47. Why is the wedge type combustion chamber not usually machined?
48. State the advantages of an inverted bath tub type combustion chamber.
49. What is the advantage of having combustion chamber inside the piston crown? What is the disadvantage?
50. What is a stratified charge combustion chamber? State its advantage.
51. Give any two examples of a stratified charge type combustion chamber.
52. What is the advantage of having a multivalve combustion chamber?
53. How does a split level combustion chamber help avoid detonation?
54. State the advantages of twin spark plug type combustion chamber.
55. Make a list of desirable factors for combustion chamber design of S.I. engines.
56. List various stages of combustion in C.I. engines.
57. State the duration of uncontrolled combustion state in a diesel engine.
58. How do you define diesel knock?
59. Enlist factors encouraging the occurrence of diesel knock.
60. What is cetane number? How is it determined?
61. What is the cetane number for most diesel fuels?
62. What is aniline point?
63. Define 'diesel index'. What is its utility?
64. Explain how does ignition delay affect (*i*) detonation, (*ii*) diesel knock.
65. It is stated that fuels good from detonation view point in a petrol engine prove poor when used in a diesel engine. Is it correct?
66. How does compression ratio affect detonation and diesel knock?
67. What is the reason that petrol engines are usually small in size, whereas diesel engines are often of bigger size?
68. Name various types of combustion chambers for diesel engines?
69. What is a direct injection combustion chamber?
70. What are the merits and demerits of open type combustion chamber?
71. What is a turbulent or swirl combustion chamber?
72. What is turbulence and how is it introduced in diesel engines?
73. What advantage is obtained by introduction turbulence in diesel engines?
74. What is a pre-combustion chamber?
75. What are the advantages of a pre-combustion chamber? What are the disadvantages?
76. What is the major source of heat loss in an ante-chamber?

MULTIPLE CHOICE QUESTIONS

Mark the correct answers:

1. One effect of detonation is
   (*a*) delay in ignition
   (*b*) interruption in lubrication
   (*c*) loss of power
   (*d*) deterioration in the quality of air-fuel mixture.
2. An effective method of prevention of detonation is the
   (*a*) cooling of the charge
   (*b*) heating of the charge
   (*c*) locating spark plug at one end of the combustion chamber

(*d*) reducing the quantity of aromatics in the fuel used.

3. Octane number of Indian lead-free petrol is
   (*a*) less than octane number of leaded petrol
   (*b*) equal to octane number of leaded petrol
   (*c*) greater than octane number of leaded petrol
   (*d*) not specified
4. The use of tetraethyl lead in gasolines is being gradually discontinued because
   (*a*) it is costly (*b*) it has bad odour
   (*c*) it blocks the catalytic converter (*d*) it decreases the engine efficiency
5. The function of quench area in a wedge-shaped combustion chamber is to
   (*a*) improve the compression ratio
   (*b*) cool the end gases
   (*c*) decrease the volume of combustion chamber
   (*d*) increase the area of combustion chamber
6. An indication of ignition quality of a diesel fuel is given by
   (*a*) detonation (*b*) octane number
   (*c*) preignition (*d*) cetane number

[**Key:** 1. (*c*), 2. (*a*), 3. (*b*), 4. (*c*), 5. (*b*), 6. (*d*)]

10 Petrol Engine Fuel Supply Systems-I

1. FUEL SUPPLY SYSTEMS

The basic fuel supply system in an automobile with petrol engine consists of a fuel tank, fuel lines, fuel pump, fuel filter, air cleaner, carburettor, inlet manifold and supply & return pipelines.

Following are the types of system which have been used for the supply of fuel from the fuel tank to the engine cylinder:

1. Gravity system
2. Pressure system
3. Vacuum system
4. Pump system
5. Fuel injection system

Out of these the first four systems make use of the carburettor while in the fuel injection system the carburettor has been dispensed with altogether. The gravity system is confined to two wheelers while the pressure and the vacuum systems are obsolete now and the pump systems is being used widely on automobiles. Due to certain advantages, the use of fuel injection system is now universal in modern petrol vehicles.

1. Gravity System

In this fuel tank is mounted at the highest position from where the fuel drops into the carburettor float chamber by gravity. The system is very simple and cheap but the rigidity of placing the fuel tank necessarily over the carburettor is a disadvantage.

2. Pressure system

In the pressure system, a hermetically sealed fuel tank is used. The pressure is created in the tank by means of engine exhaust or a separate air pump. For starting, the pump is primed by hand. It is under the pressure thus produced that the fuel flows to the float chamber of the carburettor. There are chances of pressure leak but the advantage lies in the fact that the fuel tank can be placed at any suitable location.

3. Vacuum System

This system is based upon the simple fact that the engine suction can be used for sucking fuel from the main tank to the auxiliary fuel tank from where it flows by gravity to the carburettor float chamber.

4. Pump System

In this system, a steel pipe carries petrol to the fuel pump

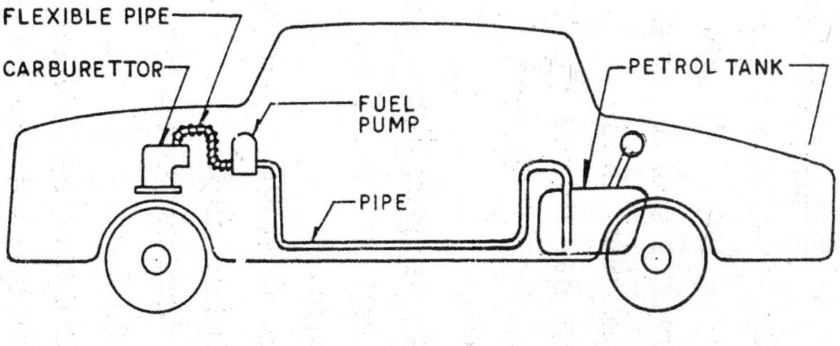

Fig. 10.1. Pump system of fuel supply.

which pumps it into the float chamber of the carburettor through a flexible pipe. If the fuel pump is mechanical, it has to be driven from the engine camshaft and hence placed on the engine itself. However, electrically operated fuel pump can be placed anywhere, the rear location (away from the hot engine) reducing the tendency of forming vapour lock. This system is used most commonly in the present day cars.

5. Fuel Injection System

The petrol injection system has been used successfully on some modern vehicles. In this carburettor is dispensed with altogether. The fuel is atomised by means of an injector nozzle and then delivered into an air stream. Separate fuel injectors are used for each cylinder while the mixture under different load and speed conditions is controlled now electronically. This is the most accurate fuel supply system.

2. FUEL TANK

It is made of steel or aluminium alloy sheet. The steel tank is usually coated on the inside with a lead-tin alloy to protect against corrosion. In case of aluminium, there is saving in weight and a high resistance to corrosion; however, the repairs are comparatively more difficult. Recently synthetic rubber compounds and flame resistant fibre reinforced plastics have also been employed to make fuel tanks by moulding.

Fuel tank is placed in the vehicle at any suitable location. For front engine vehicles, the usual locations are the underside of the luggage compartment at the rear, or directly above the rear axle. The latter location has been favoured recently because of the protection here by the rear wheels and the surrounding body structure in case of rear-end collision. For cars with engines at the rear, fuel tank may be located behind the engine compartment at the front.

Fuel tank is divided into interconnected compartments by means of baffle plates. This arrangement reduces surging of fuel on account of sudden braking or cornering. The surging if unchecked, may interrupt the fuel supply to the carburettor. Holes are made in the sides of the tank for fixing the fuel gauge sensor, the supply and the return pipes and the primary filter. The supply line from the fuel tank is taken at such a level that there is always some dead fuel storage, for the sediments to settle down. A drain plug is provided at the bottom of the tank for periodic removal of the sediments. The cap over the fuel filter tube is vented to atmosphere so that the pressure inside the tank is always atmospheric. In case this vent is chocked, a vacuum may be created in the tank with the consumption of fuel by the engine as a result of which the normal fuel delivery, which is calculated on the assumption of atmospheric pressure in the tank, may be affected. This vacuum may even cause the tank to collapse. Recently, they have started venting the fuel tank to a 'vapour recovery system' instead of to the atmosphere for reducing air pollution caused by the evaporation of fuel from the conventional fuel tank especially in hot season.

3. FUEL PUMPS

Two main types of pumps most generally used are:

1. A.C. Mechanical Pump
2. S.U. Electrical Pump

3.1. A.C. Mechanical Pump

This is a diaphragm type of pump as shown in Fig. 10.2. The diaphragm used is made out of a high-grade cotton impregnated with synthetic rubber. The valves are made of bakelite, which being lighter, keeps the inertia stresses minimum.

The drive for the pump is taken from camshaft by means of an eccentric or cam. The eccentric operates the rocker arm which in conjunction with the diaphragm return spring pushes the diaphragm up and down. Downward movement of the diaphragm causes vacuum in the chamber which causes the inlet valve to open and the fuel then goes through the strainer to the chamber. The next upward movement of the diaphragm

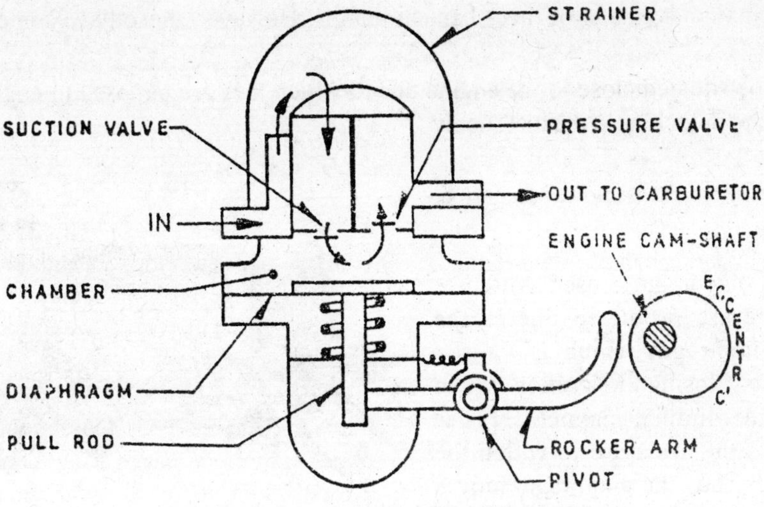

Fig. 10.2. A.C. Mechanical pump

causes the inlet valve to close while the outlet valve opens and the fuel goes out to carburettor float chamber.

In this way, the pressure in the petrol pipeline between the fuel pump and the carburettor, is kept under pressure between about 20 and 35 kPa, the exact pressure range being determined by the stiffness of the diaphragm return spring. The controlled pressure in the petrol supply line is very important to be maintained. If the pressure is low, the petrol supply will not be able to keep pace with the demand under high speed or high-load conditions. On the other hand if the pressure is excessive the needle valve of the carburettor float chamber may be forced open, causing the flooding of the carburetor and consequent stalling of the engine.

When the float chamber of the carburettor is completely filled up, there is no need of pumping more fuel till some of it is consumed. But if the engine continues to run at light load, the camshaft will be running all the time and if no other means are provided, the pump will build up excessive pressures which may damage the pump itself. This is taken care of by the diaphragm return spring, which remains partially or fully compressed depending upon the line pressure even though the rocker arm continues to be moved up and down by the eccentric. Thus the movement of the diaphragm gets restricted to partial or even zero which decreases or even stops the supply of petrol from the pump until engine consumption uses some of the petrol in the carburettor float chamber. For further surety against damage the connection between the rocker arm and the pull rod is made flexible and when the float chamber is full, the diaphragm is not operated though the camshaft is running all the time. Fig. 10.3 gives a photographic cut-away view of the pump.

Fig. 10.3. Partial cutaway view of A.C. Mechanical fuel pump

Reliability is the main advantage of mec- hanical pumps. However, these have some disadvantages also:

1. They have to be situated close to the engine due to which they are exposed to engine heat, which may result in vapour locking in the fuel supply system.
2. They operate only *after* the engine has started.

3.2. S.U. Electrical Pump

In this type also a diaphragm is used. Alternate vacuum and pressure are produced due to the movement of the diaphragm which is caused electrically in this case. Closing the ignition switch energises the solenoid winding, magnetic flux is generated which pulls the armature to which the diaphragm is attached. Thus the diaphragm moves to cause suction in the pump chamber and the fuel is drawn into the chamber. But as soon as the armature moves, it interrupts the electric supply by disconnecting the breaker points, the solenoid is de-energised and the armature falls back, causing the diaphragm to move to create pressure in the pump chamber which opens the outlet valve and the fuel goes out to the carburettor float chamber. This movement of the armature, however, completes the circuit again and the solenoid again gets energized. The whole cycle is again repeated in this way and the fuel continues to be pumped.

Fig. 10.4. Partial cutaway view of S.U. Electrical Fuel Pump.

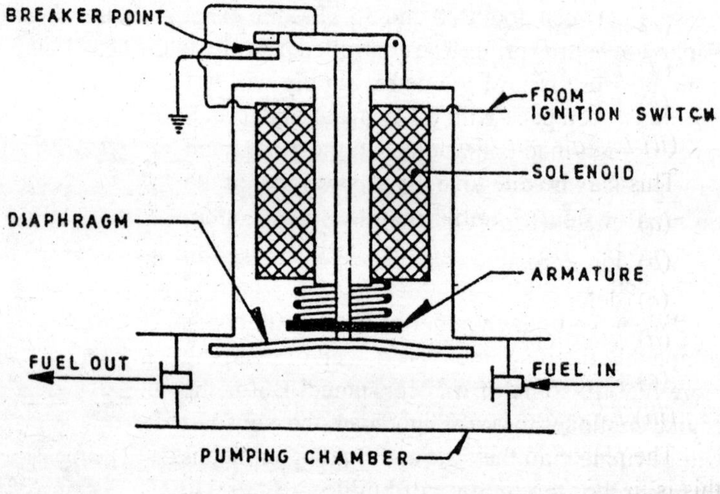

Fig. 10.5. S.U. Electrical Pump

Electric pumps need not be situated necessarily close to the engine; mostly these are located near the fuel tank. Thus these are not subjected to engine heat. Further an electrical pump need not wait for the engine to start. It starts operating immediately as the ignition is switched on.

4. FUEL PUMP TESTING

There are three main tests which are performed to judge the performance of a fuel pump. These are, pressure test, volume test and vacuum test.

Each pump must produce certain pressure on the outlet side, which is specified by the manufacturer. To test a given pump for pressure connect a suitable pressure gauge in between the pump and the carburettor and run the engine at specified speed, say 500 r.p.m. If the gauge indicates pressure which is lower or higher than the specified, even after checking that there is nothing wrong with pipes and connections etc., the pump has to be replaced or repaired and checked again to see that it gives the specified pressure reading only.

The volume test is performed by disconnecting the pump from the carburettor and measuring the discharge separately by means of a graduated container and stop watch, while the engine is running at idling speed. It is to be noted here that the engine in this case will be run only by the petrol in the carburettor float chamber. If the observed rate of flow is less than the specified one, the pump needs repairs or replacement.

The fuel pump should obviously create some vacuum on its inlet side, which is specified by the manufacturer. To test it, connect a suitable vacuum gauge in between the fuel tank and the pump and run the engine at idle speed. The reading of the vacuum gauge must conform to the specifications. Now stop the engine and see for how much time the vacuum is held in the lines. The vacuum must be retained for at least 10 seconds after closing the engine, otherwise the pump has to be taken out for repairs or replacement.

5. FUEL PUMP TROUBLE SHOOTING

1. Mechanical fuel pump

(*i*) *No fuel delivery*

The probable reasons for this defect are:

(*a*) There may be no fuel in the tank even if the fuel gauge indicates to the contrary, because the gauge may also be defective.

(*b*) The dome on the pump may be loose, thus allowing air to be drawn in.

(*c*) Air may be leaking in from the loose connections between pump and fuel tank.

(*d*) Fuel pipe may be blocked.

(*e*) Pump diaphragm may be damaged.

(*ii*) *Insufficient fuel delivery*

This may be due to:

(*a*) chocked fuel filter in the pump.

(*b*) loose dome on the pump.

(*c*) defective or dirty pump valves.

(*d*) worn out pump linkage.

(*e*) weak diaphragm spring.

(*iii*) *Excessive fuel delivery*

The reason for this may be too stiff a diaphragm spring. However, it is more likely that the cause for this is in the carburettor.

2. Electrical fuel pump

(*i*) *Pump fails to operate.*

The probable reasons for this are:

(*a*) The electric connection may be loose or partly missing even.

(*b*) Earth points may have become dirty.

(*c*) Contact points in the pump may be dirty or eroded.

(*d*) Suction pipe may be blocked.

(*ii*) *Pump operates but delivers insufficient fuel.*

This may be due to:

(*a*) chocked pump filter.

(*b*) worn out or dirty valves.

(*c*) damaged diaphragm.

6. FUEL PUMP SERVICE

1. Disconnect the fuel inlet and outlet lines. Open out the connecting bolts of the pump after rotating the engine flywheel so that the cam exerts least pressure on the rocker arm. Remove the pump.

2. Close all the external openings in the pump by plugging these temporarily and wash it thoroughly in the petrol, so that it is absolutely clean from outside.

3. With suitable tools, open out the pump. Clean and inspect each component separately. Repair or replace any part found defective.

4. Reassemble the whole pump and install it on the engine using a new gasket. Connect the fuel lines.

7. VAPOUR RETURN LINE

The petrol under the heat of the engine, tends to form into vapour. An additional factor aiding the formation of vapour in the fuel pump is the creation of vacuum during suction of fuel, which lowers its boiling point. To guard against this, a vapour return line, connecting a special outlet of the fuel pump to the fuel tank, is provided in some engines. Any vapour formed in the pump goes back to the fuel tank where it condenses. Thus the vapour does not go to the carburettor side. This arrangement also helps to decrease the formation of vapours, because in addition to vapours, the excess fuel pumped by the fuel pump is also returned to the fuel tank. This continual circulation of the fuel keeps the pump cool and thus avoids the vapour formation.

8. AIR CLEANERS

As hundreds of cubic metres of air per hour are used by the engine of an automobile, it is very important that this air should be very clean. Impurities like dust in the air cause a very rapid wear of the engine, particularly of the cylinders, pistons, rings, valves and guides. Further if the dirty air enters the crankcase, it will contaminate the lubricating oil and ultimately damage the bearings and journals and decrease the service period of the lubrication system. It is, therefore customary to install air cleaner on the intake system of automotive engines. Apart from filtering the air, air cleaner also performs other functions:

(*i*) It acts as a silencer for the carburation system *i.e.*, it reduces the engine induction noise to an acceptable level.

(*ii*) In case the engine back-fires, the air cleaner also acts as a flame arrester.

Air cleaners offer, however, a resistance to air flow which is increased as the air cleaners get clogged with dirt. Consequently air flow would get decreased, resulting in sluggish engine performance and higher consumption of fuel. These should, therefore, be cleaned regularly or replaced periodically, say, every 20,000 km.

The air cleaners generally used are of two types:

1. Heavy duty type
2. Light duty type

8.1. Heavy duty type air cleaner

This is of oil bath type (Fig. 10.6). It contains a filter element *C* saturated with oil. At the bottom there is separate oil pan *D*. The air from the atmosphere

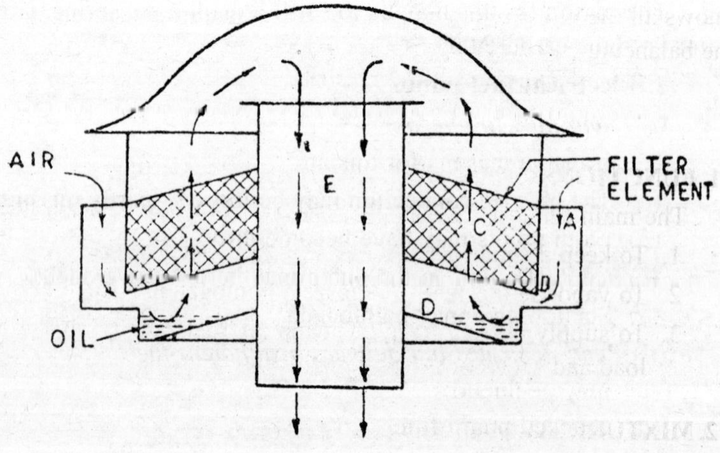

Fig. 10.6. Heavy duty air cleaner.

enters through circumferential gap A . At the corner B when the air takes a turn, it leaves large particle impurities there. Next, impinging on the surface of the oil relieves the air further of impurities. Final cleaning is done by means of filter C and the clean air passes through passage E as shown.

8.2. Light duty types air cleaner

Fig. 10.7 shows a light duty type of air cleaner. It consists of a cleaning element only. The element consists of a cylindrical cellulose fiber material, over which is put a fine mesh screen to provide strength. The element is corrugated to increase the surface area exposed to the incoming air so that the resistance offered by the air cleaner is reduced to minimum. Sides of the element are sealed against dust. The air passes through the element and any dust contained is left outside.

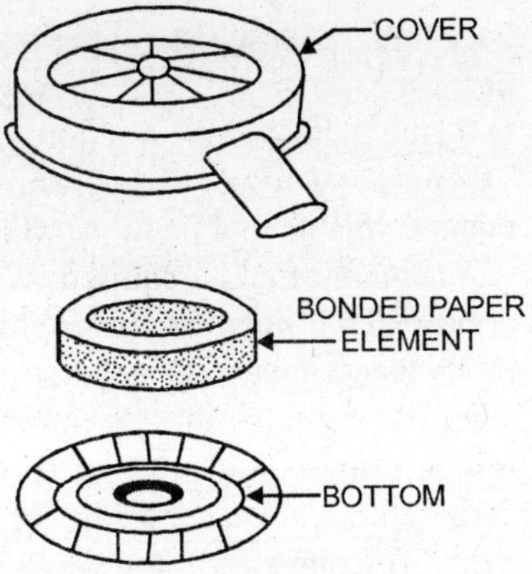

Fig. 10.7. Light duty air cleaner

8.3. Thermostatic control of air cleaner

In modern engines, lean fuel-air mixture is employed to reduce air pollution. However, such an engine would not perform efficiently when cold, *e.g.*, during cold start or idling. To remedy this incoming air is heated and then sent to the air cleaner having a thermostatic control, sensitive to a spring which controls an air bleed valve depending upon the temperature of the air entering the carburettor. When the engine is cold, the bleed valve starts opening as a result of which some cold air from the outside mixes with the hot air, thus decreasing its temperature. At about 40°C, the bleed valve opening is maximum when the hot air supply is cut off and only outside air enters the carburettor.

9. FUEL FILTERS

The most commonly used filter for cleaning petrol is the fine mesh gauge. It has worked very well, where large dust particles are involved, but has not proved much effective in preventing the fine particles and the water from going inside the cylinder. A very simple and effective device used is the ordinary chamois leather, which if first moistened with petrol will allow only the petrol to pass through it, and the water will be intercepted. Fine grit, of course, cannot pass through it.

10. FUEL GAUGE

Fuel gauge is an indispensable instrument which must be fitted on the dashboard so that the driver knows all the time how much fuel he has consumed and how much is left in the tank. The usual gauge is the balancing coil type. The construction is exactly similar to that of the oil gauge.

CARBURETTORS

11. FUNCTIONS OF A CARBURETTOR

The main functions which a carburettor is required to perform are:

1. To keep a small reserve of fuel at a constant head.
2. To vaporise the fuel to prepare a homogeneous air fuel mixture.
3. To supply correct amount of the air fuel mixture at the correct strength under all conditions of load and speed of the engine.

12. MIXTURE STRENGTH REQUIREMENT

Fig. 10.8 shows the air fuel ratio required at different engine speeds. For start the mixture required is very rich, because the engine is cold at that time due to which the fuel does not vapourize properly. Only the light, volatile fractions of the fuel would vapourize at lower temperatures. Thus to get enough volatile fuel fractions while starting a cold engine, more fuel is required. While idling there is no external load on the engine, which has to produce just the frictional horse power. However, since the engine is still not running hot at normal temperatures, and the airflow through the carburettor is slow, some liquid fuel drops out of the air-fuel mixture onto the manifold walls and condenses there. To compensate for this, a rich mixture is required, though not as rich as required for starting. For normal part throttle operation at cruising speeds a comparatively lean mixture, about 15:1 air fuel ratio, will suffice. In fact 14.6:1 mixture, *i.e.*, 14.6 parts of air and 1 part of gasoline or petrol (by mass) is the one which would give chemically complete combustion of petrol under ideal conditions. Such a mixture is called the *stoichiometric* mixture. By volume this would be about 60 parts of air to 1 part of petrol. At cruising speeds, 15:1 mixture would provide fuel economy. In fact in the present day emission-controlled engines, mixtures as lean as 18:1 have been used at cruising speeds, when operating at sea level. A rich mixture on the other hand, has more energy to produce power. Therefore, at nearly full throttle, (at high speeds or even at normal speeds when going uphill), a rich mixture will have to be supplied. Apart from this, a rich mixture is required during the acceleration period to provide the necessary power for increasing the vehicle speed.

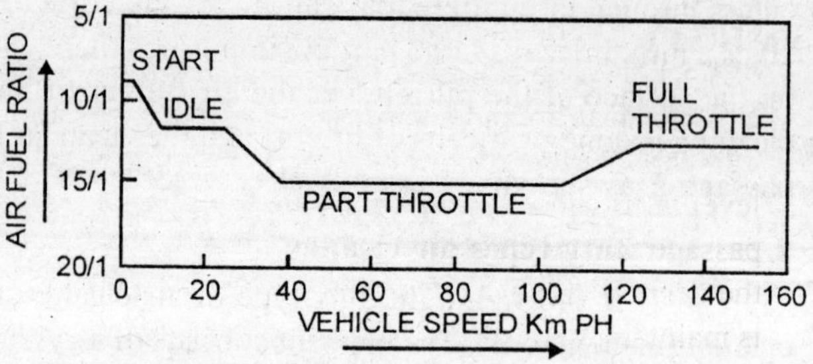

Fig. 10.8. Mixture strength requirement

Weaker mixtures than the optimum required under given conditions produce less power though affecting economy. Engine misfiring, overheating and incomplete combustion result if adequate ignition timing advance is not provided to compensate for the slow combustion of weak mixtures. Rich mixtures on the other hand provide more power but are uneconomical. Mixtures richer than the optimum required result in excess of carbon monoxide in the exhaust which is black in colour. Moreover continuous use of richer mixtures causes the carbon deposits to form on spark plug electrodes and the combustion chamber walls.

13. SIMPLE CARBURETOR

To understand the principle on which a carburettor works, consider a simple carburettor as shown in Fig. 10.9. The main parts are a float chamber, fuel jet, venturi, nozzle and a throttle valve. The float in the float chamber is made of deep drawn brass sheet and its kept hollow for lightness. Such floats have a tendency to leak along the joint seams. Due to this reason floats are now made of nylon plastic or expanded synthetic rubber. The needle valve attached to the float lever serves to close or open the fuel inlet to the float chamber

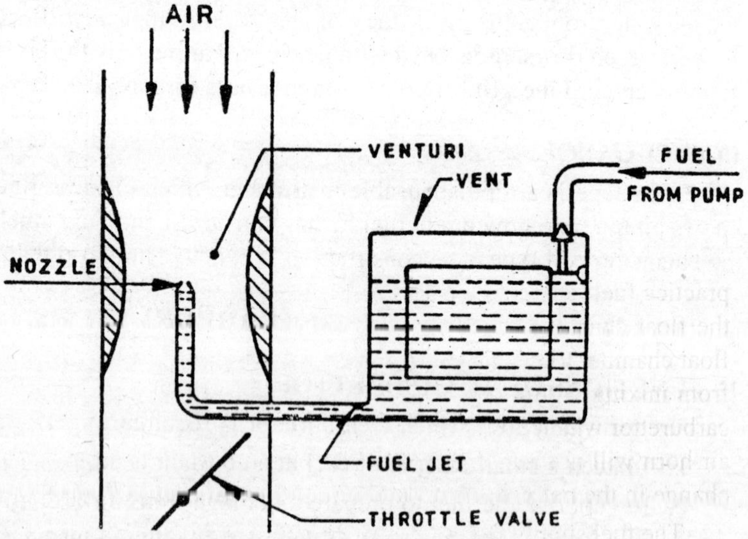

Fig. 10.9. Simple Carburettor

depending upon the requirements. The needle valve consists of a cylindrical stem with a conical tip made of steel or else a solid steel stem with a rubber seat tip. Alternatively, there may be a three-piece valve with a rubber seat tip and a spring-loaded ball in the body of the stem. The later two types maintain the liquid-tight seal during vibrations, which is not possible with ordinary single piece needle valve. When the fuel level falls below a definite predetermined value, the float also falls alongwith fuel level, thus opening the passage for the fuel supply. The fuel starts flowing in and the float rises gradually till the fuel level reaches the desired value. At this time, the float needle closes the fuel inlet passage. Thus a constant head of fuel is maintained in the float chamber. This constant level of fuel is slightly below the nozzle outlet, so that the fuel may not drop all the time from the nozzle, even when the engine is not working. This provision also prevents the fuel from spilling out when the car is tilted on account of hilly or highly cambered road. In

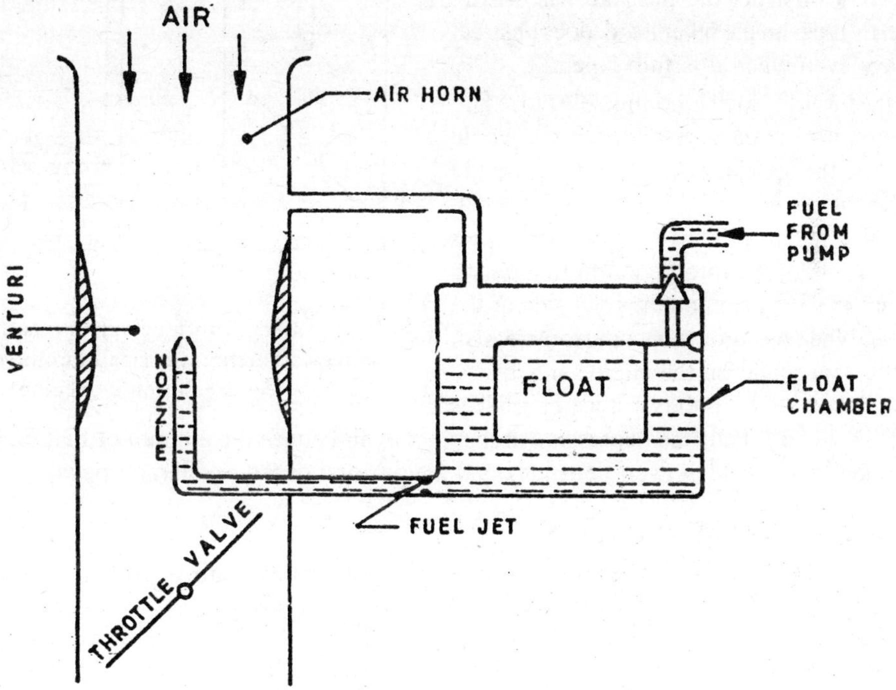

Fig. 10.10. Simple carburettor with float chamber vented to air horn.

practice fuel level in the float chamber is maintained about 5 mm below the nozzle outlet. A small vent in the float chamber keeps the pressure inside atmospheric. In modern practice it is preferred to vent the float chamber to the air intake of the carburettor (Fig. 10.10). Such an arrangement prevents dust particles from mixing with the petrol through the vent as there is always an air cleaner on the intake side of the carburettor which filters the incoming air. Moreover, in case of blocking of air cleaner, the pressure at the air horn will not remain atmospheric. In case the float chamber is vented to atmosphere, this may cause change in the net pressure causing fuel flow from the nozzle and hence faulty working of the carburettor.

The fuel supply from the float chamber at any time is metered by means of the fuel jet from where the fuel flows to the venturi through the discharge nozzle. The venturi is simply a restriction in the air passage. It is so called after the italian engineer G.B. Venturi famous for his studies on the fluid flow. Here the passage area is minimum. Thus due to less area, the air velocity increases and because of this increase in velocity, decrease in pressure is caused at the nozzle which is located in venturi itself. Due to this depression being applied at the nozzle, the fuel comes out and is vaporized by the coming air stream. The mixture then goes through the inlet manifold, to the engine cylinders. The amount of fuel delivered depends upon the jet size, float level and venturi vacuum.

The purpose of a throttle valve is to control the quantity of air fuel mixture. It is attached to the accelerator pedal by means of suitable linkage so that when the pedal is depressed, the valve opens out. The throttle valve used is of two types, the butterfly valve and the cylindrical valve. The butterfly valve is more commonly used because of its simplicity; it is merely a disc hinged at the centre. The cylindrical type, however, is advantageous in that it does not obstruct the passage when fully open. Butterfly type, on the other hand, does obstruct some passage even when it is fully open.

When the throttle valve is completely closed no engine suction is applied to the nozzle; the whole of the suction in that case, is acting below the throttle valve, in the region of the inlet manifold. A small throttle opening would still have a large depression on the engine side of the throttle permitting only a small amount of depression on the inlet side of the throttle valve. When the throttle opening is increased, however, the depression on the engine side of the throttle is reduced while in the venturi region it is increased (Fig. 10.11). Thus the depression at the venturi and hence the amount of fuel flow is controlled by the throttle opening which itself depends upon the position of the accelerator pedal.

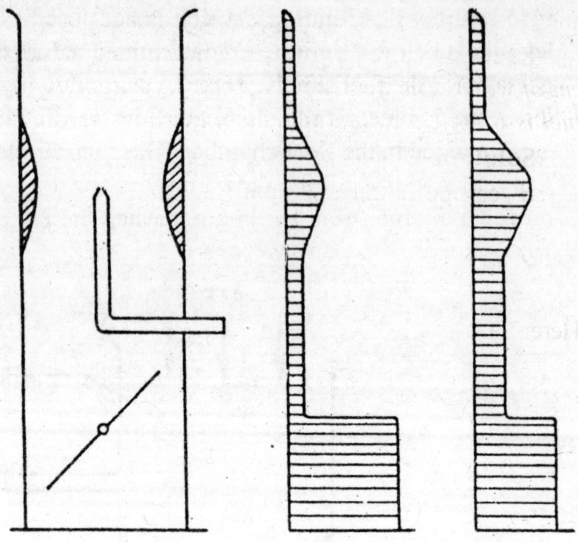

Fig. 10.11. Effect of throttle opening on the depression in the carburettor. (*a*) small throttle opening, (*b*) large throttle opening.

14. HYDRAULICS OF CARBURATION

It can be derived from the hydraulics principles that rate of flow of air W_a and of fuel W_f are given by,

$$W_a = A_v . C_a \sqrt{2\rho_a \Delta P_a} \quad ...(10.1)$$

and

$$W_f = A_f . C_f \sqrt{2\rho_f \cdot \Delta P_f} \quad ...(10.2)$$

where

A_v = Area of venturi, m^2

A_j = Area of jet, m^2

C_a = Coefficient of discharge for air

C_f = Coefficient of discharge for fuel

ρ_a = Density of air (constant, the air being assumed to be incompressible), kg/m^3

ρ_f = density of fuel, kg/m^3

ΔP_a = pressure differential causing air flow, N/m^2 (Pa)

= Atmospheric pressure – pressure at the venturi throat.

ΔP_f = pressure differential causing fuel flow, N/m^2 (Pa)

$= \Delta P_a - gh\rho_f$

h = height of nozzle above the fuel level in the float chamber, m

Note—C_a value for carburettor venturis may be assumed to be 0.83, which remains fairly constant under all conditions except that for very low velocities (below 45 m/sec.), it decreases. However, C_f depends upon various factors, *viz.*, shape, proportions of entrance to fuel jets, viscosity and temperature of fuel and rate of flow. Depending upon different factors it may vary from 0.6 to above unity.

Example 10.1. *In the carburettor of an automotive engine, the venturi depression is found to be 102 cm of water. The diameter of jet is 1 mm, coefficient of discharge for fuel is 0.65 and specific gravity of fuel is 0.78. Calculate the rate of flow of the fuel through the jet.*

Solution.

W_f can be calculated from Eq. (10.2)

$$W_f = A_f . C_f \sqrt{2\rho_f \cdot \Delta P_f} \qquad ...(10.2)$$

Here

$$A_f = \frac{\pi}{4}(1\times10^{-3})^2 = 0.785\times10^{-6}\,\text{m}^2$$

$$C_f = 0.65$$

$$\rho_f = 0.78\,\text{gm/cm}^2 = 780\,\text{kg/m}^3$$

$$h_w = 102\,\text{cm} = 1.02\,\text{m}$$

$$\rho_w = 10^3\,\text{kg/m}^3$$

$$\Delta\rho_f = gh_w\rho_w = 9.81\times1.02\times10^3 = 10006\,\text{N/m}^3$$

∴ Substituting in Eq. (10.2),

$$W_f = .785\times10^{-6}\times0.65\sqrt{2\times780\times10006}$$

$$= 2.016 \times 10^{-3}\ \text{kg/sec} = \mathbf{2.016\ gm/sec.}$$

Example 10.2. *Calculate the size of the venturi relative to the jet size in a single jet carburettor if it is to deliver an air/fuel mixture in the ratio 15:1. Assume* ρ_f = *750 kg/m³*, ρ_a = 1.3 kg/m³, *Ca* = *0.8 and* C_f = *0.65.*

Solution.

From Eq. (10.1) and (10.2)

$$\frac{W_a}{W_f} = \frac{A_v}{A_j}\cdot\frac{C_a}{C_f}\sqrt{\frac{\rho_a\cdot\Delta\rho_a}{\rho_f\cdot\Delta\rho_f}} \qquad ... (a)$$

Assuming the jet is located at the level of fuel in the float chamber,

$$\Delta P_f = \Delta P_a$$

Then the Eq. (*a*) is reduced to,

$$\frac{W_a}{W_f} = \frac{A_v}{A_j}\cdot\frac{C_a}{C_f}\sqrt{\frac{\rho_a}{\rho_f}}$$

Substituting various values,

$$15 = \frac{A_v}{A_j}\cdot\frac{0.8}{0.65}\sqrt{\frac{1.3}{750}}$$

$$\frac{A_v}{A_j} = 292.73$$

$\therefore$ Ratio of diameters $\left(\frac{d_v}{d_j}\right) = \mathbf{17.1}$

Example 10.3. *In a carburettor the petrol used has density 750 kg/m³ and the discharge nozzle is located 6 mm above the level of petrol in the float chamber. The coefficient of discharge/velocity for air is 0.85 and density of air may be assumed as 1.3 kg/m³. Calculate the critical air velocity.*

Solution.

For critical conditions *i.e.* when the fuel flow just starts.

$$\Delta P_f = 0$$

i.e.
$$\Delta P_f - gh\rho_f = 0$$

Substituting values, $\Delta P_a - 9.81 \times (6 \times 10^{-3}) \times 750 = 0$

$\therefore$
$$\Delta P_a = 44.145 \text{ N/m}^2$$

Now mass rate of air flow, $W_a = \rho_a \cdot A_v \cdot V_a$

$\therefore$
$$V_a = \frac{W_a}{\rho_a \cdot A_v}$$

$$= \frac{A_v . C_a \sqrt{2\rho_a \Delta P_a}}{\rho_a \cdot A_v} \quad \text{(From Eq. 10.1)}$$

$$= C_a \sqrt{\frac{2\Delta p_a}{\rho_a}}$$

Substituting values, the critical air velocity is given by,

$$V_a(\text{crit}) = 0.85\sqrt{\frac{2 \times 44.145}{1.3}} = \mathbf{7.005\,m/sec.}$$

15. DEFECTS IN THE SIMPLE CARBURETTOR

The simple carburettor as explained above, would work well if the engine were to run at one particular speed and load. The nozzle and the venturi sizes then can be set once and the carburettor will then work satisfactorily at that particular speed and load. But in actual practice, the engine has to run at different speeds and loads ranging from the lowest to the highest and, therefore, certain irregularities creep in the functioning of this simple carburettor. The two basic reasons for defects are:

1. If the carburettor is designed to work at high speeds at full throttle, it will not work at low speeds and part throttle, reason being that in the later case the suction created at the venturi will not be sufficient to draw fuel from the nozzle. Similarly if the carburettor is designed to work at low speeds and part throttle, it would deliver richer mixture at high speeds and full throttle.

2. The coefficients of discharge for air and fuel vary in different amounts with the change of depression

as shown in Fig. 10.12. Whereas the coefficient of discharge for air becomes almost constant at certain value of depression, the coefficient for fuel increases gradually. This means that at higher heads, *e.g.*, at higher engine speeds, the air fuel mixture goes on becoming richer and richer. Therefore, if the carburettor is set at one particular speed, it will give richer mixture at higher speeds and leaner mixture at lower speeds.

The various defects thus experienced and their remedies are discussed in the following paragraphs.

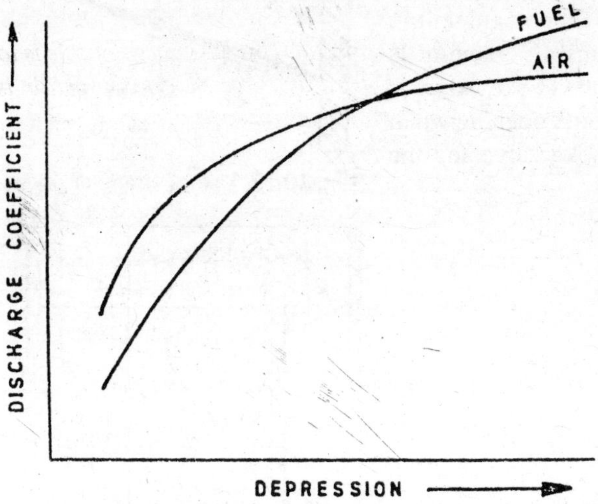

Fig. 10.12. Variation of discharge coefficient with pressure head.

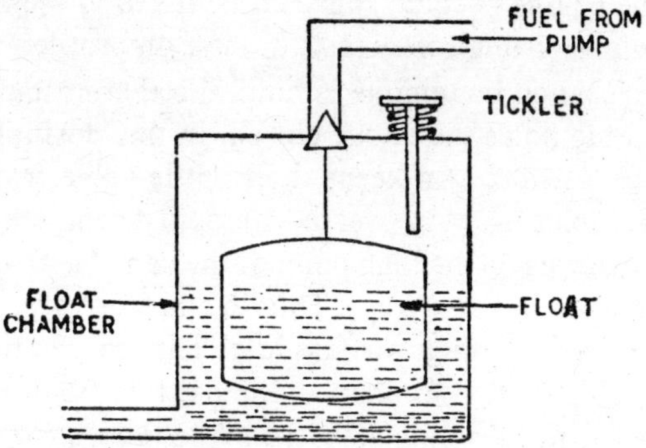

Fig. 10.13. Tickler.

1. Starting difficulty

For starting, it is seen that very rich mixture is required, but actually the mixture provided by the simple carburettor will be very lean, because of the reasons already explained. The different methods to provide for enrichment of fuel at very low speeds are:

(*a*) Ticklers. These are devices used to cause flooding of carburettor at the start (Fig. 10.13). By depressing the tickler, the float is depressed, thereby providing more fuel. These were commonly used in the old motor cycle carburettors.

(*b*) Choke. This is a simple butterfly valve fitted at the top of air horn. It may be operated by hand or it may be automatic. For starting, choke is closed so that very small amount of air gets past it and the throttle valve is open due to which whole of the suction is applied to the nozzle, which delivers sufficient fuel to provide a mixture rich in quality, though small in quantity.

The choke valve must be opened immediately when the engine starts, otherwise engine cylinder will be flooded with fuel which would cause it to stall. This is helped by any of the following methods:

(*i*) by mounting the choke eccentrically (Fig. 10.14). With this, when the engine starts, the forces due to pressure on the two sides of the choke spindle are unequal, their difference producing a turning moment to open the choke.

(*ii*) by the use of strangler valve fitted on the choke (Fig. 10.15). As soon as the engine is started, the air pressure forces open the strangler against the spring pressure and

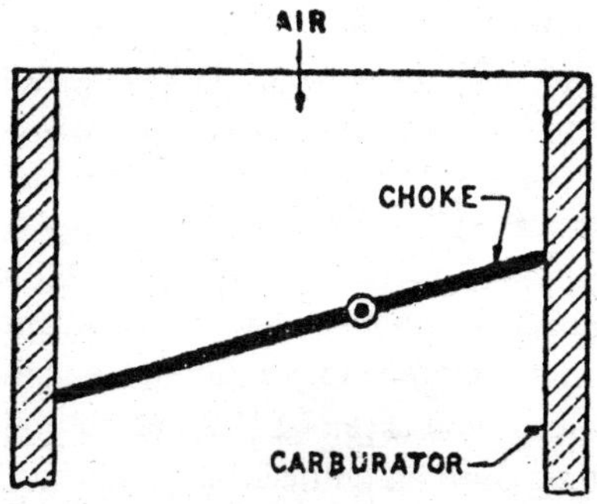

Fig. 10.14. Eccentric mounting of choke.

the air enters the carburettor avoiding the mixture being over-rich. By this time the driver also presses the choke knob to open the choke valve.

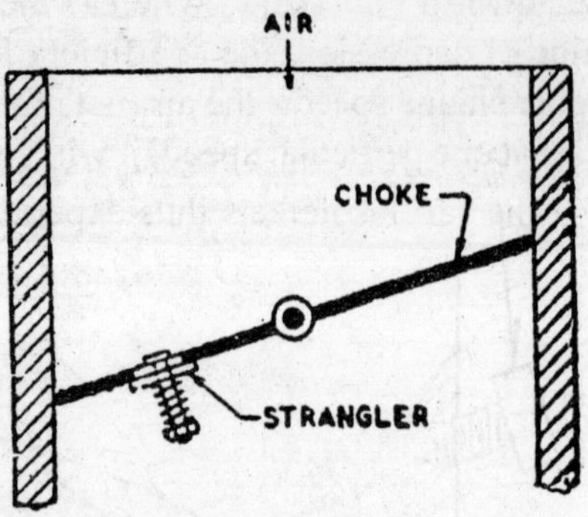

Fig. 10.15. Use of strangler in choke.

An automatic choke employed in modern automobiles is shown in Fig. 10.16. The choke valve is mounted eccentrically. A bimetal thermostatic coil located on the exhaust manifold to quickly sense the heat, is linked to the choke valve. A vacuum diaphragm unit in which the diaphragm is subjected to inlet manifold vacuum is provided. The diaphragm itself is connected to the choke valve. A fast-idle cam is provided to maintain a small throttle opening when the engine is cold. This cam is linked to the choke valve and the idle adjusting screw rests in the cam profile.

When the engine is cold, the thermostatic spring holds the choke closed. In this position, the fast-idle cam keeps the throttle valve from fully closing. Moreover the idle adjustment screw now rests on the high point of the cam due to its rotation by the linkage. Thus the throttle valve is held partly open for fast idle. This causes the engine suction to draw additional fuel from the main nozzle and also from the idle system (described later) at fast idle speed. Moreover while cranking, the manifold vacuum also opens the choke slightly to allow some air to mix with fuel.

After the engine is started the vacuum diaphragm pulls the choke valve open somewhat. This tendency is helped by eccentric mounting of the choke valve. As the engine warms, the thermostatic spring gradually winds up and allows the choke valve to open. As the choke opens the fast-idle cam is released step by step till when the engine is warm, the choke is open wide and the fast-idle cam is fully released.

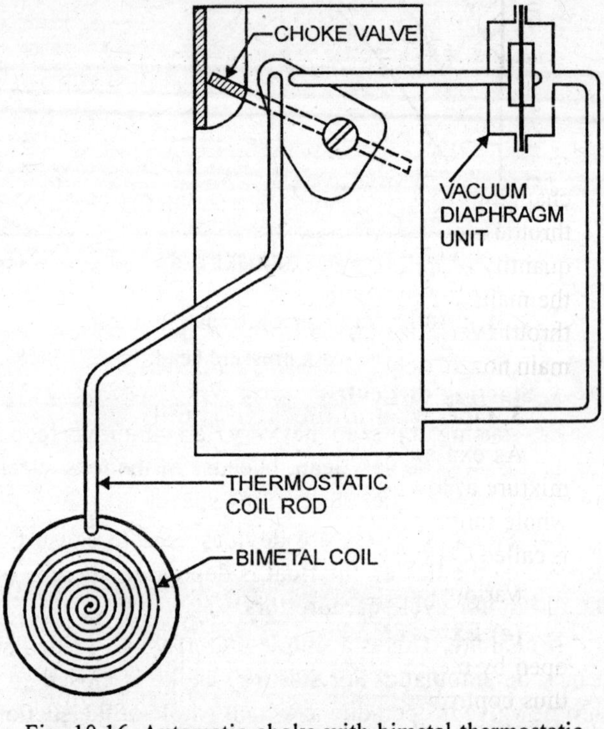

Fig. 10.16. Automatic choke with bimetal thermostatic spring and vacuum diaphragm unit.

(c) Adjustable area jet. A long tapered needle is used which is screwed into the jet as shown (Fig. 10.17). For starting, the screw is loosened, so that the jet area providing fuel is increased and thus rich mixture is provided for the start. This method has been used on some constant vacuum carburettors.

2. Idling difficulty

During idling, when the engine runs at low speed and the throttle is closed, the suction on the main nozzle is still insufficient to draw fuel from there. Therefore a separate supply circuit of fuel must be provided on the engine side of the throttle valve during idling. This is done by providing a separate idle jet and an air bleed hole (Fig. 10.18), so that metered quantities of air and fuel are drawn into the idle passage where they form an emulsion. This mixture then comes out through the idle part which is located on the engine side of the throttle. The volume of the mixture and hence the idle speed is controlled by the idle adjustment screw. To avoid a 'flat spot' when the engine is accelerated from idling to the main running condition,

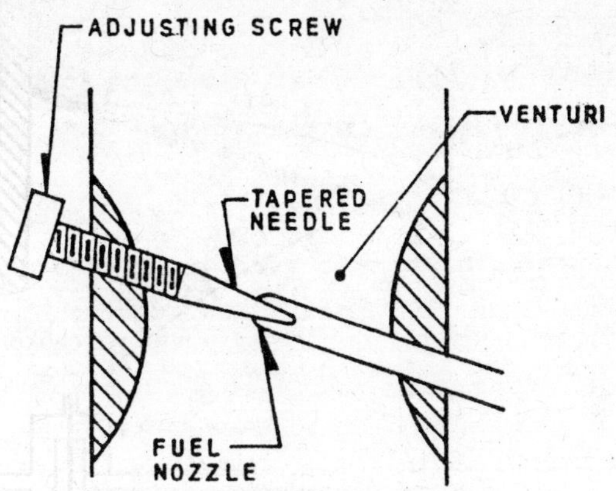

Fig. 10.17. Adjustable area jet.

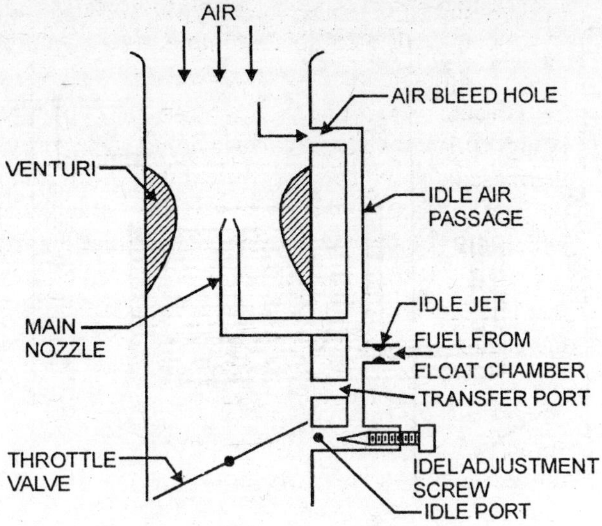

Fig. 10.18. Idling circuit.

changing from the idle circuit to the main nozzle circuit, transfer ports are provided as shown. When the throttle is opened slightly, the transfer port is also exposed to the manifold depression which causes additional quantity of mixture to enter the engine. This provides a progressive smooth transition from the idling to the main circuit. For this reason, the transfer ports are also called progression holes. When, however, the throttle valve is opened further the idle circuit stops supplying the mixture which is now supplied by the main nozzle only.

3. Operation at different speeds

As explained earlier, a simple carburettor would provide richer mixture at high speeds and weaker mixture at low speeds. The process of adjusting the mixture strength at all speeds so that throughout the whole range correct proportions of air fuel mixture are maintained is called *Compensation.*

Various devices used for providing compensation are:

(*a*) **Extra air valve.** This is a spring-loaded valve arranged to open by means of the engine suction as shown in Fig. 10.19. It is thus controlled by the spring stiffness and the engine suction and therefore, its operation has been found to be satisfactory.

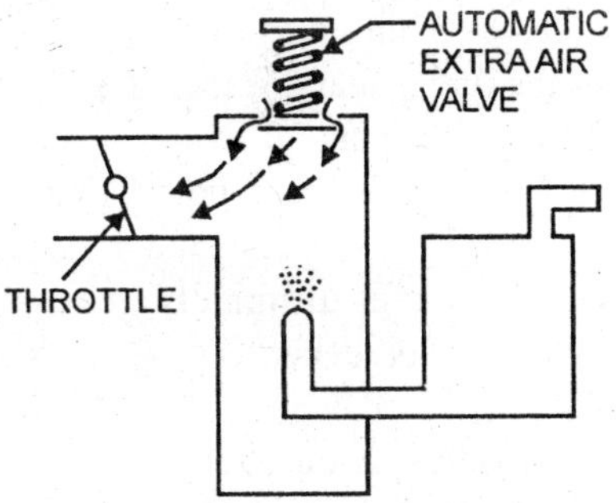

Fig. 10.19. Extra air valve

(*b*) **Compensating jet.** In this there is an additional fuel supply nozzle (Fig. 10.20). *A* is the main jet and *B* is the compensation jet. *C* is the well which is vented to atmosphere so that when the engine is not running, the fuel level in *C* is same as in the float chamber and under this condition no flow of fuel takes place through the compensating jet.

The main nozzle *D* is connected directly to float chamber and it will supply richer mixture at larger throttle openings. But the case is different of the compensating jet. When the engine is just started and the throttle valve is open a bit, small engine suction is applied, which draws fuel from the well *C* through delivery nozzle *D*, till the well *C* becomes empty. At this point, the flow through *D* is maximum. Further opening the throttle, therefore, does not increase the fuel flow through the compensating jet, it remains constant. However, the air flow is nevertheless increasing. As a result the higher speeds results in weaker mixtures through the compensating jet. The result is shown in Fig. 10.21. Thus a uniform mixture of

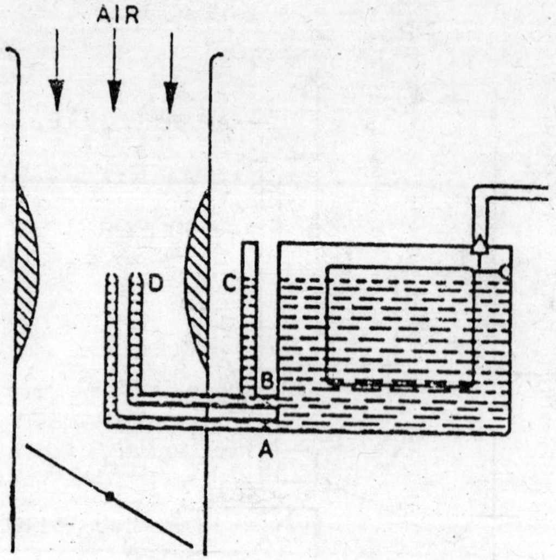

Fig. 10.20. Compensating jet.

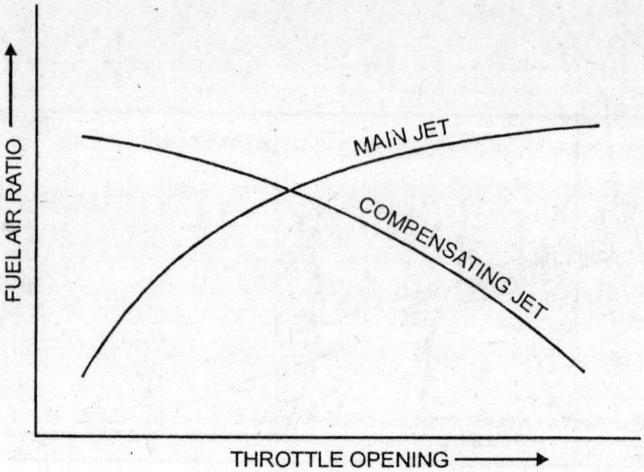

Fig. 10.21. Effect of compensating jet.

nearly constant strength is obtained at different throttle openings. The mixture strength when the petrol well is emptied depends mainly on the size of the petrol jet. This method, however, tends to overcompensate at large throttle openings. Zenith carburettor uses this method for compensation.

(*c*) Air bleed compensation. Fig. 10.22 shows the detailed construction of a carburettor nozzle with provision of air bleed compensation. *A-A* is the fuel level in the nozzle under no load conditions. *B* is the jet tube containing air bleed holes around its periphery as shown. *C* are the holes communicating the air bleed holes to the atmosphere when the nozzle is partly empty.

As the throttle is opened gradually, due to depression at the venturi, the fuel in the tube and the nozzle around is consumed, providing gradually richer mixture. Simultaneously the level of fuel in the jet tube and nozzle drops, exposing the uppermost holes in the jet tube to put them in direct communication with the atmosphere through holes *C*. This destroys the depression to some extent and the fuel flow is thus decreased. As the fuel demand is increased by the engine, the fuel level in the jet tube and around drops further, so as to expose the next row of air bleed holes, thus increasing the amount of air bleed to compensate the increased fuel flow from the jet. The fuel level ultimately gets stabilized. The degree of compensation depends upon the size and number of the air bleed holes and their location. The method has been employed in the Solex carburettor.

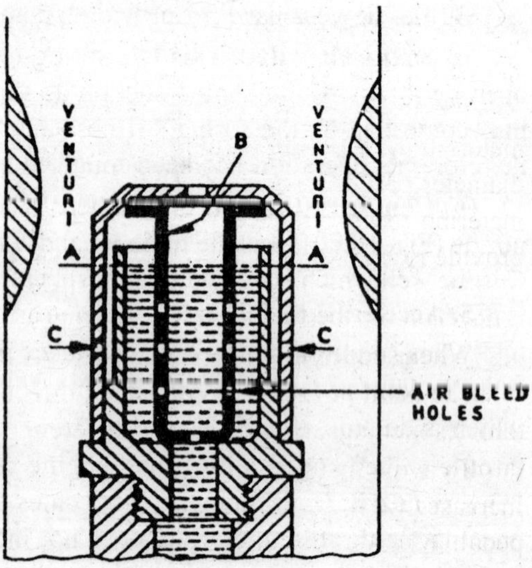

Fig. 10.22. Air bleed compensation.

(*d*) Multiple jet compensation. In this three to five jets are provided in the carburettor. A cap *E* serves to open different jet passages in turn. First when the throttle opening is small, a smaller amount of suction is applied and only nozzle 1 is then open to engine suction. The fuel is now provided by nozzle 1 only. When the throttle valve is opened further, more suction moves cap *E* still further subjecting nozzle 2 also to engine suction alongwith nozzle 1. Now both nozzles 1 and 2 will deliver the fuel. But nozzle 2 is so adjusted that it gives lesser amount of fuel than nozzle 1, thus providing compensation. Similarly, nozzles 3 and 4 which deliver still lesser amount of fuel, are put into operation by further throttle opening.

(*e*) Suction controlled devices. These devices are operated by means of engine suction, which is applied to a sliding piston. The suction effect is increased as the engine speed is increased, which controls the movement of the piston which further controls the venturi area. Sometimes the engine suction is also used to actuate a needle which decreases the effective nozzle area as the speed is increased, thus providing compensation. This method is used is the S.U. Carburettor, which has been discussed in detail in Art 23.

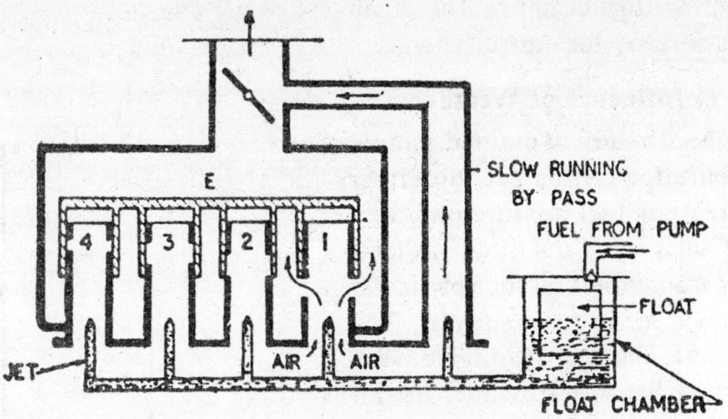

Fig. 10.23. Multiple jet compensation.

4. Difficulty at high speeds

Weak air-fuel mixtures supplied by the single jet carburettor will not give enough power at high speeds. Therefore some special system is required to enrich the mixture at high speeds. One such system uses a metering rod with stepped diameter end in the main jet (Fig. 10.24). At ordinary cruising speeds, the larger diameter part of the metering rod is in the jet which gives less fuel flow. At higher speeds, however, the metering rod is pulled up so that now smaller diameter part is in the jet. This increases the fuel flow to provide rich mixture.

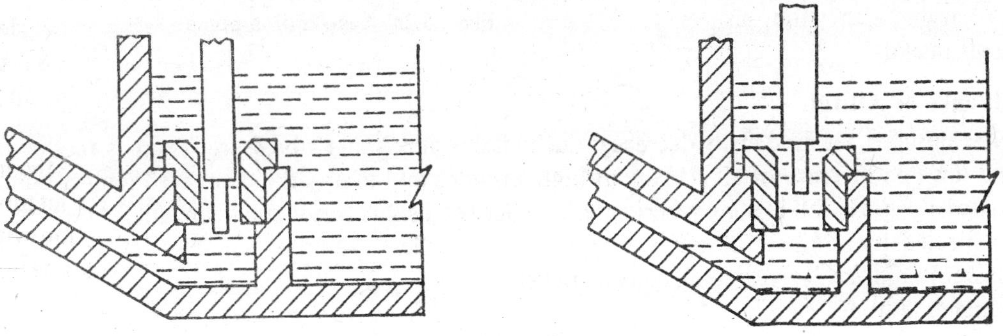

(*a*) (*b*)

Fig. 10.24. Metering rod in the jet.

5. Acceleration difficulty

When sudden acceleration is desired, the throttle is opened suddenly. This causes the maximum amount of air to come at once but the fuel supply lags thereby causing what is called *engine stumble* or *hesitation* which is due to weak mixture.

To avoid this hesitation a separate pump is used (Fig. 10.25) which provides fuel momentarily, till the increased fuel supply from the main nozzle starts. The pump is connected through linkage to the accelerator pedal. When the accelerator pedal is pressed, the outlet valve is opened and the fuel is forced out of the acceleration jet. When the pedal is released, however, the piston moves up thereby sucking the fuel in from the float chamber. The pump is thus ready for next discharge.

6. Influence of Weather

As already explained simple carburettor can be set to deliver correct air fuel mixture only at a particular speed and load. Further if the carburettor is set for a particular weather, for instance summer, then in winter months it will give weak mixture because in winter the air density increases to a large extent as compared to the increase in the fuel density. As a result the mixture becomes weak Similarly, a carburettor set for winter months will give too rich a mixture in the summer.

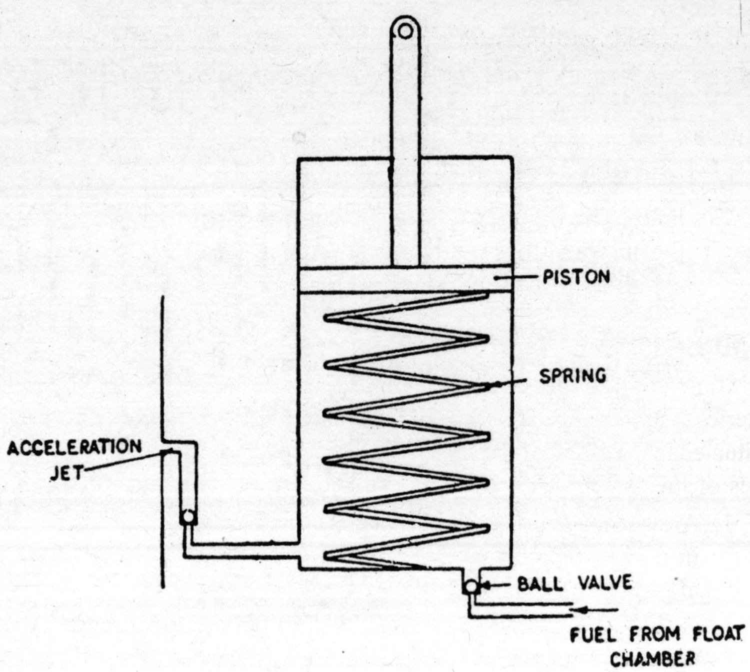

Fig. 10.25. Acceleration pump.

In the modern carburettors, climatic control devices are provided. These control the mixture strength by varying either fuel jet area or the air intake.

7. Influence of Altitute

At higher altitudes, the air density decreases due to fall in atmospheric pressure. There is another factor contributing towards increase of air density at high altitudes and that is the decreased temperature. But decrease in density due to fall in atmospheric pressure outweighs this factor and net result is an appreciable decrease in air density at higher altitudes. Due to this a simple carburettor which is set to give correct mixture at lower altitudes will give a comparatively richer mixture at higher altitudes.

A method which is used to rectify this defect in some Zenith carburettors is explained by means of Fig. 10.26.

Two pipes *AB* and *CD* connect the top of the float chamber to the air horn as shown. S is a valve in the pipe *CD*. For ordinary low altitudes, valve S is closed and thus the fuel in the float chamber is under atmoSpheric pressure because of pipe *AB*. As the altitude increases valve S is opened gradually and by so doing engine suction is applied to the float chamber, thus decreasing the pressure here which results in decreased fuel supply and hence a weaker mixture at higher altitudes, which is desired.

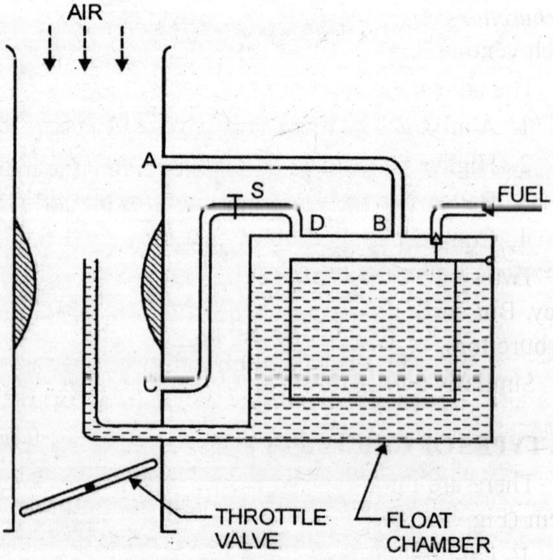

Fig. 10.26. Principle of altitude control.

8. Icing Trouble

As the fuel is atomised and evaporated in the carburettor, it causes cooling of the surrounding areas. In localities which are cold, there is thus always the danger of ice formation and choking of the venturi tube. Some of the methods used to prevent ice formation are:

(*i*) By heating the carburettor idle ports and throttle valve area by means of the engine exhaust gases.

(*ii*) By providing water passes in the carburettor body. The engine cooling system provides the hot water for heating the carburettor, thus avoiding formation of ice.

16. MULTIPLE VENTURI SYSTEM

Some carburettors employ two or three venturis in series (Fig. 10.27). The outlet of the smaller venturi is situated in the throat of the larger venturi. The advantages of the system are:

1. Air velocity and thus the depression is increased in the jet orifice portion, without adversely affecting maximum air flow capacity of the carburettor at high speeds. The higher depression increases the fuel flow.
2. There is annular blanket of air between the outer venturis which keeps the wet fuel off the walls for a greater distance. Thus more of the liquid particles of fuel reach the hot spot area of the inlet manifold.

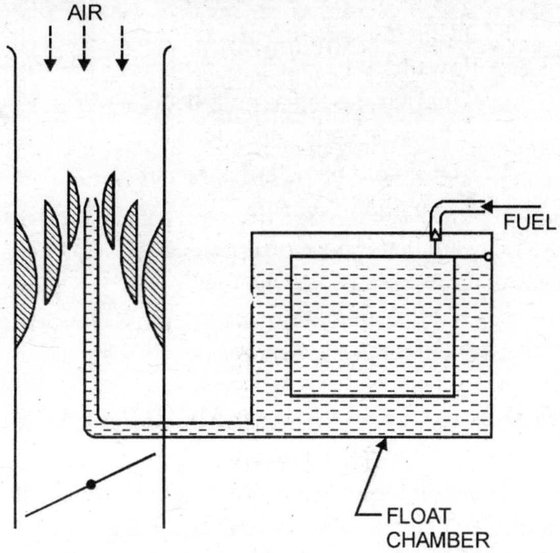

Fig. 10.27. Multi-venturi system.

The only disadvantage of the system is high manufacturing cost.

The system has been used in the Carter carburettor.

17. MULTIJET CARBURETTORS

Multijet carburettors are used in engines having a large number of cylinders *e.g.* 6 or more. A *dual carburettor* contains two sets of venturis and fuel jets. The float chamber and acceleration pump are common. Each venturi supplies mixture to half the cylinders.

The advantages of a dual carburettor over a single jet type are:

1. A mixtures of more uniform quality provided by the dual carburettor.
2. Higher volumetric efficiency.
3. Better distribution of mixture to the individual cylinders.
4. Compact design.

Two single carburettors could be used instead of a dual carburettor, providing higher volumetric efficiency. But such a system has the disadvantage that it is very difficult to synchronise the two individual carburettors.

Similarly a *quardrijet carburettor* has four venturis and four main nozzles.

18. TYPES OF CARBURETTORS

There are three type of carburettors, according to the direction in which the mixture is supplied by them (Fig. 10.28).

1. Up-drought type
2. Horizontal type

3. Down-drought type

Out of all these types, down-drought type is the one most commonly used because of the following advantages:

1. In this the gravity assists the flow of mixture. It is therefore, found that the engine pulls better at lower speeds under load.
2. Higher volumetric efficiency of the engine is achieved.
3. Carburettor position is rendered more accessible.

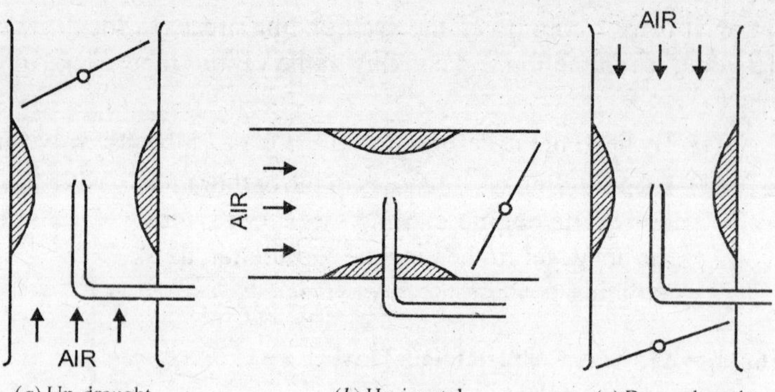

Fig. 10.28. Types of carburettors

The only disadvantage is the possibility of leakage going direct into the inlet manifold if the float is defective and the jet is overflowing.

19. SPECIAL FEATURES OF MODERN CARBURETTORS

1. Throttle return check

Closing the throttle suddenly on an engine running at very high speeds will cause a very high intake manifold vacuum, which will draw exhaust into the engine intake during valve overlap. This will dilute the intake charge to cause misfiring or even stall. To prevent this a throttle return check is connected to the throttle linkage in some modern carburettors. The return check diaphragm slows down the closing movement of the throttle valve and thus gradually reduces the speed to idle.

2. Anti-dieseling solenoid

Modern emission-controlled engines usually run hotter, which may result in some hot spots in the combustion chamber. These hot spots ignite the charge even after the ignition is turned off, which results in the engine continuing to run even after the ignition is turned off. This phenomenon is called *dieseling.* To avoid this some carburettors are equipped with an anti-dieseling solenoid. On turning off the ignition, this solenoid is de-energized, which allows the throttle valve to close completely. This keeps the air from getting into the engine which results in a mixture which is too rich to after-fire or diesel.

3. Automatic mixture control

For this, a plunger shaped valve is operated by a solenoid and a spring and controls a separate jet in the float chamber (Fig. 10.29). When the solenoid is turned on, the valve is lifted which increases the fuel supply through the jet. When the solenoid is turned off the spring pushes the valve down to decrease the fuel supply. The solenoid is operated by a computer, according to the signals received by it on engine speed, coolant temperature, manifold vacuum, throttle position and the amount of oxygen going out in the exhaust (Fig. 10.30). Due to the continual up and down movement of the

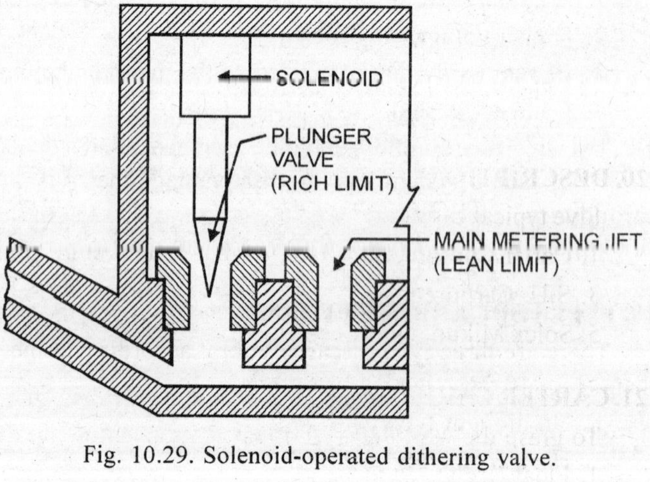

Fig. 10.29. Solenoid-operated dithering valve.

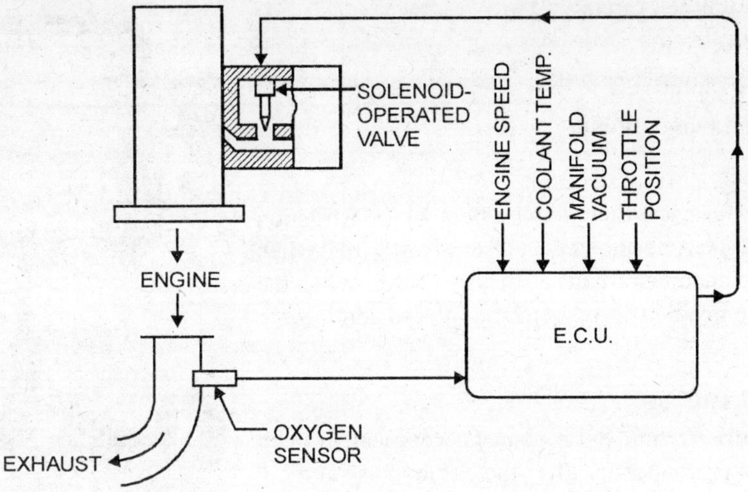

Fig. 10.30. Computer control of the solenoid valve

plunger valve, it is called a dithering valve. Thus a computer-controlled solenoid-ope rated dithering valve automatically adjusts the air-fuel ratio of the mixture in the carburettor as shown in Fig. 10.31. Such carburettors in which mixture is controlled by a computer are called feedback-controlled carburettors.

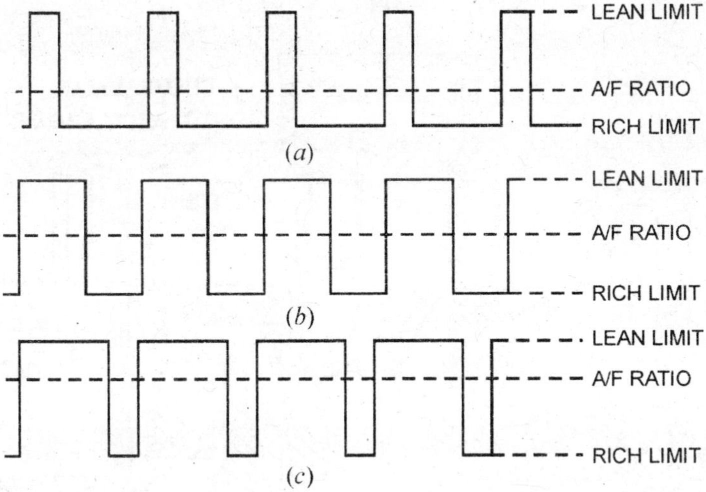

Fig. 10.31. Control of air-fuel ratio of the carburettor mixture by computer controlled, solenoid operated valve. (*a*) richer mixture, (*b*) mixture leaner than in (*a*), (*c*) still leaner mixture.

20. DESCRIPTION OF TYPICAL CARBURETTORS

Five typical carburettors will be explained in detail:

1. Carter Carburettor
2. Zenith Carburettor
3. S.U. Carburettor
4. Solex Carburettor
5. Solex Mikuni Double Venturi Carburettor

21. CARTER CARBURETTOR

To grasp the working of this carburettor, following circuits will be studied:

1. Float circuit
2. Starting circuit

3. Idle and low speed circuit
4. Part throttle circuit
5. Full throttle circuit
6. Acceleration pump circuit

1. Float circuit

The float chamber as usual contains a float, the construction is shown in Fig. 10.34. A needle valve closes the fuel inlet when the fuel in the float chamber attains a specified level. When the fuel level falls, the needle valve opens the inlet to admit more fuel.

2. Starting circuit

Choke is used for starting. It is mounted eccentrically, which facilitates its automatic opening after the engine has started.

As the choke valve is closed, the whole of the engine suction is applied at the main nozzle, which then delivers fuel (Fig. 10.35). As the air flow is quite small, the mixture supplied is very rich.

Fig. 10.32. Carter carburettor

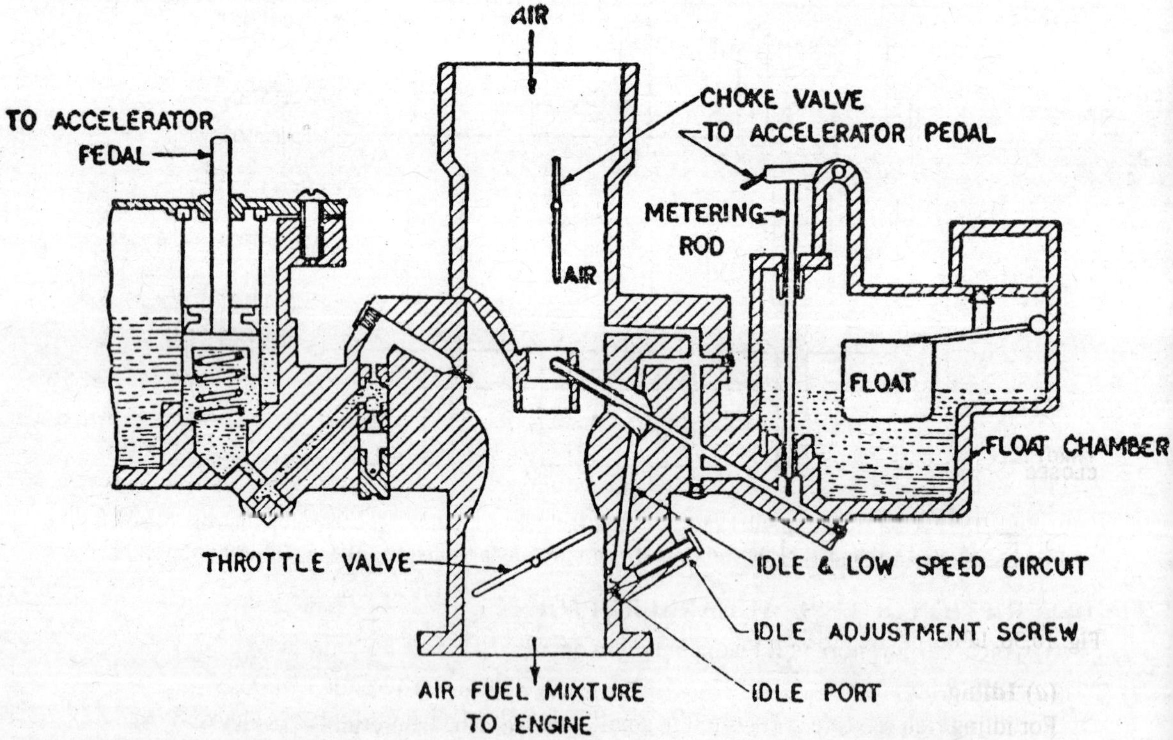

Fig. 10.33. Simplified diagram of Carter carburettor

3. Idle and low speed circuit

Figs. 10.36 and 10.37 show idling and low speed operations. Separate idling passage is provided with low speed port and idle port.

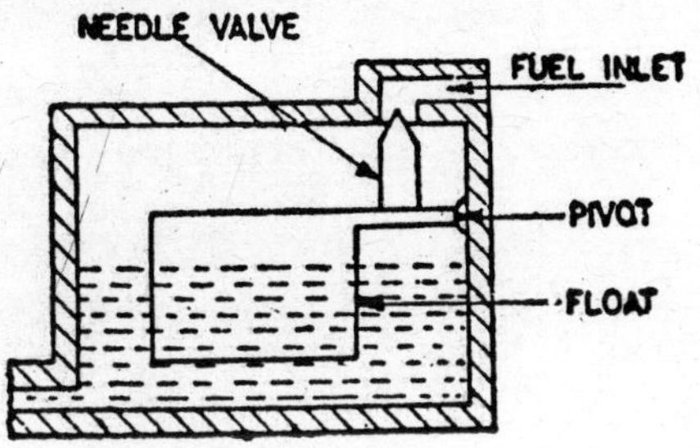

Fig. 10.34. Float

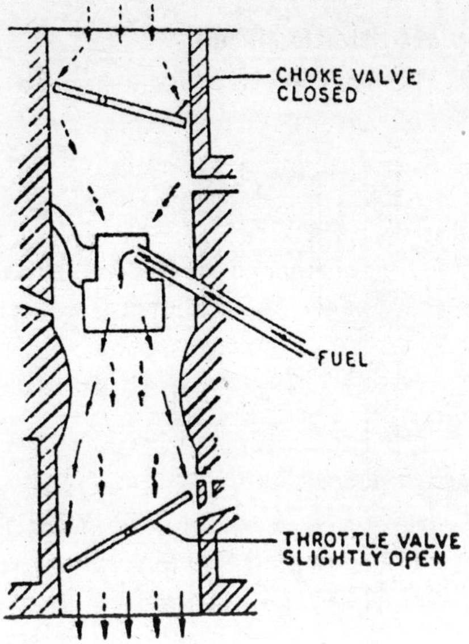

Fig. 10.35. Starting circuit. (⟶ Fuel flow, - - -▸ air flow)

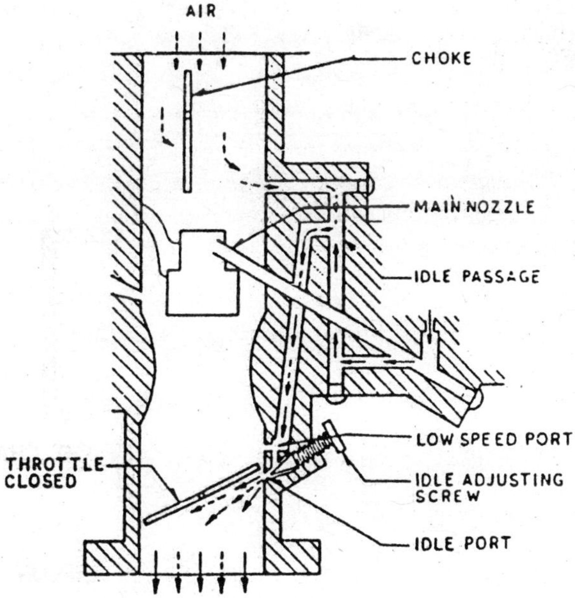

Fig. 10.36. Idling circuit (⟶ fuel flow,- - -▸ air flow)

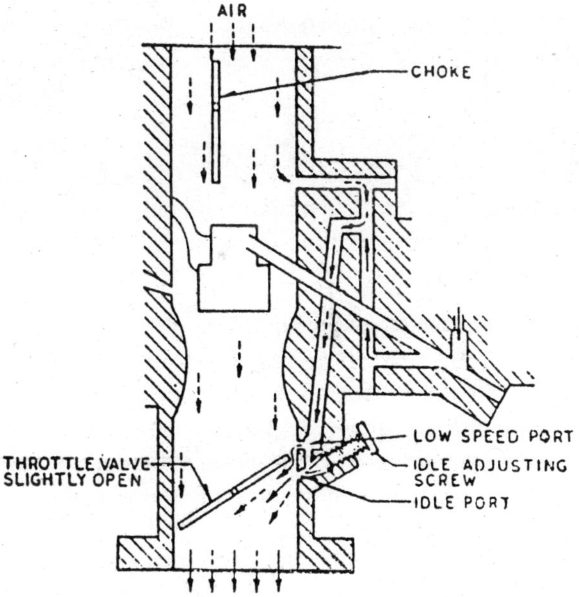

Fig. 10.37. Low speed circuit. (⟶ fuel flow,- - -▸ air flow)

(*a*) Idling.

For idling rich mixture is required in small quantity. The throttle valve is almost closed as shown in the figure. The whole of the engine suction is now applied at the idle port, through which the air and the fuel are drawn, giving rich mixture.

(*b*) Low speed operation

Throttle valve is opened further. The main nozzle also starts supplying the fuel. At this stage fuel is delivered both by the main venturi and low speed port through idle passage.

4. Part-throttle circuit

Fig. 10.38 explains clearly the part throttle circuit, where the fuel is delivered by the main nozzle only.

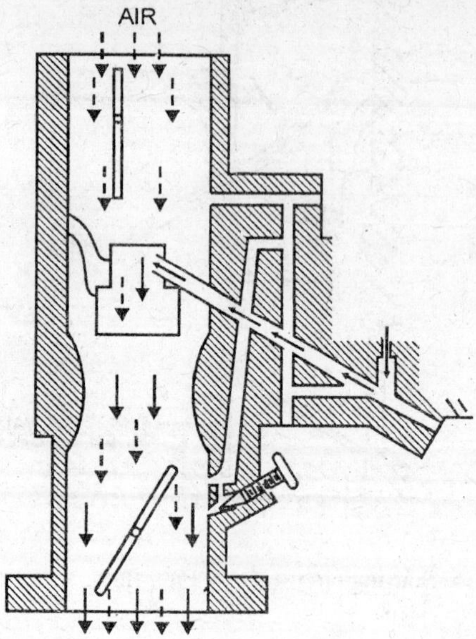

Fig. 10.38. Part throttle circuit. (⟶ fuel flow, - - -➤ air flow)

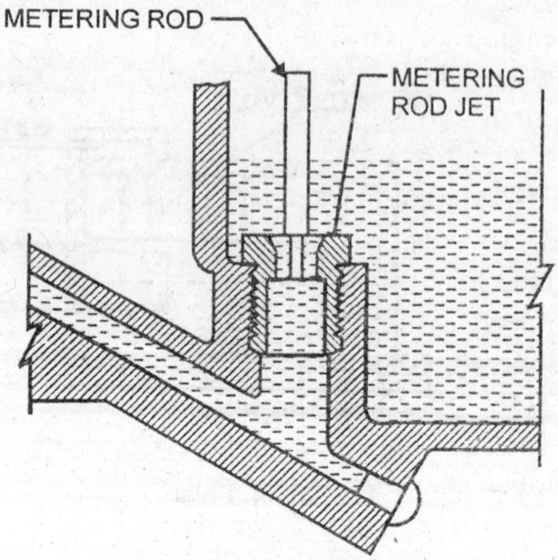

Fig. 10.39. Full throttle circuit.

5. Full throttle circuit

When the throttle is fully open, maximum amount of air is passing through the venturi. Thus a higher rate of fuel flow is desired, which is achieved by means of the metering rod (Fig. 10.39). Metering rod consists of a number of steps of diameter sizes at its bottom and is connected with the accelerator pedal through linkage. When the accelerator pedal is pressed, the throttle is held wide open and simultaneously the metering rod is lifted up. In this condition, the smaller diameter of the rod is inside the fuel-hole, providing larger flow area, thus delivering more fuel.

6. Acceleration pump circuit

The construction of the pump is clear from Fig. 10.40. When acceleration is desired, acceleration pedal is pressed, which actuates the pump, giving an extra spurt of fuel for acceleration. Leaving the accelerator pedal causes the pump piston to move up, thereby sucking more fuel from the float chamber. The pump is now ready for next operation. It must be made clear that purpose of the accelerating pump is not to provide a continuous fuel supply for acceleration, but only to provide extra spurt of fuel to avoid flat spot.

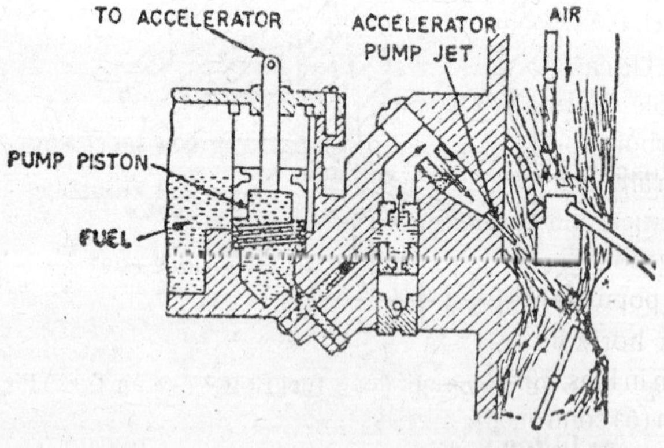

Fig. 10.40. Acceleration pump circuit.

22. ZENITH CARBURETTOR

Zenith carburettor is one of the most popular carburettors. Fig. 10.41 shows an old model of the Zenith Carburettor. The later models, however, contained economy devices and accelerating pumps.

Referring to Fig. 10.41, the carburettor contains the old type of float chamber, which design is entirely eliminated in recent years. There are three jets, the main jet, compensating jet and the idling jet. The main jet has an outer cover as shown; the annular passage between the main jet and outer cover forms a delivery tube for the compensating jet.

For starting the choke is used. For idling, the throttle valve is closed. The engine suction is applied to the idling jet, which supplies the fuel. The air for idling enters through holes. A and B and the mixture then passes out through the passage provided, as shown. A separate knob is provided for the idle adjustment. It is connected with a threaded extension tube which controls the opening *B*, thereby affecting the air intake, thus controlling the quality of mixture.

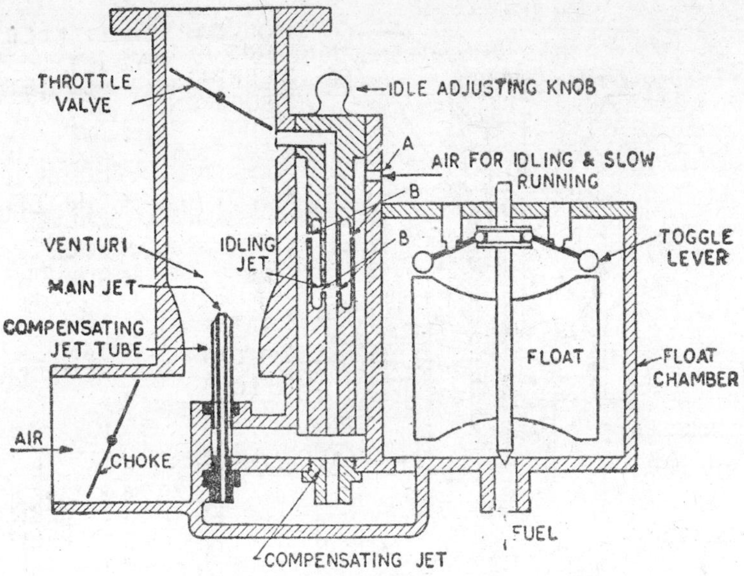

Fig. 10.41. Zenith carburettor.

As the throttle is opened a little, some air flows through the venturi. In this condition, both the main jet and the slow running jet are supplying the mixture.

With further throttle opening, the depression at the slow running jet is destroyed, the whole of the suction being applied at the main and the compensating jets. The compensating jet takes care that the correct air fuel ratio is maintained at different speeds.

23. S.U. CARBURETTOR

S.U. carburettor is a typical example of variable venturi or "constant vacuum" type of carburettor as against Carter and Zenith, which are 'constant venturi' carburettors. Both the vertical and the horizontal models are being produced. The horizontal type is, of course, more popular.

Fig. 10.42. Partial cut away view of S.U. carburettor.

A horizontal types S.U. carburettor is shown in Figs. 10.42 and 10.43. A spring loaded piston (6) controls the air passage. The throttle valve is of ordinary butterfly type. The air passage (venturi) is rectangular opening of constant width and changeable height. However, there is no separate idling or slow running device and no acceleration pump. The upper side of suction disc is connected to throttle passage through a slot cut in the piston. The lower side is exposed to atmospheric pressure. Thus

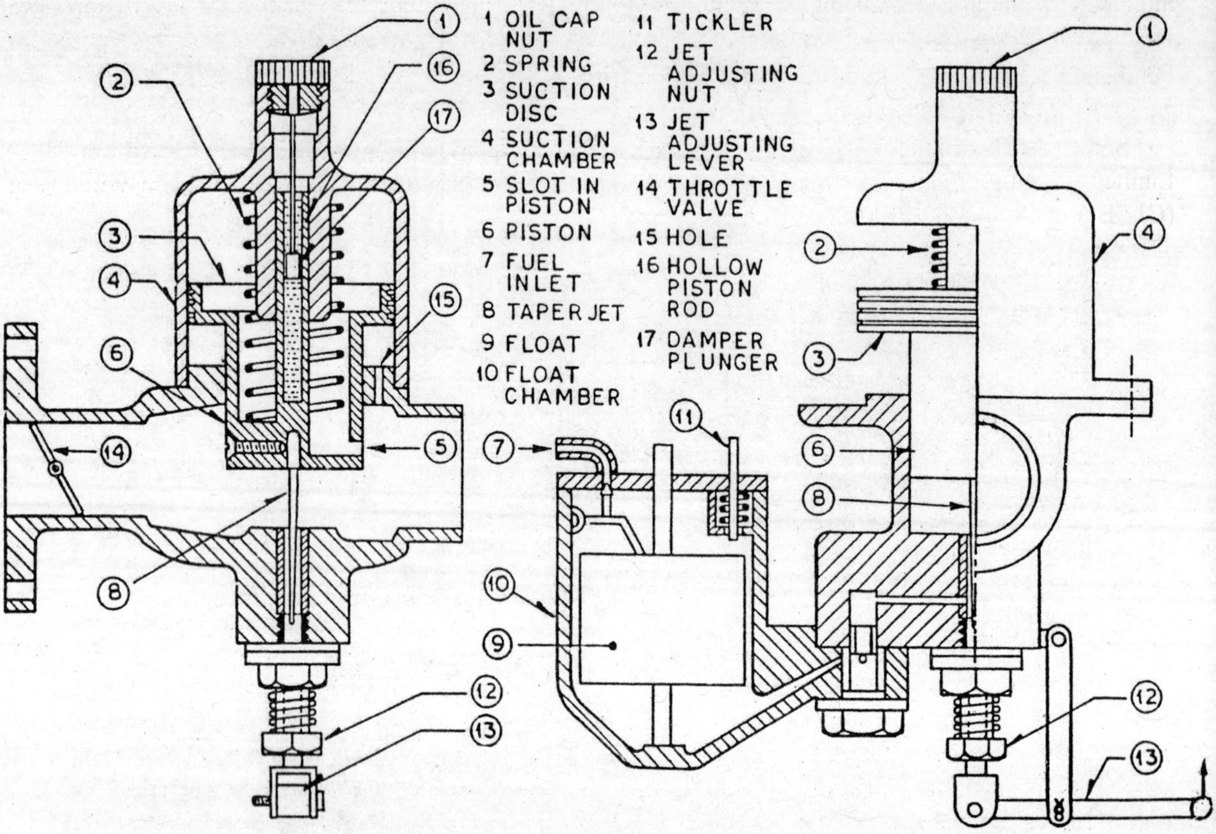

Fig. 10.43. Simplified diagram of a S.U. carburettor.

the position of piston at any instant depends upon the balance of its own weight (down) against the vacuum force (up). As the weight of piston is constant, so the vacuum also remains constant.

As the air passage varies widely through the piston movement, the size of the petrol jet should also vary. This is done with the help of a tapering needle (8) fixed to the piston. When the piston moves up and down, the needle also moves up and down. The lower end of the needle is inside the main jet. The up and down movement of the tapering needle causes the annular area for fuel flow to increase or decrease. When the needle moves up the area increases; when it moves down, the area decreases. Accurate dimensioning of this needle gives accurate petrol discharge areas for all positions of the piston.

A damper plunger (17) is installed in the hollow piston rod (16), in which oil of suitable viscosity is also filled. The damper plunger has one-way valve in it. It regulates the rate of lift of the piston, but allows the same to fall freely when throttle valve is closed. When, however, the throttle valve is opened suddenly, *e.g.*, for acceleration, the lifting speed of the piston is retarded by the damper plunger. This provides additional depression over the fuel jet which now provides more fuel. Thus no separate acceleration pump is required

Jet adjusting nut (12) is meant to adjust mixture strength. Tighening the nut will raise the jet, thereby reducing the annular area for fuel flow. Similarly loosening the nut lowers the jet and thus increases the fuel supply.

Jet adjusting lever (13) can be operated from the dashboard. For cold starting and idling this is pulled up which brings the fuel jet down and hence more fuel is supplied. Jet adjusting lever is also linked (not shown) to the throttle valve (which is otherwise operated by accelerator pedal) in such a way that as the jet lever is pulled, throttle valve is also opened gradually, applying suction to the upper side of the piston, moving it up and thus decreasing the suction at the fuel jet. Thus pulling the jet adjusting lever knob more will not enrich the fuel further even though the annular area of jet is increased. This is exactly required for idling.

24. SOLEX CARBURETTOR

Some models of Solex carburettor have been manufactured in India (under licence) by M/s Carburettor Limited, Channai. Figs. 10.44 and 10.45 show Solex carburettor model M.34 PBIC, used in Willys jeep (CJ-3B and 475-4WD engines).

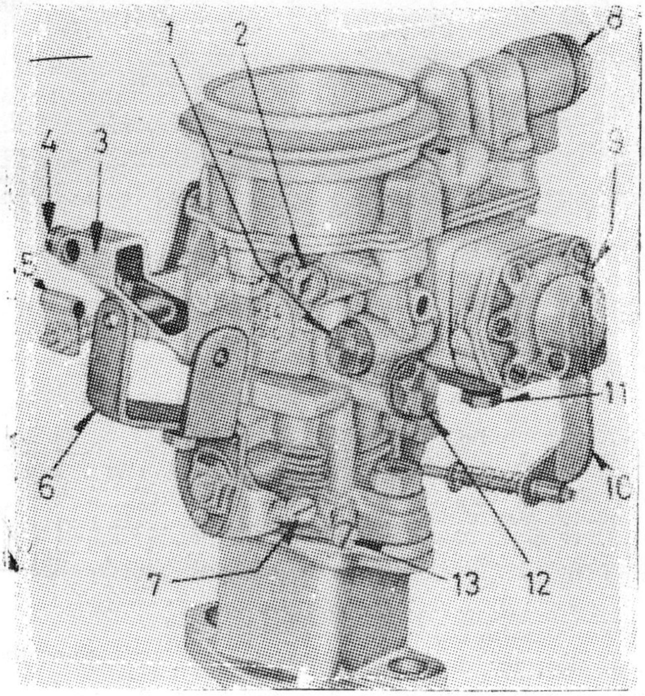

1. Pump jet
2. Pilot jet
3. Starter cover
4. Starter cable clamping screw
5. Clip for starter
6. Throttle lever
7. Siow running adjustment screw
8. Fuel inlet
9. Acceleration pump
10. Pump lever
11. Pump valve
12. Main jet holder
13. Volume control screw

Fig. 10.44. Solex carburettor model M.34 PBIC.

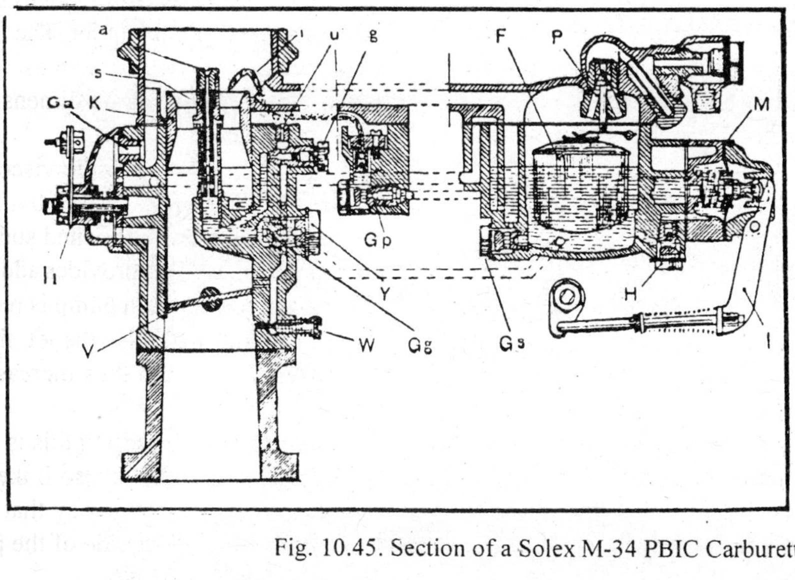

a — Air correction jet
F — Float
G_a — Starter air jet
G_g — Main jet
G_p — Pump jet
G_s — Starter petrol jet
g — Pilot jet
H — Pump valve
i — Pump injector
K — Choke tube
l — Pump lever
l_1 — Starter lever
M — Pump membrane
P — Needle valve
S — Emulsion tube
u — Pilot air bleed
W — Idling adjusting screw
Y — Main jet carrier

Fig. 10.45. Section of a Solex M-34 PBIC Carburettor

This is a down drought type of carburettor, with special provision for a progressive starter (Fig. 10.46), which supplies richer mixture for starting and then gradually weakens it till the engine has reached its normal operating temperature. The various circuits of the carburettor are explained below:

1. Starting

The starter valve is in the form of a flat disc with holes of different sizes. These holes connect the petrol jet and starter jet sides to the passage which opens into the air horn just below the throttle valve. Depending upon the position of the starter lever, which can be adjusted by the driver on the dash board, either bigger or smaller holes come opposite the passage. Initially for starting richer mixture is required and after the engine starts the richness required decreases. So in the start position, bigger holes are the connecting holes. The throttle valve being in the closed position, the whole of the engine suction is applied to the starting passage 1, due to which the petrol from the float chamber passes through the starter petrol jet and rises into passage 2, some of it comes out, and mixes with the air entering through the air jet. The jets and passages are so shaped that mixture so supplied to the carburettor is rich enough for starting.

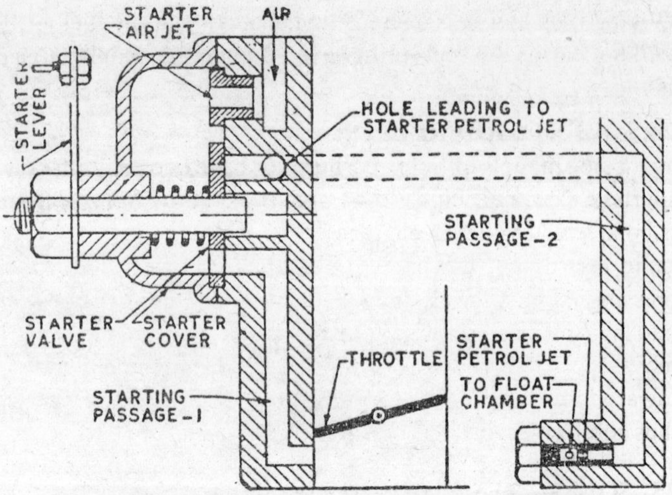

Fig. 10.46. Progressive starter for Solex carburettor

After the engine has started, the starter lever is brought to the second position, in between the initial and final positions. Now the smaller holes in the valve complete the circuit, thus reducing the amount of petrol. Also in this position, the throttle valve is partly open, so that the petrol is also coming out from the main let. The reduced mixture supply from the starter system in this situation is however sufficient to keep the engine running, till it reaches the normal running temperature, when the starter is brought to 'off' position.

2. Idling or low speed operation

An idle port controlled by idle adjusting screw is provided in the engine side of throttle valve. As the throttle is almost closed the engine suction is applied at the pilot petrol jet, which supplies petrol. The jet itself draws petrol from the main jet circuit. The air is drawn in from the pilot air jet. The petrol and the air mix in the idle passage and the mixture comes out of the idle port.

To ensure the smooth transfer from idle and low speed circuit to the main jet circuit without the occurrence of flat spot, slow running openings are provided on the venturi side of throttle valve. As the throttle is opened wide, the suction at idle port is decreased. But suction is also applied now at the slow speed openings, which more than offsets the loss of suction at the idle port and thus flat spot is avoided.

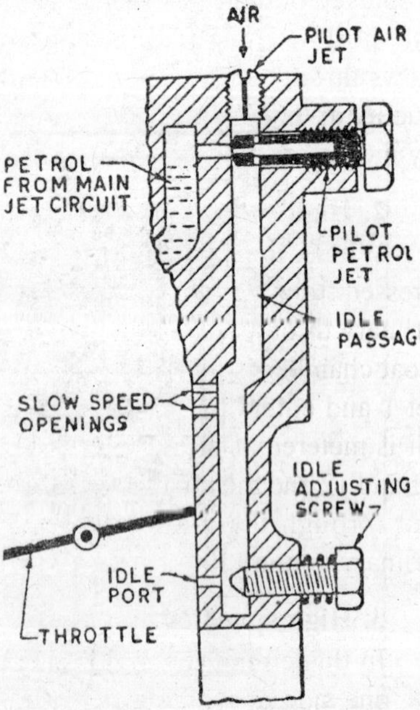

Fig. 10.47. Detail of idle circuit

3. Normal running

The throttle is held partly open, so that the engine suction is now applied at the main jet, which now supplies fuel. The air enters directly through the venturi, the quantity of mixture being governed with the throttle valve.

4. Acceleration

To avoid 'flat spot' during acceleration, a separate membrane pump is provided (Fig. 10.48), which delivers spurts of extra needed fuel for acceleration. Pump lever is connected to the accelerator pedal so that when the same is pressed, the lever moves towards left, thus pressing the membrane towards left and forcing petrol into main jet circuit. When the pedal is left free, the lever moves the membrane back towards right creating vacuum towards left which opens the ball valve provided and thus admits the petrol from chamber into the pump itself.

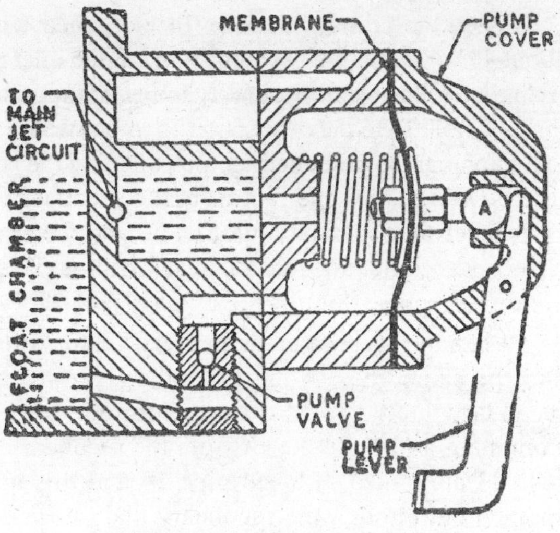

Fig. 10.48. Membrane pump

25. SOLEX MIKUNI DOUBLE VENTURI CARBURETTOR

This type of carburettor is used in maruti 800 car. Its main components are primary and secondary venturi, float chamber, acceleration pump, a fuel cut solenoid valve and a depression chamber to actuate the secondary throttle valve. Choke valve is provided in the primary air horn only. Various circuits of the carburettor are discussed here.

1. Idle and slow speed circuit

The fuel from the float chamber flows through main jet 1 and reaches slow jet 2, where it is metered and mixed with air, the air being already metered at slow air holes no. 6 and 7. This air-fuel mixture is sprayed out through bypass port 3 and idle port 4. During idling the mixture flows through mainly the idle port. The quality of this mixture can be controlled by means of the idle adjustment screw.

2. High speed circuit (primary)

When the accelerator pedal is pressed from the idling stage to open the primary throttle valve, the fuel from the float chamber is metered from the main jet 1 and enters the bleed pipe 10. The air is metered at primary main air holes 11 and 12 and the air-fuel mixture comes out through the main nozzle into the primary venturi 13.

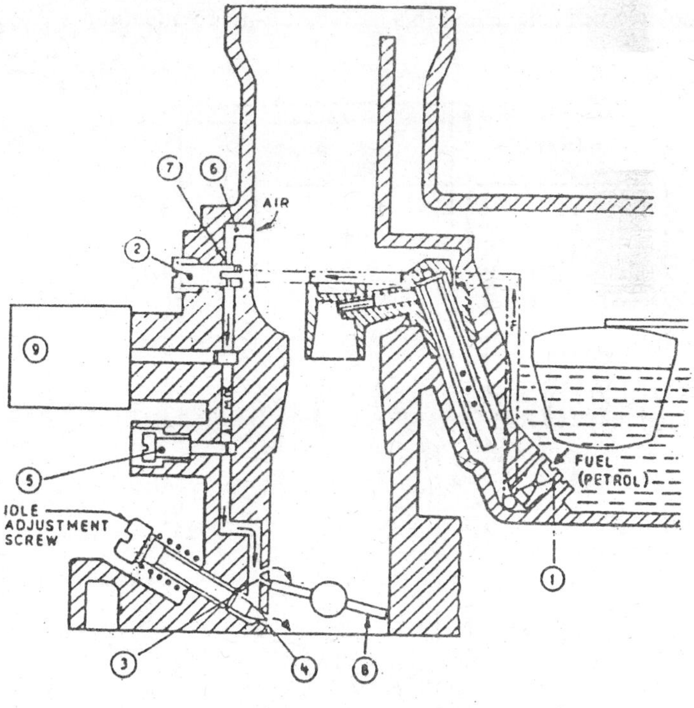

Fig. 10.49. Idle and slow speed circuit.

3. High speed circuit (secondary)

In this circuit the major part is played by a depression chamber 17 which con- sists of a diaphragm 18, one side of which is exposed to atmospheric pressure, while the other side is connected to hole 16. The diaphragm is connected to the secondary throttle valve through a rod and a lever.

When the primary throttle 14 is opened wider, *i.e.*, about 40° or more, the vacuum from engine inlet manifold is applied to the primary venturi and is further transmitted through hole 16 to the depression chamber where it causes the diaphragm to be pulled up thus opening the secondary throttle valve. Now the fuel from the secondary main jet 19 reaches the secondary pilot jet 20 and the air metered at secondary pilot air hole 21 mixes with it. This air-fuel mixture is sprayed out of the by-pass port 22.

When the vacuum in the primary venturi is increased and the same goes to the secondary venturi also, the secondary throttle valve is opened further. Then the fuel from main jet 19 and the air from the secondary main air hole 23 are mixed in bleed pipe 24 and this mixture is sprayed out through the secondary main nozzle into the secondary venturi.

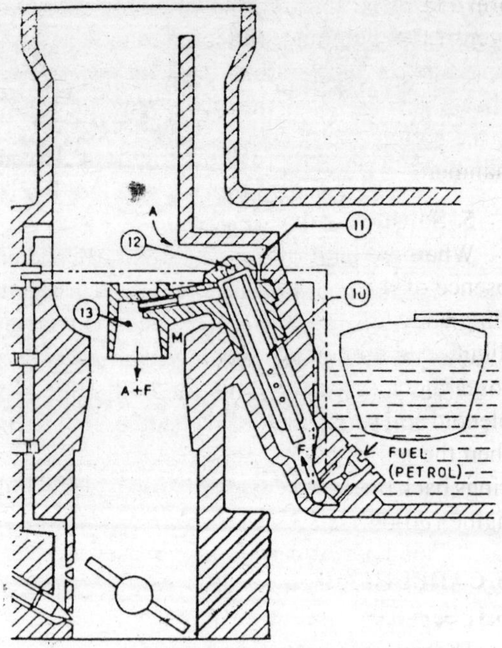

Fig. 10.50. High speed circuit (primary)

4. Acceleration power

When the throttle is opened suddenly for acceleration, while the engine is idling or is running at slow speed, a device is needed to make the carburettor respond without delay. Such a device in this carburettor is provided in the

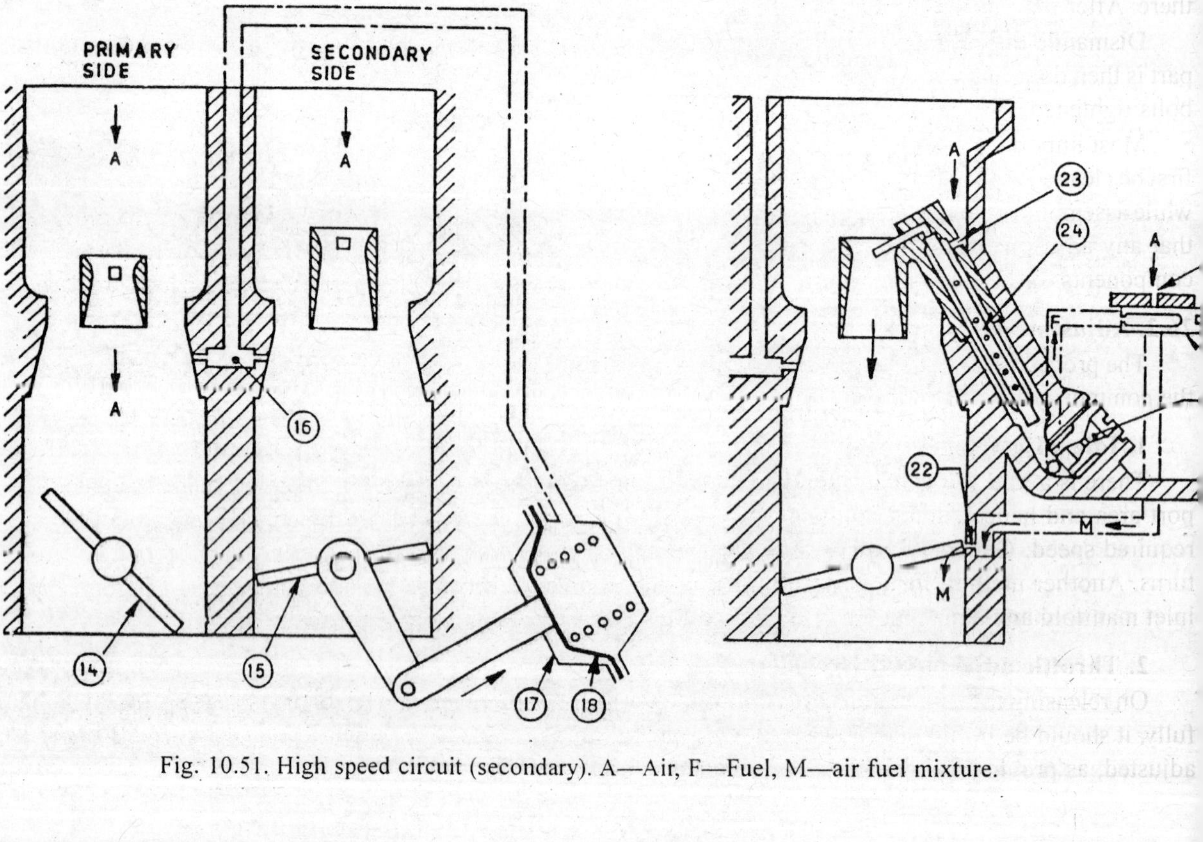

Fig. 10.51. High speed circuit (secondary). A—Air, F—Fuel, M—air fuel mixture.

form of a membrane type acceleration pump similar to one discussed already in Fig. 10.48. The operating lever of this pump is connected to tne shaft of the primary throttle valve. When the acceleration pedal is depressed for acceleration, the pump lever pushes the diaphragm forcing fuel out of pump nozzle into the primary venturi. When the accelerator pedal is released, the diaphragm is brought back to the original position by the spring, the discharge ball valve is closed and the inlet valve is opened admitted fuel in the pump chamber.

5. Solenoid valve

When the ignition key is turned off, the engine does not stop immediately with the result that in the absence of some special device, the carburettor would continue to deliver the air fuel mixture through the idling and slow speed circuit, till the engine stops. Such a run-on of the engine is prevented in the Solex Mikuni carburettor by means of a solenoid valve (Fig. 10.49). It operates a needle valve which opens or closes the passage for idling and slow speed circuit. When the ignition key is ON, current flows in the solenoid coil generating the magnetic force, which pulls the needle valve so as to open the passage. However, when the ignition key is turned OFF, the magnetic force disappears and the spring in the solenoid valve brings back the needle valve to its original position when it cuts-off the slow speed passage, thus preventing engine run-on.

26. CARBURETTOR SERVICE AND ADJUSTMENT

26.1. Service

The detailed procedure for servicing a carburettor is always specified by the manufacturers and varies in case of different types. However, the general procedure and the points of caution are given below.

To remove the carburettor from the engine, first of all remove the air cleaner and disconnect the throttle and choke linkages. Next disconnect the vacuum advance unit and any other connections that may be there. After that loosen the flange nuts and remove the carburettor gently.

Dismantle the carburettor and clean each and every component in petrol. Detailed inspection of each part is then done and the faulty ones are replaced, after which the carburettor is reassembled and the flange bolts tightened. Remake the different connections in the order reverse to that done for removing.

Most important precaution is that the jets should never be cleaned with wires etc. Rather they should first be cleaned with petrol and then with a blast of air. Secondly, you must work with clean hands, otherwise while assembling, the dirt from your hands will again go into the carburettor. Further, it must be remembered that any adjustments to the carburettor should only be done once it is first ascertained that the other components of the fuel systems are working alright; otherwise the engine performance will not improve.

26.2. Adjustments

The procedure to make various adjustments differs in case of individual carburettor models. However, the common adjustments and their procedure may be outlined as given below:

1. Idle adjustment

There is a idle adjustment screw provided in the carburettor. Screwing in the same decreases the idle port area and hence decreases the engine speed. The screw is adjusted till the engine runs smoothly at required speed. Generally this is done by screwing-in completely and then opening back by one or two turns. Another method for idle adjustment, which is precise, involves attaching a vacuum gauge to the inlet manifold and adjusting the idle screw till maximum vacuum is obtained on the gauge.

2. Throttle adjustment

On releasing the accelerator pedal throttle valve should be completely closed, while on pressing pedal fully, it should be in full open position. This may be checked and if not found correct, the linkage may be adjusted, as provided for in the manufacturer's manual.

3. Other adjustments

Apart from the idle and throttle adjustments, the other adjustments which are provided in some carburettors are for accelerating-pump, metering rod, floats etc. They may be performed as laid down in the manual.

26.3. How to recognise the mixture strength

An extra rich mixture may be recognised from the following effects:

1. High fuel consumption.
2. The colour of the exhaust smoke is black. This is due to unburnt carbon particles in exhaust.
3. Surging while idling.
4. Loud muffer explosions. They are caused by the unburnt fuel in the engine, which completes its combustion in the muffer.

An extra weak mixture is recognised from the following indications:

1. Existence of 'flat spot' while accelerating.
2. Stalling of the engine by itself.
3. Loud popping in the carburettor, which is due to engine backfiring.
4. Overheating of the engine.
5. Loss of power.

27. FUEL SUPPLY SYSTEM TROUBLE SHOOTING

The commonly occurring troubles in the fuel supply systems of automobiles (involving carburettors) are discussed here. Suitable remedies have been suggested.

1. High fuel consumption

This may be due to any or more of the following:

(*a*) Air cleaner may be choked, which may be taken out and cleaned by blowing compressed air from the inside of the element (Fig. 10.52). In case the element is too much dirty, if may be washed in a detergent, rinsed and dried.

(*b*) Fuel may be leaking from somewhere in the supply system which may be inspected and corrected if detected.

(*c*) The driver may be in the habit of pumping the accelerator pedal unnecessarily, thus wasting fuel through the accelerating pump.

(*d*) The idle adjustment may not be correct. If so, the same may be readjusted.

(*e*) The mixture supplied by the carburettor may be generally overrich which may be due to any of the following factors:

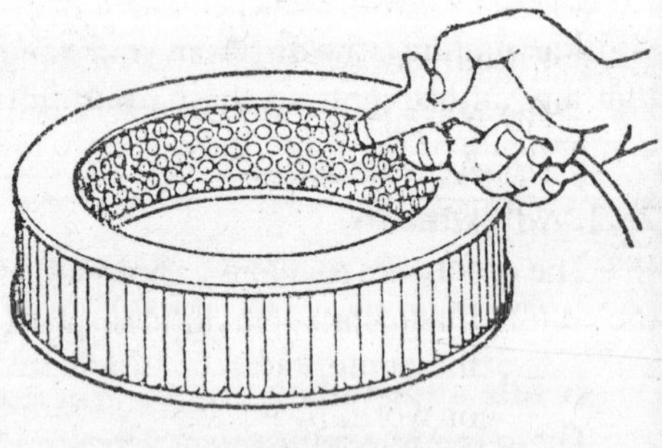

Fig. 10.52. Cleaning the air clearner element.
(Courtesy–Maruti Udyog Ltd.)

(*i*) Clogging of air bleed jets. The same may be cleaned by blowing compressed air.

(*ii*) The carburettor jets may have worn out to larger dimensions; the same may be replaced.

(*iii*) Incomplete opening of choke valve, which may be adjusted by changing the length of the choke control cable.

(*iv*) Carburettor may be flooded due to following reasons:

(α) Worn out needle valve, which may be replaced.

(β) The float may be punched, which should be replaced.

2. Difficult starting

The reasons for this may be:

(*a*) The carburettor jets may be clogged, which should be removed and serviced.

(*b*) The choke may not be closing properly when pulled up. The same may be inspected and adjusted.

(*c*) There may be some air leak into the inlet manifold, which would weaken the mixture which no longer remains sufficiently rich to start combustion. This may be inspected and remedied.

(*d*) The fuel filter may be clogged and so may not be supplying desired quantities of fuel. It may be inspected and cleaned or even replaced.

(*e*) The fuel pump pressure may not be the same as specified. The pump pressure may be tested as described and if found defective, may be repaired or replaced.

3. Poor idling

The reason pertaining to fuel system for poor idling may be listed as below:

(*a*) Incorrect adjustment of idle screw, which may be redone.

(*b*) The float level may not be correct, which may be checked and adjusted.

(*c*) The float valve needle may be sticking. This may be remedied by cleaning the carburettor inlet and valve needle thoroughly.

(*d*) The screws connecting carburettor with the induction manifold may be loose or the gasket there may be damaged. In both of these situations, the air may leak into the manifold and weaken air fuel mixture.

(*e*) The jets in carburettor may be loose, which may be inspected and if detected, tightened.

(*f*) The air cleaner may be clogged, which may be removed and overhauled.

(*g*) Throttle valve plate or spindle may be bent, which have to be replaced.

(*h*) The fuel pump may be defective, which may be tested as discussed earlier, and if found so, may be repaired or replaced.

4. Poor acceleration

The reasons for poor acceleration may be:

(*a*) Accelerator pump may not be adjusted correctly or may be working at all. This may be checked when the engine is not running. Just press the accelerator pedal and a spurt of fuel oil should be forced out of the acceleration port in the carburettor. If found defective, the pump including its linkage should be inspected and the defect remedied by repair or replacement as required. Inside the pump, the diaphragm may be punctured or the pump leather washer may have become hard or worn, or the ball check valves may be defective.

(*b*) The fuel filter may be clogged, which may be cleaned or if required, replaced.

(*c*) Fuel pump may be defective which may be inspected and repaired or replaced.

(*d*) The fuel line may be clogged, which may be cleaned.

(*e*) The fuel level in the float chamber may be low which may be corrected by readjusting the float.

5. Lack of power

This may be due to excessively rich or extremely weak air fuel mixture. The reasons of over-rich mixture have been described above under the head, "1. High fuel consumption". Weaker mixture may also be due to:

(*a*) Incorrect carburettor setting, which may be checked and rectified if necessary.

(*b*) Dirt in the carburettor jets, which should be removed, cleaned with compressed air and refixed. Look through the jets to see whether they are clear or not (before reassembling).

(*c*) Faulty inlet manifold gasket will cause the extra air to leak in and thereby weaken the mixture. The gasket has to be replaced in this case.

(*d*) Fuel filter may be choked, thereby restricting the fuel supply and weakening the mixture. The same has to be cleaned or replaced.

6. Over heating of engine

Apart from faulty ignition timing or defective cooling system, this could be also due to a weak air-fuel mixture, the reasons for which have been described above.

7. Engine spits back

Popping or banging in air cleaner is also caused due to a weak air-fuel mixture which is produced by any or all of the reasons discussed above.

TABLE 10.1. COMPARATIVE FUEL SYSTEM DATA OF SOME INDIAN AUTOMOBILES WITH PETROL ENGINES

S. No.	*Item*	*Maruti 800*	*Hindustan Ambassador Mark II*	*Fiat 1100*	*Jeep Universal*	*Dodge/Fargo Model 89 M4 (Petrol Type)*
1.	Air cleaner	Dry Type	Oil Bath Type Type	Oil Wetted Cloth	Oil Bath Type Dry Type	Heavy Duty
2.	Fuel Pump	Mechanical	S.U. Electrical	A.C. Mechanical	A.C. Mechanical	A.C. Mechanical
3.	Carburettor	Solex Mikuni 24—30 DIDS, down-draught	S.U.H.S. 2 semi-	Solex C32PBIC Down draught Down draught	Carter YF—938—SD	Carter 26815 Down Draught
4.	Fuel Tank Capacity (Litre)	27	54.6	40	40	70

THEORY QUESTIONS

1. Describe the various fuel supply systems for automotive petrol engines. Discuss in detail the pump system.
2. With the help of neat sketches, explain the construction and working of
   (*a*) A.C. mechanical fuel pump.
   (*b*) S.U. electrical fuel pump.
3. Describe various tests for checking the performance of a fuel pump. How will you service a fuel pump?
4. What are the common troubles which are likely to occur in a mechanical or an electrical fuel pump? Describe also the probable reasons for the same.
5. What are the different types of air cleaners? Discuss both heavy duty and light duty types.
6. With the help of simplified diagram, explain the construction and working of a fuel gauge.
7. Briefly mention the functions of a carburettor.
8. Discuss the working of a simple single jet carburettor. Explain various defects of such a carburettor. How are these remedied?
9. Describe clearly the requirements of air fuel mixtures for starting a petrol engine from cold. Explain various methods to achieve the same.

10. What do you understand from the term 'compensation' as applied to carburettors? Discuss various methods of compensation.
11. How does a compensation jet in a carburettor work? Explain clearly.
12. Explain briefly the triple venturi system in carburettors.
13. Write a comprehensive note on multi-jet carburettors.
14. Explain the merits and demerits of a down-drought types of carburettor over the other type.
15. Discuss in detail following special features of modern carburettors:
    (*a*) Throttle return check.
    (*b*) Anti-dieseling sloenoids.
    (*c*) Automatic mixture control.
16. Discuss extensively the construction and working of a Carter carburettor.
17. Draw a complete sketch of Carter carburettor and lebel the various parts.
18. Explain with the help of suitable sketches the working of:
    (*a*) S.U. Carburettor
    (*b*) Zenith Carburettor
19. Draw a simplified sketch of Solex carburettor and explain its working discussing clearly starting, idling, low speed operation, normal running and acceleration.
20. Describe in detail the construction and working of the carburettor used in Maruti 800 car. Discuss in particular the function of the solenoid valve in this carburettor.
21. Give a general outline of the procedure for carburettor service and adjustment. How do you recognize the mixture strength?
22. Enlist the common troubles experienced in the fuel supply of an engine. Locate their possible causes and suggest measures to remedy these.

NUMERICAL PROBLEMS

1. In a single jet carburettor, the level of fuel in the float chamber is 3 mm below the nozzle opening, which is 1.5 mm diameter. If the velocity of air at the venturi throat is 60 ms^{-1}, calculate the fuel flow, assuming the following data: density of air 1.2 kgm^{-3}, density of fuel 750 kgm^{-3}, coefficient of discharge for fuel flow 0.6 and coefficient of discharge for air flow 0.85. [**Ans.** 1.86 gm/s]

QUESTIONS FOR VIVA

1. What are the main constituents of a fuel supply system for an automobile engine running on petrol?
2. Name the various fuel supply systems.
3. What is the disadvantage of the gravity system of fuel supply?
4. Which type of gasoline fuel supply system is used most commonly in automobiles?
5. Which one is the most accurate fuel supply system for petrol cars?
6. Which material is used for fuel tank?
7. Where is the fuel tank usually located in cars?
8. Why are the baffle plates provided in the fuel tank?
9. Why is a vent necessary on a fuel tank?
10. Which engine part imparts motion to the diaphragm of mechanical fuel pump?
11. What is the material used for pump diaphragms?
12. What happens to the fuel pump operation when the carburettor float chamber becomes full?
13. Why is it important to maintain the pressure in the fuel supply line from the pump to the carburettor within specified limits? How is the same achieved?
14. What are the advantages of an electrically-operated fuel pump over a mechanical one?

15. Name the various tests to check a fuel pump.
16. What is the function of vapour return line?
17. What is the necessity of using air cleaners in the engine?
18. What happens if dusty air enters the crankcase?
19. Name two types of air cleaners in common use.
20. Why is air cleaner controlled theromstatically in some cars?
21. What happens if the air cleaner gets clogged?
22. Why are the air cleaner elements corrugated?
23. What is the special property of chamois leather as regards filtration of petrol?
24. Which types of fuel gauge is commonly used in modern automobiles?
25. Why is a richer air-fuel mixture required at the start?
26. What are the functions of a carburettor?
27. Why do we require a rich mixture even for idling?
28. What is a stoichiometric air-fuel mixture?
29. What happens if the mixture supplied to the engine is (*i*) weaker, (*ii*) richer than the optimum in a given condition?
30. What is the material used for float?
31. Why is the carburettor float made hollow?
32. Why in the modern carburettors, the float chamber is vented to the carburettor air horn instead of to the atmosphere?
33. What is the function of venturi in a carburettor?
34. What causes the air to be drawn into the carburettor?
35. What causes the fuel to flow from the main jet in the carburettor?
36. On what factors does the fuel delivery in a carburettor depend?
37. What is a butterfly valve and where is it used in the carburettor?
38. Why does the air-fuel mixture delivered by a single jet carburettor become richer at higher speed?
39. Enumerate the defects in a simple single jet carburettor.
40. What is a ticker?
41. What is the function of choke valve in the carburettor?
42. What happens if the choke valve is left closed after the engine is started?
43. Why is the choke valve usually mounted eccentrically?
44. What is the function of a bimetal spring in an automatic choke?
45. State the function of inlet manifold vacuum in the operation of automatic choke.
46. Why is the choke valve kept slightly open while cranking the engine?
47. In which type of carburettors is the adjustable area jet used instead of choke for cold start?
48. What is the function of 'idle port' in the carburettor?
49. How is the idling speed adjusted in a carburettor?
50. What is the function of transfer port in a carburettor?
51. What do you understand from the term "compensation" in case of carburettors?
52. Name any carburettor which uses compensating jet for compensation.
53. What is the disadvantage of using compensating jet for mixture control?
54. What determines the degree of compensation in case of air bleed method?
55. Which carburettor uses air bleed compensation?
56. In which carburettor is any suction control device used for compensation?
57. What is the function of a stepped metering rod in a carburettor?
58. How does the weather influence the working of a carburettor?

59. What is the effect of altitude on the mixture strength of a single jet carburettor?
60. State any method of preventing ice formation in the venturi at the atmospheric temperature.
61. What are the advantages of triple venturi system?
62. In which carburettor is the triple venturi system employed?
63. What advantages are claimed for the dual carburettor over a single jet type?
64. What is the objection to using two single jet carburettors in place of one dual jet carburettor?
65. What are the merits and demerits of a down-drought carburettor over the other two types?
66. What is the function of a 'throttle return check'?
67. What is meant by the term 'dieseling'?
68. What is an anti-dieseling solenoid?
69. What is the function of a dithering valve?
70. How is the dithering valve operated?
71. Name the various circuits in Carter carburettor.
72. What happens if the float of carburettor in a working engine gets punctured?
73. What is the necessity of providing accelerating pump in the carburettor?
74. What is the function of an anti-percolator on a carburettor?
75. What is a variable venturi carburettor? Give one example of such a carburettor.
76. Name any Indian vehicle using Solex carburettor.
77. Which carburettor is used in Maruti 800 car?
78. What is the purpose of secondary venturi in Solex Mikuni double venturi carburettor?
79. Enlist various adjustments in a carburettor.
80. How do you recognise from a running engine, whether the air-fuel mixture is extra rich or extra weak?
81. Enlist possible causes of high fuel consumption.
82. Where could the trouble possibly lie when you have difficulty in starting the engine?
83. State any five reasons for poor idling.
84. Give any five reasons for poor acceleration.
85. As regards the fuel supply system, state the reason for engine overheating.
86. What causes the 'popping' in the air cleaner?
87. What happens when:
    (*a*) the float gets punctured
    (*b*) carburettor jet is clogged
    (*c*) fuel filter is clogged
    (*d*) air cleaner is choked
    (*e*) carburettor jets have worn out

MULTIPLE CHOICE QUESTIONS

Mark the correct answers:

1. The most widely used fuel supply system for car engines is the
   (*a*) gravity system
   (*b*) pressure system
   (*c*) vacuum system
   (*d*) pump system
2. The drive for the mechanical fuel pump is taken from the
   (*a*) crankshaft
   (*b*) camshaft
   (*c*) distributor shaft
   (*d*) any the these
3. Small holes in the pump body provide a vent for
   (*a*) air
   (*b*) water
   (*c*) fuel
   (*d*) oil
4. Fuel pump outlet pressure should be approximately

(*a*) 3 kPa (*b*) 30 kPa
(*c*) 100 kPa (*d*) 300 kPa

5. Lean air-fuel mixture is required for
(*a*) starting (*b*) idling
(*c*) cruising (*d*) acceleration
6. The carburettor provides the correct quality of air-fuel mixture during
(*a*) starting (*b*) idling
(*c*) acceleration (*d*) all conditions
7. The venturi in the carburettor causes the
(*a*) increase or air velocity (*b*) decrease of air velocity
(*c*) decrease of fuel flow (*d*) decrease of manifold vacuum
8. The throttle valve controls the supply of
(*a*) air only (*b*) fuel only
(*c*) air fuel mixture (*d*) none of these
9. A single jet carburettor tends to supply richer mixture during
(*a*) starting (*b*) idling
(*c*) low speed operation (*d*) high speed operation
10. The choke is usually closed when the engine is
(*a*) hot (*b*) cold
(*c*) idling (*d*) accelerating
11. When the choke is applied the fuel come out from the
(*a*) main jet (*b*) idle port
(*c*) transfer port (*d*) progression hole
12. In a single jet carburettor the mixture tends to become richer
(*a*) during winter (*b*) at low altitudes
(*c*) at high altitudes (*d*) during idling
13. At very low temperatures the ice tends to form in the carburettor in the
(*a*) air cleaner (*b*) venturi
(*c*) idle jet (*d*) float
14. The example of a 'variable venturi' type carburettor is the
(*a*) Carter carburettor (*b*) Solex carburettor
(*c*) S.U. carburettor (*d*) Zenith carburettor
15. Anti-dieseling device is incorporated in the carburettor used in the
(*a*) Ambassador car (*b*) Premier Padmini car
(*c*) maruti 800 car (*d*) All of these
16. The carburettors in which air-fuel mixture is controlled automatically by a computer are called the
(*a*) downdraft carburettors (*b*) updraft carburettors
(*c*) Feedback carburettors (*d*) automatic carburettors

[**Key:** 1. (*d*), 2. (*b*), 3. (*a*), 4. (*b*), 5. (*c*), 6. (*d*), 7. (*a*), 8. (*c*), 9. (*d*), 10. (*b*), 11. (*a*), 12. (*c*), 13. (*b*), 14. (*c*), 15. (*c*), 16. (*c*)]

11 Petrol Engine Fuel Supply Systems-II

1. PETROL INJECTION

1.1. Introduction

There was a brief period of electrically controlled carburettors before petrol injection systems took over. In these systems also, the accelerator pedal is connected to the throttle valve, which controls the quantity of air entering the engine. When accelerator pedal is pressed to admit more air, the ECU *immediately* adjusts the fuel rate in anticipation of the air about to enter the engine. Sensors monitor the mass of air entering the engine and also the amount of oxygen in the exhaust. This information is used by the ECU to fine-tune the fuel delivery so as to achieve the just correct fuel ratio.

In the injection system, the petrol is injected by an electronically controlled valve, called *injector,* which is supplied with the petrol under high pressure by the fuel pump. When the injector is energized, an electromagnet actuates a plunger, opening the valve, allowing the pressurized fuel to squirt out through a tiny nozzle; which is designed to atomize the fuel. The amount of time the injector stays open determines the amount of fuel supplied to the engine and is called the *pulse width,* which is controlled by ECU.

Actual control algorithms are very complicated. ECU uses a formula and a large number of look-up tables to determine the pulse width for given operating conditions. All these tables are provided by the engine manufacturer from their research experience. In an actual engine, the control system may have more than 100 parameters, each with its own look-up table. However, just for illustration, a grossly simplified example with only 3 factors to calculate the pulse width is given here. These factors are, the base pulse width, the coolant temperature factor and the factor for oxygen level in the exhaust.

TABLE 1.1. BASE PULSE WIDTH

Engine speed	*Load*				
RPM	1	2	3	4	5
1,000	1	2	3	4	5
2,000	2	4	6	8	10
3,000	3	6	9	**12**	15
4,000	4	8	12	16	20

The base pulse width depends upon the engine speed and load. Let us assume the engine speed is 3000 RPM and the load is 4. From the look-up table for base pulse width, we find it to be **12** *milliseconds*. Now let us assume the coolant temperature to be 40°C and the oxygen level equals 3. From look-up tables, the respective factors for these are **0.7** and **0.9**. Therefore,

Pulse width $= 12 \times 0.7 \times 0.9 =$ **7.56 milliseconds**

TABLE 1.2. FACTORS FOR COOLANT TEMPERATURE AND OXYGEN LEVEL

Coolant temp. °C	*Factor for coolant temp.*	*Oxygen level*	*Factor for Oxygen level*
0	1.1	0	1.0
10	1.0	1	1.0
20	0.9	2	1.0
30	0.9	3	0.9
40	0.7	4	0.8

1.2. Comparison with carburettor method

An important drawback of using carburettor for metering fuel supply of a multi-cylinder engine is that it becomes difficult for a single carburettor to ensure uniformity of mixture quality in all the cylinders. Thus some may be getting a weaker, while the others may be supplied with a mixture richer than the desired. Further the venturi throat of the carburettor causes a restriction in the passage of air flow to the engine; and if throat is made larger, the problem of suitable mixture supply becomes acute at the low speeds as with a larger throat, the air velocity decreases which causes less efficient atomization and also fall in mixture strength. If to obviate the above troubles, multiple carburettors are used, their matching becomes a problem in itself during actual running. More accurate electronic closed loop control was also found not feasible for carburettors.

The solution to all these troubles has been found in the use of petrol injection. In this the petrol in the atomised form in sprayed into the air stream. Because in this system the petrol is already atomised, the mixing with the air stream is much better, giving a homogeneous air-fuel mixture, which burns more efficiently in the engine. The chief advantages obtained with this are:

1. A very high quality fuel distribution is obtained. As a result, higher compression ratios can be adopted without any danger of detonation occurring.
2. The restriction imposed by the carburettor venturi is removed in the petrol injection; this improves volumetric efficiency with the corresponding improvement in power and torque.
3. Fuel consumption is less.
4. The response of the engine to throttle control is very rapid since there is very small time lag between throttle movement and injection of the fuel, which is not directly injected into each inlet port.
5. Multi-point injection does not need time for transportation of fuel in the intake manifold. Moreover, there is no wetting of manifold walls.
6. Fuel injection equipment is inherently much more precise in metering injected fuel spray into the intake ports over the complete operating range of engine speed, load and temperature.
7. Due to the ease with which engine operating parameters can be monitored in the petrol injection system, the same can be used quite conveniently for accurate matching of air and fuel requirements under all speed and load conditions, thus improving engine performance & fuel consumption and reducing exhaust pollution.
8. Both the single-point or the multi-point systems are particularly suitable for supercharged engines.

The main drawbacks of the method are:

1. Initial cost is very high.
2. Relatively much complicated mechanism, because instead of a simple carburettor, here we need a much complicated and precise fuel injection pump, injection nozzle and pipe lines for each cylinder.

3. Increased maintenance required.
4. Due to the working tolerance of the metering and discharge components, very careful filtration is required.
5. There are more mechanical and electrical components liable to go wrong due to wearing.
6. There is more noise, mechanical and hydraulic, due to pumping and metering of the fuel.

Inspite of these drawbacks, this method was initially used extensively in the luxury and the racing cars. However, in recent times, due to gradually increasing stringent requirements of pollution laws, the petrol injection system is being used now even in case of ordinary cars.

1.3. Types

(*a*) According to location of injector:

(*i*) *Throttle body injection.* In this an injector assembly is attached to a throttle body. This injector-throttle body assembly is installed on the intake manifold in place of a carburettor. (Fig. 11.1) This method simplifies the construction of the engine block and also does not obstruct hot spots near the valves affecting

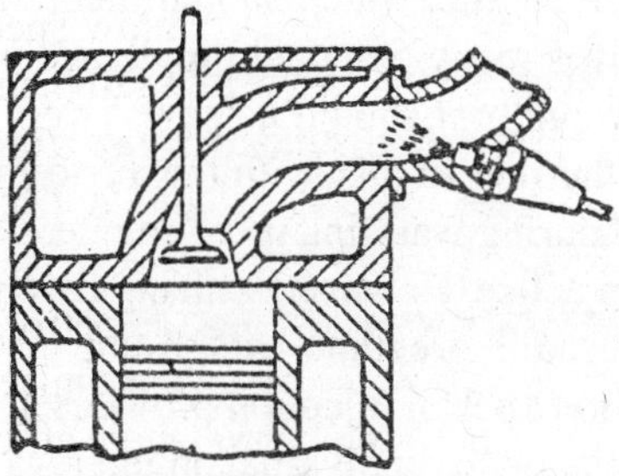

Fig. 11.1. Throttle body injection.

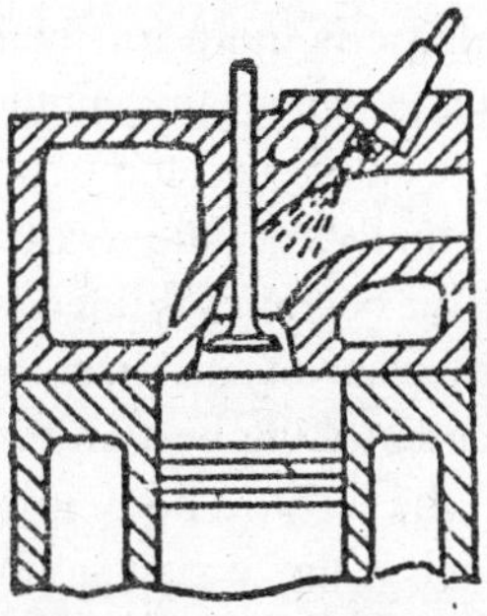

Fig. 11.2. Port injection system.

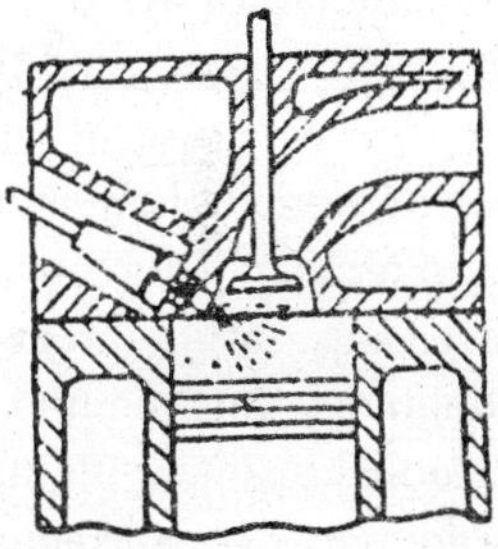

Fig. 11.3. Direct injection system.

the cooling water jacket size at that place. Moreover, it requires only one circuit in the computer to control injection. Thus this is less costly but less precise system. Nevertheless the throttle body injection controls the fuel better than carburettors. In fact at first, carburettors were replaced only by these systems that included electrically controlled fuel-injector valves into the throttle body. These were a sort of bolt-in replacement for the carburettor, without making any major changes in the engine design.

(*ii*) *Port injection.* In this the injector sprays the fuel into *each* intake port on the manifold side of the inlet valve (Fig. 11.2). As such this is also called multi-port fuel injection system.

This system has the advantage that it allows more time for the mixing of air and petrol. Moreover the system is more precise than throttle body type since in this the fuel delivery does not depend upon air to carry the fuel through the intake manifold. Therefore, gradually as the new engines were designed, throttle body injection was replaced by port injection.

(*iii*) *Direct injection.* In this the fuel injector is placed directly into the cylinder (Fig. 11.3). This system is similar to the direct injection system used in diesel engines. Mainly, the intake valve allows only the air into the combustion chamber and the fuel is sprayed- in directly through an injector. The fuel and air mix inside the combustion chamber itself. The piston crown in such engines is normally designed to create a swirling vortex to help mix the fuel and air before combustion. The piston crown also has a cavity on it for ultra-lean-burn conditions. ECU controls the amount of fuel injected based upon the airflow into the engine and demand and would operate the engine in any of the following modes:

1. **Full power mode**—Fuel-air mixture is made richer and the fuel is sprayed in during the intake stroke.

2. **Stoichiometric mode**—Fuel-air ratio is decreased a little to stoichiometric value. Fuel is still sprayed in during the intake stroke.
3. **Ultra-lean mode**—This is selected when cruising with a very small engine load, for example, when driving on a highway with a very little traffic. In this mode, the fuel is injected later on in the 4-stroke cycle, as the piston is moving up on its compression stroke. This forces the fuel-air charge right upto the spark plug points for the best burning conditions. The combustion takes place partly in the cylinder and partly in the shaped piston crown.

Since in the direct injection, petrol is injected from a tiny nozzle into a relatively large cylinder, it absorbs a high latent heat of vapourisation, thereby cooling the air within. The decrease of air temperature in the cylinder means:

1. More air charged into the combustion chamber, resulting in increased torque.
2. The engine becomes less prone to detonation, which further means increased torque and enabling the use of higher compression ratio, resulting in higher fuel efficiency. Mazda 2.0 L DISI engine employed on Mazda 5 claims 7 to 10 percent increased torque at normal engine speeds.

As in this system, the intake manifold and inlet valve do not have to carry vapourized fuel alongwith air, the amount of air going into the engine is more for the same size engine, which should theoretically result in more power and fuel economy. However this method involves many practical problems. Firstly, the injector has to work against high pressures and temperatures; secondly it should achieve a very high degree of atomization since there is virtually no time available for the fuel to mix with the air to form a uniform mixture. Moreover, the engine oil becomes diluted with fuel during warming up of engine. Due to these reasons, earlier direct injection type was not used in automotive practice. However, with improvements in technology, this method is now considered the best and is being used in premium cars. Compared to a 2001 model engine with manifold, *i.e.*, the throttle-body injection, direct petrol injection provides a mean fuel economy improvement of 15% which is the largest saving potential of any individual option regarding any aspect of engine.

Mazda has been able to achieve the desired atomisation and mixing of fuel in their MZR 2.0 L DISI engine by employing swirl injection and TSCV (Tumble Swirl Control Valves). Swirl injectors ensure that the fuel is injected in a hollow conical spray. TSCV is installed on the intake manifold. Under unfavorable combustion conditions, *e.g.*, low engine temperature, TSCV closes one port to increase the flow velocity of intake air and create an in-cylinder flow, called 'tumble swirl', which improves the fuel/air mixing, increases combustion speed and stability.

Bosch expects to achieve the ultimate in petrol injection technology for the next generation D.I. combustion processes by means of specially adopted spray diffusion mechanisms and rapid-response activators. These combined with torque-controlled engine management would result in further fuel-saving in direct injection engines. Another future strategy from Bosch is the DI-motoronic direct start. It is found that direct petrol injection allows a new starting method. The engine is started by the targeted combustion in the best possible cylinder which allows the start-stop function directly through the injection process. For this only an absolute angular sensor and a software module are required. The method can provide an additional 6% fuel economy.

Latest trends: Port injection is continuously being refined with higher-pressure injectors and also with injectors having larger number of smaller holes that increases fuel atomisation, *e.g.*, in the Ford Focus which is a Partial Zero Emission Vehicle (PZEV). Delphi Corp. is also striving to equip port injection with injectors providing better control at low fuel flow levels, better spray patterns and less dripping after the injection.

Work to improve direct injection is also at full swing. However, the fuel economy which was found to be 10-12% in experimental engines with stratified charge (rich enough around the spark plug for easy ignition, but lean elsewhere to save fuel) was reduced to 6-7% in actual engines. Moreover, since stratified charge engines do not run at stoichiometric ratio, which is necessary for efficient operation of the catalytic converter, emissions in such engines are higher which requires costly advanced catalysts which also require fuel completely free from sulphur. Such fuel is not available everywhere and even where it is available, it is quite costly. Thus ultimately a direct injection petrol engine would have a small gain in fuel economy at a much higher cost and also limited to areas where sulphur-free petrol is available.

The practical solution in case of direct injection engines is, therefore, use of stoichiometric engines which work with conventional fuel and converters. New direct injection techniques developed by Ford have resulted in 5% fuel economy in stoichiometric petrol injection engines. New Isuzu engine introduction in 2004 in their Axiom and Rodeo vehicles gives 6% fuel economy in city driving, 10% on highway and increase of 8% in maximum power on conventional 87 octane petrol.

(*b*) According to duration and timing of fuel injection.

(*i*) *Continuous type,* where the fuel injectors are always open which means fuel is being injected continuously all the time the engine is running. In this supply is controlled by the fuel pressure in the system, which in turn is determined by the position of the throttle.

(*ii*) *Intermittent type,* in which all the injector nozzles are opened only for the time period required. This is also called the jerk type or the batch type. Usually these systems fire the injectors once per engine revolution so that injectors could be sized small enough to be more easily controlled at idle.

(*iii*) *Sequential type,* in which the fuel is injected at the exact moment when it will be most useful, towards the end of the intake stroke *i.e.* just before the intake valve for its cylinder closes. This is the costliest since each injector fires separately due to which a separate control circuit is required in the computer for each injector. However, system response is very fast in this case since from the time any change is made by the driver, the system has to wait only till the next intake valve opens, instead of for the next complete engine revolution. Due to this accuracy of control is much better, resulting in improved emissions and driveability.

(*c*) According to number of injectors.

(*i*) *Single point (or central) injection.* It consists of a single injector for the entire engine mounted above the throttle butterfly valve, feeding the engine in a manner similar to a conventional carburettor system (Fig. 11.4). The mixture then passes on to inlet manifold which supplies each cylinder.

(*ii*) *Multi point (or multi-port) injection.* In this system there is a separate injector for each cylinder mounted in the inlet port. The injectors direct fuel onto the back of the inlet valves (indirect method) giving improved mixture preparation and distribution compared to the conventional single point and carburettor systems. Alternatively, injectors may spray directly into the combustion chamber (direct method Fig. 11.4). Air metering is controlled by a throttle body and butterfly valve fitted in the intake system.

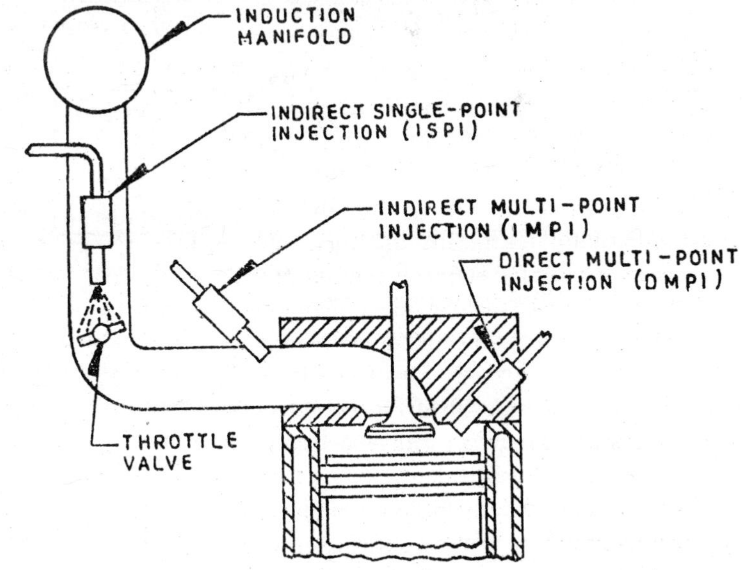

Fig. 11.4. Single point and multipoint injection systems

(*d*) According to control method.

(*i*) *Mechanical.* This is obsolete now. In this a governor was employed to control supply and a fuel distributor was used to send the fuel to the correct injector.

(*ii*) *Electronic.* A few of these systems are, L-Jetronic (introduced in 1973 and used since then on many cars), Mono-Jetronic, LH-Jetronic and Bosch Motronic system (introduced in 1980).

1.4. Mechanical Injection

An electrically driven fuel pressure pump is mounted near the fuel tank. It pumps the fuel at a specified pressure (about 700 kPa) into a metering distributor. A relief valve returns the excess fuel to the tank, thereby keeping the fuel supply to the metering distributor at constant pressure. The metering distributor supplies fuel to each injector in turn. The quantity of fuel delivered is also controlled in the distributor by engine manifold pressure. The injector is ordinarily held closed by spring until the fuel pressure opens it to deliver atomized spray of fuel. A manual control on the dash board controls the metering distributor and thereby the quantity of fuel delivered by it.

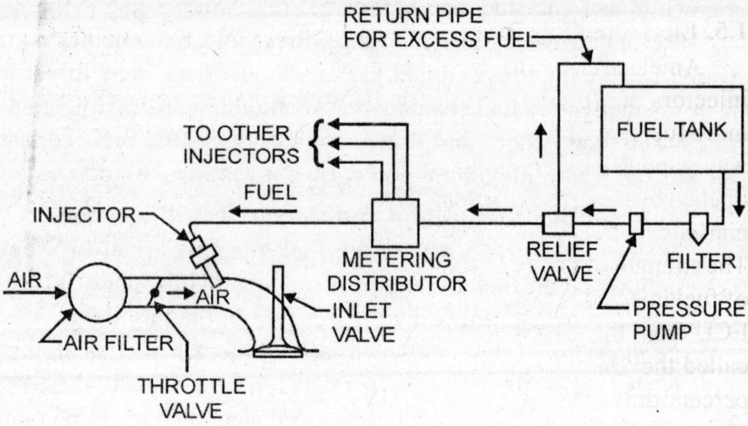

Fig. 11.5. Operation of mechanical petrol injection system.

K-jetronic mechanical fuel injection system has been developed by Bosch (Fig. 11.6). In this the controled quantity of air drawn into the engine cylinders is measured by an air flow sensor. The fuel under pressure is supplied by an electrically driven roller-type pump, through an accumulator and a filter, to the mixture control distributor unit. The accumulator maintains the fuel line pressure when the engine is not running for some time so that the injection pump system responds immediately when the engine is cranked for restarting. Moreover, it also absorbs the noise created by the charging and discharging of roller cell pump.

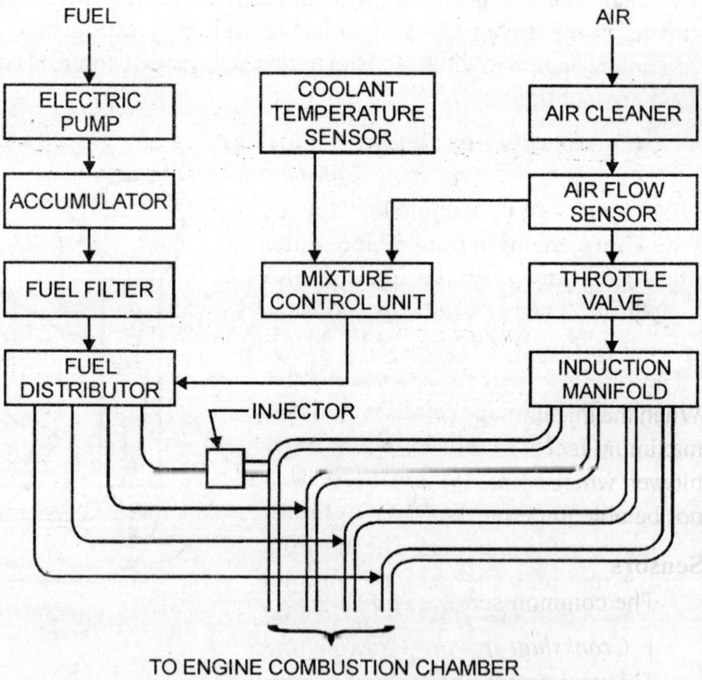

Fig. 11.6. K-jetronic fuel injection system

The fuel entering this control unit is kept at a constant pressure by means of a pressure regulator.

The mixture control unit meters the quantity of fuel discharged into the air stream before transferring it to the fuel injector valves (injectors). This metering is dependent upon the measured air flow signaled to the mixture control unit.

Fuel injection valves are thermally insulated in holders to prevent formation of fuel vapour bubbles in the fuel line due to the engine heat, which would make the starting of the engine difficult. The injection valves do not meter fuel. Their function primarily, is to open at a predeter- mined line pressure, say about 3.3 bar, and to discharge fuel in the form of spray into their respective induction ports.

1.5. Electronic injection

An electrically driven pump draws the fuel from the tank through a filter and supplies the same to the injectors at a pressure which is held constant by means of a fuel-pressure regulator. The pump draws more fuel than the required and the excess fuel is returned to the tank by the fuel pressure regulator. In this way vapour lock is prevented in the fuel lines. The injectors are held closed by means of spring and are opened by means of solenoids energized by the control signal from the electronic control unit (ECU), which consists of a small preprogrammed analog computer that translates sensor signals into command signals. The strength of the ECU control signal, which determines the open time of the injector to control the amount of fuel injected (called 'Pulse width'), depends upon the engine requirements which are determined by the ECU from the sensor signals from critical locations. The percentage of the time the injectors are open is called the '*duty cycle*'. For example if an injector remains open for half the time, it is said to have a 50 percent duty cycle. In a standard engine, at maximum power the injectors may have an 85% duty cycle.

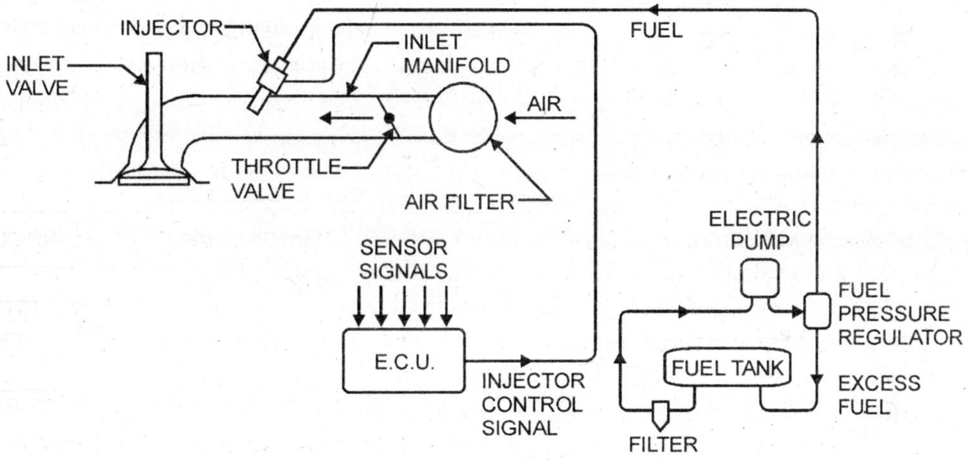

Fig. 11.7. Schematic diagram of electronic petrol injection system.

When the injectors remain open for all the time, their duty cycle is 100% which means that they are delivering maximum fuel according to their capacity. If now some accessory is to be added, for example, an aftermarket blower, which would deliver extra air, it would also require additional fuel which the earlier injectors would not be able to deliver due to the which they have to be replaced by larger injectors.

Sensors

The common sensors employed with ECU are:

1. *Crankshaft speed/position sensor*

This registers the speed and angle of crankshaft without contact. It may be inductive, differential Hall or Anisotropic Magneto Resistance (AMR) type. This allows the ECU to fire the injectors and the spark at the correct instant. These are of different types, the commonly used being the 'optical' and the 'Hall effect' types.

2. *Camshaft speed/position sensor*

Also called the phase sensor, it measures the speed and position of the camshaft without contact.

3. *Knock sensor*

This is used to recognise the onset of knocking using fuels of varying quality, thereby controlling the knock resulting in, apart from engine protection, fuel saving of about 9% and torque increase of 5%.

4. *Mass Air Flow (MAF) sensor*

It reads airflow by varying the current needed to heat a platinum wire which is cooled by engine intake air flowing around the wire (also called platinum hot wire anemometer). The air flow tenas to cool the wire while the current from ECU tries to maintain the same wire temperature at all times. More electricity needed to maintain the temperature would mean higher mass flow of the intake air, and vice versa. Such sensors do not pose any restriction in the intake air flows. MAF sensor is located in the tube between the air cleaner housing and the throttle.

5. *Manifold Air Pressure (MAP) sensor*

It measures the vacuum in the intake manifold continuously. The vacuum depends on the engine speed and the throttle opening (load on the engine). Low manifold vacuum at high engine speeds means high load, whereas high manifold vacuum at low or high speed indicates closed throttle, i.e., low load. MAP sensor is located after the throttle butterfly valve on the intake manifold and does not cause any restriction to the flow of the intake air.

6. *Barometric pressure (BARO) sensor* (correction for air density change with height).

7. *Throttle position sensor (TPS)* (correction for sluggish movement of fuel droplets during speed transition conditions).

8. *Coolant temperature sensor (CTS)* (correction for poor atomization and wall wetting). It is positioned on the thermostat housing on most cars. The sensor resistance varies with temperature so that as the temperature rises, the resistance decreases. ECU applies a regulated voltage to the sensor and from the changed voltage coming back from it, the sensor determines the coolant temperature. A cold engine needs a richer air/fuel mixture when it is first started. When the coolant reaches a certain temperature, the engine goes into closed loop operation, *i.e.*, it starts using inputs from the oxygen sensors to fine-tune the mixture. When it is operating in Open Loop (when cold or when there is no signal from the coolant sensor), the air/fuel mixture is fixed and does not change.

9. *Manifold air temperature (MAT) sensor* (correction for air density variation with atmospheric temperature). It is usually located on an intake manifold. Like the coolant temperature sensor, it is also a variable resistor, a typical sensor having a resistance of 3555 ohms at 20°C and just 475 ohms at 70°C.

10. *Exhaust oxygen sensor* (correction for emission control. It tells the ECU whether the engine is running rich or lean at part-throttle. The sensor generates its own voltage output, which is quite low (about 0.2 volt) when air/fuel ratio is lean and high (about 0.8 volt) in case of rich mixture. The ECU uses this sensor output to keep mixture at 14.7:1 strength for most of the time through a feedback control.

11. *Distributor reference pulses* (for control of open time of fuel injectors).

12. *Vehicle speed sensor (VSS).*

13. *Battery voltage sensor* (correction for supply voltage to control unit and injectors).

Basic systems

1. *Mass Airflow systems.* The MAF sensor reading is combined with other sensor inputs, *e.g.*, engine RPM, throttle position, engine load (through MAP sensor), etc., sufficient to calculate the optimum A/F desirable for that particular moment.

2. *Speed Density systems.* There is no airflow sensor. The airflow is *estimated* (not measured), using engine RPM, inlet air temperature, throttle position and engine load (from MAP sensor).

Comparison

1. Speed density system is less sophisticated than the mass airflow system, but is cheaper since no MAF sensor is needed.

2. Speed density system is less sensitive to small air-leaks occurring between the air cleaner and throttle or in the intake manifold or vacuum hoses connecting to the intake manifold. Air leaks being unmetered, cause the engine to run leaner than normal.

Desired air-fuel (A/F) ratios & their control

1. Very rich A/F ratio is needed at the time of cold start, when the engine is under load for more power and when the throttle is suddenly opened to avoid hesitation.

2. Wherever possible, The A/F ratio can be made leaner for fuel economy, e.g., when the vehicle is cruising under light load and throttle. A/F mixture may also be leaned out or even cut off completely during acceleration to save fuel.

As already explained above, A/F ratio is controlled by varying the pulse width (duration), duty cycle or on-time of the fuel injectors as per signals from oxygen sensor. The speed of the computer determines the rate at which such changes in A/F ratio can be made. Many EFI systems in late model vehicles are really very fast- they can readjust the A/F ratio between firings in individual cylinders. A/F ratio control is further explained in detail in the next article, viz., 'Fuel Trim'.

Apart from fuel and spark control, most electronically controlled fuel injection engines also control the idle speed by means of a small electric stepper motor, called the Idle Air speed Control (IAC) motor, which is controlled by ECU. When engine is subjected to additional load, *e.g.*, engaging the air-conditioning compressor or when alternator is subjected to heavy electrical loads, this IAC motor meters additional air into the engine to raise the idle speed.

Fig. 11.8 shows a typical control layout for an electronically controlled petrol injection system.

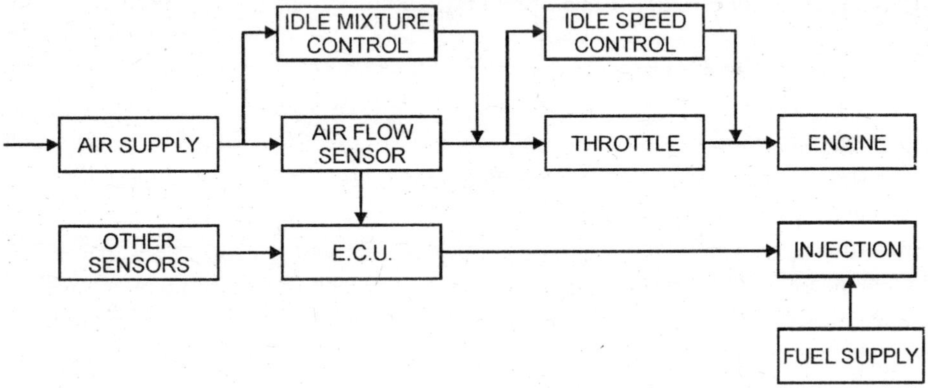

Fig. 11.8. Control layout for a petrol injection system.

Faulty inputs from any of the engine's sensors may result into the drivability, emissions or performance problems, many of which would turn on the Check Engine Light, which can be diagnosed by reading the Diagnostic Trouble Code (DTC) with a scan tool.

2. FUEL TRIM

Fuel Trim is the continuous monitoring of air/fuel mixture done by the engine computer according to the speed/load conditions under which the engine is running. It is also displayed on a scan tool as a percentage reading. Changes in fuel trim are based on the input from oxygen sensor and hence their accuracy depends directly on the accuracy of oxygen sensor inputs. The inaccuracies in the oxygen sensor readings may be caused by many conditions, *e.g.*, sensor may read leaner than the actual in case of a misfiring spark plug, a leaking exhaust valve or an air leak in the exhaust manifold gasket, allowing false air into the exhaust. Similarly, an old oxygen sensor may be sluggish, contaminated, unresponsive or dead, due to which it may give lower oxygen level in the engine exhaust, indicating a richer air/fuel mixture than its actual quality. In either of the above cases, computer would act unnecessarily to increase or decrease mixture richness, which can result in deterioration in fuel economy and exhaust emissions.

Positive fuel trim values mean the engine computer is *adding fuel* to enrich a leaner A/F mixture, whereas *negative* fuel trim values indicate that the engine computer is *subtracting* fuel to lean out a richer A/F mixture. There are two types of fuel trim values shown:

1. Short Term Fuel Trim (STFT)

This indicates what the engine computer is doing right now. This value is constantly jumping around as the computer is adjusting and readjusting the A/F ratio depending on the variations in engine load, speed temperature and other operating conditions. Values normally vary from –10% to + 10%, although may jump to as much as + or – 25% when the throttle is suddenly snap opened.

2. Long Term Fuel Trim (LTFT)

It is *not an instant value* like STFT, but is a longer term average of how the air/fuel mixture is trending over time. It is the best indicator to tell whether the engine is running too rich or too lean and is possibly experiencing an emission problem.

Ideally both STFT and LTFT should be as close to zero as possible when the engine is idling or being run at a steady speed and load. However, in practice, LTFT numbers vary between 5 and 8, depending on the condition of the engine. LTFT values more than 10 would usually indicate a problem which needs to be investigated. However in case of LTFT value going up to +20% or more, it would set a P 0171 or P 0174 lean code and turn on the 'Check Engine light'. Similarly LTFT value –20% or more would set a P 0172 or P 0175 code and turn on the 'Check Engine Light'.

3. MAIN COMPONENTS OF PETROL FUEL INJECTION SYSTEMS

1. Electric fuel pump

On engines with fuel injection, it is used to pump fuel from the petrol tank to the injectors at high pressure from 250 to 1200 bar depending upon the application. Too less pressure may starve the engine causing it to run lean, misfire, hesitate or stall. Similarly excess pressure can cause the engine to run rough, waste fuel and pollute.

Fuel pump is usually mounted inside the petrol tank, although some are mounted outside also. Some vehicles even have two pumps, one inside and the other outside the tank. The pump inside the tank raises the boiling point of petrol, avoiding vapour lock, helps muffle the noise and keep the pump cool.

In fact, the pump is generally a part of the assembly of the fuel sending unit which includes apart from the pump, the filter, the strainer and the electronic device to indicate the amount of fuel in the tank via a float attached to a sensor which sends data to the fuel gauge on the instrument panel.

Electric fuel pumps are of two types, viz., the positive displacement type and the flow type. The positive displacement types may be further sub-divided into the roller cell type or the vane type or the geroto type. The positive displacement type can generate very high pressure and flow rate is constant. However, in this type the output comes in pressure pulses to dampen which a muffler has to be mounted. In the flow pumps, the pressure build-up is continuous and pulsation-free, thus requiring no muffler. Moreover, they run at high speeds typically upto 7000 rpm and draw less current. This type is also less complicated to manufacture and has a long life. However, maximum pressure obtained is less than in the positive displacement type.

The working details of the electric fuel pump are as follows:

1. On turning the ignition key ON, the PCM (Powertrain Control Module) energizes the relay, supplying voltage to the pump, which causes the pump-motor to start spinning and run for a few seconds so as to build pressure in the fuel system. The actual time for which the pump runs until the engine starts, is controlled by a timer in the PCM.
2. When the pump starts running, the fuel is drawn into the pump through an inlet tube and mesh filter sock. The fuel exists the pump through a one-way check valve and is pushed toward the engine through the fuel line and the filter. The check valve is meant to prevent the oil to flow back, thus maintaining the pressure in the system when the pump is not working.
3. The fuel from the filter goes to the fuel supply rail on the engine and is then routed to individual fuel injectors. There is a fuel-pressure regulator on the rail which maintains the fuel pressure and returns the excess fuel back to the tank. On late model engines there is no fuel return line while the fuel pressure regulator is located in the fuel tank and is a part of the fuel pump module.
4. Once the pump starts, it runs continuously till the ignition key remains in the ON position. The speed of the pump may remain constant or vary, depending on the engine speed and load. In case of the engine stalling, the loss of the speed signal would be detected by PCM, which would turn off the fuel pump. In addition, many vehicles also have an 'inertia safety switch'. In case of an accident, the hard jolt trips the safety switches and opens the pump circuit, shutting off the pump. This is done to reduce the risk of a fire in case a fuel line is ruptured. In such an event, the safety switch has to be reset manually.

2. Air Meter

It measures the mass of air being taken in by the engine. Modern air meters use hot wire anemometry technique, wherein the output signal of the air meter is determined by the power required to keep its heated sensors at a fixed temperature. The engine control module utilizes the air meter output signal to precisely time the fuel injection, which enables it to maintain optimum air-fuel ratio, thus achieving lower emissions. Air meter is mounted between the air cleaner and the throttle body of the engine.

3. Throttle Bodies

A throttle body controls air flow to the engine. As part of the throttle body, a butterfly valve is used to open and close the passage into the intake manifold to increase or decrease the volume of the inlet air. Two types of throttle bodies are in use:

(*i*) Mechanical type, which are connected to the accelerator through a mechanical linkage. These are made out of aluminium or composite materials. Simplicity and low cost are their main advantages.

(*ii*) Electronic type, which are controlled by an electronic signal from the management system of the engine. They provide better system air control compared to mechanical throttle bodies, which enables advanced engine technologies and vehicle control features such as hybrid systems, direct petrol injection, collision avoidance, adaptive cruise control, etc., to be employed for improving engine efficiency.

4. Integrated air-fuel modules

Delphi makes integrated air-fuel modules, which allows for improvement in engine power, torque, fuel economy and emissions. Such modules include mainly the intake manifold, fuel rail and injectors, throttle body, EGR valve, PCV valve, canister purge solenoid and manifold pressure sensor.

5. Sensors

These are used to provide all the required data from the various components to achieve an efficient engine management. Important sensors used in the fuel injection system have already been listed in preceding article.

6. Fuel rail

Multi-port fuel injection fuel-rail assemblies receive fuel from the chassis fuel line and distribute the same to each injector location. These may be ordinary or damped type. In the damped type pressure pulsation levels in the rails are significantly reduced leading to improved driveability, reduced emissions and minimised fuel line noise.

7. Tuned flexible injection system

As an alternative to the fuel rail assembly, it consists of a fuel body and flexible distribution lines to the fuel injectors (Fig. 11.9). Due to its flexibility, this system does not have to be configured for each engine and vehicle, which means reduced cost. It is used on sequential fuel injection engines.

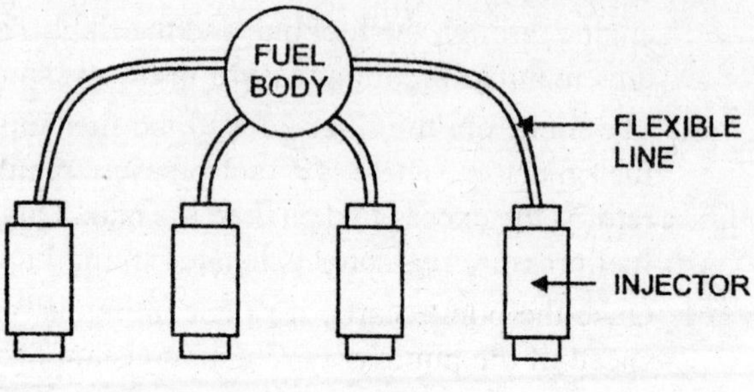

Fig. 11.9. Flexible injection system.

4. BOSCH MOTRONIC GASOLINE SYSTEM

Fig. 11.10 depicts clearly a Motoronic gasoline electronic ignition system from Mico- Bosch. Motoronic systems have microprocessor-based control systems with continuous correction of injected fuel quantity, ignition angle as well as composition of the air-fuel mixture depending on the engine and external conditions. This ensures optimum level of engine performance with minimum emission level in the exhaust under all operating conditions.

A simplified diagram of the Motoronic system for direct petrol injection engines is shown in Fig. 11.11.

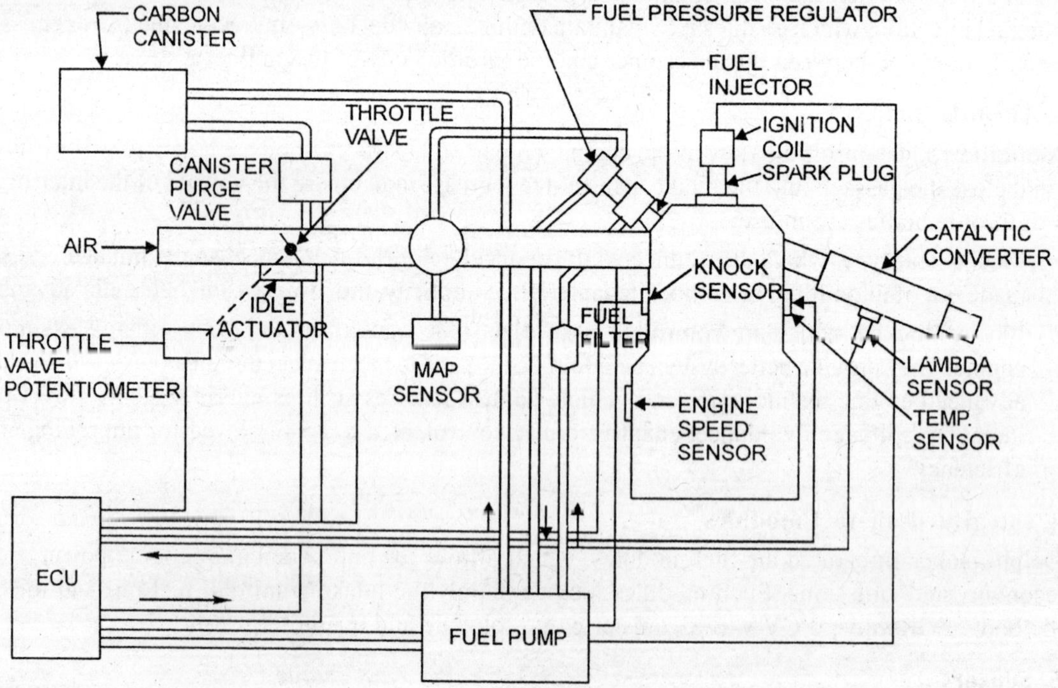

Fig. 11.11. Simplified diagram for Mico-Bosch Motoronic System

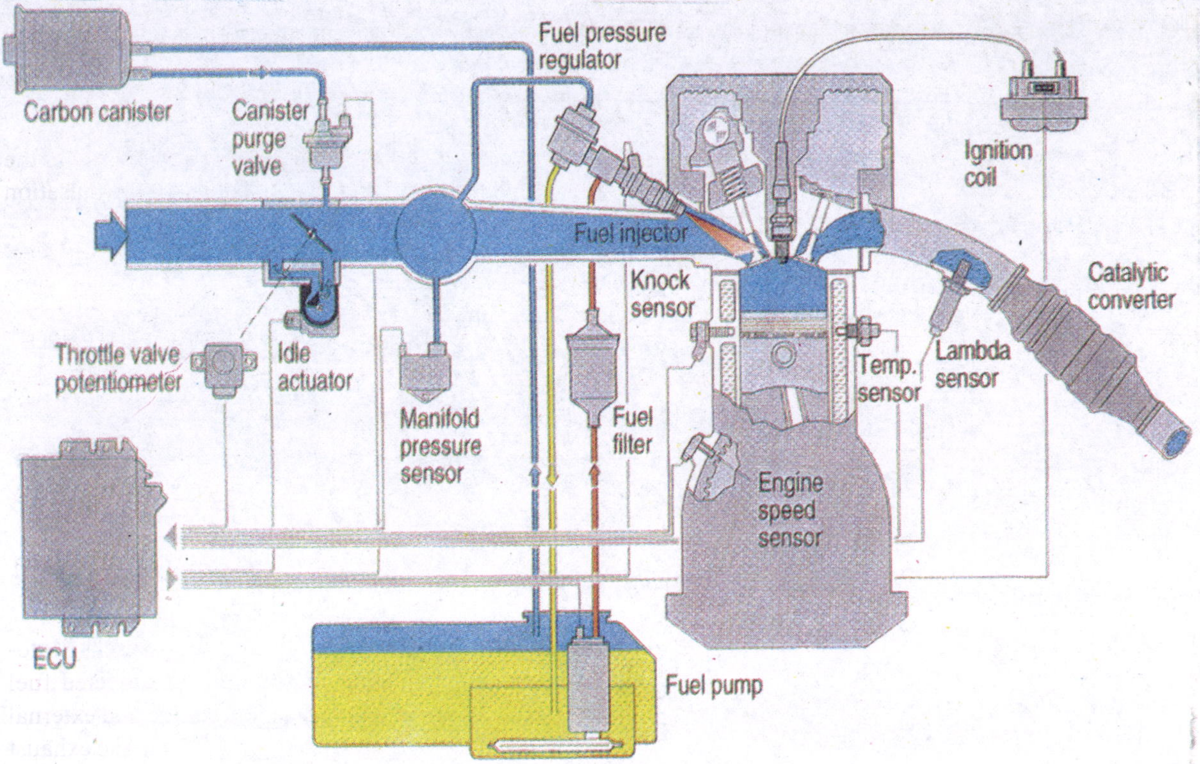

Fig. 11.10. Mico-Bosch Motoronic System. (Courtesy–MICO, India)

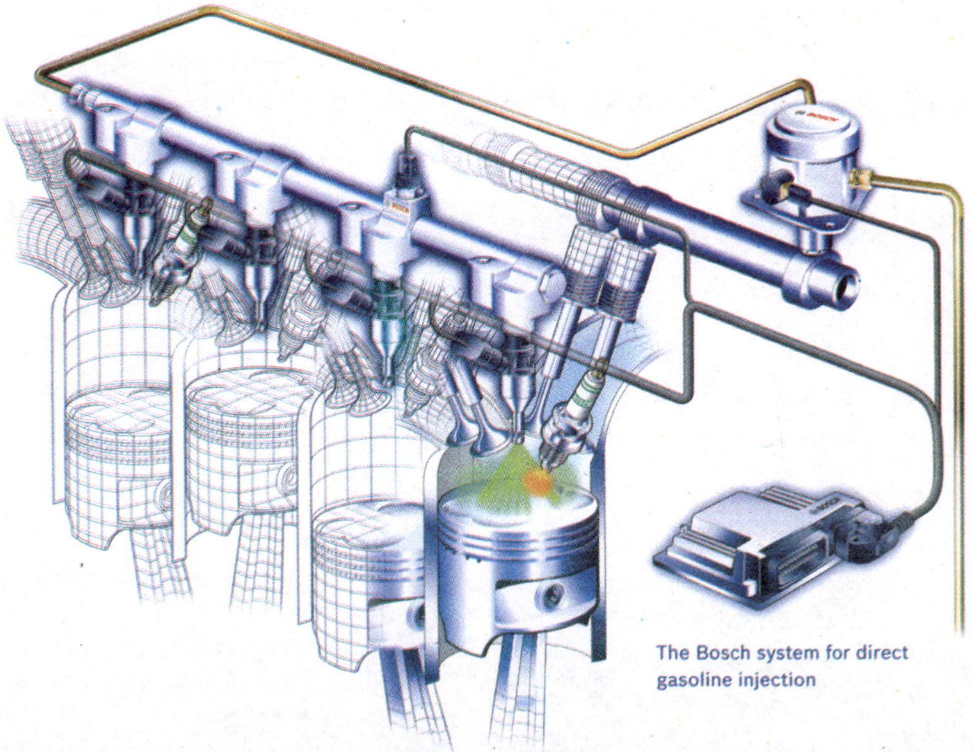

Fig. 11.12. Mico-Bosch Di-Motoronic System. (Courtesy–MICO, India)

ECU receives signals from throttle value potentiometer, MAP sensor, knock sensor, engine speed sensor, temperature sensor and lambda sensor and sends direction signals to idle actuator, canister purge valve, fuel injectors and ignition coils. The combustion takes place in different modes as follows:

1. Stratified charge mode. This occurs at low torque output upto about 3000 rpm. In this the fuel is injected during the compression stroke just before the spark plug fires, when the air flow movement in the combustion chamber transports the air/fuel mixture around the spark plug. This results in a richer mixture around the spark plug while the mixture elsewhere in the combustion chamber is relatively lean. In addition, ERG gases also enter in the rest of the chamber away from the spark plug, which results in decreased combustion chamber temperature and thereby lesser amount of NO_x emissions. In the Bosch direct injection system, the overall air/fuel ratio within the chamber can be as lean as 22:1 to 44:1, whereas in a conventional port injection system this ratio is seldom leaner than 14.7:1.

2. Homogeneous mode. This occurs at high torque output and at high engine speeds. In this mode, the injection starts on the intake stroke so that there is enough time available for the mixing and uniform distribution of the mixture throughout the combustion chamber. The Bosch system uses 14.7:1 ratio during this mode, which is the same as in case of the port injection at light loads.

3. Homogeneous lean-burn mode, which is used in the transition between the stratified and the homogeneous modes.

4. Combined Homogeneous Stratified Charge mode. In this mode, initial injection (about 75% of the total fuel injected) occurs during the intake stroke, which gets plenty of time to mix the fuel with the air. Then the remaining 25% of the fuel is injected during the compression stroke, which results in a zone of rich mixture around the spark plug. The rich mixture easily ignites which in turn helps the leaner mixture in the rest of the chamber to ignite.

5. Homogeneous anti-knock mode which occurs at full throttle.

6. Stratified charge cat-heating mode, used to heat the catalytic converter quickly to operating temperature.

7. Rich homogeneous mode, which is used to regenerate the NO_x catalytic converter in which NHO_3 nitrates are deposited. On regeneration, these nitrates and CO are reduced to N_2 and O_2.

5. DI-MOTRONIC SYSTEM

The Bosch system for direct injection in petrol engines is shown in Fig. 11.12.

Main components include a high pressure fuel pump, injectors on each cylinder, pressure accumulator or fuel rail and the E.C.U. The high pressure pump maintains fuel flow to the rail, which is connected to the injectors via hydraulic connection lines. During actual driving, ECU monitors and performs diagnosis in particular on all components which have an influence on system safety and reliability and on emission-behaviour.

During part-load operation (Fig. 11.13) only the minimum amount of petrol needed to run the engine at the corresponding torque is injected, which is achieved with the torque-guided control. This gives 15% more mileage at part-load.

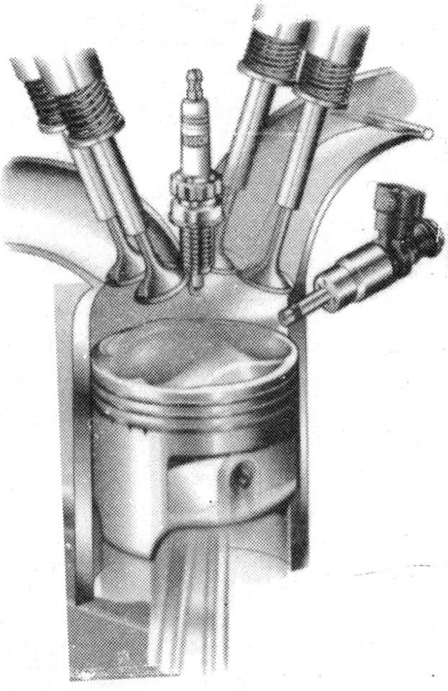

Fig. 11.13. Part-load operation with DI-Motronic (Courtesy-MICO, India)

During full-load operation (Fig. 11.14), the direct petrol injection ensures that the mixture is cooled, permitting higher compression. In this mode the engine gives 5% more mileage for the same piston displacement. Various advantages of this latest system are:

1. A number of additional engine functions have been made possible due to direct injection. These include:
   (*i*) Convenient and fuel-saving start-stop system.
   (*ii*) Intelligent cold-start combustion processes which can also feature multiple injection.
   (*iii*) Catalytic converter is heated far more quickly, reducing emissions to a minimum.

2. It supports all combustion processes and operating modes. It is ideal for implementation of stratified-charge lean burn combustion processes, as well as for homogeneous air-fuel mixture formation.

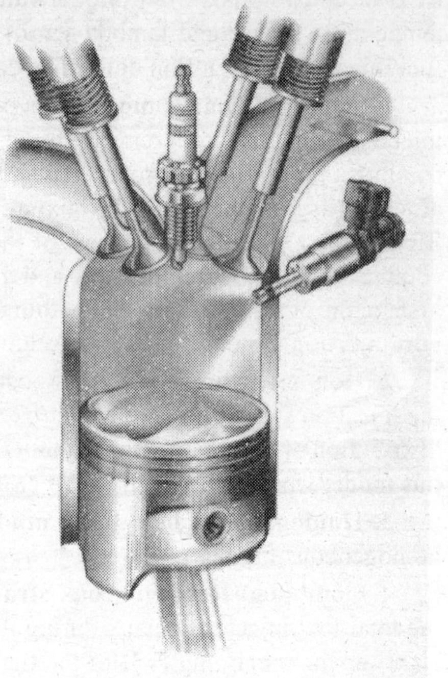

Fig. 11.14. Full-load operation with DI-Motronic (Courtesy-MICO, India)

With lean-burn combustion, fuel injection, ignition and fuel pressure are controlled so that from idle upto the medium load ranges, there is stratified charge in the cylinder. A highly economical, ignitable air-fuel mixture forms in only one small area of the cylinder, which is sufficient to generate the required power output. The remaining portion of the combustion chamber is filled with fresh air and the residual exhaust gas from fuel combustion. Since the intake air is now throttled, throttling losses are avoided resulting in higher efficiency and lower fuel consumption. For higher output, the system reverts to operation with homogeneous mixtures. Besides, other operating modes, such as dual injection, further reduce fuel consumption and exhaust emissions.

With lean-burn and homogeneous combustion processes, cooling of the combustion chamber allows the use of higher compression ratios compared to the manifold injection, resulting in potential for higher engine output, improved dynamic response and reduced fuel consumption.

3. It is very well suitable with supercharged engines and variable valve timing. It achieves increased torque and reduced fuel consumption even at low speeds due to improved air-fuel mixture formation, optimised gas exchange and use of higher compression ratios.

4. It coordinates various engine management functions, *viz.*, fuel and air delivery, mixture formation and combustion, emission control, alongwith comfort and convenience.

5. DI-Motronic also controls the exhaust gas system besides the engine management. It operates with EGR and new catalytic-converter technology. This, combined with high-pressure stratified start-up, makes it possible to keep hydrocarbon and nitrogen oxide emissions considerably below the strict SULEV (Super Ultra Low Emission Vehicle) level.

6. NATURAL GAS MOTORONIC

Fig. 11.15 shows the Bosch Motronic for CNG, installed on a car and is self-explanatory. Integrating the petrol and CNG functions in a single ECU has the following advantages:

1. At any time a coordinated change can be made between petrol and CNG operation *without any jump in torque.*
2. Lower cost.
3. The combined system as a whole has greater potential for optimization.

Fig. 11.15. Bosch NG-Motronic (Courtesy—MICO, India)

1. Pressure control module, 2. In-take unit, 3. CNG-tank, 4. CNG shut-off valve, 5. NG-Motronic ECU, 6. CNG fuel rail, 7. Petrol rail, 8. Ignition coils

7. SUPERCHARGING

Supercharging is the process of supplying to the engine, air fuel mixture at a pressure above the atmospheric. On an ordinary engine without a supercharger, the downward piston movement during the intake stroke creates a vacuum in the inlet manifold, which is used to draw in the air-fuel mixture through the carburettor into the cylinders. With supercharging, however, due to higher pressure, the density of charge increases and, therefore, its weight per stroke is increased for the same swept volume. It is seen that the power output of an engine is almost directly proportional to the weight of charge per minute; therefore, the supercharged engine gives more output. Upto 40% increase in power and torque may be obtained by super-charging.

Further at higher altitude, the air gets thin thus reducing the weight rate of air flow in the engine leading to fall in the power output. It is here also that the supercharging comes to our aid by increasing the density and hence the weight rate of air flow in the engine, thus offsetting any fall in power output due to altitude.

Theoretical operating cycles of a naturally aspirated and a supercharged petrol engine have been shown in Fig. 11.16 and 11.17 respectively. In Fig. 11.16, loop 2-3-4-5-6 represents the useful power developed, whereas loop 0127

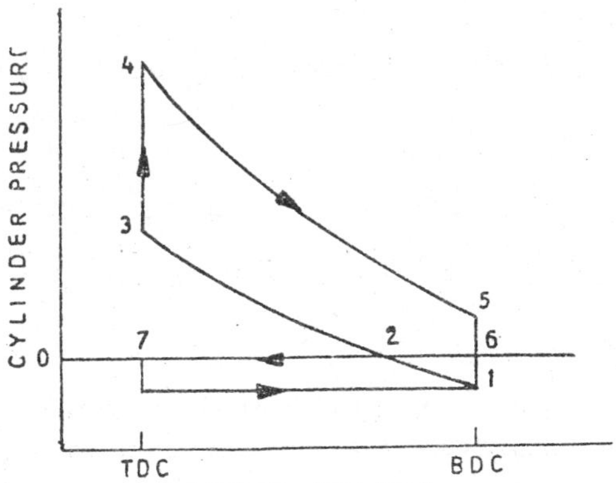

Fig. 11.16. Theoretical P-V diagram of a naturally aspirated engine

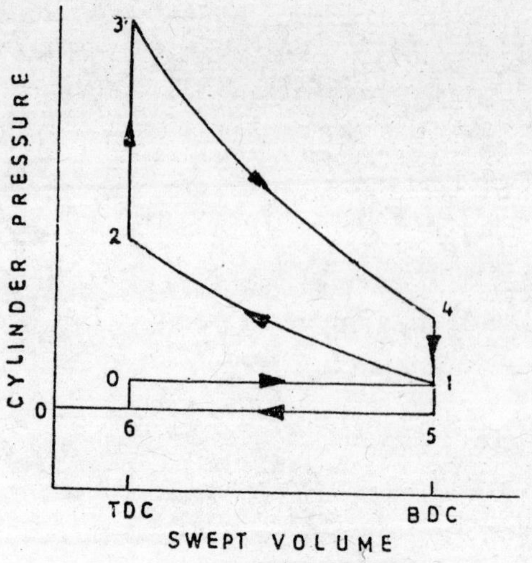

Fig. 11.17. Theoretical P-V diagram of a aspirated and supercharged engine

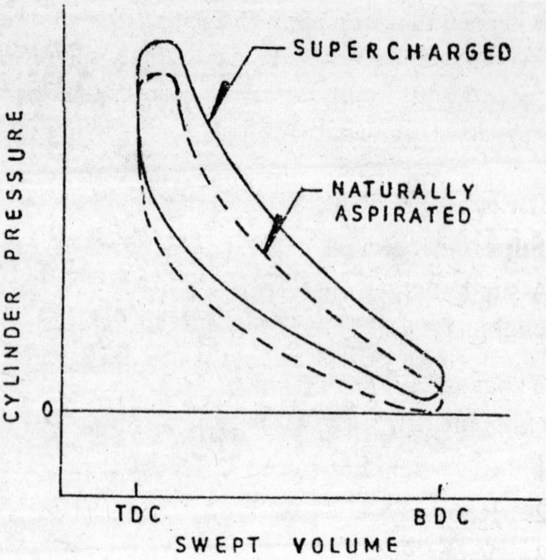

Fig. 11.18. Actual P-V diagrams of naturally the supercharged engines.

represents the negative power, *i.e.*, the power required to induct the fresh charge in the cylinder. In the same way, in Fig. 11.68, the useful work is represented by the loop 1-2-3-4, while the lower loop 0-1-5-6 corresponds to the work done in pumping the fresh charge into the cylinder. A comparison of the actual P-V diagrams of the two type of engines has been shown in Fig. 11.18. It is seen that:

1. The mean effective pressure, represented by the vertical distance between the upper and the lower curves for each engine, is greater for the supercharged engine throughout the cylinder swept volume.
2. The loop area enclosed by the supercharged engine and hence the power developed in the same is much larger than in the naturally aspirated engine.

However, it is seen that:

1. Alongwith increase of power output, the rate of fuel consumption is also increased, but then leaving aside the extremes, the fuel consumption per brake kW hour does not increase, because the power increase keeps pace with the increased fuel consumption.

2. The power consumed in compressing the air apart from the increased power losses in the engine itself, has to come from the engine power itself. Thus the net gain of power is obtained only after deducting this power consumption from the gross B.P. But it is seen that even this net power output is more than the power developed by the engine of the same size, but without supercharging (Fig. 11.19).

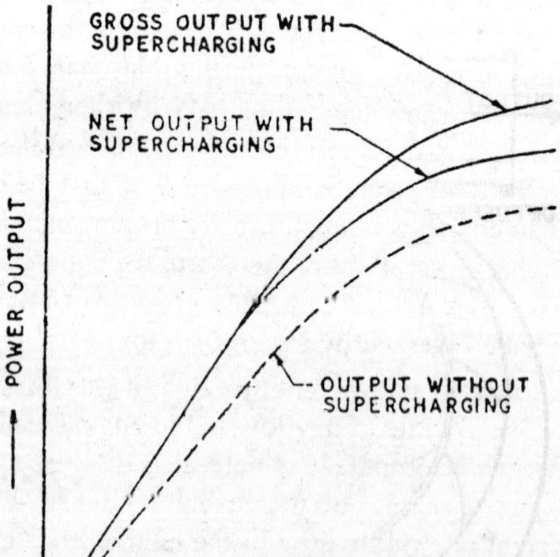

Fig. 11.19. Curves showing relative power output with and without supercharging.

3. If supercharger is used alongwith an existing engine, it demands certain precautions. Firstly it must first be accesertained that the engine will be able to take the higher forces due to supercharging and secondly the fuel with better anti-knock characteristics has to be used to avoid occurrence of detonation due to increased pressure and temperature.

It is seen that in petrol engines, if the charge density is increased (due to supercharging) and if same octane number petrol is to be used, then to avoid detonation, compression ratio must be decreased. A lower compression ratio will, however, result in decreased air standard and hence lower thermal efficiency. *That is the reason, super-charging was not being commonly used in automotive petrol engines till recently.*

8. SUPERCHARGERS AND TURBOCHARGERS

8.1. Supercharger

A supercharger, also called a blower, is a gas compressor used to pump air into the cylinders of an I.C. engine. In petrol engines, the charge is petrol/air mixture.

8.2. Types of Superchargers

Generally, four types of superchargers are in use:

1. Centrifugal type
2. Root's type
3. Vane type
4. Screw type

1. Centrifugal type supercharger

This is the most common type of supercharging on the automotive engines. It is run from the engine pulley by means of a V-belt. The air/fuel mixture enters the impeller at the centre and after passing through the impeller and the diffuser vanes, it enters the volute casing, where a part of its kinetic energy is recovered and the mixture at supercharged pressure goes out of the casing to the engine (Fig. 11.20). Generally due to the high pressure, about 30 per cent more air-fuel mixture is forced into the combustion chamber.

The impeller of this supercharger is required to run at very high speeds, say upto 80,000 r.p.m., which dictate that it should be made out of some metal, which should be able to take the accompanying stresses at such high speeds. Generally, impellers are made of duralumin, or alloy steels.

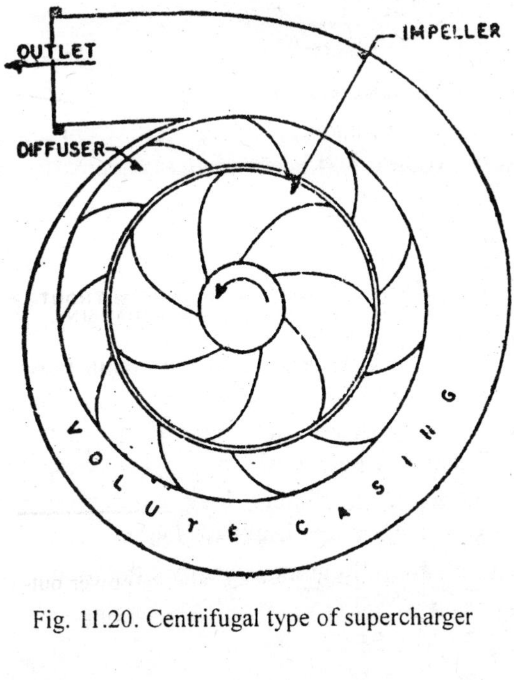

Fig. 11.20. Centrifugal type of supercharger

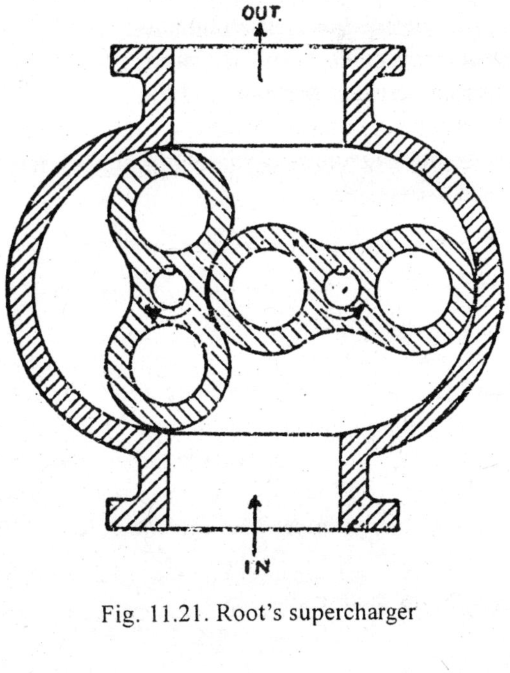

Fig. 11.21. Root's supercharger

2. Root's supercharger

It consists of two rotors of epicycloid shape (Fig. 11.21). An epicycloid is the locus of a point on the circumference of a circle which rolls on another circle. Each rotor is keyed to its shaft, which are connected together by means of gears of equal size, so that the two rotors revolve at the same speed. The working action of this supercharger is just like a gear pump, so that we get the mixture at the outlet at a high pressure.

Eaton 5th generation supercharger (2008) is a Roots-type positive displacement pump with no internal compression. Each of the two tri-lobed rotors have been twisted 60 degrees to reduce noise and improve air handling efficiency. The rotors are coated with Eaton's patented APC (Abradable Power Coat) to improve intermesh, reduce leakage and enhance efficiency. The twisted rotors also allow for axial air flow, which makes packaging in the limited engine compartment easy.

Another roots-type supercharger from Eaton (2008) is the Twin Vertices Series (TVS), which has four-lobe rotors and high-flow inlet and outlet ports which increase thermal efficiency, enhance volumetric capacity and enable higher operating speeds. The twin four-lobe rotors have 160 degree twists. The TVS has improved noise and vibration characteristics and 2.4 pressure ratio capability.

3. Vane type supercharger

Fig. 11.22 shows the construction of a vane type supercharger, in a simplified manner. A number of vanes are mounted on the drum is such a manner that they can slide in or out against some spring force, so that all the time they are in contact with the inner surface of the supercharger body.

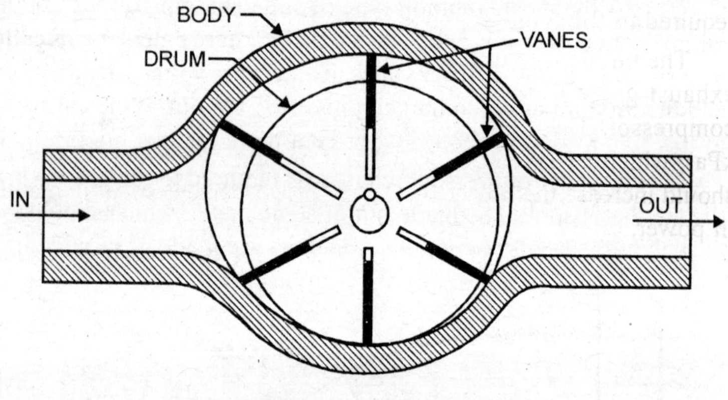

Fig. 11.22. Vane type supercharger

The vanes are usually made from laminates of linen impregnated with phenolic resin. However, Tufnol, because of its similar qualities, has also been used. The properties required for a vane material are, quiet running, low friction, low coefficient of thermal expansion and resistance to continuous exposure to oil and petrol. It is seen that space between the body and the drum goes on decreasing from the inlet to the outlet side as the drum rotates. Thus the mixture entrapped between any two vanes at the inlet will experience decrease of volume and hence increase of pressure as it reaches the outlet.

4. Screw type supercharger

These compress the air between two intermeshing screws, taking it at one end and compressing the air as it travels along the screw axis.

Out of the various types of superchargers described above, all except the centrifugal one, are the positive displacement types, *i.e.*, they have about the same output volume of air for each rotation. On the other hand, the volume output rate in case of centrifugal compressors increases with increase of speed. The centrifugal type has poor working characteristics at lower speeds, whereas the other types work well even at lower speed ranges. Further, Root's type is simple in construction, requires least maintenance and has comparatively longer life. In vane type the problem of wear of vane tips is always there.

The centrifugal type supercharger has been used mainly on aircraft engines, whereas the roots type on automobile engines.

8.3. Turbocharger

The turbocharger is bolted to the exhaust manifold of the engine. The exhaust from the engine cylinders spins the turbine which runs the compressor which is connected to the turbine through a shaft and is located between the air filter and the intake manifold. Obviously, a turbocharger is similar to the supercharger except that here the power required to run the compressor comes from the engine itself-from the exhaust gases.

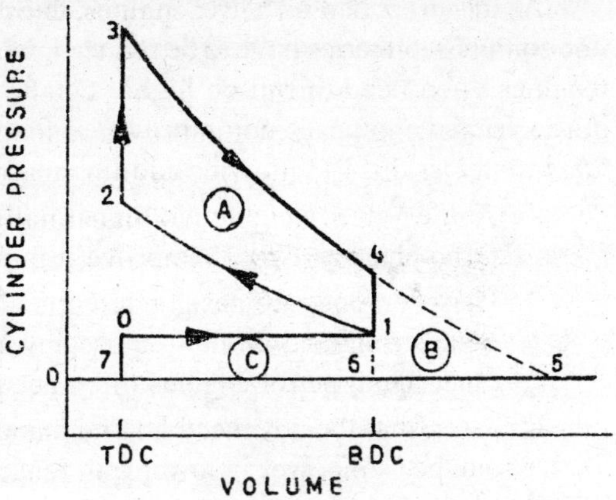

Fig. 11.23. Available exhaust gas energy in a petrol engine.

In a petrol engine, about 35-40% of the energy contained in the fuel goes waste in the outgoing exhaust gases. A turbocharger, which employs a turbine wheel and a centrifugal compressor as a supercharger and makes use of a part of the energy contained in the exhaust gases. The energy extracted from the exhaust gases is also utilized to drive the compressor. Consider PV diagram of a petrol engine (Fig. 11.23). Area A depicts the useful power developed in the engine, while area B represents the blow-down energy in the exhaust gas which is available to drive turbocharger. Likewise area C is the energy required to fill cylinder under increased air pressure, *i.e.*, the boost pressure.

The turbocharger includes a turbine and a centrifugal compressor coupled together (Fig. 11.24). The exhaust gases from the engine enter the turbocharger and run the turbine, which further drives the compressor. The atmospheric air enters the compressor which raises its pressure by about 0.5 bar (50 kPa), which means about 50 percent more air by weight (for the same volume, *i.e.*, engine size), which should increase the power by 50% from the same engine. In practice, we obtain 30 to 40 percent increase in power.

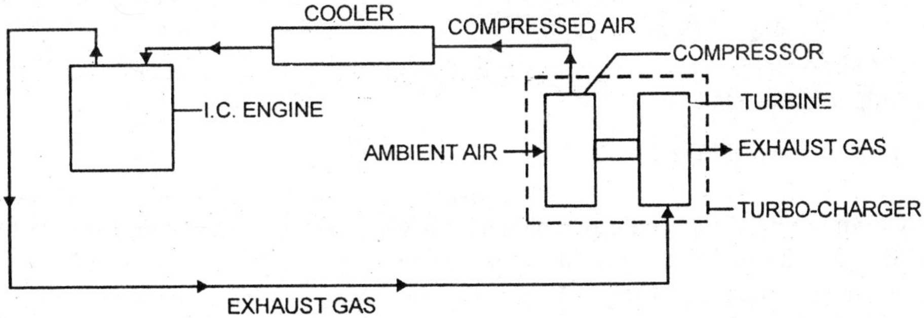

Fig. 11.24. Working of a turbochager

In the older engines using carburettors, the fuel rate is automatically increased to match the increased airflow going to the engine cylinders. Modern fuel-injection engines have oxygen sensor in the exhaust to determine air fuel ratio; thus these systems would automatically increase the fuel flow if the turbo is added.

Thus the advantages of using turbochargers in automotive engines may be summarized as:

1. Power output of a given displacement engine is increased.
2. Torque characteristic of the turbocharged engines is better.
3. Turbocharged engines have more B.P./weight ratio, compared to naturally aspirated engines.
4. Power loss due to decrease in air density at higher attitudes is reduced by using turbocharger.
5. Reduced fuel consumption.

6. Reduced noxious exhaust gas emissions.

As the turbocharged petrol engines, due to higher peak cylinder pressure, are more prone to detonation, the compression ratios have to be reduced, which reduces air standard efficiency. Besides, the turbocharged engines often need to run on higher octane fuel to avoid knocking. Moreover, the performance of the turbocharged engines is not improved at low engine speeds. However, they have been *extensively used in diesel engines* due to following additional advantages:

1. As the power/weight ratio for naturally aspirated diesel engines is lower compared to petrol engines, turbo-charging would improve it more.
2. Diesel engines are designed to operate at higher compression ratios due to which they are already quite strong and do not require any additional reinforcement for adding the turbocharger. On the other hand, petrol engines often need extensive modification for turbo-charging.

It is seen that the advanced engine management techniques available now lets turbocharged engines deliver pumped-up power in lower rpm range, by applying extra turbo boost, often called 'over boost' for brief periods. Thus over boost provides extra turbo pressure at lower speeds for periods of time long enough to accelerate the vehicle to highway speed, which means that now turbocharger engines can provide muscular response normally provided by larger displacement engines. Earlier this technique could not be used due to want of proper engine management, because running over boost for longer periods results in high temperatures, which could destroy the engine in case the temperature overshoots due to any reason.

Ford Motor Co. of USA has introduced in 2008, a new series of engines called 'EcoBoost', which is a boosting concept for direct injection petrol engines. It is claimed to provide 10-20% fuel economy with CO_2 emission-reduction upto 15%, depending upon the application. Using turbocharger with direct injection, com-pared to port injection, delivers a cooler, denser charge which reduces detonation and turbo lag. So much so that you can maintain compression ratios for better fuel economy, without the detonation occurring.

As at present (2017), use of turbochargers is no more restricted to performance cars. They are currently being used for all segments of automobiles— from city cars to luxury cars to full size pick-up trucks. Honeywell, a leading turbo-supplier, estimates that by 2020, 39 percent of all vehicles sold in North America would have turbochargers, up from 23 percent in 2015.

However, a turbo-charged engine requires good maintenance of engine oil, oil and air filters, as any dirt in oil and air can cause failures.

8.4. Comparison of supercharged and turbocharged engines

Although the supercharger and turbocharger perform the same function, yet the difference lies in the power supply to drive the same. The supercharger is driven directly by the engine at a fixed ratio to the engine speed, due to which it is able to provide the desired boost irrespective of the engine operating condition. Besides, being independent of exhaust gas energy, is always able to work regardless of exhaust temperature and pressure. However, the useful engine power gets reduced by the amount of power consumed by the supercharger. On the other hand, there is no such power loss in case of a turbocharger which gets its driving power from exhaust gases only. However, in practice a turbocharger does cause some back pressure in the exhaust system resulting in some power loss. Since there is no exhaust restriction in a supercharged engine the catalyst light off temperature (page 575) is reached rapidly during warm-up and is also able to increase compression ratio or advance spark timing due to more complete combustion chamber scavenging. Superchargers can also pump EGR (page 568) gases to benefit petrol and diesel engines both. Besides, though superchargers are easier to install, yet may prove more expansive overall, due to their inherent power consumption.

Centrifugal superchargers and turbochargers develop very less boost at very low engine speeds and full boost is available from about one-third of the engine speed range (*e.g.*, from 2000 rpm in a 6000 rpm engine). Moreover, a lag is felt by the driver of a turbocharged vehicle as a delay between pressing the accelerator pedal and feeling the turbo action. This may be due to the time taken for the exhaust system driving the turbine to come to high pressure and for the turbine rotor to overcome its rotational inertia and reach the speed required to supply boost pressure. Obviously this lag is not there in a supercharger where the compressor is driven directly from the engine. Seen from a different perspective, at lower speeds and light loads, a turbocharger supplies less boost and the engine is more efficient than a supercharged one.

The 'turbo lag' is especially noticeable at low engine speeds and can be reduced by the following:

1. By running the engine at higher speeds and with automatic transmissions which allow the engine speed and the exhaust gas flow rate to rise quickly.
2. By reducing the rotational inertia of the turbine by using lighter parts *e.g.*, by using ceramic turbine blades which are lighter than the steel blades used in most turbochargers.
3. By having a *waste gate,* which allows the use of a smaller turbocharger to reduce lag, while preventing it from spinning too quickly at higher engine speeds. A waste gate is a valve which allows the exhaust to bypass the turbine blades. It senses the boost pressure and if that is too high, which would mean that the turbine speed is too high, the waste gate bypasses some of the exhaust around the turbine blades, allowing them to slow down.
4. By changing the aspect ratio of the turbine by reducing its diameter and increasing the length of the gas-flow path.
5. By using a precision bearing, which reduces friction, thereby leading to faster acceleration of the turbine rotor.
6. By using two turbochargers of different sizes. The smaller one spins up to speed very quickly reducing lag, while the bigger one takes over at higher engine speeds to provide more boost.

8.5. Location

A supercharger is mounted on the engine where it can be driven from the crankshaft, which in most cases means that the pulley of the supercharger is in-line with the crankshaft damper and the main drive pulley. The supercharger can be located either in between the carburettor and the engine or alternately the carburettor is placed in between the supercharger and the engine.

Out of these first scheme is generally preferred, because then the carburettor can function in the normal way and there is no need to make any adjustment therein on this account. In the second case the pressure at the float chamber of carburettor has

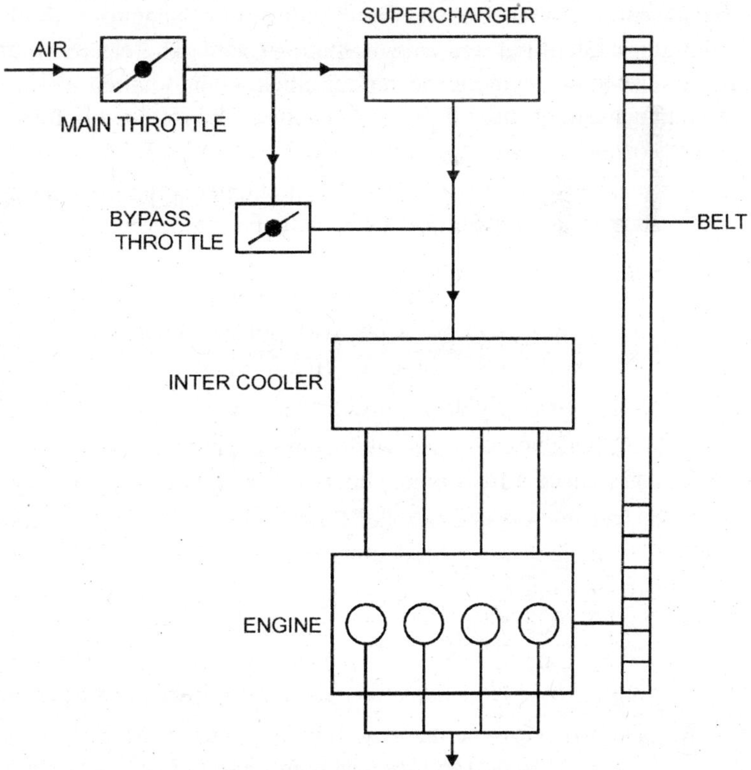

Fig. 11.25. Recommended supercharger location

carburettor has to be made equal to the supercharged pressure and to do that an equilising passage has to handle air fuel mixture (instead of pure air) which is explosive, there is always a risk of explosion in the induction manifold due to the engine backfiring. To release the high pressure so generated on account of the explosion, a safety valve has to be provided in the induction manifold. This is an ordinary valve which opens against the spring pressure.

Sometimes an inter-cooler is also used in between the supercharger and the engine. The purpose is to cool the mixture which has become hot on account of compression in the supercharger. The volumetric efficiency obviously is decreased at higher temperature which is again increased by cooling of the mixture in the inter cooler.

Mike & Daniel (1) have recommended that the supercharger should be located downstream of the main throttle (Fig. 11.25) to ensure that the air flow to the supercharger would be at sub-atmospheric pressure depending upon the throttle position. Moreover, instead of pumping excess air through the supercharger and bypassing the same afterwards under atmospheric condition, the suggested location would result in reduced air and the bypass mass flow, which would reduce the work input significantly down to almost the level of pure mechanical losses, which are very low.

A turbocharger, on the other hand, is mounted on the exhaust side of the engine and a return line is provided back to the sump. Besides, water cooling lines have to be connected to the engine cooling system.

8.6. Improvements

Some of the recent (2009-11) improvements in superchargers are:

1. Electric superchargers—Compared to a conventional centrifugal supercharger or turbocharger, an electric supercharger is driven with an electric motor, rather than the engine power or the exhaust gases. The electric motor gets power from batteries or capacitors which can be charged through regenerative braking or a generator and a 48-volt sub-system. The electric motor runs the compressor and is placed within or external to the turbocharger. Such a combination of small electric super charger and the usual turbocharger either directly boosts the engine or rotates the turbocharger faster to reduce the turbo lag without losing the benefit of waste-heat recovery.

The main advantages of electric super charger is flexible power and response, especially at low engine speeds since its operation is not tied up with the engine rpm or the exhaust energy. This results into following advantages:

1. With engine downsizing and pressurizing, e-supercharging enables the supercharging downsizing also without compressing low-rpm drivability, by filling-in the torque-less void below the turbocharger's threshold for creating boost.
2. Deactivated cylinders with electric supercharger can remain dormant longer.
3. In Miller-cycle engines, which have a longer-than-normal expansion ratio, an electric supercharger can replace a traditional supercharger to reduce parasitic losses.

Developed by U.K.'s 'Controlled Power Technologies' as 'Variable Torque Enhancement System' (VTES), which uses an electric powered supercharger, is claimed to have the following advantages:

(*i*) Being independent of engine speed, it is well suited to maintain transient performance and drivability in downsized engines, which means no boost-lag, *i.e.*, instant torque response the moment you step on the throttle, since the electric supercharger increases the air charge density significantly during first 10 combustion cycles of a low speed transient, which are critical.

(*ii*) Due to the low inertia of the compressor used, the supercharger accelerates from idle to its maximum speed of 70,000 rpm in less than one-third of a second. This enables the engine to achieve full load torque in less than a second at very low engine speeds.

(*iii*) Relatively simpler and cheaper in integrating super charging technology into an existing engine design, both for petrol as well as diesel engines.

(*iv*) Electric superchargers can be a cheaper alternative to the electric motors between the engine and the transmission in parallel hybrid electric vehicles. Both provide additional torque for acceleration and hill climbing. However, unlike electric motors, the electric supercharger can run with the existing 12 volt electrical system and does not require expensive additional battery pack.

(*v*) Easier packing in front-wheel drive vehicles.

Audi's SQ7 TDI SUV is the first production vehicle which pairs an electric supercharger with sequential turbos on its 4.0 litre diesel V-8.

2. Lontra's (U.K.) Blade supercharger—It is a small, rotary double-acting positive displacement compressor, which has a variable-flow mechanism to enable accurate matching of demand and is said to have been designed to satisfy the boosting requirements of heavily downsized petrol and diesel engine.

3. Rotrak, a joint venture of Rotrex (developer of centrifugal compressors) and Torotrak (Traction drive specialist) have developed a mechanical supercharger system which is claimed to solve problems inherent in conventional supercharger and turbocharger systems. The Rotrak system connects the supercharger with the engine through a compact variable drive and allows efficient operation throughout the entire engine speed range.

9. PROBLEMS IN D.I. ENGINES

1. Intake valve deposits

Since the fuel is injected directly into the combustion chamber and not the intake port, the fuel cannot provide any cleaning effect to prevent the build-up of carbon on the intake valves. The deposits on the valve face prevent the same from sealing resulting in compression leak, engine misfire and loss of power. Heavy deposits on intake valves can also restrict flow, less power at higher engine speeds and a drop in fuel economy and performance. Besides, intake valve carbon deposits may also flake off, pass through the combustion chamber into the engine exhaust and, if the engine has a turbocharger, this carbon could damage the turbine fins in the turbocharger. The problem of deposit build-up gets worse in direct injection engines used mostly for short trips since the intake valves never get hot enough to burn off the deposits. Moreover if the valve seals are loose or otherwise defective, too much oil would dribble down the valve stems, causing a fast build-up of carbon deposits there. Various methods to clean the intake valves in such cases are:

(*i*) Cleaning the valves with a chemical cleaner sprayed into the throttle body, intake manifold or directly into the intake ports.

(*ii*) Removing the intake manifold and spray solvent directly into the intake ports in the cylinder head.

(*iii*) Blast-cleaning the back side of the intake valves with a soft media such as walnut shells, baking soda or plastic beads.

(*iv*) In case of very heavy carbons deposits, it may be necessary to remove the cylinder head for cleaning the valves.

2. Stratified charge effect

Like diesel engines, in direct injection petrol engines also, less time is available for the fuel to mix with the incoming air before it ignites, which causes to produce richer mixtures near the spark plug and injector while leaner mixtures are produced away from the spark plug and the injector. The result is the formation by the combustion process, of larger particles of soot similar to that of untreated diesel exhaust. The particle size and quantity of the particulates produced varies, depending on the volatility of the fuel and engine operating conditions.

Current U.S. regulations permit upto 10 mg/mile of particulates, which do not require any further treatment for such exhaust particulates in petrol engines. However, if in future, the regulations are made more stringent, some form of exhaust after-treatments as in case of diesel engine particulates, may have to be required here also.

10. CLEANING OF FUEL INJECTORS

The deposits in injectors come mostly from petrol itself, which is a mixture of many hydrocarbons including olefins which are heavy, waxy compounds. The heavier the hydrocarbon, the more energy it yields on combustion. When the engine is shut off, injectors undergo 'heat soak'— fuel residue in the injector evaporates, leaving the waxy olefins behind. However, since the engine is not working, there is no cooling air flow through the ports and no fuel flow through the injectors to wash it away, heat bakes the olefins into hard varnish deposits. With time such deposits build up and ultimately clog the injectors. The middle cylinders, say number 2 and 3 in 4-cylidnr engines are in the hottest location and the injectors there tend to clog-up faster than injectors on no. 1 and 4. The formation of such deposits being normal in the engine, detergents are added to the petrol so that the injectors remain clean.

However, the detergents in the fuel may not be sufficient to keep the injectors clean for long, particularly in the vehicles with engines mostly for short trip, stop-and-go driving. In such cases the various procedures used are:

1. **Cleaning the injectors in-place.** There are many advantages:
   (*i*) Easier and cheaper since the injectors do not have to be removed.
   (*ii*) Running cleaner through the injectors with the engine running also removes many deposits on the valves and inside the combustion chamber.
   (*iii*) It takes only 10 to 15 minutes to clean.
   (*iv*) It is known usually immediately if the treatment was successful, as the engine would then run smoother without any idle misfiring, etc.

However, in-place cleaning also has some limitations:

(*i*) In case of badly clogged injectors, it may not be possible to pass enough solvent cleaner for thorough cleaning.

(*ii*) Some additional tests e.g., a test drive, a power balance test, etc. would need to be conducted to confirm that the injectors have been fully cleaned.

(*iii*) There is some risk to the vehicle's fuel system components, due to in-place cleaning. The solvent cleaning agents attack rubber and plastic components in the fuel pump, regulator and fuel lines and in case the mechanic forgets to disconnect and plug the fuel-return line before cleaning, in-place cleaning may cause additional problems.

2. **Off the vehicle injector cleaning.** Various advantages of this method are:
   (*i*) It restores heavily clogged injectors which fail to respond to in-place cleaning.
   (*ii*) Injectors can be flow-tested after cleaning, to verify their performance.
   (*iii*) Observing the spray patterns of a set of injectors, it can be seen if any of them are misshapen

The disadvantage of off-the vehicle cleaning is that it is more expensive due to:

(*i*) labour to remove the injectors, and (*ii*) cost of the special fuel injector cleaning equipment.

REFERENCES

1. Mike Lau and Daniel R. Ouwenga, "Superchargers for Downsizing", ATZ Autotechnology, Vol. 11, No. 05, 2011, pp 33-37.

THEORY QUESTIONS

1. Discuss the classification of petrol injection systems. Describe the working fundamentals and details of such a system.
2. Discuss the chief advantages and disadvantages of petrol injection in automotive engines. Describe with the help of simple schematic diagrams, the mechanical and the electronic injection methods.
3. What is `fuel trim'? Discuss in detail the short term and the long term fuel trim.
4. Write a detailed note on the electrical fuel pumps for the engines with petrol injection.
5. Discuss the latest trends in the design of petrol injection systems.
6. Draw a diagram showing the main components and layout of Bosch Motoronic petrol injection system and describe its working, including various modes of its operation.
7. What is the reason that the use of superchargers for automotive use is not common although when supercharged, the engine gives more H.P.? Discuss various types of supercharges with the help of neat sketches. Where is the supercharger located on the engine?
8. Explain briefly the function and advantages of a turbocharged engine. Discuss various methods of reducing turbo-lag.
9. Discuss thoroughly various problems and their solutions in direct injection petrol engines.
10. Describe the source of deposits in fuel injectors. Explain also the methods of cleaning the clogged injectors

QUESTIONS FOR VIVA

1. What are the advantages of petrol injection?
2. What are the drawbacks of petrol injection system?
3. What are the various types of petrol injection systems?
4. Name possible locations for injection nozzle in case of petrol injection.
5. What are the advantages of port injection?
6. State any advantages of throttle body injection.
7. How is fuel injected controlled in continuous type injection system?
8. How is the air-fuel ratio controlled in the intermittent or jerk type petrol injection system?
9. Name the main components of a mechanically controlled petrol injection system.
10. Name a few of the typical electronic injection systems used on modern cars.
11. Which component keeps the fuel supply at the specified pressure in the petrol injection system?
12. How is the injector kept closed?
13. How is the injector opened in an electronic injection system?
14. What determines the open time of the injector in the electronic injection system?
15. Name various sensors used in petrol injection systems.
16. What do you understand from the terms 'MPP' and 'MAF' sensors?
17. What is 'duty cycle' of an injector?
18. Compare the speed density system and the mass air flow system.
19. What is STFT?
20. What is LTFT?
21. What do the positive and the negative fuel trim values mean?
22. What is normal variation for STFT?
23. What happens if LTFT value exceeds + 20% or – 20%?
24. Name different modes of operation of a modern petrol injection system.
25. Define 'stratified charge' and 'homogeneous' modes.
26. What is the purpose of supercharging an engine? What are its disadvantages, if any?

27. What is a supercharger?
28. Name any four types of superchargers. Which one out of these has comparatively longer life?
29. Which is the most common type of supercharger for automotive engines?
30. Why were the automotive petrol engines rarely supercharged till recently?
31. Where is the supercharger located on engine?
32. What is the purpose of using intercooler with a supercharger?
33. What is a turbocharger?
34. State the differences in the functioning of a supercharger and a turbocharger.
35. What do you understand from the term, 'turbo-boost'?
36. What is meant by the term 'turbo-lag'?
37. Enlist the advantages of using turbocharger in automotive engines.
38. What is 'EcoBoost'?
39. At what engine speeds, the turbo lag is especially noticeable?
40. How is the turbo lag reduced?
41. What is a waste gate with reference to a turbocharger?
42. Why do we sometimes use a small supercharger with a bigger turbocharger?
43. What are the main advantages of an electric supercharger?
44. Enlist various problems in direct injection petrol engines.
45. State main four advantages of cleaning the fuel injectors in-place.
46. What is the advantage of off-the vehicle injector cleaning?

MULTIPLE CHOICE QUESTIONS

Mark the correct answers:

1. The most accurate petrol injection system is the
   (*a*) direct injection (*b*) port injection
   (*c*) manifold injection (*d*) throttle body injection
2. The cheapest yet reasonably precise gasoline injection system is the
   (*a*) direct injection (*b*) port injection
   (*c*) sequential injection (*d*) throttle body injection
3. The amount of fuel delivered by the injector depends upon the
   (*a*) size of injector nozzle (*b*) pressure pushing fuel through the injector
   (*c*) length of time the injector is open (*d*) all the above factors.

[Key: 1. (*b*), 2. (*d*), 3. (*d*)**]**

12 Diesel Engine Fuel Supply Systems

1. INTRODUCTION

Fuel supply system in a diesel engine has to perform certain functions. These functions along with the names of the components which perform the same are given below:

1. Storing of fuel

Fuel tank is usually positioned along the side of the vehicle chassis.

2. Filtering

Water and dirt must be removed from the diesel for which two filters are employed. Primary filter is usually in the form of a coarse wire gauze and is often optional. It prevents large solid particles and water from going to the fuel feed pump. Secondary filter is used after the fuel feed pump and is meant to remove fine particles of dust, dirt etc. from the diesel which is to go to the injection pump.

3. Delivery of fuel to injection pump

From the fuel tank the fuel is delivered to the fuel injection pump by means of fuel feed pump. The rate of fuel delivery depends upon the engine requirements.

4. Injecting the fuel into engine cylinders

Exact amount of fuel is *metered, atomized and injected* under high pressure to each cylinder *in correct sequence* and *at the correct moment* according to the engine requirements. This is done by means of a fuel injection pump in conjunction with injectors for each cylinder. Extra strong steel pipes transmit the metered, pressurized and timed fuel from the fuel injection pump to each injector.

5. Controlling the engine speed

Diesel engine speeds tend to overshoot to dangerous values on reduction of load. This is controlled by means of a governor, which besides limiting maximum speed also regulates the fuel supply under all conditions.

2. FUEL INJECTION SYSTEM

The function of a fuel injection system is to inject proper quantity of fuel into the engine cylinders at the correct time and at a predetermined rate. The fuel injection systems may be broadly classified into the solid injection system and the air injection system. In the solid injection system, only the liquid fuel is injected, whereas in the air injection system liquid fuel is injected alongwith compressed air. The air injection system is less reliable, less efficient and requires an air compressor for supplying air at 7 MPa or higher pressure (which consumes upto 10% of the power output of the engine), due to which reasons it has become obsolete. The solid injection system will, therefore, be discussed here in detail.

Two types of solid injection systems are in use:

1. Common rail fuel injection system
2. Individual pump fuel injection system.

2.1. Common rail fuel injection system

Layout of a conventional mechanical common rail fuel injection system is shown in Fig. 12.1. This type of fuel supply system is used in the Detroit diesel engine, commonly known as Jimmy diesel. In this a single injection pump with injector, called as unit injector is employed on each cylinder. The unit injectors are operated by rocker arms and springs similar to the engine valves. A linkage connects the control racks of all the unit injectors, so that fuel injection in all the cylinders may be equal and simultaneously controlled.

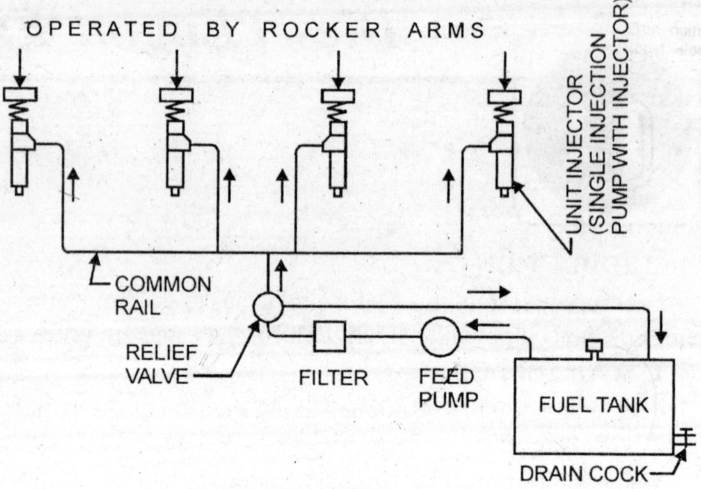

Fig. 12.1. Layout of common rail fuel injection system.

The fuel is taken from the fuel tank by the feed pump and is supplied at low pressure through a filter, to the low pressure common rail and therefore, to all the unit injectors. This avoids the high pressure fuel lines necessary in the individual pump system. Any excess fuel from the relief valve is returned to the fuel tank.

2.2. Individual pump fuel injection system

Individual pump fuel injection system using in-line pump is shown in Fig. 12.2. Fuel is drawn from the fuel tank by means of a fuel feed pump which is operated from the injection pump camshaft. Generally, the plunger type or the diaphragm type of fuel feed pumps are employed in automobiles. The pump is provided with hand-priming lever so that the diesel oil can be forced into the system and the air bled out without turning the engine. The fuel is then passed through a filter and thence to the fuel injection pump. Without the filter or with a poor quality filter, abrasive matter would reach the fuel injection pump and injectors, resulting in poor starting, irregular idling and deterioration in performance due to decreased fuel delivery from the injection pump. The abrasive matter would also cause faulty spraying and leakage in the injectors thus resulting in increased fuel consumption and heavy exhaust smoke.

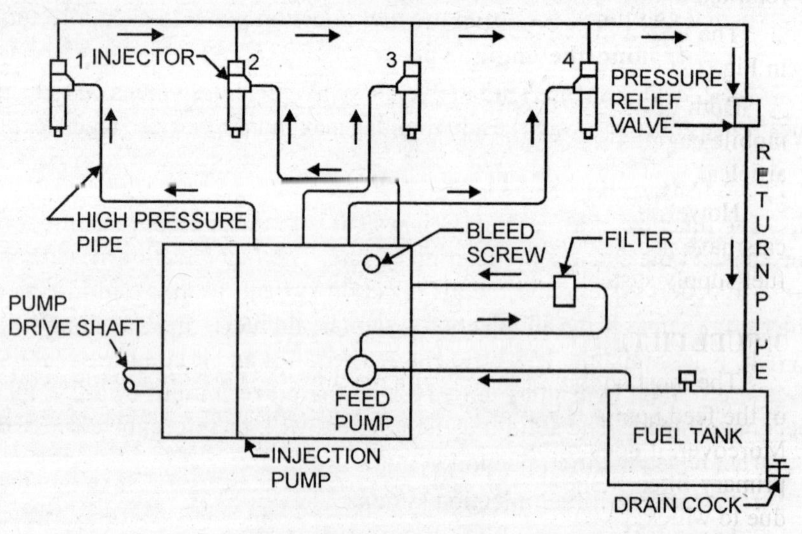

Fig. 12.2. Individual pump fuel injection system using in-line injection pump (firing order 1342)

The fuel injection pump then, injects definite quantity of fuel into individual cylinders in turn according to firing order, through injectors fitted on them. The injection pump is gear driven from the engine camshaft so that it is driven at half the engine

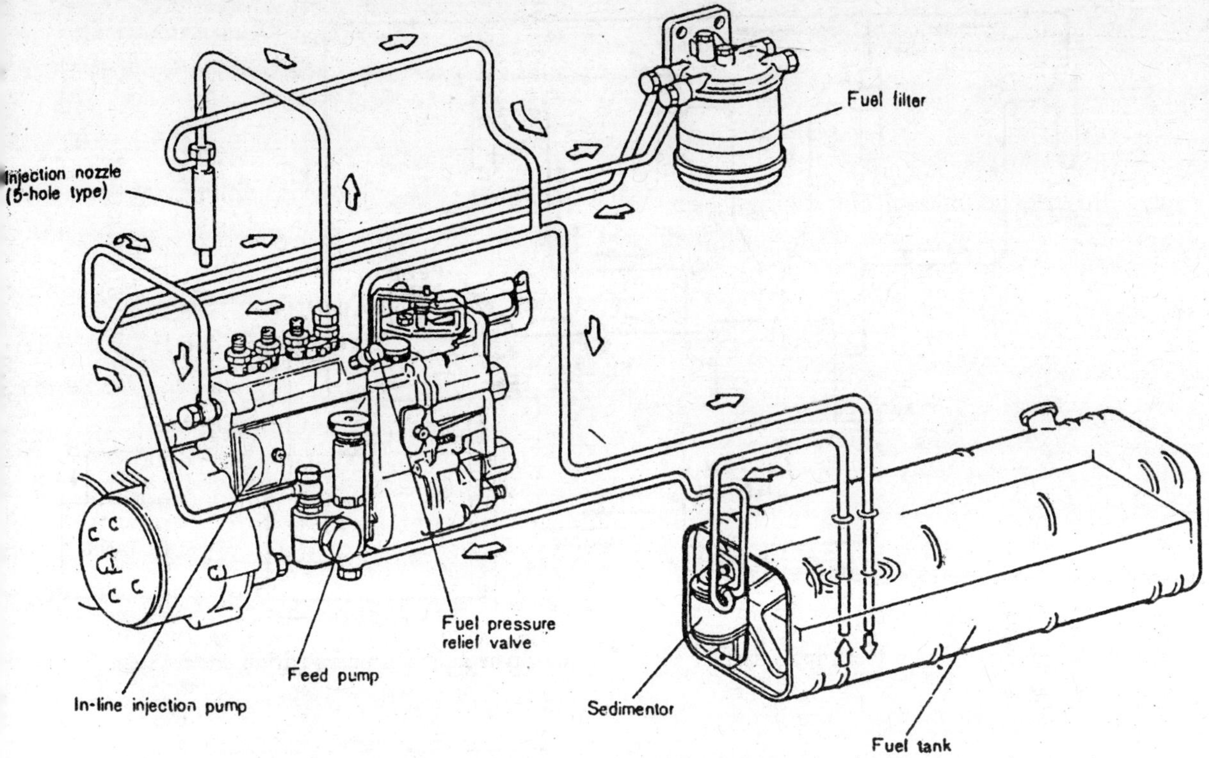

Fig. 12.3. Fuel system of direct injection SL engine for Swaraj Mazda vehicles.
(Courtesy—Swaraj Mazda Ltd.)

speed. Contained in the injection pump on its side, is a governor which provides automatic speed control, relative to any set position of the accelerator pedal. Any excess fuel after lubrication of injector nozzle is returned to the fuel tank. Such a fuel system is used in Swaraj Mazda SL engine and is shown in Fig. 12.3.

The layout of the individual pump fuel supply system using distributor type injection pump is shown in Fig. 12.4.

Both of the above conventional, mechanically operated systems have been used extensively in automobile engines. Therefore the main components of the same have been described in detail in the following articles.

However, stringent emission requirements of automotive diesel engines, coupled with their use in passenger cars have necessitated the use of very high injection pressure and more precise electronic control of their fuel supply system, which will also be discussed later in this chapter.

3. FUEL FILTERS

The most commonly used primary filter for diesel fuel is the wire gauge installed on the suction side of the feed pump. However, this type may become chocked in winter due to the formation of wax and ice. Moreover, it does not absorb any moisture in the fuel. A better alternative is to use a sedimentation type primary filter. It should be noted that diesel fuel is lighter than the impurities (water and solid particles), due to which reason the later have higher inertia and tend to gravitate quickly to the lowest position, if the fuel is allowed to become static. One such filter is shown in Fig. 12.5. The fuel from the tank enters the filter and flows around the conical diffuser funnel to accelerate downwards to the sediments chamber.

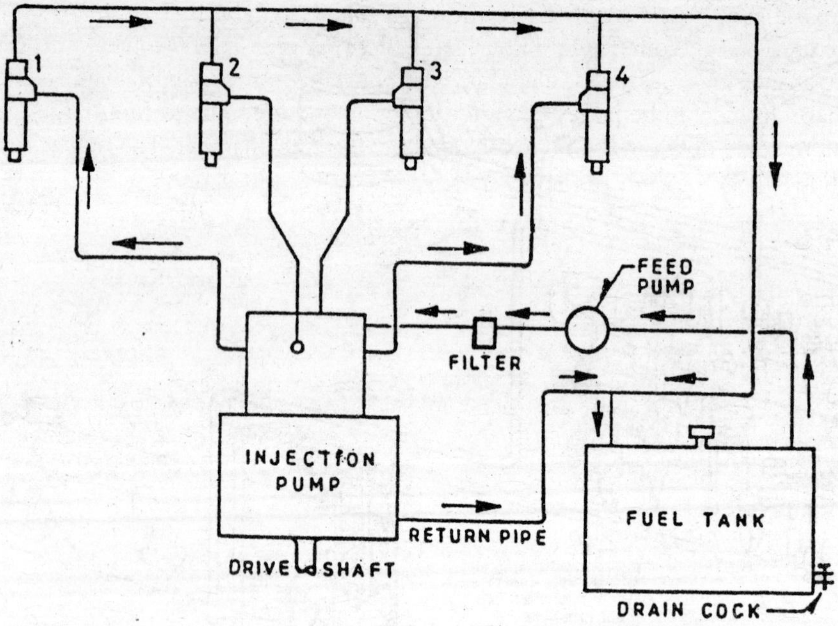

Fig. 12.4. Individual pump fuel supply system with distributor type injection pump (firing order 1342)

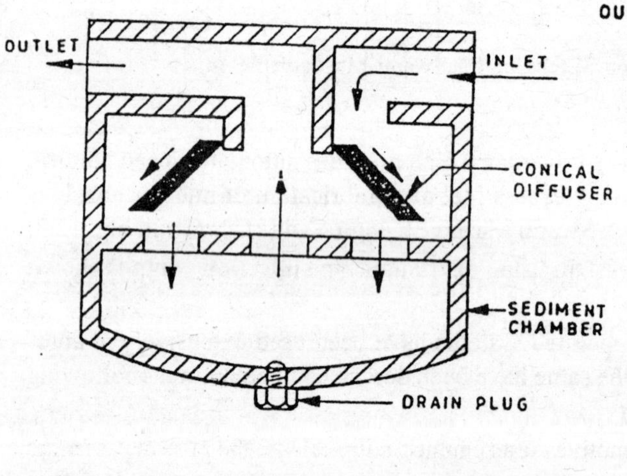

Fig. 12.5. Sedimentation type primary filter.

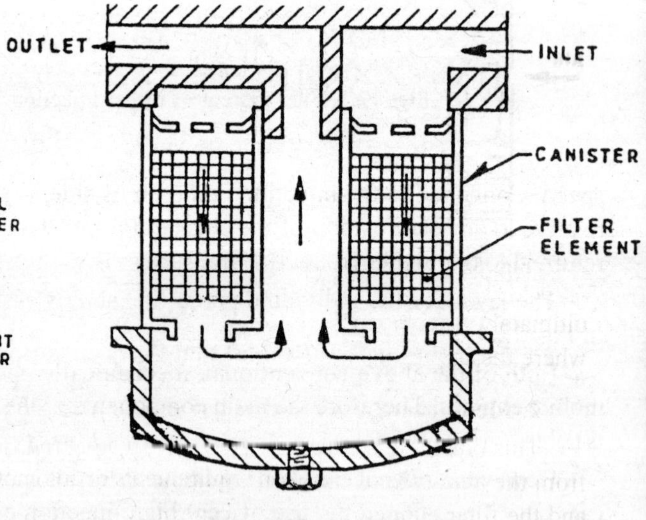

Fig. 12.6. Secondary filter

The impurities being heavier, tend to move further down towards the bottom and settle there on account of the slow movement of the fuel through the chamber, its size being relatively large. The impurities thus settled down can be drained off periodically. The clean oil passes to the outlet of the filter. Such filters will also hold back wax crystals formed within the fuel at low atmospheric temperatures, without clogging the filter element.

Resin-impregnated paper elements are now being used in most filters. On an average, such elements filter out about 95% of abrasive particles. These can also filter water content upto 3% of fuel.

The secondary filter is installed after the feed pump and it separates out abrasive particles and any water in the fuel. One such secondary filter is shown in Fig. 12.6. The fine pores of the filter element retain the abrasive particles and other solid impurities, except water droplets which are forced through the element. However, as they have passed the element, electronic attraction produced by the resin-impregnated element causes these droplets to be pulled toward each other and continue to form larger droplets which being heavier tend to settle down in the relatively larger sediments chamber. The lighter fuel then passes onto the filter outlet.

4. AIR CLEANERS

To clean air entering the engine air cleaners are used, which also act as silencers for the air-intake. A few of these are described here.

1. Light duty air cleaner (Fig. 12.7)

It consists of wire mesh element and oil reservoir at the bottom. The atmospheric air enters the air cleaner through the windows, strikes the oil surface, where heavier impurities are retained by the oil and

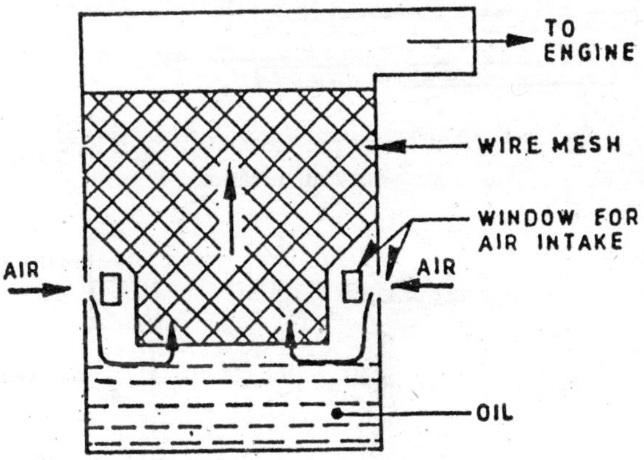

Fig. 12.7. Light duty oil bath air cleaner

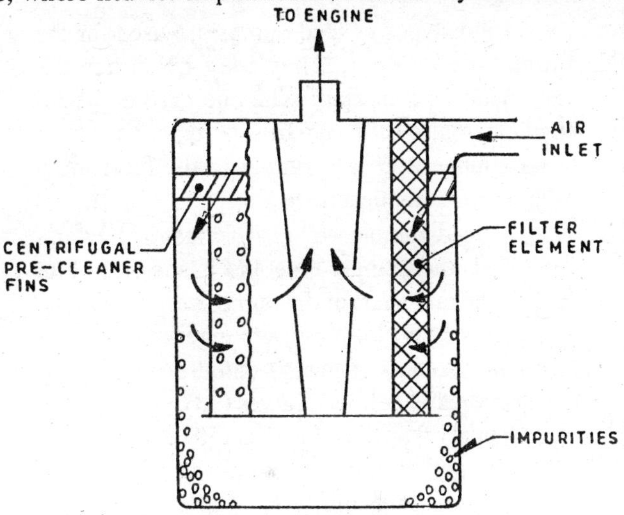

Fig. 12.8. Medium duty dry air cleaner

ultimately gravitate to the bottom, whereas air with lighter impurities passes through the wire mesh element, where the impurities are retained and the clean air passes out to the engine.

2. Medium duty air cleaner (Fig. 12.8)

This type of air cleaner consists of a paper filter element with a row of plastic fins around it. As the air from the atmosphere enters the cleaner, the plastic fins give it a high rotational speed between the casing and the filter element. This causes impurities to separate out from air due to centrifugal action, which are thrown out to the casing walls from where these flow down. Air without these dust particles then passes through the paper element, which removes any further impurities and clean air then goes to the engine.

3. Heavy duty air cleaner (Fig. 12.9)

It consists of a centrifugal pre-cleaner and two filtering elements as shown, alongwith oil reservoir at the bottom. The pre-cleaner gives a whirling motion to the incoming air, which causes the impurity particles in the air to be thrown out through the slots provided. The pre-cleaned air then impinges on the oil surface where some of the impurities are left while an emulsion of air and oil passes through the wire-wool mesh where most of the dirty oil is absorbed. This dirty oil then condenses and falls back into the oil reservoir, where the dirt settles down. The relatively clean air then passes through the second wire mesh, which retains any impurities still left and the clean air is then passed on to the engine.

5. FUEL FEED PUMP

This feeds the fuel from the fuel tank to the injection pump at a rate which varies with the engine requirements. It is driven from an eccentric or cam on the injection pump camshaft or the engine camshaft. In some systems, this pump is called a transfer pump or else, a lift pump. The feed pump is usually a part of the injection pump assembly. Some are mounted on the outside of the injection pump housing, while others are enclosed in the housing.

Various types of feed pumps are used for the automobile engines. They may be either the diaphragm type similar to the one used on petrol engines, or else they may be gear type, vane type or the plunger type *i.e.* similar to the lubricating oil pump already described.

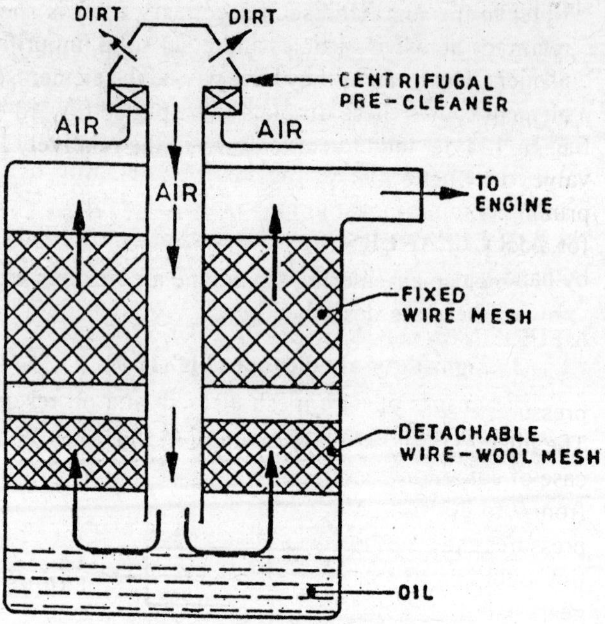

Fig. 12.9. Centrifugal pre-cleaner oil bath type heavy duty air cleaner.

The feed pump used in earlier Ashok Leyland vehicles is the plunger type and is shown in Fig. 12.10. The main part of the pump are a barrel inside the body (1), a plunger (2), plunger return spring (4), spindle (3), suction and delivery valves (13), roller tappet (7) and hand primer (50). The plunger is driven from the eccentric on the camshaft of the injection pump, through the roller tappet and the spindle. The return stroke of the piston is effected by means of the plunger return spring.

With the eccentric at the minimum throw, the plunger moves inwards by the pressure of the return spring. This creates a suction which lifts the suction valve, drawing fuel from the fuel tank to the front side of the plunger. When the eccentric rotates and moves the plunger against the spring force, the suction valve is closed and the fuel is forced through the delivery valve, to the back side of the plunger. In the next inward stroke of the plunger the fuel behind plunger is forced out to the fuel filter and

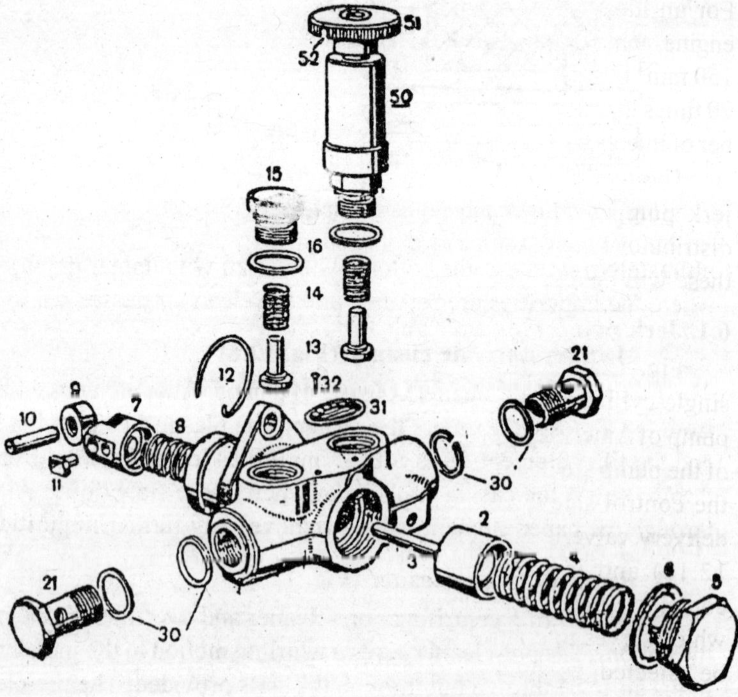

Fig. 12.10. Fuel feed pump.
(Courtesy—Ashok Leyland Ltd., India)

simultaneously, fresh fuel is also drawn through the suction valve to the front side of the plunger.

Since the fuel delivered is very much in excess of the engine requirements, a back pressure builds up on the plunger by the increased fuel pressure, against the force of the return spring thereby rendering the pump inoperative till the fuel is consumed and pressure behind the plunger falls. Thus it is seen that the fuel feed pump line pressure is determined by the stiffness of the plunger return spring. A pressure relief valve in the system keeps the diesel oil delivered by the fuel feed pump between 55 and 85 kPa. The hand priming pump is meant to bleed the fuel supply system when required. If the engine has not been working for a considerable period, the fuel can be transferred rapidly from the fuel tank to the fuel injection pump by hand-priming pump.

6. FUEL INJECTION PUMP

The function of a fuel injection pump is to deliver accurately, metered quantity of fuel under high pressure, at the correct instant and in the correct sequence, to the injector fitted on each engine cylinder. The injection pressures generally employed in case of automotive engine fuel injection range from 7 to 30 MPa. In some systems injection pressures can be as high as 200 MPa. The injection pump is driven from the engine's timing gears and its output is controlled by the driver through accelerator pedal. As the volume of fuel to be metered for each injection is very small and frequency of injection quite high, the pump has to be manufactured to very high precision. For an idea, in a 4-stroke 4-cylinder diesel engine, at maximum speed of 6000 r.p.m., about 150 mm^3 of fuel has to be metered and injected 20 times in a second. In a 2-stroke engine, number of injections per second are twice this value.

The fuel injection pumps are generally of jerk pump type. However, in many cases, distributor type pumps are also used. Both of these will be discussed here.

6.1. Jerk pump type fuel injection pump

Fig. 12.11 shows the construction of a single cylinder jerk pump type fuel injection pump of American Bosch type. The main parts of the pump are the delivery valve, the plunger, the control sleeve and the control rack. The delivery valve is of special construction (Fig. 12.12) and is spring loaded. The plunger contains a helix at its upper end (Fig. 12.13), which serves to control the quantity of fuel to be injected, as shall be discussed later. The plunger is operated by means of cam and tappet.

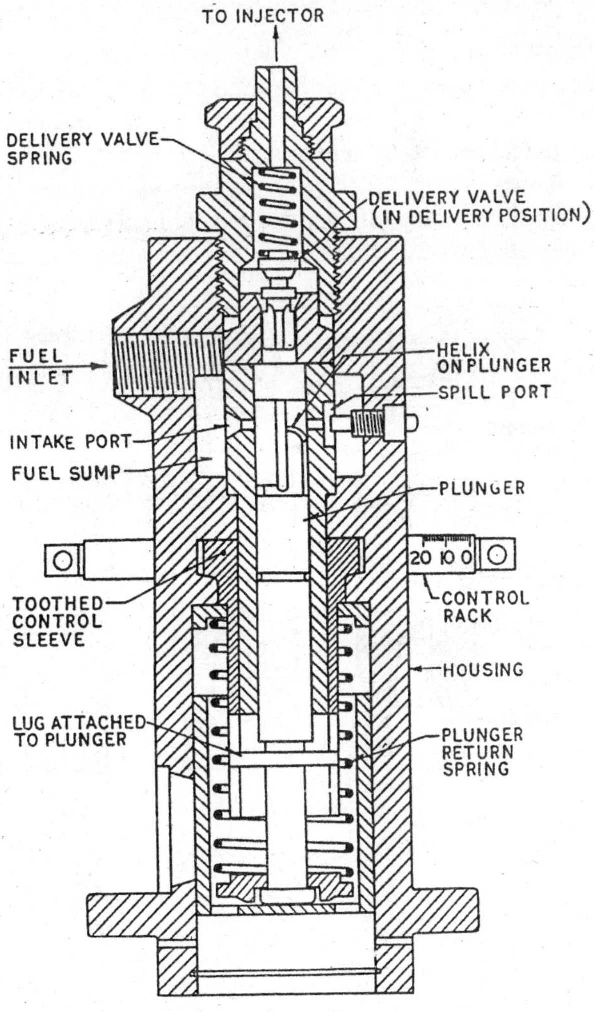

Fig. 12.11. Single cylinder jerk pump type fuel injection pump

To understand the working of the injection pump, consider Fig. 12.14. [Fig. 12.14 (*A*)]

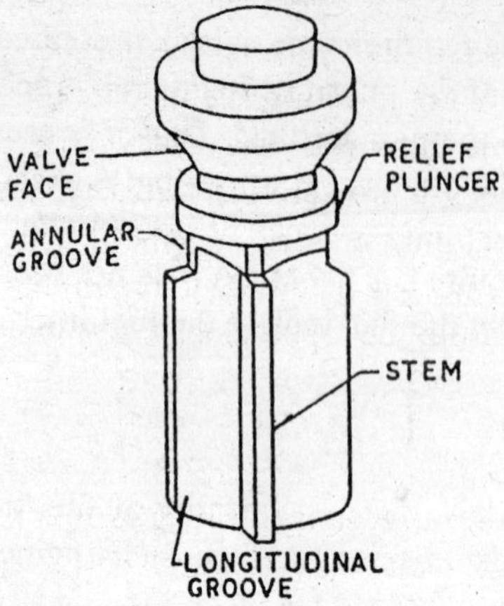

Fig. 12.12. Delivery valve

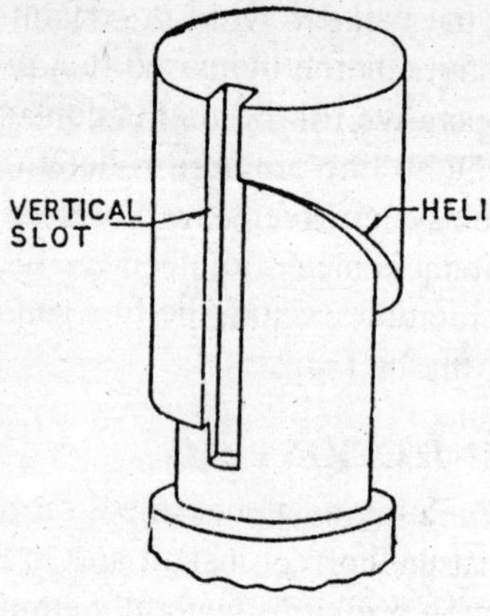

Fig. 12.13. Injection pump plunger

shows the position of the plunger at the bottom of stroke. In this position, both the intake and the spill ports are uncovered. As the plunger moves up, it covers the two ports after which the upward plunger movement exerts pressure on the fuel at the top and thence to the delivery valve which opens against the spring force, thereby delivering the fuel under pressure [Fig. 12.14 (*B*)]. The extreme position of the plunger is shown in [Fig. 12.14 (*C*)]. As soon as the helix uncovers the spill port, the fuel escapes through the vertical slot in the plunger and the spill port and pressure is released, shutting the valve down on its seat. The delivery valve has an annular groove and four longitudinal grooves (Fig. 12.12). The lowering of the

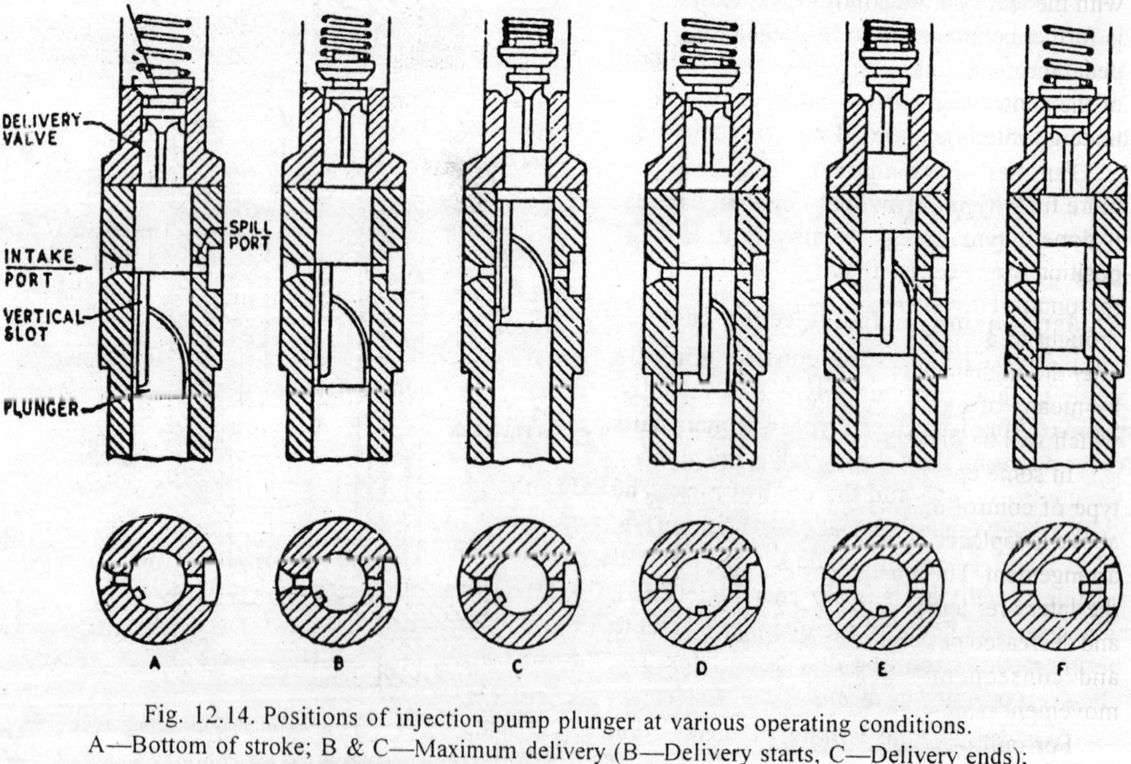

Fig. 12.14. Positions of injection pump plunger at various operating conditions.
A—Bottom of stroke; B & C—Maximum delivery (B—Delivery starts, C—Delivery ends);
D & E—Normal delivery at part load (D—Delivery starts, E—Delivery ends); F—No delivery.

delivery valve at the end of delivery increases the space on delivery side by an amount equal to that of the relief plunger, due to which a sudden decrease of pressure occurs in the delivery line causing the injector to close immediately. Thus any tendency to 'dribbling' or secondary injection is eliminated.

We have seen so far that fuel delivery starts when the plunger just covers the intake and the spill ports and ends when the helix just uncovers the spill port so as to release pressure via the vertical slot. Out of these events the first one, *i.e.*, start of delivery cannot be varied, but the later, *i.e.* end of delivery can be timed to occur earlier or later, thus varying the quantity of fuel injected according to requirements. This is done by rotating the plunger in the desired direction. Looking from the top, a clockwise plunger rotation will delay the end of delivery, increasing the effective plunger stoke and consequently the fuel supply injected. An anti-clockwise rotation, on the other hand, decreases the fuel injected per stroke. Fig. 12.14 (*B*) and (*C*) show the plunger position for maximum delivery. In this position, the vertical slot is in such a position so that the lowermost point on the helix is opposite spill port. Once the plunger is rotated in anti-clockwise direction as in Fig. 12.14 (*D*) and (*E*), the lowermost helix point is no more opposite spill port. Consequently, the effective stroke is decreased. In the no delivery position [Fig. 12.14 (*F*)], the vertical slot is opposite spill port. Thus no pressure can be built up at any vertical plunger position, reducing the effective stroke to zero. From the above it is clear that though the plunger stroke is always constant, yet the part of it which actually pumps can be varied.

The arrangement used to control the angular position of the plunger and consequently the amount of fuel injected per stroke is shown in Fig. 12.15. A lug on the plunger fits into a slot at the bottom of a toothed sleeve whose teeth engage with the teeth on the control rack, which is further connected with the accelerator pedal through linkage. To and fro motion of the control rack causes the plunger to undergo rotational movement as desired.

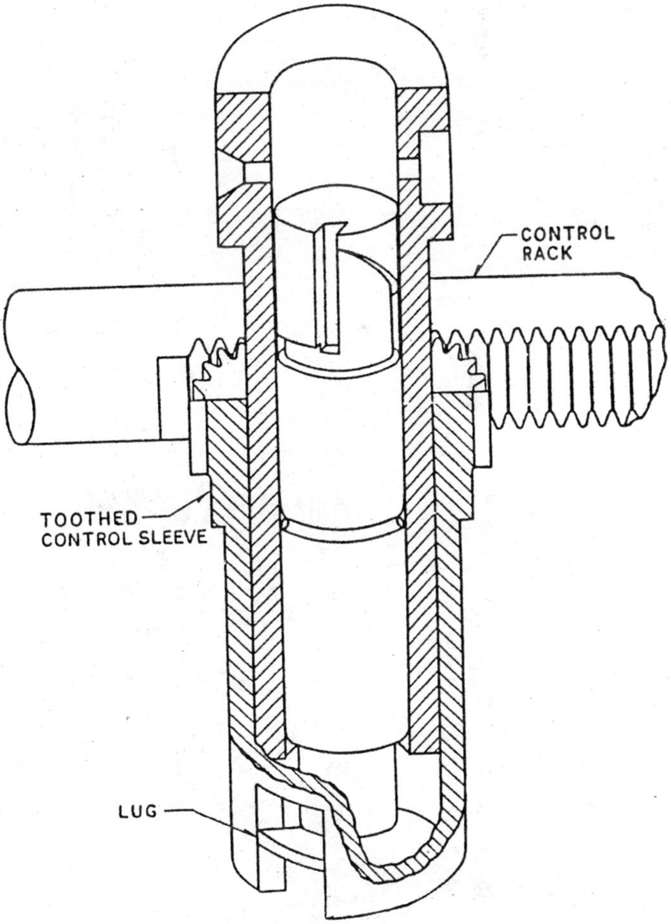

Fig. 12.15. Arrangement showing the control of fuel delivery.

For starting the engine from cold, more fuel is required to be injected. This is done by moving the control rack to a position where excess fuel is delivered by the pump. However, the rack has to be brought to normal position immediately after the engine starts. All this is achieved by means of a governor, the details of which will be discussed in Art. 8.

In some designs the rack and pinion type of control mechanism as discussed above is replaced by the 'fork-and-lever' arrangement. The advantages claimed for the later are, lesser sources of backlash and decreased possibility of misalignment and consequent sluggish control-rod movement occurring.

For multi-cylinder engines, usually many such pump assembles as explained above, one for each cylinder, are assembled into one unit to give a compact construction (Fig. 12.16). High pressure delivery lines are then connected to different fuel injectors on various cylinders. Such pumps are then referred to as 'in-line' type. The main parts of an in-line pump are pump housing, governor housing, camshaft-tappet assemblies and the pumping elements. The function of the camshaft is to produce the plunger movement for each pumping unit which occurs in the prescribed firing order and at the correct instant. The tappets are means to transfer the cam-lobe lift to the pump plungers.

Fig. 12.16. Bosch fuel injection pump for Daimler-Benz 5 cylinder engine type 617-912.

The pump housing is the main housing that contains cambox (which supports camshaft-tappet assemblies, control mechanism and governor) and the pumping head (contains pumping elements). It is made either as a single aluminium alloy casting or aluminium alloy cambox is bolted on to a pumping

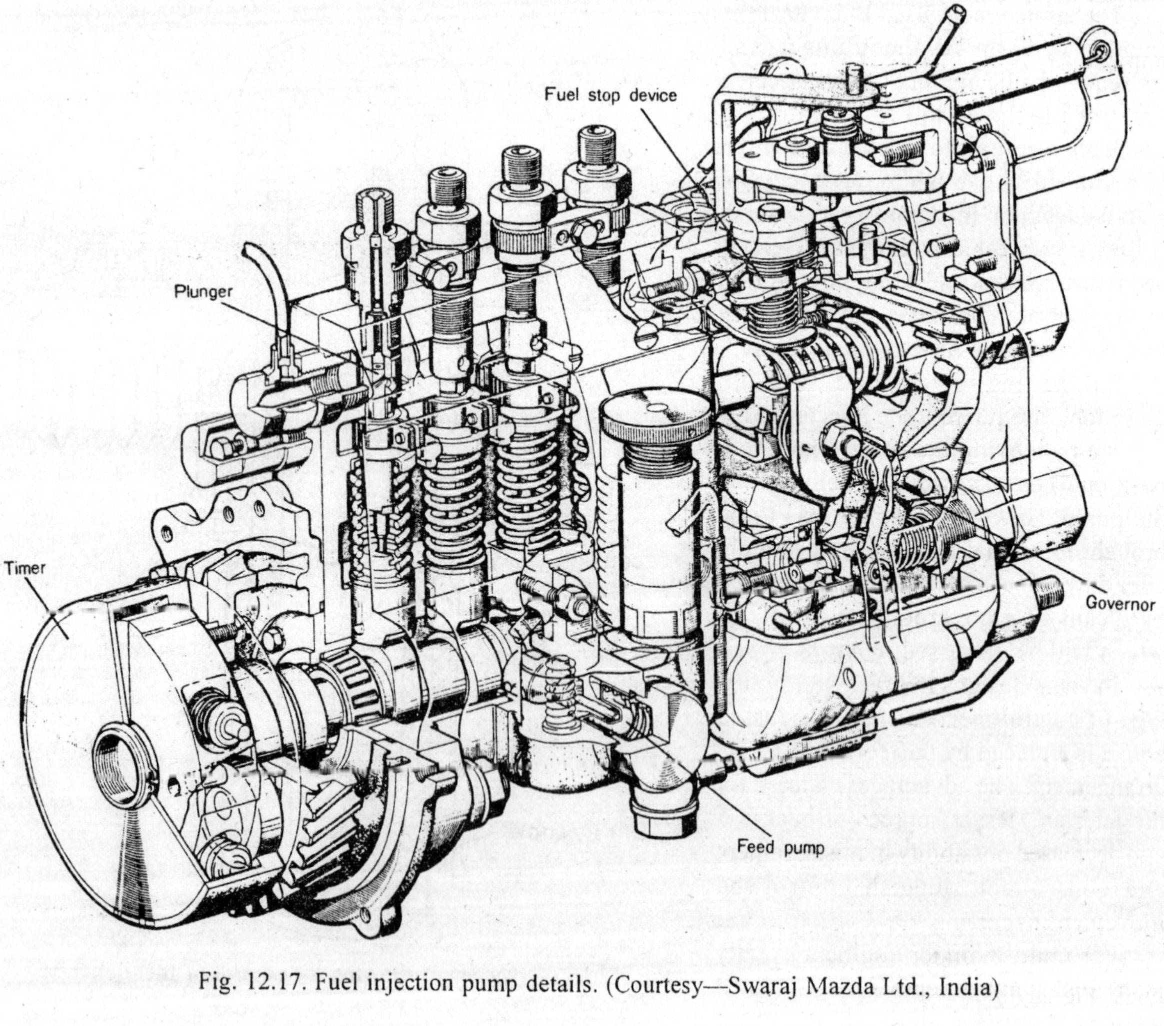

Fig. 12.17. Fuel injection pump details. (Courtesy—Swaraj Mazda Ltd., India)

head of steel. The governor housing is an extension of the cambox.

An in-line fuel injection pump used in Swaraj Mazda vehicles in India is shown in Fig. 12.17.

To get the maximum pressure in the diesel engine at the correct instant with respect to the piston-cylinder position, it is very necessary that the injection of the fuel is timed very accurately. For instance in Tata 1210 vehicle engines the beginning of injection is made to occur at 23° before T.D.C. of the compression stroke. This timing would require to be advanced with increase in engine speed. This is done automatically by means of an injection timer. In the 692 D.I. engine for Tata 1210 vehicles, the injection timer provides a maximum of 8° of advance. As shown in Fig. 12.17 for Swaraj Mazda vehicle engine, the timer is mounted on the front end of the injection pump and works on the principle of centrifugal force. The drive to the injection pump is connected to the segments on the input side of the timer, whereas the output segment plates of the injection timer are keyed to the injection pump. Two centrifugal weights connect the input and the output sides of the timer. When the speed increases, the weights fly apart causing the output segment plates and thereby the camshaft of the fuel injection pump to turn in the direction of rotation, thus providing injection advance at higher speeds. When the speed decreases, the pressure springs pull the flyweights inward, reducing the injection advance.

6.2. Distributor type fuel injection pump

The principle of working of a distributor type fuel injection pump has been illustrated by means of Fig. 12.18 and 12.19. Unlike the jerk pump explained earlier, there is a single pumping element in this type of pump and the fuel is distributed to each cylinder by means of a rotor. The rotor has a central longitudinal

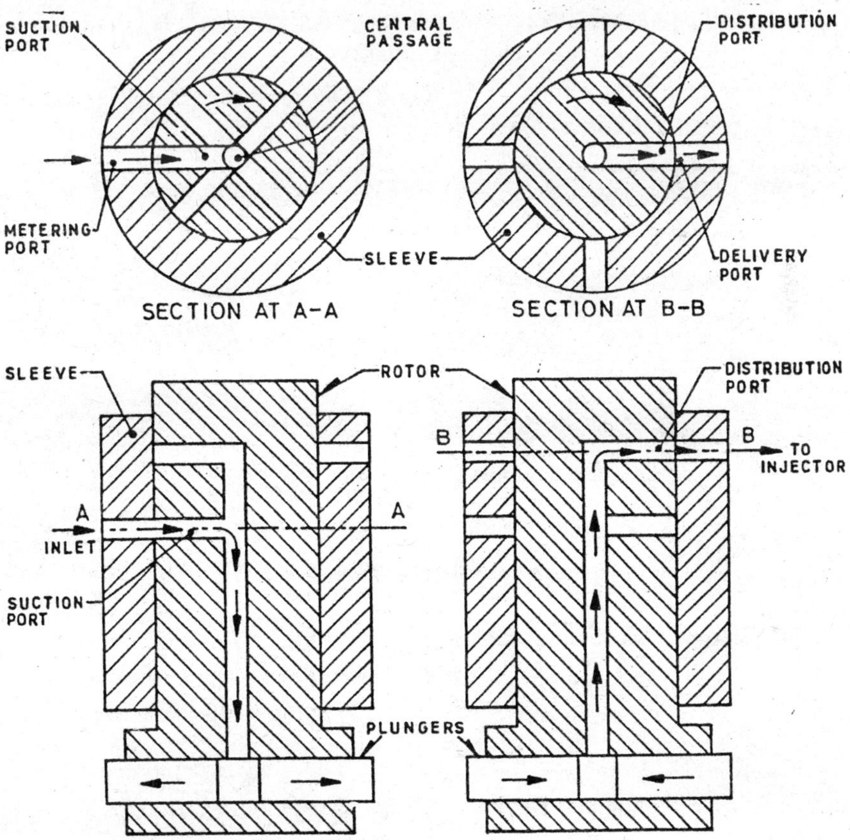

Fig. 12.18. Principle of working of distributor type fuel injection pump. (*a*) Suction, (*b*) Delivery.

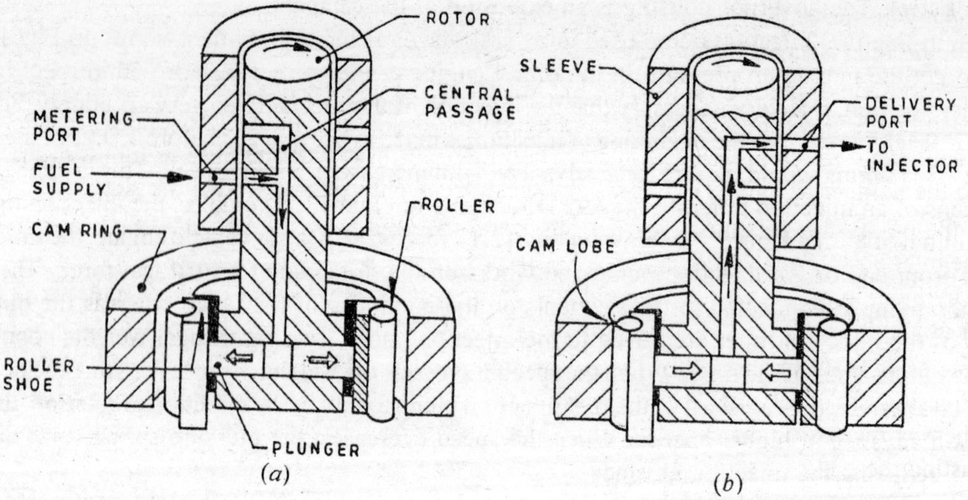

Fig. 12.19. Working of distributor type fuel injection pump. (*a*) Suction, (*b*) Delivery.

passage and a set of radial holes (suction ports) equal to the number of engine cylinders, four in the figure shown. Similarly the outer sleeve also has a set of equal number of holes (delivery ports) at a different level BB, which are offset from the suction ports in top view also. Besides there is a metering port in the sleeve for the fuel intake at level AA and a distribution port in the rotor at level BB. This distribution port is connected to the central passage in the rotor. Each of the delivery ports is connected to the high pressure delivery lines leading to injectors on the engine cylinders.

As the rotor revolves, the suction ports align with the intake metering port one by one, while the distribution

Fig. 12.20. Bosch distributor type fuel injection pump. (Courtesy—MICO India)

port aligns with the delivery ports in turn, though these alignments of suction ports and the distribution port with the relevant ports take place at different instants.

The lower end of the central passage in the rotor opens into a chamber in which two opposed pumping plungers are housed. As the rotor rotates, a stationary ring with internal cams operates the plungers through rollers and shoes which are placed in slots into the rotor base. The number of lobes on the cam ring is equal to the number of engine cylinders and these are evenly spaced around the ring.

As the pump plungers move away from each other, the fuel is drawn into the central rotor passage from the inlet port through suction ports [Fig. 12.18 (*a*) and 12.19 (*a*)]. The fuel thus charged is delivered to each cylinder in turn at high pressure, when the distribution port in the rotor coincides with the delivery port for any cylinder [Fig. 12.18 (*b*) and 12.19 (*b*)].

A metering valve controlled by the governor or hand throttle regulates the amount of fuel entering the rotor. Pressing the accelerator pedal increases the area of the metering port allowing more fuel to enter rotor, thereby increasing the engine speed. The maximum amount of fuel intake to the pump is controlled by adjusting the outward travel of the pump plungers.

Each cam on the stationary ring has usually two peaks. When the roller reaches first peak, which is higher, fuel injection stops. The valley next causes a rapid reduction in pressure in the injector pipe, preventing dribbling. The second peak on the cam prevents the pipeline pressure from collapsing completely for a time long enough for the distribution port to be cut off from the delivery port. This ensures a residual pressure in the delivery port and the injector to be maintained during cut off.

The fuel injection timing in this type of pump is controlled by rotating the internal-cam ring. As the engine speed increases, the fuel feed pump delivery pressure increases. This increase in the pump pressure is utilized to move the cam in an advance direction, thus providing more timing advance with the increase of engine speed.

Small size and less weight are the main advantages of this type of pump. Bosch distributor pump of the VE type used for passenger cars is shown in Fig. 12.20.

7. FUEL INJECTOR

This is also known as nozzle, atomizer or fuel valve. Its function is to inject the fuel in the cylinder in properly atomised form and in proper quantity. An American Bosch type fuel injector is shown in Fig. 12.21. It consists of mainly two parts, *i.e.* the nozzle and the nozzle holder, the former being connected to the later by means of a screwed cap. This facilitates the change of nozzle valve whenever required. A spring-loaded spindle in the nozzle holder keeps the nozzle valve pressed against its seat in the nozzle body, till the fuel supplied by fuel injection pump through inlet passage exerts sufficient pressure so as to lift the nozzle valve against the spring force, when a spray of atomised fuel is fed into the combustion chamber.

The fuel spray continues till the delivery from injection pump is exhausted when the spring pressure again suddenly closes the nozzle valve back on its seat. The actual opening fuel-line pressure can be varied by adjusting the initial degree

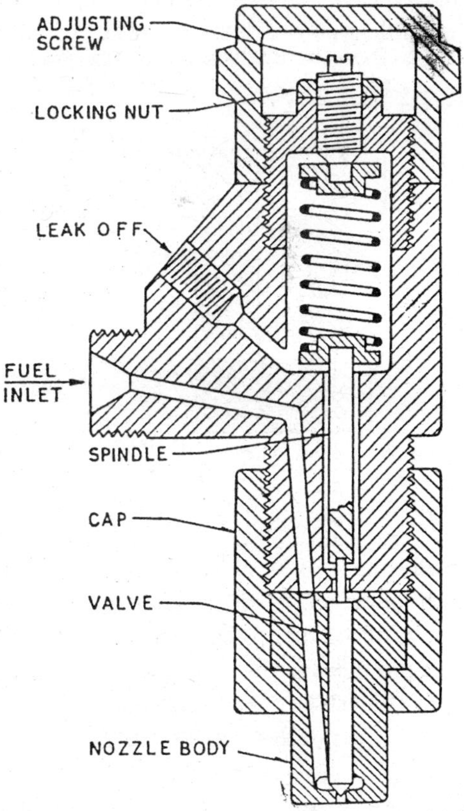

Fig. 12.21. Fuel injector (Bosch type).

of spring compression. A small quantity of fuel is purposely allowed to leak between the nozzle valve and its guide for lubrication purpose. The fuel accumulated around the spindle in this way is drained back to the fuel tank through the leak off connection. Adjusting screw provided at the top serves to adjust the tension in the spring, and hence the pressure at which the nozzle valve opens.

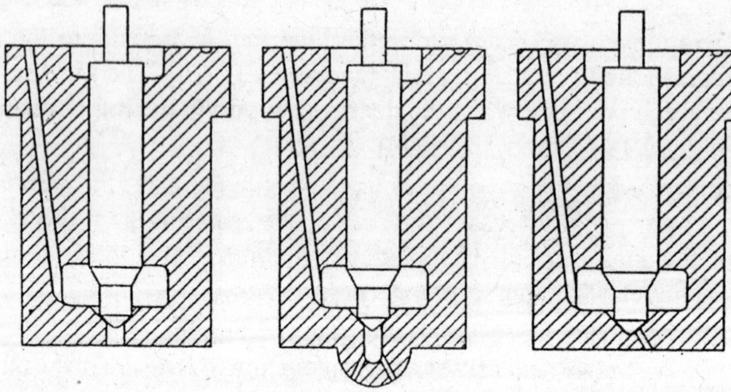

Fig. 12.22. Hole type nozzles.

The nozzles used may be classified broadly into the hole type [Fig. 12.22) and the pintle type (Fig. 12.23). Hole type nozzles are generally used in engines with open type combustion chambers, whereas pintle type nozzles are common in engines with pre-combustion chambers and some special swirl chambers. The pintle type nozzle carry an extension, which produces a hollow cone type spray. Such nozzles have

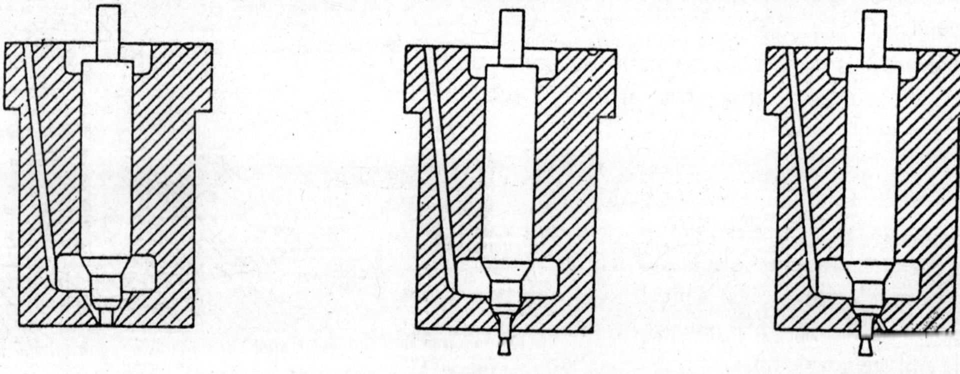

(*a*) (*b*) (*c*)

Fig. 12.23. Pintle type nozzles.

the advantage of being self-cleaning. In the throttling pintle type of nozzle [Fig. 12.23 (*b*)], the pintle is much longer and is also shaped like a truncated cone at its lower end. Such a shape causes only a small amount of the fuel to be injected as the injection starts, the rate of injection increasing gradually as the pintle protrudes further from the nozzle end. The pintaux nozzle [Fig. 12.23 (*c*)] has been developed specially for the comet type combustion chamber. This helps easy starting under cold conditions. The opening pressure of hole type nozzles varies from 17 to 34 MPa, whereas that of pintle type nozzles varies from 7 to 15 MPa.

A good nozzle should atomise the fuel uniformly so as to maintain proper injection angle and direction (Fig. 12.24).

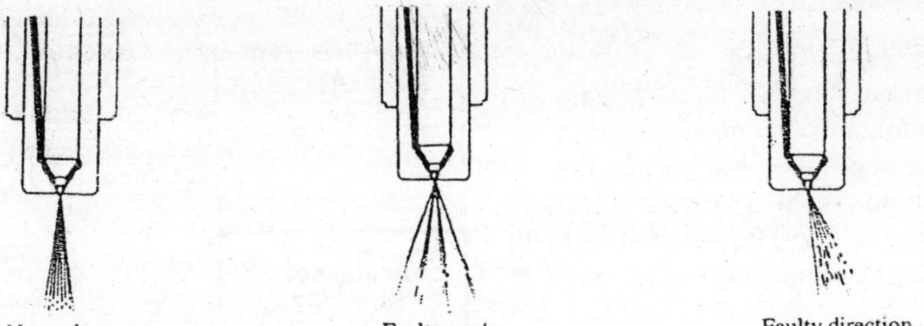

Fig. 12.24. Proper nozzle angle and direction (Courtesy—Swaraj Mazda Ltd.)

A variety of Bosch fuel injectors and nozzles meeting Euro I and Euro II emission norms are shown in Fig. 12.25.

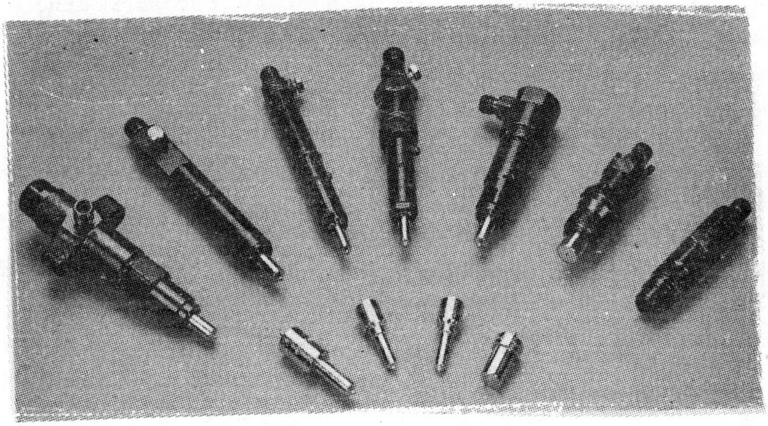

Fig. 12.25. Bosch fuel injectors and nozzles.
(Courtesy—MICO, India)

8. GOVERNOR

8.1. Necessity

Unlike petrol engines where only carburettor controls both air and fuel delivery at various speed and load conditions, in diesel engines the fuel delivered depends independently on the injection pump characteristic, and the air intake of the engine. Generally, the air intake decreases with speed of the engine, whereas the pump having a rising characteristic, results in over injection at higher speeds. On the other hand, at idling when the engine speed is less, the fuel delivered is also less when actually more fuel is required; as a result the engine will stop. Similarly with increased load on the engine, the fuel delivered by the pump also increases, causing excessive carbon deposits and high exhaust temperature. On the other hand, a reduction of engine load will cause the speed to over-shoot to dangerous values. A governor is, therefore, a necessity in case of diesel engines to control the fuel injected to ensure optimum conditions at all speeds and loads within the range specified.

8.2. Terminology

1. Governor cut-in speed

This is the speed of the fuel injection pump at which the control rod just starts to move away from the maximum fuel stop with the accelerator lever set for maximum output.

2. Governor cut-off speed

This is the speed of the fuel injection pump at which the control rod, after moving beyond the governor cut-in position, just reaches the fuel cut-off position, which means that at this point fuel delivery just stops. During this check the accelerator lever is set for maximum output.

3. Governor overrun

It is the difference between the governor cut-off speed and the governor cut-in speed.

4. Governor hysterisis

This is the difference between the governor cut-off speed at which point fuel delivery just stops with the increasing injection pump speed and the cut-off speed at which point fuel delivery just restarts with the decreasing injection pump speed.

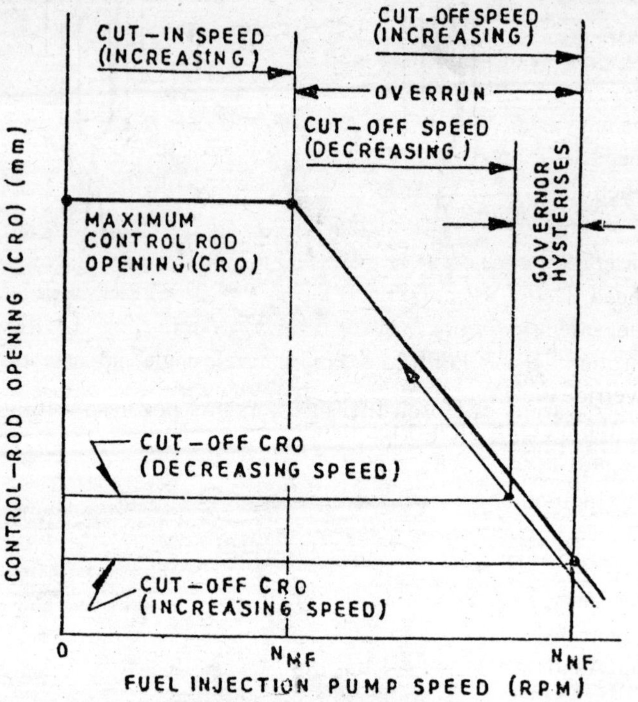

Fig. 12.26. Governor terminology.

5. Governor run-off

It is defined as the percentage speed change when the control rod is moved by the governor from the maximum fuel delivery position to the no-fuel delivery position, *i.e.*

$$\text{Governor run out} = \frac{\text{No fuel speed} - \text{maximum fuel speed}}{\text{maximum fuel speed}} \times 100$$

$$= \frac{N_{NF} - N_{MF}}{N_{MF}} \times 100$$

8.3. Types

From the point of view of method of operation, the governors may be classified as:

(*i*) Mechanical governor
(*ii*) Pneumatic governor
(*iii*) Hydraulic governor

However, from the point of view of principle of governing, they may be the following types:

(*i*) Maximum speed governor
(*ii*) All speed governor

The principles of working of all the above type will be discussed in the following articles:

8.4. Mechanical governor

The principle of working of a mechanical governor may be explained with the help of Fig. 12.27. Two spring-loaded weights are mounted on the governor shaft which gets drive from the engine. At one end, the bell crank levers carry balls whereas their other ends touch the lower surface of the flange of a sleeve on the governor shaft. As the engine speed increases, the centrifugal force due to the weights acts against the spring tension. Once the former exceeds the later, the weights fly apart, causing the other ends of the bell crank levers to raise the sleeve and hence operating the control lever in the downward direction which further actuates the control rack on the fuel injection pump in a direction which reduces the amount of fuel delivered and hence decreases the engine speed. In the same way, the amount of fuel delivery is increased when the engine speed tends to decrease.

Thus, the spring tension can be so adjusted as to actuate the control rack for decrease in fuel delivery at some maximum speed. Below this maximum speed, the control rack on the fuel injection pump may be directly operated by the driver through accelerator pedal and linkage, with a provision to override the driver control at maximum speed which is limited by the amount of spring tension. Such a governor is called the *maximum speed governor.*

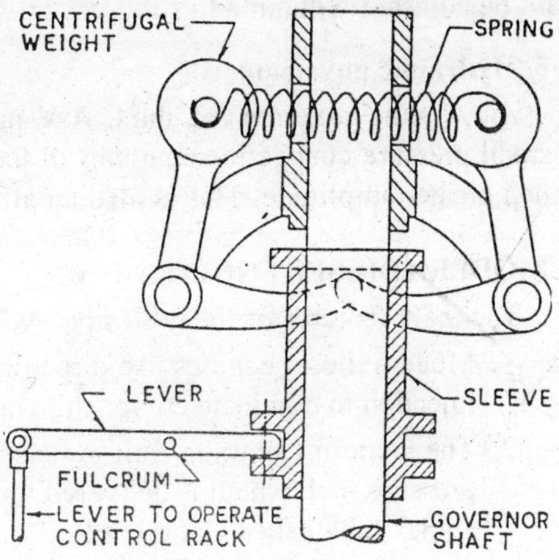

Fig. 12.27. Principle of mechanical governor

In another type of governor, however, there is no direct connection between the accelerator pedal and the pump control rack. The two are linked only through the governor. The driver can, then, change spring tension with the help of the accelerator pedal and hence fix the speed of the engine, *i.e.* of the vehicle. This type of governor is called *all speed governor.* Such a governor is used in India in Tata and Ashok Leyland vehicles.

8.5. Pneumatic governor

A pneumatic governor shown in Fig. 12.28, consists of two main parts, the venturi unit and the diaphragm unit. The venturi unit is connected to the engine inlet manifold and the diaphragm unit is fitted on the fuel injection pump. The two units are connected by a vacuum pipe.

Accelerator pedal controls the position of the butterfly valve in the venturi unit and hence the amount of vacuum from the inlet manifold, which is applied to the diaphragm via the vacuum pipe. As the diaphragm is connected to the fuel pump control rack, the rack is operated left or right depending upon the amount of vacuum applied. Thus the position of the accelerator pedal determines the position of the pump control rack and hence the amount of fuel

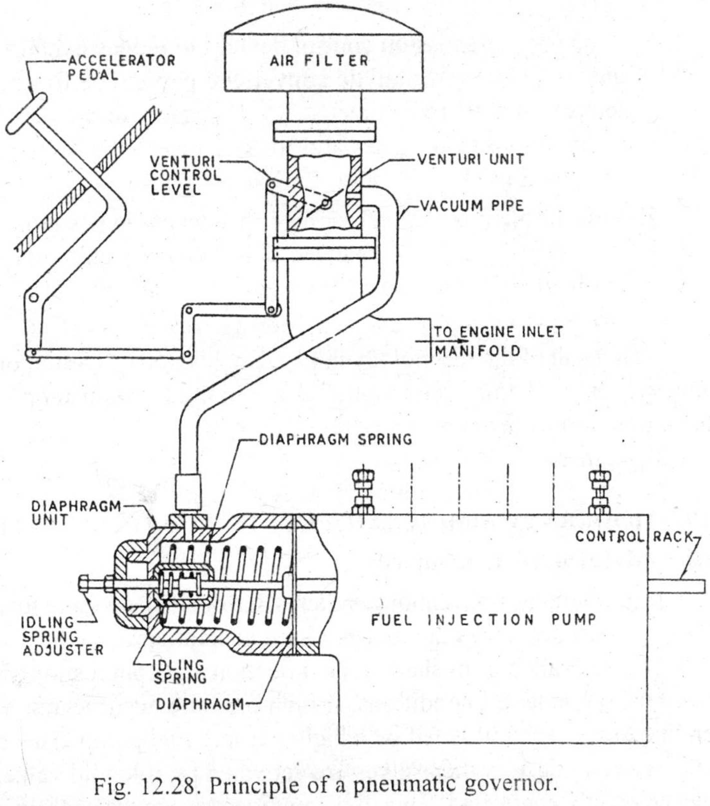

Fig. 12.28. Principle of a pneumatic governor.

injected. Idling spring adjustment is also provided by means of a separate spring and lever; the spring tension cam be adjusted with an adjusting nut provided as shown. Thus this is an all speed type of governor.

8.6. Hydraulic governor

A hydraulic governor *e.g.* the C.A.V. make works on the principle of *inverted hydraulic amplifier, i.e.* a small pressure change in some part of the system producing in some other part, an opposite change of much greater amplitude. This is also an all speed type governor.

9. MODERN DIESEL ENGINES

The main features of today's improved diesel engine technology are:

1. Modern diesel engines use direct injection, common rail fuel distribution, unit injectors and pre-injection to eliminate combustion noise.
2. The modern Common Rail systems employ engine-driven fuel pump to produce extremely high pressure fuel, which is delivered to the electrically-operated unit injector at each cylinder through a thick tube called Common Rail.
3. High pressure fuel supply means better fuel atomisation leading to higher engine efficiency, providing more power, fuel economy and improved emissions.
4. Modern diesel engines are often turbocharged and mostly have intercoolers to cool the hot compressed air coming out of the turbocharger, thereby increasing its mass density so that more air can be delivered to the cylinders to produce more power.
5. Sophisticated electronic control systems are used to regulate the injection under various conditions. This is done with the help of an Electronic Control Unit (UCU) and various accurate sensors.
6. A number of emission control devices to reduce exhaust emissions. Most commonly used devices are the oxidation catalytic converters, particulate filters, exhaust gas recirculation (EGR) and selective catalytic reduction (SCR). The catalytic converters treat CO and Hydrocarbons; particulate filters trap root particles and SCR reduces NOx in the exhaust emission. The details have been discussed in chapter 18 on Emission Control.
7. Most of the modern emission control technologies require very low sulphur content in the engine fuel; their life span is reduced drastically if significant amount of sulphur is present in fuel. Accordingly in USA, as per EPA regulations, sulphur content in automotive diesel is limited to 15 ppm (parts per million).

As a result of the technology improvements, today's automotive engines provide about three times the specific torque output and about 60% fall in the fuel consumption during the last 25 years. Besides, they have transformed from noisy, rattling engines to today's silent running breed, almost of the same level as petrol engines.

10. MODERN COMMON RAIL FUEL INJECTION SYSTEM

10.1. Main concepts involved

1. In contrast to conventional systems that build up pressure for each injection individually, CRS (Common Rail System) decouples the fuel-pressure generation from the actual fuel injection process, which extends the latitude available for shaping the injection and combustion processes individually, resulting in optimal combustion under all conditions. A high pressure accumulator element, called rail, is mounted along the engine block. The rail is fed by a high pressure fuel pump. This common rail feeds all the injectors on the engine. The injectors themselves are activated by solenoid valves. Solenoid valves and the fuel pump are electronically controlled. Thus the common rail, serves as the high pressure reservoir where the injection

pressure is independent of the engine speed and load, due to which the injection parameters can be freely controlled.

2. With CRS, high injection pressure are possible even at low engine speeds which leads to more complete combustion of the fuel. In particular, it reduces black-smoke emissions. Besides, at the same time, it becomes possible to increase engine torque in the lower engine speed ranges.

3. The combustion behaviour in diesel engines is affected by 3 variables, which are individually controlled in a CRS engine as follows:

(i) Timing of injection start

The optimal timing depends upon the engine load. In car engines at no load, the optimal injection start time is from 2° Before TDC to 4° after TDC. At part load, the optimal timing window is from 6° before TDC to 4° after TDC, while at full load it varies from 6° - 15° before TDC to 34° - 45° after TDC.

(ii) Duration of injection

Since the fuel is to be directly injected into the combustion chamber, the fuel mass flow would depend upon the difference between the injection and combustion chamber pressures, the density of fuel (which depends on temperature) and the dynamic compressibility of the fuel. These factors have to be taken into account while controlling the injector open time duration.

(iii) Discharge curve

In modern automotive diesel engines, the fuel is not injected once in a cycle. Instead fuel injection is done upto 4 or 5 times in one cycle. The first such mode is the *pilot injection* (also called *pre-injection*), which is short burst of fuel before the main injection during each combustion cycle. It helps reduce combustion noise and NO_x emission. Thus a double pre-injection would result in less noise compared to a single pre-

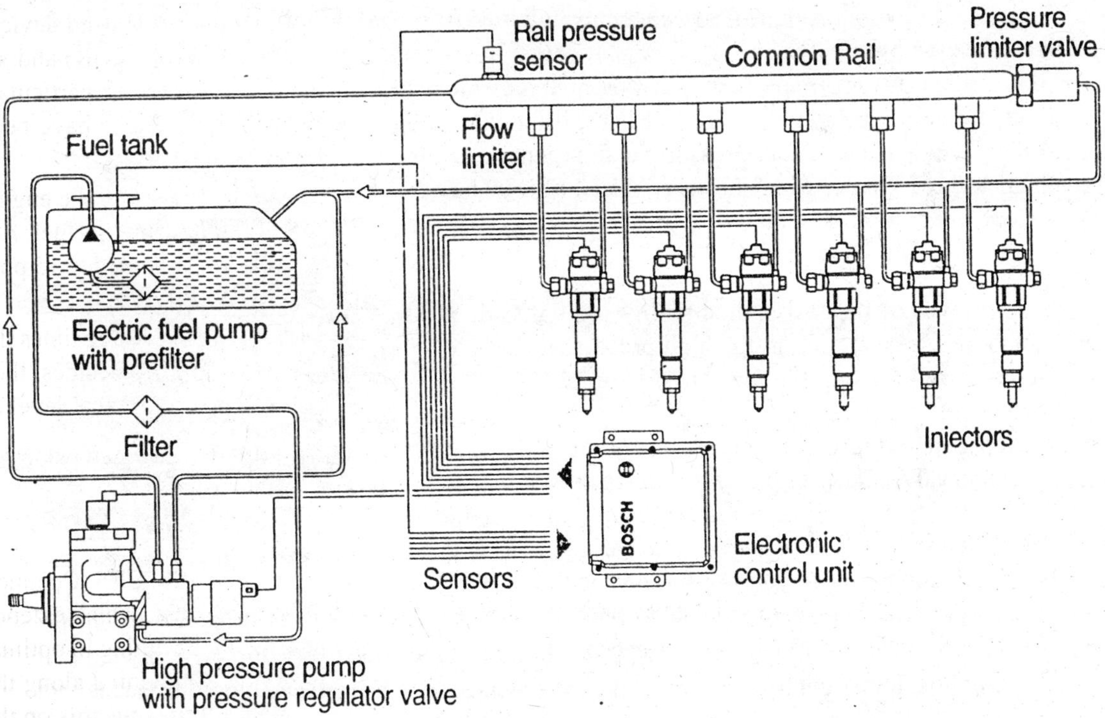

Fig. 12.29. Layout of Bosch K5-common rail system for passenger car engine.
(Courtesy—MICO, India)

injection. The bulk of the fuel is then added in the *main injection* phase, after which *post-injection* spurt of fuel is added, which reduces soot emissions. Finally, after upto 180° crankshaft degrees a *retarded post-ignition* occurs, which acts as a reducing agent for NO_x catalytic converter and/or increases the exhaust gas temperature for the regeneration of the particulate filter. The fuel injected is about 1 mm^3 for pilot injection to 50 mm^3 for full-load main injection. The injection duration is about 1 to 2 milli-seconds.

10.2. System details

Fig. 12.29 shows the layout of Bosch KS common rail fuel injection system, which is their first generation system. The layout of their second generation system is shown in Fig. 12.30.

Ref. Fig. 12.29, the fuel from the fuel tank is pumped by a low pressure electric fuel pump through a filter, to the high pressure pump, which builds up the high pressure in the common rail with the help of a pressure regulator valve which is controlled by an ECU (Electronic Control Unit) through pressure sensor. Thus the fuel pressure in the rail is independent of engine speed and the injected fuel quantity. Flow limiters protect the system against consequences of a potential leak of the injector. A pressure limiter valve guards the system against excess pressure. The fuel is injected in the engine combustion chamber by the injection nozzle of an injector controlled by a solenoid-operated valve. The injector receives its operation signals from the ECU. The system pressure, the duration of injection and the size and shape of the injection nozzle determine the fuel quantity injected.

The main components of the system are the low pressure pump, the high-pressure pump, the common rail ECU, injection lines, the injectors, sensors and actuators, the details of which are as follows:

Low-pressure pump

It is an electrical fuel pump for pre-supply to the high pressure pump. In the more advanced systems, a gear pump integrated with the high-pressure pump takes over pre-supply, making additional electrical fuel pump obsolete.

High pressure pump

This is a radial-piston pump which generates high pressure in the accumulator, *i.e.*, the rail, independent of engine speed and injected fuel quantity. Engine drives the pump through coupling, gear wheel, chain or toothed belt. The maximum fuel pressure in first generation Bosch pumps was 1350 bar which has been raised to 1600 bar is the second generation systems.

Fig. 12.31. High-pressure pump.

Common rail

Pressure level of the rail is electrically regulated by a combination of inlet metering to the high-pressure pump and fuel discharge by bleeding the pressure off quickly *via* the injector solenoid return. Fuel-pressure level is set by a closed control loop consisting of rail-pressure sensor, ECU and pressure-control valve. The pressure sensor continually measures the rail pressure. Using various programme maps stored in the ECU, the system adapts the *pressure, start* of injection and *duration* of injection to suit engine's operating conditions. Rail pressure is independent of engine speed/load so that high injection pressure can be produced even at low engine speeds, if required.

A series of injectors is connected to the rail and each injector is opened and closed by a solenoid driven by ECU.

The main functions of ECU are:

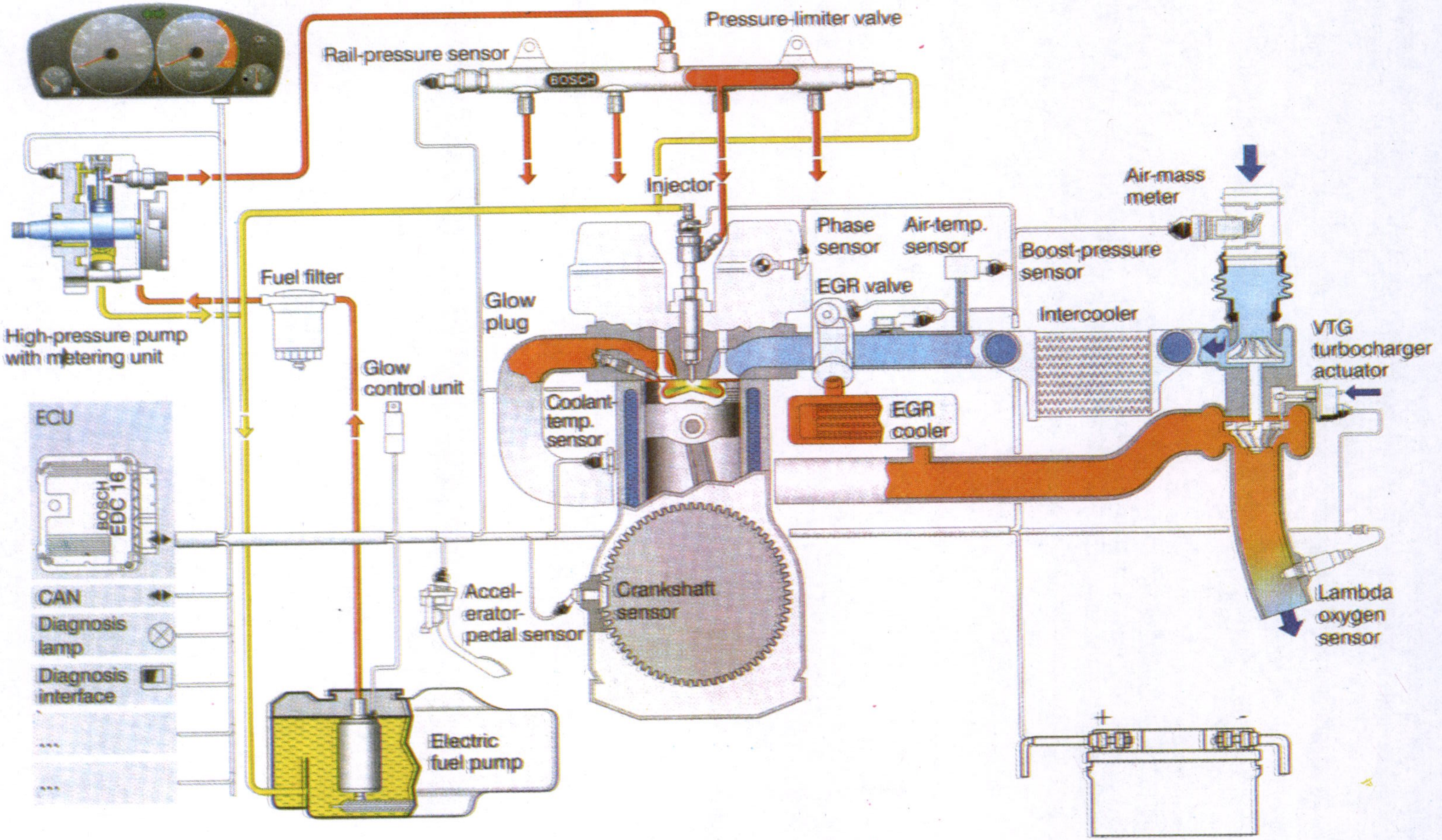

Fig. 12.30. Bosch second generation CRS for Passenger car engines. (Courtesy–MICO, India)

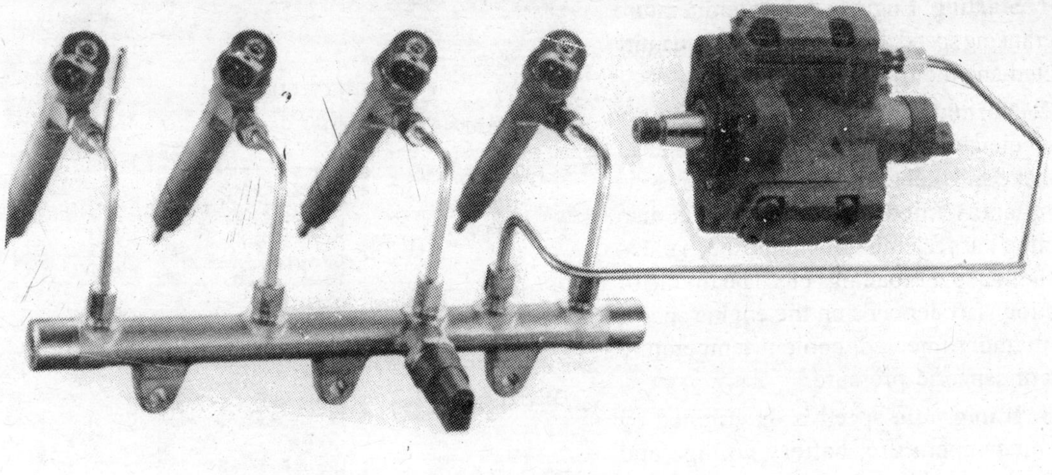

Fig. 12.32. Common rail alongwith injectors and high-pressure pump (Courtesy—MICO, India)

1. Provides map-based control of pilot injection, secondary injection, injection timing and delivery duration, providing thereby, minimal exhaust emissions and maximum fuel economy in all engine load and speed ranges.
2. Controls the fuel pressures in the common rail.
3. Integrated diagnosis-sensor monitoring and malfunction recognition.
4. Supplementary functions, such as:
   (*i*) Glow control.
   (*ii*) Exhaust gas Recirculation.
   (*iii*) Selective cylinder shut-off.
   (*iv*) Vehicle speed governing.

With the help of CAN (Controller Area Network), it is easy to network the ECU with other vehicle systems, *e.g.*, electronic transmission-shift control, thc ABS, the air conditioning system, etc.

Injectors

High-speed solenoid-controlled injectors spray fuel into the engine's combustion chambers through the integrated injection nozzles. A single such injector is fitted to each cylinder. The injection process is controlled by ECU by transmitting an optimally- timed trigger pulse to the injector solenoid. Actual flow of fuel depends upon (*i*) injector nozzle's spray aperture, (*ii*) solenoid's opening duration, and (*iii*) rail pressure.

10.3. Engine management in CRS

(*a*) Requirements

1. Very high fuel injection pressures.

2. Control of quantity of injected fuel, intake manifold pressure and timing of injection start according to different engine operating conditions.

3. Pilot-injection and post-injection.

4. Provision of rich mixture for starting, also considering the temperature.

5. Control of idle speed irrespective of the engine load.

(*b*) **Functions**

1. Starting. Engine coolant temperature and cranking speed determine the fuel quantity injected and the timing of injection start.

2. Normal operation. For this mainly the accelerator pedal position, engine speed, air intake and fuel temperature determine the fuel quantity injection. Other factors also affecting this are emission limits, mechanical and thermal over loadings etc. The timing of injection start depends on the engine speed, fuel quantity injected, coolant temperature and atmospheric pressure.

3. Idling. Idle speed is determined by coolant temperature, battery voltage and mechanical load like air conditioner. This is achieved by closed loop control by the ECU by continual monitoring of the actual speed and adjustment of the fuel quantity till the desired idle speed is achieved.

4. Limiting the engine speed. This is done by progressively decreasing the fuel quantity injected as the engine speed exceeds the engine maximum power speed, till the later value is achieved, by which time zero fuel quantity is injected.

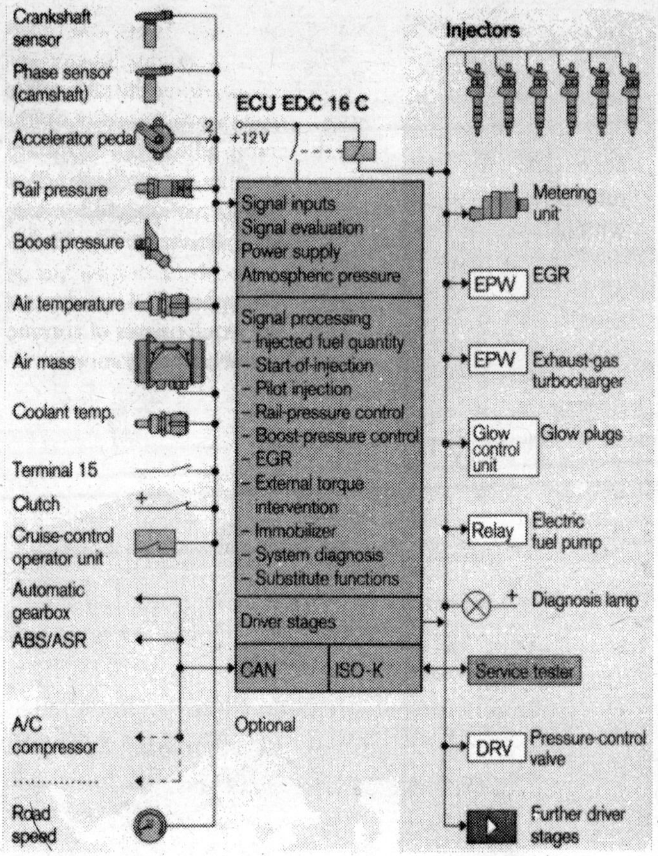

Fig. 12.33. Schematic diagram of ECU for second generation CRS for passenger cars.

5. Surge control. Sudden changes in the torque output of the engine is controlled by the ECU detecting the surging first and then by increasing the fuel quantity when engine speed decreases, and vice versa.

6. Smooth running control. In an actual engine, the torque output by various cylinders may differ slightly due to mechanical differences, which is detected from the variations in engine speed. Immediately after the fuel injection to a cylinder, the engine speed is compared with the average engine speed and depending on the former being more or less than the later, the fuel quantity to that cylinder is decreased or increased respectively.

7. Closed loop lambda control. A broadband oxygen sensor which can measure very lean air/fuel ratios such as 60:1, is employed for closed loop control in diesel engines. This control is used for short- and long-term adaptation learning of the fuel quantity injected, which facilitates limiting the smoke output and to determine whether the desired EGR is being achieved or not.

8. Fuel pressure and flow control. Fuel pressure in the common rail is regularly monitored by a pressured sensor sending its output to ECU which modulates the pulse width of the control current, thereby modulating the fuel pressure control valve through closed loop control.

10.4. Bosch Third Generation CRS

This has been available in the market since 2003. The main improvements of this are: (*i*) Higher maximum injection pressure of 1600 bar, (*ii*) Piezo injectors instead of solenoid-controlled injectors.

Piezo injectors use piezoelectric material to produce motion when excited by an electrical voltage. Piezoelectric materials are crystals which stretch when an electric voltage is applied. Motion created by the piezo device opens the injector. Switching speed of piezo-injector is twice as high as that of a solenoid valve. It takes less than 100 micro-second to open and close the needle valve of the injector and spray the high-pressure diesel fuel into the engine combustion chamber, allowing five or even possibly more injections per engine cycle, which leads to quieter running, improved fuel consumption and reduced emissions. Quieter running arises from decreased combustion noise on account of higher number of pilot and secondary injections which make the combustion process still smoother. Lesser fuel consumption and emissions result from

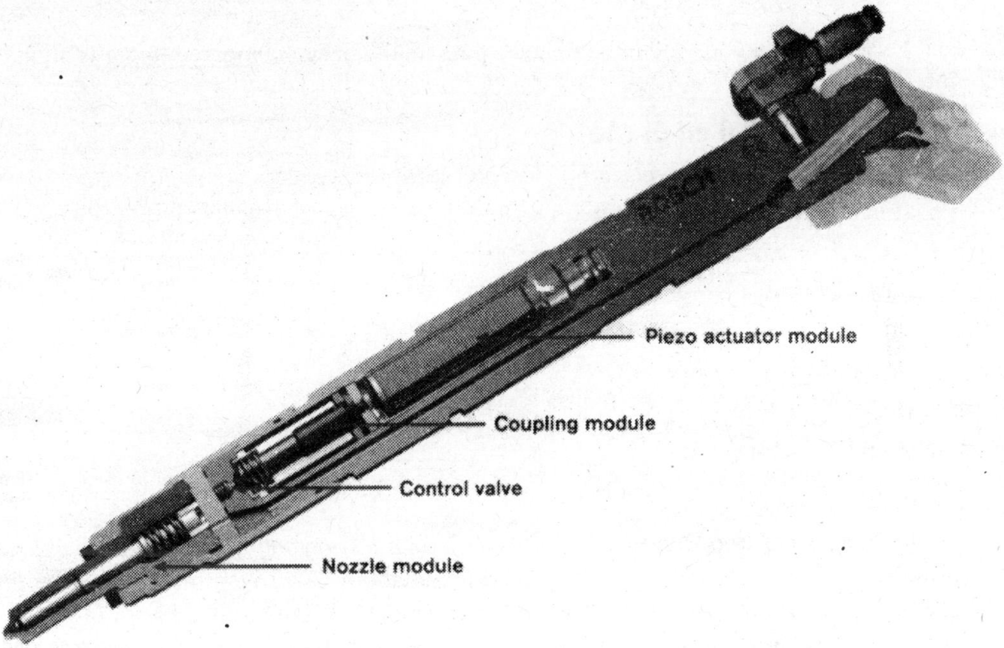

Fig. 12.34. Piezo inline injector. (Courlesy—MICO, India)

the increased switching speed of the injector which enables injection start, duration and rate-of- discharge curves to be modelled so that the programme-map configuration at the engine can be improved still further. This means that the engine design concept can be focused in any of the following directions as per requirements:

(*i*) Upto 3% lower fuel consumption.

(*ii*) 15% to 20% lower emissions.

(*iii*) 3 db less engine noise, compared to the solenoid injectors.

In comparison to this, Delphi's Direct Acting Diesel Common Rail System operates at 1800 bar maximum rail pressure. Moreover, their injectors do not need return line to operate. That means all fuel is injected and no high pressure fuel is wasted into a return flow. It also allows downsizing of the high pressure pump, helping provide an improvement in engine fuel economy. As such it is claimed to provide 10% more torque and power, upto 30% lower emissions and better fuel economy compared to servo-piezo systems.

Recent piezo ceramic injector launched by Delphi allows injection speeds three times faster than previous systems for seven or more injections per cycle. This system can raise injection pressures by 200 bar so that 2000 bar Delphi injector would give performance equivalent to a 2200 bar injector using servos. The technology would help meet Euro 6 emission standards.

10.5. Advantages of CRS

1. Due to separation of the pressure generation from the actual fuel injection process, the Common Rail System is able to provide increased fuel economy and lower emission alongwith less noisy operation, almost comparable to the petrol engines. Even the torque and power curves have been improved, in particular, at lower engine speeds.

2. The system can be easily integrated in a wide variety of engines; it can even adapt to already existing direct injection diesel engines.

3. It also provides a variety of extended functions and further freedom in the design of combustion process.

4. It is relatively simple in construction and needs less maintenance compared to conventional diesel fuel injection system.

10.6. Technical specifications of diesel CRS

Typical specifications of a *third generation* Bosch CRS for passenger cars are:

Number of cylinders	≤ 8
Injection pressure at nozzle	1600 bar
Pilot injection quantity	$= 0.5$ m^3/stroke
Number of partial injections	variable
Emission reduction	15% to 20%
Noise reduction	$\leq$ 3dB (A)
Increase in engine output	5-7%

10.7. Application in automobiles

Ford's 1.8 litre Duratorq Turbo Diesel *Common-rail direct injection engine* employed on their 'Focus' provides peak power output of 85 kW at 3800 rpm which is about 28 percent higher than the previous 'Focus', 'Endura DI,' direct injection engine (66 kW), whereas steady state peak torque (250 Nm) is 25% higher than Endura DI's 200 Nm. The emissions have also reduced markedly. Inspite of all these advantages, the fuel consumption has remained same at 18 km/litre. All this has been made possible with the second-generation common-rail technology, which is the only diesel injection system with full flexibility of injection timing, measured in crank angle degrees, between the pilot and main fuel injection. The Duratorq also employs an electronic knock sensor attached to the cylinder block. This sensor monitors the engine's vibrations and commands the injection system's ECM to modify the combustion process at optimum level.

In India, Hyundai has first employed the *Common Rail Direct Engine* (CRD) engine in 2004. Mahindra has soon followed suit with their CRDe engine for Scorpio. The system used is 'Bosch Common Rail System CRS Generation 2.2', wherein the common rail pressure is 1600 bar. The high fuel delivery pressure counters the pressure waves caused by the rising piston, which results in better engine response to throttle inputs. Further, the 'pilot injection', microseconds before the main injection, helps steady engine warm-up, which means less abrupt combustion and hence smoother operation. Both these engines meet Bharat III emission norms.

10.8. Volvo's 'i-ART' technology

Volvo has launched its new 'world-first', 'i-ART' technology in 2013, which helps to reduce fuel consumption in its new diesel engines. According to Derek Crabb, vice president of power train engineering at Volvo, "it is a break through comparable to when we invented the Lambda sensor for the catalytic converter in 1976."

In this i-Art technology, each injector has a small computer on top, which monitors injection pressure.

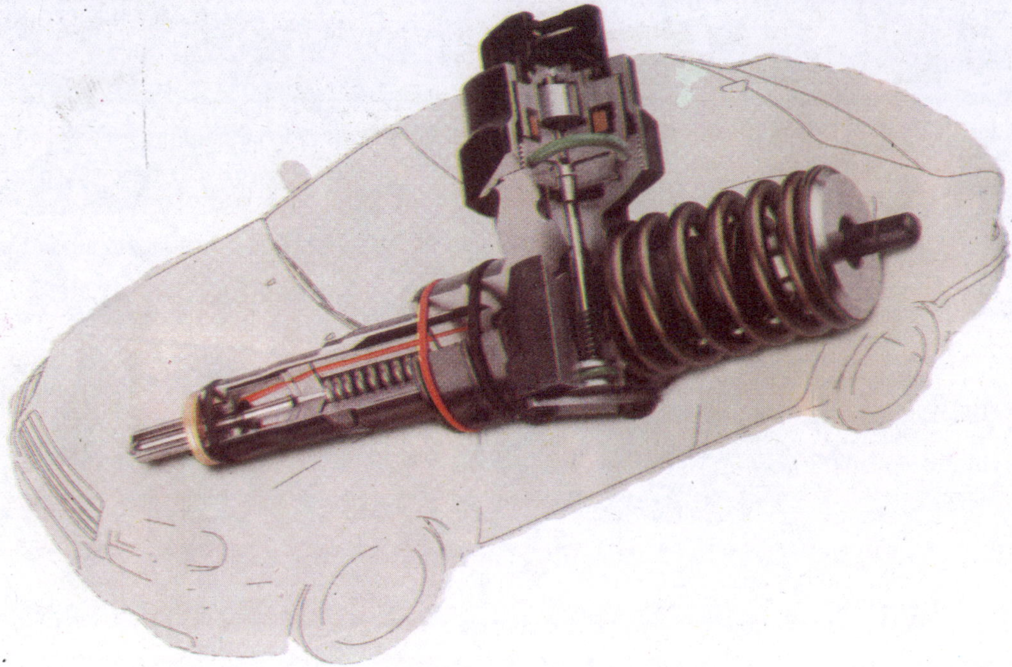

Fig. 12.35. Bosch Unit Injector. (Courtesy–MICO, India)

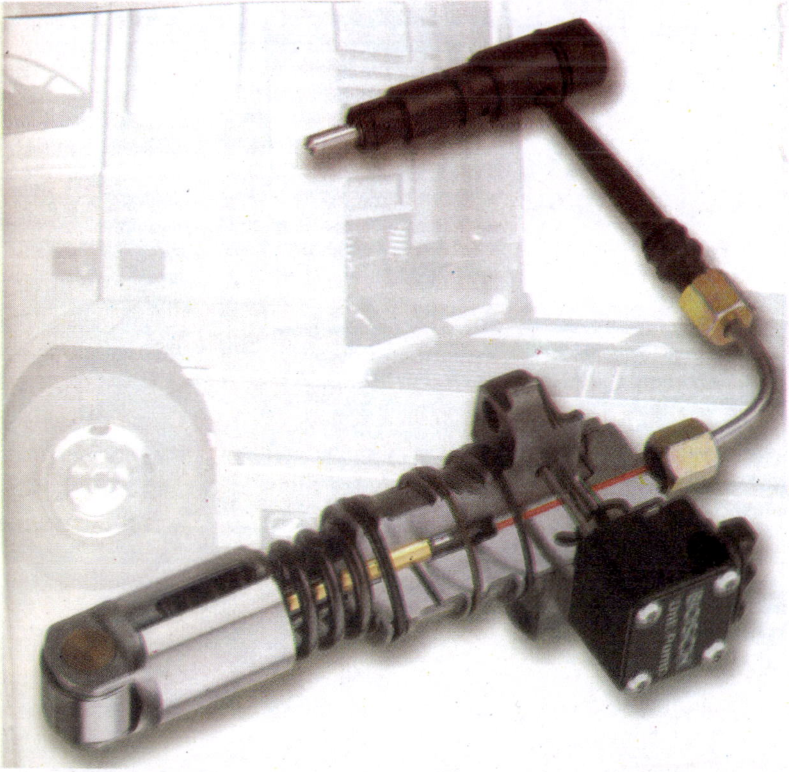

Fig. 12.36. Bosch Unit Pump. (Courtesy–MICO, India)

Using this information, the self-adapting i-ART system operates to inject the ideal amount of fuel during each combustion cycle. In this higher rail pressure of 2500 is used and pressure feedback is taken from each fuel injector instead of using single pressure sensor in the common rail in the conventional systems. The advantages claimed are lesser fuel consumption, lower emissions and higher performance output.

11. UNIT INJECTOR SYSTEM

The unit injector comprises a pumping element, injection nozzle and solenoid valve (Fig. 12.35). Each such unit injector is mounted on each engine cylinder head, thus dispensing with a high pressure fuel line. The maximum injection pressures in this systems are currently 2050 bar in passenger cars and 1800 bar in commercial vehicles.

The unit injector pumps are operated mechanically by means of overhead camshaft of the engine.

When the high-speed solenoid valve is closed, the fuel flow starts immediately towards the injectors nozzle's spray orifice, thus marking the injection timing. The duration of the solenoid valve closure determines the fuel quantity injected. When the solenoid valve opens, fuel from the injector starts flowing into an overflow line. This entire fuel injection process is defined by the electronic closed-loop control of the solenoid. The ECU also takes care of many auxiliary functions, *e.g.*, temperature-sensitive injection timing, smooth running control and active surge suppression, besides providing the diagnostic capabilities.

This system also reduces the combustion noise by means of pilot injection.

12. UNIT PUMP SYSTEM

This system is similar to the unit Injector system, except that the injector pumps in this are operated by means of a block-mounted camshaft and the pump is linked to the injector nozzle through a short pressure line as shown in Fig. 12.36. Such systems are available for commercial vehicle application and operate at maximum injection pressure of 1800 bar.

13. COMPARISON OF VARIOUS FUEL INJECTION SYSTEMS

1. In-line injection pumps alongwith mechanical governors have proved their reliability and precision till recently. However, finer requirements in the diesel engine fuel supply and combustion systems have been possible only with electronic control.

2. The distributor type injection pump is excellent for small, high speed diesel engines, on account of its compact design and low weight.

3. Single-plunger pump system (Unit Injector and Unit Pump Systems) are ideal for precision metering ... i to each cylinder.

4. Common Rail System separates the pressure generation and injection functions, thus giving more flexibility to the designer for improving engine's combustion process.

4. SUPERCHARGING OF DIESEL ENGINES

As has been seen already while discussing combustion process in diesel engines, diesel knock is not restricted by the compression ratio. Therefore, in supercharged diesel engines, there is no limitation of compression ratio. Thus in such engines, there is no apprehension of diesel knock occurring. The limit to supercharging is rather on account of the higher thermal loads on piston and valves. That is the reason, superchargers are used extensively in case of diesel engines.

The latest type of supercharger for automotive use is the G-charger developed by Volkswagenwerk of Germany. It consists of 4 main parts (Fig. 12.37), *viz.*, channel housing, secondary shaft, compressor and eccentric shaft. In the two-piece channel housing with two-spiral shaped channels, a double-acting displacer is moved on the eccentric shaft via the secondary shaft. The eccentric shaft is driven by a ribbed belt from the engine crankshaft. The circular motion of the displacer reduces the size of the compression space. The compressor conveys the intake air (white arrow) smoothly towards the centre of the housing. During his process, the charge pressure is built-up. The compressed air (shaded arrow) is forced out through the opening around the eccentric shaft bearing, through a charge air cooler where it is cooled off and then passes onto the cylinders.

The designation 'G charger' was derived from the shape of the displacer which resembles the latter 'G'. It has got the advantage that even at low engine speeds the torque is increased considerably. Moreover, since the compressed air from the working chamber of the G-charger overlaps as it is transported to the pressure side, there is no sudden interruption of the air flow. It is therefore, very uniform and the noise level is correspondingly low.

Fig. 12.37. G-charger
A. Channel housing, B. Secondary shaft, C. Compressor, D. Eccentric shaft

15. COLD STARTING DEVICES

Ignition of the cylinder charge in a diesel engine will occur only if the following conditions are met with at the time of fuel injection:

1. The temperature of the air is sufficiently higher than the self-ignition temperature of the fuel, which for most diesels is from 220° to 250°C.

2. The pressure must be high enough to ensure adequate closeness of the air and fuel mixture on injection, so that heat may be rapidly transferred from the air molecules to the liquid fuel particles.

Besides above, the engine must be cranked at the optimum speed. In case of very low cranking speed, the heat loss to the cold cylinder walls and the leakage of air past the piston rings may prevent the compression temperature reaching the desired value. On the other hand, at high cranking speed, the time available for ignition to be established will be too short. The optimum cranking speed depends upon the following factors:

1. Surface/volume ratio
2. Intensity of air swirl in the cylinder and combustion chamber.
3. Mechanical condition of the engine.
4. Type of combustion chamber. It is seen that open direct- injection chambers have the easier cold starting since they have the smallest surface/volume ratio and also the lowest intensity of air swirl compared with indirect injection chambers. Thus special devices are required to start from cold the indirect injection engines though such devices may also be used to facilitate easy start in case of direct injection engines also. Various devices used for this purpose are:

1. Decompression devices.
2. Heater plugs.
3. Inlet manifold heaters.
4. Chemical sprays.

1. Decompression devices

A lever on the cylinder head causes the compression in the cylinder to be partially released, which makes it easier to crank the engine, till it gains sufficient momentum to start the engine by itself when the lever is brought back again to restore full compression.

2. Heater plugs

A heater plug also called *glow plug* (Fig. 12.38), having low voltage (mostly 12 V) heating element at its end is screwed into the swirl chamber of each cylinder. Pushing a heater button on the dash board heats the elements which heat the air in the chamber. The heater push button is then released and the starter is pressed when the engine is started because of compression of the hot air.

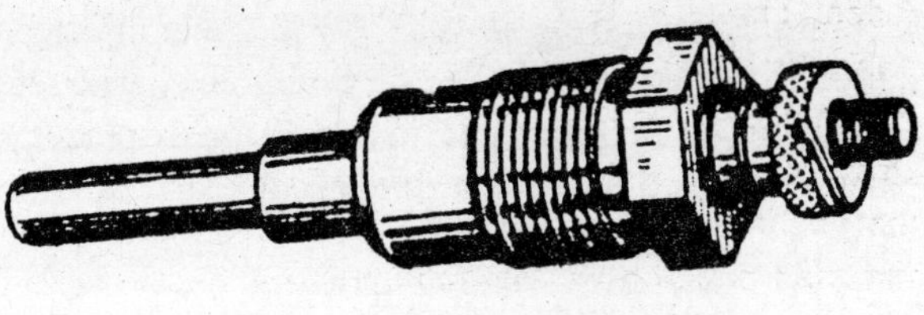

Fig. 12.38. Glow plug.
(Courtesy—Mahindra and Mahindra Ltd.)

Due to extensive development in the diesel engines the requirements of plugs for modern diesel engines have also increased manifold, which may be listed as follows:

1. Fastest heat-up speeds (1000°C per second) even with a voltage drop in the vehicle's electrical system.
2. Higher permanent-glow temperatures upto 1150°C.
3. Extended post-and intermediate-glow times in the minute range.
4. Long life, equal to the engine life.

Bosch's recent low-voltage (12 V) glow system consists of ceramic Rapiterm sheathed-element glow plugs, an electronic glow control unit and a software module integrated into the engine ECU (Fig. 12.39). These glow plugs are triggered by the control unit, thus making it possible to control the pre-glow, start, post-flow and intermediate-glow positions as per engine requirements. These glow plugs have sheathes of highly temperature- resistant ceramic composite material with adjustable electrical conductivity. Due to their very high resistance to oxidation and thermal shock, they allow immediate starting, maximum glow temperatures of 1300°C and minutes of post-and intermediate-glow at 1150°C. They consume very small power, work precisely even in case of system voltage drop and also have long life. A self-diagnosis unit and a monitoring unit are also integrated into the glow control unit, enabling on-board diagnosis as per European and U.S. norms.

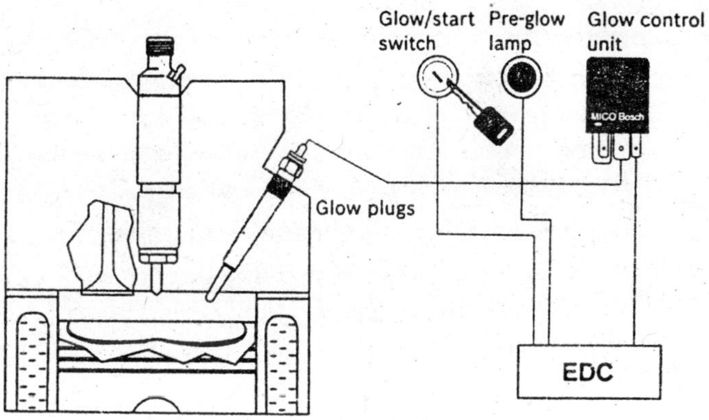

Fig. 12.39. Glow equipment in a D.I. engine
(Courtesy—MICO, India)

3. Inlet manifold heaters

Instead of heater plugs, the air in the inlet manifold can be heated by means of a heating coil, which helps the engine to start. However, the coil takes heavy current from the battery and sometimes diesel oil is burnt in the inlet manifold using special thermostatic device. A crude method of heating inlet manifold is by a burning rag dipped in the diesel oil and placing the same close to the manifold.

4. Chemical sprays

Certain chemicals are available which when sprayed into the inlet manifold vapourize readily and may be ignited at lower temperature than diesel oil. Special pumps are available to inject such chemicals. One such chemical is diethyl ether [$(C_2H_5)_2O$]. It has a boiling point of 35°C, cetane number of 53, latent heat of 384 kJ/kg. It reduces the delay period during the combustion process. The ignition point of a typical ether compound fluid is as low as 180°C. Ether injected into the air stream considerably reduces the cranking time for firing the engine. The optimum discharge rate of ether in a four stroke engine in cold weather is usually from 0.1 to 0.2 ml per litre of piston displacement per second of air intake. Ether cold-start devices are being installed in the manifold as standard equipment in cold climate countries.

16. DIESEL ENGINE TROUBLE SHOOTING

The major faults occurring in diesel engines and their remedies are discussed here.

1. Engine does not start or stalls just after starting. This may be due to any one or more of the following causes:

(*i*) Starting motor may not be turning, which may be either due to faulty starting motor, discharged battery, sulphated terminals or loose connections. Check these causes and rectify the defect if any.

(*ii*) If the engine still does not start, it may be due to air lock in the fuel system. Priming the fuel system will drive the air out, which may be done as per the instructions of the vehicle manufacturer.

(*iii*) Lack of fuel in the system. Inspect the fuel tank and ensure that fuel supply is in the 'on' position. In case these are alright, filter may be choked which should be inspected and cleaned if required. If the fuel does not reach the engine still, take out the injectors one by one, crank the engine at normal speed and check if by so doing the injectors eject the fuel out into the air. If some of the injectors do not eject, they may be opened and cleaned. If however, all of the injectors do not spray out the fuel, the defect may be in the fuel injection pump which may be opened out and cleaned.

(*iv*) Incorrect injection timing, which may be checked and adjusted, if required.

(*v*) Clogged air cleaner, which may be cleaned.

(*vi*) The fuel may be containing water. This may be checked by taking out some fuel from the bottom of the fuel tank after allowing it to stand. If the water is there it will settle down to bottom.

(*vii*) Loss of compression may also cause difficult starting because in such a case the temperature of air after compression may not be sufficiently high so as to start ignition of the fuel on injection. The compression may be checked and suitable steps taken to restore it to correct value. This may involve just changing the gasket or even overhauling the engine.

2. Faulty running. This may also be due to any or more of causes: (*ii*) to (*vii*) above. Apart from these, faulty running may also be due to the following:

(*i*) Defective governor. Check it thoroughly and repair or replace any component or the entire unit as required.

(*ii*) Defective accelerator pedal linkage, which may be checked and adjusted if required.

3. Excessive fuel consumption. This may be due to any or more of the following:

(*i*) Leak in the fuel system, which may be inspected and rectified if required.

(*ii*) Fuel injection pump timing may not be correct, which may be tested and adjusted properly.

(*iii*) Loss of compression in the engine. This may be checked and restored by suitable replacements or repairs, which may require engine overhaul.

(*iv*) Governor setting may be defective, which may be tested and adjusted suitably.

(*v*) Clogged air cleaner, which may not be correct, which may be adjusted.

(*vi*) The valve tappet clearance may not be correct, which may be adjusted.

4. Lack of power. The reasons for this may be any or more of the following:

(*i*) Loss of compression.

(*ii*) Incorrect injection timing, which may be tested and adjusted properly.

(*iii*) Injection pump plunger may have worn out. Inspect the same and if found so, the entire pump will have to be replaced.

(*iv*) Clogged air cleaner.

(*v*) Leak in the fuel system.

(*vi*) Clogged injector spray holes.

(*vii*) Clogged fuel filter.

(*viii*) Defective governor setting.

5. Excessive knocking. This may be mainly due to the following causes:

(*i*) Defective fuel. If the fuel used is not of the required cetane value, the knocking will occur.

(*ii*) Excessive carbon deposits will also cause knocking. To remove this defect, the engine has to be opened up for decarbonisation.

(*iii*) Fuel injection may be excessively advanced, which may be tested and adjusted properly.

17. SERVICE PRECAUTIONS

Following precautions must be followed while servicing diesel engines;

1. Never try to remove injection pump or injector or injector line when the engine is running.

2. Always prime the injection pump fully before starting the engine after service. An injection pump should never be run dry.

3. Never try to stop a diesel engine by covering the intake manifold opening with hand; it may cause serious injury.

4. Fuel injection pump and injectors are highly precise pieces of equipment. As such:

(*i*) When removing injection pump or injectors, cover all openings and put various components in a clean plastic bag to avoid dust.

(*ii*) Never do grinding work and avoid any such activity as may raise dust near any injection pump, injectors or open engine. In case of unavoidable situation cover the engine and any loose components with a plastic sheet.

(*iii*) Always use proper tools while working on injection system components. Never use any hammer, wrench or handle to tap injector or other components.

5. As fuel line pressures are very high (42 to 55 MPa), utmost caution should be observed for connecting lines to injectors; otherwise air may be drawn into the fuel supply system or result in fuel leaks leading to poor engine performance or even fire.

6. Always wear safety glasses while working with high pressure injector testing equipment.

TABLE 12.1. COMPARATIVE DIESEL ENGINE FUEL SYSTEM DATA OF SOME INDIAN AUTOMOBILES

S. No.	*Item*	*Ashok Leyland Viking Passenger*	*Tata Diesel Vehicles 1210*	*Swaraj Mazda (T3500 series SL engine)*
1.	Fuel injection pump	Hindustan Bosch inline Type HB-PE 6A 80B 320RS 2154	Hindustan Bosch inline Type HB-PES 6A 80B 410RS 2405	Diesel Kiki Co. Ltd. PE(S)-A (in-line type)
2.	Injection pressure, kg/cm^2 (MPa)	160 (15.7)	180 (17.6)	170 (16.7)
3.	Nozzle holder (injector)	Hindustan Bosch Type HB-KBAL 83S 7/4	Hindustan Bosch Type HB-KDAL 74 S 3/19	Bracket Type
4.	Nozzle	Hindustan Bosch Type HB-DLL 140 S 215	Hindustan Bosch Type HB-DLLA 150 S 187	Hole type
5.	Feed pump	HB-FP/KE 22 AD 223/2	HB-FP/KE 22 AD 280/2	KS (Single-acting type) Feeding pressure 3.1 kg/cm^2 (0.3 MPa)
6.	Injector Timer	—	Centrifugal HB-EP/SF 500-1400 B4R1 (Max. advance angle 8°)	SCD M (Max. advance angle 7°)
7.	Governor	Centrifugal variable speed type HB-EP/RSV 200-1200 A1A41R	Centrifugal variable speed type HB-EP/RSV 300-1400 A2B1062DL	RLD-K (all speed mechanical type with torque cam)
8.	Idling speed, r.p.m.	400	600	620-700
9.	Fuel filter	Micro dual filter, paper element type	Micro dual filter, cloth insert as primary stage filter and paper element as final stage filter	Paper element type
10.	Air cleaner	Oil bath type with filter mesh element	Oil bath type with filter mesh element	Oil bath type with filter paper element
11.	Fuel tank capacity, litres	N.A.	160	90

REFERENCES

1. 'Volvo launches world-first injection technology in new VEA engine family', http:// www.enginetechnology international.com/news.

THEORY QUESTIONS

1. Discuss in detail the function of fuel supply system in an automotive diesel engine. State also the names of various components used to perform these functions.
2. Draw diagrams showing the layout of various types of fuel supply systems for automotive diesel engines and discuss them extensively.
3. Explain the functions of a primary and a secondary diesel fuel filter. Describe any filter of each type.
4. Discuss the construction and working of various types of air cleaners for automotive diesel engines.

5. What types of fuel feed pumps are used in automotive diesel engines? Describe the construction and working of any such pump.
6. Explain the construction and working of any type of fuel injection pump. Illustrate your answer with suitable sketches.
7. Describe in detail the construction and working of an in-line fuel injection pump. Explain clearly how fuel supply is controlled with such an injection pump.
8. Draw a sketch showing the details of a distributor type fuel injection pump and explain its working. Describe clearly the function of internal-cam ring.
9. What is an injection timer? Explain its function and describe clearly how it performs the same.
10. Sketch any fuel injector and explain its working.
11. Draw sketches showing the constructional features of different types of nozzles used in automotive diesel engines, explaining the function of each.
12. Discuss in detail various terms relating to governors.
13. Why is a governor absolutely necessary in case of diesel engines unlike petrol engines? Discuss the working principles of various types of governors.
14. Discuss various technological features of modern automotive diesel engines.
15. Explain in detail modern Common Rail fuel injection system, discussing the main system components. Discuss also their various advantages over the conventional fuel supply systems for diesel engines.
16. Explain in detail the effect and control of the following in a Common Rail System in diesel engines:
    (*i*) Injection stat timing.
    (*ii*) Duration of injection.
    (*iii*) Fuel discharge curve.
17. Discuss in detail the role of engine management system in a CRS automotive diesel engine.
18. Describe in detail automotive applications of the Common Rail System.
19. What are the functions of Unit Injector and Unit Pump systems? Describe these systems and also compare them with each other.
20. Name various fuel supply systems that have been used for automotive diesel engines and describe salient features of each.
21. Explain clearly why superchargers are used more often in diesel engines. Describe the working of any type of supercharger for a automotive diesel engine.
22. What is the necessity of a cold starting device in a diesel engine? Discuss in detail various types of cold starting devices for automotive use.
23. Enlist various requirements of modern glow plugs and describe one such system in detail, mentioning also its main advantages.
24. Discuss major faults occurring in automotive diesel engines and suggest also suitable remedies.
25. Describe in detail various precautions which should be observed especially while servicing diesel engines.

QUESTIONS FOR VIVA

1. What is the function of fuel injection system in an automotive diesel engine?
2. Name the various types of fuel injection systems for diesel engines.
3. Name the main components of fuel injection system for automotive diesel engines.
4. What is air injection system?
5. What is solid injection system?
6. What is a common rail fuel injection system?
7. What is 'pilot injection'?
8. Name any Indian vehicle using common rail.

9. What is a unit injector?
10. What is individual pump fuel injection system?
11. What is a primary fuel filter?
12. Name the most common primary filter used in diesel engines.
13. What is a primary fuel filter?
14. What are the functions of an air cleaner?
15. How does oil contribute to remove impurities from air in a an oil-bath air cleaner?
16. What is the function of a centrifugal pre-cleaner in a heavy-duty air cleaner?
17. What is the function of a fuel feed pump?
18. Where is feed pump usually located?
19. From where does the feed pump get its drive?
20. What is the function of a hand primer? Where is it located?
21. What types of fuel feed pumps are employed in automotive diesel engines?
22. Describe the function of a fuel injection pump.
23. What are the usual injection pressures in use?
24. Why the fuel injection pump has to be manufactured with great precision?
25. In a 4 stroke, 4 cylinder diesel engine running at 4500 r.p.m., how many times the fuel will be injected per second?
26. What is approximately the magnitude of fuel injected per stroke?
27. What is an in-line type of fuel injection pump?
28. What is the function of helix in the pump plunger?
29. How does the injection pump plunger get its motion?
30. How does the special construction of the delivery valve in the injection pump serve to eliminate dribbling?
31. How is the effective plunger stroke varied in the injection pump?
32. At some position of the plunger rotation, there is no fuel delivery even when the injection pump is running? How does that happen?
33. What is the function of a toothed sleeve on the pump plunger?
34. What is contained in the pump housing of an in-line injection pump?
35. What material is used for the pump housing?
36. How is the injection advanced with increase in speed in an in-line injection pump?
37. What is a distributor type fuel injection pump?
38. How is the amount of fuel entering the rotor of a distributor type injection pump controlled?
39. How is the maximum amount of fuel intake controlled in a distributor type injection pump?
40. What is the function of the internal-cam ring? Is it stationary or rotating?
41. From where do the plungers in the distributor type pump get their motion?
42. How many cams are there in the internal-cam ring of the distributor type pump?
43. Why do the cams on the internal-cam ring in a distributor type pump have usually two peaks?
44. What prevents dribbling in a distributor type pump?
45. What ensures adequate residual pressure in the delivery port of the distributor type pump during cut-off?
46. How is the fuel injection timing controlled in a distributor type pump?
47. What are the advantages of a distributor type fuel injection pump?
48. What is the function of a fuel injector?
49. What causes the injector nozzle to deliver fuel?
50. What causes the injector nozzle to stop fuel delivery?
51. How can the opening fuel-line pressure be varied in the injector?
52. What is the function of nozzle?
53. Name commonly used nozzles in automotive diesel engines.
54. What is a hole type nozzle?

55. In what type of engines are the hole type nozzles generally used?
56. What is a pintle type nozzle?
57. What type of engines usually employ pintle type nozzles?
58. Why is governor necessary in diesel engines, whereas it is not so in case of petrol engines?
59. Name various types of diesel engine governors.
60. What is an "all speed governor".
61. What is a maximum speed governor?
62. What is governor cut-in speed?
63. What is governor cut-off speed?
64. What do you understand from the term 'governor overrun'?
65. Define governor hysterisis.
66. What is governor runout?
67. What is the principle of working of a mechanical governor?
68. Narrate the working principle of a pneumatic governor.
69. On what principle does the hydraulic governor work?
70. State main features of modern automotive diesel engines.
71. What is the approximate value of fuel injection pressure in a modern CRS engine?
72. Why low-sulphur fuel is a necessity in modern automotive engines?
73. What do you understand from (*i*) pre-injection, (*ii*) post-injection?
74. State any five main functions of engine management system in modern automotive engines.
75. Explain why use of supercharger is more common in diesel engines as compared to petrol engines.
76. State the two conditions which must be met for ignition of the cylinder charge in a diesel engine, to occur.
77. What is the necessity of installing a cold-starting device in a diesel engine?
78. Name various cold-stating devices?
79. What is a glow plug? Where is it used?
80. Which type of combustion chamber is best suited for easy cold starting of a diesel engine?
81. Why does a diesel engine supply richer mixture during acceleration?
82. What happens if an air lock occurs in the fuel system of a diesel engine?
83. How is the engine performance effected if the injection timing is defective?
84. What will happen if the air cleaner of an automotive engine is clogged?
85. What defects are caused in case of loss of engine compression?
86. What may be the main reasons for excessive fuel consumption of a diesel engine?
87. What may be the reasons if a diesel engine fails to start or stops immediately if starts at all?
88. What may happen if diesel oil of lower cetane number than the one specified is used?
89. What is the effect of worn out injection pump plungers on the performance of a diesel engine?
90. Why should one never attempt to remove injector with the engine running?
91. Why should the injection pump never be run dry?
92. Why should one not try to stop a diesel engine by covering its intake manifold with hand?
93. Why should safety glasses be worn while testing fuel injectors?

MULTIPLE CHOICE QUESTIONS

Mark the correct answers:

1. The compression ratio in an automotive diesel engine is usually
   (*a*) 7:1 (*b*) 10:1
   (*c*) 15:1 (*d*) 22:1

2. The cranking compression pressure in a diesel engine is about
   (*a*) 1000 kPa (*b*) 2000 kpa
   (*c*) 3000 kPa (*d*) 10000 kPa
3. The cylinder temperature after compression in a diesel engine is approximately
   (*a*) 300°C (*b*) 450°C
   (*c*) 800°C (*d*) 1000°C
4. The ignition temperature of diesel fuel is about
   (*a*) 300°C (*b*) 400°C
   (*c*) 600°C (*d*) 800°C
5. The fuel feed pump in a diesel engine is mounted
   (*a*) on the fuel tank (*b*) on the injection pump
   (*c*) on the injector (*d*) inside the fuel tank
6. In some diesel engines injection pressures can reach as high as
   (*a*) 100 MPa (*b*) 200 MPa
   (*c*) 300 MPa (*d*) 400 MPa
7. The fuel injection timing in a distributor type pump is controlled by
   (*a*) changing plunger stoke (*b*) changing speed of rotor
   (*c*) rotating the cam ring (*d*) changing the number of cams on the ring
8. The opening pressure of hole type nozzles varies from
   (*a*) 5-10 MPa (*b*) 10-20 MPa
   (*c*) 17-34 MPa (*d*) 27-54 MPa
9. The opening pressure of pintle type nozzle varies from
   (*a*) 7-15 MPa (*b*) 11-22 MPa
   (*c*) 17-34 MPa (*d*) 35-45 MPa
10. Secondary filter in a diesel engine is
   (*a*) a must (*b*) installed between the fuel tank and feed pump
   (*c*) optional (*d*) not capable of removing water from the fuel
11. A glow plug is
   (*a*) a decompression device (*b*) installed in the inlet manifold
   (*c*) screwed into the combustion chamber (*d*) used to plug the leakage of fuel
12. It is common for the diesel engine to emit excessive smoke during
   (*a*) starting (*b*) acceleration
   (*c*) deceleration (*d*) idling

[**Key:** 1. (*d*), 2. (*c*), 3. (*b*), 4. (*b*), 5. (*b*), 6. (*b*), 7. (*c*), 8. (*c*), 9. (*a*), 10. (*a*), 11. (*c*), 12. (*b*)]

13 Engine Performance

1. INTRODUCTION

Performance of an engine means how its output varies over the entire range of its operation. To study engine performance, both the power and the torque characteristics are considered. Further to see how effectively the conversion from fuel energy to engine power is carried out, the efficiency and specific fuel consumption curves are also studied. The variables to be considered are many, out of which speed, load, air-fuel ratio and mean effective pressure are important.

2. LOSSES IN THE ENGINES

The actual processes in an engine differ from the ones assumed to occur therein theoretically, which causes certain losses and thereby deteriorates the engine performance. The main sources of losses in the engine as a result of such deviation from theoretical processes, as well as due to other reasons are:

1. Loss due to ignition not being instantaneous. As the ignition of the charge in an actual engine does not occur instantaneously as assumed in a theoretical air cycle, and is spread over a definite interval of time, a loss in power occurs because of the rounding off of the peak. The loss is shown as area I in Fig. 13.1. This shape change of the pressure-volume diagram depends upon the instant at which the ignition occurs (Fig. 13.2). Diagram '1' is the optimum curve when ignition takes place at point O. When however, ignition takes place earlier than the optimum, diagram 2 is obtained. Likewise, shape 3 is obtained in case of the ignition occurring later than the optimum. Accordingly, the change in area of the pressure-volume diagram and hence the power loss will depend upon the instant when the ignition occurs.

2. Loss due to heat conducted to cylinder walls. During the power stroke, the walls of the cylinder being cooler than the products of combustion inside, transfer of heat takes place to the walls, and thence to the atmosphere through the coolant. This results in power loss shown by area II in Fig. 13.1.

3. Loss due to exhaust valve opening not being instantaneous. As it takes definite time for the exhaust valves in an actual engine to open and let out the exhaust gases, unlike theoretical air cycle, where the exhaust is assumed to occur instantaneously at constant volume, power loss represented by area III in Fig. 13.1, takes place.

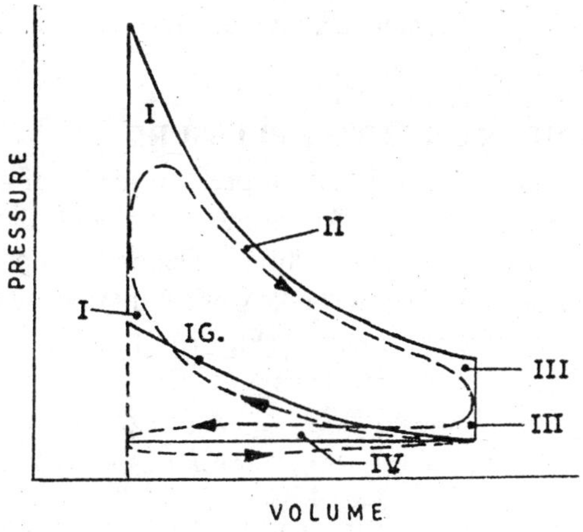

Fig. 13.1. Actual and theoretical pV-diagram (dotted lines represent actual diagram)

4. Pumping loss. Some work is required to be done for inducting the fresh charge into the cylinder and exhausting the products of combustion. This is called pumping work, which is a direct loss. As such, a finite difference of pressures is required both for the inflow of the charge as well as outflow of the exhaust gases. This results in loss of power represented by area IV in Fig. 13.1.

The design of the carburettor and inlet manifold affects the flow of charge to the cylinder. For instance, small flow areas cause high velocities and hence higher loss. Sharp bends and rough passages result in higher losses on account of high fluid friction. Likewise, fluid friction losses in case of flow of exhaust gases also depend upon design of exhaust ports, exhaust manifold and muffler. However, in all cases, pumping losses increase with increase in engine speed.

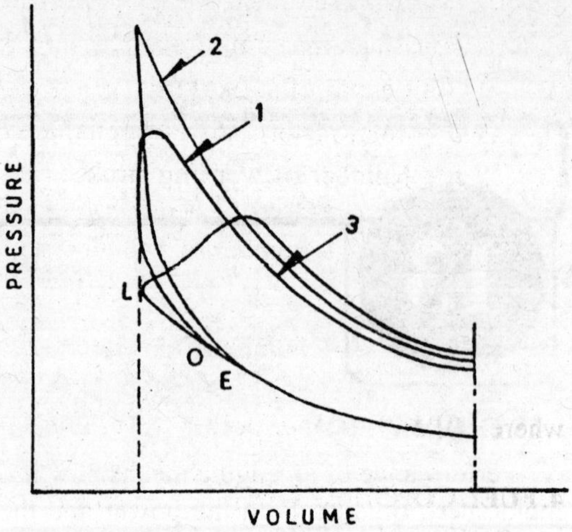

Fig. 13.2. Variation of pV diagram due to advancing or retarding of the ignition.

It is seen from Fig. 13.1 that pumping loss may be decreased by maintaining high induction pressure or low exhaust pressure. However, it is seen in practice that the former condition is much more important than the later one.

5. Loss due to incomplete combustion. In theoretical calculations it is assumed that fuel undergoes complete combustion in the cylinder to form CO_2 and H_2O. However, in actual practice there is always incomplete combustion, as a result of which the energy in the fuel goes waste.

6. Friction losses. Some power is lost in an actual engine to overcome friction of its various running parts. Friction losses are caused mainly by the shearing of the lubricant film, since no metal to metal contact can be allowed in the engine to avoid early wear.

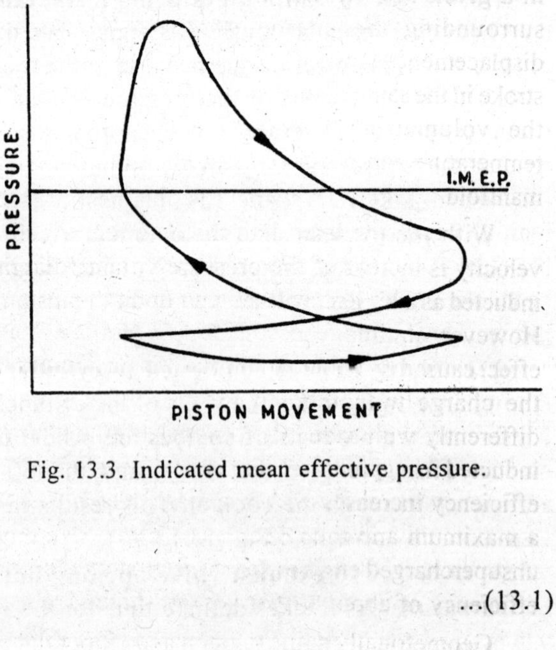

Fig. 13.3. Indicated mean effective pressure.

3. MEAN EFFECTIVE PRESSURE

This is a hypothetical pressure which if acting on the engine piston during one stroke would produce the total work of the cylinder. Depending upon the indicated work or the brake work, it may be indicated mean effective pressure (Fig. 13.3) or brake mean effective pressure. Thus

$$IMEP = \frac{IP}{lan} \quad ...(13.1)$$

and

$$BMEP = \frac{BP}{lan} \quad ...(13.2)$$

where IP = Indicated power, kW

BP = Brake power, kW

$IMEP$ = Indicated mean effective pressure, kPa

$BMEP$ = Brake mean effective pressure, kPa

l = Length of stroke, m

a = Total piston cross-sectional area, m^2

n = Number of working strokes per second

$$= \frac{RPM}{120} \times \text{number of cylinders, for } \textit{4-stroke engine}$$

$$= \frac{RPM}{60} \times \text{number of cylinders, for } \textit{2-stroke engine}$$

where RPM = engine speed in revolutions per minute.

4. FUEL CONSUMPTION

Fuel consumption in an engine under given test conditions in a given time may be measured in terms of weight or volume. Specific fuel consumption is determined by dividing the total fuel consumption per hour divided by the power developed. Depending upon the value of power used, *viz.*, indicated or brake, the consumption is termed as Indicated Specific Fuel Consumption (ISFC) or Brake Specific Fuel Consumption (BSFC).

5. VOLUMETRIC EFFICIENCY

For a normally aspirated engine, it is defined as the ratio of the volume of air or gaseous mixture inducted in a given time, at the atmospheric conditions surrounding the engine, to the total piston displacement of all the cylinders during induction stroke in the same time. For a supercharged engine, the volumetric efficiency is based on the temperature and pressure conditions in the intake manifold.

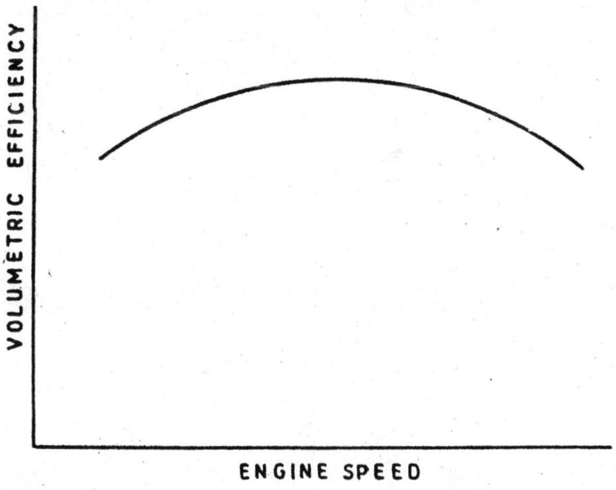

Fig. 13.4. Variation of volumetric efficiency with engine speed.

With the increase in engine speed, charge velocity is increased, which decreases the charge inducted as a higher pressure drop is then required. However, simultaneously with this the ramming effect caused by the inertia of the charge, increases the charge inducted. As the two effects vary differently with speed, the volume of the charge inducted in a given time and hence the volumetric efficiency increases with increased speed, attains a maximum and then decreases (Fig. 13.4). For unsupercharged engines the maximum volumetric efficiency of about 80 per cent is attained at a specific speed.

Geometrically similar engines are the engines which have the same ratio of their manifold, valve port and piston areas. Such engines have almost same volumetric efficiency at same engine speed.

Volumetric efficiency depends upon the engine speed and load as well as layout of the inlet and the exhaust systems, sizes of the ports and valve timing.

6. PERFORMANCE CURVES

Variation of various engine output parameters, *i.e.* torque, I.P. (Indicated Power) and B.P. (Brake Power) with engine speed are shown in Fig. 13.5. Now the Frictional Power (F.P.) is the difference between I.P. and B.P. As frictional power becomes a larger proportion of the IP with the increase of speed, BP curve drops more rapidly compared to IP curve. It is observed that engine speeds for maximum torque and for maximum power are not the same. It must be noted that in practice, engine curves are always given for full throttle conditions, which rarely occur actually on the road.

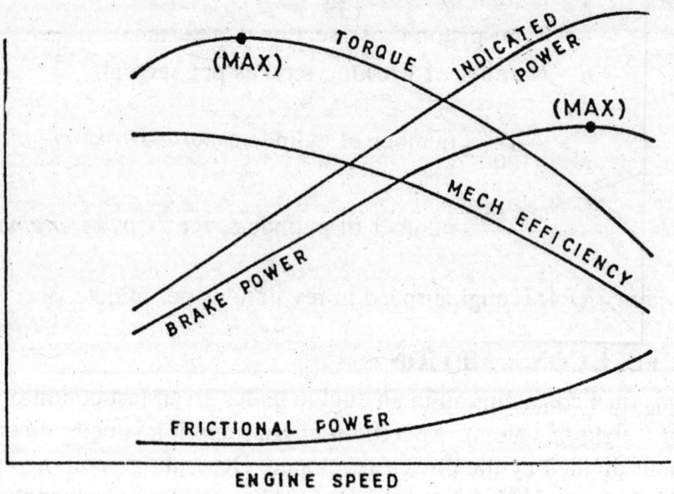

Fig. 13.5. Variation of indicated torque, indicated power, brake power and frictional power with engine speed.

In Fig. 13.6, are shown the specific fuel consumption and total fuel consumption at different engine speeds. It is seen that the specific fuel consumption, *i.e.* fuel consumed per kW hour, decreases at first and reaches some minimum value after which it again rises. Thus, the best engine speed is the one corresponding to minimum specific fuel consumption, where we get power at lowest expense. At any particular engine speed, the variation of mechanical efficiency

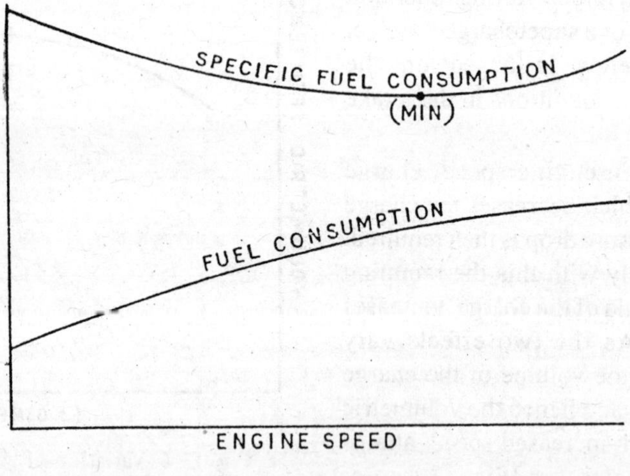

Fig. 13.6. Variation of specific fuel consumption and fuel consumption with engine speed.

and brake specific fuel consumption with load is shown in Fig. 13.7, whereas the effect of air-fuel ratio on indicated power and specific fuel consumption may be seen in Fig. 13.8. From practical view point, the manufacturers report these curves from which the suitability of the engine for the desired purpose for which it is to be used, may be ascertained. These are, the power curve, the torque curve and brake specific fuel consumption curve. These will, there- fore, be studied here in greater detail.

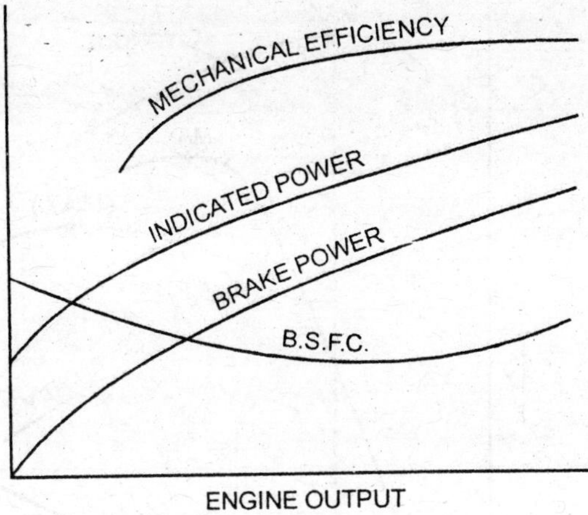

Fig. 13.7. Variation of mechanical efficiency and B.S.F.C. (brake specific fuel consumption) with load.

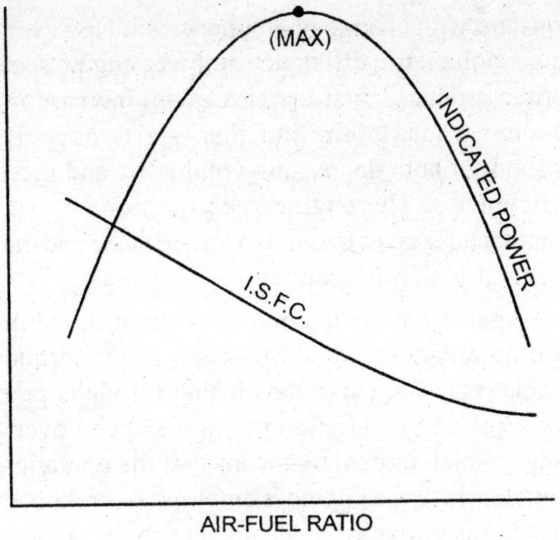

Fig. 13.8. Variation of I.P. and I.S.F.C. (indicated specific fuel consumption) with air-fuel ratio.

1. Power Curve

This shows the variation of engine brake power with engine speed. It is seen that the engine output, *i.e.* the brake power first increases with the increase in engine speed since at higher revolution per minute there would be higher number of working cycles and hence more power. However, with the increase in engine speed, the friction losses increase, while the volumetric efficiency decreases; thus both these factors contributing to decrease the output with the increase in speed. Thus the combined effect is that the slope of the power curve goes on decreasing with increase in speed till the power reaches a maximum after which it starts decreasing. The maximum value of the engine output and the engine speed at which it occurs are invariably given in the engine specifications.

Fig. 13.9 shows power curves for two different engines. It is seen that:

(*i*) Although the maximum power is same in both cases, yet the power at almost all engine speeds is lower in case of engine '2'.

(*ii*) The maximum output in engine '2' occurs at higher speed (N_2), compared to engine '1' (N_1).

Moreover the output of engine '1' remains nearer its maximum value for a larger speed range compared to engine '2' whose curve falls more sharply. These differences help to determine the exact application where the engine is going to be used.

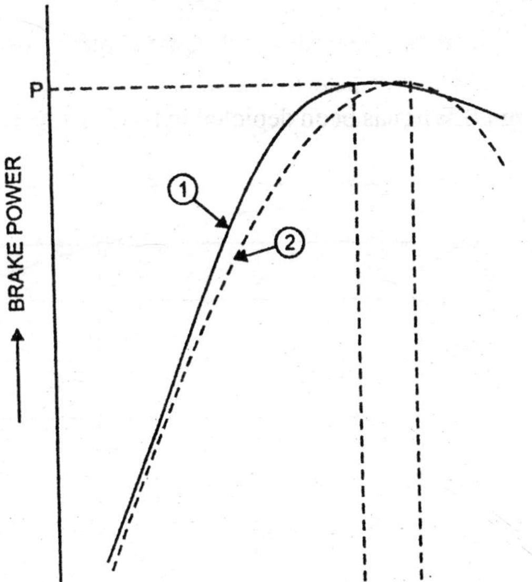

Fig. 13.9. Power Curve.

2. Torque Curve

This curve is generally flatter than the power curve. Actually in theory, the torque output should remain constant with change in engine speed. However, due to lower volumetric efficiency at lower engine speeds, the torque produced first increases with increase in speed, reaches a maximum and then starts decreasing on account of both decreasing volumetric and mechanical efficiencies. The engine specifications given by the manufacturer gives maximum torque value and the engine speed at which it occurs.

Apart from the maximum value of the torque, it is also important to know the shape of the torque curve. A relatively flat curve means that torque is practically not affected by variation in engine speed over a large range, which ultimately means that the operation of the vehicle where the engine is employed, is relatively easy. In case the torque curve drops sharply from its maxima, constant shifting of gears would be required to negotiate hills or city roads, not only making it tiresome to the driver, but also resulting in loss of power. Similarly, if the maximum torque point occurs at lower speed, starting the vehicle from a stop and operating it at lower speeds may be easy, but above this lower maximum torque speed, a rapid shifting of gears would be required. On the other hand if the maximum occurs at too high an engine speed, a small drop in rpm would require shifting of gears. Thus it is seen that besides a high maximum torque value, shape of the torque curve and the speed at which maximum torque occurs are also equally important.

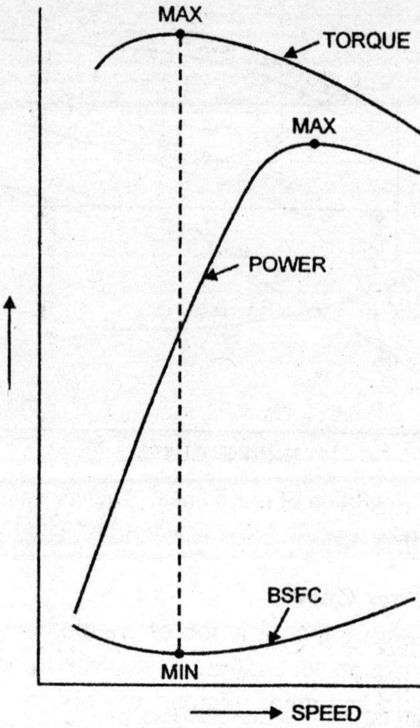

Fig. 13.10. Main engine performance curves

3. Brake Specific Fuel consumption (bsfc) curve

The curve showing the variation of fuel consumption per unit brake power in a unit time, generally in gm/kWhr has been depicted in Fig. 13.10. It may be seen that in general, this curve has an inverse relationship

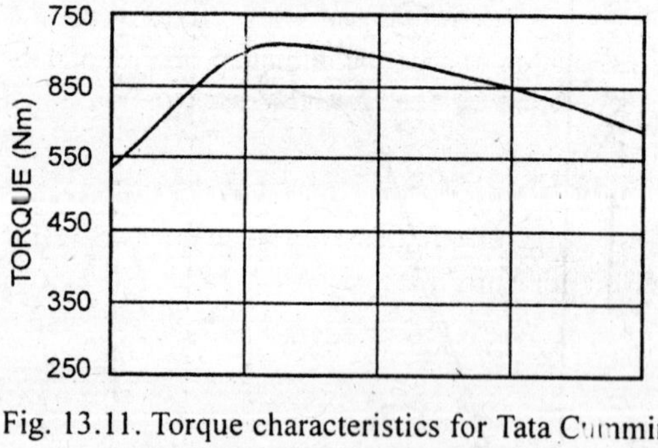

Fig. 13.11. Torque characteristics for Tata Cummins B5.9–210–10 diesel engine.

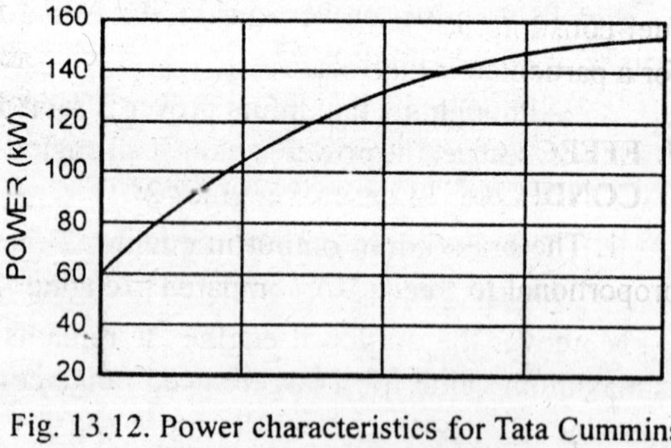

Fig. 13.12. Power characteristics for Tata Cummins B5.9-210-10 diesel engine.

with the torque curve, *i.e.* it first decreases, reaches a minimum and then starts increasing as the engine speed is increased. Also the minimum of this curve occurs nearly at the same engine rpm as the maximum of the torque curve.

As a practical example, variation of torque, BP and S.F.C. for Tata Cummins B 5.9-210-10 diesel engine

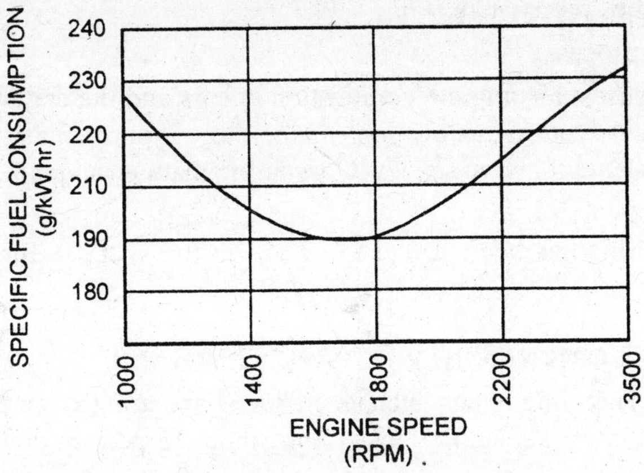

Fig. 13.13. S.F.C. for Tata cummins B5.9-210-10 diesel engine.

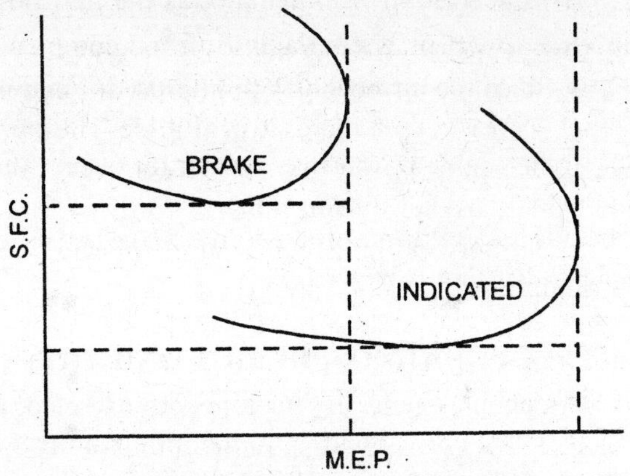

Fig. 13.14. Specific Fuel consumption at different values of indicated mean effective pressure for different A/F ratios

meeting Euro-I norms for automotive emission is shown in Fig. 13.11 to 11.13.

The effect of varying indicated mean effective pressure (I.M.E.P.) on I.S.F.C. is shown in Fig. 13.14. The horizontal tangent to these curves gives specific fuel consumption for *maximum economy,* whereas the vertical tangent point gives the specific fuel consumption for *maximum power.* If such curves are plotted for different air-fuel ratios, minimum specific fuel consumption and the corresponding air-fuel ratio may be obtained from the envelope of fuel-consumption loops (Fig. 13.15) for a particular engine speed.

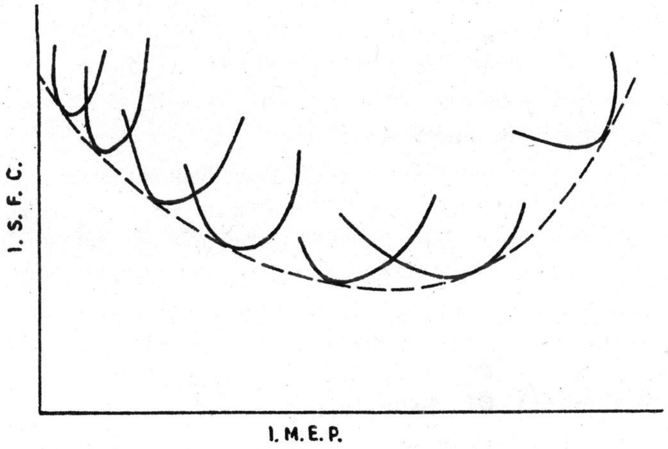

Fig. 13.15. Determination of absolute minimum I.S.F.C. and the corresponding air-fuel at a given speed.

7. EFFECT OF ATMOSPHERIC CONDITIONS ON CARBURETTED SPARK IGNITION ENGINES

1. The brake power of engine depends directly on the atmospheric pressure of dry air and inversely proportional to the square root of atmospheric absolute temperature, *i.e.*

$$B.P. \propto p$$

and

$$B.P. \propto \frac{1}{\sqrt{T}}$$

Now, dry air pressure, p = total barometric pressure − partial pressure of water vapour in the air.

Therefore, humidity condition around the cylinder has a pronounced effect on the engine performance. The loss due to humidity may be caused by:

(*i*) Reduction in the amount of air inducted.

(*ii*) Increase in fuel-air ratio, as the fuel flow does not change.

(*iii*) Effect of water vapour on the combustion process.

2. Since the air becomes thinner as altitude increases, incomplete combustion occurs and the engine output decreases with increase in altitude. The empirical relations are available in literature. Approximately 3% of the engine power may be lost for every 300 m/1000 ft above sea level, assuming that petrol quality and vehicle weight remain same.

3. No noticeable effect of atmospheric conditions has been found to occur on the specific fuel consumption.

8. EFFECT OF ATMOSPHERIC CONDITIONS ON COMPRESSION-IGNITION ENGINES

C.I. engines generally operate with excess of air. Therefore, when such an engine is operating at loads less than that corresponding to the minimum B.S.F.C., a change in air density upto about 15 percent does not affect the fuel rate and hence its performance.

According to Elliott, the performance of a diesel engine under actual test conditions can be converted to the standard test conditions by means of the formula,

$$\frac{(B.P.)_s}{B.P.} = 1 + \left(1 - \frac{\rho}{\rho_s}\right)\left(\frac{F.P.}{B.P.} - 1\right)\Big/(k-1) \qquad \text{...(13.3)}$$

where s denotes standard test conditions

ρ = air density

k = 0.189/fuel-air ratio

The power and torque curves given by the manufacturers are at standard test conditions. From practical user's view point, the power loss due to higher altitude in C.I. engines is about 10% per 1000 meters, which means, for example, that at 1000 meters only 90% and at 2000 meters only 80% of the rated power can be expected. If the amount of fuel injected at high altitude is kept at sea-level standard, combustion will be incomplete and the exhaust will include black smoke. Thus the fuel pump has to be pre-adjusted adequately for optimum combustion at the intended altitude of vehicle operation.

9. PERFORMANCE MAPS

In general, most of the performance curves, described above are not convenient for comparison with other engines as a quantity criterion. Thus it has been found more practicable to represent the engine performance with parameters like brake mean effective pressure (BMEP), piston speed, brake specific fuel consumption (BSFC) and specific power output (kW per unit piston area), which do not vary with size of the engine and hence can be used to compare different engines. Fig. 13.16 and 13.17 depict such performance curves. These are usually called *performance maps.*

10. HOW TO IMPROVE ENGINE PERFORMANCE

There are basically two methods to improve the engine performance:

1. By increasing the input energy.
2. By increasing the engine efficiency.

The energy input can be increased by increasing the fuel intake per unit time. This may be done by increasing the engine speed so long as the accompanying decrease of volumetric efficiency at higher speeds does not offset this gain. Design parameters that increase the volumetric efficiency (*e.g.,* by using more number of inlet valves in each cylinder, two being quite common) could also be helpful upto certain limits beyond which the detonation or overheating may occur. Supercharging and turbocharging are other methods

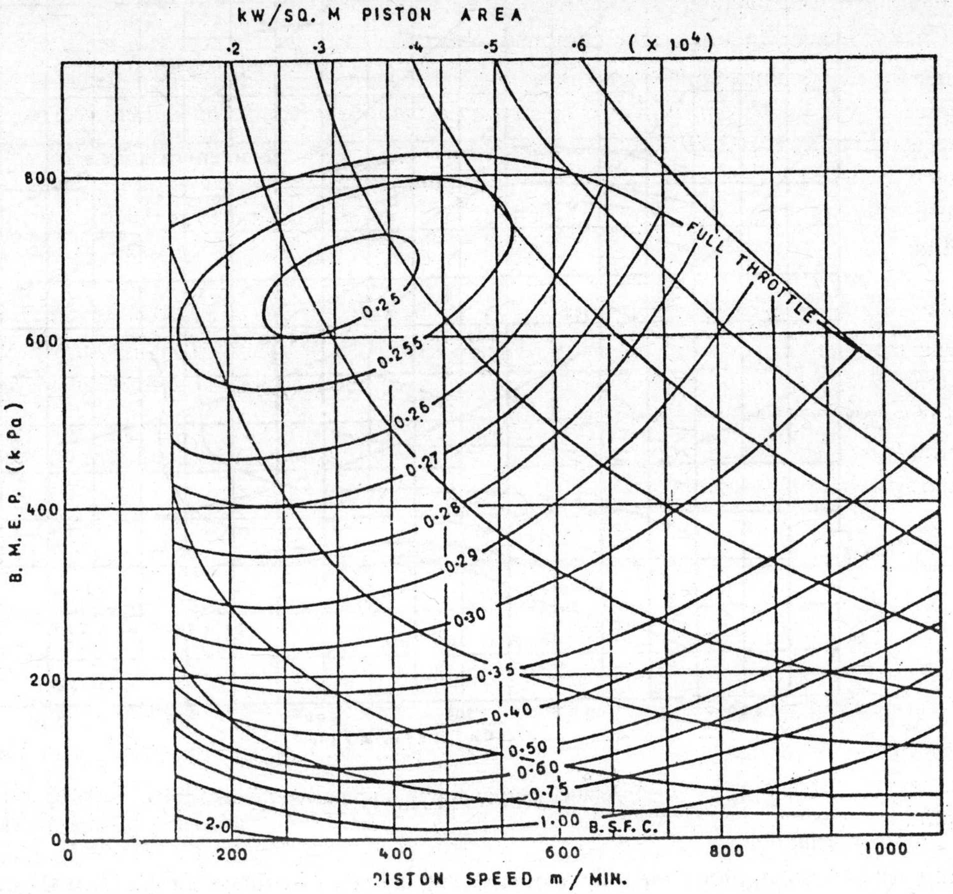

Fig. 13.16. Performance curves for an automotive petrol engine.

of increasing the intake of fuel air mixture thereby increasing the energy input in unit time. The improvement of output by supercharger or turbocharger may be further increased by means of intercooler, which is a special type of radiator through which the compressed air from the supercharger or turbocharger passes to cool off before entering the engine cylinder.

Increased efficiency is obtained by increasing compression ratio. However, beyond limits this will cause detonation, to avoid which high-octane petrol will have to be used. Further, increase in compression ratio may increase the friction losses more rapidly than the increase of indicated thermal efficiency, thus offsetting the improvement. Thus an optimum value of compression ratio has to be adopted. Reducing various losses as described in Art. 2 will also increase the efficiency, thereby improving the engine performance. Some of the methods are:

1. Use of polished intake manifolds, which minimises resistance to intake air. The cheaper method involves manual polishing using different grades of sanding and polishing tools, which are manually run around the passages, smoothing off rough edges and polishing the air passages. The more expensive method involves use of a specialized machine which pumps superheating plasma through the air passages, which melts a thin layer off these passages to make them smooth.
2. Using two exhaust valves in each cylinder, thereby decreasing the resistance to flow of exhaust gases.
3. Use of bigger tail pipes and free-flowing mufflers. Sometimes two exhaust pipes are used instead

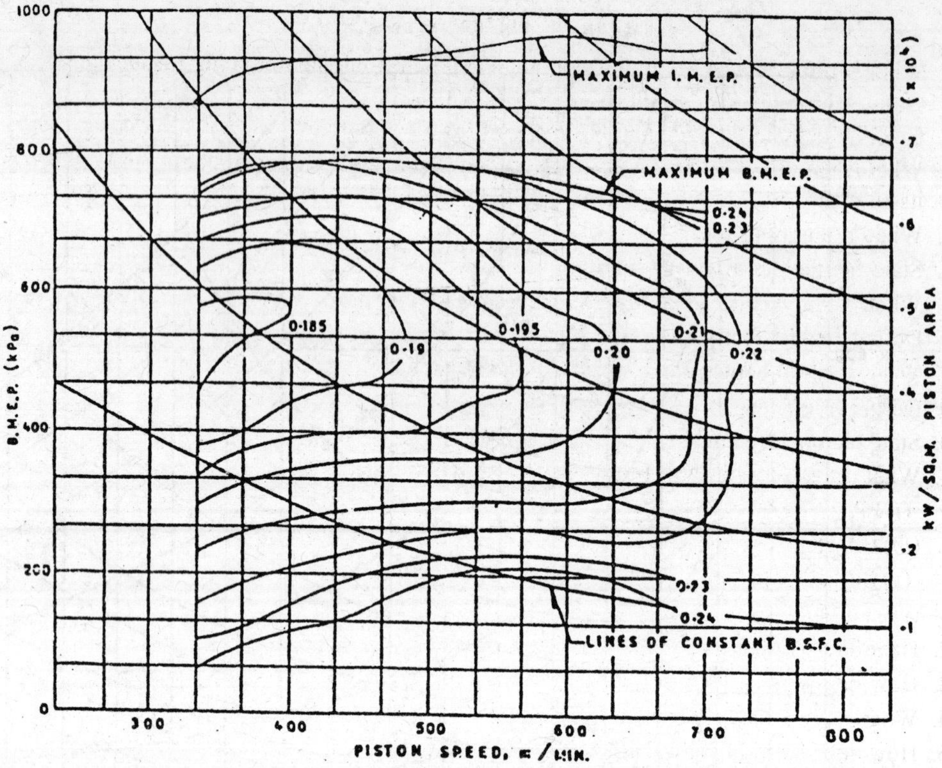

Fig. 13.17. Performance curves for an automotive diesel engine.

of conventional one.

4. Screwing the spark plug so that the electrode gap is facing the intake valves. This may be done by using a marker pen to put a straight-line mark on the plug insulator, which is aligned with the electrode gap at the bottom of the plug. A replacement crush washer which may be thinner or thicker than the standard washer may then be used to tighten the spark plug so that the alignment mark faces the intake valves.
5. Remapping, *i.e.* changing the engine management software, which may result in about 5-7% increase in power and torque. Some manufacturers do provide bolt-on upgrades to their vehicles. For example, Dodge introduced a bolt-on upgrade to their SRT-4 Noon in 2003 which includes a modified engine management computer and high-flow fuel injectors.

THEORY QUESTIONS

1. Discuss various types of power losses which occur in an actual engine as compared to its theoretical air cycle.
2. What do you understand from the term 'engine' performance'? Illustrate your answer by means of important performance curves.
3. Define volumetric efficiency and discuss the effect of different variables on it.
4. Discuss in detail the effect of atmospheric conditions on the engine performance.
5. Sketch and explain performance maps for automotive petrol and diesel engines.
6. Explain clearly various methods to improve engine performance.

QUESTIONS FOR VIVA

1. What is the effect of the ignition in an actual petrol engine not being instantaneous, as assumed in the Otto cycle?
2. On what depend the area and the shape of the p-v diagram?
3. What is the effect of the exhaust valve opening in actual engine not being instantaneous, as assumed in its air cycle?
4. What is pumping loss?
5. State the factors affecting pumping loss.
6. What is that basically causes friction loss in an engine?
7. Define mean effective pressure.
8. Write relation between engine power and mean effective pressure.
9. Define volumetric efficiency.
10. State factors affecting volumetric efficiency.
11. What do you understand from the following abbreviations?

    (*i*) I.P. (*ii*) B.P.
    (*iii*) B.S.F.C. (*iv*) I.S.F.C.
    (*v*) B.M.E.P. (*vi*) I.M.E.P.
    (*vii*) F.P.
12. How does engine power output vary with speed?
13. How does engine torque vary with speed?
14. What is the effect of ambient temperature and pressure on petrol engine performance?
15. How does humidity in the atmosphere affect the performance of a carburated engine?
16. State the effect of atmospheric conditions on diesel engine performance.
17. What is a performance map?

14 Testing of Automobile Engines

1. CLASSIFICATION OF TESTS

The tests for automobile engines may be classified either with a view to the objective for performing the test or depending upon the method or sequence of these tests. From the former view point, the tests may be categorised as:

1. **Routine tests.** These tests are performed by the manufacturer to determine or demonstrate that the engines manufactured give a performance of the specified standard.
2. **Fault finding tests.** The purpose of these tests is to assess the condition of an existing engine to diagnose any faults which may have developed with its use over time. These may be due to natural wear or poor servicing or even to bad driving habits.
3. **Tests of new or modified types of engines.** The object of these tests is to determine the effect of certain changes on the engine performance, so as to design a new engine which is better in performance and more compact than the existing ones in the market. Obviously such tests require comparatively more apparatus than used for the routine tests.
4. **Research tests.** These tests are of advanced nature and are undertaken to investigate new phenomena observed in the running engines and to determine the suitability of various theories explaining such phenomena. Such knowledge helps in the better design of new engines.

From the method or sequence point of view, the engine tests can be classified as:

1. **Laboratory tests.** The engines, whether for routine or research investigation, are first tested in the laboratory on a standard or improvised test rig.
2. **Shop tests.** The simple fault finding tests which require only simple testing equipment can be conducted in the automobile shop or garage itself. Such tests do not require any elaborate instrumentation or any highly specialized equipment.
3. **Road tests.** The laboratory tests, though very elaborate and reliable, have the limitation that these are performed only under simulated laboratory conditions. Therefore, after the laboratory tests, the engine is mounted on to a special chassis fitted with required instrumentation and then tested on road under actual running conditions.
4. **Road tests in the laboratory.** With the advance of technology it has been made possible to do 'road testing' in the laboratory itself, such tests being very common these days. A special testing rig gives drive to the chassis through two large diameter drums under each of the two rear wheels of the automobile. These drums are provided with peculiar irregularities on their surfaces, so as to reproduce road irregularities. Various parameters are thus measured with usual instruments and apparatuses.

2. FAULT FINDING TESTS

The following general fault-finding tests shall be described here:

1. Vacuum tests
2. Cylinder compression test

These will now be discussed in detail

2.1. Vacuum tests

These are performed only with the help of a vacuum gauge, which is attached to an adapter on the induction manifold (Fig. 14.1). For these tests it must be remembered that an idling engine develops power just sufficient to overcome frictional and pumping losses, while the small surplus power determines the idling speed. Various vacuum tests usually performed are:

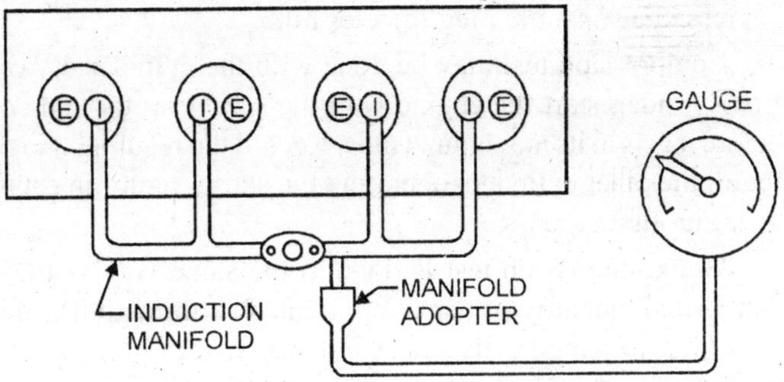

Fig. 14.1. Connecting vacuum gauge to the induction manifold.

1. Mechanical condition of the engine. This may be determined by a number of alternative tests:

(*i*) **Irregular gauge reading:** The induction vacuum in an engine will fluctuate over the operating four-stroke cycle. There will be a suction stroke every 720° in case of a single cylinder engine.

Similarly in case of four and six cylinder engines, the suction stroke will occur every 180° and 120° respectively. If the mechanical condition of the engine is O.K., the successive vacuum fluctuations seen on the vacuum gauge for different cylinders will be equal. However, if any one or more cylinders have some mechanical defect, *e.g.*, a leaking valve, broken or sticking piston rings, a damaged cylinder head gasket, the vacuum of that/those cylinder(s) would be less which will be seen as a momentary drop in vacuum indicated by the irregular gauge reading.

(*ii*) **Cranking method:** Short the 'CB' terminal of the ignition coil with the earth, which will not allow the spark to occur at the spark plug terminals. Now crank the engine by means of the starter motor for about 15 seconds, with the throttle completely closed. If the mechanical condition of the engine is good, a high manifold vacuum of the order of 350 mm Hg or more will be produced. However, if the vacuum gauge readings show lesser vacuum, it would indicate mechanical defects such as worn piston rings or cylinder bores, poor valve sealing or damaged manifold gasket.

(*iii*) **Throttle method:** When the engine is idling, note the gauge reading and quickly open the throttle so that the engine reaches its maximum speed. Then immediately close the throttle. If by so doing, the vacuum gauge reading rises immediately above the normal vacuum to 125 mm Hg or more and then comes backs to its original value, the engine is in good mechanical condition. However, if the rise in vacuum is small, it would be due to loss of compression which may be on account of worn cylinder bores, pistons and/or rings.

(*iv*) **Idling method:** In this, the engine is run at idling speed and vacuum gauge reading is noted. If the engine is in good mechanical condition, it would give, steady reading, 460-510 mm Hg in case of a 4-cylinder engine and 480-540 mm Hg in a six-cylinder engine. An intermittent drop of about 80-100 mm Hg would indicate leaking valve(s).

2. Cylinder balance. To determine this, the engine is run at idling speed and spark plug leads are removed in turn, while carefully observing the gauge reading in each case. If there is equal fall in vacuum of about 20 to 50 mm Hg in each case, it would indicate that each cylinder is contributing same power. In case there is less vacuum drop in a particular cylinder, it would mean that particular cylinder is producing lesser power, which may be due to defect in the cylinder or even due to defective fuel supply and/or ignition systems. Further investigations shall be needed to pin-point the exact cause.

2.2. Cylinder compression test

This is done with the cylinder pressure gauge. The gauge has a non-return valve so that after taking a reading, the air pressure in the gauge has to be released by means of the pressure release valve, to take another reading. Besides, the gauge is connected at the spark plug hole or injector hole by means of a screw type adopter (which is screwed into the spark plug/injector hole) or a rubber nose adapter (which is pressed against the plug/injector hole.

Compression test may be done with the cylinder dry or wet. For the test, connect the gauge with the first cylinder, start the engine with the starter motor and count the number of pulses of the gauge needle before it gets to its maximum value. Record the readings and release the gauge pressure. Repeat the procedure for all the other cylinders operating the starter motor in each case to give the same number of gauge pulses as in the first case.

Wet compression test is done in the same way as the dry compression test described above except that a small quantity of heavy oil is injected through the plug/injection hole into the cylinder to be tested. This oil will temporarily seal the piston assembly when it is reciprocating in the cylinder. The cylinder compression test is done for each cylinder, the readings recorded and the results compared with the recommended data. The individual cylirder readings should not vary by more than one bar from each other.

3. ROUTINE TESTS

The following are the important quantities to be determined from the routine engine tests:

1. Indicated power
2. Brake power
3. Mechanical efficiency
4. Fuel consumption
5. Air consumption
6. Lubricating oil consumption

The variables involved for the determination of the above quantities are:

1. Engine speed
2. Air fuel ratio
3. Throttle opening
4. Spark timing (injection timing in case of C.I. engines)
5. Temperature of coolant (water or air)
6. Temperature of exhaust gases

Various methods used for the determination of the quantities listed above shall now be discussed in the following articles:

4. MEASUREMENT OF INDICATED POWER

The indicated power (I.P. in kW) of a single cylinder engine is determined by calculation, from the following formula:

$$I.P. = \frac{P_m.L.A.N.}{60{,}000} \quad \ldots(14.1)$$

where P_m = Indicated Mean Effective Pressure (I.M.E.P.) (N/m^2 *i.e.* Pa)

L = Length of stroke (m)

A = X-Area of the cylinder (m^2)

= Revolutions per minute in 2-stroke engines

$= \frac{1}{2}$ R.P.M. in 4 stroke engines.

Note: In case of multi-cylinder engine, A = total cross-sectional area of all cylinders.

In the above formula, R.P.M. is found with the help of a tachometer, while I.M.E.P. is calculated from the P-V diagram which is obtained experimentally with an engine indicator. P-V diagram is divided into equal parts (say x) by means of lines parallel to pressure axis. The mean ordinate of each part is then measured. All these ordinates are then added and divided by the number x, which gives the I.M.E.P. Alternatively, the area of the P-V diagram is measured with a planimeter and this gives the amount of work done in one stroke, from which I.M.E.P. may be calculated as:

$$I.M.E.P. = \frac{\text{Area of indicator (P–V) diagram}}{\text{Length of indicator diagram}} \times \text{spring scale} \quad \text{...(14.2)}$$

The first method is the one, which is more commonly used.

The indicators to obtain the P-V diagram are of different types; a brief account of these being given below:

1. Mechanical indicators. These employ oscillating drum carrying a sheet of paper to reproduce the piston stroke, while a small piston working in its own cylinder and connected with the engine cylinder, operates through suitable linkage, a stylus pen on the paper. Such indicators are used only for lower speeds, well below 100 r.p.m., and there too with reservations, because of the following disadvantages:

(*i*) inertia effect of the moving parts.

(*ii*) friction between the stylus and the paper.

(*iii*) backlash in the mechanical linkage.

(*iv*) temperature effects.

2. Optical indicators. In these the piston and link mechanism is replaced by beam of light, reflected from one or more mirrors which rock to and fro corresponding to the engine pressure and piston stroke. The beam of light, after deflection, is focussed on a ground glass screen to obtain P-V diagram, or else on a photographic paper if a permanent record of the diagram is to be obtained. The inertia effects of the mechanical indicators are, thus, not present in the optical indicators. Besides, friction between the stylus and the paper is also eliminated. But, inspite of its accuracy at speeds upto about 2000 r.p.m., it has the disadvantage of being bulky, delicate and requiring a skilled operator. Thus it is not very much suitable for commercial use. Further, inertia of the moving mirrors becomes appreciable at higher speeds.

3. Cathode ray indicators. In this type of indicators, the system and linkage of the mechanical type is replaced by a beam of electrons, in the cathode ray oscillograph. The cylinder pressure variations are converted into voltage changes which are amplified and fed to plates X through which the electron beam passes. Thus the electron beam is deflected proportional to the pressure variations in the cylinder. The beam is finally projected on a fluorescent screen causing it to glow at the point of projection. Similarly another voltage representing the piston displacement is amplified and applied to another system of plates Y (at right angles to X) through which the same electron beam is passed and gets deflected proportional to the piston displacement. In this way, the electron beam is so deflected by the two voltages, as to make a P-V diagram on the fluorescent screen. As cathode ray indicators do not contain any moving mechanical parts, they are free of inertia effects.

5. MEASUREMENT OF BRAKE POWER

The brake power (B.P.) *i.e.* the power obtained at the engine flywheel is measured with the help of dynamometers, a brief description of some of the types being given here. The dynamometers actually give the engine torque, from which the B.P. (in kW) can be calculated with the formula:

$$B.P. = \frac{2\pi NT}{60,000} \quad \text{...(14.3)}$$

where N = Engine speed in R.P.M.

T = Torque in Newton meters (obtained from dynamometer test)

The dynamometers can be broadly classified as the absorption and the transmission types. In the absorption types, the whole of the engine power is absorbed by the dynamometer, whereas in the later, almost whole of the engine power is transmitted through the dynamometer and may be employed for some useful purpose. Thus, though the absorption type dynamometers are more accurate and cheaper and for these reasons almost universally employed for engine testing, yet transmission dynamometers are economical in case of longer duration tests and also more useful where the engine output is not to be disturbed, *e.g.* while during road tests under actual conditions in case of automobile engines. The principles of a few of the absorption type dynamometers will be discussed here.

1. Solid friction dynamometers. These are the most simple type in construction as well as in operation. The prony brake and the rope brake dynamometers are the typical examples of the solid friction type. In these the friction between the wooden blocks or the rope and the flywheel rim is used to absorb the engine output and convert into heat.

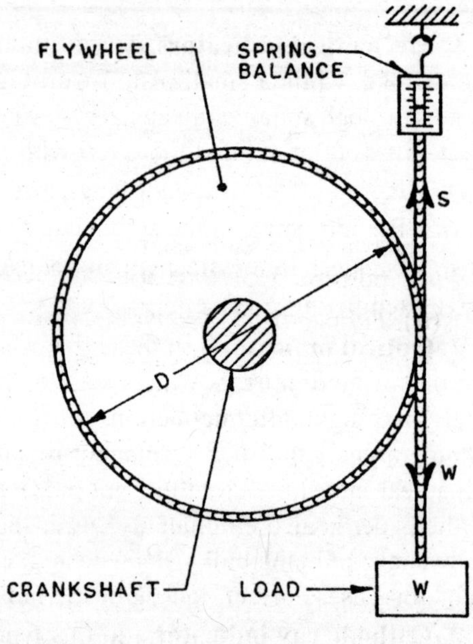

Fig. 14.2. Prony brake dynamometer

Fig. 14.2 is a schematic diagram showing various components of a prony brake dynamometer.

Let S — spring balance reading, N

W — dead load required just to stall engine, N

T — engine torque, Nm

D — diameter of flywheel, m.

Then $T = (W - S)\ D/2$, Nm.

The main disadvantage of this type is that the coefficient of friction does not remain constant which causes erratic action. Moreover the flywheel rim may become very hot in case of higher load and prolonged running.

2. Air dynamometers. This type does not find much use in the present day automotive engine industry. It consists of two radial arms fitted to a shaft and carrying light metal plates whose distance from the shaft may be changed. When it is driven by the engine, the power is absorbed in the eddying and thus heating of the air. This type is not very accurate.

3. Hydraulic dynamometers. These consist of a rotor containing a number of radial vanes and rotating inside a casing which is water tight and filled with water circulated therein. The mechanical energy of the engine, which drives the rotor is absorbed due to eddying motion set up in the casing water and is thus converted into heat. This type of dynamometer is fairly accurate and unlike the solid friction types, has a relatively smooth motion.

Fig. 14.3 shows an SAJ -Froude hydraulic dynamometer in which rotor and stators are precision castings of alloy resistant to erosion, corrosion and cavitation. The shaft assembly, power absorption elements, bearings and seals are all housed in a cast iron casing, which is supported on the cast iron bed plate by lubricated self-aligning ball-bearings. Water leakage from the vortex chamber is restricted by multi-rope glands. Dynamometer controls are installed on a separate module. Torque is measured by strain gauge load cell or a spring balance. Dynamometer shaft speed is measured by a pulse pick-up. The power absorbed by the dynamometer is controlled by the varying quantity of water circulating in the vortex of the rotor and stator elements. This is achieved by a moving slice-gate in the dynamometer casing.

Fig. 14.3. AWM hydraulic dynamometer
(Courtesy—SAJ-Froude Test Plant Pvt. Ltd. India)

4. Electrical dynamometers. These may be either the swinging field type or the eddy current type.

(*i*) *Swinging field dynamometer.* This is one of the most accurate torque measuring devices today. It consists of an electrical machine which may be run either as generator or motor. When running as generator, it absorbs the engine power and acts as a dynamometer. Running as a motor, it can be used to start the engine. Besides accuracy, it has wide range of power and speed like the hydraulic type. Further the power absorbed by the dynamometer is not lost in this case, because the current generated may be utilized for any useful purpose. However, it is complicated in construction and requires a skilled operator to run it.

(*ii*) *Eddy current dynamometer.* It consists of a stator and a rotor. A circular field coil in the stater

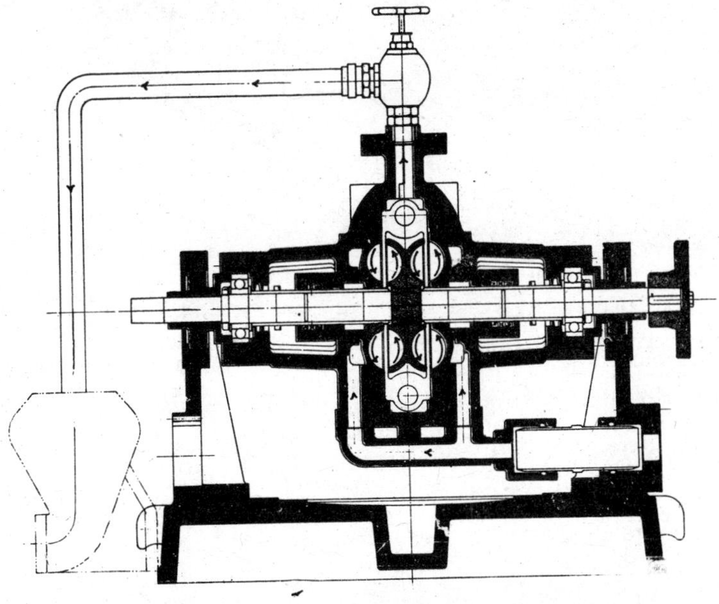

Fig. 14.4. Cross-section of hydraulic dynamometer.
(Courtesy—SAJ-Froude Test Plant Pvt. Ltd. India)

provides the magnetic flux. The rotor is connected to the engine shaft. When the engine is running, the rotating rotor in the stator field generates eddies, thus absorbing the engine power. As large amount of heat is generated in eddying, water cooling of stator is necessary in this case. This type of dynamometer obviously cannot be used for motoring the engine. However, it is quite compact and has wider range of speed than the swinging field type electric dynamometer. Another drawback is that the engine power is wasted and cannot be utilised as in case of swinging field type.

Fig. 14.5 shows an SAJ eddy current dynamometer. Power absorbed by the dynamometer is thermally dissipated through the water-cooled loss members positioned on both sides of the rotor and providing a dry gap between the rotating/stationary elements. The dual flow cooling passage in the loss members makes

Fig. 14.5. Eddy current dynamometer
(Courtesy—SAJ-Froude Test Plant Pvt. Ltd. India)

special cooling water magnetic filters redundant. Torque is measured in both directions by strain gauge load cell, while the shaft speed is measured by a pulse pick-up head. This dynamometer may be operated in any of the three modes, *i.e.*, constant torque, constant speed and the power law (representing approximately the power absorption characteristic of a propeller). Multistage test programming/computer interfacing are optional features Fig. 14.6 depicts control panel for the eddy current dynamometer.

Like indicated power, brake power may also be expressed as

$$B.P. = \frac{P_{mb}.LAn}{60,000} \qquad \ldots(14.4)$$

where B.P.= Brake power due to one cylinder, kW

P_{mb} = Brake means effective pressure (B.M.E.P.), Pa
L = Length of stroke, m
A = X-area of the cylinder, m^2
n = Number of working strokes per minute = N in case of 2-stroke engines and = N/2 in case of 4-stroke engines.

Also from Eq. (14.3),

$$B.P. = \frac{2\pi NT}{60{,}000} \quad ...(14.5)$$

where T is the torque in Newton metres and N is the engine R.P.M.

Equating Eq. (14.4) and (14.5),

$$P_{mb} = kT \quad ...(14.6)$$

where $k = \dfrac{2\pi}{LA}$ for 2-stroke engines

and $= \dfrac{4\pi}{LA}$ for 4-stroke engines.

Thus, in any case, it is constant for the engine. It is also seen that B.M.E.P. is independent of engine speed and is directly proportional to the engine torque.

Fig. 14.6. Control panel for eddy current dynamometer.
(*a*) Control system cabinet, (*b*) Details of control and indication pannels.
(Courtesy—SAJ-Froude Test Plant Pvt. Ltd., India)

6. MECHANICAL EFFICIENCY

If the I.P. and B.P. of the engine are known form the tests discussed in the preceding articles, the mechanical efficiency can be found out from the formula,

$$\text{Mechanical efficiency} \quad \eta_m = \frac{B.P.}{I.P.} \qquad ...(14.7)$$

Besides this, the mechanical efficiency can also be found out directly from Morse test. Its value usually varies from 70 to 90 percent for different types of engines.

6.1. Morse test

This simple, quick and quite accurate test is used for determining the mechanical efficiency of the engine. The equipment required is only a dynamometer and a tachometer, as used for B.P. determination.

The morse test consists of determining B.P. of the engine at any particular speed, then cutting one cylinder at a time and measuring B.P. of the rest. The desired cylinder can be cut out by removing spark plug in case of S.I. engines and removing fuel injector in case of C.I. engines.

Now I.P. = B.P. + Engine losses. Therefore as one cylinder is cut out, the losses in that cylinder must be supplied by the other cylinders. Thus, the difference between the B.P. measured for the whole engine and for the engine with one cylinder cut out gives the I.P. of that cylinder. Adding up the I.P. for all cylinders in this way gives total I.P. of the engine. Then mechanical efficiency may be determined by dividing total engine B.P. with engine I.P.

As mechanical efficiency varies with speed and load, care should be taken that they do not vary during the test.

7. MEASUREMENT OF FUEL CONSUMPTION

The fuel consumption in a given interval of time may be measured either in terms of volume or in terms of weight. In the former a vessel of suitable capacity is connected to the main fuel tank and engine is supplied with fuel from the vessel during test. The vessel is filled to its capacity and then disconnected from the fuel tank by means of a valve. The time required in emptying out the vessel is then noted with a stop watch. The volume of the vessel divided by the time, then, gives the fuel consumption of the engine. In the weight method, the fuel tank is placed on a weighing platform, which directly reads the amount of fuel consumed. The time taken to consume a specified quantity of fuel is then noted and the fuel consumption is calculated. For accurate determination of mass consumption of the fuel from the volume consumption, knowledge of the temperature at the time of measurement is required since the fuel density changes with temperature.

Specific fuel consumption can be calculated by dividing the fuel consumed per hour by the horse power. Thus if M_f kg/minute is the fuel consumption, indicated specific fuel consumption (I.S.F.C.)

$$= \frac{60\,M_f}{I.P.}, \text{kg/I.P. Hour} \qquad ...(14.8)$$

and brake specific fuel consumption (B.S.F.C.)

$$= \frac{60\,M_f}{B.P.}, \text{kg/B.P. Hour} \qquad ...(14.9)$$

The above devices give fuel consumption only for the period of duration of test. If, however, a continuous knowledge of fuel consumption rate at any moment is required, the same can be obtained with any flowmeter. These flowmeters are either of the venturi type or the variable orifice type.

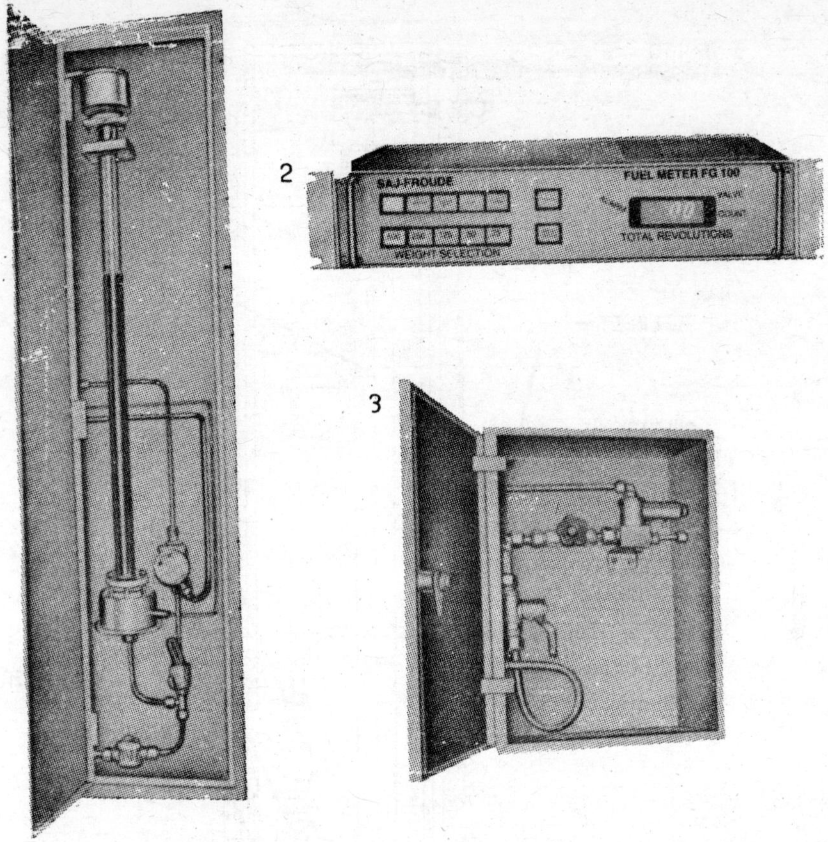

Fig. 14.7. SAJ-Froude type FG 100 fuel meter; 1. Fuel tube assembly, 2. Controller, 3. Calibration rig (Courtesy—SAJ-Froude Test Plant Pvt. Ltd., India)

Fig. 14.7 and 14.8 show a typical gravimetric fuel meter. Compared to the volumetric measurement method described above, the gravimetric method has an advantage that being based on the weight of the fuel consumed, changes in fuel density due to temperature variations do not have to be taken into account for calculating specific fuel consumption.

The operation of the gravimetric fuel meter is based on the change in head of fuel in a vertical precision bore glass tube (Fig. 14.9). Let

ΔM = fuel mass change in the glass tube

ΔW = fuel weight change in the glass tube

ΔV = fuel volume change in the glass tube

ρ = specific gravity of the fuel

Δh = fuel height change in the glass tube

ΔP = fuel head variation for change of height Δh

g = acceleration due to gravity

A = cross-sectional area of the glass tube

Then $$\Delta M = \rho \Delta V = \rho A \Delta h$$

Now $$\Delta P = \rho g \Delta h$$

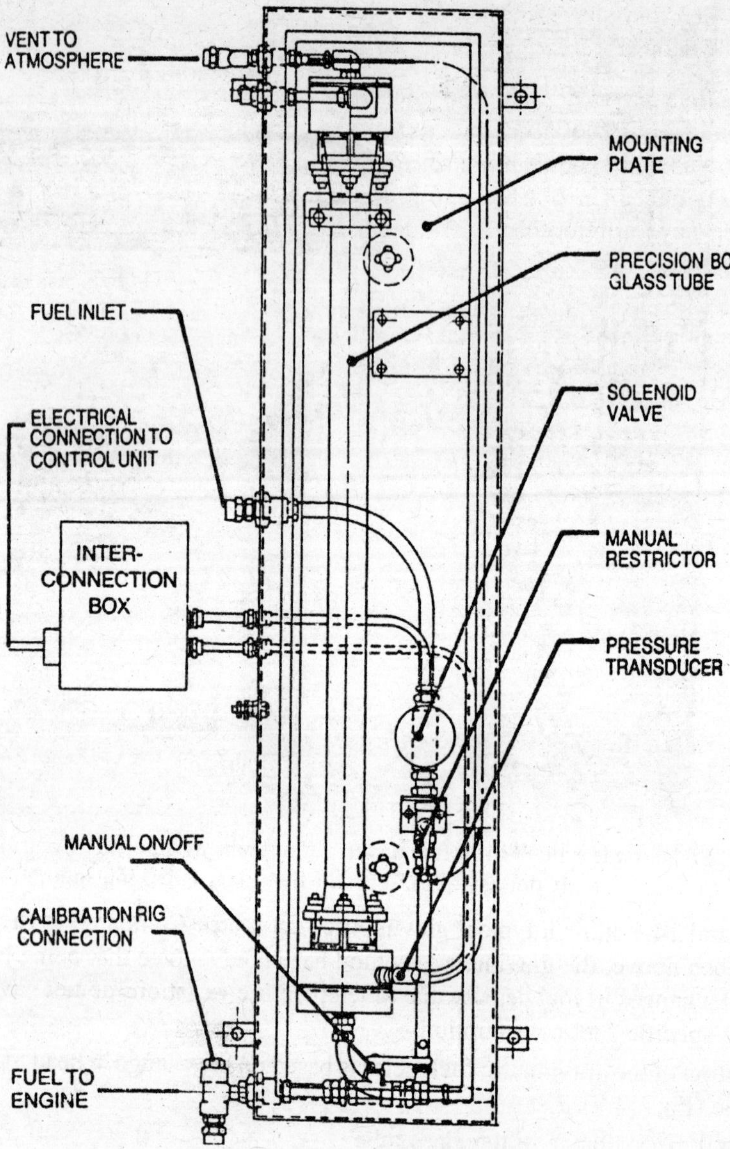

Fig. 14.8. Details of FG 100 fuel meter.
(Courtesy—SAJ-Froude Test Plant Pvt. Ltd., India)

$\therefore \quad \Delta h = \Delta P / \rho g$

$\Delta M = \rho A(\Delta P / \rho g)$

$\therefore \quad = A\Delta P / g$

$\therefore$ Weight change, $\Delta W = g\Delta M = A\Delta P$ ...(14.10)

Fuel pressure P in the tube is monitored by a high accuracy pressure transducer. The *total* engine revolutions are measured over the measurement range, from which, given the engine torque (to be measured separately by a dynamometer), the specific fuel consumption (SFC) can be calculated from the equation

$$SFC = \frac{\text{constant} \times \text{selected weight range}}{\text{total revolutions} \times \text{engine torque}} \quad ...(14.11)$$

Compared to the method of measuring speed and time, this method of noting total revolutions has the advantage that in this there is only one measurement and hence one less source of error which means higher accuracy. Further, any variation in the engine speed during the measurement is automatically taken care of in this method.

Optional accessories include repeat display output, computer interface, instantaneous fuel consumption rate, calibrating rig, header tank, weighing apparatus and spillback cooling.

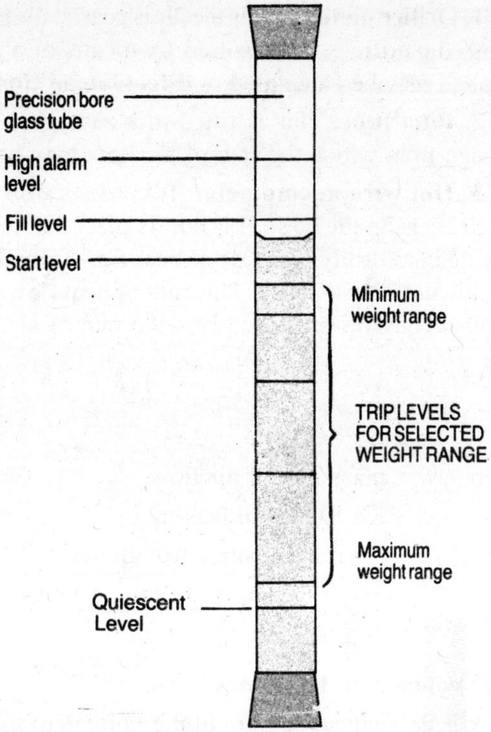

Fig. 14.9. Fuel measurement tube of FG 100 fuel meter. (Courtesy—SAJ-Froude Test Plant Pvt. Ltd., India)

8. THERMAL EFFICIENCY

It is the ratio of the work output and the heat input to the engine (in the form of fuel). Depending upon the type of the work output, it may be called indicated thermal efficiency or brake thermal efficiency. If M_f (kg/minute) is the fuel consumption, thermal efficiencies are given by the following expression:

Indicated thermal efficiency,

$$\eta_{it} = \frac{I.P.(kW) \times 60}{M_f \times \text{Calorific value of fuel}} \quad ...(14.12)$$

Brake thermal efficiency, $$\eta_{bt} = \frac{B.P.(kW) \times 60}{M_f \times \text{Calorific value of fuel}} \quad ...(14.13)$$

where calorific value of fuel is in kJ/kg

$$\therefore \quad \frac{\eta_{bt}}{\eta_{it}} = \frac{B.P.}{I.P.} = \eta_m$$

i.e. $$\eta_{bt} = \eta_m \times \eta_{it} \quad ...(14.14)$$

Value of brake thermal efficiency usually varies between 25-30% for S.I. engines and 30-40% for C.I. engines.

9. RELATIVE EFFICIENCY

It is the ratio of the indicated thermal efficiency, η_{it} of the engine to the air standard efficiency, η of the theoretical cycle on which the engine works, *i.e.*

$$\eta_r = \frac{\eta_{it}}{\eta} \quad ...(14.15)$$

10. AIR CONSUMPTION MEASUREMENT

Three methods to measure the air supply are commonly used:

1. Orifice meter. In this the air is passed through an orifice of known dimensions and pressure difference across the orifices is measured by means of a manometer. From this pressure difference the rate of air supplied may be calculated or directly read from the graph for the orifice.

2. Pitot tube. This is fitted in a passage of known dimensions and gives the velocity of fuel in the passage from which the rate of air flow may be calculated.

3. Hot wire anemometer. It consists of a resistance coil through which electric current is passed which heats up the coil. This coil is placed in the air passage in such a way that the air passing through it is heated uniformly. The rate of heat flow from the coil to the air is found from the current passing through the coil and its resistance. This rate of heat flow divided by the temperature rise of air and the specific heat of air at constant pressure gives the rate of air flow, *i.e.*

$$M_a = \frac{I^2 R}{C_p \times \Delta T} \quad ...(14.16)$$

where M_a = mass rate of air flow.

R = Resistance of heating coil.

I = Current passing through coil.

C_p = Specific heat of air at constant pressure.

ΔT = Temperature rise of air.

10.1. Volumetric Efficiency

It is defined as the ratio of the volume of the air actually inducted in a given time, at the atmospheric conditions surrounding the engine, to the total piston displacement of the induction strokes during the same time. Having known the mass rate of air consumption M_a, actual volume rate of flow, V, can be found out from the characteristic equation,

$$PV = M_a RT$$

$$\therefore \quad V = \frac{M_a RT}{P}, m^3/\text{minute} \quad ... (14.17)$$

where R is the gas constant for air = 287 J/kg°C and P, T represent the engine inlet conditions.

$$\text{Then Volumetric efficiency } \eta_v = \frac{\text{Volume rate of air consumption}}{\text{Engine swept volume}} \times n \quad ... (14.18)$$

where n = no. of induction strokes per minute.

$= \frac{\text{R.P.M.}}{2}$ in 4-strokes engines.

= R.P.M. in 2-stroke engines.

Sometimes, volumetric efficiency is also defined at N.T.P. conditions (P = 101.3 kPa, T instead of engine inlet conditions.

11. MEASUREMENT OF LUBRICATING OIL CONSUMPTION

The consumption of lubricating oil is determined by means of a depth gauge. The engine sump is filled upto a certain level before starting with the test. After the test, more oil is added to bring the oil level in the sump to the same level at which it stood earlier before the commencement of the test. The amount of the oil added is then the quantity consumed during the duration of the test.

12. HEAT BALANCE

Heat is supplied to the engine in the form of fuel, where a portion of it is converted into work, while the rest is lost to the cooling water and in the exhaust gases. Some of the heat, yet, is lost due to radiation etc. It is a common practice to prepare a heat balance sheet for the engine while performing test on it. Various quantities are measured as explained in the preceding articles, while the method of calculations and preparation of heat balance sheet is described here.

1. Heat supplied in fuel per minute. For this following quantities are required:

(*i*) Mass rate of fuel consumption per minute, M_f.

(*ii*) Calorific value of fuel, *C*, which may be found out with the help of a calorimeter.

Then heat supplied in fuel per minute $M_f \times C = H_f$, Say.

2. Heat equivalent of I.P. In heat units, this is equal to I.P. × 60 per minute (say H_i).

3. Heat lost to cooling water. To determine this, following quantities are required:

(*i*) Weight rate of cooling water per minute, M_w. This may be determined by direct measurement.

(*ii*) Temperatures of cooling water at inlet (T_i) and outlet (T_o) which are measured from thermometers installed in pockets at the inlet and outlet.

Then heat lost to cooling water per minute $= M_w \times (T_o - T_i)$

$= H_w$, say.

4. Heat lost in exhaust gases. For this the following quantities have to be determined first:

(*i*) Mass rate of exhaust gases per minute, M_e, which is equal to the sum of mass rates of fuel consumption and air supplied per minute.

(*ii*) Mean specific heat of exhaust gases C_e, which can be calculated from the results of exhaust gas analysis.

The heat lost in exhaust gases $= M_c \times C_e$

$= H_e$, say

From the above results, heat balance sheet may be prepared as follows:

Heat input per minute		*Heat utilization per minute*	
Source	*Amount (kJ)*	*Factor*	*Amount*
Fuel	H_f	(*i*) Heat equivalent of I.P. (*ii*) Heat lost to cooling water (*iii*) Heat lost in exhaust gases (*iv*) Heat lost due to radiation or unaccounted	H_i H_w H_e $H_f - (H_i + H_w + H_e)$
Total	H_f	Total	H_f

13. OPTICAL ENGINE

This is essentially similar to a standard all-metal engine; only certain parts are made of quartz and sapphire. This has been used for engine research since 1980s. With its transparent walls, it is being used as a tool to get visual and quantitative data, which makes us understand better, the physical processes that are occurring in a standard metal engine.

Researchers can observe various operating conditions inside the engine, including the time span from the injection of fuel into the cylinder on through the entire combustion. Thus a better understanding of the combustion process enables us to optimise the design of the combustion chamber. The data can also be used to validate the three-dimensional simulation models.

This engine has been and is being extensively used for making production model of the HCCI (Homogeneous Charge Compression Ignition) diesel engine, which is conceived to be the ideal engine for vehicles in the near future.

Example 14.1. *A six-cylinder four-stroke engine develops 50 kW at 5000 R.P.M. The engine bore and stroke are 70 mm each. Determine the B.M.E.P.*

Solution. From Eq. (14.4),

$$\frac{50}{6} = \frac{P_{mb} \times (70 \times 10^{-3}) \times \left\{\frac{\pi}{4} \times (60 \times 10^{-3})^2\right\} \times \frac{5000}{2}}{60,000}$$

where P_{mb} is the BMEP

$$P_{mb} = \frac{5 \times 10^5 \times 4}{70 \times 10^{-3} \times \pi \times 4900 \times 10^{-6} \times 2500}$$

$$= 742413\ Pa = \mathbf{742.4\ kPa}$$

Example 14.2. *A four-cylinder engine has a total swept volume of 1.084 litres and develops a B.P. of 23.4 kW at 3200 RPM. Calculate the B.M.E.P.*

Solution. From Eq. (14.4),

$$B.P. = \frac{P_{mb}.LAn}{60,000}$$

where $$B.P. = \frac{23.4}{4} = 5.85\,\text{kW}$$

$$LA = \frac{1.084}{4} \times 10^{-3}\,\text{m}^3$$

$$= .271 \times 10^{-3}\,\text{m}^3$$

$$n = \frac{3200}{2} = 1600$$ per minute, assuming the engine to be of 4-stroke type.

$$5.85 = \frac{P_{mb}(.271 \times 10^{-3}) \times 1600}{60,000}$$

$$P_{mb} = 809500\ Pa$$

$$= \mathbf{809.5\ kPa}$$

Example 14.3. *A car engine develops maximum B.P. of 40 kW at 3500 R.P.M. and maximum torque of 120 Nm at 2400 R.P.M. Determine (i) Percentage of maximum B.P. developed at maximum torque and (ii) Percentage of maximum torque developed at maximum B.P.*

Solution.

(*i*) Maximum torque of 120 Nm occurs at 2400 R.P.M. From Eq. (14.3),

$$\text{B.P. at 2400 R.P.M.} = \frac{2\pi \times 2400 \times 120}{60,000} = 30.16$$

$\therefore$ Percentage of maximum B.P. $= \frac{30.16}{40,0} \times 100 = 75\%$

(*ii*) Maximum B.P. of 40 kW occurs at 3500 R.P.M. From Eq. (14.3),

$$40 = \frac{2\pi \times 3500 \times T}{60,000}$$

where T is the torque developed at 3500 R.P.M.

$$T = 109.13 \text{ Nm}$$

$$\text{Percentage of maximum torque} = \frac{109.13}{120.0} \times 100 = \mathbf{90.09\%}$$

Example 14.4. *A 4-cylinder petrol engine develops 18 kW at 1200 r.p.m. When, however, one cylinder was made inoperative, average torque was found to be 100 Nm. If the engine consumes 330 gms of petrol per kW hour, calculate the indicated thermal efficiency.*

Assume calorific value of petrol as 42,000 kJ/kg.

Solution.

Let η_{it} = indicated thermal efficiency.

and M_f = fuel consumption per minute

Then from Eq. (14.12), $\eta_{it} = \dfrac{I.P. \times 60}{\text{Calorific value} \times M_f}$

Average B.P. per cylinder (with one cylinder inoperative)

$$= \frac{2\pi \times 1200 \times 100}{60,000} \times 12.56 \text{kW}$$

$$I.P. = 4(18 - 12.56) = \mathbf{21.76 kW}$$

Example 14.5. *A single cylinder engine working on four-stroke cycle was tested with a rope brake and the following data were obtained:*

Rope diameter, d = 25 mm
Diameter of brake wheel, D = 600 mm
Dead weight on the brake, W = 200 N
Reading on spring balance, S = 40 N
Engine speed, n = 500 rpm
Area of indicator diagram, a = 400 mm²
Length of indicator diagram, l = 60 mm
Spring scale, δ = 0.1 N/mm²/mm
Cylinder bore, B = 100 mm
Stroke, L = 150 mm

Calculate (i) B.P., (ii) I.P., (iii) Mechanical efficiency.

Solution.

(*i*) Torque, $T = (W - S)\left(D + \frac{d}{2}\right)$

$$= (200-40)\left(600+\frac{25}{2}\right) = 160\times 312.5\,\text{Nmm} = 50\,\text{Nm}$$

$$\therefore \qquad B.P. = \frac{2\pi\times 500\times 50}{60,000} = \mathbf{2.62\,kW}$$

(*ii*) Indicated mean effective pressure,

$$P_m = \frac{a}{l}\times\delta = \frac{400}{60}\times 0.1 = \frac{4}{6}\,\text{N/mm}^2 = \frac{4}{6}\times 10^6\,\text{N/m}^2$$

Then from Eq. (14.11),

$$\therefore \qquad I.P. = \frac{p_m LAN}{60,000} = \frac{\left(\frac{4}{6}\times 10^6\right)\times\left(\frac{150}{1000}\right)\times\frac{\pi}{4}\left(\frac{100}{1000}\right)^2\times\frac{500}{2}}{60,000} = \mathbf{3.27\,kW}$$

$$(iii)\ \text{Mechanical efficiency} = \frac{B.P.}{I.P.} = \frac{2.62}{3.27} = .80 = \mathbf{80\%}$$

Example 14.6. *In a test of a four-cylinder four-stroke petrol engine with 80 mm bore and 100 mm stroke the following results were obtained at a particular speed:*

Fuel consumption = 120 gm/minute

B.P. with all cylinders working = 21.0 kW

B.P. with cylinder no 1 cut out = 14.5 kW

B.P. with cylinder no 2 cut out = 14.2 kW

B.P. with cylinder no 3 cut out = 14.0 kW

B.P. with cylinder no 4 cut out = 14.5 kW

Calculate (i) I.P. of the Engine

(ii) Indicated thermal efficiency of the engine, if the calorific value of fuel is 42,000 kJ/kg.

(iii) Relative efficiency, if the clearance volume of each cylinder is 110 × 10^3mm^3.

Solution.

(*i*) I.P. of cylinder no. 1 = 210 – 14.5 = 6.5

I.P. of cylinder no. 2 = 21.0 – 14.2 = 6.8

I.P. of cylinder no. 3 = 21.0 – 14.0 = 7.0

I.P. of cylinder no. 4 = 21.0 – 14.5 = 6.5

∴ I.P. of the engine = 6.5 + 6.8 + 7.0 + 6.5 = **26.8 kW**

(*ii*) From Eq. (14.12),

$$\text{Indicated thermal efficiency, } \eta_{it} = \frac{I.P.\times 60}{M_f\times\text{cal. value of fuel}}$$

$$= \frac{26.8\times 60}{.12\times 42,000} = .319 = \mathbf{31.9\%}$$

(*iii*) Clearance volume of each cylinder = 110 × 10^3 mm^3

$$\text{Swept volume} \qquad = \frac{\pi}{4}(80)^2\times(100) = 502.65\times 10^3\,\text{mm}^3$$

$\therefore$ Compression ratio, $r = \dfrac{(110+502.65)}{110\times10^3} = 5.478$

$\therefore$ Air standard efficiency, $\eta = 1 - \left(\dfrac{1}{r}\right)^{\gamma-1}$

$$= 1 - \left(\frac{1}{5.478}\right)^{1.4-1} = .506$$

$\therefore$ Relative efficiency $= \dfrac{.319}{.506} = .63 = \mathbf{63\%}$

Example 14.7. *An engine develops brake power of 19.5 kW and consumes 4.5 litres of petrol during testing for 30 minutes. Specific density nd calorific value of petrol are 0.78 and 46 MJ/kg respectively. Determine (i) brake thermal efficiency (ii) specific fuel consumption.*

Solution.

(*i*) Fuel consumed per second $= \dfrac{4.5\times0.78}{30} = 1.95\times10^{-3}$ kg

Heat supplied per second $= 1.95 \times 10^{-3} \times 46$

$= 0.0897$ MJ $= 89.7$ kJ

Work done per second in kJ = Brake power developed in kW

$= 19.5$ kJ

$\therefore$ Brake thermal efficiency $= \dfrac{19.5}{89.7}\times100 = \mathbf{21.7\%}$

(*ii*) Volume of fuel used/hr. $= \dfrac{4.5}{30}\times60 = $ **9.0 litres**

$\therefore$ Specific brake fuel consumption

$= \dfrac{9.0}{19.5} = $ **0.46 litres/kWh**

Example 14.8. *In a trial an a four cylinder engine of 100 mm bore, 150 mm stroke and working on a four-stroke cycle, the following observations were made:*

Speed = 2500 r.p.m.

Net Dynamometer load at 500 mm radius = 200 N

Power required to rotor with ignition off = 4.5 kW

Petrol consumption = 200 g/m minute

Cooling water circulated = 7.5 kg/minute

Temperature rise of cooling water = 50°C

If the calorific value of petrol is 46,000 kJ/kg,

(*i*) *Calculate mechanical efficiency and indicated mean effective pressure.*

(*ii*) *Draw heat balance chart for the test in kJ/minute.*

Solution.

(*i*) $$B.P. = \frac{2\pi NT}{60,000} = \frac{2\pi \times 2500\,(200 \times 5)}{60,000} = 26.2\,\text{kW}$$

$$F.P. = 4.5\,\text{kW (given)}$$

$$I.P. = 26.2 + 4.5 = 30.7\,\text{kW}$$

∴ Mechanical efficiency $= \frac{26.2}{30.7} = .853$

$= \mathbf{85.3\%}$

I.M.E.P., $$P_m = \frac{I.P. \times 60,000}{L \times A \times N}$$ (From Eq. 14.1)

$$= \frac{30.7 \times 60,000}{.15 \times \left(\frac{\pi}{4} \times .01 \times 4\right) \times \frac{2500}{2}}$$

$= 312.7 \times 10^3$ N/m^3

$=$ **312.7 kPa**

(*ii*) (*a*) Heat supplied in fuel/minute $= M_f \times$ Cal. value

$= .2 \times 46,000 = 9200$ kJ

(*b*) Heat equivalent of I.P. $= I.P. \times 60 = 30.7 \times 60$

$= 1842$ kJ/mt.

(*c*) Heat lost to cooling water $= M_w \times 4.2$

$= 7.5 \times 4.2 \times 50$

$= 1575$ kJ/mt.

(*d*) Heat lost to exhaust, radiation etc. $= 9200 - (1842 + 1575)$

$= 5783$ kJ/mt.

Heat Balance Chart

Heat input per minute		*Heat utilized/lost per minute*	
Source	*Amount (kJ)*	*Factor*	*Amount (kJ)*
Fuel	9200	(*i*) Heat equivalent of I.P.	1842
		(*ii*) Heat lost to cooling water	1575
		(*iii*) Heat lost to exhaust due to radiation or unaccounted	5783
Total	9200	Total	9200

THEORY QUESTIONS

1. How do you classify tests for automotive engines? Give a brief description of each.
2. Describe in detail the purpose and procedure of vacuum and cylinder compression tests for auto engines.
3. Make a list of the important quantities to be measured during the testing of an engine. What are the main variables involved?

4. How is the indicated power measured? Briefly describe the principles of operation and comparative merits-demerits of various types of engine indicators.
5. How do you obtain brake power from the torque measured by a dynamometer? List various types of dynamometers and explain the principle of each. Discuss also the comparative merits and limitations of different types of dynamometers.
6. Describe in detail the construction and working of:
   (*i*) hydraulic dynamometer.
   (*ii*) eddy current dynamometer.
7. Explain in detail the procedure to perform Morse test. What is the object of this test?
8. Discuss briefly various methods of measuring fuel consumption and air consumption of automotive engines. How do you measure the consumption of the lubricating oil?
9. Explain the principle of working of a gravimetric type of fuel measurement device. Discuss the construction and working of any such device.
10. What is a heat balance sheet? Explain in detail the procedure for the preparation of heat balance sheet for an engine.

NUMERICAL PROBLEMS

1. While performing Morse test on a four-cylinder, four stroke petrol engine, the following results were obtained at a particular throttle setting and speed:
   B.P. with all cylinders working = 32.2 kW
   B.P. with cylinder no 1 cut out = 22.0 kW
   B.P. with cylinder no 2 cut out = 21.8 kW
   B.P. with cylinder no 3 cut out = 22.2 kW
   B.P. with cylinder no 4 cut out = 22.8 kW **[Ans.** 40.0 kW, 80.5%]
   Determine I.P. of the engine and is mechanical efficiency.
2. A four stroke engine with four cylinders, bore 80 mm and stroke 100 mm was tested at 3500 r.p.m. and following data were recorded:
   Fuel consumption = 300 gm/minute
   Indicated mean effective pressure = 1 MPa
   Engine torque developed = 140 Nm
   (*i*) I.P. of the engine.
   (*ii*) Mechanical efficiency. **[Ans.** (*i*) 58.5, (*ii*) 51.3, (*iii*) 87.6%, (*iv*) 24.4%]
   (*iii*) Brake thermal efficiency of the engine.
3. A four-cylinder four-stroke petrol engine with bore 60 mm and stroke 100 mm, when tested develops a torque of 65 Nm. at 3000 r.p.m. If the relative efficiency based on brake thermal efficiency is 0.5, clearance volume of each cylinder is 63,000 mm^3 and calorific value of fuel is 42,000 kJ/kg, calculate (*i*) fuel consumption, (*ii*) brake mean effective pressure. **[Ans.** (*i*) 118 gm/minute, (*ii*) 721.5 kPa]

QUESTIONS FOR VIVA

1. List different categories of engine tests from point of view of objectives of such tests.
2. How do you perform 'road tests' on automotive engines in the laboratory?
3. List the quantities to be measured during engine testing.
4. List main parameters encountered in engine testing.
5. What is indicated power?
6. Define indicated mean effective pressure.
7. What is an engine indicator?

8. What are the units of spring scale of the engine indicator?
9. What are the limitations of a mechanical indicator?
10. What is the main disadvantage of an optical indicator?
11. What is brake power?
12. What is a dynamometer?
13. What is absorption type dynamometer?
14. Give an example of absorption type dynamometer.
15. What is a transmission type dynamometer?
16. State the advantage and disadvantage of a prony brake dynamometer.
17. Name an electrical dynamometer which is highly accurate.
18. What are the merits and demerits of an eddy current dynamometer?
19. Define brake mean effective pressure. On what factors does it depend?
20. Define mechanical efficiency of an engine. What is its usual range?
21. Why should we perform the Morse test on an engine at the same conditions of load and speed throughout?
22. State the advantage of gravimetric method of fuel measurement over the volumetric method.
23. Define specific fuel consumption. What are its units?
24. What is thermal efficiency of an engine?
25. How are the brake thermal efficiency, indicated thermal efficiency and mechanical efficiency related?
26. State the usual range of brake thermal efficiency for S.I. and C.I. engines.
27. Define relative efficiency.
28. What is the principle of hot wire anemometer?
29. Define volumetric efficiency of an engine.
30. What is a heat balance sheet for an engine?

MULTIPLE CHOICE QUESTIONS

Mark the correct answers:

1. From the engine indicator diagram, we obtain
   (*a*) I.M.E.P. (*b*) B.M.E.P.
   (*c*) Mechanical efficiency (*d*) Relative efficiency
2. The units of spring scale for the engine indicator are
   (*a*) kW (*b*) N
   (*c*) Nm (*d*) $N/m^2/m$
3. Engine indicators not having any inertia effects are
   (*a*) mechanical type (*b*) electrical type
   (*c*) optical type (*d*) none of these
4. The most accurate dynamometer is the
   (*a*) prony brake type (*b*) hydraulic type
   (*c*) swinging field type (*d*) eddy current type
5. Most commonly used dynamometer on account of its simplicity of construction is the
   (*a*) prony brake type (*b*) hydraulic type
   (*c*) swinging field type (*d*) eddy current type
6. Brake mean effective pressure in an engine depends upon its
   (*a*) speed only (*b*) torque only
   (*c*) speed and torque (*d*) speed and power

7. Mechanical efficiency of an automobile engine usually varies in the range
   (*a*) 50-60% (*b*) 60-70%
   (*c*) 70-90% (*d*) more than 90%
8. The brake thermal efficiency, ς_{bt}, is given by
   (*a*) $\eta_m \times \eta_{it}$ (*b*) $\eta_m \times \eta_r$
   (*c*) $\eta_{it} \times \eta_r$ (*d*) $\eta_v \times \eta_{it}$
9. Brake thermal efficiency for S.I. engines usually varies between
   (*a*) 25% and 30% (*b*) 30% to 60%
   (*c*) 60% and 80% (*d*) more than 80%

[**Key.** 1. (*a*), 2. (*d*), 3. (*c*), 4. (*c*), 5. (*a*), 6. (*b*), 7. (*c*), 8. (*a*), 9. (*a*)]

15 Conventional Ignition Systems

1. FUNCTION

The function of the ignition system is to produce a spark in the engine cylinder towards the end of the compression stroke. In a four-stroke engine, a spark should occur in each cylinder after two revolutions of the crankshaft, whereas in a two-stroke engine a spark in each cylinder is required every revolution of the crankshaft. Thus, for instance, in a 6-cylinder 4 stroke engine running at 5000 r.p.m., the number of sparks required per minute will be 15000 and these have to be timed very accurately.

2. REQUIREMENTS OF AN IGNITION SYSTEM

1. Spark at the plug electrodes must be regular and synchronously timed with respect to the cylinder-piston position at all speeds and loads on the engine.

2. The spark should be sufficiently strong so as to start ignition of the charge.

Since lean air fuel mixtures are less conductive, they require higher ignition voltages. Thus on modern emission-controlled engines that use lean mixtures, higher-voltage ignition system is required. Moreover, due to higher secondary voltage in these systems, it becomes necessary to increase the electrical insulation.

3. The spark duration should be sufficient to establish burning of the air-fuel mixture under all conditions.

4. It should have the ability to produce spark even when a shunt is established over the spark plug electrodes, possibly due to carbon, oil or lead deposits, liquid fuel or water condensation.

5. The power consumed to produce spark should be minimum.

6. It should have service life at least equal to that of the engine.

7. The system must be easy to maintain.

8. It must not cause radio interference.

9. It should be light and compact.

10. It should be adaptable to mass production.

3. TYPES OF IGNITION SYSTEM

A battery of 12 volts is generally employed. However, a very high voltage surge (of about 25,000 volts in modern engines) is required for ignition purposes. The actual high voltage at which the spark occurs is called breakdown voltage and it depends upon so many factors, *viz.*, the gap, polarity and condition of plug electrodes, condition of plug insulation, moisture on the distributor cap and spark plugs and pressure, temperature and type of mixture in the combustion chamber. Used spark plugs may require fairly higher voltage to fire than the new plugs because of increase in the gap and rounding off of the centre electrode. In a given engine the highest ignition voltage is required during part-throttle acceleration because the fuel system is then supplying a lean mixture with increased combustion pressure.

To produce such a high voltage, a special ignition system has to be employed. Two types of conventional ignition systems (called point type ignition systems) have been used, *viz*, the 'battery ignition system' and the 'magneto ignition system'. However, both these systems have the major drawbacks of inertia of mechanical components like contact breaker and insufficient dwell period to build up adequate coil field at high engine speeds. Though for a long time, both these systems were in use and giving satisfactory performance, yet since mid nineteen seventees, these are being fast replaced by 'electronic ignition systems'. Point-type and the solid-state (electronic) ignition systems operate in the same way except that they differ in the way the primary current is stopped and restarted. The conventional point-type systems will be discussed in detail in this chapter, while the electronic systems will be explained in the next chapter.

4. BATTERY IGNITION SYSTEM

A battery of 12 volts is generally employed. Battery ignition system for a 4-cylinder engine is shown in Fig. 15.1 and 15.2. As is clear from Fig. 15.2, there are two basic circuits in the system *viz*, the primary and the secondary circuits. The battery, primary winding of the ignition coil, condenser and the contact breaker form the primary circuit, whereas the secondary winding of the ignition coil, the distributor and the spark plugs constitute the secondary circuit.

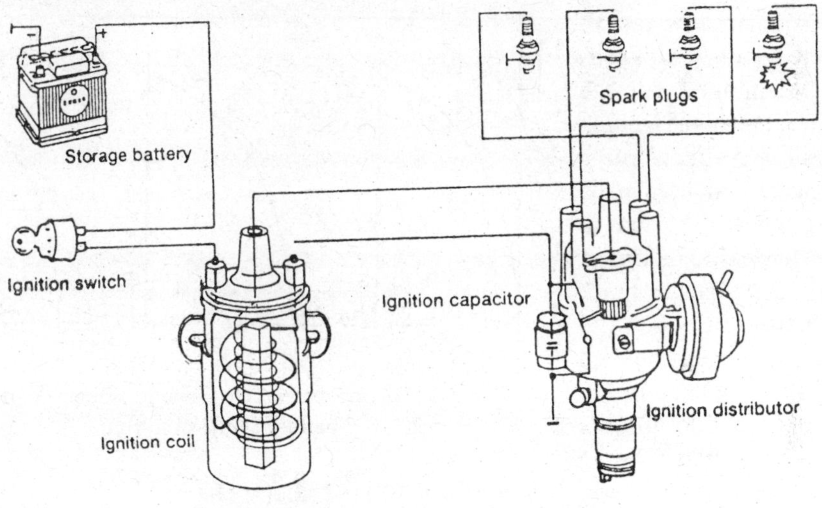

Fig. 15.1. Schematic diagram showing battery ignition system for a 4-cylinder engine. (Courtesy—MICO, India)

When the ignition switch is in the 'on' position, the current flowing in the primary circuit will grow exponentially during the make period of the contact breaker (Fig. 15.3). If V is the battery voltage and R & L the resistance and self-induction (reluctance) respectively of the primary winding, the instantaneous value of primary current I, at time t seconds after the close of the contact breaker points is given by,

$$I = \frac{V}{R}\left(1 - e^{\frac{Rt}{L}}\right) \qquad \ldots(15.1)$$

The electromagnetic energy on account of this build up of current in the primary circuit is stored in the laminated iron core of the ignition coil. As the contact breaker points open, the magnetic field built up by the growth of the current collapses and the energy stored during the make period is projected into the secondary circuit. If I_B = the value of I when break just occurred, the energy stored at break $\frac{1}{2} LI_B$. It is this energy which is suddenly passed onto the secondary, inducing an e.m.f. there. As the break is made

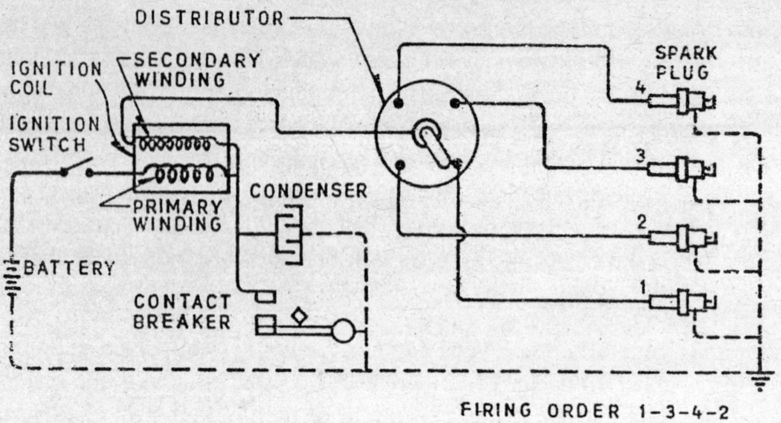

Fig. 15.2. Battery ignition system for 4-cylinder engine.

very sharp, the induced e.m.f. is very large being proportional to the rate of change of flux in the winding. This self-induction effect is combined with the transformer step-up action of the ignition coil. As a result a sudden voltage surge of very high amplitude is produced in the secondary, causing a spark to occur at the spark plug electrodes. Thus it is at points '*S*' in Fig. 15.3 that the spark occurs. As an illustration, when the breaker points open, if a self-induced e.m.f. of 250 volts is produced in the primary winding and the ratio of turns in the secondary and the primary winding is 100, the secondary voltage produced will be 250 × 100 = 25,000 volts.

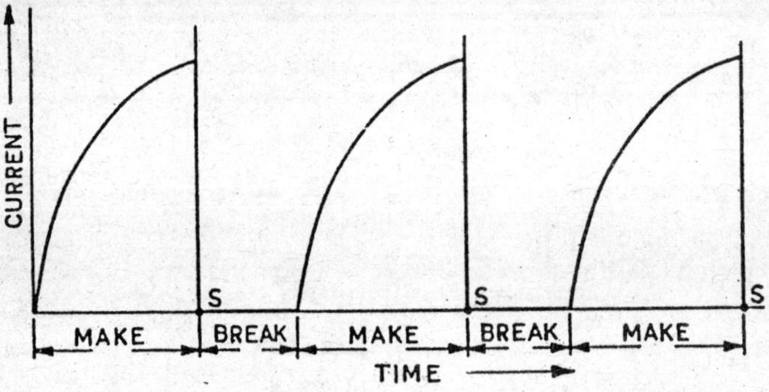

Fig. 15.3. Growth and Collapse of primary current in battery ignition system.

From the secondary winding, the high voltage surge is led to each spark lug in turn with the help of h.t. cables and the distributor.

The ignition circuit may be either positive-earth type or negative earth type. In the position earth type, the ignition coil and the spark plug centre electrode are positive with respect to the ground, whereas in the negative earth type, these are negative with respect to the ground. The advantage of negative earth type is that comparatively less voltage is required for the electrons to jump from the central electrode which is hotter than from the outer electrode which is comparatively colder. Further, larger erosion due to sparking occurs from the positive electrode. Therefore, *if the centre electrode is made negative with respect to the earth, the plug life is increased.* Moreover, since the polarity can be reversed by reversing the primary coil leads, care must be taken to connect the primary leads correctly. Otherwise, higher ignition voltage will be required for causing the spark to jump across the spark plugs.

5. COMPONENTS OF BATTERY IGNITION SYSTEM

The main components of a battery ignition system are:

1. Battery
2. Ignition coil

3. Contact Breaker
4. Condenser
5. Distributor
6. Spark Plug

The batteries are explained separately in chapter 17 while the other components are described here.

6. IGNITION COIL

The ignition coil is simply a transformer with certain characteristics making it suitable for its special use but impairing to some extent its efficiency as a transformer. It serves to convert the relatively low battery voltage into high voltage.

The battery winding consists of 200-30 turns of thick wire (about 20 S.W.G) of total resistance about 3 ohms while the secondary is made up of large number of turns (about 15,000-20,000) of fine wire (about 40 S.W.G.). As the voltage difference between adjacent turns is small, only thin insulation is required and enamelled wire is used for this purpose. However, there is sufficient voltage drop between the layers and for this paper insulation is used. The whole assembly is then impregnated in wax under vacuum to remove any air pockets or submerged in thin insulating mineral oil and hermetically sealed and in an aluminium case. The oil acts as an insulator to prevent high voltage arcing within the coil. Hermetic sealing prevents the entry of moisture in the coil, which if not prevented, would cause coil failure.

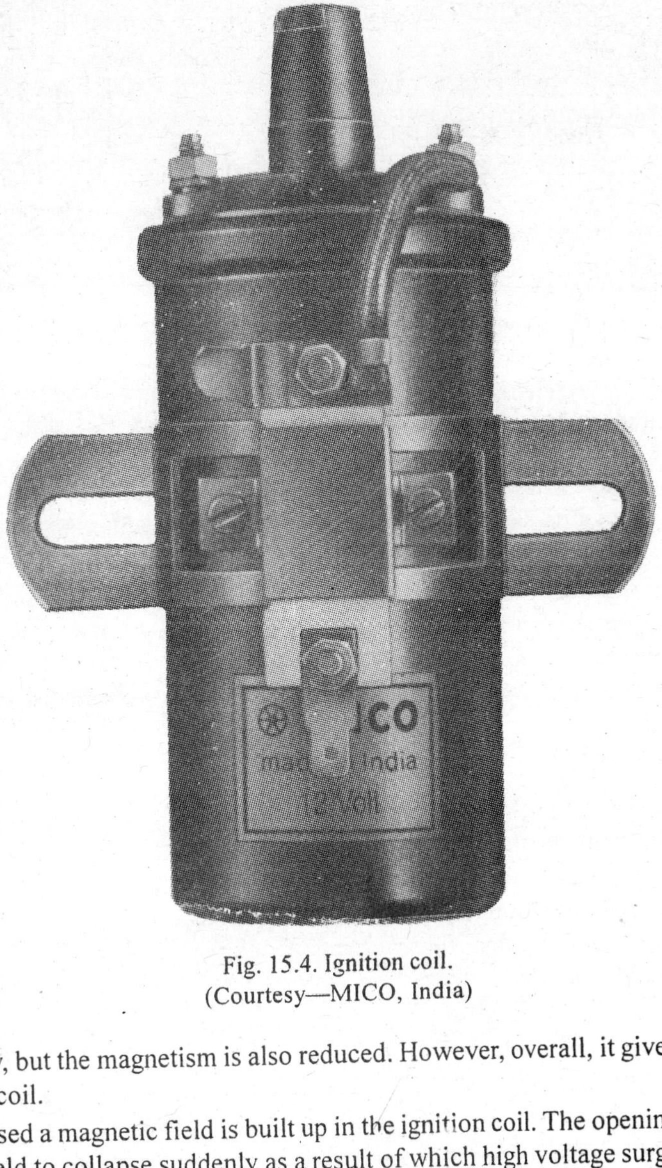

Fig. 15.4. Ignition coil.
(Courtesy—MICO, India)

The core is formed of either laminations of silicon steel or annealed iron wire insulated by varnish from each other. The laminated core has lower hysteresis loss. However, it is not easily produced in circular cross-section and because of this, square section is often used, although when square, for the same area it has a greater perimeter than that of a core formed of iron wires stacked in a circular tube. Only open cores are used; closed cores tend to cause excessive damping.

With the advancement in technology it is now possible to use powder metal as coil cores. This reduces eddy currents drastically, but the magnetism is also reduced. However, overall, it gives a more efficient and higher output ignition coil.

When the contact breaker points are closed a magnetic field is built up in the ignition coil. The opening of the breaker points causes the magnetic field to collapse suddenly as a result of which high voltage surge is induced in the secondary winding and this voltage cause a spark to jump across the spark plug gap. This collapse of magnetic field must be very rapid to produce the desired voltage. This is achieved with the help of condenser. Without the use of a condenser, the energy stored in the coil in the form of magnetic flux would be dissipated in an arc across the points and no high voltage surge would be induced in the secondary winding. The condenser momentarily absorbs the current and brings the flow to a quick stop causing the magnetic field to collapse rapidly (For details refer Art. 10).

There are two types of ignition coil:

1. Core type.
2. Metal clad or can type.

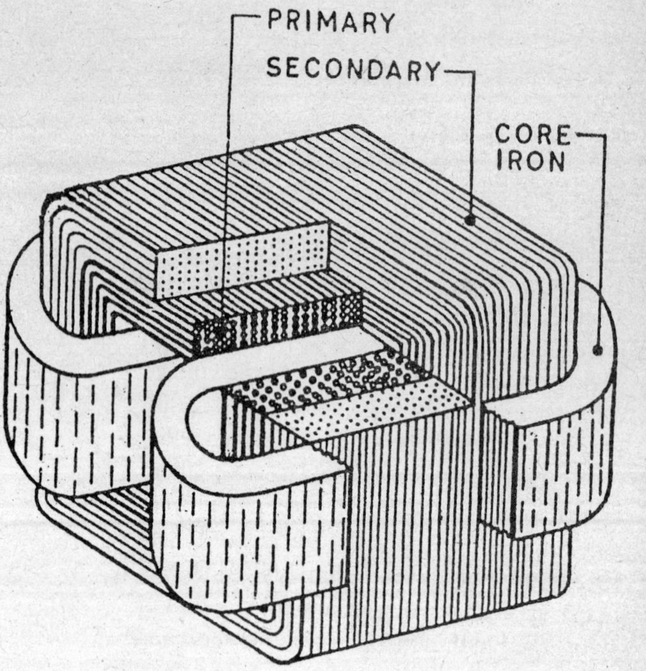

Fig. 15.5. Core type ignition coil.

Core Type Coil (Fig. 15.5)

In this type, primary is wound first on the core and then outside it the secondary is wound, the proper insulation, of course, being provided between the two.

Metal Clad or Can Type Coil (Fig. 15.6)

In this, the secondary is first wound over the core and then the primary over this. The inner end of the secondary winding is connected to the core, while the other end of the secondary is connected to the primary winding. The ends of the primary winding are connected to the L.T. terminals of the coil, one of which is connected further to the contact breaker and the other to the ignition switch.

Between the primary and the case, iron strips are sometimes placed. These increase the inductance and prevent losses that would occur otheiwise if the flux is linked with the metal case.

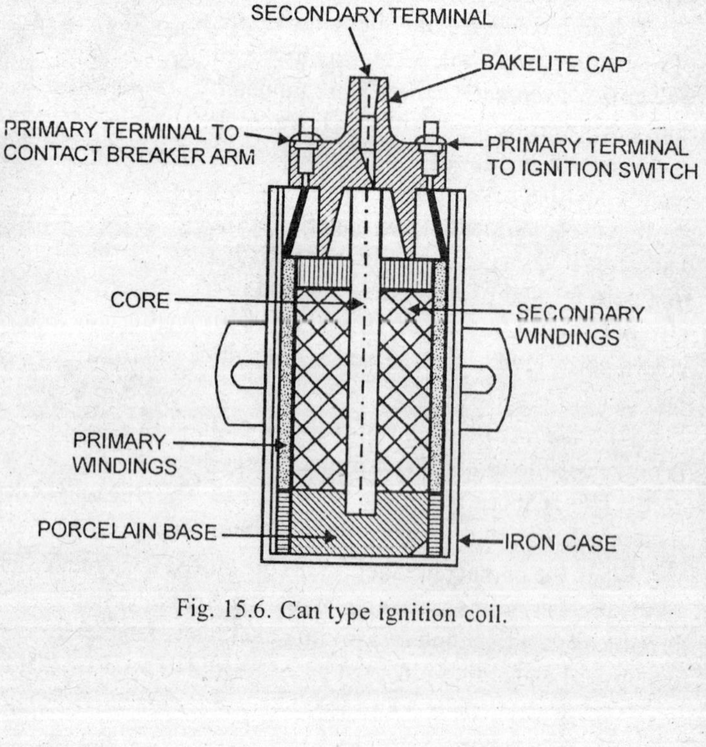

Fig. 15.6. Can type ignition coil.

Comparison of Core and Can types

1. The heat losses in the primary winding in the core type have to pass through the secondary, whereas in the can type, the primary being situated immediately under the metal case, can readily part with its heat without affecting secondary.

2. In the core type, the primary being inside, its mean turn is relatively short, Therefore, it is necessary to add external resistance called "ballast resistance", in order to limit the battery current in the oil. In the can type, however, no ballast resistance is necessary.

7. BALLAST RESISTOR

When the engine is running at slow speed, the contact breaker points are closed for a comparatively longer period of time. This causes overheating of the induction coil. To avoid this a ballast resistance is sometimes inserted in the primary circuit of the coil (Fig. 15.7). This increased resistance decreases the amount of current in the primary circuit and thus avoids overheating of the coil at low engine speeds.

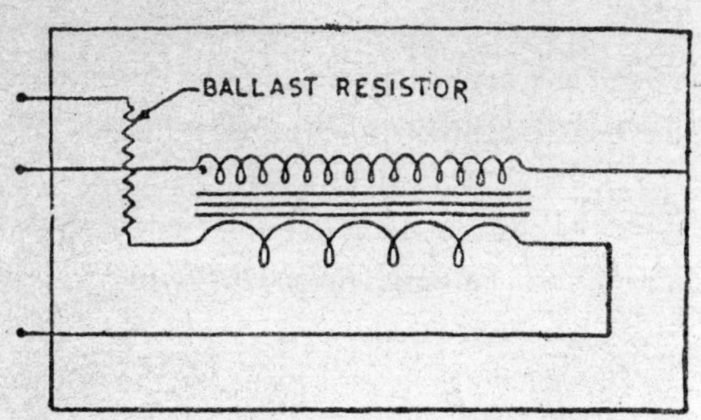

Fig. 15.7. Ignition coil with ballast resistor.

At higher engine speeds, however, due to smaller closing time of the contact breaker points, the current in the primary circuit and thus its temperature decreases, due to which the resistance value of the ballast resistor decreases. Thus it guards against any drop of secondary voltage at higher speeds, which if occurred would be quite opposite to what is desirable.

Further during starting of the engine, the resistance is bypassed from the ignition circuit to provide correct voltage to the coil when the starting motor is drawing a very heavy current, thereby drawing a very heavy current, thereby draining the battery voltage. This is done by the solenoid switch which short circuits the resistance, thus putting it out of the primary circuit completely. The value of the ballast resistance varies generally from 1 to 4 ohms. The ignition coil in Maruti 800 employs such a resistance.

8. CONTACT BREAKER

The function of a contact breaker is to make and break the primary circuit. This is probably the weakest member of an ignition system. This is clear from the fact that a four-cylinder engine operating at 4000 r.p.m. must make and break the circuit 8,000 times a minute. The essential requirements of a good contact breaker are:

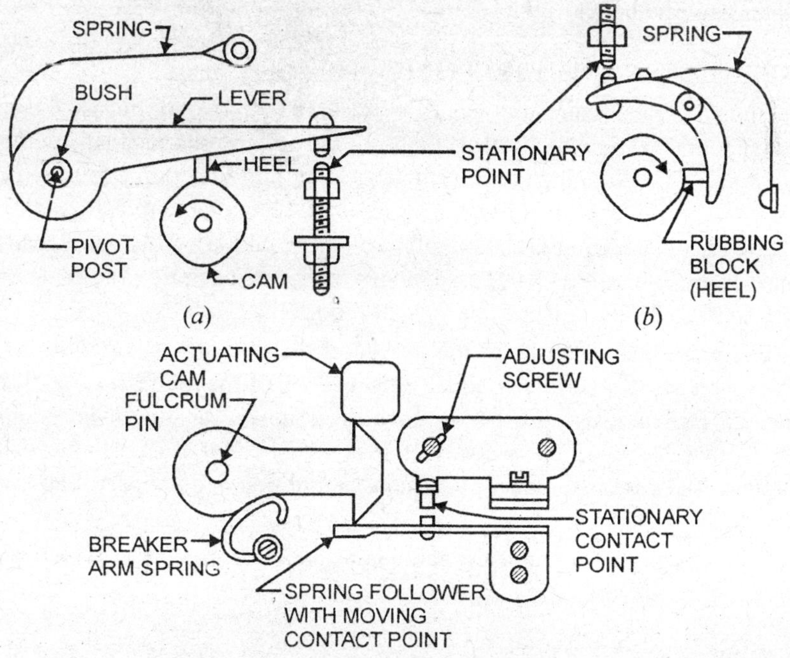

Fig. 15.8. Types of contact breakers.

1. The contacts must open and close at the correct time.
2. The contacts must close without bounce.
3. The contacts must open without "Fling". Flinging occurs when the spring force is too low. In this, heel is flung clear of the cam surface.
4. Excessive oxidation of the contacts should be avoided. The oxide film that may form on the contact faces is particularly detrimental in a low-resistance circuit of the primary winding.
5. The corrosion of the points should be minimum.

The contact breaker is placed in the distributor housing itself. The cam is fitted to the distributor spindle.

Various types of contact breakers are shown in Fig. 15.8. A hardened steel cam attached to the end of the driving spindle actuates the lever through the heel. The speed of cam is always one half of the engine speed. However, the number of cam faces is always equal to the number of cylinders except in case of the double lever type contact breakers.

The main components of a contact breaker are the lever, heel, bush and the contacts. One contact point is mounted on a steel pressing fixed by a screw or screws to the distributor base plate, thus making it adjustable. The other contact is mounted on the pivoted lever which is made of steel. The heel and the bush are generally of plastic. For the contact tungsten or alloy of platinum and iridium (with iridium content varying from 10 to 25 per cent, depending upon the degree of hardness required) is used. The tungsten is quite hard and has high melting point, but it has the disadvantage that it is liable to oxidation. Platinum-iridium contacts, however, are costlier and therefore not used so frequently. Generally their use is restricted to aircraft magnetos. Compared with tungsten they are relatively free from oxida- tion, but do not resist impact deformation so well.

The double arm type has certain advantages over the single arm type. In this for the same contact force, the force and wear on the heel are smaller, and because of smaller moment of inertia, contact bounce is less. This is why the double arm contact breaker is widely used in magnetos when operating speeds are normally higher than with the battery ignition.

9. CONTACT BREAKER GAP AND ITS SETTING

Normally the contact breaker gap should be between 0.35 to 0.45 mm. In case the gap is too small the points will not open for the period which is sufficient for the coil to function. This will cause the points to burn rapidly (since these would conduct current for too long) and the engine to misfire at all the speeds. On the other hand if the gap is too large, the closing time of the contact breaker, and hence for the e.m.f. to build, will be less, thus decreasing the maximum voltage produced. This will cause ignition failure at higher speeds Apart from this, excessive gap advances the spark because the points open earlier in this case. Similarly, the spark is retarded if the gap is kept too small.

Dwell is the time expressed in terms of the degree of distributor cam rotation for which the points remain *closed.* The dwell angle is normally equal to 55 to 70% of the total cam angle depending upon the contact breaker design and cam lobe profile. Typical dwell angles for 4-cylinder engines are 48° to 52° and for 6-cylinder engines 36° to 42°. The dwell angle is independent of camshaft speed. However, it can be changed by changing the contact-breaker point gap. A reduction in point gap of approximately .025 mm would increase the dwell angle by 1°.

The checking and adjustment of contact breaker gap should be done at intervals of 5000 km. Various methods of setting breaker gap are:

1. By feeler gauge

This is used when the breaker points are perfect and not pitted which is generally the case in the new distributors.

The Distributor cap is removed and the engine cranked (this may also be done by rotating fan blades) till the breaker gap is maximum. The gap is then checked by means of a feeler gauge. If it is found to be not correct, the screws on the plate carrying stationary point are loosened by means of a screw driver and the plate adjusted so that the gap as measured by the feeler gauge is correct. The screws are then tightened. In some cases, an offset screw is provided so that the gap can be changed simply by screwing or unscrewing.

2. By dial gauge

As stated already the feeler gauge cannot be used if the contact points are pitted, because the feeler gauge in this case will provide wrong conclusions. In such a case the dial gauge may be used (Fig. 15.9). The dial gauge is clamped on the body of the distributor and the lever arm is made to rest on the moving arm at the back of the contact point. The engine is cranked so that the gap is reduced to zero *i.e.* the contact points are touching. The dial of the gauge is then adjusted to bring the gauge pointer to zero position. The engine is now cranked again to open the contact points to the maximum gap size. The reading of a dial gauge, then directly gives the gap size. It is adjusted, as already explained above till the dial gauge reading is satisfactory as provided in the manufacturer's manual.

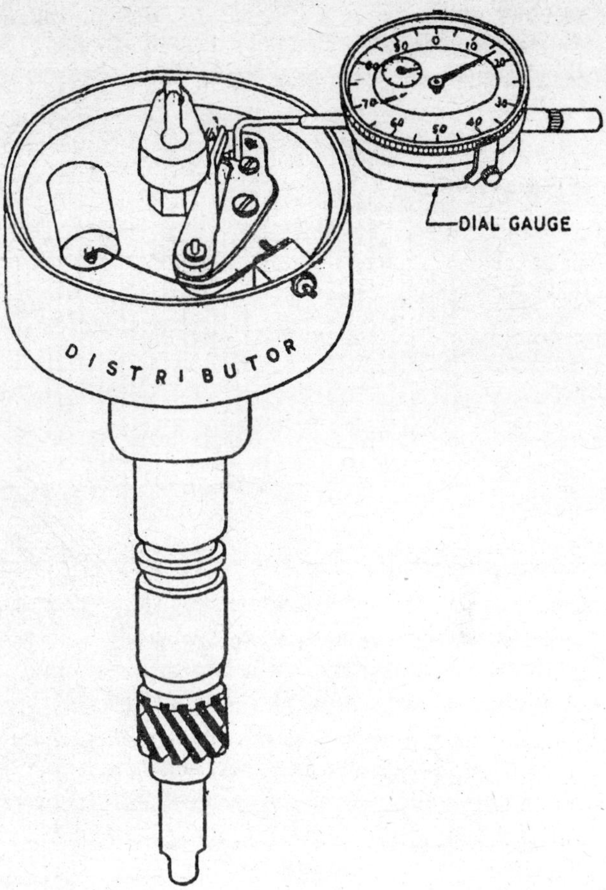

Fig. 15.9. Setting of contact breaker gap with a dial gauge

3. By dwell meter (synchronoscope)

Dwell meter is an electric instrument for measuring contact breaker point gap in terms of dwell. Dwell meters are connected to the negative side of the ignition coil to sense the opening and closing of the points. The other lead of dwell meter is usually connected to some good engine ground. A dwell meter is found to be more accurate than a feeler gauge or dial gauge for setting the point gap.

10. CONDENSER

Condenser is connected across the correct breaker. It may be considered as a kind of elastic container in which the energy due to the inertia of the current flowing during the contact period is stored. The functions of a condenser are:

1. To minimise arcing and pitting of contact breaker points.
2. To intensity the spark.

It absorbs the excess energy during the 'break' period and gives out the same at the 'make' thereby intensifying the spark and protecting the contact points as explained later.

The condenser consists of metallic plates usually of tin foil or aluminium foil separated by thin sheets of insulating paper or mica (Fig. 15.10). These insulating sheets are kept larger than the metal plates, to prevent leakage. Recently, metalised paper on which a very thin film of metal is deposited has been used instead of aluminium foil. Such a condenser is very small in size.

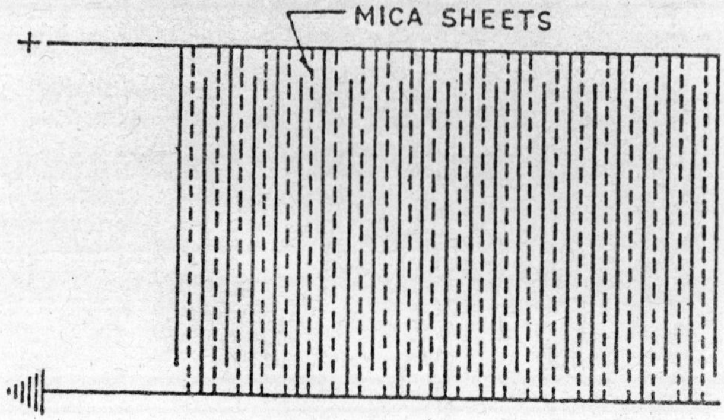

Fig. 15.10. Condenser

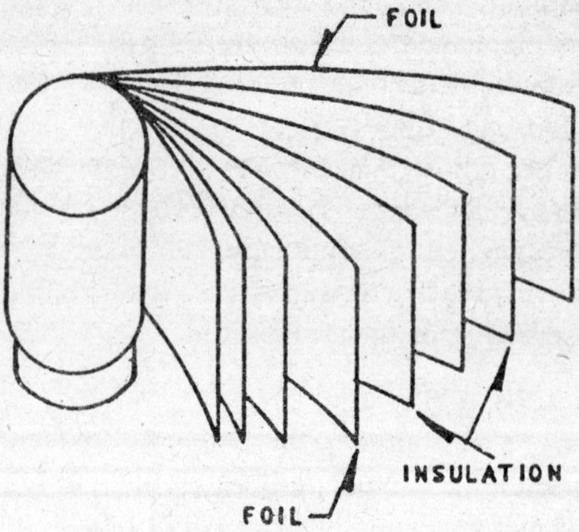

Fig. 15.11. Cylindrical condenser

In the automobile ignition systems generally the condenser used is of cylindrical form (Fig. 15.11). The principle, however, remains the same. The capacity of the condensers used in the automobiles ranges generally between 0.20 to 0.25 microfarads.

It is seen from Fig. 15.10 that no current will flow if a voltage is applied across condenser points because the two terminals of the condenser are insulated from each other. However, it is seen that when a voltage is suddenly applied to the condenser, current does flow which may either die down exponentially (Fig. 15.12) or oscillate at rate depending upon the value of the capacity of the condenser. The voltage across the condenser will also reach its full value in some time. To understand how the condenser performs its functions as given above, consider a circuit given in Fig. 15.13 (*a*). When the switch is opened, the current in the circuit will not die down immediately, but because of the self-inductance, it will take some time to do so. This means that an arc will be present across the switch points till the current due to self-induced voltage in the coil actually dies down. In other words the energy stored in the coil $\left(\frac{1}{2}I^2L\right)$ appears in the form of spark at the contact points.

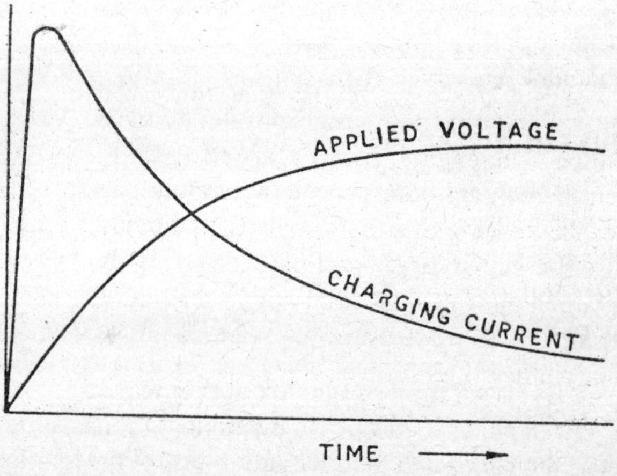

Fig. 15.12. Charging current due to voltage applied to condenser.

When, however, a condenser is connected across the contact points (switch) as shown in Fig. 15.13 (*b*), the current flowing through the coil will now flow through condenser. If the capacity of the condenser is sufficient, it will slow down the building of full voltage across the points with the result that points will

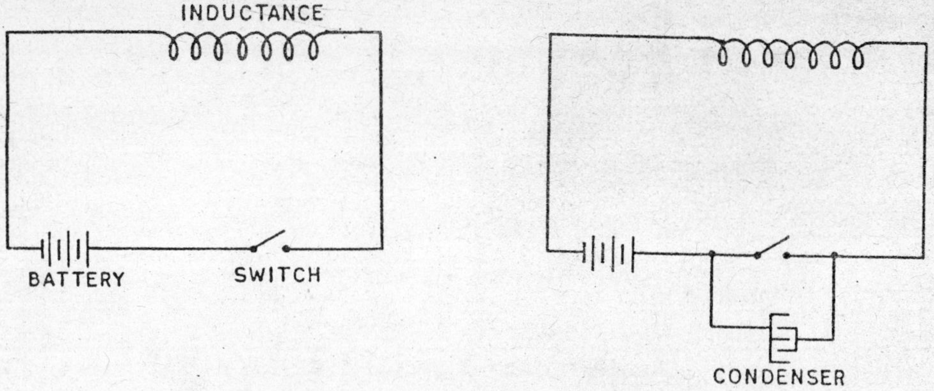

(*a*) (*b*)

Fig. 15.13. Effect of connecting the condenser.
(*a*) without condenser, (*b*) with condenser.

have separated by a distance enough for the sparking not to occur there, before the voltage has built up to dangerous value. Thus, connecting the condenser across the points avoids sparking and hence pitting. Further since the current in the coil dies down earlier a more rapid flux change is produced in the coil, which means a voltage surge of higher value in the secondary.

The best results are obtained when the condenser is placed as near the contacts as possible so that the inductance and conductance of the leads is minimum. *The condenser is therefore, always mounted in the distributor housing.*

11. DISTRIBUTOR

Situated in the same housing as the contact breaker, is the ignition distributor. However, in practice, the whole housing is called the distributor which contains the contact breaker, condenser, ignition advance mechanism and the distributor proper. The function of the distributor is to distribute the high voltage impulses to each of the sparking plugs at regularly timed intervals in the sequence of the engine's firing order. The common firing orders are:

4-cylinder in-line engine	1-3-4-2 or 1-2-4-3
6-cylinder in-line engine	1-5-3-6-2-4
8-cylinder in-line engine	1-4-7-3-8-5-2-6
8-cylinder v-type engine	1-5-4-8-6-3-7-2

The distributor proper consists of a rotor and a cap; both of these are made of bakelite. Apart from bakelite, the other materials used are alkyl plastic phenol resin, fiber glass-reinforced polyester resin plastic. All these are high dielectric strength materials. For correct location on the distributor housing, every cap has a locating tab or notch on its bottom which matches with a corresponding area on the housing. The cap is held firmly in place by spring-type clips or screws, the housing is attached to the engine block or head by means of a screw and retainer plate with a gasket in between the housing and the engine. This ensures that the housing will not move. The rotor point is made of nickel and is inserted in the bakelite casting. Rotor is fitted on the top of the shaft carrying the breaker cam. The distributor cap contains the same number of contacts as the number of cylinders. These contacts are made either from brass or nickel. The outer distributor body is usually formed of either zinc alloy or cast iron. The spindle bearing hole is bored directly along the body axis. The external surface of the distributor stem is also machined. The

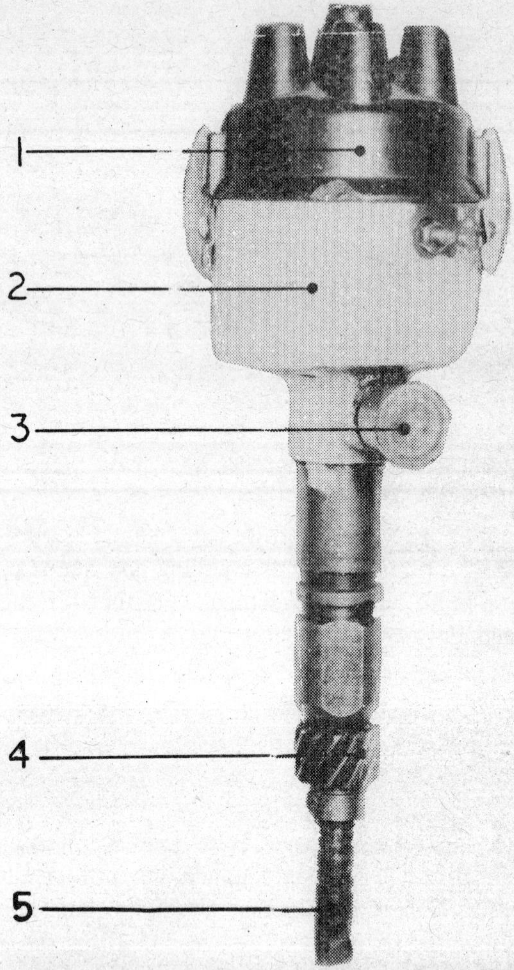

Fig. 15.14. Distributor. 1. Cap, 2. Housing, 3. Grease cup, 4. Gear for drive. 5. Extension for oil pump drive.

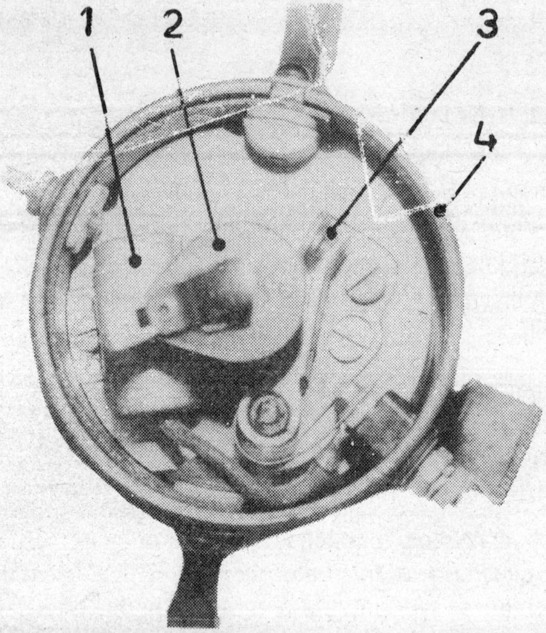

Fig. 15.15. View from top of distributor with cap removed. 1. Condenser, 2. Rotor, 3. Contact breaker points, 4. Distributor housing.

distributor is so installed on the engine that the drive connection aligns the rotor tip with the cap terminal for cylinder number 1 when the piston is at T.D.C. After this initial setting, the ignition timing still has to be set according to manufacturer's specifications. Each cap terminal is connected to a spark plug on the engine by means of h.t. cables and according to the firing order of the engine. The h.t. cable cores were previously of copper or aluminium but recent practice is to use carbon impregnated linen cores or graphite-saturated fibre-glass cores which reduce the radio and television interference.

Fig. 15.14 shows the complete outside view of distributor whereas the inner details have been shown in Fig. 15.15.

Each of the metal inserts in the cap is connected to a separate terminal from where H.T. cables lead the high voltage to the spark plugs. The central terminal is connected to the secondary of the ignition coil. As the distributor spindle rotates, the rotor point comes into contact with each terminal in the cap, thereby sending high voltage to each cylinder in turn. Alternatively, the rotor point does not actually contact the terminals, but there remains a small gap between them. This arrangement avoids mechanical wear of the contacts and also prevents the misfiring due to fused spark plugs. The role of an air gap in the spark plug circuit has been discussed in Art. 17.5.

As the distributor spindle has to rotate at half the crankshaft speed, the drive for it is taken from the camshaft through spiral gears (Fig. 15.16). Most often the distributor spindle is extended further to provide drive for oil pump also, as has been shown in the figure.

On the inside of the cap at the centre, there is a small rounded insert that contacts the rotor when the cap is on. The centre insert is made of carbon, which is a good conductor of electricity. Thus all the current coming from the ignition coil through the central tower on the distributor and carbon insert goes to the rotor.

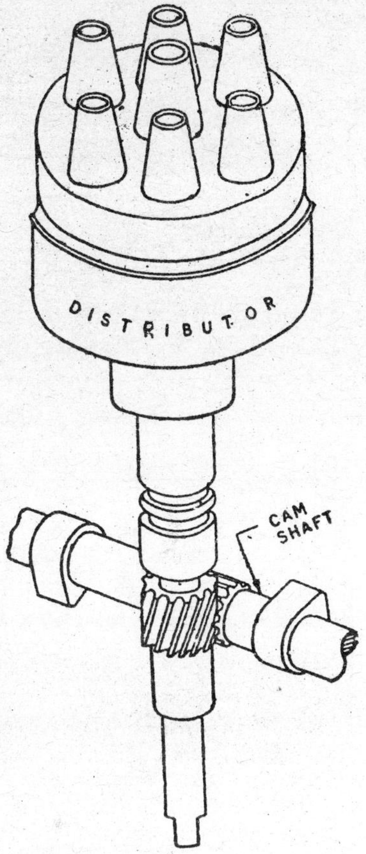

Fig. 15.16. Distributor spindle drive.

12. IGNITION ADVANCE

It is of great importance that the ignition should occur at the correct moment if the engine is to give best results. Expressed in terms of the position of the crankshaft, this moment should be such that maximum pressure occurs in the cylinder about 12° after T.D.C. Since it takes certain time called ignition delay to start the combustion after the spark, it is clear that the spark must occur prior to the moment at which maximum pressure is desired. The difference between this moment and the occurrence of spark in terms of degrees of crankshaft rotation, is termed the ignition advance. It is defined/specified with respect to the cylinder top dead position (TDC). The various factors affecting the ignition advance are:

1. Nature of fuel

Different fuels have different rates of burning and therefore the nature of the fuel determines the ignition advance required for the engine.

2. Engine speed

Obviously at higher speeds, less time is available for the combustion to start, for the same degree of crankshaft rotation. So at higher speeds more advance is required. But the turbulence increases with speed, which improves the rate of combustion, requiring lesser advance. The net requirement, however, remains the increased advance at higher speed.

3. Load

Most engines idle smoothing with an ignition advance of 15 to 32 degrees. Since other factors, apart from load, are also affecting, optimum timing is best determined by trial and error within this range. More advanced timing at speeds slightly lower than idle is sometimes used to get stable idling. This works since when the engine starts to slow down, the greater ignition advance makes the engine produce higher torque, thus increasing the engine speed. Many modern electronic systems use ignition timing mainly to smoothen idling, with increase or decrease in idle speed responded by variation of timing advance.

At smaller load, *i.e.*, at part throttle opening, lesser amount of air/fuel mixture which is also lean is drawn into the engine cylinders. On compression, this mixture produces less combustion pressure, consequently the combustion rate is slower, thus requiring more ignition advance. An ignition advance of 40 degrees or more is generally right at light loads. The hotter the camshafts, the lesser the advance required

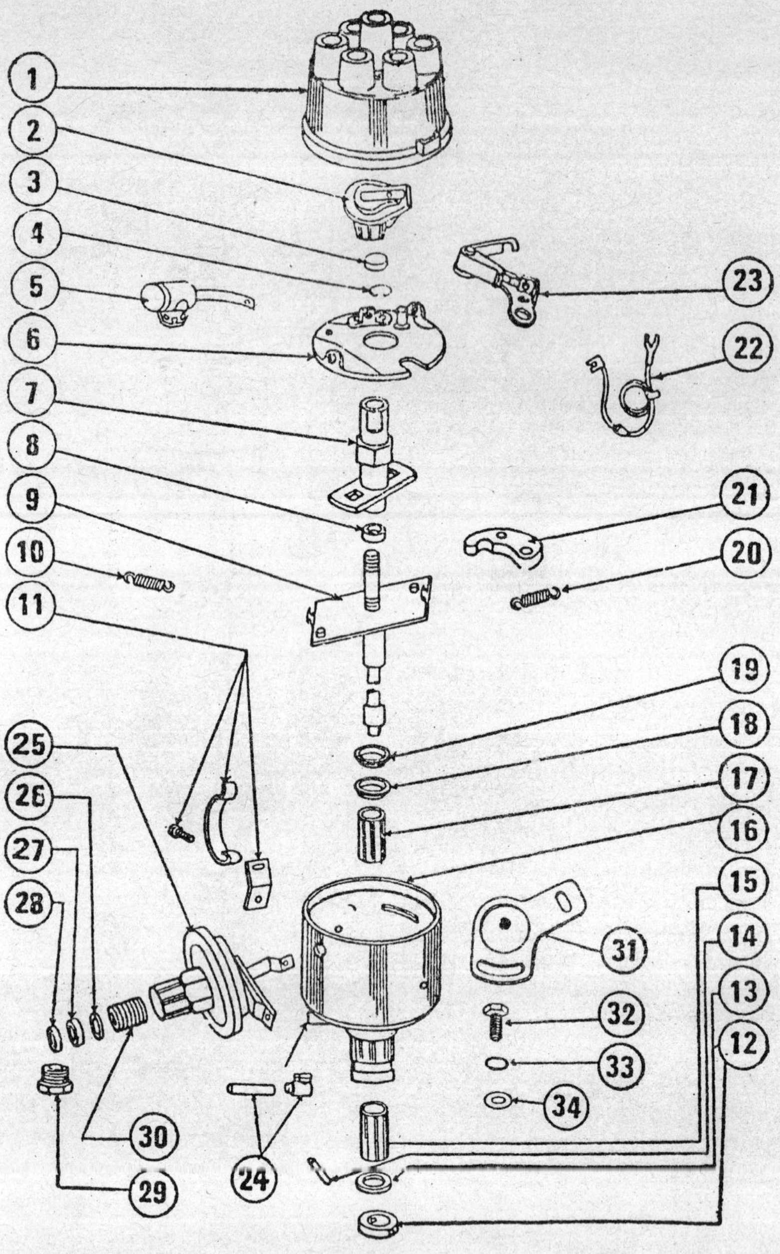

1. Distributor Cap
2. Rotor assembly
3. Cam wick
4. Cam retaining spring
5. Condenser assembly
6. Base plate assembly
7. Cam assembly
8. Step washer
9. Shaft assembly
10. Weight spring
11. Cap spring and bracket assembly
12. Shaft collar
13. Thrust washer
14. Spring pin
15, 16, 17. Base assembly with bush
18. Thrust washer
19. Thrust washer
20. Weight spring
21. Weight assembly
22. Lead wire assembly
23. Contact assembly
24. Oiler assembly

Fig. 15.17. Exploded view of Prestolite distributor design 4020 used in Fiat/Premier President cars. (Courtesy—Prestolite of India Ltd.)

at light loads; the timing in the range 35-40 degrees, is still used. Engines with good combustion chamber design will need up to 45 degrees in these conditions. However, under heavy loads, the throttle is opened fully, drawing in larger mass of air/fuel mixture which is also richer. This results in higher combustion pressure and consequently higher combustion rate requiring lesser ignition advance.

Full throttle ignition timing advance relates to the torque curve rather than the power curve. The maximum

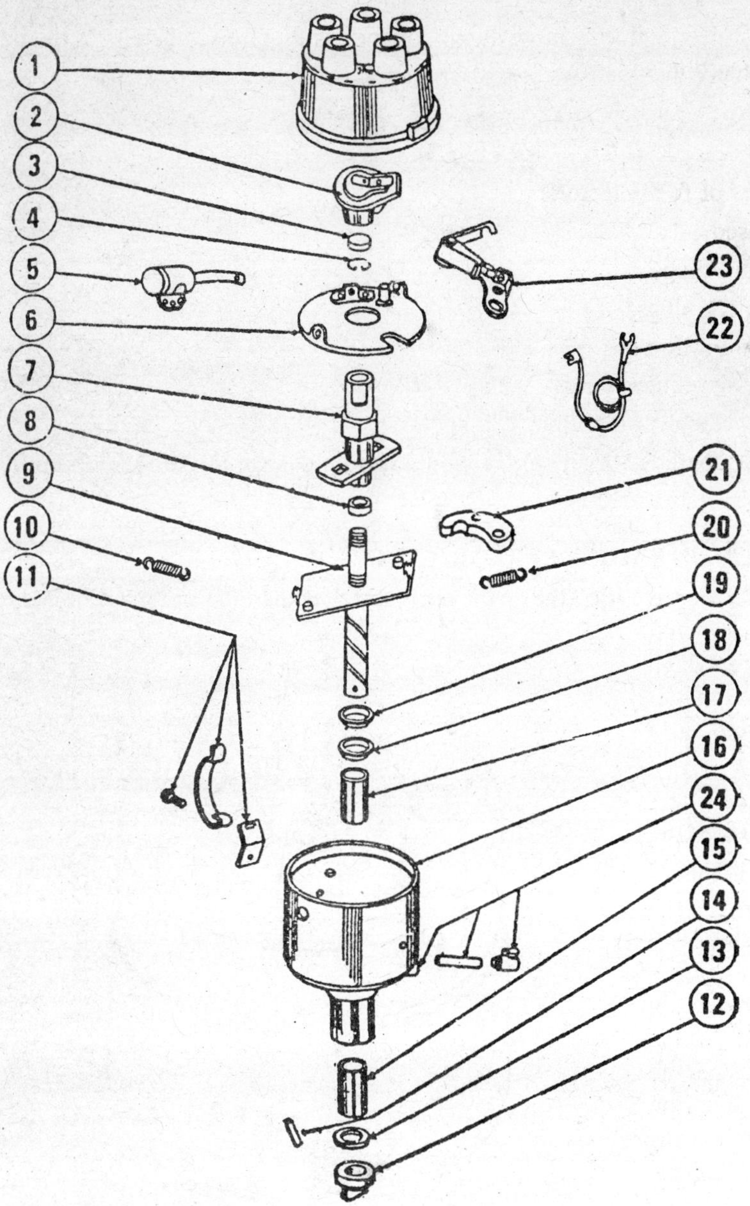

1. Distributor cap
2. Rotary assembly
3. Cam wick
4. Cam retaining spring
5. Condenser assembly
6. Base Plate assembly
7. Cam assembly
8. Step washer
9. Shaft assembly
10. Weight spring
11. Cap spring and bracket assembly
12. Shaft collar
13. Thrust washer
14. Spring pin
15, 16, 17. Base assembly with bush
18. Thrust washer
19. Thrust washer
20. Weight spring
21. Weight assembly
22. Lead wire assembly
23. Contact set assembly
24. Oiler assembly
25. Vacuum chamber
26, 27, 28. Plain washer
29. Vacuum seat plug
30. Vacuum chamber spring
31. Mounting bracket
32. Bolt
33. Spring washer
34. Plain washer

Fig. 15.18. Exploded view of Prestolite distributor ISR-402 A used in the Dodge/Fargo vehicles. (Courtesy—Prestolite of India Ltd.)

ignition timing which can be employed at peak torque is *usually* limited by the onset of detonation in naturally aspirated engines, but it is *always* the case in case of forced aspirated engines.

The ignition timing during acceleration should have sufficient attack rate to keep up with the timing requirements. *Attack rate* is the maximum permissible number of degrees per second, change, which may be as high as 650 degrees a second in some high performance engines. In case the attack rate is less, mid-range detonation may occur during transition from light load (with about 45° advance) to full throttle at peak torque (with about 15° advance only).

4. Engine Temperature

In a cold engine, combustion is slower and thus more spark advance is required, whereas in the hot engine, rapid combustion needs less ignition advance.

5. Other factors

(*i*) Compression ratio.

(*ii*) Timing of the intake valves or fuel injectors.

(*iii*) Type of ignition system used.

(*iv*) Type and condition of spark plugs.

(*v*) Position and number of spark plugs.

(*vi*) Design and size of the combustion chamber.

(*vii*) Engine coolant and intake air temperature.

(*viii*) Turbo boost pressure or intake air pressure.

Generally, any major chambers or upgrades of an engine would need a change in the ignition timing settings of the engine.

13. EFFECT OF IGNITION ADVANCE ON ENGINE PERFORMANCE

1. In case the ignition advance is excessive, *i.e.*, the spark occurs earlier than the optimum moment, it may actually push against the piston when it is still compressing the air/fuel mixture, which may result in detonation, power loss, higher combustion temperatures, backfire and early engine failure.

2. If, however, the ignition advance is less than the optimum, *i.e.*, the spark occurs too late, the ignition would occur when the piston is actually moving down during the expansion stroke. This would result in unburnt fuel, wasted energy, and loss of power.

3. NO_x emission increases with increase in ignition advance. However, emission of CO is not much affected by ignition timing; rather the same is influenced by air/fuel ratio. Emission of hydrocarbons at stoichiometric and rich air/fuel ratios increases with increase in ignition advance. However, there is very little effect of ignition advance on hydrocarbon emission in case of very lean air/fuel ratios, *e.g.*, 19:1 and higher.

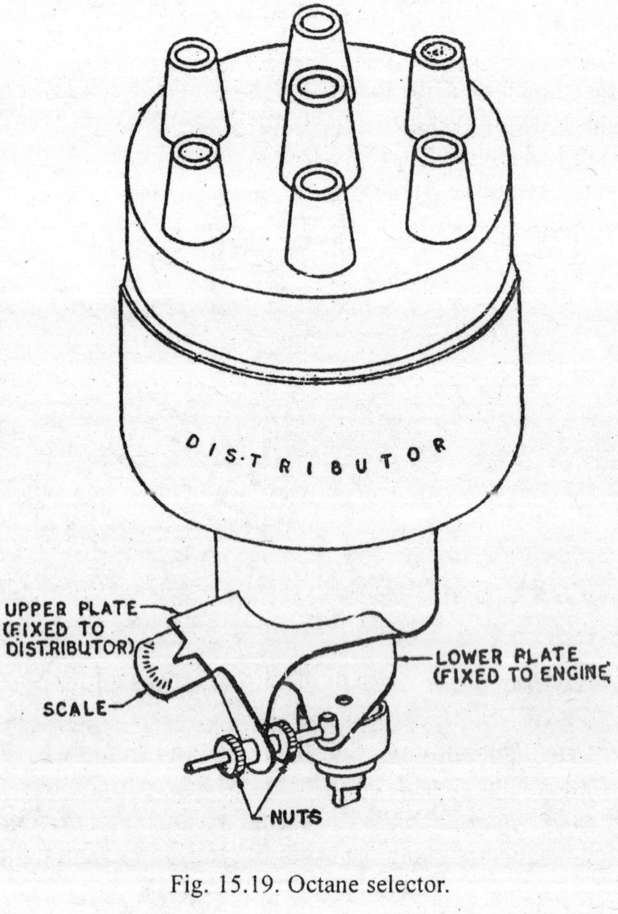

Fig. 15.19. Octane selector.

14. IGNITION ADVANCE METHODS

There are basically two methods available for adjusting ignition advance:

1. Manual. The ignition timing is controlled by hand only by means of a lever at the dash-board. The distributor has a provision that contact breaker base plate can be rotated within a certain degree. Turning the breaker plate in cam's direc- tion of rotation results in retarded ignition, turning in opposite direction in advanced ignition. With manual control, however, the timing of ignition is not satisfactory, because adjustment is normally done *after* the engine gives poor performance and not *at* the proper time. But the manual control is useful for the 'presetting of ignition advance when different fuels may be used. On actual auto- mobiles, however, this method is now rarely employed.

An octane selector to preset the ignition advance is shown in Fig. 15.19. The upper plate is fixed to the distributor while the lower one is attached to the engine block. When the nuts (shown in figure) are rotated, the upper plate is also rotated along with. The pointer attached to the upper plate then shows the rotation of the upper plate and hence of the distributor. Depending upon the requirements of a particular fuel, the ignition advance is thus preset.

2. Automatic advance mechanisms.

15. AUTOMATIC IGNITION ADVANCE METHODS

1. Centrifugal Advance

This method provides the ignition timing control according to engine speed. The distributor shaft, from its gear (lower) end to the rotor-carrying (upper) end is not a single piece. It is actually in two pieces connected together through the centrifugal advance arrangement, which consists of two centrifugal weights. *A* and *B* as shown in Fig. 15.20. These weights are pinned and connected with the breaker plate *C* by means of springs. As the engine speed increases, the weights move out due to the centrifugal force. Thus the weights turn about their fulcrums and by so doing they turn the breaker plate and cam and thus the upper shaft piece in the anticlockwise direction with respect to the lower portion of the distributor shaft. This causes the cam to open and close the contact points earlier thus providing the desired advance. As the speed increases, the centrifugal force on the weights increases, resulting in increased advance, till the weights touch some post which limits their movement thus limiting the maximum advance. The centrifugal advance variation with speed is shown in Fig. 15.21. This is true when both the springs loading the centrifugal weights have the same spring stiffness.

In some designs, however, the springs of unequal stiffness are used. The spring of higher stiffness is

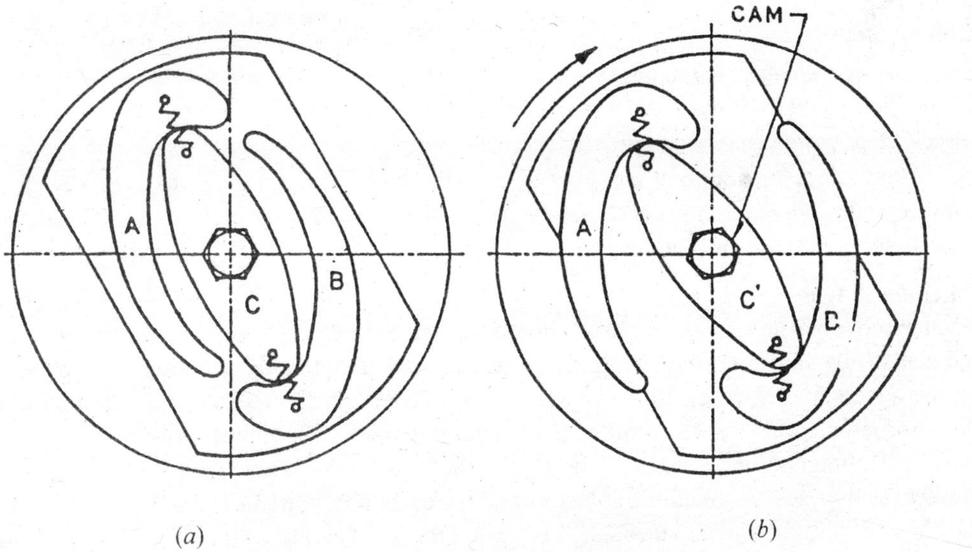

Fig. 15.20. Centrifugal advance mechanism.
(*a*) No advance, (*b*) Full advance

kept loose so that at lower speeds it does not pull the centrifugal weight to which it is attached. Only at a higher speed when the spring looseness is taken up, it starts loading the weight. With such an arrangement, therefore, a differential rate of ignition advance is provided (Fig. 15.22). The need to have reduced advance rate at higher speeds arises on account of increased heat and turbulence generated within the combustion chamber at higher speeds.

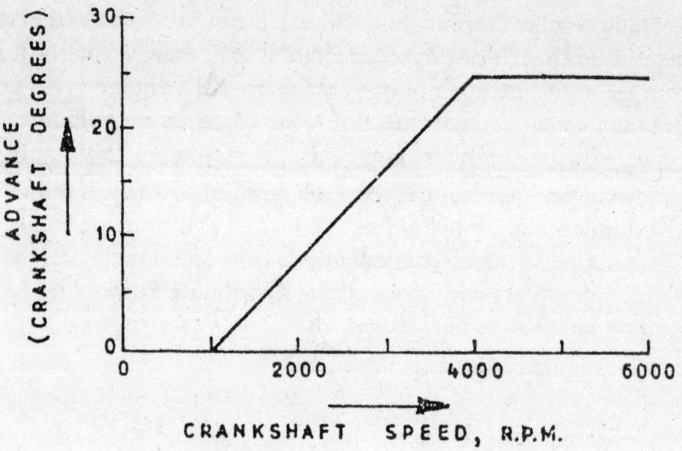

Fig. 15.21. Variation of centrifugal advance with speed.

The distributor shaft drive is timed to a specific crankshaft angle when the advance mechanisms are in full retard. This is called *initial (or basic) timing.* The exact centrifugal advance varies with different distributor designs. However, in general, it varies between 34 and 38 crankshaft degrees, including the initial timing, which is usually about 8° BTDC. For most engines, the maximum advance is reached at about 4000 rpm.

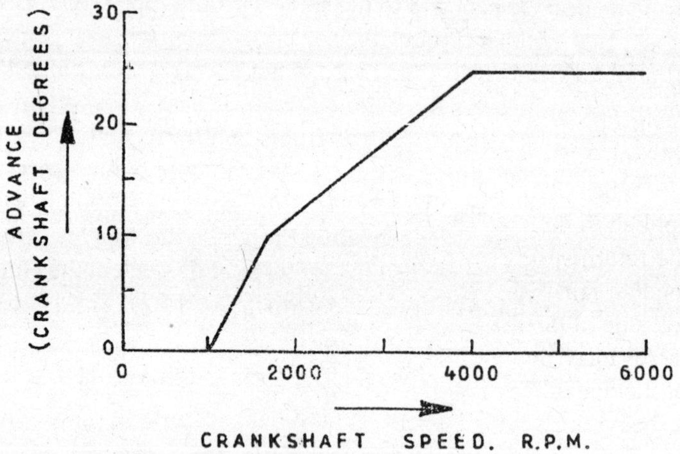

Fig. 15.22. Differential rate centrifugal advance.

It is obvious that if due to any reason (*e.g.* rusting or sludge built-up), the centrifugal advance mechanism is not working, poor engine performance will result. If the weight springs have become less stiff or either or both springs are broken, excessive advance will result causing detonation to occur.

If the centrifugal advance mechanism does not function properly it may result in the following:

1. Lack of power during acceleration.
2. Constant knocking sound at low speeds and slow, jerky cranking due to over-advanced ignition.
3. Increase in fuel consumption.
4. Increase in exhaust emissions.

2. Vacuum Advance

A vacuum advance unit senses any variation of ported or manifold vacuum, which are direct measure of engine load. Thus this unit provides ignition advance according to engine load and entirely independent of the engine speed. Manifold vacuum is the depression in the intake manifold and is available at a port on the carburettor situated below the throttle valve closed position. The manifold vacuum during light load conditions like idling and deceleration is highest while it is lowest under heavy engine load. The ported vacuum refers to the vacuum available at the port on the carburettor situated above the fully closed throttle position of the carburettor and is thus controlled by the throttle valve position. Thus it is zero during idling and deceleration when the throttle valve is closed. As the throttle valve is opened, the ported vacuum increases gradually to the same value as the manifold vacuum.

The vacuum advance unit contains a rubber diaphragm which is connected to the base plate of the distributor. One side of the diaphragm is open to atmosphere while the other side is subjected to the vacuum from the carburettor. One such unit operating on the ported vacuum is shown in Fig. 15.23. A return spring

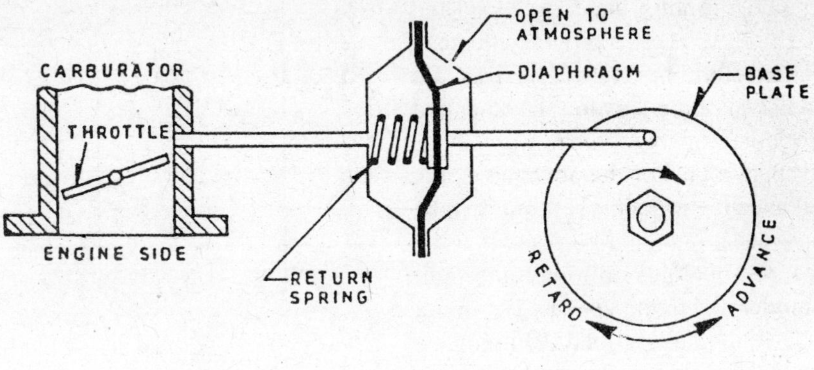

Fig. 15.23. Vacuum advance mechanism

on the vacuum side of the diaphragm keeps the advance unit at zero advance if no vacuum is applied as during idling. When, however, the throttle is opened the vacuum is applied which moves the diaphragm against the return spring to rotate the distributor base plate in the direction opposite to the usual distributor rotation, causing the ignition to advance.

When the engine is idling, the throttle valve is closed, thus preventing the engine vacuum from being applied to the diaphragm. Both sides of the diaphragm are then subjected to atmospheric pressure so that the return spring keeps the ignition setting fully retarded. When however, the engine is running at light load, at fairly high speed, the throttle is partly open with the fair amount of engine vacuum applied to the left side of the diaphragm. Due to this spring compression force would be overcome and the timing would be advanced proportional to the vacuum applied. A stop is, however, employed to keep the maximum movement of the diaphragm and hence the ignition advance at the desired value. If, however, the engine is subjected to heavy load with throttle wide open, the vacuum in the induction manifold will be very small, which would not be sufficient to press the return spring. Thus no advance would be provided under this condition.

Such units provide ignition advance of about 14 to 30 crankshaft degrees.

If the vacuum advance mechanism connected to ported vacuum is not working properly it will result in increased fuel consumption. However, if the vacuum advance controlled by manifold vacuum fails to work properly, apart from increa- sed fuel consumption, the following may also result:

1. Rough idle, sometimes stalling.
2. Knocking sound only at part-throttle.

Besides, if the vacuum leaks on account of a punctured diaphragm, it may results in weak air-fuel mixture, poor engine performance and engine overheating.

A distributor with a vacuum advance unit is shown in Fig. 15.24.

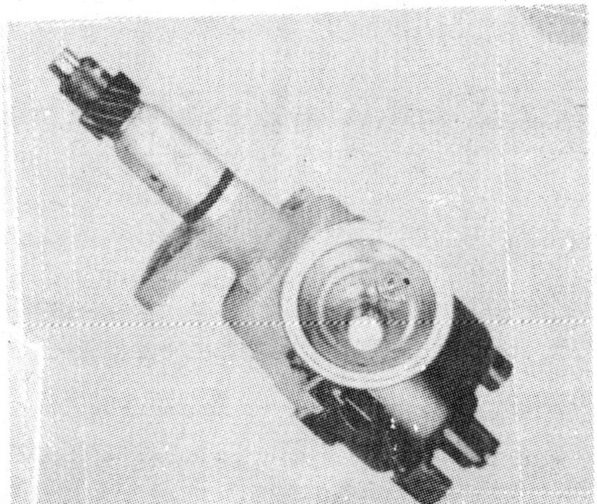

Fig. 15.24. Distributor with vacuum advance unit. (Courtesy—Lucas-TVS, India)

3. Centrifugal and vacuum advance combination

It has been observed that the centrifugal advance mechanism takes care of speed only and not the load conditions, which are looked after by vacuum advance mechanism. Further, the vacuum advance is independent of centrifugal advance and this mechanism operates only when the engine is running under light load and at moderately high speed so as to provide adequate engine vacuum. Therefore the two mechanisms are independent of and complimentary to each other and are provided together on the distributors. At a particular speed then, a part of the advance is due to the centrifugal advance while the rest of it will be due to vacuum advance. It should be noted that the vacuum advance unit operates only with the light engine load at moderate engine speeds. Beyond a certain engine speed, the pulling-load requirements increase and for that the throttle has to be opened gradually which causes the vacuum advance to decrease from its maximum value until it becomes zero (Fig. 15.25). In most of the distributors, the combined maximum advance including the basic timing, centrifugal advance and vacuum advance varies between 55° and 60° B.T.D.C. It must be remembered, however, that this combined maximum advance is not likely to actually occur since the maximum centrifugal and maximum vacuum advances may not occur at the same time.

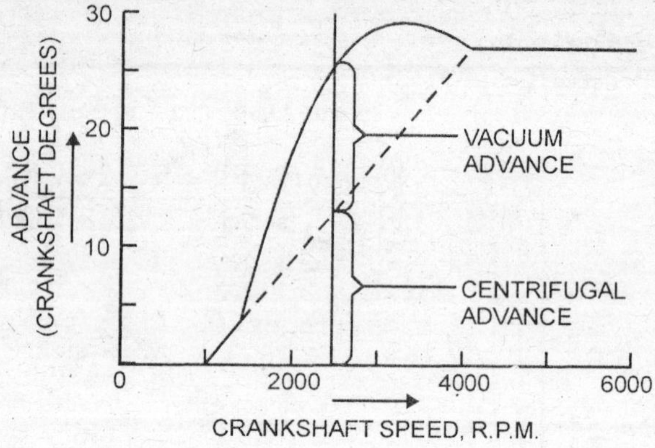

Fig. 15.25. Combined centrifugal and vacuum advance.

4. Complete vacuum control

Distributors have also been designed in which no centrifugal advance mechanism is provided and the entire ignition advance is by means of engine vacuum as described above. In such systems there are two vacuum openings in the carburettor: one is at the venturi and the other one is just on the atmospheric side of the throttle when the same is closed. Both openings are joined through passages and further connected to one side of the diaphragm in the vacuum advance mechanism of the distributor which is similar in construction to the one described in Fig. 15.23. The vacuum in the venturi increases with the increase of engine speed and thus provides the ignition advance proportional to speed. Thus it performs the same function as the centrifugal advance mechanism. The advance due to load is provided by the second opening at the throttle valve as usual. Thus the system eliminates the need for separate centrifugal advance mechanism.

5. Computerised advance control

Modern engines use computerised ignition system. The computer has in its memory, a timing map, which contains spark advance values for all combinations of speed, load and other conditions. Based upon the signals received from various sensors, the computer ECU sends a signal to the ignition coil at the precise moment indicated in the timing map, which fires the spark plug at the right instant. In most of the original systems, the timing map and hence the timing advance curve cannot be modified. However, aftermarket ECUs mostly allow the changes to be made later in the timing map.

16. IGNITION TIMING

Ignition timing is the exact instant a spark is sent to the spark plug and is expressed in degrees of crankshaft rotation relative to TDC. It is set initially by positioning the distributor body relative to one of the contact breaker cam lobes. It has already been explained that the spark occurs at the instant the primary circuit of the ignition system starts to open. Invariably cylinder number 1 is used as a reference, *i.e.*, the engine design is such that if the timing of cylinder number 1 is correct, it is correct for all the other cylinders also. In most engines the spark is designed to occur slightly before the top dead centre (abbreviated as so any degrees BTDC) to take into consideration the time taken for the fuel to burn.

If the ignition of an engine is not adjusted correctly, the spark may occur either before or after it should. In the former case it would result in difficult starting, poor performance, increased fuel consumption and slow, rough idling. On the other hand in the later case, slow and jerky cranking with the warm engine may occur. Detonation may also be experienced while accelerating.

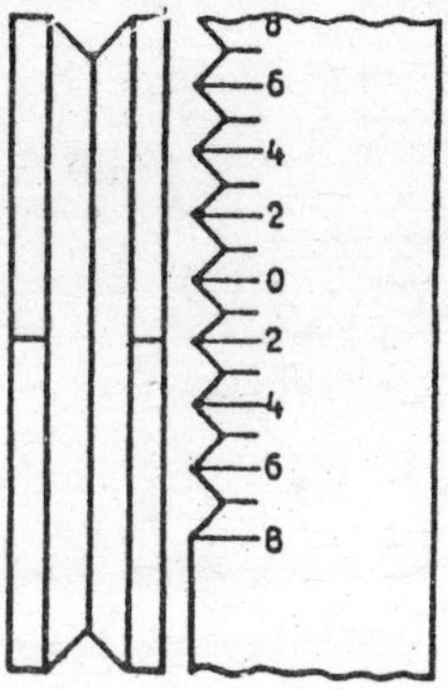

Fig. 15.26. Timing mark location on crankshaft pulley

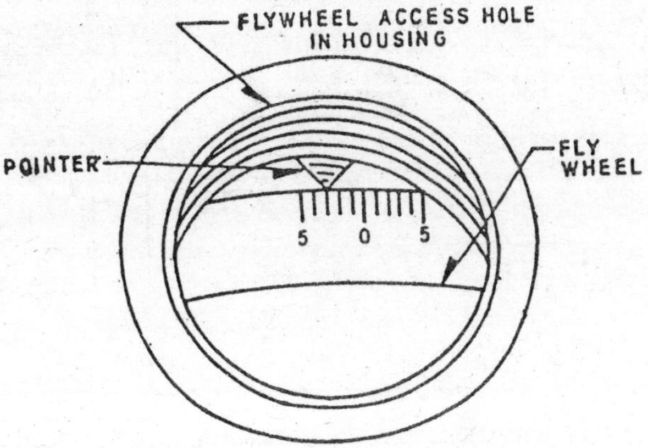

Fig. 15.27. Flywheel timing marks.

For adjusting ignition timing, a timing mark is provided on the crankshaft pulley, which must be opposite to the specified degree marked on some stationary part of the engine (Fig. 15.26). Typical examples of these are engines for Audi, Ford, Lincoln, Mercedes-Benz, Mazda, Toyota, Volvo. Alternatively the crankshaft degrees are marked on the flywheel while a stationary pointer is attached to the engine block (Fig. 15.27). The typical examples of these are the engines of BMW, Chrysler, Dodge, Honda, Renault.

At the instant the timing mark is correctly aligned, the spark must be sent to the spark plug of cylinder number 1, which means that the distributor contact points must start to open at that instant.

Before checking or setting ignition timing, it must be ensured that:

(*i*) the dwell must be correct.

(*ii*) the engine is running at its normal temperature.

(*iii*) the engine is running at specified timing speed.

(*iv*) the vacuum hose is disconnected from the vacuum advance unit.

Timing light is an important accessory for checking or setting ignition timing (Fig. 15.28). The power for a timing light is usually taken from the engine battery (Fig. 15.29).

Fig. 15.28. Model PTL-51 portable timing light. (Courtesy—Sun Electric Corp., U.S.A.)

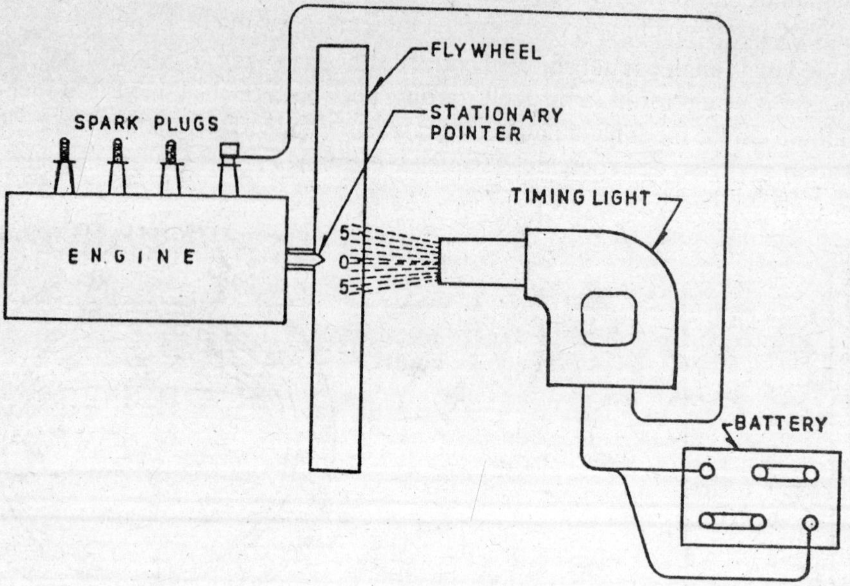

Fig. 15.29. Use of Stroboscopic light for setting ignition timing.

One terminal of the light is also connected to the spark plug number 1. The light goes off whenever the plug fires. Thus the light flashes off at half the engine rpm. If this light is now flashed onto the flywheel marks, they will appear to be stationary due to stroboscopic effect.

Thus the following procedure for checking or adjusting ignition timing may be followed:

1. Start the engine and see that the same has reached the normal operating temperature.
2. Connect the timing light to cylinder number 1 as explained above and aiming it on the crankshaft pulley or the flywheel, as the case may be, check whether the stationary pointer is opposite the specific mark.
3. If not, loosen the distributor locking bolt or nut and turn the distributor in the desired direction so that the stationary pointer comes opposite the specified timing mark. (For advancing, rotate the distributor in the direction opposite to that to the crankshaft rotation and for retarding, rotate the distributor in the direction of crankshaft rotation).
4. Lock the distributor bolt/nut and recheck the timing with the timing light. If required, readjust and check to ensure that it has been set correct.

17. SPARK PLUG

Next to contact breaker, spark plug is probably the weakest part of the ignition system. It is mounted in the combustion chamber of the engine, where working conditions are severe. During peak combustion conditions the temperature of gases in the combustion chamber in a modern car engine may be around 2500°C and the pressure about 7 MPa. Moreover, a spark plug is also exposed to thermal and load cycling fatigue due to sudden changes in temperature and pressure-from the high temperature of burnt gas to the relatively low temperature of the air/fuel mixture and from the high pressure at the time of explosion of the air/fuel mixture to low pressure during induction. In addition, the spark plug has to endure high voltage, mechanical vibration and the corrosive atmosphere of combustion gases. A modern spark plug has an economical life of about 40,000-100,000 km.

The minimum voltage required to make the spark jump across the air gap depends upon the following:

1. Compression pressure– With the increase of compression pressure, the air density in the cylinder is increased so that greater voltage is required for the spark to jump across the plug electrodes.
2. Mixture strength– Minimum voltage is required to make the spark jump in case of stoichiometric air-fuel ratio. If the mixture is weaker or richer than this, a higher voltage is required.
3. Air gap– As the gap between the plug electrodes increases, the voltage required to jump the gap is increased considerably.
4. Electrode temperature– With the increase of temperature of the central electrode, the voltage required to produce the spark decreases rapidly between 200° and 600°C and a slower rate thereafter.
5. Vehicle speed and load– Plug voltage requirement curve is like a typical engine torque curve (Fig. 15.30). The voltage required rises initially with the increase of speed as the engine's valve timings improve the cylinder filling, thereby raising the compression pressure. However, at higher speeds, the voltage required decreases due to cylinder breathing becoming more difficult, resulting in less dense charge induced. Voltage requirement also depends upon the load, *i.e.*, the throttle opening. For example, when the throttle is suddenly opened wide to accelerate the vehicle, the extra mixture charge entering the cylinder briefly raises the cylinder pressure so that more voltage is now required for the spark to jump at the pump electrodes.

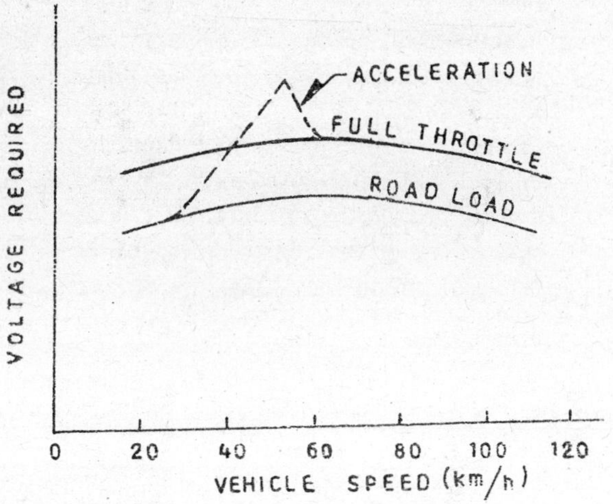

Fig. 15.30. Effect of vehicle speed and load on the voltage.

17.1. Requirements

The requirements of a good sparking plug are:

1. Very high resistance to current leakage.
2. Continued maintenance of the proper gap under all conditions.
3. Gas tightness—Any leakage of hot gas will upset the normal steady state of the plug, causing such high temperatures to be reached that the insulator and the electrode disintegrate and fall into the cylinder.
4. Resistance to corrosion—Corrosion is detrimental because it increases gap length and thus raises the sparking potential required.
5. Sufficient 'reach' into the combustion chamber—Reach is defined as the length of the threaded portion that enters the cylinder. If a long-reach plug is fitted into a short-reach hole, combustion chamber space is reduced which increases the compression ratio, more heat is produced and the exposed edges of the threaded portion may be heated to incandescence, resulting in preignition. Moreover the protruding end of a long reach plug in a short reach hole may also damage the piston or valve. If a short-reach plug, on the other hand, is fitted into a long reach hole the combustion-chamber space is increased, compression ratio is reduced and carbon is deposited in the threads of the hole which makes difficult the fitting of a correct reach plug at a later date. This has been made clear in Fig. 15.31.
6. Electrode temperature must be maintained between certain limits. At higher temperatures (900°C or above) there is danger of preignition and also corrosion of electrodes. At temperatures lower

than 350°C, however, carbon or even oil deposits are formed causing the leakage of electricity to earth instead of sparking across the plug electrodes. The optimum temperature range is 500-800°C, which is maintained by proper heat range of the plug. The heat range is defined as the ability of the spark plug to dissipate heat by transferring the same to the engine cooling system. It depends on (*i*) the insulator nose length, (*ii*) gas volume around the insulator nose and (*iii*) the materials/ construction of the centre electrode and the insulator.

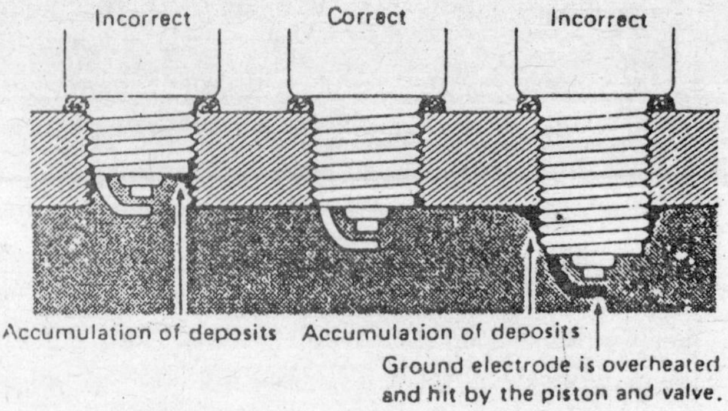

Fig. 15.31. Use of correct reach plug.
(Courtesy—Maruti Udyog Ltd., India)

17.2. Construction

Figs. 15.32 and 15.33 show clearly the construction of a spark plug. The plug has three main parts, the centre electrode, the ground electrode and the insulator separating them. Besides these, there are the body shell, the sealing ring and the gasket washer. The upper end of the centre electrode is connected to the spark plug terminal, where H.T. cable from the ignition coil in case of single cylinder engines (or from distributor in case of multi-cylinder engines) is connected. The lower end of the centre electrode projects

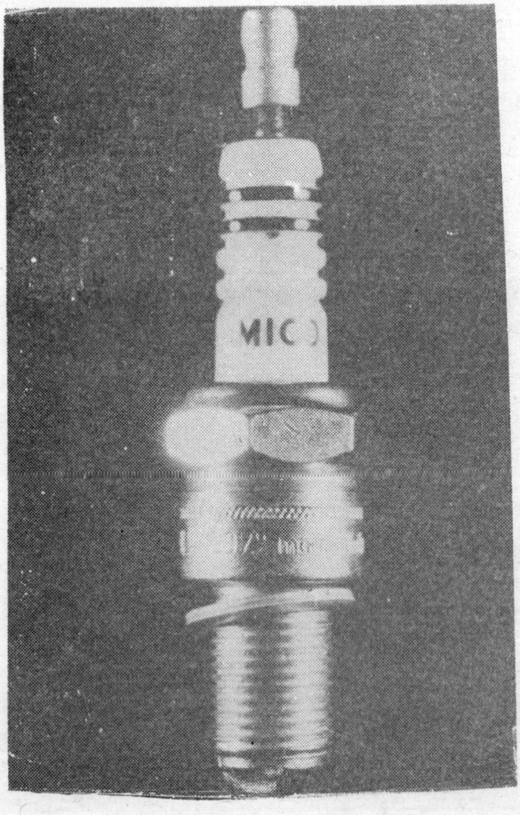

Fig. 15.32. Mico-Bosch spark plug.

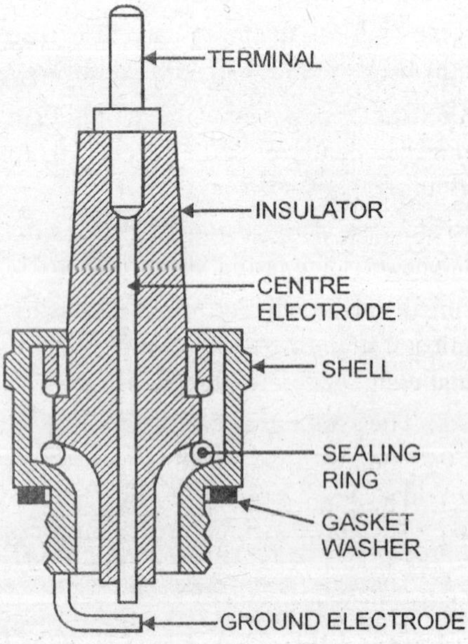

Fig. 15.33. Spark plug.

beyond the insulator to form a gap with the ground electrode. The insulator is meant to fulfill the following functions:

(*i*) To insulate the centre electrode from body shell, thereby preventing the leakage of high voltage surge from leaking to earth within the shell.

(*ii*) To control the working temperature of the centre electrode by suitably adopting the thermal conductivity of the insulating material, its shape and the length of the heat path, while designing the spark plug for a given engine.

The *insulator nose length* is the distance from the firing end of the insulator to the point where insulator meets the metal shell. The insulator tip is the hottest part of the spark plug, which makes the tip temperature, the main factor in pre-ignition and fouling.

The body shell serves to house the electrodes and the insulator. Gas-tight seal is necessary to prevent the hot gas from leaking between the insulator and the body shell and between the insulator and the centre electrode. Such seals may be of different types, *e.g.* solid ring, dry powder, metal powder fused into glass, etc. Besides above the hot gas from the combustion chamber may also leak between the plug and the cylinder head. To seal this a flat ring gasket washer is commonly used.

In some modern spark plugs, the centre electrode is made in two pieces. By doing so the designer can use different metals to suit best the different requirements of the upper piece which has to be connected to the h.t. cable and the lower piece which has to go into the combustion chamber.

The main dimensions of a spark plug are shown in Fig. 15.34.

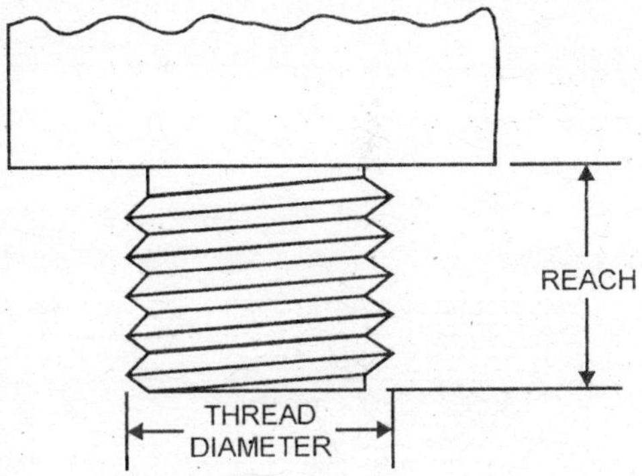

Fig. 15.34 Main dimensions of a spark plug.

In some spark plugs, centre electrode contains a built-in resistance in series with the electrode. This resistance reduces the number of high voltage surges accompanying the full voltage surge for the spark, thereby reducing the radio and television interference due to ignition system and also providing protection to the on-board computer. The resistance also increases the plug life by cutting down peak current that would burn the electrodes. However, the resistance increases the voltage required to jump across the spark plug gap. Such plugs are called *resistor spark plugs* and their life is about double the life of ordinary plugs.

17.3. Materials

The body shell is generally made of low carbon steel. The body is formed by impact extrusion. The threads are cold rolled to provide a low-friction profile. Porcelain was used initially for insulation, but this has the disadvantages of brittleness and low resistance to thermal shock. Porcelain was therefore replaced by mica. The use of mica, too, was restricted by leaded fuels which attached it. Present day insulators are almost exclusively of sintered alumina. These possess some distinct advantages:

1. They are much stronger than porcelain.
2. Their thermal conductivity is higher than in a case of porcelain.
3. These possess high electrical resistivity at operating temperatures.
4. These can be moulded accurately and with ease, to the desired shape.
5. These have a high resistance to abrasion, erosion and chemical attack from the products of combustion. Thus with these materials, the plug can be cleaned by sand blasting in contrast to mica and porcelain plugs which are eroded by sand blasting.

The ground electrodes are usually formed of nickel or sometime an alloy of nickel and manganese.

The addition of manganese (2.5—3.5 per cent) improves the tensile strength and the resistance to sulphur attack at high temperatures. For centre electrodes silicon-manganese-nickel alloys are used. These are particularly resistant to the high temperature effects of leaded fuels. Copper-cored central electrodes are also used which increase the heat range of a plug. A nickel alloy jacket around the copper core, due to its very good anti-corrosion and anti-erosion properties, protects it from the aggressive combustion environment.

Platinum tips are also used for some spark plugs due to the very high burn-off resistance of this material. This property also makes it possible to use electrodes of much smaller diameter, thus increasing mixture accessibility. Platinum is also found to have a catalytic effect which further accelerates the combustion process. However, higher cost inhibits their use in common automobiles.

Modern spark plugs use rare-earth element yttrium for the centre electrode. Yttrium provides excellent wear resistance of the electrodes at higher temperatures. The Nickel-yttrium alloy developed by Bosch increases appreciably the service life of the spark plug. Moreover, since in this case, the spark gap remains nearly constant for entire service life of the plug, it also results in consistent performance and increased life of the catalytic converter.

17.4. Types

The spark plugs may be long reach or short reach type depending upon the length of threaded portion (Fig. 15.35) and should be used only in the corresponding hole in the combustion chamber as described earlier.

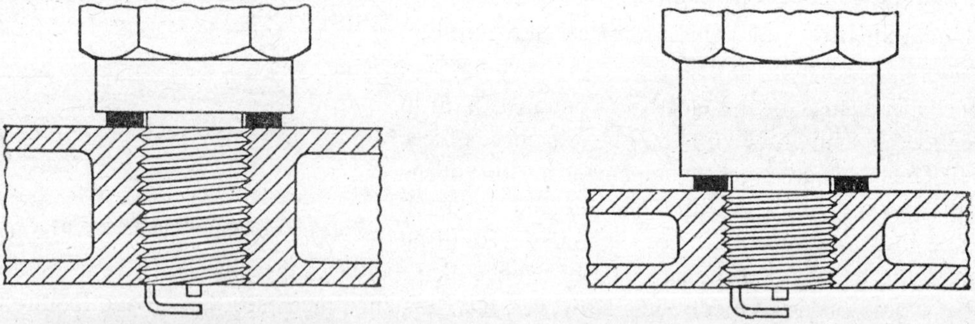

(*a*) (*b*)

Fig. 15.35. (*a*) Long reach plug, (*b*) Short reach plug.

From heat dissipation point of view also the spark plugs are also divided into two classes, *viz.*, 'hot' or 'hard' and 'cold' or 'soft'. Hot plug runs hotter than the cold plug because the path of heat dissipation to the coolant in the jackets is longer in the hot plug than in the cold plug (Fig. 15.36). Some cold spark plugs have a copper core in the centre electrode to help carry heat from the tip of the electrode (Fig. 15.37). Cold plugs are generally used in heavy duty, high speed engines where high temperatures are encountered. Lower speed, medium duty and colder operating conditions require a hot plug.

17.5. Air gap

The correct spark plug gap size is specified by the manufacturers but in general it varies from 0.6 mm to 1 mm. For the new plug gap is generally 0.6 mm, but with use it is increased, thus requiring higher voltage for the spark to jump across it. On the other hand the carbon deposits on the plug electrodes provide alternative paths for the current to pass, thereby limiting the maximum voltage applied across the gap to a lower value; the result is voltage surge of lower intensity which may not be sufficient to make the spark jump across the plug points, thus causing the plug to mis.

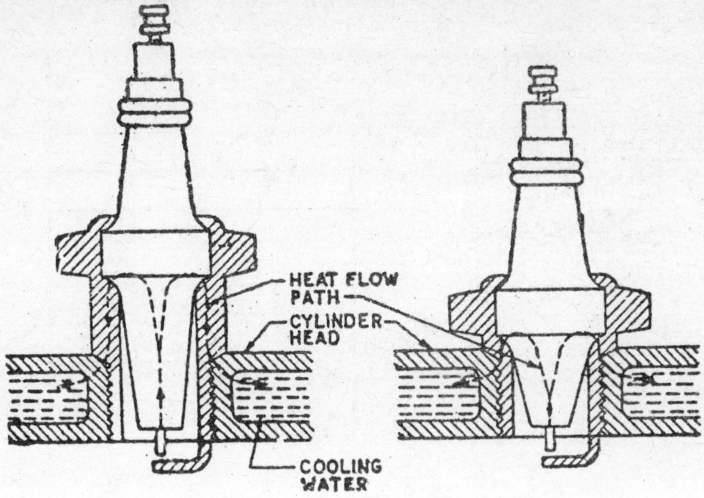

Fig. 15.36. (*a*) Hot plug, (*b*) cold plug.

If an auxiliary air gap is introduced in the plug circuit, the same will help the voltage to build up to a high value sufficient to jump across the air gap. If the gap size is sufficient, obviously the voltage would be enough to cause sparking at the plug points. It is this basic principle of air gap only, which is made use of in the so called 'spark intensifiers', which contain simply two screws providing an adjustable gap.

17.6. Surface discharge plug

This is a special type of spark plug where the space between the plug electrodes is filled with some semi-conductor material. Thus the spark has to jump across this material instead of air in the ordinary plugs.

The advantages of this types of plug are the more intense and efficient spark which results in fuel economy, increased output and improved starting. The plug is claimed to be not effected by carbon deposits on the electrodes. The only disadvantage is its higher cost.

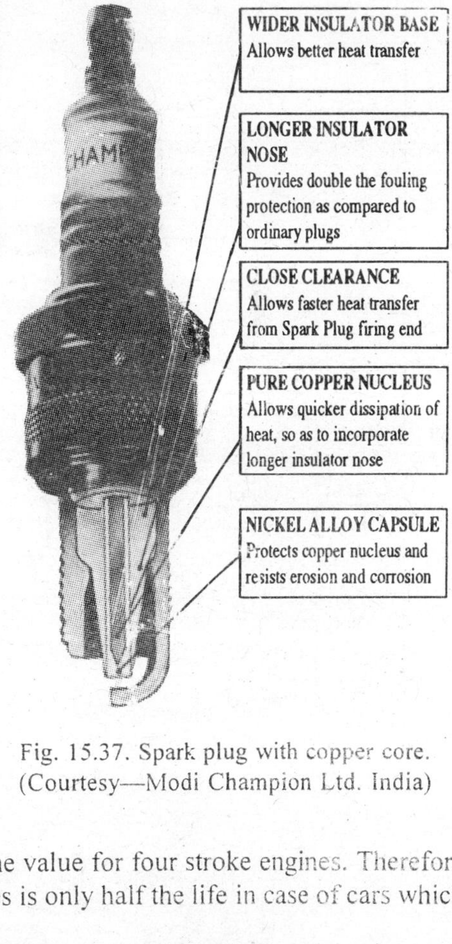

Fig. 15.37. Spark plug with copper core. (Courtesy—Modi Champion Ltd. India)

17.7. General

1. The spark plug sizes have been standardized. 14 mm diameter thread size is most commonly used. 18 mm diameter plugs are also used, however.

2. Higher temperatures are encountered in air cooled engines than the water cooled ones, due to which spark plugs for the motor cycles and scooters have to be of higher heat value than the comparable car engines.

3. The frequency of firing in two stroke engines is twice the value for four stroke engines. Therefore expected spark plug life in motor cycles with two stroke engines is only half the life in case of cars which

New Type Designation—Spark Plug

W 8 D C

M	M18 × 1.5
W	M14 × 1.25
U	M10 × 1

THREAD DIAMETER AND DESIGN FEATURES

SPECIAL TYPES

New	Old
13	25
12	50
11	75
10	100
9	125
8	150
7	175
6	200
5	225
4	250
3	275
2	300
09	325
08	350
07	375
06	400
HEAT RANGE	CODE NO.

SPARK POSITION (Insulator nose) AND THREADED LENGTH

Code	Spark position	Thread length mm
A	Normal	12.7
B	Projected	
C	Normal	19
D	Projected	
E	Normal	9.5
F	Projected	
H	Extremely Projected	19

CENTRE ELECTRODE MATERIAL

C	Bimetal compound centre electrode
—	Standard Cr-Ni alloy solid electrode
S	Silver centre electrode
P	Platinum electrode

OTHER FEATURES AND ELECTRODE GAP

X	Electrode gap 1.1 mm
Y	Electrode gap 1.5 mm
O	Deviation from basic
1, 3, 5, ...	Heat range deviation towards "cold"
2, 4, 6, ...	Heat range deviation towards "hot"

Fig. 15.38. Spark plug designation.

always have four stroke engines. In actual practice it is even less due to other reasons, *e.g.* less time available for cooling by the fresh charge.

4. Spark plugs are designated in a prescribed manner. One such method (courtesy Mico, India) is shown in Fig. 15.38.

18. FACTORS AFFECTING SPARK PLUG OPERATION

1. Air/fuel mixture quality

1. Richer mixtures result in lowering of tip temperature, causing fouling and poor engine running.

2. Leaner mixtures increase the tip and cylinder temperature to rise, resulting in pre-ignition and/or detonation. Such mixture may also cause serious damage to the plug and the engine.

2. Compression

1. Forced induction, *e.g.,* by supercharging, turbocharging etc.
2. Reduction of combustion chamber volume, *e.g.,* using domed pistons, smaller chamber heads, etc.
3. Changing of camshaft to increase compression ratio.

Any of the above would raise the temperatures of the spark plug tip and inside of the cylinder. Therefore, under these conditions a colder plug and higher octane fuel have to be used alongwith appropriate modification of ignition timing and mixture quality, failing which may result in damage to the spark plug and engine.

3. Ignition advance

Increase of ignition advance by 10° may result in increase of plug-tip temperature by 70°-100°C.

4. Engine speed and load

Both these factors result in proportional increase in the cylinder and plug tip temperatures. Thus an engine which needs to run consistently at high speeds or under heavy loads, requires to be fitted with colder spark plugs.

19. SPARK PLUG DEFECTS

If insulator tip of the spark plug is greyish-yellow to light brown (Fig. 15.39), it indicates normal condition of the spark plug, which means the engine running is correct and the correct spark plug has been used to give optimum engine performance.

Fig. 15.39. Normal spark plug condition. (Courtesy—MICO, India)

Fig. 15.40. Spark plug fouled by oil. (Courtesy—MICO, India)

Fig. 15.41. Spark plug fouled by petrol (Courtesy—MICO, India)

1. Erosion of electrodes

The spark plug gap increases as a result of erosion of electrodes, which is caused by the corrosive combustion gases. The increased gap size requires higher voltage to cause the spark to jump across it. If the gap size increases so much that the voltage surge produced is not sufficient for the spark to jump, the plug will misfire.

The gap may be readjusted till it is found that electrodes are so much eroded away that the plug has to be replaced.

2. Fouling of plug

Spark plug fouling provides alternative paths for the current to pass, due to which the voltage surge is not able to attain its maximum value. This may lead to misfiring.

This can be detected by any of the two methods:

(*i*) Stop the running engine and touch the spark plugs. The plug which was misfiring would be cool.

(*ii*) While the engine is running, disconnect the distributor lead of the plug which is suspected to misfire. If there is a change of sound, the plug in question was not missing, but if there is no change in the sound the plug was missing.

The fouling of spark plug may be due to oil (Fig. 15.40) or petrol (Fig. 15.41). The oil fouling is visible in the form of shining black, oily deposits. The plug may be cleaned with a cloth and refitted. But the causes of the oil fouling have also to be removed. The possible causes may be worn out cylinder, piston rings or defective valve seals (most common).

The carbon deposits due to burning of petrol on the spark plug are recognised by their black colour and dry, fluffy nature. The possible causes for the same are, defective fuel system delivering extra rich mixture, defective choke in the carburettor which does not open automatically when engine starts, clogged air cleaner, poor engine compression, continuous low speed/light load operation, insufficient high tension voltage or use of a plug colder than that which is specified by the manufacturer. The exact fault may be traced out and remedied. The spark plug one grade hotter may be tried if necessary. However the foulded plug may be cleaned and reused.

The lead deposits are caused by the use of tetraethyl lead in the high octane value fuels and by continuous running of the engine at high speeds. The deposits are easily recognised by the colour which may be red, white or yellow. The plug may be cleaned with usual methods.

3. Plug overheating

Metal particle on insulator tip, eroded electrodes, insulator burnt white with pearly deposits of metallic enamel beads indicate that the plug was running overheated (Fig. 15.42).

This defect may arise on account of any of the following reasons:

(*i*) Ignition may be too much advanced.

(*ii*) The plug used may be "hotter" than the specified one.

(*iii*) Weak mixture.

(*iv*) Defective cooling system.

(*v*) Loose fitting causing the exhaust gases to leak.

(*vi*) Improper closing of valve.

The reason may be traced out and corrected.

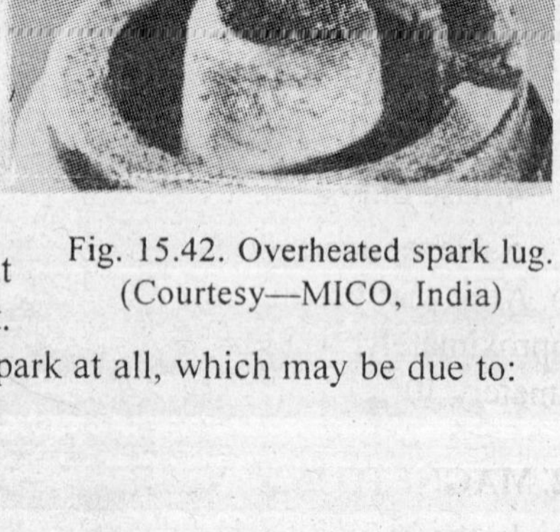

Fig. 15.42. Overheated spark lug.
(Courtesy—MICO, India)

4. Plug misfiring

A plug is said to misfire when the spark intensity is insufficient to ignite all the fuel in the combustion chamber at the proper moment. A misfiring spark plug may be causing a weak spark or even no spark at all, which may be due to:

(*i*) Defective ignition coil(s).

(*ii*) Incorrect plug gap.

(*iii*) Fouled spark plug(s).

(*iv*) Incorrect ignition timing.

20. SERVICING THE SPARK PLUG

1. First of all the spark plug is cleaned. This may be best done in a spark plug cleaning machine (Fig. 15.43). One simply has to insert the firing end of the spark plug into the rubber adapter of the cleaner,

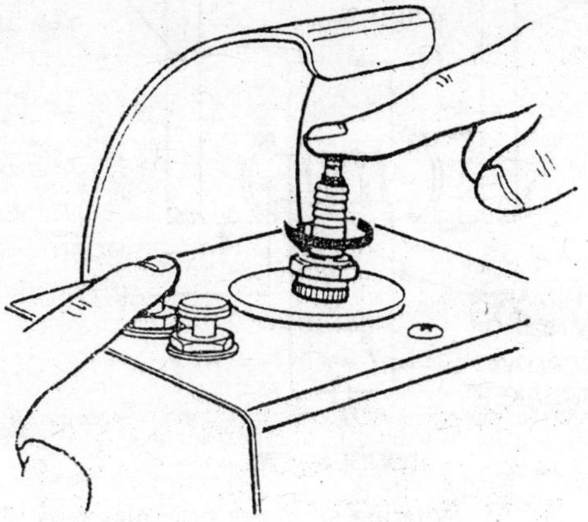

Fig. 15.43. Cleaning spark plug. (Courtesy—Maruti Udyog Ltd.)

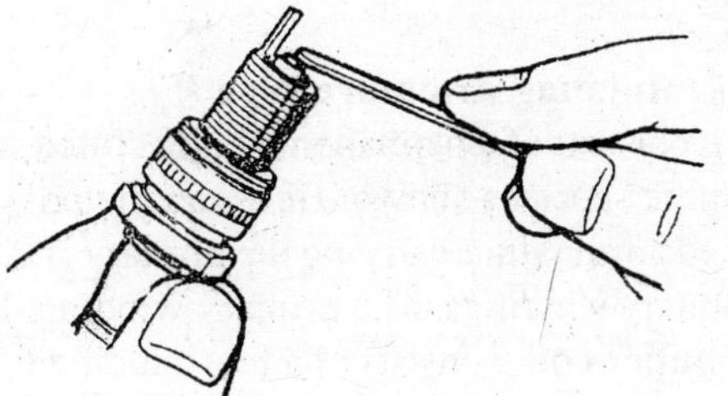

Fig. 15.44. Filing centre electrode. (Courtesy—Maruti Udyog, Ltd.)

lightly hold down with the finger tip and press the switch to blast the cleaning compounds on to the plug. The plug is rotated slowly with the finger so that it is cleaned uniformly on all sides. After cleaning, the air blast switch is pressed to remove the cleaning compounds from the firing end of the spark plug.

2. After cleaning in the cleaner:

(*a*) Inspect the plug for any cracks on the firing end of the insulator. If discovered, the plug should be discarded.

(*b*) Clean the metal shell and the thread with a wire bush and wipe the insulator with a dry and clean cloth.

(*c*) Inspect the centre electrode. If it is worn out, file its tip evenly to make a sharp edge (Fig. 15.44). Again clean the plug.

(*d*) Adjust the spark plug gap according to the specifications.

Replace the gasket in case the same has been damaged or is missing.

21. PRECAUTIONS REGARDING FITTING OF SPARK PLUGS

1. Select the spark plug type as specified by the engine manufacturer. The plug gap is also to be kept as specified.

2. Ensure that the gasket on the plug is not damaged.

3. Make sure that the surfaces are clean on both the plug as well as the engine head.

4. Use only a suitable box spanner for tightening.

5. Never overtighten the spark plug. As a rule of thumb, first screw in the plug by hand as far as it can go. After that use the correct box spanner. If it is a new plug, tighten after first resistance by further 90° approximately, whereas in case of already used spark plug, tighten further with box spanner, by approximately 30°.

22. MAGNETO IGNITION SYSTEM

In the battery ignition system, the current in the primary winding is supplied by the battery, whereas in the magneto ignition system it is magneto which produces and supplies the current in the primary winding.

23. MAGNETO

Basically magneto is of two types, the rotating armature type and the rotating magnet type.

1. Rotating armature type (Fig. 15.45)

It consists of a permanent magnet fitted with two pole shoes as shown. Between the poles is rotated an armature carrying the primary and the secondary windings. The primary winding, like the ignition coil, consists of a few hundred turns of thick wire, whereas secondary winding comprises some thousands turns of thin wire. The contact breaker and condenser are also inserted into the primary circuit and rotated with the armature.

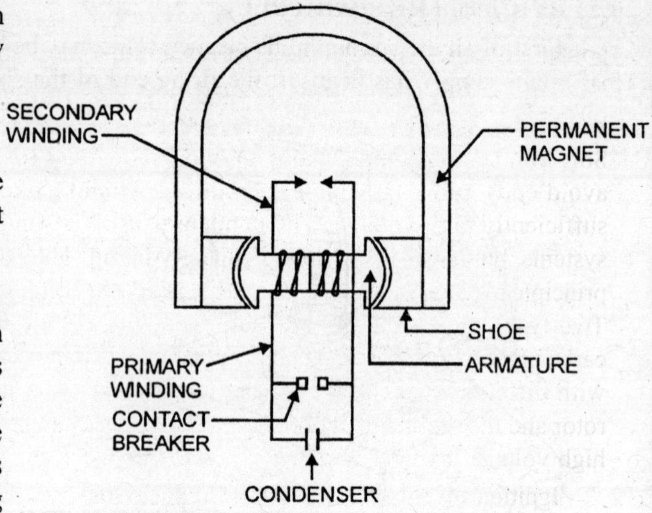

Fig. 15.45. Rotating armature type magneto.

To understand the working of magneto, consider Fig. 15.46 showing the positions of armature after every 90° of its rotation. When the armature is in the position shown in Fig. 15.46 (*a*), the flux in the armature core is zero. In Fig. 15.46 (*b*) the flux passing through the armature core is maximum. In Fig. 15.46 (*c*), the flux through the core is again zero, whereas in Fig. 15.46 (*d*), the flux through the core is in the opposite direction and of course maximum. Thus it is seen that in the positions (*a*) and (*c*), the flux changes sign and hence the rate of change of flux is greatest here. Obviously, rate of change of flux is zero at (*b*) and (*d*). This has been shown more clearly in Fig. 15.47.

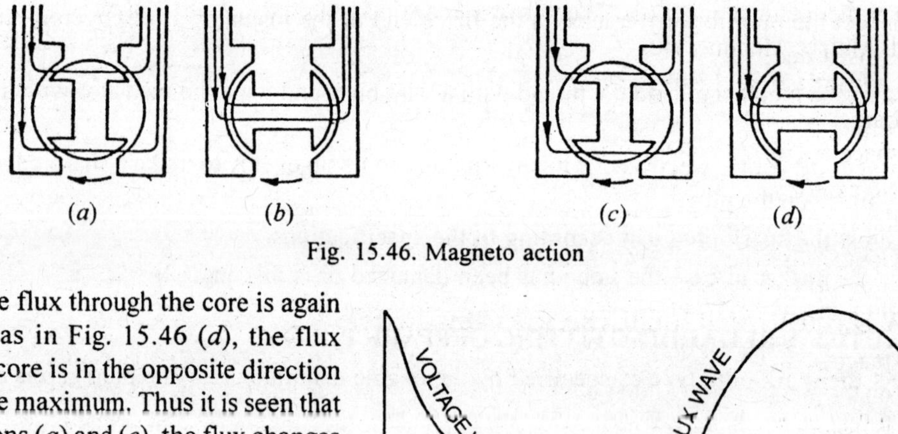

(*a*) (*b*) (*c*) (*d*)

Fig. 15.46. Magneto action

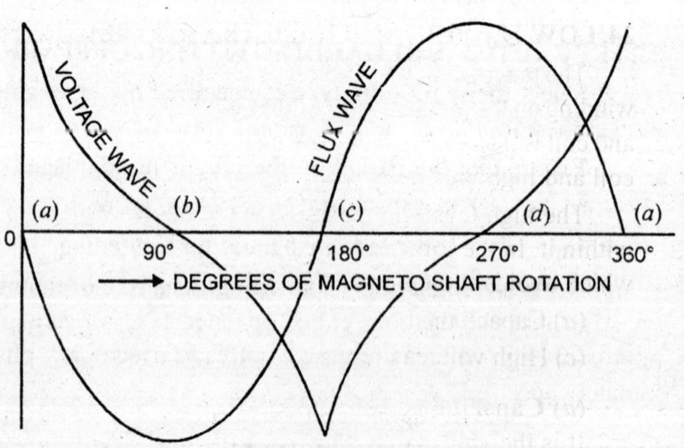

Fig. 15.47. Flux changes and voltage wave generated in magneto

The contact breaker contacts are closed when the armature reaches position (*b*). When it is at (*c*), the primary current and rate of change of flux are maximum. At (*c*) the contacts are suddenly broken. This produces a very rapid change of flux and high voltage surge is experienced in the secondary winding which is sufficient to establish a spark at the spark plug points. Fig. 15.47 depicts clearly how these voltage surges are produced in the magneto.

The various components of a magneto are, the magneto housing, rotor, windings, condenser, contact breaker and the distributor. The housing is usually a one-piece die casting made of aluminium or zinc alloy. Soft-iron laminated pole shoes are mounted within the housing. Armature rotor is also made laminated to avoid eddy currents, which if produced, would make the break less sharp and hence will not produce sufficiently high voltage. The condenser used is similar in construction to that used on battery ignition systems, but is usually of lower capacity rating. The contact breaker and the distributor are also similar in principle to battery ignition system, but generally differ in their mechanical construction and relative location. Two types of distributors used on magnetos are the carbon brush type and the jump spark type. In the carbon brush type there is actual contact between the distributor rotor and the segments which are connected with different spark plugs. In the jump spark type, however, there is no physical contact between the rotor and the segments: there is a small gap between them. The high voltage jumps across the gap to reach the spark plug.

Ignition advance mechanism is similar to that used in battery ignition systems.

2. Rotating Magnet Type (Fig. 15.48)

The principle of this is virtually the same as that of the rotating armature type explained previously. However, in this it is the magnet and not the armature winding which rotates. This type was developed after the rotating armature type because of the difficulty to produce a rotating magnet of reasonable dimensions from the earlier magnet materials. The alloys used for magnets are Nifal, Alnico, Alcomax etc.

The advantages of this type are:

1. Larger armature may be provided, which means more space for insulation.
2. No centrifugal stresses occur in the windings, because these remain stationary.
3. Contact breaker and condenser are also stationary.

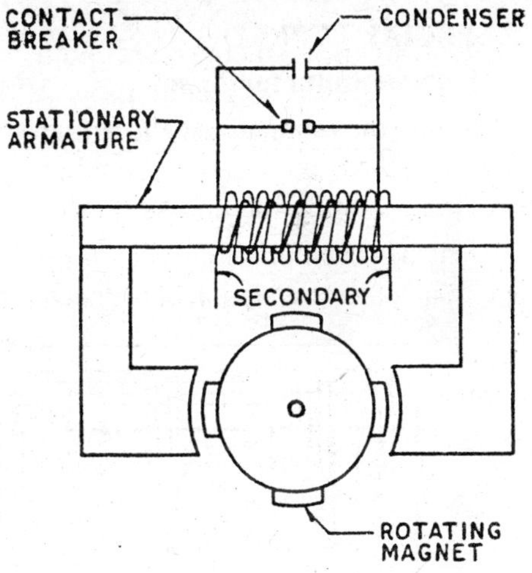

Fig. 15.48. Rotating magnet type magneto.

24. LOW TENSION AND HIGH TENSION MAGNETOS

The magneto in which there is separate ignition coil for stepping up the voltage, there being no secondary winding on the magneto armature itself, is called the low tension magneto. In this only low voltage is produced and coil is installed somewhere near the spark plug. In the high tension magneto there is no separate ignition coil and high voltage is produced in the magneto itself.

The high tension magneto has the advantage that it is a compact unit which contain the ignition coil within it. In the low tension magneto, however, it is the low voltage which is present in most of the circuit which avoids the defects of a high voltage. These defects are:

(*a*) Capacitance (*b*) Moisture
(*c*) High voltage corona

(*a*) Capacitance

It is the ability of a conductor to store electrons. In the high tension ignition system, the capacitance of high tension leads from the magneto to the spark plug causes the leads to store a portion of the electrical charge until the voltage is built up sufficiently to cause a spark to jump across the gap of a spark plug.

When the spark has jumped and established a path across the gap, the energy stored in the leads during the rise of voltage is dissipated in heat at the spark plug electrodes. Since this discharge of energy is in the form of a relatively low voltage and high current, it causes burning of the electrodes and shortens the life of the spark plug.

(*b*) Moisture

The existence of moisture results in increased electrical conductivity, thus providing new and unforeseen routes for the escape of high voltage.

(*c*) High voltage corona

When a high voltage is impressed between the conductor of an insulated lead and any metallic mass near the lead, an electrical stress is set up in the insulation. This effect is called high voltage corona. Repeated application of this stress to the insulation will eventually cause its failure.

In automobiles, however, H.T. magnetos are used.

25. SPECIAL TYPES OF MAGNETOS

1. Polar Inductor Type

In this the magnet and the armature are stationary. The change of flux is effected by means of an inductor, which is virtually a magnetic switch that periodically switches the flux to and from armature core, thus creating the necessary flux changes.

This has been made clear by means of Fig. 15.49. The two inductors N are connected together while two inductors S are similarly connected together. When the inductor N is conveying flux to the N pole of

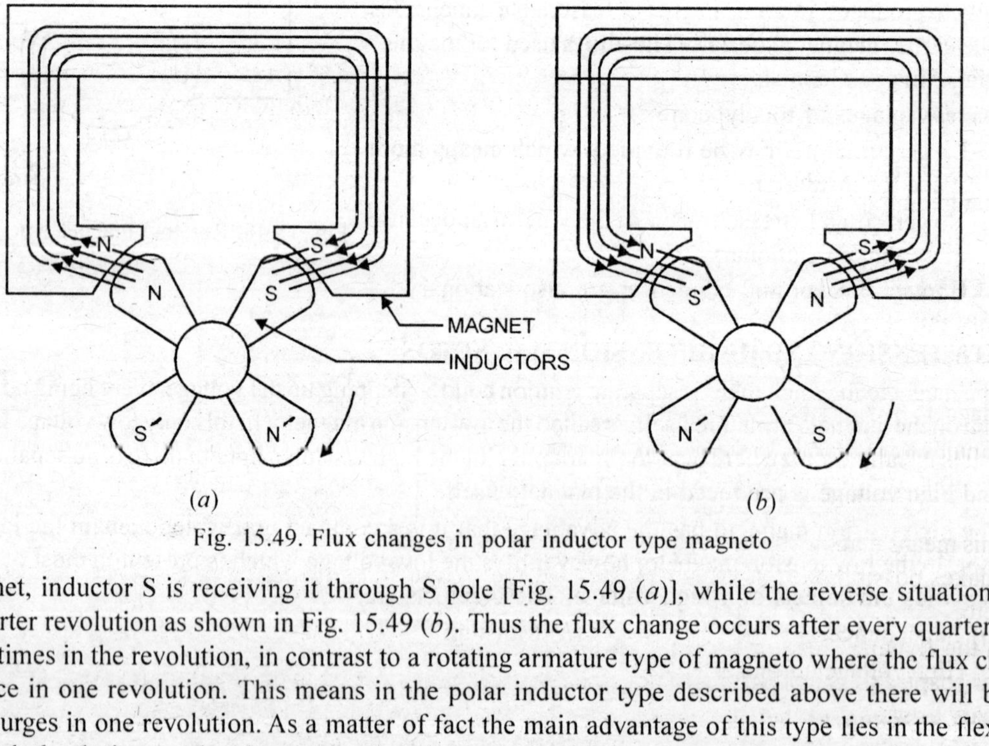

Fig. 15.49. Flux changes in polar inductor type magneto

the magnet, inductor S is receiving it through S pole [Fig. 15.49 (*a*)], while the reverse situation arises after quarter revolution as shown in Fig. 15.49 (*b*). Thus the flux change occurs after every quarter cycle, *i.e.* four-times in the revolution, in contrast to a rotating armature type of magneto where the flux changes only twice in one revolution. This means in the polar inductor type described above there will be four voltage surges in one revolution. As a matter of fact the main advantage of this type lies in the flexibility with which the design can be changed easily to have desired even number of sparks, *e.g.* 2, 4, 6, 8... per revolution. For example with only two inductors spaced 180° apart, the magneto gives two sparks in one revolution.

2. Sleeve Inductor Type

It consists of two 90° iron-segments built into the form of a sleeve which can rotate, while the armature and the magnet remain stationary. The flux changes occurring in half a revolution of the sleeve inductor

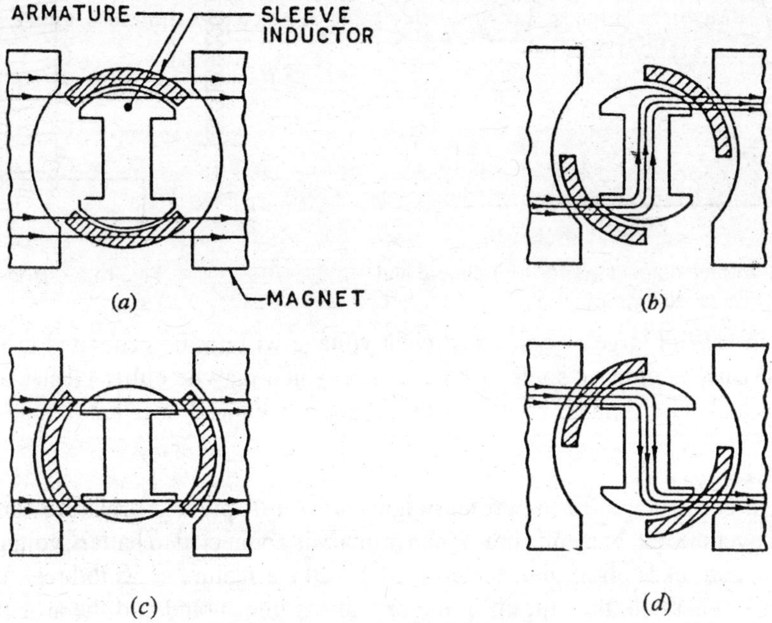

Fig. 15.50. Flux changes in sleeve inductor magneto.
(*a*) zero flux-0°, (*b*) maximum flux-45°, (*c*) zero flux 90°, (*d*) maximum flux-135°

i.e. magneto shaft, are shown in Fig. 15.50. It is seen that flux change in this type of magneto occurs four times in one revolution. Thus it gives four sparks in each revolution.

3. Flywheel Magneto

Only one spark per revolution is needed in two-stroke engines having single cylinder only. In particular,

$$\frac{\text{Magneto Speed}}{\text{Engine Speed}} = \frac{\text{No. of cylinders}}{\text{No. of sparks per magneto revolution}}$$

$$= 1$$

This means that the magneto runs at the crankshaft speed. This makes possible the use of magneto called the flywheel magneto.

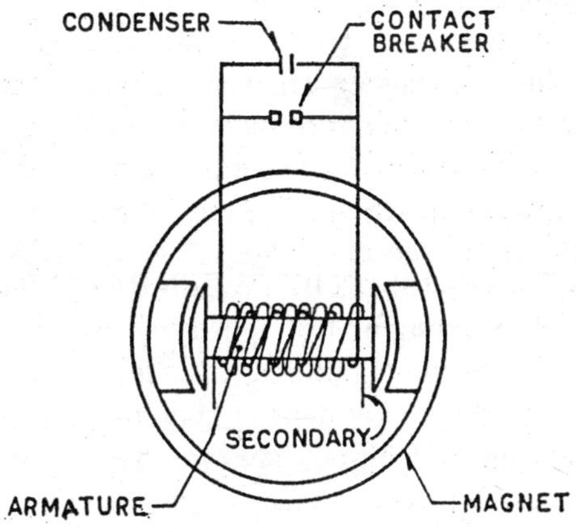

Fig. 15.51. Flywheel magneto.

In the flywheel magneto, (Fig. 15.51) the magnet is of circular shape and is attached to the engine flywheel, thus rotating at the crankshaft speed. Inside there is stationary armature containing primary and the secondary windings. Flywheel magneto is mainly used in single cylinder engines but is also employed in 2 and 4 cylinder engines.

26. AUXILIARY STARTING UNIT

The magneto has a rising characteristics shown in Fig. 15.52. Thus it is observed that at low cranking

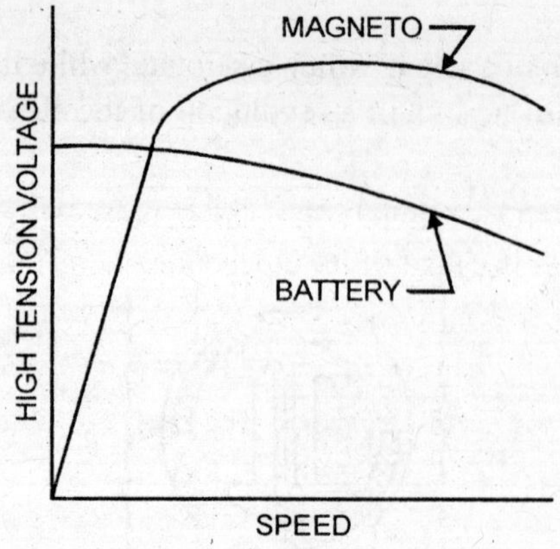

Fig. 15.52. H.T. voltage characteristics of magneto and battery ignition system

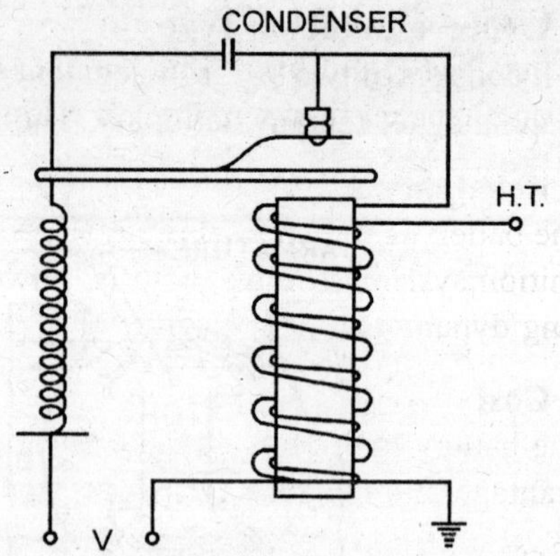

Fig. 15.53. Booster coil.

speeds particularly in case of large engines, sufficient voltage will not be generated to effect a spark. In such situations it is usual to employ some auxiliary unit, which may be either electrical or mechanical. Electrical auxiliaries are usually called booster coils and the mechanical ones are termed impulse starters.

27. BOOSTER COIL

This is similar in construction to the ordinary ignition coil. A pivoted armature carrying one of the contacts is mounted over the end of a coil core. When primary is connected to battery voltage *V*, exponential current rise in this circuit takes place until the core attracts the armature so as to break the contacts. Just when the contacts are separated, the current in the primary is interrupted and the armature again moves up to close the contacts. This way the oscillations involving make and break of the contacts start. Thus a booster coil produces a series of sparks so long as primary is connected to the supply source.

28. IMPULSE STARTER

The impulse starter is a purely mechanical auxiliary and is placed between the magneto shaft and the engine driving member. The impulse starter is so designed that at low cranking speeds it automatically gives the magneto rotor shaft, the quick turn so that a hot spark is produced. It also provides for automatic retard of the spark during the time the engine is being started. There are two main types of impulse starters: the pivoted pawl type and the sliding pawl type.

29. COMPARISON BETWEEN BATTERY AND MAGNETO IGNITION SYSTEMS

1. Starting or low speed operation

The battery ignition gives adequate voltage even at very low engine speeds because the high voltage produced does not depend upon the engine speed. Thus this system has ideal starting characteristics. In the magneto ignition, however, the engine speed affects the voltage produced and during low cranking speeds sufficient voltage may not be created to produce a spark at the plug terminals, for which a separate booster coil or impulse starter has to be used.

2. Strength of Spark

The spark is quite adequate in battery ignition though the voltage curve drops slightly with increase of speed. Magneto spark, however, is even more powerful.

3. Maintenance

It is easier in case of battery ignition, because the items requiring maintenance attention are located in more accessible position than in the magnetos.

4. Circuit Wiring

For magneto ignition system, circuit wiring is very simple. In the battery system, there is more circuit wiring and hence it is more liable to wiring faults.

5. Reliability

The battery is indispensable for a battery system. A run-down battery will result in complete failure of the ignition system. The magneto is, however, more reliable because of its independence of battery and charging dynamo.

6. Cost

The battery ignition is cheaper because of less amount of precision work. The high cost is the chief disadvantage of a magneto ignition system.

7. General

Both the battery and the magneto ignition systems have proved satisfactory in practice. Particularly for automobiles, there has been complete satisfaction over the performance of these two systems. However, where reliability and not the cost is the main consideration as in racing cars, aeroplanes, the magneto system has been frequently employed.

30. TESTING IGNITION CIRCUIT

To locate the fault in the ignition circuit is quite easy. This can be done by means of simple tests as has been explained here. The detailed testing of the various components of the ignition system however will be discussed in the next article (Art. 31).

30.1. Testing Primary Circuit

Remove the distributor cap and take out the rotor. See that contact breaker points are opening and closing freely. Check the contact breaker gap which should be around 0.35-0.40 mm. Adjust the gap, if necessary. Switch on the ignition. With the help of the cranking handle, rotate the engine and observe the ammeter on the dash board.

If the ammeter gives a reading which fluctuates, the primary circuit is in order. If the needle shows a reading which remains stationary, it indicates that either the condenser is short- circuited or grounded or the contact breaker points are not separating. However, if the ammeter does not give any reading it clearly means that the primary circuit is open.

If from the above it is established that the fault lies in the primary circuit, the following test may be conducted to locate the fault.

With the ignition ON, crank the engine slowly so that the contact breaker points are open. Take a voltmeter of range 0-20 volts and connect it in turn, across the various terminals in the primary circuit and the earth. If the component to whose terminal the voltmeter is connected reads nearly the battery voltage, it is alright and you may proceed to test the next one. However, in case the voltmeter does not give any reading, it means the concerned component is defective. For example, connect the voltmeter across the ammeter and the earth. If the voltmeter reads about 12 volts (equal to the battery voltage), the ammeter is in order, but if the voltmeter does not indicate anything, it means the ammeter is defective. In the same way all the main components of the primary circuit, *viz.*, the ignition switch, ignition coil, breaker etc. may be tested and the fault spotted.

30.2. Testing secondary circuit

The major defects in the secondary circuit are likely to occur in spark plugs, distributor cap and rotor, high voltage leads and the ignition coil.

To spot the fault, remove the lead from one of the spark plugs, place its end at about 10 mm from the engine block or head (ground). With the ignition on, crank the engine. It the spark jumps across the gap regularly and with good intensity, it indicates that the entire secondary circuit upto and including the distributor is working alright and the fault, if any, lies in the spark plug. Remove that spark plug and instead fit a new one, connect the high tension lead from the distributor to this plug and crank the engine. If the engine runs now alright, the spark plug was defective which should be replaced. If the engine does not run smoothly even now, repeat the test with other spark plugs to locate the defective spark plug.

If, however, on cranking the engine with the h.t. leads form the distributor to spark plug removed, a spark of good intensity is not obtained, with the removed end held about 5 mm from the cylinder head, then the fault must lie either in the distributor or the coil.

Once any defective components is indicated by the above simple tests, it may be removed from the engine and tested vigorously, to detect the nature of defect, whether the same can be repaired or the replacement would be desirable.

It must be noted, however, that the smooth starting and running of the engine also depends upon efficient working of the fuel system of engine. This must be kept in mind while testing the ignition circuit.

31. TESTING OF IGNITION SYSTEM COMPONENTS

The various tests meant to check the different components of the ignition system are discussed below.

1. Testing of the Ignition Coil

Ignition coil is tested with special coil test meter. The following are the main tests conducted:

(*i*) Resistance of primary winding.

(*ii*) Resistance of secondary winding.

(*iii*) Internal short circuit test.

(*iv*) Coil performance test.

Out of these the first three tests can also be performed with the help of a multimeter, while the coil performance may be tested by putting all other new and tested components in the ignition system and observing the spark. A weak spark would then indicate the faulty coil.

A special oscilloscope, called ignition scope is also used in the modern automobile shops to observe the coil output waveform, which indicates clearly where the faulty lies in the ignition system. For example, uneven firing voltage in various cylinders indicates that either the spark plugs have worn out unevenly or some of the H.T. leads from distributor to spark plugs may be defective. Similarly absence of oscillations in the ignition surface pattern indicates defective ignition coil. In the same way, other typical patterns have been observed which indicates defects in contact breaker, condenser, plugs etc.

2. Testing of condenser

Following are the tests usually carried out for the condenser:

(*i*) *Short-circuit test.* D.C. current is passed through the condenser with a lamp of suitable resistance

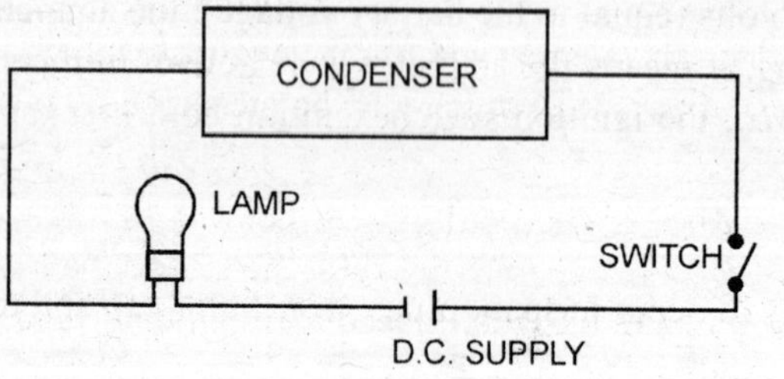

Fig. 15.54. Short circuit test of the condenser.

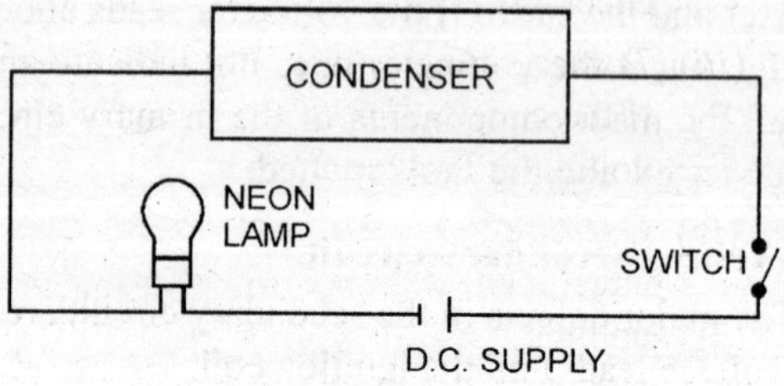

Fig. 15.55. Open circuit test of condenser.

in series (Fig. 15.54). If the condenser is short circuited, the lamp will light when the switch is closed.

(*ii*) *Open-circuit test.* D.C. current of suitable voltage (above 100 volts) is passed through the condenser which is in series with a neon lamp (Fig. 15.55). If no flashing of the lamp occurs, the open circuit exists in the condenser which must be replaced.

(*iii*) *Capacity test.* Test the capacity of the condenser and compare it with the value given by the manufacturer.

3. Contact breaker testing.

The method of testing the gap between contact points has alright been discussed while studying the contact breaker gap setting. The spring tension is measured with the help of a spring balance.

4. Distributor Testing.

The main test for a distributor is the checking of ignition timing which has already been discussed. The detailed testing of distributor, including the condenser and the contact breaker can be carried out on special testers. The following tests can be performed with the help of these testers:

(*i*) Electrical resistance of leads.

(*ii*) Condenser tests.

(*iii*) Contact breaker testing involving.

(*a*) stiffness of the contact breaker spring.

(*b*) alignment of the contact breaker points.

(*c*) dwell period of contact breaker cam.

(*iv*) Centrifugal advance tests.

(*v*) Vacuum advance test.

5. Spark plug testing.

(*i*) *Visual inspection.* If the insulator tip is brownish or light grey in colour, it indicates that the engine conditions have been good, proper fuel has been used. If, however, a dry layer of carbon is formed on the insulator tip, electrodes and plug points, it indicates that suitable fuel has not been used, carburettor adjustment is not correct or/and the ignition system was not correctly tuned. Further, a thick layer of oil on the body and electrodes of the plug means badly worn out piston rings, excessively rich air/fuel mixture and improperly adjusted ignition system. This may also be due to the use of a spark plug of higher heat value than required. If, however, the insulator tip is white in colour and carbon is deposited on the plug body, the probable cause may be a weak air-fuel mixture, excessively advanced ignition, inefficient cooling system or/and spark plug used may be of lower heat value.

(*ii*) *Gap checking.* This may be done with a feeler gauge. A flat gauge will give wrong indication because of uneven wear of the points. So round wire type of feeler gauge should be used for gap checking of spark plugs.

(*iii*) *Testing for current leakage.* Increase the gap to say, 1.5 mm and run the engine to medium speed. If the current is leaking, the terminal voltage at the plug (which may be measured by means of a voltmeter) will be zero.

(*iv*) *Gas tightness test.* Compressed air is passed through the spark plug which is screwed to special attachment at the end of the air tubing. The plug is immersed in the kerosene oil and the air bubbles coming out of the spark plug are collected in a measuring jar placed inverted over plug. The air collected should not exceed 5 cm^3 in one minute.

32. IGNITION SYSTEM SERVICING

The servicing of the ignition system involves systematic testing of all the components of the ignition system, putting them in proper order or replacing the defective components and lubricating various components so that the system works at maximum efficiency. The steps for the systematic service of the ignition systems are given below:

1. Test the battery and the leads. The detailed procedure of battery testing has been discussed in the chapter on storage batteries. Take suitable steps so that the battery is in proper condition.
2. Test the primary and secondary circuits as explained in Art. 28. Test also various components individually to make sure that each one of these is working properly. If the contact breaker points are dirty or pitted they should be cleaned by polishing them with a fine carborundum stone, and then wiping them with a petrol-moistened cloth. Adjust the C.B. points after cleaning.
3. Check the ignition timing. If found defective, set it properly.
4. Lubricate the distributor at the distributor shaft, contact breaker cam and advance mechanism. Take care to use lubricant of correct specifications.
5. Inspect and test the spark plugs. Clean the plugs thoroughly and set the gap properly. For refitting the plugs, consult the stop manual to see if the plug gaskets have to be replaced.

33. IGNITION SYSTEM TROUBLE SHOOTING

The various troubles encountered in the engine due to faults in the ignition system are mentioned below. Of course, it must be kept in mind that the troubles mentioned are not caused only because of the defects in the ignition system which have been mentioned here. Defects in other engine components may also be the sources of these troubles.

1. Engine missing

If it is only one cylinder which is missing, the spark plug is defective which may be taken out and serviced or replaced as required.

Defective distributor cap or the high tension lead may be another reason for this trouble. Replacement of the defective part is the only solution available.

If many cylinders are missing in an irregular manner, the ignition timing may be wrongly set, which should be tested and reset.

The other reasons could be:

(*i*) Defective condenser or coil which may be tested separately. If defect is established, the component should be replaced by a new one of same specifications.

(*ii*) The contact breaker points may be worn out or out of adjustment. They may be cleaned and inspected. Consequently they are properly set or replaced if found beyond adjustment.

(*iii*) The spark plugs may be fouled or otherwise defective which may be cleaned and their gap properly set or even replaced if necessary.

2. Engine overheating

Too much retarded ignition timing may be the cause of this trouble, which should be checked and reset correctly.

3. Engine backfiring

There may be any of the two causes:

(*i*) The spark plugs used may not be of the correct heat range. The only remedy is to replace the same with new plugs of suitable range.

(*ii*) The ignition timing may be too much retarded, which should be checked and reset.

4. Engine making noise when accelerating

The possible reasons in this case also may be the use of plug hotter than that specified and too much retarded ignition timing.

5. Starting trouble

If engine does not start at all (assuming that the defect is already traced to ignition system, *i.e.* no spark is occurring), proceed as follows to locate the source of trouble.

(*i*) The battery may be discharged or the battery connections may be loose or the battery leads may be broken or disconnected. A quick check is to switch on the head lamps. If these light with adequate brilliance, it is sure there is no trouble on account of battery.

(*ii*) Check whether all the external connections are tightly secured.

(*iii*) Remove the distributor cap and inspect. The rotor should be in order and the C.B. points should not have struck in open or closed position. If C.B. points open and close properly, remove the H.T. lead from the distributor, hold it at about 5 mm from any convenient metallic part and rotate the engine shaft by hand. If this does not produce a spark the C.B. points, surfaces are oxidised or otherwise contaminated with lubricating oil. In such an eventuality, the C.B. points have to be opened, cleaned, reassembled and readjusted. It must be remembered while cleaning that too much rubbing will also damage the points by removing completely the thin layer of the rare metal from them. If still negative result is obtained, the fault may lie in the wiring, contact breaker, ignition switch or the coil. Test all these components separately and locate the cause. However, if the test described above does produce a spark, it indicates that the primary circuit is in order.

(*iv*) Hold the H.T. cable at (distributor side) about 5 mm from the distributor rotor points while the C.B. points are opened. If a spark is produced, it indicates that rotor point is short-circuited with the spindle. In such case the rotor has to be replaced.

(*v*) Proceeding as above, by now it is ensured that everything is in order upto the distributor. There are very remote chances of any defect further in this case because it is highly unlikely that all the leads from distributor to plugs or all the spark plugs are defective.

In case the engine does start, but only with difficulty, probably it needs a complete tune up of the entire ignition circuit.

If the starting difficulty is experienced only in damp weather, the same could be due to moisture entering the distributor through a cracked cap or otherwise and short-circuiting the ignition circuit. The moisture may also accumulate on the terminals of spark plugs or coil secondary. The remedy is to wipe out the moisture from the distributor components, spark plug terminals and the coil with a clean cloth. Replace cap if it is cracked.

6. Rapid burning of contact breaker points

This may be due to one or more of the following reasons:

(*i*) Contact breaker points may not be set properly (the gap may be too small). Set the gap as specified in the shop manual of the vehicle.

(*ii*) The spring on the contact breaker arm may have lost its stiffness, which should be replaced.

(*iii*) The condenser may be defective. This will be evident from observation of contact points where one point will be having a pit, whereas the other point will have a large build up of transferred material from the pitted point. The condenser may be tested for leakage, open contacts and capacity. A good condenser should have insulation resistance of at least 100 MÙ when measured at 500 volts D.C. The capacity should match with the value given in the shop manual. If found defective, the new condenser should be installed.

(*iv*) Too much primary current also results in rapid burning of breaker points. Check the current and if found excessive, test the individual primary circuit components. The ignition coil may be defective in which case it has to be replaced by a new one. Even the operating voltage may be excessive due to malfunctioning of the voltage regulator, which may be suitably adjusted.

(*v*) The lubricating oil may be creeping to the contact breaker points. With oil on the points an arc is formed which causes pitting. The source of the oil seepage may be traced and remedied.

7. Engine lacks power

If due to ignition, the cause is usually retarded timing. This may also be on account of dirty plugs and wrongly adjusted C.B. points.

34. IGNITION TESTERS

Electronic ignition testers are now available that not only detect any faults in ignition system, but also locate the same. Apart from this, these also indicate the general condition of the ignition system. Earlier ignition testers were audio devices indicating the system condition by means of peculiar sounds in the headphones. However, modern ignition testers make use of the cathode ray oscilloscope, where time base is synchronized with the engine cycle, usually by an electrical impulse corresponding to one of the spark plugs. The vertical image on the cathode ray screen is proportional to the voltage across primary or secondary of the ignition circuit. The ignition cycles of all the cylinders in the engine may be displayed simultaneously on the screen in their firing order sequence.

Investigation of ignition system with a cathode ray oscilloscope is illustrated with Figs. 15.56 to 15.64. Normal

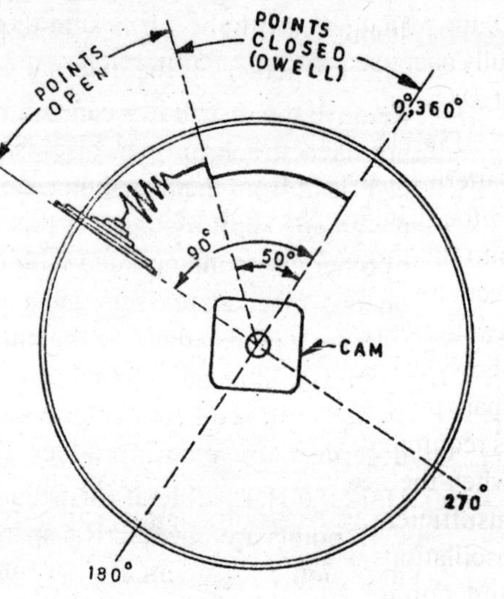

Fig. 15.56. Relation between distributor working and oscilloscope pattern for a 4-cylinderengine.

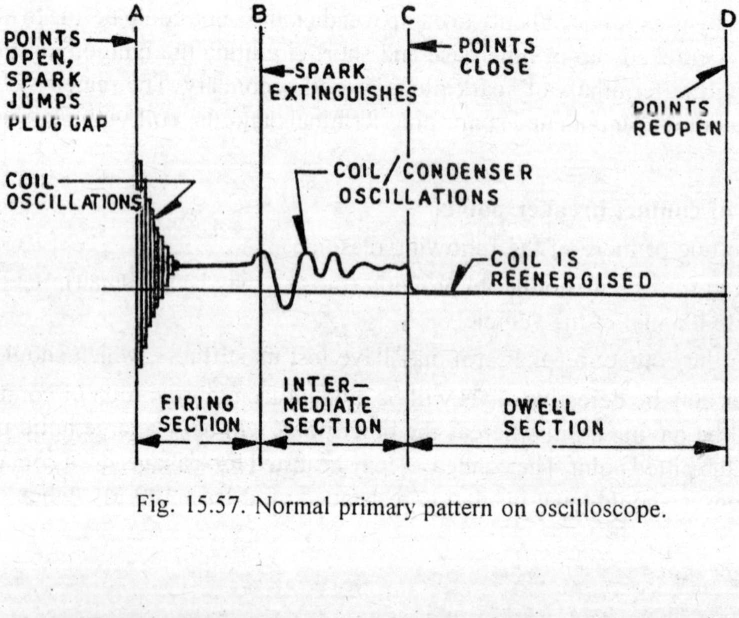

Fig. 15.57. Normal primary pattern on oscilloscope.

primary pattern on oscilloscope is shown in Fig. 15.56 and 15.57. The C.B. points open at *A* and the spark jumps the gap between the plug electrodes. The spark extinguishes at *B*, while the C.B. points are still open. Thus in the intermediate section shown when there is inadequate voltage to maintain the spark at the plug, coil-condenser oscillations decreasing in amplitude occur. These oscillations stop when the C.B. points close at C. During the dwell time between *C* and *D*, the primary winding of the coil becomes fully energised and the pattern starts repeating at *D*.

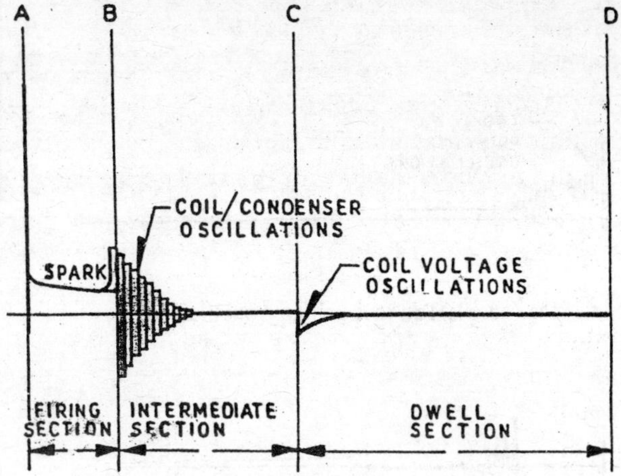

Fig. 15.58. Normal secondary pattern on oscilloscope.

Fig. 15.58 depicts normal secondary pattern on the oscilloscope. This is similar to the primary pattern and may be divided into the three sections, *viz.*, the firing section, the intermediate section and the dwell section. The length of spike at *A* shows the voltage required to fire the spark plug. Subsequently lesser voltage is required to continue the spark. At *B*, when the secondary voltage becomes insufficient to continue the spark, the oscillations of energy between the coil and condenser take place till they decrease to disappear at *C* when the points close accompanied by some minor oscillations. The cycle of events start repeating itself when the primary winding is fully energised.

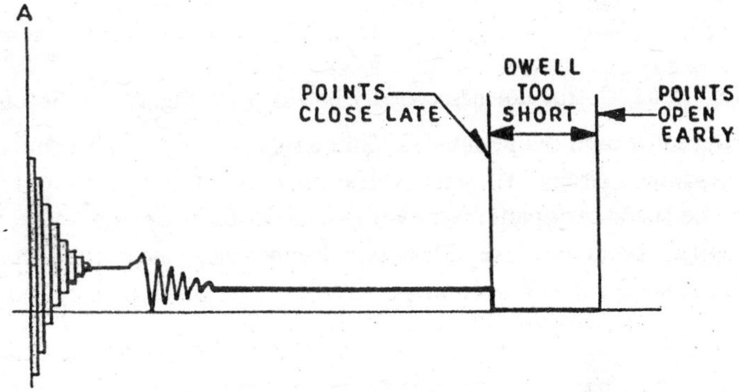

Fig. 15.59. Primary pattern—dwell too short.

Variation of the characteristic features of the primary and the secondary patterns of the actual system from the normal as described above indicates the typical fault and its location. Some examples of such faults in the primary as well as in the secondary circuits are discussed below.

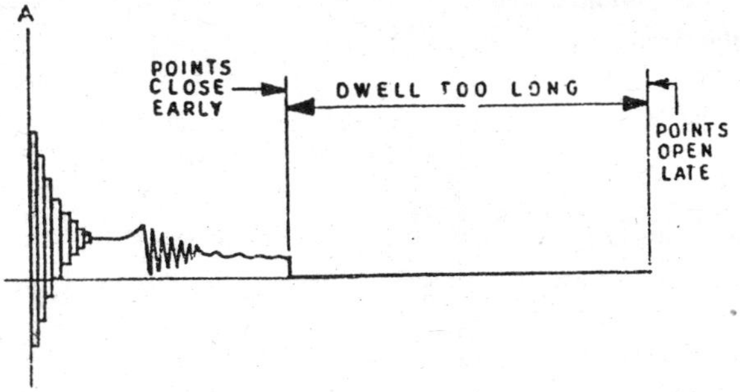

Fig. 15.60. Primary pattern—dwell too long.

Fig. 15.59 shows a primary pattern with dwell too short. This may occur on account of C.B. points closing too late and/or opening too early and/or C.B. gap being too wide. As the coil gets less time for energising, the height of voltage spike at *A* is less. Similarly Fig. 15.60 depicts primary pattern with excessive dwell, which may be due to smaller gap, C.B. points

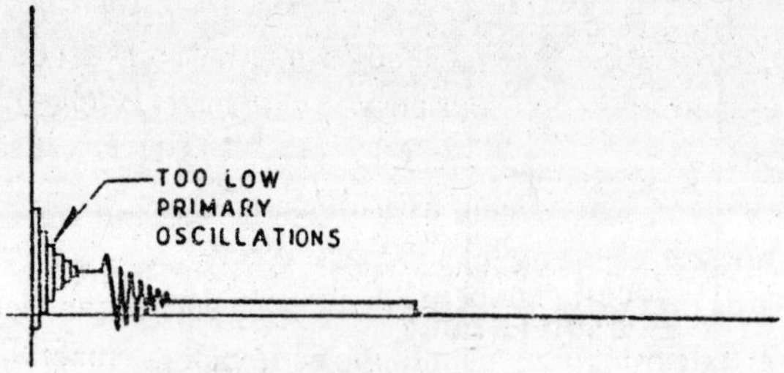

Fig. 15.61. Too low primary oscillations.

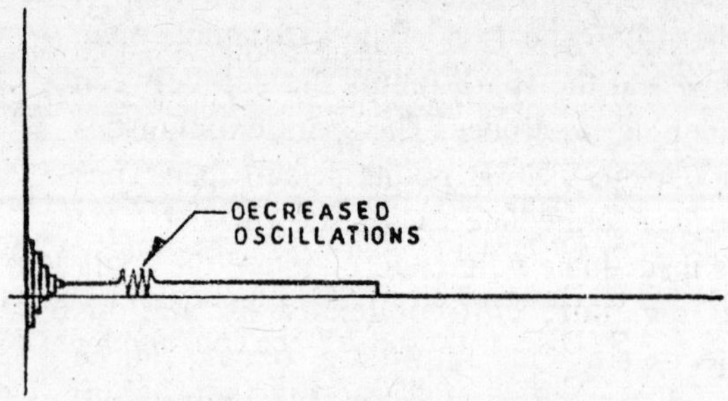

Fig. 15.62. Less primary oscillations.

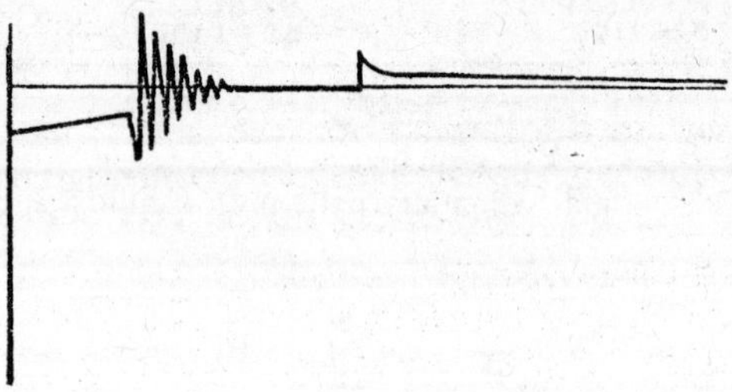

Fig. 15.63. Upside down secondary pattern.

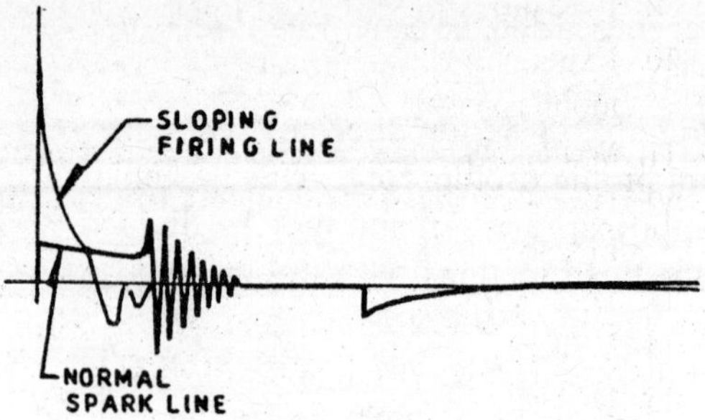

Fig. 15.64. Secondary pattery with downsloping firing line.

closing earlier and/or opening late. Likewise too low primary oscillations (Fig. 15.61) show excessive resistance in the primary circuit, whereas too less primary oscillations (Fig. 15.62) indicate short-circuit in the condenser.

The secondary pattern shown in Fig. 15.63 is upside down, which means that primary connections of the coil are reversed. Likewise downsloping firing line of the secondary pattern shown in Fig. 12.64 indicates high resistance in the secondary circuit of the particular cylinder for which this pattern was obtained.

TABLE 15.1. COMPARATIVE IGNITION SYSTEM DATA OF SOME INDIAN AUTOMOBILES

S. No.	*Item*	*Maruti 800*	*Hindustan Ambassador Mark II*	*Fiat 1100*	*Jeep Universal*	*Dodge/Fargo Model 89M4 (Petrol Type)*
1.	Type of Ignition System	Battery	Battery	Battery	Battery	Battery
2.	Ignition Coil	Suzuki	Lucal Model L.A. 12	Prestolite make	Prestolite	Prestolite
3.	Distributor	Suzuki	Lucau DM2/ Prestolite Dist. 4030	Lucas make	Prestolite IAY-4401	Prestolite IBR_4002
4.	Contact breaker gap (mm)	0.4—0.5	0.36—0.40	0.45 ± 0.03	0.5	0.457—0.559
5.	Type of Automatic Ignition advance	Centrifugal and Vacuum	Centrifugal and Vacuum	Centrifugal	Centrifugal	Centrifugal and Vacuum
6.	Centrifugal advance	20°—24°	16°—18°	30°	11°	8°
7.	Spark Plug	NGK BP 5 ES (14 mm)	Mico Bosch H-W 145T 2 or KLG 14 mm	M 14-12/225 14 mm	Mico Bosch HW 145 T3 or KLG TF-550. 14mm	Mico Bosch HW-145T3 or KLG-TFS-50
8.	Spark Plug gap (mm)	0.7—0.8	0.64	0.5—0.6	0.76	0.71-0.8

THEORY QUESTIONS

1. Discuss in detail the requirements and the functions of an ignition system of an internal combustion engine.
2. Give detailed account of the battery ignition system. Illustrate your answer with neat sketches.
3. Discuss in detail the construction of an ignition coil.
4. Explain the requirements of a contact breaker and discuss in detail its various types.
5. Discuss in detail various methods of contact breaker gap setting.
6. Describe clearly the function of a condenser in the ignition system. Explain its constructional details.
7. Discuss thoroughly the constructional details of a distributor, explaining in particular the functions of its various components and the materials used for the same. Describe also the drive for the distributor.
8. What is ignition advance? On what factors does it depend? Explain clearly different methods for providing automatically the desired ignition advance.
9. Explain with the help of suitable sketch, the operation of an octance selector.
10. Explain the term 'ignition timing' and describe in detail the procedure for adjusting the same in an automobile engine.
11. Write a brief note on the effect of ignition advance on engine performance.
12. Discuss in detail the requirements of a good sparking plug. Explain with the help of a sketch the construction of a spark plug, describing in particular the materials used for its different parts.
13. Discuss in detail various types of spark plugs for an automotive engine.
14. Write a brief note on surface discharge plug.
15. Discuss in detail various factors affecting the operation of spark plug.
16. Enlist various spark plug defects. Explain their probable causes and suitable remedies in each case.
17. Describe in detail the procedure to service a spark plug. Discuss also important precautions for fitting the same.
18. Discuss the construction and working of the rotating armature type of magneto.
19. Draw a diagram showing rotating magnet type of magneto and discuss its working.
20. Discuss the defects in a high tension magneto.
21. Give a brief account of special types of magnetos with the help of simple diagrams.
22. Explain the function of a booster coil in the ignition system.
23. Write a note on impulse starter.
24. Discuss in detail the comparison between the battery and magneto ignition systems.
25. Discuss clearly the procedure to check ignition circuits.
26. Describe in detail various tests which are performed to check different components of the ignition system.
27. How will you service complete ignition system? Discuss in detail.
28. Describe the various causes of the troubles most likely to occur due to faults in the ignition system of an automobile engine. Suggest also appropriate remedies in each case.
29. Discuss the use of cathode ray oscilloscope for trouble shooting the ignition system. Illustrate your answer with patterns indicating typical defects.

QUESTIONS FOR VIVA

1. What is the voltage of the batteries generally used in the automotive ignition systems?
2. What is the voltage causing the spark at the plug electrodes? How does this depend upon the strength of the air-fuel mixture?
3. Why do the used spark plugs require higher voltage to fire than the new plugs?
4. Name the various components of battery ignition system.
5. What are the 'positive earth' and 'negative earth' ignition circuits? Which out of these is preferred and why?

6. What is the function of an ignition coil?
7. Which of the primary and the secondary windings is thicker and which contains higher number of turns?
8. How are the different turns of the windings in the ignition coil insulated from each other?
9. What is the material used for core of ignition coil?
10. To which component of the ignition system is the secondary terminal of the ignition coil connected?
11. Name two types of ignition coils.
12. What is the function of oil in the ignition coil?
13. Why is the ignition coil hermetically sealed?
14. What is the function of a ballast resistor? What is it approximate value?
15. What is the purpose behind using a contact breaker in the ignition system?
16. What are the requirements of a good contact breaker?
17. Where is the contact breaker located on the engine?
18. Name the different components of a contact breaker?
19. What is the material used for contact breaker points?
20. State the advantages of a double arm type contact breaker.
21. What is approximately the correct contact breaker gap?
22. What happens if the contact breaker gap is (*i*) less (*ii*) excessive?
23. Name various methods of setting contact breaker gap.
24. What is dwell?
25. What are the commonly used values of dwell angle?
26. How can be dwell angle be changed?
27. What is a dwell meter?
28. What does the condenser do in the ignition system?
29. How is it ensured that no spark is produced at the contact breaker points?
30. What is the approximate value of capacity of a condenser used in the automotive ignition system?
31. Why is the condenser always mounted within the distributor housing?
32. What are the firing orders commonly used for 4 and 6 cylinder engines?
33. Name the various components of the ignition system housed in a distributor.
34. State the materials used for a distributor rotor and cap.
35. How is the distributor housing attached to the engine?
36. Which material is used for distributor rotor point?
37. What is the function of a carbon inset in the distributor?
38. Where is the distributor located on the engine?
39. Which engine part gives the drive to the distributor spindle?
40. At what speed relative to the crankshaft speed does the distributor shaft rotate?
41. How is the rotor positioned while initially installing the distributor?
42. What is 'ignition advance'?
43. What is the purpose of providing ignition advance?
44. What are the factors on which the ignition advance requirements of an engine depend?
45. Name various methods of ignition advance.
46. Where are the centrifugal weights located in the distributor?
47. On which engine parameter does the centrifugal advance depend?
48. Why are sometimes springs of unequal stiffness used in centrifugal advance mechanism?
49. Why is a reduced rate of ignition advance required at higher engine speeds?
50. What is basic timing of a distributor?
51. What is approximately the maximum value of centrifugal advance and at what engine speed does it occur?

52. What happens if due to any reason the centrifugal advance mechanism is not working?
53. On which engine factor does the vacuum advance depend?
54. What is manifold vacuum and where on the engine is it available? When is it maximum?
55. What is 'ported vacuum'? How does it vary with different engine conditions?
56. What is the approximate maximum value of vacuum advance in automobile engines?
57. What happens if the vacuum advance mechanism fails to work properly.
58. What is 'ignition timing'? How is it expressed?
59. Which cylinder is used as reference cylinder for setting ignition timing of an engine?
60. What happens if the ignition timing is advanced
    (*i*) excessively,
    (*ii*) less than optimum.
61. What should be ensured before actually checking or setting the ignition timing?
62. Name a few vehicles on the engines of which the timing mark is provided on the
    (*i*) crankshaft pulley,
    (*ii*) flywheel.
63. What is a 'timing light'?
64. In which direction do you rotate the distributor if the timing is to be (*i*) advanced, (*ii*) retarded?
65. What are the approximate values of temperature and pressure in the combustion chamber of an automotive engine under peak combustion conditions?
66. State various factors which determine the minimum voltage required for the spark to occur at the spark plug electrodes.
67. What is approximately the life of a spark plug?
68. State the requirements of a good spark plug.
69. What is 'reach' of a spark plug?
70. What happens if a plug with short reach is fitted into a combustion chamber hole with long reach?
71. What happens if the spark plug electrode temperature is
    (*i*) below 300°C,
    (*ii*) above about 800°C?
72. Name main parts of a spark plug.
73. What is 'insulator nose length' in a spark plug?
74. Define 'heat range' of a plug.
75. What are the functions of insulator in a spark plug?
76. What is the necessity of a sealing ring in a spark plug?
77. What advantage is obtained by having a 2-piece centre electrode in a spark plug?
78. What is a 'resistor spark plug'? What is its life compared to ordinary spark plugs?
79. What are the main dimensions of a spark plug?
80. What is the material used for body shell in a spark plug?
81. For what purpose is the porcelain used in spark plugs?
82. What is the material used for the spark plug electrodes?
83. What is the objection to the use of platinum alloys as the material for spark plug points of ordinary automobiles?
84. How do you compare the detachable and the non-detachable types of spark plug?
85. What are the hard and the soft spark plugs?
86. What is the function of air gap in a spark plug?
87. What is a spark intensifier?
88. What is a 'surface discharge plug'? What are its advantages and disadvantages?

89. What is the most commonly used size of spark plugs?
90. Why should the spark plug of two-stroke engines, in general, be of higher heat values than the comparable four-stroke engines?
91. Why is the life of spark plugs of two-stroke engines far less than those of four-stroke engines?
92. How do you designate a spark plug?
93. Name various spark plug defects.
94. What is the general range of spark plug gap size?
95. What is the cause of spark plug fouling?
96. What is the reason for the overheating of the spark plug?
97. What may be the possible reasons for:
    (*i*) oiled spark plug,
    (*ii*) over heated plug?
98. Name various magneto types.
99. What is the advantage of a rotating magnet type of magneto compared to a rotating armature type?
100. Enumerate the defects of a high tension magneto.
101. What is high voltage corona?
102. What is a polar inductor type of magneto?
103. What is a flywheel magneto? Where is it used mainly?
104. What is the function of booster coil?
105. What is an impulse starter?
106. Out of the coil and the magneto ignition systems, which do you consider better for an ordinary automobile and which for a racing car?
107. Name different tests generally conducted to check ignition coils.
108. What are the tests usually performed to check the condenser?
109. Enlist various spark plug tests.
110. What is the colour of the insulator tip of a spark plug that
    (*i*) has been working in normal engine conditions,
    (*ii*) is of higher heat value than required?
111. What ignition defect would cause the engine to overheat?
112. What are the causes of engine back-firing?
113. How is a cathode ray oscilloscope used for investigating an ignition system?

MULTIPLE CHOICE QUESTIONS

Mark the correct answers:

1. The three components of primary ignition circuit are
   (*a*) the contact breaker, condenser and distributor cap
   (*b*) the contact breaker, ignition coil and spark plugs
   (*c*) the contact breaker, ignition switch and condenser
   (*d*) the contact breaker, ignition switch switch and rotor
2. The components of secondary ignition circuit include the secondary winding of ignition coil, distributor rotor, distributor cap and
   (*a*) condenser (*b*) spark plugs
   (*c*) ignition switch (*d*) distributor drive gear
3. The primary winding of ignition coil consists of
   (*a*) a few turns of thin wire (*b*) many turns of thin wire
   (*c*) a few turns of thick wire (*d*) many turns of thick wire

4. Contact breaker points are generally made of
   (*a*) plastic (*b*) steel
   (*c*) copper (*d*) tungsten
5. The 'dwell' is
   (*a*) the time for which the points remain closed
   (*b*) the distance between the cam lobes
   (*c*) the angle at which the heal contacts the cam
   (*d*) none of the above
6. Excessive contact breaker gap results in
   (*a*) advanced timing (*b*) increased dwell
   (*c*) rapid burning of points (*d*) all of the above
7. The spark occurs when the
   (*a*) points close (*b*) increased dwell
   (*c*) ignition switch is on (*d*) none of the above
8. Contact breaker points are opened by the cam and closed by the
   (*a*) same cam (*b*) centrifugal force
   (*c*) magnetic force (*d*) spring tension
9. The dwell angle on a six-cylinder engine compared to a four- cylinder engine is
   (*a*) more (*b*) less
   (*c*) equal (*d*) sometimes less and sometimes more
10. The contact breaker gap is set most accurately by means of a
   (*a*) feeler gauge (*b*) steel rule
   (*c*) dial gauge (*d*) dwell meter
11. Contact breaker gap should be set
   (*a*) before adjusting dwell angle (*b*) after adjusting dwell angle
   (*c*) first after starting the engine (*d*) after setting spark plug gap.
12. The capacity of an automotive engine condenser is approximately
   (*a*) 0.2 microfarads (*b*) 2 microfarads
   (*c*) 20 microfarads (*d*) 2 millifarads
13. The ratio of the distributor shaft and the crankshaft speeds is
   (*a*) one (*b*) one-half
   (*c*) one-fourth (*d*) two
14. The centrifugal advance mechanism provides ignition advance proportional to
   (*a*) engine load (*b*) engine speed
   (*c*) both (*a*) and (*b*) (*d*) none of the above
15. Vacuum advance mechanism operates at
   (*a*) light engine load (*b*) heavy engine load
   (*c*) high engine speed (*d*) any of the above
16. Ignition timing can be adjusted by a
   (*a*) accurate clock (*b*) stop watch
   (*c*) stroboscopic light (*d*) vacuum gauge
17. The commonly used material for insulator of spark plug is
   (*a*) bakelite (*b*) asbestos
   (*c*) alumina (*d*) copper
18. Compared to the life of a spark plug of two stroke engine, the spark plug life of a four-stoke engine is approximately

(*a*) same (*b*) twice
(*c*) one-half (*d*) none of the above

19. A spark may be fouled by
(*a*) petrol (*b*) oil
(*c*) lead (*d*) all of the above

20. Spark plug having white insulator tip indicates
(*a*) over advanced ignition (*b*) retarded ignition
(*c*) leaded fuel (*d*) gas leak

21. A spark plug with a black centre insulator indicates
(*a*) over advanced ignition (*b*) retarded ignition
(*c*) stuck-up choke valve in the carburettor (*d*) excessive dwell

[**Key:** 1. (*c*), 2. (*b*), 3. (*c*), 4. (*d*), 5. (*a*), 6. (*a*), 7. (*b*), 8. (*d*), 9. (*b*), 10. (*d*), 11. (*b*), 12. (*a*), 13. (*b*), 14. (*b*), 15. (*a*), 16. (*c*), 17. (*c*), 18. (*b*), 19. (*d*), 20. (*a*), 21. (*c*)]

16 Electronic Ignition Systems

1. INTRODUCTION

Although the conventional mechanical contact breaker point type electrical ignition system has been very common in use because of its simplicity, yet due to its following drawbacks and limitations the same has now been replaced by the electronic ignition system:

1. The contact breaker points have to operate with a heavy current which causes them to get pitted, burnt or worn out, thus necessitating their servicing and resetting at smaller periods.
2. Because the contact breaker is only a mechanical device, it cannot operate precisely at higher speeds because of inertia.
3. At higher speeds, the dwell period is not sufficient for building up the magnetic field to its full value, due to which spark is weakened. It is found that the satisfactory performance of ordinary contact breakers has been limited to about 400 sparks per second, which corresponds to an engine speed of 8000 r.p.m. (in case of 6-cylinder engine).
4. Compared to the secondary output voltage of 10-15 kV at the spark plug points used in older engines, modern high speed engines need much stronger sparking voltage of 15-30 kV required to ignite leaner air-fuel mixtures for improved fuel economy and emission. The conventional ignition systems have to employ a resister to reduce the operating voltage of the primary circuit, to increase the life of the breaker points. In contrast in case of electronic systems, the primary circuit operates on full battery voltage.
5. Ignition timing also becomes inaccurate soon due to wear of the contact heel, cam and spindle.
6. Due to contact bounce, the heel cannot follow the cam at high speeds.

Thus it is seen that for modern high speed car engines the electric ignition system cannot work satisfactorily. The solution of the problem has been found in the use of electronic ignition which can provide upto 1000 sparks per second, which is the requirement for an eight cylinder engine running at 16,000 r.p.m. Besides this there are other disadvantages also:

1. The electronic ignition is easy to control with a computer to give accurate timing for all operating conditions.
2. Except spark plugs, the other components of electronic ignition system have a much higher life (about 80,000 km).

In the electronic ignition, the semi-conductor transistors are used. Apart from semi-conductors, the use of piezoelectric materials has also been made in the modern electronic systems. Piezoelectric effect, which has been known for quite some time, is the property of certain ceramic crystals to produce electrical energy on being subjected to pressure. The materials exhibiting the piezoelectric effect are called piezoelectric materials. The advantages claimed for such ignition systems are simplicity (no ignition coil is required) and reliability (no outside electrical source like battery is needed).

Electronic ignition systems may be broadly divided into two categories, viz., the induction type and the capacitor discharge type. Earlier induction type systems were derived from mechanical systems. As such these were still using breaker points, but instead of switching the primary current, these were used to trigger a transistor, which switched the heavy primary current. These were, however, soon replaced by the contact less, or all-electronic type in which the breaker points are completely eliminated. There are two types of such electronic systems in current use:

1. Basic distributor type.
2. Distributorless type.

Until 1984, only the basic distributor type electronic systems were used, However, since then, distributorless type ignition systems are being widely used due to reduced emissions, improved fuel economy and greater reliability. Product reliability was further improved by electrically controlling high-voltage distribution and ignition timing. By mid 1990s, the stick coils appeared, featuring smaller igniter. Over recent years the trend has been for combining the ignition and fuel supply systems into integrated engine management. The beginning of twenty first century saw the use of the capacitor discharge system, claiming to have better ignition system characteristics.

An introduction to basic electronics will be given first in a few articles before the details of the automotive electronic ignition systems are discussed.

2. SEMI CONDUCTORS

These are the materials having exactly four electrons in their outermost orbit, due to which reason these are neither good conductors nor good insulators. *Germanium and silicon* are two common examples of semi-conductor materials. These cannot provide a current flow when pure. However, when another material is added to a semi-conductor material in very small amount (the process being called *doping*), the necessary conditions are provided for the movement of electrons. The doping elements are called *impurities* and are added to pure semi-conductor materials in very small quantity, say one atom of impurity for every 100 million atoms of the pure semi-conductor material. These impurities are of two types depending upon the number of electrons in their outermost orbits. In one type, for example, phosphorus, arsenic, antimony or bismuth. there are five electrons in their outer must orbit, while in the second type like boron, indium, aluminium or gallium, there are three electrons in their outermost orbit. Depending upon the type of impurity added, the semi-conductor material obtained is called as N-type or P-type.

N-type semi-conductor material is produced by doping silicon or germanium with an element having five electrons in their outer orbit, which combine with four electrons of silicon or germanium to make a total of nine electrons. As a maximum of eight electrons can be accommodated in the outermost orbit, this leaves one spare electron. Even though the material is electrically neutral, yet these free electrons repel other electrons outside the material. Being free, these electrons are mobile and move from the negative end of the semi-conductor to opposite positive terminal if the semi-conductor is connected to a battery.

When silicon or germanium is, however doped with an element having three electrons in its outer orbit, it results in a material with seven electrons in the outer orbit, which enables it to attract electrons even though the material is electrically neutral. The one vacancy so produced in commonly called a *hole*. These positively charged holes attract and accept electrons from other neighbouring atoms, which correspondingly form their own holes and accept electrons from other atoms, and this goes on. Thus it is seen that in such a material, the conduction is due to continuous hole movement from one atom to the next. Such a material is called P-type semi-conductor material.

In N-type semi-conductors, the conduction of electrical current is through excess of electrons, whereas in P-type semi-conductors, electrical conduction in through holes *i.e.*, absence of electrons. When a voltage is applied to a semi-conductor, electrons travel towards the positive terminal while the holes travel towards the negative terminal. It is thus seen that holes move in the conventional direction of current.

3. DIODES

A diode is formed by combining a N-type and a P-type semi-conductor material. The joint of the two materials is called junction (Fig. 16.1). If a battery is connected across a diode in such a way that positive terminal is connected to the P-type material and the negative terminal is connected to the N-type material

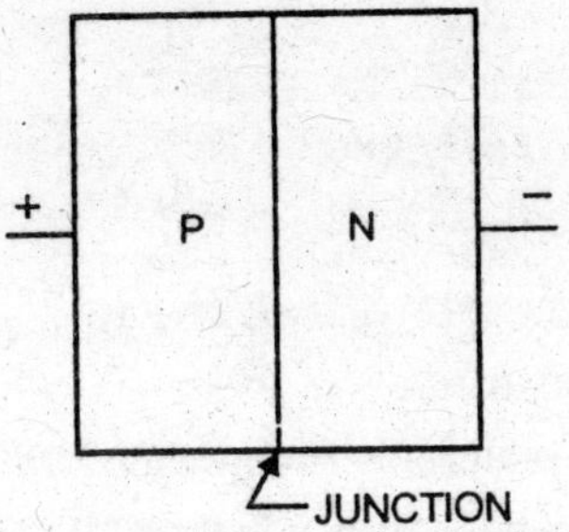

Fig. 16.1. Construction of a diode

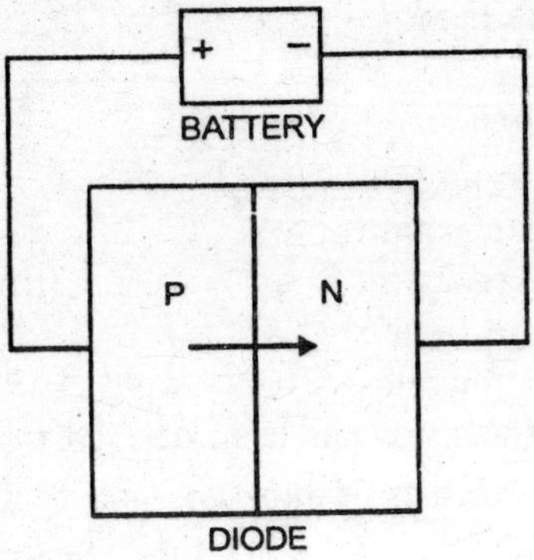

Fig. 16.2. Diode-battery connection for forward bias

(Fig. 16.2), the excess electrons leaving the N-type material to flow into the P-type material to fill the holes there, would be quickly replaced by electrons from the battery. Therefore, a current flow through diode would be maintained. Such a condition is termed *forward bias*. However, if the battery connections are reversed (Fig. 16.3), the electrons will be attracted toward positive terminal of the battery, which means away from the diode junction. Obviously, there would not be any electron (*i.e.* current) flow through the diode in this condition, called the *reversed bias*. Thus it is seen that the current would flow only when the diode is connected in forward bias. Due to this reason, diodes are used as rectifiers in alternators to convert a.c. produced into d.c. These are also used in many other circuits to prevent any possible damage due to reverse current flow in the circuit. The conventional representation of the diode has been shown in Fig. 16.4. It must be remembered, however, that excessive reverse voltage would break down the bond structure, allowing the reverse current. Excessive reverse current would destroy the diode on account of excessive heat. Such a blown diode will not let current to pass through resulting in an open circuit and has to be

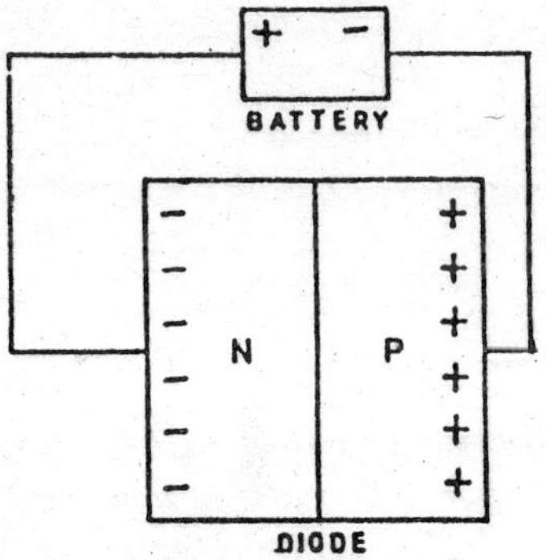

Fig. 16.3. Diode-battery connection for reverse bias.

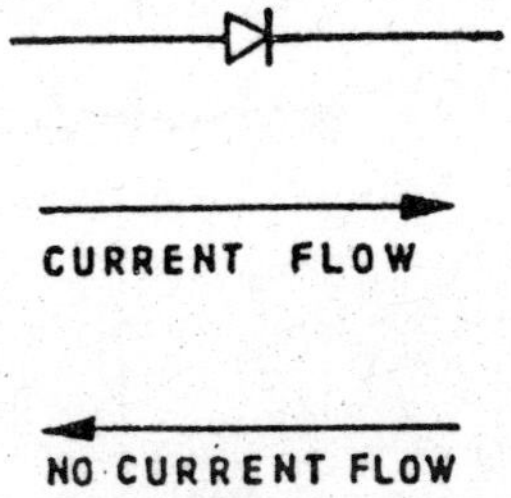

Fig. 16.4. Conventional representation of a diode

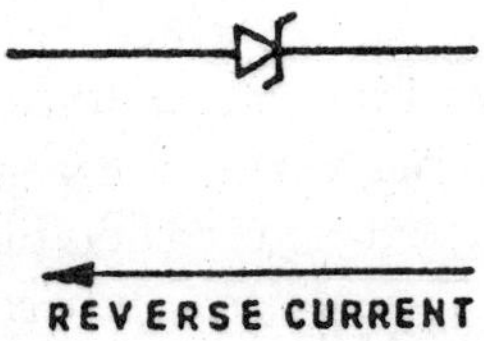

Fig. 16.5. Conventional representation of a Zener diode

replaced. Similarly a shorted diode will conduct current in both the directions and has to be replaced. A special types of diode, called *Zener diode* (Fig. 16.5) acts as ordinary diode upto a certain voltage called the breakdown voltage, above which it allows the reverse-bias current to flow. Such a diode is heavily doped and the reverse-bias current does not harm the material. A Zener diode is thus very suitably used in control circuits for various automotive applications, such as to regulate field current in an alternator.

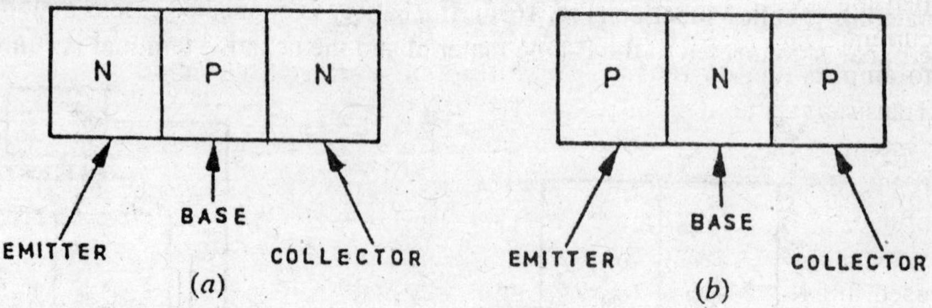

Fig. 16.6. Types of transistors (*a*) NPN type, (*b*) PNP type.

4. TRANSISTORS

A transistor is a semiconductor device which can either be used as an electronic solid-state switch or a current amplifier. The name transistor has been derived from 'transfer' and 'resistor'.

A transistor consists of three alternate sections of N-and P-type materials. It may be a NPN transistor or a PNP transistor (Fig. 16.6). In either type the central section is called the *base* and it controls the current through the transistor. Again the section at on end is called the *emitter* and at the other end is called the *collector*. Conventionally, a transistor is represented as shown in Fig. 16.7. The arrow on emitter points in the direction of current flow according to the conventional theory.

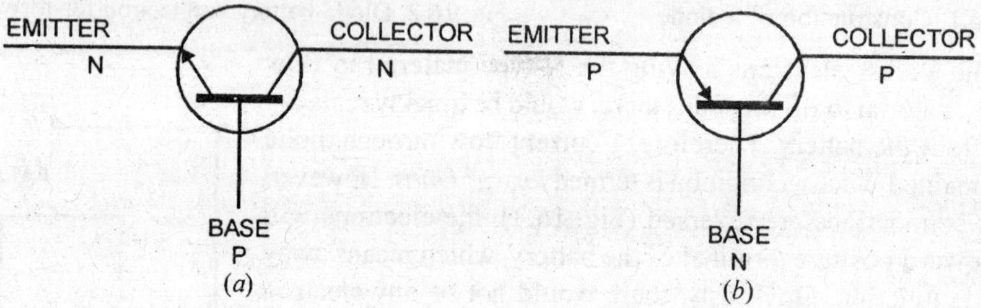

Fig. 16.7. Conventional representation of a transistor.

A transistor may be seen as two diodes attached back to back. Thus here also, N-type material would conduct electricity by supplying its free electrons, while P-type material would conduct electricity by supplying holes. A small current is carried by the base when the proper voltage and polarity is applied to it. This current through the base is called *control current* and should be high enough to switch the transistor on or off. When this base or control current is turned on, the main circuit current starts flowing through the emitter and the collector (Fig. 16.8). Similarly when the base current is turned off, the main circuit current through the emitter and the collector is also turned off. In this way, the transistor acts as an electrical switch in a circuit. It operates like a relay except that it does not have any moving

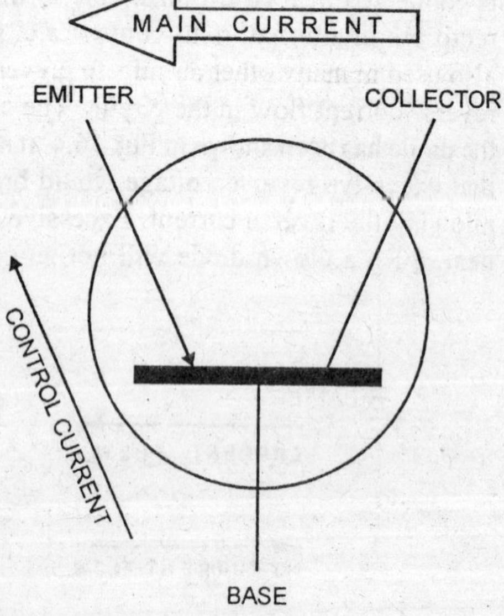

Fig. 16.8. Operation of transistor

parts. Moreover if a signal is strong enough to trigger on or off the base current of a transistor, the resulting emitter-collector current will be magnified and exactly duplicated as the base current. Thus a transistor is also used to amplify the current in a circuit. Transistors are used to control charging system voltage in electric voltage regulators for alternators, to control primary current in electronic ignition systems and in computers. Transistors used in automotive components are usually of the PNP type.

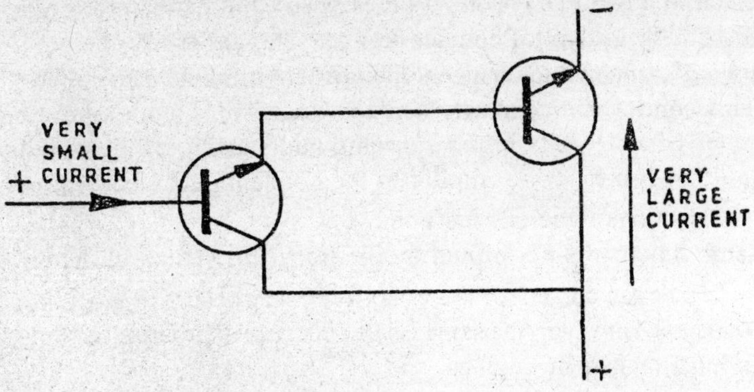

Fig. 16.9. Use of Darlington pair

Two transistors wired together are called the *Darlington pair,* which allows a very small current to control large current (Fig. 16.9). Such pairs are commonly used in electronic ignition systems and computer control circuits for engines.

5. THYRISTOR

It is a semiconductor rectifier which can control the output power supplied to an electrical load, with the minimum wastage of energy. The word 'thyristor' has been derived from 'thyratron' and 'resistor', the thyratron-tube being the predecessor of the thyristor. The thyristor has four layers of alternative P-type and N-type semi-conductors forming three P-N junctions. The two outer layers have terminals, forming anode and cathode, whereas one of the inner regions also has a terminal, known as 'Gate'.

6. PRINCIPLE OF DISTRIBUTOR TYPE INDUCTIVE ELECTRONIC IGNITION

Working of basic distributor type electronic system is similar to the conventional electrical ignition system, in that both are inductive type systems, *i.e.*, the energy for the system is stored in the coil. In this

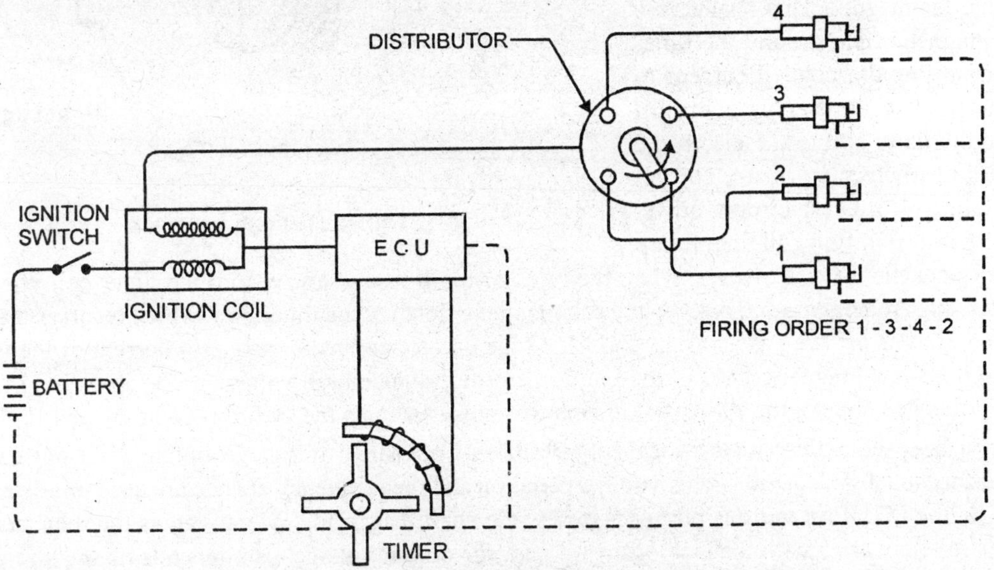

Fig. 16.10. Simplified diagram for an electronic ignition system.

electronic ignition system, a timer (also called armature, trigger wheel, reluctor, etc.) is employed in the distributor instead of contact breaker. This timer may be a *pulse generator* or a *Hall-effect switch* or an *optical switch* which triggers the ignition module, also called the electronic ignition control unit (I.C.U.) This control unit primarily contains transistor circuit whose base current is triggered off and on by the timer which results in the stopping and starting of the primary current. Other than this, the electronic ignition system works similar to the conventional electrical point-type system (Fig. 16.10).

However, whereas the point-type system employed ballast resistor to limit the primary current, the same function is performed by the transistor switch in the electronic systems.

Peak gap current of about 40 to 60 mA occurs in most inductive type systems, which is sufficient for road cars. However, the same is not adequate to produce the long arc discharge needed for peak performance in most racing car engines.

A few types of commonly used timers are discussed in the following articles.

7. PULSE GENERATOR

A pulse generator is used to generate an alternating voltage, which is used, instead of contact breaker points, to control the make and break of the current build-up in the primary winding of the ignition coil.

A magnetic (inductive) pulse generator consists of three main components, *viz.*, a permanent magnet, a timer coil and a reluctor. Out of these the first two are stationary, while the reluctor (also called timer core or armature) which is in the form of a toothed wheel, is mounted on the distributor shaft. It has the same number of teeth as the number of engine cylinders. As the reluctor wheel rotates, its teeth come very close to the pole plates of the permanent magnet as shown in Fig. 16.11. This reduces the reluctance of the air gap between the reluctor tooth and the timer coil and the other reluctor tooth and the magnet. This results in a strong magnetic field around the timer coil, permitting the control current to flow across the emitter-base circuit through the timer coil to the electronic control unit where the primary current in the emitter-collector circuit flows. This allows the ignition coil to build up a strong magnetic field. However, when the reluctor tooth passes away from the timer coil, the wide air gap offers high reluctance and results in weak magnetic field for the timer coil, which reverses the induced voltage and turns off the base current and hence the emitter-collector (primary) current of the transistor. This collapses the magnetic field in the ignition coil to produce high voltage at the spark plug. The mid-point of change between the positive and negative pulses signals the occurrence of the spark.

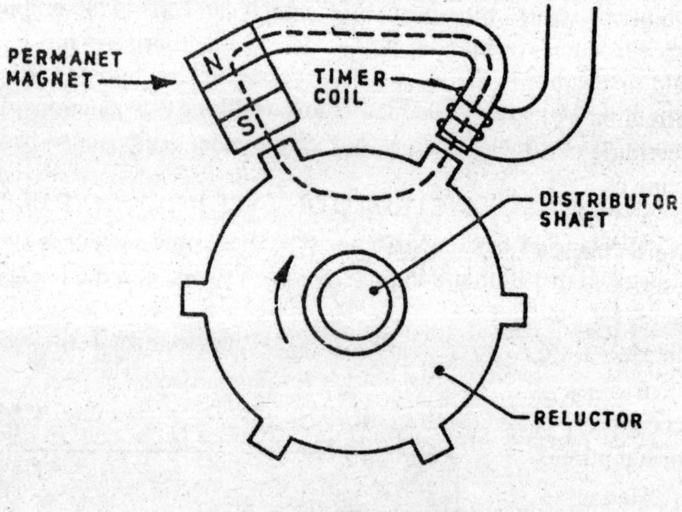

Fig. 16.11. Pulse generator.

Since the speed of the reluctor wheel controls the rate of change of flux, the output of the pulse generator varies from about 0.5V to 100V. This voltage variation and the frequency change are used jointly as sensing signals by the I.C.U. for various purposes except the spark-triggering. Moreover, as the reluctance of the magnetic circuit varies with the size of the air gap, the output voltage also depends on the size of the air gap.

8. HALL-EFFECT SWITCH

According to the Hall Effect, when a thin semi-conductor chip (usually silicon) carrying current is crossed at right angles by a magnetic field, a potential difference is produced at the edges of the chip at 90° to the path taken by the signal current (Fig. 16.12). The Hall Effect is very small in metals, but is considerably increased when applied to semiconductors such as germanium and silicon, where that effect has been used as a very accurate electronic switch. The Hall voltage depends on the change of strength of the magnetic field, due to which this effect can be used as a switching device to trigger the ignition point by varying the Hall current.

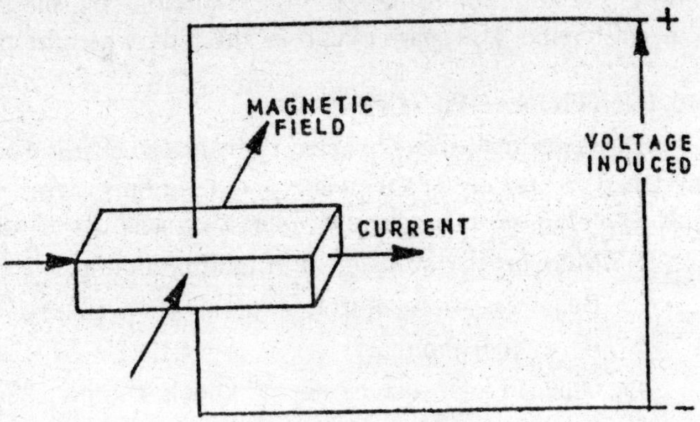

Fig. 16.12. Hall effect.

A semi-conductor chip, retained in a ceramic support is fixed opposite a permanent magnet. The input and the output (Hall) currents flow in the chip at right angles to each other. There is a steel shutter with windows between the chip and the magnet.

The steel shutter is connected to the distributor rotor and the number of windows is equal to the number of engine cylinders. As the distributor rotates, the shutter and the window alternately covers and uncovers the semi-conductor chip from the permanent magnet, thereby disallowing and allowing the magnetic field to strike the semi-conductor sensor. As the shutter covers the sensor from the magnetic field, the current stops flowing from the sensor. This stoppage of the sensor current stops the base current in the electronic control unit transistor thereby turning off the primary (emitter-collector) current there, thus acting as a timer to switch off the primary current, producing the spark. The dwell period is governed by the angular spacing of the windows. If the space between the windows decreases, the closing time for the primary circuit increases. This type of timer switch is more accurate than the magnetic pulse generator described earlier. Further, it is also possible to produce a spark with a Hall-effect switch even with the engine stationary, which is not possible with the inductive pulse generator. Thus it needs care while handling this system due to the risk of getting an electric shock.

Besides ignition, Hall-effect switch is commonly used on engines having electronic fuel injection on account of its high accuracy and long term reliability, which results in optimum performance, lowest fuel consumption and lowest emissions.

Hall effect switch can be tested with a DC voltmeter. However, an ohmmeter should never be used for testing the same, as the Hall chip can be damaged by the voltage from the ohmmeter.

9. OPTICAL SWITCH

It replaces the contact breaker points and is assembled inside the distributor cap. It consists of a radiation source, a photo-cell and a rotating disc attached to the distributor drive & having as many slits at regular intervals as the number of cylinders. The radiation source is a gallium arsenic cell operating at a constant level by means of a zener diode stabilizer, whereas the photocell is a silicon photo-transistor directly coupled to a second transistor, both forming a Darlington amplifier.

The gallium arsenide lamp provides infra-red rays. When these radiations fall upon a semi-conductor material, valance electrons are released, creating holes. As a result, a base current flows in the P-type region which triggers the transistor's collector-emitter circuit. The spark occurs at the plug when the primary current is broken.

The radiation source is placed above the disc whereas the photo-diodes are below it. When the engine is cranked or running, the slits in the disc periodically interrupt the radiation rays reaching the photo-cell which cause the base-current there to stop, thereby stopping the primary current. However, when the slit allows the radiation to pass on to the photodiode, the base current is 'on', thus triggering the primary current circuit. The spark occurs at the plug when the primary current is broken.

10. IGNITION ADVANCE

The centrifugal and the vacuum advance mechanism as described in the conventional point type ignition systems are also used in the electronic ignition systems. However, computer-controlled advance is being increasingly used in electronic systems on account of following advantages.

1. More precise adjustment of ignition timing.
2. Better emission control.
3. Better fuel economy.
4. Other inputs, such as engine knock, can be taken into account.
5. Improved starting and better idle control.
6. Other control inputs can also be used, *e.g.*, the coolant temperature and ambient air temperature.
7. Number of components subject to wear considerably reduced.

Computer-controlled advance system may or may not have the conventional centrifugal and the vacuum mechanisms. In general, such a system has sensors for engine coolant temperature, engine speed, exhaust temperature, manifold absolute pressure, atmospheric pressure and throttle position. An engine speed sensor obviates the necessity to have centrifugal advance mechanism, whereas manifold pressure sensor avoids the requirement of vacuum advance mechanism.

11. CAPACITOR DISCHARGE SYSTEM

In this system, electrical energy is stored in a charging capacitor. When ignition is timed to occur, a thruster power switch completes the capacitor-primary winding circuit of the ignition coil. Due to this, the capacitor will discharge through the primary winding instantaneously, causing a sudden flow of current their which induces a very high voltage surge in the secondary winding, causing the spark to occur at the spark plug terminals This voltage has a very fast rise time compared with a conventional inductive system.

Fig. 16.13 shows the layout of a capacitor discharge system. In this a 475 volt power supply is used

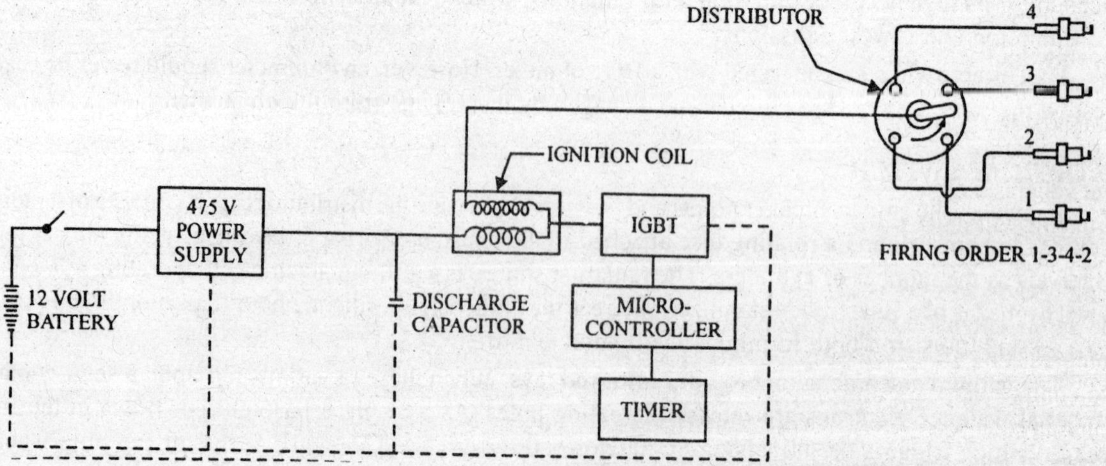

Fig. 16.13. Capacitor discharge ignition system

to charge the capacitor. The energy stored in the capacitor = V^2C, where V is the voltage and C is the capacitance of the capacitor used. An electronic timer such as an IGBT (Insulated Gate Bipolar Transistor) then discharges this stored energy into the primary winding of the ignition coil instantaneously, when a high voltage pulse appears on the secondary winding of the coil, firing the spark plug.

The rise time for a typical capacitor discharge ignition system is 3-10 kV/ìs compared with 300-500 V/ìs in case of pure inductive system. This super fast rise time and high voltage ensures that even a carbon- or oil-fouled plug will be fired.

The characteristics of this system are much better than the induction type system. However, the duration of the spark, about 0.1 millisecond, being very small, can cause problems particularly during starting and idling and even during normal running in modern high speed engines for igniting reliably the lean mixtures used therein. This short duration can be sometimes remedied by taking the advantage of high spark voltage and increasing the spark plug gap which provides a larger spark, or alternatively, this problem is also overcome by providing multi-sparking facility. However, even this very small spark duration works very well with a *direct ignition* (*i.e.* separate coil for each plug) system.

Multi-sparking systems use a specially designed power supply that quickly recharges the storage capacitor, allowing multiple sparks to be fired upto about 3000 rpm, above which all multiple spark systems revert back to single spark system.

Various advantages of the capacitor discharge system are:

1. It produces a very high secondary voltage.
2. The secondary voltage output is constant throughout the speed range since the primary winding of the ignition coil always receives same energy discharge from the capacitor.
3. The build-up of the output voltage is very fast, about ten times the speed in case of the inductive type of ignition. Due to this there is much smaller risk of the h.t. current shorting to earth, e.g., through a fouled plug-insulator.
4. There is no need to control dwell angle since the charging time is independent of engine speed.

Capacitor discharge ignition has been successfully used on some models of Porsche 911 and Ferrari.

12. DISTRIBUTORLESS IGNITION

This type of electronic ignition system was introduced in mid 1980s. In this there is no distributor and the spark plugs are fired directly from the coils. The spark timing is controlled by an Ignition Control Unit (ICU) and the Engine Control Unit (ECU). Therefore, the only maintenance required is to replace the spark plugs. The distributor of spark to various cylinders according to firing order is controlled by the Electronic Control Unit and/or the vehicle computer. Besides, instead of a single ignition coil for all cylinders, there is separate ignition coil for each cylinder (*direct ignition system*), or two cylinders may share one coil (*waste spark method*). Usually the number of ignition coils used is equal to half the number of engine cylinders for which 'waste spark method' is used. To understand how this method works, one has to remember that in a four stroke cycle engine, both valves are closed during compression and power strokes whereas at least one valve is open during the intake and exhaust strokes. If a spark is made to fire at the end of the exhaust stroke, it will have no effect on the engine (lost spark) since there are only burnt gases there. However, a spark at the end of the compression stroke would ignite the charge there. Moreover, because of the low compression and the exhaust gases in the 'lost spark' cylinder, the voltage required for the spark to jump the gap is only about 3 kV, which is quite low compared to the voltage required for spark to occur in the compression cylinder. Therefore, there is no much consumption of energy in the 'lost spark', due to which the compression cylinder spark is not much effected. Thus if spark plugs are made to fire at the same point on each revolution, it would ignite the charge in one cylinder and would not consume any noticeable energy in the other cylinder (Fig. 16.14). Thus two cylinders are paired with one ignition coil. The system is generally used only on four-or six-cylinder engines as the control system becomes too complex for higher number of cylinders. On a four-cylinder engine with firing order 1342 cylinders 1 and 4 are paired and cylinders 3 and 2 are paired (Fig. 16.15), while on a six cylinder engine with firing order 153624, cylinders 1-4, 5-2 and 6-3 are paired. Each ignition coil is a true transformer. To trigger the coils at the correct time, a crankshaft Hall-effect sensor is usually employed.

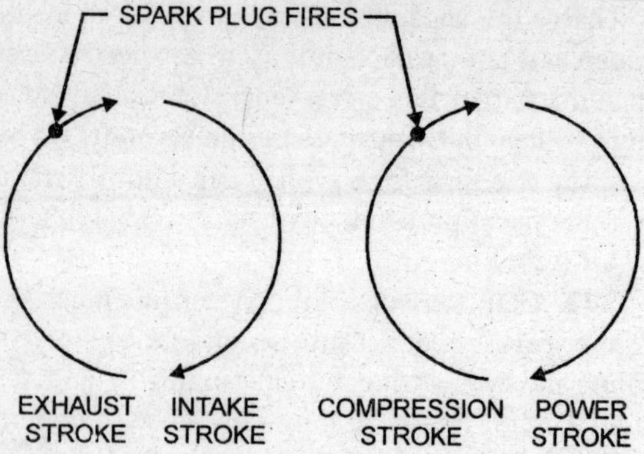

Fig. 16.14. Spark plugs firing at the same point of each revolution in two cylinders.

It may also be noted that the spark on one of the paired cylindes jumps from the spark plug centre electrode to the earth electrode, whereas in the other cylinder of the pairs, it jumps from the earth electrode to the centre electrode. The disadvantage of this is reduced life of the spark plugs on account of increased wear.

In the distributorless ignition system, the ignition module takes the place of the distributor. It uses crankshaft/camshaft sensor data to control the timing of the primary circuit in the coils. Primary current is controlled by transistors in the control module. The dwell time is also controlled by the control module.

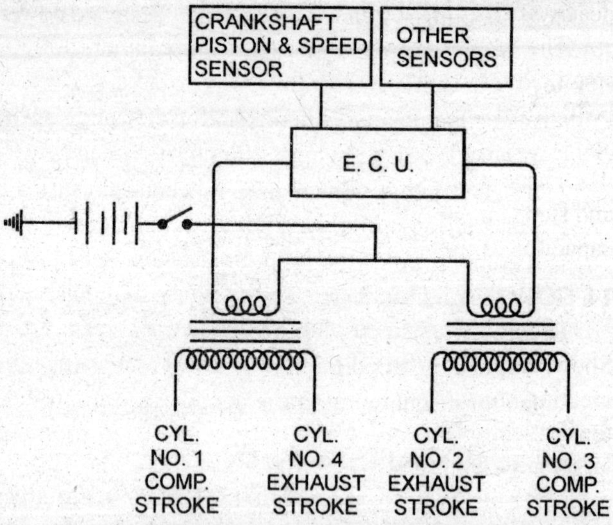

Fig. 16.15. Paired cylinders in a distributorless 4-cylinder

The distributorless ignition system has the following distinct advantages due to which it is being increasingly used:

1. More stable ignition timing resulting in better performance, saving in fuel and reduced emissions.
2. Due to fewer moving parts, less friction and wear; consequently less maintenance required.
3. No mechanical timing adjustments required.
4. No mechanical load on the engine for operation of distributor.
5. Less ratio frequency interference due to absence of rotor-to-cap gap in the case of a distributor.

Direct ignition system has been discussed in the next article. In a way, this can be considered an extension of the distributorless system

13. COIL ON PLUG (COP) OR DIRECT IGNITION

In this, individual ignition coils are used on the spark plugs, thus eliminating the spark plug wires, thereby avoiding the following disadvantages of the wires:

1. Voltage leaks and misfires due to bad plug wires.
2. Even the toughest insulation can burn if a wire rubs up against a hot exhaust manifold.
3. Connection between the wire and the plug terminal inside the plug boot can also be damaged if someone jerks the wire to remove the plug boot when changing spark plugs.
4. Plug wires can also radiate magnetic fields which may affect the nearby sensor wires or other electronic circuits.

Thus the use of coil on spark plug top gives the following benefits due to which it is being increasingly used in new engines:

1. Reliability and durability of the ignition system is improved.
2. Individual coil for each plug means more time (compared to single coil) between each firing, thus increasing the coil saturation time (the time required to build up its magnetic field), thereby increasing the coil output voltage at high engine speeds, where misfire is likely to occur in case of a single ignition coil.
3. COP delivers about 30 percent more spark energy compared to single coil system, thereby improving combustion and avoiding misfire in case of lean air/fuel mixtures exceedingly being used in modern engines.
4. Easy to mount.

In a typical ignition system using coils on top of spark plugs, a crankshaft position sensor generates a basic timing signal by reading notches on the crankshaft, flywheel or harmonic balances. This signal goes to the PCM (Powertrain Control Module), where it is used to determine firing order and activate/deactivate the individual ignition coils.

The 'direct ignition' method ensures that the rise time for the low inductance primary winding is very small, which results in a very high voltage spark to occur at the plug electrodes. This voltage, which may be higher than 40 kW, provides efficient initiation of the combustion process under cold starting conditions and with weak mixtures. Some direct ignition systems use capacitor discharge ignition.

These coils have low primary resistance (0.4 to 0.6 ohm in case of Chrysler 2.7, 3.2 & 3.5 L engines) and steps up the primary voltage of 13 volts to 40,000 volts. On Chrysler COP systems, there is also a coil capacitor for each bank of coils to suppress radio noise.

14. CORONA IGNITION SYSTEM

Federal-Mogul's ACIS (Advanced Corona Ignition System) introduced in 2011 at the Frankfurt Auto Show, employs a high-energy, high frequency electrical field to produce repeatable, controlled ionization, creating multiple streams of ions to ignite the fuel mixture throughout the combustion chamber. In contrast the conventional spark ignition creates only a small arc in the gap of the spark plug. ACIS creates a significantly larger high-intensity plasma ignition source spread throughout the combustion chamber which promotes more thorough combustion and enables advanced combustion strategies, e.g., lean-burn, very high EGR to be used, resulting in further increase of fuel efficiency. The advantages claimed for ACIS over the conventional coil and plug system are:

1. Better combustion resulting in fuel economy and CO_2 emission reduction. Up to 10 percent improvement of fuel recorded by the Company on a 1.6 litre turbocharged direct ignition petrol engine.
2. Compatible with existing petrol engine architecture which means ready replacement of the conventional system on the existing engines, by the manufactures.
3. Allows for optimised timing of the start of combustion.
4. Prolonged system life and extended service intervals due to elimination of the spark plug and its arc, which are the source of electrode erosion in the conventional ignition system.

15. COMPARISON OF CHARACTERISTICS

1. In the inductive system, the build-up and collapse of magnetic field is slow compared to the capacitor discharge system which causes the spark to jump the spark plug gap for a longer time, thus helping the better ignition and hence combustion of the air/fuel mixture. On the other hand the rapid production of the secondary voltage in case of capacitor discharge system does not give any time for the secondary voltage to leak off and even an oil-fouled plug would also fire.

2. It is seen from Fig. 16.16 that in all inductive type ignition systems, the power consumption decreases

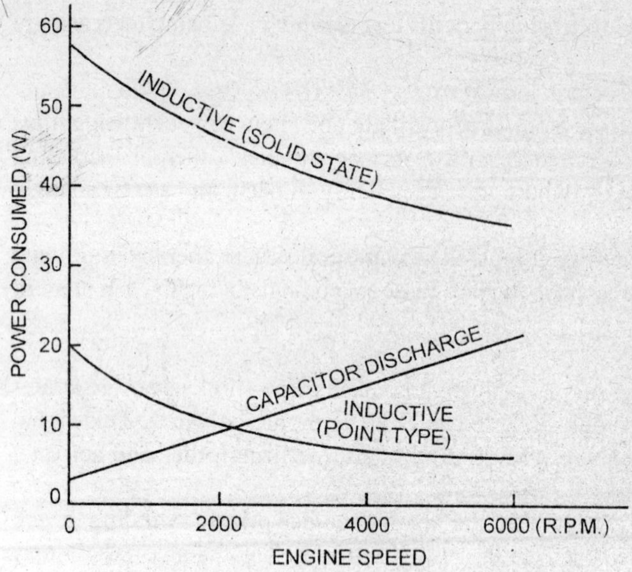

Fig. 16.16. Comparison of power consumed

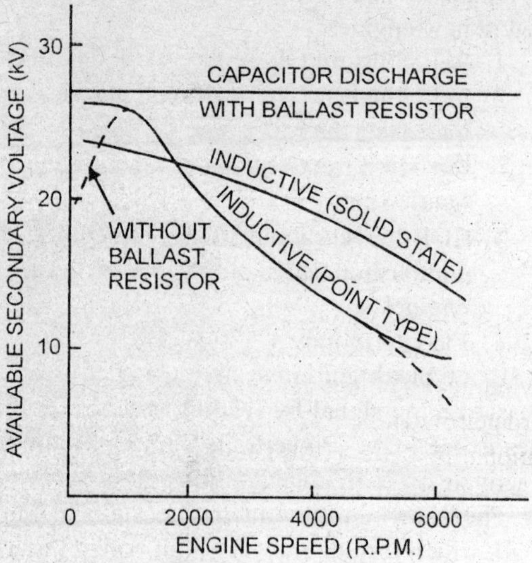

Fig. 16.17. Comparison of secondary voltage

with engine speed, whereas the solid state inductive system consumes almost three times the power consumed by the conventional point type system. In case of capacitor discharge system, there is linear power consumption increase with speed.

3. Regarding the available secondary voltage (Fig. 16.17), in case of all inductive systems, it decreases with increase of engine speed, the rate of decrease being more with the conventional point type system, wherein further inaccuracies may arise due to bouncing of contact breaker points at higher speeds. Further in case of point type, the low rate of points opening at idling speed may cause arcing at the points. The secondary voltage provided in case of the capacitor discharge system is both highest as well as constant with speed. The sparks thus have higher energy content at all speeds. Moreover, since the capacitor discharge is very fast, the voltage rise is extremely rapid so that even the conducting deposits over spark plug electrodes do not cause any hindrance to the production of a high energy spark. An inherent limitation of this type is the extremely short spark duration, which makes it unsuitable for certain types of combustion chambers.

16. LASER IGNITION

The latest development in the automotive ignition systems is the replacement of conventional spark plugs with lasers for creating sparks. The advantages claimed for laser ignition are:

1. Laser ignition provides multipoint, multipulse ignition.
2. It provides flexibility of positioning the ignition.
3. It does not have metal electrodes which erode with use, thus providing a much longer life.
4. With this type of ignition, the spark is going to be stronger which allows leaner air-fuel mixture to be used in the engine, causing more complete combustion of the fuel, resulting in lower consumption and reduced hydrocarbon emissions.
5. The lower idle speeds and better cold engine performance compared to engines with conventional spark plugs.
6. Lower idle speeds and better cold engine performance.
7. Faster and more stable combustion.
8. Some of the laser can be reflected back from inside the cylinder to provide information on the level

of ignition and type of the air/fuel mixture being used, thus serving as a sensor to control the air and fuel quantities automatically resulting in optimised performance.

However, as yet there are still many obstacles to make it viable for production. These are:

1. Downsizing of lasers to the size of a normal plug.
2. Delivering the laser beam through optical fibre is quite difficult and bending the optical fibre tightly causes the fibre to fail after some time.
3. Higher density of laser energy may cause degradation of fibre with time, resulting in loss of beam transmission and thus in loss of ignition.

However, this work is still at an early stage, with engineers from Ford Motor Co. through their U.K. engine R & D centre and researchers from the University of Liverpool and GSI Group in England already having achieved significant emission reductions with experimental laser ignition systems on production Ford petrol engines. The impressive work in this connection is also being done by Takunori Taira of Japan's National Institute of Natural Sciences.

THEORY QUESTIONS

1. Discuss the advantages of electronic ignition system compared to the conventional electrical ignition systems.
2. Describe briefly various types of electronic ignition systems.
3. Describe in detail (*a*) semi-conductors, (*b*) diodes, (*c*) thyristor.
4. Explain the functions and operation of a transistor. Describe also how a transistor is represented conventionally.
5. Discuss in detail the construction and use of an optical switch.
6. Explain clearly the principle of an electronic ignition system. Discuss in detail the pulse generator and the Hall-effect types of timers used in electronic ignition systems.
7. Describe the capacitor discharge method of ignition and discuss its merits and demerits.
8. Describe in detail the working of a distributorless ignition system. Explain also its advantage over the distributor type ignition systems.
9. Write a brief note on Corona ignition system.
10. Compare in detail the characteristics of the point type, pointless inductive type and the capacitor discharge type ignition systems.
11. Write a detailed note on 'Laser ignition in automotive engines'.

QUESTIONS FOR VIVA

1. Why does an electronic ignition system have a longer life than the conventional electrical systems?
2. Why is the electronic ignition system better suited especially for modern automotive engines?
3. What is a semi-conductor?
4. What is 'doping'?
5. What are N-type and P-type semi-conductor materials?
6. What is a 'hole'?
7. What is a diode? What is its function?
8. How does the current flow across the junction of a diode?
9. What is 'reverse bias' in case of a diode?
10. What is a Zener diode? For what purpose is it used?
11. What is a transistor?
12. State the functions of a transistor.

13. Name the three parts of transistor.
14. How is the conventional representation of a (*i*) diode, (*ii*) zener diode, (*iii*) transistor done?
15. What is a thyristor? Where is it used?
16. What is a Darlington pair?
17. What is essentially different in case of electronic ignition system from the conventional electrical systems?
18. Name the three parts of a pulse generator.
19. What is the function of a pulse generator?
20. State the Hall effect as used in electronic ignition.
21. What is the function of a Hall effect switch?
22. What is the advantage of a Hall effect switch over a pulse generator?
23. Why should an ohmmeter never be used for testing a Hall-effect switch?
24. What does an optical switch consists of?
25. Which type of radiation source is used in an optical switch?
26. How many ignition coils are used in a 4-cylinder engine with distributorless ignition system?
27. How are the ignition coils triggered in a distributorless ignition system?
28. What are the advantages of a distributorless ignition system?
29. What is the advantage of placing ignition coils in a distributorless ignition system, directly on the spark plugs?
30. What are the main advantage and disadvantage of capacitor discharge ignition system?
31. What is the duration of spark in a capacitor discharge ignition system?
32. What does 'ACIS? stand for?
33. State the advantages claimed for laser ignition.

MULTIPLE CHOICE QUESTIONS

Mark the correct answers:

1. A semi-conductor is a material with
   (*a*) more than four electrons in its outermost orbit
   (*b*) less than four electrons in its outermost orbit
   (*c*) exactly four electrons in its outermost orbit
   (*d*) none of the above
2. N-type semi-conductor material is obtained by doping with
   (*a*) antimony (*b*) iron
   (*c*) boron (*d*) silicon
3. Two types of semi-conductors are
   (*a*) doped and undoped (*b*) P and N
   (*c*) hole and electron (*d*) while and grey
4. Current will flow through a diode when it is connected
   (*a*) in forward bias (*b*) in reversed bias
   (*c*) to any suitable battery in any manner (*d*) none of the above
5. A zener diode is used for
   (*a*) rectification of current (*b*) blocking the current flow in reverse direction
   (*c*) voltage regulation (*d*) all of the above
6. A transistor consists of a base,
   (*a*) diode and emitter (*b*) diode and collector
   (*c*) emitter and collector (*d*) diode and thyristor

7. A transistor is controlled by the current at
   (*a*) the base (*b*) the emitter
   (*c*) the collector (*d*) both the emitter and the collector
8. The ignition coil in an inductive electronic ignition system is triggered on and off by means of a
   (*a*) contact breaker (*b*) diode
   (*c*) permanent magnet (*d*) timer
9. A pulse generator consists of a
   (*a*) permanent magnet, ignition coil and electronic control unit
   (*b*) permanent magnet, reluctor and electronic control unit
   (*c*) ignition coil, reluctor and electronic control unit
   (*d*) permanent magnet, reluctor and timer coil
10. The most accurate timer for electronic ignition system is the
    (*a*) diode (*b*) transistor
    (*c*) Hall effect switch (*d*) pulse generator
11. The number of ignition coils in a distributorless ignition system for a 6-cylinder engine is
    (*a*) 1 (*b*) 2
    (*c*) 3 (*d*) 6

[**Key:** 1. (*c*), 2. (*a*), 3. (*b*), 4. (*a*), 5. (*c*), 6. (*c*), 7. (*a*), 8. (*d*), 9. (*d*), 10. (*c*), 11. (*c*)]

17

Storage Batteries

1. FUNCTION

A battery is a device for storing chemical energy and converting the same into direct current in a controlled manner. Storage battery is perhaps the most important component of the entire electrical system of a vehicle. It supplies the current to operate the lighting and various accessories, when the engine is not running, for the starting of the vehicle as well as to ignition system, lighting system and to various electrical/electronic components and accessories when the engine is running and over a wide temperature range, say –30°C to 70°C. Besides this the battery also serves to stabilize the system voltage by absorbing any abnormal, transient voltages in the electrical system of the entire vehicle, thereby protecting many components against excessive voltage surges. Thus it is seen that the storage battery is a foundation block in the structure of the vehicle electrical system.

2. TYPES

The most widely used battery is the lead-acid type. Lead-acid batteries are secondary batteries, *i.e.*, they can be recharged after having been discharged. Primary batteries can be discharged only once after which they have to be thrown away.

The following types of batteries shall also be discussed here:

(*i*) Alkaline Battery
(*ii*) Zinc-air Battery
(*iii*) Nickel-Metal hydride battery
(*iv*) Lithium-ion Battery

3. LEAD-ACID BATTERY

3. 1. Construction

The major components of the battery are:

1. Container
2. Plates
3. Separators
4. Cell covers
5. Electrolyte.

Fig. 17.1. 6 Volt Lead-acid battery. 1. Separator, 2. Battery Plate, 3. Cell connector, 4. Plate connector

A 6-volt 'Exide' lead-acid battery is shown in Fig. 17.1. Different components are shown separately.

3.2. Container

Battery containers are of single-piece construction and were at one time made of either hard rubber (moulded ebonite) or of a bituminous material. However, modern car batteries are invariably made of polypropylene, which is a strong, translucent and light-weight plastic. Unlike some plastics, it does not become brittle when cold due to which it can resist jerks and knocks during handling. It is not attacked by acid and can withstand the fluids (petrol, diesel, anti-freeze, brake fluid) usually found on a vehicle.

There are partitions inside the container, which divide it into compartments for different cells. At the bottom of each compartment there are bridges formed on which the battery plates rest. The spaces between the bridge ribs serve to contain sediment resulting from normal chemical action. Without these ribs the sediment would short the positive and the negative plates.

3.3. Plates

There are two types of battery plates, the positive and the negative. For each plate there is a supporting frame work or grid made of an alloy of lead and antimony. The percentage of antimony is usually upto 1.5. It give strength to otherwise soft lead, has a good electrical conductivity and is not affected by the action of the electrolyte. However, the greater the percentage of antimony, the greater is the amount of gassing which requires more water to be added to the battery cells periodically. To reduce the evolution of hydrogen, alloys with a low antimony content (less than 2 percent) were developed. It was found that addition of very small (0.02 percent) formed fine lead selenide (PbSe) particles in the molten low-antimony lead alloy act as nucleants during the solidification process, forming a large number of crystals resulting in the desired fine-grain structure which improves the grid strength inspite of the percentage of antimony being low.

Batteries using these types of alloy lose water very slowly, which increases the refilling intervals, under normal conditions to about 3 years. That is why these are called "maintenance-free" as per DIN. Besides this, the antimony-content is sufficient to stabilize the cyclic performance of the battery so that more than 1000 charging/discharging cycles are possible. 'Maintenance-free' batteries use about 0.1% calcium instead of antimony. Such batteries generally do not require the addition of water during normal use. Moreover in their case there is less corrosion of battery cables and terminals, relatively longer shelf life (about three times longer than a lead-antimony battery) and resistance to overcharging due to higher internal resistance of lead-calcium plates. However, due to the same higher internal resistance lead-calcium grid batteries are difficult to jump-start and cannot tolerate deep cycling as compared to lead-antimony grid batteries. Moreover, due to increased shedding of active plate material, they are more prone to deep discharge failures (dead battery).

The function of the grid is to hold the active materials and to carry current from the electrodes to the external load. One such battery grid is shown in Fig. 17.2, which consists of horizontal and vertical ribs. This design has been used universally. However, recently radial design battery grids have been used (Fig. 17.3) which has lower resistance for the current generated anywhere in the grid to reach the connector lug compared to the conventional design grids.

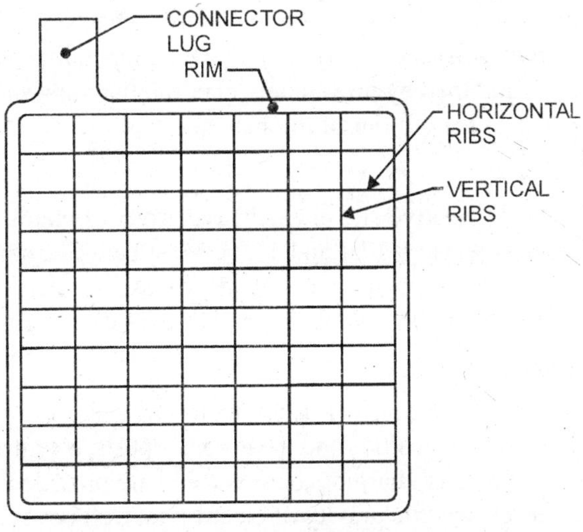

Fig. 17.2. Battery grid

The active material in the positive plate grid is

red lead (Pb_3O_4) and in the negative plate grid it is litharge (PbO) in the form of paste. After the 'forming' process (which consists of electrolysis after immersing the plates with the paste in a dilute solution of sulphuric acid), they are converted into *lead peroxide* (chocolate brown in colour) for the positive plate and *spongy lead* (grey in colour) for the negative plates. The negative plate also contains small amounts of additives to give the battery good discharge-performance at low temperatures, which helps to improve starting. A number of positive plates are lead burnt to a post strap to form a positive plate group, while the negative plate group contains one plate more than the positive plate group so that both sides of the positive plates can be utilized on which greater electron chemical activity takes place. Positive terminal is identified either by + sign or is coloured red. Also the positive post is usually larger in diameter than the negative post.

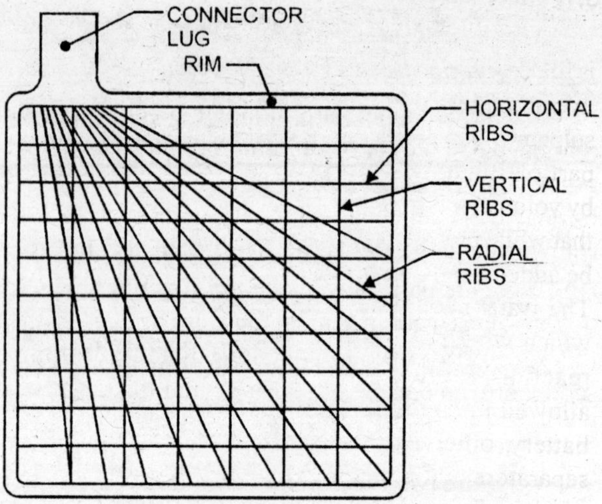

Fig. 17.3. Radial design battery grid

3.4. Separators

To avoid direct contact and thus short-circuiting of positive and negative plates, thin sheets of some non-conducting material are inserted between them. These sheets are called separators. These are made from treated wood, ebonite, porous rubber, fibre glass, resin-bonded cellulose and sintered P.V.C. (polyvinyl chloride), which is perhaps the most widely used separator material.

The material of the separators, apart from being non-conducting, must be sufficiently porous or perforated to allow diffusion of the acid. It should also be able to resist the high temperatures and strongly acidic oxidizing conditions occurring in the battery. Most modern separators are made of micro-porous polyethylene, which has the right properties to withstand the conditions in the batteries.

The separators must be slightly thinner than the space into which they fit. There are vertical grooves on one face of the separator. They are made as wide as possible to facilitate free circulation of electrolyte over the surface of the active material. The grooved sides are arranged adjacent to the positive plates to provide free circulation of acid there, because it is on the positive plates that the greatest electro-chemical activity exists. Recently envelope type separators have been used in some batteries which enclose the entire plate and thereby prevent any material that may shed from the plates from causing a short circuit between plates at the bottom of the battery.

3.5. Cell

One positive and one negative group of plates are slid over each other, with separators in between, to form a cell (Fig. 17.1 and 17.4). Most batteries have one more negative plate than positive plates in each cell. However, many recent batteries use the same number of positive and negative plates in the cell. The size of the plates and their number per cell determine the capacity of the battery.

3.6. Cell Covers

Cell covers provide a ceiling for the cells. They are moulded from hard rubber. Each cell cover contains holes for the positive and the negative posts, a vent and a filler opening. The vents allow the gases produced during battery charging to escape and are provided in the plugs to cover filler opening which are also used to check specific gravity of electrolyte. Cell coves are sealed in the container by running pitch around the edges to form a leak proof joint.

3.7. Electrolyte

After assembling completely, the battery is filled with electrolyte. The electrolyte used in the lead-acid batteries is a solution of sulphuric acid. It contains approximately one part of sulphuric acid to 2 parts of water by volume. It is very important to remember that while preparing electrolyte, acid should be added very slowly to the distilled water. The water should never be added to acid, which would result in a violent chemical reaction. Further, electrolyte should be allowed to cool before it is poured in the battery, otherwise it may damage plates, separators and container. Specific gravity of the electrolyte is a very important property, which has to be kept under consistent observation to have a proper watch over the state of the battery. The state of charge, temperature and the self-discharge are the factors affecting the gravity value. (Specific gravity is generally referred to as only 'gravity' in the usage related to batteries).

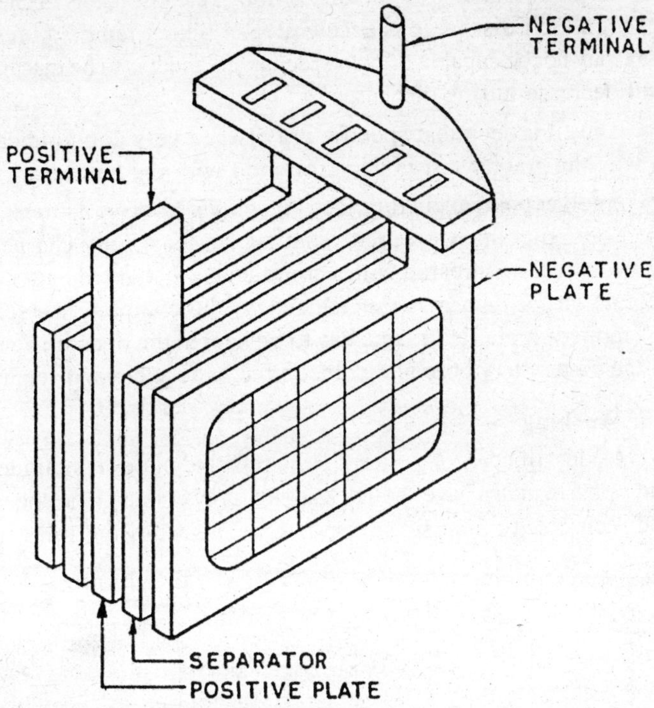

Fig. 17.4. Battery cell.

In the discharge process, water is produced, which dilutes the electrolyte, reducing its gravity. In the fully charged condition, the gravity is 1.290 at 15°C, while it is reduced to about 1.110 in the completely discharged state. In common usage the decimal points are omitted; for instance gravity value 1.290 will usually be referred to as 'twelve ninety.')

The electrolyte gravity also varies with temperature. It increases with decrease of temperature and *vice-versa.* The gravity value, therefore, is always mentioned at a specified standard temperature, which is usually 15°C. The readings noted at the other temperatures are applied correction to reduce them to gravity values at standard temperature. A table showing gravity values at different temperatures for the battery in fully charged, half charged and fully discharged conditions or else a diagram depicting graphically these values, is invariably given by the manufacturer (Fig. 17.5).

As a rough guide 18 points are added for every 25°C excess of temperature above 15°C and the same number is subtracted for every 25°C reduction of temperature below 15°C. For instance, if the gravity reading at 40°C is 1.250, the corrected reading at 15°C would be

$$1.250 + \frac{0.018}{25}(40 - 15)$$

$$= 1.250 + 0.018 = 1.268$$

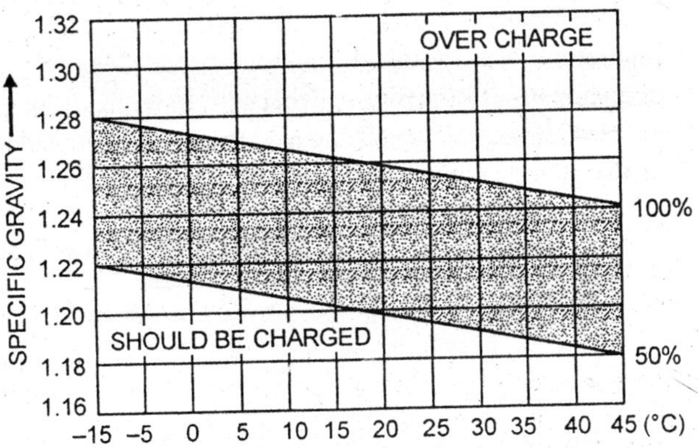

Fig. 17.5. Variation of specific gravity of battery electrolyte with temperature (Courtesy—Maruti Udyog Ltd.)

When the battery is kept idle for long periods, it gets discharged by itself. This is due to internal chemical reaction. Self discharge is accelerated at higher temperatures. A battery which is badly discharged in this way may not be capable of being recharged and has to be discarded. Obviously, the gravity value of electrolyte will decrease during discharge.

Thus it is seen that specific gravity is a very good indicator of the state of charge to the battery. That is why the gravity test is very common with the batteries.

Another point about the electrolyte is variation of its freezing point with gravity. As the gravity increases, the freezing point is lowered. That means for a fully charged battery, the freezing point of electrolyte is much lower as compared with a completely discharged battery. This is a pointer towards one very important precaution for battery maintenance; *i.e.* the completely discharged batteries, in particular, should not be left in the vehicle if the same has to be kept in the open on a very cold night. That may cause the electrolyte in the battery to freeze and damage the battery beyond repairs.

3.8. Working

As described earlier initially the positive plates contain lead peroxide PbO_2 (Chocolate brown colour) and negative plates have spongy lead, Pb (grey colour). The chemical changes that take place during discharging and recharging processes are shown by means of the following equation:

$$\underset{\left[\begin{array}{c}\text{Positive}\\\text{plate}\end{array}\right]}{PbO_2} + \underset{\left[\begin{array}{c}\text{Electro}\\\text{lyte}\end{array}\right]}{2H_2SO_4} + \underset{\left[\begin{array}{c}\text{Negative}\\\text{plate}\end{array}\right]}{Pb} \underset{\text{charge}}{\overset{\text{discharge}}{\rightleftarrows}} \underset{\left[\begin{array}{c}\text{Positive}\\\text{plate}\end{array}\right]}{PbSO_4} + \underset{\left[\begin{array}{c}\text{Electro}\\\text{lyte}\end{array}\right]}{2H_2O} + \underset{\left[\begin{array}{c}\text{Negative}\\\text{plate}\end{array}\right]}{PbSO_4}$$

On discharging, both PbO_2 and Pb are converted to lead sulphate, $PbSO_4$ (while colour). Water is produced, which dilutes the electrolyte solution. During recharging, lead sulphate on positive plates is converted into lead peroxide, PbO_2 while on negative plates lead sulphate is converted to spongy lead. Water is split up into oxygen and hydrogen gases during charging process. The hydrogen combines with sulphate ions to reform sulphuric acid. In this way the specific gravity of electrolyte is increased during charging process.

However, as the recharging process is complete the chemical action stops and hydrogen and oxygen gases are liberated at negative and positive plates respectively. Thus the gassing phenomenon is an indication that battery has been fully charged. Any further charging will only result in the decomposition of water in the electrolyte.

The chemical activity inside the battery is dependent upon the temperature of electrolyte. At higher temperatures the chemical reactions proceed more vigorously while at lower temperatures they become sluggish. This is opposite to what would be desirable; because at lower temperatures, the oil viscosity is increased and hence the starting load is increased considerably. The following table by Exide gives an indication of the effect of temperature on the engine requirement and battery efficiency.

TABLE 17.1. EFFECT OF TEMPERATURE

S. No.	*Temperature °F (°C)*	*Power required to crank engine (percent)*	*Power available from battery (percent)*
1.	80 (26.6)	100	100
2.	32 (0)	155	65
3.	0 (17.8)	210	40
4.	– 32 (– 35.5)	350	25

3.9. Cell Voltage

The open circuit voltage of a fully charged battery cell is about 2.1. A six-volt battery would, therefore, be made of 3 cells, while a 12-volt battery would contain 6 cells connected in series as shown in Fig. 17.6. With the exception of some old cars, all cars use 12 volt negative earth systems. The cell voltage should not fall below 1.7 volts during service. It may be noted here that cell voltage is not affected by the cell size and the number of plates.

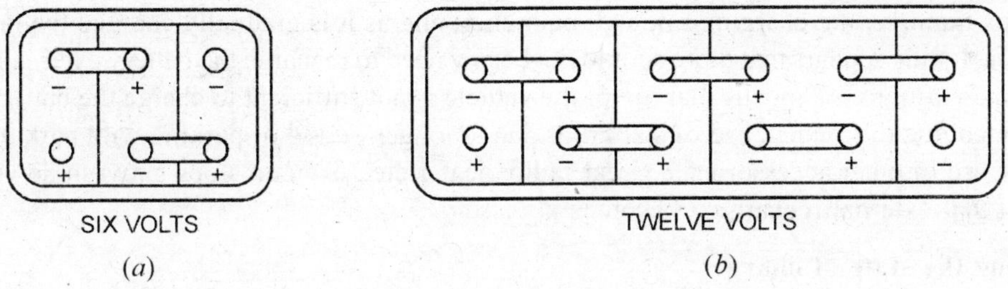

Fig. 17.6. Arrangement of cells in a battery (*a*) 6 volt battery, (*b*) 12-volt battery

4. BATTERY CAPACITY

The capacity of a battery is defined as the amount of current it can deliver when fully charged. It depends upon the number and area of plates in the cell and the quality of the electrolyte present. As mentioned earlier, the battery capacity decreases with decrease of temperature.

5. BATTERY RATINGS

Rating of a battery is determined by the current it can produce and the time for which it can sustain this current. Batteries are rated in many ways, the most important of which are given below:

1. **Twenty-hour rate (Ampere-hour capacity).** It indicates the lasting power of a battery on small load. It represents the rate of current a battery can deliver continuously for 20 hours after which the cell voltage should not drop below 1.75. During the test the temperature of the battery should be approximately 80°F (26.7°C). This is usually given in ampere hours (Ah). This rating is important for powering accessories when the engine is not running.
2. **Twenty minute rate.** It represents the rate of current a battery can deliver continuously for 20 minutes, the cell voltage not dropping below 1.5. The temperature of the battery at the start of the test is brought to 80°F (26°C).
3. **Reserve capacity.** It is an important measurement for the total capacity of the battery. It indicates how long a battery can keep the engine running if the charging system fails. It is the time in minutes for which the battery can supply 25 amp at 80°F (26.7°C), without the cell voltage falling below 1.75. A 44 Amp hour battery has a reserve capacity (RC) of about 60 minutes.
4. **Cold rate (zero test or cranking ability test).** It is a measure of the worst conditions under which a battery is expected to deliver current, *e.g.*, during starting the engine and operating under extreme cold. It represents the current in amperes which the battery can supply continuously for 30 seconds (DIN) or 1 minute (BS) without the cell voltage dropping below 1.2 V. The temperature of the battery at the beginning of discharge is kept 0°F (– 18°C). The cold rating is also abbreviated as CCA (*Cold Cranking Amperes*). The CCA rating of an automotive battery must equal or exceed the number of cubic inches of engine displacement. In metric terms 1 ampere of CCA is required for every 16 cm^3 of engine displacement. Thus, a 2.4 litre engine must have a battery with CCA rating of 2400/16 = 150 A.

A battery for an ordinary car may be rated as 44 Ah, 60 RC and 170A CCA. A 'heavy duty' battery will have the same Ah rating as equivalent 'standard duty' battery, but shall have higher RC and CCA.

6. BATTERY CHARGING

6.1. Need for recharging

A generator is always provided which is constantly charging the battery. The rate of charging current flowing from the generator to the battery depends upon the state of charge of the battery. If the battery is run down, for example, the charging rate will be higher; and as it is gradually charged the rate tapers off to lower value. It thus appears that there should not be any need to recharge the battery externally. However, under certain conditions it happens that run of the vehicle is not sufficient to charge the battery, which has been discharged due to frequent use of starter, or due to longer period of parking with parking lights ON. Similarly the use of other accessories, *e.g.* car radio, heater etc., even on stops may run down the battery to the extent that external recharging becomes necessary.

6.2. Checking the state of charge

The state of battery charge is checked most commonly by noting the specific gravity (the details have been discussed in Art. 3.7 and 7.1). The temperature is also recorded and the gravity reading is corrected accordingly. This is then compared with the chart supplied by the battery manufacturer to ascertain whether the battery needs recharge or not. However, the following are the general indications:

TABLE 17.2. STATE OF BATTERY CHARGE

S. No.	*Condition of battery cell*	*Specific Gravity*	
		climates usually below 32°C	*climates usually above 32°C*
1.	Fully charged	1.290	1.230
2.	Half discharged	1.200	1.140
3.	Completely discharged	1.110	1.050

The cell voltage is also measured to indicate the state of charge. The details of these tests have been explained in Art 7.2 and 7.3.

6.3. Charging details

1. First of all, check the electrolyte level in the battery. If it is below the top edge of plates, put more electrolyte so that correct level is attained. Also, note the gravity and temperature readings.

2. If the battery needs recharging connect its positive and negative terminals to the respective terminals of the battery charger, which is simply a source of direct current supply. The battery terminals are usually marked, but even if they are not, their polarity can be established by immersing the ends of the wires connected to them in the electrolyte solution contained in a separate vessel. More intense formation of hydrogen bubbles will take place at the wire connected to negative terminal.

3. Adjust the value of the charging current. This is generally kept as half (in amperes) of the number of plates in the cell. For example, the charging current for a 17-plate battery would be 8.5 amperes. Alternatively, the recharging current should be set to one tenth of the battery's twenty hour rating. For example, in case of a battery with 30 ampere-hour capacity, the charging current should be set at 3 amperes. Continue the charging till the gassing beings; then decrease the charging current and continue till there is no further increase in the gravity and cell voltage readings for three hours.

4. Keep a watch over the temperature of the electrolyte during charging process. If the temperature exceeds 55°C, discontinue charging till the battery cools down, when the charging may be restarted.

5. During the charging process the gravity and the temperature readings should be taken hourly reguiarly.

6. Battery overcharging must be avoided because it damages the battery, particularly the positive plates.

7. In case several batteries are to be charged simultaneously, the same may be connected in series.

Generally it takes from 12 to 20 hours for recharging a battery depending upon the condition of the battery and the charging rate. However, batteries with sulphated plates require low charging current and hence take longer time for charging. In these, initial charging is done at lower value of current and it may take 40-60 hours. It must be ensured that no flame is brought near the battery which is being charged as due to evolution of hydrogen and oxygen gases during charging an explosive mixture is being formed.

6.4. Slow rate and quick rate charging

Ordinarily batteries are charged by slow rate charging method. As mentioned already it takes about 12 to 20 hours for charging a battery with this method. It is no doubt a safe method for charging but the long charging period is a disadvantage. In the quick rate charging method, a high charging current is employed till the battery comes to about 80 per cent charge level, when the current is reduced to low value. With this method, a battery can be charged in about one hour only.

The quick charging method demands more rigorous observation and watch during charging and if used with care it does not damage batteries which are in good condition. However, if a battery is not in good condition, the same must not be charged by this method.

6.5. Trickle charging

As the batteries left in storage for longer periods are subjected to self discharge, continuous charging at very low charging rate of about 5 percent of the normal charging current has been employed for keeping in proper condition, the batteries for display purpose for vehicles doing little running. However, continuous trickle charging should not be used and it is recommended that only during day time the battery should be connected to trickle charge while during the night it should be disconnected.

6.6. Factors affecting charging

1. Temperature

Charging rate decreases with decrease of temperature, and vice versa. That means a battery, at higher temperature would take less time for charging, all other conditions remaining same.

2. Area of plates

Battery with small plates is charged faster compared to a similar battery with large plates.

3. State of the charge

A battery which remains long in severely discharged condition will have sulphate crystals on its plate which increases the internal resistance and hence the decrease in charging rate.

4. Gassing

Excessive hydrogen and oxygen bubbles due to overcharging may wash away active material on the plates, decreasing the charging rate.

5. Impurities

Electrolyte impurities decrease the charging rate.

6. Corrosion of terminals and clamps

This may be caused due to spilled electrolyte and condensation from gases and would increase internal resistance of the battery, resulting in decrease of charging rate.

6.7. Dry charged batteries

These days dry charged batteries are also available, which do not contain any electrolyte, but their plates are already charged. The vent holes are closed by plugs so that no moisture enters the battery. Thus the battery may be stored without damage for 2 to 3 years. The electrolyte is supplied, separate, packed in an acid proof carton. When the battery is intended to be put into service, all that has to be done is to open up the vent holes, and put the electrolyte from the carton in the battery to half-fill each cell. Allow it to stand for one to two hours, then top it with additional electrolyte to bring it upto proper level; the battery is ready for use. Alternatively the electrolyte may be prepared and filled in.

7. BATTERY TESTING

Following are the important tests conducted to ascertain the conditions of a battery.

1. Specific Gravity test
2. Open volt test
3. High discharge test
4. Cadmium test.

7.1. Specific gravity test

This is conducted with the help of a hydrometer (Fig. 17.7). The hydrometer is in the form of a glass tubular body with a rubber bulb at the top and a sampler tube at the bottom. There is a glass float inside the glass body. This float has a vertical density scale. To test the specific gravity of the electrolyte, immerse the sampler tube in the cell electrolyte, squeeze the rubber bulb and release the same which would cause a sample of the electrolyte to be drawn inside the glass body. Let the float inside rise and then read off the scale at the surface level of the sample drawn in.

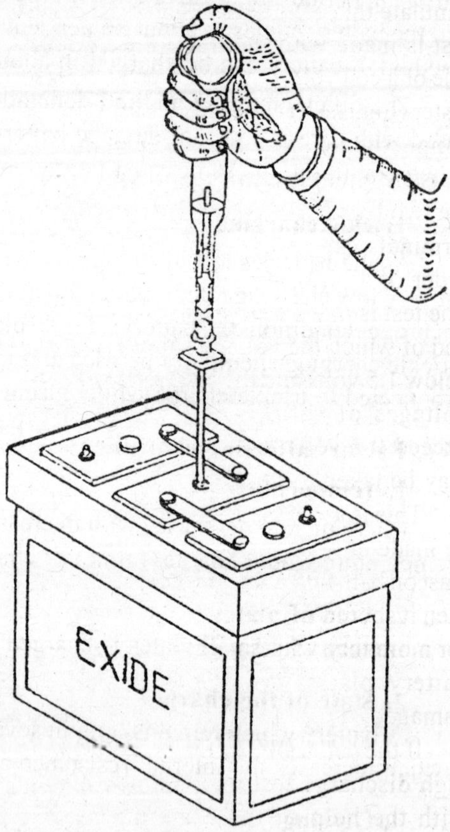

Fig. 17.7. Specific gravity test with a hydrometer.

As already discussed, the gravity of electrolyte is effected by its temperature. The same, therefore, has also to be recorded with the help of a thermometer which is either built-in within the hydrometer or is separate. The correction is then calculated and the specific gravity reading corrected accordingly.

There are certain precautions regarding the test. First is that gravity test should be carried out while the battery is gassing as this would affect the reading. Secondly, the test should also not be made immediately after the water has been added to adjust the electrolyte level, because the reading will be erroneous and will not be representative of the bulk of the electrolyte. So either the test should be made before adding any water or if it very necessary to add the water, the battery, should be operated for sufficient time afterwards so that the water mixes thoroughly with the electrolyte.

7.2. Open volt test

More and more vehicles are now fitted with maintenance-free (sealed) batteries, where the hydrometer cannot be used to determine the state of charge. In such a case, the same can be determined from the open volt test of the battery. The open circuit voltage of the battery cell is measured with the help of an ordinary voltmeter. The open circuit voltage of a fully charged battery cell is about 2.1 volts. If the car has been recently driven or the battery has been just charged, the 'surface charge' must be removed from the battery by running on the headlights for one minute, turning off and waiting for about two minutes. Then with the engine and all electrical accessories shut off, connect the voltmeter to battery terminals. In case of a 12-V battery, if the voltmeter reads 12.6 V, it is fully charged, if it reads 12.2 V it is half-full charged while below 11.9 V it may be taken as fully discharged. It has been observed that a change of 0.01 volt of open-circuit voltage is equivalent to a change of 0.010 in the specific gravity of the electrolyte. Thus the gravity measurement can be indirectly made with an accurate voltmeter also. There is another way of expressing relation between the open circuit voltage of a battery cell and its specific gravity. This is:

Voltage of the cell = Specific gravity + .840

7.3. High discharge test

Open circuit voltage is not representative of the cell voltage under actual operating conditions. The cranking motor at the time of starting draws a very heavy current which causes the cell voltage to fall. To simulate this condition a high discharge test is made with the help of a carbon pile or a cell voltage tester. A cell voltage tester (Fig. 17.8) consists of a voltmeter connected to two legs with a high resistance placed across these. Pressing the two legs of the cell tester on the cell terminals causes heavy current (of the order of 150 to 200 amperes) to flow. The test is made for 5-10 seconds at the end of which the voltage should not fall below 1.5 volts and the difference of cell voltages of various cells should not exceed 0.2 volts; otherwise the battery may be taken to be defective.

This is a very severe test and should be made only on the fully charged or at least on half-fully charged batteries. Even then it should not be made continuously for more than 15 seconds, otherwise the battery plates will get permanently damaged.

If the cell tester is not available, the high discharge test may also be made with the help of the cranking motor when the ignition switch is off. The voltage readings are taken with the help of an accurate voltmeter.

Fig. 17.8. Cell voltage tester.

7.4. Cadmium test

This test is conducted to know the chemical conditions of the positive and negative plates and is performed only when the battery is either on charge or discharge.

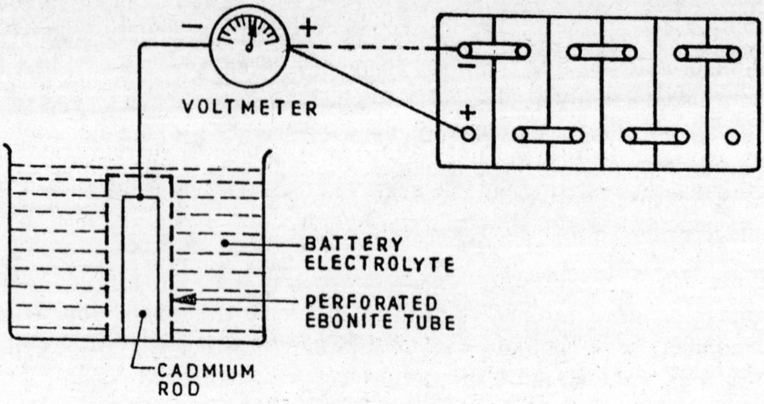

Fig. 17.9. Cadmium test

A cadmium rod is enclosed in a perforated ebonite tube and is immersed in the battery electrolyte. The rod is then connected to the negative terminal of a high resistance voltmeter whose positive terminal is connected alternately to the positive and negative terminals of the battery cell. In case of positive terminal, the voltage reading should not be less than 2.5 volts if the battery is nearing completion of its charge or 2.0 volts if the battery is being discharged at normal rate. A lower reading indicates defective positive plates. On the other hand, reading in case of connection with negative battery terminal should not be more than 0.2 volt when the battery is being discharged at normal rate. A higher reading in this case will indicate defect in negative plates.

8. BATTERY PROBLEMS

1. Self discharge

A battery, if left standing gets discharged by itself; this phenomenon is called self-discharge. It is common for the new batteries even to get self discharge at the rate of one percent per day of storage. The rapid self-discharge (3 percent or above) is considered, a serious defect which must be remedied.

The causes of self-discharge are contamination of electrolyte, damaged separators and long term storage. Care must be taken, therefore, to ensure always to use pure sulphuric acid (not the commercial one) and distilled water for preparing electrolyte. The use of tap water should never be resorted to. Only in case of emergency pure clean drinking water may be used. While inspecting the battery also, it must be ensured that no dirt, dust etc. enters inside.

Long term storage of the battery causes stratification of the electrolyte, *i.e.* the bottom layers become heavier than the top layers. Such stratification sets up local equalising currents which increase the self-discharge rate.

The first thing to be done about a battery experiencing self discharge is to discharge it to 1.2 volt per cell. This will transfer metallic impurities from negative plates to the electrolyte. Then the electrolyte is drained and the battery is washed thoroughly first with clean water and then with distilled water. Fresh electrolyte is then put in and the battery recharged.

2. Sulphation

As discussed already the battery plates in discharged condition contain lead sulphate which gets converted into active materials on being recharged. However, if a battery is kept standing in discharged condition, large lead sulphate crystals, whitish in colour, are formed on the plates. These crystals are difficult to be converted back into the active materials on recharging. If the sulphated battery is kept standing for long time, the lead sulphate crystals grow and rupture the plate grid, thus rendering the battery useless. Further, sulphation of the battery plates reduces the capacity and increases the internal resistance of the cells.

The other prominent causes of sulphation are, very high specific gravity of electrolyte and insufficient electrolyte in the battery.

The batteries not excessive sulphated can be recharged keeping the charging rate low. The precaution to avoid sulphation are that the battery should be inspected regularly and water should be added whenever required to keep the electrolyte level and gravity at proper value. Further, heavily discharged batteries should not be used.

3. Overcharge

Normally when a battery is fully charged, a very small amount of charging current is allowed to flow into the battery. However, in case of a defect in the alternator, a much higher current may pass through the battery all the time the vehicle is running, which would damage the positive plates and would cause excessive water-loss from the battery, which no longer remains maintenance-free. A dark brown or black colour on the bottoms of the vent plugs is a sure indication of overcharge.

If the alternator, in case of non Start-Stop vehicle, has voltage more than 14.8 volts at normal temperature, it indicates the faulty charging system. Defective rectifier diode will cause the charging voltage of 16.0 V at the battery. This alternator must be repaired immediately to avoid permanent damage to the battery.

In latest 'start-stop' vehicles with regenerating brakes, higher voltages (15.2 V) are used to optimize charging efficiencies and to reduce charging periods.

4. Undercharge

This happens when the battery is not receiving sufficient charge to bring it to the full charging state. This can occur if the vehicle is being used only for occasional short trips or for start-stop city driving. Undercharging also arises if the alternator voltage is between 13.6 and 13.8 volts.

Undercharging slowly results in sulphation of battery plates.

5. Deep cycling

Ordinary lead-acid batteries are not designed for deep cycling *i.e.* continuous cycles of charge and discharge to extreme level. Special deep-cycling batteries are used in modern Start-Stop vehicles. However in case of ordinary batteries, continuous deep cycling may cause the active material on positive plates to fall gradually to the bottom of the battery, slowly reducing the ability of the plates to store electricity. A strong indication of the deep-cycled battery is the presence of a large number of small brown/black particles in the battery electrolyte.

6. Internal short-circuiting

The internal short-circuiting is caused by either the damaged separators or when the active material dropping from the plate forms itself into a heap so high as to cause short-circuiting. It results into rapid self discharge and sulphation. The remedy is to clean the battery and replace the damaged separators.

7. Deterioration of plates

This may be caused by any of the following:

(*i*) Very high charging or discharging rate. Excessive heat generated will cause the plates to buckle. For this reason, starter should not be used continuously for long periods.

(*ii*) High specific gravity of the electrolyte.

(*iii*) Freezing of electrolyte.

(*iv*) The electrolyte level not being maintained properly.

(*v*) Mechanical damage due to rough ride, particularly when the battery is not mounted properly. Their only remedy is the replacement of damaged plates.

8. Cracking of container

The container develops cracks generally due to mechanical shocks caused by rough road jolts or rough handling. If the cracks are on the outer walls, the electrolyte will leak. On the other hand, cracks on the interior walls will cause the internal short circuiting of the plates of the two adjoining cells. The cracked container has to be replaced.

9. Corrosion of battery terminals and clamps

It occurs due to oxidation and the battery terminals and clamps should be cleaned after regular intervals. After cleaning with a cloth moistened with liquid ammonia, a coat of petroleum jelly helps to retard their corrosion again.

10. Loss of water

The water in the electrolyte keeps on evaporating and periodic checks must be made to ensure that proper level and gravity is being maintained. If this is not done, the separators and the plates may be permanently damaged. For old lead-antimony grid batteries, the requirement of 30 to 60 ml of water per cell per 1600 km was considered normal. Lead-calcium grid batteries, however, do not require any water to be added during normal service, as discussed earlier. Excessive water loss may be caused due to higher than normal charging rate due to incorrect setting of the regulator. A cracked container may also be the reason of unduly high water loss.

Always use distilled water for topping up and then make sure that proper gravity value, apart from correct level, is being maintained.

11. Cell gravity variation

The gravity readings of all the cells should be same. If the reading of some cell is different from the rest by more than ten gravity points, it indicates loss of acid from this particular cell.

12. Discolouring of electrolyte

This is an indication that the plates are disintegrating and the battery has almost exhausted its life.

9. BATTERY MAINTENANCE

The life of the battery, apart from its original design and valid use, depends to a large extent on the attention which is given for its proper maintenance.

The vehicle batteries must be inspected periodically, say, every week. The following points may be taken care of:

1. The battery terminals should be clean and tight. The electrolyte may escape from the battery due to overfilling and cause the corrosion of the terminals. A loose connection at the terminals will result in a faulty starting and discharged battery. The battery top collects dust due to moisture present there and if not cleaned for long, will cause the battery to discharge slowly.

To clean the battery top and terminals, brush away the dirt first, then wipe a rag moistenend with weak ammonia or sodium bicarbonate solution and finally flush off with clean water. Care should be taken that ammonia or soda solution does not enter the battery cells. After cleaning tighten the battery terminals properly and apply vaseline or petroleum jelly thereon. This provides protection against corrosion. (Do not use grease).

2. Remove vent plugs, taking care that no flame is brought near the vents because the gas inside is highly inflammable.

Use a hydrometer to measure the specific gravity of the electrolyte to ascertain the state of charge of the battery. Apply the temperature correction to the measured gravity reading. It is important to remember that battery must be stabilized (*i.e.*, the surface charge removed if any), before ascertaining the state of charge. To stabilize a battery after charging turn on the car headlights on main beam for 15 seconds, or otherwise apply a 15 ampere load for 15 seconds. This will remove any surface charge after which the battery may be tested. If the battery is overcharged the regulator should be set properly. If the battery is found discharged, the same may be recharged and fitted again. However in case the same trouble occurs consistently, the cause for the same must be located and defect, if any, must be rectified.

3. In case the electrolyte level in the battery is not sufficient, top up with distilled water. If distilled water is not available sometime clean water may also be used, but this should always be kept in mind that any mineral content in the water shortens battery life. For topping up, use only earthen ware or a glass funnel, never use a metal funnel.

If too much water is required for topping up and particularly if this happens consistently, investigate the reason for this. This may be due to overcharging which may be avoided by setting the regulator properly. Another reason could be a cracked container allowing the electrolyte to leak out. This is rather more likely the reason if all the battery cells do not required excessive topping up simultaneously.

4. Never let the battery remain in discharged condition, otherwise the plates will become sulphated.

10. PRECAUTIONS DURING HANDLING OF A BATTERY

1. Battery electrolyte being corrosive, avoid its contact with eyes, skin and clothes. In case it goes into the eyes accidentally, wash the eyes with clean water and consult your doctor. In case of skin, however, wash with solution of bicarbonate (baking soda) to neutralise the acid and then wash clean water. Avoid the contact of battery electrolyte with painted surfaces and metal parts, which are easily attacked by the acid.

2. Do not bring any flame near the battery, as the highly explosive vapours coming out of the battery may get ignited accidentally.

For the same season never use spark-producing tools near a battery.

3. To avoid accidental arcing, always remove the negative (ground) cable first while disconnecting and connect the same last while assembling.

4. Use proper protective clothing, *i.e.*, apron, gloves and face shield, while handling batteries.

5. While making battery connections, take special care to observe proper polarities so that correct connections are made.

6. Jump start batteries only if permitted by the battery manufacturer. Some manufacturers specially prohibit jump start.

7. Do not use the ground terminal when checking for ignition sparks.

8. Never disconnect the battery with the ignition switch ON.

9. Never use an electric welder without the battery cables disconnected.

10. Never place tools on the battery.

11. As disconnecting the battery may erase the memory on electronic devices; it is a good practice to note down trouble codes and programmed settings before disconnecting the battery.

11. JUMP STARTING

In case of emergency when the battery is run down, your vehicle can be jump started with the charged battery of another vehicle, *provided* the battery manufacturer has not prohibited such a procedure to be adopted. Various steps involved are:

1. Move the vehicle with the fully charged battery closer to your vehicle so that their batteries are as near as possible. However, take care that the *two vehicles should not touch each other at any point.*
2. Apply the hand brake of your car and bring the gear lever to neutral position.
3. Attach one end of one jumper cable to the positive terminal of the charged battery, while the other end be connected to positive terminal of your discharged battery.
4. Attach one end of the second jumper cable to the negative terminal of the charged battery, while its other end be connected to some solid engine ground (such as the generator mounting bracket) at a distance of at least 500 mm from the discharged battery of your vehicle. *Do not connect directly to the negative terminal of the discharged battery.*
5. Now start the engine of the other vehicle with charged battery after turning off all electrical accessories and lights. Then turn the ignition key of your engine, which will start the same.
6. Disconnect the negative jumper cable *first* from the other vehicle and *then* from your vehicle. Similarly remove the positive jumper cable *in the reverse* order from the one it was connected.
7. Do not shut-off the engine of the vehicle that was jump started, for at least 10-15 minutes, which would give the vehicle's alternator time to recharge the battery. Ideally, the engine should not be stopped until you get home or a garage, to be sure that the battery-life is not over (which may need immediate replacement) or the alternator/regulator is not defective.

Important precaution— Before attempting to jump start, inspect both batteries for any signs of physical damage, particularly the cracks in the container and/or leaking fluid. In case either is visible, forget about jump-starting, because any such attempt could result in a catastrophic explosion caused from sparks igniting volatile gases escaping from the battery.

12. FUTURE OF LEAD-ACID BATTERY

The euphoria over the development of lithium-ion battery makes one think about the standard wet-flooded (conventional lead-acid) battery going obsolete. However thinking logically, no such possibility is seen in the near future. It is estimated that fully hybrids, including the plug-ins where Li-ion battery gets preference, would be having only 6% of the market by 2020. Moreover, 12 volt lead-acid battery still remains the choice in most hybrids for starting, lighting and some accessories. It is also used in micro-hybrids, vehicles that can stop and restart automatically.

Besides above, practical improvements are also taking place in the design of lead-acid batteries,

1. Absorbent Glass Mat (AGM) battery
2. Enhanced Flooded Battery (EFB)

These will now be discussed brefly as follows:

1. AGM battery

Johnson Controls Inc. (JCI), Yuasa of U.K. have brought out recently a new battery, 'AGM', which has far longer life, greater capacity for deep cycling and larger number of cycles possible, useful for microhybrids. This battery has two lead plates with a glass mat between them, which absorbs the electrolyte so that there is no free liquid. Any gases given off during charging are also allowed to be recombined into water, making the battery totally maintenance-free. The sandwich of lead plates and AGM (Absorbent Glass Mat) is rolled into a tight, rigid cylindrical cell. This is a sealed battery, although a relief valve is also provided. The other advantages of this battery are:

1. The AGM battery has 3-4 times the cycle life of a conventional lead-acid battery.
2. It is the most efficient and flexible of the lead-acid type of batteries. Due to ultra high pure lead and acid used, it converts upto 96% of the charge into electrical energy as compared to 80-85% for the conventional lead-acid battery.

3. It has lower internal resistance and similar life compared to the conventional design.
4. It can discharge quickly, which is a plus point for restarting.
5. Use of pure lead for plates gives it improved performance and reduced corrosion for double the life of the conventional lead-acid batteries using lead-strengthening alloys.
6. It is claimed to have 0.8 to 2.9 kg less mass than the comparable ordinary lead-acid batteries, the difference varying according to size.
7. Higher starting power even at lower temperatures. Increased reaction surface area ensures increased energy densities for faster engine rotation during starting.
8. Spill proof/leak proof.
9. Compatible with sensitive electronic equipment.
10. The acid absorption of the glass mat means that the battery packs can be operated under higher pressures than conventional flooded batteries which produce superior vibration resistance of AGM design.

The only disadvantage is that it costs double than the conventional lead-acid battery. It must be noted, however, that an AGM battery should never be opened after it leaves the factory, otherwise sulphation could occur on the plates leading to irreversible loss in performance.

2. EFB battery

Also known as AFB (Advanced Flooded Battery) or ECM (Enhanced Cycling Mat), this type of batteries are enhanced version of the standard wet-flooded type, but as a lower tier option to AGM batteries in terms of performance and durability.

Compared to 30,000 engine starts from standard flooded type, EFB type will provide 85,000 starts. The improvement comes mainly from the addition of carbon additives in the plate manufacturing process and high charge acceptance envelope type separators. EFB technology also provides improved (about 100%) cyclic durability over standard wet-flooded batteries.

This type is suitable for entry-level start-stop vehicle as a cost-effective solution. EU emission targets for these vehicles are lower than those for higher performance models.

AGM batteries are now factory-fitted to many luxury cars and 'Stop-Start' vehicles, for example, the 'Optima' brand of JCI, which is a spiral-cell AGM, is already used on Corvettes. Similarly Yuasa's AGM batteries are being used in many Japanese and European vehicles, *e.g.*, Mazda MX5 and Toyota Prius.

13. ALKALINE BATTERY

Alkaline batteries are of two types, the Edison type or the nickel-iron type, and the nickel-cadmium type. The construction of these is similar except the active materials which are different. However, only the latter type is suitable for automobile service.

The basic construction of the alkaline battery is similar to that of lead-acid battery. In both the nickel-iron and the nickel-cadmium types, the active material on positive plates is nickel hydroxide. On negative plates, it is metallic iron in nickel-iron cell and cadmium oxide in nickel-cadmium cells. The electrolyte used is potassium hydroxide solution. The active material for positive as well as negative plates is contained in finely perforated steel tubes, which combine together to form a plate. Due to this reason, even under severe jolts, no active material is lost from the plates and thus no sediment is produced.

The electrolyte does not take part in chemical reactions on charging or discharging, unlike in the lead-acid batteries. Its specific gravity remains, therefore, constant at about 1.20. For this reason, smaller quantities of the electrolyte are required. However, due to the same reason, there is no indication of the state of charge of the battery.

The advantages of the alkaline battery over the lead acid type are:

1. Alkaline battery is lighter than lead-acid battery.
2. It is mechanically very strong and can stand very rough use.
3. The rate of charging can be much higher in case of alkaline battery.
4. The discharge rate when left standing is very low as compared to lead-acid type, since the active materials are insoluble in the electrolyte. Therefore, the battery has got much longer shelf life.
5. Comparatively longer life (10-15 years).

The disadvantages, however, are:

1. The voltage of each cell in the alkaline battery is only 1.2 as compared to 2 volts for a lead-acid cell. For this reason for a 6-volts battery 5 cells will be required as against 3 cells of the lead acid type.
2. Due to relatively higher internal resistance, the efficiency of the alkaline battery is slightly lower than that of the lead accumulator.
3. Initial cost is higher compared to the lead-acid battery. However, this is amply compensated by the extra life.

14. ZINC-AIR BATTERY

This type is of recent origin and unlike the conventional battery in which energy is stored and drawn as required from the active material of the electrodes, it drawn its energy from the zinc within its cells and from oxygen of the atmosphere. It can be operated as a primary battery, that is, to be thrown away after use, and as a secondary or rechargeable battery.

The zinc-air battery cell (Fig. 17.10) consists of porous zinc oxide in the centre and a thin air-cathode on each side. Air cathodes are made from a hydrophobic material, which supports a conducting mesh and a layer of catalyst (to improve the rate of reduction of atmospheric oxygen). This hydrophobic material is impervious to electrolyte, but allows the air to pass through it : it is used as container also. The electrolyte used is potassium hydroxide During discharge, cathode remains uncharged, while anode is changed to zinc oxide. The system has an open-circuit voltage of 1.4 volts.

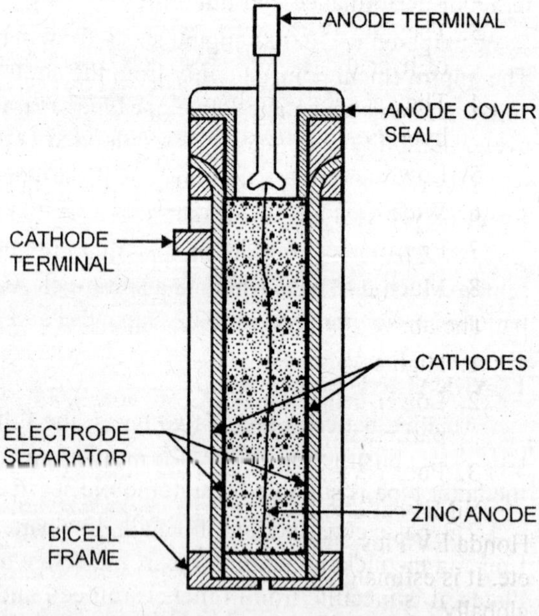

Fig. 17.10. Zinc-air battery cell.

The advantages of the zinc-air battery are:

1. High energy densities.
2. Readily available and cheap raw materials are used.
3. Good discharge characteristics over a wide range of temperature.
4. Minimum of ancillary equipment is needed.

The main disadvantage of this type of battery is that carbon dioxide from the air forms carbonate which reduces conductivity. For this reason, these batteries have to be used quickly after activation.

Zinc-air batteries have been widely used as primary batteries in the form of prismatic or cylindrical or button cells.

15. NICKEL-METAL HYDRIDE BATTERY

This type of battery is similar to the nickel-cadmium battery, but uses a hydrogen-absorbing alloy for the negative electrode instead of cadmium. As in NiCd batteries, the positive electrode is nickel oxyhydroxide (NiOOH).

NIMH cells have an alkaline electrolyte, usually potassium hydroxide. For separation hydrophilic polyolefin non-wovens are used. A NiMH battery has two to three times the capacity of an equivalent size nickel-cadmium battery. Cell voltage is 1.2V.

The reaction taking place on the positive electrode of NiMH is

$$\mathrm{Ni(OH)_2 + OH^-} \underset{\text{Discharging}}{\overset{\text{Charging}}{\rightleftarrows}} \mathrm{NiO(OH) + H_2O + e^-}$$

On negative electrode, it is

$$\mathrm{H_2O + M + e^-} \underset{\text{Discharging}}{\overset{\text{Charging}}{\rightleftarrows}} \mathrm{OH + MH}$$

Metal 'M' in the negative electrode is an intermetallic compound, the most commonly being used currently is AB_5 where A is rare earth mixture of lanthanum, cerium, neodymium, praseodymium and B is nickel, cobalt, manganese and/or aluminium.

Various advantages of NiMH batteries are:

1. Having bean patented in 1986, it is a mature technology, by now.
2. Their energy density is more than double that of lead acid and 40% higher than that of Ni Cd batteries.
3. They have very long life which has been demonstrated by the fact they are still functioning in case of hybrid cars well after, 1,60,000 km and over a decade of service.
4. They accept both higher charging and discharging rates (deep cycling) and thus are suitable for hybrid cars. Rapid charge possible in one hour.
5. Low internal resistance, although higher than in case of Nickel Cadmium type.
6. Wide range of operating temperature.
7. Environmental friendly since there is no use of lead, cadmium or mercury.
8. Much safer than lithium type batteries in case of an accident due to the use of lesser active chemicals.

The above advantages are accompanied by the following disadvantages:

1. High self-discharge rate.
2. Lower cell voltage (1.2 V) means more cells are required to make the desired voltage batteries, particularly compared to the Lithium batteries, where cell voltage is 3.2-3.7 V.
3. Poor performance in cold weather.

NiMH batteries are currently being used in all-electric and electric hybrid vehicles such as G.M. EV1, Honda EV Plus, Ford Ranger EV, Toyota Prius, Honda Insight, Ford Escape Hybrid, Honda Civic Hybrid, etc. It is estimated that these batteries will help power about a million newly built hybrid vehicles in 2010 globally.

16. LITHIUM-ION BATTERY

16.1. General

This type of battery has a promising future as rechargeable battery for the electric hybrid vehicles due to its very high storage capacity, which reaches 120 Wh/kg and 270 Wh/*l*, coupled with the fact that it operates at room temperature. Presently its use in the form of small cylindrical cells is fast increasing mainly in cellular phones, computers and camcorders.

Like other batteries, lithium-ion batteries also work by shuttling ions between their electrodes. During charging, ions travel in one direction; during discharging process, they go in the other direction (Fig. 17.11 and 17.12). Positive electrode is usually made from cobalt or manganese oxide and negative electrode is made from graphite. The electrolyte is a lithium-based gel or polymer. As is seen from Fig. 17.11 and 17.12, the Li ions move through the perforated separators during charging and discharging. Tiny bits of

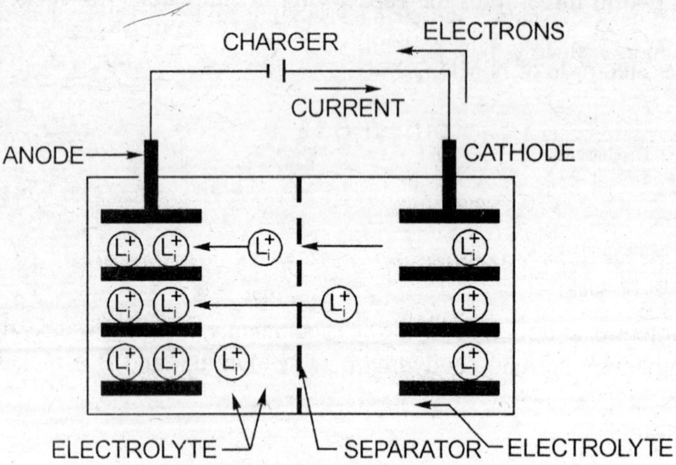

Fig. 17.11. Charging mechanism of Lithium-ion battery

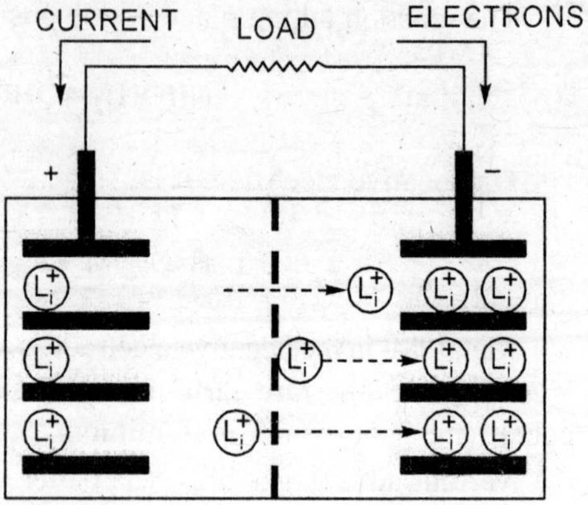

Fig. 17.12. Discharge mechanism of Lithium-ion battery

metal, which may result from the manufacturing process, have the potential to get stuck in these perforations, thus preventing ions from moving freely. If this happens, pressure and heat can build up, which may result in an explosion. Even allowing the ions to move too quickly may also result in overheating. Thus when thousands of such batteries have to be packed in a single casing for automotive use, it has to be ensured that the batteries do not overheat.

Mercedes-Benz S400 Hybrid is the first hybrid series production car equipped with Li-ion battery, launched in 2009.

These batteries having round Li-ion cells employ anodes of aluminium foils coated with graphite and cathodes of copper foils coated with lithium metal oxide. A plastic separator separates the anodes and the cathodes. The wound electrode is soaked with an organic electrolyte, which enables the transfer of lithium ions in the first place. During charging process, lithium ions are released from the metal oxide and move through the separator to the graphite electrode where lithium carbide is formed. The energy stored in the charging process can be released again by the inverse chemical reaction occurring during the discharging process.

The battery consists of a welded sheet metal housing containing a pack of cells consisting of 35 round cells alongwith an aluminium cooler that connects directly to the vehicle air conditioning system. The total weight of the battery is 24.5 kg.

The on going research points out that most promising electrode materials are nano-engineered materials like phosphates of iron, lithium-titanate spinal and bundles of silicon nanowires. Use of these materials increases the available surface area for the ions to bond to and exchange electrons, avoiding a bottleneck that used to result in deformation and hence in decreased effectiveness with traditional materials like graphite. These also decrease the internal resistance of the battery, thereby improving its ability to store and deliver energy.

16.2. Advantages and disadvantages

Compared to other rechargeable batteries, the *advantages* of lithium-ion batteries are:

1. Higher specific energy density than most other types (80 percent more compared to NiMH battery). This means, they can store more energy for their size and weight.
2. Higher specific power density (50 percent more compared to NiMH battery), which means longer cycle life for identical weight/space and/or increased payload/enlarged stowage space for identical power.
3. They can operate at higher voltages than other rechargeables, *e.g.*, the cell voltage for Li-ion is 3.6 volts compared to 1.2 volts for NiMH or NiCd. This would mean lesser number of cells for obtaining required voltage, which would ultimately mean greater reliability.
4. They have lower self-discharge rate.
5. Safety status is equivalent to NiMH batteries.

However, there are some *disadvantages* also:

1. More expensive, since these are more complex to manufacture. However, the cost is coming down fast, with continuous research.
2. These require sophisticated chargers to carefully monitor the charging process, which makes these more complex and costly.

Most of modern all-electric vehicles and plug-in hybrid vehicles use lithium-ion batteries.

16.3. Thermal management

Li-ion batteries should not be stored at temperatures above 60°C and operated at above 40°C. Likewise there is appreciable reduction in the potential charging and discharging rates below –5°C. Operating the Li-ion battery beyond this temperature range shortens the battery life due to irreversible damage occurring within the Li-ion cells. In hybrid vehicles, this would result in reduced electrical support for traction, while in electric vehicles it can even result in breakdown if the above temperature range is violated. Thus thermal management of the Li-ion batteries becomes essential, especially in case of large batteries in plug-in hybrid vehicles and electric cars.

For cooling, the heat from the battery can be dissipated only through the air conditioning circuit. The battery heat from the refrigerant circuit of the air conditioning system can be transferred, using conditioned air, a dedicated coolant circuit or the refrigerant itself. The method used depends upon the type of the vehicle. All the three methods are being currently used in different production vehicles.

Likewise battery can be heated by either an integrated direct electric heater for the battery plate or by heating the secondary fluid using fuel heater or electric heater or waste heat of the I.C. engine.

17. COMPARISON OF BATTERIES

Table 17.3 gives representative values of the specific energy and the specific cost for four types of automotive batteries.

TABLE 17.3. COMPARISON OF BATTERIES

S. No.	*Type of battery*	*Specific Energy (Wh/kg)*		*Specific Cost (U.S. $/kWh)*	
		Present	*Potential*	*Present*	*Potential*
1.	Lead-acid	35	55	125	75
2.	Nickel-Cadmium	40	60	540	115
3.	Nickel-Metal-Hydride	65	120	570	110
4.	Lithium-Polymer	—	200	—	100

18. BATTERY SAFETY TESTING

Batteries in electric and hybrid vehicles can be very dangerous in case of an accident, when they can short-circuit and catch fire. The danger depends on three factors, *viz.*,

(*i*) Chemistry and construction of battery cell.

(*ii*) Battery pack and module design.

(*iii*) Integration of the battery pack into the vehicle.

Currently there are no specific standards to test a battery pack for crashworthiness. However, some relevant industry norms do exist, such as,

(*i*) ECE-R100 relates to requirement for the construction and functional safety of battery electric vehicles.

(*ii*) SAFE J2464 for the safety and abuse testing of energy storage systems.

(*iii*) FMVSS 305 for electrolyte spillage and electrical shock protection.

However, surely coolıng helps to decrease the risk of overheating and work has been done to increase the protection of battery packs. Besides, tests are also being developed to determine failure mechanisms and hazard levels with finite element models to validate results of such a work. The related work is being done at the following places, among others:

(*i*) Virtual Vehicle Research Center in Graz, Austria.

(*ii*) Vehicle Safety Institute of Technical University of Graz, Austria.

(*iii*) Impact and Crashworthiness laboratory at Massachusetts Institute of Technology, U.S.A.

19. RECYCLNG BATTERIES

Hazardous materials have to be prevented from entering the waste stream both during its production as well as at the end of its useful life. That is why recycling of batteries is very important. However, due to small number of electric vehicles so far, the facilities for recycling are limited even in advanced countries like USA. Various types of the recycling processes are:

1. Smelting

These are meant mainly to recover basic elements or salts. These processes are being used extensively today and can be employed for all types of batteries including lead-acid, nickel-metal hydride and lithium-ion batteries. Smelting is done at high temperatures where organic materials such as the carbon anodes and the electrolyte are burnt as fuel or reductant. The precious metals are recovered which may be further refined. The slag contains other materials including lithium.

2. Direct recovery processes

These directly recover battery-grade materials. Using a number of physical and chemical processes, all metals and active materials can be recovered. These are low-temperature processes requiring minimal energy.

3. Intermediate processes

These lie between smelting and direct recovery. These are also suitable for all types of batteries. But in their case the recovery occurs later along the production line than in case of smelting.

20. SUPERCAPACITOR

Also called 'ultra-capacitor' or 'double-layer capacitor', it is a high-capacity capacitor which is rated in Farads, is used for energy storage undergoing frequent charge and discharge cycles at high current and short duration. It can withstand however, low voltage only, between 2.5 and 2.7 volts. To get higher voltage, many capacitors are connected in series, which reduces the total capacitance and increases the internal resistance.

Super-capacitors, unlike batteries, store energy through the separation of highly mobile electrical charges on the surfaces of electrodes thereby producing current very fast. A typical super-capacitor consists of a pair of identical metal plates coated with activated carbon, which has been made porous to increase surface area and charge-holding. The electrode pair is immersed in liquid organic electrolyte which speeds up the transportation of the charge. Such super-capacitors have a specific capacitance of 100 F/g. New nano-structured carbons based on rayon fibre provide specific capacitance of 135 F/g, which is higher (150 F/g) in case of high-purity, stable carbon with 'tunable porosity' made by using a sol-gel process. Graphene which is one atom-thick carbon, is also being tried as super-capacitor-coating material, which is expected to enhance the specific capacitance still further. Either propylene/ethylene carbonates or acetonitrile are used as electrolytes. The later is better, but is prone to fire hazards.

Another alternative for improvement is to boost the voltage of the complete system from 2.7 volts to 3.5 volts, which could yield a 40% improvement in cell price per energy unit, due to better energy density and consequently fewer cell requirements. However, this would need high-purity carbon, costing more. Moreover, it may also cause breakdown of the electrolyte.

20.1. Comparison with batteries

1. Batteries store energy electrochemically where chemical reactions release electrical carriers which can be extracted into a circuit. In the super capacitors, energy is stored electro-statically on the surface of the material and it does not involve any chemical reactions.

2. A battery not only produces and stores energy but also delivers energy over a relatively long time-period, thus providing an extended driving range for an electrical vehicle. On the other hand, super capacitor dispenses its energy over a much shorter period of time, making it suitable for short power boosts.

3. Compared to battery, supercapacitors can be charged rapidly. A super capacitor can be charged in 1-10 seconds whereas it takes 10-60 minutes to charge a Lithium-ion battery. Due to this, supercapacitors can repeatedly release power in quick bursts of current, thus being useful for regenerative braking and stop-start systems. Some Japanese and European micro-hybrids employ super capacitors to enable them to turn themselves off when their engines would normally be idling, and also capture the braking energy. The largest application of the super capacitors is in about 10,000 hybrid diesel buses in China.

4. Compared to cycle life of 500-1000 cycles in case of a Lithium ion battery, a super capacitor can be recycled 500,000 - 20,000,000 times.

5. A lithium ion battery has a life of 5-10 years, whereas it is much longer (10-15 years) in case of a super capacitor.

6. Higher charging temperature range (–40°C to 65°C) in case of super capacitors, as compared to batteries (– 40°C to 45°C in case of a lithium ion battery).

7. Much higher cost per kWh in case of super capacitors (10-40 times compared to a lithium-ion battery), which is the main reason for these being not widely used so far.

8. Low specific energy; holds a fraction of a regular battery.

9. High self-discharge; higher than most batteries.

10. Lower cell voltage needs series connections with voltage balancing.

20.2. General

A battery can be used to store a large amount of electricity over a long period of time while supercapacitors are ideal when a quick charge is needed to fill a short-term power requirement. This avoids too severe demands, on battery. Thus combining the two into a 'hybrid battery' satisfies both long term and quick charge needs and also reduces battery stress, resulting into a longer service life.

TABLE 17.4. COMPARATIVE STORAGE BATTERIES DATA OF SOME INDIAN AUTOMOBILES

S. No.	*Make*	*Type*	*Capacity (amp. hr. at 20 hour rating)*
1.	Maruti 800	NS-405	30
2.	Hindustan Ambassador Mark II	Exide 6XCV9R-12V	62
3.	Fiat 1100	Exide 6XCV9R-12V	62
4.	Jeep Universal	Exide 6XTA9R-12V	50
5.	Dodge/Fargo Model 89M4 (Petrol Type)	Exide 6XCV9L-12V	62
6.	Ashok Leyland Comet Passenger	Exide 6XCS19L-24V	139
7.	Standard 20	11 plates, 12 volts	75
8.	Tata Diesel vehicles	12 volts battery	180

REFERENCES

1. Rachel Evans, Electric & Hybrid Vehicle Technology International, January 2014, pp 42-46.

THEORY QUESTIONS

1. With the help of suitable sketches, explain the construction and working of lead-acid battery.
2. Describe different methods of battery rating.
3. Describe in detail the method of battery charging.
4. Discuss various factors affecting the charging of lead-acid batteries.
5. Explain in detail how you would perform specific gravity tests.
6. Discuss in detail various tests for ascertaining the fitness of a battery to be used in a vehicle.
7. What are the different troubles experienced in an automotive battery? Discuss their causes, effect on the battery and suggest suitable methods to remedy these.
8. Write a comprehensive note on the maintenance of batteries. Describe in detail various precautions to be observed.
9. Describe step by step procedure to jump-start a vehicle battery. Mention also important precautions to be observed.
10. Discuss in detail the future of lead-acid battery, considering the extensive development and use of other types, like the lithium-ion batteries.
11. What are the advantages claimed for an alkaline battery? Describe briefly its construction.
12. Describe clearly the working of the Nickel-metal hydride batteries.
13. Discuss the working of a zinc-air battery and its advantages over the conventional types.
14. Discuss in detail the lithium-ion batteries for automotive use. State also their advantages and disadvantages.
15. Write a comprehensive note on the necessity and methods for thermal management of Li-ion batteries.
16. Discuss briefly the comparison of different types of batteries used in automotive practice.
17. Discuss thoroughly the sources of danger and relevant industrial norms regarding safety of batteries for electric vehicles.
18. Describe various types of recycling processes for modern automotive batteries.
19. Explain clearly how a super-capacitor performs its function.
20. Write a detailed note comparing super-capacitor and batteries for automotive use.

QUESTIONS FOR VIVA

1. What is a battery?
2. What is the function of an automotive storage battery?
3. Name three types of storage batteries.
4. Enumerate main components of a lead-acid battery.
5. What is the material used for battery containers?
6. What is a 'maintenance-free' battery?
7. What is the function of separators in a battery?
8. What does a battery cell consist of?
9. What is cell cover?
10. What is the electrolyte used in a lead-acid battery?
11. What should be the specific gravity of the electrolyte in a fully charged battery?
12. What is the open circuit voltage of a fully charged lead acid- battery cell?
13. How do you define capacity of a storage battery?
14. Name different battery ratings.
15. What are the factors affecting the specific gravity of battery electrolyte?
16. How is the electrolyte gravity reduced when the battery gets discharged?
17. What do you infer from the statement that gravity value is twelve ninety?
18. Approximately, what is the relative variation of gravity with temperature of the battery electrolyte?
19. What is "battery self-discharge"?
20. Why should the battery, particularly a discharged one should not be left outside in a cold night?
21. A generator is always fitted for charging the battery. Why should then the need arise for recharging?
22. How do you check the state of charge of a battery?
23. To what approximate value would you set the charging current, while recharging a battery?
24. Why should the battery be not overcharged?
25. In case several batteries are to be charged simultaneously, how would you connect them?
26. What are the 'slow rate' and the 'quick rate' charging method?
27. What is trickle charging?
28. What is a dry charged battery?
29. What is advantage of a dry charged battery?
30. Enumerate different tests for ascertaining the condition of a battery.
31. What is high discharge test? What is open volt test?
32. What do you understand from the term 'sulphation' and 'internal short-circuiting'?
33. What is the precaution to be observed, if any, while recharging a sulphated battery?
34. What is deep cycling?
35. What indication points to a deep-cycled battery?
36. How is internal short circuiting to the battery caused? What are the possible causes of deterioration of battery plates?
37. What is the precaution to be taken against corrosion and rusting of battery terminals and clamps?
38. Why should only the distilled water be used in the battery?
39. What is to be inferred if the gravity reading of one cell in a battery is different from the rest by more than 10 points?
40. Will the rate of self-discharge increase or decrease with the increase of temperature?
41. Does the concentration of sulphuric acid in the battery electrolyte effect the rate of self-discharge? If so, in what way?
42. What do you understand from the term 'jump start' in case of batteries?

43. Mention important precautions during jump start of batteries.
44. What is an AGM battery?
45. What does 'AGM' stand for?
46. What is an EFB battery?
47. What does 'EFB' stand for?
48. For which type of vehicles does the EFB type battery suit?
49. What electrolyte is used in an alkaline battery?
50. What is the active material on positive plates in a nickel cadmium battery? What material is there on negative plates?
51. What is the open circuit voltage of a cell in the alkaline battery?
52. Out of the lead-acid and the alkaline types of storage batteries which has the higher charging rate and which has the higher discharge rate when left standing?
53. Name the two types of alkaline batteries.
54. How does the specific gravity of electrolyte in case of nickel- cadmium battery vary with the state of battery charge?
55. Why does the alkaline battery have longer shelf life?
56. Why is alkaline battery cleaner compared to the lead-acid type?
57. State the materials used for Ni-metal hydride batteries.
58. Write down the reaction equations for charging and discharging of NiMH batteries.
59. State the advantages of a Nickel Metal Hydride battery.
60. What is a primary battery?
61. For which reasons the danger arises in case of accident of a vehicle with deep-cycling batteries?
62. Name any two places where research work regarding the safety of automotive batteries is being carried out.
63. Name various processes for recycling of batteries for electric vehicles.
64. What is a super-capacitor? Where is it used in automobiles?
65. How does a super-capacitor differ from a battery regarding their functioning?
66. How does a super-capacitor compliments the function of battery for automotive use?
67. What is the electrolyte in a zinc-air battery?
68. What is the open circuit voltage of a zinc-air battery cell?
69. Which materials are used for the cathode and the anode of a zinc- air battery?
70. State advantages and disadvantages of zinc-air battery.
71. What are the merits and demerits of lithium-ion battery?

MULTIPLE CHOICE QUESTIONS

Mark the correct answer:

1. The colour of positive plate of a lead-acid battery is
   (*a*) brown (*b*) grey
   (*c*) white (*d*) black
2. The colour of negative plate of a lead-acid battery is
   (*a*) brown (*b*) grey
   (*c*) white (*d*) black
3. The positive plate of a lead-acid battery has
   (*a*) PbO_2 (*b*) Pb
   (*c*) $PbSO_4$ (*d*) H_2SO_4

4. The negative plate of a lead-acid battery has
   (*a*) PbO_2 (*b*) Pb
   (*c*) $PbSO_4$ (*d*) H_2SO_4
5. Battery electrolyte consists of approximately (by volume)
   (*a*) 35 percent sulphuric acid and 65 percent water
   (*b*) 65 percent sulphuric acid and 35 percent water
   (*c*) 65 percent nitric acid and 35 percent water
   (*d*) 35 percent nitric acid and 65 percent water
6. To increase output voltage, battery cells are connected in
   (*a*) series (*b*) parallel
   (*c*) series-parallel (*d*) None of the above
7. The capacity of a battery is determined by the number of plates per cell and
   (*a*) number of cells (*b*) shape of plates
   (*c*) size of plates (*d*) number of separators
8. A maintenance-free battery
   (*a*) has lead-antimony plate grid (*b*) has lead-calcium plate grid
   (*c*) does not contain acid (*d*) does not contain water
9. The number of positive plates in a battery cell is
   (*a*) one more than the negative plates (*b*) one less than the negative plates
   (*c*) two less than the negative plates (*d*) two more than the negative plates
10. A discharged lead-acid battery has on its plates
    (*a*) PbO_2 (*b*) Pb
    (*c*) Pb_3O_4 (*d*) $PbSO_4$
11. A fully-charged 6 cell automotive battery should indicate
    (*a*) 12 V (*b*) 12.6 V
    (*c*) sp. gravity of 1.290 at 32°C (*d*) Both (*b*) and (*c*)
12. With the increase of battery temperature, the specific gravity of electrolyte
    (*a*) increases (*b*) decreases
    (*c*) stays the same (*d*) none of these
13. An example of improved form of lead-acid batteries is
    (*a*) Absorbent Glass Mat battery (*b*) Zinc-air battery
    (*c*) Nickel-cadmium battery (*d*) Lithium ion battery.
14. The battery which cannot be deep-cycled is
    (*a*) Ordinary lead-acid battery (*b*) Advanced Flooded Battery
    (*c*) Absorbent Glass Mat battery (*d*) Lithium ion battery
15. Out of the following terms which is not a process for recycling of batteries?
    (*a*) Smelting (*b*) Direct recovery
    (*c*) Deep cycling (*d*) Intermediate processes
16. The time required to charge a modern automotive super-capacitor is about
    (*a*) one hour (*b*) 30 minutes
    (*c*) 10 minutes (*d*) 10 seconds

[**Key:** 1. (*a*), 2. (*b*), 3. (*a*), 4. (*b*), 5. (*a*), 6. (*a*), 7. (*c*), 8. (*b*), 9. (*b*), 10. (*d*), 11. (*d*), 12. (*b*), 13. (*a*), 14. (*a*), 15. (*c*), 16. (*d*)]

18 Charging System

1. INTRODUCTION

The function of the charging system in an automobile is to generate, regulate and supply the electrical energy for charging the battery. Main requirements of a charging system, when the vehicle is running, are:

1. To supply the current demands of all the loads.
2. To supply the current required for charging of battery under all conditions.
3. To supply constant voltage under all conditions.
4. Should include a provision to indicate proper working.
5. Should have highest power to weight ratio.
6. Should be reliable and quiet.
7. Should require least maintenance.

The charging system consists of a generator for converting mechanical energy from the engine into electrical energy, a regulator to control the amount of electrical energy so produced, a relay to regulate the flow of the charging current from the generator to the battery relevant to the state of charge of the battery and a voltmeter, an ammeter or indicating lamp to indicate whether the system is operating or not. In the modern cars, the charging system includes an alternator only instead of a generator.

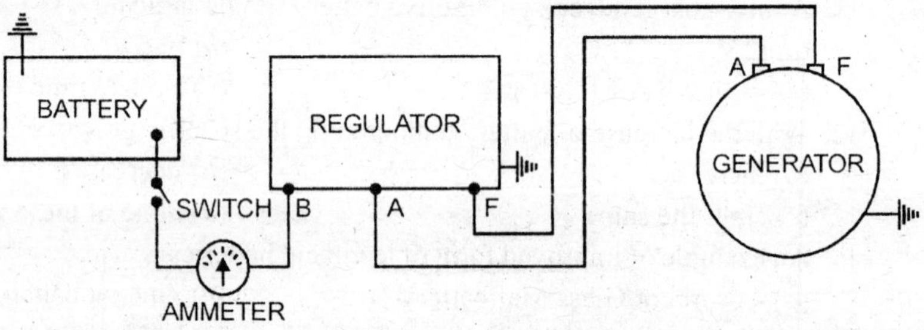

Fig. 18.1. Charging system

Here a detailed description of both the older as well as the modern charging systems will be discussed.

2. GENERATOR

A generator in an automobile supplies the electrical energy for charging the battery in the vehicle. The generator is driven from the engine by means of a belt. As the engine speed is subject to large variations, so will be the generator speed. But the necessity demands that the voltage should remain nearly constant. Further automobile generator is a low voltage generator producing direct current only because the use of battery requires direct current. These requirements make the automobile generator or dynamo unique in itself.

Shunt wound type of d.c. generator or dynamo is the one that has been used for automotive purposes till recently when it has given way to the a.c. generator or the alternator whose output is rectified by in-built semi-conductor rectifier diodes. In this chapter, the d.c. generator still being used in many older Indian vehicles, will be discussed in detail. Besides, a.c. generator will also be described and its comparison with d.c. generator studied.

3. PRINCIPLE OF D.C. GENERATOR

When a conductor is moved in a magnetic field, a current is produced in it, the direction of which is determined by Fleming's right hand rule. Referring to Fig. 18.2 it is seen that if the conductor moves downward,

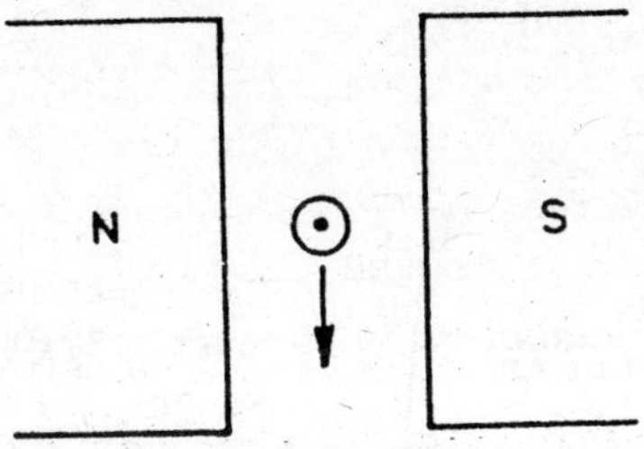

Fig. 18.2. Direction of current of downward conductor movement

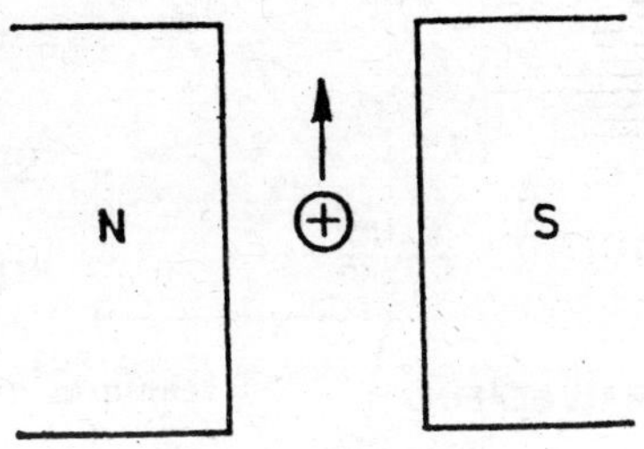

Fig. 18.3. Direction of current for upward movement of conductor

the direction of current in the conductor is outward, shown by a dot. Similarly upward movement of the conductor produces a current in the inward direction as shown by a cross (Fig. 18.3).

Similarly, if instead of a straight conductor a U-shaped conductor (Fig. 18.4) is taken and rotated in the magnetic field, the direction of current in the two legs of U will be opposite, because one is moving up

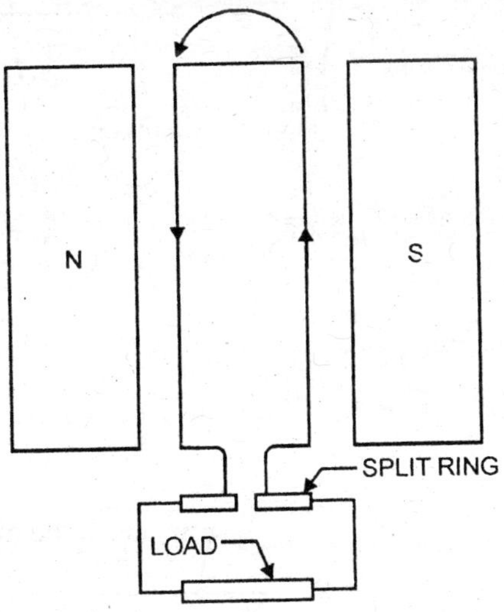

Fig. 18.4. Rotation of U-shaped conductor in the magnetic field

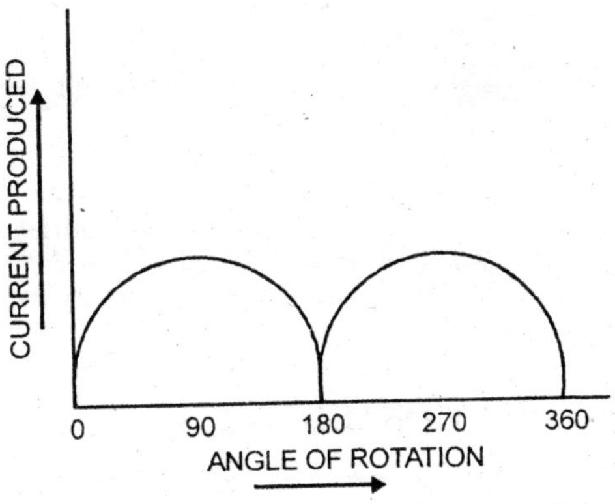

Fig. 18.5. Current produced by a single conductor

while the second is moving down. If at the ends of this conductor two split rings are attached the current produced will be led through the wiring to the load attached. However, the current produced in this manner will not be uniform; its shape will be as given in Fig. 18.5. To make it uniform, many such conductors are used, as a result of which a relatively smooth current is produced (Fig. 18.6).

In the above discussion it is assumed that the magnets used are permanent, but in actual practice the

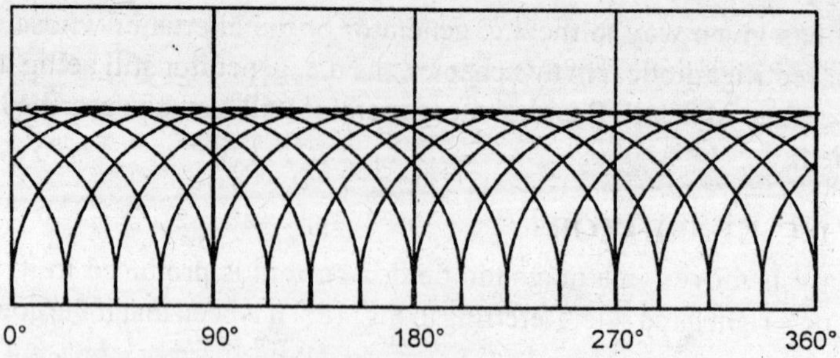

Fig. 18.6. Production of smooth current wave

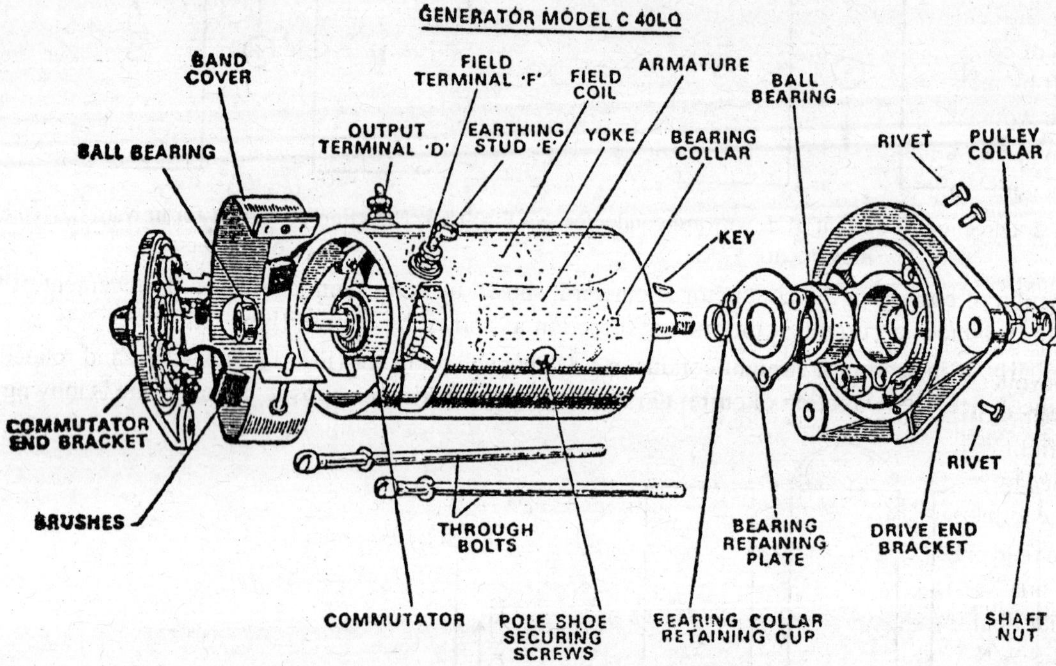

Fig. 18.7. Exploded view of Lucas-TVS generator model C40LQ
(Courtesy—Lucas-TVS Ltd.)

magnets used are electromagnets, the energy for which is taken from the current produced by the generator itself.

4. CONSTRUCTIONAL DETAILS

The main components of a generator are:

1. Frame
2. Armature
3. Field coils

These will now be discussed in detail.

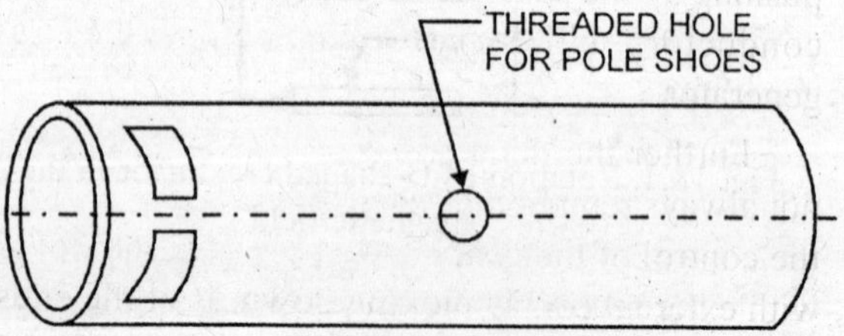

Fig. 18.8. Generator frame.

4.1. Frame

Generator frame is made of steel and is of cylindrical form (Fig. 18.8). It is machined on the inner surface to accommodate pole shoes. The end covers support the bearings for the armature shaft. The end cover on the side of the commutator carries the brushes and the generator output terminals. Openings are provided for inspection as well for dissipation of the heat produced while the generator is running.

4.2. Armature

Fig. 18.9 shows a typical automotive generator armature. The major parts are - shaft, core, commutator and coil windings. The shaft is made of mild steel, whereas soft iron laminations are used to make up the core. The core has longitudinal slots over it which contain the coil windings. The windings are secured in position in the slots by wedges of insulation (mica). Sometimes a band or steel wire is tightened around the armature to protect the armature windings against the centrifugal force which tries to fly these out. The windings are coated with insulating varnish and dried. Commutator consists of copper segments insulated from each other and from the shaft by mica. It is pressed on the armature shaft. Armature coils are wound in two distinct ways, the lap winding and the wave winding.

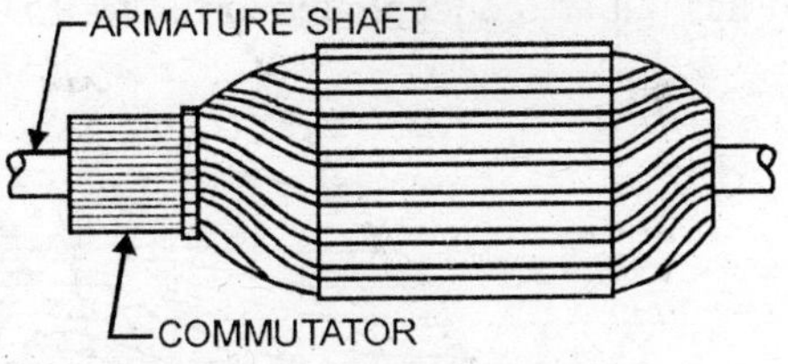

Fig. 18.9. Generator armature.

Two spring-loaded carbon brushes held in brush holders are employed to connect the armature coils to the outside circuit.

4.3. Field coils

As mentioned previously the field magnets used are not permanent, but they are electromagnets energized by the generator current itself. The generator field coils are wound as shunt windings (Fig. 18.10), as this method is best suited for the attainment of nearly constant voltage with varying speed, and for current regulation. About one fifth of total output is consumed as the field current and the generator output is controlled by the field circuit regulation. Field coils are in the form of many turns of insulated fine wire.

The field magnets, though not permanent ones, have a small residual magnetism which provides the initial field to start the generator operation.

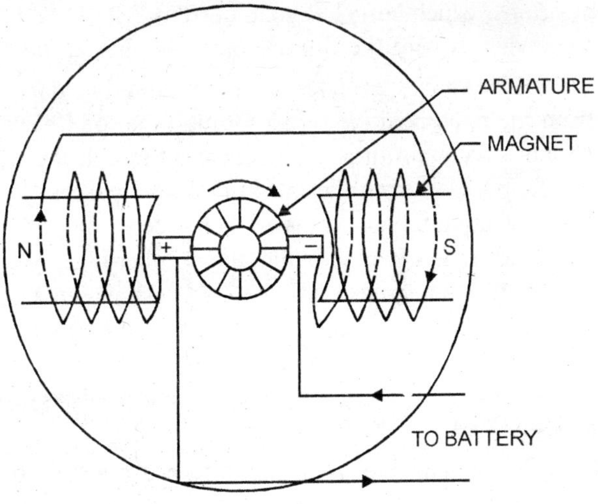

Fig. 18.10. Shunt wound generator

5. GENERATOR OUTPUT CONTROL

The shunt generator has a rising output versus speed characteristic curve. Thus if there is no control, the output will go on rising, speeding up the generator, producing still larger output and pushing up the speed still further, till the armature conductors will fly away, thus destroying the generator.

Further the demands on the generator, too are varying, because the state of the battery charge does not always remain the same and different accessories are needed at different times. Keeping in view these the control of the generator output becomes a necessity. This may be achieved either with a third brush or with external control. Both of these methods will be discussed here, although external regulation is almost universally being used these days.

An important consideration of the charging system control is the charging voltage. If this voltage is set equal to the battery terminal voltage when fully charged, there will not be any possibility of over charging the battery. However, keeping in view the voltage drops in wiring and operating temperature of the charging system and the battery, the charging voltage, which is kept constant is kept somewhat higher. For a 12-volt system, the commonly accepted voltage is 14.2 ± 0.2 V.

5.1. Third Brush Control

When an armature containing a number of conductors is rotated within a magnetic field, distortion of the field occurs as shown in Fig. 18.11. At higher speeds, lines of magnetic force are no longer straight lines but are concentrated at the trailing edges. This effect is termed *armature reaction.*

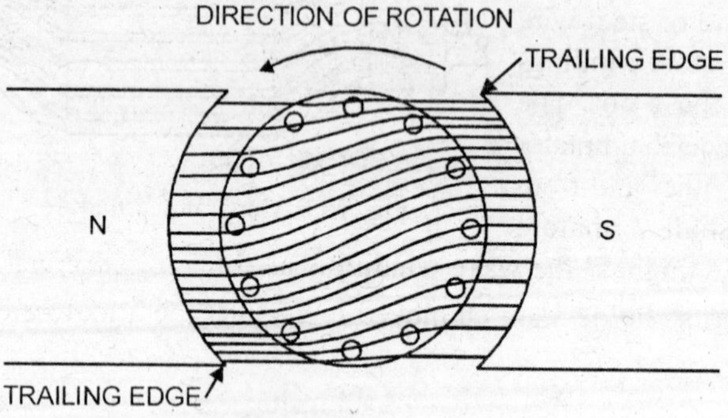

Fig. 18.11. Magnetic field distortion by conductor rotation.

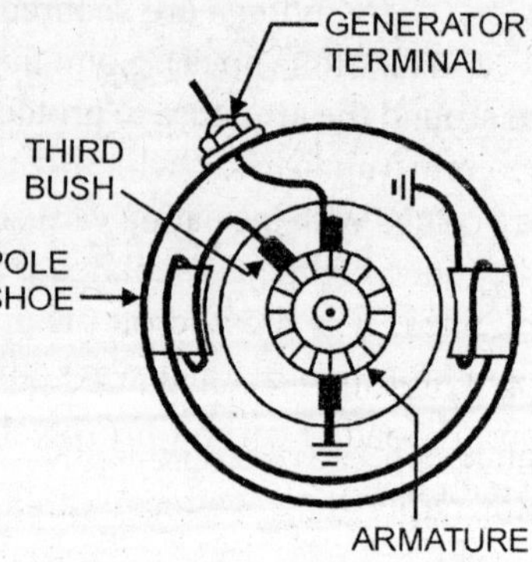

Fig. 18.12. Third brush control.

In a third-brush generator, there are three brushes. The two main brushes are attached at the usual positions, which carry the load current. The third brush conveys current to field windings and is placed somewhere where the flux becomes weaker at the higher speeds due to armature reaction (Fig. 18.12).

As the armature rotates, the current produced passes from the main positive brush through the load back to the negative main brush. The current through the field windings is led from the third brush and passes to the main negative brush. At the higher speeds, the voltage output tends to increase while, the number of magnetic lines of flux is decreased at the third brush which means decreased voltage applied to field windings and consequently less voltage output. Thus a third brush generator will have the output at different speeds as given in Fig. 18.13.

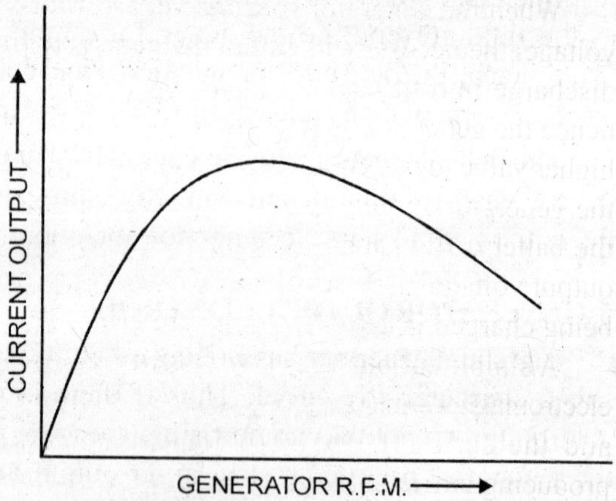

Fig. 18.13. Output from a third brush generator

The main advantage of the third brush generator is its simplicity in construction. However, this type of regulation does not take into account the requirements of the electrical system. The state of the charge of the battery and the fact that different accessories are in use at different times, are the important considerations on which the output requirements of the generator depends. But these considerations are not taken care of by the third brush regulation. Rather at higher battery voltage it supplies more current which is just the reverse of what is desirable. Further the regulator comes into operation *after* the high voltage already resulted. Due to these reasons, this type of generator output regulation is not in much use these days.

5.2. External Output Control

Principally, external regulation comprises inserting a resistance in a generator field circuit, whenever the output rises to a predetermined value. There are two methods of achieving it. The generators in which

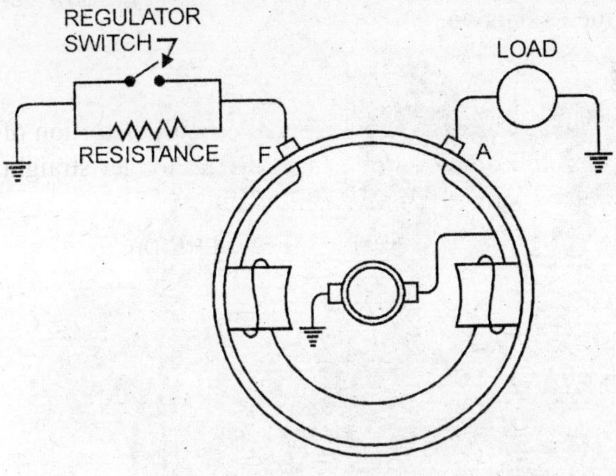

Fig. 18.14. Externally grounded field type generator

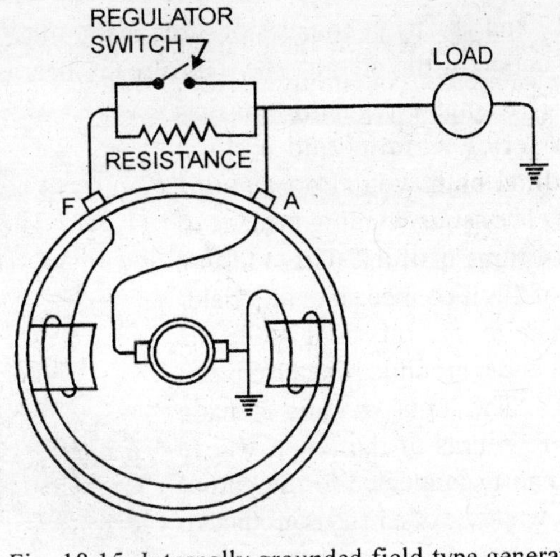

Fig. 18.15. Internally grounded field type generator

this resistance is inserted between the ground and the field windings, and the field circuit is grounded in the regulator rather than the generator itself are called the externally grounded type (Fig. 18.14). In the internally grounded type, the field circuit is connected to the grounded brush inside the generator and the resistance is connected between the field windings and the insulated brush (Fig. 18.15). In actual practice, action of insertion and taking out the external resistance from the circuit, is achieved by means of some automatic regulator, the principal features of which will be discussed in the Article 5.4.

5.3. Cut Out Relay

When the generator speed is very low, due to which the output is not sufficient to balance the battery voltage, the necessity to cutout the generator from the battery arises, because otherwise the battery would discharge into the generator. When the engine and hence the generator speed has reached a sufficiently higher value to match its output to the battery voltage the generator should be automatically connected to the battery. The speed of the generator at which its output voltage just rises above voltage of the battery being charged is called *cutting in speed.*

A simple cut-out relay is shown in Fig. 18.16. The electromagnet consists of two coils, the shunt (voltage) and the series (current). When the generator is producing sufficient voltage so that the field due to both the current and the voltage coils support each other, the electromagnet pulls down the armature, the contacts are closed and the generator-battery circuit is completed. However, when due to low engine speeds the generator voltage falls below the battery voltage, the current flows from the battery to the generator. Then the fields due to current and voltage windings become opposed to each other. Hence the pull on the armature decrease, the contacts open out and the battery is cut from the generator.

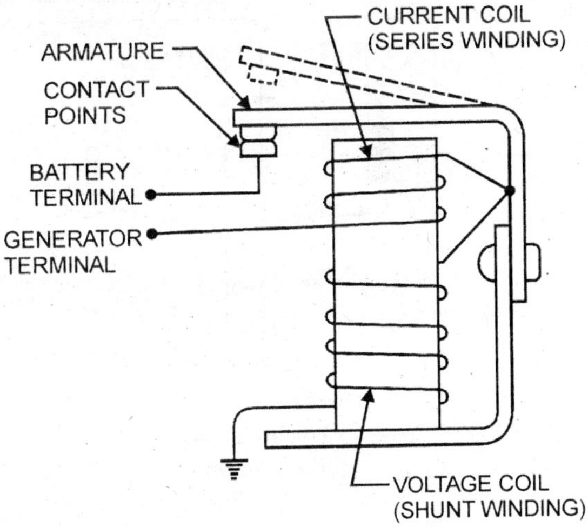

Fig. 18.16. Cut out relay.

5.4. Regulators

The most widely used type of regulator is the combined current and voltage vibrating regulator (Figs.

18.17 and 18.18). It consists of three main units, *viz,* cutout relay, current regulator and voltage regulator. The action of the cutout relay has already been described as above.

The voltage regulator consists of a series winding and a shunt winding, both wound on a single core. The series winding is made of a few turns of thick wire and one end of this is connected to the 'field' terminal of the regulator while the other end is grounded via the contact points. The shunt winding is made of a few turns of thin wire, whose one end is connected to the cutout relay while the other end is grounded. Both the series and the shunt windings are wound in such a manner that the fields due to these support each other.

The working of the *voltage regulator* is as follows: When the generator is producing higher voltage than for which the regulator is set,

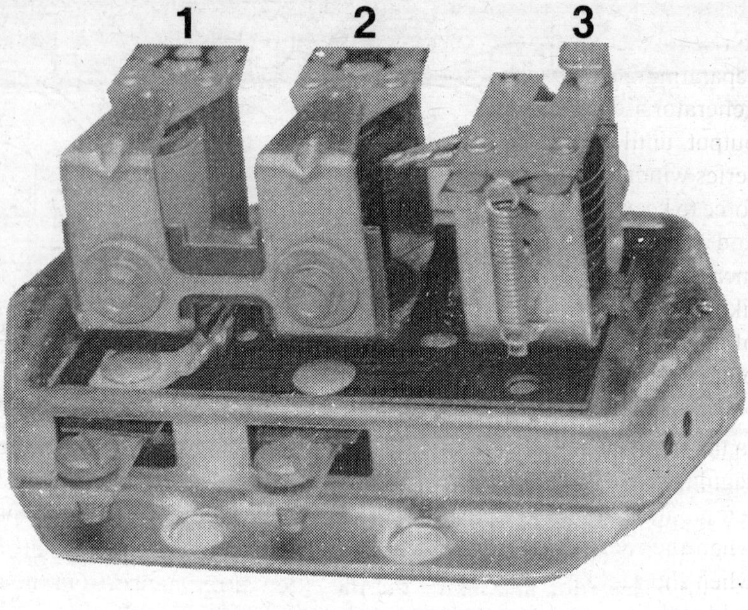

Fig. 18.17. Freedco Regulator.
1. Voltage regulator, 2. Current regulator, 3. Cut out relay.

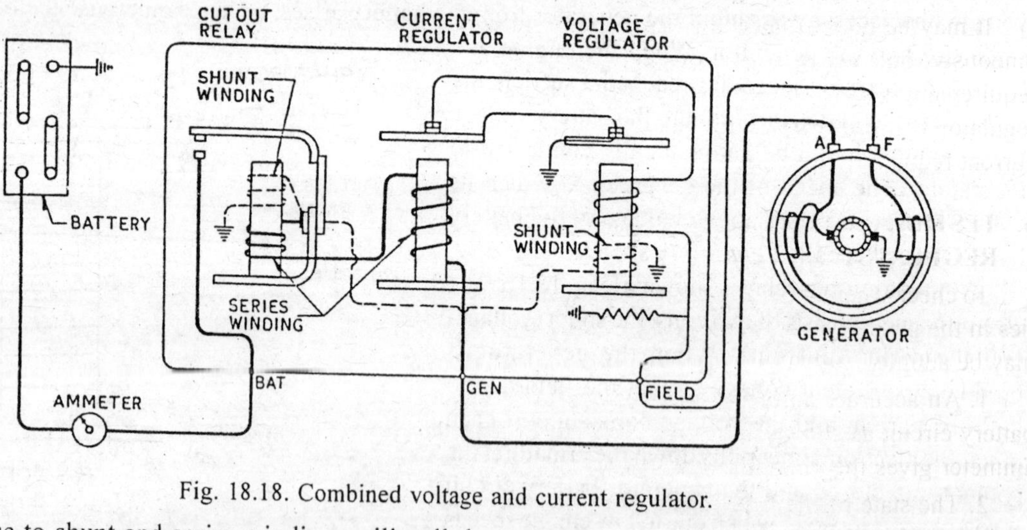

Fig. 18.18. Combined voltage and current regulator.

the force due to shunt and series windings will pull the armature down, thereby separating the contact points. This inserts a resistance in series in the generator field circuit as a result of which the voltage drops. When the voltage has decreased so that magnetic force is not sufficient to pull the regulator armature down, it goes up again making the contacts and thereby grounding the generator field circuit. Thus the regulator goes on vibrating at the preset mean generator field voltage, at a frequency generally in the range 150-200 times per second. In this way it guards the circuit against high voltage and the battery against overcharging.

The current regulator consists of a heavy series winding. When the regulator is not operating, the contact points are closed and the generator field circuit is grounded. When the load on the generator increases and the generator voltage is not sufficient to operate the voltage regulator, the generator current continues to rise till the stage is reached where the current in the regulator coil is sufficient to pull the armature down separating the contacts, thus inserting the resistance in the generator field circuit which brings down the generator output, until the current decreases to a value such that the series winding in the current regulator cannot exert enough force to keep the contacts separated. The contacts then close and the generator field circuit is grounded again thus increasing its output. In this way the current regulator also, like the voltage regulator continues vibrating at a frequency of about 200 times per second, and maintains the current output at a preset constant value.

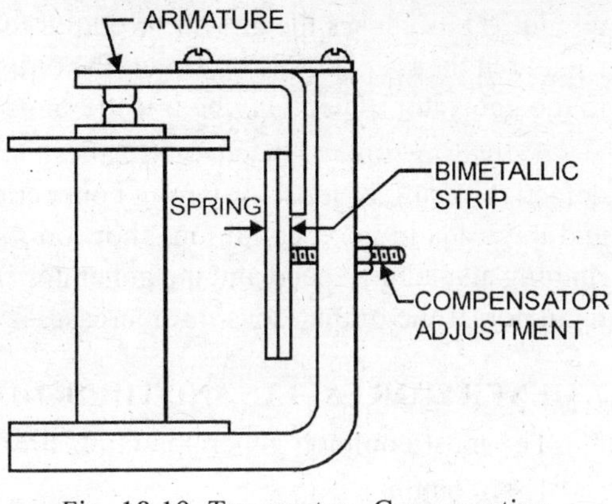

Fig. 18.19. Temperature Compensation.

A temperature compensator is also provided in regulator in the shape of a bimetallic strip on the armature of voltage regulator (Fig. 18.19). The bimetallic strip bends as the temperature rises thus weakening the spring. Therefore, when the regulator is cold, it requires more force, *i.e.*, higher voltage, to open the contacts. However, when the regulator temperature increases, the voltage required decreases. Thus, when the regulator is cold at the start of the engine, it will maintain a higher generator output, which will fall off to normal with the attaining of normal operating temperature of the regulator. In this way, higher charging rate is attained when the engine is started from cold, which is desired. This arrangement takes care of seasonal temperature variation and will provide higher charging rate in winter than in summer.

It may be noted that at any time, both the voltage and the current regulator do not operate simultaneously; only one of these would be working at a time. When the battery charge is low, the current requirement is high, whereas the voltage required for charging such a battery is low. Thus only the current regulator will operate to keep the current within the safe limits. However as the battery is charged, the current required decreases but the voltage regulator will now operate to limit the voltage.

6. TESTING PROCEDURE FOR GENERATOR REGULATOR SYSTEM

To check whether the source of trouble for faulty output lies in the generator, or the regulator, following procedure may be adopted:

1. An accurate ammeter is connected in series in the battery circuit as shown in Fig. 18.20. The reading on the ammeter gives the charging rate.

2. The state of the battery is tested with a hydrometer as explained in the chapter on "storage batteries".

3. From the above two observations, any of the following four conditions will be found to exist:

(*a*) Discharged battery and a high charging rate.

(*b*) Charged battery and low charging rate.

(*c*) Charged battery and high charging rate.

(*d*) Discharged battery and low charging rate.

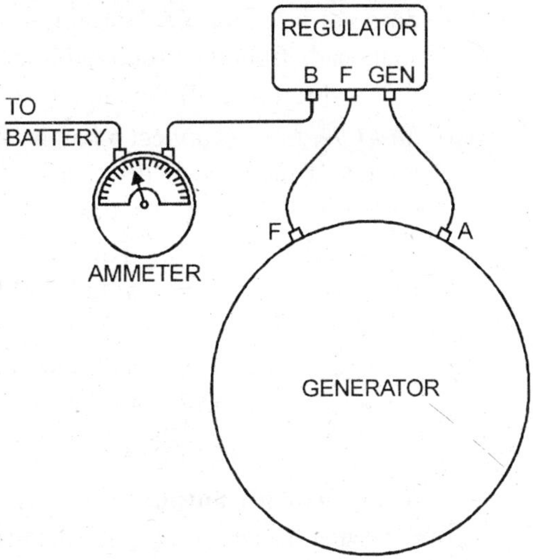

Fig. 18.20. Circuit for testing of generator-regulator system.

Out of these (*a*) and (*b*) are the normal conditions. The remaining two conditions may be investigated further to locate the source of trouble.

4. Investigation of charged battery and a high charging rate—Run the generator at medium speed with ammeter in circuit. Disconnect the *F* lead at the regulator terminals, thereby opening the generator field circuit. If this causes the generator output to drop, it indicates that generator is alright and so look for the trouble in the regulator. If however, the output does not fall on disconnecting *F* lead, there is some defect in the generator which may be traced out as detailed in Art. 7.

5. Investigation of discharged battery and a low charging rate—The possible causes for this may be defective wiring or loose or wrong connections. If on inspection, the connections are found to be correct and the leads in good condition, short out the generator field circuit in the regulator. With the generator running at medium speed and the generator field circuit shorted, if the output increases, the fault lies in the regulator; if the output does not increase, it indicates a faulty generator.

7. GENERATOR FAULTS AND THEIR DIAGNOSIS

The most common generator faults are:

1. No output
2. Low or unsteady output
3. Excessive output
4. Excessive noise.

1. No output

(*a*) *Visual Inspection.* If the generator is not giving any output, the first step is to remove the cover band, if any, and have a visual inspection. Check for sticking brushes, burnt commutator and loose connections. Inspect the inner surface of the cover band. Dots of solder on it point to the melting away of commutator bar soldering because of overloading. If everything looks alright, disconnect the generator leads and proceed with further tests with the help of test lamp.

(*b*) *Ground Test.* Insert paper or some other insulation between the commutator and the ground brush. Check for ground with a lamp between the generator main brush and the frame. If the lamp lights, the generator is grounded. In case of internally grounded field type, disconnect the field ground wire before making the tests.

(*c*) *Open field test.* Place test lamp between the armature and the generator field terminal. If the lamp does not light, the field circuit is open. In case of internally grounded field type, disconnect the field leads from the grounded brush and check with test lamp between the disconnected lead and the field terminal. If the lamp does not light, the field is open.

(*d*) *Short field test.* Connect a battery of the specific voltage and an ammeter in series with the field circuit. If the current flowing is higher than the value specified by the manufacturer, the field coils are shorted and should be replaced.

(*e*) *Open armature circuit.* This is generally obvious as in such a case the commutator bars will be found to be burnt and pitted. Solder thrown on inside the cover band is another sign of open armature circuit.

(*f*) *Short armature circuit.* Short circuits in the armature are tested by means of a growler. The armature is placed on the growler, with a strip of steel held above the armature core. When the armature is rotated slowly, any vibration of steel strip would indicate the shorted armature circuit.

2. Low or unsteady output

The following may be the possible causes of low or unsteady generator output:

(*a*) Generator drive belt may be loose and be slipping.

(*b*) Worn out or sticking brushes.

(*c*) Defective brush springs.

(*d*) Out-of round, dirty or worn out commutators.

3. Excessive output

In the externally grounded field type of generator, excessive output is caused because of internal grounding of the field circuit, which prevents excessive regulation. This may be tested by connecting the test lamp between the '*F*' terminal and the generator frame, with the '*F*' terminal leads disconnected and some insulation placed between the commutator and the field brush. If the test lamp lights, the field is internally grounded.

In the internally grounded field type, however, the excessive output may be caused by a short between the field and the insulated brushes.

4. Excessive Noise

Excessive generator noise may be due to loose drive pulley, worn out bearing, worn commutator, or improperly seated brushes.

8. POLARIZING THE GENERATOR

The generator when newly built or after repairs must be polarized to correct polarity. Failure to do this after the generator is assembled after repair may result in burnt relay contact and a run down battery.

The generators of externally grounded field type are polarized by connecting jumper wire momentarily between the *A* (or GEN) and 'BAT' terminals of the regulator. In case of internally grounded field type, lead from the 'FIELD' terminal of the regulator is disconnected and temporarily touched to the 'BAT' terminal momentarily. This is sufficient to polarize the generator correctly.

9. GENERATOR MAINTENANCE

For exact description of maintenance schedule the shop manual for the generator installed or the vehicle should be consulted. However, the general guidelines are given here.

The maintenance of a generator in fact includes mainly its lubrication, brush gear inspection and adjustment of belt. Certain generators (*e.g.* Lucas C40LQ) do not need lubrication as their bearings are prepacked with grease, whereas other models (*e.g.* Lucas C40Q) should be lubricated at the appropriate holes provided for this purpose after intervals of 10,000 km. The brushes should be inspected after every 30,000 km and should be replaced if their wear is found to be more than the specified value, which may be known from the manual. The inspection of the driving belt should be about 12 mm. If the belt is too tight, it is liable to damage the water pump gland and the generator bearings. On the other hand a very loose belt will not drive the generator efficiently. Also make sure that the generator belt or pulley has not become oily, as this will cause the belt to slip.

10. REGULATOR SERVICING

Inspect for burning of various regulator contacts. The contacts are burnt due to excessive current caused by faulty connections. Even in normal operation, a small arc occurs each time the regulator contacts open, forming deposits of oxide in the contacts which necessitates their cleaning or replacement. If the oxide formation is quite small, the contacts may be cleaned by means of a clean rag wetted in a suitable cleaning compound *e.g.* carbon tetrachloride, which is inserted between the points and moved back and forth. After this, contacts are wiped dry. However, if there are large oxide deposits, the points may be filed with a fine file and then cleaned thoroughly so as to remove filed metal particles from the contacts. Emery paper or cloth should not be used for this purpose. In case the points are damaged beyond repair, they should be replaced. Generally both the contact points should be replaced together as a combination. Air gaps of the contact points should also be checked which may be done by means of a feeler gauge and set to the prescribed value, which is nearly 1.25 mm.

After inspecting the contacts, examine the series coils of the cutout relay and the current regulator, which might have been damaged on account of heavy current which may be caused by:

1. grounding of the field in the generator regulator due to which the generator output may remain uncontrolled causing high current output.
2. grounding of the armature lead between regulator and generator and contacts of the output relay sticking closed. Due to this the battery discharges through the windings.

If any coil is found to be defective, the same has to be replaced.

The regulator setting may be performed finally, as described in the next article.

11. REGULATOR SETTING

Before proceeding to perform regulator setting, make sure that the generator belt is not loose and there is no fault in the battery or the generator. The adjustment may be carried out either with the regulator on the vehicle, or on the special test bench. Though basically the connections in the two cases remain same, yet the discussion here will be oriented towards 'on the vehicle' regulator setting.

1. Setting of cutout relay

(*a*) *Cutout relay closing*—Connect a moving coil voltmeter between 'GEN' terminal of the regulator and earth. Switch on some electrical load *e.g.*, head lights and increase the engine speed gradually. The volt meter should indicate a slight drop during closing of contacts which should be within the prescribed limits. If it does not happen, make suitable adjustment and check again. In the Prestolite (India) regulators, the spring tension is increased by pressing downwards the hinge, for increasing the closing voltage and vice versa for decreasing.

(*b*) *Cutout relay opening*—Disconnect the battery from the regulator, connect a voltmeter between 'BAT' terminal of the regulator and the earth, start the engine and increase the speed to 300 r.p.m. Slowly decrease the speed and observe the voltmeter which should indicate zero reading at the opening of the cutout relay. This should happen at the specified voltage. If it does not, make adjustment and check again. Decrease the gap between the relay contacts for increasing the opening voltage and vice versa for decreasing it.

2. Setting of current regulator

Short circuit the voltage regulator contacts and connect a moving coil ammeter between BAT terminal and the battery. Switch on lights and accessories. Start the engine and run the generator at approximately 4500 r.p.m. The ammeter should indicate current equal to the maximum rated output of the generator. If it does not, make suitable adjustments and test again. Increasing the spring tension in the Prestolite (India) regulator increases the current value and vice-versa.

3. Setting of voltage regulator

Run the generator for about fifteen minutes, then stop it and connect a moving coil voltmeter of 0.40 A range between 'GEN' terminal and the earth. Start the engine and run the generator at about 3000 r.p.m., at a load of about 6-7 amperes. The voltmeter reading will increase with the increase of generator speed and will become steady when the generator speed is no more varied. This steady reading is the open-circuit voltage at which the generator is running. This should be within the specified limits which can be determined from the shop manual. If it is beyond limits, stop the generator, make adjustment as provided in the regulator and retest. In Prestolite (India) regulators, increasing the spring tension by pressing the hinge down increases the voltage and vice-versa.

12. ALTERNATOR

The alternator or a.c. generator is the generator which produces alternating current instead of the direct current as produced by a commutator type generator already discussed in detail in this chapter. This has replaced the d.c. generator since early 1960's because of its distinct advantages which will be discussed later. However, as the automotive electrical system requires only direct current, the alternating current produced in the a.c. generator has to be converted first into the direct current, which is easily done with the help of diodes.

12.1. Principle

As shown in Fig. 18.21, an electromagnet (rotor) is mounted on a shaft and is supplied current for its energisation from a battery though the slip rings and brushes. The rotator is turned by a belt and pulley by

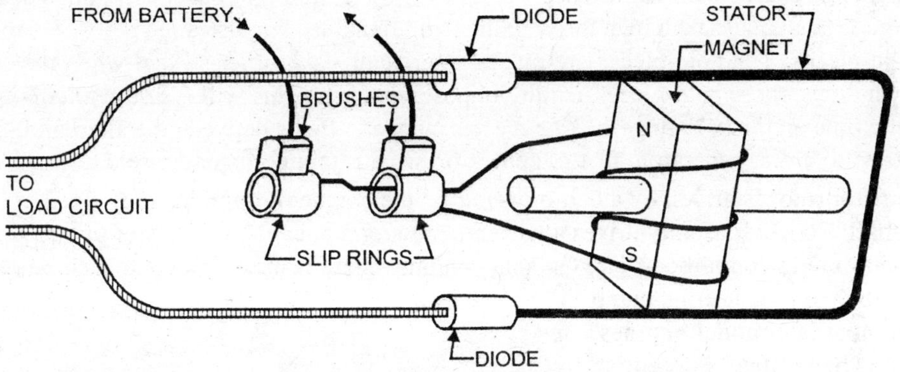

Fig. 18.21. Principle of alternator

the engine. In series with the stator winding are two diodes as shown which are further connected to outside electric circuit.

When the electromagnet is rotated, the magnetic lines of force cut the stationary stator loop to produce an induced current. However, as after every half revolution, the magnet reverses its polarities, the current produced in the stator alternates in sign in a revolution. To convert this alternating current into the direct current, diodes are employed, which allow the current to pass through them only in one direction. Thus it it the d.c. current in the a.c. generator also which is fed to the electrical circuit.

12.2. Constructional Details

(*a*) *Frame or housing*—This encloses he entire alternator assembly and is made of cast aluminium in two pieces. Aluminium is light weight, non-

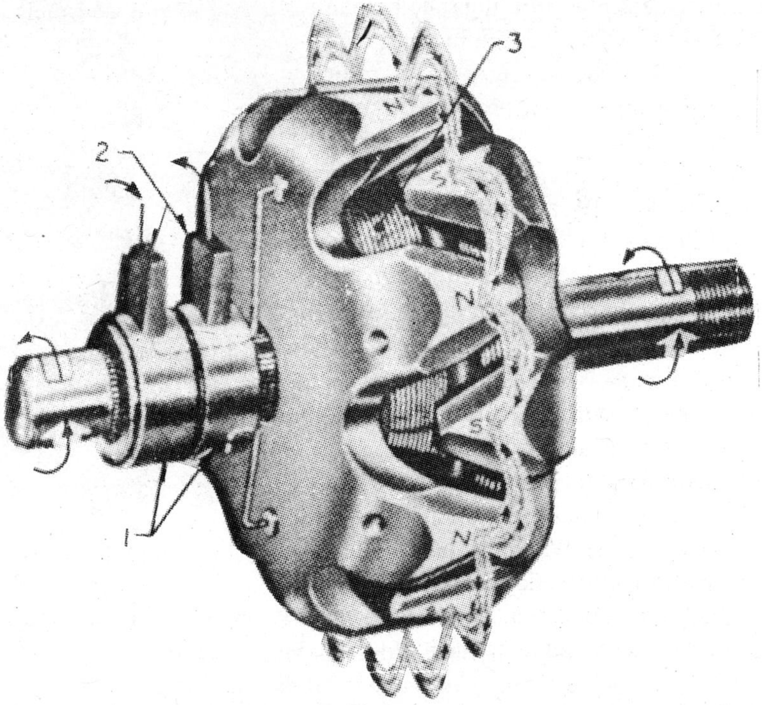

Fig. 18.22. Rotor of a Delco-Remy alternator.
1. Slip rings, 2. Brushes, 3. Winding.

magnetic and has higher thermal conductivity so as to keep the alternator assembly cool by more efficient transfer of heat. The front part of the frame (called drive-end housing) has ball bearings, while the rear part (called brush-end housing) contains a roller bearing.

(*b*) *Rotor*—In practice the rotor is not a bar magnet as stated above while explaining the principle of working of the alternator. It is actually a specially shaped electromagnet and not a permanent magnet. This is done so as to be able to control the field current and hence the strength of the rotor magnetic field. This enables us to control the alternator output according to the requirements.

The most common is the claw pole rotor which consists of an iron core around the rotor shaft. Many turns of copper wire coated with varnish are wound over the core (Fig. 18.22). On both sides of the rotor winding are thick metal plates bent over the winding with triangular fingers called poles. These metal plates are called pole-pieces. The pole-pieces are placed interlacing each other so that when the current flows through the windings, these pole-pieces acquire opposite polarity. This will result in alternating north and south magnetic pole on the pole fingers. This creates magnetic fields between the alternating pole fingers. Thus in contrast to a d.c. generator, an a.c. generator has a rotating magnetic field.

The alternator rotor is driven by a belt powered by the engine through a pulley fixed at the front end of the crankshaft. This belt usually drives also some other components, viz., water pump, power steering pump, air conditioning compressor, etc. On some engines there is more than one such belt, e.g., the fan belt, alternator belt, power steering pump belt, etc. On most late model engines, there is only one belt, called 'Surpentine belt', which drives all the components.

(*c*) *Slip rings and brushes*— The current to the rotor winding is carried through the copper slip rings and carbon brushes (Fig. 18.22). The brushes ride the surface of the slip rings on the rotor under spring tension provided by the brush holders. As these brushes carry only the field current (about 2- 5 amp) in contrast to the carbon brushes in d.c. generator where all the current produced in the generator has to pass through the brushes, the former have a much longer life.

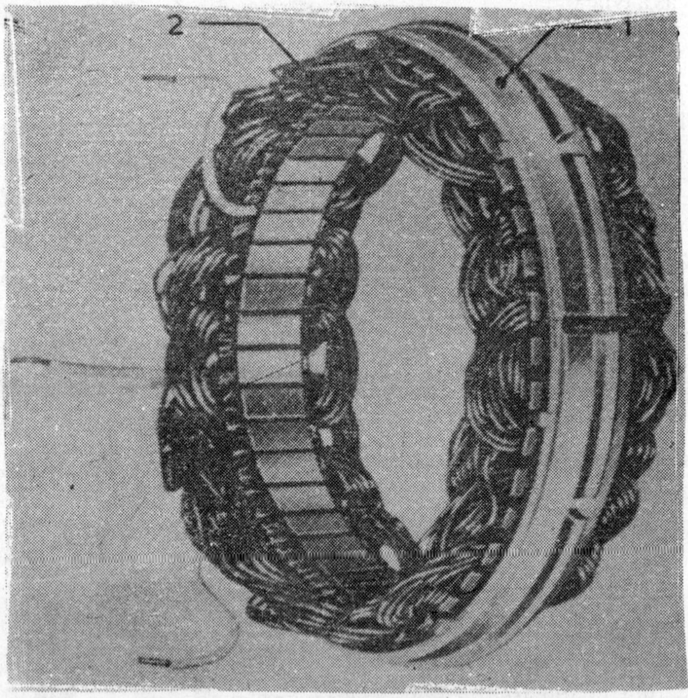

Fig. 18.23. Stator of a Delco-Remy alternator.
1. Stator, 2. Stator winding

(*d*) *Stator*—Three-phase alternator is commonly used in automobiles. Between two halves of the alternator casing, is situated the stator consisting of three sets of winding wound over a laminated iron core (Fig. 18.23). When the rotor rotates, its moving magnetic field induces current in all the three stator windings. All alternators contain at least six diodes, one pair of a positive and a negative diode for each winding to have full-wave rectifi- cation of the a.c. current produced in the alternator. One diode would achieve only half-wave rectification. Further, three positive diodes are usually included in the rectifier circuit. These are known as the auxiliary, field or excitation diodes. These are smaller than the main diodes and serve to supply a small current back to the field windings in the rotor.

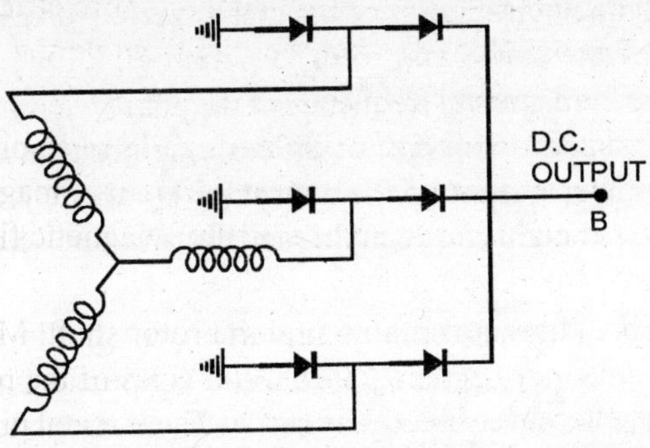

Fig. 18.24. Star connection for stator winding

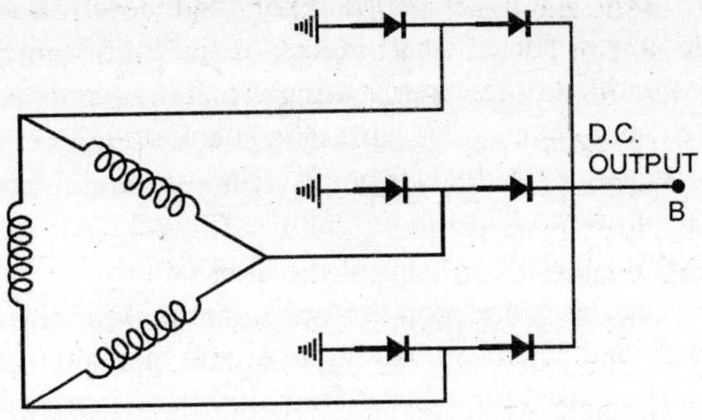

Fig. 18.25. Delta connection for stator winding

The three windings of the stator are connected together. The two commonly used connections are the star (or Y) connection (Fig. 18.24) and the delta (or) connection (Fig. 18.25).

The current and voltage characteristics of the star and delta wound rotors are quite different. Star connection can be considered as a series combination of phases. As such, the output voltage across any two phases will be the vector sum of the phase voltage, while the current output will be the same as the phase current. Thus star-wound stators produce higher voltage. Likewise, a delta connection can be taken as a type of parallel circuit, which means that the output current will be the vector sum of the phase currents, while output voltage would be the same as the phase voltage.

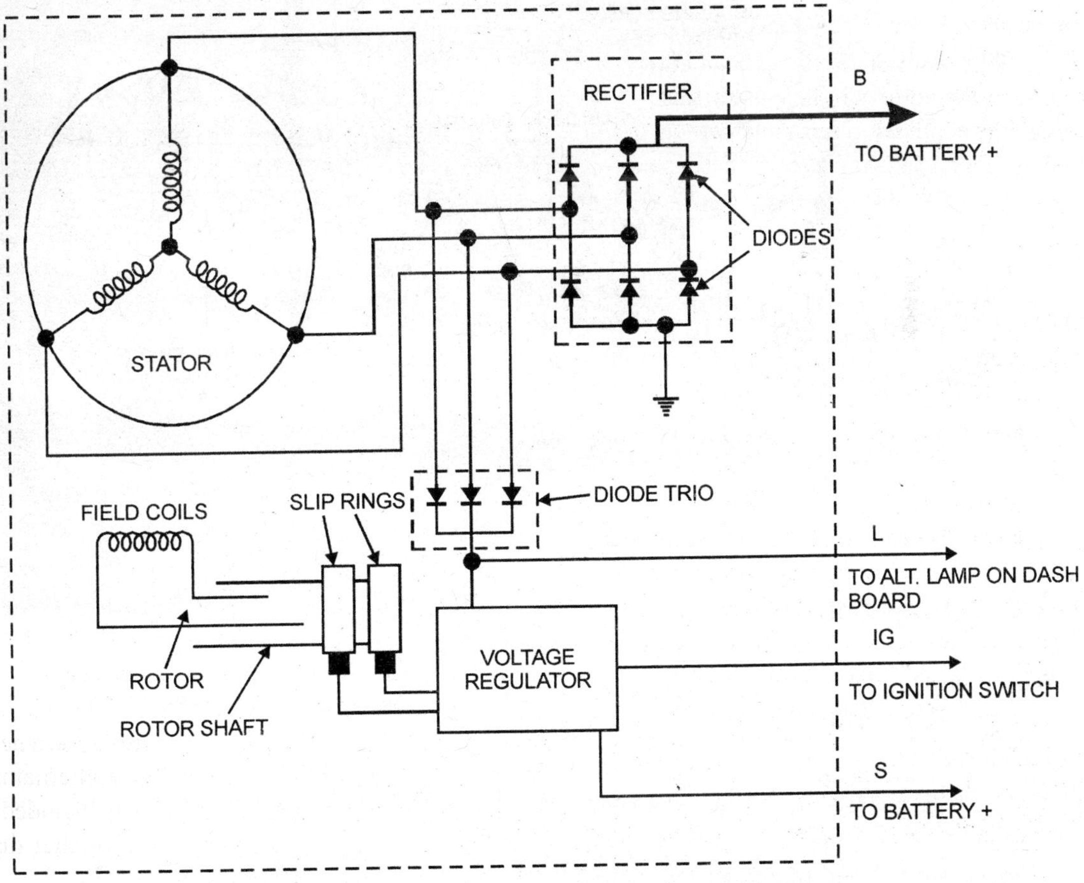

Fig. 18.26. A typical electrical circuit for an alternator

In case of a star-wound stator, the theoretical voltage at the neutral point of the star is zero. However, in practice, due to small inaccuracies in the construction of the stator and rotor, a potential, known as *third harmonic,* develops at this point. Its frequency is thrice the fundamental frequency of the phase windings. The energy due to this harmonic is collected by employing an additional pair of diodes, one positive and one negative, at the star point. This results in increase of power output of the alternator by about 15%. Thus it is seen that full circuit of an actual alternator consists of eight-diode main rectifier and three field diodes.

The star connection is more common. With this connection, alternator is able to start charging at lower speeds and has lower maximum output than alternator with delta connection stator. Delta connections for stators are used for those alternators where high output at high alternator speed is required. Such alternators can produce about 73% more current than the same alternator with star connections. These, however, produce lower current at low speeds and, therefore, have to be operated at high speed to obtain maximum output.

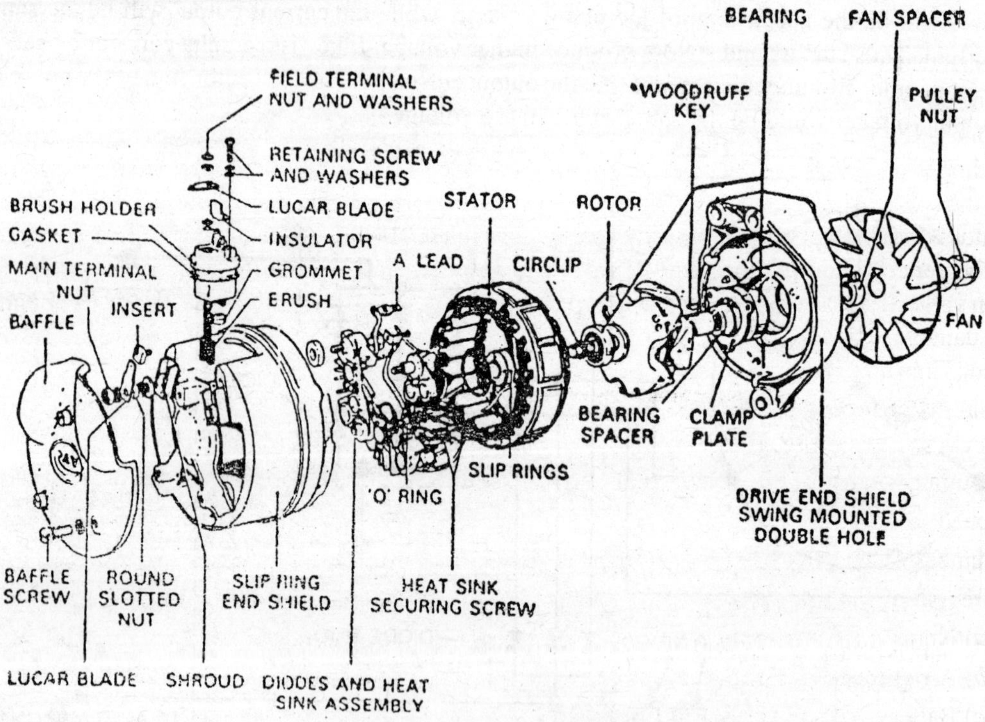

Fig. 18.27. Lucas-TVS alternator
(Courtesy—Lucas-TVS Ltd.)

A typical electrical circuit for an alternator is shown in Fig. 18.26. The voltage regulator can be mounted inside or outside the alternator housing.

An exploded in view of a Lucas-TVS AC-5 alternator is shown in Fig. 18.27, whereas Fig. 18.28 and 18.29 depict the external views of Lucas-TVS type SA 108 alternator for cars and A 115 for LCV's and trucks respectively.

Fig. 18.28. Lucas-TVS type SA 108 alternator.

Fig. 18.29. Lucas-TVS type A 115 alternator.

12.3. Alternator regulation

In case of alternators, diodes are employed in the stator output circuit, which allow the current to flow in one direction only. Thus the cutout relay used in d.c. generator regulators is not required here because diodes do not allow the current to flow from the battery to the alternator when it stops or slows, although the connection between the two still remains. Moreover, the magnetic field produced by the current in the stator windings opposes in polarity and approaches in magnitude the magnetic field provided by the rotor as the generator output increases. This causes the alternator to limit its own current output to a safe maximum value. Thus alternators also do not require current limiters as they do not produce enough current to cause damage to the alternator itself. However, an alternator could produce more than 250 V if not controlled. Thus *alternator regulators require only the voltage to be controlled.*

An alternator regulator should keep the charging voltage between 13.5 and 14.5 volts to protect the electrical components throughout the vehicle.

The output voltage of an alternator depends upon:

1. Speed of rotation of the rotor.
2. Number of turns in the stator windings.
3. Strength of the rotor magnetic field which itself is dependent on the following:
   (*a*) Number of turns and type of wire used for the rotor winding.
   (*b*) Air gap between the stator and the rotor poles. Increasing the gap deceases the field strength.
   (*c*) Battery voltage applied to the rotor.

Out of the above, it is not practical to keep the rotor rpm constant at different engine speeds. Moreover, the number of turns in the stator and the rotor windings are fixed in a given alternator. Therefore, the maximum output voltage of the alternator is controlled by the control of the field (rotor) current. This is done by means of regulator. However, whenever the output voltage is below the maximum limit, the voltage regulator does not come into action to limit the field current. Rather at low alternator speeds, the field current is maintained at its maximum so as to allow the alternator output to be as high as possible. Further, when the output voltage is below the regulator setting, it is controlled by the battery. If the battery is discharged, its voltage will be lower than that of the fully charged battery. Therefore, the alternator will charge it at lower voltage but high current. As the battery becomes charged, the battery voltage would increase till the regulator setting is reached when the regulator will start operating to limit the maximum charging voltage. Thus below the maximum limit, the battery acts as the voltage regulator of the alternator. That is the reason why testing of the charging system must be done with the battery in fairly good condition, say at least 75 percent charged, so as to obtain accurate results. Moreover all connecting wires and grounds must be in good condition and tightly secured, otherwise the voltage regulator may sense a battery voltage which is lower than the actual and hence may allow the alternator output to exceed the normal charging voltage, thus causing overcharging of the battery.

The alternator regulators may be of two types:

1. Mechanical, *i.e.*, the Vibrating contact point type.
2. Electronic, *i.e.*, the Transistorised type.

The vibrating contact point regulators are similar to the point type voltage regulator for d.c. generator discussed earlier. The transistorised regulators do not employ any vibrating contacts which increases their life. Due to this, these are now almost universal on modern cars. A simplified basic circuit for such a transistorised regulator is shown in Fig. 18.30. In the circuit, there are two transistors T_1 and T_2, one Zener diode D_z, three fixed resistances R_1, R_2 and R_3, one resistance R_A whose connection point can be changed and one resistance R_T whose resistance decreases with increase in temperature. This resistance is called a *thermister.*

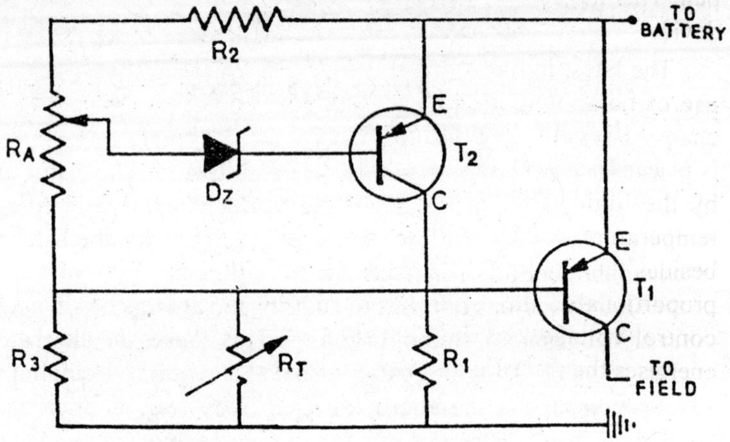

Fig. 18.30. Simplified circuit of a transistorised regulator

From the battery, the current comes to emitter of T_1. As the base of T_1 is conducting to the ground through R_1, current will flow from emitter to the collector of T_1 and hence full current goes to the alternator field.

Battery voltage is also applied to T_2, but the reverse biased zener diode D_z would not allow the base current to flow so long as voltage applied to it is below the zener point with the result that T_2 remains off. A small current would then flow through R_2, R_A and R_3 to the ground. When the reverse-bias voltage at D_z reaches the zener point, it would start the reverse current to flow. The voltage across D_z at zener point is the voltage drop through R_2 and part of R_A. Thus zener point and hence the regulator can be adjusted by setting resistor R_A. When reverse current starts to flow through D_z, T_2 also comes 'on' so that the current flows from collector of T_2 to the ground via R_1, R_T and R_3. Further the battery voltage is applied by T_2 at the base of T_1 with the result that there is no voltage drop from emitter to base of T_1 which causes T_1 to be turned off and thus stops completely the flow of rotor field current. With the stoppage of field current, the alternator voltage drops which reduces the reverse-bias voltage on the zener diode causing the same to stop conducting base current of T_2 thereby turning T_2 off. With this T_1 again starts working and supplies the field current raising the alternator voltage. This cycle is repeated at very high frequency (upto 7000 cycles per second) to keep the output voltage of the stabilizer at the set value. The resistance of thermister R_T decreases with increase of temperature, so that current flow through R_2 and R_A is increased. This means a lower voltage is required to start the reverse flow of current at zener point. Thus the regulator operates at a lower voltage when warm compared to when it is cold. This is what is exactly required from the battery consideration where chemical activity is less when cold requiring higher charging voltage.

In practice the above basic circuit is improved by using various resistors and capacitors. Besides alternator regulators using *integrated circuits* (IC's) are being increasingly used. 'Integrated' implies that the transistors and circuits are made together in a single part. Such a regulator has the following advantages:

1. Outside wiring of the alternator is simplified.
2. Number of moving parts are reduced.
3. Better resistance to vibration.
4. Voltage fluctuations are reduced.

A hybrid IC voltage regulator circuit is shown in Fig. 18.31. The hybrid system uses both the IC chip as well as the connection of discreet components on a ceramic plate using film techniques. The I.C. forms the main part of the regulator and contains the sensing elements and temperature compensation components. The IC controls an output stage such as a Darlington pair. This method results in a compact and reliable device.

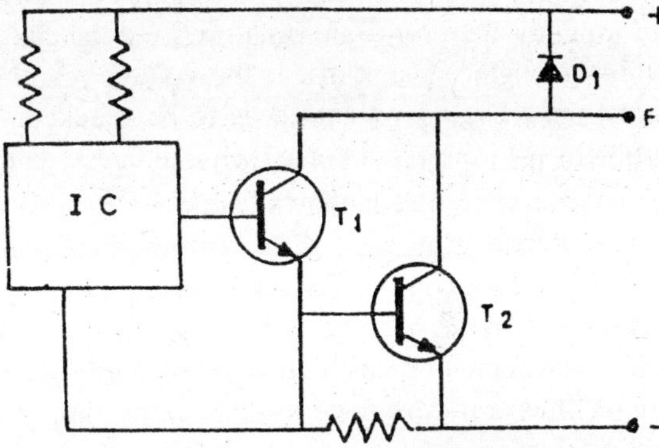

Fig. 18.31. Hybrid IC voltage regulator circuit.

The latest in the charging system control is the use of computer. In such a computer-controlled charging system, (Fig. 18.32) the voltage regulator is placed inside the power module and is controlled by the logic module, which also senses battery temperature and battery voltage at the alternator, besides the alternator output voltage. A duty cycle proportional to the difference between the desired control voltage and the alternator output voltage energises the rotor field.

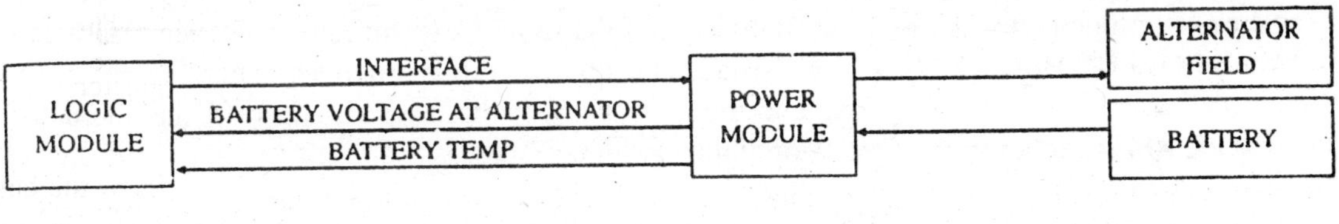

Fig. 18.32. Schematic diagram of computer-controlled charging system.

In simple words, the computer switches the rotor (field) current on and off at a fixed frequency of about 400 cycles per second and the correct average field current is produced by varying the on-off periods, which provides the desired alternative output. For instance, at high engine speeds with negligible electrical load, on-time for field circuit may be only 10% whereas at low engine speeds with high electronic load, the computer may have the field circuit charging on 75% of the time to produce the higher average field current to meet the increased demand. Such an accurate control permits the use of smaller and lighter storage batteries. It also reduces the magnetic drag of the alternator, thereby increasing considerably the engine power output. In other words the fuel consumption is reduced. Moreover, the diagnosis of the charging system becomes easier, being computerised.

13. CHARGING SYSTEM WARNING

It is necessary to have some method of warning the driver in case something goes wrong with the charging system so that vehicle is not stranded on the way. Three types of such devices found on the vehicle dashboard are:

1. Warning lamp

2. Voltmeter
3. Ammeter

The warning lamp is often labeled as 'Gen', 'Bat' or 'Alt'. This lamp lights up in case of any problem occurring in the charging system. The most common cause is that the alternator stops working due to broken drive belt.

A voltmeter is a much better indicator of the state of the charging system. When the engine is not running, a fully charged battery would read about 12.5 volts. However, when the engine is running, the charging system takes over so that the voltmeter would read 14 to 14.5 volts all the time unless there is a heavy load on the electrical system, e.g., lights, heater, wipers, rear defogger all working together at idling, when the voltmeter reading would drop. If the voltage drops below 12.5 volts, it would mean that some of the current is being provided by the battery also; dash lights may also dim in such a situation. If this continues for long, the battery would run down and the charge left therein may not be sufficient to restart the vehicle after the engine is shut off. If the voltage reading with the running engine remains constantly below 14 volts, the charging system needs to be checked. Again, the voltage ever exceeding 15 volts would mean problem with the regulator which can cause damage to the entire electrical system.

It case of vehicle having an ammeter as a warning device, a negative ammeter reading would mean the battery is providing most of the current, thereby discharging itself while a positive reading would indicate that most of the load-current is being supplied by the generator. Normally the ammeter should read close to zero but on the positive side, if the battery is fully charged and with only minimum electrical load. It is OK if the ammeter reads high while recharging the battery after starting, but it should gradually taper down soon. In case it continues to read more than 10-20 amp with no main electrical devices ON, it would indicate a weak battery; in which case, the charging system requires investigation.

14. ALTERNATOR TESTS

1. Rotor tests

(*a*) Connect ohmmeter between the slip ring and the rotor shaft. Infinite resistance indicates the rotor is O.K. However, very low resistance means the rotor is grounded. Alternatively, connect test lamp between the slip ring and the rotor shaft. If the lamp remains off, the alternator is O.K. On the other hand if the lamp lights, grounded rotor is indicated.

(*b*) Connect test lamp across slip rings. If the lamp lights, it is normal indication. However 'no light' indicates open circuit.

2. Stator tests

(*a*) Connect ohmmeter between any stator lead and the frame. In normal condition resistance should be infinite, whereas very low resistance indicates ground circuit. The same test may also be done with a test lamp. Normal condition would be indicated by no light, whereas if the lamp lights, it indicates grounded circuit.

(*b*) Connect ohmmeter across any pair of stator leads. Less than 0.5 ohm resistance indicates rotor is O.K. whereas any high reading would mean open circuits.

3. Diode tests

Connect ohmmeter across the diode to be tested and then reverse leads. A normal diode would show low reading one way and high reading the other way. However, in case both readings are low, it means the diode is short-circuited; whereas both high readings would indicate the open circuit.

Alternatively, a 12 volt test lamp can be used instead of ohmmeter. Lamp lighting one way but remaining off the other way means the diode is O.K. However, open-circuit is indicated if the lamp does not light either way. The lamp lighting both ways means short circuit.

15. ALTERNATOR TROUBLE SHOOTING

The most common alternator faults are:

1. Noisy alternator　　　　2. Incorrect charging

1. Noisy alternator

Before proceeding any further, it must be made sure that the noise is coming actually from the alternator because quite often a noisy water pump, power steering pump, fan or the other drive belt is confused with the noise from the alternator. Once the source is established, check the following causes preferably in the given sequence.

(*i*) Worn belt, which has to be replaced.

(*ii*) Loose alternator mounting. The same may be tightened.

(*iii*) Rotor fan may be touching stator leads, which may be easily corrected.

(*iv*) Damaged rotor or rotor fan. Any defective component has to be replaced.

(*v*) Worn or defective alternator bearings; the only alternative is their replacement.

(*vi*) One or more rectifiers may be open or short-circuited. The same may be tested as described earlier in detail. The defective diode, if any, must be replaced.

(*vii*) Defective stator wiring. The same may be tested as described in the above article and appropriate remedial action taken.

2. Incorrect charging

Excessive gassing from the battery and the need to add water more often definitely indicates high voltage of the charging system, whereas a weak battery may mean either the defective battery or faulty charging system. To pin-point the source(s) of trouble, proceed as follows in the sequence recommended.

(*i*) Check condition of the battery, cables and terminals. If found O.K., check the tension of the drive belt; adjust if required.

(*ii*) Check for loose or corroded connections at the alternator and regulator. Remedy appropriately if needed.

(*iii*) Check whether the regulator and alternator are grounded. If not, ground the same.

(*iv*) With ignition switch 'on', the voltage at field input brush terminal should be equal to the battery voltage, *i.e.* 12 volts in case of cars. If it is O.K. proceed separately in case the charging rate is low or excessive.

(*a*) *If the charging is low or steady.* Test resistance of charging circuit. If found O.K., perform alternator bench test as per manual. However, if charging circuit resistance is too high, look for shorted or corroded cables or high resistance of the fuse.

(*b*) *If the charging is excessive.* Check the alternator field wire and field terminal. If grounded, rectify the same. If found O.K. check for internal grounding of the alternator field with the help of the bench test. If the defect still persists, it may be due to voltage regulator sensing circuit being open, which may be set right.

16. COMPARISON OF ALTERNATOR WITH D.C. GENERATOR

1. The output characteristics of the alternator shows that its performance is better than the commutator type d.c. generator with external control which itself is an improvement over the third brush type d.c. generator. At low speeds it produces more current than the commutator type generator avoiding at the same time more output than desired at high speeds. It is thus self-regulating in this respect. No external current regulator is, therefore required for the control of maximum current.

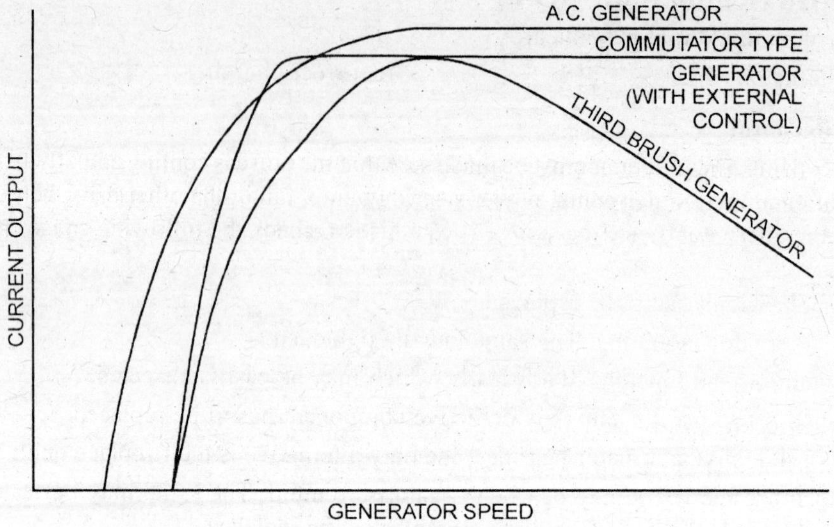

Fig. 18.33. Output characteristics of various generators.

2. As the rectifiers installed in the alternator itself serve as one-way valves for the current flow, a separate cutout relay is not required.

3. In a d.c. generator the output windings are in the armature which rotates, whereas in alternator the same are on the stator. Due to this the cooling is comparatively easy in alternators and hence their maximum output is more (*e.g.* 50 amp compared with 30 amp. in case of d.c. generator).

4. The maximum permissible speed in case of alternators is generally about 30 percent higher than the d.c. generator which is restricted to about 9000 r.p.m. due to commutator operation. In fact speed of a properly designed alternator can be as high as 20,000 r.p.m. or even more. However, in automobiles, belt and pulley sizes restrict the speed to a maximum of 2.5 times the engine speed.

5. The output/weight ratio of alternators is higher than that of generators. This is mainly on account of higher permissible maximum speeds in case of alternators. For alternators with built-in rectifiers the output/weight ratio is about 100% higher than d.c. generator.

6. In the d.c. generator, the entire current produced has to pass through the carbon brushes, whereas in the alternator the carbon brushes carry only the field current (approximately, 2 to 5A). Moreover an alternator is simpler and sturdier in construction compared to the d.c. generator. Therefore, it requires less maintenance and is more reliable.

7. The only disadvantage with alternator system is its higher initial cost than the d.c. generator. This is mainly on account of rectifiers, which are quite expensive.

17. MODERN ALTERNATORS

The second generation of alternators from Bosch, Germany are the Bosch Compact Series LIC-B (Fig. 18.34) which weigh less, generate more power, have higher efficiency and extended range of voltage-regulator functions. Their later series LI-E give still better performance, are safe against corrosion and withstand extreme temperatures and dynamic loads.

These compact alternators are equipped with the single-chip multi-function regulator (Fig. 18.35), whose capabilities include:

1. Battery sensing.
2. Defect diagnosis

Fig. 18.34. Bosch compact alternator,
(Courtesy-MICO India Ltd.)

Fig. 18.35. Bosch Single-chip regulator
(Courtesy-MICO India Ltd.)

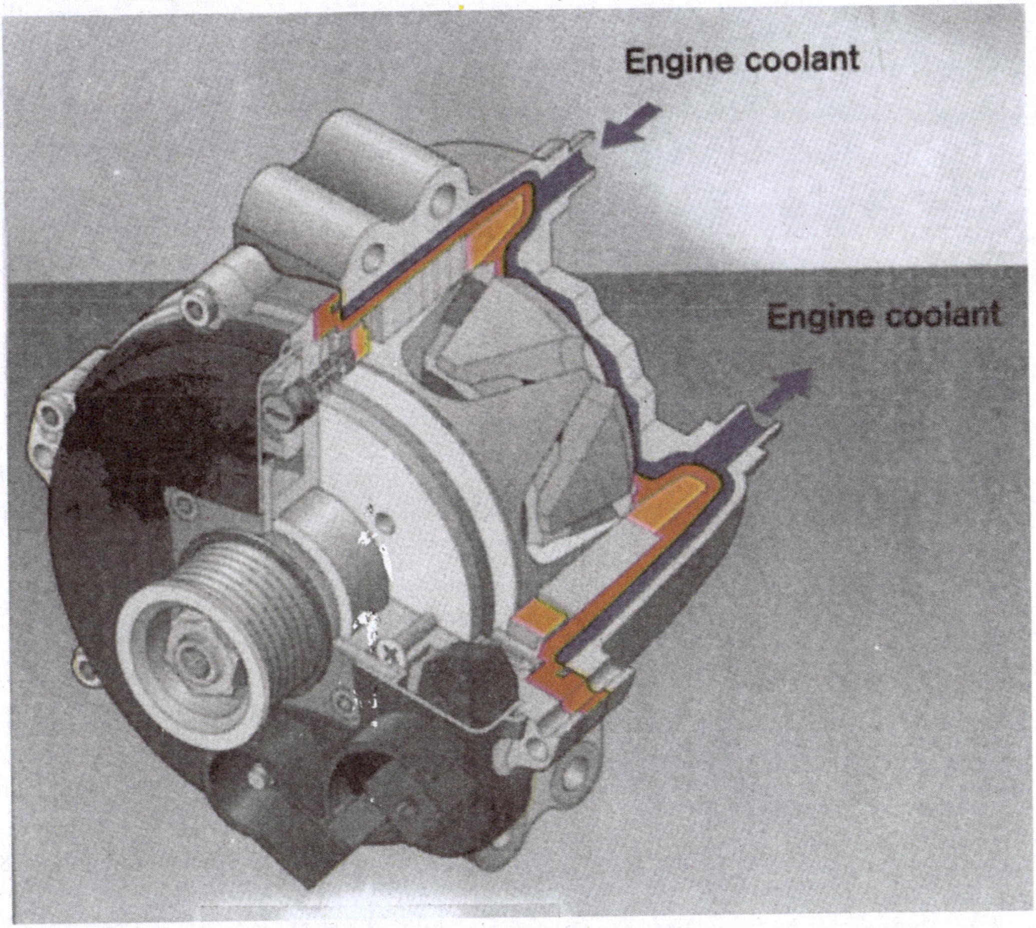

Fig. 18.36. Bosch fluid-cooled generator
(Courtesy-MICO India Ltd.)

3. Various load-response functions

This regulator works in tandem with Zener power diodes to guard both alternator as well as on-board electrical system against abnormal voltage peaks.

Bigger alternators are also fluid-cooled such as Bosch-alternator generating currents of 150 A and more (Fig. 18.36). This fully-encapsulated alternator is directly linked to the engine's cooling circuit. The advantage is early warming up the engine by transferring heat to the coolant. Moreover, it is maintenance-free, being completely enclosed and having dispensed with wearing parts, *e.g.*, slip sings and brushes.

Bosch LI-L Long Life alternators have long life of 8 lakh kilometers. These generate upto 130 A at 28 volts and upto 200 A at 14V.

18. WIRELESS VEHICLE CHARGING

Conductive charging systems using plug and wires for electric cars have been standardised by now. 'J 1772' is the SAE standard for the plug and receptacle systems, including fast charging in a single 'combo connector' for all types of charging systems using cables, for electric and electric-hybrid cars.

Wireless charging systems are basically of 2 types:

1. Inductive charging technique, wherein the energy between two objects in close proximity is transferred through electromagnetic field. An electrical device on the vehicle receives energy from the charging station through an inductive coupling and transfers the same to charge the battery. 3-7 kW of energy can be transferred by this technology as demonstrated by research models.

2. Magnetic resonance technique, in which the energy is exchanged between two objects through their changing or oscillating magnetic fields. Resonant coupling occurs when the natural frequencies of the energy-exchanging objects are almost equal. 'WiTricity corp.' of USA claim that their system being developed based on this technology can deliver upto 3.3 kW over a distance of 200 mm at acceptable efficiency rate, *i.e.*, at more than 90% efficiency.

Various Companies currently involved with the development of Wireless charging systems include Ford, Pike, General Motors, Evatran and WiTricity of USA and Conductix-Wampfler of Germany in partnership with Daimler AG.

The first wireless charging guidelines (SAE TIR J2954 Wireless Power Transfer for Light-Duty Plug-in/Electric Vehicles and Alignment Methodology) have been approved by SAE. The relevant paper was published online on May 31, 2016 by the chairman of the SAE Wireless Power Transfer Committee, at an electric vehicle conference in Logan, Utah (USA). SAE TIR J2954 establishes 85 kHz as a common frequency band for light duty vehicle charging systems. Four levels of plug-in and electric WPT are defined in the guidelines. Specifying the frequency band makes it possible for the manufacturers to build and install wireless charging receivers in vehicles, which are compatible with different wireless charging stations from various manufacturers. This standard has already been co-signed by many automotive manufacturers (BMW, Daimler, Fiat Chrysler, Ford, GM, Honda, Jaguar, Karma, Mitsubishi, Nissan and Toyota) and OEM Tier 1, *i.e.*, Delphi, Lear, LG, Magna, Panasonic, TDK and Toshiba, besides a large number of Bus OEMs and Technology Suppliers. Bench testing and live in-vehicle testing of WPT1 and WPT 2 are expected to be completed by 2018.

REFERENCES

1. Charles Ofria, ''A short course on charging systems'', carparts.com/classroom/charging.htm.

TABLE 18.1. COMPARATIVE GENERATOR DATA OF SOME INDIAN AUTOMOBILES

S. No.	*Make*	*Generator Type*	*Regulator Type*
1.	Maruti 800	Alternator	Vibrating voltage type
2.	Hindustan Ambassador Mark II	Lucas, Motolite C-40	Lucas Regulator vibration type
3.	Fiat 1100	Fiat Generator R90-180/12-2500	A/4-180/12 vibrator type
4.	Jeep Universal	Perstolite GJP-7432	Prestolite-VBO 4231 E-5A vibrator type
5.	Dodge/Fargo Model 89M4 (Petrol type)	Prestolite GJM-7001S	Prestolite-VRX 6301A vibrator type
6.	Ashok Leyland Comet Passenger	Lucas GGS 524/3	Lucas 3GC-24V vibrator type
7.	Tata Diesel Vehicles	Lucas-TVS Type C-40 LQ 22740B (20 Amps) (For SE/SK/SA/SD vehicles) or type GGS-12-1 (35 amps) (for LP/LPT/LPO vehicles)	Lucas-TVS Type 3GC 37367 (20 amps) (For Se/SK/SA/SD vehicles) or Type 3GC-37379 (35 amps) (For IP/LPT/LPO vehicles)

THEORY QUESTIONS

1. Draw a diagram showing the charging circuit and explain the functions of its various components.
2. Explain the principle of a generator. Give its constructional details.
3. How do you control the generator output in the automobiles? Explain the third brush control.
4. Write a note on the cut-out relay as used in the battery generator circuit.
5. Discuss the construction and working of a combined vibrating voltage and current regulator. How is the temperature compensation achieved in such a regulator?
6. What is the necessity of generator output control? Discuss various methods of achieving the same.
7. Explain in detail the procedure adopted to test the working of a generator-regulator system.
8. Describe the common faults occurring in the generators. Explain fully how these faults may be detected and remedied.
9. Described in detail the procedure for servicing of a generator regulator mentioning clearly how the correct regulator setting is obtained.
10. Discuss the construction and principle of operation of an alternator for automobile charging system. Illustrate your answer with suitable sketches. Compare the alternator with the d.c. generator for automotive use.
11. Explain clearly the type of regulation required in case of alternators and describe various types of regulators for the same. Discuss in particular the principle of an electronic regulator for alternators.
12. Write a comprehensive note on various warning devices for malfunctioning of charging system in an automobile.
13. Describe in detail various tests to ascertain the fitness of various components of an alternator.
14. Discuss in detail the procedure for trouble shooting in case of an alternator.
15. Write a detailed note on wireless systems for charging electric cars including hybrids.

QUESTIONS FOR VIVA

1. What are the main requirements of the vehicle charging system?
2. What is the function of a generator in an automobile?
3. Wherefrom does the generator get its drive?
4. Where is the generator located on the engine?
5. What is the Fleming's left hand rule?
6. What is the material used for:

(*a*) generator frame
(*b*) armature?

7. What is commutator?
8. Name the two types of armature windings.
9. Does the generator output vary with the engine speed?
10. What is the ideal charging voltage for a 12-volt battery?
11. How do you define armature reaction?
12. What is the main advantage of third brush control generator?
13. What is the purpose of using a cut out relay in the battery generator circuit?
14. What are the objections in the use of third brush generator?
15. What is the type of generator used in modern cars?
16. How do you provide for generator cooling?
17. What is the necessity of controlling generator output?
18. What are the two types of external output control of generator?
19. What is the necessity of employing a temperature compensator in the generator regulators?
20. In a generator-regulator system which of the following conditions indicate defect:
    (*a*) Discharged battery and a high charging rate.
    (*b*) Charged battery and a high charging rate.
    (*c*) Discharged battery and a low charging rate.
    (*d*) Charged battery and a low charging rate.
21. What is indicated by the dots of solder on the inner surface of the generator cover band?
22. Test lamp is connected across the armature and the generator field terminal but the same does not light. Is the field circuit open or closed?
23. What happens if the generator drive belt is loose?
24. What are the possible causes of excessive generator noise?
25. What is an alternator?
26. Where is the field winding located in the alternator?
27. What is the material used for alternator frame?
28. State the material used for alternator windings?
29. What is the purpose of carbon brushes in an alternator?
30. How is an automotive alternator rotor driven?
31. What is a serpentine belt?
32. How many windings does the stator of an alternator contain?
33. How many diodes are used in an alternator stator circuit?
34. Name the two types of connections for the stator windings of an alternator. Which type is used more commonly?
35. State the advantages of the star connection and the delta connection.
36. What is 'third harmonic' in a star-wound stator? What is its significance?
37. Why is a cutout relay not required in case of an automotive charging system with alternator?
38. Why do the alternator regulators not require maximum current regulator?
39. State the factors on which the output of an alternator depends.
40. Name the one factor that is used to control the maximum voltage output of an alternator.
41. Why is an alternator regulator not required to function at low alternator speed?
42. How is the alternator output voltage controlled when it is below the regulator setting?
43. Name various warning devices for malfunctioning of charging system in a vehicle.
44. What does a negative reading on dashboard-mounted ammeter mean?

45. What indication is given by a positive reading on the ammeter on the dash board?
46. Why must a battery in good condition be used for testing of the charging system?
47. Name the two types of alternator regulators.
48. What is a thermister? What is its function in an alternator regulator?
49. What is an integrated circuit?
50. State the advantages of an I.C. type alternator regulator.
51. State the advantages of an alternator over d.c. generator.
52. What is inductive charging?
53. What is magnetic resonance technique of charging?

MULTIPLE CHOICE QUESTIONS

Mark the correct answers:

1. In a d.c. generator magnetic field is produced in the
   (*a*) armature (*b*) commutator
   (*c*) carbon brushes (*d*) stator
2. The three units contained in a regulator for automobile d.c. generator are:
   (*a*) voltage regulator, current regulator and Zener diode
   (*b*) voltage regulator, current regulator and temperature compensator
   (*c*) voltage regulator, current regulator and cutout relay
   (*d*) any three of the above
3. The frequency of a vibrating contact type regulator for d.c. generator is
   (*a*) 20 per second (*b*) 200 per second
   (*c*) 20 per minute (*d*) 200 per minute
4. The charging voltage at the start of the engine is ________ than when the engine is running on load.
   (*a*) lower (*b*) higher
   (*c*) same (*d*) any of the above
5. In an alternator the magnetic field is produced in the
   (*a*) rotor (*b*) stator
   (*c*) frame (*d*) regulator
6. An alternator frame is made of
   (*a*) cast iron (*b*) brass
   (*c*) aluminium (*d*) copper
7. The number of windings in the stator of an alternator is
   (*a*) 1 (*b*) 2
   (*c*) 3 (*d*) 4
8. The number of windings in the stator of an alternator is
   (*a*) 1 (*b*) 2
   (*c*) 4 (*d*) 6
9. The output of an alternator is controlled by
   (*a*) voltage regulator (*b*) cutout relay
   (*c*) current regulator (*d*) all of the above
10. A thermister is used in an alternator regulator to
   (*a*) control maximum voltage (*b*) control maximum current
   (*c*) compensate for temperature change (*d*) all of the above

[**Key:** 1. (*d*), 2. (*c*), 3. (*b*), 4. (*b*), 5. (*a*), 6. (*c*), 7. (*c*), 8. (*d*), 9. (*a*), 10. (*c*)]

19 Starting System

1. GENERAL

To start an internal combustion engine, the same has to be rotated at a certain minimum speed, which is about 100 rpm in case of automotive engines. Achieving this minimum speed depends upon the following factors:

1. Rated voltage of the system, which is 12 V in case of car engines and 24 V for heavier vehicle engines.
2. Lowest temperature at which the engine may be started. This is usually taken as –18°C to –25°C for cars and –15°C to –20°C for transport vehicles.
3. The torque required to crank the engine at the lowest temperature.
4. Minimum cranking speed of the engine at the lowest temperature.
5. Characteristics of the battery.
6. Voltage drop between the battery and the stator motor.
7. Characteristics of the starter motor.

Table 19.1. gives thc minimum cranking speed ranges for various types of engines.

TABLE 19.1. MINIMUM CRANKING SPEEDS

Type of Engine	*Speed Range (RPM)*
Reciprocating S.I.	60–90
Rotary S.I.	150–180
C.I. with glow plugs	60–140
C.I. without glow plugs	100–200

2. STARTING MOTOR

2.1. Introduction

The starting motor is driven by means of the current taken from the battery. It is usually mounted on the side of the engine.

The motor must be powerful enough to turn the engine at a speed such that the carburettor will give a proper air-fuel mixture. It must be capable of exerting a very heavy torque when starting and at low speeds. For this purpose the starter motor is usually series wound. In a series motor, the field coils are connected in series with the armature windings so that field and armature currents are equal.

Present day starting motors are also sometimes series-shunt wound. The advantage of this type of winding system is the lower internal resistance. Due to this motor output for a given current consumption is higher compared to a series wound motor.

2.2. Construction

The construction of the starter motor is basically similar to the d.c. generator construction, except that it is series wound and the generator is shunt wound. The main components are the body, the armature, the commutator and the field windings. Most automotive starting motors have four field poles and four brushes. Brushes are held against the commutator by means of springs.

Fig. 19.1. Partial cutaway view of a starting motor alongwith Bendix drive

The armature consists of a segmented copper commutator and heavy duty copper windings. Wave winding technique is generally used for starter motors as the game gives most appropriate torque and speed characteristic for a four-pole system. At the end of the armature shaft there is some form of drive mechanism by means of which the motor starts the engine (Fig. 19.1). For light and medium cars, starter motors of stalling torque 10 to 30 Nm are employed, whereas for lager cars and heavy vehicles, the starting torques required are 50 to 150 Nm.

The starter motor is directly connected to the battery from where it draws the current for its operation. The amount of current drawn for starting may be as high as 500 amperes. Thus motor winding of standard copper cable of larger diameter is used. To keep the resistance in the circuit low, the cable of shortest possible length is employed. It may also be noted that due to very heavy current discharge during staring , the battery voltage falls by considerable amount. Therefore unless battery of adequate capacity is provided and is maintained in good condition, the excessive battery voltage drop will not allow the maximum torque to be developed.

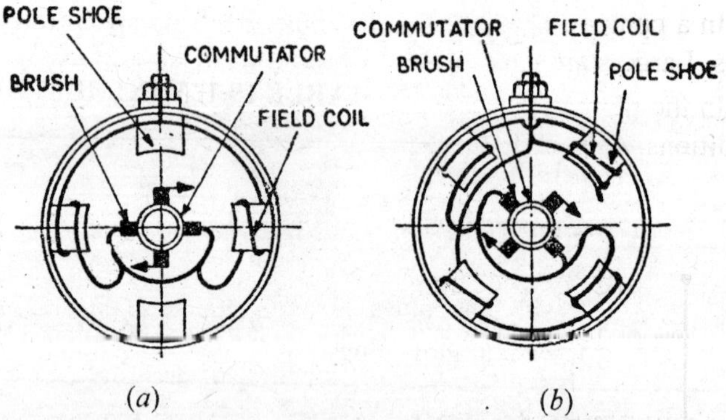

Fig. 19.2. Starting motors (*a*) 2-pole, (*b*) 4-pole.

Fig. 19.2 gives schematic diagram for 2-pole and 4-pole starting motors. In the series wound starter motor having 2-poles, the opposite field coils are wound in the same direction, so that they have the same polarity. The intermediate unwound poles then adopt opposite polarity by induction. In case of 4-pole motors, the adjacent coils are wound in the opposite direction.

2.3. Considerations affecting size of the motor

1. Torque required at the crankshaft to start rotation and the power required to maintain at the cranking speed. Starting torque depends upon piston displacement per cylinder, number of cylinders and the compression pressure. The power required for cranking depends upon the weather conditions. In winter, it may be three to four times the power required in summer or when the engine is otherwise warm.

2. Crankshaft speed required for starting the engine. It also depends upon the weather. It is found that on modern car engines a cranking speed of 100 r.p.m. is adequate even in adverse winter conditions.

3. Speed ratio between the starting motor and the crankshaft. This depends upon the size of the flywheel, which is further dictated by engine design considerations.

3. CHARACTERISTICS OF D.C. MOTORS

There are four types of D.C. motors, *viz.*, shunt wound, series wound, compound wound and with permanent magnet excitation type. The speed-torque characteristics of each of these will be discussed briefly with an object to determine the suitably of each as a starter motor.

A shunt wound motor is shown in Fig. 19.4. In this the field winding is connected in parallel with the armature. As the field excitation in this remains constant, the speed also remains practically constant, *i.e.* independent of torque. Thus the stalling torque in this case is low, due to which it is not suitable as starter motor.

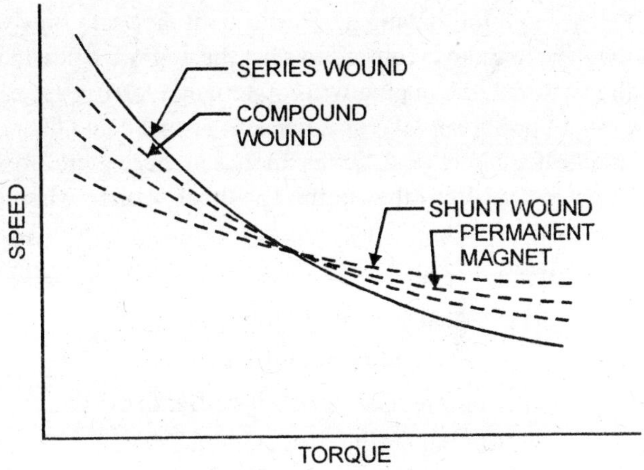

Fig. 19.3. Torque-speed characteristics of various types of D.C. motors.

In a series motor (Fig. 19.5), the field and armature are in series, which means the armature and field currents are same. Due to high field current, low resistance field windings, *viz.*, a few turns of heavy wire, are used. Due to high initial current, low resistance and no back EMF, this type of motor produces a very strong magnetic field and hence very high initial torque (Fig. 19.3). That is why series wound motors are considered most suitable as starter motors.

In a compound wound motor (Fig. 19.6), the field windings are a combination of series and shunt types. Large starter motors are quite often compound wound.

In the permanent magnet types (Fig. 19.7), the field excitation remains constant under all operating conditions. As such their characteristics are quite similar to the shunt wound motors. However due to fall

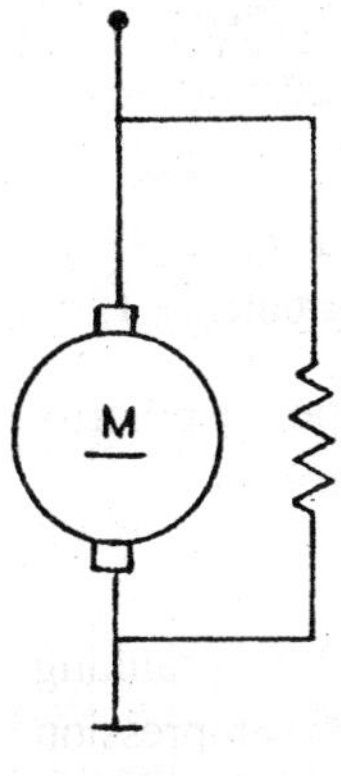

Fig. 19.4. Shunt wound motor.

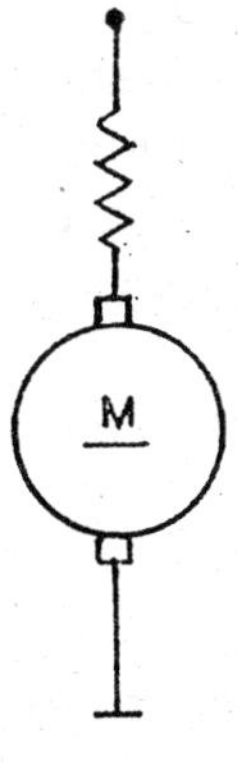

Fig. 19.5. Series wound motor.

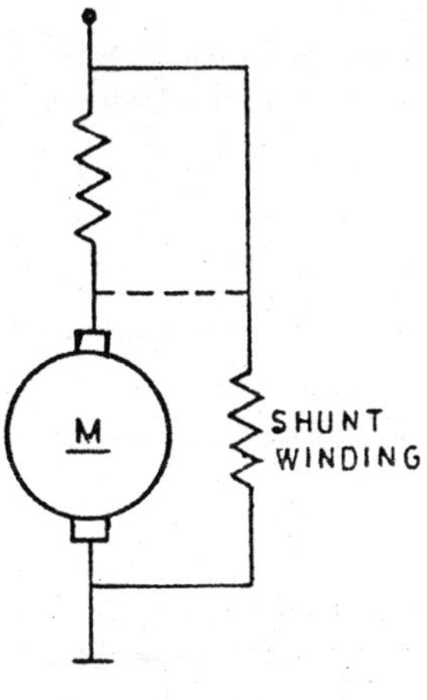

Fig. 19.6. Compound wound motor.

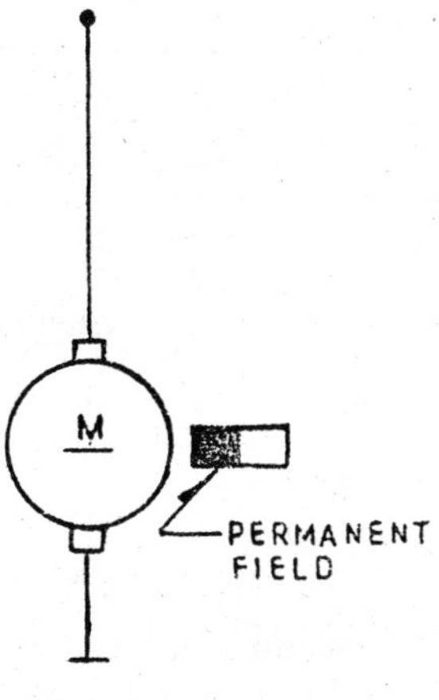

Fig. 19.7. Permanent magnet motor.

in battery voltage under load and low resistance of armature windings, their characteristics are comparable to series wound motors. Coupled with their lesser weight and smaller size compared to the conventional wound motors, their use as starter motors is on the increase. For use with bigger engines, permanent magnet motors with intermediate transmission have been developed. Epicyclic type of transmission of about 5:1 reduction is used, which increases the output toque. Such motors with about 1.7 kW power output are suitable for spark ignition engines of upto about 3 litres and compression ignition engines of upto about 1.5 litres capacity.

4. STARTING DRIVES

For staring the engine it is sufficient to rotate it at about 100 R.P.M. If a 4-pole motor is used at 1500 R.P.M. a pinion having gear ratio with the teeth on flywheel of 15:1 can be used to drive the flywheel so that less torque is required to start the engine. But when the engine starts, if the pinion is still in mesh with the flywheel, the engine will rotate it at a terrific speed equal to 15 times engine speed. The starting motor will not be able to take up such a large speed and the commutator segments and the windings will fly away due to the centrifugal force. Such a device, therefore, is necessary which should disconnect the starting motor pinion from the engine flywheel at once when the engine is started.

Main types of starting devices in present day use are:

1. Bendix drive
   - (*i*) Standard Bendix drive.
   - (*ii*) 'Folo-thru' Bendix drive.
   - (*iii*) Compression spring type Bendix drive.
   - (*iv*) Rubber spring type Bendix drive.
2. Overrunning clutch or the pre-engaged type drive.
3. Dyer drive.

Out of these Bendix drives are the inertia drives, whereas the other two are pre-engaged drives. In case of inertia type drives the starter motor is allowed to attain sufficient speed and hence power before it engages with the engine flywheel starter-ring gear. Inertia type drives are commonly used for cars, whereas for heavy vehicles, particularly with the diesel engine, the pre-engaged drives are employed. Because of large inertia of the flywheel and crankshaft in pre-engaged drives, it is imperative that the starter pinion must engage with the flywheel teeth before the starter motor develops full torque, as otherwise gear teeth would be subjected to heavy shock due to engagement and to excessive wear.

Problems have been experienced with the inertia type drives as explained below, due to which reason these are being gradually replaced by the pre-engaged type:

1. Excessive wear of the pinion and the ring gear due to aggressive nature of the engagement.
2. Pinion prone to seizure quite frequently due to the contamination of dust from the clutch. Oiling the pinion to remedy the same tended to attract more dust and hence more frequent seizure.
3. Premature disengagement of the pinion due to a false start. Methods to prevent the same as in case of 'folo-thru' drive were not found very satisfactory.

The construction and principle of working of each of these will be discussed in detail in the following few articles.

5. BENDIX DRIVES

Bendix dries are the inertia drives in which the starter motor pinion is made to engage or disengage with the toothed ring on the periphery of the engine flywheel. These may be either the *outboard* type or the *inboard* type. In the outboard type pinion moves away from the starting motor to engage the flywheel whereas in the inboard type it moves towards the starting motor. An inertia type Bendix drive for cars is shown in Fig. 19.8.

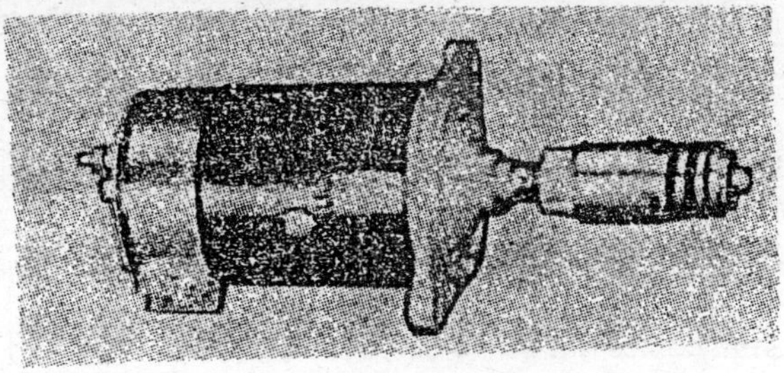

Fig. 19.8. Type M35G (in board)/M35G SCOB (outboard) starter motor drive. (Courtesy—Lucas-TVS, India)

5.1. Standard Bendix Drive

Fig. 19.9 shows a standard Bendix drive (inboard type) for the starter motor. There is a threaded sleeve on the armature shaft. The sleeve can slide or turn freely over the shaft. The shaft is keyed to the fixed drive head which is connected torsionally to the sleeve through a coil spring (which takes up the shock of engagement) and the sliding dog. On the sleeve there is a pinion to which an unbalance weight is attached, the purpose of the weight being to prevent the rotation of the pinion on the sleeve threads.

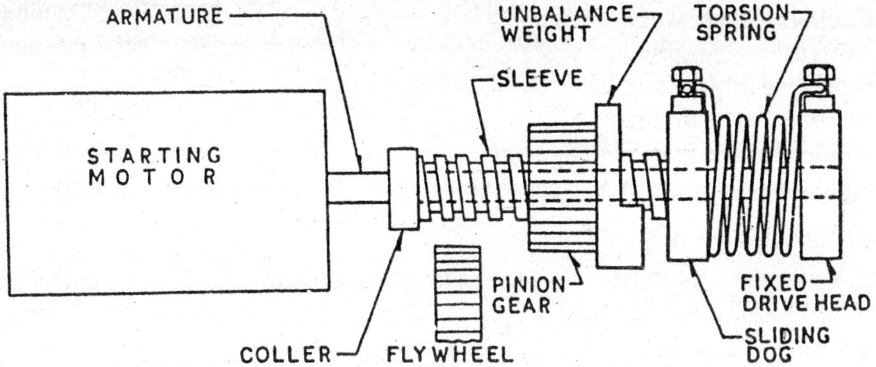

Fig. 19.9. Standard Bendix Drive (inboard type)

When the motor starts, the armature shaft rotates causing the sleeve to rotate and because the pinion cannot rotate due to the unbalance weight, it moves axially towards the motor till it is engaged with flywheel. Further movement of the pinion is prevented by the collar attached on the sleeve and because of this pinion has to start rotating. As it is also in mesh with the engine flywheel, the flywheel is rotated and the engine starts.

When the engine starts, it is the flywheel that rotates the pinion and because of its bigger size, the flywheel rotates the pinion much faster than the armature (which by now, has slowed down due to releasing of the self-starter switch) with the result that the pinion is backed out of mesh with the flywheel.

An important precaution with this and other types of inertia drives is that the pinion and the sleeve should not be lubricated. Lubrication causes the dirt to accumulate on the sleeve and pinion teeth which causes sticking of pinion.

5.2. 'Folo-thru' drive

This is similar in construction to the standard Bendix drive. The armature shaft is connected to the threaded sleeve through a spring and overrunning clutch. The inside of the pinion barrel fits into the sleeve threads. Towards the end of the sleeve is provided a detent as shown (Fig. 19.10). Two spring-loaded pins, *viz.*, the lock pin and the anti-drift pin are also provided. The anti-drift pin provides only a frictional contact with the sleeve teeth, whereas the lock pin action is in connection with the detent as is further explained below.

The engagement of the pinion with the flywheel takes place in the similar manner as in case of standard Bendix drive. However, towards the end of the pinion travel, the lockpin drops into the detent and would not let the pinion disengage prematurely due to false start *e.g.* weak or intermittent firing. The pinion thus continues driving the flywheel till the engine really gets stated and attains a speed of about 400 r.p.m., when the lockpin comes out of the detent due to the centrifugal force which overcomes the spring force. The pinion then gets disengaged as in case of the standard Bendix drive.

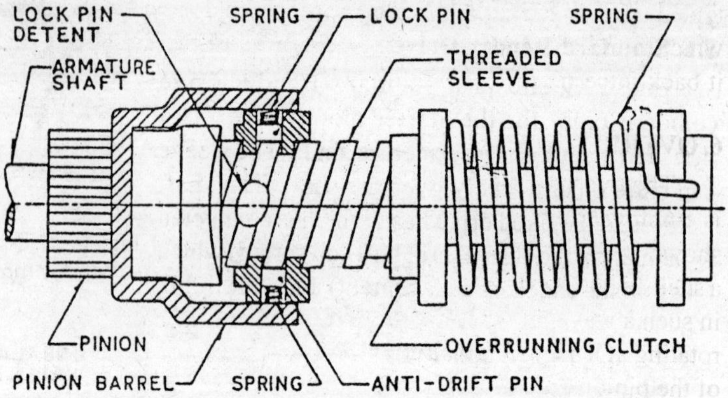

Fig. 19.10. 'Folo-thru' starting drive.

The anti-drift pin is loaded with a spring which is stiffer than the spring of lock-pin. This prevents the drifting of the pinion to engage with the flywheel teeth accidently. The over-running clutch serves to avoid damage to the starting motor in case due to any reason, the pinion remains engaged with the flywheel even after starting of the engine.

5.3. Compression spring type Bendix drive

This differs from the standard Bendix drive in that the threaded sleeve is mounted directly on the splined armature shaft (Fig. 19.11). A spring under compression is employed in between the sleeve and the nut

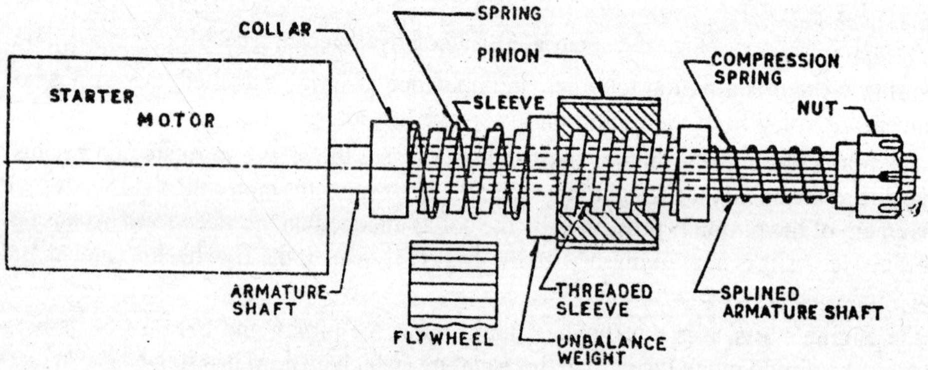

Fig. 19.11. Compression type Bendix drive.

which is fixed on the end of the armature shaft. Another spring of lower stiffness is there between the pinion and the collar on the shaft.

When the motor is started, the sleeve starts rotating alongwith the armature shaft. This causes the pinion to travel towards the motor till its teeth engage completely with the teeth on the periphery of the flywheel. By this time the pinion also strikes against the collar which stop its travel further (the spring on the sleeve over which the pinion moves, serves to avoid shock due to striking of the pinion with the collar). The pinion, therefore, tries to rotate but is offered initial resistance by the flywheel which is stationary by now. The torque of the shaft then has a tendency to force the threaded sleeve further out against the spring tension, till the flywheel starts rotating. The spring tension is consequently relieved. When the engine starts, the flywheel rotating at a faster speed causes the pinion to be backed out of mesh.

5.4. Rubber spring type Bendix drive

The principle of operation of this type of starting drive is similar to that of the compression-spring type described earlier. Only the spring is replaced here with a rubber bush. This serve to avoid the damage which is caused when due to false engine starting the pinion gets disengaged and attempt is made to bring it back into mech.

6. OVERRUNNING CLUTCH DRIVE

The overrunning clutch, which is employed in this type of drive is shown in Fig. 19.12. It consists of a shell and a pinion collar connected in such a way that when the shell is rotating at a speed greater than that of the pinion collar, the former will drive the later. However, whenever the speed of the collar becomes more than the speed of the shell, there is no more any connection between the two. Such a connection is provided by the spring-loaded rollers, the shell is connected with the armature shaft through splines, whereas the collar is attached to the pinion.

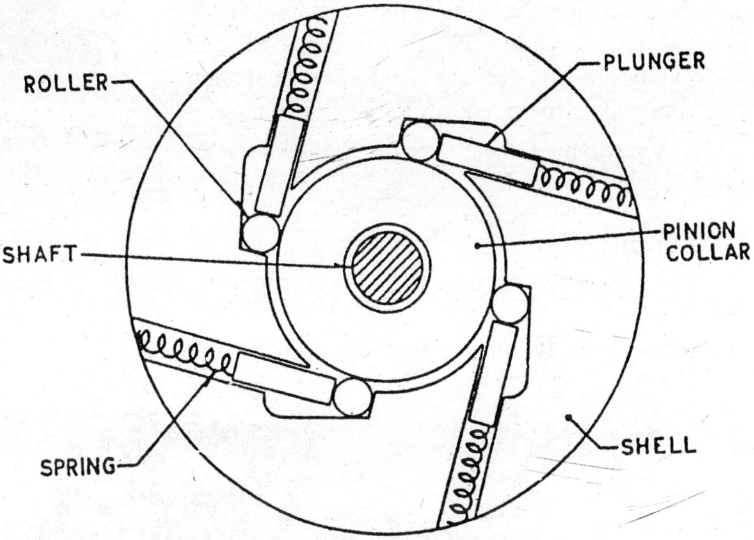

Fig. 19.12. Overrunning clutch for starting drive.

A starting drive employing the overrunning clutch is shown in Fig. 19.13. The shift lever is attached to the starting pedal in case of mechanical starters as shown, or may be operated by means of solenoid switches, to be described later. When the starting pedal is pressed by the drive, the shift lever moves about its pivot, thereby pushing the switch of starting motor and moving the overrunning clutch and the pinion assembly through the spring

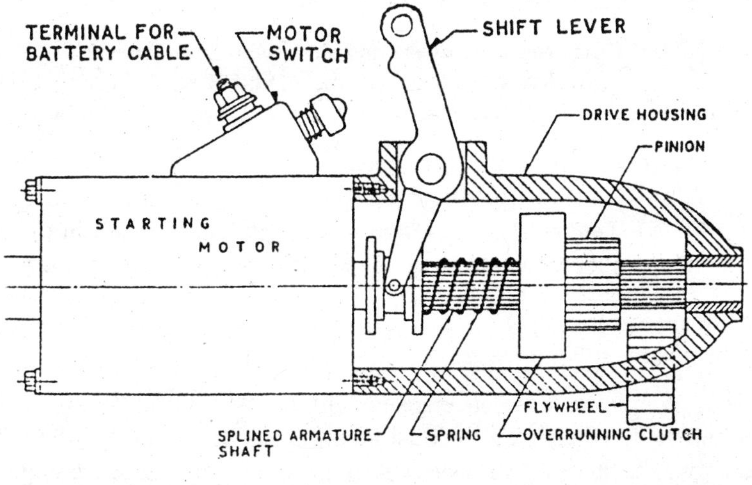

Fig. 19.13. Overrunning clutch type starting drive.

simultaneously. This causes the pinion to get engaged with teeth on the flywheel and also get the armature shaft and therefore the pinion to rotate, which starts the engine. However, as soon as the engine gets started, the flywheel and hence the pinion rotates at much faster speed, causing the overrunning clutch to disconnect the pinion side from the motor side. Further the motor also stops because as soon as the engine starts, the driver leaves the starting pedal or key which causes the shift lever to come back to its previous position, leaving the motor switch open and disengaging the pinion from the flywheel.

An overrunning clutch type pre-engaged starter motor drive for cars is shown in Fig. 19.14.

Fig. 19.14. Type SM82 pre-engaged starter motor drive. (Courtesy—Lucas-TVS, India)

7. DYER DRIVE

This type of drive (Fig. 19.15) is particularly suitable for heavy duty engines. It is so designed that complete engagement of the pinion with the flywheel ring gear is achieved before the starter motor switch is operated, thus avoiding the possibility of gear damage.

Fig. 19.15. Delco-Remy starting motor with Dyer drive.
1. Oiler, 2. Bearing, 3. Washer, 4. Oiler, 5. Solenoid, 6. Shift lever, 7. Bearing, 8. Drive, 9. Bearing, 10. Field coil, 11. Armature, 12. Brush, 13. Brush spring, 14. End cover.

The main components of the Dyer drive have been shown in Fig. 19.16. The shift sleeve is free to move on the armature shaft, in which are provided the spiral teeth. The shift sleeve is operated by means of the shift lever (Fig. 19.15). The snug on the pinion guide fits into the slot in the pinion, which has internal splines corresponding to the armature shaft splines. However the pinion fits on the armature shaft rather loosely.

When the shift lever is pressed, the shift sleeve is pushed to the right and consequently the pinion is also moved in that direction. However, because of the spiral teeth, there is angular motion also. With further pushing of the shift lever the pinion thus gets engaged with the flywheel. However, the chances are that the teeth in the pinion and the flywheel may not align for engagement when the pinon is moved adjacent to the flywheel. In such a case the shift sleeve continues to move the pinion guide along the armature shaft, which, because of snug fitting on the spiral splines of the shaft, rotates without any linear movement,

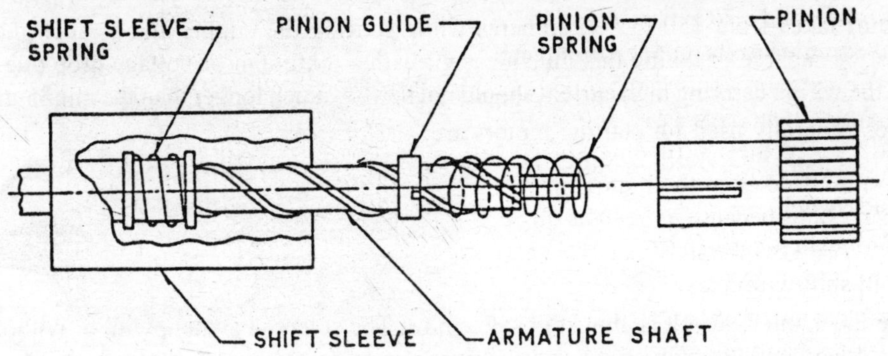

Fig. 19.16. Components of Dyer drive.

taking with it the pinion also, till the teeth on the pinion and the flywheel mesh. After the complete engagement, the shift lever is pushed further to press the starting motor switch.

On starting the motor, the shift lever is pushed further to press the starting motor switch.

On starting the motor, the shift sleeve is immediately rotated back in its initial position. Further as soon as the engine gets started, the flywheel rotates faster than the pinion and thus disengages the latter as in case of Bendix drive.

The exploded view of Lucas-TVS pre-engaged type starter motor with the drive is shown in Fig. 19.17.

EXPLODED VIEW

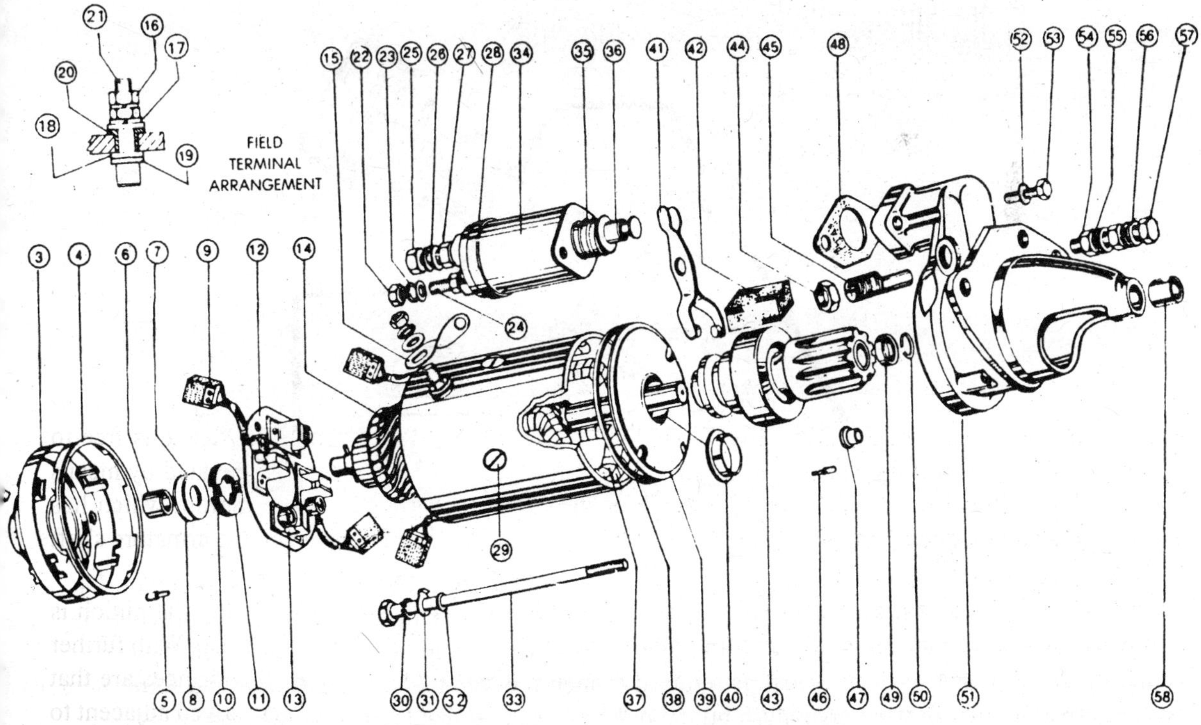

Fig. 19.17. Exploded view of SM114 starter motor
(Courtesy—Lucas-TVS Ltd.)

8. STARTING MOTOR SWITCHES AND CONTROL CIRCUIT

Starting motor takes from 150 to 500 amperes while starting the engine and as such the switches operating it must be capable of taking this much current with a minimum of voltage drop due to contact resistance. Also the wiring carrying high current should not be very much longer than the minimum required.

The switches generally used for starting motors are:

(*i*) Manual switch.

(*ii*) Solenoid switch.

(*iii*) Solenoid switch-cum-shaft.

(*iv*) Solenoid shift with relay.

Out of these the manual switch is almost obsolete now. The solenoid switch, with or without relay, is used with Bendix drives. For overrunning clutch type drives, apart from starting the motor itself, the mechanism for obtaining the engagement of pinion with the flywheel also has to be actuated. This is done with a solenoid switch-cum-shift, commonly referred to as solenoid shift only. To obtain more positive operation of the solenoid, a relay is sometimes used alongwith the solenoid shift. All these types of switches shall now be explained in detail.

8.1. Manual Switch (Fig. 19.18)

The construction is very simple. As the plunger is pressed the contacts are made. It may be operated either with hand or with foot. It is sometimes operated by means of accelerator lever also.

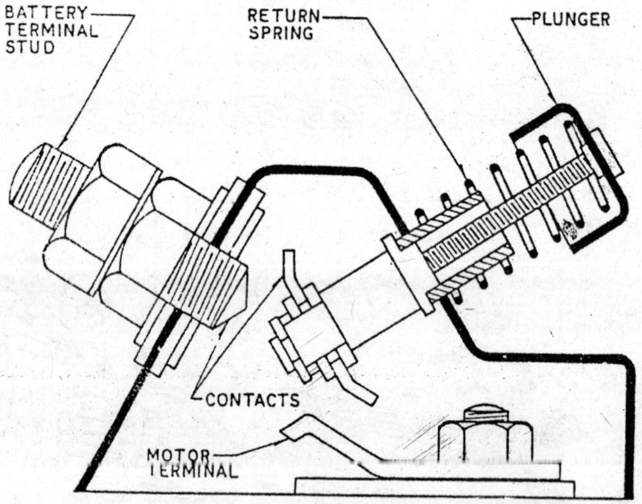

Fig. 19.18. Manual switch.

8.2. Solenoid Switch (Fig. 19.19)

It is also called magnetic switch. As the button is pressed, the current flows from the battery to the winding which produces a magnetic field, resulting in the movement of the plunger to close the switch connecting the battery and the starting motor. The advantage of solenoid switch (compared to the manual) is that the heavy current wiring length is reduced (which reduces the voltage drop in the starter circuit) and the driver has to operate only a push button or key switch (at the dashboard) which carries a nominal amount of current only. This driver's switch is only of light duty type and and its wiring is also thin and light.

The solenoid switch here has got one winding only. Some solenoid switches have two windings also, *viz.*, the pull-in winding and the hold-in winding. Both the windings exert combined force to pull the armature so as to make the contacts of the switch when the pull-in winding is short-circuited and only hold-in winding remains to hold the contacts closed. The advantage of this type is that less current is drawn from the battery during the period of hold-in. Further only the leads from the battery to the solenoid terminal and from the other solenoid terminal to the starting motor need only be thick, other wires including the one going to the driver switch may be thin as explained above.

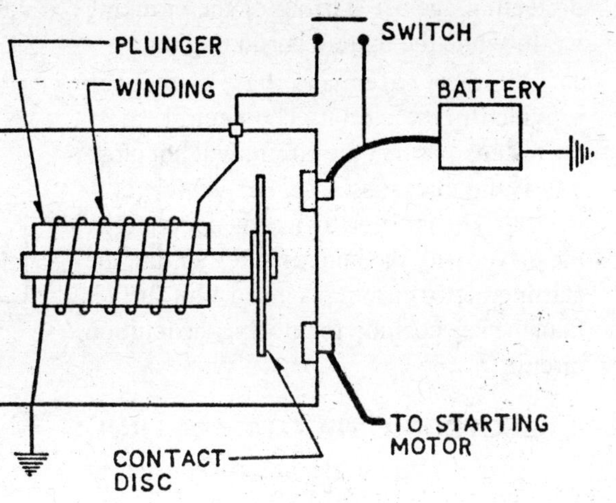

Fig. 19.19. Solenoid switch.

8.3. Solenoid switch cum-shift

Solenoid switch cum-shift consists of a solenoid switch which also performs the function of actuating the shift lever in an overrunning clutch type of Dyer's drive. The solenoid shift shown in Fig. 19.20 is similar in construction to the solenoid switch (Fig. 19.19), except that it has got a different plunger construction suited for operating the shift lever. Further this is of two-winding type. It is not necessary, however, that a solenoid shift must have always two windings.

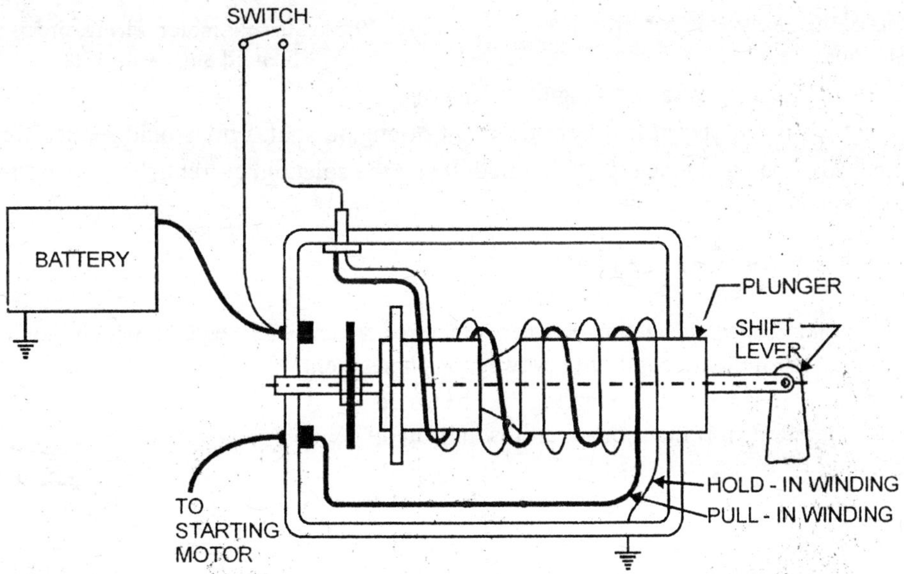

Fig. 19.20. Solenoid switch-cum-shift.

8.4. Solenoid shift with relay (Fig. 19.21)

The wiring diagram shown in Fig. 19.21 is for a fully automatic circuit for the starting motor. The relay used serves to control the solenoid currents. Further, completing the relay circuit through the generator provides protection against operating of the starter while the engine is running, because then the generator voltage is approximately equal to the battery voltage. As such no voltage is applied through the relay even when the switch contacts are closed. A vacuum switch is also sometimes employed in the circuit, which provides added protection against starting of the cranking motor while the engine is running.

Another safeguard has also been provided in the circuit. The neutral safety switch ensures that the circuit will not close unless the transmission of the vehicle is in neutral. The ignition switch is operated by the driver with the help of a key so that the starting motor circuit is closed a bit further than the key position for closing the ignition circuit.

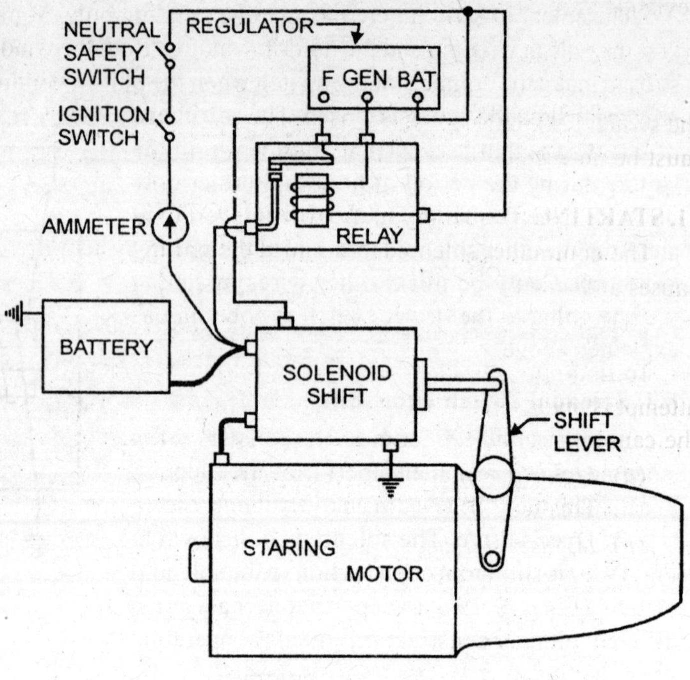

Fig. 19.21. Starting motor wiring circuit using a solenoid shift with relay.

9. ELECTRONIC STARTER CONTROL

An electronic starter control consists of a static relay on a circuit integrated with the solenoid switch.

Such a control, with necessary improvements of the circuit can perform the following functions:

1. To prevent cranking when the engine is running.
2. To shunt off the starter at the precise instant of engine start. This would reduce wear and noise.
3. To Redesign the mechanical parts by modulating the solenoid current. This would result in softer operation and eight reduction.

10. TESTING THE STARTING SYSTEM

1. Armature

Use a multimeter to check insulation between armature core and each segment which should be complete. The meter should also indicate continuity between each segment.

Any short circuit in the armature winding may be tested with a growler (Fig. 19.22). If the short-circuit is indicated, the winding has to be replaced.

2. Field coil

Continuity should exist in the field windings which may be tested with a multimeter. There should also be perfect insulation between the yoke and the field terminal.

3. Brushes

Examine the brushes visually for wear and damage. If the wear or damage is more than the permissible limits, replace the concerned brush.

4. Brush holders

Use a multimeter to check the insulation between the positive brush holder and the holder frame. In case of

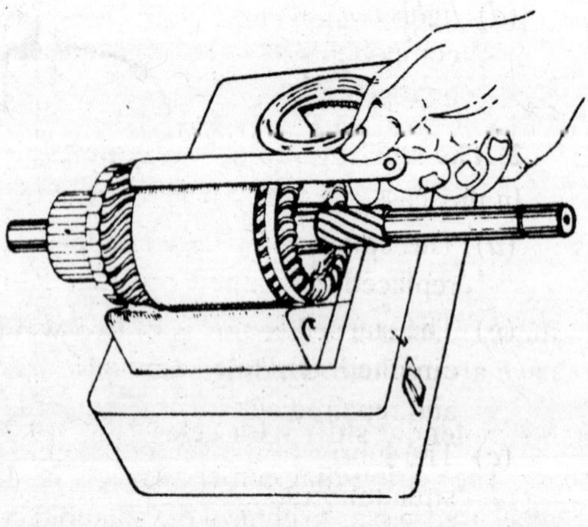

Fig. 19.22. Armature winding test with growler. (Courtesy—Lucas-TVS Ltd.)

deviation from complete insulation, rectify the fault suitably by repairing or replacing the damaged component.

5. Solenoid switch or shift

Disconnect the lead wire connecting the switch to the starter motor and apply battery voltage between the switch terminals. If correct, the plunger is attracted. In case of shift, the pinion would also project. It must be noted that the test should not be done continuously for more than 10-15 seconds.

11. STARTING SYSTEM TROUBLE SHOOTING

The common faults occurring in the starting system are described below, alongwith their possible causes and suitable remedies.

1. The starting motor does not operate

To find out specific cause for this trouble it is quite helpful if the car lights are turned on before an attempt is made to start cranking motor. The lights will then be affected in different ways depending upon the cause of trouble, as soon as the starter key (or button) is operated:

(*a*) *Lights go off complete.* This is due to some loose connection in the circuit, most likely at battery. This may be located and rectified accordingly.

(*b*) *Lights dim heavily.* In this case the possible causes could be any or more of the following:

- (*i*) Weak battery. It may be checked and if found discharged may be removed for recharging.
- (*ii*) The starting motor field or armature may be grounded. It may be checked by means of a circuit tester (ordinary multimeter will do) and rectified suitably.
- (*iii*) The pinion of the Bendix drive may be jammed into engagement with the flywheel. This may be freed by placing the transmission in high gear and rocking the vehicle to and fro. If this action does not free the pinion then losen the mounting bolts of the starter motor, pull the motor slightly till the pinion is disengaged, and retighten the bolts.
- (*iv*) There may be some defect in the engine, *e.g.* excessive friction, which may be properly tested.

(*c*) *Lights dim, but only slightly.* The reasons for this may be:

- (*i*) The end-clearance for the pinion may not be sufficient, so that it reaches the end of its travel before the starting motor switch may be operated by the solenoid shift.
- (*ii*) Open-circuit in the field windings or the armature of the starting motor, which may be suitably tested.

(*d*) *lights remain unaffected.* This means that the circuit is open somewhere. The wiring an the individual components have to be tested separately for open-circuit. Any component found defective has to be replaced.

2. The starting motor does operate, but slowly

In this case possible causes for the trouble may be the following:

(*a*) The battery may be weak or defective. The same should be tested therefore and recharged or replaced with a new one depending upon the fault.

(*b*) The starting motor may be defective; the field windings or the armature may be open or short-circuited. Or the commutator bars may be dirty or burnt. The motor should be tested thoroughly and repaired suitably as required.

(*c*) The lubricating oil in the engine may be of viscosity, higher than the one prescribed. In such a situation, replace the oil with the one of proper specifications.

(*d*) The sleeve or the pinion threads in the starting drive may be dirty and thus be sticking. The same should be cleaned thoroughly with kerosene oil.

3. The starting motor operates, but the engine is not cranked

This may be due to any or more of the following reasons:

(*a*) Some component (more commonly the spring) in the starting drive may be broken, which should be replaced.

(*b*) The pinion gear may not be aligned properly which may be checked and rectified.

(*c*) The overrunning clutch may be slipping, which should be replaced.

(*d*) The teeth in the flywheel ring may be broken. The ring gear has to be replaced in such a case.

4. The starting drive does not disengage, after the engine starts

The possible causes for this defect may be:

(*a*) the mounting bolts of the starting motor may be loose, which may be tightened.

(*b*) The armature shaft may be bent, which may be straightened or replaced.

(*c*) The circuit-switches may be defective. They should be checked thoroughly. Any defective switch detected has to be replaced.

(*d*) The pinion gear may not be aligned properly which may be suitably rectified.

(*e*) Dirty threads on sleeve or the pinion gear, which should be thoroughly cleaned with kerosene oil.

(*f*) There may be short-circuit in the solenoid, which may be tested. If found defective, the same should be repaired or replaced.

(*g*) The return-spring of the shift lever may be very weak or even broken. In either case the same has to be replaced with a new spring.

5. Noisy operation

This may be due to:

(*a*) Teeth on flywheel ring may have worn out. The same has to be replaced.

(*b*) Pinion spring may be broken, which has to be replaced.

(*c*) Starter motor bearings may have worn out. These may be inspected and if required, replaced.

(*d*) Lack of lubrication of the starting motor bearings may also be a source of undue noise.

(*e*) Motor shaft may be bent, which may be checked and replaced if required.

12. MODERN STARTERS

These are lighter and more compact compared to earlier starters. An example is the Bosch DW starter (Fig. 19.23), which is 40% less in weight, its diameter being only 70 mm. On units in the 0.7 to 2.2 kW range, the conventional field windings and pole shoes have been replaced by permanent magnets. These are meant for petrol engines of upto 6.0 litre size and diesel engines of upto 3.0 litre size. For diesel engines upto 4.0 litre size, electrically excited 2.3 kW starters are being made by Bosch.

13. START-STOP SYSTEM

13.1. General

It is a system installed on many modern vehicles cars; in particular, in all hybrid cars. When the vehicle comes to a stop, *e.g.*, at traffic lights, or even when the driver starts braking with an intention to stop, this system stops the fuel delivery and ignition spark to the engine. The engine restarts immediately when the brake pedal is released or the clutch is engaged or even when the driver has decided to stop, but then changes his mind suddenly depending on the traffic movement. There may be some special situations also when the system is designed not to work, *e.g.*, when the engine is cold so that it may warm up quickly, or if the battery charge is less than a specified threshold or if the driver unfastens his/her seat belt. Although all this stopping and restarting happens automatically, yet the driver can choose whether to activate or disable the system.

The system operates with a combined starter and alternator, with a battery which can sustain deep cycling. The combined starter-generator is similar to a conventional starter, except that in this the motor is more powerful and faster acting. Moreover, the automatic action is due to control by an ECU and a number of sensors.

The system results in fuel saving and emission-reduction. But how much fuel is saved, would depend on the type of driving. For example, the improvement in fuel consumption and emissions is maximum in crowded urban areas where traffic jams are frequent.

13.2. Effect on the engine

A vehicle without the start-stop system normally experiences about 50,000 start-stop events during its lifetime, which number increases to 500,000, *i.e.*, ten times in case of a vehicle with automatic start-stop. Due to this, the engine and in particular, main bearings, experience increased wear, requiring special remedial measures. The main bearings are only plain bearings on which the crankshaft (which is one of the heaviest engine components) journals are supported directly. When the engine is running, the journals and the bearing surfaces do not touch on account of hydrodynamic lubrication conditions maintained by the lubricating oil fed under pressure and pumped around the bearing surfaces due to the action of the spinning crankshaft. However, when the engine stops, the crankshaft journals settle onto the bearings and the two metal surfaces come into contact. When the engine restarts, the hydrodynamic lubrication conditions are not achieved immediately. Till the engine crankshaft achieves a certain threshold speed, the 'boundary lubrication' prevails, where the metal-to-metal contact continues. Similar action happens when the engine stops. Since the start-stop cycles increase to ten times, the wear in such vehicles is bound to be very high.

The problem of wear has been dealt with a two-pronged strategy:

1. By improving bearing materials. Federal Mogul has developed a new material called 'Irox' with a polymer-coating containing particles of iron oxide (rust), which is so slippery in microscopic form that coefficient of friction of an Irox bearing is reduced to only 50% that of a conventional aluminium bearing.
2. By improvements in lubricating oils. A modern engine oil contains additive package consisting of a complex chemical cocktail. 'Millers Oils' of U.K. claims that its new triple ester nano-technology-based 'Nanodrive' contains tiny nano-particles like microscopic ball bearings exfoliate under high pressure, the polymer 'flakes' adhering to the engine surfaces, which can reduce the friction by 50 percent.

14. INTEGRATED STARTER-GENERATOR (ISG)

With the progress in electrical and electronics fields, a highly efficient and integrated system has been developed, which is called 'Integrated Starter Alternator Damper (ISAD).

Functions

1. To perform the functions of both an alternator and a starter motor.
2. To provide an automatic vehicle start-stop system. The ECU switches off the engine at no load, *e.g.,* during stop at traffic lights and restarts the engine instantly when accelerator pedal is pressed to start the vehicle. The combustion process in the engine starts only when it is accelerated to the specified cranking speed. It can produce upto 2 kW at idling speed.
3. To convert the vehicle's kinetic energy into electrical energy when the brakes are applied and then, to feed this energy back into the electrical system. This helps stabilize the battery voltage during start-stop operations.

4. It transfers the drive from the engine to the transmission and also acts as a vibration absorber unit. The damping effect is achieved by a rotation capacitor. A change in relative speed between the engine and the rotor due to vibration, causes one pole of the capacitor to be charged, thus absorbing the vibration energy.

Advantages

1. As the ISG saves engine power due to 'start-stop' function and also provides regenerative braking, a smaller engine can be used. Compared to a conventional generator and starting motor system, ISG saves upto 20% fuel.
2. As there is no need to enrich fuel for starting with an ISG, lower start-emissions occur, in particular during the cold start.
3. Longer life and reduced maintenance for the engine since the same is not used at all when the car is stationary.
4. Reduced noise and vibration due to built-in damping.
5. ISG can be integrated with most of the existing car systems. Thus there is no necessity for any modification of the existing car, which makes it highly cost-effective.
6. It can also start the engine at very low temperatures.

Disadvantages

ISG needs specialised power systems. Quite severe design requirements are:

1. High starting torque under severe operating conditions.
2. Large range of speed in generator mode.
3. High efficiency over a wide speed range, from say, 600 to 8000 rpm.
4. High cycle life, say, more than 2.5 lakh cycles in 10 years.
5. High reliability at reasonable cost.

Types

They may be classified either according to the drive system or the electrical design.

1. According to the drive system. It may be the belt driven or directly connected to the crankshaft between the engine and the gear box. The former needs lesser amount of copper and iron, but is otherwise complex and costlier due to the drive, which is eliminated in the later case.

2. According to electrical design. Four types are currently in use, which have their own merits and demerits:

(*i*) **Induction Machine (IM).** This design gives robust construction and is also less costly. The disadvantages are, lower efficiency, cooling problems, costly starter windings and greater complexity of control system.

(*ii*) **Permanent Magnet Synchronous Machine (PMSM).** This has the merits of simpler construction and higher efficiency. The negative considerations are costly permanent magnets and the inability of the ISG to be switched off electrically with the engine running, thus needing field weakening even at low speeds.

(*iii*) **Brushless DC Machine (BDCM).** It has constant torque and power operation due to reluctance torque. Provides larger torque for the same peak current and voltage compared to the other types. Moreover, there are no heat producing components on the rotor, the inertia is quite low and no maintenance is required. However, higher costs, corrosion and demagnetisation of the permanent magnets with time are its main demerits.

(*iv*) **Variable Reluctance Machine (VRM).** These are cheap and robust. Moreover, these can be switched off electrically even at high speeds. Higher temperatures around this type of machine also have negligible effect on its performance.

Battery and Sensors

Battery used with the ISG must be able to withstand large number of deep cycling. For this purpose AGM (Absorbent Glass Mat) type lead-acid batteries may be used.

The ISG system also has to include a 12V, DC to DC converter to stabilize voltage levels during deep cycling required for starting operation.

For ECU control, speed sensors already being used for antilock braking are used to transmit speed data to the ECU. A crankshaft sensor registers the attitude of the crankshaft and positions of the individual cylinders so that immediately when the driver releases the clutch or the brake pedal to restart the vehicle, ECU can determine in which cylinder fuel is to be injected and in which cylinder ignition should take place. Direct fuel injection is obviously more suitable for quick acceleration. Such engines can be restarted and accelerated to the idling speed in half the time required for conventional engines. Another important sensor is the one to measure the charge and condition of the battery. The battery current, voltage and temperature along with integrated software are used to calculate battery charge and to predict future battery behaviour. This information is then conveyed to the ECU which ensures that the engine is stopped only if the battery has adequate capacity to restart the engine quickly and not otherwise.

Practical Applications

1. DENSO in 2001, launched the world's first belt-driven Integrated Starter Generator.

2. The Siemens ISG works on the pulse-start technology and accelerates the engine to the idling speed; only then the combustion process for instant ignition starts. It automatically restarts the engine in less than 0.1 second when the accelerator pedal is pressed. It is very compact and is placed directly on the crankshaft between the engine and the transmission. In the 'boost mode', the starter briefly serves a second engine to provide acceleration in the low speed range. In the 'retard mode'. It converts the vehicle kinetic energy into storable electric energy. This ISG gives a peak generator output of 8 kW with 80% efficiency over the entire speed range as compared to 1.5 kW at 70% maximum efficiency, dropping to only 30% at high speeds in case of conventional generators

3. Delphi is working on a device that allows an engine to start itself by only injecting fuel into a cylinder that stopped under compression. It has been found that in such a situation, there is sufficient torque available for starting the engine. However, this device will not be useful for cold starts; it would serve only to provide automatic engine stop/start in traffic. Due to this reason, conventional starter would have to be retained alongwith this device.

4. Visteon's 'Speed Start 12' is a belt driven combined starter/alternator, which provides 5% fuel economy. However, it costs more.

5. Valeo's belt-driven starter alternator combines the functions of the starter and the alternator and performs the start-stop function, *i.e.* it automatically stops the engine when the vehicle is at a standstill and instantaneously restarts it when the driver engages a gear or releases a brake. The advantages claimed are:

(*i*) Upto 10 percent reduction in fuel consumption in urban traffic conditions.

(*ii*) Zero noise, zero consumption and zero emission at standstill. Currently standstill phases may amount to 30 percent of travel time in urban driving.

Another next-generation belt-driven starter-alternator from Valeo combines four functions:

1. Starter function starts the engine by electronic control over the alternator's reverse motor mode.
2. In the alternator mode, the unit generates a high electrical output.

3. 'Green-Boost' function, where additional torque is delivered to the engine.
4. Regenerative braking, which makes it possible to regenerate the energy of braking phases into useful electrical energy.

Besides environmental benefits at standstill, this unit is claimed to reduce upto 20% fuel consumption in urban traffic conditions.

6. Bosch also produces the starter-alternators of two types, *viz.*, the crankshaft-integrated (ISG) type and the belt-driven (RSG) type. Both the types can perform start- stop function. ISG is a permanently excited synchronous electrical motor installed between the IC engines and the transmission. RSG (Fig. 19.24) is a built-in generator that can also function as starter and is meant for small to medium displacement engines.

7. European Research and Development Centre of Mazda has produced a Smart Idle Start Stop System (SISS) for Mazda's petrol engines. The system restarts the engine by injecting fuel directly into the cylinder when the engine is stopped and then igniting it to force the piston down. This direct-injection enables a fast and quiet restart. Combining DI engine with SISS is said to improve fuel economy by upto 9%. For an automatic transmission vehicle, the system can restart in 0.35 second.

The main thing necessary for this type of restart by combustion is the stopping of the compression-stroke pistons and expansion stroke pistons at exactly correct positions so as to have right balance of air volumes. Mazda's SISS controls precisely the piston positions when the engine is shutting down so that all the pistons are stopped in their optimum position. For starting the system identifies the initial cylinder to fire, injecting fuel into it and then igniting it. Cylinders continue to be selected for ignition even at very low engine speeds and the engine quickly runs at the idle speed.

TABLE 19.2. COMPARATIVE STARTING SYSTEM DATA OF SOME INDIAN AUTOMOBILES

S. No.	*Make*	*Starting Motor*	*Starting Drive*
1.	Maruti 800	N.A.	Over-running clutch with solenoid shift
2.	Hindustan Ambassador Mark II	Lucas M35G	Bendix
3.	Fiat 1100	E76-0, 5/12	Over-running clutch
4.	Jeep Universal	Prestolite MDU-7004	Bendix
5.	Dodge/Fargo Model 89M4 (Petrol type)	Prestolite MDU-5003	Bendix
6.	Ashok Leyland Comet Passenger	Lucas GBS-524P190	Axial starter
7.	Tata Vehicles	Lucas-TVS Type GSPS-12B-25	Beacon Type 512STR16 Axial starter

REFERENCES

1. Viorel I.A., Szabo L. and others, "Integrated starter-generators for automotive applications", Acta Electrotehnica, Vol. 45, no. 3, 2004, pp. 255-260.

THEORY QUESTIONS

1. Discuss various factors on which the minimum cranking speed of an engine depends.
2. Describe the construction and working of starter motor for automobiles.
3. Discuss in detail the characteristics of various types of D.C. motors.
4. What are the different type of starting drives in current practice? Describe all of them in detail.

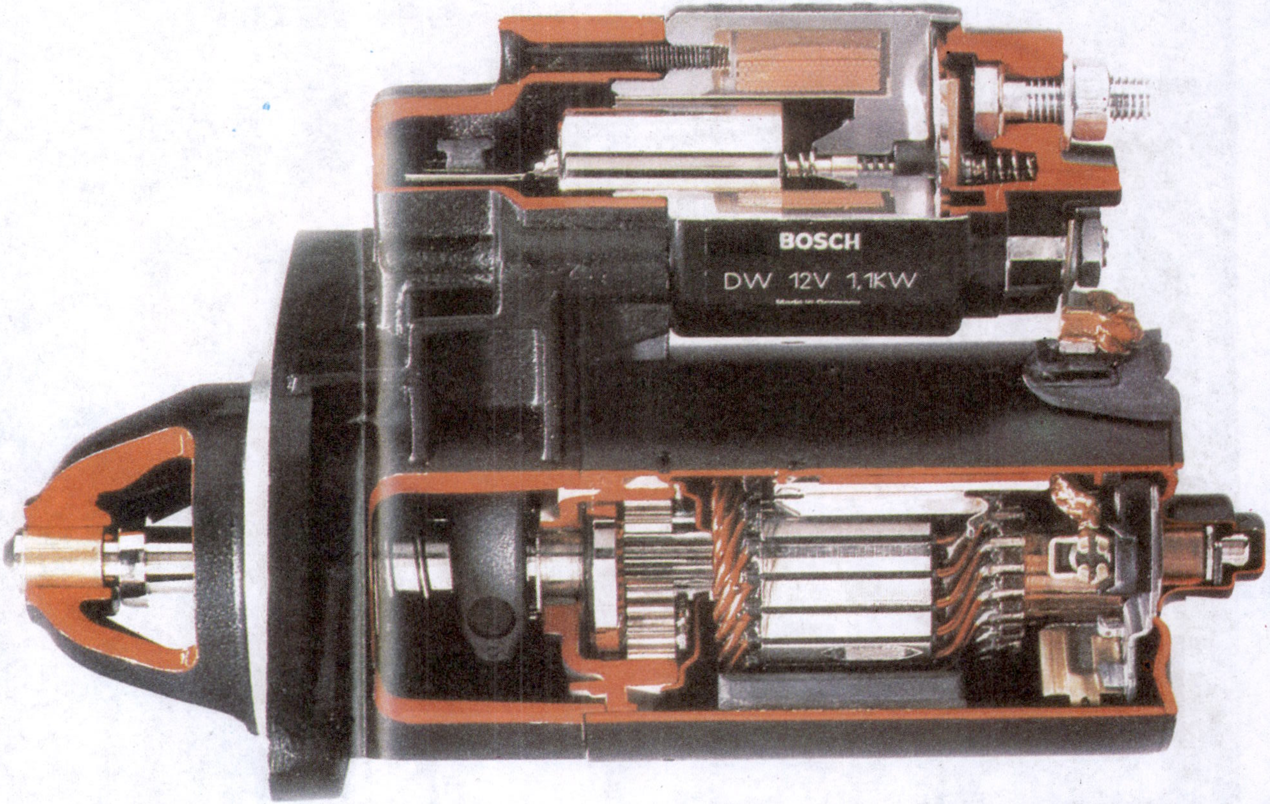

Fig. 19.23. Bosch DW reduction-gear starter with permanent-magnet technology (Courtesy-MICO India Ltd.)

Fig. 19.24. Bosch belt - driven alternator RSG (Courtesy-MICO India Ltd.)

5. Discuss thoroughly an integrated starter.
6. Write a comprehensive note on starting motor control switches.
7. Describe in detail the general procedure for testing the starting system of an automobile engine.
8. Discuss the common troubles occurring in the starting system of an automobile engine. Suggest also suitable remedies.
9. Describe the function and advantages of start-stop system in automobiles. Discuss also its effect on the vehicle engine.
10. What is an I.S.G.? Discuss in detail its functions, contents and types. Describe also its advantages and disadvantages compared to the conventional alternator and starting motor.
11. Write a detailed note on the latest trends in the I.S.G.

QUESTIONS FOR VIVA

1. What is the minimum cranking speed for a car engine?
2. Where is the starting motor located on the engine?
3. Wherefrom the current for the starting motor is taken?
4. What are the requirements of a starting motor?
5. Name different types of d.c. motors.
6. "The starting motor is series wound and the generator is shunt wound". Is the statement true?
7. Which type of d.c. motor has the highest stalling torque?
8. What is the approximate stalling torque of an automotive starting motor?
9. What is the magnitude of current taken by the starting motor?
10. On what considerations does the size of starting motor depend?
11. Why do we provide heavy leads for the starting motor?
12. What is the maximum approximate time for which the starting motor may be run continuously?
13. While servicing starting motor why should we not lubricate the commutator?
14. What is the necessity of incorporation a separate starting drive?
15. How much approximately is the gear reduction for starting the engine?
16. Name three types of starting drives.
17. What is an inertia type drive? In which type of automobiles is it commonly used?
18. State problems with inertia type drive.
19. What is a pre-engaged starter? In which type of automobiles is it usually employed?
20. How does the bendix drive separate the starting motor from the engine flywheel after the engine starts?
21. What is the function of an overrunning clutch in the starting drive?
22. What does the abbreviation ISAD stand for?
23. State the functions performed by ISAD.
24. Name various types of starting motor switches?
25. What is the advantage of having two windings in a solenoid switch?
26. What is the advantage of a solenoid switch compared to the manual type?
27. How does the vacuum switch in the starting motor control circuit operate?
28. Why do we incorporate neutral safety switch in the starting motor control circuit?
29. What improvements in the starting system are possible with an electric control?
30. What may be the possible reasons if:
    (*a*) the starter motor operates, but the engine is not cranked.
    (*b*) the starting drive does not disengage after the engine starts.
31. What is the main characteristic required for a micro-hybrid battery?
32. What is a start-stop system in modern cars?

33. How does the start-stop system function?
34. What type of battery is a must with a start-stop system?
35. Can ordinary lead-acid battery work with a start-stop system?
36. What are the advantages of the start-stop system?
37. What is the disadvantage, if any, of the start-stop system?
38. Name the two strategies to mitigate the problem of wear in a vehicle engine with start-stop system.
39. What do you mean by the term ISG?
40. State the functions of an ISG.
41. Enumerate the advantages of ISG.
42. Name different types of electrical machines for ISG.
43. State the full form of (*i*) IM, (*ii*) PMSM, (*iii*) BDCM and (*iv*) VRM.
44. Which type of lead-acid battery is most suitable for ISG?
45. What is Visteon's 'Speedstart 12'?
46. What do you understand from the term 'SISS'?

MULTIPLE CHOICE QUESTIONS

Mark the correct answer:

1. Gear reduction between the starting motor and the flywheel is about
   (*a*) 5 (*b*) 10
   (*c*) 15 (*d*) 50
2. In a starter motor the field windings are wound around
   (*a*) armature (*b*) commutator
   (*c*) brush (*d*) pole shoes
3. Automotive starting motors are
   (*a*) series wound (*b*) series-shunt wound
   (*c*) shunt wound (*d*) both (*a*) and (*b*) above
4. The stalling torques of starting motors for cars vary between
   (*a*) 10 to 30 Nm (*b*) 30 to 60 Nm
   (*c*) 60 to 100 Nm (*d*) 100 to 200 Nm
5. Inertia type drives are commonly employed in
   (*a*) cars (*b*) light transport vehicles
   (*c*) heavy transport vehicles (*d*) cross-country vehicles
6. Any short-circuit in the armature winding may be tested with a
   (*a*) multimeter (*b*) growler
   (*c*) voltmeter (*d*) ammeter
7. Continuity in the field windings can be tested with a
   (*a*) multimeter (*b*) growler
   (*c*) voltmeter (*d*) none of the above

[Key: 1. (*c*), 2. (*d*), 3. (*d*), 4. (*a*), 5. (*a*), 6. (*b*), 7. (*a*)**]**

Emission Control

20

1. INTRODUCTION

There are four areas in the automobile, which can emit pollutants into the atmosphere. These are, the fuel tank, the carburettor, the crankcase and the exhaust system. The fuel tank and the carburettor emit fuel vapours, the crankcase gives out the partly burnt air-fuel mixture blown off though the piston rings, while the emissions from the exhaust system include unburnt hydrocarbons, carbon monoxide, nitrogen oxides, sulphur oxides and carbon dioxide. Besides, in diesel engines, particulate matter (PM) forms a considerable portion of the exhaust pollutants. Particulate matter consists of substances other than water that can be collected by filtering the exhaust gas. Basically, the particulate matter is carbon or soot particles on which some organic compounds, such as unburnt hydrocarbons, get absorbed. A portion of the organic matter may also come from lubricating oil.

As per US Environmental Defense Fund, the amalgam of gases comprising NOx contribute to the formation of fine particles and zone smog which result in serious diseases and deaths; NOx also gives year-round environmental problems, from acid rain to eutrophication which causes oxygen depletion in coastal areas, lakes, etc., resulting in harm to animal life.

Particulate matter consists of airborne particles which include dirt, dust, smoke and liquid droplets. A portion of the organic matter, such as hydrocarbons, may also come from lubricating oil. PM from diesel engines typically consists of unburnt carbon (soot) from the fuel serving as nucleus with other constituents like sulphuric acid, condensed water and soluble hydrocarbons collecting on the outside of the nucleus particle. PM particles range from very fine to coarse. Particles less than 2.5 ìm in size cannot be seen with eyes and are designated as $PM_{2.5}$ as per U.S. Department of Energy; whereas PM_{10} are the particles between 2.5 and 10 ìm size which can be seen. PM_{10} particles are relatively less dangerous since you can see and avoid the same just like the smoke coming out of the exhaust of an old diesel bus. The finer particles can accumulate in the respiratory track over time which may lead to worsening of the condition in case of people afflicted with cardiopulmonary conditions, *e.g.*, heart disease, asthma, etc., which may cause worsening heart and lung symptoms, reduced lung function and even death. Other effects of PM are, limited visibility, reduced crop yield, damage to buildings and ecosystems, etc.

Since it is possible for even the unsuspecting persons to inhale fine PM, the toxic organic particles condensed on the same, may harm the body. It has been concluded that long-term inhalation of diesel exhaust may cause lung cancer.

Table 20.1 shows the serious diseases caused by various exhaust constituents.

Various exhaust constituents cause many serious diseases as given in Table 20.1.

Further, although carbon dioxide is essential to plant life and has no immediate harmful effects to human beings, yet it is also now being considered a pollutant because it is a greenhouse gas significantly contributing to global warming by trapping earth's heat.

TABLE 20.1. DISEASES CAUSED BY AUTO EXHAUST

S. No.	*Constituent*	*Disease*
1.	Carbon Monoxide (CO)	Heart/Blood circulatory problems, particularly angina; may also affect foetuses, sickle cell anaemics and young children; can affect central nervous system, impairing physical coordination, vision and judgement, producing nausea and headaches.
2.	Hydrocarbons (HC)	Lung diseases
3.	Nitrogen Oxides (NO_x)	Asthma/Bronchitis; increased susceptibility to infections, pulmonary disease and eye, nose and throat irritations.
4.	Lead (Pb)	Blood cancer, brain damage
5.	Suspended Particulate Matter (P.M.)	Asthma/Bronchitis

Obviously, the atmospheric pollution can be decreased by controlling these areas of the automobile. To do this, two different approaches have been followed:

1. To reduce the formation of pollutants in the emissions by redesigning the engine ventilating system, carburettor and fuel tank. The combustion chamber, alongwith fuel system, cooling system, ignition system and the exhaust system are also redesigned, thus improving upon the combustion efficiency which reduces the emissions.
2. To destroy the pollutants after these have been formed.

The details of both these approaches have been discussed in the following articles.

To give an idea about the level of pollutants that may be safely allowed in the exhaust of passenger cars, the emission standards of U.S.A., Europe and India are discussed here.

U.S.A.

In 1990 US 'Clean Air Act' was amended to phase-in 'Tier 1' Federal emission norms from 1990 to 1996. These norms remained in force till their phasing out from 2004 to 2007, for Tier 2 norms to become effective. Under the Tier 2 regulation, the same emission standards apply to all types of vehicles, *i.e.*, cars, minivans, light duty trucks and SUVs (upto 10,000 lbs of GVWR), irrespective of the type of fuel used also.

Tier 2 regulates the following 5 tail pipe emissions:

1. Non-Methane Organic Gases (NMOG)
2. Carbon Monoxide (CO)
3. Nitrogen Oxides (NO_x)
4. Particulate Matter (PM)
5. Formaldehyde (HCHO).

However, Tier 2 does not regulate any green house gas, *viz.*, carbon dioxide (CO_2), methane (CH_4), Nitrous oxide (N_2O) or water vapour.

Tier 2 regulations consist of 8 certification levels, called 'bins' to control each of the above five pollutants. Bins 5 to 8 have two standards each at 50,000 miles and at 120,000 miles while bins 1 to 4 have only 120,000 miles standard. Bin 8 is most lenient which permits highest level to emissions. The bins 7 to 1 are graded as per the severity of the permitted values, so that bin 1 allows zero quantity of all of the five pollutants. The permitted values are shown in Table 20.2.

An automobile manufacturer's fleet must average at least 0.07 gm/mile of NO_x. Of course only manufacturer can sell as many of relatively dirty vehicles of Bin 8 norms, provided the desired average of 0.07 gm/mile of NOx is achieved over the entire lot sold in a y

TABLE 20.2. TIER 2 EMISSION LIMITS FOR CARS AND LIGHT DUTY TRUCKS

Pollutant	*Distance (km)*	*Bin 8 (g/mile)*	*Bin 7 (g/mile)*	*Bin 6 (g/mile)*	*Bin 5 (g/mile)*	*Bin 4 (g/mile)*	*Bin 3 (g/mile)*	*Bin 2 (g/mile)*	*Bin 1 (g/mile)*
NMOG	50 K	0.100	0.075	0.075	0.075				
	120 K	0.125	0.090	0.090	0.090	0.070	0.055	0.010	0.000
CO	50 K	3.4	3.4	3.4	3.4				
	120 K	4.2	4.2	4.2	4.2	2.1	2.1	2.1	0.0
NOx	50 K	0.14	0.11	0.08	0.05				
	120 K	0.20	0.15	0.10	0.07	0.04	0.03	0.02	0.00
PM	120 K	0.02	0.02	0.01	0.01	0.01	0.01	0.01	0.00
HCHO	50 K	0.015	0.015	0.015	0.015				
	120 K	0.018	0.018	0.018	0.018	0.011	0.011	0.004	0.000

Tier 2 norms have also set restrictions on the maximum amount of sulphur in petrol and diesel, which is 30 ppm (parts per million) with maximum cap of 80 ppm for petrol and 15 ppm (maximum) for diesel.

A second set of emission norms, called 'Low Emission Vehicle II (LEV 2) standards also exist in USA in parallel with the Tier 2 norms of EPA (Environmental Protection Agency). Tier 2 rules are also called 'Federal standards' and 'LEV 2' are also known as 'California/CARB (California Air Resources Board)' standards. The later are generally more stringent than the former and states can decide which standards out of these two are to be adopted. Table 20.3 gives LEV 2 standards.

TBALE 20.3. LEV 2 EMISSION LIMITS FOR CARS AND LIGHT DUTY TRUCKS

Pollutant	*Distance (km)*	*Maximum Limits (gm/mile)*			
		LEV	*ULEV*	*SULEV*	*ZEN*
NMOG	50 k	0.075	0.040		0.000
	120 k	0.090	0.055	0.010	0.000
CO	50 k	3.4	1.7		0.0
	120 k	4.2	2.1	1.0	0.0
NOx	50 k	0.05	0.05		0.0
	120 k	0.07	0.07	0.02	0.0
PM	50 k	0.01	0.01	0.01	0.00
	120 k				
HCHO	50 k	0.015	0.008		0.000
	120 k	0.018	0.011	0.004	0.000

It is seen from Table 20.3 that there are 4 categories of vehicles for LEV 2:

1. LEV (Low Emission Vehicle)
2. ULEV (Ultra Low Emission Vehicle)
3. SULEV (Super Ultra Low Emission Vehicle)
4. ZEV (Zero Emission Vehicle). This has following subcategories:
   (*i*) PZEV (Patrol Zero Emission Vehicle)
   (*ii*) AT-PZEV (Advanced Technology Partial Zero Emission Vehicle)

(*iii*) NEV (Neighborhood Electric Vehicle)

(*iv*) FFEV (Full-Function Electric Vehicle)

(*v*) CEV (City Electric Vehicle)

A manufacturer can choose any of the above LEV 2 categories, provided their entire fleet meets the prescribed average NMOG limits, which are currently 0.035 g/mile, which is much stricter than Bin 5 of Tier 2.

Tier 3 rules have now been proposed which would be phased in from 2017 to 2025.

Europe

Emission norms for the European Countries are gives in Table 20.4 and 20.5.

TABLE 20.4. EURO NORMS FOR PETROL (INCLUDING CNG/LPG)/DIESEL PASSENGER VEHICLES

Standard	*CO gm/km*	*HC/NOx gm/km*	*PM gm/km*
Euro I (1992/93)			
Petrol	2.72	0.97	N.A.
Diesel	2.72	0.97	0.14
To 1994 diesel DI	2.72	1.36	0.19
Euro II (1995/97)			
Petrol	2.2	0.5	N.A.
Diesel	1.0	0.7	0.08
To 1999 diesel DI	1.0	0.9	0.10
Euro III (1999/2000/2001)			
Petrol	2.3	0.2/0.15	N.A.
Diesel	0.64	0.56/0.5	0.05
Euro IV (2005)			
Petrol	1.0	0.1/0.08	N.A.
Diesel	0.5	0.3/0.25	0.025

TABLE 20.5. EURO NORMS FOR PETROL (INCLUDING CNG/LPG)/ DIESEL PASSENGER VEHICLES

Standard	*CO (mg/km)*	*THC+NO$_x$ (mg/km)*	*THC (mg/km)*	*NMHC (mg/km)*	*NO$_x$ (mg/km)*	*PM (mg/km)*	*PN (x/km)*
Euro 5 (2009)							
Petrol	1000	—	100	68	60	*5.0/4.5@	—
Diesel	500	230	—	—	180	5.0/4.5@	6.0×10^{11}
Euro 6 (2014)							
Petrol	1000	—	100	68	60	*5.0/4.5@	—
Diesel	500	170	—	—	80	5.0/4.5@	6.9×10^{11}

* For D.I. engines only,

@ 4.5 mg/km standard wef. Sept 2011.

TABLE 20.6. B.S. VI EMISSION STANDARDS FOR M & N CATEGORY VEHICLES WITH G.V.W. LESS THAN 3.5 TONS

		Reference Mass (RM) (kg)	Mass of Carbon Monoxide (CO)		Mass of Total Hydrocarbons (THC)		Mass of Non-Methane Hydrocarbons (NMHC)		Mass of Oxides of Nitrogen (NOx)		Combined mass of Hydrocarbons and Oxides of Nitrogen (THC + NOx)		Mass of Particulate Matter (PM)		Number of Particles (PN)	
			L1 (mg/km)		*L2* (mg/km)		*L3* (mg/km)		*L4* (mg/km)		*L2 + L3* (mg/km)		*L5* (mg/km)		*L6* (numbers/km)	
Category	Class		PI	CI	PI	CI	PI	CI	PI	CI	PI	CI	PI(1)	CI	PI(1), (2)	CI
M (M1 & M2)	—	All	1000	500	100	—	68	—	60	80	—	170	4.5	4.5	6.0×10^{11}	6.0×10^{11}
N1	I	RM ≤ 1305	1000	500	100	—	68	—	60	80	—	170	4.5	4.5	6.0×10^{11}	6.0×10^{11}
	II	1305<RM ≤ 1760	1810	630	130	—	90	—	75	105	—	195	4.5	4.5	6.0×10^{11}	6.0×10^{11}
	III	1760 < RM	2270	740	160	—	108	—	82	125	—	215	4.5	4.5	6.0×10^{11}	6.0×10^{11}
N2	—	All	2270	740	160	—	108	—	82	125	—	215	4.5	4.5	6.0×10^{11}	6.0×10^{11}

PI = Positive Ignition, CI = Compression Ignition

1. For positive ignition, particulate mass and number of particles limit shall apply only to vehicles with direct injection engines.
2. Until three years after date of implementation for new type approvals and new vehicles, particle number emission limit of 6.0×10^{12} /km shall apply to BS VI gasoline direct injection vehicles upon choice of the manufacturer.

India

In comparison to USA, emission norms in India are not as stringent. Although Indian norms are equivalent to Euro norms, yet their implementation in India is lagging behind European implementation.

Bharat Stage II (equivalent of Euro II) norms for two-wheelers, cars and commercial vehicles were enforced in the entire country w.e.f. April 1, 2005, whereas Bharat Stage III (equivalent of Euro III) were made mandatory for two wheelers, cars and commercial vehicles in the entire country and Bharat Stage IV norms in the 13 Metros were enforced with effect from April 1, 2010. Ultimately, Bharat Stage IV norms (equivalent to Euro 4 norms) were enforced for all types of road vehicles in the entire country with effect from April 1, 2017. BS IV norms require low sulphur fuel due to which the implementation was gradual according to supply provision of such a fuel by the oil companies.

Now, to keep pace with the advanced countries, the Govt. of India has decided to skip Euro 5 norms and enforce directly Bharat Stage VI (Euro 6) norms with effect from April 1, 2020. Table 20.6 gives BS-VI emission standards for M & N category vehicles with gross vehicle weight less than 3.5 tons, as per the booklet 'Indian Emission Regulations' Published by ARAI.

Phased implementation of the exhaust emission standards in some of the countries have been given in Table 20.7.

TABLE 20.7. EXHAUST EMISSION STANDARDS IMPLEMENTATION SCHEDULE WORLDWIDE

Country	2005	2006	2007	2008	2009	2010	2011	2012	2013	2014	2015
U.S.A.	Tier 2, Bin 8		Tier 2, Bin 5						Tier 2, Bin 4		Tier 2, Bin 2
Argentina	Euro 3				Euro 4			Euro 5			
Europe	Euro 4					Euro 5			Euro 6		
Chine (Bejing)	Euro 2	Euro 3		Euro 4			Euro 5				
China (Nation wide)	Euro 2			Euro 3			Euro 4				
Thailand	Euro 2						Euro 3		Euro 4		
South Korea	Euro 3		Euro 4				Euro 5				
India (Metros)	Euro 3						Euro 4		Euro 5		
India (Nation wide)	Euro 2						Euro 3		Euro 4		

Process of prescribing standards for CO_2 in the exhaust emission of vehicles is also actively on, for protection from global warming.

The European Parliament issued on April 23, 2009, regulation **(EC) 443/2009** for new passenger cars. The main points therein are as follows:

1. Community target of 120 g/km of CO_2 with effect from 2012.
2. As per phase-in scheme, the percentage of manufacturer's car fleet must meet the targets as follows:

Year	2012	2013	2014	2015
% of fleet must meet target	65	75	80	100

3. Apart from the above, there is also a scheme for super credits during the phase-in period. According to this, if the CO_2 emission is less than 50 g/km, super-credits would be as follows:

Year	2012	2013	2014	2015	2016
One car counts as	3.5 cars	3.5 cars	2.5 cars	1.5 cars	1 car

2. REDUCTION OF FORMATION OF POLLUTANTS

This is achieved by the following:

1. Closed crankcase ventilation.
2. Reducing evaporative emissions.
3. Redesigning the engine:
   (*i*) Compression ratio
   (*ii*) Combustion chamber
   (*iii*) Cooling system
   (*iv*) Valve timing
   (*v*) Fuel supply system
   (*vi*) Ignition system
4. Improving vehicle efficiency
5. Use of HCCI engine
6. Use of hybrid vehicles
7. Improving driving efficiency and traffic conditions.
8. Improving on-board diagnostic systems.

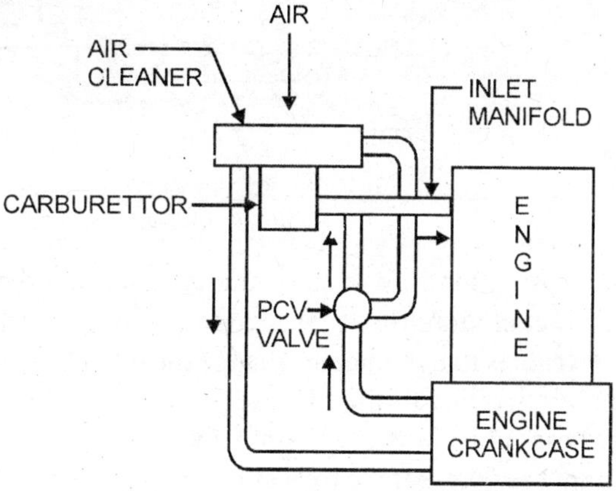

Fig. 20.1. Layout of Positive Crankcase Ventilation.

3. CLOSED CRANKCASE VENTILATION

There are two variations of this system, *viz.*, the positive crankcase ventilation and the Fixed orifice system. In the former, filtered air from the carburettor air cleaner is introduced in the crankcase from where it carries away the blow-by gases and the gasoline vapours to the engine inlet manifold through a special PCV valve. Thus the blow-by gases and the gasoline vapours enter the engine combustion chamber along-with fresh charge and are burnt there.

The construction of the special PCV valve is shown in Fig. 20.2 (*a*). It consists of a spring-loaded tapered valve for flow control. The crankcase pressure and the manifold vacuum act together to close the valve whereas the spring pressure tends to keep it open.

At idle and low speed crankcase emissions are very less due to lower cylinder pressure. Moreover, manifold vacuum is high. Therefore, only a small flow through the PCV valve would be sufficient to keep the crankcase clean. If the air flow through PCV at idle and low speed is high, it would cause the air-fuel mixture to become excessively lean resulting in poor drivability. High manifold vacuum at idle and low speed would pull the valve to the right so as to maintain a small air flow [Fig. 20.2 (*b*)].

At normal speeds, blow-by increases and manifold vacuum decreases, due to which the valve moves towards left, increasing the flow. At high speeds or heavy loads when the throttle is held wide open, there is practically no manifold vacuum acting on the valve, due to which the valve opens maximum [Fig. 20.2 (*c*)],

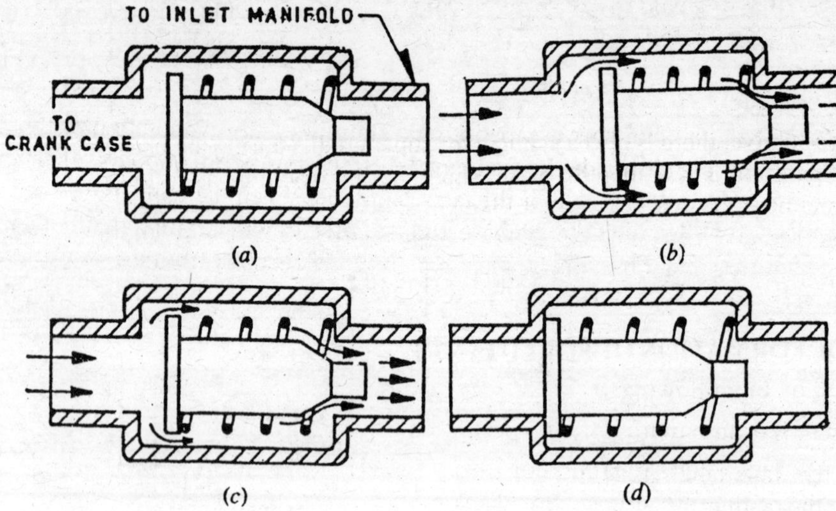

Fig. 20.2. P.C.V. Valve (*a*) Construction, (*b*) Operation at idle and low speed, (*c*) High speed/load operation, (*d*) Preventing backfire during cranking.

increasing the flow to maximum capacity of the PCV valve. However, at higher speeds a large volume of air-fuel mixture flows through the carburettor. Therefore, greater flow of crankcase vapours through PCV valve does not practically change the quality of the mixture. However, when the crankcase emissions exceed the flow rating of the PCV valve, the excess emissions are forced by the crankcase pressure to bypass the PCV valve into the air cleaner.

In case the engine backfires during cranking, a high pressure will be produced in the intake manifold, which would cause the valve to 'back-seat', thereby sealing the inlet [Fig. 20.2 (*d*)]. Thus the crankcase is protected from the backfire.

In the Fixed Orifice system, a fixed orifice in the base of the carburettor takes the place of the PCV valve. The blow-by gases are diverted to the inlet manifold through this orifice (Fig. 20.3).

In both these types of closed crankcase ventilation systems, the sludge and the other similar elements during the course of time, clog the PCV valve or the orifice which may decrease and ultimately stop the flow of the blow-by gases altogether, which would affect the engine performance

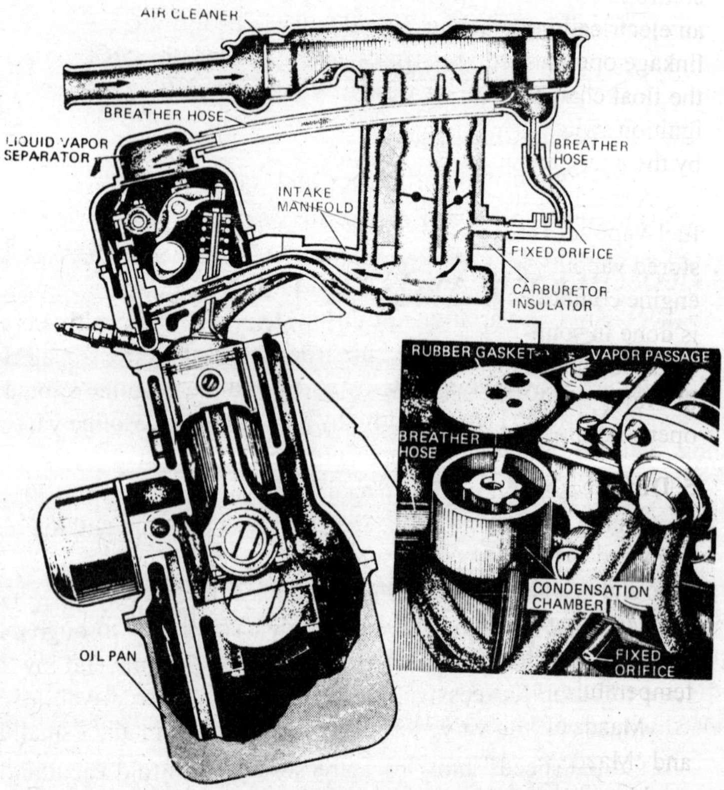

Fig. 20.3. Honda Fixed Orifice closed-crankcase ventilation system.

very seriously. A coating of oil on the air cleaner filter, apart from various engine troubles, indicates that service of the system is needed urgently, which mainly requires the cleaning of the P.C.V. valve or the orifice, the air-cleaner filter, hoses, etc.

4. REDUCING EVAPORATIVE EMISSIONS

The petrol vapours from the fuel tank escape into the atmosphere by evaporation or 'breathing'. With the increase of temperature, the air inside the fuel tank which carries petrol vapours, expands and is thus forced out through either the filter cap vent or the tank vent tube. When the temperature decreases the air from outside enters the tank. Thus with the increase and decrease of temperature, the fuel tank 'breathes', thereby discharging petrol vapours into the atmosphere. Petrol is also lost from the carburettor float bowl by evaporation when the engine is not running. It is estimated that the losses from the fuel tank and carburettor are about 10% of the total hydrocarbon emission of the engine into the atmosphere.

A simple system to control these losses is shown in Fig. 20.4. The fuel tank and the carburettor float chamber are not vented into the atmosphere. They are, instead, vented to a carbon canister, which adsorbs the vapour and stores it. The vapour-liquid separator is a chamber on the fuel tank. Vapours go to the top of the separator while liquid petrol is returned to the tank. Flow of vapours from the fuel tank is controlled by a mechanically operated vent valve or an electrically operated solenoid valve. The throttle linkage operates the mechanical valve so that it is open during idling which makes the vapour flow from the float chamber to the canister. When the throttle is opened, the vent valve is closed. Likewise, with the ignition switch off, the electrical vent-valve is open. When the ignition switch is on, the vent-valve is closed by the energization of the solenoid.

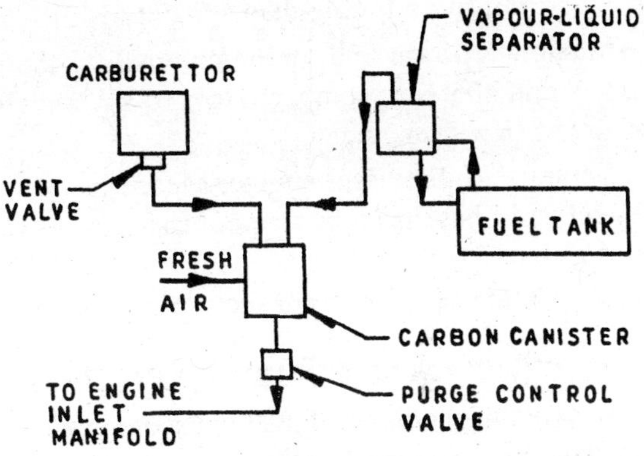

Fig. 20.4. Layout of a Vapour Recovery System

A special filler cap is employed on the fuel tank which maintains some pressure in the tank, that forces fuel vapours form the tank to the carbon canister. When the engine is operated, fresh air removes the stored vapours from the carbon in the carbon canister and carries them through the inlet manifold into the engine combustion chamber to be burnt there. The control of flow of fuel vapours from the carbon canister is done in some cases by a special purge control valve which is operated usually by EGR port (discussed later) vacuum. The purge control valve is closed when the engine is not running. When the engine is idling, the valve remains closed. However, when the throttle is opened, vacuum applied to the purge control valve opens it, allowing the vapours to be purged from the canister into the intake manifold.

5. DECREASING COMPRESSION RATIO

As the air standard efficiencies of both auto and diesel cycles increases with increase of compression ratio and the power developed directly depends on the cylinder pressure, higher values of compression ratio and cylinder pressures have been traditionally used for I.C. engines. However, with the gradual awareness of decreasing exhaust emissions it was seen that NO_x and soot both increase with increase of compression ratio. On the other hand, reduced compression ratio in diesel engines causes difficulty in starting and low temperature running.

Mazda of Japan have took up the challenge and developed 'Skyactive D' engine (used in Mazda 'Demco' and 'Mazda z' vehicles) with compression ratio of only 14.0:1, which is claimed to be the lowest of any modern (2013) production diesel engine in the world. This low ratio allows more complete air/fuel mixing in the combustion chamber, thus helping in preventing the formation of NO_x and soot. Maximum combustion pressure in this are also lower by 20%, which helps the charge to be ignited at TDC, *i.e.*, some what advanced compared to high-pressure diesels. This enables a longer power stroke, improving efficiency.

Apart from improved NOx/soot emissions, this engine also has lower fuel consumption and also reduced CO in the exhaust. Additional advantages are:

1. Lower mechanical stresses in the engine structure and rotating assembly.
2. Reduced weight, which is about 20 kg less than the comparative standard engine.
3. Extended power band. 'Skyactive D' produces useful power till 5000 rpm in contrast to the standard Mazda MZR engine, where power curve drops at 3500 rpm only.
4. More torque is also produced at lower speeds. At 1200 rpm, 'Skyactive D' is said to produce 68 Nm of more torque than standard MZR-CD.

Maximum power and torque claimed for the lower compression, 2.2 litre 'Skyactive D' engine are 126 kW and 405 Nm, compared to 110 kW and 360 Nm for the standard high compression ratio (16.3) 2.2 litre MZR-CD engine.

Contrary to the diesel engine, Mazda has gone in opposite direction in case of petrol engine. Their 'Sky active G' has the highest compression ratio (14.0:1) compared to any current production petrol engine and it gives 15% more torque and 15% increase in fuel economy over their current MZR 2-litre petrol engine, without any adverse effect on exhaust emissions.

6. REDESIGNING THE COMBUSTION CHAMBER

To avoid the addition of lead in gasoline, the engine combustion chambers have been redesigned with lower compression ratios (from approximately 10.5: 1 to 8.5 : 1). This loss of efficiency has been made good by increasing the temperature in the combustion chamber which results in improved combustion of the air-fuel mixture and consequently in the decrease of pollutants in the exhaust.

The shape of the combustion chamber also has a bearing on the combustion efficiency. It is seen that for a given volume, the hemispheric combustion chamber has the least surface area, which means least surface to cool the air-fuel mixture and hence lower percentage of unburnt hydrocarbons in the exhaust.

The other modern trends in the combustion chamber design include reduction or elimination of quench area and causing controlled turbulence in chamber. Such turbulence improves the mixing of air/fuel mixture and consequently improved combustion, reducing the *HC* and *CO* emissions.

In a chamber designed for the stratified charge, the richer part of the mixture is kept away from the surfaces of the chamber, whereas the weaker part is first introduced. Thus the percentage of unburnt hydrocarbons is decreased.

In a new system being developed in Britain, first only a charge of air is injected after which the fuel-air mixture is injected (Fig. 20.5). A specially designed piston on its compression stroke sets the combustion charge oscillating as ignition takes place. The system attains the complete combustion effect by utilizing the natural frequencies of the system and the resonance effect.

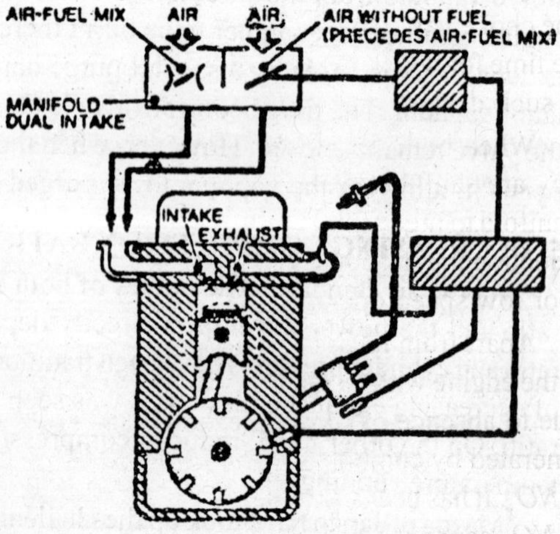

Fig. 20.5. Sonex system for emission control.

7. CHANGES IN COOLING SYSTEM

High engine temperatures result in reduction of *HC* and

CO emissions. However, beyond a certain limit, they also tend to increase the formation of the NO_x. The engine cooling systems have been redesigned to allow higher engine temperatures, but at the same time these are not allowed to be too high.

8. VARIABLE VALVE TIMING

In conventional engines, opening and closing of inlet and exhaust valves is restricted by the piston movement. Providing an independent valve control is now possible, wherein valve timings and stroke can be varied depending upon the engine working conditions due to which more efficient combustion of conventional engine fuels, *viz.,* petrol and diesel and also of alternative fuels such as ethanol, bio-diesel, etc. takes place, which ultimately leads to lesser exhaust emissions. Details of various variable valve timing technologies have already been discussed in Art 12 to 16 of Chapter 4.

9. CHANGES IN FUEL SUPPLY SYSTEM

Suitable modifications have been made in the intake manifolds so that rapid vaporization of fuel takes place during warm up of the engine. Further, crossover of the inlet and the exhaust manifolds is also redesigned so as to reduce the time required for the transfer of heat from the exhaust gases to the inlet charge. Moreover, in case of engines with petrol injection in particular, a tuned intake manifold is employed, wherein all runners are of equal length with very little restriction so that equal air flow takes place to all the cylinders. This combination of petrol injection and tuned intake manifold helps reduce emission of *HC* and NO_x.

Lean air-fuel mixtures ignite easier when they are preheated. In emission-controlled engines, the air cleaner is controlled thermostatically so that the temperature of the air entering the carburettor is about 30-40°C. The heating is done by utilizing the heat of the exhaust gases only. The supply of air to the engine at higher temperature results in decrease of exhaust emissions due to increase in combustion efficiency. Fig. 20.6 shows the exhaust controlled heating system employed in SL engines of Mazda vehicles.

The carburettors have also been redesigned so as to increase idling speeds. Higher idling speed allows the engine to run with a lean mixture. Further suitable designs are now adopted so that the idle setting cannot be disturbed later by mechanics, as it has been found that too rich an idle mixture results in increase of exhaust emissions (Fig. 20.7).

When the choke valve is closed, the mixture supplied is very rich, emitting higher quantities of exhaust. The choke valves have, therefore, been redesigned so that these open and close more quickly, thus reducing the time for which the engine is producing higher emissions. An electric-assisted choke is a typical example of such design.

When both the inlet and the exhaust valves are partly open between the intake and exhaust strokes, they are said to overlap. Increasing the overlap causes some of the exhaust gases to return to the inlet manifold to mix with the fresh charge. This decreases the maximum combustion temperature with consequent reduction in NO_x emission. However, this advantage is accompanied by disadvantages such as rough idling, poor low speed operation, etc.

Apart from increasing the valve overlap discussed above a better method of diluting the intake charge of the engine with exhaust gas is the *Exhaust Gas Recirculation* (EGR) system, which reduces NO_x emission. Due to absence of oxygen, exhaust gas itself will not burn. However, it will absorb a portion of the heat generated by combustion, thus only lowering the combustion temperature, thereby reducing the formation of NO_x. It has been estimated that a 16% lowering of peak temperature would result in roughly 85% reduction of NO_x concentration in the ejected exhaust gas.

The effect of exhaust gas recirculation on the engine power, specific fuel consumption and NO_x emissions in case of an engine with fixed throttle setting, constant ignition timing and constant air-fuel ratio are shown in Figs. 20.8 to 20.10. It is seen that with the increase in percentage of exhaust gas recirculation, the

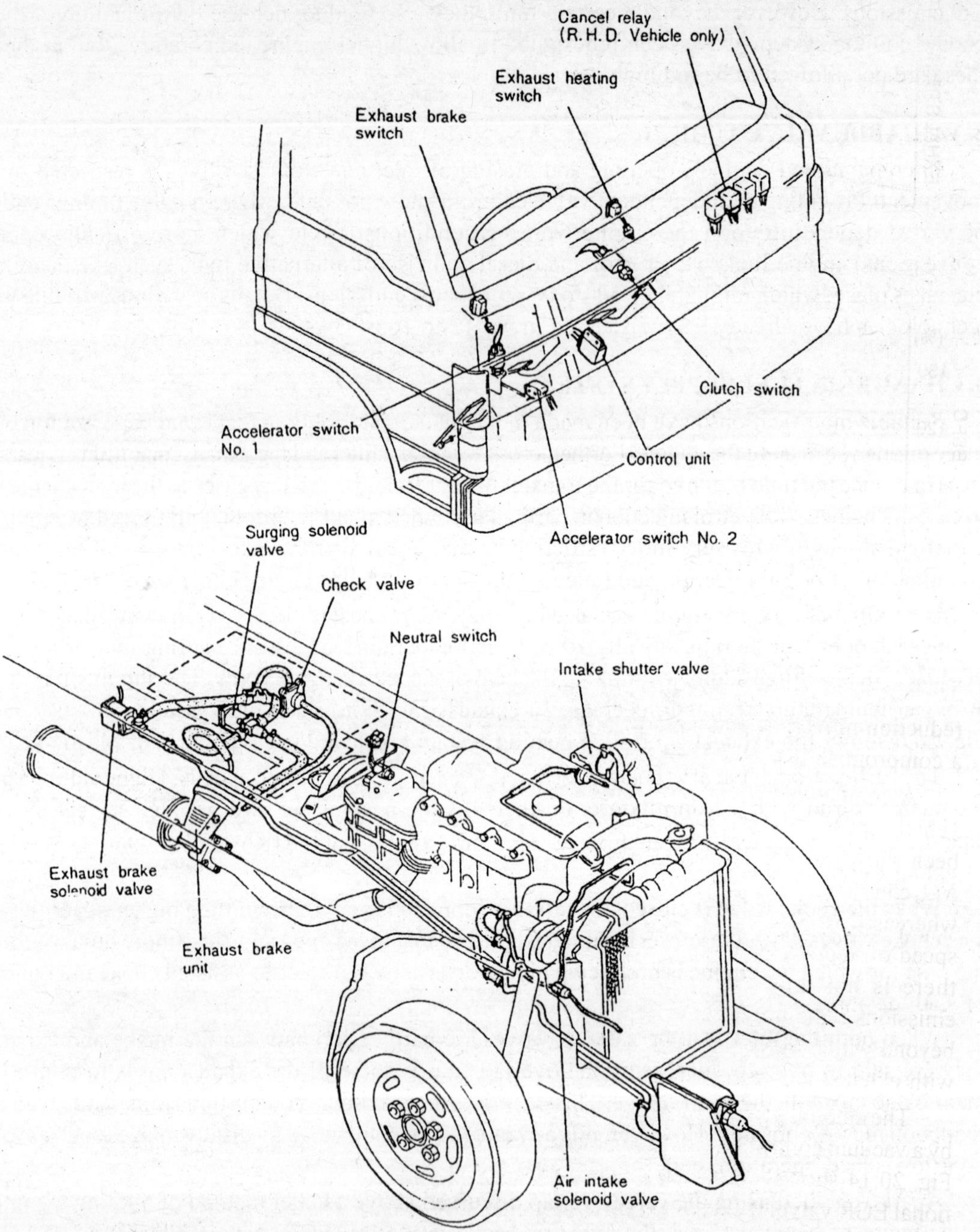

Fig. 20.6. Exhaust-controlled heating system. (Courtesy—Swaraj Mazda Ltd.)

engine power decreases while brake specific fuel consumption increases; the nitrogen oxide emissions also decrease upto 15% E.G.R., beyond which they tend to level out. At 15% E.G.R., the reduction in NO_x amounted to 88% and, while the power loss was 16% and rise in BSFC was 14%. This deterioration in power and fuel consumption can, however, be offset to a large extent by optimising the ignition advance and the air-fuel ratio for different quantities of E.G.R. under operating conditions. But then instead of 88%

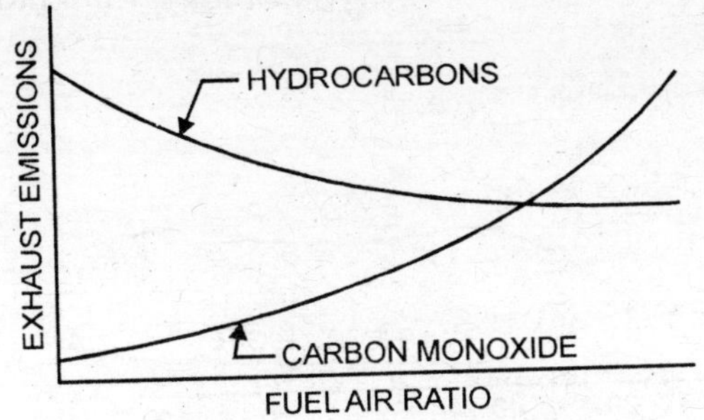

Fig. 20.7. Effect of fuel-air ratio on exhaust emissions at idle.

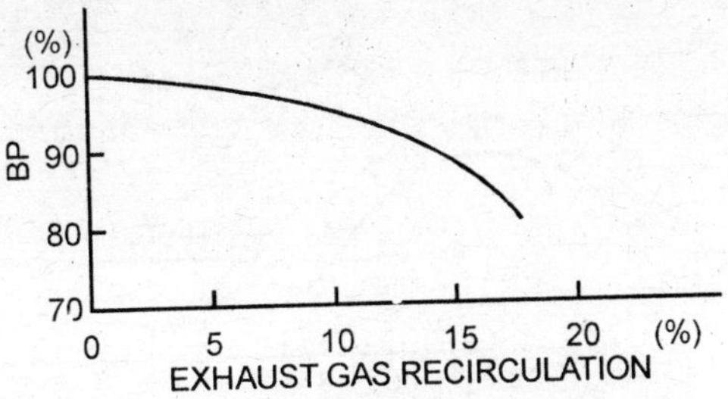

Fig. 20.8. Effect of exhaust gas recirculation on engine brake power.

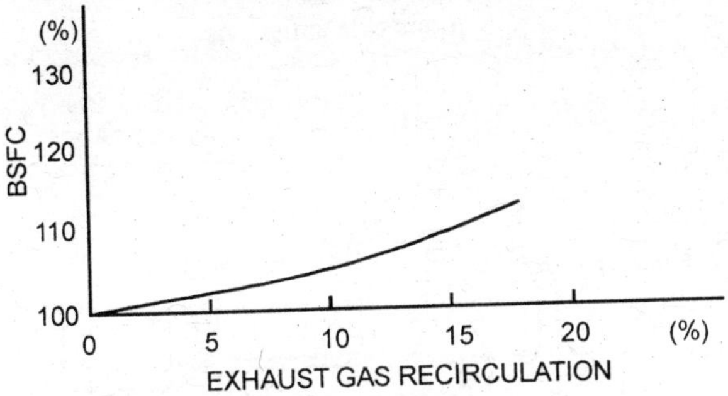

Fig. 20.9. Effect of exhaust gas recirculation Specific Fuel consumption.

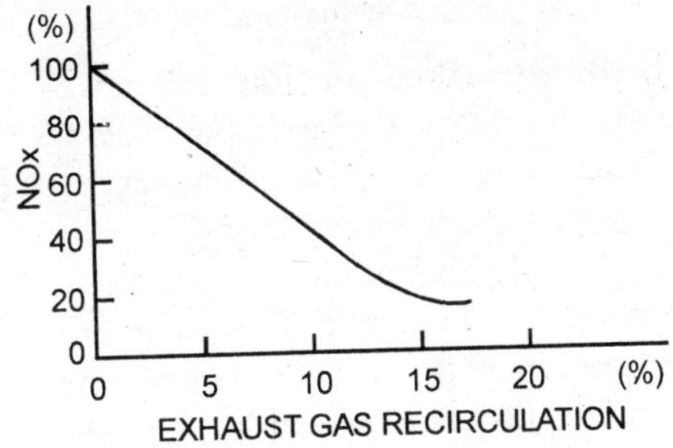

Fig. 20.10. Effect of exhaust gas recirculation on nitrogen oxide emissions.

reduction in NO_x as above, there would be about 60% decrease in nitrogen oxide emissions. In this way, a compromise is worked out and EGR system is accordingly designed.

A comparison of exhaust emissions at various vehicle speeds with and without EGR systems has been shown in Fig. 20.11 to 20.13. It is seen that NO_x emissions are less in the EGR case at all speeds, whereas hydrocarbon emissions are lesser upto a speed of about 60 kmph; even beyond that speed, there is not much deterioration. However CO emissions remain unaffected upto about 15 kmph, beyond which they increased in the case of engine with exhaust gas recirculation.

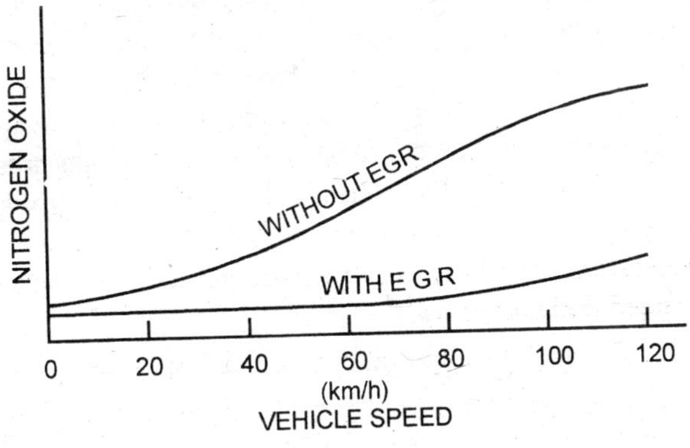

Fig. 20.11. Effect of EGR on NO_x emissions

The amount of exhaust gas admitted is regulated by a vacuum controlled valve, called the EGR valve. Fig. 20.14 shows simplified diagram of a conventional EGR valve. It consists of a spring-loaded vacuum diaphragm linked to a tapered valve which controls the passage for the exhaust gas. Ported vacuum from a calibrated signal port just above the throttle valve is connected to the EGR valve vacuum chamber. The EGR valve is closed at idle on account of spring pressure and less ported vacuum. This is required also because in case exhaust gas recirculates at idle, it would cause rough idling. When the throttle is opened from the idle position, ported vacuum applied will gradually open the tapered valve, causing the exhaust gas to flow into the intake manifold. However when the throttle valve is fully open, there is no manifold vacuum, so the tapered valve is closed by the spring.

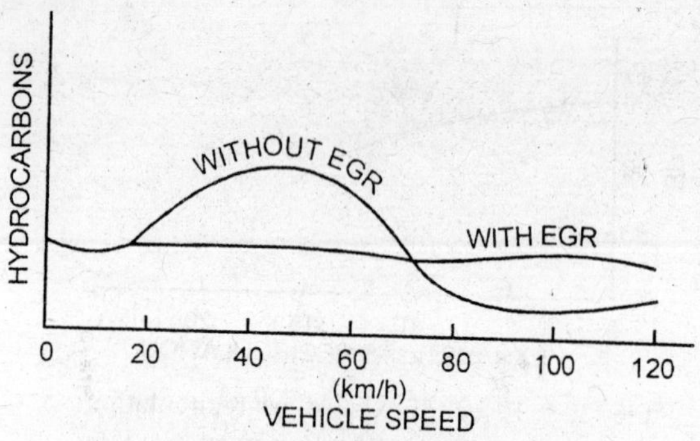

Fig. 20.12. Effect of EGR on hydrocarbon emissions.

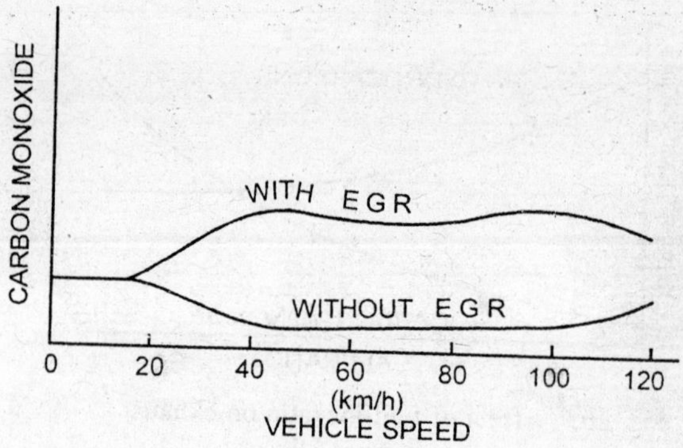

Fig. 20.13. Effect of EGR on carbon monoxide emissions.

Thus EGR system does not affect full power operation. It is thus seen that in this system, exhaust gas is recirculated only when the engine operating conditions form NO_x. The thermal control valve in the vacuum line is installed in the engine cooling system and serves to prevent EGR operation at low engine temperatures. Such a system is especially useful in diesel engines where catalytic converter cannot promote chemical reduction reaction due to the high oxygen excess, the NO_x emissions thus remaining same. EGR has also been achieved in some cases by exhaust back-pressure- modulated system which employs a transducer located inside the EGR valve for control of the operating vacuum signal, which is generated in the same way as described above. The transducer controls an air bleed within the valve by means of the exhaust gas pressure to modify the vacuum signal from the carburettor. The latest in this direction is the computer-controlled EGR. One such system consists of a pulse-width-modulated solenoid to control EGR supply vacuum, the duty cycle of the solenoid being controlled by the electronic control module depending upon the transducer signals of engine coolant temperature, engine speed and engine load.

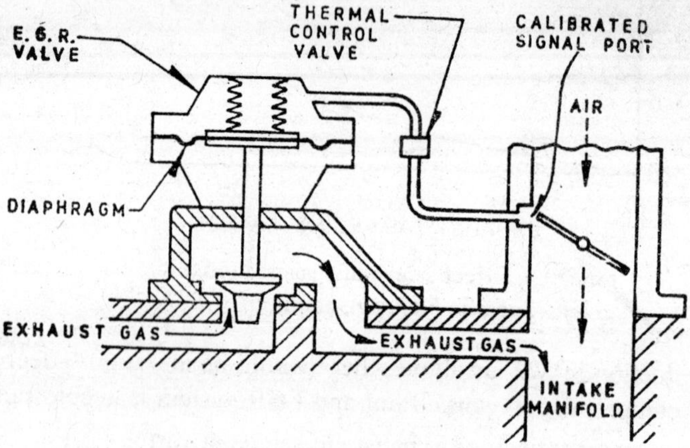

Fig. 20.14. Conventional EGR valve.

Exhaust gas recirculation as used on a Mercedes diesel engine is shown in Fig. 20.15.

Some engines have diaphragm-operated *jet valve* in each combustion chamber. Jet valve is a small air valve that allows air flow through a small passage to a separate small third valve at cruising speeds This air injected into the cylinder increases the turbulence and combustion efficiency, thereby reducing *HC* and *CO* emissions.

Coasting air valve (CAV) is employed in some engines to decrease *HC* emissions during vehicle deceleration. When the engine is decelerating above the preset speed, say, 1800 r.p.m., CAV is opened by the manifold vacuum, allowing additional air flow into the intake manifold to reduce *HC* emissions. However, when the engine speed has decreased to the above preset value, the speed sensor energises the solenoid valve to vent out the manifold vacuum, thereby closing the CAV and consequently providing smooth engine operation.

Air switching valve (ASV) is also operated by manifold vacuum. It increases air flow through the

Fig. 20.15. Exhaust recirculation system on diesel engine.
(Courtesy—Mercedes-Benz A.G., Germany)

idling circuit during deceleration which weakens the air-fuel mixture so that the idle circuit becomes inoperative. This improves HC emissions and fuel economy. When the speed has decreased to the preset value, the ASV closes, the idle circuit starts functioning providing smooth engine operation avoiding stalling.

Sequential fuel injection system with electronic control and a three way catalytic converter is found so far to be the best yet costliest method to optimize the fuel supply and to keep *HC*, *CO* and NO_x within the stringent limits imposed by current pollution control regulations in advanced counties.

10. MODIFICATIONS IN IGNITION SYSTEM

As the exhaust emissions at idle increase with the increase of ignition advance (Fig. 20.16) the distributor is redesigned to provide retarded ignition at idling speeds, which are also kept higher. At higher idling speeds the

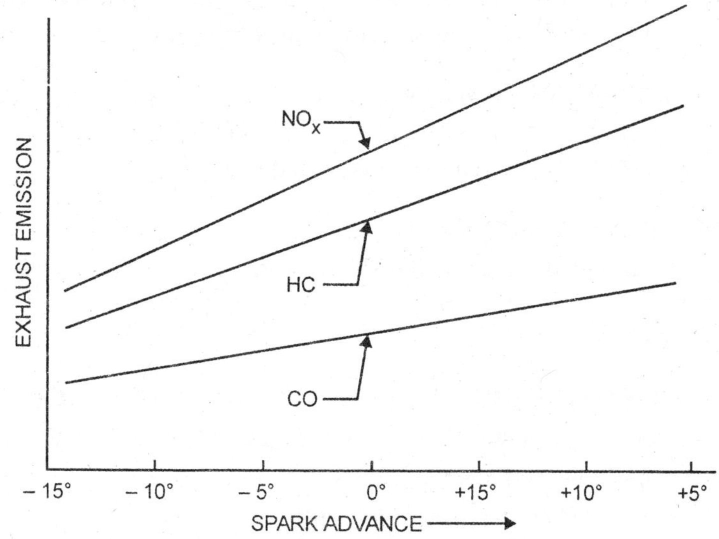

Fig. 20.16. Effect of ignition timing on exhaust emission at idle.

engine runs smoothly with a leaner mixture. Thus lean mixtures and higher temperatures resulting from retarded timing reduce hydrocarbon and carbon monoxide emissions.

The vacuum advance units have also been modified so as not to operate under certain conditions, *e.g.*, on deceleration, in low gear etc. This decreases the exhaust emissions under these conditions. A Deceleration Spark Advance System (DSAS) is shown in Fig. 20.17. Ported vacuum from carburettor and intake manifold vacuum are both applied to a solenoid valve, which is operated by a speed sensor. The solenoid valve is further connected to the vacuum advance unit of the distributor. At engine speed higher than the specified, say 1300 r.p.m., the solenoid valve is de- energized which causes the manifold vacuum to be applied to the distributor vacuum advance unit, through the solenoid valve. In case the engine decelerates, the manifold vacuum keeps the vacuum advance in fully advanced position, which reduces hydrocarbon emissions. However, when the engine speed decreases below the preset value, the solenoid is energized which shuts off the manifold vacuum and applies ported vacuum to the vacuum advance unit.

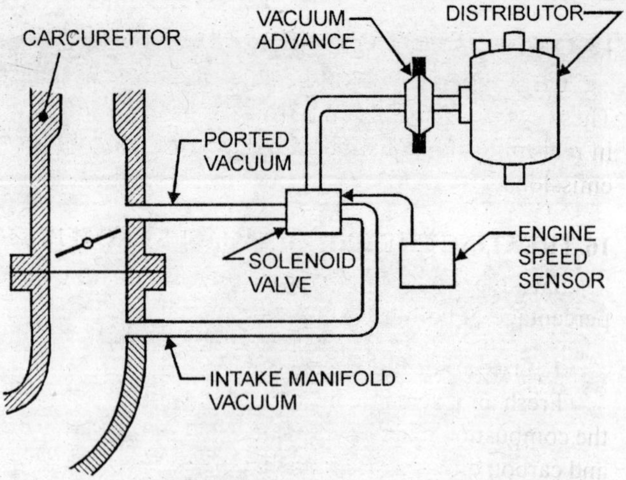

Fig. 20.17. Deceleration Spark Advance System

In computer-controlled systems, the computer combines the signals from the sensors for manifold air pressure (MAP), speed, inlet air temperature, manifold air temperature and throttle position to provide the correct ignition advance for the operating conditions.

11. IMPROVING VEHICLE EFFICIENCY

This would result in lesser fuel consumption and consequently lower emissions. This can be done by:

1. Lightweight vehicle design.
2. Minimised air resistance by better streamlining vehicle body.
3. Reduced rolling resistance with improved suspension, tyre and steering design.
4. Higher transmission and drive line efficiency.

12. USE OF HCCI ENGINE

As already discussed in Art 16 of Chapter 1, use of HCCI engine results in significant reduction of exhaust emissions, in particular NO_x and CO_2.

13. USE OF HYBRID VEHICLES

It has been described in detail in Chapter 2, how the use of hybrid vehicle results in significant reduction in fuel consumption and exhaust emissions. In particular, fuel-cell hybrid would be able to achieve almost zero pollution.

14. IMPROVING DRIVING EFFICIENCY AND TRAFFIC CONDITIONS

Significant reduction of emission can be achieved by optimising driving and traffic as follows:

1. Driving techniques.
2. Regulated traffic without jams.
3. Reducing the number of cold starts.

15. IMPROVING ON-BOARD DIAGNOSTIC (OBD) SYSTEMS

OBD systems monitor vehicle emissions & controls and inform the driver in case of a malfunction. These systems have become quite sophisticated in recent years so that these are able to monitor conditions in real-time and adjust vehicle operation accordingly, thus increasing operating efficiency and reducing emissions.

16. TREATING THE EXHAUST GASES FROM PETROL ENGINES

Two types of techniques are used for treating the exhaust gases from the petrol engine to reduce the percentage of pollutants in them:

1. Oxidation methods

Fresh air is sent into the exhaust manifold. The fresh air supplies additional oxygen required to complete the combustion of any hydrocarbons or carbon monoxide in the exhaust gases producing instead, the water and carbon dioxide which are harmless. *However, by this method, nitrogen oxides in the exhaust cannot be removed.* A few such systems are described here.

(*i*) Air injection system

In this system, compressed air is injected into each exhaust port slightly downstream of the exhaust valves. The injected air mixes with very hot, fully and partially burnt exhaust gases and causes oxidation of the *CO* and hydrocarbons. The effectiveness of this system, however depends upon the following:

1. Strength of the air-fuel mixture.
2. Temperature and pressure of exhaust gas and mixture reaction.
3. Time available for exhaust gas-air mixture
4. Heat losses from the exhaust-reaction zone.

The effect of injected air flow as percentage of the engine air flow on the hydrocarbon emission is shown in Fig. 20.18. It is seen that hydrocarbon emissions decrease upto injected air flow of about 20% of the engine air flow, beyond which these start rising moderately. On the other hand CO emission rate

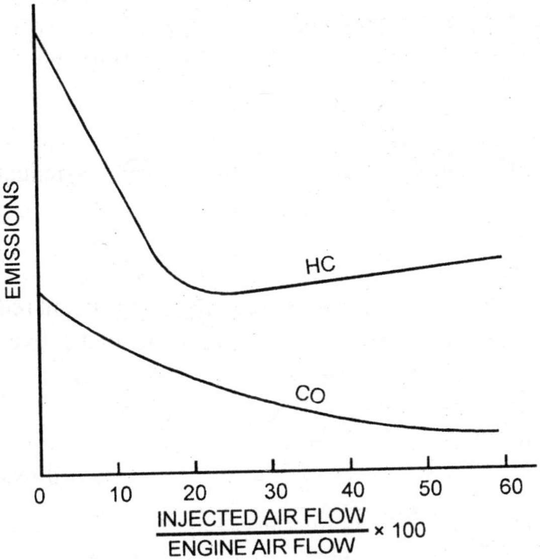

Fig. 20.18. Effect of injected air flow on the emission

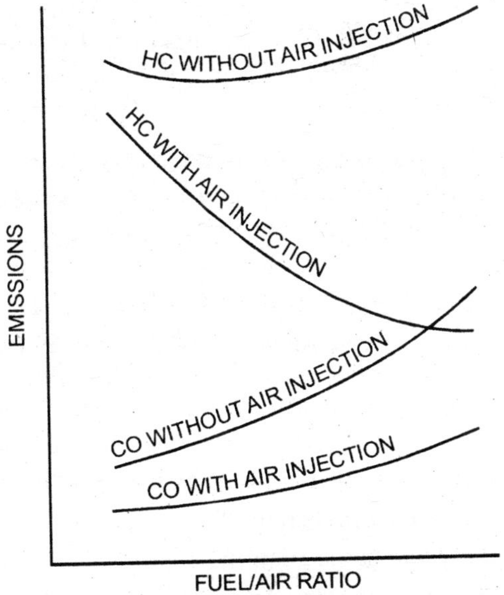

Fig. 20.19. Effect of idle fuel-ratio on CO and HC emissions.

goes on decreasing with increase of injected air. Thus, at about 20% injected air flow, overall emission effect may be minimised.

Effect of idle air-fuel mixture ratio on CO and HC emission is shown in Fig. 20.19. It is seen that air injection markedly reduces both these emissions. However, reduction in HC with air injection is much more as the air-fuel mixture is enriched. However, with CO emissions, the trend is different. These tend to increase, whether with or without air injection, as the air-fuel mixture is enriched.

On such air injection system used to reduce the hydrocarbons and carbon monoxide is shown in Fig. 20.20. Fresh air from the belt driven vane pump goes through a diverter valve, one-way check valve and piping into the exhaust manifold. This fresh air supplies excess oxygen to oxidize HC and CO into CO_2 and water vapours. Check valve is meant to prevent back flow whenever exhaust pressure is greater than air pump pressure. This may happen in the event of pump belt failure, excessively high exhaust system pressure or rupture of air hose. The diverter valve is operated by a vacuum diaphragm. When idling or in part or full throttle conditions, the manifold vacuum is not enough to move the diverter valve diaphragm. However, on sudden deceleration, manifold vacuum increases to lift the diverter valve diaphragm thereby diverting the air to the atmosphere or air cleaner. High manifold vacuum draws rich mixture from the carburettor which is too difficult to ignite in the combustion chamber, due to which reason it goes out in the exhaust manifold. When air is pumped into exhaust manifold the mixture becomes lean and would ignite when hot exhaust gases are released next from the cylinder. This may result in a dangerous explosion into the exhaust system. The diverter valve, by diverting the air flow under conditions prevents this backfiring. Air injection systems work more precisely when computer control is used. This method of exhaust cleaning does not affect engine efficiency.

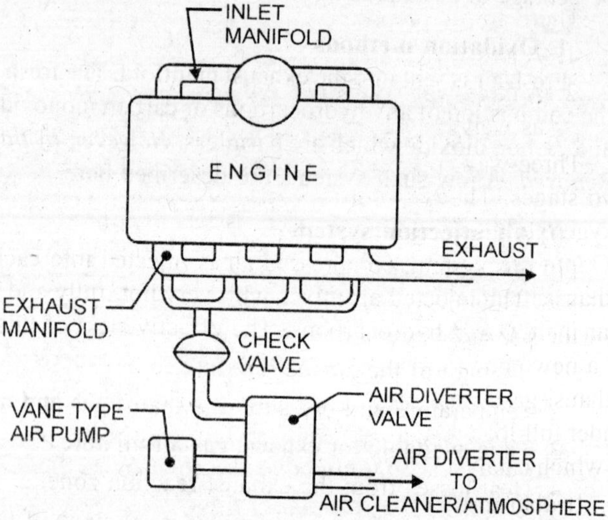

Fig. 20.20. Air injection system with computer control.

(*ii*) Pulse air-injection reactor (PAIR) system

Some engines do not require all the air supplied by the pump in the air injection system. Such engines may be provided with pulse air system. The system employs the pressure pulses of the exhaust to pull air into the exhaust system. When exhaust valve opens, exhaust gases rush out at high speed. Due to their inertia, more exhaust is pulled from the cylinders as the exhaust valve closes, causing thereby a momentary partial vacuum in the exhaust ports. Negative pressure in the exhaust manifold moves air from the air cleaner through the check valves and pipes into the exhaust ports. This system has a advantage over the air injection system described above in that no engine power is consumed to run a belt-driven pump. Such systems are usually employed on engines with catalytic converters (described later). Apart from burning HC in the exhaust manifolds, these also supply oxygen to the catalytic converter.

2. Catalytic converters

The exhaust gases from the engine are passed through catalytic converters (Fig. 20.21). A catalytic converter is a cylindrical unit about the size of a small silencer and is installed into the exhaust system of a vehicle such as a car, scooter, moped, motor cycle or auto-rikshaw. It is placed between the exhaust manifold and the silencer. Inside the converter there is a honeycomb structure of a ceramic or metal, which

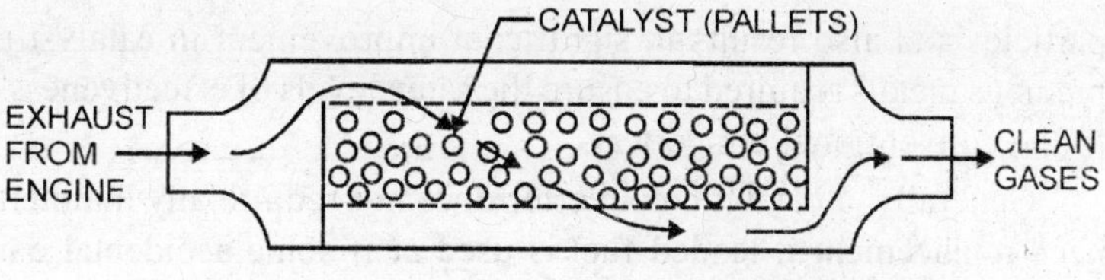

Fig. 20.21. Flow of gas through catalytic converter.

is coated with alumina base materials and thereafter a second coating of precious metals platinum, palladium or rhodium or combinations of the same. This second coating serves as a catalyst. A catalyst is a substance which causes a chemical reaction that normally does not happen in the given conditions. It promotes the required chemical reaction, but itself does not participate in that reaction. As a result of catalytic reaction, as the exhaust gases pass over the converter substrate, toxic gases such as CO, HC and NO_X are converted into harmless CO_2, H_2 and N_2.

There are two types of catalytic converters:

1. A two-way converter, which is used to control only CO and HC emissions by oxidation.
2. A three-way converter, which is used almost in all petrol cars. It controls CO and HC by oxidation as well as NO_X by reduction.

Three-way converters (TWC) are now standard on all vehicles using petrol engines and operate in two stages. The first converter stage uses rodium to reduce the NO_x in the exhaust into nitrogen and oxygen. In the second converter stage, platinum or palladium acts as oxidation catalyst to change HC and CO into harmless water and CO_2. For supplying the oxygen required in the second stage, air is fed into the exhaust after the first stage. The catalyst allows the oxidation of the exhaust gases at a much lower temperature than in the combustion chamber. With temperatures exceeding 300°C, the steady-state conversion efficiencies of a new converter are about 98% for carbon monoxide and about 95% for hydrocarbons. The outgoing exhaust gases in a petrol engine are at a temperature of 300 -400°C during idling and may reach 900°C under full-load conditions. Converters should be best operated between 400°C and 800° C. The temperature at which catalyst is 50% effective, is known as the *light-off temperature,* which should be as low possible.

It also means that relatively high emissions occur initially during the warm-up period. Recent developments have reduced the necessary operating temperatures and also made catalytic converters more robust so that they can be mounted closer to the engine and thus get warmed up quickly, thus reducing the warm-up time to a few seconds and consequently start reducing the emissions earlier.

Apart from converting hydrocarbons into CO_2 and water, oxidation converter also reduces the mass of the particulate emissions by oxidizing some of the hydrocarbons on the carbon particles.

Any impurities in the fuel shorten the life of the catalyst. As such the automobiles using catalytic converter have to be run on *unleaded* gasoline. Such a converter for well maintained car engines and run only with lead-free petrol has a service life from 80,000 to 100,000 km. Besides, metallic additives in the petrol may lead to deposits on exhaust system components (such as oxygen sensor and catalyst) under some driving conditions. In diesel fuel metallic additions or other ash-forming materials will increase the amount of ash to be captured by the particulate filters, which would have to be bigger in size. Sulphur in the fuel decreases the efficiency of catalytic converter. However, this effect is reversible and the efficiency is restored on reverting to low-sulphur fuel. Due to this reason, sulphur content has been restricted to 10 parts per million in Europe for the automotive fuels with effect from 2009. The detergent additives, on the other hand, help keep the fuel injection and the combustion system clean, thus helping prolong the maintenance period for emission control.

Conventionally, the precious metal catalyst particles are adhered to a base material. Heat from the exhaust gases causes these particles to collect together and agglomerate to form larger particles, which reduces the surface area of the precious metals, thereby deteriorating their performance as catalysts. Modern nanotechnology enables the manufactures to individually embed the precious metal particles into the base materials. Such a system prevents the thermal deterioration caused by the agglomeration of precious metal particles and also results in significant improvement in catalyst performance. As a result, the amount of precious metals required to ensure the same levels of effectiveness is reduced by 70 to 90 percent compared to the conventional converters.

Generally, a catalytic converter does not require any maintenance or servicing. The converter would need replacement if leaded fuel is used or if some accidental damage occurs. It may be noted, however that overrich mixtures and malfunctions can produce hotter catalytic reactions, causing over heating of the converter and heat damage to vehicle components nearby. To prevent this happening, avoid running an engine which misfires or is having overrich mixture. Overheating from overrich mixtures, late timing or misfiring can melt the catalyst element. In case of excessive overheating even lumps can be formed on melting, blocking the flow of exhaust gases which may lead to major problems such as reduced power, backfiring and overheating.

The catalyst in a converter should have low thermal inertia and should warm up rapidly. The warm-up time should be preferably 30 seconds, but not more than one minute. Placing the converter very close to the exhaust manifold shall help in rapid warm-up, but too close a placement can cause the exhaust gases under certain conditions, to exceed the safe working temperature, thereby reducing drastically, the converter life. Alternatively, electrically heated catalyst systems have been used to reduce the time to attain light- off temperature. These employ a small converter ahead of the main converter. The substrate onto which the catalyst is deposited, is made of metal so that, when electric current is passed, it would heat-up quickly.

It is seen that sulphur in petrol and sulphur or phosphorus in engine oil, delays light-off of the converters and reduces their efficiency. On the other hand, a major fraction of the pollutants that come out of the tailpipe are produced during the first two minutes after the start of the engine, since the catalyst in the converter is yet cold and has not reached its light-off point. The Corona Discharge Device from Litex claims to make the catalyst more active at lower temperature allowing the converter to start operating effectively more quickly. The device is inserted in the exhaust, upstream of converter and generates a non- thermal plasma (partially ionised gas) in the exhaust stream, which reduces CO emissions by more than 80% and both HC and NO_x emissions by more than 50%, depending, of course, on the sulphur content of the fuel. This means use of this device also reduces the size of the catalytic converter and hence significant saving in cost. Moreover, since the device reduces the amount of toxic exhaust that enters the converter, it results in increased converter life.

Toyota Motor Corporation has, in February 2017, announced the commercial availability of a new improved smaller catalytic converter, which uses 20 percent less precious metal in about 20 percent less volume, still maintaining the same performance in exhaust purification. It will be first used in Lexus LC 500 h towards the end of 2017 and would be gradually installed in new vehicle models.

The new catalytic converter is based on integrally moulded FLAD (Flow Adjustable Design Cell) substrate

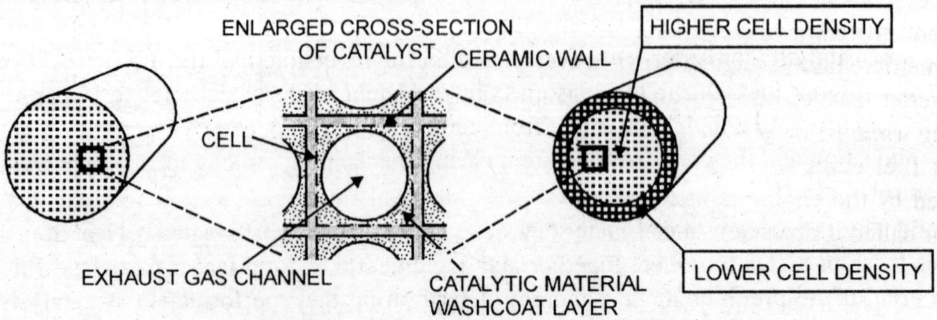

Fig. 20.22. Structure of FLAD substrate
(Courtesy–Toyota Motor Corporation, Japan)

jointly developed by Toyota and Denso Corporation. The FLAD substrate has smaller cell cross-sectional area at the inner portion (narrower channels for exhaust gas), with higher cell density compared to that at the outer portion, which has relatively larger cell-cross-sectional area, *i.e.* lower cell density, providing wider exhaust-flow channels (Fig. 20.22).

The walls of cells within this substrate are wash coated with catalytic materials like platinum, rhodium, palladium, etc. Due to the different cell densities at the inner and the outer portions of the new catalytic converter, the uniformity of the exhaust gas flow through the converter is improved (Fig. 20.23). In case of a conventional converter, there is higher volume of gas flow towards the centre than on the outside, due to which more catalytic precious metal is required at the inner portion of the converter, but since current technologies allow only uniform wash coat, higher amount of precious material has to be coated all over. Making flow relatively more uniform allows relatively lesser overall amount of precious material to be coated.

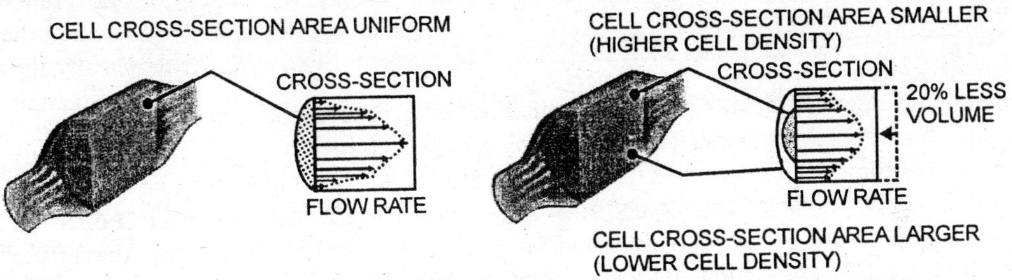

Fig. 20.23. Flow of exhaust gas through the converter (*a*) Conventional converter with uniform cell cross-sectional area, (*b*) FLAD design converter with lesser cross-sectional area of the inner cells compared to the outer ones.

It is seen from Fig. 20.24 that to operate at highest efficiency, a three- way converter requires stoichiometric, *i.e.*, about 14.6:1, air-fuel ratio. Lean mixtures will provide lower HC and CO levels and converter is quite efficient in oxidising these, but these ratios produce higher cylinder temperatures and consequently higher NO_x level to reduce which converter efficiency is very low. On the other hand rich air-fuel ratios would provide low NO_x emission levels, which are reduced quite efficiently by the converter. Yet these also cause high emission of HC and CO and extremely low converter efficiency for oxidising these. Thus, for the catalyst to operate at nearly 90 percent efficiency, the air-fuel ratio should be exactly stoichiometric. Slight change in the air-fuel ratio within which the converter may be allowed to work is called the *catalyst operating window.* Incidentally, this is also air-fuel ratio for the best fuel economy. Any deficiency or excess of air consumed by the engine is usually expressed by an excess air factor called ë (lambda), which is defined as

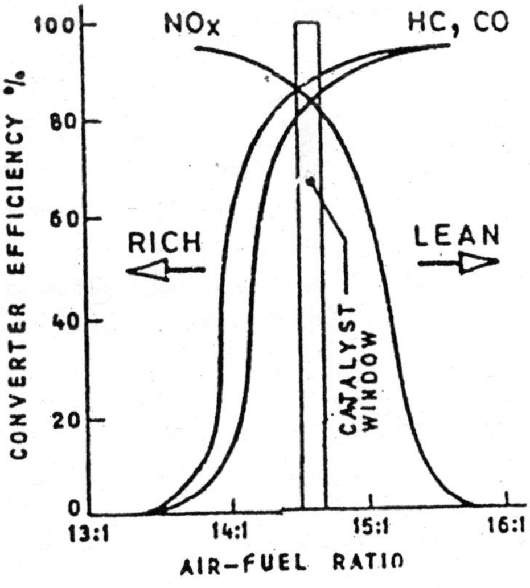

Fig. 20.24. Catalytic converter efficiency

$$\lambda = \frac{\text{actual air consumed}}{\text{theoretical air required}}$$

For a stoichiometric air-fuel mixture $\lambda = 1.0$. Similarly, for a rich mixture having 5% deficient air, $\lambda = 0.95$.

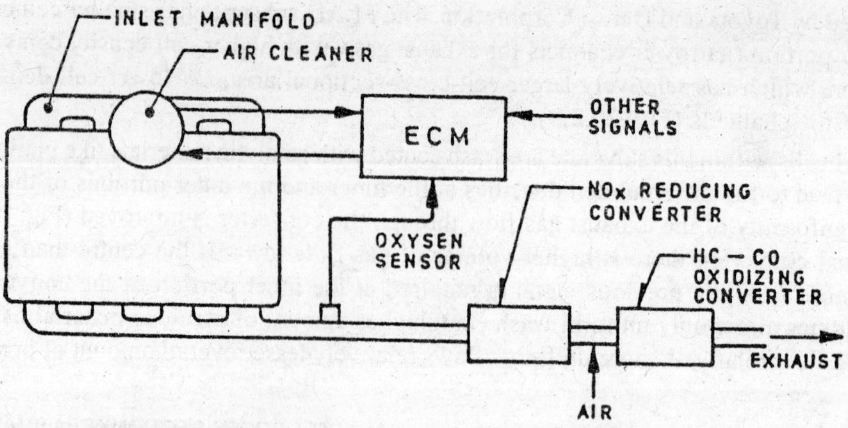

Fig. 20.25. Feedback computer control of air-fuel ratio for efficiency converter operation.

To achieve the described air-fuel ratio at all operating conditions, a feedback system is used that determines the correct air-fuel ratio of the intake charge by measuring the amount of oxygen remaining in the exhaust (Fig. 20.25). Electronic control module (ECM) is the control centre of the Computer Command Control (3C) system. It receives signals from oxygen sensor (λ-probe), throttle position sensor, coolant temperature sensor, manifold pressure sensor, vehicle speed sensor etc., and accordingly adjusts the air-fuel ratio to remain within the catalyst operating window. Lambda probe (or sensor) as used in modern petrol engines is shown in Fig. 20.26. It is fitted in the exhaust pipe where it measures the exhaust gas composition continuously. The information signal is fed to the E.C.M.

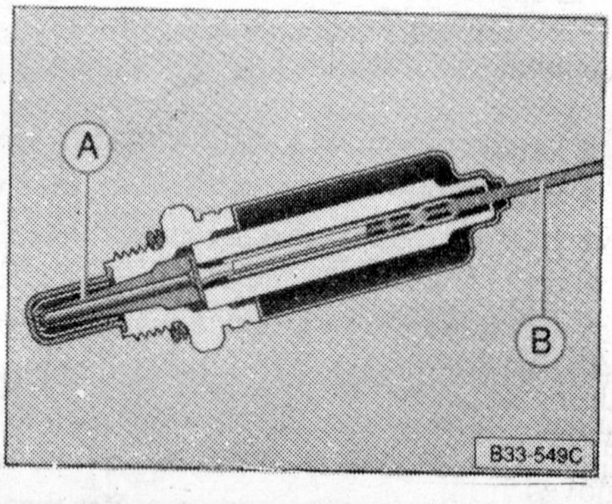

Fig. 20.26. Lambda probe: (A) sensor with precious metal coating, (B) connecting wire. (Courtesy—Volkswagenwerk A.G., Germany)

Lambda probe *i.e.* sensor may be unheated or heated type. An unheated type is shown in Fig. 20.27. It is shaped like an extended thimble. The solid state electrolyte is made of ceramic zirconium oxide ($Z_1 O_2$) stabilized with yttrium oxide (Y_2O_3). The external and internal surfaces of the solid ceramic electrolyte are coated with gas-permeable microporous platinum. These surfaces represent the electrodes involving electrical activity. The inner surface is exposed to the atmosphere, while the outer surface is surrounded by the hot exhaust gases. For physical support, steel housing is provided, which is threaded on the outside, for fitting into the exhaust system. The lower half is further protected by a slotted shield made of steel. The electrical cable is taken out through an insulator. The heated type sensor is similar in construction, except that in this a ceramic heating rod-element projects into the thimble. With this type, the warm-up time after starting the engine is reduced and the sensor characteristics are also better. However, principle of working remains the same in the two types.

Lambda sensor is very sensitive to the smallest variation of air supply and oxygen concentration in the exhaust gas at the stoichiometric ratio ($\lambda = 1$). When the air-fuel mixture composition moves to the rich side of the stoichiometric ratio, very small quantity of oxygen is available to be fed to the outer platinum electrode catalyst; then the oxygen is made available by depleting the outer boundary region of the electrolyte. This causes a large number of negatively charged oxygen ions to travel from the oxygen rich inner electrode to the outer electrode, due to which a voltage is produced between the two electrodes in proportion to the rate of travel of oxygen ions through the electrolyte. This e.m.f. is about 800 mV and is available in the form of a relatively large stepped voltage change, when the air-fuel mixture tends towards a rich mixture. On the other hand when the air-fuel mixture is on the leaner side of the stoichiometric ratio, there is excess oxygen available for oxidation of CO and HC around the outer electrode with the result that very few ions diffuse through the electrolyte and thus producing a very small voltage across the electrolyte. In this way, whenever the air-fuel mixture goes to the lower side ($\lambda > 1$), a voltage step of only about 50 mV is generated.

As the magnitude of the generated voltage in the lambda sensor depends upon the difference in oxygen concentration around the two electrodes, it has been effectively used to measure the oxygen content in the exhaust gas at any instant. Voltage signal from the lambda sensor goes to the lambda regulator, wherein it is amplified and finally integrated after which it is sent to the fuel metering control, which is operated by an E.C.M., which takes information from other sensors also, apart from the lambda regulator voltage control output.

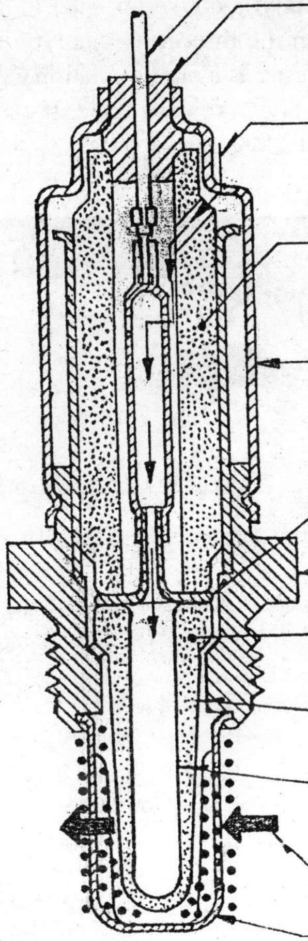

Fig. 20.27. Unheated lambda probe.

17. EXHAUST TREATMENT FOR DIESEL ENGINES

17.1. Introduction

As the diesel engines operate with excess air, problem of incomplete combustion resulting in emission of hydrocarbons is not there. However, in these engines, apart from NO_x, particulate matter (PM) also has to be removed from exhaust. Various technologies used are:

(*a*) *For controlling PM*

- Diesel oxidation catalysts (DOCs)
- Diesel particulate filters (DPF)
- Closed crankcase ventilation (CCV)

(*b*) For *controlling* NO_x

- Exhaust gas recirculation (EGR)
- Selective catalytic reduction (SCR)
- Lean NOx catalysts (LNCs)
- Adsorbers

Out of the above, CCV and EGR have already been discussed. The remaining technologies will now be discussed in detail.

17.2. Oxidation Catalysts

These have been found to reduce 20 to 50 percent PM on mass basis.

The oxidation catalytic converter used in diesel engines is a *two-way converter* which operates in a different manner than the three-way catalytic converter for petrol engines described above. The diesel engine catalytic converter is a pure oxidation catalytic converter, which oxidises hydrocarbons and carbon monoxide into water and carbon dioxide. It cannot reduce nitrogen oxides. Such a catalytic converter as used on an old model Mercedes-Benz car is shown in Fig. 20.28.

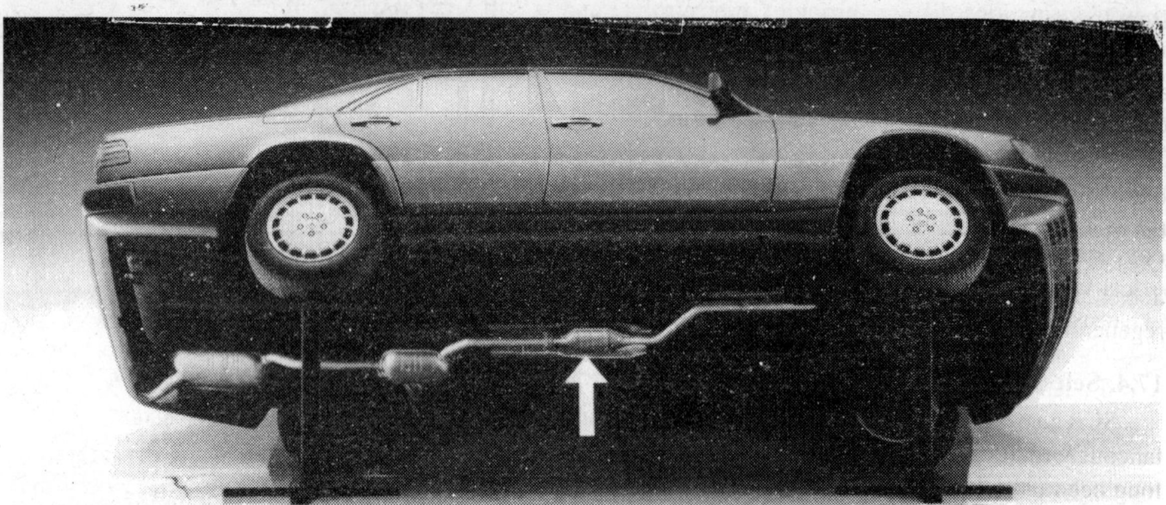

Fig. 20.28. Catalytic converter mounted on a diesel car.
(Courtesy—Mercedes-Benz A.G., Germany)

Earlier it was thought that since the diesel engine basically operates with excess air, lambda control is not feasible. However, modern diesel engines do have a lambda closed-loop control, wherein signals from lambda sensor can be used for precise adaptation of EGR, boost pressure and start-of-injection according to various operating modes of the engine, leading to reduction of emissions by about 20%. Besides, full-load smoke limitation is also achieved with the help of lambda sensor. This determines the maximum fuel quantity which can be injected at full load without exceeding the smoke limit.

17.3. Particulate filters (Traps)

The particulate matter can be removed by passing the exhaust gas through a filter (usually made of ceramic honeycomb materials, *e.g.*, cordierite, silicon carbide or aluminium titanate) wherein the particulates

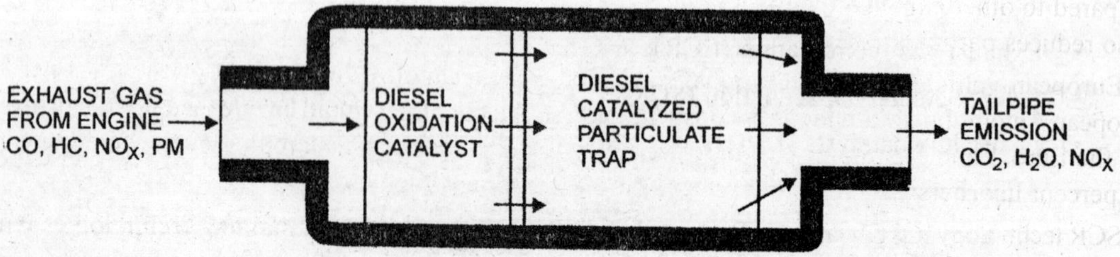

Fig. 20.29. Diesel catalytic convert-cum-particulate trap.

are retained on the filter element, which are periodically removed by their oxidation by means of hot exhaust gases, either at elevated temperature or in the pressure of a catalyst. In this way, the filter is kept clean automatically. It must be noted that diesel particulate matter ignites at 500°C to 600 °C, whereas the normal operating temperature range of exhaust gases is from 200°C to 500°C from idling to full load. There are two methods available to make the two temperature ranges compatible. One method is to use an electric heating element in the filter passage. The other method is to use a catalyst, which brings down the ignition temperature of the particulates by about 200°C. It is seen that such filter traps are effective for small diesel engines wherein exhaust temperatures are high. In case of large diesel engines, wherein exhaust gases are at lower temperature and the particulate matter is also large, the process of removal of particulate matter from the exhaust is relatively more difficult.

Particulate filters may be of flow-through or wall flow designs. While the former type are capable of achieving 30-75 percent PM reduction, the later design can remove over 90 percent PM. Both these types can trap sub-micron, ultrafine particles.

Conventional particulate filters get clogged by sulphur, which is usually present as an impurity in diesel fuel. It is, therefore, necessary that sulphur-free (sulphur less than 10 ppm) diesel fuel is used in the diesel vehicle engines. AEA Technology in the U.K. have developed a special filter named 'Electrostat Diesel Particulate Filter (DPF)', which is claimed to operate at low exhaust temperature, is self-cleaning, does not need fuel additives and works unaffected by sulphur content of the fuel. DPF used plasma filter regeneration to remove all types of particulates.

17.4. Selective catalytic reduction

SCR uses urea as a reducing agent. It has been found to reduce 75-90 percent NOx, while simultaneously reducing HC emissions upto 80 percent and PM emission by 20-30 percent. Such systems have long been used on most Euro IV and V heavy-duty trucks in Europe.

Bosch of Germany have also developed 'Denoxtronic', which is a Selective Catalytic Reduction (SCR) technique. The SCR operates with 'Ad Blue' (urea/water solution), which is a reduction agent and is injected into the exhaust gas. The toxic NO_x in the exhaust gas is converted into a harmless Nitrogen gas and water vapours. Obviously, it is necessary to meter accurately the quantity of Ad Blue depending upon the operating condition of the engine.

$$NO + NO_2 \quad + \quad \underset{\text{(from Ad Blue)}}{2NH_3} \longrightarrow 2N_2 \quad + \quad 3H_2O$$

Fig. 20.30 shows the details of the Denoxtronic. The ECU processes all the relevant data form the engine management together with the measured data from the exhaust system sensors. The Ad Blue metering is done precisely according to the characteristics of the engine and the catalytic converter in all operating modes. The system also fulfills the requirements of the on-board diagnosis which is now mandatory as per Euro – 4 norms. The system is claimed to be maintenance-free for its entire life.

In test vehicles, Denoxtronic has been able to reduce NO_x emissions by 80% alongwith fuel economy. Compared to other exhaust treatment systems, it is claimed to achieve 5% to 8% fuel savings. Moreover, it also reduces particulate emissions.

European vehicle manufacturers have selected SCR technology as the best method. As per ACEA (European Automobile Manufacturers Association), SCR technology would enable their members to comply with the Euro 4 and 5 emission standards and also to achieve fuel consumption levels which would be 5 to 6 percent lower than those of equivalent Euro 5 compliant engines.

SCR technology has already worked well in stationary engines. For automotive applications, the following additional monitoring systems have been designed.

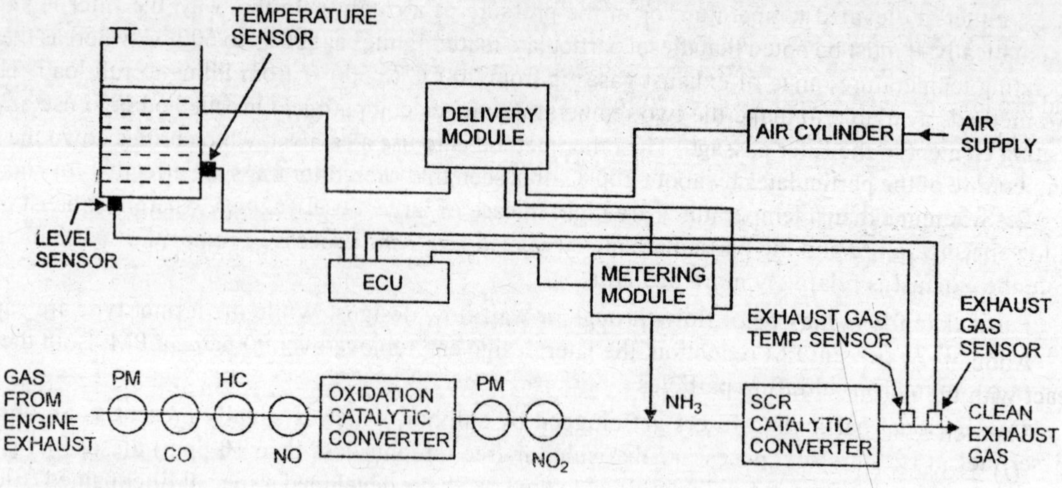

Fig. 20.30. Schematic diagram of diesel engine exhaust gas treatment with Denoxtronic.

1. Gas sensors, particularly *No_x sensors* are used downstream of the catalytic converter system. The use of a NO_x sensor gives closed loop control of the complete system since any failure or tempering (*e.g.* disabling of the urea injector or use of wrong fluid in the Ad Blue tank) would result in sensor detecting a sudden change in NO_x levels.
2. *Urea sensors* have been employed to detect the presence, concentration and level of Ad Blue in the tank.
3. *Flow sensors* have been used to monitor the delivery of the reducing agent to the injectors.

With Euro 6 requirements in mind, Delphi has developed an SCR dosing system with a sensor to detect and help control the level of ammonia produced in the exhaust when urea is introduced in the system. Delphi system uses higher pressure to provide better distribution, improved mixing and more unform distribution of urea across the catalyst which results in better performance. Besides, Delphi system does not need a separate feed pump in the urea tank. Flow of urea is caused by the pump in the dosing nozzle, which, apart from reducing complexity, would also reduce system cost. The urea flows into a pumping chamber, where the pressure is taken upto about 40 bar, then across a poppet nozzle which breaks at 20 bar. So the urea is injected at 20 plus bar. The increased pressure and the ammonia sensor are able to reduce the size of the SCR and catalyst.

17.5. Lean NOx catalysts

Lean Nox catalysts often include a porous material made of zeolite having a microporous, open framework structure. However, such systems are not much in use due to relatively low NO_x conversions (20 to 30%) and also corresponding fuel economy penalties due to the injection of small amount of diesel fuel into the exhaust upstream of the catalyst, to serve as a reductant.

17.6. Adsorbers

We have seen that, outside the catalyst operating window, the engine's operating conditions are not ideal for conventional catalysts to achieve their maximum efficiency to reduce pollutants. Besides, it does take about 10 seconds for the best catalytic converter to reach 'light off' temperature during which period there will be some emissions going uncontrolled. Adsorbers serve to collect certain pollutants, especially NO_x or HC during the engine operating condition which are not ideal, to store these pollutants and to treat the same when conditions become suitable. Currently two types of adsorbers are mainly in use:

1. NOx adsorbers or Lean NOx traps (LNTs)

These adsorb and store NO_x under lean conditions in lean-burn petrol engines and from diesel engines. A brief return to nearly stoichiometric conditions for a few seconds is sufficient to desorb the stored NO_x and provide the conditions for a conventional three-way catalystic converter downstream, to destroy NO_x.

This is done in following steps:

1. Oxidising NO to NO_2 over a precious metal catalyst, which is platinum.
2. Storing NO_2 in an adjacent alkaline earth oxide (usually barium oxide) trapping site as a nitrate.
3. The stored NO_x is then periodically removed by two-step regeneration, which includes temporarily inducing a rich exhaust condition (for a short period, from 3 to 9 seconds) followed by reduction to nitrogen by an ordinary three-way catalytic converter.

Another result of running rich is higher levels of CO and HC in the exhaust stream. These CO and HC react with the released NO_x reducing it to N_2. The engine then runs lean again and the cycle repeats.

The rich exhaust condition, also called *'lean-rich modulation'* is achieved by any of the following:

(*i*) Exhaust Gas Recirculation.

(*ii*) Post-combustion fuel injection in the cylinder.

(*iii*) In-exhaust fuel injection.

The platinum catalyst also promotes the oxidation of hydrocarbons to help eliminate completely, the oxygen at the trap, which is absolutely necessary if NO_x is to be released. In addition, during the rich phase, platinum enables the trap to regenerate by assisting the NO_x release and reduction steps.

GM has recently found that 'perovskite oxide catalysts' with the addition of a 'little bit of palladium' can replace platinum. New catalytic system also resists degradation of catalytic action due to heat exposure, even better than platinum. However, the use of the new catalyst in production system is expected to take some time.

It may be noted that NO_x adsorbers also adsorb sulphur oxides, due to which reason fuels with very low sulphur content need to be used. Besides, sulphur compound being relatively more difficult to desorb, the system has to run automatically periodically a short 'desulphation cycle' to remove them.

2. Hydrocarbon adsorbers

These are used with conventional three-way converters to reduce the HC emissions from rich fuel that has to be used to start the engine from cold. Such adsorbers use special materials, *e.g.*, zeolites into or ahead of the converter. HC emissions collected during cold start are desorbed at higher temperatures when the converter has reached its operating temperatures.

17.7. Combined technologies

Basically, due to the thermodynamics of the diesel combustion process, engine design can only be opti-mized either for NO_x or for PM, but not simultaneously for both. NO_x-optimized engines offer less opportunity to optimise fuel efficiency and CO_2 emissions. Likewise, a highly fuel efficient engine optimized for low PM will have higher engine-out NO_x levels. Therefore, to meet the more stringent levels beyond Euro 3, special techniques have been/are being developed to simultaneously optimize PM and NO_x levels.

1. Toyota DPNR

Toyota of Japan have also developed a system called DPNR (Diesel Particulate-NO_x Reduction) system, which can simultaneously and continuously reduce PM (Particulate Matter) and NO_x in diesel exhaust gases with a new simple and compact catalytic converter. DPNR consists of a fine porous ceramic filter coated with a NO_x storage-reduction three-way catalyst which was designed for lean- burn petrol engines. During lean-mixture operation, PM is oxidised by active oxygen released in the NO_x storage process and by excess oxygen in exhaust gas. When, however, the engine momentarily switches to stoichio-metric operation, PM is also oxidised by active oxygen released in the process of reducing the stored NO_x. By switching the direction of the exhaust gas flow in the catalytic converter, PM-oxidation capability is increased. However, there are limitations, firstly it needs low-sulphur fuel to maintain high conversion effi-ciency for a long time and, generally it can be fitted only on common rail, direct injection engine vehicles, since it needs the precise control of the injected fuel quantity, which is not available with other vehicles.

2. BASF's 'two-brick' system

BASF has unveiled in March 2010, a new system for treatment of diesel engine exhaust. This simpler system has been demonstrated on Hyundai's i-flow concept car at Geneva Auto Salon.

The new system consists of two blocks:

1. DOC/LNT block
2. SCR catalyst integrated with the soot filter substrate

The DOC/LNT block is a ceramic monolith with a catalyst coating, which traps NO_x under lean operating conditions. This NO_x is released by switching to rich conditions for a few seconds, which is then converted to N2 and ammonia (NH_3) is generated in the DOC/LNT block. This generated NH_3 is then used by the SCR catalyst in the second block to remove excess NO_x, while the filter removes the PM. Thus the real advantage of this system is that urea does not have to be stored on board; the same is produced 'in situ' only in the first block. The added advantages are reduced mass, simplicity and lower overall cost. Reduced system mass also results in lower fuel consumption.

18. PRACTICAL EXAMPLES

Recently (2006), Volvo introduced PZEV (Partial Zero Emission Vehicle) certified petrol sedans S40 and S60 and wagons V50 and V70. The salient features of these vehicles are the computerized emission control during the start-up process and a modified fuel system to reduce evaporative emissions. Various techniques used are:

1. Volvo uses new software to precisely control the start-up sequence and minimize engine-out emissions during this period. Moreover, computerized variable valve timing is used to manage emissions during the cold-start by:

(*i*) delaying the opening of the intake valve, which improves the air/fuel mixing and increases stability during combustion.

(*ii*) running the combustion process very late, burning up the fuel on the cylinder wall and reducing HC and NO_x emissions. and

(*iii*) delaying the opening of the exhaust valve which brings the hydrocarbon-rich exhaust gas "residual" back into the cylinder, further reducing HC emissions.

2. Catalyst warm-up time is shortened. During start-up the car is run with excess air in the cylinders. The resulting lean mixture heats the gas in the cylinder quickly, hastening the warming-up of the catalyst. Besides, PZEV-certified volvo cars use two catalytic converters instead of one. The first catalytic converter is placed very near the engine exhaust manifold so as to get heated quickly.

3. These vehicle also achieve almost-zero levels of evaporative emissions. Their evaporative emission control system employs two different canisters. The first canister with three chambers filled with activated carbon is fitted near the fuel tank, while the second one containing a honeycomb structure coated with activated carbon (called Hydro Carbon Scrubber, *i.e.*, HCS) is employed downstream. Hydrocarbons are trapped by the carbon in these canisters. When the car is running, these trapped hydrocarbons are sucked out of the canister and are burnt in the engine.

Further, it has been stated already that cooler engine conditions reduce NO_x emissions, but simultaneously increase the emission of PM, which cannot be oxidized fully at lower temperatures. Likewise, high engine temperatures oxidize more particulate matter, but generate greater amounts of NO_x.

Rolf D. Reltz and Matthew P. Thial have invented a multiple fuel injection scheme that can simultaneously reduce both NO_x and PM emission in diesel engines. In this several fuel charges are injected, instead of a single charge during a combustion cycle. Each charge is injected at a pressure higher than the previous one, but the total fuel injected during one combustion cycle remains the same as in case of conventional single charge. This spacing of fuel injection into several charges enlarges the surface area of the soot (PM) clouds exposed to oxidation and also maintains more uniform combustion temperatures leading to cooler engine conditions and thus lower NO_x emissions. The method may be used simultaneously with EGR to further reduce NO_x emissions. Thus this method eliminates the need for cumbersome and expensive particulate traps.

19. FUTURE SCENARIO FOR DIESEL ENGINES EMISSIONS CONTROL

Currently (2017) diesel engine vehicles in Europe and India make up about 50 percent of the total market. Recently, diesel engine vehicles came prominently in news due to their very high NO_x emissions. In Europe, diesel powered vehicles were found in 2016 to contribute 80% of the total NO_x emissions released by transportation vehicle engines only. Moreover although NO_x emissions from diesel cars were reduced by 68% from Euro 4 to Euro 6, yet it has been found recently that real-world performance, under normal conditions on the road, are much worse than the laboratory results using the NEDC driving cycle- not only for NO_x but also for CO_2, which was found to be 20-30% higher than the laboratory results and the difference is even greater for NO_x emissions, particularly in case of diesel engines, where the real-world NO_x emissions are seen to be 4-5 times the value obtained from laboratory measurements. Thus NO_x emissions are a major issue in case of diesel engines. Therefore, various future strategies for automotive engines, in particular for diesel engines would be briefly as follows:

1. Tightening of regulations

Test procedures for emissions and fuel consumption of light-duty vehicles to be improved:

(*i*) Real Driving Emissions (RDE) for measuring pollutants on the road by a Portable Emission Measurement System (PEMS), and

(*ii*) Worldwide Harmonized Light-Duty Vehicles Testing Procedure (WLTP) for measuring CO_2 emissions.

2. Technology Improvement

(*i*) *Exhaust after-Treatments*– Various treatments as discussed in Art 17 would be combined and used.

(*ii*) *Variable Compression Ratio*– Higher compression ratio for cold ambient start and engine warm up would result in more complete combustion and less HC/CO emissions for starting, warm-up and part-load conditions. For increasing loads, compression ratio would be gradually reduced.

(*iii*) *Cylinder deactivation*– For diesel engines, Tula Technology has developed Dynamic Skip Fire (DSF) which uses audio electronics-based digital signal processing to vary continuously the number of cylinders firing at any one time, thus controlling diesel exhaust temperature irrespective of engine load.

In case of NOx traps and SCR catalysts, there is problem at low loads, viz., the exhaust at low loads is too cold for the after-treatment devices to be fully effective. DSF allows control of the exhaust temperature, so the after-treatment systems would be working all the time at or near their peak efficiency.

(*iv*) *Electric super chargers*– These would be used to supplement conventional turbochargers for improving or even eliminating turbo lag at low engine speeds and loads. Besides improving engine efficiency at low loads, electric supercharger helps in bringing the SCR to the operating temperature quickly resulting in decreased emissions and improved fuel economy. Electric Air Charger (eAC) from Pierburg is to be used on a small diesel engine in 2017-18.

20. USE OF ALTERNATIVE FUEL

Vehicle have been made with compressed Natural Gas (CNG) as fuel. Bifuel vehicles with petrol/CNG and diesel/CNG are also available, wherein either fuel can be used with the flick of a switch. Even the existing vehicles can be converted to pure CNG type or bifuel type simply by means of a conversion kit. The advantages of CNG as alternative fuel are:

1. Reduced pollution. With CNG, CO emissions are reduced by 70-90%, HC emissions by 40-60%, No_x emissions by 30-40%, and CO_2 emissions (which cause global warming) by about 10% compared with the petrol. Moreover, being lighter than air, CNG dissipates quickly in the air & moves up.
2. Economical. In terms of calorific value, the cost of CNG is almost a third of the cost of petrol.
3. Reduced maintenance. Since CNG burns cleanly, vehicle maintenance is reduced.
4. Easy availability. Natural gas it available in India in large quantities. Moreover, it does not require refining as in case of petrol and diesel.
5. Better anti-knock property. CNG can be safely used in engines with compression ratio as high as 12:1. Its octane number is higher than that of petrol.

However, there are accompanying disadvantages also:

1. Safe compression ratio of CNG engines are lower compared with diesel engines.
2. CNG cylinder takes a considerable portion of the luggage space.
3. Special safe fuel storage cylinders are expensive to design and build, which increases the overall cost of a CNG vehicle.
4. The range is only half that of a petrol car with same size of fuel tank.

We have already discussed in the previous articles that catalytic converter cannot reduce NO_x and particulate matter from the exhaust of diesel engines. The trap technology to remove particulate matter satisfactorily is as yet in the development stage. Under these conditions, the use of CNG as fuel in place of diesel is a rational and viable means of reducing pollution. That is why in the Indian metropolitan cities, the diesel vehicles have been converted to CNG vehicles.

The CNG conversion kit consists of the following main components:

1. Cylinder. It stores CNG at a working pressure of 200 bar. It is fitted with a shut-off valve and a safety burst disc.
2. Vapour bag. This encloses the cylinder and the pipes connecting the cylinder valve. It is vented out of the car.
3. High pressure pipe. It connects the refuelling valve to the CNG cylinder and pressure regulator.
4. Refuelling valve. it is used to refuel the CNG cylinder.
5. Pressure regulator. It has a solenoid valve to shut off gas supply to the engine. The CNG stored at high pressure in the cylinder is reduced to just below atmospheric pressure by the pressure regulator. The negative pressure does not allow the gas to pass through when the engine is not running.
6. Gas-air mixer. It meters accurately the gas fed into the engine.
7. Fuel-solenoid valve. This is used to cut off petrol or diesel supply to the engine when it is run on CNG.
8. Selector switch. It is fitted on the dashboard, enabling the driver to choose whether the engine is

to be run on petrol (or diesel) or CNG. This is so designed that when the engine is switched off, the gas supply is also switched off. It also indicates the quantity of CNG available in the cylinder.

However, these advantages are also accompanied by the following disadvantages:

1. Safe compression ratios of CNG engines are lower compared with diesel engines.
2. CNG cylinder takes a considerable portion of the luggage space.
3. Dearth of refuelling stations.
4. Special safe CNG cylinders are expensive to design and build.
5. Driving range is only nearly half that of a similar petrol car.
6. High cost of conversion kit, which may be of the order of Rs. 30,000/-.
7. Like petrol, natural gas is also a fossil fuel and cannot be considered a renewable source.

21. SOURCES OF EXCESSIVE EXHAUST EMISSIONS

1. Hydrocarbons

(*i*) Improper ignition timing.
(*ii*) Ignition system misfiring.
(*iii*) Improper quantity of air/fuel mixture.
(*iv*) Defective piston & rings and/or worn out cylinder.
(*v*) Faulty valves or components of valve actuating mechanism.
(*vi*) Vacuum leaks.

2. Carbon Monoxide

(*i*) Clogged air cleaner.
(*ii*) Defective choke.
(*iii*) Incorrect idle setting.
(*iv*) Excessively rich Carburettor setting.
(*v*) Faculty injectors.

3. Both Hydrocarbons as well as Carbon Monoxide

(*i*) Non-functioning PCV system.
(*ii*) Sticking air cleaner preheat door.
(*iii*) Excessive crankcase dilution with fuel.
(*iv*) Defective carbon canister purging system.
(*v*) Excessive rich air/fuel ratio.

4. Nitrogen Oxides

(*i*) Excessively lean air/fuel ratio.
(*ii*) Excessively advanced ignition timing.
(*iii*) Defective cooling system.
(*iv*) Defective EGR system.
(*v*) Vacuum leaks.

22. EXHAUST GAS ANALYSIS

The analysis of exhaust gases can be done with the help of infrared analyzers. One such analyzer, capable of measuring the hydrocarbons and carbon monoxide is shown in Fig. 20.31.

Fig. 20.31. Infra red exhaust analyzer.
(Courtesy—Sun Electric Corp., U.S.A.)

The working of the same can be understood by the schematic diagram shown in Fig. 20.32. The probe is inserted into the vehicle's tail pipe and the analyzer is switched on. The exhaust sample is moved by an internal pump, through the sample line and filter, to the sample cell from where it is vented to atmosphere.

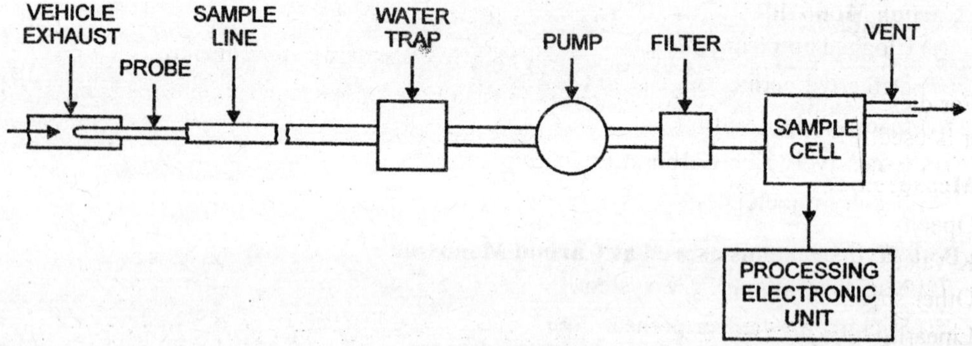

Fig. 20.32. Operation of exhaust gas analyzer.

In the sample cell, a beam of infrared light is made to pass through the exhaust sample. The analyzer then determines the quantities of Hydrocarbons and Carbon Monoxide with the help of the processing electronics and indicates the same by moving the respective indicators. The quantities indicated are in parts per million (PPM). In some analyzers four gases *viz.*, HC, CO, CO_2 and O_2 can be analyzed, whereas in still others, called five-gas analyzers, NO_x is also measured. From the analysis of exhaust gases, we can come to know the possible sources of engine malfunction and take remedial action to bring pollution within permissible limits.

Specifications of a typical modern four gas infrared emission analyser are as follows:

CO	Ranges	0—9.9%
	Accuracy	± 0.2%
	Repeatability	± 2% fs.

HC	Ranges	0—5000 ppm
	Accuracy	± 40 ppm at 0—2000 ppm
	Repeatability	± 150 ppm at 2000—5000 ppm
O_2	Ranges	0—20.9%
	Accuracy	± 0.4% at 0-5%
		± 1.0% at 5-20.9%
	Repeatability	± 2% fs
CO_2	Ranges	0-20.0%
	Accuracy	± 0.9% at 0-10%
		± 0.48% at 10-16%
		± 1.44% at 16-20%
	Repeatability	± 2% fs
	Warm up time	10 minutes at 20°C
	Response time	14 seconds for 90% of final reading (10 m pipe)
	Operating temperature	2°C to 43°C
	Dimensions	315 mm (W) × 282 mm (H) × 270 mm (D)
	Weight	14.3 kg
	Power	200-240, V, 50-60 Hz, 100 W
	Options	(*i*) Printer (*ii*) R.P.M./Oil temperature display.

For diesel engines, it is also required to measure the smoke capacity and K-value for which a smoke meter is used. Main specifications of a typical Smoke Meter (Manatec-DSM 2000) are as follows:

Measurement Parameters	**Range**	**Resolution**
Opacity	0-99.9%	0.10%
K-Value	0-99.9%	0.10%
Other specifications		
Linearity		± 0.1 m^{-1}
Repeatability		± 0.1 m^{-1}
Response time-Physical		< 0.4 sec.
Response time-Electrical		< 1 ms
Warm-up time		< 3 minutes
Smoke measuring cell length		430 mm

23. REMOTE SENSING EMISSION LEVELS

While it is mandatory to get the vehicle tested for 'emissions under control' certificate periodically, yet this is no guarantee that even if the certificate was obtained after prescribed testing procedure, the emission control system in the vehicle would not become defective during the period of the certificate. The remote sensing of emission level on the road helps to detect the actual emission levels of the vehicle any time and have been found to be a valuable means to keep the vehicle emissions under effective check.

Fig. 20.33 shows a simplified layout of the on-road remote sensing emission measurement system. IR (Infra Red) beam is used for CO and HC, whereas UV (Ultra Violet) beam is meant for NO_x emissions. As

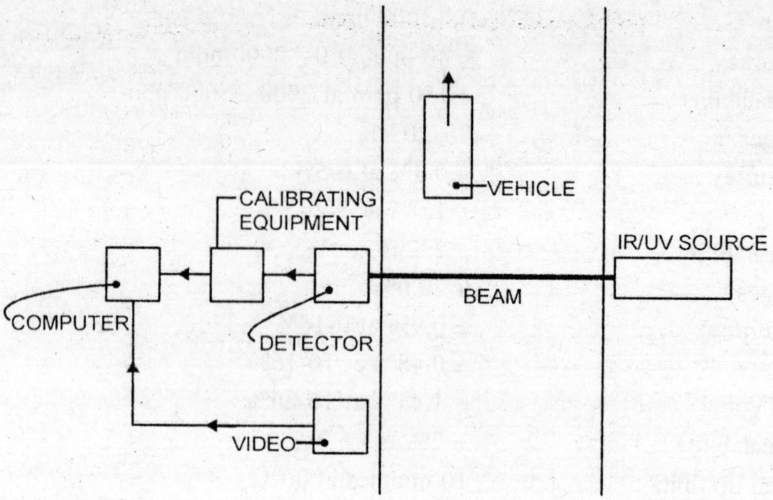

Fig. 20.33. Remote sensing system for emission measurement

the vehicle is driven through the specified spot, the beam is absorbed by CO, HC and NO_x and by CO_2, in proportion to the amount of these gases present. CO_2 serves to provide the calibration since its concentration in S.I. engine exhaust gases is fixed. The registration plate of the vehicle is recorded by the video camera.

24. FUTURE EMISSION STRATEGIES

Automotive manufacturers world-over seem to be ready for Euro 6 emission regulations to be enforced around 2020. However, perceptions differ about the strategies that would be adopted after that. Views of various industry analysts are quite varied and listed here:

1. Manufacturers will choose the right after-treatment system based on vehicle segments. Luxury sedans will use costly technologies like SCR, while some compact vehicles will employ NO_x traps or even variable valve train technology.
2. Efforts will be made to reduce transient emissions by means of faster cam-phasing response using electromechanical cam phasers and variable valve timing, either by switching or by using continuous systems such as UniAir.
3. The new NO_x limits will need a reduction in engine-out emissions alongwith exhaust gas after treatment. Engine-out NO_x can be reduced by optimizing the combustion system, (the combustion bowl, compression ratio, injection pressure and injector nozzles) as well as with advanced EGR. For instance, low-pressure EGR may be achieved by recirculating air at the turbocharger or compressor inlet and may be quite effective to cool down the air-EGR mixture for a better soot/NO_x trade off.
4. No further restriction is expected on PM and PN. Hence particle filter technology based on silicon carbide, most likely in a closed couple position, will become the mainstream in Europe. But for NO_x a further reduction to match the 60 mg/km gasoline engine value is possible. Also pollutants like NO_2 could be included in post Euro 6 emission limits, requiring the development of new and more complex NO_x after-treatment systems.
5. Another key driver post Euro 6 would be the compliance with the EC's 95 g/km CO2 limits under the CAFÉ agreement. As this also affects gasoline engines, advanced combustion technologies, friction-reduction, turbo charging, thermal management, downsizing and optimized after-treatment systems will all have to be considered.
6. Hybridization will continue to progress driven in part by increasingly high-performance engine

stop/start solutions. Micro-hybrids will become the standard by 2020 with 95% of European cars featuring some form of electric engine boost.

7. Harmonizing all regional standards into one global test cycle (the Worldwide harmonized Light Vehicles Test Procedure-WLTP) will further drive down emissions and fuel consumption across various drive trains, as they would be based on real-world drive conditions, rather than laboratory test cycles. However, for WLTP to become a reality, it would take quite a long time since there are many stake holders like governments and regional regulatory authorities, etc.

REFERENCES

1. AECC News Letter, April 2017.
2. 'Indian Emission Regulations', ARAI, January 2017.
3. 'What next after Euro 6?' Engine Technology International.com, June 2013, pp 52-6.
4. AECC News letter, Nov.-Dec. 2007.
5. Bruce & Bartalsen, "The U.S. Motor Vehicle Emission Control Programme", Platinum Metals Rev, Vol. 45, 2001, pp. 50-59.
6. 'Mazda Skyactive', Engine Technology International.com September 2011, pp 5-7.

THEORY QUESTIONS

1. How does an automobile contribute towards pollution of the atmosphere? Discuss various methods to reduce the emission of pollutants in exhaust gases.
2. Discuss the Federal and the CARB emission standards in U.S.A.
3. What is closed crankcase ventilation? Discuss in detail the two types of crankcase ventilation systems used in automobiles.
4. Discuss in detail the PCV system of emission control, describing the construction and working of the PCV valve under various conditions.
5. Explain clearly how the atmospheric pollution from the carburettor and the fuel tank is prevented.
6. Describe how proper design of combustion chamber helps to reduce exhaust emissions.
7. Explain clearly how the decrease in compression ratio of a diesel engine is used to reduce emissions, without decrease of engine efficiency.
8. What is Sonex system for emission control? Describe briefly the principle of its working?
9. Discuss in detail various changes in the fuel supply system of engines for controlling emissions.
10. Describe clearly the function and working details of the following:

(*i*) Jet valve

(*ii*) Coasting air valve

(*iii*) Air switching valve

11. How does 'Deceleration Spark Advance System' help to control engine emission? Describe its working in detail.
12. Describe clearly how you treat the engine exhaust gases to convert the pollutions therein into harmless constituents.
13. Discuss in detail the following techniques of reducing the pollutions in the exhaust gases:
    (*i*) Air injection system.
    (*ii*) Pulse air injection reactor system.
14. What is a 'Three-way converter'? Explain clearly how it is used to convert the pollutants in the engine exhaust into harmless constituents. Describe also why it is desirable to have computer feedback control of air-fuel ratio while using catalytic converter.

15. Explain the function of a lambda probe. Describe the working of any type of this probe with the help of a simple sketch.
16. Explain how Carona Discharge Device is used in an engine. Describe also its various advantages.
17. Discuss in detail various methods for exhaust treatment of automotive diesel engines.
18. Describe clearly how NO_x emission is diesel engines is controlled by an LNT.
19. What is Diesel Particulate-NO_x reduction system of Toyota? Explain how it works. Discuss also its utility and limitations.
20. Explain the function of an adsorber in the automotive emission control. Describe different types of adsorbers in use.
21. Discuss in detail the future scenario regarding diesel engines emission control.
22. Compare CNG with conventional automotive fuels and describe various components of the kit for converting a petrol/diesel vehicle into a bifuel vehicle with CNG as one of the fuels, the other fuel being petrol or diesel.
23. Describe clearly any method of remote sensing emission level of a moving vehicle.
24. Write a comprehensive note on future emission strategies for automobiles.

QUESTIONS FOR VIVA

1. Name the areas of automobile emitting pollutants.
2. Name the main pollutants in the engine exhaust.
3. Name various diseases caused by engine exhaust.
4. State the safe level of pollutants for cars according to (*i*) American Federal standard, (*ii*) India 2000 norms, (*iii*) Euro stage IV norms.
5. Name the current emission standard being used in (*i*) USA, (*ii*) Europe, (*iii*) India
6. What do you understand from the abbreviations (*i*) LEV, (*ii*) ULEV, (*iii*) SULEV, (*iv*) ZEV, (*v*) NMOG?
7. What is the maximum limit of CO_2 in the engine exhaust as per the current European standard?
8. When shall Bharat Stage VI emission norms come into effect in India?
9. What are the two basic approaches to decrease pollution from automobiles?
10. What is PCV? What purpose does it serve?
11. State the forces that control the PCV valve.
12. What is the function of PCV valve?
13. What happens when at higher speeds the crankcase emissions exceed the flow rating of the PCV valve?
14. How does PCV valve protect crankcase from engine backfiring?
15. What does a coating of oil on the air cleaner filter of an engine with PCV indicate?
16. What do you understand from the 'breathing' of the fuel tank? How does it cause pollution?
17. What is a vapour recovery system?
18. How is the flow of vapours from fuel tank to the carbon canister controlled?
19. What is the function of a vapour-liquid separator?
20. What is the function of a carbon canister in the vapour recovery system?
21. State the function of a purge control valve.
22. How is the purge control valve operated?
23. What is the advantage of reducing compression ratio of the engine?
24. How is the loss of combustion efficiency due to lower compression ratio compensated?
25. How does the shape of combustion chamber effect the combustion efficiency?
26. What is the 'stratified charge' system?
27. What is the effect of increasing valve overlap on the engine emission?
28. If the opening temperature for the thermostat valve in the engine cooling system is raised, how does it effect the pollution?

29. What is the effect of heating the supply air on the engine emission?
30. How does an electric-assist type of choke help decrease the emission of pollutants?
31. How does the fuel-air ratio affect he exhaust emission at idle?
32. What does the abbreviation 'EGR' stand for?
33. What is the function of EGR?
34. Why should the EGR valve remain closed at idle and at full throttle?
35. What is the purpose of installing thermal control valve in the EGR system?
36. What is a 'jet valve'? Which pollutants does it control?
37. State the function of a 'coasting air valve' (CAV).
38. How is 'air switching valve' (ASV) operated?
39. How does the fuel injection help to reduce automobile pollution?
40. What is the effect of ignition timing on exhaust emission at idle?
41. What is the effect of ignition of a 'declaration spark advance system' (DSAS)?
42. Which pollutants in the engine exhaust are controlled by the 'air injection system?
43. What is a pulse air injection reactor (PAIR) system?
44. What is the advantage of PAIR system over the air injection system?
45. Why is PAIR system usually employed on engines with catalytic converters?
46. What is a catalytic converter?
47. What is a catalyst?
48. Define 'light-off temperature' of a catalytic converter.
49. What are the advantage and the drawbacks of the catalytic converter method compared to blowing of air only into the exhaust manifold?
50. What is three-way converter (TWC)?
51. Which catalyst is used for reduction of NO_x?
52. Which catalyst is employed for oxidising HC and CO?
53. Why should unleaded petrol be used for engines employing catalytic converters?
54. At what air-fuel ratio does the three-way converter operate at maximum efficiency? How is this ratio achieved?
55. What does abbreviation FLAD stand for?
56. What is the advantage claimed for Toyota's new catalytic converter based on FLAD?
57. What air-fuel ratio gives best fuel economy?
58. What is lambda?
59. What does the amount of oxygen in the exhaust indicate?
60. At what value of λ, is the lambda probe most sensitive?
61. Name the electrolyte used in a lambda probe.
62. What is 'catalyst operating window'?
63. Why does the three-way catalytic converter not work in case of diesel engines?
64. Which two methods can be used for self-cleaning of particulate trap in diesel engines?
65. What is the effect of sulphur in petrol or lubricating oil on the performance of Catalytic Converter?
66. Why should sulphur-containing diesel not be used in conventional particulate filters?
67. What is the function of 'Electrostat Diesel Particulate Filter'?
68. Name various technologies for controlling PM in diesel engine exhaust.
69. Name various technologies for controlling NO_2 in the exhaust of diesel engine.
70. What do you understand from the following abbreviations?
    (*i*) PM, (*ii*) DOC, (*iii*) DPF,
    (*iv*) LNC, (*v*) LNT, (*vi*) SCR.
71. How are future emission regulations for automotive diesel engines proposed to be tightened?

72. How is CNG better than diesel from pollution view point?
73. Name various components of a CNG conversion kit.
74. Name the main technologies likely to be used in future for improving emissions in automotive diesel engines?

MULTIPLE CHOICE QUESTIONS

Mark the correct answers:

1. Apart from hydrocarbons, the main pollutants in the engine exhaust are
(*a*) CO and CO_2 (*b*) CO_2 and NO_x
(*c*) CO and NO_x (*d*) CO_2 and H_2O
2. Maximum allowable hydrocarbons in the car emission are approximately
(*a*) 10 ppm (*b*) 100 ppm
(*c*) 1000 ppm (*d*) 5000 ppm
3. CO in the exhaust of a car engine should not be more than
(*a*) 0.5% (*b*) 2%
(*c*) 5% (*d*) 10%
4. The PCV valve is located between the
(*a*) air cleaner and the carburettor (*b*) carburettor and the intake manifold
(*c*) intake manifold and air cleaner (*d*) intake manifold and crankcase
5. In case of engines backfiring, the PCV valve
(*a*) opens maximum (*b*) closes
(*c*) disintegrates (*d*) sticks in the centre position
6. The position of the PCV valve plunger at idle speed is
(*a*) towards the valve outlet side (*b*) towards the valve inlet side
(*c*) towards the crankcase (*d*) near the valve centre
7. The function of charcoal granules is to absorb
(*a*) water vapour (*b*) carbon monoxide
(*c*) petrol vapour (*d*) liquid petrol
8. The liquid-vapour separator is located near the
(*a*) fuel tank (*b*) carburettor
(*c*) charcoal canister (*d*) purge valve
9. EGR system is employed for controlling emissions of
(*a*) HC (*b*) CO
(*c*) NOx (*d*) HC and CO
10. The pump for the air injection system is of
(*a*) gear type (*b*) eccentric type
(*c*) vane type (*d*) reciprocating type
11. The main purpose of the diverter valve in the air injection system is to
(*a*) close the check valve
(*b*) prevent backfire in the exhaust system
(*c*) prevent backfire in the inlet manifold
(*d*) weaken the air-fuel mixture
12. The catalyst used in the reduction converter is
(*a*) copper (*b*) charcoal
(*c*) rodium (*d*) platinum

13. The catalyst used in the converter for oxidising HC and CO is
    (*a*) copper (*b*) charcoal
    (*c*) rodium (*d*) platinum
14. In a three-way converter, the first converter controls
    (*a*) HC (*b*) CO
    (*c*) NO_x (*d*) all of these
15. The air fuel ratio required for the efficient operation of a three-way converter is approximately
    (*a*) 5:1 (*b*) 10:1
    (*c*) 15:1 (*d*) 20:1

[**Key:** 1 (*c*), 2. (*b*), 3. (*a*), 4. (*d*), 5. (*b*), 6. (*a*), 7. (*c*), 8. (*a*), 9. (*c*), 10. (*c*), 11. (*b*), 12. (*c*), 13. (*d*), 14. (*c*), 15. (*c*)]

21 Miscellaneous Topics

1. TECHNOLOGIES TO IMPROVE ENGINE FUEL ECONOMY

1. Lightweight materials

Reducing vehicle weight is an effective way to improve efficiency. However, it must be done consistent with passenger safety. Newer vehicles are making extensive use of advanced materials, *e.g.,* composite body panels, high-strength lightweight aluminium structural components, etc. Even conventional materials can improve safety while reducing weight if more sophisticated structural designs are made use of. More details can be seen in Vol. 1 on chapter 18, 'Automotive Materials'.

2. Reduced resistance

Drag and the friction between road and wheels are the two main resistances to the vehicle motion, which if reduced, improve the efficiency. Improved aerodynamic design of the vehicle body alongwith slippery body panels can decrease drag. Improving the profile of components such as side-view mirrors, radio antenna, etc. can further reduce wind resistance. Rolling friction can be reduced by using low-resistance tyres.

3. Regenerative braking

This causes some of the vehicle's kinetic energy to be recaptured as electricity, when the brakes are applied. During braking, the motor acts as a generator taking kinetic energy from the wheels and converting the same to electrical energy, which is then used to charge the batteries.

4. Variable valve timing and lift

Valve opening and closing can be fine-tuned with the help of computer-control, which would improve the engine efficiency and hence result in fuel economy. 5% average efficiency increase is possible with this.

5. Direct fuel injection

In this system, also called 'fuel stratified injection', fuel is injected directly into the cylinder so that the timing and shape of the fuel mist can be controlled accurately, due to which higher compression ratios may be used, which results in higher performance and reduced fuel consumption. When used with supercharger/ turbocharger, this gives average efficiency increase of 11-13 percent.

6. Cylinder deactivation

When less power is required by the vehicle, one or more engine cylinders can be deactivated, which can then be reactivated whenever the power requirement is increased, thus producing only optimum power. Thus considerable saving in fuel consumption is achieved. This method is especially useful in case of heavy trucks, where eight- or ten-cylinder engine may be used for hauling a trailer carrying large pay loads, but the maximum engine power is not required when only lighter loads are being carried.

Cylinder deactivation technology also helps to maximize engine efficiencies for S.I. engines by minimizing operation in the low-to-medium load range, where efficiencies are lower due to lower in-cylinder temperatures and lower volumetric efficiencies. As on today (2016), cylinder deactivation for diesel engines is still at the investigation stage. A detailed description of cylinder deactivation is given in Art. 6.

7. Integrated starter-generator

The use of integrated starter-generator in conventional vehicles reduces the fuel consumption and improves the acceleration. It is found that this may improve fuel economy of conventional vehicles by about 20 percent. However, due to heavy electrical power requirement for the integrated starter-generator, the same is being developed preferably with 42 volt systems.

8. Engine downsizing

Some of the more commonly available technologies which support engine downsizing include pressure boosting using a turbocharger or supercharger, hybrid technologies to supplement the engine, *e.g.*, batteries, electric motor or hydraulics and alternative sources of energy. A downsized engine results in lower fuel consumption and emissions.

The loss of power output in case the engine size is reduced (which means decreased volume of intake air) can be compensated by turbocharger (or supercharger). However, in a petrol engine, higher intake pressure would raise the combustion temperature, resulting in higher NO_x emissions and even detonation. This can be taken care of in modern direct injection petrol engines wherein the fuel is pumped right inside the cylinder. Thus cooling due to evaporation of petrol inside compensates to a large extent, the temperature rise due to use of turbocharger. Therefore, it is seen that the decisive factors for downsizing are the direct injection and the turbocharger. However, these have to be accompanied by:

1. Effective and efficient control such as the Bosch DI-Motronic system for petrol engines.
2. Variable valve timing of inlet and exhaust valves, including scavenging control through valve lap. Scavenging control ensures more fresh air supply to the cylinders which boosts torque and provides the desired performance level even in the downsized engine. Bosch expects to achieve the performance level of its current 2.0 litre 100 kW engine from only 1.1 litre engine without any reduction of power or torque, but with 29% reduction in fuel consumption by 2015.

Volkswagen has been able to replace their 3.2 litre, 184 kW, V6 petrol engine by downsizing to 2.0 litre TSI (Turbocharged Spark Ignition) engine with a higher output (199 kW) engine which also gives slightly higher maximum torque (350 Nm) across a broader range 2500 to 5000 rpm against 2800 to 3000 rpm for V6). There is also improvement in performance, lesser fuel consumption and reduced CO_2 emission.

9. Lubricants

Effective lubrication is essential for friction reduction at normal and extreme operating conditions and temperatures. Specially blended lubricating oils are being developed to optimize the performance and efficiency by using temperature- and pressure-sensitive nanotech additives which can adjust their performance depending on the changing operational loads.

10. Mahle Jet Ignition

Mahle's Jet Ignition technology, developed for Motorsport cars and under development for series-production vehicles, is a new technology for gasoline (petrol) engines which uses a jet-ignition chamber assembly, replacing the standard spark plug for combustion of the air/fuel mixture in the cylinders. The technology is capable of producing very high fuel efficiency as in diesel engines, through ultra lean-burn operation, raising the knock limit and permitting the use of higher compression ratios. Even engine out

NO_x and particulates are significantly reduced, with the hydrocarbon emission remaining maintained at the levels of standard spark ignition operation.

Mahle Jet Ignition employs a pre-chamber along with the main chamber, the two connected by a number of orifices to allow jets of partially burnt gases to ignite the main chamber. The pre-chamber has a spark plug and a secondary fuel injector, receiving only 3% of the total system fuel injected. Injection occurs at about 60° BTDC and the spark reaches the pre-chamber at about 20° BTDC. The main chamber gets fuel through a conventional port or direct injection injector. Depending on the engine application, 4 to 8 plasma jets shoot through the orifices between the pre-and the main chambers. Due to small orifice size, turbulence is produced in the hot gas jets which then penetrate deeper into the main combustion chamber and produce an evenly distributed ignition effect, causing extensive ignition of the main charge, generating a faster combustion and pressure build-up. The mixture in the main chamber is ignited between 12° and 5° BTDC.

Test results on a 2.4 litre production engine showed brake specific fuel consumption below 200 g/kWh with 41% peak brake thermal efficiency and significant related reduction in CO_2 emissions. The company also claims reductions in engine-out NO_x emissions by more than 99% in the ultra-lean area.

11. Divided Exhaust Boosting

DEB is a concept which uses bifurcated exhaust ports in the cylinder head, an additional integrated exhaust manifold to split the exhaust stream from the engine into two routes– to the turbocharger and to the catalytic converter. When the valve, shaped like the EGR valve, opens, all of the hot, blow-down gas goes directly to the turbocharger, while all the remaining exhaust gas which the piston has to push out, is directed by another valve to go straight to the catalytic converter, bypassing the turbocharger.

Due to the exhaust gas going directly to the catalytic converter, it lights off much faster - much faster than in case of a naturally-aspirated engine. Apart from the faster cat light-off, other benefits include pumping benefit, about 25% more low- end torque, upto 4% higher fuel economy, use of a smaller turbine and an 18° shift in the knock limit at 4000 rpm and wide-open throttle.

2. CURRENT USE OF MODERN TECHNOLOGIES IN VEHICLE ENGINES

Main modern technologies used to augment the performance and fuel saving in automotive vehicle engines are discussed briefly as follows:

1. Naturally aspirated engines

(*i*) *High compression ratios*– Mazda SkyActive engine achieves compression ratio of 13:1 in U.S. Compare this with 1.8 litre 2015 Toyota Corolla (Comp ratio = 10:1) and Formula one racing car (Comp ratio = 17:1). Combined with variable valve timing for a longer intake stroke and gasoline direct injection, the fuel economy in case of the SkyActive is improved by 10% - 15% depending on the driving conditions. Although fuel efficiency improvement from high-compression ratio engines is quite less compared to the hybrids and downsized turbocharged engines, but increase in manufacturing cost in case of high-compression engines is significantly less. For more comments on this, see next article.

(*ii*) *Atkinson cycle*– Although Toyota and Ford have been using Atkinson cycle (to increase the intake stroke and decrease the compression stroke) for more than 10 years, yet Toyota has now used this in case of their non-hybrid naturally aspirated ESTEC engine, employing variable valve timing and lift and achieving higher compression ratio. These engines have been used in Lexus, Camry and some other models with improved performance/fuel consumption.

(*iii*) *Cylinder deactivation*– Improvements in electronic engine controls has enabled us to use *dynamic* cylinder deactivation in engines with even three cylinders, wherein individual cylinders can deactivate every compression stroke automatically, to increase fuel efficiency, resulting in estimated fuel savings of 2.0% to 10.5%. GM's 'Dynamic Skip Fire' engine uses this technology to achieve an improvement of 14% in fuel consumption without any decrease in power output.

(*iv*) *Gasoline Direct Injection* (GDI)– Although by itself GDI gives only small (up to 3%) improvement in fuel consumption yet due to better fuel control and charge-cooling effect, it enables high compression ratio to be used. It is estimated that by 2022, almost 98% of new passenger vehicles would be employing GDI engines.

(*v*) *Start-Stop*– In real-world driving conditions, about 3% - 7% reduction in fuel consumption is achieved with this technology in case of naturally aspirated engines. However, on combined city-highway cycle, improvement in fuel consumption is only about 2%. Although start-stop is standard in all hybrid vehicles, yet in case of conventional I.C. engine vehicles, this system is being increasingly used, due mainly to its rapidly falling prices. GM's 2016 model 'Chevrolet Malibus' cars are employing the start-stop system achieving fuel economy of upto 15%.

2. Downsized, boosted petrol engines

(*i*) *Miller Cycle*– It is estimated that Miller cycle increases the geometric compression ratio from 10.0:1 to 12.0:1 and decreases fuel consumption by 3.9% to 5.7% as compared to a downsized, turbocharged engine with variable timing and lift. Volkswagen used the Miller cycle with a variable-geometry turbocharger and other cylinder improvements in their EA211 TSI evo engine family in 2016, claiming 10% improvement in efficiency. Audi also is going to employ Miller cycle engines in their 2017 model A4. Mazda is already using this cycle in their 2.5 litre turbocharged version of its 14 SkyActive - G, achieving a 23% decrease in fuel consumption compared to the naturally aspirated V6 engine which the SkyActive has replaced.

(*ii*) *E-boosting and 48-volt systems*– Prototypes have already been made for customer evaluation, by the main manufacturers of turbochargers, such as BorgWarner, Hitachi, Vales and Honeywell. Audi has already started production of e-boost system for their V8 diesel SQ7 engine with single-stage turbocharger instead of earlier 2-stage system, with the e-boost system consuming fuel only as a conventional V6 diesel. It is estimated that 48-V, e-boost systems, in general, achieve a minimum 60% efficiency of deeper hybridization/electrification at only 40% of cost.

(*iii*) *Variable Compression Ratios* (VCR)– This permits the use of higher compression ratio at light loads (when detonation is not a problem) and a low ratio at heavy loads and increased power. Nissan has already started production of such a system for one of their Infiniti 2017 models. Further, instead of a continuously variable compression ratio, two-step VCR system is being developed by FEV. As per FEV estimates such a two-step system would deliver more than 80% of the potential improvement in case of the fully variable system at a much lower cost than the later. FEV also estimated that 4.2% to 6.2% (on U.S. combined cycle) fuel-reduction would be achieved with the two-step VCR with a low-pressure EGR system.

3. FUTURE TRENDS FOR PETROL ENGINES

1. Higher compression ratios. At present 13:1 to 14:1 ratios for petrol engines are possible. The only limitation appears to be the higher manufacturing costs due to the necessity of using higher tolerances.

Mazda, is now producing next-gen engines 2.0 litre 'Skyactive-G' (Gasoline) and 2.2 litre 'Skyactive-D' (Diesel) using same (14:1) compression ratio. Various features that makes it possible to use such high ratios in Skyactive-G are:

(*i*) Direct injection.

(*ii*) Long, intricate 4-into-2-into 1 exhaust system, which is designed to minimize exhaust blow-back into the inhaling cylinder and reducing the percentage of residual exhaust gas from 8% in the previous MZR engine to 4% in Skyactive-G. However this new exhaust system is much costlier, since it has to be stamped and fabricated due to its complex section. Besides, it needs more space under the vehicle body, to package it.

(*iii*) A small bowl-shaped cavity in the piston crown which ensures fast and smooth flame propagation.

(*iv*) Skyactive-G runs on Atkinson cycle during idling and under low loads. Higher compression/expansion ratios, without the detonation occurring, are made possible by using late intake valve closing, for which Mazda has employed electrically actuated variable valve timing (EVVT) and camshaft phasing on both inlet and exhaust. This dual EVVT provides higher phasing velocities and accuracy with greater control flexibility than hydraulic VVT systems.

2. Lean-burn combustion. The probability of boosting the petrol engines with EGR makes it possible to achieve the brake thermal efficiencies of HCCI, which makes it less attractive for the Auto Manufactures to go for all-new HCCI engine.

3. Turbocharging. This is bound to become progressively popular for petrol engines. With the downsizing of these engines, boost pressure will increase and it is expected that these would be branched into two categories, viz., the fuel economy type with 22-25 bar BMEP and the performance type with 30 to 32 bar BMEP.

4. Cooled EGR. EGR use became redundant with the use of variable valve timing technologies. But the cooled EGR is coming again in downsized booted, engines. With this compression ratio can be increased, narrowing down the difference between the petrol and the diesel engines.

5. Camless valve trains. Renewed interest of Auto Manufacturers have been noticed in these and we may have practical engines using this by about 2020.

4. FUTURE TRENDS FOR COMMERCIAL VEHICLE DIESEL ENGINES

1. High pressure fuel injection. High injection pressures can reduce exhaust emissions, in particular the particulate emission. It is expected that in commercial vehicle diesel engines, the pump-nozzle and pump-line-nozzle systems would be replaced by high pressure (upto 2500 bar) common rail systems by 2015.

2. Supercharging. Downsizing trend is also there in commercial engines. To do this for large engines would mean increase in power output, per litre of displacement to achieve which more supercharging and higher EGR rates would be needed, which would mean necessity for higher injection pressures and consequently more robust injection system.

However, in smaller commercial vehicle engines, if the supercharging boost is increased and the valve size is reduced, losses due to EGR increase, which would result in increase of fuel consumption. This means that in such cases, downsizing, if done, would be for some reason other than the reduction in fuel consumption.

3. Treatment of exhaust gases. Majority of truck manufacturers like Mercedes Benz and Daf, have opted to use SCR method which reduces the pollutants as per emission regulations without reducing the output and reducing the fuel consumption at the same time. However, some manufacturers, such as MAN and Scania do use EGR jointly with diesel particulate filter (DPF), which does not require any additional reagent like Adblue in SCR. The later method uses high values (up to 40% of fuel load) of *cooled* EGR. Higher rate and cooling of EGR decreases the temperature of the mixture of fresh air and exhaust gas in the combustion chamber, consequently producing less NO_2 during combustion. However, this requires:

(*i*) Higher charging pressure, for which two-stage supercharging with inter cooling is necessary.

(*ii*) Sturdier mechanical design due to (*i*) above.

(*iii*) Higher cooling capacity with two-stage cooling for the supercharging as well as EGR cooling.

(*iv*) High-pressure common rail injections with more than 2500 bar pressure have to be used.

4. Use of hybrid systems. Hybridisation is feasible in engines of all types of commercial vehicles, except for the fact that so far any battery to give long range suitable for these type of vehicles and also with a long service life and reasonable price is not available.

The direction of current development of hybrid system for commercial vehicles is towards parallel hybrids, which is a cost-effective solution and can also be integrated into existing manual or automatic transmissions.

5. INFINITY VARIABLE COMPRESSION RATIO ENGINE

Infiniti (Hongkong) presented at the 2016 Paris Motor Show, the world's first production-ready VC-T (Variable Compression-Turbocharged) engine. Depending on the driving situation demands, the engine control logic automatically applies the optimum compression ratio seamlessly from 8:1 (for high performance) to 14:1 (for high efficiency).

The technology uses an advanced multi-link system to seamlessly raise or lower the reach of the pistons, instantly selecting the most suitable compression ratio.

The Company claims significantly reduced fuel consumption and improved emissions, besides lower levels of noise and vibration. The infinity VC-T is a 2.0 litre 4-cylinder turbocharged petrol engine.

6. CYLINDER DEACTIVATION

6.1. Necessity

To satisfy demands for quicker acceleration and/or heavy towing capacity, high-powered multi-cylinder engines are necessary. But it is seen that in a typical light load driving, only about 30 percent of the engine's maximum power is used. In this condition, the throttle valve is almost closed due to which the engine has to do more work to draw in air. This results in higher pumping loss. Moreover, under part-throttle conditions, the cylinder pressure also decreases. Mercedes has found that some large capacity engines have to be throttled so much at light loads that the cylinder pressure at TDC is almost half that of a small 4-cyinder engine. Obviously, low cylinder pressure results in lower thermal efficiency.

Thus, if at lighter loads, some cylinders are stopped from working so that the remaining fewer cylinders would now have to operate at higher load, which means throttle valve can be opened further to produce the same power. This results in reduced pumping loss and increased cylinder pressure, leading to higher fuel efficiency. It has been found that in highway conditions, fuel consumption reduction of upto 20 percent, is possible to achieve with this method.

6.2. Principle of working

In simple words, cylinder deactivation is done by keeping the intake and the exhaust valves closed through all cycles for a particular set of cylinders in the engine.

To deactivate a cylinder, the exhaust valve is prevented from opening after the power stroke and the exhaust gas charge is retained in the cylinder, which is compressed during the exhaust stroke. After the exhaust stroke, the intake valve is prevented from opening, which causes the exhaust gas in the cylinder to be expanded and compressed gain and again, acting like a gas spring in the combustion chamber. The work required for compression of the gas in one cylinder is compensated by the work done by the expansion of the gas in another cylinder. That is why even number of cylinders is deactivated at a time.

Depending upon the engine design, valve actuation is controlled by any of the following two methods:

1. *For push rod designs*– For cylinder deactivation, the hydraulic tappets are collapsed by using solenoids to change the oil pressure delivered to the tappets. When collapsed, tappets cannot elevate their companion push rods under the rocker arms, so that the valves cannot be actuated and thus remain closed.

2. *For overhead cam designs*– In such engines, a pair of locked-together rocker arms is used for each valve, out of which one rocker follows the cam profile while the other one actuates the valve. For deactivation, solenoid-controlled oil pressure releases a locking pin between the two rocker arms. The result is that one arm continues to follow the cam profile, while the other (unlocked) arm, being disconnected, remains motionless and thus is unable to activate the valve.

In the latest cylinder deactivation systems, the engine management system also cuts off fuel supply to the disabled cylinders, while the transition between the normal operation and deactivation is smoothened by means of changing ignition timing, cam timing and throttle position, controlled electronically.

Mostly, cylinder deactivation is applied to large displacement engines, which are particularly inefficient at light loads. For example, in case of a V-12, upto 6 cylinders and in case of a V-8, upto 4 cylinders can be disabled.

6.3. Typical examples

General Motors was the first to apply cylinder deactivation to their V-8 engine for Cadillac, in 1981. However, due to their throttle-body fuel injection system, which continued to supply fuel to the deactivated cylinders, GM had to discontinue deactivation after one year only. In 2004, GM re-introduced the deactivation under the name 'Displacement On Demand (DOD)' in their 'Envoy' and 'Trail Blazer' SUVs, which employ a V-8 engine. The engine runs on all 8 cylinders during start-up, idle and heavy acceleration, but switches to 4-cylinder mode on light loads.

After 'Cadillac', Daimler Chrysler was the first company to have cylinder deactivation on mass-produced cars in 1999 model Mercedes CL 600, S600 and CL500, which were powered by 6.0 litre V-12 or 5.0 litre V-8. Their system was named as 'Active Cylinder Control (ACC). Mercedes claim improved fuel economy of about 7 percent in city driving and upto 20 percent at a steady cruise.

In 2004, Daimler Chrysler employed the deactivation in 5.7 litre HEMI V-8 engine in their Chrysler 300C, Dodge Magnum, Dodge Charger, Jeep Grand Cherokee, Dodge Durango, Dodge Ram and Jeep Commander. This system operates in completely different manner compared to their Mercedes system and is named as 'Multi-Displacement System (MDS)'. In this, 4 of 8 cylinders are disabled at road speeds above 30km/h and engine speed less than 3000 rpm. It uses a special set of lifters called 'lost motion devices', which are operated hydraulically (controlled by ECU-operated solenoids). For cylinder deactivation, oil under pressure pushes a locking pin which causes the lifter to collapse, disengaging the camshaft from the push rod and the valve, resulting in cylinder deactivation. MDS can switch between 8 and 4 cylinder operation in just 40 milliseconds. Fuel is also cut off to the deactivated cylinders and electronically controlled throttle is used to maintain engine power during transition.

Honda employed the deactivation under their brand name 'Variable Cylinder Management' (VCM) on their 2005 model 3.5 litre 'Odyssey' and 3.0 litre 'Accord Hybrid'. Their system is similar to the Daimler Chrysler Hemi MDS.

Volkswagen's 'Cylinder On Demand (COD)' is employed on 2012 Audi and Bentley models, with new 4.0 litre turbocharged direct fuel injection V-8. The Company also is the world's first in applying cylinder deactivation in a 4-cylinder turbocharged, direct-injected, volume-production engine (1.4 TSI with ZAS)' in 2012. In this engine, two cylinders are temporarily deactivated at low and medium loads resulting in fuel saving of 0.4 litre per 100 km.

The Audi COD system is triggered at low loads, low to moderate engine speeds and when the transmission is in the third gear or higher. The upper limit for COD initiation is from 25% to 40% of maximum torque. It closes the inlet and exhaust valves of cylinders 2, 3, 5 and 8, by means of sliding sleeves on the camshafts. Immediately before the valves are closed, the combustion chambers fill with fresh air, while the fuel injection and ignition cease. COD is disabled whenever the control logic monitoring the vehicle's accelerator, brake pedal and/or steering input detects an irregular pattern of operation, such as on bad, rural roads. The driver can also deactivate COD manually in 0.3 second by firmly pushing the accelerator pedal.

It case of Volkswagen's 1.4 litre 4-cylinder engine, cylinders 2 and 3 are deactivated. In this COD operates from 1400 to 4000 rpm and torque range of 25 to 75 Nm.

6.4. Cylinder deactivation for diesel engines

It is being investigated jointly by Eaton, Cummins Engine and Purdue University's Herrick Laboratory. The research work is being done on a Cummins 6-cylinder test rig at Purdue, equipped with a variable geometry turbocharger and cooled EGR. The test rig also has camless VVA. The positive findings so far are:

1. Deactivation increases the rate at which particulate filter heats up.
2. Deactivating two of the six cylinders, by shutting off valve motion and fuel injection, raises the engine outlet temperatures to 520°C (maximum) due to the reduced air/fuel ratios.
3. Deactivation of three out of six cylinders at loaded idle causes a rise of 90°-310°C at the turbine outlet. At 310°C, a 39% reduction in fuel consumption was achieved compared to the six-cylinder operation.

The research work on transient operation and control-algorithm development for seamless transitioning groups of cylinders in and out of activation, is still continueing.

7. 48-VOLT ELECTRICAL SYSTEM

7.1. Introduction

Dual-voltage 48/12 volt electrical system has come into prominence due to the decision of the European Union regulators to replace the current NEDC test protocol with the new WLTP (Worldwide harmonized light vehicle Test Procedure) in 2017, which would be much near the real world driving conditions and ever-expanding electronic features in the new vehicle as demanded by customers. More than a decade earlier, there was also a talk about pure 42-volt electrical system, but that could not be employed mainly due to the high cost of the full high-volt system. The duel 48/12-volt system would deliver performance and functionality of the high-voltage system at a fraction of the cost in case of the later. The system is so designed that higher voltage is applied only to the subsystems or components which can benefit more from the higher available current, still retaining the most of the 12-V system for which the reliable and inexpensive lead-acid battery would suffice. However, for the higher voltage, lithium-ion would be the preferred medium for energy storage.

7.2. Advantages and disadvantages

1. It is able to stop and start the engine in mild hybrids even more frequently than 12-V start-stop systems. With this it is possible to stop the engine while the vehicle is still in motion and to restart it seamlessly if the driver slows down but does not stop completely and then presses the accelerator pedal.

2. A stronger electrical motor can provide some torque-assistance to the IC engine of the hybrid in case of more acceleration required.

3. An electrical supercharger powered by the 48-V system not only eliminates completely the lag in case of turbochargers but is also able to enable the booster to provide extra performance right at the instant the IC engine restarts.

4. Four times voltage means four times power. The additional power enables to install additional accessories, especially the ones requiring high torque, such as ABS pumps, stability control systems with high-load actuators and adjustable stabilizer bars.

5. 48-V systems operate below the safe vicinity of the threshold of the danger of mortal electrocution,

providing the maximum safe power possible, also simultaneously enabling the use of lighter, smaller-gauge wiring which contributes to both the mass as well as cost reduction.

6. The system can be installed easily in all types of vehicles- petrol or diesel, with manual or automatic transmission.

The only disadvantage is that space has to be found in the engine compartment, for the 48-V battery, which can be different for different vehicles and models.

7.3. Applications

1. Audi announced the use of 48-V system for their SQT sport SUV, 2017 model, for electrical boosting and active roll-control chassis system.

2. Robert Bosch is to launch the mild-hybrid module in next 2-3 system.

3. Continental will start production soon, of a 48-V system.

4. Valeo's 48-V package with engine e-boosting is under development.

8. SCUDERI ENGINE

Scuderi Group is based in West Springfield, Mass., USA. In the Scuderi Engine (Fig. 21.1), the four strokes of the thermodynamic cycle are divided between two cylinders, viz., the intake/compression cylinder and the power/exhaust cylinder, which are interconnected by a crossover Passage. By separating the compression and the power strokes in two different cylinders, Scuderi has devised a means to reduce the negative compression work and increase the positive power work by simply decreasing the size of the compression cylinder and increasing the size of the power cylinder. A compressed air tank is also added to the engine to make it an 'air hybrid' in the Scuderi terminology. This compressed air tank serves to store energy when the car decelerates, by the air flowing under pressure from the engine to the air tank. During acceleration, the compressed air from the air tank is released to power the engine. A turbo charger is also added without increasing the cylinder size. There are five modes of engine operation:

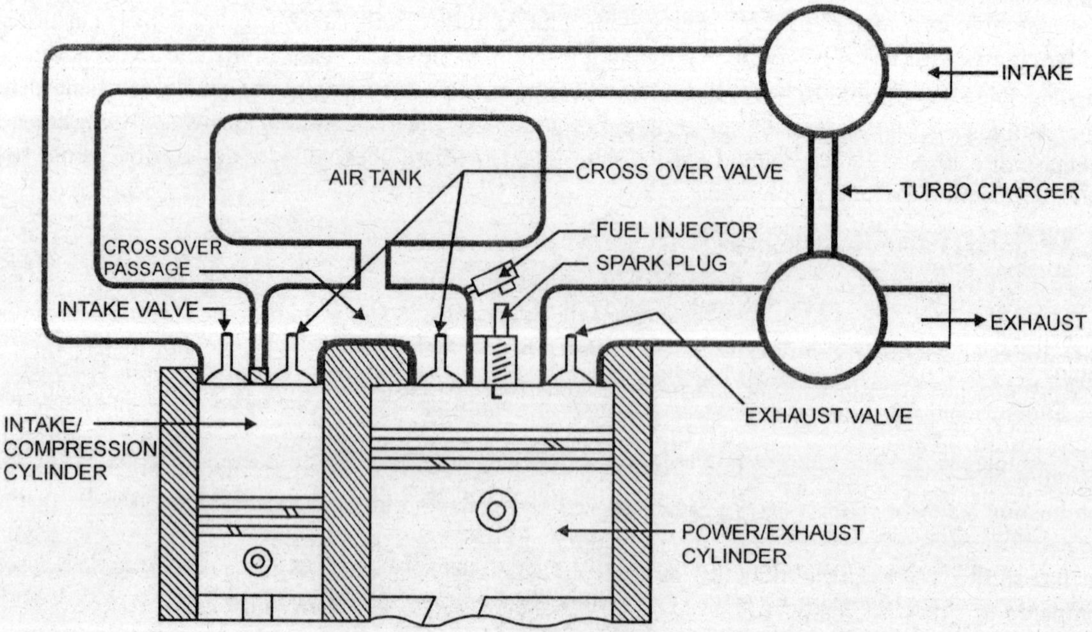

Fig. 21.1. Scuderi Engine

1. *Normal engine firing mode,* wherein both the cylinders are enabled and high pressure air from the cylinder is used for firing. In this mode air does not flow either from or to the air tank.
2. *Air compressor mode,* in which power cylinder is disabled and the compression cylinder recharges the air storage tank, without fuel injection or firing.
3. *Air expander mode,* in which compression cylinder is disabled and the high pressure air from the air tank goes to the power cylinder to produce power, without fuel injection or firing.
4. *Air expander and charging mode,* wherein the compression cylinder is disabled and high pressure air from the tank flows to the power cylinder, the fuel injected and the mixture fired to produce power.
5. *Firing and charging mode,* in which both the compression as well as the power cylinders are working and the air-fuel mixture is fired in the power cylinder, producing power and also recharging the air storage tank.

The advantages claimed are:

1. The company expects to improve fuel efficiency in the present petrol and diesel engines by 25 to 50% and reduce NO_x emissions by 80% while offering greater torque and power than traditional engines. Lower NO_x emissions occur, because higher cylinder pressures allow the engine to run at lower *peak* tem-peratures. The average temperature is higher, but the peak is lower, due to which the chemical reaction to create NO_x become sluggish. The engine also recaptures some of the post-compression air pressure and feeds it back into the compression cylinder to pre-compress the intake air, which results in improved efficiency.

Scuderi has so far been able to assemble and run only a 'proof-of-concept' engine in 2009. In May 2011, the company announced results from a simulation testing with a Nissan Sentra sedan and reported 36 percent reduction in fuel consumption.

2. It is also claimed to be cost-effective due to minimal manufacturing and retooling costs, by using similar or common components.

9. TRANSONIC COMBUSTION TECHNOLOGY

The Transonic Combustion injection-ignition (TSCi) system involves direct injection of fuel as a supercritical fluid, later in the compression stroke. The principle of this process is that the delay in the injection of fuel may be controlled such that the heat release mainly takes place after TDC of the engine power stroke. For this the combustion process must have a short delay period, followed by rapid fuel-air mixing and combustion, which are achieved by injecting the fuel as a supercritical fluid.

A *supercritical fluid* is any substance at a temperature and pressure above its critical point. Such fluids have properties between those of a gas and a liquid and have no surface tension, have the ability to solvate other liquids and solids and form small particles with a narrow size distribution during phase change to liquid.

Supercritical injection of fuel promotes rapid mixing with the air in the cylinder, which, after a short delay, results in spontaneous ignition at multiple locations. Rapid combustion at multiple sites results in optimum heat release and high cycle efficiency, alongwith elimination of droplet burning and increased combustion stability. All these result in improved fuel consumption and decreased emissions. Testing of the petrol engine with a diesel engine architecture using TSCi system, employed on a production passenger car has shown fuel consumption reduction of 5 to 21% under a number of speed and load conditions, compared to a production, port fuel injection, SI engine.

TSCi fuel injection system consists of special fuel injectors, fuel heaters, an electronic control unit and a high-efficiency fuel pump. Like diesel and HCCI engines, Transonic engine also does not require a spark plug. At present, the system is being developed for petrol engines, but is also claimed to be compatible with low carbon bio-fuels in the future.

10. OBD-II AND CHECK ENGINE LIGHT

OBD-II stands for Second Generation On-Board Diagnostic System which has been made mandatory on all cars and light trucks in USA since 1996. It monitors most engine functions while the vehicle is running. This is achieved with the help of a PCM (Power train Control Module) and various sensors. It automatically detects almost any problem which can cause emissions to exceed to Federal limit by 1.5 times. For detailed discussion on OBD, refer ''Automobile Engineering Vol. I, 14th Edition, 2018'' by this Author.

The 'Check Engine Light', also sometimes called, the 'Malfunction Indicator Lamp' (MIL) alerts the driver when any problem regarding emissions or sensors occurs. When the light comes ON, a problem is indicated; the mechanic plugs a code reader or scan tool into the vehicle's diagnostic connector and reads out the code which indicates the cause of the problem which can then be rectified.

However, in practice, the procedure is not so simple. After getting the code, someone still has to check out the various components in the affected circuit to determine the exact cause of the problem and for this we need a lengthy diagnostic chart to isolate the fault. For example, OBD-II code P0 130 is for the oxygen sensor circuit, which might indicate a bad sensor or a loose connector or some wiring problem, which necessitates further detailed investigation of each possible cause indicated. Similarly, if the OBD generates a misfire code for a single cylinder, say P0 301 for number 1 cylinder, it only tells you that cylinder no. 1 is misfiring, *but not why,* which could have been caused by a defective spark plug, a bad plug cable, a weak coil in the ignition system, a dirty or dead fuel injector or a compression problem (which may mean a leaking head gasket, a bad valve, rounded cam lobe, etc.). Obviously, there are numerous possibilities; so it would need a lot of diagnostic expertise to isolate the exact fault before any parts can be replaced.

Diagnosis is further complicated by the fact that all OBD-II codes are not standardized and uniform for all vehicles. There are two types of codes:

1. *Generic or basic codes,* which are used by all vehicle manufacturers to indicate the same type of problems. All these codes start with 'P0'.
2. *Special codes* which cover problems not included in the basic code list for the engine management system and also include codes for problems beyond the engine, such as ABS codes, air bag codes, climate control codes, etc. These special codes are often vehicle-specific and may need special software or a dealer scan-tool to read. Such OEM special enhanced codes all start with 'P1'.

11. BOSCH WATER BOOST INJECTION TECHNOLOGY

The system propagated by Bosch is quite simple. Water is drawn from a small tank and a fine mist is sprayed directly into the intake part where it evaporates, thereby cooling the petrol-air mixture in the cylinder, resulting in lower combustion temperatures within the engine, allowing a higher compression ratio without any risk of detonation occurring. As claimed by Bosch, this translates into improved performance and about 4% less CO_2 under test conditions. Under the real world conditions, there may be fuel savings up to 13%.

It is estimated that 5 litres of demineralized water is sufficient for about 3000 km of regular driving. Even after the tank dries up, the car would be able to continue operating as normally.

12. ETHANOL-BOOSTED DIRECT INJECTION

Ricardo is collaborating with Behr, Bosch, Delphi, Federal Mogul, GW Castings and Honeywell to develop technology which optimizes ethanol-fuelled engines to achieve efficiency better than that of petrol engines and to approach that of diesel engines. The technology, called ethanol-boosted direct injection (EBDI) is based on ethanol's best properties, *viz.*, higher octane value and higher heat of vaporization and is expected to create a renewable fuel scenario, which does not depend on the oil price. EBDI would mean that a higher level of turbocharging can be used to achieve higher cylinder pressure enabled by ethanol. EBOF would also use other technologies, *viz.*, direct injection, variable valve timing, optimized ignition and advanced gas re-circulation.

13. INCREASING ENGINE POWER

The following are the commonly used techniques to increase the available engine power:

1. Increasing displacement, which can be achieved by increasing the cylinder size or the number of cylinders.
2. Increasing compression ratio. This works practically upto a limit, beyond which higher octane fuel would have to be used to avoid detonation.
3. Increasing air input. This may be done by using supercharger or turbocharger, which compresses the inlet air to increase its mass density, resulting in more power.
4. Cooling the inlet air, especially after supercharging or turbocharing, where an intercooler is employed which cools the hot air from the turbocharger/supercharger to increase its density and hence the power produced.
5. Smooth air entry means less power loss due to air resistance. This is achieved by using polished air inlet manifolds and using two inlet air valves per cylinder instead of one.
6. Easy exhaust, which may be obtained by using (*i*) two exhaust valves per cylinder, (*ii*) bigger tail pipes, (*iii*) free-flowing mufflers, and (*iv*) using two exhaust pipes instead of one.
7. Fuel injection gives more precise fuel metering compared to carbuettors. Besides, direct injection is more efficient than port injection system.
8. Decreased mass of engine components reduces power loss due to inertia which means higher net power available at the crankshaft.

14. HINTS TO SAVE FUEL

1. Keep the engine properly tuned. A poorly tuned engine can increase fuel consumption by upto 50%.
2. Get your vehicle serviced regularly by authorized service centre.
3. While driving, avoid excessive use of clutch.
4. Switch off the engine if wait at the red lights is more than 30 seconds.
5. Drive in gear corresponding to the vehicle speed. For this engine RPM would be quite useful.
6. Drive the vehicle at medium speeds as far as possible. Increasing speed from 80 kph to 90 kph may increase the fuel consumption by 10%.
7. Maintain steady speed. Fast acceleration and sudden braking results in increase of fuel consumption, apart from damage to tyres. Switching off the engine and using only the rolling resistance to slow down is the bad deceleration technique.
8. Keep the tyres properly inflated always. Under-inflated tyres may increase fuel consumption by 5%.
9. Do not use the roof luggage rack unless absolutely necessary. A loaded roof rack may increase fuel consumption on highway by 5%. Even an empty roof rack would increase the fuel consumption by 1%.
10. Minimise the use of air conditioner, particularly at lower speeds.
11. Keep the windows closed on the highway, to reduce the wind resistance.

15. WHAT DOES COLOUR OF ENGINE SMOKE MEAN?

The exhaust emissions of a perfectly tuned engine should be invisible. However, in case of a visible smoke, it is its colour that gives a clue as to the possible engine defects.

1. Thin white vapours

This is due to condensation of water vapours inside the exhaust system and may be accompanied by a slow drip of water. It is perfectly harmless and is often seen during the starting of the vehicle from cold, in the morning.

2. Thick blue or grey smoke

This may be due to the lubricating oil in the crank case reaching and burning in the combustion chamber. Apart from loss of oil, it can result in fouled spark plugs, rough idle, hard starting and sluggish acceleration. On turbo-charged engines, bluish-grey smoke may mean failure of the turbocharger.

3. Black sooty smoke

Such a smoke is usually due to the air-fuel mixture being excessively rich. This may be due to many factors, *e.g.*, a clogged air cleaner, the choke stuck closed, a faulty sensor, a clogged component in the fuel injection system. The exact cause will need systematic investigation by a trained mechanic.

4. Thick white smoke

This type of smoke indicates serious trouble and needs immediate attention. It may be caused by the burning of coolant in the engine, a damaged cylinder head gasket, crack in the cylinder block or a damaged cylinder head. Any of these would require the complete engine overhaul. It is better and safer in such a case to tow the vehicle to the garage.

16. COMPREHENSIVE TROUBLE SHOOTING & FAULT RECTIFICATION IN PETROL ENGINES

Trouble shooting of various systems of automotive engines has already been discussed at appropriate places. However, in practice any trouble may be caused due to malfunctioning of any one or more related engine systems. As such a detailed view of engine diagnosis involving all the engine systems is very important and necessary from practical view point. The same is, therefore, discussed here.

1. Starting difficulty

The brief step-by-step procedure for determining the source of trouble and their rectification is:

1. Check all fuses. If these are OK, proceed ahead.
2. Test electrical system voltage. Connect positive lead of the multimeter with the positive battery terminal and the negative lead with the negative battery terminal. The battery voltage should be between 13.4 and 12.6 volts.

   In case the multimeter or voltmeter is not available, put the headlights ON. If the electrical system is OK, the headlights should illuminate brightly. If these are dim or not working, either the battery is discharged or is damaged and needs replacing.
3. Turn the headlights on and activate the ignition key to crank the engine. If the headlights stay bright, the electrical circuit is not connected which may be due to shorting of the starter motor winding or to the worn out armature bushes, which would have to be replaced. However, if the headlights dim too much, it would indicate that the battery is weak and needs recharging or replacement.
4. In case of automatic transmission, check whether the gear selector is either in park or neutral. If not, bring it into either of these positives. Only then the engine would crank over.
5. Defective anti-theft system would also not allow the starter to function. Check it and repair or replace the defective component or system.
6. In case the above is O.K., check for any defects in the fuel supply system (Art. 34, Ch. 9) or the

ignition system (Art 31, Ch. 13). Take appropriate steps to rectify in case any defect is found out.

7. In modern vehicles, there is OBD (On Board Diagnostic) system installed. In case of any engine trouble, 'check engine light' will be ON. Then you may use a scanner tool to retrieve the trouble codes and proceed accordingly. However, sometimes conditions do occur which are not detected by the computer. For example any valve failure would not be detected by the engine, since the problem is not sensor-related in this case.

2. The engine stalls at idle

Idling allows the engine to operate at the lowest speed without your foot on the accelerator pedal. If the engine starts, but stalls at idle, follow the following steps:

1. Check the electrically controlled components to identify any system trouble. Use a scanner tool to retrieve trouble codes and check if they relate to this problem, *e.g.*, injection driver or ignition coil failure. First repair these problems and retest the system.
2. Inspect all engine and accessory vacuum lines. Vacuum hoses are generally connected to the engine intake manifold and would supply engine vacuum to various accessories, *e.g.*, power brakes. In some cars, a larger vacuum hose connects the intake manifold to the IAC (Idle Air Control) motor. Check in particular the integrity of all vacuum hoses at each end of the hose, where a vacuum hose usually fails. Replace any broken or damaged hose and then proceed to the next step.
3. In the modern cars, there is an Idle Air Control motor alongwith the throttle body, which control engine speed by regulating volume of the idle air. After a long use, this system develops a light tar condition, also called *coking*, which may result in poor idling condition. To remedy this, it requires servicing.
4. IAC motor may also have failed, either due to short-circuiting or by developing high resistance and thereby becoming very slow. To check the motor, remove it and with the wires connected, turn the key to ON, but without stating the engine. If the motor does not move in or out, it has failed. Replace the motor and check the system again.
5. The EGR valve is so designed that it should remain open only when the engine is loaded and not at idle. Any carbon from the exhaust system may cause the EGR valve to stick open, allowing the exhaust gases leak into the engine, which is similar to vacuum leak. To check the valve, remove the same and blow air from its intake part. If the air is able to flow and come out of the outlet port, the valve is struck open, which may be cleaned or replaced as required.

3. Engine surging during idling

The following are the probable sources of this trouble:

1. Check the fuel pump for proper pressure and working; rectify as required.
2. Check idling speed; adjust as required.
3. Check proper working of the choke mechanism; if sticking, repair or replace as necessary.
4. Inspect the spark plugs for correct gap and proper working; adjust or replace as necessary.
5. Check working of the ignition system, especially the ignition timing; adjust as required.
6. Check the head gasket; replace if necessary.
7. Check for wear of camshaft, chain or sprockets; replace as required.
8. Check the PCV valve for proper working; rectify as necessary.

4. Engine misses during idling

To diagnose this, proceed as follows:

1. Check the spark plug; adjust gap or replace as necessary.
2. Check ignition wires; if loose, tighten and if broken, replace.

3. Check the working of ignition coil. If found weak or defective, replace the same.
4. Look for cracked distributor cap and damaged or worn out distributor rotor. Replace if necessary.
5. Check ignition timing; readjust if required.
6. Check ignition advance mechanism. Repair or replace as required.
7. Check the idling speed. Adjust if necessary.
8. Check if the choke mechanism is sticking. Any defective component, if detected, may be repaired or replaced.
9. Check fuel filter. Replace if found choked with impurities.
10. Inspect the condition of gasket between the carburettor and the inlet manifold. Replace if found defective.
11. Check engine compression. Perform necessary repairs if compression found low.
12. Check the working of valve train. Any damaged or worn out component may be replaced and tappet clearance may be readjusted if required.

5. The engine hesitates

If the engine hesitates, it means interruption of continuous power during acceleration. This may be either due to the engine shutting off for a very small period or due to the engine simply losing power temporarily. When an engine hesitates it would give a sudden jerk to the vehicle motion. Besides, it may also cause greater stress in the engine drive train components, thereby resulting in their premature failure. To remedy this proceed as follows:

1. Scan for trouble codes and look for the faults in any directly related systems. Make rectifications, if required and check the air and fuel filters, which should not be blocked even partially.
2. Check for damaged or broken vacuum hoses or air intake boot on or near the engine. Check also their connections to be tight. If OK, proceed to check the fuel pump pressure and the pressure regulator as described earlier in case of the previous trouble.
3. Next check the ignition timing with the help of a timing light. If the timing is not correct, check the timing belt or chain as applicable. If OK, reset the timing.

6. The engine runs rough

1. If the fuses are OK and all the electronically controlled components have been fused to have no specifically selected problem like ignition coil failure, remove all the spark plugs and inspect them. Clean or replace as needed.
2. Look for any damaged or broken vacuum hoses or their integrity at each end. If OK, proceed next.
3. The deposits on the spark plug indicate the conditions inside the engine. Remove spark plugs after marking the plug cables for their identification regarding the firing order. Inspect all the plugs and if the condition of one plug is considerably different from the others, that cylinder may be having a problem. Take appropriate action.
4. For smooth running, all the engine cylinders should have equal compression. A poor compression in one or more cylinders may result in rough idle and low power output. Use engine compression gauge for testing compression in each cylinder. Readings in all cylinders should not vary more than 5 percent of each other. If low compression is found in any or more cylinders, further troubleshooting would have to be done to determine the cause, which may be a burnt intake or exhaust valve, broken piston or ring, broken valve spring or a damaged head gasket. Take appropriate steps to rectify and when the compression is OK, go to the next step.
5. Test the ignition system as described in the relevant chapter of this volume.
6. Test the fuel pressure at the common rail or anywhere else in the fuel line. It should be between 13

and 17 psi for most throttle body injection systems, while for most direct port injection systems, it should be between 40 and 55 psi. If the pressure is O.K., proceed to the next step. However, if no or less fuel pressure is there, check the fuel pump fuse and fuel pump control relay, usually placed in the fuse box and ensure that these are OK.

7. Next step is to test the injectors. Usually the sound of the injectors clicking when the electronic valve opens and closes is audible. To facilitate hearing this, a stethoscope can be used, touching it against each injector one by one. However, if no such sound is heard test fuel injector pulse and supply voltage output, which would indicate whether the computer has the operating voltage and injector trigger signal or not. If not, make further tests to find the cause and rectify the same. Test all the injectors in this way. If all the injectors are OK, check the fuel pressure regulator. Ruptured diaphragm inside the regulator may feed the raw fuel into the engine resulting in its rough running.
8. A broken valve spring or worn out cam lobe would also cause the engine to misfire and run rough due to loss of compression in the engine cylinder. These conditions do not always show up in a compression test, which is performed at the cranking speed. The effect of a worn out cam would be evident only when the engine is running. The extent of damage to the valve spring would determine whether the effect is visible when the engine is running under load or both during idling and under load. To test this disconnect the power to the ignition or the fuel system and remove the valve covers for visual inspection. Ask some helper to crank the engine while you are watching the rocker arms or cam lobes. If any one or more cam lobes are travelling less than the others, they are worn out and in that case you would have to change the camshaft. A visual inspection would also reveal any broken valve springs, which may be replaced.
9. Next check the head gasket. A damaged head gasket may cause the coolant or exhaust to enter the combustion chamber and causing the affected cylinder(s) to misfire.
10. Check the engine mounts for integrity, look for any cracks or separation or the mount. Remove the damaged mount. Preferably replace all the engine mounts.

7. Low power output

If the engine is running smoothly but giving low power output proceed as follows:

1. Scan for trouble codes using a scanner tool and check whether these trouble codes relate to the specific problem, such as a fuel pump circuit failure code. If so, repair the related problems and retest. If now no trouble codes are present, you may proceed to the next step.
2. Check the air filter. If it were clogged, adequate quantity of air would not enter the engine, restricting its output. Clean or replace as required.
3. Check the fuel filter. For this remove the filter and wipe out the excess fuel from it. Then blow through the filter. If you cannot blow freely, it is clogged, which would need to be replaced.
4. Check the fuel pump pressure and pressure regulator. Most vehicles have a fuel pressure test port at some place on the fuel rail, which has to be located visually. Vehicle repair manual may be referred for help to locate this. Fuel pressure regulator should also be inspected. If should not be leaking. In case of any fault, take the necessary steps and proceed to the next step once fuel system pressure is OK.
5. Use the timing light. If it is OK, check the main engine sensors, *viz.*, the MAF (Mass Air Flow) and MAP (Manifold Absolute Pressure) sensors. OBD diagnosis computer may detect this if the sensor is shorted or too much out of range. Slightly off the range would not be detected. So these sensors should be checked as described in the vehicle repair manual.
6. Next check the spark plugs. If not adjusted correctly or are badly worn, the same should be readjusted or replaced as required. For this check you would have to wait till the engine is cold.

7. Check the timing chain or belt setting by looking at the timing marks. If found to be correct, go to the next step.
8. Check the exhaust system. The catalytic converter should not be blocked. To check this, warm up the engine and measure the temperature at the entry and exit of the catalytic converter, using an Infrared/laser thermometer. If the temperature at the exit of the converter is higher than at the entry, it is OK, otherwise it would mean a blocked converter, which would need to be replaced.
9. Check the engine compression by means of the engine cylinder compression test. If OK, check whether the clutch is slipping. If this is also right, proceed to test the crankshaft harmonic balancer. For this, remove the balancer and inspect. Replace if required.

8. Engine overheating

The cause of this trouble may be determined as follows:

1. Check coolant level. Repair the leaks if any and bring up the coolant to correct level.
2. Check coolant quality. Replace if needed.
3. Check the fan belt for tightness; adjust if necessary.
4. Check the thermostat for correct operation. Replace if found stuck.
5. Check the thermostat switch for radiator fan. Replace if found defective.
6. Inspect the radiator core for any blocked passages. Repair or replace as necessary.
7. Check the working of coolant pump. Repair as required.
8. Inspect the condition of the hoses and their connections. Refix or replace as necessary.
9. Check for blocked engine cooling water passage. Flush the system in the reverse direction if necessary to open up the blocked passages.
10. Check up the quality of the air-fuel mixture delivered by the carburettor; too lean a mixture would cause engine overheating. Adjust as required.
11. Too much retarded ignition may be another cause of overheating. Check up and adjust if necessary.

9. Detonation

For diagnosis, proceed as follows:

1. Inspect the ignition system for proper working. Correct as required.
2. Check the ignition timing. Adjust if found excessively advanced.
3. Check the quality of the fuel used. Replace if found substandard.
4. Check the fuel pump for proper working. Replace if necessary.

10. Excessive fuel consumption

The probable causes are:

1. Leakage in fuel lines which may be checked and rectified.
2. Carburettor defects:
   (*i*) flooded; the cause has to be rectified.
   (*ii*) too rich mixture; adjust.
   (*iii*) idle speed setting too high; adjust.
   (*iv*) choke sticking; rectify.
3. Air filter clogged: clean or replace.
4. Valve system defect:
   (*i*) sticking valves; rectify.
   (*ii*) incorrect valve timing; replace.

(*iii*) valve spring broken; replace.
(*iv*) incorrect tappet clearance; readjust.

5. Spark plug faults:
   (*i*) incorrect gap; readjust.
   (*ii*) fouled plug; clean/use correct plug.
   (*iii*) wornout plug electrodes; replace.
6. Ignition coil defective, which will have to be replaced.
7. Distributor defects:
   (*i*) contact breaker points burnt or the gap increased; replace.
   (*ii*) defective advance mechanism; rectify/replace.
   (*iii*) ignition timing incorrect; readjust.
8. Miscellaneous:
   (*i*) clogged silencer; clean/replace.
   (*ii*) low engine compression; overhaul the engine.
   (*iii*) Apart from the above engine defects, excessive fuel consumption may also be caused by slipping/wornout clutch, dragging brakes, under-inflated tyres, incorrect geometry and unsteady driving.

11. Smoky exhaust

The probable causes may be:

1. Low engine compression; overhaul the engine.
2. Broken valve springs; replace.
3. Fuel supply system defects:
   (*i*) Clogged air filter; clean or replace.
   (*ii*) Adulterated fuel; change the fuel.
   (*iii*) Flooded carburettor; rectify defect.
   (*iv*) Excessively rich mixture; adjust the carburettor.
   (*v*) Sticky choke; rectify.
4. Ignition system defects:
   (*i*) incorrect ignition timing; readjust.
   (*ii*) defective advance mechanism; rectify or replace.

REFERENCES

1. ''Downsized Boosted Gasoline Engines'', ICCT Technical Brief no 5, December 2016, pp. 1-6.
2. Scott Collie, ''Bosch Water Boost injection opens door to better power and economy'', new atlas.com/bosch-water-injection/45205, August 31, 2016.
3. ''Naturally Aspirated Engines'', ICCT Technical Brief no. 2, June 2016, pp. 1-5.
4. Lindsey Brooke, ''Pushing the ICE forward gradually'', Automotive Engineering, June 2016, pp. 24-26.
5. Dan Carney, ''Preparing for a 48-volt revival'', Automotive Engineering, April 2016, pp. 26-28.
6. Thomas ESCH, "Trends in Commercial Vehicle Power Trains", ATZ Auto technology, May 2010, Vol. 10, No. 5, pp. 26-31.

THEORY QUESTIONS

1. Discuss briefly various technologies which can be used to improve vehicle fuel economy.

2. Write a detailed note on current use of modern technologies in automobile engines.
3. Describe in detail the Jet Ignition technology developed by Mahle for petrol engines.
4. Write a brief note on 'divided exhaust boosting'.
5. Discuss in detail, the necessity and principle of operation of cylinder deactivation. Describe also various such technologies in use.
6. Describe in detail future trends for (*i*) Petrol engines, (*ii*) Commercial diesel engines.
7. Describe salient featuri's of Infiniti's variable compression ratio engine.
8. Describe the main positive results of Eaton, Cummins Engine and Purdue University research project on cylinder deactivation for diesel engines.
9. Explain thoroughly the advantages of the 48-V electrical system for automotive use. Describe also its practical application so far.
10. Write detailed notes on:
    (*i*) Scuderi engine
    (*ii*) Transonic combustion technology.
    (*iii*) Opposed piston diesel engine.
11. Write a comprehensive note on 'OBD-II and check Engine Light'.
12. Explain briefly Bosch's 'Water boost injection technology'.
13. Write a comprehensive note on ethanol-boosted direct injection.
14. Describe commonly used techniques briefly, to increase engine power.
15. Write down various measure switch the drivers/owners of vehicles can take to keep the fuel consumption minimum.
16. Describe in detail various steps involved in complete trouble shooting and fault rectification in an automotive petrol engine.

QUESTIONS FOR VIVA

1. What is cylinder deactivation?
2. Why was cylinder deactivation not commonly used earlier in 4-cylinder engines?
3. Why can the compression ratio not be increased beyond a limit?
4. Why do we use two intake air valves per cylinder instead of one?
5. What is the reason for using two exhaust valves per cylinder instead of one?
6. Why is it preferred to use an inter cooler with a turbocharger?
7. Why sometimes two exhaust pipes are used instead of one?
8. Name any two cars using cylinder deactivation.
9. Name the engine wherein four engine strokes are divided between two cylinders.
10. State the advantages claimed for a Scuderi engine.
11. What is the advantage claimed for the 'Transonic combustion technology?'
12. What is indicated if the engine exhaust smoke is (*i*) thick blue or grey, (*ii*) black sooty, (*iii*) thick white?
13. What is Mahle's Jet Ignition technology?
14. What is 'divided exhaust boosting'?
15. What is the range claimed for seamless compression ratio change in Infiniti's variable compression ratio engine?
16. State the main advantages of 48-V electrical system.
17. What is the disadvantage, if any, for the 48-V electrical system?
18. What does OBD-II stand for?
19. What is the function of 'Check Engine Light'? What is a (*i*) Generic, (*ii*) special OBD code?

22 Automotive Engine Specifications

1. INTRODUCTION

In this chapter the main specifications of the automobile engines, both indian as well as foreign, have been given. Apart from the use of data for design engineers, the same is helpful to develop a practical insight in a serious student. With a close scrutiny of data for various vehicles one may easily be aware of the effect of various engine parameters on the output. The engine data of the following automobiles have been given here:

A. INDIAN AUTOMOBILES

1. Small cars (upto 1000 cc)

Chevrolet Spark; Ford Ecosport; Hyundai Eon; Maruti Alto 800, Celerio, Estilo, A-Star, Wagon R; Tata Nano Twist.

2. Medium Cars (1000-2000 cc)

Audi A4 2.0 TDI, A3; BMW 320d, Mini Cooper S, 1-series; Fiat Palio, Palio 1.9D ELX, Stile 1.1, Linea T-Jet, 500, Grand Punto; Ford Fiesta (Petrol), Ecosport Diesel, Figo; G.M. Chevrolet Gruze, Beat, Sail, Enjoy; Hindustan Ambassador; Honda City VTEC, Brio, Amaze, CRV 2.0, Jazz BR-V; Hyundai Tucsox, Eon, Verna, i10, i20, Elantra, Grand i10, Creta, i20 Elite, Xcent; Mahindra Verito, Quanto, KUV100 K8, Nuvosport N8, TUV 300, Maruti Swift, Desire, Ritz (Petrol/Diesel), Eeco, Ertiga, S-Cross, Baleno, Ciaz; Mercedes A180 (Petrol), B 180; Mitsubishi Lancer (Petrol/Diesel), Cedia Sports; Nissan Micra (Petrol/ Diesel), Evalia, Datsun Go, Radigo S, Terrano; Premier Rio; Renault Scala, Duster, Lodgy; Skoda Superb (Petrol/Diesel), Laura (Petrol/Diesel), Yeti, Superb, Octavia; Tata Vista, Bolt, Tiago; Toyota Corolla Altis, Etios (Liva, D40), Prius; Volkswagen Passat, Polo, Jetta, Vento.

3. Large Cars (more than 2000 cc)

Audi A6 3.0 T FST Quattro, Q5 3.0 TDI Quattro, Q73.0 TDI Quattro, R8 Spyder 5.2 FSI Quattro, RS 5 Coupe; BMW M5, X5; Ford Endeavour; Lamborghini Aventador; Land Rover Discovery 4; Mahindra Scorpio CRDe, Xylo, XUV 500, Ssangyang Rexton; Mercedes A180 (Diesel), S 500, GL 350 CDI; Tata Aria, Safari Storme; Toyota Camry, Fortuner, Innova MT/AT; Volvo XC60, S60.

4. Transport Vehicles

Tata Ace, Xenon 4X4, LPT-2518/62 TC truck; Ashok Leyland Comet, Newgen 2526; Force One Truck; Volvo City Bus.

5. Electric/Hybrid Vehicles

Mahindra e20

B. FOREIGN AUTOMOBILES

1. Small cars (upto 1000 cc)

Citroen C1, C3; Kia Picanto 1.0; Vauxhall New Viva, Wuling Minivan.

2. Medium cars (1000-2000 cc)

Citroen C4, C5; Kia Picanto 1.25; Ssangyong Korando; Peugeot 508, 3008, 5008; Vauxhall Meriva; Audi A4 2006, Volkswagen Golf GTI, Caddy, Polo; Mercedes Benz 200, 200 D; Smart for Four; Skoda Octavia Petrol; BMW 318i; Alfa Romeo 2000 Spider Veloce, Alfa 75 2.0 T, Spark; Chery A3; FAW Vita N3; Toyota Yaris.

3. Large cars (more than 2000 cc)

Ssangyong Rexton; Rolls Royee Phantom, Ghost Series II; Audi 100 SE, A6 (2006); Porsche 911 Carrera, 718 Boxster S; Ferrari 458 Spider, SA Alberta; Volvo 740 GL/GLE/Turbo/GL Diesel; Hyundai Genesis; Ford Mustang, Escape hybrid; Chevrolette Corvette, Cadillac Escalade; Suzuki Kizashi.

2. ENGINE FOR 'AMBASSADOR' CAR

Manufacturer	: Hindustan Motors Limited, India	
Model	: 2011	
	Petrol	**Diesel**
Type	: 4-cylinder in line	4-cylinder in line
Bore	: N.A.	73.0 mm
Stroke	: N.A.	88.9 mm
Displacement	: 1817 cc	1489 cc
Maximum Power	: 53 kW at 4800 rpm	26.5 kW at 4000 rpm
Maximum Torque	: 134 Nm at 2250 rpm	73 Nm at 2250 rpm
Power to Weight ratio	: 46 kW/tonne	22 km/tonne
Power to Wegiht ratio	: 117 Nm/tonne	97 Nm/tonne
Specific output	: 29 kW/litre	14.6 kW/litre
Valve Gear	: SOHC	SOHC
Fuel system	: MPFI	Diesel
Fuel tank capacity	: 54 litres	54 litres

3. ENGINE FOR AUDI A6 3.0 T FSI QUATTRO (2011 MODEL)

Manufacturer	: Audi India Ltd.
Engine Type	: 6-cylinder, V-type, front, longitudinal, petrol
Bore	: 84.5 mm
Stroke	: 89 mm
Displacement	: 2995 cc
Compression ratio	: 10.3 : 1
Maximum Power	: 220 kW at 5250-6500 rpm
Maximum Torque	: 440 Nm at 2900-4500 rpm
Power to Weight ratio	: 121.2 kW/tonne
Torque to Weight ratio	: 235 Nm/tonne
Specific Output	: 73.5 kW/litre

Valve Gear	: 4 valves per cylinder, DOHC per bank
Fuel System	: Direct injection, supercharged
Fuel Tank Capacity	: 75 litres
Fuel consumption	: City: 7.0 kpl, Highway: 10.0 kpl

4. ENGINE FOR AUDI Q5 3.0 TDI

Manufacturer	: Audi India Ltd.
Engine Type	: 6-cylinder, V-type
Bore	: 83.0 mm
Stroke	: 91.4 mm
Displacement	: 2967 cc
Compression ratio	: 16.8 : 1
Maximum Power	: 176 kW at 4000-4400 rpm
Maximum Torque	: 500 Nm at 1500-3000 rpm
Power to Weight ratio	: 90 kW/tonne
Torque to Weight ratio	: 256 Nm/tonne
Specific Output	: 59 kW/litre
Valve Gear	: 4 valves per cylinder, DOHC
Fuel System	: Common rail direct injection, turbocharged diesel
Fuel Tank Capacity	: 75 litres
Fuel consumption	: City: 8.0 kpl, Highway: 11.1 kpl

5. ENGINE FOR AUDI Q7 3.0 TDI QUATTRO

Manufacturer	: Audi India Ltd.
Engine Type	: 6-cylinder, V-type, front, longitudinal
Bore	: 83.0 mm
Stroke	: 91.4 mm
Displacement	: 2967 cc
Compression ratio	: 16.8 : 1
Maximum Power	: 180 kW at 3800-4400 rpm
Maximum Torque	: 550 Nm at 1750-2500 rpm
Power to Weight ratio	: 77 kW/tonne
Torque to Weight ratio	: 235 Nm/tonne
Specific Output	: 60.6 kW/litre
Valve Gear	: 4 valves per cylinder, DOHC for each bank
Fuel System	: Common rail, exhaust gas turbo-diesel
Fuel Tank Capacity	: 100 litres
Fuel consumption	: City: 5.5 kpl, Highway: 9.1 kpl

6. ENGINE FOR R8 SPYDER 5.2 FSI QUATTRO

Manufacturer	: Audi India Ltd.
Model	: 2012
Type	: V-10, petrol

Bore : N.A.
Stroke : N.A.
Displacement : 5204 cc
Compression ratio : N.A.
Maximum Power : 386 kW at 8000 rpm
Maximum Torque : 530 Nm at 6500 rpm
Power to Weight ratio : 215 kW/tonne
Torque to Weight ratio : 295 Nm/tonne
Specific Output : 61.6 kW/litre
Valve Gear : 4 valves per cylinder, DOHC
Fuel System : Direct fuel injection
Fuel Tank Capacity : 80 litres
Fuel consumption : City: 4.5 kpl, Highway: 9.3 kpl

7. ENGINE FOR RS 5 COUPE

Manufacturer : Audi India Ltd.
Model : 2012
Type : V-8, petrol
Bore : N.A.
Stroke : N.A.
Displacement : 4163 cc
Compression ratio : N.A.
Maximum Power : 331 kW at 8250 rpm
Maximum Torque : 430 Nm at 4000-6000 rpm
Power to Weight ratio : 184 kW/tonne
Torque to Weight ratio : 239 N/mtonne
Specific Output : 79.5 kW/litre
Valve Gear : 4 valves per cylinder, DOHC
Fuel System : Direct fuel injection
Fuel Tank Capacity : 64 litres
Fuel consumption : City: 6.7 kpl, Highway: 11.7 kpl

8. ENGINE FOR AUDI A3

(Figures in brackets are for diesel engine)

Manufacturer : Audi India Ltd.
Engine Type : 4-cylinder, inline, front, transverse, turbo-petrol (turbo-diesel)
Bore : 82.5 (81.0) mm
Stroke : 84.1 (95.5) mm
Displacement : 1798 (1988) cc
Compression ratio : 9.6 : 1 (16.2:1)
Maximum Power : 132 (105) kW at 5100-6200 (3500-4000) rpm
Maximum Torque : 250 (320) Nm at 1250-5000 (1750-3000) rpm

Power to Weight ratio	: 101.9 (78.3) kW/tonne
Torque to Weight ratio	: 193 (239) Nm/tonne
Specific Output	: 73.4 (58.4) kW/litre
Valve Gear	: 4 valves per cylinder, DOHC
Fuel System	: Direct injection, turbo
Fuel Tank Capacity	: 50 litres
Fuel consumption	: City: 9.8 (12.3) kpl, Highway: 14.2 (17.5) kpl

9. ENGINE FOR BMW D 320 d

Manufacturer/Supplier	: BMW India Pvt. Ltd.
Model	: 2012
Engine Type	: 4-cylinder, in-line, diesel, light alloy cylinder head
Bore	: 90.0 mm
Stroke	: 84.0 mm
Displacement	: 1995 cc
Compression ratio	: 16.5 : 1
Maximum Power	: 135 kW at 4000 rpm
Maximum Torque	: 380 Nm at 1750-2750 rpm
Power to Weight ratio	: 95 kW/tonne
Torque to Weight ratio	: 267 Nm/tonne
Specific Output	: 67.6 kW/litre
Valve Gear	: 4 valves per cylinder, DOHC
Fuel System	: Common-rail, direct injection, turbo-charged
Fuel Tank Capacity	: 60 litres
Fuel consumption	: City: 11.1 kpl, Highway: 15.5 kpl

10. ENGINE FOR MINI COOPER S

Manufacturer	: BMW India Pvt. Ltd.
Model	: 2012
Type	: 4-cylinder, inline, aluminium alloy, petrol
Bore	: 77.0 mm
Stroke	: 85.8 mm
Displacement	: 1598 cc
Compression ratio	: 10.5 : 1
Maximum Power	: 135 kW at 5500 rpm
Maximum Torque	: 240 Nm at 1600-5000 rpm
Power to Weight ratio	: 109 kW/tonne
Torque to Weight ratio	: 194 kW/tonne
Specific Output	: 84.5 kW/litre
Valve Gear	: DOHC, 4 valves per cylinder
Fuel System	: Direct injection, with twin-scroll turbocharger
Fuel Tank Capacity	: 50 litres
Fuel consumption	: City: 9.5 kpl, Highway: 13.5 kpl

11. ENGINE FOR M5 CAR

Manufacturer	: BMW India Ltd.
Model	: 2012
Type	: All-aluminium; 90° V-8 petrol engine, front, longitudinal
Bore	: 88.3 mm
Stroke	: 89.0 mm
Displacement	: 4395 cc
Compression ratio	: 10 : 1
Maximum Power	: 412 kW at 6000-7000 rpm
Maximum Torque	: 680 Nm at 1500-5750 rpm
Power to Weight ratio	: 220 kW/tonne
Torque to Weight ratio	: 363 Nm/tonne
Specific Output	: 93.7 kW/litre
Valve Gear	: 4 valves per cylinder, with valvetronic variable lift system with doublevanos variable camshaft timing
Fuel System	: Direct injection with Twin-turbo; 200 bar injection pressure
Fuel Tank Capacity	: 80 litres
Fuel consumption	: City: 4.7 kpl, Highway: 7.1 kpl

12. ENGINE FOR 1-SERIES CAR

Manufacturer	: BMW India Pvt. Ltd.
	(Figures in brackets are for diesel engine)
Engine Type	: 4-cylinder, inline, turbo-petrol (turbo-diesel), front, longitudinal
Bore	: 85.8 (90) mm
Stroke	: 77 (84) mm
Displacement	: 1598 (1995) cc
Compression ratio	: 10.5 : 1 (16.5)
Maximum Power	: 100 (105) kW at 4400 (4000) rpm
Maximum Torque	: 220 (320) Nm at 1350-4300 (1750-2500) rpm
Power to Weight ratio	: 76.3 (80.3) kW/tonne
Torque to Weight ratio	: 163.5 (238) Nm/tonne
Specific Output	: 62.5 (52.7) kW/litre
Valve Gear	: 4 valves per cylinder, DOHC
Fuel Tank Capacity	: 52 litres
Fuel consumption	: City: 9.7 (14.2) kpl, Highway: 11.5 (18) kpl

13. ENGINE FOR X5 CAR

Manufacturer	: BMW India Pvt. Ltd.
Model	: 201.4
Engine Type	: 6-cylinder, inline, turbo-diesel, front, longitudinal
Bore	: 90 mm
Stroke	: 84 mm

Displacement	: 2993 cc
Compression ratio	: 16.5:1
Maximum Power	: 190 kW at 4000 rpm
Maximum Torque	: 560 Nm at 1500-3000 rpm
Power to Weight ratio	: 91.8 kW/tonne
Torque to Weight ratio	: 270.6 Nm/tonne
Specific Output	: 63.5 kW/litre
Valve Gear	: 4 valves per cylinder, DOHC

14. ENGINE FOR PALIO CAR

Manufacturer : Fiat India Automobiles Ltd.

Engine Type : 4-cylinder, in line, petrol, transverse, iron block and alloy head

	1.2	**1.6**
Bore	: 70.8 mm	N.A.
Stroke	: 78.86 mm	78.4 mm
Displacement	: 1242 cc	1596 cc
Compression ratio	: 9.1	9.5 ± 0.15
Maximum Power	: 52 kW (72 PS) at 6000 rpm	72 kW (100 PS) at 5500 rpm
Maximum Torque	: 102 Nm (10.4 kgm) at 3250 rpm	137 NM (14.1 kgm) at 4250 rpm
Power to Weight ratio	: 59.75 PS/tonne	83 PS/tonne
Torque to Weight ratio	: 8.63 kmg/tonne	11.62 kgm/tonne
Specific output	: 58 PS/litre	62.7 PS/litre
Valve Gear	: 2 valves per cylinder	4 valves per cylinder DOHC
Fuel Consumption (kpl)	: 9.2 (city), 14.0 (highway)	9.1 (city), 14.1 (highway)

Fuel system : Electronic MPI integrated with injection

Ignition : Electronic integrated with injection

Fuel tank capacity : 47 litres

15. ENGINE FOR PALIO 1.9 D ELX CAR

Manufacturer	: Fiat India Auto Ltd.
Engine Type	: 4-cylinder, inline, diesel, traverse
Bore	: N.A.
Stroke	: N.A.
Displacement	: 1910 cc
Compression ratio	: N.A.
Maximum Power	: 45.5 kW (63 PS) at 4500 rpm
Maximum Torque	: 120 Nm (12.2 kgm) at 2500 rpm
Power to Weight ratio	: 55.2 BHP/tonne
Torque to Weight ratio	: 10.7 kgm/tonne
Specific output	: 32.9 BHP/litre

Valve Gear	: 2 valves per cylinder
Fuel system	: Electronic MPFI
Fuel tank capacity	: 47 litres
Fuel consumption	: City: 11.8 kpl, Highway: 15.7 kpl

16. ENGINE FOR PALIO STILE 1.1

Manufacturer	: Fiat India Auto Ltd.
Type	: 4-cylinder, inline, front transverse
Bore	: 70 mm
Stroke	: 72 mm
Displacement	:1108 cc
Compression ratio	: 9.6 : 1
Maximum Power	: 42.5 kW at 5250 rpm
Maximum Torque	: 92.0 Nm at 2750 rpm
Power to Weight ratio	: 42.94 kW/tonne
Torque to Weight ratio	: 92.9 Nm/tonne
Valve Gear	: SOHC, 2 valves per cylinder
Fuel system	: Petrol
Fuel tank capacity	: 47 litres
Fuel consumption	: City : 10.2 kpl, Highway : 15.05 kpl

17. ENGINE FOR FIAT LINEA T-JET

Manufacturer	: Fiat India Auto Ltd.
Model	: 2010
Type	: 4-cylinder, inline, front, transverse
Bore	: 72 mm
Stroke	: 84 mm
Displacement	:1368 cc
Compression ratio	: 9.8 : 1
Maximum Power	: 83.5 kW at 5000 rpm
Maximum Torque	: 207 Nm at 2200 rpm
Power to Weight ratio	: 67.9 kW/tonne
Torque to Weight ratio	: 168 Nm/tonne
Specific output	: 61 kW/litre
Valve Gear	: DOHC, 4-valves per cylinder
Fuel system	: MPFI, Turbo-petrol
Fuel tank capacity	: 45 litres
Fuel consumption	: City : 10.3 kpl, Highway : 15.4 kpl

18. ENGINE FOR FIAT 500 1.3/LINEA MULTI-JET

Manufacturer	: Fiat India Automobiles Ltd.
Type	: 4-cylinder, inline, turbocharged, diesel

Bore	: 69.6 mm
Stroke	: 82.0 mm
Displacement	:1248 cc
Compression ratio	: 17.6 : 1
Maximum Power	: 56 kW at 4000 rpm
Maximum Torque	: 145 Nm at 1500 rpm
Power to Weight ratio	: 57.1 kW/tonne
Torque to Weight ratio	: 147.9 Nm/tonne
Specific output	: 44.8 kW/litre
Valve Gear	: DOHC, 4-valves per cylinder
Fuel system	: Common-rail, direct injection
Fuel tank capacity	: 35 litres
Fuel consumption	: City : 15.1 kpl, Highway : 20.2 kpl

19. ENGINE FOR FIAT GRANDE PUNTO

(Data inside brackets is for diesel engine)

Manufacturer	: Fiat India Auto Ltd.
Type	: 4-cylinder, inline, alloy head, iron block
Bore	: 72.0 (69.6) mm
Stroke	: 84.0 (82) mm
Displacement	: 1368 (1248) cc
Compression ratio	: 10.8 : 1 (18.0)
Maximum Power	: 66.4 kW at 6000 rpm (56 kW at 4000 rpm)
Maximum Torque	: 115 Nm at 4000 rpm (197 Nm at 4500 rpm)
Power to Weight ratio	: 58 (47.0) kW/tonne
Torque to Weight ratio	: 100 (165) Nm/tonne
Specific output	: 48.5 (44.8) kW/litre
Valve Gear	: DOHC, 4-valves per cylinder, DOHC
Fuel system	: Petrol (Diesel)
Fuel tank capacity	: 45 litres
Fuel consumption	: City :10.2 (13.7) kpl, Highway : 14.6 (17.5) kpl

20. ENGINE FOR FORD NW 'FIESTA' (PETROL)

Type	: 4-cylinder, in-line, all-aluminium-alloy, front, transverse petrol engine (1.5 Ti VCT)
Bore	: 79.0 mm
Stroke	: 76.4 mm
Capacity	: 1499
Compression ratio	: 11.0 : 1
Maximum Power	: 81 kW, at 6000 rpm
Maximum Torque	: 140 Nm at 4000 rpm
Power to Weight ratio	: 72.8 kW

Torque to Weight ratio	: 12.5 Nm/tonne
Specific Output	: 54.2 kW/litre
Valve Gear	: 4 valves per cylinder, DOHC
Fuel tank capacity	: 43 litres
Fuel Consumption	: City: 10.8 kpl; Highway: 16.2 kpl

21. ENGINE FOR ECO SPORT

Manufacturer	: Ford India Ltd. (Fig. in brackets are for diesel engine)
Engine Type	: 4-cylinder, inline, front transverse petrol (turbo-diesel)
Bore	: 71.9 (73.5) mm
Stroke	: 82.0 (88.3) mm
Displacement	: 999 (1498) cc
Compression ratio	: 10.0 (16) : 1
Maximum Power	: 91.7 (67.1) at 6000 (3750) rpm
Maximum Torque	: 170 (204) Nm at 1400-4500 (2000-2750) rpm
Power to Weight ratio	: 72.9 (52.0) kw/tonne
Torque to Weight ratio	: 135 (158) Nm/tonne
Specific Output	: 91.8 (44.8) kW/litre
Valve Gear	: 4 valves per cylinder, DOHC (2 valves per cylinder, SOHC)
Fuel Tank Capacity	: 52 litres
Fuel consumption	: City: 11.8 (13.3) kpl, Highway: 17 (18.6) kpl

22. ENGINE FOR FIGO

Manufacturer	: Ford India Ltd. (Data in brackets are for diesel model)
Engine Type	: 4-cylinder, inline, front transverse
Bore	: 79.0 (73.5) mm
Stroke	: 76.4 (88.3) mm
Displacement	: 1499 (1498) cc
Compression ratio	: 11.1 (16) : 1
Maximum Power	: 82.4 kW at 6300 rpm (73.8 kW at 3750 rpm)
Maximum Torque	: 136 Nm at 42.50 rpm (215 Nm at 1750-3000 rpm)
Power to Weight ratio	: 79.5 (79.0) kW/tonne
Torque to Weight ratio	: 141 (206) Nm/tonne
Specific Output	: 55 (49.3) kW/litre
Valve Gear	: 4 valves per cylinder, DOHC (2 valves per cylinder, DOHC)
Fuel Tank Capacity	: 42 (40) litres
Fuel consumption	: ARAI : 28.53 kpl

23. ENGINE FOR ENDEAVOUR (2016)

Manufacturer	: Ford India Ltd.
	(Data in brackets for 5-cylinder engine)
Engine Type	: 4-cylinder, inline (5-cylinder, inline), front transverse
Bore	: 86.0 (89.9) mm
Stroke	: 94.6 (100.76) mm
Displacement	: 2198 (3198) cc
Compression ratio	: 15.7 (15.5) : 1
Maximum Power	: 118 kW at 3200 rpm
	(147 kW at 3000 rpm)
Maximum Torque	: 385 Nm at 1600-2500 rpm
	(470 Nm at 1750-2500 rpm)
Power to Weight ratio	: 52.7 (65.7) kW/tonne
Torque to Weight ratio	: 172 (196) Nm/tonne
Specific Output	: 53.7 (46.0) kW/litre
Valve Gear	: 4 valves per cylinder, DOHC
Fuel Tank Capacity	: 80 litres
Fuel consumption	: City: 8.1 (7.0) kpl, Highway: 11.8 (10.3) kpl

24. ENGINE FOR CHEVROLET SPARK

Manufacturer	: General Motors India Ltd.
Type	: 4-cylinder, in-line, front transverse, Daewoo's S-Tec engine
Bore	: 68.5 mm
Stroke	: 67.5 mm
Displacement	: 995 cc
Compression ratio	: 9.3 : 1
Maximum Power	: 47 kW at 5400 rpm
Maximum Torque	: 90 Nm at 4200 rpm
Power to Weight ratio	: 56 kW/tonne
Torque to Weight ratio	: 107 Nm/tonne
Specific output	: 47.24 kW/litre
Valve gear	: SOHC, 2-valves per cylinder
Fuel system	: Petrol
Fuel tank capacity	: 38 litres
Fuel consumption	: City: 12.3 kpl, Highway: 16.9 kpl

25. ENGINE FOR CHEVROLET SRV

Manufacturer	: General Motors India Ltd.
Type	: 4-cylinder, in-line, front transverse
Bore	: 79 mm
Stroke	: 81.5 mm
Displacement	: 1598 cc

Compression ratio	: 9.5 : 1
Maximum Power	: 75.3 kW at 5800 rpm
Maximum Torque	: 140 Nm at 4500-6500 rpm
Power to Weight ratio	: 61.2 kW/tonne
Torque to Weight ratio	: 114 Nm/tonne
Specific output	: 61.2 kW/litre
Valve gear	: DOHC, 4-valves per cylinder
Fuel tank capacity	: 60 litres
Fuel consumption	: City: 9.7 kpl, Highway: 12.8 kpl

26. ENGINE FOR CHEVROLET CRUZE (M.T.)

Manufacturer	: General Motors India Ltd.
Model	: 2012
Type	: 4-cylinder, inline, front, transverse
Bore	: 86 mm
Stroke	: 86 mm
Displacement	: 1998 cc
Compression ratio	: 16.3 : 1
Maximum Power	: 122 kW at 3800 rpm
Maximum Torque	: 380 Nm at 2000 rpm
Power to Weight ratio	: 79.6 kW/tonne
Torque to Weight ratio	: 247 Nm/tonne
Specific output	: 61 kW/litre
Valve gear	: 4-valves per cylinder, DOHC
Fuel system	: Common-rail direct injection, turbocharged diesel
Fuel tank capacity	: 60 litres

27. ENGINE FOR CHEVROLET BEAT

(Figures in brackets are for 2012 diesel model)

Manufacturer	: General Motors India Ltd.
Type	: 4-cylinder (3-cylinder), inline, front, transverse
Bore	: 69.5 (69.7) mm
Stroke	: 79 (82) mm
Displacement	: 1199 (936) cc
Compression ratio	: 9.8 : 1 (16.5 : 1)
Maximum Power	: 59 (43.6) kW at 6200 (4000) rpm
Maximum Torque	: 107 (150) Nm at 4400 (1750) rpm
Power to Weight ratio	: 61 (42.5) kW/tonne
Torque to Weight ratio	: 111 (146) Nm/tonne
Specific output	: 46.6 kW/litre
Valve gear	: 4-valves per cylinder, DOHC
Fuel system	: MPFI (Petrol), Turbocharged diesel

Fuel tank capacity	: 35 litres
Fuel consumption	: City: 11.5 (16.0) kpl, Highway: 15.3 (19.1) kpl

28. ENGINE FOR CHEVROLET 'SAIL'

Manufacturer	: General Motors India
Model	: 2013
Type	: 4-cylinder, inline, front, transverse
Bore	: 69.7 mm
Stroke	: 82.0 mm
Displacement	: 1249 cc
Compression ratio	: 16.5 : 1
Maximum Power	: 57.4 kW at 4000 rpm
Maximum Torque	: 205 Nm at 1750 rpm
Power to Weight ratio	: 51.1 kW/tonne
Torque to Weight ratio	: 182 N/mtonne
Specific Output	: 46 kW/litre
Valve Gear	: DOHC, 4 valves per cylinder
Fuel System	: Turbo-diesel
Fuel Tank Capacity	: 40 litres
Fuel consumption	: City: 12.9 kpl, Highway: 18.7 kpl

29. ENGINE FOR CHEVROLET ENJOY

Manufacturer	: General Motors India
Engine Type	: 4-cylinder, inline turbo-diesel
Bore	: 69.7 mm
Stroke	: 82.0 mm
Displacement	: 1248 cc
Compression ratio	: 16.5 : 1
Maximum Power	: 57.0 kW at 4000 rpm
Maximum Torque	: 187 Nm at 1750 rpm
Power to Weight ratio	: 42.3 kW/tonne
Torque to Weight ratio	: 139 Nm/tonne
Specific Output	: 45.7 kW/litre
Valve Gear	: 4 valves per cylinder, DOHC
Fuel Tank Capacity	: 55 litres
Fuel consumption	: City: 12.3 kpl, Highway: 16.0 kpl

30. ENGINE FOR CITY i-VTEC CAR

(Data for diesel engine inside brackets)

Manufacturer	: Honda Siel Cars Ltd., India
Model	: 2013
Engine Type	: 4-cylinder in-line, front, transverse, petrol (diesel engine with turbo)
Bore	: 73 (76) mm

Stroke	: 89.4 (82.5) mm
Displacement	: 1497 (1498) cc
Compression ratio	: 10.4 : 1 (16.0:1)
Maximum Power	: 86.5 (73.5) kW at 6600 (3600) rpm
Maximum Torque	: 146.2 (200) Nm at 4600 (1750) rpm
Power to Weight ratio	: 78.7 (69) kW/tonne
Torque to Weight ratio	: 132.9 (171.7) Nm/tonne
Specific Output	: 57.8 (49.1) kW/litre
Valve Gear	: 4-valves per cylinder, SOHC (DOHC)
Fuel tank capacity	: 40 litres
Fuel consumption	: City: 11 (14.2) kpl, Highway: 17 (19.5) kpl (Manual)

31. ENGINE FOR BRIO

Manufacturer	: Honda Siel Cars, India
Model	: 2011
Type	: 4-cylinder, inline, front, transverse, alloy block and head
Bore	: 73.0 mm
Stroke	: 71.6 mm
Displacement	: 1198 cc
Compression ratio	: 10.2 : 1
Maximum Power	: 65.6 kW at 6000 rpm
Maximum Torque	: 109 Nm at 4600 rpm
Power to Weight ratio	: 71 kW/tonne
Torque to Weight ratio	: 117.8 Nm/tonne
Specific Output	: 54.8 kW/litre
Valve Gear	: 4-valves per cylinder, SOHC, i-VTEC
Fuel System	: Petrol
Fuel Tank Capacity	: 35 litres
Fuel consumption	: N.A.

32. ENGINE FOR 'AMAZE' CAR

Manufacturer	: Honda Siel Cars, India
Model	: 2013
Type	: 4-cylinder, inline, petrol/diesel (without/with turbocharger) Petrol: i-VTEC, Diesel: i-DTEC
Bore	: 73.0 (76.0) mm
Stroke	: 71.6 (82.5) mm
Displacement	: 1198 (1498) cc
Compression ratio	: 10.2 : 1 (16.0 : 1)
Maximum Power	: 64.6 (73.5) kW at 6000 (3600) rpm
Maximum Torque	: 109 (200) Nm at 4500 (1750) rpm
Power to Weight ratio	: 67.1 (64.4/68.4) kW/tonne

Torque to Weight ratio	: 112.9 (108.4/186.0) Nm/tonne
Specific Output	: 54.0 (49.1) kW/litre
Valve Gear	: 4-valves per cylinder, SOHC (DOHC)
Fuel System	: i-VTEC
Fuel Tank Capacity	: 35 litres
Fuel consumption	: City: 12.5 (11.5/15.2) kpl, Highway: 17.0 (16.1/20.8) kpl

(Figures in brackets are for diesel engine. Figures before slash are for engine without turbocharger and after slash are for engine with turbocharger)

33. ENGINE FOR CR-V 2.0

Manufacturer	: Honda Siel Cars India Ltd.
Engine Type	: 4-cylinder, inline, front transverse
Bore	: 81.0 mm
Stroke	: 96.9 mm
Displacement	: 1997 cc
Compression ratio	: 10.6 : 1
Maximum Power	: 114 kW at 6500 rpm
Maximum Torque	: 190 Nm at 4300 rpm
Power to Weight ratio	: 77.6 kW/tonne
Torque to Weight ratio	: 129 Nm/tonne
Specific Output	: 57.1 kW/litre
Valve Gear	: 4 valves per cylinder, SOHC
Fuel Tank Capacity	: 58 litres
Fuel consumption	: City: 9.0 kpl, Highway: 12.0 kpl

34. ENGINE FOR JAZZ

Manufacturer	: Honda Siel Cars India Ltd. (Data in brackets for diesel engine)
Engine Type	: 4-cylinder, inline, front, transverse
Bore	: 73.0 (76.0) mm
Stroke	: 71.6 (82.5) mm
Displacement	: 1198 (1498) cc
Compression ratio	: 10.8 (16.0) : 1
Maximum Power	: 66.0 kW at 6000 rpm (73.1 kW at 3600 rpm)
Maximum Torque	: 110 Nm at 4800 rpm (200 Nm at 1750 rpm)
Power to Weight ratio	: 63.3 (70.2) kW/tonne
Torque to Weight ratio	: 105.6 (173.3) Nm/tonne
Specific Output	: 55.1 (48.8) kW/litre
Valve Gear	: 4-valves per cylinder, SOHC (4 valves per cylinder, DOHC)

Fuel Tank Capacity	: 40 litres
Fuel consumption	: City: 12.3 (14.4) kpl, Highway: 16.9 (19.8) kpl

35. ENGINE FOR HONDA BR-V

Manufacturer	: Honda Siel Cars India Ltd. (Data in brackets for diesel engine)
Engine Type	: 4-cylinder, inline
Bore	: 73.0 (76) mm
Stroke	: 89.4 (82.5) mm
Displacement	: 1497 (1498) cc
Compression ratio	: 10.3:1 (16.0:1)
Maximum Power	: 88.8 kW at 6600 rpm (74.6 kW at 3600 rpm)
Maximum Torque	: 145 Nm at 4600 rpm (200 Nm at 1750 rpm)
Power to Weight ratio	: 71.9 (60.4) kW/tonne
Torque to Weight ratio	: 117.4 (153.1) Nm/tonne
Specific Output	: 59.3 (49.8) kW/litre
Valve Gear	: 4 valves per cylinder, SOHC (DOHC)
Fuel system	: Petrol (Diesel)
Fuel Tank Capacity	: 42 litres
Fuel consumption	: City: 11.01 (13.88) kpl, Highway: 14.01 (20.04) kpl

36. ENGINE FOR TUCSON (SUV)

Manufacturer	: Honda Motor India Ltd. (Data in brackets for diesel engine)
Engine Type	: 4-cylinder, inline
Bore	: 81 (84) mm
Stroke	: 97 (90) mm
Displacement	: 1999 (1995) cc
Compression ratio	: 10.3 : 1 (16 : 1)
Maximum Power	: 115.6 kW at 6200 rpm (138 kW at 4000 rpm)
Maximum Torque	: 192.2 Nm at 4000 rpm (400 Nm at 1750-2750 rpm)
Power to Weight ratio	: 76.3 (91.1) kW/tonne
Torque to Weight ratio	: 126.9 (241.6) Nm/tonne
Specific Output	: 57.8 (69.2) kW/litre
Valve Gear	: 4 valves per cylinder, (DOHC)
Fuel system	: Petrol (Diesel)
Fuel Tank Capacity	: 62 litres
Fuel consumption	: City: 7.0 (10.3) kpl, Highway: 11.8 (12.7) kpl

37. ENGINE FOR 'EON' CAR

Manufacturer	: Hyundai Motor India Ltd.
Model	: 2011

Type : 3-cylinder, inline, front, transverse
Bore : 67.0 mm
Stroke : 77.0 mm
Displacement : 814 cc
Compression ratio : 10.1 : 1
Maximum Power : 41 kW at 5500 rpm
Maximum Torque : 74.5 Nm at 4000 rpm
Power to Weight ratio : 53.1 kW/tonne
Torque to Weight ratio : 96.1 Nm/tonne
Specific Output : 50.4 kW/litre
Valve Gear : 3-valves per cylinder, DOHC
Fuel System : Petrol
Fuel Tank Capacity : 32 litres
Fuel consumption : City: 13.7 kpl, Highway: 17.2 kpl

38. ENGINE FOR HYUNDAI VERNA (2011)

(Data in brackets for Diesel, manual transmission model)

Manufacturer : Hyundai Motor India Ltd.
Type : 4-cylinder, inline front transverse, aluminium head and iron block
Bore : 77.0 (77.2) mm
Stroke : 85.44 (84.5) mm
Displacement : 1591 (1582) cc
Compression ratio : 10.5 : 1 (17.3 : 1)
Maximum Power : 90.5 kW at 6300 rpm (94.2 kW at 4000 rpm)
Maximum Torque : 155 Nm at 4200 rpm (260 Nm at 1900-2750 rpm)
Power to Weight ratio : 84.5 (79.1) kW/tonne
Torque to Weight ratio : 144.7 (218.3) Nm/tonne
Specific output : 56.9 (59.3) kw/litre
Valve Gear : DOHC, 4 valves per cylinder. Hyundai VTVT variable valve timing system working on the camshaft phasing
Fuel system : (Common rail)
Fuel tank capacity : 43 litres
Fuel consumption : City : 9.8 (13.8) kpl, Highway : 15.3 (17.9) kpl

39. ENGINE FOR HYUNDAI I10 (1.2 KAPPA 2)/ I20

Manufacturer : Hyundai Motor India Ltd.
Type : 4-cylinder in line, front traverse, alloy head and block
Bore : 71.0 mm
Stroke : 75.6 mm
Displacement : 1197 cc
Compression ratio : 10.5 : 1

Maximum Power	: 59.2 kW at 6000 rpm
Maximum Torque	: 111.8 Nm at 4000 rpm
Power to Weight ratio	: 63.6 kW (55.3 for i 20)/tonne
Torque to Weight ratio	: 120 Nm (105 for i 20)/tonne
Specific output	: 49.2 kw/litre
Valve Gear	: DOHC, 4 valves pe : cylinder.
Fuel tank capacity	: 35 litres
Fuel consumption	: City : 11.7 kpl, Highway : 16.0 kpl

40. ENGINE FOR ELANTRA CAR

Manufacturer	: Hyundai Motor India Ltd. (Data in brackets are for diesel engine)
Engine Type	: 4-cylinder, inline
Bore	: 81.0 (77.0) mm
Stroke	: 97.0 (84.5) mm
Displacement	: 1999 (1582) cc
Compression ratio	: 10.3:1 (17.3:1)
Maximum Power	: 113.4 kW at 6200 rpm (95.5 kW at 4000 rpm)
Maximum Torque	: 192 Nm at 4000 rpm (260 Nm at 1900-2750 rpm)
Power to Weight ratio	: 89.2 (70.7) kW/tonne
Torque to Weight ratio	: 151 (194.0) Nm/tonne
Specific Output	: 56.7 (60.4) kW/litre
Valve Gear	: 4 valves per cylinder, DOHC
Fuel System	: Diesel engine is with variable-geometry turbo (VGT)
Fuel Tank Capacity	: 56 litres
Fuel consumption	: City: 9.8 (12.8) kpl, Highway: 14.4 (17) kpl

41. ENGINE FOR GRAND I10

Manufacturer	: Hyundai Motor India Ltd.
Model	: 2013
Engine Type	: 3-cylinder, inline, common rail, diesel, front, transverse
Bore	: 75.0 mm
Stroke	: 54.5 mm
Displacement	: 1120 cc
Compression ratio	: 16.0 : 1
Maximum Power	: 52.2 kW at 4000 rpm
Maximum Torque	: 160 Nm at 1500-2750 rpm
Power to Weight ratio	: 47.8 kW/tonne
Torque to Weight ratio	: 146.3 Nm/tonne
Specific Output	: 46.6 kW/litre
Valve Gear	: 4 valves per cylinder, DOHC

Fuel Tank Capacity	: 43 litres
Fuel consumption	: City: 15.4 kpl, Highway: 19.6 kpl

42. ENGINE FOR CRETA

Manufacturer	: Hyundai Motor India Ltd. (Data in brackets for diesel model)
Engine Type	: 4-cylinder, inline, front, transverse
Bore	: 77.0 (77.2) mm
Stroke	: 85.44 (84.5) mm
Displacement	: 1591 (1582) cc
Compression ratio	: 10.5 (17.3):1
Maximum Power	: 90.5 kW at 6400 rpm (94.2 kW at 4000 rpm)
Maximum Torque	: 151 Nm at 4850 rpm (260 Nm at 1900 rpm)
Power to Weight ratio	: 71.5 (74.5) kW/tonne
Torque to Weight ratio	: 119.4 (192.6) Nm/tonne
Specific Output	: 56.9 (59.5) kW/litre
Valve Gear	: 4 valves per cylinder, DOHC (4 valves per cylinder DOHC)
Fuel systek	: Petrol, Turbo diesel
Fuel Tank Capacity	: 55 litres
Fuel consumption	: City: 9.5 (12.5) kpl, Highway: 14.5 (17.5) kpl

43. ENGINE FOR I20 ELITE

Manufacturer	: Hyundai Motor India Ltd. (Fig. in brackets are for diesel model)
Engine Type	: 4-cylinder, inline, front, transverse, petrol/turbo-diesel
Bore	: 71.0 (75.0) mm
Stroke	: 75.6 (79.0) mm
Displacement	: 1197 (1396) cc
Compression ratio	: 10.5 (17) : 1
Maximum Power	: 61 (66.4) kW at 6000 (4000) rpm
Maximum Torque	: 115 (220) Nm at 4000 (1500-2750) rpm
Power to Weight ratio	: 56.6 (55.3) kW/tonne
Torque to Weight ratio	: 95.8 (183) Nm/tonne
Specific Output	: 51 (55.5) kW/litre
Valve Gear	: 4 valves per cylinder, DOHC
Fuel Tank Capacity	: 45 litres
Fuel consumption	: City: 11.5 (13.8) kpl, Highway: 15.2 (19.5) kpl

44. ENGINE FOR XCENT CAR

(Data in brackets for diesel engine)

Manufacturer	: Hyundai Motor India Ltd.
Engine Type	: 4-cylinder, inline, front, transverse, petrol/diesel
Bore	: 71.0 (75.0) mm
Stroke	: 75.6 (54.4) mm
Displacement	: 1197 (1120) cc
Compression ratio	: 10.5 (16.0) : 1
Maximum Power	: 61 (53) kW at 6000 (4000) rpm
Maximum Torque	: 113.8 (180) Nm at 4000 (1750-2500) rpm
Power to Weight ratio	: 60 (52) kW/tonne
Specific Output	: 51 (47.3) kW/litre
Valve Gear	: 4 valves per cylinder, DOHC
Fuel Tank Capacity	: 43 litres
Fuel consumption	: City: 11.5 (16.2) kpl, Highway: 16.3 (20.3) kpl

45. ENGINE FOR 'AVENTADOR' CAR

Manufacturer	: Automobile Lamborghini S.p.A., Italy
Model	: 2012
Type	: V-12, mid, longitudinal, petrol engine
Bore	: 95.0 mm
Stroke	: 76.4 mm
Displacement	: 6498 cc
Compression ratio	: 11.8 : 1
Maximum Power	: 515 kW at 8250 rpm
Maximum Torque	: 690 Nm at 5500 rpm
Power to Weight ratio	: 327 kW/tonne
Torque to Weight ratio	: 438 Nm/tonne
Specific Output	: 79.2 kW/litre
Valve Gear	: 4-valves per cylinder, DOHC
Fuel Tank Capacity	: 90 litres
Fuel consumption	: City: 3.6 kpl, Highway: 8.8 kpl

46. ENGINE FOR 'DISCOVERY 4' SUV

Manufacturer	: Land Rover
Model	: 2012
Type	: 6-cylinder, in V, twin-turbo diesel
Bore	: 84.0 mm
Stroke	: 90.0 mm
Displacement	: 2993 cc
Compression ratio	: 16.1 : 1

Maximum Power	: 180 kW at 4000 rpm
Maximum Torque	: 6000 Nm at 2000 rpm
Power to Weight ratio	: 69.6 kW/tonne
Torque to Weight ratio	: 232 Nm/tonnes
Specific Output	: 60 kW/litre
Valve Gear	: 4-valves per cylinder, DOHC
Fuel Tank Capacity	: 82 litres
Fuel consumption	: City: 6.5 kpl, Highway: 9.5 kpl

47. ENGINE FOR SCORPIO CRDE S.U.V.

Manufacturer	: Mahindra & Mahindra Limited, India
Model	: 2014
Engine Type	: 4-cylinder, in-line, common rail turbo-diesel, iron block and aluminium alloy head
Bore	: 85.0 mm
Stroke	: 96.0 mm
Displacement	: 2179 cc
Compression ratio	: 18.5
Maximum Power	: 88 kW at 4000 rpm
Maximum Torque	: 280 Nm at 1800-2800 rpm
Power to Weight ratio	: 48.35 kW/tonne
Torque to Weight ratio	: 153.8 Nm/tonne
Specific Output	: 40.4 kW/litre
Valve Gear	: 4 valves per cylinder
Fuel System	: Common rail direct injection

48. ENGINE FOR XYLO

Manufacturer	: Mahindra & Mahindra Limited, India
Model	: 2012
Type	: 4-cylinder in-line, front longitudinal, diesel
Bore	: 85.0 mm
Stroke	: 96.0 mm
Displacement	: 2179 cc
Compression ratio	: 18.5 : 1
Maximum Power	: 91 kW at 4000 rpm
Maximum Torque	: 280 Nm at 2800 rpm
Power to Weight ratio	: 50.5 kW/tonne
Torque to Weight ratio	: 155 Nm/tonne
Specific Output	: 41.7 kW/litre
Valve Gear	: DOHC, 4 valves per cylinder
Fuel System	: Common-rail, direct injection
Fuel Tank Capacity	: 55 litres
Fuel consumption	: City: 10.1 kpl, Highway: 14.4 kpl, Turbo-diesel

49. ENGINE FOR 'VERITO' CAR

Manufacturer	: Mahindra and Mahindra
Model	: 2013
Type	: 4-cylinder, inline, front, transverse
Bore	: 76.0 mm
Stroke	: 80.5 mm
Displacement	: 1461 cc
Compression ratio	: 17.9 : 1
Maximum Power	: 48.5 kW at 4000 rpm
Maximum Torque	: 160 Nm at 2000 rpm
Power to Weight ratio	: 140 kW/tonne
Torque to Weight ratio	: 42.5 Nm/tonne
Specific Output	: 109 kW/litre
Valve Gear	: 2-valves per cylinder, SOHC
Fuel System	: Turbo-diesel
Fuel Tank Capacity	: 50 litres
Fuel consumption	: City: 13.6 kpl, Highway: 19.3 kpl

50. ENGINE FOR 'QUANTO' SUV

Manufacturer	: Mahindra and Mahindra, India
Model	: 2012
Type	: 3-cylinder, inline
Bore	: 83.0 mm
Stroke	: 92.0 mm
Displacement	: 1493 cc
Compression ratio	: 18.5 : 1
Maximum Power	: 73.5 kW at 3750 rpm
Maximum Torque	: 240 Nm at 1600-2800 rpm
Power to Weight ratio	: 44.8 kW/tonne
Torque to Weight ratio	: 146 Nm/tonne
Specific Output	: 49.2 kW/litre
Valve Gear	: DOHC, 4-valves per cylinder
Fuel System	: Common rail, turbo-diesel
Fuel Tank Capacity	: 55 litres
Fuel consumption	: City: 11.4 kpl, Highway: 16.1 kpl

51. ENGINE FOR 'E 20'

Manufacturer	: Mahindra and Mahindra Limited, India
Model	: 2013
Type	: Electric
Battery	: 48 V, Lithium-ion
Maximum Power	: 19 kW at 3750 rpm

Maximum Torque	: 53 Nm at 0-3400 rpm
Power to Weight ratio	: 22.9 kW/tonne
Torque to Weight ratio	: 63.7 Nm/tonne
Range	: 100 km in one full charge under test conditions
Charging time	: 5 hours to full charge from a standard 220 V, 15 A socket.

52. ENGINE FOR 'REXTON' SUV

Manufacturer	: Mahindra Ssangyong
Model	: 2012
Type	: 5-cylinder, inline,
Bore	: 86.2 mm
Stroke	: 92.4 mm
Displacement	: 2696 cc
Compression ratio	: 17.5 : 1
Maximum Power	: 137 kW at 4000 rpm
Maximum Torque	: 402 Nm at 1600-3100 rpm
Power to Weight ratio	: 69.1 kW/tonne
Torque to Weight ratio	: 202.5 Nm/tonne
Specific Output	: 50.8 kW/litre
Valve Gear	: DOHC, 4-valves per cylinder
Fuel System	: Common-rail, turbo-diesel
Fuel Tank Capacity	: 78 litres
Fuel consumption	: City: 7.2 kpl, Highway: 11.8 kpl

53. ENGINE FOR KUV 100 K8

Manufacturer	: Mahindra & Mahindra Ltd. (Data in brackets for diesel engine)
Engine Type	: 3-cylinder, inline, front, transverse
Bore	: 76.0 mm
Stroke	: 88.0 mm
Displacement	: 1198 cc
Compression ratio	: NA
Maximum Power	: 61 kW at 5500 rpm (57.4 kW at 3750 rpm)
Maximum Torque	: 115 Nm at 3500-3600 rpm (190 Nm at 1750-2250 rpm)
Power to Weight ratio	: 53.5 (50.3) kW/tonne
Torque to Weight ratio	: 100.9 (166.7) Nm/tonne
Specific Output	: 50.9 (47.9) kW/litre
Valve Gear	: 4 valves per cylinder, DOHC
Fuel system	: Petrol, Turbo-diesel
Fuel Tank Capacity	: 35 litres
Fuel consumption	: ARAI: 18.15 (25.32 kpl) kpl

54. ENGINE FOR NUVOSPORT N 8

Manufacturer	: Mahindra & Mahindra Ltd. (Data inside brackets for AMT model)
Engine Type	: 4-cylinder, inline, front, longitudinal
Bore	: 83.0 mm
Stroke	: 92.0 mm
Displacement	: 1493 cc
Compression ratio	: 16.5:1
Maximum Power	: 75.7 kW at 3750 rpm
Maximum Torque	: 240 Nm at 1600-2800 rpm
Power to Weight ratio	: 45.3 (44.6) kW/tonne
Torque to Weight ratio	: 143.7 (141.3) Nm/tonne
Specific Output	: 50.7 kW/litre
Valve Gear	: 4 valves per cylinder, DOHC
Fuel system	: Diesel
Fuel Tank Capacity	: 60 litres
Fuel consumption	: City: 11.0 (10.7) kpl, Highway: 15.0 (13.6) kpl

55. ENGINE FOR XUV 500

Manufacturer	: Mahindra & Mahindra Ltd.
Engine Type	: 4-cylinder, inline, front, transverse
Bore	: 85.0 mm
Stroke	: 96.0 mm
Displacement	: 2179 cc
Compression ratio	: 16.5:1
Maximum Power	: 102.9 kW at 3750 rpm
Maximum Torque	: 457 Nm at 1600-2800 rpm
Power to Weight ratio	: 56.2 kW/tonne
Torque to Weight ratio	: 250 Nm/tonne
Specific Output	: 47.3 kW/litre
Valve Gear	: 4 valves per cylinder, DOHC
Fuel system	: Diesel
Fuel Tank Capacity	: 70 litres

56. ENGINE FOR KUV 100 K8

Manufacturer	: Mahindra & Mahindra Ltd. (Data in brackets is for diesel engine)
Engine Type	: 3-cylinder, inline, front, longitudinal
Bore	: 83.0 mm
Stroke	: 92.0 mm
Displacement	: 1493 cc
Compression ratio	: 16.5 : 1

Maximum Power	: 61.5 (58.9) kW at 3750 rpm
Maximum Torque	: 230 Nm at 1500-2250 rpm
Power to Weight ratio	: 37.8 (35.6) kW/tonne
Torque to Weight ratio	: 141.5 (139.0) Nm/tonne
Specific Output	: 41.2 kW/litre
Valve Gear	: 4 valves per cylinder, DOHC
Fuel system	: Turbo diesel
Fuel Tank Capacity	: 60 litres
Fuel consumption	: City: 11.4 (NA) kpl, Highway: 15.5 (NA) kpl

57. ENGINE FOR ALTO 800

Manufacturer	: Maruti Suzuki, India
Model	: 20
Type	: 3-cylinder, inline, front, transverse petrol engine
Bore	: 68.5 mm
Stroke	: 72.0 mm
Displacement	: 796 cc
Compression ratio	: 10.3 : 1
Maximum Power	: 35 kW at 6000 rpm
Maximum Torque	: 69 Nm at 3500 rpm
Power to Weight ratio	: 48 kW/tonne
Torque to Weight ratio	: 95 Nm/tonne
Specific Output	: 44 kW/litre
Valve Gear	: 3-valves per cylinder
Fuel Tank Capacity	: 35 litres
Fuel consumption	: City: 13.3 kpl, Highway: 17.8 kpl

58. ENGINE FOR ESTILO/A-STAR

Manufacturer	: Maruti Suzuki India
Engine Type	: 3-cylinder inline, petrol, front, transverse (K10B)
Bore	: 73 mm
Stroke	: 79.5 mm
Displacement	: 996 cc (LX); 1061 cc (VX)
Compression ratio	: 10.0
Maximum Power	: 50 kW at 6200 rpm (LX)
Maximum Torque	: 90 Nm at 3500 rpm (LX)
Specific output	: 50 kW/litre
Valve Gear	: 4 valves per cylinder, DOHC
Ignition and Fuel System	: Electronic MPFI
Fuel tank capacity	: 35 litres
Fuel consumption	: City : 12.1 kpl, Highway : 17.1 kpl

59. ENGINE FOR WAGON R

Manufacturer	: Maruti Suzuki India
Model	: 2012
Engine Type	: 4-cylinder, in line, ferrous engine F10D
Bore	: N.A.
Stroke	: N.A.
Capacity	: 998 cc
Compression ratio	: N.A.
Maximum Power	: 52 kW at 6200 rpm
Maximum Torque	: 92 Km at 3500 rpm
Power to Weight ratio	: 60.4 kW/tonne
Torque to Weight ratio	: 109 Nm/tonne
Specific output	: 52 kW/litre
Valve gear	: 4-valves per cylinder
Fuel system	: Multi-point fuel injection
Fuel tank capacity	: 35 litres
Fuel consumption	: City : 13.0 kpl, Highway : 18.0 kpl
Ignition System	: Electronic, transistorised

60. MARUTI SWIFT (SWIFT DZIRE)

(Data inside brackets is for diesel model)

Manufacturer	: Maruti Suzuki India
Engine Type	: 4-cylinder, naturally aspirated petrol (K 12 M)/diesel (1.3 L DDiS) engine
Bore	: 73.0 (69.6) mm
Stroke	: 71.5 (82) mm
Displacement	: 1197 (1248) cc
Compression ratio	: 10.0 : 1 (17.6 : 1)
Maximum Power	: 62.7 kW at 6000 rpm (56 kW at 4000 rpm)
Maximum Torque	: 113 Nm at 4500 rpm (190 Nm at 2000 rpm)
Specific output	: 52.3 (52.09) kW/litre
Valve Gear	: 4 per cylinder, SOHC (4 per cylinder, DOHC)
Fuel supply system	: M.P.F.I.
Fuel tank capacity	: 43 litres
Fuel consumption	: City : 11.9 (14.4) kpl, Highway : 17.0 (19.1) kpl

61. MARUTI RITZ (PETROL/DIESEL)

(Figs. in brackets are for diesel model)

Manufacturer	: Maruti Suzuki India
Engine Type	: 4-cylinder, alloy block and head
Bore	: 73.0 (69.6) mm
Stroke	: 71.5 (82) mm
Displacement	: 1197 (1248) cc

Compression ratio	: 10.0 : 1 (17.6 : 1)
Maximum Power	: 62.7 kW at 6000 rpm (55.2 kW at 4000 rpm)
Maximum Torque	: 112.8 Nm at 4500 rpm (190 Nm at 2000 rpm)
Power to Weight ratio	: 60.8 (53.6) kW/tonne
Torque to Weight ratio	: 101.6 (171.2) Nm/tonne
Specific output	: 52.4 (44.2) kW/litre
Valve Gear	: DOHC, 4 valves per cylinder
Fuel tank capacity	: 43 litres
Fuel consumption	: City : 11.8 kpl, Highway : 16.7 kpl

62. ENGINE FOR 'EECO' VAN

Manufacturer	: Maruti Suzuki, India
Model	: 2012
Type	: 4-cylinder, inline, aluminium
Displacement	: 1196 cc
Compression ratio	: 9.9 : 1
Maximum Power	: 54.5 kW at 6000 rpm
Maximum Torque	: 101 Nm at 3000 rpm
Power to Weight ratio	: 58.7 kW/tonne
Torque to Weight ratio	: 109 kW/tonne
Specific Output	: 45.6 kW/litre
Valve Gear	: 4 valves per cylinder
Fuel System	: MPFI
Fuel Tank Capacity	: 40 litres
Fuel consumption	: ARAI: 15.1 kpl

63. ENGINE FOR ERTIGA

Manufacturer	: Maruti Suzuki India Ltd.
Engine Type	: 4-cylinder, inline, turbo diesel, front, transverse
Bore	: 69.6 mm
Stroke	: 82.0 mm
Displacement	: 1248 cc
Compression ratio	: 17.6 : 1
Maximum Power	: 66.4 kW at 4000 rpm
Maximum Torque	: 200 Nm at 1750 rpm
Power to Weight ratio	: 53.7 kW/tonne
Torque to Weight ratio	: 162 Nm/tonne
Specific Output	: 53.2 kW/litre
Valve Gear	:
Fuel Tank Capacity	: 45 litres
Fuel consumption	: City: 12.8 kpl, Highway: 16.8 kpl

64. ENGINE FOR CELERIO

Manufacturer	: Maruti Suzuki India Ltd. (Data inside brackets for diesel engine)
Engine Type	: 3-cylinder, inline (2-cylinder), front, transverse
Bore	: 73.0 (77.0) mm
Stroke	: 79.5 (85.1) mm
Displacement	: 998 (793) cc
Compression ratio	: 11.0 (15.0) : 1
Maximum Power	: 50 kW at 6000 rpm (35 kW at 3500 rpm)
Maximum Torque	: 89 Nm at 3500 rpm (125 Nm at 2000 rpm)
Power to Weight ratio	: 60 (38.9) Nm/tonne
Torque to Weight ratio	: 107 (139) Nm/tonne
Specific Output	: 50 (44.1) kW/litre
Valve Gear	: 4 valves per cylinder, DOHC
Fuel Tank Capacity	: 35 litres
Fuel consumption	: City: 16.7 kpl, Highway: 20.9 kpl)

65. ENGINE FOR S-CROSS

Manufacturer	: Maruti Suzuki India Ltd.
Engine Type	: 4-cylinder, inline, front, transverse
Bore	: 79.5 mm
Stroke	: 80.0 mm
Displacement	: 1598 cc
Compression ratio	: 16.5 : 1
Maximum Power	: 88 kW at 3750 rpm
Maximum Torque	: 320 Nm at 1750 rpm
Power to Weight ratio	: 69.0 kW/tonne
Torque to Weight ratio	: 251 Nm/tonne
Specific Output	: 55.1 kW/litre
Valve Gear	: 4 valves per cylinder, DOHC
Fuel system	: Diesel
Fuel Tank Capacity	: 48 litres

66. ENGINE FOR BALENO

Manufacturer	: Maruti Suzuki India Ltd. (Data in brackets for diesel engine)
Engine Type	: 4-cylinder, inline, front, transverse
Bore	: 73.0 (69.6)
Stroke	: 71.5 (82.0)

Displacement	: 1197 (1248) cc
Compression ratio	: 11.0 (17.6) : 1
Maximum Power	: 62.0 kW at 6000 rpm (55.2 kW at 4000 rpm)
Maximum Torque	: 115 Nm at 4000 rpm (190 Nm at 2000 rpm)
Power to Weight ratio	: 69.7 (56.0) kW/tonne
Torque to Weight ratio	: 129.2 (192.9) Nm/tonne
Specific Output	: 51.8 (44.2) kW/litre
Valve Gear	: 4 valves per cylinder, DOHC
Fuel system	: Petrol/Turbo diesel
Fuel Tank Capacity	: 37 litres
Fuel consumption	: City: 13.4 (14.9) kpl, Highway: 18.1 (21.4) kpl

67. ENGINE FOR CIAZ

Manufacturer	: Maruti Suzuki India Ltd.
Engine Type	: 4-cylinder, inline, front, transverse
Bore	: 73.0 mm
Stroke	: 82.0 mm
Displacement	: 1373 cc
Compression ratio	: 11.0 : 1
Maximum Power	: 67.9 kW at 6000 rpm
Maximum Torque	: 130 Nm at 4000 rpm
Power to Weight ratio	: 66.2 kW/tonne
Torque to Weight ratio	: 127 Nm/tonne
Specific Output	: 49.5 kW/litre
Valve Gear	: 4 valves per cylinder, DOHC
Fuel system	: Petrol
Fuel Tank Capacity	: 43 litres
Fuel consumption	: City: 11.5 kpl, Highway: 17.2 kpl

68. ENGINE FOR A 180 CAR

Manufacturer	: Mercedes-Benz India (Data in brackets are for diesel CDI engine)
Engine Type	: 4-cylinder, inline, front, transverse
Bore	: 83.0 (83.0) mm
Stroke	: 73.7 (83.0) mm
Displacement	: 1595 (2143) : 1
Compression ratio	: 10.3 (16.2) : 1
Maximum Power	: 91 (80) kW at (3200-4400) mm
Maximum Torque	: 145 (250) Nm at (1400-2800) rpm
Power to Weight ratio	: 66.2 (53.2) kW/tonne

Torque to Weight ratio : 105 (116) Nm/tonne
Specific Output : 57.0 (37.3) kW/litre
Valve Gear : 4 valves per cylinder, DOHC
Fuel system : Direct injection turbo-petrol (Common rail turbo-diesel)
Fuel Tank Capacity : 50 litres
Fuel consumption : City: NA (11.0) kpl, Highway NA (15.0) kpl

69. ENGINE FOR B 180 CAR

Manufacturer/Assembler : Mercedes-Benz India
Engine Type : 4-cylinder, inline
Bore : 83 mm
Stroke : 73.7 mm
Displacement : 1595 cc
Compression ratio : 10.3 : 1
Maximum Power : 90.2 kW at 5000 rpm
Maximum Torque : 200 Nm at 1250-4000 rpm
Power to Weight ratio : 63.3 kW/tonne
Torque to Weight ratio : 140 Nm/tonne
Specific Output : 56.5 kW/litre
Valve Gear : 4 valves per cylinder, DOHC
Fuel system : Direct injection, turbo
Fuel Tank Capacity : 50 litres
Fuel consumption : City: 9.5 kpl, Highway: 14.2 kpl

70. ENGINE FOR S 500 CAR

Manufacturer : Mercedes-Benz India
Engine Type : 4-cylinder, twin-turbo, petrol, front, longitudinal
Bore : 92.9 mm
Stroke : 86.0 mm
Displacement : 4663 cc
Compression ratio : 10.5 : 1
Maximum Power : 338 kW at 5250-5500
Maximum Torque : 700 Nm at 1800-3500 rpm
Power to Weight ratio : 153.6 kW/tonne
Torque to Weight ratio : 318 Nm/tonne
Specific Output : 72.5 kW/litre
Valve Gear : 4 valves per cylinder, DOHC
Fuel Tank Capacity : 80 litres
Fuel consumption : City: 4.95 kpl, Highway: 7.0 kpl

71. ENGINE FOR GL 350 CDI

Manufacturer : Mercedes-Benz India
Engine Type : V-6 front, longitudinal

Bore	: 83 mm
Stroke	: 92 mm
Displacement	: 2987 cc
Compression ratio	: 15.5 : 1
Maximum Power	: 192.4 kW at 3600 rpm
Maximum Torque	: 620 Nm at 1600-2400 rpm
Power to Weight ratio	: 78.4 kW/tonne
Torque to Weight ratio	: 252 Nm/tonne
Specific Output	: 64.4 kW/litre
Valve Gear	: 4 valves per cylinder, DOHC per bank
Fuel system	: Common-rail diesel
Fuel Tank Capacity	: 100 litres
Fuel consumption	: City: 6.5 kpl, Highway: 10.5 kpl

72. ENGINE FOR MITSUBISHI LANCER CAR (PETROL/DIESEL)

(Data within brackets for diesel model)

Manufacturer	: Mitsubishi motors, Japan and Hindustan Motors Ltd., India
Type	: 4-cylinder, in line, SOHC, GLXi and (GLD) engine
Capacity	: 1468 cc, (1998 cc)
Maximum Power	: 64.9 kW at 5500 RPM (50.7 kW at 4500 RPM)
Maximum Torque	: 132.4 Nm at 3300 RPM, (122.6 Nm at 3000 RPM)
Power to weight ratio	: 64.3 (50.2) kW/tonne
Torque to weight ratio	: 131 (121) Nm/tonne
Specific output	: 44.2 (25.4) kw/litre
Valve gear	: SOHC, 3 valves per cylinder, (SOHC, 2 valves per cylinder)
Fuel System	: MPFI (Distributor type injection pump)
Fuel Tank Capacity	: 50 litres

73. ENGINE FOR MITSUBISHI CEDIA SPORTS

Manufacturer	: Mitsubishi Motors, Japan and Hindustan Motors Ltd., India
Type	: 4-cylinder in line, naturally aspirated, front transverse, alloy head & iron block
Bore	: 81.5 mm
Stroke	: 95.8 mm
Displacement	: 1999 CC
Compression ratio	: 9.5 : 1 (± 0.3)
Maximum Power	: 85.8 kW at 5250 rpm
Maximum Torque	: 175 Nm at 4250 rpm
Power to Weight ratio	: 70 kW/tonne

Torque to Weight ratio	: 143 Nm/tonne
Specific output	: 42.9 kW/litre
Valve gear	: S0HC, 4 valves per cylinder
Fuel system	: Petrol/LPG, Sequential multi-point LPG injection system.
Fuel tank capacity	: 50 litre (48 litre)
Fuel consumption	: City : 9.2 (6.4) kpl, Highway : 12.42 (11.1) kpl

74. ENGINE FOR MITSUBISHI PAJERO CAR (B.S. III AND IV)

Manufacturer	: Mitsubishi Motors, Japan and Hindustan Motors Ltd., India
Engine Type	: 4-cylinder in-line diesel, longitudinal, alloy head, iron block
Bore	: 95 mm
Stroke	: 100 mm
Capacity	: 2835 cc
Compression ratio	: 21:1
Maximum Power	: 87.3 kW at 4000 rpm; BS IV: 80 kW at 4000 rpm
Maximum Torque	: 292 Nm at 2000 rpm; B.S. IV: 275 Nm at 2000 rpm
Power to Weight ratio	: 40.5 kW/tonne; BS IV 37.1 kW/tonne
Specific output	: 30.8 kW/litre; BS IV: 28.2 kW/litre
Value Gear	: 4 valves per cylinder, DOHC (2 valves per cylinder, SOHC)
Fuel system	: Electronically controlled direct diesel injection, turbocharger, intercooled
Fuel tank	: 92 litres

75. ENGINE FOR MITSUBISHI MONTERO 3.2 DI-D

Manufacturer	: Mitsubishi Motors, Japan and Hindustan Motors Ltd., India
Type	: 4-cylinder, in-line, longitudinal, diesel, intercooled turbocharger
Bore	: 98.5 mm
Stroke	: 105 mm
Displacement	: 3198 cc
Compression ratio	: 17.0
Maximum Power	: 148.6 kW at 3500 rpm
Maximum Torque	: 441 Nm at 2000 rpm
Power to Weight ratio	: 64.3 kW/tonne
Torque to Weight ratio	: 191 Nm/tonne
Specific output	: 46.5 kW/litre
Valve gear	: DOHC, 4-valves per cylinder
Fuel system	: Common-rail, direct injection
Fuel tank capacity	: 88 litres
Fuel consumption	: City : 7.3 kpl, Highway : 10.1 kpl

76. ENGINE FOR MITSUBISHI OUTLANDER

Manufacturer	: Mitsubishi Motors, Japan and Hindustan Motors Ltd., India
Engine type	: 4-cylinder, inline, front, transversse, petrol

Bore	: 88 mm
Stroke	: 97 mm
Displacement	: 2360 cc
Compression ratio	: 10.5 : 1
Maximum Power	: 124.6 kW at 6000 rpm
Maximum Torque	: 225.6 Nm at 4100 rpm
Power to Weight ratio	: 77.1 kW/tonne
Torque to Weight ratio	: 139.7 Nm/tonne
Specific output	: 52.8 kW/litre
Valve gear	: DOHC, 4-valves per cylinder
Fuel system	: MPFI
Fuel tank capacity	: 60 litres
Fuel consumption	: City : 7.2 kpl, Highway : 9.7 kpl

77. ENGINE FOR MICRA PETROL/DIESEL

(Data in brackets for Diesel model)

Manufacturer	: Nissan Motor India
Model	: 2010
Type	: 3-cylinder (4-cylinder), inline, front, transverse
Bore	: 78.0 (76.0) mm
Stroke	: 73.6 (80.5) mm
Displacement	: 1198 (1461) cc
Compression ratio	: 9.8 (17.9 : 1)
Maximum Power	: 56 kW at 6000 rpm (47 Nm at 4000 rpm)
Maximum Torque	: 104 Nm at 4000 rpm (160 Nm at 2000 rpm))
Power to Weight ratio	: N.A. (46.6) kW/tonne
Torque to Weight ratio	: N.A. (32.2) kW/litre
Specific Output	: 46.7 (32.2) kW/litre
Valve Gear	: 4 valves per cylinder, DOHC (2 valves per cylinder, SOHC)
Fuel System	: ECCS, Electronic Fuel Injection Control (Common rail)
Fuel Tank Capacity	: 41 litres
Fuel consumption	: Petrol: N.A. Diesel: City: 14.6 kpl, Highway: 19.5 kpl

78. ENGINE FOR 'EVALIA' SUV

Manufacturer	: Nissan Motor, India
Model	: 2012
Type	: 4-cylinder, inline
Bore	: 76.0 mm
Stroke	: 80.5 mm
Displacement	: 1461 cc
Compression ratio	: 18.8 : 1
Maximum Power	: 62.7 kW at 3750 rpm

Maximum Torque	: 200 Nm at 1750 rpm
Power to Weight ratio	: 43.9 kW/tonne
Torque to Weight ratio	: 140 Nm/tonne
Specific Output	: 42.9 kW/litre
Valve Gear	: SOHC, 2 valves per cylinder
Fuel System	: Common rail, turbo-diesel
Fuel Tank Capacity	: 55 litres
Fuel consumption	: City: 12.8 kpl, Highway: 17.5 kpl

79. ENGINE FOR 'DATSUN GO' CAR

Manufacturer	: Nissan Motor India
Model	: 2014
Engine Type	: 3-cylinder, inline, front, transverse, petrol
Bore	: 78.0 mm
Stroke	: 83.6 mm
Displacement	: 1198 cc
Compression ratio	: 9.8 : 1
Maximum Power	: 50 kW at 5000 rpm
Maximum Torque	: 104 Nm at 4000 rpm
Power to Weight ratio	: 63.5 kW/tonne
Torque to Weight ratio	: 132 Nm/tonne
Specific Output	: 41.7 kW/litre
Valve Gear	: 4 valves per cylinder, DOHC
Fuel Tank Capacity	: 35 litres
Fuel consumption	: City: 12.8 kpl, Highway: 17.9 kpl

80. ENGINE FOR DATSUN REDIGO S

Manufacturer	: Nissan Motor India
Engine Type	: 3-cylinder, inline
Bore	: 69.0 mm
Stroke	: 71.2 mm
Displacement	: 799 cc
Compression ratio	: 10.5 : 1
Maximum Power	: 40.3 kW at 5678 rpm
Maximum Torque	: 72 Nm at 4386 rpm
Power to Weight ratio	: 60.1 kW/tonne
Torque to Weight ratio	: 107.5 Nm/tonne
Specific Output	: 50.4 kW/litre
Valve Gear	: 4 valves per cylinder, SOHC
Fuel system	: Petrol
Fuel Tank Capacity	: 28 litres
Fuel consumption	: City: 15.8 kpl, Highway: 19.97 kpl

81. ENGINE FOR TERRANO

Manufacturer	: Nissan Motor India
Engine Type	: 4-cylinder, inline, front, transverse
Bore	: 76.0 mm
Stroke	: 80.5 mm
Displacement	: 1461 cc
Compression ratio	: 15.3 : 1
Maximum Power	: 80.9 kW at 3900 rpm
Maximum Torque	: 248 Nm at 2250 rpm
Power to Weight ratio	: 61.8 kW/tonne
Torque to Weight ratio	: 190 km/tonne
Specific Output	: 55.4 kW/litre
Valve Gear	: 2 valves per cylinder, SOHC
Fuel system	: Diesel
Fuel Tank Capacity	: 50 litres

82. ENGINE FOR RIO

(Data in brackets for Diesel Model)

Manufacturer	: The Premier Automobiles Ltd.
Model	: 2012
Type	: 4-cylinder, inline, petrol/diesel
Bore	: N.A. (69.6) mm
Stroke	: N.A. (82.0) mm
Displacement	: 1173 (1248) cc
Compression ratio	: N.A.
Maximum Power	: 56.3 kW at 5800 rpm (53 kW at 4000 rpm)
Maximum Torque	: 103.9 Nm at 3500-4250 rpm (183 Nm at 1800-2400 rpm)
Power to Weight ratio	: 52.13 (46.3) kW/tonne
Torque to Weight ratio	: 96.2 (159.9) Nm/tonne
Specific Output	: 48 (42.5) kW/litre
Fuel System	: Petrol: Naturally aspirated, Diesel: Fixed geometry turbo charger
Fuel Tank Capacity	: 46 litres
Fuel consumption	: City: N.A. (13.8) kpl, Highway: N.A. (17.0) kpl

83. ENGINE FOR 'SCALA AUTO' CAR

Manufacturer	: Renault India Pvt. Ltd.
Model	: 2012
Type	: 4-cylinder, inline, petrol
Bore	: 78.0 mm
Stroke	: 78.4 mm
Displacement	: 1498 cc

Compression ratio	: 10.1 : 1
Maximum Power	: 72.8 kW at 5600 rpm
Maximum Torque	: 133.4 Nm at 4000 rpm
Power to Weight ratio	: 70 kW/tonne
Torque to Weight ratio	: 128 Nm/tonne
Specific Output	: 48.6 kW/litre
Valve Gear	: 4-valves per cylinder, DOHC
Fuel Tank Capacity	: 41 litres
Fuel consumption	: City: 10.7 kpl, Highway: 16.0 kpl

84. ENGINE FOR DUSTER AWD

Manufacturer	: Renault India Pvt. Ltd.
Engine Type	: 4-cylinder, inline, front, transverse
Bore	: 76.0 mm
Stroke	: 80.5 mm
Displacement	: 1461 cc
Compression ratio	: 15.3 : 1
Maximum Power	: 80.6 kW at 4000 rpm
Maximum Torque	: 152 Nm at 1750 rpm
Power to Weight ratio	: 56.7 kW/tonne
Torque to Weight ratio	: 107 Nm/tonne
Specific Output	: 55.2 kW/litre
Valve Gear	: 2 valves per cylinder, SOHC
Fuel system	: Diesel
Fuel Tank Capacity	: 50 litres

85. ENGINE FOR LODGY DCI 110

Manufacturer	: Renault India Pvt. Ltd.
Engine Type	: 4-cylinder, inline, front, transverse
Bore	: 76.0 mm
Stroke	: 80.5 mm
Displacement	: 1461 cc
Compression ratio	: 15.3 : 1
Maximum Power	: 81.0 kW at 4000 rpm
Maximum Torque	: 245 Nm at 1750 rpm
Power to Weight ratio	: 59.2 kW/tonne
Torque to Weight ratio	: 179 Nm/tonne
Specific Output	: 55.4 kW/litre
Valve Gear	: 2 valves per cylinder, SOHC
Fuel system	: Diesel
Fuel Tank Capacity	: 50 litres
Fuel consumption	: City: 12.3 kpl, Highway: 17.5 kpl

86. ENGINE FOR KWID

Manufacturer	: Renault India Pvt. Ltd.
Engine Type	: 3-cylinder, inline, transverse
Bore	: 69.0 mm
Stroke	: 71.2 mm
Displacement	: 799 cc
Compression ratio	: 10.5 : 1
Maximum Power	: 39.7 kW at 5678 rpm
Maximum Torque	: 72 Nm at 4386 rpm
Power to Weight ratio	: 59.3 kW/tonne
Torque to Weight ratio	: 107.6 Nm/tonne
Specific Output	: 49.7 kW/litre
Valve Gear	: 4 valves per cylinder, DOHC
Fuel system	: Petrol
Fuel Tank Capacity	: 28 litres
Fuel consumption	: City: 15.3 kpl, Highway: 21.5 kpl

87. ENGINE FOR SKODA LAURA

Manufacturer	: Skoda India Ltd.
Model	: 2011
Engine Type	: 4-cylinder, inline
Bore	: 82.5 mm
Stroke	: 84.2 mm
Displacement	: 1798 cc
Compression ratio	: 9.6 : 1
Maximum Power	: 118 kW at 4500-6200 rpm
Maximum Torque	: 250 Nm at 1500-4500 rpm
Power to Weight ratio	: 91.5 kW/tonne
Torque to Weight ratio	: 192 Nm/tonne
Specific output	: 66.4 kW/litre
Valve Gear	: 4-valves per cylinder, DOHC
Fuel supply system	: Direct injection, turbocharged
Fuel Tank Capacity	: 55 litres
Fuel Consumption	: City : 9.8 km/litre, Highway : 14 km/litre

88. ENGINE FOR SUPERB (3RD GEN) (2016)

Manufacturer	: Skoda India Ltd. (Data inside brackets for diesel model)
Engine Type	: 4-cylinder, inline, front, transverse
Bore	: 82.5 (81.0) mm
Stroke	: 84.1 (95.5) mm
Displacement	: 1798 (1968) cc

Compression ratio	: 9.6 (15.8) : 1
Maximum Power	: 134.3 kW at 4000-6200 rpm (132 kW at 3600-4000 rpm)
Maximum Torque	: 320 Nm at 1450-3900 rpm (350 Nm at 1500-3500)
Power to Weight ratio	: 89.9 (84.3) NW/tonne
Torque to Weight ratio	: 214.2 (227.3) Nm/tonne
Specific Output	: 74.7 (67.1) kW/litre
Valve Gear	: 4 valves per cylinder, DOHC
Fuel system	: Petrol/Diesel
Fuel Tank Capacity	: 66 litres
Fuel consumption	: City: 8.9 (10.1) kpl, Highway: 14.2 kpl (14.6)

89. ENGINE FOR OCTAVIA

(Figs. In brackets are for diesel engine)

Manufacturer	: Skoda India Ltd.
Model	: 2013
Engine Type	: 4-cylinder, inline, turbo-petrol (turbo-diesel), front, transverse
Bore	: 82.5 (81.0) mm
Stroke	: 84.1 (95.5) mm
Displacement	: 1798 (1968) cc
Compression ratio	: 9.6 (16.1) : 1
Maximum Power	: 132.4 (105.1) kW at 5100-6200 (4000) rpm
Maximum Torque	: 250 (320) Nm at 1250-5000 (1750-3000) rpm
Power to Weight ratio	: 96.2 (75.4) kW/tonne
Torque to Weight ratio	: 181.7 (229.4) Nm/tonne
Specific Output	: 73.6 (53.4) kW/litre
Valve Gear	: 4 valves per cylinder, DOHC
Fuel Tank Capacity	: 50 litres
Fuel consumption	: City: 9.5 (12) kpl, Highway: 14 (17) kpl

90. ENGINE FOR INDICA VISTA

Manufacturer	: Tata Motors Ltd., India
Engine Type	: 4-cylinder, inline
Bore	: 72 mm
Stroke	: 72 mm
Displacement	: 1172 cc
Compression ratio	: 9.7 : 1
Maximum Power	: 47.7 kW at 5500 rpm
Maximum Torque	: 96 Nm at 3000 rpm
Power to Weight ratio	: 44.4 kW/tonne
Torque to Weight ratio	: 89.4 Nm/tonne
Specific output	: 40.7 kW/litre

Valve Gear : 2-valves per cylinder, SOHC
Fuel supply system : MPFI
Fuel Tank Capacity : 37 litres
Fuel Consumption : City : 10.3 km/litre, Highway : 15.5 km/litre

91. ENGINE FOR ARIA

Manufacturer : Tata Motors, India
Model : 2010
Type : 4-cylinder, inline
Bore : 85.0 mm
Stroke : 96.0 mm
Displacement : 2179 cc
Compression ratio : 17.2 : 1
Maximum Power : 103 kW at 4000 rpm
Maximum Torque : 320 Nm at 1700-2700 rpm
Power to Weight ratio : 46.3 kW/tonne
Torque to Weight ratio : 143 Nm/tonne
Specific Output : 47.2 kW/litre
Valve Gear : 4 valves per cylinder, DOHC
Fuel System : Common rail, turbo-diesel
Fuel Tank Capacity : 60 litres
Fuel consumption : City: 9.8 kpl, Highway: 13.8 kpl

92. ENGINE FOR SAFARI STORME (4/2, 4 X 4)

Manufacturer : Tata Motors
Model : 2.2 L VTT Dicor (2012)
Type : 4-cylinder, inline
Bore : 85 mm
Stroke : 96 mm
Displacement : 2179 cc
Maximum Power : 104 kW at 4000 rpm
Maximum Torque : 320 Nm at 1700-2700 rpm
Power to Weight ratio : 49.8 kW/tonne
Torque to Weight ratio : 152.7 Nm/tonne
Specific Output : 47.7 kW/litre
Valve Gear : 4 valves per cylinder, DOHC
Fuel System : Variable Turbine Technology, 32 Bit Electronically controlled, Direct Injection Common rail diesel
Fuel Tank Capacity : 55 litres
Emission standard : BS IV

93. ENGINE FOR NANO TWIST

Manufacturer	: Tata Motors
Engine Type	: 2-cylinder, inline, rear, transverse
Bore	: 73.5 mm
Stroke	: 73.5 mm
Displacement	: 624 cc
Compression ratio	: 10.3 : 1
Maximum Power	: 28 kW at 5500 rpm
Maximum Torque	: 51 Nm at 4000 rpm
Power to Weight ratio	: 42.4 kW/tonne
Torque to Weight ratio	: 77 Nm/tonne
Specific Output	: 44.9 kW/litre
Valve Gear	: 2 valves per cylinder, SOHC
Fuel Tank Capacity	: 15 litres
Fuel consumption	: City: 15.1 kpl, Highway: 20.6 kpl

94. ENGINE FOR BOLT XT

Manufacturer	: Tata Motors (Data in brackets for diesel engine)
Engine Type	: 4-cylinder, inline, front, transverse
Bore	: 75.0 (69.6) mm
Stroke	: 67.5 (82.8) mm
Displacement	: 1193 (1248) cc
Compression ratio	: 9.0 (17.6) : 1
Maximum Power	: 66.2 kW at 5000 rpm (55.2 kW at 4000 rpm)
Maximum Torque	: 140 Nm t 1500-400 rpm (190 Nm 1750-3000 rpm)
Power to Weight ratio	: 58.8 (47.6) kW/tonne
Torque to Weight ratio	: 124 (164) Nm/tonne
Specific Output	: 55.5 (44.2) kW/litre
Valve Gear	: 2 valves per cylinder, SOHC (4 valves per cylinder, DOHC)
Fuel system	: Petrol/Diesel
Fuel Tank Capacity	: 44 litres
Fuel consumption	: City: 11.4 (N.A.) kpl, Highway: 16.1 (N.A.) kpl

95. ENGINE FOR TIAGO

Manufacturer	: Tata Motors (Data in brackets are for diesel engine)
Engine Type	: 3-cylinder, inline, front, transverse
Bore	: 77.0 (75.0) mm

Stroke	: 85.8 (79.0) mm
Displacement	: 1199 (1047) cc
Compression ratio	: 10.7 (16.5) : 1
Maximum Power	: 62.5 kW at 6000 rpm (51.5 kW at 4000 rpm)
Maximum Torque	: 114 Nm at 1800-3000 rpm)
Power to Weight ratio	: 61.8 (50.9) kW/tonne
Torque to Weight ratio	: 112.7 (129.6) Nm/tonne
Specific Output	: 52.1 (49.2) kW/litre
Valve Gear	: 4 valves per cylinder, DOHC
Fuel system	: Petrol/Diesel

96. ENGINE FOR TATA "ACE" MINI-TRUCK

Model	: Tata 275 IDI NA BS-II (and BS-III)
Engine Type	: 4-stroke, naturally aspirated, indirect injection engine
Displacement	: 700 cc
Maximum Power	: 11.94 kW at 3200 rpm
Maximum Torque	: 37.3 Nm at 2000 rpm
Special Item	: Catalytic converter in BS-III model
Fuel tank capacity	: 30 litres

97. ENGINE FOR TATA XENON 4 X 4

Manufacturer	: Tata Motors Ltd. India
Engine Type	: 4-cylinder, inline, front, longitudinal, diesel
Bore	: 85 mm
Stroke	: 96 mm
Displacement	: 2179 cc
Compression ratio	: 17.2 : 1
Maximum Power	: 103 kW at 4000 rpm
Maximum Torque	: 320 Nm at 1700-2700 rpm
Power to Weight ratio	: 51.7 kW/tonne
Torque to Weight ratio	: 160.7 Nm/tonne
Specific output	: 47.3 kW/litre
Valve Gear	: 4 valves per cylinder, DOHC
Fuel supply system	: Common rail, direct injection
Fuel Tank Capacity	: 65 litres
Fuel Consumption	: City : 9.2 km/litre, Highway : 12.9 km/litre

98. ENGINE FOR TATA LPT-2518/62 TC TRUCK

Manufacturer	: Tata Motors, India
Model and Type	: Cummins B 5.9-180 BS-III, 6-cylinder, inline, direct injection, turbocharged intercooled diesel engine

Bore	: 102 mm
Stroke	: 120 mm
Displacement	: 5833 cc
Maximum Power	: 133.5 kW at 2500 rpm
Maximum Torque	: 675 Nm at 1400-1600 rpm
Power to Weight ratio	: 20.1 kW/tonne
Torque to Weight ratio	: 101.5 Nm/tonne
Specific Output	: 22.9 kW/litre
Valve Gear	: 4 valves per cylinder
Fuel System	: Turbocharged, intercooled diesel engine
Fuel Tank Capacity	: 350 litres

99. ENGINE FOR CAMRY AUTOMATIC CAR

Manufacturer	: Toyota Kirloskar Motor Ltd., India
Model	: 2012
Engine Type	: 4-cylinder VVT-i, aluminium alloy
Bore	: 90.0 mm
Stroke	: 98.0 mm
Displacement	: 2494 cc
Compression ratio	: 10.4 : 1
Maximum Power	: 132.8 kW at 6000 rpm
Maximum Torque	: 233 Nm at 4100 rpm
Power to Weight ratio	: 90 kW/tonne
Torque to Weight ratio	: 158 Nm/tonne
Specific output	: 53.2 kW/litre
Valve Gear	: 4 valves per cylinder, DOHC
Fuel System	: Electronic, MPFI
Fuel Tank Capacity	: 70 litre
Fuel consumption	: 8.2 kpl (city), 12.0 kpl (highway)

100. ENGINE FOR TOYOTA FORTUNER

Manufacturer	: Toyota Kirloskar Motor
Engine Type	: 4-cylinder inline
Bore	: 96.0 mm
Stroke	: 103.0 mm
Displacement	: 2982 cc
Compression ratio	: 17.9 : 1
Maximum Power	: 125.3 kW at 3600 rpm
Maximum Torque	: 343 Nm at 1400-3400 rpm
Power to Weight ratio	: 64 kW/tonne
Torque to Weight ratio	: 175 Nm/tonne

Specific output	: 42 kW/litre
Valve Gear	: 4 valves per cylinder, DOHC
Ignition and fuel system	: Common-rail, direct injection, turbo-diesel
Fuel tank capacity	: 80 litres
Fuel Consumption	: 9.3 kpl (city), 12.9 kpl (highway)

101. ENGINE FOR TOYOTA COROLLA ALTIS

Manufacturer	: Toyota Kirloskar Motor
Engine Type	: 4-cylinder inline, front, transverse
Bore	: 79.0 mm
Stroke	: 91.5 mm
Displacement	: 1794 cc
Compression ratio	: 10.0 : 1
Maximum Power	: 97 kW at 6200 rpm
Maximum Torque	: 148 Nm at 4200 rpm
Power to Weight ratio	: 79.8 kW/tonne
Torque to Weight ratio	: 122 Nm/tonne
Specific output	: 54 kW/litre
Valve Gear	: 4 valves per cylinder, DOHC
Fuel system	: MPFI
Fuel tank capacity	: 55 litres
Fuel Consumption	: City: 19.5 kpl, HIghway: 15 kpl (Manual Transmission)

102. ENGINE FOR ETIOS LIVA

Manufacturer	: Toyota Kirloskar Motors
Model	: 2011
Type	: 4-cylinder, inline, front, transverse
Bore	: 72.5 mm
Stroke	: 72.5 mm
Displacement	: 1197 cc
Compression ratio	: 10.5 : 1
Maximum Power	: 58.9 kW at 5600 rpm
Maximum Torque	: 104 Nm at 3100 rpm
Power to Weight ratio	: 64 kW/tonne
Torque to Weight ratio	: 113 Nm/tonne
Specific Output	: 49 kW/litre
Valve Gear	: DOHC, 4 valves per cylinder
Fuel System	: Petrol
Fuel Tank Capacity	: 45 litres
Fuel consumption	: City: 12.1 kpl, Highway: 16.7 kpl

103. ENGINE FOR ETIOS D-4D

Manufacturer	: Toyota Kirloskar Motors
Model	: 2011
Type	: 4-cylinder, inline, front, transverse
Bore	: 73.0 mm
Stroke	: 81.5 mm
Displacement	: 1364 cc
Compression ratio	: 16.5 : 1
Maximum Power	: 50.7 kW at 3800 rpm
Maximum Torque	: 170 Nm at 1800-2400 rpm
Power to Weight ratio	: 50.5 kW/tonne
Torque to Weight ratio	: 169 Nm/tonne
Specific Output	: 37.2 kW/litre
Valve Gear	: 2 valves per cylinder, SOHC
Fuel System	: Diesel engine, Common Rail Direct Injection with turbocharger
Fuel Tank Capacity	: 45 litres
Fuel consumption	: City:14.2 kpl, Highway: 19.0 kpl

104. ENGINE FOR PRIUS HYBRID CAR

Manufacturer	: Toyota Kirloskar Motors
Model	: 2010
Type	: 4-cylinder, inline (petrol) plus electric motor, front, transverse
Bore	: 80.5 mm
Stroke	: 88.3 mm
Displacement	: 1798 cc
Compression ratio	: 13.0 : 1
Maximum Power	: Petrol: 73.1 kW at 5200 rpm
Maximum Torque	: Petrol: 142 Nm at 4000 rpm
Power (Electric)	: 59.7 kW
Torque (Electric)	: 207.5 Nm
Combined power	: 100 kW
Battery type	: Sealed Nickel-metal hydride
Fuel Tank Capacity	: 45 litres
Fuel consumption	: City:15.1 kpl, Highway:18.7 kpl

105. ENGINE FOR INNOVA MT/AT (2016)

Manufacturer	: Toyota Kirloskar Motors Ltd. (Data within brackets for AT model engine)
Engine Type	: 4-cylinder, inline, front, transverse
Bore	: 92.0 (92.0) mm
Stroke	: 90.0 (103.6) mm
Displacement	: 2393 (2755) cc

Compression ratio : 15.6 (15.6) : 1
Maximum Power : 112 kW at 3400 rpm
(130 kW at 3400 rpm)
Maximum Torque : 343 Nm at 1400-2800 rpm
(360 Nm at 1200-3400 rpm)
Power to Weight ratio : 60.4 (69.5) kW/tonne
Torque to Weight ratio : 185 (192.5) Nm/tonne
Specific Output : 46.8 (47.2) kW/litre
Valve Gear : 4 valves per cylinder, DOHC
Fuel system : Turbo-diesel
Fuel Tank Capacity : 55 litres
Fuel consumption : ARAI 15.1 (14.29) kpl

106. ENGINE FOR VW PASSAT 2.0 TDI (2011 MODEL)

Manufacturer : Volkswagen India Ltd.
Engine Type : 4-cylinder in line, front traverse, common rail, turbo-diesel
Bore : 81 mm
Stroke : 95.5 mm
Displacement : 1968 cc
Compression ratio : 16.5 : 1
Maximum Power : 127 kW at 4200 rpm
Maximum Torque : 350 Nm at 1750-2500 rpm
Power to Weight ratio : 79 kW/tonne
Torque to Weight ratio : 218 Nm/tonne
Specific Output : 53.06 kW/litre
Valve Gear : DOHC, 4 valves per cylinder
Fuel System : Direct injection with turbo-charger (Pumpe Duse technology); fuel injection at 2200 bar (220 MPa)
Fuel Tank Capacity : 70 litres
Fuel consumption : City: 10.8 kpl, Highway: 15.2 kpl
Emission compliance : Euro 4

107. ENGINE FOR VW POLO 1.2 TDI

Manufacturer : Volkswagen India Ltd.
Model : 2010
Engine Type : 3-cylinder, inline, front, transverse, EA 189 engine
Bore : 79.5 mm
Stroke : 80.5 mm
Displacement : 1199 cc
Compression ratio : 16.5 : 1
Maximum Power : 56 kW at 4200 rpm
Maximum Torque : 180 Nm at 2000 rpm

Power to Weight ratio	: 49.7 kW/tonne
Torque to Weight ratio	: 160 Nm/tonne
Specific Output	: 46.7 kW/litre
Valve Gear	: 4 valves per cylinder, DOHC
Fuel System	: Turbocharged diesel
Fuel Tank Capacity	: 45 litres
Fuel consumption	: City: 14.0 kpl, Highway: 18.9 kpl

108. ENGINE FOR JETTA HIGHLINE TDI

Manufacturer	: Volkswagen India Ltd.
Engine Type	: 4-cylinder, inline, diesel engine
Displacement	: 1968 cc
Kerb weight	: 1445 kg
Maximum Power	: 103 kW at 4250 rpm
Maximum Torque	: 320 Nm at 1750 rpm
Power to Weight ratio	: 71.3 kW/tonne
Torque to Weight ratio	: 221 kW/tonne
Specific Output	: 52.3 kW/litre
Valve Gear	: 4 valves per cylinder, DOHC
Fuel system	: Turbo-diesel, variable geometry turbo charger
Fuel Tank Capacity	: 55 litres
Fuel consumption	: ARAI-19.33 kpl

109. ENGINE FOR VENTO

Manufacturer	: Volkswagen India Ltd. (Data inside brackets for diesel engine)
Engine Type	: 4-cylinder, inline
Bore	: NA
Stroke	: NA
Displacement	: 1598 (1498) cc
Compression ratio	: NA
Kerb weight	: 1137 (1211) kg
Maximum Power	: 76.8 kW at 5250 rpm (76.8 kW at 4400 rpm)
Maximum Torque	: 153 Nm at 3800 rpm (250 Nm at 1500 rpm)
Power to Weight ratio	: 67.5 (63.4) kW/tonne
Torque to Weight ratio	: 134.6 (206.4) Nm/tonne
Specific Output	: 48.06 (51.26) kW/litre
Valve Gear	: 4 valves per cylinder, DOHC
Fuel system	: MPI (TDI-variable Geometry turbocharger)
Fuel Tank Capacity	: 55 litres
Fuel consumption	: ARAI-15.04 (20.34) kpl

110. ENGINE FOR XC 60 CAR

Manufacturer	: Volvo Car India Ltd.
Model	: 2010
Type	: 5-cylinder, inline, front, transverse
Bore	: 81.0 mm
Stroke	: 93.15 mm
Displacement	: 2400 cc
Compression ratio	: 16.5 : 1
Maximum Power	: 153 kW at 4000 rpm
Maximum Torque	: 420 Nm at 1500-3250 rpm
Power to Weight ratio	: 81.8 kW/tonne
Torque to Weight ratio	: 224.5 Nm/tonne
Specific Output	: 63.7 kW/litre
Valve Gear	: 4 valves per cylinder, DOHC
Fuel System	: Common rail direct injection diesel, turbocharged
Fuel Tank Capacity	: 70 litres
Fuel consumption	: City: 9.0 kpl, Highway: 12.2 kpl

111. ENGINE FOR S60 SUMMUM D5

Manufacturer	: Volvo Car India
Engine Type	: 5-cylinder, inline, diesel
Bore	: 81.0 mm
Stroke	: 93.15 mm
Displacement	: 2400 cc
Compression ratio	: 16.5 : 1
Kerb weight	: 1551 kg
Maximum Power	: 160.4 kW at 4000 rpm
Maximum Torque	: 440 Nm at 3250 rpm
Power to Weight ratio	: 103.4 kW/tonne
Torque to Weight ratio	: 283.6 Nm/tonne
Specific Output	: 66.8 kW/litre
Valve Gear	: 4 valves per cylinder, DOHC
Fuel system	: Turbo-diesel
Fuel Tank Capacity	: 67.5 litres
Fuel consumption	: ARAI: 20.4 kpl

112. ENGINE FOR ASHOK LEYLAND COMET

Manufacturer	: Ashok Leyland Limited Limited, India
Type	: Overhead valve, vertical diesel engine, 6 cylinder in line
Bore	: 103.38 mm
Stroke	: 120.7 mm
Cubic capacity	: 6079 cc

Fig. 22.1. Engine for Ashok Leyland Comet.
(Courtesy–Ashok Leyland Ltd.)

Compression ratio	: 16.1
Brake Power	: 82.1 kW at 2400 r.p.m.
Torque	: 369 Nm at 1600 r.p.m.
Lubrication	: Wet sump, gear-type pump. Full flow, cloth element filter.
Cylinder head	: Detachable one piece casting
Cylinder block	: One piece casting; liners pre-finished, press fit, flange located.
Piston	: Spheroidal cavity
Fuel injection	: Mico-Bosch fuel injection equipment (Injection pump, governor, fuel feed pump, primary and secondary fuel filters and fuel injector).

113. ENGINE FOR ASHOK LEYLAND 'NEWGEN 2526 J'

Model	: JO8CTI-BS2 (upgradeable to BS3/BS4)
Type	: 6-cylinder, turbo-charged, intercooled diesel engine
Capacity	: 8 litres
Maximum Power	: 191 kW at 2500 rpm
Maximum Torque	: 746 Nm at 1500 rpm
Valve system	: 4 valves per cylinder
Fuel system	: In-line fuel injection pump MD TICS with timing and injection-rate electronically controlled
Electrical system	: 24 V negative earthed system. Battery—2 ×12 V, 120 Ah rating

114. ENGINE FOR FORCE ONE

Manufacturer	: Force Motors Ltd. India
Model	: 2011
Type	: 4-cylinder, inline (2.2 FM Tech)
Bore	: 88.0 mm
Stroke	: 88.34 mm
Displacement	: 2199 cc
Compression ratio	: 16.1 : 1
Maximum Power	: 103.7 kW at 3800 rpm
Maximum Torque	: 321 Nm at 1600-2400 rpm
Power to Weight ratio	: 55.7 kW/tonne
Torque to Weight ratio	: 172.5 Nm/tonne
Specific Output	: 48.2 kW/litre
Valve Gear	: 4 valves per cylinder, DOHC
Fuel System	: Diesel, Common rail direct injection
Fuel Tank Capacity	: 70 litres
Fuel consumption	: City: 9.8 kpl, Highway: 11.9 kpl
Emission status	: BS-IV

115. ENGINE FOR VOLVO 8700 CITY BUS

Model	: D7C 275 HP
Type	: 6-cylinder, 4-stroke diesel engine
Bore	: 107 mm
Stroke	: 135 mm
Capacity	: 7.3 litre
Compression ratio	: 19.5 : 1
Maximum Power	: 202 kW at 2200 rpm (ISO 1585)
Maximum Torque	: 1200 Nm at 1200 rpm (ISO 1585)
Valve System	: Overhead valves
Fuel System	: Direct Injection
Emission standard	: Euro III compliant
Fuel tank Capacity	: 2 × 155 litres
Electrical System	: 24 V system, 2 batteries of 165 Ah Capacity. Max. alternators output : 2 × 110 A. Starter motor : 4.5 kW

116. ENGINE FOR MODEL C1

Manufacturer	: Citroen Cars, U.K.
Engine Type	: 3-cylinder, inline, petrol
Bore	: 71.0 mm
Stroke	: 84.0 mm
Displacement	: 998 cc
Kerb weight	: 855 kg

Maximum Power	: 51 kW at 6000 rpm
Maximum Torque	: 95.5 Nm at 4800 rpm
Power to Weight ratio	: 59.6 kW/tonne
Torque to Weight ratio	: 111.7 Nm/tonne
Specific Output	: 51.1 kW/litre
Fuel system	: Mid-point electronic fuel injection
Fuel Tank Capacity	: 35 litres
Fuel consumption (litres/100 km)	: Urban cycle = 5.0 Extra urban cycle = 3.6 Combined cycle = 4.1
Emissions	: CO_2 = 95 g/km

117. ENGINE FOR MODEL C3, PURE TECH 68

Manufacturer	: Citroen Cars, U.K.
Engine Type	: 3-cylinder, inline, petrol
Displacement	: 999 cc
Kerb weight	: 1115 kg
Maximum Power	: 50 kW at 6000 rpm
Maximum Torque	: 95 Nm at 3000 rpm
Power to Weight ratio	: 44.8 kW/tonne
Torque to Weight ratio	: 85.2 Nm/tonne
Specific Output	: 50.1 kW/litre
Fuel system	: Mid-point electronic fuel injection
Fuel Tank Capacity	: 50 litres
Fuel consumption (litres/100 km)	: Urban cycle = 5.4 Extra urban cycle = 7.2 Combined cycle = 6.4
Emissions	: Complies Euro 6 CO_2 emissions, g/km = 102

118. ENGINE FOR MODEL C4 PURE TECH 110

Manufacturer	: Citroen Cars, U.K.
Engine Type	: 3-cylinder, inline
Bore	: 75.0 mm
Stroke	: 90.5 mm
Displacement	: 1199 cc
Kerb weight	: 1200 kg
Maximum Power	: 81 kW at 5500 rpm
Maximum Torque	: 205 Nm at 1500 rpm
Power to Weight ratio	: 67.5 kW/tonne

Torque to Weight ratio	: 170.8 Nm/tonne
Specific Output	: 67.6 kW/litre
Fuel system	: Turbocharged diesel fuel injection
Fuel Tank Capacity	: 60 litres
Fuel consumption	: Urban cycle = NA Extra urban cycle = NA Combined cycle = NA
Emissions Standard	: Complies Euro 6 CO_2 emission = 110 g/km

119. ENGINE FOR MODEL C5 e-HDi 115, ETG 6

Manufacturer	: Citroen Cars, U.K.
Engine Type	: 4-cylinder, inline
Bore	: 75.0 mm
Stroke	: 88.3 mm
Displacement	: 1560 cc
Kerb weight	: 1705 kg
Maximum Power	: 84 kW at 3600 rpm
Maximum Torque	: 279 Nm at 1750 rpm
Power to Weight ratio	: 49.3 kW/tonne
Torque to Weight ratio	: 163.6 Nm/tonne
Specific Output	: 53.8 kW/litre
Fuel system	: High Pressure diesel injection
Fuel Tank Capacity	: 71 litres
Fuel consumption	: Urban cycle = NA Extra urban cycle = NA Combined cycle = NA
Emissions Standard	: Complies Euro 5 CO_2 emission = 111 g/km

120. ENGINE FOR MODEL 508 SALOON

Manufacturer	: Peugeot, U.K.
Engine Type	: 4-cylinder, inline
Displacement	: 1997 cc
Maximum Power	: 120 kW at 3850 rpm (Diesel) 27 kW at 2000 to 7500 rpm (Electric)
Maximum Torque	: 300 Nm at 1750 rpm (Diesel) 200 Nm at 2000 rpm (Electric)
Kerb weight	: 1660 kg
Power to Weight ratio	: 72.3 kW/tonne (Diesel), 16.3 kW/tonne (Electric)
Torque to Weight ratio	: 180.7 Nm/tonne (Diesel), 120.5 Nm/tonne (Electric)
Specific Output	: 60.1 kW/litre (Diesel), 13.5 kW (Electric)
Fuel system	: Diesel/Electric

Fuel Tank Capacity	: 72 litres
Fuel consumption	: Urban cycle = 3.9 litres/100 km Extra urban cycle = 4.1 litres/100 km Combined drive cycle = 4.4 litres/100 km
Emissions Standard	: Complies Euro 6 CO_2 emission = 95 g/km

121. ENGINE FOR MODEL 3008 AND MODEL 5008 ACTIVE 1.2

Manufacturer	: Peugeot, U.K.
Engine Type	: 3-cylinder, inline
Displacement	: 1199 cc
Kerb weight	: 1400 kg
Maximum Power	: 96 kW at 5500 rpm
Maximum Torque	: 230 Nm at 1750 rpm
Power to Weight ratio	: 68.6 kW/tone
Torque to Weight ratio	: 164.3 Nm/tone
Specific Output	: 80.1 kW/litre
Fuel system	: Petrol
Fuel Tank Capacity	: NA
Fuel consumption	: Urban drive cycle = 6.1 litres/100 km Extra urban cycle = 4.3 litres/100 km Combined drive cycle = 4.9 litres/100 km

122. ENGINE FOR MODEL 5008, 'ACTIVE' ENGINE, 1.6 LITRE BLUE HDI

Manufacturer	: Peugeot, U.K.
Engine Type	: 4-cylinder, inline
Bore	: 75 mm
Stroke	: 88.3 mm
Displacement	: 1560 cc
Kerb weight	: 1490 kg
Maximum Power	: 88 kW at 3500 rpm
Maximum Torque	: 300 Nm at 1750 rpm
Power to Weight ratio	: 59.1 kW/tone
Torque to Weight ratio	: 201 Nm/tone
Specific Output	: 56.4 kW/litre
Valve gear	: 8 valves
Fuel system	: Petrol
Fuel Tank Capacity	: 60 litres
Fuel consumption **(Standard 17" tyre)**	: Urban drive cycle = 4.9 litres/100 km Extra urban cycle = 4.0 litres/100 km Combined drive cycle = 4.3 litres/100 km

Emissions Standard	: Euro 6 Carbon dioxide emission = 113 g/km (17´´ tyre)

123. ENGINES FOR PICANTO CAR

Manufacturer	: Kia Motors, U.K.	
Model	: 2012	
	1.0	**1.25**
Bore	: 71 mm	71 mm
Stroke	: 84 mm	78.8 mm
Displacement	: 998 cc	1248 cc
Compression ratio	: 11.0 : 1	11.0 : 1
Maximum Power	: 51 kW at 6200 rpm	62 kW at 6000 rpm
Maximum Torque	: 95 Nm at 3500 rpm	120 Nm at 4000 rpm
Power to Weight ratio	: 52.5 kW/tonne	63 kW/tonne
Torque to Weight ratio	: 98 Nm/tonne	122 Nm/tonne
Specific Output	: 51.1 kW/litre	49.7 kW/litre
Fuel System	: MPI	MPI
Fuel Tank Capacity	: 35 litres	35 litres
Fuel consumption	: Urban: 18.5 kpl Extra urban: 27.8 kpl Combind: 23.8 kpl	Urban: 16.9 kpl Extra urban: 25.6 kpl Combined: 21.3 kpl
CO_2 emission	: 99 g/km	109 g/km

124. ENGINE FOR PHANTOM CAR

Manufacturer	: Rolls Royce Motors, U.K.
Engine Type	: 8-cylinder, V-type, petrol engine
Bore	: 92.0 mm
Stroke	: 84.6 mm
Displacement	: 6749 cc
Kerb weight	: 2560 kg
Maximum Power	: 338 kW at 5300 rpm
Maximum Torque	: 720 Nm at 3500 rpm
Power to Weight ratio	: 132 kW/tone
Torque to Weight ratio	: 281.2 Nm/tone
Specific Output	: 50.1 kW/litre
Valve gear	: 4-valves per cylinder
Fuel system	: Direct injection
Fuel Tank Capacity	: 100 litres
Fuel consumption (litres/100 km)	: Urban drive cycle = 22.8 Extra urban cycle = 10.2 Combined drive cycle = 14.8

Emissions Standard	: Euro 5
CO_2 emission	: 347 g/km

125. ENGINE FOR GHOST SERIES II CAR

Manufacturer	: Rolls Royce Motors, U.K.
Engine Type	: 12-cylinder, V-type, petrol engine
Bore	: 89.0 mm
Stroke	: 88.3 mm
Displacement	: 6592 cc
Compression ratio	: 10.1 : 1
Unladen weight (DIN)	: 2360 kg
Maximum Power	: 420 kW at 5250 rpm
Maximum Torque	: 780 Nm at 1500 rpm
Power to Weight ratio	: 178 kW/tone
Torque to Weight ratio	: 330 Nm/tone
Specific Output	: 63.7 kW/litre
Valve gear	: 4-valves per cylinder
Fuel system	: Direct injection
Fuel Tank Capacity	: 82.5 litres
Fuel consumption (litres/100 km)	: Urban cycle = 21.2 Extra-urban cycle = 9.8 Combined cycle = 14
Emissions Standard	: Euro 6
CO_2 emission	: 327 g/km

126. ENGINE FOR REXTON

Manufacturer	: Ssangyong Motor, U.K.
Model	: 2012
Type	: 5-cylinder, inline, diesel
Displacement	: 2696 cc
Maximum Power	: 123 kW at 4000 rpm
Maximum Torque	: 340 Nm at 1800-3250 rpm
Power to Weight ratio	: 59 kW/tonne
Torque to Weight ratio	: 163 Nm/tonne
Specific Output	: 45.6 kW/litre
Valve Gear	: DOHC
Fuel System	: Common rail
Fuel Tank Capacity	: 80 litres
Fuel consumption	: Urban: 8.2 kpl, extra urban: 14.9 kpl, Combined: 11.6 kpl
CO_2 emission	: 229 g/km
Emission standard	: Euro 5

127. ENGINE FOR 'KORANDO'

Manufacturer	: Ssangyong Motor, U.K.
Model	: 2012
Type	: 4-cylinder, inline
Displacement	: 1998 cc
Maximum Power	: 130 kW at 4000 rpm
Maximum Torque	: 360 Nm at 2000-3000 rpm
Power to Weight ratio	: 78 kW/tonne
Torque to Weight ratio	: 216 Nm/tonne
Specific Output	: 65 kW/litre
Fuel System	: Diesel
Fuel Tank Capacity	: 57 litres
Fuel consumption	: Urban:13.3 kpl, Extra urban: 19.6 kpl, Combined: 16.66 kpl
CO_2 emission	: 157 g/km
Emission standard	: Euro 5

128. ENGINE FOR NEW VIVA 1.0i 75 PS

Manufacturer	: Vauxhall Motors, U.K.
Engine Type	: 3-cylinder, inline
Displacement	: 999 cc
G.V.W.	: 1353 kg
Maximum Power	: 55 kW at 6500 rpm
Maximum Torque	: 95 Nm at 4500 rpm
Power to Weight ratio	: 40.6 kW/tonne
Torque to Weight ratio	: 70.2 Nm/tonne
Specific Output	: 55.1 kW/litre
Valve Gear	: 4 valves per cylinder, DOHC
Fuel system	: Naturally aspirated with Variable Valve Timing, Multi-point sequential electronic fuel injection
Fuel Tank Capacity	: 32 litres
Fuel consumption (litres/100 km)	: Urban cycle = 5.6 Extra urban cycle = 3.9 Combined cycle = 4.5
Emissions Standard	: Complies Euro 6 CO_2 emission = 104 g/km

129. ENGINE FOR MERIVA, 1.4i 100 PS ECOTEC-4

Manufacturer	: Vauxhall Motors, U.K.
Engine Type	: 4-cylinder, inline, petrol
Displacement	: 1398 cc
G.V.W.	: 1905 kg

Maximum Power	: 74 kW at 6000 rpm
Maximum Torque	: 96 Nm at 4000 rpm
Power to Weight ratio	: 38.8 kW/tonne
Torque to Weight ratio	: 50.9 Nm/tonne
Specific Output	: 52.9 kW/litre
Valve Gear	: 4 valves per cylinder, DOHC
Fuel system	: Naturally aspirated VVT Multi-point sequential fuel injection
Fuel Tank Capacity	: 54 litres
Fuel consumption (litres/100 km)	: Urban cycle = 7.6 Extra-urban cycle = 5.0 Combined cycle = 6.0
Emissions Standard	: Complies Euro 6 CO_2 emission = 140 g/km

130. ENGINE FOR AUDI 100 SE CAR

Manufacturer	: Audi AG, Germany
Type	: Five-cylinder in-line engine
Bore	: 79.5 mm
Stroke	: 86.4 mm
Cubic capacity	: 2144 c.c.
Compression ratio	: 9.3 : 1
Brake Power	: 100 kW at 5700 r.p.m. (DIN/SAE net)
Torque	: 185 Nm at 4200 r.p.m. (DIN/SAE net)
Valve Gear	: Overhead valve camshaft
Cooling system	: Water cooling with electrically driven thermostatically controlled fan
Oil filter	: Full-flow oil filter
Fuel System	: K-Jetronic injection system

131. ENGINE FOR A4 CAR 2006 MODEL

Manufacturer	: Audi, Germany
Type	: 4-cylinder, in-line, front, longitudinal diesel engine
Bore	: 81.0 mm
Stroke	: 95.5 mm
Capacity	: 1968 cc
Compression ratio	: 18.0 : 1
Maximum Power	: 104.4 kW at 4000 rpm
Maximum Torque	: 320 Nm at 1750-2000 rpm
Power to weight	: 92.1 bhp/tonne
Torque to weight	: 21.46 kgm/tonne
Specific Output	: 71.3 bhp/litre
Valve Gear	: 4-valves per cylinder

132. ENGINE FOR A6 CAR 2006 MODEL

Manufacturer	: Audi, Germany
Type	: 6-cylinder, in-line, aluminium head and block, front, longitudinal petrol engine
Bore	: 82.5 mm
Stroke	: 92.8 mm
Capacity	: 2976 cc
Compression ratio	: 10.5 : 1
Maximum Power	: 162.6 kW at 6300 rpm
Maximum Torque	: 300 Nm at 3200 rpm
Power to weight	: 138.85 bhp/tonne
Torque to weight	: 19.48 kgm/tonne
Specific Output	: 73.25 bhp/litre
Valve Gear	: 5-valves per cylinder, DOHC

133. ENGINE FOR GOLF GTI 16 V CAR

Manufacturer	: Volkswagenwerk AG, Germany
Type	: 4-cylinder in-line engine mounted traversed at front
Bore	: 81 mm
Stroke	: 86.4 mm
Cubic capacity	: 1781 c.c.
Compression ratio	: 10.0 : 1
Max. Brake Power	: 102 kW at 6100 r.p.m. (DIN)
Max. Torque	: 168 Nm at 4600 r.p.m.
Mixture preparation	: Mechanical fuel injection system
Valve Gear	: Four valves cylinder. Valve gear driven by toothed belt from double overhead camshafts and bucket type tappets with maintenance-free hydraulic valve clearance adjustment
Cooling system	: Closed, pressurised liquid coolant circuit with pump, radiator, coolant reservoir and electrically driven thermostat controlled fan.
Lubrication system	: Pressure lubrication system with oil pump, renewable filter element and oil cooler.
Fuel system	: Mechanical fuel injection system.
Ignition system	: Electronic characteristic data ignition system
Electrical system	: 12-volt electrical system with alternator (max. 65 Amp current) and 45 Amp/hour battery.

134. ENGINE FOR 'CADDY' CAR

Manufacturer	: Volkswagenwerk AG, Germany
Type	: 4-stroke, 4-cylinder in-line, installed transversely. Three models:

(*i*) Carburettor engine (55 kW).

(*ii*) Fuel injection engine (70 kW)

(*iii*) Diesel engine (39 kW)

Valve system : Valves operated by toothed belt and overhead camshaft, hydraulic tappets.

Cooling system : Filled for life of vehicle, electric radiator fan controlled by thermoswitch.

	Petrol engine with catalytic converter	*Petrol engine without catalytic converter*	*Diesel engine*
Bore, mm	81.0	81.0	76.5
Stroke, mm	86.4	77.4	86.4
Capacity, cm³	1781	1595	1589
Compression ratio	8.5	9.0	23.0
Output, kW at rpm	70/5500	55/5000	39/4500
Maximum torque, Nm at rpm	142/3000	125/2500	95/2600
Mixture formation	Fuel injection	Carburettor	Fuel injection
Fuel	91 RON unleaded	90 RON unleaded	Diesel

Carburettor engine : Downdraft twin barrel carburettor, automatic choke, thermostatically controlled intake air preheating, electrical mixture preheating.

Fuel injection engine : Electronic ignition, mechanically controlled petrol injection, exhaust emission system with Lamda probe.

Diesel engine : Distributor type injection pump with cold starting aid, filter preheating, self bleeding fuel system.

135. ENGINE FOR 'POLO' CAR

Manufacturer : Volkswagenwerk AG, Germany

Type : Four stroke 4-cylinder, in-line petrol or diesel engine, installed traversely. Light alloy cylinder head

	Petrol engine with catalytic converter	*Petrol engine without catalytic converter*	*Diesel engine*
Bore, mm	75	75	75
Stroke, mm	72	72	78.7
Capacity, cm³	1272	1272	1391
Compression ratio	10.0	10.0	22.0
Fuel	95 RON unleaded	95 RON unleaded 98 RON unleaded/leaded	Diesel
Output, kW at rpm	55/5900	57/6100	35/4500
Maximum torque, Nm at rpm	99/3200-4000	99/3200-4000	85/2700-3500
Fuel consumption, litres/100 km (DIN)	5.4 (at 90 km/h) 7.4 (at 120 km/h) 8.6 (urban driving)	5.3 (at 90 km/h) 7.2 (at 120 km/h) 8.5 (urban driving)	4.2 (at 90 km/h) 5.9 (at 120 km/h) 5.7 (urban driving)

Valve system : Valves operated by toothed belt and 5 bearing overhead camshaft. Hydraulic tappets.

Cooling system : Filled for life or vehicle, Radiator with separate expansion tank. Radiator lectric fan controlled by thermoswitch.

Air supply : Dry air-cleaner with paper element. Load-and temperature-sensitive intake air preheating (on petrol only), mechanically driven, G-supercharger with charge air cooling (on 83 and 85 kW engines only) which boosts the output by about 50%.

Fuel supply : (*i*) Carburettor engines

Downdraft twin choke carburettor with automatic choke (40 kW engine).

(*ii*) Fuel injection engines

Digitally controlled fuel injection system called 'Digifant' on 83kW engines. Overrun cut-off.

(*iii*) Diesel engines

Distributor type injection pump with cold starting aid. Preheated diesel filter. Self-bleeding system.

Ignition : Electronic Ignition System.

136. ENGINE FOR MERCEDES BENZ 200 CAR

Manufacturer : Mercedes Benz AG, Germany

Model and type : Engine model 115.938, 4-cylinder petrol engine

Bore : 87 mm

Stroke : 83.6 mm

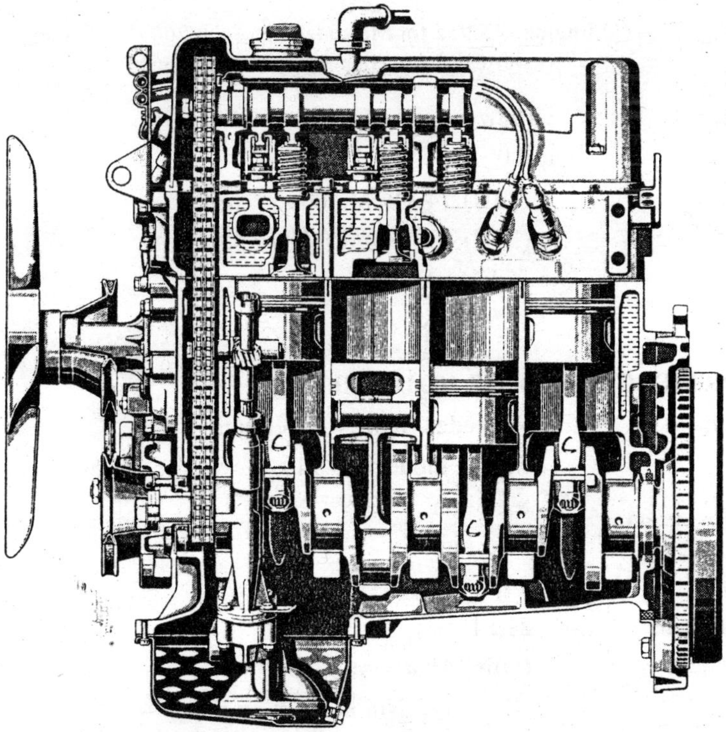

Fig. 22.2. (*a*) Engine 115.938 for Mercedes Benz Car 200 (Front Section)
(Courtesy–Mercedes Benz A.G. Germany)

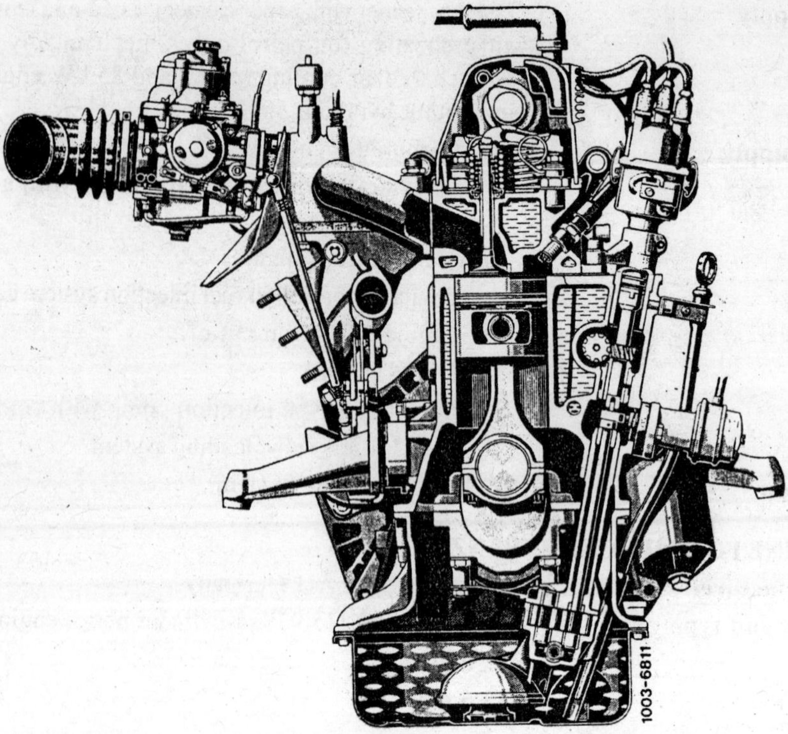

Fig. 22.2. (*b*) Engine 115.938 for Mercedes Benz Car 200 (left Sectional view)
(Courtesy–Mercedes Benz A.G. Germany)

Cubic capacity	:1988 c.c.
Compression ratio	: 9.0 : 1
Brake Power	: (DIN) 74 kW at 4800 r.p.m.
Torque	: (DIN) 158 Nm at 3000 r.p.m.
Battery	: 12 V/44 Ah
Generator	: 14 V/35 A

137. ENGINE FOR MERCEDES-BENZ 200 D CAR

Manufacturer	: Mercedes Benz AG, Germany
Model type	: Engine model 615.940, 4-cylinder diesel
Bore	: 87.0 mm
Stroke	: 83.6 mm
Cubic capacity	: 1988 c.c.
Compression ratio	: 21 : 1
Brake Power	: (DIN) 41 kW at 4200 r.p.m.
Torque	: (DIN) 112 Nm at 2400 r.p.m.
Battery	: 12 V/69 Ah
Generator	: 14 V/36 A

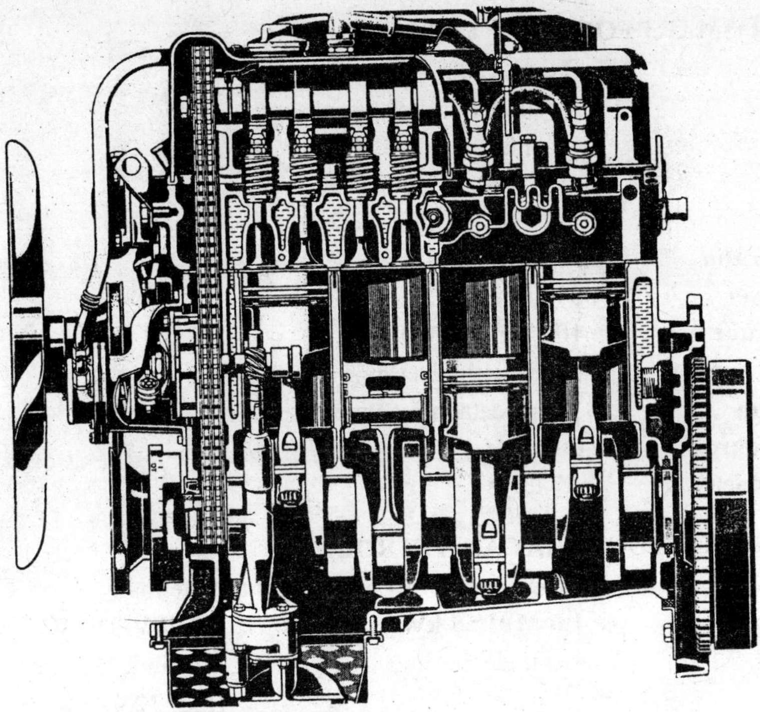

Fig. 22.3. (*a*) Engine 615.940 for Mercedes Benz Car 200D (Front Section)
(Courtesy–Mercedes Benz A.G. Germany)

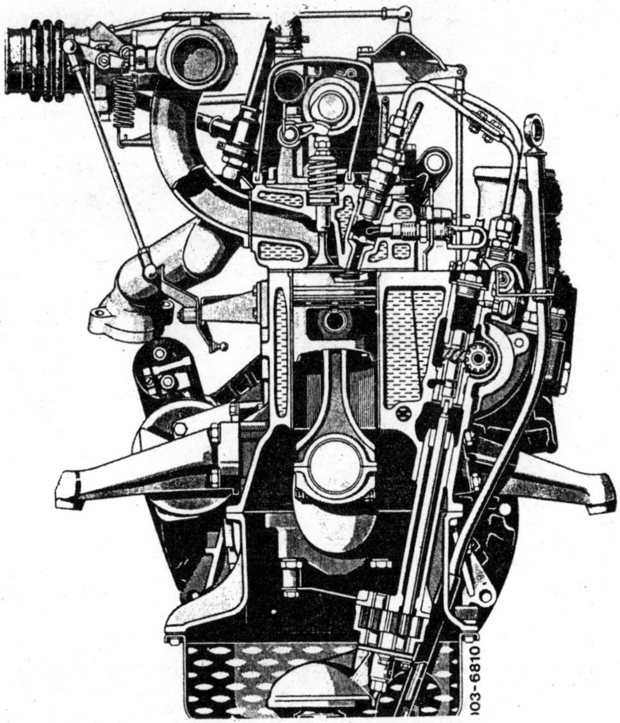

Fig. 22.3. (*b*) Engine 615.940 for Mercedes Benz Car 200D (Left Sectional view)
(Courtesy–Mercedes Benz A.G. Germany)

138. ENGINE FOR 'SMART FOR FOUR' CAR

Manufacturer	: Daimler Chrysler Germany
Type	: 3-cylinder, in-line petrol engine
Bore	: 75 mm
Stroke	: 84.8 mm
Capacity	: 1124 cc
Compression ratio	: 10.5 : 1
Maximum Power	: 55 kW at 6000 rpm
Maximum Torque	: 100 Nm at 3500 rpm
Fuel delivery	: MPFI with electronic accelerator
Exhaust system	: 3-way catalytic converter
Emission Standard	: Euro IV
Fuel Tank Capacity	: 47 litres

139. ENGINE FOR SKODA OCTAVIA CAR (PETROL)

Manufacturer : Skoda Auto (Volkswagen group), Germany

	1.6 MPI/55 kW	**1.8 20V/92 kW**
Type	: 4-cylinder, in-line	4-cylinder in-line with 20 valves (Fig. 22.4)
Capacity	: 1598 cc	1781 cc
Fuel	: Unleaded 95 RON	Unleaded 95 RON
Fuel supply system	: Multipoint fuel injection	Sequential multipoint fuel injection

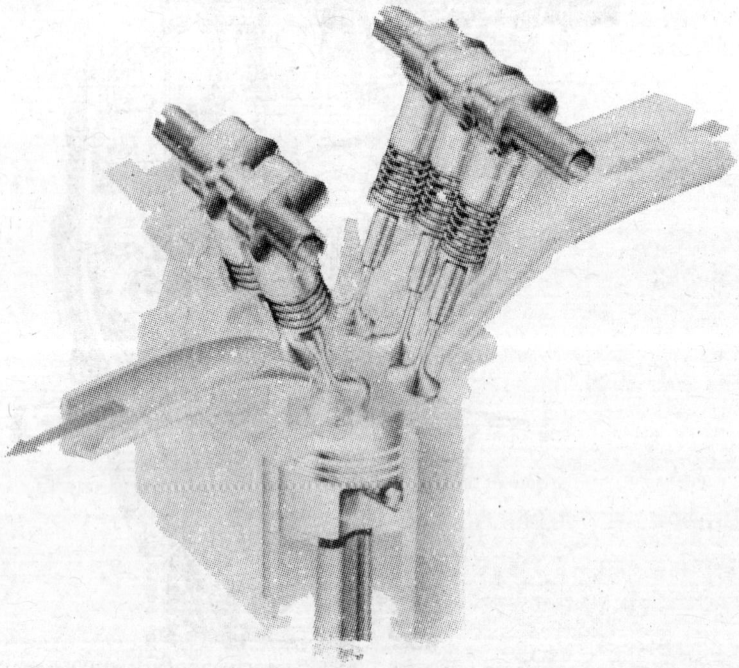

Fig. 22.4. 20-valve technology of Octavia car

Maximum Power	: 55 kW at 4600 RPM	92 kW at 6000 RPM
Maximum Torque	: 135 Nm at 3200 PM	170 Nm at 4200 RPM

140. ENGINE FOR PORSCHE 911 CARRERA CAR

Manufacturer	: Porsche Aktiengesellschaft, Germany
Type	: Air-cooled, six-cylinder, 4-stroke boxer engine, dry-sump lubrication, two controlled, three-way, metal catalytic converters, twin ignition, two valves per cylinder, hydraulic valve-play adjustment
Bore	: 100 mm
Stroke	: 76.4 mm
Displacement	: 3600 cc
Compression ratio	: 11.3
Fuel type	: Super plus unleaded, 98 RON
Mixture preparation	: Sequential, multi-point, manifold injection
Engine management	: Engine-up controlled, electronic engine management, load-monitoring by hot film, knock regulation
Electrics	: Battery 75 Ah, Generator 115 amp/1610 W
Maximum Power	: 200 kW at 6100 RPM
Maximum Torque	: 330 Nm at 5000 RPM
Fuel tank capacity	: 74.5 litres (92 litres optional)

141. ENGINE FOR 718 BOXSTER S

Manufacturer	: Dr. Ing. H.C.F. Porsche AG
Engine Type	: 4-cylinder, horizontally opposed
Bore	: NA
Stroke	: NA
Displacement	: 2497 cc
Compression ratio	: 9.5 : 1
Unladen weight (DIN)	: 1355 kg
Maximum Power	: 257 kW at 6500 rpm
Maximum Torque	: 420 Nm at 1900-4500 rpm
Power to weight ratio	: 189.7 kW/litre
Torque weight ratio	: 310 Nm/tonne
Specific output	: 103 kW/litre
Fuel system	: Petrol engine
Fuel tank capacity	: 64 litres
Fuel consumption (litres/100 km)	: Urban : 10.7 Non-urban : 6.5 Combined : 8.1
CO_2 emissions	: 184 g/km

142. ENGINE FOR BMW 318I CAR

Manufacturer	: BMW, Germany
Engine Type	: 4-cylinder, in-line petrol engine. Alloy head and block
Bore	: 84 mm
Stroke	: 90 mm
Displacement	: 1995 cc
Compression ratio	: 10.2 : 1
Maximum Power	: 106.6 kW at 6000 rpm
Maximum Torque	: 200 Nm at 3750 rpm
Specific Output	: 53.46 kW/litre
Power to weight ratio	: 74.86 kW/tonne
Torque to weight ratio	: 140.28 Nm/tonne
Valve Gear	: 4 valves per cylinder, DOHC, Valvetronic
Ignition and Fuel system	: Bosch ME 9.2 engine management, MPFI
Fuel tank capacity	: 63 litres
Fuel consumption	: 6.8 kpl (city), 10.2 kpl (highway)

143. ENGINE FOR ALFA ROMEO CAR (2000 SPIDER VELOCE)

Manufacturer	: Alfa Romeo s.p.a., Italy
Type	: 4-cylinder in-line petrol engine
Bore	: 84 mm
Stroke	: 88.5 mm
Cubic capacity	: 1962 c.c.
Compression ratio	: 9 : 1
Brake Power	: Maximum B.P. (SAE)-109.7 kW at 5300 r.p.m.
Torque	: Maximum (SAE)-205 Nm at 4400 r.p.m.
Valve	: V-Overhead valves directly operated by two overhead camshaft acting through oil bath cups (Fig. 20.6): sodium cooled valves
Fuel System	: Fuel tank capacity-51 litres, two horizontal twin- choke carburettors
Electric system	: Battery-12 V, 50 Ah; Alternator.

144. ENGINE FOR ALFA 75 2.0 T. SPARK "USA 83" CAR

Manufacturer	: Alfa Romeo s.p.a., Italy
Type	: 4-cylinder in-line
Bore	: 84 mm
Stroke	: 88.5 mm
Displacement	: 1962
Compression ratio	: 10.0
Max. Power output	: 104 kW at 5800 r.p.m.
Max. Torque	: 185.8 Nm at 4800 r.p.m.
Timing control	: Double overhead camshaft (chain) variable advance
Fuel supply	: Bosch ML 4.1 Motronic electronic injection combined with ignition

Fig. 22.5. Engine for Alfa Romeo 'Spider car
(Courtesy–Alta Romeo s.p.a.)

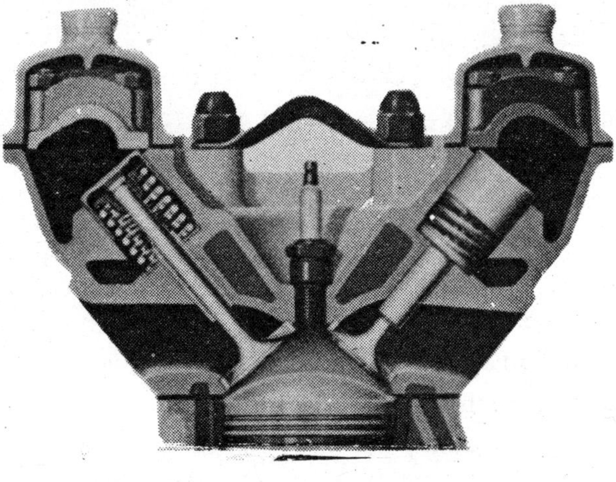

Fig. 22.6. Valve arrangement for Alfa Romeo for 'Spider car
(Courtesy–Alta Romeo s.p.a.)

Emission control	: Three-way catalytic converter with lambda sensor
Fuel consumption	: 6.3 litres/100 km (at 90 km/h)
	: 8.6 litres/100 km (at 120 km/h)
	: 10.2 litres/100 km (urban driving)

145. ENGINE FOR '458 SPIDER' SPORTS CAR

Manufacturer	: Farrari S.P.A., Italy
Model	: 2012
Type	: V-8, 90° petrol engine, dry sump
Bore	: 94.0 mm
Stroke	: 81.3 mm
Displacement	: 4500 cc
Compression ratio	: 12.5 : 1
Maximum Power	: 419 kW at 9000 rpm
Maximum Torque	: 550 Nm at 6000 rpm
Power to Weight ratio	: 273 kW/tonne
Torque to Weight ratio	: 358 Nm/tonne
Specific Output	: 93 kW/litre:
Fuel System	: Direct injection
Fuel Tank Capacity	: 86 litres
Fuel consumption	: Combined cycle with HELE system: 8.46 km/litre

146. ENGINE FOR 'SA APERTA' SPORTS CAR

Manufacturer	: Farrari SpA, Italy
Model	: 2012
Type	: 65° V 12 petrol engine
Bore	: 92.0 mm
Stroke	: 75.2 mm
Displacement	: 5999 cc
Compression ratio	: 11.9 : 1
Maximum Power	: 493 kW at 8250 rpm
Maximum Torque	: 620 Nm at 6500 rpm
Power to Weight ratio	: 289 kW/tonne
Torque to Weight ratio	: 363 Nm/tonne
Specific Output	: 82 kW/litre
Fuel Tank Capacity	: 105 litres
Fuel consumption	: Combined cycle: 5.7 kpl
CO_2 emission	: 411 g/km

147. ENGINES FOR VOLVO CARS

Manufacturer : AB Volvo, Sweden

The specification are given in Table 22.1.

TABLE 22.1. ENGINE SPECIFICATIONS FOR VOLVO CARS

Specification	*Volvo 740 GL*	*Volvo 740 GLE*	*Volvo 740 Turbo*	*Volvo 740 GL diesel*
Type	4-cylinder in-line	4-cylinder in-line	4-cylinder in-line	4-cylinder in-line
Bore/Stroke (mm)	96/80	96/80	96/80	76.5/86.4
Displacement (cc)	2316	2316	2316	2383
Compression ratio	10.3 : 1	10.3 : 1	9.0 : 1	23.0 : 1
Max. output (kW DIN/rpm)	84/5000	98/5400	132/5700	61/4800
Max. Torque (Nm DIN/rpm)	192/2500	190/3600	253/3400	140/2800
Engine block	Cast iron	Cast iron	Cast iron	Cast iron
Cylinder head	Light alloy	Light alloy	Light alloy	Light alloy
Valve system	Overhead camshaft	Overhead camshaft	Overhead camshaft	Overhead camshaft
Fuel system	Carburettor	CL-injection	Intermittent injection	Indirect injection
Ignition system	Transistorized/ Breakerless	Microprocessor controlled	Microprocessor controlled	—

148. ENGINE FOR 'GENESIS' SEDAN

Manufacturer	: Hyundari Motor Co., Korea	
Model	: 2012	
	Lambda 3.8 Engine	**Tau 4.6 Engine**
Type	: V-6 petrol	V-8 petrol
Displacement	: 3778 cc	4627 cc
Maximum Power	: 248 kW at 6400 rpm	273 kW at 6500 rpm
Maximum Torque	: 492 Nm at 5100 rpm	439 Nm at 3500 rpm
Power to Weight ratio	: 143 kW/tonne	149 kW/tonne
Torque to Weight ratio	: 283 Nm/tonne	240 Nm/tonne
Specific Output	: 65.6 kW/litre	59 kW/litre
Valve Gear	: DOHC with dual CVVT	DOHC with dual CVVT
Fuel System	: MPFI, Petrol, Direct Injection	MPFI, Direct Injection, Petrol
Fuel Tank Capacity	: 73 litres	77 litres
Fuel consumption	: City: 8.2 kpl, Highway: 12.5 kpl	City: 7.3 kpl Highway: 11.2 kpl

149. ENGINE FOR 'MUSTANG V6' COUPE

Manufacturer	: Ford Motor Co., U.S.A.
Model	: 2012
Type	: Ti-VCT V-6 petrol engine, all aluminium
Bore	: 95.5 mm
Stroke	: 86.7 mm
Displacement	: 3720 cc
Compression ratio	: 10.5 : 1

Maximum Power	: 227 kW at 6500 rpm
Maximum Torque	: 388 Nm at 4250 rpm
Power to Weight ratio	: 144 kW/tonne
Torque to Weight ratio	: 247 Nm/tonne
Specific Output	: 61 kW/litre
Valve Gear	: Direct acting mechanical bucket type valve lifters
Fuel System	: Sequential Multiport Electronic fuel injection
Fuel consumption (Estimated)	: City: 8.2 kpl, Highway: 12.5 kpl

150. ENGINE FOR ESCAPE HYBRID

Manufacturer	: Ford Motor Co., U.S.A.
Model	: 2012
Type	: Atkinson cycle I-4 petrol engine and Permanent magnet electric motor. Engine block and head made of aluminium
Displacement	: 2500 cc
Compression ratio	: 12.3 : 1
Maximum Power	: 115 kW at 6000 rpm (Engine); 70 kW at 5000 rpm (Motor)
Net power	: 127 kW
Maximum Torque	: 188 Nm at 4500 rpm
Power to Weight ratio	: 31.5 kW/tonne
Torque to Weight ratio	: 51.5 Nm/tonne
Specific Output	: 46 kW/litre
Valve Gear	: Mechanical bucket type valve lifter
Fuel System	: Sequential Multi-port electronic fuel injection
Fuel Tank Capacity	: 57 litres
Fuel consumption (Estimated)	: City: 14.6 kpl, Highway: 13.3 kpl, Combined: 13.8

151. ENGINE FOR CHEVROLET CORVETTE

Manufacturer	: General Motors, USA
Model	: 2012
Type	: V 8, 90° push rod, aluminium block and heads
Bore	: 10.3 mm
Stroke	: 92.0 mm
Displacement	: 6200 cc
Compression ratio	: 10.7 : 1
Maximum Power	: 320 kW at 5900 rpm
Maximum Torque	: 588 Nm at 4600 rpm
Power to Weight ratio	: 218 kW/tonne
Torque to Weight ratio	: 401 Nm/tonne
Specific Output	: 51.6 kW/litre
Valve Gear	: 16 valves, hydraulic lifters

Fuel System	: Petrol
Fuel Tank Capacity	: 42 litres
Fuel consumption	: City 6.9 kpl, Highway: 11.2 kpl

152. ENGINE FOR CADILLAC 'ESCALADE' SUV

Manufacturer	: General Motor, USA
Model	: 2012
Type	: V-8, 'Vortec' petrol/ethanol engine
Displacement	: 6200 cc
Maximum Power	: 300 kW at 5700 rpm
Maximum Torque	: 577 Nm at 4300 rpm
Power to Weight ratio	: 119 kW/tonne (RWD), 115 kW/tonne (AWD)
Torque to Weight ratio	: 229 Nm/tonne (RWD), 222 Nm/tonne (AWD)
Specific Output	: 48.4 kW/litre
Valve Gear	: Variable Valve Timing
Fuel System	: SFI Petrol/ Ethanol
Alternator	: 160 Amp

153. ENGINE FOR 'KIZASHI' CAR

Manufacturer	: Suzuki Motor of America
Model	: 2013
Type	: 4-cylinder, inline
Bore	: 92 mm
Stroke	: 90 mm
Displacement	: 2393 cc
Maximum Power	: 134 kW at 6000 rpm
Maximum Torque	: 230 Nm at 4000 rpm
Power to Weight ratio	: 89 kW/tonne
Torque to Weight ratio	: 157 Nm/tonne
Specific Output	: 54.3 kW/litre
Valve Gear	: DOHC with variable timing, 4 valves per cylinder
Fuel System	: MPFI
Fuel Tank Capacity	: 45 litres
Fuel consumption	: City: 10.7 kpl, Highway:14.7 kpl

154. ENGINE FOR A3 HATCHBACK

Manufacturer	: Chery Automobile Co. Ltd. China
Model	: 2012
Type	: 4-cylinder, inline, SQR 481 FC engine
Displacement	: 1597 cc
Compression ratio	: 10.5 : 1
Maximum Power	: 87.5 kW at 5150 rpm

Maximum Torque	: 147 Nm at 4300-4500 rpm
Power to Weight ratio	: 64.8 kW/tonne
Torque to Weight ratio	: 109 Nm/tonne
Specific Output	: 55 kW/litre
Fuel System	: Petrol
Fuel Tank Capacity	: 57 litres
Emission standard	: Euro 4

155. ENGINE FOR VITA N3 SEDAN

(Figures in brackets for 1.3 litre engine)

Manufacturer	: FAW, China
Model	: 2012
Type	: 3-cylinder, inline 1.1 litre (4-cylinder inline 1.3 litre), water cooled
Bore	: 78.0 mm (78.7 mm)
Stroke	: 73.4 mm (69.0 mm)
Displacement	: 1052 cc (1342 cc)
Compression ratio	: 9.5 : 1 (9.3 : 1)
Maximum Power	: 42.5 kW at 6000 rpm (63 kW at 6000 rpm)
Maximum Torque	: 87 Nm at 3600 rpm (110 Nm at 5200 rpm)
Power to Weight ratio	: 50.3 kW/tonne (70 kW/tonne)
Torque to Weight ratio	: 103 Nm/tonne (122 Nm/tonne)
Specific Output	: 40.4 kW/litre (46.9 kW/litre)
Fuel System	: Petrol
Fuel Tank Capacity	: 37 litres
Fuel consumption	: 20 kpl (18.2 kpl) at 60 km/hr with full load

156. ENGINE FOR XING WANG MINI VAN

Manufacturer	: Wuling Motors, China
Model	: 2012
Type	: 4-cylinder, inline, LJ 465Q3E2
Displacement	: 970 cc
Compression ratio	: 8.8 : 1
Maximum Power	: 34.7 kW at 5200 rpm
Maximum Torque	: 71 Nm at 3000-3500 rpm
Power to Weight ratio	: 38.5 kW/tonne
Torque to Weight ratio	: 78.9 Nm/tonne
Specific Output	: 35.8 kW/litre
Fuel System	: Petrol
Fuel Tank Capacity	: 32 litres
Fuel consumption	: 17.2 kpl at 40 km/hr.

157. ENGINE FOR YARIS

Manufacturer	: Toyota Motor Co. Japan
Model	: 2012
Type	: 4-cylinder, inline
Displacement	: 1500 cc
Maximum Power	: 79 kW at 6000 rpm
Maximum Torque	: 480 Nm at 4200 rpm
Power to Weight ratio	: 75.8 kW/tonne
Torque to Weight ratio	: 460 Nm/tonne
Specific Output	: 52.7 kW/litre
Valve Gear	: 16 valves DOHC, variable timing with intelligence (VVTi)
Ignition System	: Electronic, with Toyota Direct Ignition (TDI)
Emission rating	: Ultra Low Emission Vehicle II (ULEV-II)

QUESTIONS FOR VIVA

1. State the compression ratios of engines of (*i*) Tata Nano (*ii*) Nissan Teana, (*iii*) Hyundai Santro, (*iv*) Tata Indica (petrol and diesel).
2. What is the values of B.P. for (*i*) Alto 800, (*ii*) Mahindra Xylo, (*iii*) Tata Indica Diesel, (*iv*) Honda City, (*v*) Mercedes Benz E-220 engines?
3. What fuel is used in Honda Tazz engine?
4. What is the firing order for Alto 800 engine?
5. What is the cubic capacity for engine of (*i*) Scorpio mHawk, (*ii*) Ford Fiesta, (*iii*) Tata Indica, (*iv*) Hyundai i10 and, (*v*) Skoda Fabia?
6. Which type of fuel supply system is used in (*i*) Santro, (*ii*) Indica, (*iii*) Skoda Laura, (*iv*) Wagon R, (*v*) Mercedes Benz, E 220 and (*vi*) Audi A-4 (1.8T)?
7. State the maximum torque of Tata Axe truck.
8. Which type of fuel injection system is used in Tata LPT-2518 truck?
9. What is the compression ratio of (*i*) Force One, (*ii*) Ashok Leyland Comet engines?
10. State the type of engine for H.M. 'Winner' Truck vehicles.
11. Which type of lubrication system is employed in Swaraj Mazda engine?
12. Which type of ignition system is used in Vauxhall 'Agila' car engine?
13. Give the compression ratio for Ferrari '458 spider' sports car engine.
14. Which type of valve gear is employed in Hyundai 'Genesis' engine?
15. Which type of fuel supply system is used in Suzuki 'Kizashi' car engine?
16. Which type of fuel system is used in Renault 'Fluence' diesel car engine?
17. State the type of engine used in Honda 'Insight SE Hybrid' car.
18. Which type of fuel supply system is used in Ford 'Mustang' car engine?
19. Which type of camshaft is used in Audi 100 SE car engine?
20. State the type of valve gear employed in Volkswagen Golf 16 V car engine.
21. Which main types of engines are employed in Volkswagen 'Polo' car?
22. Which type of fuel supply systems are used in a typical 'Polo' car engine of each of the types referred to at S. No. 21 above.
23. Which type of valves are used in Alfa Romeo 2000 spider Veloce car engine?
24. Which type of fuel supply system is used in engine of (*i*) Volvo 740 GLE, (*ii*) Volvo 740 Turbo, (*iii*) Volvo 740 G.L. diesel?

25. Which type of ignition system is used in engine of Volvo model (*i*) 740 GLe, (*ii*) 740 GLE, (*iii*) 740 Turbo?
26. Name five Indian and five foreign car manufacturers.
27. Name the car with engine having (*i*) 4 valves, (*ii*) 5 valves per cylinder.

MULTIPLE CHOICE QUESTIONS

Mark the correct answers

1. Compression ratio of 17.0 is used in the engine of
   (*a*) Force One Truck (*b*) Tata 207 DI Pickup
   (*c*) Mahindra Scorpio (*d*) Ashok Leyland Comet
2. V-eight engine was employed in
   (*a*) Ferrari '458 Spider' (*b*) Chery A3 Hatchback
   (*c*) Citroen C4 (*d*) Audi 100 SE
3. Five cylinder in-line engine is used in
   (*a*) Chrysler Delta (*b*) Mercedes Benz 200
   (*c*) Fiat 500 (*d*) Audi 100 SE
4. V-12 engine is used in
   (*a*) Chevrolet Corvette (*b*) Ssangyong Korando
   (*c*) Rolls Royce, Phantom (*d*) Alfa Romeo 2000 Spider Veloce
5. Electronic fuel injection system is used in engine of
   (*a*) Chrysler Delta (*b*) Audi 100 SE
   (*c*) Volvo 740 GLE (*d*) All of these
6. Valve-actuating mechanism with two overhead camshafts is employed in the engine of
   (*a*) Volvo 740 (*b*) Alfa Romeo 2000 Spider Veloce
   (*c*) Volkswagen Caddy (*d*) All of these
7. Catalytic converter for controlling exhaust emission is used in engine of
   (*a*) Volkswagen Caddy (*b*) Mercedes Benz 200
   (*c*) Audi 100 SE (*d*) All of these
8. Microprocessor-controlled ignition system is used in engine of
   (*a*) Audi 100 SE (*b*) FAW 'Vita N3'
   (*c*) Ssangyong Rexton (*d*) Volvo 740 Turbo

[Key: 1. (*d*), 2. (*a*), 3. (*d*), 4. (*c*), 5. (*d*), 6. (*d*), 7. (*d*), 8. (*d*)**]**

Appendix – A

BS IV REFERENCE FUEL SPECIFICATIONS

1. BS-IV Petrol

Parameter	*Unit*	*Limit*[1]		*Test method*
		Min.	*Max.*	
Research Octane Number, RON		95.0	–	EN 25164
Motor Octane Number, MON		85.0	–	EN 25163
Density at 15°C	kg/m^3	740.0	754.0	ISO 3675
Reid Vapour Pressure	kPa	56.0	60.0	Pr EN ISO 13016-1 (DVPE)
Distillation				
Evaporated at 70°C	% v/v	24.0	40.0	EN-ISO 3405
Evaporated at 100°C	% v/v	50.0	58.0	EN-ISO 3405
Evaporated at 150°C	% v/v	83.0	89.0	EN-ISO 3405
Final boiling point	°C	190.0	210.0	EN-ISO 3405
Residue	% v/v	–	2.0	EN-ISO 3405
Hydrocarbon Analysis Olefins	% v/v	–	10.0	ASTM D 1319
Aromatics	% v/v	29.0	35.0	ASTM D 1319
Benzene	% v/v	–	1.0	ASTM D 1319
Saturates	% v/v	Report	EN 12177	
Carbon/Hydrogen ratio		Report	ASTM D 1319	
Induction Period[2]	minutes	480	–	EN ISO 7536
Oxygen content	% min	–	0.1	EN 1601
Existent gum	mg/ml	–	0.04	EN ISO 6246
Sulphur content[3]	mg/kg	–	10	ASTM D 5453
Copper corrosion		–	Class 1	EN ISO 2160
Lead content	mg/l	–	5	EN 237
Phosphorus content	mg/l	–	1.3	ASTM D 3231

1. The values quoted in the specifications are "true values". In establishment of their limit values the terms of ISO4 259 'Petroleum products-Determination and application of precision data in relation to methods of test' have been applied and infixing a minimum value, a minimum difference of 2R above zero has been taken in to account in fixing a maximum and minimum value, the minimum difference is 4R (R = reproducibility).
2. Not with standing this measure, which is necessary for technical reasons, the manufacturer of fuels should never the less aim at a zero value where the slipulated maximum value is 2R and at the mean value in the case of quotations of maximum and minimum limits. Should it be necessary to clarity the questions as to whether a fuel meets the requirements of the specifications, the terms of ISO 4259 should be applied.
3. The range for cetane number is not in accordance with the requirements of a minimum range of 4R. However, in the case of a dispute between fuel supplier and fuel user, the terms of ISO 4259 may be used to resolve such disputes provided replicate measurements, of sufficient number to archive the necessary precision, are made in preference to single determinations.
4. The actual sulpur content of the fuel used for the Type I test shall be reported.

2. BS-IV Diesel

Parameter	*Unit*	*Limit*(1)		*Test method*
		Min.	*Max.*	
Catane number(2)		52.0	54.0	EN-ISO 5165
Density at 15°C	kg/m3	833	837	NE-ISO 3675
Distillation:				
– 50% point	°C	245	–	EN-ISO 3405
– 95% point	°C	345	350	EN-ISO 3405
– Final boiling point	°C	–	370	EN-ISO 3405
Flash point	°C	55	–	EN 22719
CFPP	°C	–	–5	EN 116
Viscosity at 40°C	mm^2/s	2.3	3.3	EN-ISO 3104
Polycyclic Aromatic Hydrocarbons	% m/m	3.0	6.0	IP 391
Sulphur content(3)	mg/kg	–	0.2	EN-ISO 10370
Copper corrosion		–	Class 1	EN ISO 2160
Conradson carbon residue (10% DR)	% m/m	–	0.2	EN-ISO 10370
Ash content	% m/m	–	0.01	EN-ISO 6245
Water content	% m/m	–	0.02	en-iso 12937
Neutralisation (strong acid) number	mg KOH/g	–	0.02	ASTM D 974
Oxidation stability(4)	mg/ml	–	0.025	EN-ISO 12206
Lubricity (HFRR wear scan diameter at 60°C)	µm	–	400	CEC F-06-A-96
FAME		PROHIBITED		

1. The values quoted in the specifications are "true values". In establishment of their limit values the terms of ISO4259 "Petroleum products-Determination and application of precision data in relation to methods of test" have been applied and infixing a minimum value, a minimum difference of 2R above zero has been taken in to account, in fixing a maximum and minimum value, the minimum difference is 4R (R = reproducibility).
2. Not with standing this measure, which is necessary for technical reasons, the manufacturer of fuels should never the less air at a zero value where the stipulated maximum value is 2R and at the mean value in the case of quotations of maximum and minimum limits. Should it be necessary to clarify the questions as to whether a fuel meets the requirements of the specifications, the terms of ISO 4259 should be applied.
3. The range for cetane number is not in accordance with the requirements of a minimum range of 4R. However, in the case of a dispute between fuel supplier and fuel user, the terms of ISO 4259 may be used to resolve such disputes provided replicate measurements, of sufficient number to archive the necessary precision, are made in preference to single determinations.
4. The actual sulphur content of the fuel used for the type test shall be reported.
5. Even though oxidation stability is controlled, it is likely that shelf life will be limited. Advice should be sought from the supplier as to storage conditions and life.

3. BS IV - CNG

For Category M and Category N Vehicles not exceeding 3,500 Kgs GVW

Parameter	*Unit*	*Basis*	*Limits*		*Test method*
			Min.	*Max.*	
Reference fuel G20					
Composition:					
Methane	% mole	100	99	100	ISO 6974
Balance(1)	% mole	–	–	1	ISO 6974
N_2	% mole				ISO 6974
Sulphur content	mg/m$^{3(2)}$	–	–	10	ISO 6326-5
Wobbe index (net)	MJ/m$^{3(3)}$	48.2	47.2	49.2	
Reference fuel G25					
Composition:					
Methane	% mole	86	84	88	ISO 6974
Balance(1)	% mole	–	–	1	ISO 6974
N_2	% mole	14	12	10	ISO 6974
Sulphur content	mg/m$^{3(2)}$	39.2	38.2	40.6	ISO 6326-5
Wobbe index (net)	MJ/m$^{3(3)}$	39.4	38.2	40.6	

(1) Inerts (different from N_2) + C_2 + C_2.
(2) Value to be determined at 293, 2K (20°C) and 101.3 kPa.
(3) Value to be determined at 273, 2K (0°C) and 101.3 kPa.

For Vehicles above 3,500 Kgs GVW

Parameter	*Unit*	*Basis*	*Limits*		*Test method*
			Min.	*Max.*	
Reference fuel GR					
Composition:					
Methane		87	84	89	
Ethane		13	11	15	
Balance(1)	% mole	–	–	1	ISO 6974
Sulphur content	mg/m$^{3(2)}$	–	–	10	ISO 6326-5
(1) Inerts (different from N_2) + C_2					
Reference fuel G23					
Composition:					
Methane		92.5	91.5	93.5	ISO 6974
Balance(1)	% mole	–	–	1	ISO 6974
N_2		7.5	6.5	8.5	ISO 6974
Sulphur content	mg/m$^{3(2)}$	–	–	10	ISO 6326-5
Reference fuel G25					
Composition:					
Methane		86	84	89	
Balance(1)	% mole	–	–	1	ISO 6974
N_2		14	12	16	
Sulphur content	mg/m$^{3(2)}$	–	–	10	ISO 6326-5

(1) Inerts (different from N_2) + C_2 + C_2.
(2) Value to be determined Standard conditions 2K (20°C) and 101.3 kPa.

Note: Commercial CNG fuel shall be used for testing till availability of reference fuel.

4. BS - IV LPG

For Category M and Category N Vehicles not exceeding 3,500 Kgs GVW

Parameter	*Unit*	*Fuel A*	*Fuel B*	*Test method*
Composition:				ISO 7941
C_3-content	% vol.	30 ± 2	85 ± 2	
C_4-content	% vol.	balance	balance	
$<C_3, >C_4$	% vol.	max. 2	max. 2	
Olefins	% vol.	max. 12	max. 15	
Evaporation residue	mg/kg	max. 50	max. 50	ISO 13757
Water at 0°C		free	free	Visual inspection
Total sulphur content	mg/kg	max. 10	max. 10	EN 24260
Hydrogen sulphide		none	none	ISO 8819
Copper strip corrosion	Rating	Class 1	Class 1	ISO 6251(1)
Odour		Characteristic	Characteristic	
Motor Octane Number		min. 89	min. 89	EN 589 Annex B

(1) This method may not accurately determine the presence of corrosive materials if the sample contains corrosion inhibitors or other chemicals which diminish the corrosivity of the sample to the copper strip. Therefore, the addition of such compounds for the sole purpose of biasing the test method is prohibited.

For Category M and Category N Vehicles above 3,500 Kgs GVW

Parameter	*Unit*	*Fuel A*	*Fuel B*	*Test method*
Composition:				ISO 7941
C_3-content	% vol.	50 ± 2	85 ± 2	
C_4-content	% vol.	balance	balance	
$<C_3, >C_4$	% vol.	max. 2	max. 2	
Olefins	% vol.	max. 12	max. 14	
Evaporation residue	mg/kg	max. 50	max. 50	ISO 13757
Water at 0°C		free	free	Visual inspection
Total sulphur content	mg/kg	max. 10	max. 10	EN 24260
Hydrogen sulphide		none	none	ISO 8819
Copper strip corrosion	Rating	Class 1	Class 1	ISO 6251(1)
Odour		Characteristic	Characteristic	
Motor Octane Number		min. 92.5	min. 92.5	EN 589 Annex B

(1) This method may not accurately determine the presence of corrosive materials if the sample contains corrosion inhibitors or other chemicals which diminish the corrosivity of the sample to the copper strip. Therefore, the addition of such compounds for the sole purpose of biasing the test method is prohibited.

Reference: "Indian Emissions Regulations", ARAI (The Automotive Research Association of India)

Appendix – B

BS VI REFERENCE FUEL SPECIFICATION

1. BS-VI Petrol (E5)

Parameter	Unit	Limit[1]		Test method
		Min.	Max.	
Research Octane Number, RON		95.0	–	EN25164/prENISO5164
Motor Octane Number, MON		85.0	–	EN25163/prENISO5163
Density at 15°C	kg/m3	743	756	ENISO 3675/ENISO12185
Reid Vapour Pressure	kPa	56.0	60.0	ENISO 13016-(DVPE)
Water content	% v/v		0.015	ASTME 1064
Distillation:				
– Evaporated at 70°C	% v/v	24.0	44.0	ENISO 3405
– Evaporated at 100°C	% v/v	48.0	60.0	ENISO 3406
– Evaporated at 150°C	% v/v	82.0	90.0	ENISO 3405
– Final boiling point	°C	190	210	ENISO 3405
Residue	% v/v	–	1.0	ENISO 3405
Hydro-carbon Analysis:				
– Olefins	% v/v	3.0	13.0	ASTMD 1319
– Aromatics	% v/v	29.0	35.0	ASTMD 1319
– Benzene	% v/v	–	1.0	EN 12177
– Saturates	% v/v	Report		ASTM 1319
Carbon/Hydrogen ratio		Report		
Carbon/oxygen ratio		Report		
Induction Period[2]	Minutes	480		ENISO 7536
Oxygen content[4]	% m/m	Report		EN1601
Existent gum	mg/ml	–	0.04	ENISO 6246
Sulphur content[3]	mg/kg	–	10	ENISO 20846/ENISO20884
Copper corrosion		–	Class 1	ENISO 2160
Lead content	mg/l	–	5	EN 237
Phosphorus content	mg/l	–	1.3	ASTMD 3231
Ethanol[5]	% v/v	4.7	5.3	EN 1601/EN 13132

1. The values quoted in the specifications are "true values". For establishing the limit values, the terms of ISO4259:2006 (Petroleum products–Determination and application of precision data in relation to method of test) have been applied and for fixing a minimum value, a minimum difference of 2R above zero has been taken into account; for fixing a maximum and minimum value, the minimum difference is 4R (R=reproducibility). Not with standing this measure, which is necessary for technical reasons, the fuel manufacturer shall nevertheless aim at a zero value where the stipulated maximum value is 2R at the mean value when quoting maximum and minimum limits. Should it be necessary to clarify whether a fuel meets the requirements of the specitications, the terms of ISO 4259:2006 shall be applied.
2. The fuel may contain oxidation inhibitors and metal deactivators normally used to stabilize refinery petrol streams, but detergent/dispersive additives and solvent soils shall not be added.
3. The actual sulphur content of the fuel used for the type test shall be reported.
4. Ethanol metting the specification of prEN15376 is the only oxygenate that shall be intentionally added to the reference fuel.
5. There shall be no intentional addition to this reference fuel of compounds containing phosphorus, iron, manganese or lead.

2. BS VI REFERENCE FUEL (E85)

Parameter	*Unit*	*Limit*[1]		*Test method*[2]
		Min.	*Max.*	
Research octane number, RON		95.0	–	EN ISO 5164
Motor octane number, MON		85.0		EN ISO 5163
Density at 15°C	kg/m^3	Report		ISO 3675
Vapour pressure	kPa	40.0	60.0	EN ISO 13016-1 (DVPE)
Sulphur content[3,4]	mg/kg	–	10.0	EN ISO 20846 EN ISO 20884
Oxidation stability	minutes	360		EN IEO 7536
Existent gum content (solvent washed)	mg/(100ml)	–	5	EN-ISO 6246
Appearance This shall be determined at ambient temperature or 15°C whichever is higher		Clear and bright, visible free of suspended or precipitated contaminants		Visual inspection
Ethanol and higher alcohols[7]	% v/v	83	85	EN 1601 EN 13132, EN 14517
Higher alcohols (C3-C8)	%v/V		2.0	
Methanol	% v/v		0.5	
Petrol[5]	% v/v	Balance		EN 228
Phosphorus	mg/l		0.3[6]	ASTM D 3231
Water content	%v/v		0.3	ASTM D 1064
Inorganic chloride content	mg/l		1	ISO 6227
pHe		6.5	9.0	ASTM D 6423
Copper strip corrosion (3h at 50°C)	Rating	Class 1		EN ISO 2160
Acidity (as acetic acid CH_2COOH)	% m/m, mg/l		0.005(40)	ASTM D 1613
Carbon/hydrogen ratio		Report		
Carbon/oxygen ratio		Report		

1. The values quoted in the specifications are "true values". In establishment of limit values the terms of ISO 4259 Petroleum products-Determination and application of precision data in relation to methods of test have been applied and in fixing a minimum value, a minimum difference of 2R above zero has been taken into account; in fixing a maximum and minimum value, the minimum difference is 4R (R=reproducibility). Not with standing this measure, which is necessary for technical reasons, the manufacturer of fuels shall nevertheless aim at a zero value where the stipulated maximum value is 2R and at the mean value in the case of quotations of maximum and minimum limits. Should it be necessary to clarify whether a fuel meets the requirements of the specifications, the terms of ISO 4259 shall be applied.
2. In case of dispute, the procedure for dissolving the dispute and interpretations of the results based on test method precision, describe, in EN ISO 4259 shall be used.
3. In case of national dispute concerning sulphur content, either EN ISO 20846 or EN ISO 20884 shall be called up similar to the reference in the national annex of EN 228.
4. The actual sulphur content of the fuel used for the Type I Test shall be reported.
5. The unleaded petrol content can be determined as 100-minus the sum of the percent content of water and alcohol.
6. There shall be no intentional addition of compounds containing phosphorus, iron, manganese, or lead to this reference fuel.
7. Ethanol to meet the specification of EN 15376 is the only oxygenate that shall be intentionally added to this reference fuel.

3. BS VI Reference Fuel (ED95)[1]

Parameter	*Unit*	*Limit*[1]		*Test method*[2]
		Min.	*Max.*	
Total alcohol (Ethanol including content on higher saturated alcohols)	% m/m	92.4		EN 15721
Other higher saturated Mono-alcohols (C3-C5)	% m/m		2.0	EN 15721
Methanol	% m/m		0.3	EN 15721
Density 15°C	kg/m3	793.0	815.0	EN ISO 12185
Acidity, calculated as acetic acid	% m/m		0.0025	EN 15491
Appearance		Bright and clear		
Flash point	°C	10		EN 3679
Dry residue	mg/kg		15	EN 15691
Water content	% m/m		6.5	EN 15489[4] EN-ISO 12937 EN 15692
Aldehydes calculated as acetaldehyde	%m/m		0.0050	ISO 1388-4
Esters calculated as ethylacetat	% m/m		0.1	ASTM D 1617
Sulphur content	mg/kg		10.0	EN 15485 EN 15486
Sulphates	mg/kg		4.0	EN 15492
Particulate contamination			24	EN 12662
Phosphorus	mg/l		0.20	EN 15487
Inorganic chloride	mg/kg		1.0	EN 15484 or EN 15492
Copper	mg/kg		0.100	EN 15488
Electrical conductivity	µS/cm		2.50	DIN 51628-4 or prEN 15938

1. Additives are necessary to fulfil the ED95 fuel specification such as cetane improver as specified by the engine manufacturer, shall be added t the ethanol fuel, as long as no negative side effects are known. If these conditions are satisfied, the maximum allowed amount is 10% m/m.
2. Denaturants for ED95 fuel, if required, shall be approved by the vehicle manuf.acturer or shall be either Ethyl Tert Butyl Ether (ETBE); or Iso butanol; or Methyl Ethyl Ketone (MEK); or Tert butanol or Iso Propanol.

   The values quoted in the specifications are "true values". In establishment of their limit values the terms of ISO 4259 Petroleum products-Determination and application of precision data in relation to methods of test have been applied and in fixing a minimum value, a minimum difference of 2R above zero has been taken into account; in fixing a maximum and minimum value, the minimum difference is 4R ® = reproducibility). Not with standing this measure, which is necessary for technical reasons, the manufacturer of fuels shall nevertheless aim at a zero value where the stipulated maximum value is 2R and at the mean value in the case of quotations of maximum and minimum limits. Should it be necessary to clarify whether a fuel meets the requirements of the specifications, the terms of ISO 4259 shall be applied.
3. Equivalent EN/ISO methods will be adopted when issued for properties listed above.
4. Should it be necessary to clarify whether a fuel meets the requirements of the specifications, the terms of EN 15489 shall be applied.

4. BS VI Ethanol Fuel: E100 as per IS 15464:2004

Sr. No.	*Characteristic*	*Requirement*
1.	Relative density at 15.6/15/6 OC, max	0.7961
2.	Ethanol content percent by volume at (excluding denaturant)	99.5
3.	Miscibility with water	Miscible
4.	Alkalinity	Nil
5.	Acidity (as CH_3COOH) mg/l, max	30
6.	Residue on evaporation percent by mass, Max	0.005
7.	Aldehyde content (as CH_3CHO) mg/l, max	60
8.	Cooper mg/kg, max	0.1
9.	Conductivity, μ S/m, max	300
10.	Methyl alcohol, mg/litre, max	300
11.	Appearance	Clear and bright

5. BS VI Hydrogen Fuel

Characteristics	*Units*	*Limits*		*Test Method*
		minimum	*maximum*	
Hydrogen Purity	% mole	98	100	ISO 14687-1
Total Hydrocarbon	μmol/mol	0	100	ISO 14687-1
Water[1]	μmol/mol	0	2	ISO 14687-1
Oxygen	μmol/mol	0	2	ISO 14687-1
Argon	μmol/mol	0	2	ISO 14687-1
Nitrogen	μmol/mol	0	2	ISO 14687-1
CO	μmol/mol	0	1	ISO 14687-1
Sulphur	μmol/mol	0	2	ISO 14687-1
Permanent Particulates[3]				ISO 14687-1

1. Not to be condensed
2. Combined water, oxygen, nitrogen, argon: 1.900 mmol/mol.
3. The hydrogen shall not contain dust, sand, dirt, gums, oils or other substances in an amount sufficient to damage the fuelling station equipment of the vehicle (engine) being fuelled.

6. BS VI Diesel Fuel (B7)

Parameter	*Units*	*Limits*[1]		*Test Method*
		minimum	*maximum*	
Cetane index		46.0		EN ISO 4264
Cetane number[2]		52.0	56.0	EN ISO 5165
Density at 15°C	kg/m^3	833.0	837.0	EN ISO 12185
Distillation:				
– 50% point	°C	245.0	–	EN ISO 3405
– 95% point	°C	345.0	360.0	EN ISO 3405
– Final boiling point	°C	–	370.0	EN ISO 3405
Flash point	°C	55	–	EN ISO 2719
Cloud point	°C	–	–10	EN 23015
Viscosity at 40°C	mm^2/s	2.30	3.30	EN ISO 3104
Polycyclic aromatic hydrocarbons	%m/m	2.0	4.0	EN 112916
Sulphur content	mg/kg	–	10.0	EN ISO 20846 EN ISO 20884
Copper corrosion 3 hrs 50°C		–	Class 1	EN ISO 2160
Conradson carbon residue (10% DR)	% m/m	–	0.20	EN ISO 10370
Ash content	% m/m	–	0.010	EN ISO 6245
Total contamination	mg/kg	–	24	EN 12662
Water content	mg/kg	–	200	EN ISO 12937
Acid number	mg KOH/g	–	0.10	EN ISO 6618
Lubricity (HFRR wear scan diameter at 60°C)	μm	–	400	EN ISO 12156
Oxidation stability @ 110°C[3]	h	20.0		EN 15751
FEME[4]	% v/v	6.0	7.0	EN 14078

1. The value quoted in the specifications are "true values". In establishment of their limit values the terms of ISO 4259 Petroleum products-Determination and application of precision data in relation to methods of test have been applied and in fixing a minimum value, a minimum difference of 2R above zero has been taken into account; in fixing a maximum and minimum value, the minimum difference is 4R (R = reproducibility). Not with standing this measure, which is necessary for technical reasons, the manufacture of fuels shall nevertheless aim at a zero value where the stipulated maximum value is 2R and at the mean value in the case of quotations of maximum and minimum limits. Should it be necessary to clarify whether a fuel meets the requirements of the specifications, the terms of ISO 4259 shall be applied.
2. The range for cetane number is not accordance with the requirements of a minimum range of 4R. However, in the case of a dispute between fuel supplier and fuel user, the terms of ISO 4259 may be used to revolve such disputes provided replicate measurements, of sufficient number to archive the necessary precision are made in preference to single determinations.
3. Even though oxidation stability is controlled, it is likely that shelf life will be limited. Advice shall be sought from the supplier as to storage conditions and life.
4. FAME content to meet the specification of EN 14214.

7. BIO Diesel

Biogas (Biomethane) As per IS 16087: 2013

Sr. No.	*Characteristic*	*Requirement*	*Method of Test*
1.	CH_4 percent Min	90	IS 15130 (part 3)
2.	Moisture, mg/m^3 Max	16	IS 15641 (part 2)
3.	H_2S, mg/m^3 Max	30.3	ISO 6326-3
4.	$CO_2 + N_2 + O_2$, percent Max (v/v)	10	IS 15130 (part 3)
5.	CO_2 Prcent Max (v/v) (when intended for filling in cylinders)	4	IS 15130 (part 3)
6.	O_2, Percent Max (v/v)	0.5	IS 15130 (part 3)

Reference: "Indian Emissions Regulations", ARAI (The Automotive Research Association of India)

Appendix – C

FUEL CONSUMPTION OF SOME INDIAN PETROL VEHICLES

S. No.	Make and Model	Engine c.c.	City (km/h)	Highway (km/h)	S. No.	Make and Model	Engine c.c.	City (km/h)	Highway (km/h)
1.	Audi	1798	9.8	14.2	23.	Hyundai Creta	1591	9.5	14.5
2.	Audi RS5 Coupe	41.63	6.7	11.7	24.	Hyundai i20 Elite	1197	11.5	16.3
3.	BMW i-series	1598	9.7	11.5	25.	Hyundai Xcent	1197	11.5	17.2
4.	BMW Mini Cooper S	1598	9.5	13.5	26.	Maruti Alto	796	12.1	17.1
5.	BMW M5	4395	4.7	7.1	27.	Maruti Wagon R	998	13.0	18.0
6.	Chevrolet Spark	995	12.3	16.9	28.	Maruti Swift	1197	11.9	17.0
7.	Chevrolet SRV	1598	9.7	12.8	29.	Maruti Ritz	1197	11.8	16.7
8.	Chevrolet Beat	1199	11.5	15.3	30.	Maruti Baleno	1197	13.4	18.1
9.	Fiat Palio Stile	1108	10.2	15.0	31.	maruti Ciaz	1373	11.5	17.2
10.	Fiat Grand Punto	1368	10.2	14.6	32.	Mercedes B180	1595	9.5	14.2
11.	Ford Fiesta	1499	10.8	16.2	33.	Mercedes S500	4663	4.9	7.0
12.	Ford Ecosport	999	11.8	17.0	34.	Nissan Datsun GO	1198	12.8	17.9
13.	Ford Endeavour	2198	8.1	11.8	35.	Nissan Datsun Redigo	799	15.8	20.0
14.	Honda CR-V 2.0	1997	9.0	12.0	36.	Renault Kwid	799	15.3	21.5
15.	Honda Jazz	1198	12.3	16.9	37.	Renault Scala Auto	1498	10.7	16.0
16.	Honda Br-V	1497	11.01	14.01	38.	Skoda Superb	1198	8.9	14.2
17.	Honda City i-VTEC	1497	11.0	17.0	39.	Skoda Octavia	1798	9.5	14.0
18.	Honda Amaze	1198	12.5	17.0	40.	Tata Nano Twist	624	15.1	20.6
19.	Hyundai Eon	814	13.7	17.2	41.	Tata Bolt	1193	11.4	16.1
20.	Hyundai Tucson	1999	7.0	11.8	42.	Tata Indica Vista	1172	10.3	15.5
21.	Hyundai Elantra	1999	9.8	14.4	43.	Toyota Altis	1794	9.8	14.5
22.	Hyundai Grand i10	1120	15.4	19.6	44.	Toyota Etios Liva	1197	12.1	16.7

Appendix – D

FUEL CONSUMPTION OF SOME INDIAN DIESEL VEHICLES

S. No.	Make and Model	Engine c.c.	City (km/h)	Highway (km/h)	S. No.	Make and Model	Engine c.c.	City (km/h)	Highway (km/h)
1.	Audi A3	1968	12.3	17.5	23.	Mahindra Xylo	2179	10.1	14.4
2.	Audi Q5 3.0 TDI	2967	8.0	11.1	24.	Mahindra XUV 500	2179	10.2	14.3
3.	Audi Q 73.0 TDI Quattro	2967	5.5	9.1	25.	Mahindra Verito	1461	13.6	19.3
4.	BMW X5 3.0d	2993	5.7	9.8	26.	Mahindra Quanto	1493	11.4	16.1
5.	BMW i-Series	1195	14.2	18.0	27.	Mahindra Nuvosport	1493	11.0	15.0
6.	Chevrolet Enjoy	1248	12.3	16.0	28.	Mahindra TUV300MT	1493	11.4	15.5
7.	Chevrolet Beat	936	16.0	19.1	29.	Mahindra Rexton	2696	7.2	11.8
8.	Chevrolet Sail	1249	12.9	18.7	30.	Maruti Ertiga	1248	12.8	16.8
9.	Fiat Grand Punto	1248	13.7	17.5	31.	Maruti Celerio	793	16.7	20.9
10.	Ford Ecosport	1498	13.3	18.6	32.	Maruti Baleno	1248	14.9	21.4
11.	Ford Endeavour	3198	7.0	10.3	33.	Maruti Swift Desire	1248	14.4	19.1
12.	Honda Jazz	1498	14.4	19.8	34.	Maruti Ritz	1248	11.8	16.7
13.	Honda BR-V	1498	13.9	20.0	35.	Mercedes A180	2143	11.0	15.0
14.	Honda City i-VTEC	1498	14.2	19.5	36.	Mercedes GL 350CDI	2987	6.5	10.5
15.	Honda Amaze (Turbocharger)	1498	15.2	20.8	37.	Nissan Micra	1461	14.6	19.5
16.	Hyundai Tucson	1995	10.3	12.7	38.	Nissan Evalia	1461	12.8	17.5
17.	Hyundai Elantra	1582	12.8	17.0	39.	Renault Lodgy	1461	12.3	17.5
18.	Hyundai Creta	1582	12.5	17.5	40.	Skoda Superb	1968	10.1	14.6
19.	Hyundai i20 Elite	1396	13.8	19.5	41.	Skoda Octavia	1968	12.0	17.0
20.	Hyundai Xcent	1120	16.2	20.3	42.	Tata Aria	2179	9.8	13.8
21.	Land Rover Discovery 4	2993	6.5	9.5	43.	Toyota Altis	1364	12.1	17.1
22.	Mahindra Scorpio mHawk	2179	10.6	13.6	44.	Toyota Etios D-4D	1364	14.2	19.0
					45.	Volkswagen Passat	1968	10.8	15.2
					46.	Volkswagen Polo	1199	14.0	18.9
					47.	Volvo XC60	2400	9.0	12.2

Appendix – E

SELECTED INDIAN STANDARDS CONCERNING AUTOMOTIVE ENGINES

S. No.	IS Number	Reaffirmed	Title
1.	IS 810: 1991	2015	Internal Combustion Engines-Engine Valves-Specification
2.	IS 1062: 2005		Road vehicles-Spark Plugs-Test methods and requirements
3.	IS 1063: 1997		Automotive Vehilces-M14 X 1.25 Spark Plugs with Flat Seating and Their Cylinder Head Housing-Specification (Amendment No. 1)
4.	IS 1543: 1964	2014	Specification for Single cylinder Fuel Injection Pumps
5.	IS 2077: 1966		Automotive vehicles-Electric relays-Specification.
6.	IS 2081: 1998		Automotive Vehicles-Taper Terminal Cable Connectors for Batteries-Specification for brass type connectors
7.	IS 2325: 1981		Ignition coils for battery-coil ignition systems for automobiles (Amendment No. 3)
8.	IS 2577: 1974		Specification for Cartridge Fuse-links for Automobiles
9.	IS 2765: 1982	2015	Specification for Radiator hose
10.	IS 3141: 2007		Starter Motors for internal combustion engines used for automotive and other applications-Specification.
11.	IS 3169: 1991	2015	Internal Combustion Engines-Two Stage, One Litre, Diesel Fuel Filters
12.	IS 3170: Part 1, 2006	2015	Internal combustion Engines-Fuel Injection Nozzles Part 1 Injection Nozzles-Size 'S'
13.	IS 3170: Part 2, 2006	2015	Internal Combustion Engines-Fuel Injection Nozzles Part 2: Calibrating Nozzle, Delay Pintle Type
14.	IS 3171: Part 1, 1997	2016	Internal combustion Engines-Fuel Injection Nozzle Holders - Part 1: Flange Mounted Fuel Injectors Size 'S' Types 2, 3, 4, 5 and 6
15.	IS 3171: Part 2, 2006	2016	Internal combustion Engines-Fuel Injection Nozzle Holders - Part 2: Screw-in Injection Nozzle Holders, Types 20, 21, 21.1 and 27 for Pintle Nozzle Sized 'S', Type B
16.	IS 3171: Part 3, 1997	2016	Internal combustion Engines - Fuel Injection Nozzle Holders - Part 3: Screw Mounted Injection Nozzle Holders, Types 12, 13, 14, 15, 16, 17, 18 and 19.
17.	IS 3172: 1997		Internal Combustion Engines - Fuel Injection Equipment - Single and Double Ended Pipe Unions (single and double ended banjo)- Specification (Amendment No. 1)
18.	IS 3173: 1965	2017	Specification for High pressure connections for fuel injection equipment for diesel engines.
19.	IS 3351: 2006	2015	Road vehicles-Fuel filters for diesel engines Test methods
20.	IS 3649: 1966	2016	Ciutch facing for Automotive Transmission
21.	IS 4086: 1983		Specification for Distributors for Spark Ignition Engines
22.	IS 5439: 1978		Specification for Oil Pressure Switches for Automobiles
23.	IS 5546: 1981		Specification for Distribution Caps and Rotor Arms for Automobiles.

S. No.	IS Number	Reaffirmed	Title
24.	IS 5562: 2000		Automotive Vehicles-Fuel Gauges-Cross Coil Movement Type-Specification
25.	IS 5577: 2000		Automotive Vehicles-Ammeters-Specification
26.	IS 5791: 2006	2015	Internal combustion engines-Piston rings-Material specifications
27.	IS/ISO 6621: Part 1: 2007	2014	Internal Combustion Engines- Piston Rings, Part 1, Vocabulary
28.	IS/ISO 6621: Part 2: 2003		Internal Combustion Engines- Piston Rings, Part 2, Inspection Measuring Principles
29.	IS 6740: 1985	2015	Specification for Gudgeon pins for internal combustion engines (Amendment No. 1)
30.	IS 6750: 1986	2015	Specification for Cylinder liners for internal combustion engines (Amendment No. 1)
31.	IS 7065: 1973		Steering Wheels
32.	IS 7449: Part 1, 1974	2014	Glossary of terms for IC engines: Part 1 Fuel injection equipment
33.	IS 7451: Part 1, 2007	2012	Reciprocating internal combustion engines-Vocabulary : Part 1 Terms for engine design and operation
34.	IS 7451: Part 2, 2006	2016	Reciprocating Internal combustion Engines: Part II Designation of the direction of rotation and of cylinders and valves in cylinder heads, and definition of right-hand and left-hand in-line engines and locations on an engine (Amalgamating Part 3, 4 and 5)
35.	IS 7451: Part 6, 2007	2017	Reciprocating Internal Combustion Engines- Part VI: Hand Operated Control Devices - Standard Direction of Motion
36.	IS 7611, 1993	2014	Internal Combustion Engines - Radiators-Specification
37.	IS 7657 Part 1, 1975	2014	Specification for Starter Ring Gears for Internal Combustion Engines - Part 1: Gears for Inertia and Solenoid Pre-engaged Starters (Amendment No. 1)
38.	IS 7657: Part 2, 1975	2014	Specification for Starter Ring Gears for Internal Combustion Engines - Part II: Gears for Axial and Coaxial Starters
39.	IS 7998: 1986		Specification for Contact Breakers for Two Wheelers, Three Wheelers and Stationary Spark Ignition Engines
40.	8118: 2008	2014	Automotive Vehicles - Opacity (Smoke) of Exhaust Gas from Vehicles Equipped with compression Ignition Engines Operating Under Free Acceleration - Method of Measurement
41.	IS/ISO 8178: Part 3: 1994	–	Reciprocating Internal Combustion Engines-Exhaust Emission Measurement. Part 3: Definitions and Methods of Measurement of Exhaust Gas Smoke under Steady-state conditions.
42.	IS/ISO 8178: Part 4: 2007	2017	Reciprocating Internal Combustion Engines-Exhaust Emission Measurement.
43.	IS/ISO 8178: Part 7: 1996	2017	Reciprocating Internal Combustion Engines- Exhaust Emission Measurement, Part 7: Engine Family Determination.
44.	IS 8422: Part 1, 1977	2014	Specification for Piston Rings for IC Engines - Part 1: Plain Compression Rings from 30 up to 200 mm Nominal Diameter R-Rings
45.	IS 8422: Part 2, 1977	2014	Specification for Piston Rings for IC Engines - Part II: Taper Faced Compression Rings from 30 up to 200 mm Nominal Diameter M-rings
46.	IS 8422: Part 3, 1977	2014	Specification for Piston Rings for IC Engines - Part III: Keystone Rings from 82 up to 200 mm Nominal Diameter T-rings 15

S. No.	IS Number	Reaffirmed	Title
47.	IS 8422: Part 4, 1977	2014	Specification for Piston rings for IC engine: Part 4 Napier oil scraper rings from 30 up to 200 mm nominal diameter N-Rings
48.	IS 8422: Part 5, 1977	2014	Specification for Piston rings for IC engines: Part 5 Stepped oil scraper rings from 30 up 200 mm nominal diameter Z-Rings
49.	IS 8422: Part 6, 1977	2014	Specification for Piston rings for IC engines: Part 6 Slotted oil control rings from 50 to 200 mm nominal diameter S-Rings
50.	IS 8422: Part 7, 1977	2014	Specification for Piston rings for IC engines: part 7 Double beveled slotted oil control rings from 50 up to 200 mm nominal diameter G-Rings
51.	IS 8422: Part 8, 1977	2014	Specification for Piston rings for IC engines: Part 8 Narrow land slotted oil control rings from 50 up to 200 mm nominal diameter D-Rings
52.	IS 8422: Part 9, 1986		Specification for Piston rings for IC engines: Part 9 Terms and Definitions
53.	IS 8503: 1986	2015	Technical supply conditions for aluminium alloy pistons for internal combustion engines
54.	IS 8925: 1978		Specification for Alternators for Automobiles (Amendment No. 2)
55.	IS 9057: 1998	2012	Automotive Vehicles - Exhaust Emissions-Carbon Monoxide Concentration at Idling for Vehicles Equipped with Spark Ignition Engines-Method of Measurement
56.	IS 9418: 1980	2015	Dimensions for Mounting Flanges for In-line Fuel Injection Pumps for Multi-cylinder Compression Ignition Engines
57.	IS 9420: Part 1, 1988	2014	Specification for Feed pumps for diesel fuel injection equipment: Part 1 External dimensions
58.	IS 9420: Part 2, 1988	2014	Specification for Feed pumps for diesel fuel injection equipments: Part 2 Types of drive
59.	IS 9432: 1980		Ignition switches for automobiles (Amendment No. 2)
60.	IS 9465: 1980	2015	Mounting dimensions for inline injection pump assemblies for fuel injection equipment for multi-cylinder compression ignition engines
61.	IS 10000: Part 1, 1980	2015	Method of tests for internal combustion engines: Part 1 Glossary of terms relating to test methods
62.	IS 10000: Part 2, 1980	2015	Methods of tests for internal combustion engines: Part 2 Standard reference conditions (Amendment No. 1)
63.	IS 10000: Part 3, 1980	2015	Methods of tests for internal combustion engines: Part 3 Measurements for testing-units and limits of accuracy
64.	IS 10000: Part 4, 1980	2015	Methods of Tests for Internal Combustion Engines: Part 4: Declaration of Power, Efficiency, Fuel Consumption and Lubricating Oil Consumption (Amendment No. 1)
65.	IS 10000: Part 5, 1980	2015	Methods of tests for internal combustion engines: Part 5 Preparation for tests and measurements for wear
66.	IS 10000: Part 6, 1980	2015	Methods of tests for internal combustion engines: Part 6 Recording of test results
67.	IS 10000: Part 7, 1980	2015	Methods of tests for internal combustion engines: Part 7 Governing tests for constant speed engines and selection of engines for use with electrical generators
68.	IS 10000: Part 8, 1980	2015	Methods of tests for internal combustion engines: Part 8 Performance tests
69.	IS 10000: Part 9, 1980	2015	Methods of tests for internal combustion engines: Part 9 Endurance tests

S. No.	IS Number	Reaffirmed	Title
70.	IS 10000: Part 10, 1980	2015	Methods of Tests for Internal Combustion Engines; Part 10 Test for Smoke Levels, Limits and Corrections for Smoke Levels for Variable Speed compression Ignition Engines
71.	IS 10000: Part 11, 1980	2015	Methods of tests for internal combustion engines: Part 11 Information to be supplied by the purchaser to the manufacturer and information to be supplied by the manufacture alongwith the engine
72.	IS 10000: Part 12, 1980	2015	Methods of Tests for Internal Combustion Engines Part 12: Specimen Test Certificates
73.	IS 10000: Part 13, 1980	2013	Methods of Tests for Internal Combustion Engines Part 13: Recommendations of Nature of Tests Required for Functional Changes in Critical Components
74.	IS 10250: 1982		Specification for Severities for Environmental Test for Automotive Electrical Equipment (Amendment No. 2)
75.	IS 10478: 2007	2012	Diesel engines - Fuel injection pumps - Tapers for shaft ends and hubs
76.	IS 10533: 1983	2014	Specification for Valve guides for internal combustion engines
77.	IS 11509: Part 1, 2012		Methods of Test for Full-flow Lubricating Oil Filters for Internal combustion Engines - Part 1: Pressure Drop/Flow Characteristics
78.	IS 11509: Part 2, 2012		Part 2 Element by-pass characteristics
79.	IS 11509: Part 3, 2012		Part 3 Resistance to high pressure drop and to elevated temperature
80.	IS 11509: Part 6, 2006	2016	Methods of Test for Full-flow Lubricating Oil Filters for Internal Combustion Engines Part 6 Cold start simulation and Hydraulic Pulse Durability Test
81.	IS 11838: 1986	2016	Recommendations for measurement of quality characteristics of gudgeon pins
82.	IS 12159: 2007		Method of evaluation of accuracy of parameters of steering geometry of automotive vehicles.
83.	IS 12022: 1987		Methods of testing of quality characteristics for piston rings (Amendment No. 1)
84.	IS 12025: 1987	2012	Recommendations for Measurement of Quality Characteristics for Piston
85.	IS 12352: 1988	2014	Specification for Fuel injection pumps mounting bolts
86.	IS 12404: 1988	2014	Specification for Coupling for driving fuel injection pumps
87.	IS 12455: 1988	2014	Performance requirements of air cooled spark ignition automotive engines
88.	IS 12460: 1988	2014	Automotive Vehicles–Transmission Systems—Manual Control Sequence in Automatic Transmissions—Recommendations
89.	IS 12499: 1988	2014	Specification for Two stages, 0.5 litre diesel fuel filters
90.	IS 12500: 1988	2014	Specification for 0.2 litre diesel fuel filters
91.	IS 12508: 1988		Specification for Electric Tachometer System for Automotive Vehicles
92.	IS 12535: Part 1, 1988	2014	Automotive Vehicles—Transmission Systems-Glossary-Part 1 General definitions
93.	IS 12535: Part 2, 1991	2016	Automotive Vehicles—Transmission Systems-Glossary-Part 2 Universal Joints and Drive shafts
94.	IS 12535: Part 3, 1991	2016	Automotive Vehicles—Transmission Systems-Glossary-Part 3 Drive Axles Definition
95.	IS 12587: 1989	2014	Automotive Vehicles—Transmission Systems-Gear Arrangement-Recommen-dations.

S. No.	IS Number	Reaffirmed	Title
96.	IS 12588: 1988		General Requirements for Automotive Oil and Air Assisted Pressure Gauges
97.	IS 12902: 1990	2016	Commercial Vehicles and Buses-Gearbox flanges-Type A
98.	IS 12903: 1990	2016	Commercial Vehicles and Buses-Gearbox flanges-Type S
99.	IS 12904: 1990	2016	Fly wheel housing for reciprocating internal combustion engines - Nominal dimensions and tolerances
100.	IS 12969: 1990		Internal Combustion Engines - Methods of Test for Quality Characteristics of Valves
101.	IS 12971: 1990	2016	Road Vehicles-Commercial Vehicles-Clearance Envelope for Power Take-offs (PTO)
102	IS 12972: 1990	2016	Commercial Road Vehicles-Connections for Rear-mounted Power Take-offs (PTO)
103.	Is 12973: 1990	2016	Commercial Road Vehicles-Side openings for Truck Power Take-offs (PTO)
104.	IS 12978: 2006	2015	Commercial Vehicles and Buses—Cross-tooth Gearbox Flanges-Type T
105.	IS 12996: 1990	2014	Internal Combustion Engines - Radiator Pressure Caps - Specification
106.	IS 13016: 1991	2015	Internal Combustion Engines - Radiator Drain Cocks - Specification
107.	IS 13018: 1990	2014	Internal Combustion Engines - Method of Test for Pressure Charged Engines
108.	IS 13029: 1991	2015	Internal combustion Engines - Setting Ignition timing in Spark Ignition Engines-code of Practice
109.	IS 13050: 1991	2015	Internal Combustion Engines - Valve Seat Inserts - Specification
110.	IS 13051: 1991	2015	Internal Combustion Engines - Radiator Fans - Specification
111.	IS 13052: 1991	2015	Internal Combustion Engines - Cylinder Liners - Quality Characteristics
112.	IS 13090: 1991	2016	Automotive Vehicles— Commercial Vehicles-Clutch housings-Dimensions
113.	IS 13226: 1991		Automotive vehicles - Resistances in starter motor circuits - Methods of measurement
114.	IS 13313: 1991		Automotive vehicles - Electrical wiring - Colour coding
115.	IS 13437: 1992		Automotive Vehicles - Recommendation for Voltages for Electrical Systems
116.	IS 13460: 2005		Road vehicles - Connectors for the electrical connection of towing and towed vehicles 7-pole connector type 12 S (supplementary) for vehicles with 12 V nominal supply voltage
117.	IS 13461: 2005		Road vehicles - Connectors for the electrical connections of towing and towed vechiles-7-pole connector type 12 N (Normal) for vehicles with 12 V Nominal supply voltage
118.	IS 13476: 2006		Automotive Vehicles-Mechanical steering gear-Method of test
119.	IS 13491: 2005		Road Vehicles - Connectors for the Electrical Connection of Towing and Towed Vehicles-7-Pole Connector Type 24 N (Normal) for Vehicles with 24 V Nominal Supply voltage
120.	IS 13499: 2005		Road Vehicles - Connectors for the Electrical Connection of Towing and Towed Vehicles-7-pole Connector Type 24 S (Supplementary) for Vehicles with 24 Nominal supply voltage
121.	IS 13509: 1992		Automotive vehicles - Towing vehicles- Mounting of electrical connections on rear cross members
122.	IS 13599: 1993		Automotive vehicles-Instruments systems-Temperature gauges electrical type-Specification
123.	IS 13602: 1992		Automotive vehicles-Instrument systems-Temperature gauges mechanical types-Specification

S. No.	IS Number	Reaffirmed	Title
124.	IS 13646: 1993		Automotive Vehicles-Instrument Systems-Pressure Gauges Electrical Type-Specification
125.	IS 13686: 1993	2014	Internal combustion engines-Radiators-Method of test
126.	IS 13687: 1993	2014	Internal combustion engines-Radiators-Heat dissipation performance-Method of test
127.	IS 13824: 1993	2014	Internal Combustion Engines-Method of Verification of Emission of Crankcase Gases for Vehicles Powered with Four Stroke Spark Ignition Engines
128.	IS 14269: 1995		Automotive Vehicles-Integral Power Steering Gear-Method of test
129.	IS 14273: 1999	2014	Automotive Vehicles-Exhaust Emissions-Gaseous Pollutants from Vehicles Fitted with Compression Ignition Engines-Method of Measurement
130.	IS 14380: 1996		Automotive vehicles-Ignition coils for magneto ignition systems-pecification (Amendment No. 1)
131.	IS 14381: 1996		Automotive vehicles-Electrical circuits-Recommended graphical symbols
132.	IS 14382: 1996		Automotive Vehicles-Speed Limiters-Installed Requirements
133.	IS 14511: Part 1, 2007	2012	Diesel engines-Testing of fuel injectors: Part 1 Hand-lever-operated testing and setting apparatus
134.	IS 14511: Part 2, 2008	2012	Diesel engines-Testing of fuel injectors: Part 2 Test methods
135.	IS 14553: 2008	2014	Automotive vehicles-Apparatus for the measurement of opacity (smoke) of exhaust gas from vehicles equipped with compression ignition engines-Specification
136.	IS 14554: 1998	2012	Automotive vehicles-Exhaust emissions-Apparatus for the measurement of carbon monoxide concentration from vehicles equipped with spark ignition engines at idling-Specification
137.	IS 14555: 1998		Automotive Vehicles-Evaporative Emissions from Vehicles Equipped with Spark Ignition Engines-Method of Measurement
138.	IS 14556: 1998		Automotive Vehicles - Catalytic Converters for Spark ignition Engines - Installed Requirements
139.	IS 14599: 1999	2014	Automotive Vehicles - Performance Requirements (measurement of Power, SFC, Opacity) of Positive and Compression Ignition Engines - Method of Test
140.	IS 14600: 1999	2014	Automotive Vehicles - Exhaust Emissions - Gaseous Pollutants from Vehicles Equipped with Internal Combustion Engines - Method of Measurement
141.	IS 15350:.2003		Automotive Vehicles - Ignition Systems - Methods of Tests
142.	IS 15364: 2003		Automotive Vehicles - Spark-Pluge M10 X 1 with Flat Seating
143.	IS 15653: 2006	2016	Reciprocating internal combustion engines - Vocabulary - Terms for engine maintenance
144.	IS 16057:2013	–	LPG-Operated Internal Combustion Engines-Safety and Performance Requirements.

Appendix – F

ELEMENTS OF ELECTRICAL AND ELECTRONIC CIRCUITS

1. ELECTRICAL QUANTITIES

1. Charge (Q) - It is the quantity of electricity residing in an electrostatically charged body. Alternatively, it may also be defined as the quantity of electricity conveyed by a given current in the given time.

$$Q = It \qquad \text{...(F.1)}$$

where I is the current in amperes and t is the time of current flow in seconds.

Charge is measured in coulombs. A coulomb (C) is the quantity of electricity conveyed by a current of 1 ampere in 1 second.

2. Current (I) - The number of electrons having passed a fixed point in 1 second. It is measured in amperes (A)

$$I = V \text{ D } R \qquad \text{...(F.2)}$$

where V is the voltage in volts and R is the resistance in ohms.

3. Resistance (R) - It is the opposition to current flow in a material or circuit when a voltage is applied across it. It is measured in ohms.

$$R = V/I \qquad \text{...(F.3)}$$

4. Resistivity ρ) - This is measure of the ability of a material to resist the flow of an electric current. Its unit is ohm-meter.

$$R = \rho L/A \qquad \text{...(F.4)}$$

where L and A are the length and cross-sectional area of the material sample in metres and square meters respectively.

5. Conductivity (σ) - It is the reciprocal of resistivity

$$\sigma = 1/\rho \qquad \text{...(F.5)}$$

6. Power (P) - The rate at which electrical energy is converted into another form such as heat, motion, electromagnetic field. When a voltage of 1 volt produces a current of 1 ampere in a material or circuit, the power developed is one **Watt** (W)

$$P = IV \qquad \text{...(F.6)}$$

$$P = I^2R \qquad \text{...(F.7)}$$

7. Inductance (L) - That property of an electric circuit element by which energy is capable of being stored in a magnetic flux field. Inductance makes itself felt in a circuit only when there is a changing current. It unit is Henry (H)

$$i = \frac{V}{R}\left[1 - e^{[Rt/L]}\right] \qquad \text{...(F.8)}$$

8. Capacitance (C) - It is the property of a capacitor that determines how much charge can be stored in it for a given potential difference across its terminals. It is measured in Farads (F).

$$C = Q/V \qquad \text{...(F.9)}$$

2. PRINCIPLES OF ELECTROMAGNETIC INDUCTION

1. When a conductor cuts or is cut by magnetic lines of force, a voltage is induced in the conductor, the magnitude of which is proportional to the rate at which the conductor cuts or is but by the magnetic lines of force.

2. The direction of the induced voltage depends upon the direction of the magnetic field and the direction of the conductor movement relative to the field.

3. PRINCIPLE OF MUTUAL INDUCTION

If the two coils are wound around the same iron core, then any change in magnetism of one coil will induce a voltage in the other. The magnitude of this induced voltage would depend upon the primary current, the speed at which the magnetism is changed and the turn ratio between the two coils.

4. RESISTOR

Resistors are used to limit current values and to provide definite voltage drops in a circuit. Most resistors in the electronic circuits are made from small carbon rods. Variable forms of resistors in linear or logarithmic forms are also available.

5. INDUCTOR

These are often used in oscillator or amplifier circuits. Basically, it is in the form of a coil of wire wound on a former. Iron cores are used to increase the inductance value due to change in the permeability of the core. Tuning a circuit by change in the inductance value by changing the position of the core is also possible. Inductors of high value in particular are usually known as chokes. These are used to smoothen the voltage in DC circuits.

6. CAPACITOR

It is a device for storing an electric charge.

7. CIRCUIT SYMBOLS

Various symbols used in electrical/electronic circuits are shown in Fig. F.1.

8. INTEGRATED CIRCUITS

These are made on a single slice of silicon called substrate. A single slice can contain thousands of components. Apart from very small size, the bigger advantage is the high speed at which these circuits can be made to work due to very small distance between components.

9. AMPLIFIERS

One resistor with one transistor can form the simplest amplifier. However, the gain of such as amplifier is variable and non-linear. As such other components, *i.e.*, resistors and capacitors are added to obtain a practical amplifier.

10. SCHMITT TRIGGER

It is employed to change variable signals into a pure square wave type signals for use in digital or switching circuits. These are available as ready-made I.C.'s.

11. FILTERS

The filter is used to block undesired frequencies in electronic circuits. A low pass filter blocks high frequencies, while a high pass filter blocks low frequencies. If the circuit includes amplification, the filter

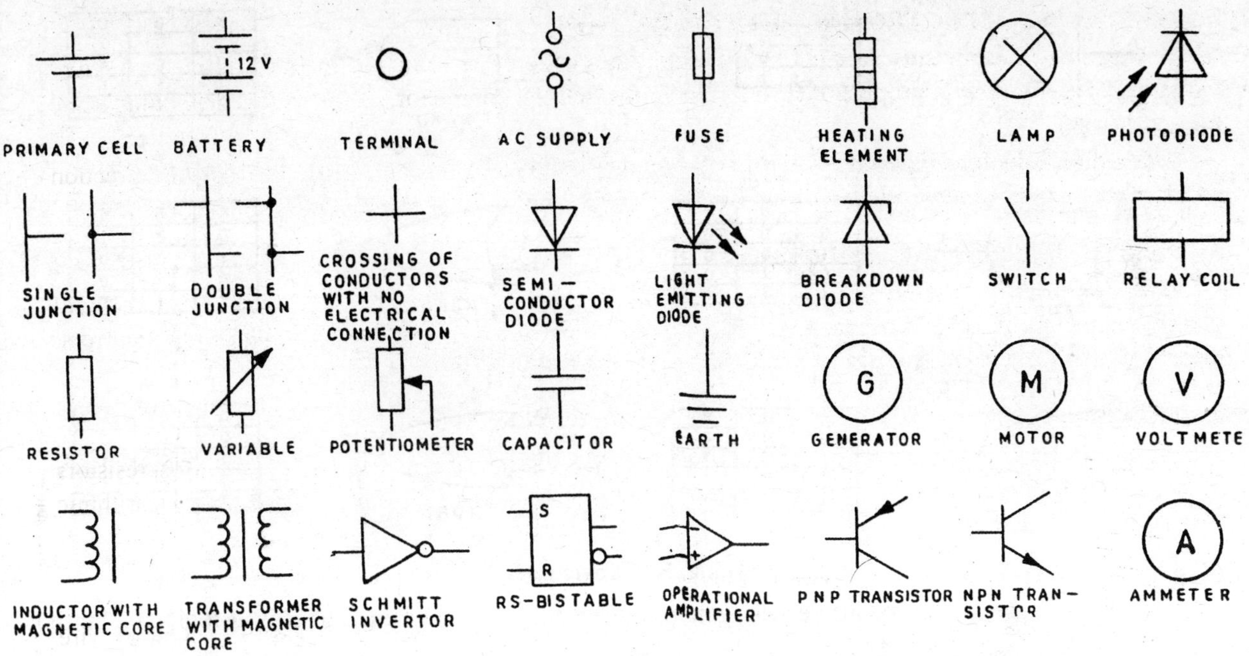

Fig. F-1. Circuit Symbols

is called active; without amplification the filter is termed passive. The basic principle of the filter circuits is that the reactance of the capacitors changes with frequency. When the frequency increases, the reactance decreases and vice-versa.

12. STEPPER MOTOR

These are being increasingly used as actuators in automobiles due to the ease with which these can be controlled by electronic systems. These are of three type, *viz.*, variable reluctance type, permanent magnet type and the hybrid type.

13. SYNCHRONOUS MOTOR

These are employed when a drive is required to be time-synchronised. They always rotate at a constant speed, N given by $N = 120 \times f'/p$, where f is the system frequency and p is the number of poles in the motor.

14. LOGIC GATES

Digital device to carry out simple task as per the required logic is called a logic gate. In digital circuits, '1' means 'yes' and '0' means 'no'. Accordingly, a truth table is used to describe what combination of inputs will produce a particular output. Six basic logic gates are commonly used (Fig. F.2)

(*i*) NOT gate. Here the output is always the logically opposite of the input.

(*ii*) AND gate. In this the output will be '1' only if both A and B inputs are '1'.

(*iii*) OR gate. This will produce output '1' only if either A or B input (or both) are '1'.

(*iv*) NAND gate. This is a combination of NOT and AND gates. Here the output is the inverse of the AND gate.

(*v*) NOR gate. This is a combination of NOT and OR gates. Here the output is the inverse of the OR gate.

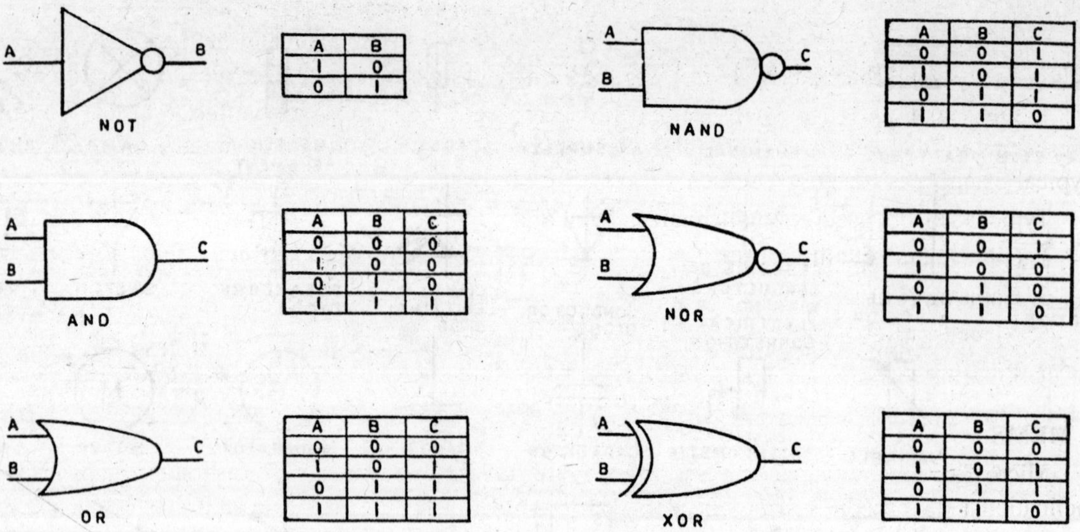

A	B
1	0
0	1

A	B	C
0	0	1
1	0	1
0	1	1
1	1	0

A	B	C
0	0	0
1	0	0
0	1	0
1	1	1

A	B	C
0	0	1
1	0	0
0	1	0
1	1	0

A	B	C
0	0	0
1	0	1
0	1	1
1	1	1

A	B	C
0	0	0
1	0	1
0	1	1
1	1	0

Fig. F. 2. Logic gates

(*vi*) XOR gate. The 'Exclusive OR' gate produces output only when *A* or *B* is at logic '1', but not when both are '1' or '0'.

15. COMBINATIONAL AND SEQUENTIAL LOGIC

Circuits consisting of a number of logic gates as given in the previous section and having no memory or counter circuits are called combinational logic circuits. In these the output of the system is determined by the present inputs.

Circuits having the ability to memorize previous inputs or logic states are called sequential logic circuits. In these the output is determined by the sequence of previous inputs and they store information after the inputs have been removed. Thus the computer memories are built upon the sequential circuits as building blocks.

16. FLIP-FLOPS

These are the basic memory circuits and are also called bistables since they have two steady states. These are of three types, *viz.,* RS, D and JK types. The RS type is the simplest and stands for set-reset. The D-type has two inputs CK for clock and D and two outputs' Q' and 'not Q'. Q takes on the logic state of D when clock pulse is applied. The JK type is combination of the RS and the D types.

17. SENSORS

These may be of different types:

1. Hall-effect sensors. Used for non-contacting measurements of angles, position or displacement and also rotational speeds.

2. Microsystem Technology sensors. Microsystems combine one or more micro-electronic, micro-mechanical or micro-optical functions. The miniaturised systems feature structures only a few millimeters in size.

3. AMR (Anisotropic Magneto Resistance) sensors. These are also used for non-contacting measurement of angles and rotational speed.

4. GMR (Giant Magneto Resistance) sensors. These consist of a number of thin layers stacked one on top of the other. These layers react to even extremely weak magnetic field and have thickness of only a few atoms. These are also used for the contactless measurement of angles and rotational speed.

5. Piezoceramic sensors. These are ceramic sensors using piezoelectric effect, wherein a mechanical stress applied to a piezocrystal produces electrical polarisation resulting in a measurable electric current. Typical example is the knock sensor.

Typical sensors used in engine management are:

1. Oil level and quality sensor	2. Crankshaft speed sensor
3. Camshaft phase sensor	4. Temperature sensor
5. Air-mass-flow sensor	6. Knock sensor
7. Pressure sensors	8. Exhaust gas sensor for lambda control

6. MEMS

Micro Electro-Mechanical Systems (MEMS) are very popular sensors for measuring motion, acceleration, inclination and vibration. Technologically, MEMS are different from other Silicon ICs, which mainly employ the electrical properties of silicon. However, the inside of MEMS has actually miniature mechanical structure made up completely of silicon. Inside a MEMS structure, there are actual moving mass, springs, cavities, dampers, etc. Thus MEMS combine the electrical and mechanical properties of silicon.

Main automotive applications of MEMS include vehicle security systems, black box and infotainment.

Appendix – G

TECHNICAL TERMS

Adsorber. Engine component meant to collect certain pollutants, *e.g.* NO_x and HC during the unfavourable engine operating conditions to store these and treat the same when operating conditions become suitable.

Air cleaner. A device connected to the engine air intake, which filters dirt and dust from all the air entering the engine.

Air injection. A system of emission control in which fresh air is supplied to the engine exhaust which facilitates the conversion of CO to CO_2 and partially burnt hydrocarbons to burn completely.

Alternative fuels. Automotive fuels other than those derived from petroleum, *e.g.*, alcohol, CNG, LPG, etc.

Alternator. A generator producing alternating current, which is converted to direct current, for recharging the vehicle battery.

Ampere hour rating. Measurement of the ability of a battery to deliver a stated amount of current for a given period of time.

Aniline Point of a fuel is the temperature at which equal parts of the fuel and pure aniline dissolve in each other.

Antechamber. A small auxiliary combustion chamber employed in compression-ignition engines, in which partial combustion of the fuel is used to force the burning mixture into the cylinder, thus causing more perfect combustion.

Antifreeze. A chemical such as alcohol or glycerine, added to the engine coolant, to lower its freezing point.

Armature reaction. When an armature containing a number of conductors is rotated within a magnetic field, distortion of the field occurs. This effect is termed armature reaction, which was earlier used to control the output of a D.C. generator.

BDC (Bottom Dead centre). The position of the crankshaft when the piston is in its lowest position, *i.e.*, the position farthest from the cylinder head.

BMEP (Brake Mean Effective Pressure). The mean pressure in the I.C. engine cylinder during the working stroke.

Backfire. An engine is said to backfire when fresh unburnt charge of fuel burns in the exhaust system on account of leakage from the exhaust valve.

Back pressure. It is a form of resistance to the free flow of exhaust from the cylinder head. It may be due to badly designed or fabricated or choked exhaust manifold, catalytic converter and/or muffler.

Battery capacity. Amount of current the battery can deliver. It depends upon the number and area of plates, quality of the electrolyte and the temperature.

Battery reserve capacity. The time in minutes of which the battery can supply 25 amp and 25°C, without the cell voltage falling below 1.75.

Bendix drive. Inertia type of starter drive used in smaller vehicles.

Bio-diesel. Fuel oil obtained from bio-sources like soya bean, jetrophe, etc. Having properties similar to diesel, it can be used in existing diesel engines without major modification, either as pure bio-diesel or as blend of bio-diesel and ordinary diesel.

Blow-by is said to occur when air-fuel mixture or the burnt gases from the combustion chamber leak past the piston rings to the sump. This occurs due to wear of cylinder bore and/or piston rings or due to excessive piston ring gaps.

Bore. Diameter of engine cylinder.

Boxer engine. Horizontally opposed engine, where opposing pistons are attached to different crankpins.

Brake Points (B.P.). It is the actual power delivered at the crankshaft.

Breaker points (C.B. points). A set of contact points for opening and closing the primary circuit of an electrical ignition system.

Breakerless ignition. Ignition system where the mechanical switching device, such as contact breaker is replaced by an electronic switching device through the use of transistors.

Breathing. Ability of an engine to take in air-fuel mixture and to discharge exhaust gases efficiency.Many of the modern car engines are provided with 4 valves per cylinder for better breathing.

Butterfly valve. A valve consisting of a flat circular pivoted along a diameter; used as throttle valve in carburettor engines.

Bypass filter. A filter through which only a part of the total engine lubricating oil flows.

CAFE (Corporate Average Fuel Economy. U.S. Federal regulation according to which a vehicle manufacturer has to achieve a minimum average fuel economy across their entire range of cars and light vehicles.

Camshaft. A shaft having a number of cams at appropriate angular positions for operating the valves at timings relative to the piston movement and in a sequence as per the firing order. It rotates at half the crankshaft speed.

Camshaft position sensor (CMP). A sensor which sends signals about the rotational position of the camshaft to the ECU, which enables the computer to time the fuel injection and ignition systems more accurately for faster starting of the engine.

Capacitor Discharge Ignition (CDI). It is a type of automotive electronic ignition system which uses capacitor discharge current output to fire the spark plugs. It has been widely used in motor cycles and recently in some cars also.

Carburettor. A device in an engine which supplies correct amount of the air-fuel mixture of correct strength under all operating conditions.

Catalyst. A material that can cause a chemical reaction of other substances, without being a part of that reaction. Examples in the engine catalytic converter are platinum, palladium, etc.

Catalytic converter. It is a cylindrical unit installed into the exhaust system of a vehicle, between the exhaust manifold and the silencer. Due to the presence of catalyst in the converter toxic gases in the exhaust, *i.e.*, CO, HC and NO_x are converted into harmless CO_2, H_2O and N_2. Commonly used catalysts are platinum and rodium. A three way catalytic converter can be used in case of a petrol engine only. In diesel engines, only a two-way converter is used.

Cetane. *n*-hexadecane ($C_{16}H_{34}$) is a straight chain paraffin, which is assigned the cetane number of 100.

Cetane rating. A measure of the auto-ignition value of diesel fuel. It is defined as the percentage of cetane in a mixture of cetane and alpha methylnaphthalene, which has the same diesel knock characteristics as fuel under test.

Check engine light. The malfunction indicator light that the vehicle's on-board computer illuminates when it senses a fault in the engine.

Choke. A butterfly valve fitted at the top of the air horn of a carburettor, to provide rich mixture during starting of engine.

Circuit breaker. A switch which protects an electrical circuit from overload by opening the circuit when the current flow exceeds a pre-determined level.

Clearance volume. The volume of the engine cylinder (including the combustion chamber volume) above the piston when it is in the TDC position.

CNG (Compressed Natural Gas). Alternative fuel consisting of natural gas compressed from 160 to 200 bar.

Combustion chamber. The small space in the engine cylinder head and/or piston into which air-fuel mixture (in petrol engines) or air (in diesel engines) is compressed and is burnt.

Common rail injection. A diesel engine fuel supply system in which a single pump delivers fuel at very high pressure, say, 1400 bar to a common reservoir from which supply goes to various injectors.

Compression ratio. It is the ratio of the cylinder volume above the piston at BDC to the volume above the piston at TDC.

Condenser. An electrical device which acts to store electrical charge at the break of circuit and supplies the same during the make period of circuit.

Connecting rod. It converts the linear motion of the piston into the rotary motion of the crankshaft. Its small end is connected to the piston and big and to the crankshaft.

Controller. Electronic power control device to regulate the d.c. voltage supplied by the batteries to the d.c. motor; usually done by chopping the d.c. current up to a different duty cycle, which is called pulse width modulation and is done at a very high frequency.

Coolant. A mixture of antifreeze and water which is circulated the engine to cool the same.

Coolant-controlled EGR. A system which prevents exhaust gas recirculation until engine coolant temperature reaches a specific value.

Crankcase. The lower part of an engine in which the crankshaft and related parts operate.

Crankshaft. The main engine shaft from which power is taken. It converts the reciprocating motion of the engine pistons into rotary motion for automobile use.

Crankshaft counter-balances. A series of weights attached to or forged integrally with the crankshaft, placed in such a manner as to offset the reciprocating weight of each piston-connecting rod assembly.

Crankshaft position sensor. A sensor ring with lobes, which sends information related to the exact position of the crankshaft so as to achieve accurate ignition/injection timing.

Crankshaft pulley. A pulley attached to the front end of the crankshaft, which drives the fan, the alternator and other devices by means of a fan belt.

Cubic capacity. It is the sum of the displacement of all the engine cylinders.

Cylinder. A circular tubular opening in the cylinder block in which the piston moves in a reciprocating manner.

Cylinder block. The basic engine casting containing the cylinders, coolant jackets, crankshaft main bearings as well as machined surfaces for mounting other components such as the cylinder head, oil pan, transmission, etc.

Cylinder head. The detachable portion of the engine, usually fastened to the top of the cylinder block and containing all or most of the combustion chambers. On overhead valve engines, it contains the valves and their operating parts. On overhead camshaft engines, it contains the camshaft as well.

Darlington pair. Two transistors wired together are called the Darlington pair, which allows a very small current to control large current.

Dead centre. The extreme top or bottom of the piston stroke.

Deep cycling. The process of discharging a battery almost completely before recharging.

De NOx catalyst. A catalyst which, unlike a conventional catalyst used in a three-way catalytic converter, is able to remove NOx from the exhaust gases even in the presence of oxygen.

Detonation. An uncontrolled explosion of the unburnt air-fuel mixture in the engine cylinder occurring after the regular combustion of some of the charge caused by the spark at the spark-plug. It becomes audible through vibration of the combustion chamber walls.

Detonation sensor. A sensor, which is often piezoelectric, mounted near the cylinders which can detect engine knocking or frequencies of detonation so as to send a message to the ECU to retard the ignition timing to prevent damage to the engine.

Diagnostic code. Code number obtained by accessing the diagnostic mode of the engine management computer. This code can be used to determine the area in the system where a malfunction may be there.

Diagnostics. Process of identifying the cause or nature of a condition or problem to determine the appropriate corrective action to take for repairing an automotive system.

Diagnostic socket. A socket on the vehicle, usually located in the engine compartment, which is part of the onboard electronic sensor system.

Diesel index. It is defined as:

$$\text{Diesel Index} = \frac{\text{Aniline point}\,(^\circ\text{F}) \times \text{API gravity}\,(\text{deg})}{100}$$

Dieseling. A condition in which an S.I. engine continues to run for some time after the ignition is turned off; caused by the hot spots in the combustion chamber.

Diesel Particulate Filter (DPF). A filter which removes small particles from exhaust of a diesel engine.

Diesel Particulate Matter (DPM). Sub-micron size particles found in the exhaust of a diesel engine. DPM consists of both solid and liquid particles and is usually classified into three fractions, viz., inorganic carbon (Soot), organic fraction and sulphate fraction.

Digital fuel injection (DFI). A system, identical to earlier electronic fuel injection system, but with digital microprocessors. In this analog inputs from various engine sensors are converted to digital signals before processing. The system is self-diagnosing and self-monitoring.

Diode. It is a basic electronic component formed by combining a N-type and a P-type semi-conductor back-to-back. Diodes are commonly used as rectifiers.

Direct ignition. An ignition system without any distributor, wherein the spark distribution is controlled by the computer.

Direct injection. A fuel injection system for diesel engines, wherein the fuel is injected directly into the combustion chamber, onto the top of the piston.

Displacement. The volume of the cylinder between the top dead centre and the bottom dead centre positions of the piston.

Distributor. A mechanically driven device on an engine which is responsible for electrically firing the spark plugs at a pre-determined point of the piston stroke and in the desired firing order.

Distributorless Ignition System (DIS). An electronic ignition system which does not have a conventional rotating distributor. It employs multi-spark ignition coils or one ignition coil for each spark plug.

D-Jetronic. Fuel injection system by Bosch, where 'D' stands for 'druck', which means 'pressure' in German language. The system is controlled by manifold pressure, indicating engine load, *i.e.*, the amount of air engine is using, which is fed as input signal to the engine ECU for calculating the correct amount of fuel delivery.

Doping. The process of adding a very small quantity of impurity element to a semi-conductor material. Depending upon the type of impurity added, the semi-conductor is N-type or P-type.

Double Overhead Camshaft (DOHC). The engine has two camshafts mounted in one cylinder head. One camshaft operates the intake valves, while the other operates the exhaust valves.

Dwell is the time expressed in terms of the degrees of distributor cam rotation for which the contact breaker points remain closed. It is usually 55 to 70 percent of the total cam angle.

Dynamic timing meter. A General Motors diesel engine tool used to measure timing while the engine is running by using a quartz sensor in the combustion chamber which measures the point of combustion and converts the same to timing in degrees of crankshaft rotation by using a magnetic crankshaft pick-up and microprocessor.

Dynamo. A device driven by the engine and producing direct current (d.c.)

Dynamometer. An equipment used to measure engine power.

ECU. Electronic Control Unit, also called Electronic Control Module (ECM) is the central controlling unit of a mechanical system, *e.g.*, the fuel injection system or the ignition system.

EGR system. Exhaust Gas Re-circulation is one of the methods of diluting the intake charge of the engine with exhaust gas, which reduces NO_x emission.

Electronic Ignition. A system in which the timing and firing of the spark plugs is controlled by an electronic control unit, usually called a module. These systems do not have contact points or breaker.

Engine capacity. This is the total piston displacement of all the engine cylinders.

Exhaust manifold. Set of pipes carrying exhaust gases from the cylinder head to the exhaust system.

Feeler gauge. A blade, usually metallic, of precisely predetermined thickness, used to measure the clearance between two parts.

Flywheel. A heavy metallic wheel attached to the engine crankshaft to smoothen out the power surges from the engine power strokes. This also serves as part of the clutch and the starting system.

Fuel cell. A power source for the future automobiles. Fuel cells produce electricity through chemical reaction with no harmful exhaust emission.

Fuel injector. A component of the fuel injection system. Its function is to inject the fuel in the cylinder in properly atomised form of spray.

Fuel rail. Metallic or plastic pipes carrying fuel from the fuel injection pump to the fuel injectors.

Gasket. A thin laminate sandwiched between two mating surfaces to improve sealing between them.

Gear pump. Positive displacement pump using two meshing gears. Commonly used in engine lubrication system.

Glow plug. An electrically heated device placed in the pre-combustion chamber of a diesel engine to assist in starting from cold.

Governor. A mechanical or electronic device to restrict the performance of an engine usually for reasons of safety.

Governor cut-in speed. The speed of the fuel injection pump at which the control rack just starts to move away from the maximum fuel step, with the accelerator lever set for maximum output.

Governor cut-off speed. The speed of the fuel injection pump at which the governor just stops the fuel delivery.

Governor Hysterisis. The difference between the governor cut-off speed at which point fuel delivery just stops with the increasing injection pump speed and the cut-off speed at which fuel delivery just restarts with the decreasing injection pump speed.

Governor overrun. The difference between the governor cut-off speed and the governor cut-in speed.

Governor run-out. The percentage speed change when the control rod is moved by the governor from the maximum fuel delivery position to the no-fuel delivery position.

Gudgeon pin. A hollow cylindrical shaped pin connecting the piston to the small end of the connecting rod.

Hall effect. When a thin semi-conductor chip (usually silicon) carrying current is crossed at right angles by a magnetic field, a potential difference is produced at the edges of the chip.

Hall effect switch. A device for switching off or on the supplied voltage with the absence or presence of the magnetic field. Also called Hall generator, it consists of a rotor with vanes, a conductive element with a permanent magnet and the Hall IC.

Halogen lamp. A type of incandescent lamp which is more efficient and has a much longer life than the common incandescent lamp. It uses a halogen gas, usually bromine or iodine, which causes the evaporating tungsten to be redeposited on the filament, thereby increasing its life.

HCCI engine. An internal combustion engine in which well-mixed fuel and air mixture is compressed to the point of auto-ignition. Thus it has the characteristics of combustion of both the spark ignition as well as the compression ignition engine.

Heat gasket. The gasket located between the cylinder head and the cylinder block to prevent leakage of compression pressure, gases, and coolant.

High Intensity Discharge (HID) lamp. A lamp which produces light by passing electricity through gas, which causes the gas to glow. Examples are mercury vapour lamps, metal halide lamps, high-pressure sodium lamps, etc. Such lamps emit very high intensity light compared to fluorescent lights and also have a very long life.

Honing. Final process used in the machining of cylinder bores, either during manufacture or in re-boring. Honing is used to put cross-hatch lines on the cylinder bore which will give good life span and aid lubrication and oil consumption characteristics in use.

Honing stones. The grinding tool, *i.e.,* stones used to execute the honing process.

Horizontally opposed engine. Also called a *boxer engine*, it has two banks of cylinders placed flat, which gives it a lower center of gravity, thereby improving handling. Besides, lower height would mean lower hood height, which improves aerodynamics.

Hybrid vehicles. The vehicles propelled by a combination of an internal combustion engine and an electric motor either in series or in parallel arrangement.

Idle screw. A screw on the carburettor, for adjusting the idling speed of the engine.

Ignition Coil. A device, basically a transformer, used to step up the battery voltage to very high value of about 25 kV or even higher for causing spark at the spark plug electrodes for ignition of the charge.

Ignition lag. In S.I., engines. It is the duration between the occurrence of spark at the spark plug and deviation of the combustion curve from the motor curve; also called the delay period. In C.I. engines, it is the time elapsed between the start of fuel injection and first appearance of flame or pressure rise.

Ignition module. The electronic control unit for opening the closing and opening of the primary circuit in the electronic ignition system.

Ignition timing. It is the exact instant when a spark is sent to the spark plug and is expressed in degrees of crankshaft rotation relative to TDC.

Indicated Power (I.P.). The power developed within the engine cylinders, expressed in kW.

Inlet manifold. Group of tubes carrying air-fuel mixture from the carburettor to intake ports of various engine cylinders.

Jump start. The process of starting the engine of a car with discharged battery with the help of a charged battery by connecting the two batteries.

Jumper Cables. The leads that are used to connect the discharged battery of one vehicle to the charged battery of another vehicle, to jump-start the engine of the former.

Knocking. See *detonation.*

LNG (Liquefied Natural Gas). Alternative fuel comprising of natural gas liquefied at –160°C.

LPG (Liquefied Petroleum Gas). Comprises of petroleum gases (mainly propane and butane) liquefied under pressure and used as an alternative fuel.

Lambda (λ). It is the excess air factor defined as the ratio of the actual air consumed in the engine to the theoretical air required. For a stoichiometric air-fuel mixture, $\lambda = 1.0$. For a rich mixture, $\lambda < 1.0$ and for a lean mixture, $\lambda > 1.0$.

Lambda sensor or probe is an oxygen sensor which is an essential part of a closed loop computer-controlled fuel injection system. The sensor is placed in the exhaust manifold. Lambda control is used only in spark ignition engines. It is not possible in a diesel engine.

Magneto. A self-contained device which generates and distributes electricity to the spark plugs from the mechanical energy of the engine. Now obsolete in four-wheelers, but still used in many two-wheelers.

Main bearings. The bearings supporting the crankshaft. These are located in the lower portion of the engine block.

Manifolds. Sets of pipes carrying the air-fuel mixture and the exhaust gases to and from the engine.

Mean effective pressure. This is the average effective pressure throughout the whole power stroke. It is expressed in bars or kilo Pascals (1 bar = 100 kPa).

Mechanical Efficiency of an engine is the ratio of its B.P. and I.P.

Micro-hybrid. A vehicle using a "start-stop system", where regenerative braking technology helps to stop the engine when the vehicle pulls to a stop, is in neutral or is idling and to restart it when the driver presses the accelerator pedal.

Multi-Point Fuel Injection (MPFI). A petrol fuel injection system using one injector per cylinder. Only air enters the inlet manifold. As this air approaches the inlet valve, the injector opens in the valve port, spraying fuel into the air stream.

Misfire. Absence of or inadequate firing of the air-fuel mixture in an engine due to fault in the electrical and/or fuel systems.

Morse Test. Simple test for determining the mechanical efficiency of an engine.

Muffler. A component to reduce the noise of engine exhaust; also called silencer.

Naturally aspirated engine. An engine using only the atmospheric pressure and vacuum in the intake manifold to fill the cylinder. Supercharged and turbo-charged engines are not naturally aspirated.

NOx. A general chemical formula to represent a number of nitrogen oxides, *e.g.*, NO and NO_2. These tend to form in exhaust gases of an engine when combustion temperatures are high or the air-fuel mixture is lean.

Octane number of any fuel is the percentage of isooctane by volume in mixture of isooctane and normal heptane which gives the same anti-knock characteristics as the fuel under standard test conditions.

Oil additive. Chemical substances added to the lubricating oil either to reinforce some of its natural properties or to provide it with certain new properties which it does not possess originally.

Oil galleries are the passages inside the engine cylinder block which carry lubricating oil under pressure to engine bearings.

Oilpan. A metal bowl attached to the bottom of the crankcase into which the lubricating oil drains when the engine is not running.

On-Board Diagnostic (OBD) system. A system which monitors some or all computer input and control signals. Any signal outside predetermined limits means a fault in the system, in which case the driver is alerted.

Opposed engine. An engine that has its cylinders parallel to the ground on either side of the crankshaft.

Otto cycle. It is a 4-stroke engine cycle named after Niokolaus August Otto who patented the four-stroke engine in 1876. All petrol engines work on this cycle.

Overhead camshaft. A camshaft that actuates the valve from over the cylinder head rather than in the cylinder block.

Oversquare engine. An engine with stroke less than bore.

PCV Valve. A valve usually located in the rocker cover that vents crankcase vapours back into the engine cylinder to be reburnt.

Particulates. Solid particles in the exhaust, mainly soot or unburnt carbon particles, usually in a diesel engine.

Piezoelectric ignition. A system of ignition which employs a small section of piezoelectric material. When such a material is compressed even a very small amount, it emits a high voltage sufficient to fire a plug. Such a system does not need a coil, condenser or breaker points.

Pilot injection. Small amount of fuel injected into the combustion chamber before the main fuel delivery. Usually employed in diesel engines.

Piston. A hollow round cylindrical plug in the engine cylinder that transmits the force of combustion in the combustion chamber to the crankshaft via the connecting rod.

Piston displacement. The volume of cylinder swept by the piston in moving from TDC to BDC. It is also called swept volume.

Piston skirt. The lower part of the piston containing the gudgeon pin and the small end of the connecting rod.

Piston slap. Slapping sound produced by a worn out piston moving up and down in the engine cylinder.

Plastigage. A special gauge to measure bearing clearance.

Plug-in hybrid. A hybrid vehicle with batteries which can be recharged from ordinary household electric power plug.

Poppet valve. Engine valve deriving its name from its motion of popping up and down. It is also called 'mushroom valve' due to its shape like a mushroom.

Port fuel injection. Also called multi-port fuel injection, it is the system used in petrol engines wherein there is a fuel injector at each intake port to spray fuel into the intake air as it approaches the intake valve.

Power. Work done by the engine in a given period of time.

Preignition. The phenomenon of a hot spot such as a glowing spark or deposit igniting the air-fuel mixture earlier than the spark plug.

Pressure cap. Radiator cap containing a pressure relief valve that operates at a pre-determined pressure, thus maintaining the cooling system at the desired pressure. This raises the boiling point of the coolant, thus allowing the engine to operate at higher temperature, increasing its efficiency.

Primary cell/battery. The battery/cell in which the chemical reaction completely destroys one of the metals after some time. The example is the small cells for flashlights and toys.

Primary filter. Also called 'surface filter', it is installed on the inlet side of the oil pump. It uses coarse filter material. More commonly known as 'strainer'.

P-series. A new fuel recently added to the definition of 'alternative fuel'. It consists of certain blends of MTHF (Methyl tetrahydrofuran), ethanol and hydrocarbons (pentane plus).

Push rod. A rod meant to transmit the reciprocating motion of the tappet to the rocker arm.

Quench area. The area of the flat surface of the piston crown that opposes the flat surface of the cylinder head as the piston approaches TDC.

Radial engine. An engine in which the cylinders are arranged radially in a circle around the crankshaft.

Radiator. A component of the engine cooling system, which is in fact a heat exchanger. Its functions to dissipate the engine heat from the coolant to the atmosphere.

Regulator. A device which maintains the current and/or voltage levels of a circuit within predetermined range.

Relative Efficiency of an engine is the ratio of its indicated thermal efficiency and its air standard efficiency.

Relay. An electrical switch whose movement is caused by the action of an electromagnet inside the box. When power is applied to the relay's coil, the electromagnet comes alive and pulls across the switching contacts. With a relay, a small current can energise a device consuming large current.

Reluctor. A wheel that rotates inside the distributor and triggers the release of voltage in an electronic ignition system.

Rocker arm. A valve train component that reverses the upward motion of the push rod to downward motion of the valve and vice-versa.

Rotary engine. An engine with a triangular shaped rotary piston moving inside an epitrochoidal housing. Invented by Dr. Wankel engine, it is also called Wankel engine.

Run on. Tendency of an engine to continue firing even after the ignition is switched off; also called 'dieseling'.

Secondary cell/battery. In this the metals and the electrolyte change when the battery supplies the current, the process being called 'discharging'. All these changes can be reversed by applying current to the battery in the opposite direction, the process being known as 'charging'. Lead-acid battery is the most commn example of the secondary cell.

Semi conductors. Materials having exactly four electrons in their outermost orbit. Germanium and silicon are two examples.

Sensor. A device to record data for measurement and/or control.

Shim. Thin metallic sheets used as spacers between two parts.

Sludge. Thick, black deposits in engine formed from dirt, oil, water, etc. It is usually formed in engines when oil changes are neglected.

Smart coolant pump. A coolant pump which is controlled directly by the engine management system so that precise engine temperatures can be maintained by varying the rate of coolant flow as per requirements.

Smog. A mixture of smoke and fog.

Solid state ignition. The modern ignition system with no mechanical contact breakers and conventional distributor replaced by a crankshaft-actuated electronic trigger, making the computer fire high voltage ignition coils in a predetermined sequence.

Spark plug. A device screwed into the combustion chamber of a spark ignition engine wherein electrical charge from the distributor jumps a preset air gap to a grounding point(s). The resultant spark ignites the fuel/air mixture in the combustion chamber.

Square engine. This is an I.C. engine having stroke equal to its bore size.

Starter. A high-torque electric motor used for starting the engine, typically through a high ratio geared drive connected to the flywheel ring gear.

Stoichiometric mixture. The air-fuel mixture which would achieve complete combustion of fuel under standard conditions. It is 14.6 parts of air and 1 part of petrol by mass. By volume, this would be about 60 parts of air to 1 part of petrol.

Stratified charge. A charge containing stratas or layers progressively becoming leaner from one layer to the other. With such a charge very lean air-fuel mixtures can be used, but the NO_x emissions are increased.

Supercharging. The process of supplying to the engine, air-fuel mixture at a pressure above the atmospheric.

Surface discharge plug. This is a special type of spark plug where the space between the plug electrodes is filled with some semi-conductor material.

Surge. A driving problem where the vehicle engine on its own appears to gain power for a small moment and then falls back down to the previous speed. The reason for this is very lean air-fuel mixture (which may be due to wrong carburettor setting, clogged air filter, blocked main jet, etc.) or retarded ignition timing.

Swept volume. Also known as 'displacement', it is the maximum volume of air-fuel mixture which can be induced into a cylinder in a single intake stroke.

Swirl. Rotational flow of charge within the engine cylinder, about its axis.

Swirl ratio. The ratio of the angular rotational speed of air about the cylinder axis to the angular rotational speed of the crankshaft.

TDC (Top Dead Centre). The highest point to which the piston travels in the cylinder of an I.C. engine.

Tailpipe. Component of the exhaust system after the silencer, which allows the exhaust gases to go to the atmosphere.

Tappets. The components to convert the rotary motion of the camshaft into reciprocating up and down motion of the push rods.

Thermal Efficiency of an engine is the ratio of the work output and the heat input to the engine. Depending upon the type of work, it may be called indicated thermal efficiency or brake thermal efficiency.

Thermosyphon system. A simple engine cooling system based on the natural convection caused by the difference in densities of the hot and the cold regions of the cooling water; is now obsolete.

Three-way converter. A catalytic converter employed to control the levels of all the three main pollutants in the engine exhaust, viz., unburnt hydrocarbons, carbon monoxide and nitrogen oxides.

Throttle body. An engine housing containing the throttle butterfly valve for regulating the air flow through the inlet manifold.

Throttle body injection. The type of fuel supply system in which the injectors are located on the engine throttle body, thereby supplying fuel to more than one cylinder. It is also known as *single point* or *central injection.*

Thyristor. A semiconductor rectifier which can control the output power supplied to an electrical load, with the minimum wastage of energy. The word 'thyristor' has been derived from 'thyratron' and 'resistor', the thyratron being the predecessor of the thyristor.

Timing belt. A square-toothed, reinforced rubber belt which is driven by the crankshaft and operates the camshaft.

Timing chain. A roller chain driven by the crankshaft for operating the camshaft.

Torque. It is the force of rotation acting along the crankshaft axis at any instant of time. From the engine it goes through the transmission, to the road wheels. It is responsible for rotation of the wheels and hence for the pulling of the vehicle.

Trailing throttle. The driving condition when the vehicle is being driven with accelerator pedal released.

Transistor. It consists of three alternate sections of N- and P- materials. It may be NPN or PNP type. Its central section is called base which controls the main current through the emitter and the collector. The current through the base is called control current.

Transverse engine. An engine is which the crankshaft is placed side-to-side relative to the wheels of the vehicle. Also called an *east-west engine.* Most modern front wheel drive vehicles use this engine orientation.

Tune-up. A regular maintenance function, usually associated with the replacement and adjustment of components in the electrical and fuel systems of a vehicle for the purpose of achieving optimum performance.

Turbocharger. A centrifugal supercharger which is driven by the exhaust gases. It forces additional quantity of air or air-fuel mixture in the engine intake manifold.

Turbo-charging. A method of increasing engine power and decreasing emissions by rerouting hot exhaust gases through a turbine which drives an air compressor which forces more air into the engine cylinders.

Turbulence. A type of flow in which randomly dispersed vortices of different sizes become superimposed into the charge (air or air-fuel mixture) flow stream. These vortices are carried alongwith the flow stream and form breakways having concentric spiral motion. The whirling of these vortices causes interaction with adjacent vortices, increasing the fuel mixing and heat transfer rate.

ULEV (Ultra Low Emission Vehicle). A vehicle which emits emission less than the limits specified by the California Air Resource Board, U.S.A.

Undersquare engine. An engine with the stroke to bore ratio more than one.

Valve overlap. The duration when both the inlet and the exhaust valves are open simultaneously, in terms of degrees of crankshaft rotation

Valve train. The system which operates intake and exhaust valves, consisting of camshaft(s), valves springs, lifters, push rods and rocker arms.

VLIM (Variable Length Intake Manifold). An automobile engine manifold technology, which can vary the length of the intake tract to optimize power and torque as well as to improve fuel efficiency. Also called VRIS (Variable Resonance Induction System), it has two main effects: 1. *Swirl,* which helps form a homogeneous air-fuel mixture which ignites without detonation. At low engine speeds, air is directed through a longer path with limited capacity, increasing the speed of an flow. At higher loads, shorter and higher capacity path opens to increase the amount of air flow. 2. *Pressurisation* by means of a tuned intake path which has a high pressurizing effect similar to a low-pressure super charger, due to Helmholtz resonance. This effect occurs only over a narrow engine speed band, increasing engine output.

VTEC (Variable Timing Electronic Control). Honda's variable valve timing system controlled electronically.

Venturi. A restriction in the entrance passage of a carburettor where the main fuel jet is placed.

Viscosity. Resistance of a fluid to flow; most important property of a lubricating oil.

Volumetric Efficiency of an engine is the ratio of the volume of the air actually inducted in a given time at the atmospheric conditions surrounding the engine, to the total piston displacement or swept volume of the engine.

Wankel engine. A rotary engine working on Otto cycle.

Water jacket. Passage around the engine cylinder for the circulation of coolant. These are cast as integral parts of the cylinder block and head.

Wheel motor. An electric motor which is installed into a wheel and drives it directly.

Zener diode. A special type of diode that acts as ordinary diode upon a certain voltage called the breakdown voltage above which it allows the reverse-bias current to flow. It is used in control circuit, such as to regulate field current in an alternator.

ZEV. A vehicle not emitting any pollutants through exhaust, is called ZEV or Zero Emission Vehicle. As at present only electric vehicles and some fuel cell vehicles may be rightly called ZEV.

Index

A

B

C

D

E

G

H

I

J

L

M

N

O

P

W

Z